Automotive and Small Truck Fuel Injection Systems

Gas and Diesel

Automotive and Small Truck Fuel Injection Systems

Gas and Diesel

ROBERT N. BRADY
Department Head,
Diesel Mechanic/Technician Program
Vancouver Community College
King Edward Campus
Vancouver, British Columbia

A Reston Book
Prentice-Hall, Inc.
Englewood Cliffs, New Jersey 07632

Library of Congress Cataloging-in-Publication Data

Brady, Robert N.
Automotive and small truck fuel injection systems.

Includes index.
1. Automobiles—Motors—Fuel injection systems.
2. Automobiles—Motors (Diesel)—Fuel injection systems. 3. Trucks—Motors—Fuel injection systems.
4. Trucks—Motors (Diesel)—Fuel injection systems.
I. Title.
TL214.F78B73 1986 629.2′53 85-14368
ISBN 0-8359-0315-X

Editorial/production supervision and
interior design by Camelia Townsend

A Reston Book
Published by Prentice-Hall, Inc.
A Division of Simon & Schuster, Inc.
Englewood Cliffs, NJ 07632

10 9 8 7 6 5 4 3 2 1

PRINTED IN THE UNITED STATES OF AMERICA

To all of those terrific and dedicated mechanics at the Scottish Gas Board, and others through the years who inspired and instilled in me the desire and ability to continue lifelong learning techniques and to maintain always continued levels of excellence in all my endeavors.

Contents

Preface

The information contained within this textbook is designed as both a teaching aid and self learning guide for automotive mechanics who aspire to broaden their working knowledge of both gasoline and diesel fuel injection systems now in wide use in passenger cars and light trucks and vans worldwide.

Although carburetors are still in use and will continue to be used for some years to come in specific types of vehicle engines, gasoline fuel injection systems are now the dominant choice on not only high performance vehicles, but also in mass produced family type sedans. North American, Japanese and European car manufacturers all employ either throttle body or ported gasoline fuel injection systems with the trend indicating that by 1987–88, that at least 70% of all passenger cars will be equipped with one of these forms of fuel injection. Several manufacturers have indicated that by as early as the 1986–87 model year, that all of their passenger car engines will be equipped with gasoline fuel injection.

The use of electronics will allow carburetion to be used in certain applications, however with the increased EPA (Environmental Protection Agency) exhaust emissions limits, fuel injection has proven that it has the capability to meet, and in many cases exceed these tough emissions standards as well as providing superior engine/vehicle performance characteristics and improved fuel economy. This is very important to vehicle manufacturers who must also contend with the legislated North American CAFE (Corporate Average Fuel Economy) standards. It is the general industry feeling and opinion that the carburetor as we have known it, has reached its limit of improvement and will no longer be capable of providing the desired results in all areas of automotive performance.

Current and future automotive mechanics must have a thorough knowledge and understanding of not only how a fuel injected system operates, but must also be capable of diagnosing and using the multitude of special test gauges and instruments available to them from both the vehicle and aftermarket manufacturers, so that they can effectively and efficiently service/maintain these fuel systems.

In order to analyze each type of fuel injection system, it is important that the theory of combustion be well understood, and within this book both this subject along with various types and designs of high performance combustion chamber styles are shown and discussed so that the reader has an appreciation of the similarities between many of the major engine makes and fuel injection systems.

A basic understanding of both electricity and electronics is also a prerequisite to mastering the intricacies of both gasoline and diesel fuel injection systems, since both systems now employ electronic sensors and control modules with access to a memory system for interpretation and analysis of vehicle/engine problems.

The one thing in the mechanics favor however, is that the majority of gasoline fuel injected systems in use today employ technology that is very similar in both layout/arrangement and also general operation.

The largest supplier of gasoline fuel injection systems for passenger car use is Robert Bosch Corporation who supply these various systems to every major North American, Japanese and European manufacturer.

Several manufacturers have modified these systems slightly to suit their particular engines and conditions, but basically speaking they all are of Bosch design. Various Bosch licencees worldwide produce these fuel injection systems for use in a wide model range of passenger cars.

This factor alone simplifies the mechanic/technicians job since each system is not a total opposite to another in use by a different vehicle manufacturer.

Although the success of the diesel automobile in North America has not met with the same success rate that it has in other areas of the world, both the European and Japanese manufacturers continue to enjoy a high success rate with their diesel powered passenger cars and light trucks and vans. Many of these vehicles are imported into North America in many forms and therefore the well trained auto mechanic should have a working knowledge and an appreciation of this alternate power plant.

The greatest success for the automotive diesel in North America appears destined for the light pickup truck and recreational vehicle market with Ford/IHC, General Motors, Chrysler/Mitsubishi, Isuzu, AMC/Renault, Peugeot, Mercedes-Benz, BMW, Volkswagen/Porsche/Audi, Nissan, Toyota, and Mazda all offering a diesel engine in one or more cars or light trucks of their own or in concert with another major manufacturer.

This book is not intended to make a diesel fuel injection specialist out of the automotive mechanic since this is a specialized field in its own right, but it will allow you to gain a close insight and thorough understanding into just how and what the differences are between these powerplants and their opposite counterpart the gasoline engine. Any automotive mechanic who genuinely desires to master the basic maintenance, testing and adjusting of passenger car and light truck diesel engines can do so with the aid of this book and the application of hands-on exposure to the various types of diesel engines available to them. Many of the basic concepts that you will absorb related to gasoline fuel injection can be used as a foundation on which to build and increase your knowledge of diesel fuel injection.

As far as the actual gasoline versus diesel engine designs are concerned, little difference exists other than the fact that the diesel due to its higher compression ratio and firing pressures is usually more robust and slightly heavier than its gasoline counterpart. Both engines however are available with overhead camshafts and turbochargers, and the internal components that together provide a smooth running and reliable powerplant are similar in both design and operation.

Rather than viewing the diesel engine as a noisy, sluggish and foul smelling competitor, accept it for what it is and appreciate that it is a horse of a different color with many advantages to its credit, and one that will be around in one form or another for many years to come especially in mid-range and heavy-duty trucks and many other applications. Your ability to understand and service/troubleshoot both gasoline and diesel fuel injection systems will make you a very valuable addition to many employers and will provide you with a rewarding and fullfilling career for many years to come.

Automotive and Small Truck Fuel Injection Systems

Gas and Diesel

1

Introduction

The internal combustion engine has been based upon the same concept of operation since 1794 when an inventor named Street developed this idea. Thirty years later a French engineer named Sadi Carnot came up with ideas that were later incorporated into the engine concept named after Rudolf Diesel; Diesel patented a compression ignition engine in 1892. Carnot did not build such an engine; however, he initiated the compression ignition concept.

The first four-stroke cycle internal combustion engine that ran on gasoline using the flame ignition principle was constructed by Nickolaus Otto in 1876, hence the term "Otto Cycle" that we often hear used today to distinguish it from that of the diesel cycle.

The basic historical dates for piston-type engines are as follows:

Type	*Inventor*	*Date*
1. Coal gas four cycle	Otto	1877
2. Automobile four cycle	Benz	1879
3. Compression ignition	Diamler	1883
4. Diesel engine	Diesel	1895

Sweeping changes have taken place in the automotive industry over the years that have led to engineering and design improvements that would not have been thought possible when the original idea was proposed.

For years, carburation was accepted as an adequate form of fuel control for the regular family sedan and did in fact serve admirably well. In later years it was only racing and high performance imported European vehicles that offered the possibility of anything as technologically advanced as "fuel injection."

Even today, many manufacturers still produce vehicles with two- and four-barrel carburetors; however, these units are far more sophisticated than they were in the past and are equipped with electronic feedback controls that are required in order to provide suitable performance and yet still meet EPA (Environmental Protection Agency) pollution control standards.

With the strict EPA standards now in effect, vehicle manufacturers are also faced with the CAFE (Corporate Average Fuel Economy) standards whereby the average fuel consumption in miles per gallon or liters per 100 kilometers must meet a minimum standard.

The basic function of a carburetor is to perform the function of metering, atomizing, and mixing fuel with the air flowing into the engine. In addition, the carburetor regulates the volume of the air/fuel mixture that enters the engine by the driver/operator manipulating the throttle to control the position of the butterfly valve inside the venturi of the carburetor throat.

Although the carburetor does its job reasonably well in terms of mixing the air/fuel mixture, the firing order of the engine creates a high air turbulence inside the engine's intake manifold whereby the result is that the engine cylinders closest to the carburetor mounting position invariably receive a richer air/fuel mixture than those cylinders that are located further away. This problem is more acute on a 6-cylinder engine for example that employs only one centrally mounted carburetor, than on a 4- or 6-cylinder engine that employs dual carburetors.

On a 6-cylinder engine that employs only one carburetor with an engine firing order of 1-5-3-6-2-4, you can appreciate how this internal air/fuel mixture is constantly switching back and forward due to the uneven firing order. This turbulence will automatically favor the cylinders closest to the carburetor mounting position as far as the ratio of the air/fuel mixture is concerned. The faster the engine speed, the greater the spread will be between the correct air/fuel ratio distribution to all engine cylinders.

With the need for engine/vehicle manufacturers to comply with the EPA regulations, a better system was required. However, in addition to meeting exhaust emissions regulations and the CAFE limits, the general public also demanded performance from smaller, lower compression engines that were available at the time.

With the first Arab oil embargo in the early seventies, gasoline prices rose very steeply, causing consumers to demand smaller and more fuel-efficient automobiles. Car manufacturers were put into a position where they had to downscale both the size and horsepower of their existing vehicles. Consumers demanded not only the luxury that they had been used to for many years in the larger style automobiles, but also vehicle performance along with improved fuel economy.

Under the Energy Policy and Conservation Act of 1975, the United States established projected standards of vehicle fuel economy and emissions limits that the automobile manufacturers were faced with meeting under the risk of paying hundreds of millions of dollars in penalties.

A variety of engineering improvements have been incorporated over the years to meet not only the desires of the general car buying public, but also to improve the performance/fuel economy as well as meeting the EPA regulations, which have tended to become stricter every year. Standards for 1983 were put at 26 miles per gallon; 1984 standards were placed at 27 miles per gallon while the statutory requirement for 1985 and beyond is 27.5 miles per gallon. These fuel economy limitations can be averaged over a manufacturer's product line so that some vehicles may achieve less than the standard while others would obviously have to obtain more in order to maintain what is commonly called a Corporate Average Fuel Economy (CAFE) standard. Manufacturers that exceed the CAFE requirements for each year of the standards can accumulate credits that may be used to offset shortfalls during a three-year time period immediately preceding or following the year in which they fail to meet the minimum standard.

If a manufacturer is placed into a negative position with respect to the CAFE standards, then the manufacturer is subject to a penalty of $5.00 per vehicle sold for each one-tenth of a mile per gallon that their CAFE falls short of the standard. Considerable lobbying is now under way by the automobile manufacturers to have the CAFE standards reduced to 26 miles per gallon or less.

Although tremendous improvements have been made to engines by way of various monitoring devices, the carburetor has reached its limit of improvement from both an economic point of view and a performance/EPA level standard. The Buick Division of General Motors Corporation expects fuel injection to completely supplant carburetors by 1986 or shortly thereafter.

Due to the inherent design characteristics of the carburetor, a percentage of fuel can flow into the engine when decelerating where it is then blown out of the exhaust system in an unburnt state. The result is that the engine efficiency is lowered while at the same time poisonous hydrocarbon emissions are released into the atmosphere. Due to this fault of the carbureted engine, it generally requires a greater degree of emission control equipment than does a fuel injected engine.

A natural successor to the carbureted gasoline engine is the electronically controlled fuel injection system that is now being employed in ever-increasing numbers by major automobile manufacturers. The carburetor engine cannot guarantee the type of precision control that a central-point or individual port-type electronic fuel injection system can; therefore, the fuel injection system is here to stay. Various forms of fuel injection systems are now in use by vehicle manufacturers; however, regardless of the type employed, the function of each is similar, namely to perform the function of injecting gasoline into the intake manifold in a finely atomized state in the area of each cylinder's intake valve so that each cylinder's air/fuel ratio will be the same regardless of engine speed. In this way, engine performance is greatly improved along with a decrease in fuel consumption and the ability to meet both the EPA exhaust emissions regulations and the CAFE requirements.

The gasoline fuel injection system was once very costly to produce and required special skills and knowledge to service because it was a detailed technical system that could prove to be unreliable under many conditions. Advances in electronics technology have now raised the design and operation of the average gasoline fuel injection system to a plateau whereby the average service technician with minimal special tools can service and troubleshoot the system in a short period of time. In addition, little or no adjustment is required with the gasoline fuel injection system between major service intervals.

Within this book, the various types of gasoline fuel injection systems are discussed in detail from their basic operating conditions as well as the service requirements of each with respect to all current automobiles now employing this type of fuel delivery system.

Diesel Powered Vehicles

The spark ignition gasoline engine since its inception has emerged as the single best configuration although many changes have taken place over the years. The diesel engine, however, due to its slower ignition characteristics was not given as much attention to improvement as was its gasoline counterpart. Indeed, not until 1927, when Robert Bosch developed a successful high pressure fuel injection system for the diesel engine, did it start to gain wider acceptance as a viable alternate power source to the gasoline engine.

Although companies such as Benz did produce a diesel-powered passenger car option as early as 1936, it wasn't until 1948 that they started to effectively mass produce diesel-powered passenger cars, with Peugeot following suit in 1958 with their 403 model car. In reality then, the high speed diesel-powered passenger car did not come into being until 1948. Since then the diesel engine has not produced a "one best" configuration because today many differences exist between makes, such as four cycle versus two cycle, direct injection versus pre-combustion chamber indirect injection, unit fuel injectors, in-line injection pumps and nozzles and rotary/distributor injection pumps.

Between 1945 to the mid-1960s, the diesel made tremendous inroads against its gasoline counterpart, especially in heavy trucks and equipment. Today very few heavy trucks or equipment use anything other than a diesel engine for power.

The diesel engine, although successful in many areas of the world as a passenger car powerplant, has been slow to gain acceptance in North America due to a number of factors. Contrary to U.S. disenchantment with diesel cars due to the slower acceleration characteristics, and the noise, smell, and smoke that they tend to emit, European sales of diesel-powered cars and light trucks continue to increase. The reason for this can probably be credited to the fact that gasoline prices range anywhere from 70 to 125% more in Europe than they do in North America and that the average weekly salary in Europe is about ⅓ to ¼ of what it is in North America.

European production of automotive diesels in 1981 was 1,177,000 units compared to 297,500 in North America. Diesel car sales now account for 15% of total car sales in West Germany, 10% in France, and 7% in Italy. Projected total sales for 1985 in these countries are 23%, 18%, and 13% respectively. Peugeot diesel car sales in France are actually about 24% of their total car production with their world diesel car sales running at about 20% of their total vehicle production.

Diesel car sales, although not as dramatic in the U.S., have increased their penetration of the marketplace. In 1975, diesel car sales were a mere 0.24%, which climbed to 3.4% in 1980 and 4.5% in 1982; however, the 1983 figures showed a drop to about half of the 1982 penetration probably due in part to the depressed economy and the improvement of gasoline engine fuel mileage and performance. These lower sales, however, were only in domestic-built cars, since the sales of imported diesels showed an increase.

Both Mercedes-Benz and the Audi Division of VWA (Volkswagen America) increased sales of diesel cars in the 1983 model year over that in 1982. According to some optimistic estimates, the proportion of diesel engine-equipped passenger cars will never exceed 10% of that of the gasoline engine up to the year 2000 and that only 25% of all fuel required will be diesel fuel to that same time.

The strongest penetration for diesel power in the North American market will most likely take place in the pickup truck field rather than in passenger cars. Already the trend in North America is away from large V8 diesel engines in cars and towards their use in pickup and light trucks. Cars are being equipped with 4- and 6-cylinder diesels with some offering turbocharging for improved vehicle performance.

Both General Motors and Ford offer V8 diesels in pickup trucks, RV's and mid-range trucks as well as having access to in-line 4- and 6-cylinder engines. American Motors and Renault have also signed an agreement whereby Renault will supply diesel engines to AMC. One such example is the current use of the turbocharged diesel engine now available in the Jeep range of vehicles produced by AMC in North America.

Automotive diesels are also currently available in RV's and motor homes such as Winnebago offering Renault's 2-OL diesel as an option in its 20 ft. Class C (see Figure 1–1) and Chevrolet's 6.2L V8 diesel in their 22–27 ft. Class A rear-wheel-drive units. The Centauri, LeSharo, and Phasar chassis are available with Renault's 4-cylinder J85-234 turbocharged 2L diesel as the standard powerplant. These vehicles are 19.7 ft. long and have a GVWR of 5830 lb. Claimed fuel consumption is about 24 mpg (US) or 28.5 mpg Imperial for the Centauri model and 22 mpg (US) or

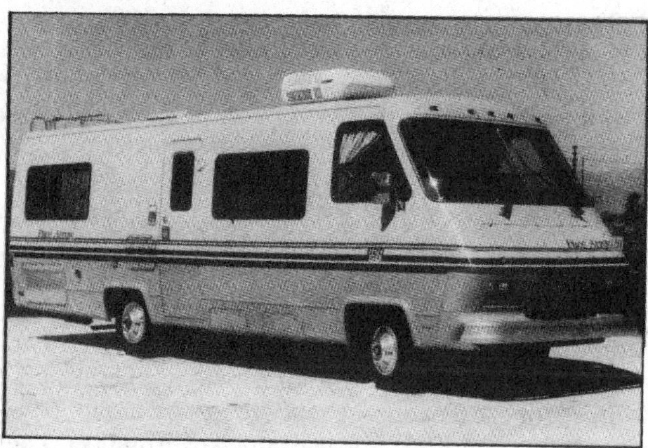

Figure 1–1. Chevrolet 6.2L or Ford/IHC 6.9L V8 diesel-powered Fleetwood Class A Pace Arrow 27 foot motor home. [Courtesy of Fleetwood Enterprises Inc.]

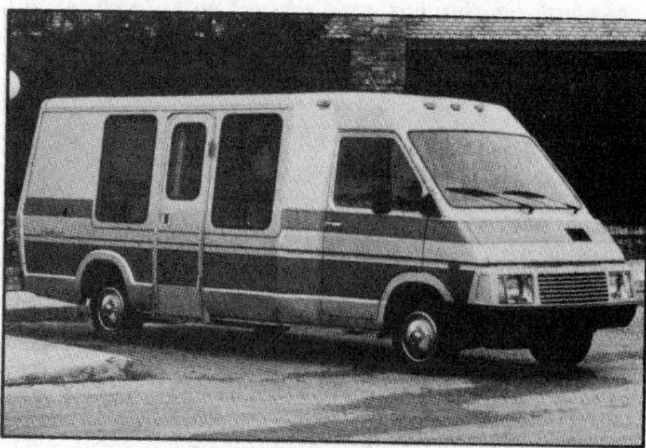

Figure 1–3. Renault turbocharged 2.0L diesel-powered Winnebago Class C Centauri 19.7 foot motor home. [Courtesy Winnebago Industries]

26 mpg Imperial for the LeSharo and Phasar models. This is a fuel performance increase of almost 100% over the gasoline counterpart.

Coachmen Industries, Inc., offer the Chevrolet 6.2L V8 diesel as an option in their 23 ft., 11,100 lb. GVWR Class C mini-motor homes built on a Chevrolet chassis similar to the Ford unit shown in Figure 1–2. Fleetwood Enterprises, Inc., also offer the Chevrolet 6.2L diesel on both Pace Arrow (see Figure 1–3) and Southwind Class A Models. This engine is also offered as an option in the 20–23 ft., 10,200–11,000 lb. GVWR Tioga Arrow and Jamboree Rallye built on a Chevrolet chassis. The 20–26 ft. versions of these mini-motor homes with a Ford chassis offer the optional International Harvester 6.9L V8 diesel engine.

British Ford recently entered the small automotive diesel market with the release of a 1.6L (97.5 cu.in.) engine for use in their small Fiesta and Escort cars. At this time there are no plans to import this engine into the U.S. because Ford USA is committed to using a Toyo Kogyo (Mazda) 2.0L (122 cu.in.) diesel engine as an option for the Ford Tempo/Mercury Topaz twins as well as in the Escort and Lynx models. Ford USA and Mazda recently signed an agreement to jointly produce a new small car that will be built in Mexico for the North American market.

Figure 1–2. Chevrolet 6.2L V8 diesel-powered option Class C motor home. [Courtesy Coachmen Industries]

Ford in the U.S. introduced in 1984 a 2.4L (146 cu.in.) 6-cylinder turbocharged diesel engine produced by BMW/Steyr as an option for use in their Lincoln Continental Mark VII, Thunderbird, and Cougar. In addition, the 1984 model Ford F350 pickup trucks and E350 Econoline vans are offered with the International Harvester 6.9L (421 cu.in.) pre-combustion chamber V8 diesel engine. Ford arranged to have 72,000 of these engines produced in the 1984 model year.

General Motors through the Oldsmobile Division have of course offered their V8-350 cu.in. (5.7L) diesel engine as an option in other GMC division cars and light pickup trucks since the 1977 model year and now also offer a V6-263 cu.in. (4.3L) diesel (1982 and up). General Motors announced in late 1984 that it will drop the optional diesel engine in some 1986 mid-size, large and luxury car lines. Chevrolet offers a 6.2L V8-378 cu.in. diesel (1982 and up) for use in their pickup trucks and vans and in GMC RV's and Motorhome chassis. This engine is now being manufactured by the Detroit Diesel Allison Division of GMC. The U.S. Armed Forces has also arranged to purchase 53,000 of the 6.2L (378 cu.in.) V8 pre-combustion chamber diesels in Chevrolet Blazers and pickup trucks. Delivery started in August of 1983 and will continue until April 1986.

Also offered as an option by GMC in the Chevette and Pontiac 1000 models since 1981 is the Isuzu 4-cylinder 1.8L 111 cu.in. diesel engine. General Motors currently has a 40% interest in Isuzu and a 5%

interest in Suzuki. GM is now importing a 3-cylinder minicar built by Suzuki and a small front-wheel-drive car built by Isuzu known as the 'Sprint.'

General Motors and Toyota Motor Corp. began joint production of 250,000 cars a year in late 1984 from a Fremont, California, plant. Look for some possible Toyota diesel options in GMC vehicles and light trucks. The S-15 GMC pickup truck is now available (1984) with a 2.2L 134 cu.in. 4-cylinder Isuzu diesel engine. The Dodge Ram 50 is available with a Mitsubishi 4-cylinder 2.3L turbocharged silent-shaft diesel engine.

In addition to the North American manufacturers' tendency to diesel power in some of their product lines, such notable European manufacturers as Volkswagen (with their highly successful Rabbit/Golf), Mercedes-Benz, Peugeot, Renault, Opel, Fiat, Audi, BMW, and Volvo (VW) believe that there is merit in the diesel engine for passenger car and light truck applications.

Although it appears that the diesel engine is a viable alternate for automotive power, it also has problems with regards to emissions regulations standards in the U.S. The EPA limit for diesel particulates (solid carbon particulates impregnated with high molecular weight hydrocarbons and lower hydrocarbon aerosols) for 1985 will go to 0.2 gram per mile from the 1982 standard of 0.6 gram per mile. Various approaches are being followed, and research is under way to attempt to meet this standard by way of engine design modifications, fuel system modifications, and exhaust cleaning devices. Nitric oxides in the exhaust gases is another area that both the diesel and gasoline engines have to improve upon, based on current EPA proposed standards; therefore, further technological breakthroughs are on the horizon for both types of motive power.

At the present time it would appear that both the gasoline and diesel automotive engines will continue to predominate as the main means of motive power; however, alternate automotive fuels are now being actively pursued by many manufacturers.

The initial success of the diesel-powered car and light truck in North America was a direct result of the Arab oil embargo of the seventies. The diesel offered an average fuel economy improvement of 50 to 100% over its gasoline counterpart depending upon the driving cycle that it was subjected to.

However, North American manufacturers in their haste to capitalize on the apparent demand for automotive diesel power simply converted their existing gasoline engines to diesel. Moderate success was obtained at a minimum expense to the manufacturer; however, the consumer, being unfamiliar with the fact that operating characteristics of a diesel-powered vehicle differ from a gasoline unit, simply expected too much.

The diesel is an inherently slower accelerating engine; it is noisier, harder to start especially in cold weather, and emits more exhaust smoke and odor than its gasoline counterpart. In addition, the diesel engine requires more frequent maintenance (such as oil change, and fuel and water filter changes). Another initial problem with the introduction of the diesel-powered car in North America was one of a lack of adequately trained qualified servicemen. This problem has now been improved by the manufacturers, and with the general education of the consumer to the diesel's different operating characteristics, the current production diesel engines now available in cars and light trucks seem destined for moderate success in North America.

It is doubtful, however, that the diesel engine will ever be able to rival its gasoline counterpart from the standpoint of performance—acceleration, quietness, and vibration. The best that can probably be expected from the diesel is that it will gain some added success with advanced technological breakthroughs such as turbocharging and the possible advent of direct injection versus the present pre-combustion chamber design now used by all manufacturers of passenger car units. However, additional penetration of the light van and pickup truck market will more than likely be where the diesel excels in the final analysis in North America.

Peugeot recently completed a fuel economy run with their 305 model car equipped with a 1360 cc inline 4-cylinder turbocharged diesel that went from Pau to Paris at 92.9 mpg. This same test was conducted in the U.S. from Chicago to Knoxville driving at an average speed of 33.1 mph for 516.1 miles and a fuel economy of 91.2 mpg. This type of a result is typical of the average speed that might be encountered in city/suburban driving cycles, therefore the diesel can still be considered as a logical alternative to the gasoline engine depending on your choice of driving mode.

Daihatsu in Japan has what is the world's smallest passenger car diesel. It is a 993 cc (60.5 cu.in.) unit that produces 38 bhp JIS (32 bhp SAE) at 4800 rpm. Used in the Charade model of economy hatchbacks, the diesel unit has attained 88 mpg, while the best the gasoline version could achieve was 75 mpg with the aid of an electronically controlled carburetor and a 3-way catalytic converter system.

Although the two examples above may not be typical of the type of driving cycle experienced by the average driver, consider that producers of the 1984

VW Rabbit Weltmeister diesel model have claimed about 65 mpg (4.3L/100 km) average, which is excellent fuel economy.

As mentioned earlier, all current automotive diesel engines are of the pre-combustion chamber design. Several manufacturers are now testing the possibility of direct injection to improve the overall performance of automotive diesel engines. One such manufacturer is British Leyland who has joined forces with Massey-Ferguson (Perkins) Holdings Ltd., and who, in conjunction with Austin-Rover, will develop what is to be the world's first high output direct-injection passenger car diesel.

Ford of Germany offered as an option starting in the 1984 European Granada, a diesel engine supplied by Peugeot. This vehicle may show up in North America eventually. In 1986 Chrysler and Peugeot may jointly produce a car to replace the Omni/Horizon with a possible optional powerplant being a 1.9L Peugeot diesel, however this diesel option has been dropped for the North American market. Chrysler might also build a turbo-diesel version of its venerable slant-6 engine for 1985 and 1986 with a 2.2L 4-cylinder version of the same engine to follow a year later. Both turbo boost and EGR will be controlled electronically. Chrysler claims that the turbodiesel 3.7L slant-6 will provide V8 performance with exceptional fuel economy. The larger 6-cylinder unit will most likely be for pickup and light trucks while the 4-cylinder engine may be offered for use in the company's front-wheel-drive products such as the T-115 mini-van (1984).

Figure 1–4 illustrates the present trend toward international cooperation between automobile manufacturers. This trend will create a situation whereby many of the engines, both gasoline and diesel, that will be used in future vehicles will be manufactured by one particular company, but used by one or more manufacturers. As you can see, many major carmakers are closely inter-related.

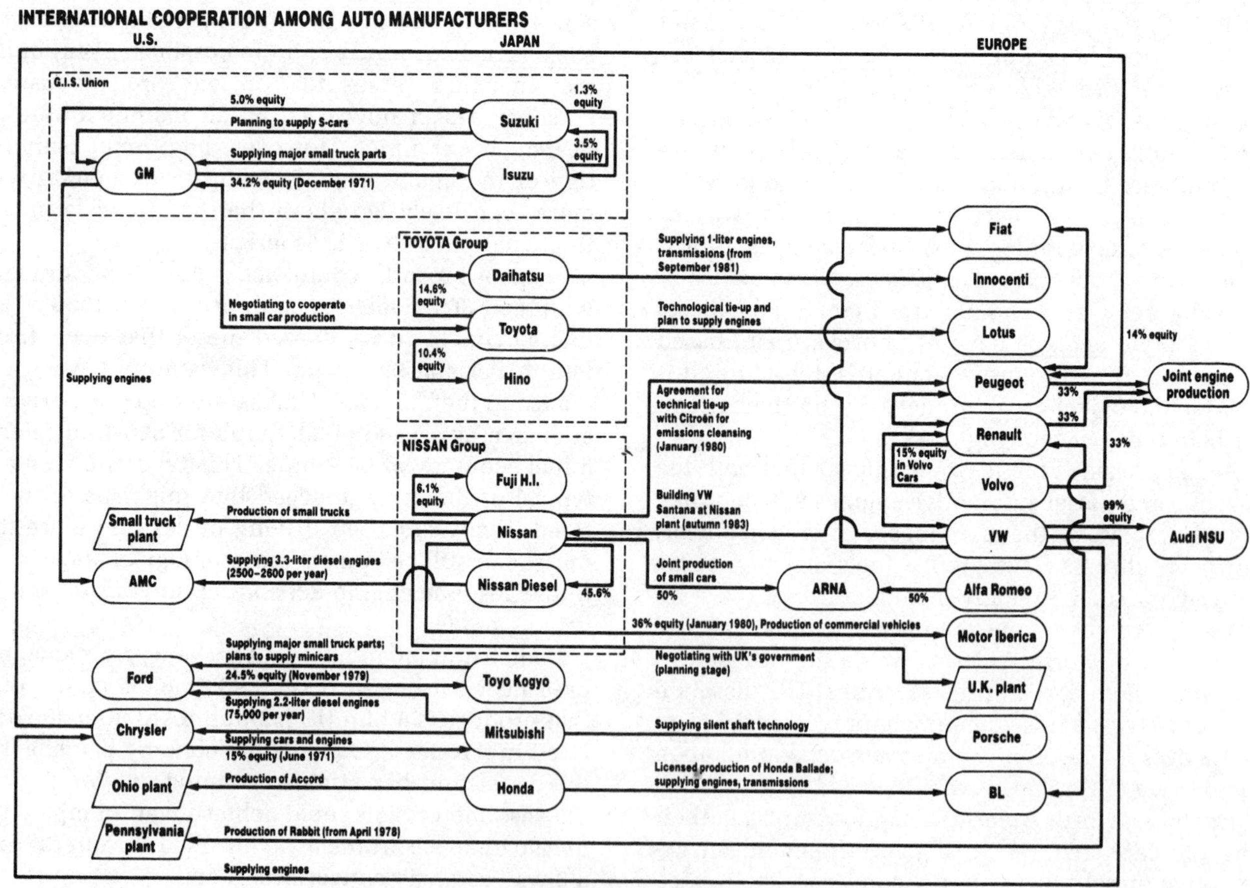

Figure 1–4. Chart showing international cooperation among automotive manufacturers. (Courtesy of *Road and Track Magazine*)

2

Diesel Operating Principles

The operating principle of a 4-stroke cycle diesel engine is very similar to that of a 4-stroke gasoline engine in that it requires two complete revolutions of the engine crankshaft (720 degrees) in order to complete the four strokes of the piston, namely the intake, compression, power, and exhaust.

In a 4-stroke cycle gasoline or Otto cycle engine, these events occur as illustrated in Figure 2–1. If we now compare the same four piston movements of the 4-stroke gasoline engine with its diesel counterpart shown in Figure 2–2, we see that the main difference occurs during the intake and compression strokes.

On the intake stroke of the gasoline engine (carburetor or fuel injected), a mixture of air and gasoline is drawn into the engine cylinder at a ratio of approximately 14 parts air to one part gasoline. When the piston begins its upward stroke (compression) with both the intake and exhaust valves closed, the air/fuel mixture is continually placed under compression, which raises its pressure and temperature. A number of degrees before the piston reaches TDC (top dead center), the air/fuel mixture is ignited by a high tension spark from the spark plug. The rapid increase in cylinder pressure reaches its peak just as the piston attains TDC, and the piston is driven down the cylinder on its power stroke, which rotates the engine crankshaft and flywheel.

In the 4-stroke diesel cycle, however, only air is forced into the engine cylinder during the intake stroke. During the compression stroke, air only is placed under compression, which increases its pressure and temperature. NOTE: The diesel engines compression ratio is generally much higher than that of the gasoline engine. A gasoline engine usually will have a CR (compression ratio) of between 8 and 9:1 in most engines today, although some can be higher. The diesel requires a considerably higher CR in order to promote ignition of the diesel fuel once it is injected into the cylinder. Therefore all current automotive diesel engines that use an indirect injection-type combustion chamber normally have CR's of between 21 and 23:1.

As the piston moves up the cylinder on its compression stroke, the actual point of injection of fuel will vary between engines; however, once the diesel fuel is injected, it penetrates the air mass in the cylinder. The high pressure atomized diesel fuel, forcing its way into the hot compressed air, creates friction between the atomized fuel droplets, which causes combustion to take place as they mix with the oxygen in the combustion chamber. The fuel is injected for a number of crankshaft degrees depending upon the design of the particular engine.

In summary then, the only difference between the 4-stroke gasoline engine and the 4-stroke diesel engine operating cycles occurs during the intake and compression strokes.

Polar Valve Timing

A polar valve timing diagram is shown in Figure 2–3, which is a graphical representation of the sequence of events in four piston movements (two crankshaft revolutions = 720 degrees) in a typical 4-stroke cycle internal combustion engine. The actual degrees of each stroke, namely intake, compression, power, and exhaust will vary between engines. Figure 2–3 is simply an illustration of such a sequence showing each stroke's relative duration in crankshaft degrees. The diagram can be considered common to both a gasoline and a diesel engine for our discussion here. Neither the start of diesel fuel injection nor the point before TDC that the spark plug fires is shown because this can vary in relation to the type of fuel system used and other engine design characteristics.

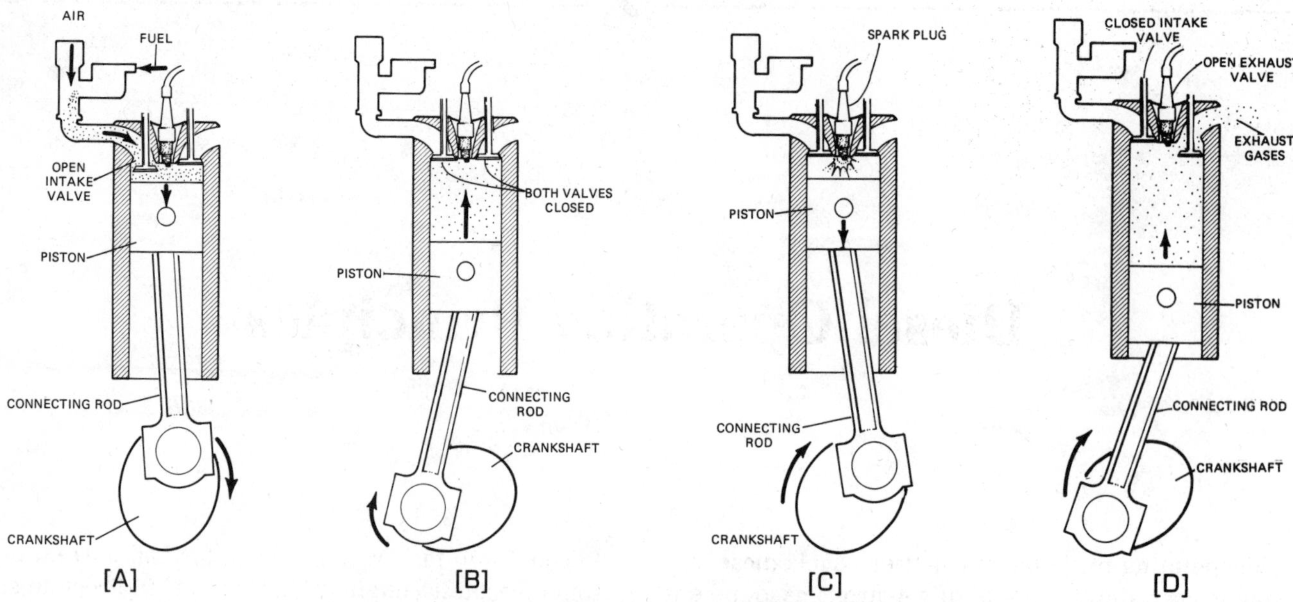

Figure 2-1. Four-stroke cycle gasoline engine. (a) Intake stroke; (b) Compression stroke; (c) Power stroke; (d) Exhaust stroke. (Courtesy of Reston Publishing Company, Inc.)

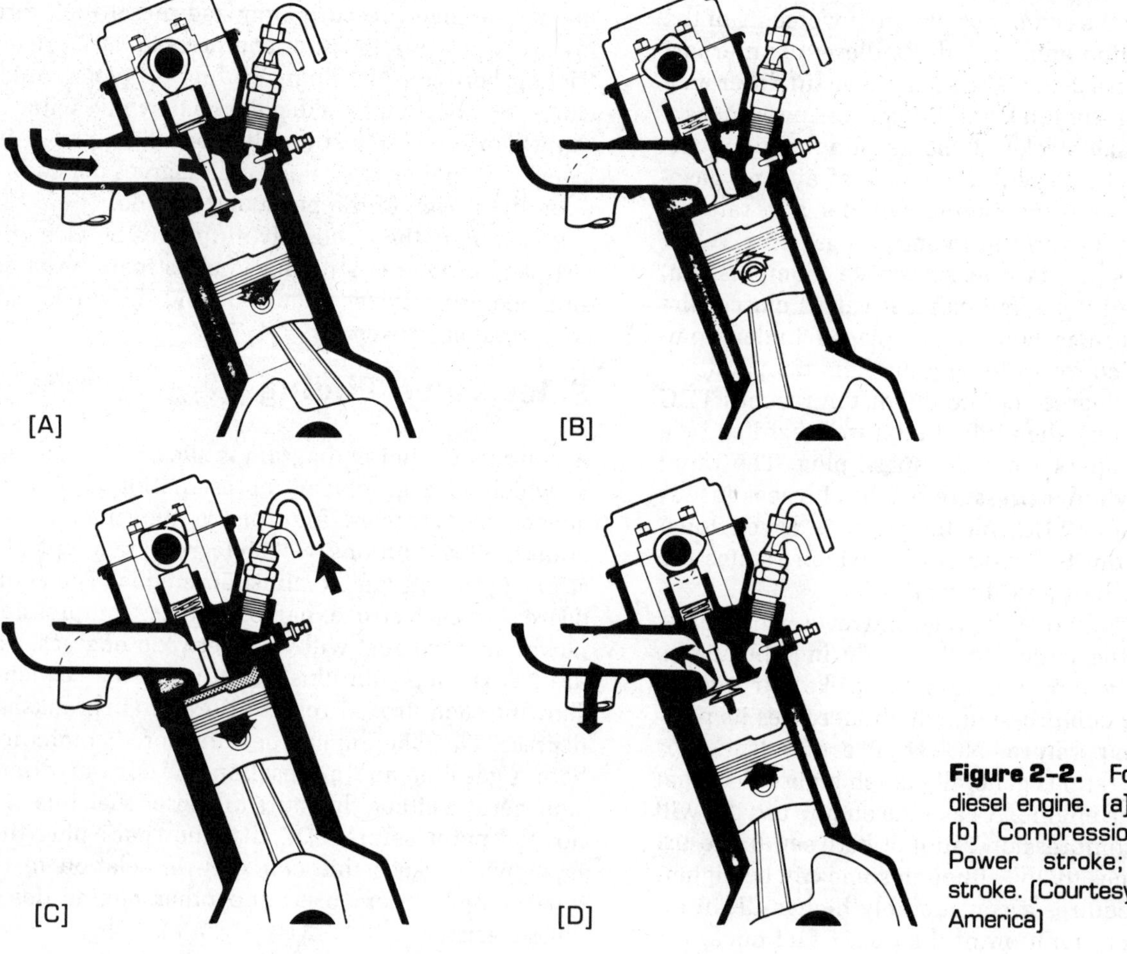

Figure 2-2. Four-stroke cycle diesel engine. (a) Intake stroke; (b) Compression stroke; (c) Power stroke; (d) Exhaust stroke. (Courtesy Volkswagen of America)

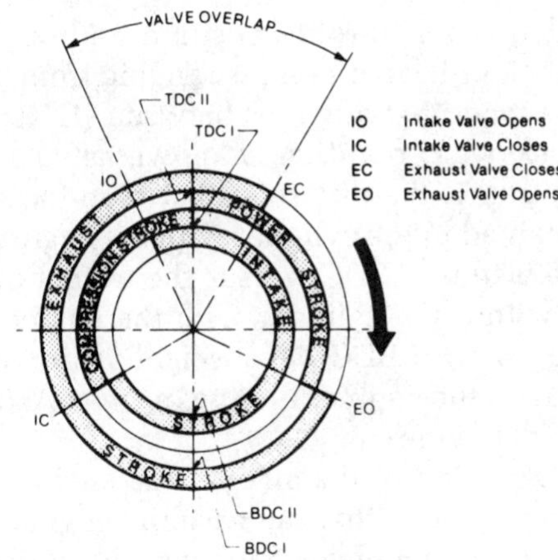

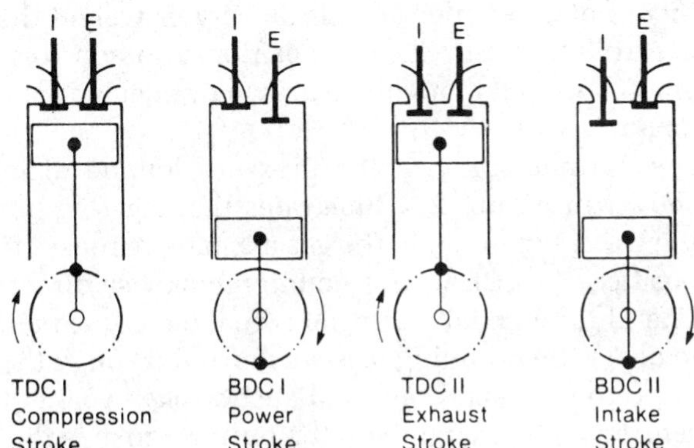

Figure 2-3. Polar valve timing diagram for a 4-stroke cycle diesel engine. (Courtesy of Mercedes-Benz Trucks, USA)

Diesel versus Gasoline Engine

Although the diesel engine has gained increased popularity over its gasoline engine counterpart in the fuel economy race, the diesel still lacks the performance characteristics of the gasoline engine—which will restrict its overall impact and influence as an alternate automotive powerplant.

The following discussion looks at the advantages and disadvantages of both the gasoline and diesel engines in relation to their use in the passenger car. Prior to this comparison, however, it will assist us if we consider the operating characteristics of both in some detail.

Since the main advantage of the diesel engine over its gasoline counterpart is in the area of fuel economy, it would be helpful to study and interpret a cylinder pressure-volume diagram of both types of engines. Such a diagram is illustrated in Figures 2–4 and 2–5.

The key information that can be interpreted from a pressure-volume diagram as far as we are concerned is the energy required to compress the cylinder charge versus the energy returned on the power stroke of both engines since herein lies the basic clue to the difference in the fuel economy of the gasoline versus the diesel engine.

Gasoline Pressure-Volume Curve Interpretation

In Figure 2–4 (gasoline engine), we start our study with the actual compression stroke at position "A," which is the piston position at BDC (bottom dead center). Position "B" represents the piston at TDC (top dead center) having moved up the cylinder while compressing the air/fuel charge. The actual degrees

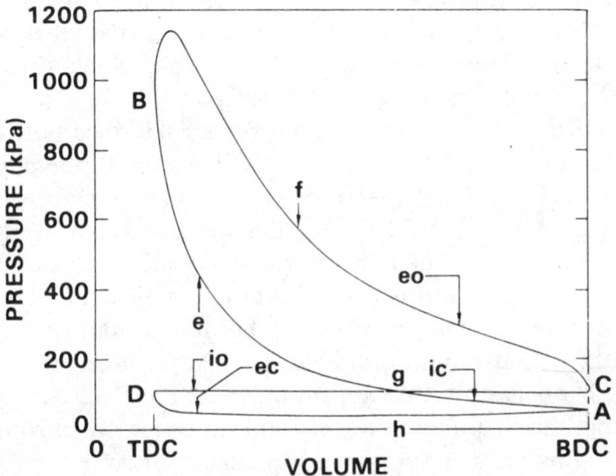

Figure 2-4. Pressure-volume diagram for a throttle controlled 4-stroke cycle spark ignition engine at part load. (Courtesy of Society of Automotive Engineers)

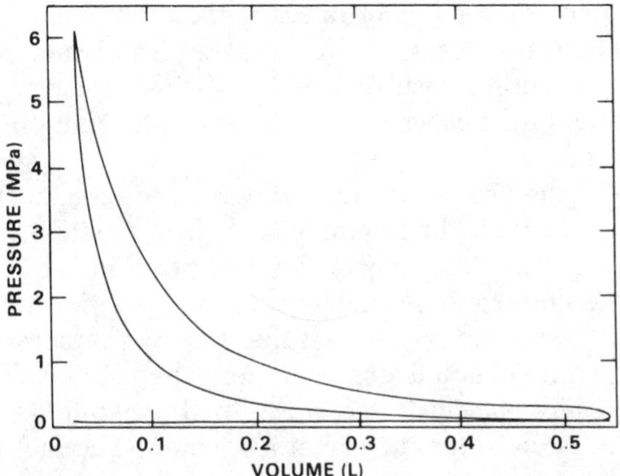

Figure 2-5. Pressure-volume diagram for an indirect injection 4-stroke cycle diesel engine at part load. (Courtesy of Society of Automotive Engineers)

BTDC (before top dead center) at which the spark plug fires will vary between engines and also with the speed of the engine. However, in this diagram the spark plug fires and ignites the air/fuel charge at position "e" to increase the pressure and temperature within the combustion chamber. This ignition, or flame front, is completed at point "f," and the energy of the expanding gases (power stroke) starts at point "B" (peak cylinder pressure), which pushes the piston down the cylinder rotating the crankshaft until it reaches BDC at point "C."

If we now subtract the energy expended (negative work) in pushing or forcing the piston up the cylinder to compress the air/fuel mixture from the energy returned to drive the piston down the cylinder, we are left with what is commonly referred to as the "indicated work of the cycle." This is shown in the pressure-volume diagram as that area under the curve "AeB" (negative work) versus the positive work under the curve at positions "BfC" in a clockwise direction on the diagram "AeBfC."

Keep in mind that the power stroke does not last from TDC all the way to BDC because the exhaust valve(s) will be opened by the camshaft before the piston reaches BDC. On the pressure-volume diagram shown, the exhaust valve actually opens at position "eo" and a loss of pressure will occur since part of the gas pressure will be moving out through the exhaust manifold. From BDC at position "C," the piston is now moving up on its exhaust stroke, expelling or pushing the remaining gases out through the exhaust manifold at a pressure greater than atmospheric from position "C" to position "D." All high speed automotive gasoline engines in use today have a camshaft timed to open the cylinder intake valve BTDC and close the exhaust valve after TDC.

This condition allows the flow of exhaust gases, which are at a higher pressure than the intake (non-turbocharged engine), to cause a partial suction to the incoming fresh air, providing a better scavenging of the burnt gases and more complete filling of the cylinder.

On the pressure-volume diagram, the intake valve is opened by the engine camshaft at position "io," while the exhaust valve closes at position "ec."

The intake stroke that occurs from position "D" to "A" is caused by the fact that the air pressure in a non-turbocharged engine is less than atmospheric pressure (14.7 psi); therefore, intake manifold pressure is considered to be at a vacuum. The fact that atmospheric pressure exists above the throttle control valve is the reason that air is forced into the cylinder through the open intake valve. Generally, the pressure of air in the intake manifold will vary, being dependent upon throttle position and engine load. Most engines have pressures ranging from 85 to 90% of atmospheric—which is between 12.49 psi and 13.23 psi. This condition is known as *volumetric efficiency* and is the difference between the weight of air contained in the cylinder with the engine stopped and the piston at BDC, versus the weight of air contained within the cylinder with the engine running and the piston at BDC. This weight of air will always be less with the engine running on an N.A. (naturally aspirated) engine.

The throttling of the air into the engine is one of the major reasons that the gasoline engine does not provide as good a fuel economy as the diesel engine, which does not restrict the air flow into the engine by use of a throttle. The air density or weight that is controlled by the throttle position also controls the amount of fuel that will be drawn or injected into the engine.

A further study of the pressure-volume diagram shown in Figure 2–4 illustrates the negative pumping work expended in the gas exchange process from positions "CgDhA" in a counterclockwise direction. The difference between the actual energy expended to cover the pumping losses incurred through the intake, compression, and exhaust strokes versus that returned to the engine on the power stroke is the net work done by the expanding gases on the piston during one engine cycle (four strokes—intake, compression, power, and exhaust) and is represented by the points indicated on the pressure-volume diagram as positions "AgBCgDhA."

We now have to take into account the energy or work lost in overcoming the frictional resistance in the piston and rings and the bearings and valve train components in order to arrive at the actual brake horsepower (useable power at the flywheel) delivered by the engine. These pumping losses are greatest at part-load conditions on the gasoline engine therefore affecting its fuel economy.

Diesel Pressure-Volume Curve Interpretation

The diesel engine does not use a throttle to control the speed of the engine as the gasoline engine does. The throttle on the diesel engine is connected to a fuel control mechanism to vary the amount of fuel that is injected into the cylinder. No throttling of air takes place, which means that the diesel operates with a stratified charge of air/fuel in the cylinder under all operating conditions.

The net result of the unthrottled air in the diesel engine is that at idle rpm and light loads the air/fuel ratio in the cylinder is very lean. Because of the excess amount of air (oxygen) contained in the cylinder of a diesel engine during idle and light load situations, the average specific heat of the cylinder gases is lowered, which increases the indicated work obtained from a given amount of fuel. This condition along with that of an unthrottled air flow means that the diesel engine's efficiency increases as load is reduced over its gasoline counterpart, a condition that occurs much of the time in city-type driving modes.

Another condition that favors the diesel engine is that it is not limited by octane requirements such as in the gasoline engine. The gasoline engine's compression ratio is limited by the availability of sufficiently high octane fuel, which at present in North America is about 91 octane unleaded. This limits the gasoline engine's CR (compression ratio) to about a maximum of 9:1 while the diesel, which is not dependent upon a fuel with a high octane rating, can have CR's of 23:1. If you compare the pressure-volume (PV) diagram of the diesel engine with that of the gasoline engine, you will notice that the negative pumping area is much less than that of the gasoline engine (see Figure 2–5).

Another advantage that the diesel engine enjoys over its gasoline counterpart is that diesel fuel contains about 11% more BTU's (British Thermal Units) per unit volume than gasoline; therefore, even if both engines produced the same thermal efficiency (heat efficiency, or how much energy is returned at the flywheel for each dollar of fuel expended), the diesel would still have a better return per dollar spent on fuel.

An example of the brake specific fuel consumption of a gasoline versus a diesel engine is illustrated in Figure 2–6.

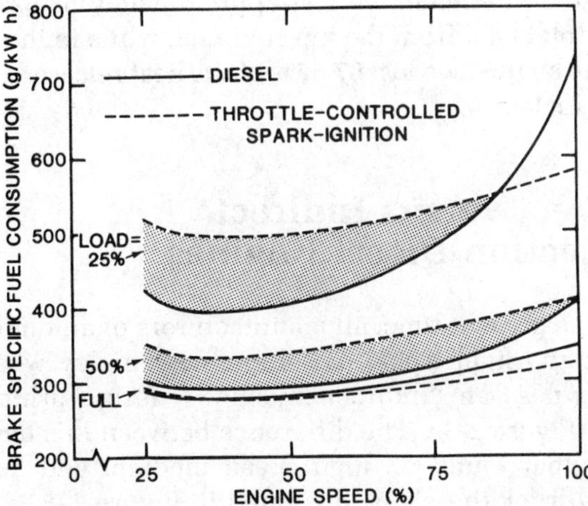

Figure 2–6. Brake specific fuel consumption comparison chart of a 4-stroke cycle diesel versus a 4-stroke cycle spark ignition engine. (Courtesy of Society of Automotive Engineers)

Thermal Efficiency

The heat value of diesel fuel is approximately 11% greater than gasoline per unit volume. Consequently this factor, along with the others mentioned above, will allow the diesel engine to be more heat efficient than its gasoline counterpart.

The term used to describe the heat efficiency of an internal combustion engine is *thermal efficiency*. In its simplest form this means how much energy the engine can extract from a given volume of fuel and turn into useable power at the flywheel.

The heating value of a fuel is measured in terms of BTU's (British Thermal Units) with one BTU equal to 778 foot pounds of work; or looking at it another way, the energy of one BTU would raise a weight of 778 pounds one foot.

It has also been established that 33,000 ft.lb./min. is equivalent to one horsepower; therefore if we divide 33,000 by 778, we find that it is necessary to release 42.416 BTU's in one minute to produce one horsepower.

The accepted practice when computing horsepower is to calculate it over a one-hour period (horsepower/hour); therefore, it is now necessary to multiply 42.416 BTU's times 60, which will therefore represent the total number of BTU's that would be required to produce one horsepower/hour in a perfect internal combustion engine. We would therefore find that $60 \times 42.416 = 2544.96$ BTU's; or for general practice, we accept 2545 BTU's as being the amount of heat required to produce one horsepower per hour.

Since we know that there is no perfect engine and that we lose a considerable portion of this heat to the cooling system, the exhaust system, friction, and radiation, then we can reason that all engines have to be supplied with more heat energy than 2545 BTU's per horsepower/hour in order to account for the heat losses mentioned above. The ratio of the number of BTU's that is actually used to develop one horsepower/hour to the number of BTU's that must be used in the engine to produce this one horsepower/hour is known as the thermal efficiency.

In other words, brake thermal efficiency is the ratio of the heat equivalent of one horsepower/hour to the heat units actually supplied per brake horsepower/hour. An example illustrating this condition is as follows: Determine the thermal efficiency (TE) of a 250-horsepower engine that consumes 0.350 pound of fuel/brake horsepower/hour. This fuel consumption figure is always shown on the engine/vehicle manufacturer's sales literature.

Alternately, the amount of fuel consumed in an

hour by an engine at a steady load and speed can be measured and the thermal efficiency (TE) determined by calculation.

To determine the TE, we must also know the engine horsepower and the heat value of the fuel being used. Typical heat values of the fuel can be found in Chapter 4. Since TE is calculated on an hourly basis, it is necessary to determine just how much fuel is consumed in one hour by the engine. In this example, the engine consumes 0.350 lb./hp/hr.; therefore, it would burn 0.350 × 250 = 87.5 lb./hr.

The volume of fuel consumed in gallons/hour can be determined by dividing the 87.5 pounds by the specific gravity of the fuel. These specific gravities average about 7.1 pounds/US gallon for diesel fuel with a heat value of approximately 124,800 BTU's/US gallon. Gasoline weighs about 5.8 pounds/US gallon with an average heat content of approximately 113,800 BTU's/gallon.

This sample engine would therefore burn about 12.32 US gallons/hour if running on diesel fuel and about 15.08 US gallons/hour if running on gasoline.

NOTE: The gasoline engine would burn at least 10–15% more fuel than the .350 lb./bhp/hr. figure stated in this example. Therefore, it is very important that you remember this when dealing with this oversimplified example. However for simplification, we will assume that both engines do burn the same 0.350 lb./bhp/hr.

To determine its TE, we can either multiply the pounds of fuel used by the heat value/pound or multiply the total gallons used by the BTU's/gallon. An example is as follows:

Diesel Engine . . . 12.32 gallons × 124,800 BTU's = 1,537,536 BTU's

Gasoline Engine . 15.08 gallons × 113,800 BTU's = 1,716,104 BTU's

In the diesel engine, it actually required 1,537,536 BTU's to produce 250 brake horsepower in one hour, while in the gasoline engine it required more heat (fuel), namely 1,716,104 BTU's to produce the same horsepower in one hour.

Each horsepower developed in the diesel engine therefore required 6150 BTU's (1,537,536 divided by 250 horsepower), while the gasoline engine required 6864.4 BTU's or 714.4 BTU's more than was required in the diesel engine to produce the 250 horsepower over a one-hour period.

We already know that a perfect engine should use 2545 BTU's/hp/hr., so to determine the engine's thermal efficiencies, we simply divide 2545 by the actual amount that the engine did use. For the diesel engine, this would be 2545 divided by 6150 times 100% equals 41.38%, while in the gasoline engine, this TE would be 2545 divided by 6864.4 times 100% equals 37%.

In this example, the diesel engine was 4.38% more thermally efficient than its gasoline counterpart; however, this difference in actual operation between vehicles in actual road use can vary anywhere between 10% to 50% in favor of the diesel engine in a city/urban driving cycle due to the very lean mixtures that the diesel engine enjoys over its gasoline counterpart at low engine speeds. Additional benefits for the diesel are discussed throughout Chapters 2 and 3.

Remember that we indicated earlier in this explanation that the gasoline engine would use about 10–15% more fuel/bhp/hr. than its diesel counterpart. If we assume, therefore, that the gasoline engine consumed 0.400 lb./bhp/hr., then it would use 250 hp × 0.4 lb./bhp/hr. = 100 pounds of fuel in one hour.

100 pounds divided by 5.8 pounds/US gallon equals 17.24 gallons/hr. fuel consumption. 17.24 × 113,800 BTU's (heat per gallon) = 1,961,912 BTU's of fuel used to produce the 250 horsepower/hour. Dividing 1,961,912 BTU's by 250 horsepower equals 7847.64 BTU's to produce one bhp/hr. The TE is therefore 2545 divided by 7847.64 times 100% equals 32.43%, for a difference in TE between the diesel and the gasoline engine of 8.95 or roughly 9% with the engine running at a steady load and speed.

This TE indicates the amount of heat lost in both engines as a consequence of friction, radiation, jacket water coolant, and the exhaust system. Therefore in the diesel engine, we lost approximately 58.62% of the total heat from the injected fuel, while in the gasoline engine, we lost 67.57% of the heat released from the fuel.

Direct versus Indirect Injection Diesel Engines

At the present time, all manufacturers of automotive (car and light truck) diesel engines employ what is known as an "indirect injection" design principle (see Figure 2–7). The difference between this design and that found on high speed medium and heavy duty truck diesels is illustrated in Figure 2–8.

You will note from Figure 2–8 that on the direct injection diesel engine design the fuel is injected directly into the open combustion chamber, while in the indirect injection design the fuel is injected into

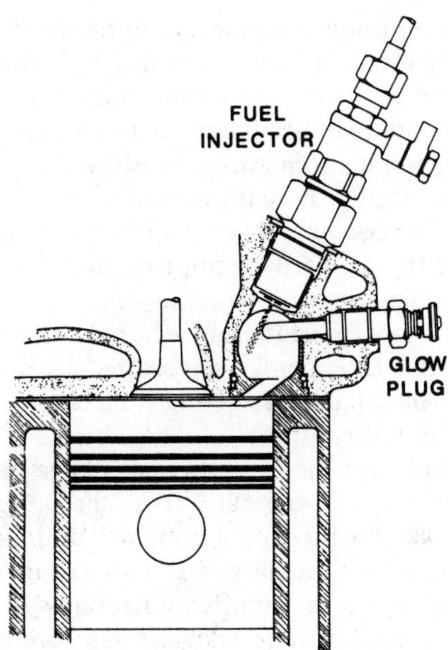

Figure 2-7. Cross-sectional view through a 4-stroke cycle indirect injection (pre-combustion chamber) type diesel engine. (Courtesy of Society of Automotive Engineers)

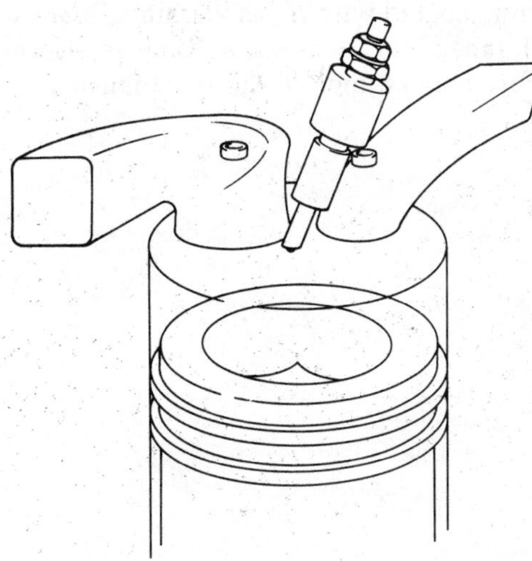

Figure 2-8. Basic design of a 4-stroke cycle direct injection type diesel engine. (Courtesy of Society of Automotive Engineers)

a small antechamber or pre-combustion chamber above the main cylinder.

There are several reasons why the indirect-injection (IDI) diesel engine has been favored to this time over the direct-injection design in passenger cars and light duty pickup trucks. The main reasons are that the IDI unit is generally quieter, it operates well over a broad speed range, and it is less sensitive to the grade of fuel used and can be used with a less expensive fuel system. The IDI engine has also been easier to tune to the stringent U.S. EPA emissions standards.

In addition, the IDI engine through its pre-combustion chamber design allows intense air turbulence to be developed, which assists greatly in mixing the injected diesel fuel with the compressed air. Cylinder pressures with the IDI engine can usually be maintained around 85 bar (1234 psi) for a typical naturally aspirated engine. This feature allows the same basic engine block to be used while the cylinder head layout can also be reasonably simple. Since the combustion chamber design eliminates the need for a special chamber in the piston crown, the piston height and cylinder block can be kept on a par with that used in the gasoline engine. Companies such as VW and General Motors have both adapted existing gasoline engine designs to IDI diesel engines, namely introduction to the VW Rabbit and Oldsmobile V8–350 cu.in. engine where the adaptation of identical tooling and sharing of some components saved millions of dollars in total redesign.

The injection system used with the IDI diesel engine is considered to be "low pressure" (around 300 bar or 4356 psi) with the use of a variable orifice nozzle (injector unit).

The direct-injection (DI) diesel engine for passenger cars has the following problem areas as opposed to that of the IDI engine:

1. Less high speed potential in terms of smoke-limited BMEP (brake mean effective pressure)
2. Higher exhaust emissions
3. Higher combustion noise

The direct-injection engine does, however, offer one major advantage over its IDI counterpart and that is in the area of fuel consumption. A similar sized DI engine can show a fuel economy improvement of 10 to 15% over the IDI engine.

The reason for this fuel economy difference is the high surface to volume ratio of the IDI engine (greater overall combustion chamber area) exposed to possible heat losses (radiation). In addition, there is a percentage of throttling loss in the IDI engine caused by the flow of gases into and out of the antechamber or pre-combustion chamber. Also, because the ignition and release of the pressure gases from the antechamber is slower than that developed within the DI engine, the IDI engine suffers some loss of efficiency but lowers the combustion chamber noise and NOx (nitric oxide) emissions.

In addition, the IDI diesel requires a higher compression ratio than its DI counterpart because of greater heat loss (radiation) from the larger exposed surface area of the combustion chamber area. The IDI diesel also requires the use of a glow plug in the pre-combustion chamber to aid starting in cold weather. As well, the increased compression ratio of the IDI engine (22 to 23:1) versus the DI (16 to 18:1) causes an increase in friction because of the higher effort required to compress the air on the compression stroke.

At the present time, several manufacturers are conducting extensive research and development to perfect the DI diesel engine. As mentioned in the introductory comments in this book, British Leyland (Austin-Rover) in concert with Massey-Ferguson Holdings, Ltd. (Perkins), are developing a direct-injection diesel engine for passenger cars. Ricardo Consulting Engineers Ltd. (England), M.A.N, BMW, and VW in Germany are others who are actively working on this design.

Any increase in engine displacement of the DI unit to offset power losses at a given speed range would carry a severe fuel economy penalty.

M.A.N. in Germany has shown, through extensive testing, that in addition to the superior fuel economy of the direct injection engine, their patented "M" combustion design system (combustion chamber contained within the head of the piston) also enables low emission and noise levels to be obtained. This system was tested using an existing Robert Bosch VE distributor pump similar to that used on the VW Rabbit. This pump is produced in large quantities for diesel engine manufacturers worldwide.

Turbocharging, as with an IDI engine, would allow an increase in power requirements if desired along with a similar improvement in vehicle performance.

The DI diesel must therefore be capable of not only meeting the EPA smoke emissions regulations but must also be capable of starting from cold in the same way as the IDI engine.

EPA limits for diesel particulates have been set at 0.2 gram per mile for 1985 from the 0.6 gram per mile in 1982. Further reductions will undoubtedly be sought in the following years. To meet these standards, engine design modifications, fuel system modifications, and exhaust devices are some of the concepts presently under study. Exhaust particulate traps not unlike the catalytic converter concept used for gasoline engines is a possible addition to the diesel engine in order to meet these proposed EPA particulate standards. The diesel trap may take the form of alumina coated wire mesh, ceramic fibers, and ceramic honeycomb. A ceramic foam is also another possibility for automotive diesel exhausts.

3

Combustion Chamber Designs

The design of the actual combustion chamber in a gasoline and a diesel engine differ in that the gasoline engine operates on the Otto cycle while the diesel engine operates on the compression ignition (CI) cycle.

Simply put, this means that the gasoline engine requires a spark plug to ignite the air/fuel mixture, while the diesel engine relies on the heat of the compressed air within the cylinder or combustion chamber to initiate burning of the injected diesel fuel.

Several terms have been used to describe the combustion concept used in both types of engines; however, the gasoline engine is said to operate on a "constant volume" cycle, while the diesel engine is considered to operate on a "constant pressure" cycle. It would be helpful to study just what actually transpires within the engine cylinder of each engine before considering the various types of combustion chamber designs for a thorough understanding of these differences in operation.

Constant Volume (Gasoline) Cycle

In a gasoline engine, when the piston moves up the cylinder on its compression stroke, the spark plug is timed to fire the air/fuel mixture at a point just before top dead center (BTDC). When the spark plug fires, we have almost instantaneous combustion because the high tension electrical spark causes flame propagation to begin and progress through the combustion chamber. The speed at which this action takes place is so sudden that a flame front travelling at anywhere from 20 to 150 feet/second (13.6–102.3 mph or 22–165 km/hr) expands through the combustion chamber. The upward moving piston has not moved any appreciable distance from the point where the spark plug fired until the point where ignition occurred. The net result is that the volume above the piston has, for all intents and purposes, remained constant. Generally, the time taken for the air/fuel mixture compressed pressure once ignited to increase to its peak cylinder pressure (firing) is timed to occur just as the piston arrives at the top dead center (TDC) position where a constant volume exists, so that the greatest work effort can be transferred to the piston crown for the power stroke.

In summation, therefore, in the Otto cycle, combustion is completed during the instant that the piston is at TDC (pressure increases, volume remains constant).

In a gasoline engine, the theoretical ratio for complete fuel combustion is known as the "stoichiometric ratio," which simply means that in order to promote complete combustion of fuel in an engine cylinder we require approximately a ratio of 14 parts air to 1 part fuel. This ratio of 14:1 is actually 14.7:1, which requires that more air than fuel is needed.

For example we can consider it in two ways: first, that 14.7 lb. of air are required to burn 1 lb. of fuel. Second, since the word stoichiometric is explained in the dictionary as "the process or art of determining or calculating the equivalent and atomic weights of the elements participating in any chemical reaction," if we were to consider the atomic weights of the elements contained in gasoline and air, we would arrive at a figure in terms of volume that would require about 9800 cubic feet of air for every cubic foot of gasoline in order to attain complete combustion.

Some people bump this figure of 14.7:1 up to a round figure of 15:1 as a rule of thumb, which would require approximately 11,500 liters of air for every 1 liter of fuel to maintain a stoichiometric situation.

In both a gasoline and diesel engine, the stoichiometric air/fuel ratio for both is the same for complete combustion of 1 lb. of fuel.

Gasoline engines attain their maximum output with an air deficiency of 0–10%, i.e., Lambda = 1–0.9 (rich mixture) and maximum thermal efficiency and fuel economy with 10% excess air, i.e., Lambda = 1.1 (lean mixture).

On the other hand, diesel engines tend to obtain maximum horsepower with 10–15% excess air, i.e., Lambda = 1.1–1.15 so as to remain within the smoke limit; however, at an idle rpm, the diesel engine operates with an excess air value of between 600 and 1000%.

In a gasoline engine, if there is an excess of fuel, then the fuel will not be completely burned due to air starvation and the exhaust gases will contain unburnt hydrocarbons and high carbon monoxide content. On the other hand, if there is less fuel than necessary for complete combustion, the engine will run "lean." The engine power will decrease and because of the slower combustion, the engine exhaust gas temperatures will be higher, resulting in high concentrations of NOx (nitric oxides), which are the particles in the exhaust stream that actually produce a yellowish smog in areas like Los Angeles.

Engine power, specific fuel consumption, and exhaust gas composition are all affected by an excess air factor. There is no single ideal excess air factor at which these conditions will be at an optimum value. Engine manufacturers have found that an excess air factor anywhere between 0.9 and 1.1 of Lambda have proven to be the best compromise.

$$\text{NOTE: Lambda} = \frac{\text{amount of air supplied}}{\text{theoretical air requirement}}$$

When Lambda = 1, then the amount of air supplied to the engine corresponds to the theoretical amount of air required or the "stoichiometric" air/fuel ratio of 14 to 14.5:1.

Power produced by an engine is related to its Volumetric Efficiency (see turbocharger chapter) or the amount of air that it can take in on the intake stroke.

The density (weight) of the air charge in the cylinder affects the horsepower of the engine. The heavier (denser or colder) the air charge, the lower will be the combustion efficiency as a result of reduced evaporation and lower turbulence. In a diesel engine, a reduction in air temperature from 80°F above zero to one of 20°F below zero will generally result in a reduction in air temperature at the end of compression of between 230–300°F (110–149°C), depending upon the engine compression ratio, piston and combustion chamber design, and speed of rotation, which all affect air turbulence.

Humid air, on the other hand, contains less oxygen than dry air and will therefore reduce engine power; this effect is quite noticeable when operating engines in the tropics.

Gasoline Combustion

Under normal conditions of combustion within the cylinder, the air/fuel charge is ignited by a high tension spark that is timed to occur at a predetermined number of crankshaft degrees BTDC.

If the ignition timing is early (advanced) or fuel with too low an "octane" rating/value is used in the engine, pre-ignition or detonation will result. This condition is sometimes referred to as combustion/spark knock or ping by a mechanic/technician.

This condition is caused by a portion of the air/fuel mixture igniting before the normal flame front reaches it. The result is that an uncontrolled burning takes place (explosion), and pressure is created within the combustion chamber that results in a pronounced knock or pinging sound. This noise is a result of pressure sound waves emanating through the combustion chamber against the piston and cylinder components, which tend to resonate. Extremely high pressures are exerted upon the piston crown and engine bearings, which can lead to internal mechanical problems. Severe combustion knock can lead to melted piston crowns and internal engine damage.

Under normal combustion chamber situations, the pressure rise of the air/fuel mixture is as shown in Figure 3–1; the condition is known as controlled burning. Combustion chamber design, compression ratio, octane value of the fuel, and combustion chamber deposits (carbon) can affect the air/fuel burning rate.

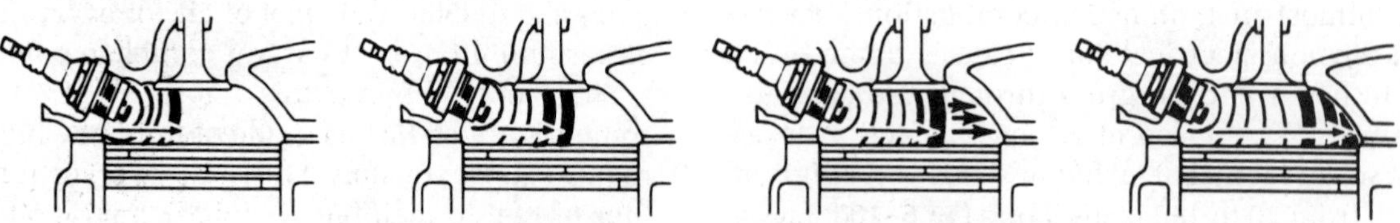

Figure 3–1. Flame front during normal combustion is a fast moving wall of flame that propagates across the combustion chamber. (Courtesy of Reston Publishing Co., Inc.)

When the spark plug fires, the air/fuel mixture is ignited and a wall of flame will spread across the combustion chamber from the area closest to the spark plug. The speed at which the flame front travels is commonly called the "rate of flame propagation."

A steady, controlled burn therefore occurs under normal conditions within the combustion chamber. If, however, ignition of the air/fuel charge should occur along with a secondary explosion or uncontrolled burning prior to or following the spark plug firing, then burning has already started when the spark plug does ignite the mixture. The result is that an uncontrolled burning or dual flame front condition exists. The air/fuel mixture therefore tends to explode rather than burning rapidly in a controlled state; the end result is that the explosion causes high pressure sound waves to emanate throughout the combustion chamber. The design of the combustion chamber affects the actual intensity of the sound waves with the result that a pinging noise radiates from the engine (see Figure 3–2).

Another similar condition that can cause this noise is when carbon formation creates hot spots within the combustion chamber and the air/fuel mixture ignites before the spark plug fires (pre-ignition). This causes extremely high pressures to oppose the upward moving piston with the result that high bearing loads are encountered leading to a bottom end rumble from the engine. Other serious problems are holes actually burned through the piston crown. Low octane fuel can also cause pre-ignition to occur.

Pressure within the combustion chamber will vary depending upon compression ratio; however, most engines in use today have CR's ranging from a low of about 8:1 up to 9.5:1 in the average production vehicle. This results in compression pressures in the range of 100–190 psi (689.5–1310 kPa) and temperatures of between 550–1100°F (288–593°C).

Considering the various design changes between engine combustion chambers, valve arrangement, piston shape, etc., average compression pressures can be said to vary between 116–217 psi (8–15 bar) with temperatures running between 752–1112°F (400–600°C).

When the air/fuel charge is ignited, however, both a pressure and temperature rise will occur in the cylinder with a resultant pressure increase usually between 3 to 4 times greater than the compression pressure. Firing temperatures can increase rapidly to a peak of several thousand degrees F with some high compression engines recording pressures up to 800–900 psi (5516–6205 kPa) and peak temperatures (extremely short duration) as high as 3500–4500°F (1927–2482°C).

The quality of combustion in a gasoline engine is dependent on how the flame front advances once the spark plug fires in terms of both space and time. The flame front is basically the ignition of the air/fuel mixture closest to the spark plug tip. The speed at which the burning fuel (flame front) will advance is affected by the heat entrapment, heat conduction, and radiation in the combustion chamber but specifically by how well and how fast the gas flow and turbulence occurs within the combustion chamber. This action is affected by compression ratio, piston and combustion chamber design, engine speed, load, and whether an excess or air deficiency exists.

Once the air/fuel mixture is ignited by the spark plug, the flame front starts at a slow rate and will reach its maximum once combustion chamber pressure peaks usually between 30–40 bar (435–580 psi) and temperatures of 3632–4532°F (2000–2500°C). Towards the end of the combustion phase, the flame front will slow down due to a lack of available oxygen and the flame coming into contact with the cooler cylinder walls. The actual speed of combustion (flame front propagation) is normally between 10–25 m/sec. (33–82 ft./sec. = 22.5–56 mph) or 36–90 km/hr.

Engine knocking is caused by a condition whereby the unburnt air/fuel mixture self-ignites towards the end of combustion thereby burning in an uncontrolled or violent state. This causes the flame front to advance much faster than in a controlled burning state (initiated by spark plug firing). Speed of the flame front during uncontrolled burning can vary between 250–300 m/sec. (820–983 ft./sec.) or between 559–670 mph (899–1078 km/hr).

In contrast to knocking in a diesel engine, the

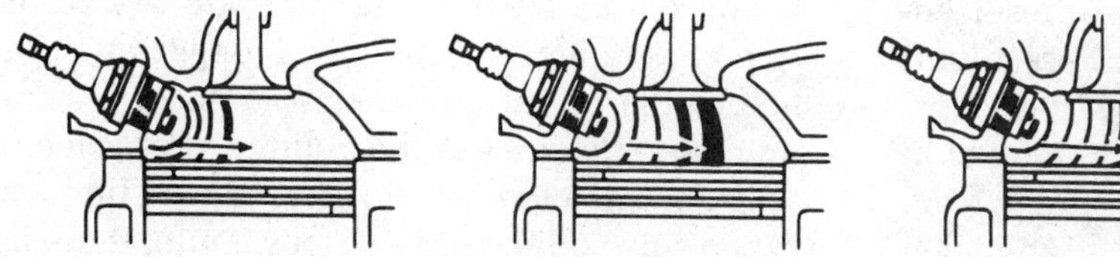

Figure 3–2. Detonation produces an explosion as the last portion of the compressed air/fuel charge (the gas) ignites. (Courtesy of Reston Publishing Co., Inc.)

knock in the gasoline engine occurs at the end of combustion rather than at the start of combustion as in the diesel cycle. Knocking during the combustion phase in a gasoline engine can be influenced by various factors such as too low a grade of fuel or by improperly adjusted ignition timing (advanced). However, the single factor that most affects the "knocking" in an engine is the compression ratio. A reduction by 1 in the compression ratio will lower the fuel octane requirement by as much as 10–15 octane numbers. In addition, in a non-turbocharged engine, an increase in altitude will decrease the air pressure taken into the combustion chamber. For every 1000 meters (3278 feet) of altitude, the fuel octane requirement will decrease by approximately 8 numbers. For every 100 meters (324 feet) increase in altitude the general rule of thumb is that horsepower will decrease by about 1%.

Constant Pressure (Diesel) Cycle

In the diesel engine, air only is drawn into the cylinder and subsequently compressed during the upward moving piston compression stroke. The diesel engine always operates with an excess air over fuel ratio due to the unthrottled entry of air. A diesel engine mechanically or electronically regulates the fuel flow and is therefore a leaner burning engine than its gasoline counterpart. At an idle rpm, the diesel engine tends to operate at an extremely lean air/fuel ratio with the excess air running between 600–1000%; at the high speed end of the operating range, the diesel still has an excess air/fuel ratio of about 10–15% over its gasoline counterpart when producing its maximum horsepower. This excess air = Lambda + 1.1–1.15 ratio so as to remain within the acceptable smoke limits. The point at which fuel is injected directly into the compressed air will vary between engines and also with the load and speed on the engine similar to that of the spark plug firing in a gasoline engine through the advance mechanism.

Compression ratios in the automotive diesel engine average between 20 and 23:1 with resultant compression pressures from as low as 275–490 psi (1896–3378 kPa). Temperatures in the combustion chamber before fuel injection will usually range from 800–1200°F (427–649°C). Heavy-duty high-speed diesel engines used in highway trucks, etc., can obtain pressures between 435–800 psi (30–55 bar) and compression temperatures before fuel is injected of between 700–900°C (1292–1652°F).

Compression pressures and temperatures after the fuel is injected will range from 1000–1200 psi (6895–8274 kPa), providing peak temperatures of around 4000°F (2204°C). Higher pressures and temperatures can occur in some engines; however, these figures are a mean average.

Fuel is injected into the diesel engine from a nozzle of an injector in a finely atomized state. This atomization of the fuel occurs as a direct result of the high pressure created within the injector assembly (average of 1800–2300 psi) and the small holes through which the fuel is forced (sprayed) into the combustion chamber resulting in fuel penetration pressures (spray-in pressures) of between 14,000–22,000 psi, depending upon whether the engine is of the direct-injection (DI) or the indirect-injection (IDI) design.

In addition, the size and number of the fuel spray holes in the injector spray tip can also vary from as small as 0.005″ in multiple hole nozzles to a single hole of 0.020″ to 0.050″ on a pintle-type nozzle, which is the type used on current IDI type passenger car diesels.

Since all passenger car diesels in use at this time employ the IDI design, these injector spray-in pressures will be on the lower side of the figures given. The speed of penetration of the fuel leaving the injector tip or nozzle can reach velocities of as high as 780 mph, which is a little faster than the speed of sound.

This fuel (atomized) is basically in a liquid state; therefore, in order for ignition to take place, the fuel must "vaporize" (known as distillation temperature—see fuels section). This means that the fuel must penetrate the air mass (high pressure air/high temperature) in order to allow the fuel molecules to mix with the oxygen molecules within the combustion chamber. Unlike a gasoline engine where the air/fuel mix has already taken place during the intake and compression strokes, the diesel fuel must achieve this after injection. In order for the fuel to actually reach a state of ignition, there is a time delay from the point of injection to the point of ignition. This time delay is approximately 0.001 second and therefore results in a slower igniting fuel. The longer this time delay before the initial fuel that was injected takes to ignite, then the greater the volume of injected fuel that will be collected within the combustion chamber. When this volume of fuel does ignite, there is a pressure increase within the combustion chamber.

A time delay of longer than approximately 0.002

second would be an excessively long ignition delay period and would therefore result in a rough running engine (knocking). This knocking occurs at the start of combustion in a diesel engine versus at the end of combustion in a gasoline engine.

Ignition lag will increase if the injection timing is either very late or very early because the fuel will be injected into an air mass that has lost a lot of its compression heat (late timing) or not yet attained it through early injection timing. Since the injector will continue to inject fuel into this already burning mass, the pressure will rise to a peak pressure as the piston attains the TDC position. As the piston starts down into its power stroke, this additional injected fuel maintains a steady pressure as it starts to burn, thereby providing the diesel engine with the term "constant pressure cycle." In some engines, the fuel will be cut off just BTDC, others may cut off fuel at TDC while still others may not cut fuel off until after TDC.

Because of the fact that diesel fuel continues to be injected into the already burning fuel of the combustion chamber as the piston moves down the cylinder on its power stroke, the cylinder pressure is said to remain constant during the combustion process.

With the gasoline engine the instantaneous ignition concept produces a very rapid rise in cylinder pressure with a very fast burn rate resulting in a hammer-like blow on the piston crown. In the diesel cycle, the pressure rise is sustained for a longer time period resulting in a more gradual and longer push on the piston crown than that in the gasoline engine. Diesel's original concept was that his engine would continue to have fuel injected during the power stroke and that no heat losses would occur in his uncooled engine. This concept was known as an "adiabatic diesel engine," which in the true sense of the word meant that there would be no loss of heat to the cylinder walls while the piston moved up on its compression stroke. In addition, no cooling system would be used, resulting in the transfer of waste heat to the exhaust for a gain in thermal efficiency. Since no cooling system would be required, no frictional losses would occur through having to use a gear-driven water pump, etc. We know this was impossible to achieve; however, his original idea of producing a true constant pressure cycle, although never achieved, did attain some measure of success in the engines that now bear his name.

There is no internal combustion engine today that operates on either a constant pressure or constant volume cycle since they all require a few degrees of crankshaft rotation in order to complete combustion with a consequent rise in cylinder pressure.

Comparison of Diesel versus Gasoline Engines

The following chart will allow you to compare the major differences between standard production gas and diesel engine vehicles' combustion phases in automotive applications. Higher compression ratios in high performance or racing engines will increase both the temperatures and pressures at the end of the compression stroke and once the spark plug fires.

Stroke/Condition	*Gasoline*	*Diesel*
1. Intake stroke	Air/fuel	Air only
2. Compression ratios	8–9.5:1 Ratio	20–23.5:1
3. Compression pressures	115–200 psi (793–1379 kPa)	435–600 psi (2999–4137 kPa)
4. Compression temperatures	850–1100°F (454–593°C)	1400–1650°F (760–899°C)
5. Power stroke pressures	725–815 psi (5000–5619 kPa)	1100–1370 psi (7584–9446 kPa)
6. Power stroke peak average temperatures	2912–3767°F (1600–2075°C)	3000–4000°F (1649–2204°C)
7. Exhaust temperatures average full load	1300–1800°F (704–982°C)	800–1100°F (426–593°C)
8. Exhaust temperatures average at idle rpm	392–482°F (200–250°C)	212–392°F (100–200°C)
9. Exhaust temperatures at part load	1022–1202°F (550–650°C)	482–1022°F (250–550°C)
10. Mean effective pressure (typical averages)	116–145 psi (800–1000 kPa)	109–174 psi (751–1199 kPa)
11. Thermal Efficiency*	23–28%	32–42%

*Thermal Efficiency is the amount of work effort returned at the flywheel for every dollar in fuel burnt in the engine; more specifically it is the ratio of the work done by the engine to the mechanical equivalent of the heat supplied in igniting the fuel.

Advantages of Diesel Engine

1. Better fuel economy (Thermal Efficiency)
2. Reduced risk of fire
3. Higher torque at a lower engine speed
4. Longer life between overhauls (especially on heavy duty equipment)
5. Lower carbon monoxide content
6. Lower carbon dioxide content
7. Higher exhaust oxygen content
8. No lead content in exhaust

Disadvantages of Diesel Engines

1. Noisier
2. Slower acceleration characteristics
3. More exhaust smoke
4. Lower power to weight ratio (heavier)
5. Unfavorable fuel odor
6. More expensive to produce
7. Generally more expensive to repair (overhaul)

Gasoline Combustion Chamber Designs

At the present time, gasoline fuel injected automotive engines employ IDI (indirect injection) because the fuel is injected outside of the combustion chamber. The injected fuel can take place in the throat of a carburetor as shown in Figure 3–3. This is called throttle body injection (TBI), shown in Figure 3–4. TBI is currently employed by General Motors Corporation, Ford Motor Co., and Chrysler Corporation on many of their vehicles.

The other method of gasoline fuel injection employs individual injectors (one for each engine cylinder), such as shown in Figure 3–5. In this system, commonly known as manifold injection, the injector is located several inches from the inlet valve port area so as to protect it from the heat of combustion (see Figure 3–6). Manifold, or port injection as it is sometimes referred to, has been perfected by Robert Bosch and is currently available on many of the world's major makes of automobiles. (See Chapter 6, Gasoline Fuel Injection, dealing with Robert Bosch systems.)

Examples of combustion chamber designs using gasoline fuel injection are illustrated in the following section.

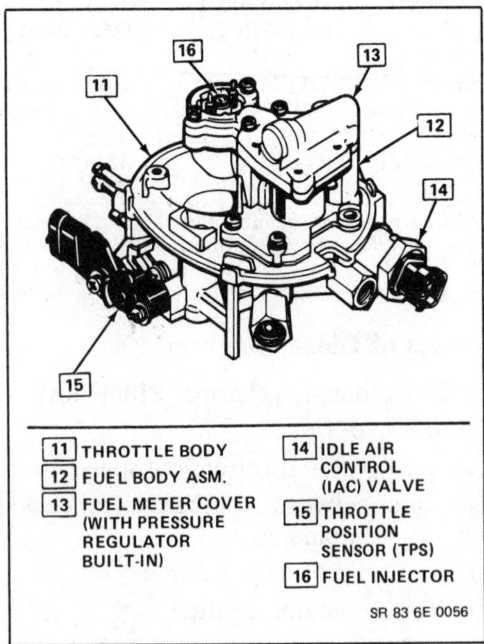

Figure 3–3. Typical throttle body injection (TBI) unit. (Courtesy of GMC Service Research)

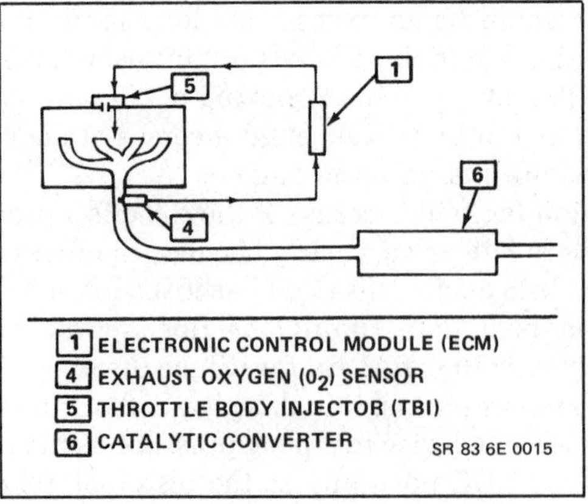

Figure 3–4. Closed-loop control system (TBI). (Courtesy of GMC Service Research)

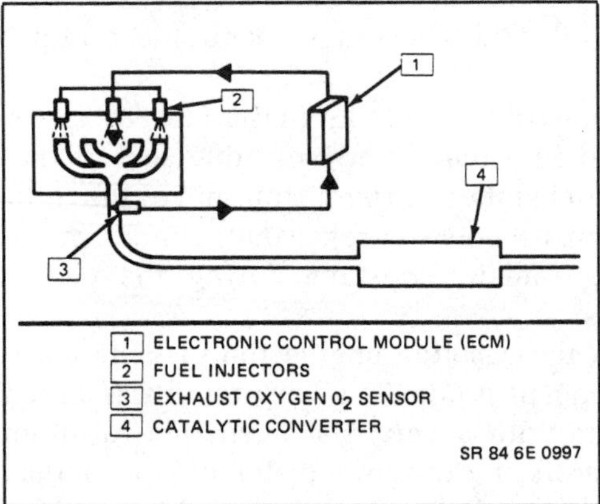

Figure 3–5. Closed-loop system, multi-port fuel injection. (Courtesy of GMC Service Research)

Current High Performance Gasoline Engine Designs

For years, the North American type mass-produced passenger car employed large displacement V8's that were capable of delivering an average horsepower of between 250 and 375. However, with the first oil embargo from the Arab states, manufacturers were forced to redesign their existing engines so that they

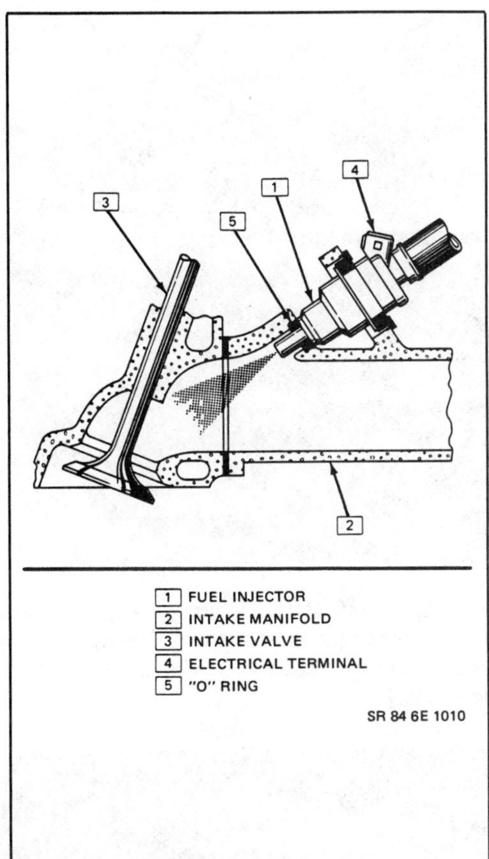

Figure 3-6. Location of fuel injector on a port-type fuel injection system. (Courtesy of GMC Service Research)

would provide much better fuel economy as well as meeting the EPA exhaust emission regulations.

Although large displacement V8's are still available in the larger North American cars and trucks, most of the mass-produced vehicles are now offered with in-line four or V6 engines.

The European and Japanese producers on the other hand have always manufactured either 4- or 6-cylinder powerplants due to their better fuel economy, weight, and financial penalties imposed by the government on large displacement, gas-guzzling engines.

The trend is now fairly well established towards producing engines of small displacement that will provide both good performance as well as good fuel economy. The consumer, however, has been looking increasingly towards a powerplant that offers better performance and yet will still return good fuel economy.

To this end, European and Japanese manufacturers now offer small displacement, high horsepower engines in quite a few of their models. Undoubtedly this trend will continue, and the North American manufacturers will not be far behind.

In order to obtain this high performance from a small displacement engine, the manufacturers have chosen to employ a four-valve head concept whereby each cylinder is equipped with two exhaust as well as two intake valves for better breathing and expulsion of exhaust gases. Four valve heads are not a new idea because they were employed as long ago as 1912 by Peugeot in a racing car and have also been widely used in diesel engines for many years.

Many of these new style engines also are available with turbocharging and aftercooling in addition to the use of fuel injection. Some examples of the basic engine design and combustion chamber shapes used with these high performance small displacement gasoline engines are shown in this chapter.

Figure 3–7 illustrates the design being used by Jaguar in their new 6-cylinder 3.6L (219.5 cu.in.) engine, which produces 225 horsepower (168 kW) at 5300 rpm. This engine employs a four-valve head with a "pentroof" design combustion chamber, so named because of the straight sides that peak at the single spark plug location between the four valves. The inlet valves are mounted at 24 degrees, while the exhaust valves are mounted at 22.5 degrees from the vertical position in order to form a pentroof shape to the combustion chamber. Figure 3–8 also shows the engine in cutaway form. This engine will be used to replace the older in-line six XK model in the XJ40 sedan as well as being offered in the latest XJS coupe and Cabriolet models.

Figure 3-7. Jaguar's pentroof combustion chamber with four-valve head. (Reprinted with Permission © 1984 Society of Automotive Engineers, Inc.)

Figure 3–8. Cutaway view of Jaguar AJ6 3.6L engine which produces 225 hp (168 kW) at 5300 rpm showing wide angled valves operated by twin chain driven camshafts. (Reprinted with Permission © 1984 Society of Automotive Engineers, Inc.)

Domed pistons used with this engine provide a compression ratio of 9.6:1. This engine uses a fuel injection system produced by Lucas/Bosch which is known as the P-Type Digital system. Jaguar intends to offer an economy version of this engine using the standard two-valve head concept with a May Fireball design combustion chamber similar to that used on the Jaguar V12 engine.

Featured in Figure 3–9 is the 5-cylinder Audi Quattro Sport 2.1L (128 cu.in.) engine also employing four-valve head design. This engine with turbocharging and intercooling can produce a staggering 300 horsepower (220 kW) at 6500 rpm. This is 50% more than that now available from the two-valve turbocharged engine. In "rally-trim," Audi expects that the engine is capable of about 450 horsepower (335 kW) which is a phenomenal power-to-displacement ratio.

The Audi Quattro four-valve head uses straight (upright) exhaust valves, while the inlets are mounted at a 25-degree angle. The high performance four-valve head engine attains a turbocharger boost of 32 psi (220 kPa) and has an 8:1 compression ratio.

The Volkswagen Scirocco 1.8L (110 cu.in.) 4-cylinder engine is now available with a 16-valve head that employs straight mounted exhaust valves while the inlets are mounted at 25 degrees from the vertical axis similar to that used in the Audi Quattro engine. This arrangement provides what is commonly called a "semi-pentroof" design combustion chamber (see Figures 3–10 and 3–11). Hydraulic bucket self-adjusting tappets are used with the VW engine as well as sodium filled exhaust valves. This engine will produce 139 horsepower (102 kW) at 6300 rpm using a 10:1 compression ratio.

Figure 3–9. Audi 2.1L 5-cylinder engine with closely spaced camshafts for operation of 4 valves per cylinder. (Reprinted with Permission © 1984 Society of Automotive Engineers, Inc.)

Figure 3–10. VW Scirocco 16-valve 1.8L engine uses a semi-pentroof design combustion chamber and belt driven counter-rotating camshafts. (Reprinted with Permission © 1984 Society of Automotive Engineers, Inc.)

Mercedes-Benz recently introduced their "Baby Benz" to the marketplace; it is designated the 190 Series and is available in either a gasoline or diesel version with a 4-cylinder engine. In the gasoline version, Diamler-Benz decided to pursue a four-valve head design in order to improve the overall engine per-

Figure 3–11. Close-up view of valve location in VW Scirocco 16-valve head using hydraulic tappets (lifters). (Reprinted with Permission © 1984 Society of Automotive Engineers, Inc.)

formance. The engine is a 2.3L (140 cu.in.) displacement unit designated the 190E, and the four-valve head design was done by Cosworth engineering in Britain who of course has done high performance engine designs for many racing engines. The result is that the 190E engine produces 185 horsepower (136 kW) at 6000 rpm from a compression ratio of 10.5:1.

Figures 3–12 and 3–13 illustrate that the design is similar to that used by Jaguar in that the 190E has a pentroof design combustion chamber with 45-degree inclined valves that are sodium cooled and employ bucket tappets. The 190E aluminum cylinder heads are actually produced in Britain by Cosworth.

Figure 3–13. Mercedes Baby Benz uses ram air intake pipes from a plenum chamber to feed air to mate with twin inlet ports on 4-valve head engine. Also note that the exhaust manifold handles paired cylinders with each section. (Reprinted with Permission © 1984 Society of Automotive Engineers, Inc.)

BMW has redesigned their 6-cylinder 3453 cc (210.5 cu.in.) engine with a four-valve layout similar to that used by Mercedes (see Figure 3–14). Flat-topped pistons with four recessed crescents for valve head clearance provide a 10.5:1 compression ratio, and the engine now produces 286 hp (210 kW) at 6500

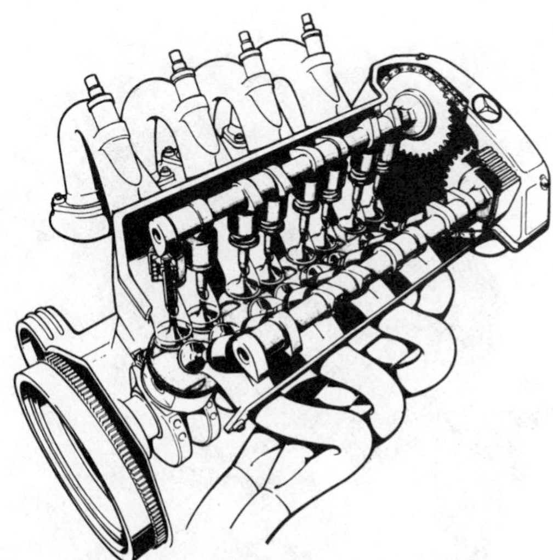

Figure 3–12. Mercedes-Benz (Baby Benz) 16-valve 190E 2.3L engine. Cylinder head is produced by Cosworth in Britain. Engine output is 185 hp (136 kW) at 6000 rpm. (Reprinted with Permission © 1984 Society of Automotive Engineers, Inc.)

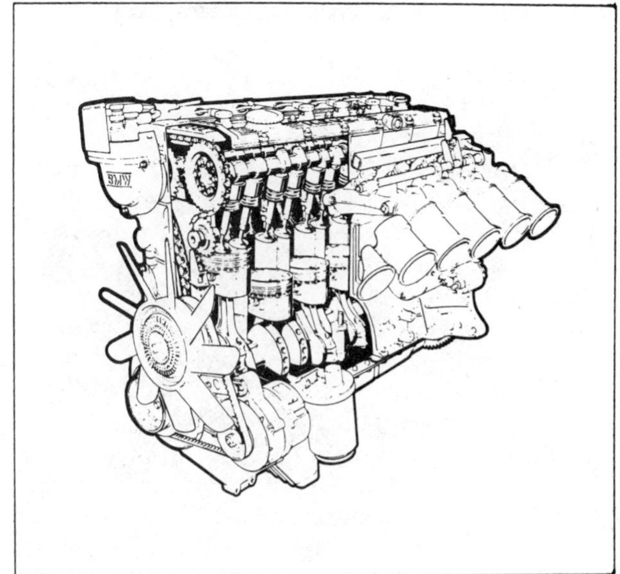

Figure 3–14. BMW's 3.5L 6-cylinder 24-valve engine produces 286 bhp (210 kW) at 6500 rpm. (Reprinted with Permission © 1984 Society of Automotive Engineers, Inc.)

rpm (see Figure 3–15). Robert Bosch Motronic digital management fuel injection is used with this unit.

Saab in Sweden has also gone to the four-valve per cylinder design for use in their third generation turbocharged engine. Saab feels that this design will dominate their engines throughout the 1980s. Saab claims a 10% fuel saving plus a substantial horsepower increase with this design in their 1985 900 model car. Using the four-valve head design with a turbocharger and an 8.5:1 compression ratio, the 4-cylinder 2.0L (122 cu.in.) engine will produce 160 hp (118 kW) and 180 hp (132 kW) with turbocharger intercooling. This design is shown in Figures 3–16 and 3–17.

Figure 3–15. BMW M 635 CSi Coupe is adapted from mid-engined M1 Road Racer; twin overhead camshafts are driven by a simple roller chain. (Reprinted with Permission © 1984 Society of Automotive Engineers, Inc.)

Figure 3–16. Saab 2.0L turbocharged 16-valve engine develops 160 hp (118 kW) in non-intercooled version. (Reprinted with Permission © 1984 Society of Automotive Engineers, Inc.)

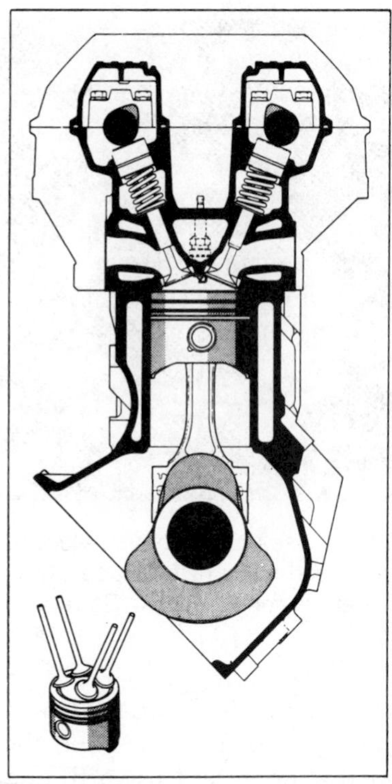

Figure 3–17. Saab 2.0L 16-valve engine uses hydraulic tappets thereby eliminating the need for periodic valve adjustment. (Reprinted with Permission © 1984 Society of Automotive Engineers, Inc.)

Volvo also has a four-valve head design for their 2.3L engine with a 38-degree symmetrical valve spread and a quoted 230 hp (170 kW) output at 7200 rpm. This engine is shown in Figure 3–18. Peugeot

Figure 3–18. Volvo DOHC 2.3L 16-valve engine produces 230 hp (170 kW) at 7200 rpm. (Reprinted with Permission © 1984 Society of Automotive Engineers, Inc.)

has also developed a lightweight 1775-cc turbocharged competition engine that develops an outstanding 320 hp (230 kW) at 8000 rpm. This engine, which uses a light alloy wet cylinder block, special forged pistons, conrods, and crankshaft, also employs twin overhead camshafts with the four valves per cylinder inclined at a 45-degree angle. The compression ratio is a moderate 8:1; however, the turbocharger can deliver up to 25 psi (170 kPa) of boost pressure. A similar engine to this may very well appear in future 80s Peugeot production vehicles. This engine is shown in cutaway form in Figure 3–19 and on a dynomometer test bed in Figure 3–20.

Although many of these four-valve head designs are still being used for competition rather than full stock production, look for some of these designs to surface in the 80s.

Diesel Combustion Chamber Design

Diesel combustion chambers (see Figure 3–21) are either of the direct-injection (DI) design or of the indirect-injection (IDI) design. Since we are mainly concerned here with the combustion chamber design as it applies to passenger car and light trucks, we will concentrate on the IDI engine. For comparison and general information, the DI type is also shown in Figure 3–21 although there are no cars using the DI design concept as of this writing. The DI design is prevalent in most high speed diesel engines employed in heavy duty trucks and equipment due to their superior fuel economy over the IDI engine.

Figure 3–20. View of Peugeot XU9T engine on a test stand will be transversely mounted in the rear of a 4 × 4 rally car. [Reprinted with Permission © 1984 Society of Automotive Engineers, Inc.]

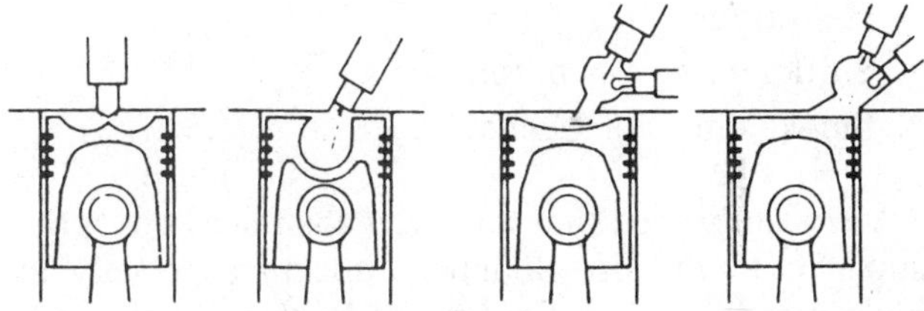

Figure 3–21. Typical diesel engine combustion chamber designs from left to right: direct injection, M.A.N. system; pre-combustion chamber; swirl chamber. [Reprinted with Permission © 1984 Society of Automotive Engineers, Inc.]

Figure 3–19. Peugeot XU9T 16-valve head engine with intercooler produces 320 hp [240 kW] from 1775 cc. [Reprinted with Permission © 1984 Society of Automotive Engineers, Inc.]

Extensive research is currently under way by many diesel engine and vehicle manufacturers to perfect the DI diesel for passenger car application since this design offers potential fuel economy savings of between 10 to 15% over its currently popular IDI-type diesel counterpart.

Problems to date with the DI engine lie in the following areas:

1. Noisier than the IDI engine
2. More sensitive to the grade of fuel used
3. Greater exhaust nitric oxides (NOx) and hydro-

carbons (HC) emissions than the IDI engine (about 15% worse than the IDI engine)
4. Requires more expensive injection equipment (smaller hole sizes in the nozzle tips—multiple hole versus single hole)

Figure 3–21 illustrates the various types of typical diesel combustion systems.

One of the first successful passenger car diesels in North America was the VW Rabbit. When VW decided to produce this engine, the IDI combustion chamber was selected for a number of reasons such as lower noise levels, less susceptibility to the grade of fuel used over its DI counterpart, and easier cold weather starting when a glow plug is employed. In addition, the IDI diesel was easier to design and make comply with U.S. EPA smoke and emissions regulations than the DI unit with respect to control of CO (carbon monoxide), hydrocarbons, and nitric oxides (NOx).

VW decided in favor of the swirl chamber concept rather than the pre-combustion chamber concept because of the following conditions that were desirable in the finished product:

1. Satisfactory fuel consumption under all operating conditions
2. Sufficient power output
3. Smoke emissions behavior

VW also proved that the swirl chamber concept is superior to the pre-chamber concept especially at high speeds, and since the VW Rabbit diesel is designed for engine speeds in excess of 5000 rpm, this was a logical choice. Further tests indicated that black smoke emissions at high altitudes were also more favorable with the swirl chamber design than with the pre-chamber, and finally it was also found that with the exception of engine noise at idle rpm, the noise level in the swirl chamber design could be brought down to that of a gasoline engine.

Specific Combustion Chamber Designs

Now that we have studied the difference between a DI engine and IDI engine, let's look at some actual combustion chamber designs (see Figure 3–22) in use by various diesel engine manufacturers in their passenger cars and light trucks.

Mercedes-Benz uses a pre-combustion chamber design in both their naturally aspirated and turbocharged diesel passenger cars; this arrangement is shown in Figures 3–22 and 3–23. Volkswagen, on the other hand, chose the swirl chamber concept rather than the pre-combustion chamber arrangement; the VW setup is illustrated in Figures 3–22 and 3–24. Both systems used by VW and Mercedes-Benz offer advantages that both feel suit their particular engine design concept and provide the best performance and exhaust emissions control over the engines speed range.

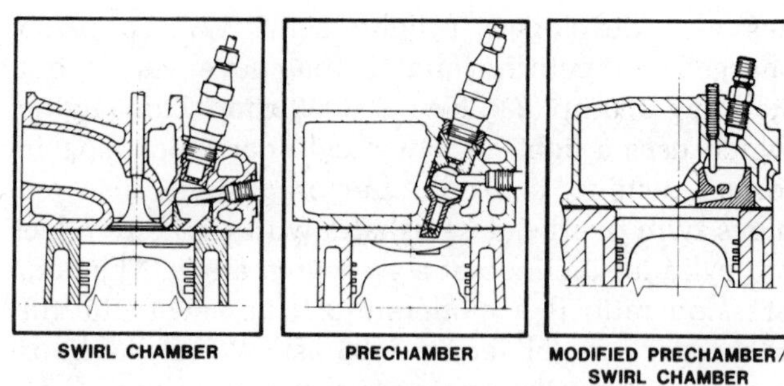

Figure 3–22. Light duty diesel indirect injection type of combustion chambers. [Reprinted with Permission © 1984 Society of Automotive Engineers, Inc.]

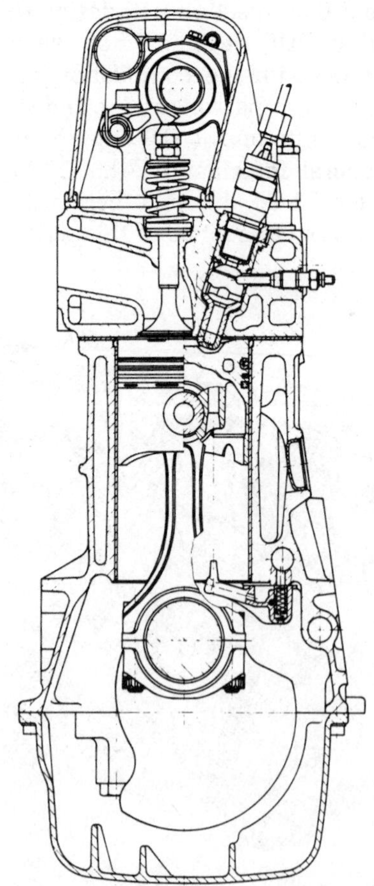

Figure 3–23. Mercedes-Benz 5-cylinder 300D engine combustion chamber design. [Reprinted with Permission © 1984 Society of Automotive Engineers, Inc.]

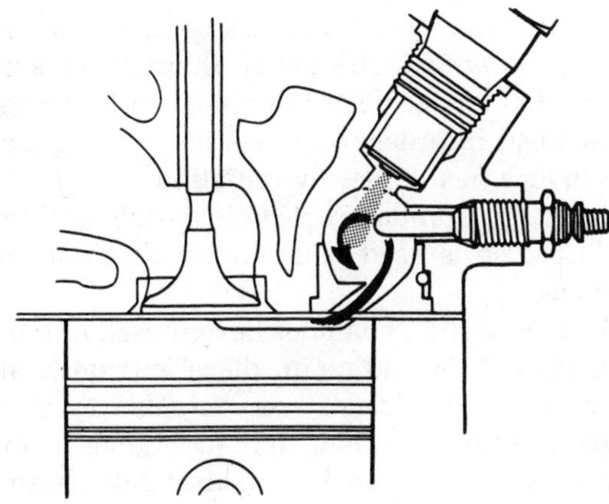

The engine has indirect fuel injection. The fuel is injected into the swirl chambers.
The swirl chambers are located in the cylinder head. They comprise approx. 50% of the total combustion volume.
During the compression stroke, air is forced into the swirl chamber and forced to rotate with high velocity. This air speed influences the combustion speed and therefore the engine can reach high engine speeds (= wide rpm range).
The fuel is injected just before the piston reaches top dead center. It mixes with the turbulent air within a short period of time.
During compression, the air/fuel mixture is heated to approx. 800° C = 1400° F which causes it to ignite. Combustion starts in the swirl chamber and spreads to the cylinder. The pressure increase is thus controlled resulting in a smoother running engine.

Figure 3-24. Diesel engine swirl chamber design commonly used in automotive diesels. (Courtesy of Volvo of America Corp.)

Pre-Combustion Chamber versus Turbulence/ Swirl Chamber

As shown in Figure 3-23, this system incorporates a small separate combustion chamber above the main cylinder and piston. Mercedes-Benz have continued to use this type of combustion chamber design over the turbulence/swirl chamber design, which is being used by a number of other diesel vehicle manufacturers.

Mercedes-Benz claims that the pre-combustion chamber design offers lower peak cylinder pressures, lower pressure rise rate and shorter engine combustion chamber warm-up time, as well as multifuel capability. In addition, the availability of more design parameters which influence the combustion process and which can be optimized with this system have made Mercedes continue with this type of combustion chamber rather than the swirl chamber design, which is now in use by many other automotive diesel engine manufacturers.

Mercedes (Diamler) Benz has always regarded the pre-combustion chamber system as the best compromise. Mercedes-Benz claim that the pre-combustion chamber engine exhibits the lowest hydrocarbon emissions, but it does have marginally higher NOx (nitric oxides) emissions.

Initial combustion takes place within the pre-combustion chamber, which generally contains up to 40% of the total combustion chamber area above the piston when at TDC. The nozzle usually directs its injected fuel towards the outlet or throat of the pre-combustion chamber so as to oppose the force of the incoming pressurized air and to assist the atomization of the fuel droplets.

This injected fuel will actually burn initially in a rich zone (pre-combustion chamber) after which time the burning fuel forces its way through the restricted throat and into the main combustion chamber area under high velocity to complete the combustion process.

Mercedes-Benz engines employ several radial holes around the circumference of the pre-combustion chamber so that the velocity of the gases leaving the throat are increased as they enter the main combustion chamber.

Volkswagen, Oldsmobile DDA (6.2L), Fiat, and BMW are some of the producers of diesel engines that have chosen to go with what is technically referred to as a turbulence chamber design on their engines. However, the common term used for these combustion chamber designs is that of the swirl chamber, patented and designed by Ricardo and Company Engineers Ltd. in Britain who actually refer to it as the Ricardo Comet Mark V combustion chamber.

Figure 3-24 VW, BMW; Figure 3-25 Olds GMC; and Figure 3-26, Fiat illustrate this type of a combustion chamber. Its actual design, although similar to that of the pre-combustion chamber used by

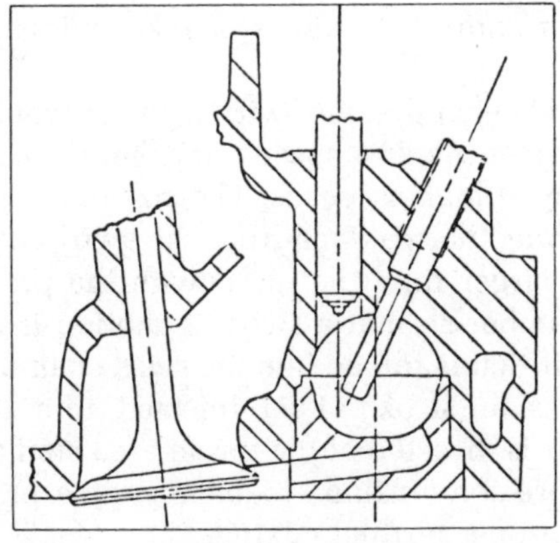

Figure 3-25. Revised swirl chamber design used with 5.7L (350 cu.in.) Oldsmobile diesel engine. (Reprinted with Permission © 1984 Society of Automotive Engineers, Inc.)

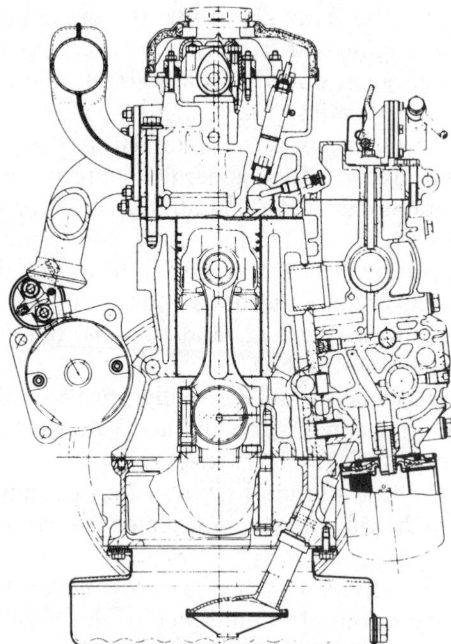

Figure 3–26. Cross section of a 4-cylinder Fiat diesel engine. (Reprinted with Permission © 1984 Society of Automotive Engineers, Inc.)

Mercedes-Benz, differs in that the swirl chamber generally contains about 50 to 80% of the combustion chamber volume above the piston when it is at TDC versus the 40% of the pre-combustion chamber unit.

Volkswagen and GMC (6.2L engine), for example, chose the swirl chamber design over the pre-combustion chamber unit because they felt that with the swirl chamber they could achieve:

1. Satisfactory fuel consumption under all operating conditions
2. Sufficient power output
3. Better control of exhaust smoke emissions

They also felt that the swirl chamber would permit higher rated engine speeds and better overall fuel economy at higher speeds. The size of the throat, or interconnecting passage from the swirl chamber, is usually larger than that used with the pre-combustion chamber so that a swirling motion is imparted to the air that is forced into the swirl chamber on the compression stroke. Fuel injected into the swirl chamber is directed at the throat area similar to that in the pre-combustion chamber engine in order to ensure proper air/fuel mixing.

Prior to the end of injection, the latter portion of the fuel causes the burning gases to pour into the main chamber (it can be a figure 8 depression on the piston crown), which creates complete mixing of the remaining fuel with the air there. Often a part of the swirl chamber where the injected fuel impinges is uncooled in order to reduce ignition lag and further reduce heat losses by radiation.

Generally, a pintle (single hole) nozzle or throttling pintle nozzle is used with this type of combustion chamber.

The combustion chamber design used in the Oldsmobile/GMC V8–350 cu.in. diesel engine is also of the prechamber design, very similar to that employed by Mercedes-Benz in their engines. However, as illustrated in Figure 3–25, Oldsmobile chose, after extensive testing, what is known as a "side outlet prechamber" with inboard pencil type nozzles in their pre-1980 5.7L V8 diesel. Before selecting this design of combustion chamber, Oldsmobile evaluated nearly 300 combinations of prechambers and injection nozzles.

This side outlet design is very similar to that of the swirl chamber Ricardo Comet Mark V design found on a variety of other automotive diesel engines.

The injector nozzles that were used by Oldsmobile initially were Stanadyne fixed orifice (hole) type with two 0.017″ diameter holes in all engines except those used in Chevrolet and GMC trucks sold in California. The California truck nozzles have three 0.014″ diameter holes in order to comply with the more stringent California exhaust emissions regulations, 1980 and later diesel engines use a screw-in type microjector (see chapter 22).

The Peugeot diesels also employ the IDI (indirect-injection) combustion chamber system using the Ricardo Comet V design, which is similar to that employed by VW and others. Fiat passenger car and light truck diesels employ a system as shown in Figure 3–26.

The diesel engine used by Datsun/Nissan in their pickup trucks employs the swirl chamber design similar to that employed by Volkwagen. This system is used on the 2.2 and 2.5 liter models.

M.A.N. Direct-Injection Diesel Engine

The advantage of a DI (direct-injection) automotive diesel engine is the potential of a 15 to 20% fuel economy improvement over its IDI (indirect-injection) counterpart. To date, no manufacturer has been able to perfect the DI engine option and yet still meet exhaust emissions regulations as well as the quietness of operation of the IDI engine.

Recent tests however have indicated that the M.A.N. controlled direct-injection system (W. Germany) has the potential with further research to possibly meet these regulations and also be capable of 5000 rpm engine speed. The concept of operation of the M.A.N. system is one whereby the injected fuel is deposited upon the combustion chamber wall, which is shown in Figure 3–27 and is actually cast within the crown of the piston.

Theory of Combustion

Since both gasoline and diesel engines employ internal combustion to produce power, the ongoing legislation by the EPA in the United States and the ECE in Europe to reduce exhaust emissions makes it important for the automotive technician to understand the actual theory of combustion in both these types of engines.

Successful performance of any internal combustion engine is related to a quick change from chemical energy (fluid, i.e., gas/diesel) of the fuel into heat energy in the combustion chamber so that the piston will receive maximum effort from the expanding gases as it reaches TDC.

In a gasoline engine, the vaporization of the fuel can be done either in a carburetor or by fuel injection. This fuel injection can be done through carburetor TBI (throttle body injection) or by intake manifold port injection towards each cylinder's intake valve.

In a diesel engine, air only is taken in on the intake stroke; therefore, vaporization of the diesel fuel occurs at the point of actual injection into the compressed air mass in the cylinder during the compression stroke as it mixes with the hot air in the cylinder.

Combustion can only take place when the carbon and hydrogen in the fuel mixes with the oxygen (air) to provide combustion.

The actual horsepower (kilowatts) that any engine is capable of producing depends on a number of factors, but in the final analysis, the amount of air that the engine is capable of inhaling or swallowing on each intake stroke determines the final output from each cylinder. Once the air is trapped in the engine cylinder, the remaining power factor is dependent upon just how much of the trapped air (oxygen) can be consumed (burned).

Let's look briefly at the chemistry of combustion within an engine cylinder to establish specifically what requirements constitute a successful combustion chamber burn. Since in our discussion of internal combustion engines we are concerned with liquid fuels that are made up of two main elements, namely hydrogen and carbon, let's look at both the atomic and molecular weights of these elements in relation to oxygen.

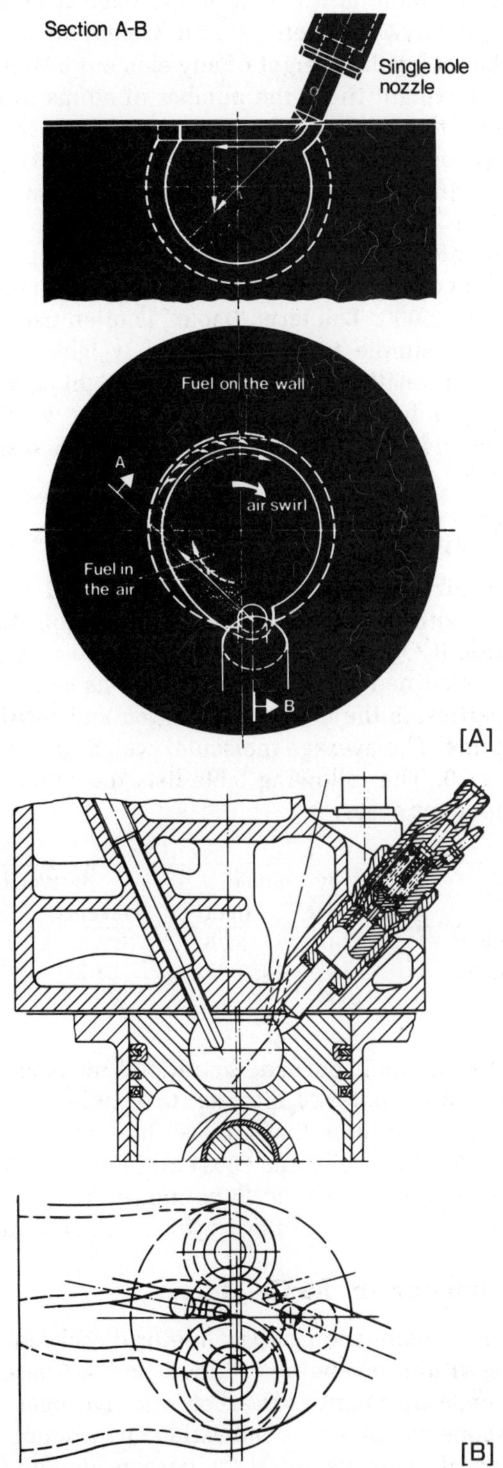

Figure 3–27. M.A.N. systems. (a) Layout of M.A.N. piston combustion chamber "M" type; (b) M.A.N. "M" type-combustion system with controlled direct injection and glow plug location in combustion chamber. (Reprinted with Permission © 1984 Society of Automotive Engineers, Inc.)

Atomic weights are generally compared to that of oxygen, which has an atomic weight of 16, while the lightest atomic unit is that of hydrogen which is approximately 1. Carbon's atomic weight is 12.

The molecular weight of any element is simply its atomic weight times the number of atoms in a molecule of the element. Therefore the molecular weight of hydrogen (H_2) is 1.008 times 2 = 2.016, oxygen (O_2) is 16 times 2 or 32, while carbon (C) would be 12 times 1 = 12.

When considering the molecular weight of hydrogen for combustion calculations, we round it off from 2.016 to 2.000. The term "mole" is often used to express in simple terms the actual weight of a substance in relation to its molecular weight. Therefore hydrogen, for example, with a molecular weight of 2 would indicate that a mole of hydrogen weighs 2 pounds.

Makeup of Air

Since air consists mainly of oxygen and nitrogen along with very small amounts of argon, carbon dioxide (CO_2), hydrogen, and other gases, what we are concerned with in relation to its combustion properties, is the amount of oxygen and nitrogen it contains. The average molecular weight of dry air is about 29. The following table lists the ratios of the component gases:

	By Volume		*By Weight*	
	Percent	*Ratio*	*Percent*	*Ratio*
Nitrogen	79	3.76	76.8	3.32
Oxygen	21	1.00	23.2	1.00
Total	100		100	

When a fuel is burned, the ratio of nitrogen to oxygen is 3.76 moles of nitrogen to 1 mole of oxygen. Also, if we assume that a diesel fuel oil consists of some 15% hydrogen and 85% carbon by weight, this means that there will be 15 pounds of hydrogen and 85 pounds of carbon in every 100 pounds of fuel oil.

Combustion in the Cylinder

Internal combustion of gasoline or diesel fuel takes place within the engine cylinder. Both of these fuels are made up mainly of carbon and hydrogen. As a consequence of this combustion, byproducts consisting of water vapor (H_2O), carbon dioxide (CO_2), and carbon monoxide (CO) are formed.

All diesel engines operate throughout their speed range with an excess of air in the combustion chamber. Depending upon the load and speed range of the engine, the ratio of air to fuel can vary from as low as 18/20:1 to as high as 90/100:1 at an idle rpm.

Another way to consider this excess air is to consider that the required weight of air for each pound of fuel burned would be 120% of 14.86 = 17.83 lb. One mole of a gas generally occupies 379 cu.ft. at 60°F at 14.7 psi (atmospheric pressure) with the molecular weight of air being 29. Therefore the volume of air required for the combustion of 1 pound of diesel fuel would be 379 × (17.83/29) = 233 cu.ft.

From the above information, a simplified chemistry of combustion can be calculated. The general ratio by weight of carbon-hydrogen in diesel fuel varies between 6 to 8. If we were to assume that the fuel burns with no excess air, then the reactions for complete combustion could be equated as follows:-

Chemical equations:

$$C + O_2 = CO_2 \text{ and } 2H_2 + O_2 = 2H_2O$$

By Molecular Weight:

$$12 + 32 = 44 \text{ and } 4 + 32 = 36$$

Therefore, 1 pound of hydrogen requires 32 ÷ 4 or 8 lb. of oxygen, and 1 pound of carbon requires 32 ÷ 12 or 2.667 lb. of oxygen for complete combustion. Since 1 lb. of air contains 0.2315 lb. of oxygen, 1 lb. of hydrogen requires 8 ÷ 0.2315 or 34.56 lb. of air while 1 lb. of carbon requires 2.667 ÷ 0.2315 or 11.52 lb. of air for complete combustion. For complete combustion of an average fuel containing one part hydrogen and seven parts carbon, the weight of air required is

$$\frac{34.56 + (7 \times 11.52)}{8} = 14.4 \text{ lb.}$$

From the above equations, under normal temperature/pressure conditions, 1 lb. of air occupies 12.4 cu.ft. and 1 lb. of fuel with a specific gravity of 0.85 would occupy a volume of 0.01888 cu.ft. Therefore, in order to completely burn 1 cu.ft. of diesel fuel, we would require (14.4 × 12.4) ÷ 0.01888 or 9450 cu.ft. of air.

Generally speaking then, it requires approximately 10,000 times the volume of air than fuel oil to burn it to completion.

Processes of Combustion—Diesel Engine

The general air/fuel mixture discussed under the heading combustion in the cylinder undergoes four basic chemical stages in order to provide complete burning of the injected fuel.

These stages would be as follows:

1. Ignition delay—this is the time delay period from the actual beginning of fuel injection to the actual start of ignition of this injected fuel.
2. A period or stage of rapid combustion of the fuel that was injected during the delay period in (1) above.
3. A combustion period during which the continually injected fuel burns as rapidly as it enters the combustion chamber.
4. An afterburning period during which late combustion of previously unburnt fuel or intermediate products of combustion is completed.

In the diesel combustion process, air only is compressed on the compression stroke. The final pressure and temperature of this air charge depends upon such items as:

1. Compression ratio
2. Engine speed and load
3. Piston/combustion chamber shape (air turbulence)
4. Inlet air temperature

The final cylinder pressure before the diesel fuel is injected will of course be related to the four items listed above; however, generally speaking the air pressure would be between 400–600 psi (2758–4137 kPa) and at a temperature of 600–1000°F (315.5–538°C).

The actual number of degrees BTDC that the fuel will be injected depends upon the engine rpm since at a faster speed the fuel is injected earlier in the piston's upward moving stroke to enable sufficient time for burning (larger fuel volume injected).

The type of combustion chamber and DI (direct injection) versus IDI (indirect injection) also have a bearing on the actual start of injection.

Ignition Delay Period. This time period involves both a physical and a chemical delay. The length of the ignition delay is basically controlled by engine design and the fuel characteristics such as:

1. The Cetane number or rating of the fuel—a higher number has a shorter delay period (quicker vaporization of fuel).
2. The size of the fuel droplets leaving the injector nozzle or spray tip (number of holes and diameter plus compression pressure).
3. Combustion chamber air pressure/temperature at the start of injection.
4. Air turbulence within the combustion chamber.

The shorter the ignition delay period, the less volume of injected fuel there will be and the lower will be the cylinder pressure rise after the fuel ignites. This ignition delay is very important because the longer the delay period, the greater will be the volume of fuel prior to ignition; therefore, when ignition does occur, the accumulated fuel charge will burn very violently. This causes a rapid rise in cylinder pressure, which causes gas vibrations throughout the combustion chamber to occur very quickly and emit a noise that is known as combustion knock. All diesel engines suffer from this occurrence, some more so than others.

This physical delay period exists because the injected fuel has to be atomized; it has to mix adequately with the air (oxygen) as well as vaporizing in order that combustion can be started.

Parallel with or immediately behind the physical delay is a period of pre-flame reaction (chemical delay). The fuel is injected at speeds approaching 780 mph, which is faster than the speed of sound. This fuel is ignited by contact with hot metal surfaces and the air temperature, with burning usually starting with ignition of the vaporized fuel droplets in the first portion of the injected fuel.

One of the most important factors that establishes the time delay period is the Cetane number of the fuel because a fuel with a higher Cetane number will vaporize more readily and therefore propagate combustion quicker. Most high speed diesel engines require a fuel with at least a Cetane number of 40.

Once the combustion period starts, a flame front is established; therefore, fuel that is still being injected will burn almost immediately. Peak flame temperatures will occur when the combustion pressures are at maximum and while the fuel is still being injected.

Combustion is generally considered smooth if the rate of cylinder pressure rise is less than 30 psi/degree of crankshaft rotation. The maximum pressure in the engine cylinder is usually limited to about 900–1200 psi (6205–8274 kPa) in small high speed automotive type engines; however, this pressure can be as high as 2000 psi (13,790 kPa) in larger diesel engines. Normally, the peak temperature of combustion can run as high as 3500–4500°F (1926–2482°C), depending on the peak cylinder pressure that is developed.

Because of ignition lag and combustion time, high

speed engines cannot attain constant pressure combustion. It is therefore necessary to start injection earlier with the result that much of the combustion occurs at a constant volume condition similar to a gasoline engine while the remainder of combustion will occur at or nearly constant pressure. Studies indicate that a more efficient burn would occur at a constant volume condition than at a constant pressure condition. However, if all of the injected fuel were to burn at a constant volume condition, the peak cylinder pressures would become excessively high (small volume and large fuel mass).

Because of incomplete mixing of the injected fuel with the air in the combustion chamber, some fuel droplets will not burn until late in the cycle. This means that the combustion chamber temperature is lower and there is also less oxygen to sustain the remaining burn with the result that incomplete combustion will occur. This situation is reflected as smoke in the exhaust stack or pipe. White smoke will result from incomplete combustion in overlean combustion chamber areas, by fuel spray impingement on metal surfaces, and also with low temperatures in the cylinder such as when starting an engine (more so on cold days).

Fuel with too high a Cetane rating can also cause white smoke when used in high ambient temperature conditions.

Gray or black smoke is the result of incomplete combustion in rich combustion chamber areas caused by such conditions as engine overload, insufficient fuel injector spray penetration, or late ignition due to retarded injection timing. Air starvation can also cause black smoke.

Exhaust Emissions

Both gasoline and diesel engines today are regulated by government agencies regarding exhaust emissions. Gasoline vehicles have been designed for use on unleaded fuel and have also been designed to use catalytic converters. Lowering of engine compression ratios along with ignition and valve timing changes have resulted in a reduction in exhaust emissions.

The three main emissions that both gasoline and diesel engines have to comply with are:

1. Hydrocarbons (HC)
2. Carbon Monoxide (CO)
3. Nitric Oxides (NOx)

Other emissions limits mandated by EPA (Environmental Protection Agency) in the United States are sulfate emissions and visible smoke. Both gasoline- and diesel-powered vehicles now employ EGR (exhaust gas recirculation) as an aid to reducing combustion chamber temperatures, which assists in lowering the NOx. EGR up to about 20% is successful in reducing NOx at low engine loads, while anywhere from 0 to 50% (modulated) is required under full-load conditions.

Hydrocarbon emissions result from unburned fuel particles or the decomposed fuel molecules. Carbon monoxide formation is a direct function of the air/fuel ratio. CO is formed at an intermediate stage of combustion in the air/fuel region where either the local oxygen concentration or the combustion chamber temperature is not high enough for the oxidation of carbon monoxide to carbon dioxide. Both hydrocarbon and carbon monoxide emissions are affected by engine/fuel injection timing during full-load operation.

Nitric oxide emission is related to the concentration of local oxygen and local high temperature. It is formed in a slightly lean combustion region where the combustion temperature is high and oxygen is available for its combination with nitrogen.

Sulfate emissions are a direct result of the sulphur content of the fuel, either gasoline or diesel. Current gasoline fuels contain approximately 0.03% sulphur content by weight, while diesel fuel contains anywhere from 0.1 to 0.5% by weight. Of the total amount of sulphur contained in diesel fuel, a fraction ranging from 1 to 2% is converted into sulfates at the exhaust, which is equivalent to the conversion rate for spark ignition engines without catalytic converters. Approximately 80 to 90% of the sulphur is burned into sulphur dioxide (SO_2) with the remaining 10 to 20% being shared by several other sulphur compounds. Both gasoline and diesel engines emit about the same concentration of aldehydes.

CO emissions in a diesel engine are generally lower than in a non-emission controlled gasoline engine due to the lean air/fuel ratio of the diesel, which ranges from 16 to 90:1 in comparison to the stoichiometric (14.5:1) air/fuel ratio of the gasoline engine.

NOx emissions from the diesel engine are also lower than for a non-emissions controlled gasoline engine because of the locally rich fuel droplet combustion that tends to suppress NOx formation due to the lack of oxygen availability. Diesel emissions are also lower because of the two-stage combustion consisting of the rich pre-combustion chamber air/fuel and lean main chamber. This suppresses NOx formation due to the limited oxygen availability and moderate temperature levels along with the large exhaust residual in the prechamber, which reduces oxygen availability in the prechamber.

Proposed standards on HC/CO/NOx for 1985 and beyond are as follows:

PASSENGER CAR REGULATORY EMISSION CONTROL REQUIREMENTS

		1960 No Control	1968 (66 CA)	1970	1971	1972	1973	1974	1975-76	[1] 1977	1978-79	1980	[1] 1981-82	[1,7] 1983-85 and Later
		75 FTP gpm	1970 FTP (Tailpipe Conc.)			1972 FTP gpm (CVS)			1975 FTP gpm (CVS)					
Exhaust Emissions	HC	10.6	275 ppm [8] [6.3]	2.2 gpm [8] [4.1]		3.4 [8] [3.0] CA 3.2			1.5 CA 0.9	1.5 CA .41		.41 [2] CA .39	.41 [2] CA .39	.41 [2] CA .39
	CO	84	1.5% [8] [51]	23 gpm [8] [34]		39 [8] [28]			15 CA 9.0	15 CA 9.0		7.0 CA 9.0	[3] 3.4 [5] CA 7.0	3.4 CA 7.0
	NOx	4.1			CA 4 gpm	CA 3 gpm on 70 FTP	3	CA 2	3.1 CA 2.0	2.0 CA 1.5		2.0 CA 1.0	[4] 1.0 [5] CA 0.7	[4] 1.0 [6] CA 0.7
Evap. g/Test	Trap			CA 6	6	2 →				2				
	SHED	46.6*									6	6 CA 2	2	
Crankcase		4.1 gpm												

*Uncontrolled Evap. 46.6 g/Test - 3.7 gpm.

NOTES: FTP - Federal Test Procedure
ppm - parts per million
gpm - grams/mile
CVS - constant volume sampler (true mass meas.)
CA - California only
GVW - Gross Vehicle Weight, IW - Inertia Wt.
HFET - Highway Fuel Economy Test
LDV - Light Duty Vehicle
LDT - Light Duty Trucks
MDT - Medium Duty Trucks
HDT - Heavy Duty Trucks

1 High Altitude Standards:
1977 Same as Sea Level
1978-81 None
1982-83 Exh. HC 0.57, CO 7.8 (waiver 11), NOx 1.0 (Waiver 1.5) gpm, Evap. 2.6 g/test
1984 All cars meet Stds. at all altitudes.

2 Non-CH_4 Std.; .41 total HC w/CH_4 correction allowed 1980. Post 1980 std. must be met as measured: 0.39 Non CH_4 or 0.41 total HC.

3 Possible 2 yr. waiver to 7 gpm.

4 Possible waiver to 1.5 gpm for diesel or innovative technology through 1984.

5 Federal stds. optional in CA for '81, but selection of '81 Fed. option requires 7.0/0.4 for '82.

6 0.7 NOx std. optional in CA for '83 and later, but requires limited recall authority for 7 yrs./75,000 mi. Primary std. 0.4 g/mi.

7 1983 exhaust standards represent the following reductions from uncontrolled levels: HC 96%; CO 96%; NOx 76%.

8 Numbers in [] are 1975 FTP gpm equivalents of listed standards.

ADDITIONAL CAR REQUIREMENTS

1985 No Crankcase Emissions Allowed
Tampering by Service Industry, Dealers, etc. Prohibited
Fuel Filler Must Exclude Leaded Fuel Nozzles (Catalyst Veh.)
Exhaust Standards Apply to Diesel w/Test Modif.
Assembly Line Test Requirement—SEA
Parameter Adjustment Requirements; Idle Mixture and Choke
Diesel particulate std. 0.6 gpm.

1987 Diesel Particulate Std. 0.2 g/mi.

CALIFORNIA—IN ADDITION TO FED. CAR REQ'TS.

1984 End-of-Line Exhaust Test.
Fuel Filler Specs. for Vapor Recovery
Higher optional Std. for 100,000 mi. Cert.
Restrictions on Allowable Maintenance
HFET NOx Std. 1.33 x FTP Std.
Must "meet" CO Standard to 6000 ft.
Limited sale of some Federal vehicles permitted

Diesel Particulate Std. (gpm)
1984 – 0.6, 1985 – 0.4
1986 – 0.2, 1989 – 0.08

LIGHT DUTY TRUCK REGULATORY EMISSION CONTROL REQUIREMENTS

		[1]1975	[3]1976-77	1978	Vehicle Weight, # GVW	IW	[2, 3] 1979 GVW	1979 IW	[3, 4] 1980 GVW	1980 IW	[3, 4] 1981-82 GVW	1981-82 IW	[3, 4] 1983 GVW	1983 IW	[3, 4] 1984 GVW	1984 IW
Exhaust Emissions, 1975 FTP gpm	HC	2.0	2.0	2.0		– 4000 –		CA .41 – –		CA .39 – –		CA .39 – –		CA .39 – –		CA .39 – –
			CA 0.9	CA 0.9	6000 –		1.7	CA 0.5	1.7	CA 0.5	1.7	CA 0.5	1.7	CA 0.5	0.80	CA 0.5
		Classed as HDT	Classed as HDT	*CA 0.9		– 6000 –	[5] CA 0.9	– –	[5] CA 0.9	– –		– – CA 0.6		– – CA 0.6		– – CA 0.6
	CO	20	20	20		– 4000 –		CA 9.0 – –		CA 9.0 – –		CA 9.0 – –		CA 9.0 – –		CA 9.0 – –
			CA 17	CA 17	6000 –		18	CA 9.0	18	CA 9.0	18	CA 9.0	18	CA 9.0	10	CA 9.0
		Classed as HDT	Classed as HDT	*CA 17		– 6000 –	[5] CA 17	– –	[5] CA 17	– –		– – CA 9.0		– – CA 9.0		– – CA 9.0
	NOx	3.1	3.1	3.1		– 4000 –		CA 1.5 – –		CA 1.5 – –		CA 1.0 – –		[6] CA 0.4 – –		[6] CA 0.4 – –
		CA 2.0	CA 2.0	CA 2.0	6000 –		2.3	CA 2.0	2.3	CA 2.0	2.3	CA 1.5	2.3	CA 1.0	2.3	CA 1.0
		Classed as HDT	Classed as HDT	*CA 2.3		– 6000 –	[5] CA 2.3	– –	[5] CA 2.3	– –		– – CA 2.0		– – CA 1.5		– – CA 1.5
Evap. g/Test by SHED post 1977		2	2	6		– 4000 –	6		6		2		2		2	
				[5] CA 6	6000 –	– 6000 –			CA 2							

NOTES:

1 Prior to 1975, Passenger Car requirements apply to trucks equal to or less than 6000# GVW. Heavy Duty Truck (HDT) Requirements apply to trucks greater than 6000# GVW.

2 Federal Definition of Light Duty Trucks (LDT) effectively changed in 1979 from equal to or less than 6000# GVW to equal to or less than 8500# GVW (10,000# optional); except veh. greater than 6000# Curb or greater than 45 sq ft Frontal Area are HDT.

3 High Altitude Standards :
1977 Same as Sea Level
1978-81 None
1982-83 Exh HC 2.0, CO 26, NOx 2.3 gpm, Evap. 2.6 g/test
1984 Exh. HC 1.0, CO 14, NOx 2.3 gpm, Evap. 2.6 g/test

4 Post 1979 CA HC Standards are non-CH_4 ; See Passenger Car Note

5 CA MDV category through 1980 defined as 6000–8500# GVW (10,000# opt.); IW only definition applies after 1980; 1979-80 LDT (equal to or less than 6000 GVW) are sub divided by IW as shown.

6 1.0 NOx Standard optional in CA for 1983 and later, but requires limited recall authority for 7 yrs./75,000 miles

ADDITIONAL LDT REQUIREMENTS

1984 No Crankcase Emissions Allowed (Gasoline and Diesel)
Tampering by Service Industry, Dealers, etc. Prohibited
Fuel Filler must exclude Leaded Fuel Nozzles (Catalyst Veh.)
Federal LDT Exh. Stds. apply to Diesel w/test modific.
Assembly Line Test Req't.—SEA 40% AQL
Parameter adj. rules same as passenger car
Diesel Exhaust Particulate Std. 0.6 gpm
Optional certif. test procedures
Restrictions on allowable maintenance
Idle Standard CO 0.50% (gasoline only)
Full-life Req. 130 Kmi.; Half-life Opt. 50 Kmi. '83 dura/d.f. proc.
Full-life Opt.- limited warranty mileage
1985 Full Life Req. 120 Kmi Relaxed Cert. Procedure Warranty Limited to 60Kmi
1987 Diesel Particulate Std. 0.26 g/mi with Averaging
198? Possible 1.2 g/mi NOx Std. Plus NOx Averaging

CALIFORNIA—IN ADDITION TO FED.REQ'TS.

1983 Car End-of-Line Tests apply through 8500# GVW
Fuel Filler Specs. for Vapor Recovery—see pass. car
Higher Optional Stds. for 100,000 mi. Cert.
Restrictions on Allowable Maintenance
HFET NOx Std. 2.00 x FTP Std.
Must "meet" CO std. to 6000 ft.
Limited sale of some federal vehicles permitted
Diesel Particulate Std. (gpm:)
1984–0.6, 1985–0.4,
1986–0.2, 1989–0.08

CANADIAN EMISSION REQUIREMENTS DIFFERING FROM U.S.

1975–85 Passenger car and LD truck exhaust:
HC 2, CO 25, NOx 3.1 gpm by 1975 FTP.
Evap.: 2.0 g/test by Trap.
Post 1985 req'ts to be announced.
Trucks greater than 6,000 lb GVW exhaust:
HC+NOx 16 g/bhp-h, CO 40 g/bhp-h
(LDV Stds. Opt., 6000–8500 lb GVW)

4

Gasoline and Diesel Fuels

Introduction

Crude petroleum oil and natural gas are found in many areas of the world under pressures of up to several hundred psi, and both crude oil and natural gas are often obtained from the same well.

The crude oil must then be subjected to a refining process in order to remove the impurities; special additives are then blended with the fuel to make it commercially acceptable for use in both gasoline and diesel engines. Crude oils vary widely in composition and processing requirements with Alaskan North Slope crude producing less gasoline than Arabian light crude. However from a quality standpoint, Arabian light crude contains more sulphur than Indonesian Minas crude.

Products suitable for automotive engine fuels are blended using straight distilled fractions and processed components. Petroleum distillates normally used as fuel for diesel engines are composed of heavy hydrocarbons. The more volatile portions of the crude oil are used for gasoline, while the heavier distillates are used as diesel fuel and home heating oil.

The refining process for diesel fuel is not as complex as that for gasoline; therefore, it is less expensive to produce.

Because of its heavier hydrocarbon content, diesel fuel contains more energy or BTU's per gallon than gasoline or any other fuel now used in internal combustion engines, including LPG and CNG. This greater heat value is one of the reasons that the diesel engine is more efficient than the gasoline engine. Other specific reasons can be found in Chapter 3, dealing with combustion chamber design.

The heat value of both diesel and gasoline will vary slightly depending upon the actual crude used, the refining process and the grade of fuel desired; however, gasoline and diesel fuel generally contain the following weights and heat values per U.S. gallon.

Diesel. Average of 7.1 pounds/U.S. gallon and an average of around 128,400 BTU's per U.S. gallon.

Gasoline. Average of 5.8 pounds/U.S. gallon and an average of around 113,800 BTU's per U.S. gallon.

Distribution of products from an average barrel of 1977 crude oil is shown in Figure 4–1.

Diesel Fuels

Diesel fuel is refined from crude petroleum oil, which is basically a mixture of many types of molecules that contain only carbon (C) and hydrogen (H) atoms and is referred to as a mixture of hydrocarbons. These are mixed with some sulphur and nitrogen compounds, small amounts of soluble organic compounds, and a few impurities such as water and sediment.

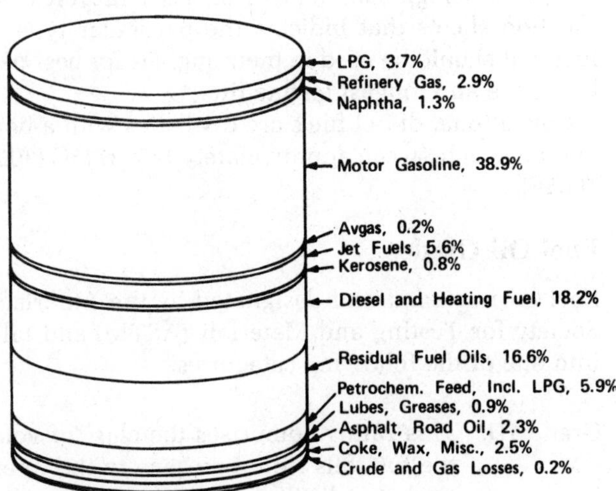

Figure 4–1. Typical end products from a barrel of crude oil. (Courtesy of U.S. Department of Energy)

Hydrocarbon molecules found in petroleum consist of a practically unlimited number of combinations of carbon and hydrogen atoms. These molecules are classified into three general groups:

1. Paraffin
2. Naphthene
3. Aromatic

Varying proportions of these hydrocarbon molecules are found in all petroleum crude oils. These variations have caused a designation to be given to the type of crude oil in terms of the predominating hydrocarbon group, namely paraffinic, mixed, naphthenic, or aromatic.

Refining residue from the paraffinic crudes is mostly "paraffin wax," which is the cause of fuel filter plugging when ambient temperatures drop to the cloud point of the fuel oil. Naphthenic and aromatic crudes produce a residue that is comprised mostly of asphalt.

When crude oil is refined, it is heated and most of it is vaporized so that the gases pass to a fractionating cooling tower where the various portions such as gasoline, kerosene, gas oil (fuel oil), and other distillates are separated from each other as they condense at different temperatures in the tower.

All crude oils contain varying amounts of soluble organic compounds of vanadium, nickel, iron, sodium, and other elements. Fuels with high concentrations of these elements can create rapid wear of internal engine components, although diesel fuels containing small amounts of vanadium in the region of about 70 ppm (parts per million) will generally give satisfactory results.

All diesel engine manufacturers issue fuel oil specification sheets that indicate the particular type of fuel that should be used in their engines for best performance and longest engine life.

Coventional diesel fuels are distillates with a boiling range of between approximately 149–371°C (300–700°F).

Fuel Oil Grades

Fuel oil is graded and designated by the American Society for Testing and Materials (ASTM) and falls into one of the following categories:

Grade 1D. This rating comprises the class of volatile fuel oils from kerosene to the intermediate distillates. Fuels within this classification are applicable for use in high speed engines in services involving frequent and relatively wide variations in loads and speeds and also in cases where abnormally low fuel temperatures are encountered (for easier starting in cold weather).

Grade 2D. This rating includes the class of distillate gas oils of lower volatility; they are applicable for use in high speed engines in services involving relatively high loads and uniform speeds, or in engines not requiring fuels having the higher volatility or other properties specified for Grades 1D.

Grade 3D. This fuel oil has been proposed but has not as yet been accepted since it does contain a higher degree of sulphur (up to 0.7%) and its Cetane rating is lower than existing 1D and 2D diesel fuels now in use. This type of fuel usually manifests poor cold weather properties (wax formation tendencies) as well as poor ignition quality, which adversely affects both engine noise and exhaust emissions levels.

Other classifications of diesel fuels below these grades are not considered acceptable for use in high speed automotive or truck type engines, therefore they will not be discussed here.

On a volume basis, typical No. 2 diesel fuel has about 13% more heating value in BTU's per gallon than does gasoline; while No. 1 diesel fuel, which is a lighter distillate and therefore less dense than No. 2, has approximately 10% more BTU content per gallon than gasoline.

Fuel Grade Versus Engine Performance

Selection of the correct diesel fuel is a must if the engine is to perform to its rated specifications. Generally the factors that must be considered in the selection of a fuel oil are:

1. Starting characteristics
2. Fuel handling
3. Wear on injection equipment
4. Wear on pistons
5. Wear on rings, valves, and cylinder liners
6. Engine maintenance
7. Fuel cost and availability

Other considerations in the selection of a fuel oil are:

1. Engine size and design
2. Speed and load range
3. Frequency of load and speed changes
4. Atmospheric conditions

Fuel Properties

The major functional properties of diesel fuel and how important they are are listed below.

Diesel Fuel Property	*Why Is It Important*
Heating value (BTU/gal.)	Affects power and economy
Viscosity	Affects atomization
Cetane number	Affects cold starting, smoke, and combustion roughness
Sulphur content	Affects wear and deposits
Cloud and pour point	Affects low temperature handling
Cleanliness	Affects fuel filter life and injector life

Heating Value

Heating value of a fuel correlates with its API (American Petroleum Institute) gravity. If we considered that an engine has been tuned to run on the heaviest regular fuel of about 30 degrees API and the fuel grade is changed to a 40-degree API rating, then there will be a heat loss of about 3.5%. Unit injector engines will not lose any more power than this; however, engines using separate pumps and nozzles can show a loss of horsepower up to 10% due to actual flow reduction because of increased fuel compressibility.

The 3.5% loss could also be attributed to an air inlet temperature increase from 16°C (61°F) to 27°C (81°F) or an increase in altitude from sea level to 300 meters (985 feet). A 10% power loss could be caused by a temperature increase from 16°C (61°F) to 50°C (122°F) or an altitude increase of about 850 meters (2800 feet).

The following properties of a diesel fuel should be understood as to how these properties can affect the engine's operation.

Cetane Number

This is a measure of the ignition quality of the fuel, and it influences both the ease of starting and combustion roughness of an engine because the ignition delay period is lengthened with a decrease in Cetane number. A low Cetane fuel permits a lot of the injected fuel to evaporate before the flame front actually begins. When the flame front begins, this previously injected fuel burns very rapidly causing cylinder pressure to rise to very high peaks with the resultant diesel knock.

Ignition delay is discussed in detail under Chapter 3, dealing with combustion chambers. The duration of this delay is expressed in terms of Cetane number (rating). Rapidly ignited fuels have high Cetane numbers (50 or above), while slowly ignited fuels have low Cetane numbers (40 or below). The lower the ambient temperature, the greater the need for a fuel of a higher Cetane rating so that it will ignite rapidly.

Difficult starting can be experienced if the Cetane number of the fuel is too low. This can be accompanied by engine knock and puffs of white smoke during engine warm-up in cold weather when using a low Cetane fuel.

High altitudes and low ambient temperatures require the use of a diesel fuel with an increased (higher) Cetane number.

Low temperature starting is enhanced by the use of high Cetane fuel oil in the proportion of 1.5°F lower starting temperature for each Cetane number increase in the fuel.

Current 1D and 2D diesel fuels have a Cetane rating between 40 and 45. Cetane rating is actually a measure of the fuel oil's volatility; the higher the rating, the easier the engine will start and the smoother will be the combustion process within the ratings specified by the engine manufacturer.

Cetane rating differs from the octane rating that is used for gasoline in that the higher the number of gasoline on the octane scale, the greater that fuel's resistance to self-ignition, which is a desirable property in gasoline engines with a high compression ratio. Using a low octane fuel will cause pre-ignition in a high compression engine.

The higher the Cetane rating, the easier the fuel will ignite once injected into the diesel combustion chamber. However, in engines that are more sensitive to Cetane number, the tendency towards black smoke is greater as the Cetane number increases. This is due to the short ignition delay, which assures that some raw fuel is sprayed into an established flame—which is why soot is produced. On the other hand, the use of a low Cetane number fuel will cause a longer ignition delay to occur once the fuel is injected into the combustion chamber as well as harder starting in cold weather.

Hexyl nitrate or amyl nitrate is often added to the fuel at the refinery stage in order to increase the Cetane number of the fuel, especially in areas that are subjected to cold weather conditions.

Volatility (ASTM) Designation

Fuel volatility requirements depend on the same factors as Cetane number stated earlier. The more volatile fuels are best for engines in cars, buses, and trucks where rapidly changing loads and speeds are encountered. Low end point fuels tend to give better fuel economy where their characteristics are needed for complete combustion and will definitely produce less exhaust smoke, odor, deposits, crankcase dilution, and engine wear.

The importance of fuel end point increases with an increase in engine rpm.

The "end point" of a fuel is established by a distillation test whereby a given volume of fuel is placed into a container which is then heated gradually. As the fuel boils, vapors pass through a tube located in an ice bath where the vapors are condensed and collected in a graduated container. Due to the many types of hydrocarbons in fuel oils, different boiling temperatures will occur; therefore, during the test boiling temperatures will keep rising.

The temperature at which the first 10% of the fuel is recovered in the container is known as the 10% point; similarly the temperature corresponding to 90% recovered is called the 90% point, and the highest boiling temperature reached at the end of the test is called the "end point." These temperatures at which the fuel vaporizes are more commonly called the "distillation temperature." Typical distillation temperatures of both a No. 1D and No. 2D diesel fuel are as follows:

Distillation Range	*No. 1D*	*No. 2D*
10% Point	350–475°F (177–246°C)	400–500°F (204–260°C)
90% Point	450–600°F (232–315°C)	550–650°F (288–343°C)
End Point (100%)	500–625°F (260–329°C)	625–700°F (329–371°C)

Viscosity

This is a measure of the fuel's resistance to flow, and it will decrease as the fuel oil temperature increases. A high viscosity (thick) fuel oil may cause extreme pressures in the injection system and will cause reduced atomization and vaporization of the fuel spray.

Recommended fuel oil viscosity for high speed diesel engines is generally in the region of 39 SSU (Seconds Saybolt Universal) derived from a test that measures the time taken for a given quantity of fuel to flow through an orifice (restricted hole) in a tube.

This vicosity rating of 39 SSU maximum will provide good penetration into the combustion chamber and atomization of the fuel. The viscosity of diesel fuel also acts to lubricate the internal components of the injection system. The viscosity of diesel fuel is normally specified at 40°C (104°F) with the Canadian standard being between 1.2–4.1 centistokes, while the limits are 1.3–2.4 for ASTM grade 1D and 1.9–4.1 for grade 2D.

Carbon Residue

The amount of carbon residue left within the combustion chamber has a direct bearing on the engine deposits and cleanliness of combustion; therefore, the smaller the amount of carbon residue at the end of the combustion process, the longer the engine life will be and the cleaner the exhaust smoke.

The amount of carbon in a fuel is determined by burning a given quantity in a sealed container until all that remains is carbon residue. Carbon residue is expressed as a percentage by weight of the original sample of the fuel oil.

Sulphur

Sulphur has a definite effect upon the wear of the internal components of the engine such as piston rings, pistons, valves, and cylinder liners. In addition, a high sulphur content fuel requires that the engine oil and filter be changed more frequently due to the corrosive effects of the hydrogen sulfide in the fuel and the sulphur dioxide or sulphur trioxide that is formed during the combustion process—which combines with water vapor to form acids. High additive lubricating oils are often used and desirable when high sulphur fuels are used.

Refer to the engine manufacturer's specifications for the correct lube oil to use when a high sulphur fuel is used.

Sulphur content can only be established by chemical analysis of the fuel oil. Fuel sulphur content above 0.4% is considered as medium or high, whereas fuel with a sulphur content below 0.4% is considered low. Summer grade diesel fuel contains between 0.2 and 0.5% sulphur, while winter grades often contain less than 0.2%.

Flash Point

This condition has nothing at all to do with the combustion phase or performance of the fuel in the engine but is rather a measure of the temperature at which the fuel oil vapors will flash when in the presence of an open flame. Safety in handling and storage are the only points warranting consideration for flash point.

Cloud Point

This is the temperature at which the wax crystals in the fuel (paraffin base) begin to settle out with the result that fuel filter plugging occurs. This condition exists when cold ambient temperatures are encountered and is the reason that a thermostatically controlled fuel heater is required on vehicles that are to operate in cold weather environments.

Failure to use a fuel heater will prevent fuel from flowing through the filter and the engine will not run. Cloud point generally occurs 5–8°C (5.5°F) above the pour point.

Pour Point

The pour point of the fuel determines the lowest temperature at which the fuel can be pumped through the system. Pour point is expressed as the temperature 5°F above the level at which the oil becomes solid or refuses to flow.

Pour point is generally about 10°F lower than the cloud point, although flow improvers can result in satisfactory fuel flow at about 9°C colder temperatures than is possible with untreated fuel.

Ash Content

Contained within the fuel oil are ash-forming materials in the form of abrasive solids or soluble metallic soaps. The solids will cause wear of injection equipment, pistons, rings, and liners and also increase engine deposits. Ash from soluble soaps will contribute to engine deposits and wear. Determination of ash content is established by burning a given weight of fuel oil in an open container until even all the carbon deposits are consumed. Weight of the remaining ash is then expressed as a percentage of the weight of the original test sample of fuel oil.

Corrosion

This is the tendency of the fuel oil to react with copper, brass, or bronze parts of the fuel system. This specification does not indicate the corrosion of steel parts of the engine, which may occur from the use of high sulphur fuels with low engine temperatures.

Corrosion is determined by immersing a strip of polished copper in the fuel for a period of three hours at 212°F; the results are interpreted as (1) slight tarnish, (2) moderate tarnish, (3) dark tarnish, or (4) corrosion.

API Gravity

The API (American Petroleum Institute) gravity of a fuel oil has no direct bearing on engine performance, but it does indicate the fuel oil's viscosity, its distillation characteristics (temperature at which the fuel vaporizes), and the heating value of the fuel in BTU's.

Heavier distillates of fuel oil contain more BTU's per gallon than the lighter distillate fuel oils; therefore, all things being equal, the engine can produce more horsepower per gallon of fuel used. However, these heavier distillates generally create more exhaust smoke and odor. The fuel economy with the heavier distillates may not be as good in some cases as with that of the lighter fuel oils.

Specific Gravity

This is the ratio of the weight of the fuel oil to that of an equal volume of water, which weighs 10 pounds per Imperial gallon; therefore, water is given a SG number of 1.

SG is commonly designated as "sp gr 60/60F" indicating that both the fuel oil and the water are weighed and measured at a temperature of 60°F.

On the other hand, the API gravity of a liquid is measured with a special hydrometer, the float of which gives a reading at the surface of the liquid in degrees API. A temperature correction must be applied to the observed reading to obtain the degrees API at 60°F (standard temperature).

Fuel Additives

Generally no fuel additives are required when a suitable fuel is used; however, certain fuel characteristics can be improved by treatment with additives which fall into the following five general types:

1. Ignition or Cetane improvers
2. Detergents or solvents
3. Oxidation inhibitors or stability improvers
4. Corrosion inhibitors or stability improvers
5. Smoke and odor control

Figure 4–2 illustrates the typical additives used in automotive diesel fuels. Although the chart shows barium being used as an antismoke additive, this is no longer used because of environmental concerns.

Heat Value of a Fuel

One of the major constituents of a diesel fuel oil is hydrogen, with water being the by-product of combustion when the hydrogen is burned. This water may remain in vapor form in the hot combustion gases or, if the gases are cooled, then the water vapor will be condensed to a liquid state, giving up its latent heat of vaporization.

AUTOMOTIVE DIESEL FUEL ADDITIVES

Additive	Type	Function
Detergents	Polyglycols, Basic Nitrogen-Containing Surfactants	Prevent Injector Deposits - Increase Injector Life
Dispersants	Nitrogen-Containing Surfactants	Peptize Soot and Products of Fuel Oxidation - Increase Filter Life
Metal Deactivators	Chelating Agents	Inhibit Gum Formation
Rust and Corrosion Inhibitors	Amines, Amine Carboxylates and Carboxylic Acids	Prevent Rust and Corrosion in Pipelines and Fuel Systems
Cetane Improvers	Nitrate Esters	Increase Cetane Number
Flow Improvers	Polymers, Wax Crystal Modifiers	Reduce Pour Point
Antismoke Additives or Smoke Suppressants	Organic Barium Compounds	Reduce Exhaust Smoke
Oxidation Inhibitors	Low Molecular Weight Amines	Minimize Deposits in Filters and Injectors
Biocides	Boron Compounds	Inhibit Growth of Bacteria and Microorganisms

Note: No Commercial Additives Reduce Cloud Point

Figure 4-2. Automotive diesel fuel additives. [Reprinted with Permission © 1984 Society of Automotive Engineers, Inc.]

The heat value of a fuel is determined in a calorimeter. The products of combustion are cooled to the original temperature, the water vapor is condensed, and total heat released is known as the gross, or high heat value (HHV) of the fuel (BTU per pound for liquid fuel or BTU per cubic foot for gaseous fuels).

If we assume that the water vapor is not condensed, then the latent heat of vaporization of the water is subtracted to give the fuel's net or low heat value (LHV).

In North America, the thermal efficiency (heat efficiency) of an engine using liquid fuel is calculated on the basis of HHV (high heat value) of fuel, but for a gas engine (natural gas, LPG, CNG, etc.), the LHV (low heat value) is used.

Since the exhaust temperature of an internal combustion chamber engine is well above the boiling point of water (212°F), no internal combustion engine is capable of utilizing the heat that would be released from the exhaust gases if the water vapor were to be condensed. Therefore, when we determine the thermal efficiency of an engine using the HHV of the fuel, the engine is being compared to a theoretical engine capable of utilizing all of the heat released by combustion of the fuel—that is, the ratio of the heat equivalent of work done to the total heat supplied.

Determination of thermal efficiency can be calculated as described in Chapter 2 under the heading "Thermal Efficiency."

Approximate values of both HHV and LHV of diesel fuels are shown in Table 4–1 in relation to API gravity. The average HHV of most fuel oils used in high speed diesel engines is between 19,000 and 19,750 BTU's per pound.

Fuel Requirements of Diesel Engines

Wide variations in the design of diesel engines requires that particular fuel oils be used in a particular engine in order to provide the best performance. The

Table 4–1. *High and Low Heat Values of Some Typical Diesel Fuel Oils**

Gravity deg API	Sp. Gravity, at 60°F	Weight Fuel, lb/gallon	High Heat Value BTU/lb	High Heat Value BTU/gallon	Low Heat Value BTU/lb	Low Heat Value BTU/gallon
44	0.8063	6.713	19,860	133,500	18,600	125,000
42	0.8155	6.790	19,810	134,700	18,560	126,200
40	0.8251	6.870	19,750	135,800	18,510	127,300
38	0.8348	6.951	19,680	137,000	18,460	128,500
36	0.8448	7.034	19,620	138,200	18,410	129,700
34	0.8550	7.119	19,560	139,400	18,360	130,900
32	0.8654	7.206	19,490	140,600	18,310	132,100
30	0.8762	7.296	19,420	141,800	18,250	133,300
28	0.8871	7.387	19,350	143,100	18,190	134,600
26	0.8984	7.481	19,270	144,300	18,130	135,800
24	0.9100	7.578	19,190	145,600	18,070	137,100
22	0.9218	7.676	19,110	146,800	18,000	138,300
20	0.9340	7.778	19,020	148,100	17,930	139,600
18	0.9465	7.882	18,930	149,400	17,860	140,900
16	0.9593	7.989	18,840	150,700	17,790	142,300
14	0.9725	8.099	18,740	152,000	17,710	143,600
12	0.9861	8.212	18,640	153,300	17,620	144,900
10	1.000	8.328	18,540	154,600	17,540	146,200

Note: It should be understood that heating values for a given gravity of fuel oil may vary somewhat from those shown in the above table.

*Bureau of Standards, Miscellaneous Publication No. 97; Thermal Properties of Petroleum Products, April 28, 1933.

engine manufacturer supplies information as to the suggested type of fuel that best meets the operating conditions of their engines.

However, since all current automotive passenger car and light pickup type truck diesel engines use the pre-combustion chamber design, the fuel specifications for these units is basically the same across the board.

For example, a direct-injection type engine used in a heavy duty Class 8 type highway truck application might require a minimum of 45 Cetane fuel with an allowable maximum of 0.5% sulphur with 2D fuel being preferred over 1D because the No. 2D diesel fuel has a greater overall BTU heat value per gallon. In this particular engine the No. 1D fuel is too volatile (vaporizes too quickly) and therefore gives poor fuel economy and performance, but it can still be used. Straight run distillates (fuels) are recommended rather than cracked blends for these types of heavy duty applications.

On the other hand, a No. 1 diesel fuel will offer easier starting in cold weather and is cleaner burning; however, it contains less BTU content per gallon than does the No. 2 since it is a more highly refined fuel and therefore weighs less per gallon.

The fuel economy or miles per gallon is affected by the weight, or API gravity of the fuel. Within the range of 32–40 deg API gravity, a lower number which indicates a heavier grade of fuel will provide more power and increased miles per gallon. A higher number API fuel indicates a lighter fuel that will result in lower power and decreased miles per gallon.

Experience has shown that many trucking companies prefer a fuel with a minimum 40 Cetane number and an API gravity of 36 maximum.

Tests have indicated that for each API gravity number above 36, a 1% lower power and 2% lower fuel mileage occurs. Therefore the difference between 32 and 40 API gravity can make a difference of up to 15% in fuel costs. (See Table 4–1.)

Gasoline has an API gravity rating of about 65; kerosene comes in at about 42; while average diesel fuel is about 35–36. Another interesting comparison between these three fuels is that gasoline has an almost non-existent lubricating equality of 0.45 Kinematic Viscosity cSt at 40°C, while kerosene is 1.2 and diesel fuel runs between 1.9–3.4 Kinematic Viscosity cSt at 40°C. From this information, you can appreciate that use of other than a diesel fuel can cause serious problems due to lack of lubrication to injection system components. Many injection system components operate with clearances as small as 0.0025 mm (0.0000984"), therefore it is imperative that the fuel used have an adequate viscosity in order to provide suitable lubrication for these finely lapped components.

Diesel Fuel Classification

Since the majority of heavy duty truck diesel engines used in North America are of domestic manufacture, the most common requirement of diesel fuel is to meet the 1D or 2D specifications. Automotive passenger cars and light pickup trucks are therefore subjected to the same basic fuel specifications.

The 1D and 2D fuel grades are specified by the American Society for Testing and Materials (ASTM) specification D975 as a guide for engine operators.

The corresponding Canadian Standard for diesel fuels is CAN 2–3.6-M82, which recognizes the extreme low temperatures (Type AA) and the low to high temperatures (Types A and B) and wide area distribution patterns in the Canadian markets.

In Canada, because the diesel grades are adjusted to meet local winter conditions by the refiner, blending with stove oil or kerosene by the customer is seldom practiced. It requires about 30% of No. 1D (Stove) in No. 2D fuels to achieve a 3°C lowering of the cloud point and if blended cold, wax crystals in the cold 2D fuels are not dissolved.

Table 4–2 represents the comparison in diesel fuel between the ASTM and Canadian specifications.

A typical No. 2 diesel fuel produced by Chevron Canada Limited would have the characteristics shown in Table 4–3.

Table 4–4 lists the typical characteristics of a Chevron No. 1 diesel fuel versus that of the No. 2 shown in Table 4–3.

Diesel Fuel Operating Problems

Fuel selection can cause operating problems in an engine—such as unacceptable exhaust smoke concentrations. White smoke in a diesel engine is caused by minute particles of unburnt fuel, generally in cold weather operation especially on initial engine start-up and is caused by low engine air temperatures. However, this condition will disappear when the engine warms up.

Black smoke is generally caused by air starvation, some mechanical defect such as a faulty injector, using a fuel with too high a boiling point, by engine overload, or overfueling the engine through maladjustment.

Table 4–2. *Canadian Government CAN 2–3.6-M82 and ASTM D975 Specifications*

	Type AA	Type A	Type B	ASTM 1-D	ASTM 2-D	ASTM Method
Flash Point °C min.	40	40	40	38	52	D93
Cloud Point °C max.	−48	−34	*	*	*	D2500
Pour Point °C max.	−51	−39	-	-	-	D97
Kinematic Viscosity 40°C cSt	min. 1.2	1.3	1.4	1.3	1.9	D445
	max. -	4.1	4.1	2.4	4.1	
Distillation:						
90% recovered °C max.	290	315	360	288	282* min to 338 max	
Water and Sediment, % vol. max.	0.05	0.05	0.05	0.05	0.05	D1769*
Total Acid Number, max.	0.10	0.10	0.10	-	-	D974
Sulphur, % mass max.	0.2	0.5	0.7	0.5*	0.5*	D1552*
Corrosion, 3 hr. @ 100°C max.	No. 1	No. 1	No. 1	No. 3	No. 3	D130
Carbon Residue (Ramsbottom) on 10% *bottoms, % mass max.	0.15	0.15	0.20	0.15	0.35	D524
Ash, % Wt., max.	0.01	0.01	0.01	0.01	0.01	D482
Ignition Quality, Cetane No., min.	40	40	40	40*	40*	D613

*Note—The Canadian Government specifications and the ASTM specifications provide for modification of these requirements appropriate for individual situations. Refer to these sources for detail.

Table 4–3. *Chevron Diesel Fuel 2: Chevron Heating Fuel 2: Typical Test Data*

Characteristic	*Winter*	*Summer*	*−40°F Pour Winter*
Gravity Degrees API	31.8	31.1	37.9
Flash Point, PMMC °C	59	57	51
Viscosity at 40°C, cSt	3.4	3.0	2.1
Pour Point, °C	−15	−17	−45
Sulphur, Wt.%	0.28	0.22	0.1
Sediment & Water, Vol.%	nil	nil	nil
Carbon Residue, 10% Botts; %	0.05	0.05	0.03
Color, ASTM	Less than 1.5	Less than 2	Less than 1
Cetane Index	45	42	42
BTU per Pound	19,480	19,450	19,675
BTU per Imperial Gallon	168,800	169,300	164,400
Distillation, °C:			
Initial Boiling Point	174°C(345°F)	183°C(361°F)	173°C(343°F)
10% Recovery	236°C(457°F)	220°C(428°F)	210°C(410°F)
50% Recovery	285°C(545°F)	278°C(532°F)	230°C(446°F)
90% Recovery	334°C(633°F)	335°C(635°F)	276°C(529°F)
100% End Point	362°C(684°F)	365°C(689°F)	327°C(621°F)

Blue/gray smoke results from oil being burned in the combustion chamber and usually will be more noticeable on cold starts rather than when the engine warms up, but in either case the engine is in need of mechanical repair.

To prevent fuel line freeze-up due to minute water particles in the fuel, a fuel line water filter can be used as well as adding methyl or isopropyl alcohol in the ratio of 0.0125% or 1 part in 8000, which equals out to about 1 pint of isopropyl alcohol (isopropanol) to every 125 gallons of diesel fuel.

Diesel Fuel Temperatures

The temperature of diesel fuel can greatly affect the power output of the engine. It was mentioned earlier that when the ambient temperature drops to the cloud point of the fuel, wax crystals start to settle

Table 4–4. *Chevron Heating Fuel 1: Typical Test Data*

Characteristics	*Test Data*
Gravity Deg API at 60°F	42.8
Flash Point, PMCC, °C	53 (127.4°F)
Viscosity at 40°C, cSt	1.53
Pour Point, °C	−46
Sulphur, Wt.%	0.08
Sediment & Water, Vol.%	nil
Carbon Residue, 10% Botts; %	0.03
Ash Weight %	nil
Color, ASTM	Less than 1.0
BTU's per Pound Gross	19,825
BTU's per Imperial Gallon Gross	160,900
Distillation Range, °C (Vaporization Point)	
Initial Boiling Point	171°C (340°F)
10% Recovery	167°C (333°F)
50% Recovery	217°C (423°F)
90% Recovery	256°C (493°F)
100% End Point	282°C (540°F)

out. This action can reduce the fuel flow through the filter leading to starvation, and in some cases, the engine will not start. Temperature controlled fuel heaters are now available from a variety of sources for protection against wax formation in cold weather operation.

In the summer when high ambient temperatures can be encountered, the temperature of the diesel fuel can also affect the engine's performance. It is generally considered that the optimum fuel temperature should run between 90–95°F (32–35°C). For every 10°F increase over this temperature, the engine will lose about 1% of its gross horsepower due to expansion of the fuel (less dense), this is 1.5% on a turbocharged engine. Maximum fuel temperatures should never be allowed to exceed 150°F (65.5°C) since fuel injection component damage can result (lack of lubrication) as well as possible flash point temperatures of the fuel when exposed to an open flame condition.

Fuel Additives

A number of commercially available fuel additives are manufactured by several different companies for both gasoline and diesel fuels. These additives are not normally required since the fuel refiner ensures that their specific fuel meets and in many cases exceeds federal specifications.

Geographical location and operating conditions along with the type of fuel storage and handling characteristics as well as maintenance all play a part in determining whether or not some of these commercially available fuel additives might help a base fuel.

These additives fall into the following categories:

1. Cold weather diesel additive. This contains a pour point depressant as well as a fuel conditioner and it allows the fuel to flow down to 40°C below zero. This will assist in preventing waxing of the fuel in cold ambients so that the fuel filters do not become clogged.
2. Diesel fuel treatment. This additive eases starting and prevents corrosion and thereby is claimed to keep the fuel lines, pumps, and injectors clean with an engine performance improvement.
3. Biocide and conditioner. This additive prevents fungi and bacteria from growing in the diesel fuel, either in bulk storage or inside the tank or fuel lines. Fuel kept in storage longer than 12 months can start to break down. Microorganisms actually metabolize fuel creating stringy gooey masses that can plug filters and corrode metal components. Rubber and tank coatings can also be damaged. Evidence of fungi growth is usually noticed on filters as slimy, unfilterable blobs that may appear black, brown, or greenish.
4. Gasoline conditioner. This additive is ideal for gasoline fuel injection systems. It cleans and lubricates pumps and injectors and removes gums and varnish, prevents rust and corrosion to prolong fuel system life and improves the overall performance of the engine with easier starting and smoother idling characteristics.

Gasolines

As mentioned earlier in this chapter, gasoline originates from crude oil and is a lighter distillate than diesel fuel. It is therfore a little bit more expensive to produce. Gasoline is a blend of a number of hydrocarbons with a boiling range of between 27–204°C (81–399°F). Different types of hydrocarbons affect the characteristics of gasolines in which they occur.

Gasoline Requirements

Gasoline is blended at the refinery to minimize driveability problems on a seasonal basis since driveability problems will increase as the ambient temperature drops. The boiling range of gasoline is lowered during the winter months to minimize driveability problems.

If gasoline is stored for a period of time and results in a summer product being used in the winter months, then poor starting and poor driveability will result; while if a winter grade gasoline is used in the summer months, vapor lock can occur, resulting in poor fuel mileage.

To avoid any of the above conditions and to ensure provision of top road performance and best fuel economy, a gasoline should provide the following:

1. Good fuel mileage under all driving conditions
2. Easy starting in all weather conditions
3. Rapid engine warm-up
4. Rapid acceleration
5. Smooth performance
6. Minimum engine maintenance

In addition to the six characteristics quoted above, there are three major factors that govern the performance of a particular grade of gasoline:

1. Fuel volatility (vaporization temperature)
2. Anti-knock quality
3. Deposit control

Volatility

This characteristic is important because it relates to the fuel's ability to change from a liquid to a vapor as it passes through the carburetor, throttle body, or fuel injector into the air stream to provide an acceptable air/fuel ratio mix.

As with diesel fuel explained earlier, gasoline has what is known as a distillation range or temperature at which it was distilled at the refinery. This distillation range indicates the temperature at which the fuel will vaporize and is shown in graph form in Figure 4–3. Distillation curves for typical summer and winter gasolines are shown in Figure 4–4.

Another test that is often used to express the volatility of a particular gasoline is its Reid Vapor Pressure (RVP) at 38°C (100.4°F) under specified conditions. The equivalent ASTM test is ASTM D323. In both cases, this test measures the butane content in the gasoline.

Insufficient fuel volatility can cause the following engine/vehicle problems:

1. Difficult cold starting
2. Vapor lock
3. Carburetor icing
4. Difficult warm-up and acceleration
5. Poor fuel economy especially on short trips

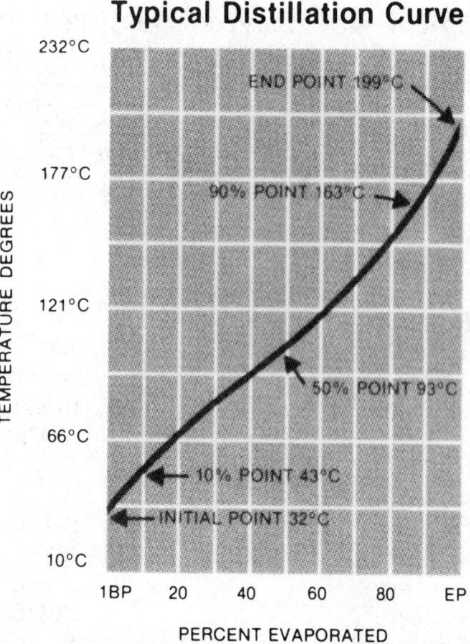

Figure 4–3. Typical gasoline fuel distillation curve. (Courtesy of Imperial Oil Limited, Toronto, Canada)

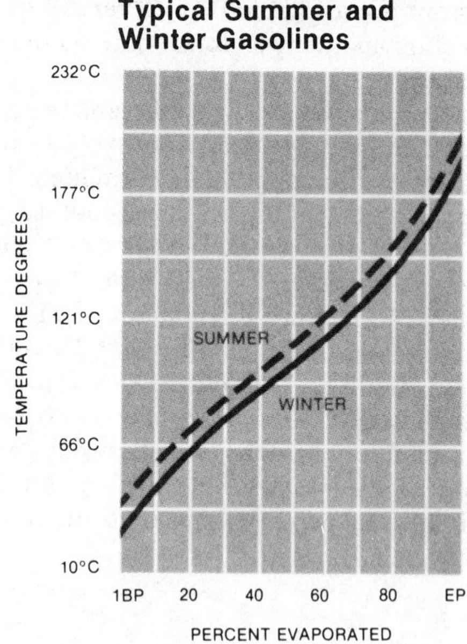

Figure 4–4. Typical distillation curve for summer and winter grades of gasoline. (Courtesy of Imperial Oil Limited, Toronto, Canada)

6. Combustion chamber deposits
7. Crankcase dilution

A typical RVP diagram illustrating the RVP in kPa (1 psi × 6.895 = kPa) and the RVP limits throughout the months of the year with the vapor lock point generally most prominent between the months of mid-April to mid-September are shown in Figure 4–5.

Starting and Vapor Lock Protection

Starting ability is controlled by the Reid Vapor Pressure and the fuel's front end of the distillation curve, or the 10–20–30% temperatures, or the % at 70°C (158°F). Startability is dependent upon the front end volatility or amount of butane blended into the gasoline. This amount is varied as much as 4 times in a year to meet seasonal requirements in various geographical areas of the country. The gasoline is blended with more butane in the winter months to improve starting and less in the summer to minimize vapor lock conditions.

The condition of vapor lock has of course almost been eliminated with the use of fuel injected systems which use a high pressure electric fuel pump to keep the fuel pressure at between 35–45 psi average on most vehicles that use fuel injection today.

Carburetor Icing Protection

In carbureted engines, icing can occur at ambient temperatures between −2 and 13°C (29–55°F)when the humidity exceeds 65%. This occurs because moisture-laden air enters the carburetor and is mixed with fuel droplets; therefore, as the fuel evaporates, it removes heat from the air and the surrounding metal parts, which lowers their temperature.

This condensing water vapor being drawn into the engine will form ice that causes the engine to stall due to lack of air flow passing between the throttle plate and the carburetor body, as shown in Figure 4–6.

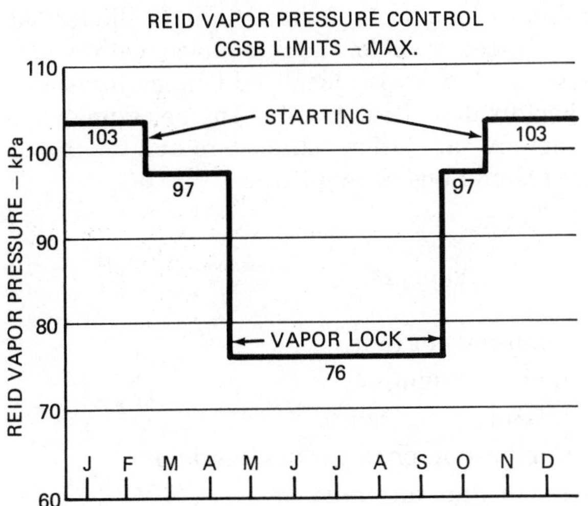

Figure 4–5. Reid vapor pressure control chart. [Courtesy of Imperial Oil Limited, Toronto, Canada]

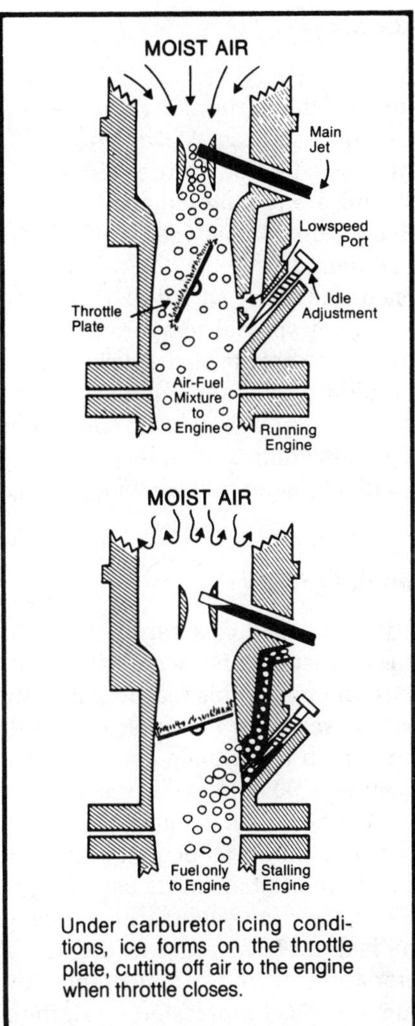

Figure 4–6. Carburetor icing problems. [Courtesy of Imperial Oil Limited, Toronto, Canada]

The percentage of gasoline evaporated up to 100°C (212°F) will reduce the icing condition because higher volatility gasolines create a much greater degree of evaporative cooling.

Engine Warm-Up and Acceleration

The lower the gasoline's evaporative temperatures related to the 50% and 90% figure or to the % at 100°C and 160°C (212°F and 320°F), the faster the engine will warm up. This is because the temperature in the intake manifold has to rise to a point where enough gasoline is vaporized to ensure good mixture distribution to all engine cylinders and also to permit full throttle acceleration without hesitation or bucking without the choke having to remain on.

One other major advantage of a low 90% evaporated temperature is the reduction of crankcase dilution from blow-by gases.

Fuel Economy

Fuel economy in any vehicle is related to driving conditions and the carburetor, throttle body, or fuel injected system setting. A rich mixture will obviously cause a reduction in fuel economy plus odor problems and a smoking exhaust. Figure 4-7 illustrates the relationship between air/fuel ratio, power, and fuel economy.

Because of the many varied driving situations and changes in geographical temperatures and weather conditions, a compromise between fuel economy, crankcase oil dilution, and combustion chamber deposits is necessary. Therefore, since most driving trips cover less than 6–10 miles, the more volatile gasoline with a greater mid-fill volatility gives greater fuel economy.

Anti-Knock Quality

Gasoline is composed of a variety of hydrocarbons; one of these is iso-octane, which is highly resistant to self-ignition and for this reason it is defined as 100 on the octane scale. At the other end of the scale is n-heptane, which is assigned a value of zero. Therefore a gasoline is 90 octane if it reacts like a blend of 90% (by volume) iso-octane and 10% n-heptane.

Gasoline is rated on an octane scale, which is exactly opposite the Cetane scale used for grading diesel fuels. In a gasoline engine, the octane number of a gasoline is a direct measure of the fuel's ability to resist detonation during combustion. Therefore the higher an engine's compression ratio, the higher the octane rating necessary to prevent detonation (spark knock or ping). This condition is explained in detail under the chapter dealing with combustion systems.

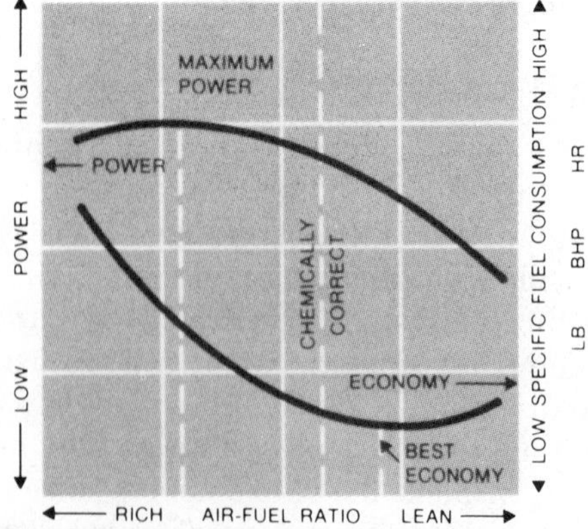

Figure 4-7. Maximum power versus best fuel economy chart. (Courtesy of Imperial Oil Limited, Toronto, Canada)

In a diesel engine, the fuel rating is a measure of the ability of the fuel to minimize ignition delay. In a gasoline engine, the important characteristic of the fuel is its ability to resist spontaneous combustion, while in a diesel engine it is desirable to have a fuel that will ignite fairly quickly once injected into the combustion chamber compressed air.

Therefore, the higher the octane rating of gasoline, the greater will be its resistance to ignite too early or before the spark plug fires. In a diesel engine, the higher Cetane rating indicates a more volatile fuel and therefore ease of ignition once it has been injected.

Prior to 1971, gasoline engine compression ratios regularly averaged about 10–10.5:1. However, post-1971 engines have had their CR's reduced to between 8–9:1 because of the use of lower octane unleaded gasolines. 1971 and later engines can usually operate satisfactorily with regular grade gasolines, but if poor performance is experienced with this grade of fuel, then premium grades can be used. Nothing is gained or lost by using gasoline of a higher octane rating than required by the engine.

Lead as an additive is an anti-knock agent; however, it is dangerous to human health and has therefore almost been totally eliminated in many gasolines today. Current unleaded gasolines use an additive known as MMT (methylcyclopentadienyl manganese tricarbonyl) in place of lead to improve the anti-knock quality of the gasoline.

Detonation can occur, of course, due to incorrect ignition timing setting and carbon deposits within the combustion chamber as well as using the wrong grade of gasoline in an engine with a high compression ratio. Detonation can cause serious engine damage because it causes severe shock pressure waves within the combustion chamber and cylinder with a pronounced audible knock coupled with a loss of power and excessive localized temperatures.

Factors that affect gasoline engine octane requirements are the design parameters and engine operating conditions shown below.

Design Parameters.

1. Compression ratio
2. Ignition timing
3. Mixture temperature
4. Combustion chamber configuration
5. Air/fuel mixture ratio
6. Heat transfer to cooling system
7. Air induction turbulence

8. Volumetric efficiency
9. Exhaust gas dilution

Operating Conditions.

1. Ambient air temperature
2. Ambient humidity
3. Barometric pressure
4. Engine speed
5. Engine load

Octane rating is established by either the Motor Method or the Research Method. Both use the same single-cylinder lab engine. The test conditions for the Research Method are less severe than that for the Motor Method; therefore, commercial gasolines have a higher octane number by the Research Method than by the Motor Method.

In heavy equipment, the Motor Octane Number is more commonly used than the Research Octane Number, while in passenger cars with automatic transmissions under full load and part throttle, the Motor Octane Number is more significant. In standard transmission equipped cars at full throttle where the engine can be loaded at low engine speed, the Research Octane Number is more significant, while at part throttle, the reverse is true.

A Road Octane Number rating by a test car under full throttle acceleration from 10–50 mph (16–80 km/hr) is assigned to the gasoline in terms of a reference fuel. Since road testing is expensive, this method is only used for special octane requirement surveys.

Calculation of the average of Research and Motor Octane Numbers can be established by using the formula

$$\frac{R + M}{2}$$

Although few gasoline retail outlets advertise their fuel's octane rating, the following chart indicates what designation number is assigned to a fuel in relation to its anti-knock index.

Anti-Knock Index

Designation	*Minimum pump octane*
1	Less than 87
2	87
3	89
4	91.5
5	95
6	97.5

The octane requirement for a particular engine is affected by several factors which are shown in Figure 4–8.

Elimination of spark knock by the use of electronic ignition control is common on many engines today with the use of a knock sensor located in the intake manifold. This sensor picks up a given frequency of vibrations/knock and relays a signal to the vehicle electronic control module which then causes spark ignition to retard until an acceptable level is obtained. This knock sensor is used on many of the newer turbocharged gasoline engines.

Gasoline Additives

A number of additives are used in current gasolines in order to alter the characteristics so that a variety of conditions are met. These additives are discussed below.

1. Oxidation Inhibitors. These inhibitors are added to assist in controlling gum and deposit formation.
2. Metal Deactivators. Metal deactivators are used to inhibit reactions between the fuel and metals in the system.
3. Detergents. These additives function to keep the carburetor, throttle body, or fuel injector parts clean and also assist in reducing carburetor icing.
4. Rust Inhibitors. Prevent rust.
5. Tetraethyl Lead (TEL). TEL is used to improve anti-knock quality. Amounts up to 0.66 gram of TEL added to each liter of gasoline (3.0 grams per Imperial gallon) will increase the octane rating by about 15 numbers on the octane scale. In Canada, current TEL content is limited to a maximum of 0.77 gram per liter (3.5 grams per Imperial gallon), while in the U.S.A. the E.P.A. has proposed

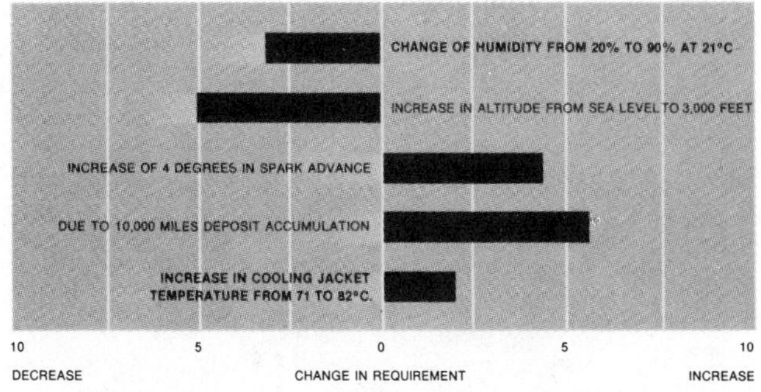

Figure 4–8. Average effect of variables on gasoline octane requirement. (Courtesy of Imperial Oil Limited, Toronto, Canada)

a new regulation that would reduce lead in gasoline to 0.10 gram per U.S. gallon beginning January 1, 1986, and to 0.29 grams/liter in Canada by January 1, 1987. Scavenger additives are used with the TEL additive to promote removal of lead salts formed in the combustion chamber after combustion.

Unleaded Gasoline

Since 1974, catalytic exhaust converters have been used extensively to minimize exhaust pollutants vented into the atmosphere. These engines cannot use leaded gasolines; otherwise, the catalytic converter will plug up rapidly. Unleaded gasolines must therefore be used in these engines; the gasolines are limited to 0.06 gram per Imperial gallon in lead content.

The required anti-knock quality of these unleaded gasolines is obtained by a special refining process that is more expensive to produce than leaded gasolines of a similar octane quality.

Small amounts of MMT (methylcyclopentadienyl manganese tricarbonyl), which is a catalyst-compatible octane improver, are used to improve the anti-knock quality of unleaded gasolines.

Typical properties of gasolines manufactured by Esso Petroleum Canada, Imperial Oil Limited are shown in Table 4–5 where the letters Res O.N. and M.O.N. stand for Research Octane Number and Motor Octane Number respectively.

Chevron Gasolines

The following charts illustrate the properties of various widely used gasolines produced by Chevron Canada Limited and can be considered typical of the gasolines used in cars today. Specific gasoline brands shown are:

1. Premium leaded gasoline
2. Supreme unleaded gasoline
3. Regular gasoline
4. Unleaded gasoline

It is interesting to note the drastic reduction in grams of lead per gallon between the premium leaded and supreme unleaded gasolines. Production costs are greater for unleaded gasolines than for a leaded gasoline of similar octane quality. Small amounts of MMT (methylcyclopentadienyl manganese tricarbonyl), which is a catalyst-compatible octane improver, can be used to improve the anti-knock qualities of unleaded gasolines.

As mentioned earlier, the use of lead as an anti-knock agent is being phased out through governmental legislation because of the danger to human health when it is emitted into the atmosphere from the exhaust system as a by-product of combustion.

You will note that the Research Octane difference between these two gasolines is as low as 0.3 in the summer months while the Motor Octane Number is as much as 2.7.

Although there is a reduction in lead content in the unleaded gasoline, there is a slight increase in the sulphur content. Also, because of various additive concentrations, the color of the gasoline is different, with the premium leaded gasoline being red and the supreme unleaded being green. A similar condition exists between the regular unleaded gasoline and the regular gasoline with the former being water white (clear) in color while the latter is bronze in color.

Research Octane of the regular unleaded is 95 in summer months while the regular gasoline is 94. Their Motor Octane Numbers are 84.9 and 86.2 in the summer months respectively.

There is a substantial reduction in lead content between the regular and unleaded gasolines from 1.4 grams per U.S. gallon in the summer months to 0.002 gram per gallon year around in the unleaded gasoline.

Sulphur content is similar in both gasolines although the Reid Vapor Pressure varies some between them as can be seen in the spec sheets.

Table 4–5. *Properties of Esso Gasolines*

	Esso 2000 (Unleaded)	*Esso (Leaded)*	*Esso Extra Unleaded*
Res. O.N.*	93	94	96
M.O.N.*	84	85	87
TEL content	NIL**	approx. 0.45 gram/L	NIL**
Approximate Compression Ratio served	8.2:1	8.2:1	8.2:1 and higher
Carburetor Icing Control		Adequate	
Volatility Adjustments		Continually with season for each market area	
Detergent		As required for carburetor deposit control	

*Lower for higher altitude market areas.
**Trace lead contents are due to pick-up in the distribution system.

Distillation temperatures (vaporization point) vary considerably between all of these four brands of gasoline, which can be seen in the following information charts.

Table 4–6. *Chevron Premium Leaded Gasoline: Typical Test Data*

	Winter	*Transition*	*Summer*
Octane, F-1 (Research)	98.5	98.3	98.7
Octane, F-2 (Motor)	89.8	89.7	89.3
(R + M)/2	94.2	94.0	94.0
Color	Red	Red	Red
Sulphur, wt. %	0.03	0.03	0.03
Gum, ASTM	2.0	2.0	2.0
Lead, grams per U.S. gal.	1.55	1.78	0.8
Gravity, Deg API @ 60°F	59.3	57.7	59.6
Reid Vapor Pressure, kPa	94.5	101	90
Doctor Test	Pass	Pass	Pass
Copper Strip, 3 hrs. @ 150°F	Pass	Pass	Pass
Distillation, °C:			
Initial Boiling Point	27	28	31
10% Evaporation	39	36	44
50% Evaporation	96	103	103
90% Evaporation	158	149	164
End Point	191	202	204
% Recovered	96.2	96.1	96.6
% Residue	1.1	1.0	1.0

Table 4–7. *Chevron Supreme Unleaded Gasoline: Typical Test Data*

	Winter	*Transition*	*Summer*
Octane, F-1 (Research)	97.6	96.8	97.4
Octane, F-2 (Motor)	86.5	85.4	86.6
(R + M)/2	92.1	91.4	92.0
Color		---------green---------	
Sulphur, wt. %	0.12	0.10	0.12
Gum, ASTM, mg per 100 ml	1.2	1.0	1.0
Lead, gram per gallon	0.002	0.002	0.002
Gravity, Deg. API	54.9	52.4	52.1
Reid Vapor Pressure, kPa	95	90	87
Doctor Test		---------negative---------	
Copper Strip, 3 hrs. @ 66°C		---------pass---------	
Distillation, °C:			
Initial Boiling Point	28	29	30
10%	41	56	45
50%	103	116	107
90%	154	162	157
EP	191	197	195
% Recovered	97.5	97.2	96.5
% Residue	1.1	1.0	0.8

Table 4–8. *Chevron Regular Gasoline: Typical Test Data*

	Winter	*Transition*	*Summer*
Octane, F-1 (Research)	93.9	93.9	94.0
Octane, F-2 (Motor)	90.0	86.3	86.2
(R + M)/2	90.0	90.1	90.1
Color	Bronze	Bronze	Bronze
Sulphur, wt. %	0.03	0.03	0.02
Gum, ASTM	2.0	2.0	2.0
Lead, grams per U.S. gallon	1.2	1.5	1.4
Gravity, Deg. API @ 60°F	65.2	64.3	61.8
Reid Vapor Pressure, kPa	97	96	89
Doctor Test	Neg.	Neg.	Neg.
Copper Strip, 3 hrs. @ 66°C	Pass	Pass	Pass
Distillation, °C:			
Initial Boiling Point	28	28	34
10% Evaporation	37	39	40
50% Evaporation	81	87	91
90% Evaporation	158	163	170
End Point	201	201	206
% Recovered	96.5	96.8	96.7
% Residue	1.0	0.8	1.1

Table 4–9. *Chevron Unleaded Gasoline: Typical Test Data*

	Winter	*Transition*	*Summer*
Octane, F-1 (Research)	94.1	93.1	95.0
Octane, F-2 (Motor)	84.4	84.1	84.9
(R + M)/2	89.3	88.6	90.0
Color		---------water-white---------	
Sulphur, wt. %	0.02	0.02	0.02
Gum, ASTM, mg per 100 ml	1.0	1.0	1.0
Lead, gram per gallon	0.002	0.002	0.002
Gravity, Deg. API	59.9	56.8	55.7
Reid Vapor Pressure, kPa	101	91	83
Doctor Test		---------negative---------	
Copper Strip, 3 hrs. @ 66° C		---------pass---------	
Distillation, °C:			
Initial Boiling Point	28	29	30
10%	37	47	42
50%	92	104	99
90%	154	160	162
EP	197	194	199
% Recovered	96.9	97.0	97.0
% Residue	1.0	0.7	0.9

5

History and Comparisons of Gasoline Fuel Injection Systems

Gasoline fuel injection systems rather than the carburetor are now the preferred type of fuel system for passenger car applications. However, fuel injection was not always so popular, and indeed it is only since the early 1970s that it has been installed as standard equipment on imported European vehicles.

North American and Japanese manufacturers of passenger cars were not committed to gasoline fuel injection in mass-produced vehicles until the late 1970s and the early 1980s; as a result, it is a field that is fairly new to the average automotive mechanic/technician.

The principle of gasoline fuel injection has been around since the invention of the internal combustion "Otto cycle" engine. Technology did not exist at that time to perfect the fuel injection system; therefore, the cheaper and less complicated carburetor method of fuel delivery was chosen.

Carburetors are still in use on a variety of gasoline engines worldwide, but it appears that the carburetor has reached its highest level of improvement as we have known it in its current design. However, recent advances in technology have allowed Robert Bosch and Pierburg in Germany to develop an electronic carburetor that presently costs approximately one-third less to manufacture than a fuel injection system. This electronic carburetor is known as the "Ecotronic" and has been adopted for use on the 316 and 518 BMW models. This carburetor costs 40% more than a conventional carburetor, but it improves fuel economy about 15% over the standard carburetor along with a 20% reduction in exhaust emissions.

The Ecotronic carburetor is used with an electronic digital control unit or ECU (electronic control unit) similar to current fuel-injected technology vehicles. On two-barrel models of this carburetor, electronic control is only used with the primary barrel; the secondary barrel, with its separate float chamber, jets, and throttle valve, operates on the same principle as a conventional carburetor.

Although detailed discussion on the operation and maintenance of carburetors will not be discussed in this book, Figures 5–1 through 5–3 indicate the con-

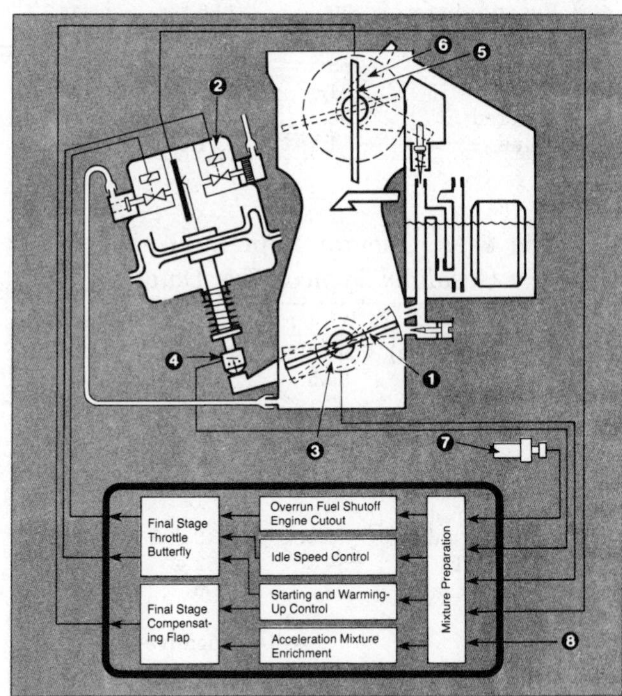

Stage 1 control layout. 1 - throttle butterfly valve; 2 - electro-pneumatic throttle actuator; 3 - throttle spindle potentiometer; 4 - idle switch contact; 5 - choke valve; 6 - choke actuator; 7 - temperature sensor; 8 - engine speed input. Lambda exhaust sensor is additional input where required.

Figure 5–1. Ecotronic carburetor, stage 1 control layout. [Reprinted with Permission © 1984 Society of Automotive Engineers, Inc.]

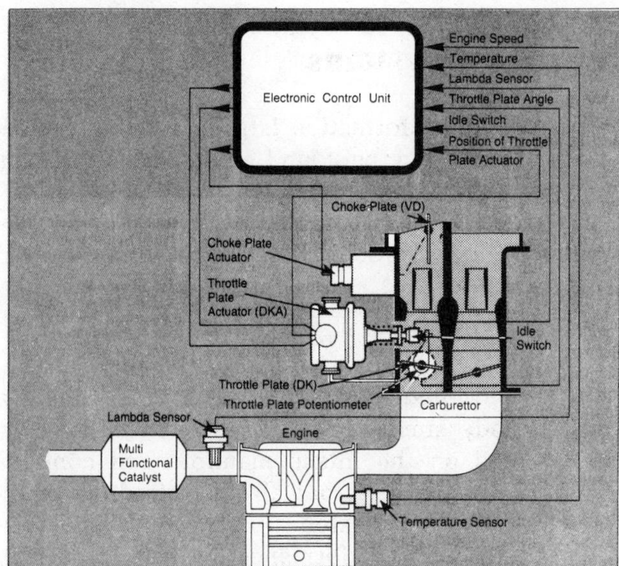

Installation diagram of electronic carburetor. ECU controls all starting, idle, and fuel cutoff functions, and eliminates usual accelerator pump. Second barrel with mechanical throttle operation comes in at moderate cruise speed.

Figure 5-2. Ecotronic carburetor installation diagram. [Reprinted with Permission © 1984 Society of Automotive Engineers, Inc.]

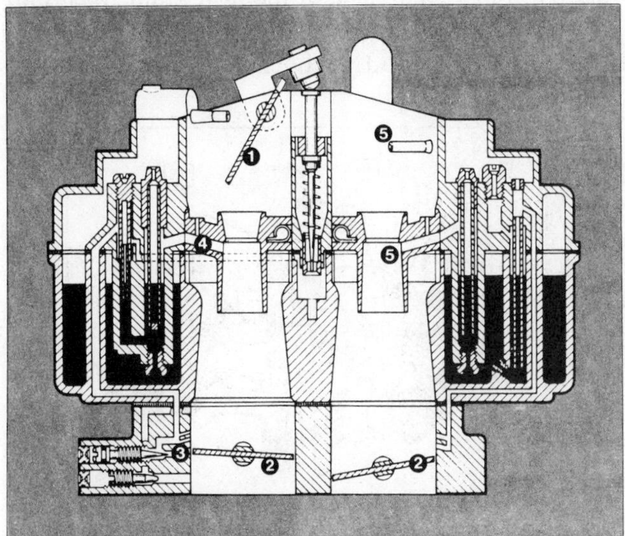

Section of two-barrel economy carburetor, with electronic control on stage 1 (left), mechanical throttle operation for stage 2 throttle valve. 1 - choke; 2 - throttles; 3 -idle jet; 4 - main jet; 5 - full-load jets (3).

Figure 5-3. Section of a two-barrel Ecotronic carburetor. [Reprinted with Permission © 1984 Society of Automotive Engineers, Inc.]

cept of operation of the Ecotronic model carburetor because it does employ electronic controls similar in some respects to those now in use with gasoline fuel injection systems.

Continuous injection of gasoline into the intake manifold was actually employed by the Wright brothers in their early airplane engines. The most prominent application of gasoline fuel injection was that which was applied to military aircraft as far back as 1925 by the United States Army/Air Force because the severe limitations of the commonly used carburetor when applied to fighter aircraft were not acceptable. The Army/Air Force then set about designing a mechanical fuel injection system, which was not perfected for military use until 1936 when injection type carburetors similar to existing throttle body injection systems were adopted.

In 1934 the Germans developed their first fighter aircraft with gasoline fuel injection and subsequently employed this system on most of their fighter aircraft during World War II. Towards the end of the war, the United States adopted direct cylinder fuel injection (similar to a diesel fuel system) rather than carburetor injection for use on their B29 bombers.

Gasoline fuel injection offers the following advantages over its carburetor counterpart:

1. Much better cylinder fuel distribution, therefore more even load distribution between cylinders with less tendency towards detonation because of lean air/fuel mixture ratios.
2. An average of a 10% increase in engine power due to better volumetric efficiency, which is achieved through the use of larger air inlet passages. This is further assisted by the cooler fuel that is delivered directly by the fuel injector and also accounts for better fuel vaporization. Increased valve overlap is generally possible as well as the elimination of air throttling and heat transfer imparted to the incoming air.
3. Leaner air/fuel ratios can be more easily obtained along with provision for fuel shutoff during coasting or deceleration, such as is now available on some imported vehicles (e.g., VW).
4. Faster acceleration is usually available because atomized fuel is delivered directly to the inlet valve port with port type fuel injection for momentary fuel enrichment.
5. A wider range of fuel can be employed because of the mechanical atomization of the fuel.
6. In aircraft applications, carburetor icing is minimized because the gasoline is vaporized at the inlet valve port or in the engine cylinder.

7. Reduction or elimination of back-firing in the inlet manifold or throttle body.
8. Higher engine torque, quicker starts and engine warm-up.
9. Closer control of air/fuel mixture and the ability to maintain stoichiometric conditions over all operating conditions.
10. Less exhaust emissions because of Item 9.
11. Elimination of many pollution control devices required with carburetor systems.
12. No requirement for mixture screw adjustments as with carburetor systems.
13. Better fuel economy.
14. No choke system requirements.

Fuel injection in gasoline engines is less critical than that for diesel engines because, in the gasoline engines, the injection pressures are very much lower and timing is less sensitive. The rate of injected fuel is usually in phase with and proportional to the volume of air taken in by the engine for any operating condition.

Since the fuel is not injected directly into the compressed air in the engine cylinder with a gasoline engine as it is with a diesel engine, the spray of gasoline has less penetration, but it still requires good atomization characteristics.

Gasoline fuel injection systems now in wide use by most of the major vehicle manufacturers are derived from the Robert Bosch concept; Robert Bosch, of course, has used their Model "D," "L," and "K" systems on passenger cars longer than anyone else in the industry. In North America, General Motors, Ford, and Chrysler all employ gasoline fuel injection systems that are derived from the Robert Bosch concept with minor design changes to suit their particular style and size of engines. Bendix also offers a gasoline fuel injection system that parallels the Robert Bosch idea; Bendix systems have been used by American Motors and Renault in the Alliance model car as well as by Cadillac in their late 1970s digital fuel injected engines.

The Robert Bosch gasoline fuel injection systems are discussed in detail in Chapter 6, dealing with that make of system, while the systems employed by GMC, Ford, and Chrysler are discussed in chapters 7, 8 and 9.

Other passenger car manufacturers using gasoline fuel injection employ Robert Bosch systems with some modifications having been made for their particular engine design. These vehicles employ the "D," "L," "K" D, L, K or Motronic Bosch fuel system; information requirements for these vehicles can be found under the Robert Bosch chapter.

Basic Types of Gasoline Injection Systems

The following information is general in nature because details can be found under each system's chapter.

The two main types of fuel injection systems now in use are the throttle body injection and individual fuel injectors.

TBI

TBI (throttle body injection) is a system where a throttle body (similar in appearance to a carburetor) is mounted on the intake manifold and contains either one or two electronically energized fuel injectors. Usually one injector is used on 4-cylinder engines; some high performance systems (especially with turbochargers) use two; and two fuel injectors are found on V-type engines. Figures 5–4 and 5–5 illustrate the concept of operation of the TBI fuel system.

The TBI system is often referred to by different names by the individual vehicle manufacturers. For example, General Motors refers to their TBI systems by either this term or as EFI (electronic fuel injection), while Ford refers to their TBI system as CFI (central fuel injection). Chrysler calls the TBI system used on their vehicles CPI (central point injection); while it is referred to as ECI (electronic controlled injection) on the 1.6L turbocharged Dodge Colt en-

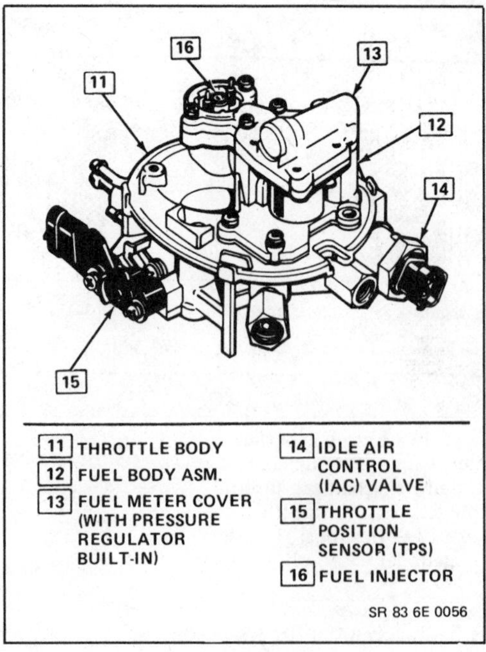

Figure 5–4. Throttle body injection (TBI) unit. (Courtesy of GMC Service Research)

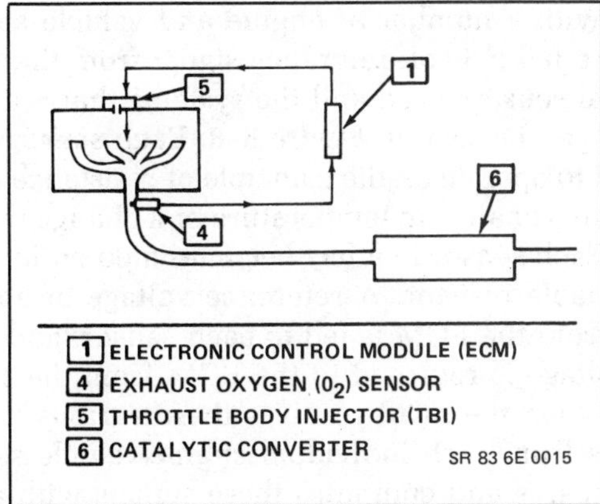

Figure 5-5. TBI closed-loop control system. (Courtesy of GMC Service Research)

gine. Chrysler also called their early TBI system, which was used on their V8's, EFM (electronic fuel metering).

TBI systems use an electric fuel pump and a spring-loaded pressure regulator to maintain fuel pressures of between 9–13 psi (62–90 kPa) average with the maximum being held to approximately 18 psi (124 kPa) on General Motors vehicles. Chrysler systems run up to about 36 psi (248 kPa), while Ford averages about 39 psi (269 kPa).

Individual Fuel Injector System

The second type of fuel injection system now in use is a system that employs individual fuel injectors for each engine cylinder. This system is actually a multi-port injection system and has the injector installed very close to the intake valve as can be seen in Figure 5-6 and 5-7.

Each injector is served by a common fuel rail that has fuel delivered to it by an electric fuel pump; pressures are maintained at a higher level than in the TBI system because a greater number of injectors are used. The higher pressure allows for the pressure drop and pulsations that occur in the system when the injectors are activated. The port-type fuel injection systems maintain average fuel pressures of between 26–28 psi (179–193 kPa) on some systems, while others run as high as 46–53 psi (317–365 kPa).

Again, the different manufacturers refer to their particular system by different terms. General Motors refers to their system as MFI (multi-port fuel injection) as well as SFI (sequential fuel injection) on some Buick models. Cadillac called their earlier system DEFI (digital electronic fuel injection). Ford calls their system EFI (electronic fuel injection). Chrysler refers to their port fuel injection system as EFI.

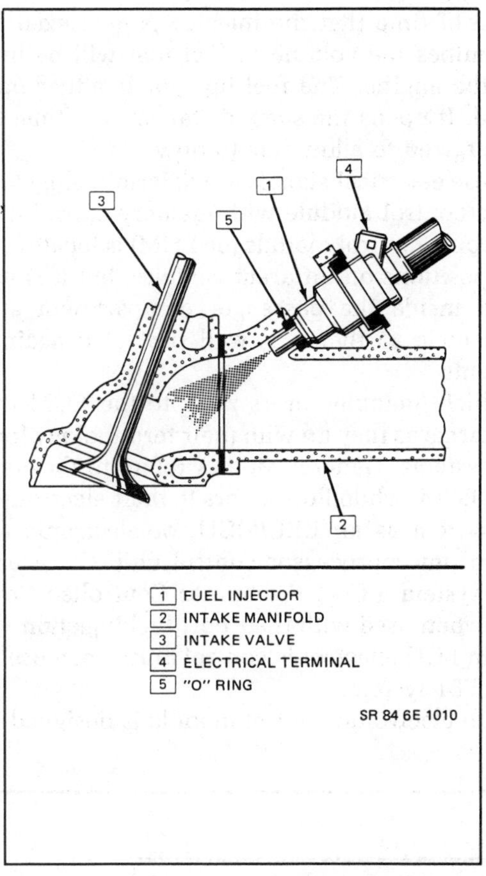

Figure 5-6. Injector used with multi-port type fuel injection. (Courtesy of GMC Service Research)

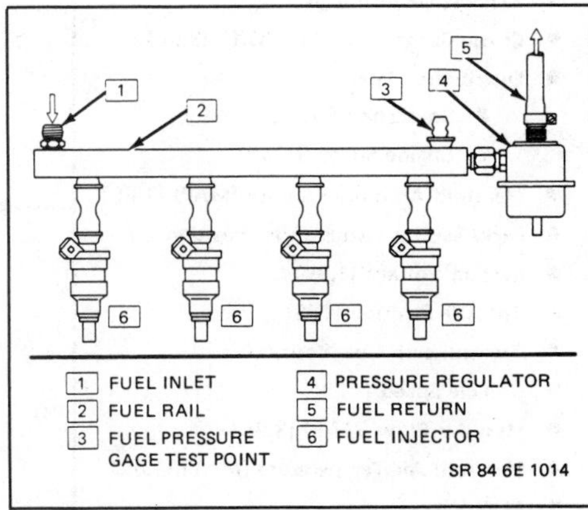

Figure 5-7. Multi-port injection fuel rail assembly. (Courtesy of GMC Service Research)

Injection System Controls

Regardless of the type of gasoline fuel injection system in use, all systems use fuel injectors that are pulsed open and closed by an electrical signal. The length of time that the injector is energized (open) determines the volume of fuel that will be injected into the engine. The fuel injector is either open or closed. It opens the same distance each time that it is energized to allow fuel to flow.

These electrical signals are determined by an electronic control module with memory capability. The electronic control module (or ECM) is located in various positions on different vehicles, but it is usually found inside the passenger compartment area of the vehicle so that it is well protected against the elements.

Vehicle manufacturers refer to the ECM by various terms as they do with their terminology for their fuel systems. General Motors refers to their system as an ECM, while Ford refers to their electronic control system as an EEC/MCU, or electronic engine control microprocessor control unit. Chrysler calls their system a CCC (Computer Controlled Combustion) when used with their port fuel injection system and an ECU (electronic control unit) when used with their TBI system.

Each electronic control module is designed to operate with a number of engine and vehicle sensors that are fed a 5-volt reference signal from the ECM. Typical sensors used and the systems that are monitored can be seen in Figure 5–8. Each sensor is designed to operate on the principle of resistance either through a change in temperature or a change in pressure. Each sensor in effect becomes a potentiometer, or variable resistor. A reference voltage of about 5 volts from the ECM is fed to each sensor, and a signal voltage is returned to the ECM from the sensor unit. When the ECM receives these various voltage signals from each individual engine/vehicle sensor, it compares and computes these signals with a preprogrammed memory system that can be changed. Based on a reference mode, the ECM then sends out voltage signals to alter the air/fuel ratio under various speeds, loads, and operating conditions. In this way, the engine's performance, fuel consumption, and exhaust emissions are held at the maximum efficiency level.

If any condition arises whereby a sensor fails or an abnormal condition exists with the engine or vehicle monitored systems, then a voltage signal from

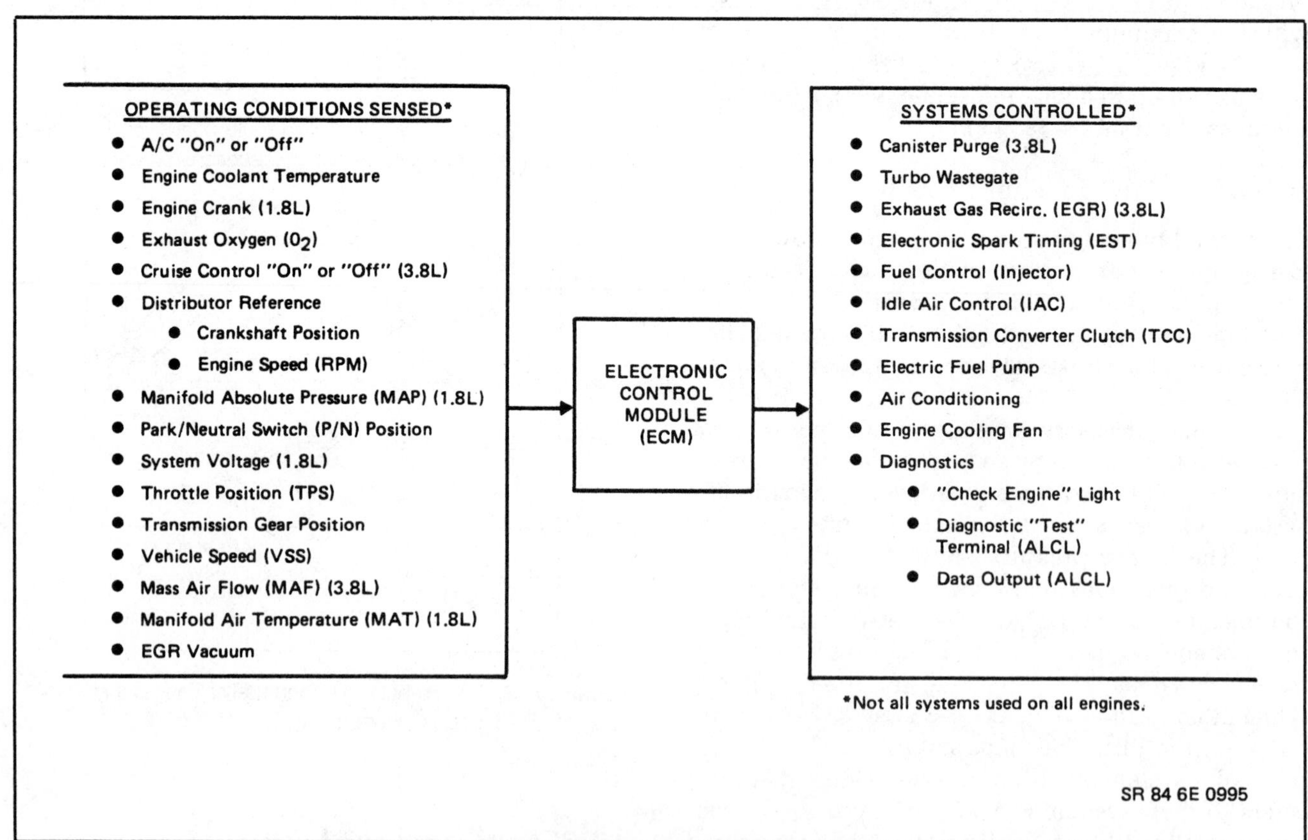

Figure 5–8. Electronic control module (ECM) operating conditions sensed and systems controlled chart. (Courtesy of GMC Service Research)

the ECM will sense the problem and will flash a *check engine light* on the instrument panel.

The mechanic/technician, by using a simple jumper wire arrangement, can then access the memory bank of the ECM by an ALCL (assembly line communication link) or similar arrangement, which is located under the dashboard/instrument panel of the vehicle inside the passenger compartment (See Figure 7–49). This initiates a sequence that will cause a series of flashing numbers to show on the instrument panel. The mechanic/technician then follows the vehicle/engine service manual diagnostic routine to isolate the given problem.

Each service manual contains a troubleshooting chart and a list that indicates where and what the problem might be for a particular flashing number. These code numbers, as they are commonly referred to, can be found in Chapters 7, 8 and 9 in this book dealing with General Motors, Ford, and Chrysler products.

Effective troubleshooting with gasoline fuel injected vehicles is no more difficult than it is for a carbureted engine. The basic rules of troubleshooting apply equally well to both types of fuel systems. Information contained in each chapter for a specific fuel system can in many cases be applied to general service/repair work for other types of fuel systems; therefore, store this information in your head just like an ECM system and use it to advantage when troubleshooting and servicing any type of fuel injection system.

Diesel engines operate on a slightly different philosophy than the gasoline engine fuel injection system; however, information within this textbook is laid out in such a fashion that you will be able to quickly and effectively isolate and service/repair the majority of the complaints that you encounter in everyday work. Troubleshooting charts and helpful hints located within each chapter will allow you to assimilate and effectively understand the system operation and differences that exist between gasoline and diesel fuel injection systems.

Open-Loop versus Closed-Loop Operation

Two terms that you will come across when working with both electronic feedback carburetors and electronic fuel injected systems are "open loop" and "closed loop." These terms refer to the situation in the engine when the air/fuel mixture is, or is not, being controlled by the ECM sensor inputs.

When the engine is cold and the exhaust temperature is below 360°C (680°F), the oxygen sensor located in the exhaust gas stream will not affect the system's operation and therefore the ECM ignores this sensor. Under this condition, the engine is considered to be in an "open-loop" control operating mode. NOTE: The condition of "open" versus "closed" loop is discussed in detail in Chapter 7—GMC Gasoline Fuel Injection Systems.

When the exhaust gases are hot enough to activate a voltage signal from the oxygen sensor, the ECM will enter the control function and alter the air/fuel ratio control to maintain efficient operation. This is the closed-loop operation.

The term "stoichiometric" is often used when discussing internal combustion engines. It refers to the theoretical ratio of the mass of air required for complete combustion of a given mass of fuel. For gasoline this is an average of 14:1 (14 pounds of air to one pound of gasoline).

With this stoichiometric air/fuel ratio, the excess air factor (greater than stoichiometric) is:

$$\text{Lambda} = \frac{\text{quantity of air supplied}}{\text{theoretical requirement}}$$

(where Lambda = 1, or 100%)

A lean mixture contains more air; a rich mixture contains less air. The air/fuel ratio in a spark-ignited engine must be between 0.7 and 1.25 (Lambda) whether it is using a carburetor or fuel injection system. With an air/fuel ratio of 1.25, the air/fuel mixture is no longer ignitable; this is known as the lean misfire limit.

Gasoline engines tend to obtain their maximum power output with a rich mixture (less air) of between 0 and 10%, while minimum fuel consumption is obtained with an excess of air (10%). Satisfactory idling and good transition occurs with an air deficiency of between 30–40%. The most efficient mixture to minimize exhaust emissions is 14.7 to 1, which allows the catalytic converter to operate most efficiently.

6

Gasoline Fuel Injection Systems

Robert Bosch Gasoline Injection Systems

The name of Robert Bosch has been synonymous with fuel injection systems, both gasoline and diesel, for many, many years. It was Robert Bosch who actually was the first company to perfect mass-produced diesel fuel injection systems back in 1927 (see Chapters 10 and 12).

With the experience that Bosch had with diesel fuel injection systems, his magneto and gasoline ignition system, and automotive electrics, this company also became involved with gasoline fuel injection many years ago.

In 1951 Robert Bosch conducted a series of tests on cars running in the Indianapolis 500 race by adapting gasoline fuel injection systems to several cars. The concept used followed Bosch's own success with diesel in-line multiple plunger injection pumps to deliver gasoline directly to a combustion chamber mounted injector.

The success of the Indy 500 cars convinced Robert Bosch that there was a market for high performance fuel injection systems in racing vehicles, and in 1954 the Mercedes-Benz W-196 was produced using a Robert Bosch gasoline fuel injection system. Direct injection of gasoline into the combustion chamber of the engine was reasonably popular on post-World War II Diamler-Benz model Silver Arrow vehicles.

This success was followed in late 1954, and early 1955 with the adoption by the British racing teams of BRM and Vanwall of gasoline fuel injection in their competition cars. Current Formula 1 racing vehicles such as Ferrari, Matra, Alfa Romeo, Ford Cosworth V8, and BRM 12-cylinder engines all run fuel injection on their vehicles.

The power of an internal combustion engine depends on its ability to inhale air—whether it is naturally aspirated or turbocharged. The use of fuel injection allows engine efficiency to be increased through a more even distribution of the air/fuel ratio throughout the engine's operating range.

The theoretical value for complete combustion in an engine is commonly referred to as stoichiometric, which means that we require 14.7 parts of air to 1 part of gasoline. This stoichiometric ratio can be more closely maintained with electronically controlled fuel injection than it can with carburetion.

Because of the greater efficiency of the engine using fuel injection, a horsepower increase of at least 10% is produced over its carburetor version. In addition, better fuel economy and less exhaust emissions are also obtained.

Bosch was the first company to mass produce and have their gasoline fuel injection systems adopted and used by many of the world's leading passenger car manufacturers. The list of users of Robert Bosch gasoline fuel injection systems at the end of this chapter is testimony to this fact, with over 100 vehicle models using Robert Bosch gasoline fuel injection systems.

Today many of the world's vehicle manufacturers that do not appear on the list as users of Bosch gasoline fuel injected systems have worked closely with the Robert Bosch Company and use systems that have been derived from one of the Bosch systems now in use. These other manufacturers have modified the Bosch fuel system to suit their own engines but have retained many of the Bosch ideas and concepts.

Nearly all European passenger car models sold in the U.S. have Bosch gasoline fuel injection systems. All Japanese models that have fuel injection have systems or components produced under Bosch license or using Bosch patents, and all U.S. manufacturers use fuel injection components or systems from Bosch.

Robert Bosch is truly an international corporation with manufacturing facilities in 16 countries, sales and service in over 130; and they employ 110,000 people with 7000 of these working in research and development alone.

Robert Bosch offers a number of gasoline fuel injection systems, which are described in this chapter; however, they currently offer five basic types of systems for use in a wide variety of engines. These systems are briefly described below, with a detailed explanation given within this chapter.

1. The Mono-Jetronic system, which is a throttle body injection (TBI) system employing a single injection valve mounted centrally in a throttle body mounted onto the intake manifold similar to a carburetor. This system will not be discussed in this chapter because it is the same basic system that is used by many North American passenger car manufacturers such as General Motors, Ford, and Chrysler. Refer to Chapters 7, 8, and 9 for an explanation of TBI.
2. The D-Jetronic system, which was Bosch's first mass-produced multi-port electronic gasoline fuel injection system, has now been replaced with the L-LE and LH-Jetronic system and a later model known as the Motronic system. The D system employs an inlet manifold pressure-sensitive (vacuum or low pressure side of throttle valve) control sensing system to control intermittent injection of fuel to the engine through electronic means.
3. The L-LE and LH-Jetronic system, which is also a multi-port system, employs an intake manifold air-flow sensor on the inlet side of the throttle valve for its operation, which is also of intermittent injection operation and is electronically controlled. The L system is more precise and reliable than the D system was.
4. The Motronic system, which is an integrated system for the dual control of both the gasoline fuel injection system as well as the electronic ignition system.
5. The K and KE multi-port Jetronic system, which is the most widely used of the five systems, is so named because the German word for "continuous" starts with the letter K; it is also referred to as a CIS system (continuous injection system). This system employs an air-flow sensor that controls fuel flow proportional to air flow or volume. It is a mechanical system rather than an electronic system such as the Motronic and L-type systems.

The easiest way to remember the difference between the Robert Bosch types of fuel injection systems is that the K system is mechanical and continually injects fuel, while the D-, Motronic, and the L-Jetronic systems are electronic and air-flow controlled with fuel injection being intermittent rather than continuous.

Carburetor/Fuel Injection Comparisons. Earlier in this book we listed the major advantages of fuel injection over its carburetor counterpart. It would be helpful to list the comparable function of the carburetor component with the part that is used in the injection system. The following list shows the comparative part and its function for these two systems.

Carburetor Part	*Fuel Injection Part*
1. Accelerator pump	Throttle switch
2. Fast idle cam	Thermo-time switch
3. Float	Fuel pressure regulator
4. Power valve metering rods	Inlet manifold pressure sensor on TBI systems or the air-flow sensor on Bosch L-Jetronic system
5. Metering jets and idle fuel system	Injector valves and electronic control unit

If you can relate the function of the above parts between the carburetor fuel system and that of the electronic fuel injection system, you will have little trouble in understanding the fuel injected system.

Detailed service information or technical assistance on Robert Bosch gasoline or diesel fuel injection systems can be readily obtained by contacting:

1. Robert Bosch Corporation, 2800 South 25th Avenue, Broadview, Illinois 60153, USA.
2. Robert Bosch Canada Limited, 6811 Century Avenue, Mississauga, Ontario. L5N 1R1, Canada.

Robert Bosch K-Jetronic Fuel Injection System

The easiest way to understand the various types of Bosch fuel injection systems is to start with an explanation of the K-type system because it probably most closely resembles the typical mechanical carburetor fuel system that you are familiar with.

As you know, the carburetor system relies on the position of a throttle (butterfly) valve to control air flow and hence fuel flow into the engine cylinders. The K system is much like a carburetor system in that it also allows a continuous fuel flow to the engine related to throttle position. In the K system, the amount of air drawn in by the engine is controlled by a throttle valve similar to that on the carburetor engine. However this air flow is measured by an air-flow sensor that mechanically controls the fuel quantity delivered to the engine's injectors. Figures 6–1 and 6–2 illustrate this control mechanism—which is the key to the system's operation.

In Figure 6–1, the position of the plate (2) is controlled by the operator's foot through mechanical linkage. This throttle plate establishes the amount of air that will be drawn into the engine cylinders at any given time.

NOTE: There is no direct mechanical connection between the throttle butterfly valve and the air-flow sensor plate.

The air-flow sensor plate (2) in Figure 6–1 is connected to a counter-balanced lever 5 (weight on the right-hand side) that is moved up or down by the air flow into the engine cylinders. In Figure 6–2, the position of the air-flow sensor plate in its funnel (1) (intake throat) transmits this up-or-down motion to the control plunger (5) within the fuel distributor. The position of the fuel control plunger establishes how much fuel (volume) can flow from the inlet passage (3) out through the fuel passage (4) to the individual fuel injection valves at each cylinder.

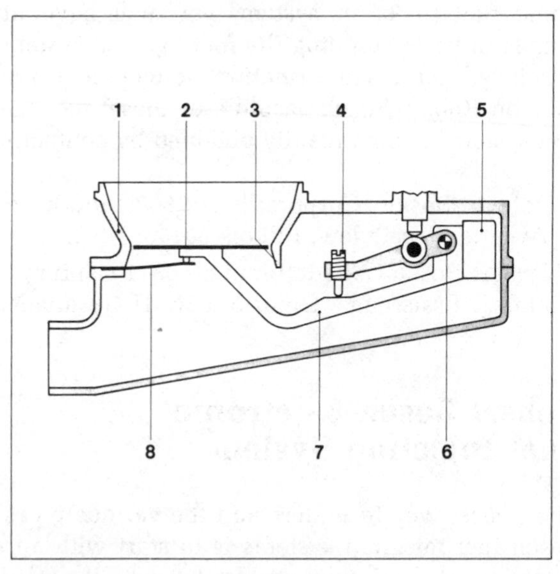

1 Air funnel
2 Sensor plate
3 Relief cross-section
4 Idle mixture adjusting screw
5 Counterweight
6 Fulcrum
7 Main lever
8 Leaf spring

Figure 6–1. K-Jetronic updraft air-flow sensor in zero position. (Courtesy of Robert Bosch Corporation, Broadview, Illinois)

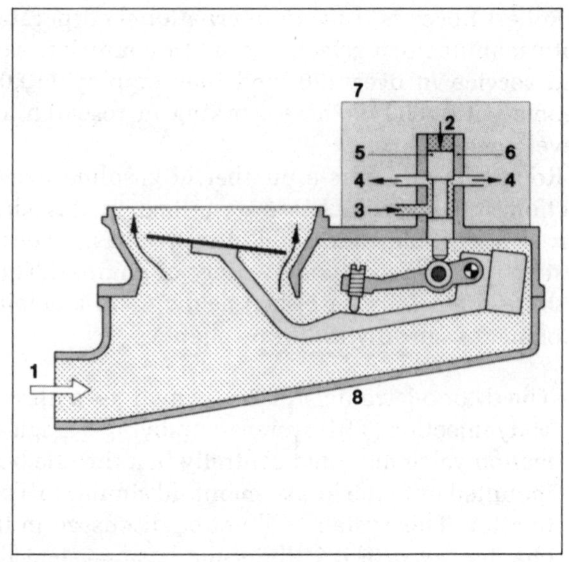

1 Intake air
2 Control pressure
3 Fuel intake
4 Fuel metered to cylinders
5 Control plunger
6 Barrel with metering slits
7 Fuel distributor
8 Air-flow sensor

Figure 6–2. K-Jetronic barrel with metering slits. (Courtesy of Robert Bosch Corporation, Broadview, Illinois)

With the throttle at an idle position, the volume of air drawn into the engine will be small; therefore, the air-flow sensor plate's movement will be small. The fuel plunger will only move up a small amount to deliver a small amount of fuel to the injectors.

If, however, the throttle valve is opened wider, the engine draws in a larger volume of air. The air-flow sensor plate will move up a greater distance in its funnel and transmit this movement to the fuel plunger, which will also move up further to allow a greater volume of fuel to be delivered to the individual cylinder injectors.

K System Schematic. A schematic diagram of the complete fuel system is illustrated in Figure 6–3. Study this diagram and familiarize yourself with the location and names of all the components in the K-Jetronic fuel system.

KE-Jetronic System

Latest improvements to the K-Jetronic system have resulted in the basic system now being known as a KE model designation, with the "E" indicating that the system is electronically controlled rather than mechanically as is the case with the K system.

The KE system differs from the K in that the KE system's air/fuel mixture has been improved, particularly during the engine warm-up phase and under conditions where a load is applied and released from the engine.

A visual interpretation of the differences between the standard K system and the KE system can best be determined by looking at Figures 6–3(a) and 6–3(b), which illustrate the model K and model KE systems in schematic form respectively.

The major differences between the K and KE systems are shown in block diagram form in Figure 6–3(c). Basically, the advances to the KE are that a number of previously mechanical/electrical controls are now fed to an ECU (electronic control unit) that controls the system functions similar to that used on the Robert Bosch model L and Motronic systems discussed later in this chapter.

The KE system alters the fuel pressure differential at the barrel metering slits thereby influencing the amount of fuel injected.

Adoption of the ECU allows the exhaust emissions to be lowered along with an improvement in fuel economy as well as accommodating closed-loop idle speed regulation. An explanation of closed- and open-loop control is discussed later in this chapter as well as in Chapter 7 dealing with General Motors gasoline fuel injection systems.

Other than the adoption of the ECU on the KE, the KE operates in a similar fashion as that described for the K system.

Description of K System Components

Each component of the K-Jetronic system that requires additional explanation is discussed below in relation to the system's operational diagram shown in Figure 6–3.

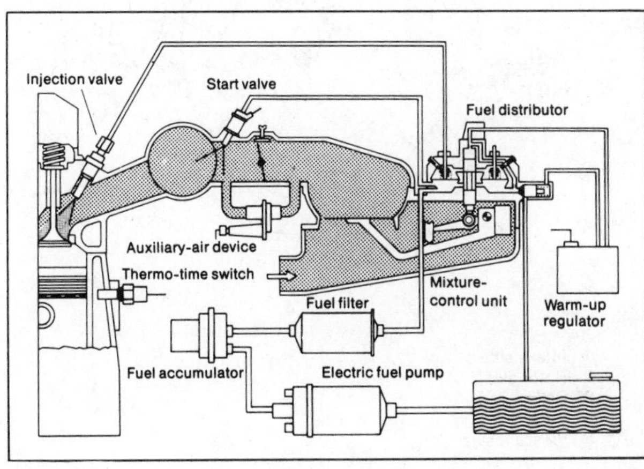

[A]

[B]

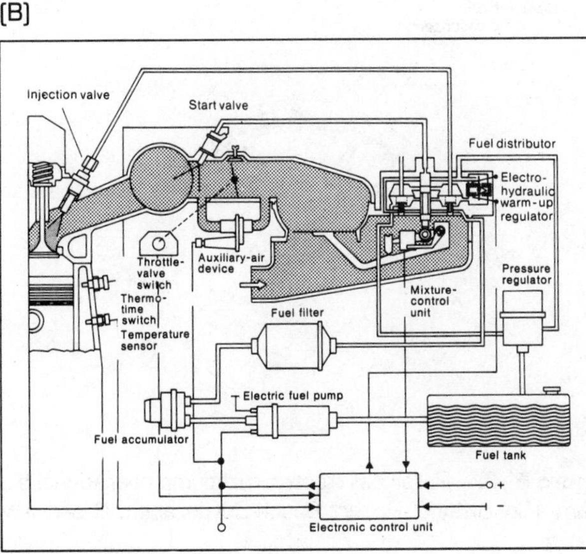

[C]

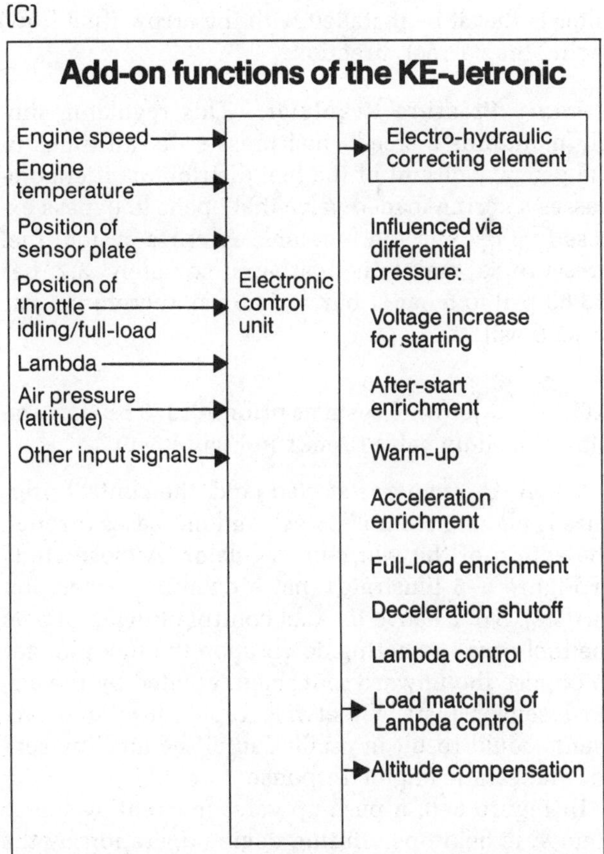

Figure 6–3. K-Jetronic (CIS) with Lambda control operational diagram. (Courtesy of Robert Bosch Corporation, Broadview, Illinois)

Electric Fuel Pump. The fuel pump is an electrically operated roller-cell design unit that may be installed inside the fuel tank or outside the fuel tank, depending on the particular make of vehicle to which it is fitted. Figure 6–4 shows the basic arrangement of the electric fuel pump, which is driven by a permanent magnet type motor.

As the pump rotates, fuel is pressurized by the action of the roller being thrown outwards by centrifugal force. The fuel is trapped in the cavities between the rollers. When the ignition key is turned on, the pump will rotate and continue to do so as long as the engine is running. However, through a safety switch circuit, the pump will stop running if the engine stops but the key is left on.

Fuel Accumulator. The fuel accumulator is simply a spring-loaded diaphragm that keeps pressure on the fuel pumped into the body of this unit to maintain fuel pressure in the system for a period of time after the engine has been switched off. It also acts to dampen the operating noise of the electric fuel pump.

Fuel Filter. The most important item with the fuel filter is that it be installed with the arrow (fuel flow) facing the correct direction.

Primary Pressure Regulator. This regulator simply maintains a steady fuel pressure at all times in the primary circuit of the fuel distributor. It encompasses a spring-loaded valve that opens to bypass excess fuel back to the fuel tank. Average system fuel pressure is maintained at approximately 3.7 bar (53.65 psi) to 5 bar (1 bar = 14.5 psi, therefore 5 bar = 72.5 psi).

NOTE: K-Jetronic systems prior to 1978 did not employ a push-up valve (Item 4 in Figure 6–6).

When the engine is started cold, the control pressure is about 0.5 bar (7.25 psi) and increases through the action of the warm-up regulator. A close study of Figure 6–5 illustrates that a damping restriction (orifice), No. 2 above the fuel control plunger, allows the fuel pressure acting down upon the fuel plunger to oppose the upward movement created by the air-flow sensor plate. Otherwise rapid throttle movements could result in oscillation of the air-flow sensor and erratic engine response.

In Figure 6–6, a push-up valve (current systems), Item 4, is held open during engine operation by the pressure regulator plunger (3). When the engine is switched off, the plunger of the primary pressure regulator returns to its zero position and the non-return valve is closed by a spring.

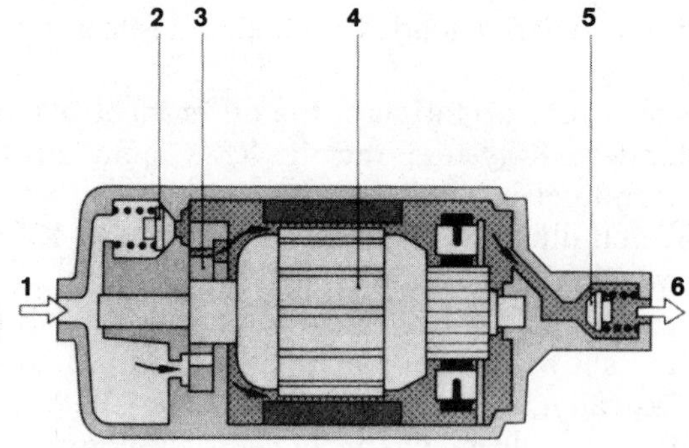

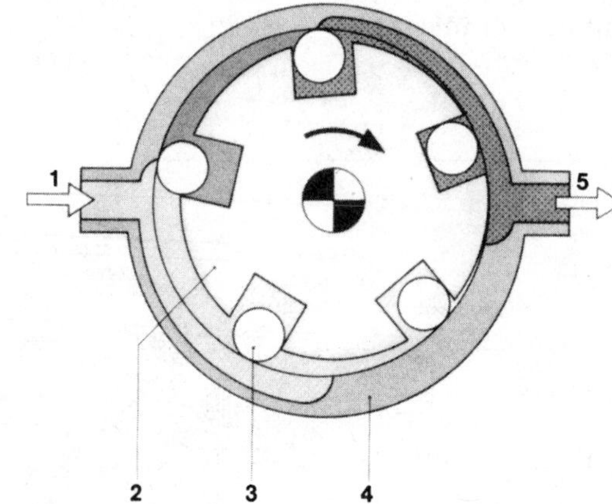

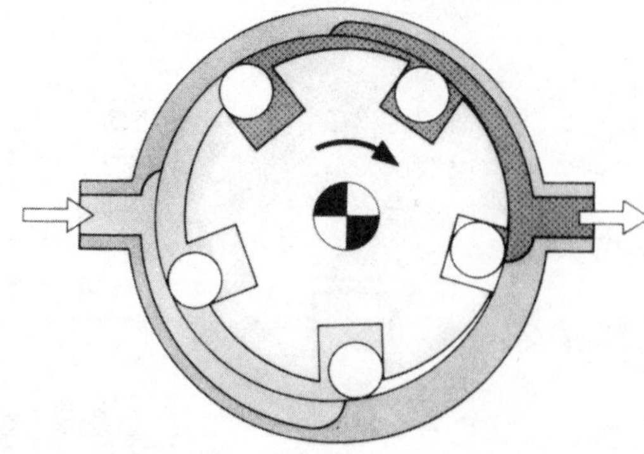

Figure 6–4. Roller cell electric fuel pump operational diagram. (Courtesy of Robert Bosch Corporation, Broadview, Illinois)

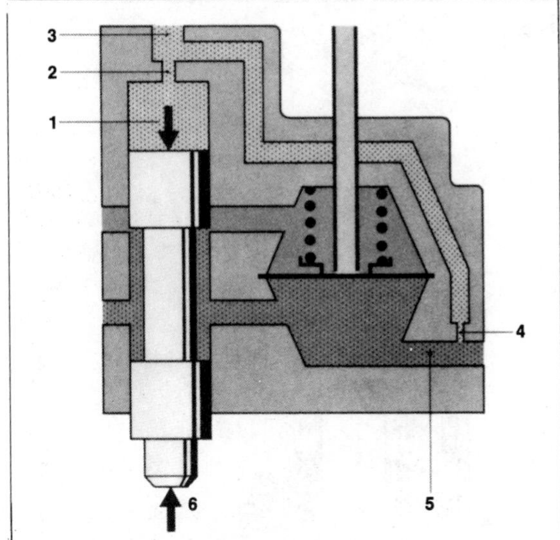

1 Control-pressure effect (hydraulic force)
2 Damping restriction
3 Line to warm-up regulator
4 Decoupling restriction bore
5 Primary pressure (delivery pressure)
6 Effect of air pressure

Figure 6-5. K System primary and control pressure circuit. [Courtesy of Robert Bosch Corporation, Broadview, Illinois]

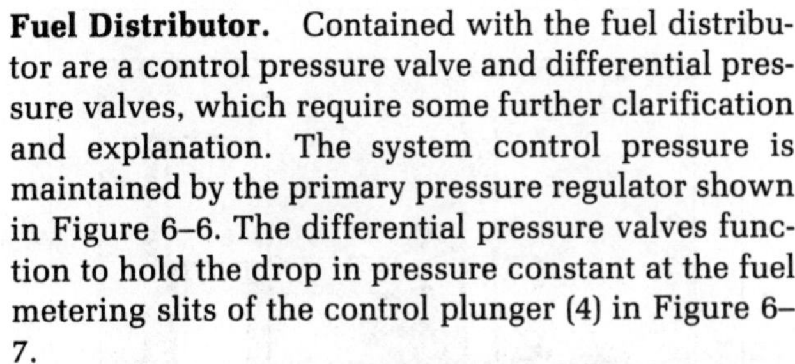

Fuel Distributor. Contained with the fuel distributor are a control pressure valve and differential pressure valves, which require some further clarification and explanation. The system control pressure is maintained by the primary pressure regulator shown in Figure 6-6. The differential pressure valves function to hold the drop in pressure constant at the fuel metering slits of the control plunger (4) in Figure 6-7.

In order to accurately control the fuel delivery in relation to the air-flow sensor in the intake manifold, the position of the fuel plunger must change with air-flow changes. The mechanical connection that exists between the air-flow sensor lever and fuel plunger is in fact capable of doing this. However, to ensure that the fuel quantity will increase or decrease in "direct proportion" to the air flow, a constant pressure drop must take place at the fuel plunger metering slits independently of the volume of fuel flowing through them. This action is obtained through the differential pressure valves illustrated in Figures 6-8 and 6-9. This pressure difference is approximately 0.1 bar (1.45 psi).

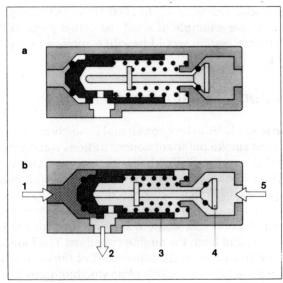

a In zero (inoperated) position
b In operating position
1 Primary-pressure intake
2 Return (to fuel tank)
3 Plunger of the primary-pressure regulator
4 Push-up valve
5 Control-pressure intake (from warm-up regulator)

Figure 6-6. Primary pressure regulator with push-up valve in the control pressure circuit. [Courtesy of Robert Bosch Corporation, Broadview, Illinois]

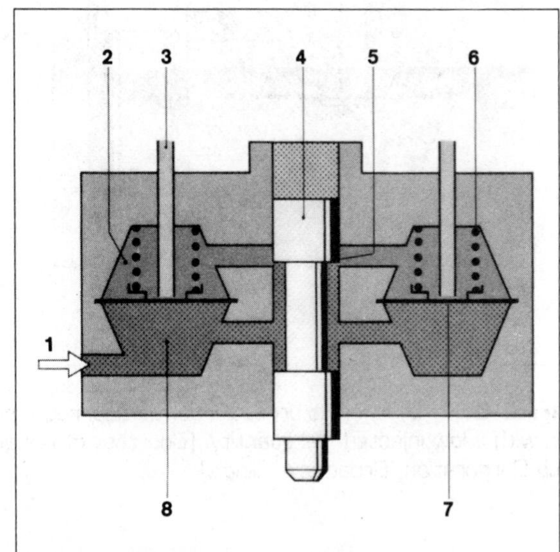

1 Fuel intake (primary pressure)
2 Upper chamber of the differential-pressure valve
3 Line to the fuel-injection valve (injection pressure)
4 Control plunger
5 Control edge and metering slit
6 Valve spring
7 Valve diaphragm
8 Lower chamber of the differential pressure valve

Figure 6-7. K Fuel distributor with differential pressure valves. [Courtesy of Robert Bosch Corporation, Broadview, Illinois]

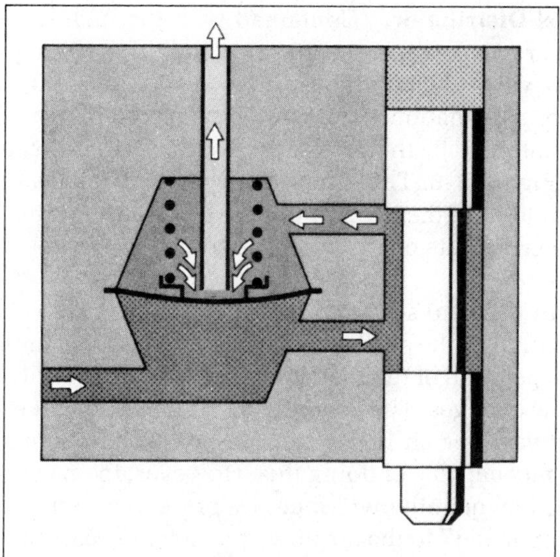

Figure 6–8. K Differential pressure valve, diaphragm position with a large injected fuel quantity. (Courtesy of Robert Bosch Corporation, Broadview, Illinois)

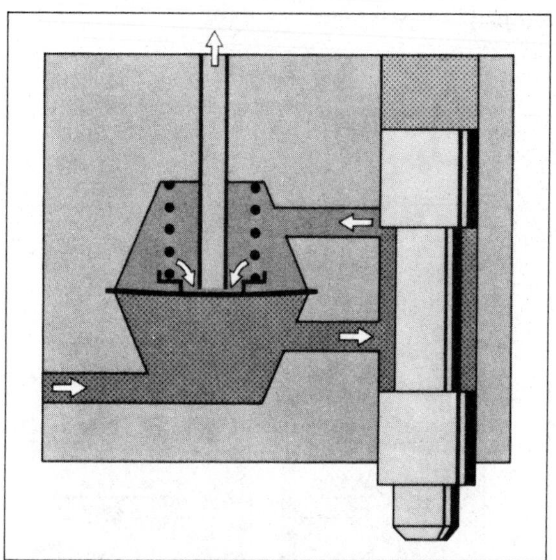

Figure 6–9. Differential pressure valve, diaphragm position with a low injected fuel quantity. (Courtesy of Robert Bosch Corporation, Broadview, Illinois)

Item 7—Fuel Injection Valve. A fuel injection valve, or injector, is fitted at each engine cylinder and is screwed into the intake manifold. Figure 6–10 illustrates the internal components of the injection valve, which is designed to open at pressures in excess of 3 bar (47.85 psi). This pressure is adjustable for various types of engines.

The injection valves do not meter (volume control) the fuel but are opened by fuel pressure. Metering of

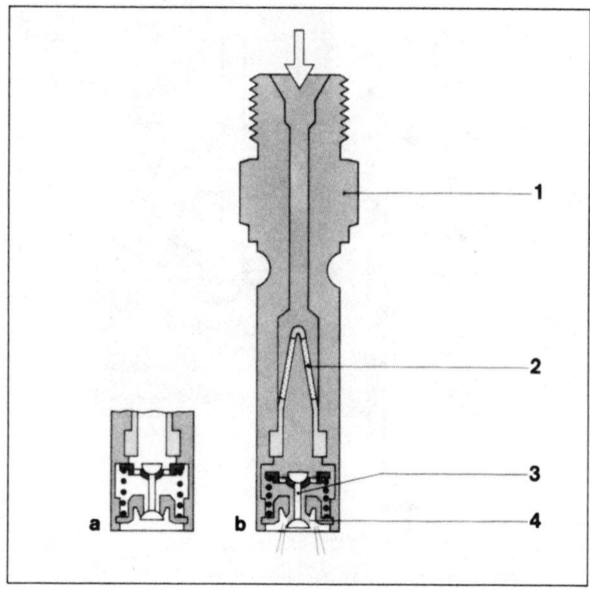

a Inoperative
b During injection
1 Valve housing
2 Filter
3 Valve needle
4 Valve seat

Figure 6–10. Fuel injection valve. (Courtesy of Robert Bosch Corporation, Broadview, Illinois)

fuel takes place at the control plunger slits in the fuel distributor. Atomization of the fuel occurs through the rapid oscillation of the injector needle valve as it opens and closes, at up to 2000 Hertz frequency.

A further example of what the actual parts of the K-Jetronic system would look like appears in Figure 6–11.

Lambda Sensor Probe

Due to the EPA's (Environmental Protection Agency) exhaust smoke pollutant concentrations standards in the U.S. and the ECE exhaust controls in Europe, all engines today are employing what is commonly known as a *Lambda sensor* in the exhaust system.

The purpose of the Lambda sensor probe is to measure the excess air content flowing through the exhaust system from the engine cylinders. The Lambda sensor allows accurate adjustment of the air/fuel ratio to a tolerance of 0.02% of its stoichiometric value in a fuel injected system up to 120 times a minute. Figure 6–12 shows the basic construction of the Lambda probe sensor, which operates on the principle of conduction (heat).

Current Lambda sensors employ both a platinum-coated outer and inner surface. However, the latest developments indicate that titanium dioxide film type exhaust gas sensors have improved transient re-

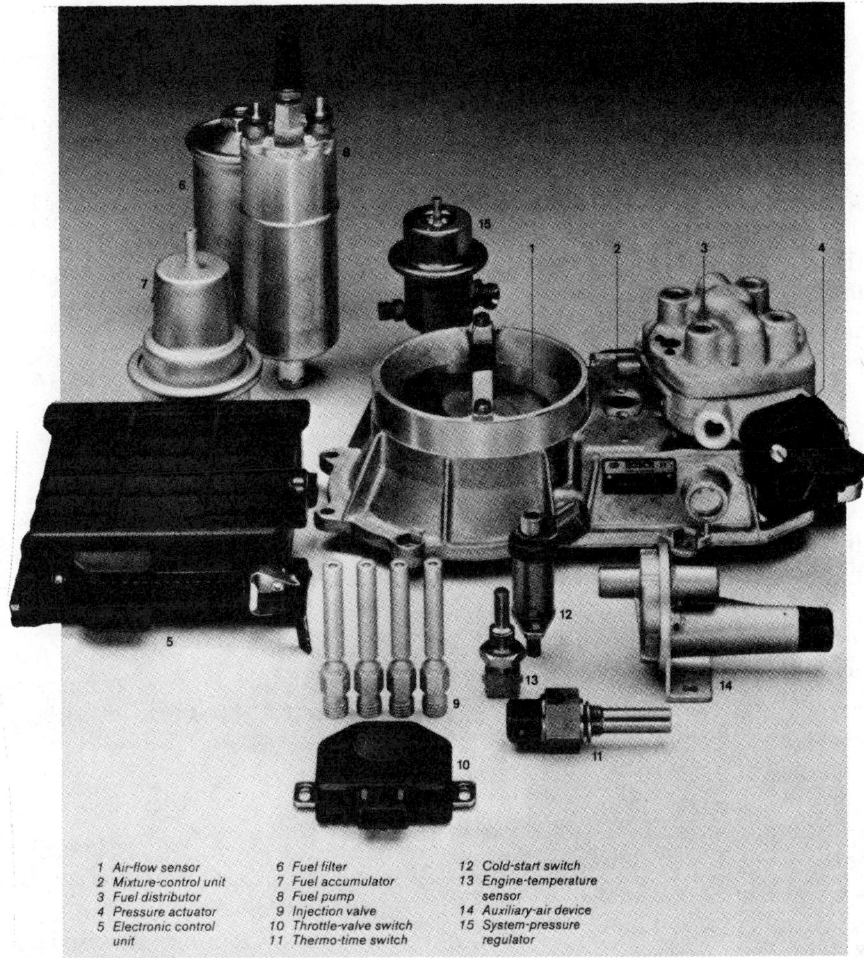

Figure 6-11. Component parts of a KE-Jetronic system. [Courtesy of Robert Bosch Corporation, Broadview, Illinois]

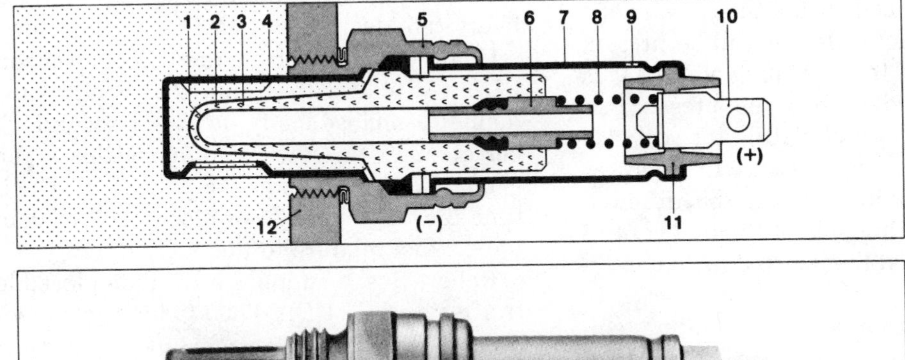

1 Electrode (+)
2 Electrode (−)
3 Sensor ceramic
4 Protective tube (exhaust-gas side)
5 Housing (−)
6 Contact bushing
7 Protective tube (atmosphere)
8 Contact spring
9 Opening to atmosphere
10 Electrical terminal (+)
11 Insulator
12 Exhaust-pipe or manifold wall

Left: Exhaust-gas side

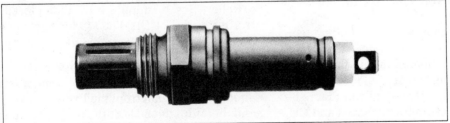

Figure 6-12. Sectional drawing through a Lambda [oxygen] sensor. [Courtesy of Robert Bosch Corporation, Broadview, Illinois]

sponse and can apparently respond faster than the titanium dioxide ceramic-type sensor and the zirconium dioxide sensor. Two basic designs of oxygen sensors are used:

1. The titanium dioxide film is deposited on an insulating substrate and is contacted with platinum electrodes. This design permits the simple incorporation of a film-type heater and/or a compensating thermistor.
2. The titanium dioxide film is deposited on a precious metal electrode and contacted with a second electrode.

The ceramic material used becomes conductive for oxygen ions at temperatures of about 572°F (300°C) and higher. Therefore, if the concentration of oxygen at the probe differs from that of the oxygen outside the probe, an electrical voltage is generated between the two surfaces, which is then sent to an ECU (electronic control unit) or equivalent to interpret and analyze. The ECU will then send out a corrective signal to the engine fuel and timing controls to rectify the situation and to reduce the concentration of nitric oxides emanating from the engine as a by-product of combustion.

In the Bosch K-Jetronic fuel system, an electromagnetic valve (timing valve No. 4) shown in Figure 6–13 is used as a variable throttle. This valve is controlled by electrical pulses from the Lambda control unit.

The correct air/fuel mixture is controlled by the Lambda sensor sending a signal to the timing valve (4) in Figure 6–13. If the timing valve (4) is open, the fuel pressure in the lower chambers of the fuel distributor unit can be reduced. If the timing valve (4) is closed, the primary fuel pressure will be present in the lower chambers. If the timing valve (4) is opened and closed rapidly, then the pressure in the lower chambers of the fuel distributor will also change rapidly. This changing pressure in the lower chambers will cause an increase in the differential pressure at the metering slits of the fuel control plunger with the result that the injected fuel quantity is also increased.

Until the Lambda sensor attains its operating temperature shortly after engine start-up, closed-loop control cannot come into operation and the system is switched to an open-loop mode.

The timing or frequency valve (Item 4 in Figure 6–13) is controlled by an electronic signal from an ECU. The temperature rise at the Lambda sensor causes it to react to the oxygen content surrounding it. If the air/fuel mixture was lean, the timing/frequency valve would be in an open position, which would therefore

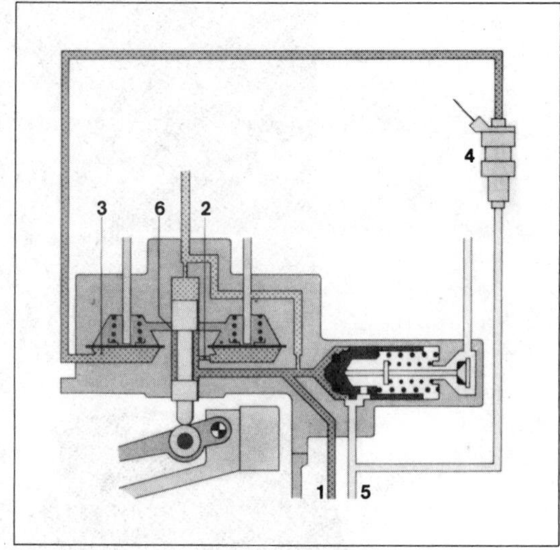

1 Fuel inlet
2 Decoupling throttle (fixed throttle)
3 Lower chambers of the differential pressure valves
4 Timing valve (variable throttle)
5 Fuel return
6 Metering slits

Figure 6–13. Fuel distributor model for a Lambda closed-loop control. (Courtesy of Robert Bosch Corporation, Broadview, Illinois)

result in reduced fuel pressure in the lower chamber of the fuel distributor.

When the Lambda sensor reacts to a rich air/fuel mixture, a voltage signal is sent first to the ECU and them to the timing/frequency valve, which would now close, causing an increase in fuel pressure in the lower chamber of the distributor. The diaphragm moves up against the return spring, and the fuel delivery rate decreases to the engine injectors.

Changing signals from the Lambda sensor and the ECU will cause the timing/frequency valve to actually open/close many times per second. The number of times that the valve opens/closes per second is commonly known as the *duty cycle*. The length of time the valve is open or closed will of course alter the air/fuel mixture to the engine.

Robert Bosch supplies a Lambda closed-loop control tester, P/N KDJE-P600 (which is shown in Figure 6–70 dealing with special tools and equipment).

NOTE: The term "open loop" refers to the situation that exists when the fuel system is not being affected by a signal from the Lambda (oxygen) sensor and therefore the ECU to the timing/frequency valve. Closed loop therefore refers to the fuel system in an operating mode whereby it does receive a corrected Lambda and ECU signal to the timing/frequency valve.

Although the use of Special Tool KDJE-P600 is the preferred method for testing the open/closed cycle of the engine fuel system control circuit, it can be tested with the use of a dwell meter capable of a gauge reading swing of at least 70 degrees. Many manufacturers supply a test socket on the vehicle for this check using the Bosch meter arrangement. The location of the test socket will vary between makes of vehicles; therefore, check the vehicle service manual or underhood emissions sticker to locate this test connection.

Procedure

1. Disconnect the thermo-time switch wiring so that you can temporarily hot wire the switch to duplicate a rich mixture on a cold engine, which will deliver about a 60% enrichment mixture.
2. With this situation as per Step 1, the dwell meter needle should register in a steady position because of the open-loop condition.
3. Reconnect the thermo-time switch.
4. When the engine attains operating temperature, disconnect the wiring from the Lambda (oxygen) sensor; the dwell meter needle should now indicate about a 50% reading.
5. Reconnect the Lambda sensor wiring and wait for a minute or two to allow the Lambda sensor to react to the temperature increase caused by the hot exhaust gases flowing around it.
6. If the Lambda sensor and timing/frequency valve are operating correctly, the needle on the dwell meter should swing back and forward thereby indicating that the Lambda sensor is operating correctly (on/off signal that happens many times per second).

Checks and Tests for the K-Jetronic

Maintenance and repairs to the K-Jetronic fuel injection system require a number of special tools and equipment in order to successfully troubleshoot these systems. Refer to the Special Tools section at the end of this chapter, which deals with Robert Bosch gasoline injection systems for part numbers (P/N) and references as to their use.

CAUTION: Many times when a problem exists on a gasoline fuel injected system, the mechanic/technician overlooks the simplest areas first and subsequently ends up performing checks and tests that do not solve the problem. The easiest way to perform checks and tests on this system is to follow a routine similar to that which you would use when attempting to troubleshoot a carburetor-type engine problem.

Ask yourself if the problem is electrical or fuel system? What are the symptoms of the complaint? If you are unfamiliar with the system, refer to the K-Jetronic Troubleshooting Chart, Figure 6–36, which will help you pinpoint the problem area.

NOTE: Before condemning the fuel injection system, make sure that the ignition system is operating correctly—and that the valve adjustment and engine compression is correct.

Preliminary Checks

Before conducting specific tests, make sure that no fuel leaks exist at any fuel lines, fuel pump, or fuel accumulator. Also check closely for signs of a vacuum leak in the air intake system, especially between the mixture control unit and the engine. In a fuel injected system, just as in a carburetor system, such leaks can cause the engine to run lean due to the fact that on the K system, any leaks would not be sensed by the air-flow sensor valve.

An example of vacuum leaks can be seen in Figure 6–14. Typical areas of leaks would be:

1. Connection dome between the mixture control unit and the common intake manifold
2. The seal at the flange of the cold start valve
3. All hose connections at the common intake manifold
4. Hose connections at the auxiliary air device
5. Injection valve seal rings
6. Intake manifold to cylinder head gasket
7. Brake booster vacuum lines
8. Vacuum Limiter
9. Evaporative canister
10. Air conditioner door actuator, if used

Air (Vacuum) Leakage Check

The quickest method to use to isolate vacuum leaks is to pressurize the intake system hoses. This can be done by removing one hose from the auxiliary valve air regulator and, while holding the throttle valve wide open, apply air pressure to it. Using a spray bottle, apply soapy water to all air hose and manifold connections to detect leaks. Alternately, start and idle the engine and spray non-flammable solvent around possible leakage areas. The idle speed will smooth out if the solvent finds a leak.

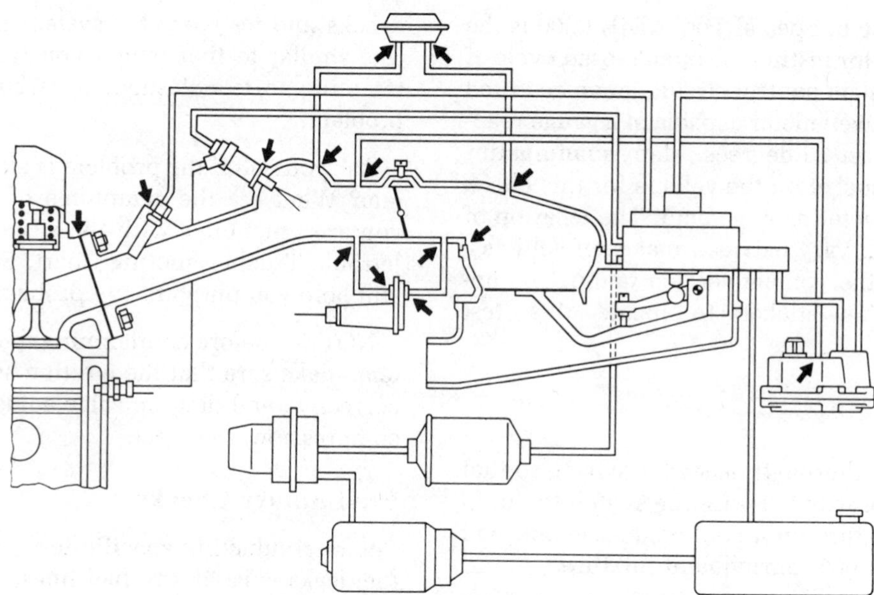

Figure 6-14. K-Jetronic fuel system, possible air leakage points. (Courtesy of Robert Bosch Corporation, Broadview, Illinois)

Water in System

If water gets into the fuel tank, then it will invariably get drawn up into the electric fuel pump. Pressures at the electric fuel pump generally will reach 65–85 psi so that any water in the fuel usually passes through the filter unless a fuel/water separator is being used. This water will be delivered to the center of the fuel distributor where it is trapped by the action of a small plastic filter. Rusting of components will take place fairly quickly, and the water will find its way into both the warm-up regulator and fuel injectors. Blockage of the fuel metering slits at the fuel control plunger will result in a misfire at one or more injectors so that you may think the cause is a bad spark plug.

If the fuel distributor is badly rusted, it should be replaced.

NOTE: Fuel system pressure should be relieved from the system before loosening any fuel lines. This can be done by placing a rag around the large connection on the fuel line at the pressure regulator. Another method is to remove the electric plug from the cold start valve and hot wire (jumper wire from 12-volt source) the cold start valve for 10 seconds, which will cause it to pump pressure into the intake manifold thereby releasing system pressure adequately to allow safe removal of any fuel lines.

SPECIAL NOTE: In areas that experience freezing weather, any water in the fuel will freeze. If this occurs within the electric fuel pump, a high resistance to initial pump rotation will exist. This resistance can cause a fuse to blow in the circuit to the pump.

CAUTION: Be sure to bridge the safety circuit with a jumper wire.

The engine should be warmer than 68°F or 20°C. The safety circuit controls the operation of the electric fuel pump, the warm-up regulator, and the auxiliary air device. If the safety circuit is not bridged, then the electric fuel pump will not operate without the engine "running" and no fuel pressure will be developed—which is necessary for many of the tests that will be conducted when checking out the fuel system.

The safety circuit is used to prevent the electric fuel pump from delivering fuel when the ignition key is turned on or left on accidentally. If the vehicle engine should stall or the vehicle was involved in an accident, then the safety circuit would stop the delivery of fuel. This prevents the possibility of a fire should there be an open fuel line condition.

The bridging of the safety circuit will vary between vehicles, but this usually involves one of two methods:

1. Pull the electric plug out of the air flow sensor.
2. Pull the electronic relay out of its plug and bridge terminals 30 and 87 in the plug.

You can make a typical bridging adaptor by assembling an 8-amp in-line fuse with two male connectors and an on/off toggle switch. This bridging adapter was necessary on all early K fuel systems. Figures 6–15 and 6–16 illustrate the two commonly used fuel pump relay circuits that can be bridged.

Air-Flow Sensor Check

The check of the air-flow sensor is similar on most units and involves a freedom of movement check. When the engine is stopped, the air-flow sensor plate rests lightly against a small leaf spring at the base of the air inlet funnel, as shown in Figure 6–19.

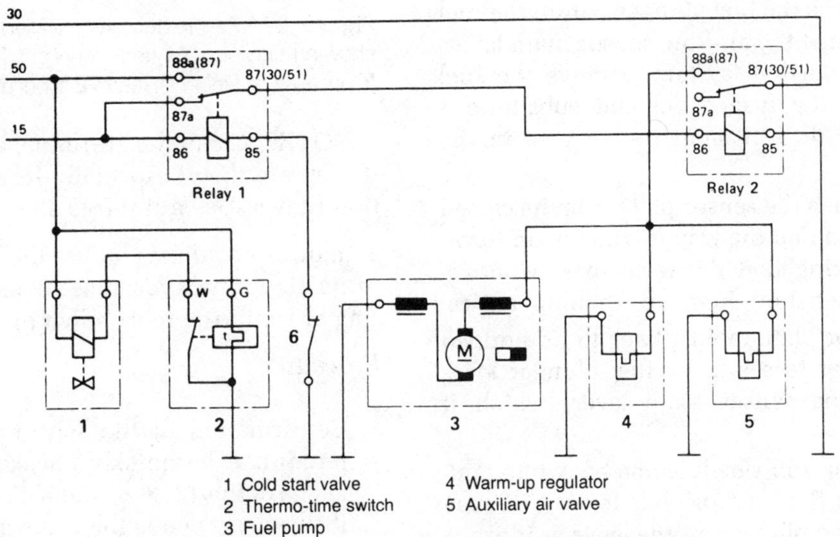

Figure 6–15. Electric wiring diagram, dual relay K system. [Courtesy of Robert Bosch Corporation, Broadview, Illinois]

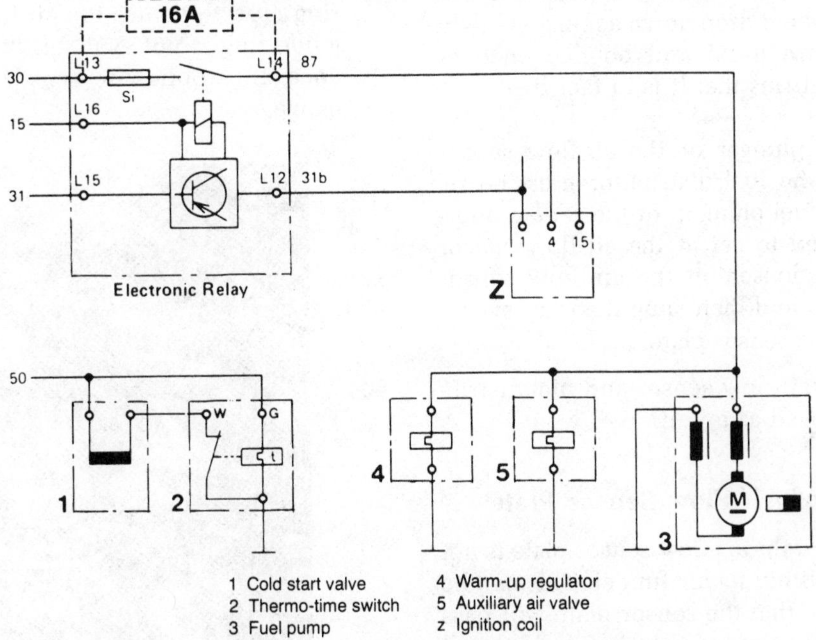

Figure 6–16. Electric wiring diagram, RPM relay 'K' system. [Courtesy of Robert Bosch Corporation, Broadview, Illinois]

Procedure

1. To check the air-flow sensor plate for freedom of movement, remove the rubber air boot on updraft versions, or the air cleaner on downdraft models, from the top of the sensor plate.
2. It will be necessary to allow the electric fuel pump to create pressure in the primary circuit. This is necessary to force the fuel plunger within the fuel distributor against the air-flow sensor main lever. To pressurize the fuel system, remove the fuel pump relay at the fuse panel and substitute a jumper wire with a 16 amp in-line fuse in the jumper.
3. Gently lift or push the sensor plate from its closed (stopped) position all the way to wide open throttle while checking that the resistance to movement remains constant. Remove the jumper wire.
4. Move the sensor plate rapidly back to its stopped position in order to see if the fuel plunger loses contact with the sensor plate lever, which it should do.
5. The fuel plunger will slowly come back into contact with the air-flow sensor plate lever. You may be able to feel the plunger hit the lever as it drops or possibly even hear it. If this is the case, then the fuel plunger is operating correctly.
6. To make sure that the air-flow sensor lever is free, simply lift or push it all the way to a wide open throttle position, then let it fall freely under its own weight. It should drop down against the light leaf spring (Figure 6–19) and bounce once or twice, which confirms that it is in fact free.

If either the fuel plunger or the air-flow sensor plate sticks, either the fuel distributor must be removed to get at the fuel plunger, or the sensor housing must be removed to get at the air-flow sensor plate. However, try loosening the air flow sensor plate mounting bolts and then snug them up evenly first, then recheck the sensor plate.

NOTE: Both the air-flow sensor and fuel distributor should be removed as a unit.

Centering Check of Air-Flow Sensor Plate

Problems will occur if the air-flow sensor plate is not correctly centered within its air funnel. This can be done by making sure that the sensor plate will pass through the narrowest part of the air funnel throat (its base) without touching the sides. An example of a correctly centered and an off-center air-flow sensor plate can be seen in Figure 6–17.

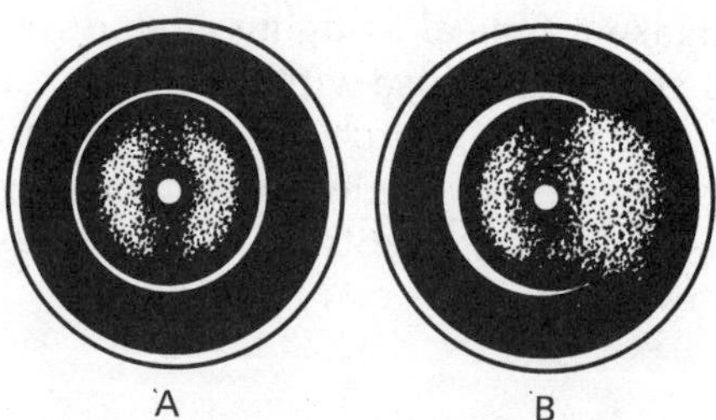

Figure 6–17. K-Jetronic system sensor plate position showing correct (A) and incorrect (B) centering in venturi. (Courtesy of Robert Bosch Corporation, Broadview, Illinois)

NOTE: Clean the air funnel and sensor plate first if they are dirty, especially from crankcase ventilation fumes. Use a lint-free shop towel.

In order to adjust (center) the air-flow sensor plate in its air funnel, you must obtain the correct guide ring or centering tool shown in Figure 6–73.

Procedure

1. Confirmation of the sensor plate position (centered) can be quickly checked by using a feeler gauge at the 12, 3, 6, and 9 o'clock positions to see if the clearance is the same at all points.
2. To adjust (center) the plate, loosen off the central bolt in the plate and, using the correct guide ring (special tool), center the plate in the venturi of the air funnel. Tighten the retaining bolt to 5–5.5 N·m (44–48 in.-lb). Figure 6–18 shows the special guide ring used to center the air-flow sensor plate. If a guide ring is not available, check for equal clearance all around the sensor plate with a feeler gauge.

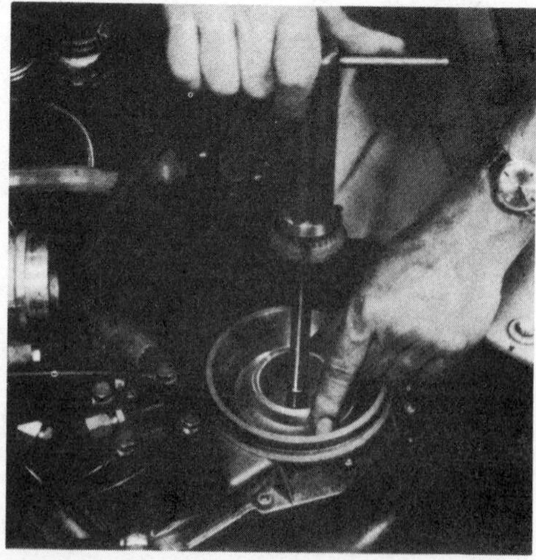

Figure 6–18. Centering adjustment of K air-flow sensor plate. (Courtesy of Robert Bosch Corporation, Broadview, Illinois)

Sensor Plate Height Adjustment

Pressurize the fuel system again by removing the fuel pump relay at the fuse panel and substitute a 16 amp fuse protected jumper wire.

Once the air-flow sensor plate has been adjusted, you should check to see that the plate also sits horizontal in the funnel on the side nearest the fulcrum of the lever arm. An example of this is shown in Figure 6–19. The edge of the plate can lie either below or above the air funnel venturi line but never above the narrow point of the cone, depending upon whether it is of the updraft or downdraft type. The allowable tolerance is 0.5 mm (0.020″).

Adjustment is made either by bending the simple wire loop at the point shown in the illustration; on some newer models, a screw may be used instead.

NOTE: Height adjustment of the air-flow sensor becomes critical to the successful operation of the engine because if the plate sits too high, the engine will tend to want to keep running due to air flow and reaction at the fuel plunger (plate lever). With the sensor plate sitting too low, the engine will exhibit both poor cold and warm start-up characteristics.

Auxiliary Air Regulator Check

The purpose of the auxiliary air regulator was explained earlier under the description of operation of the system. Briefly, the function of this valve is to ensure that the engine receives more air/fuel when started up from cold to ensure a smoother idle. When the engine warms up, this valve closes off.

The auxiliary air valve is designed somewhat like a sliding gate valve in that it should be fully open on a cold engine and closed on a warm engine.

Quick Check. Disconnect the electric plug at the body of the auxiliary valve, then start the engine at an idle speed. Squeeze shut the rubber hose between the outlet of the valve and the intake manifold at the engine. This action reduces the air flow to the engine; therefore the engine should slow down slightly.

Reconnect the electric plug to the valve and let the engine run until it is warm. Again squeeze shut the rubber hose from the engine to the intake manifold. If the auxiliary valve is operating correctly, it should have closed off the extra air flow to the engine; therefore, no reaction (change in idle rpm) should exist when this hose is squeezed shut.

Regular Check. If the quick check fails to confirm the valve condition, proceed as follows:

1. Remove the electrical plug and both hoses from the valve.
2. On some vehicles/engines, you may be able to look through one side of the auxiliary air valve directly and see the actual internal sliding valve position. However, on some engines, in order to effectively see the valve position, you may need a trouble-light or flashlight and a mirror such as shown in Figure 6–20.
3. When looking into the valve or mirror, the valve should appear to be partly open on a cold engine. If it is not, replace the valve assembly.

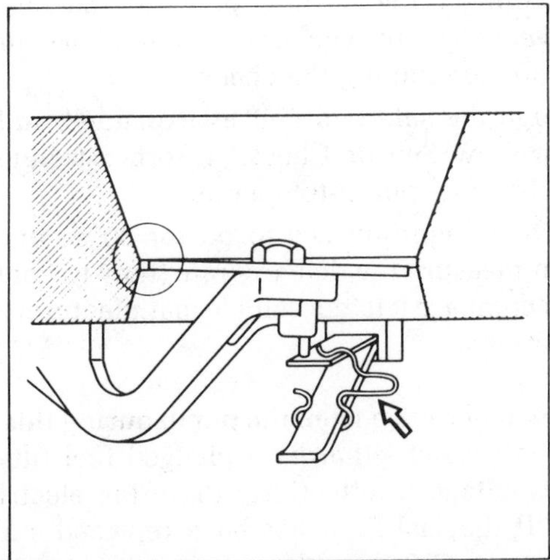

Figure 6–19. Air-flow sensor plate (K) return spring and checking of air-flow sensor plate for horizontal placement in the air funnel. (Courtesy of Robert Bosch Corporation, Broadview, Illinois)

Figure 6–20. Auxiliary air regulator valve (K) opening check using a small mirror. (Courtesy of Robert Bosch Corporation, Broadview, Illinois)

4. If the valve is open, reconnect the electric plug and two valve hoses.
5. Start and run the engine until it warms up (5 to 10 minutes).
6. Recheck the internal position of the valve to see if it is in fact closed now, which it should be. If it isn't closed, replace it.

Electric Fuel Pump Check

The major operating check of the fuel pump involves one of delivery (volume) over a given time period. The flow rate (volume) will obviously be the same in some cases for engines of the same basic displacement and power; however, because of the wide range of vehicles fitted with Bosch K-Jetronic fuel injection systems, you should refer to the vehicle service manual for the correct delivery specifications.

Vehicles before 1978 that were not equipped with a push valve in the fuel distributor require a slightly different hookup than current K systems do. On systems fitted with a push valve, you will note from Figure 6–21 that a slight extension of the fuel distributor on the right-hand side (opposite the air-flow sensor) exists.

Figures 6–21 and 6–22 illustrate the two arrangements required in order to effectively measure the delivery rate of the electric fuel pump.

NOTE: Fuel delivery will be low if the fuel filter is partially plugged.

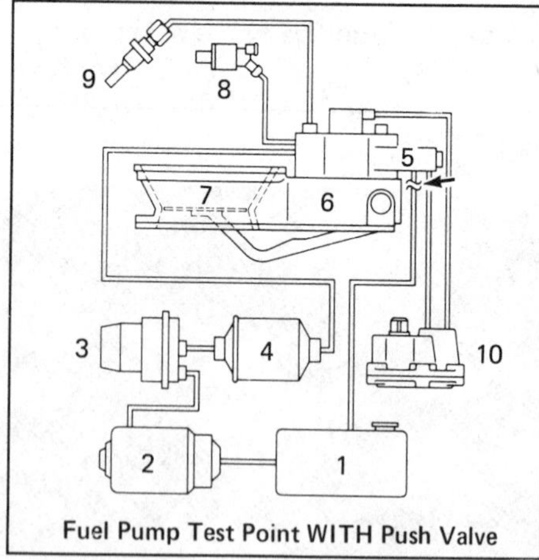

Index

1 - Fuel tank
2 - Electric fuel pump
3 - Fuel accumulator
4 - Fuel filter
5 - Push valve
6 - Fuel distributor
7 - Air-flow sensor plate
8 - Start valve
9 - Fuel injector
10 - Warm-up regulator

Figure 6–21. Fuel pump test point with a push valve-K system. (Courtesy of Robert Bosch Corporation, Broadview, Illinois)

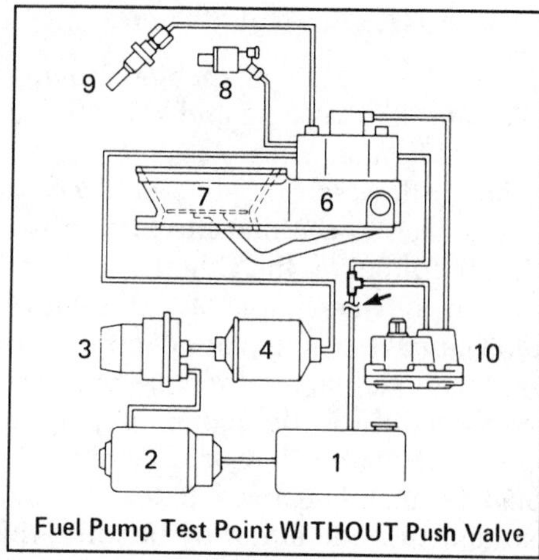

Figure 6–22. Fuel pump test point without a push valve-K system. (Courtesy of Robert Bosch Corporation, Broadview, Illinois)

Procedural Check

1. Refer to the appropriate figure for the type of system you are checking.
2. Remove the gas tank filler cap to vent the system.
3. Hold a container under the return fuel line as you disconnect it to catch any fuel spillage.
4. Use a receptacle that will hold at least 1.5 liters (0.4 U.S. gallon) to receive the return fuel (if a graduated (measurement marks) container is not available).
5. If necessary, push a piece of flexible hose over the fuel return pipe in order to route the fuel without spillage to the measuring container.
6. Disconnect the electric plugs at the cold start valve, warm-up regulator, and auxiliary air valve before conducting the check.
7. Bridge the safety circuit as explained under the "Air-Flow Sensor Check" in order to allow the electric fuel pump to operate.
8. Turn the ignition key to on for 30 seconds and then measure the delivery rate (flow to container) and compare it to the vehicle manufacturer's specifications.

A low fuel return from the pump during this check could be caused either by a plugged fuel filter or a lack of voltage to effectively drive the electric fuel pump. If the fuel filter has been replaced and or is known to be OK, then check for a minimum voltage of 11.5 volts at the electric plug to the pump. If there is no restriction in the fuel line or fuel tank strainer, you can safely assume that the electric pump is the problem.

NOTE: In freezing weather, water in the fuel can freeze and will cause a high resistance to initial pump rotation. This action can cause a safety fuse to blow, and no voltage will be available to the pump.

Thermo-Time Switch Check

This switch is used to control how long the start valve will remain on during initial start-up of the engine. The thermo-time switch is located in the engine cylinder head coolant water passage of most engines.

The thermo-time switch was shown earlier during our discussion of the basic system operation. Each Robert Bosch-supplied thermo-time switch has both the cutoff temperature and the switching time stamped on the flats of the switch hex nut.

To check the condition of the thermo-time switch, proceed as follows:

1. Ensure that the engine water temperature is lower than that stamped on the switch.
2. Disconnect the plug-in harness from the start valve at the intake manifold.
3. Connect a 12-volt test light across the plug-in harness terminals in a series hookup to the cold start valve.
4. Crank the engine for ten seconds after disconnecting the ignition coil and note how long the test light remains illuminated.
5. Generally if the test light does not stay on for the time stamped on the thermo-time switch hex nut flats, then it is defective (usually most thermo-time switches will stay activated for between two seconds with the engine coolant at room temperature, and for about five seconds at zero degrees F.

Another method of checking the thermo-time switch is by checking its resistance from each terminal (two) to ground. Each switch, depending upon the vehicle that it is designed for, will exhibit a different *ohm* reading. Check the switch part number and refer to the engine/vehicle service manual for the correct readings.

Cold Start Valve Check

The cold start valve is activated by a voltage signal that originates at the thermo-time switch. The cold start valve functions to supply excess fuel for successful engine start-up (similar to a choke arrangement on a carburetor).

Generally the cold start valve will only spray fuel at an engine coolant temperature lower than 95°F (35°C). The thermo-time switch determines the actual injection time of the cold start valve. The cold start valve does not operate when the engine is warm.

The cold start valve is located on the intake manifold downstream from the air-flow sensor and just past the throttle valve; it is retained by two small bolts. The cold start valve can be removed with its fuel line attached in order to quickly check its operating condition.

CAUTION: Do not smoke or use an open flame when testing the cold start valve for operation.

Procedure

1. Disconnect the electrical plug from the CSV (cold start valve) and remove it, with its fuel line attached, from the manifold.
2. Obtain a glass jar or receptacle that you can hold the CSV over while it is being tested.
3. Bridge the safety circuit as described in detail earlier under the air-flow sensor check.
4. Run a jumper wire from the hot (positive) side of the battery or B+ on the ignition coil to one terminal on the CSV while grounding the body of the CSV. CAUTION: Avoid creating electric sparks or running another jumper wire from the other CSV terminal to ground which would cause a bypass of the thermo-time switch.
5. Turn on the ignition switch (key) for 10 to 20 seconds and while holding the CSV in a glass jar, note that the valve sprays fuel and that its spray pattern resembles that of a cone.
6. If the valve sprays fuel, check that the thermo-time switch is operating correctly and that the CSV is receiving voltage to its electrical plug.
7. If the valve fails to spray fuel or fails to spray fuel in a cone-shaped pattern, replace it.
8. If the valve is OK, turn off the ignition key and remove the jumper wire, dry off the CSV nozzle tip, then reconnect the jumper wire.
9. With the ignition left on for about one minute, check that no fuel droplets form on the end of the nozzle; fuel leakage would cause the engine to run rich.

Warm-up Regulator (Control Pressure) Check

The function of the WUR (warm-up regulator) is to provide a fuel enrichment process after start-up when the engine is cold, similar to that of a choke on a carburetor circuit.

The WUR is electrically heated (12-V system), and its operation is explained in detail under the K-Jetronic system operation. Through the action of the WUR, control pressure is reduced when the engine is cold. This allows the metering slits of the fuel distributor control plunger to open further.

NOTE: Before checking the control pressure in the system, you must establish whether or not the engine/vehicle is fitted with a WUR that may have vacuum controlled full-load enrichment and/or high altitude compensation systems.

WUR Identification. If a hose is connected from the air intake manifold venturi back to the WUR, then it is equipped with vacuum full-load enrichment. If, on the other hand, the WUR has a hose connected between the WUR and the air intake system just before the air filter (in front of the throttle valve), then the WUR is equipped with high altitude compensation. Some WURs may have no hose but simply employ an opening in the housing to the atmosphere for high altitude compensation.

Testing WUR.

Procedure

1. Disconnect the electrical plug from the WUR and measure the voltage across the plug terminals with the ignition key turned on.
2. Minimum voltage should be 11.5 volts, otherwise the WUR will not operate properly (note: electrically energized bimetal spring).
3. Using an ohmmeter, check for continuity between the two terminals of the WUR, which will confirm that the bimetallic spring is not open. An average reading here should be between 16–22 ohms; otherwise, the WUR is faulty.
4. If the ohmmeter reads infinity between the WUR contacts, it indicates that the bimetallic spring is open; therefore the WUR should be replaced.

WUR Removal/Replacement.

CAUTION: If you have to replace the WUR, always make sure that you bleed down the fuel system pressure by energizing the start valve. This is done by first disconnecting the electrical plug at the CSV and then using a jumper wire applying voltage to the CSV terminal for about 10 seconds, or by loosening off the large nut on the side of the control pressure regulator. Make sure that no open flame is present or that no one is smoking around you when you do this because a positive fire hazard exists. Cover any line that you are loosening off with a rag and wipe up the spilled fuel immediately. Once the fuel pressure has been bled down, proceed to remove the necessary components to enable you to take the WUR off of the system. Install it in the reverse order of removal.

Control Pressure Fuel Delivery Check

Refer to Figure 6–23 and remove the fuel line at the fuel distributor. Use one of the test fuel lines from special tool kit KDJE-P100 or earlier tool kit KDEP-1034 and attach it to the fuel distributor as shown. With the other end of the fuel line placed into a suitable container or beaker graduate, bridge the fuel pump electric safety circuit as explained in Figures 6–15 and 6–16 to allow the electric fuel pump to operate for about one minute. Measure the quantity of fuel delivered into the container or beaker graduate after this time; it should be between 160–240 cc's. Otherwise the fuel distributor is defective and should be replaced.

K-Jetronic Fuel System Pressure Tests

A number of fuel pressure tests are required on the K-Jetronic fuel system to determine its condition. These tests should be done in the following sequence:

1. System cold control pressure check
2. System warm control pressure check
3. System primary pressure check
4. System "at rest" pressure check

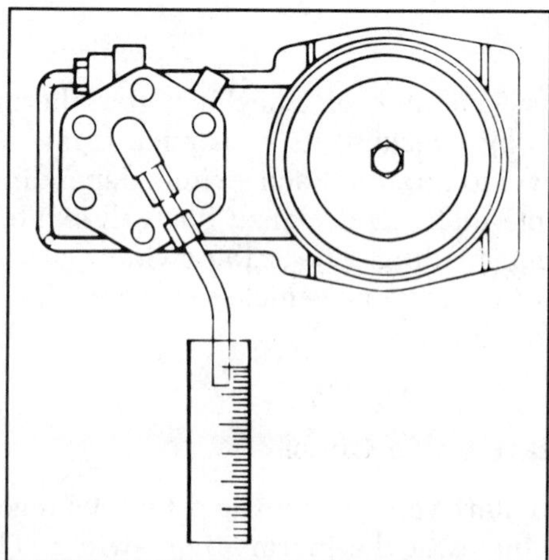

Figure 6–23. Control pressure fuel delivery check. (Courtesy of Robert Bosch Corporation, Broadview, Illinois)

Fuel system pressures developed within the system will vary between different makes of vehicles because not all systems employ components with the same part number. For this reason, these pressure readings may be the same on some engines, while on others they may be quite different. Refer to either the vehicle service manual or *Robert Bosch K-Jetronic Service Guide* booklet P/N 251204683 for the specific pressures for 1979 and later vehicles and booklet P/N 251250883 for 1973–78 vehicles. An example of what K-Jetronic system pressures would be monitored can be seen in Figure 6–24.

Approximate fuel system pressures will vary between makes of vehicles, with many being very close to one another. Most K-Jetronic vehicle equipped engines will have fuel pump delivery rates varying from a low of 750 cc/30 seconds on the Audi 4000 and Fox; the BMW 320i; Peugeot 505; Porsche 924; Saab 99 and 900; VW Dasher, Rabbit, Jetta, Scirroco, and Pickup; Volvo 240 DL, GL, GT, GLT, and GLT turbocharged engines up to the 1500 cc/30 seconds in the Porsche turbo of 1979.

Vehicles delivering 850 cc/30 seconds are the Audi 79-82 Fox, the 4000 and 4000 + 5, and the 5000 units.

VW Rabbit (Cal), Scirocco and Dasher (Cal) 1980
VW Rabbit, Rabbit Pickup, Rabbit CNA Scirocco, Jetta & Jetta CNA 1981
VW Pickup (Cal) and Scirocco 1982

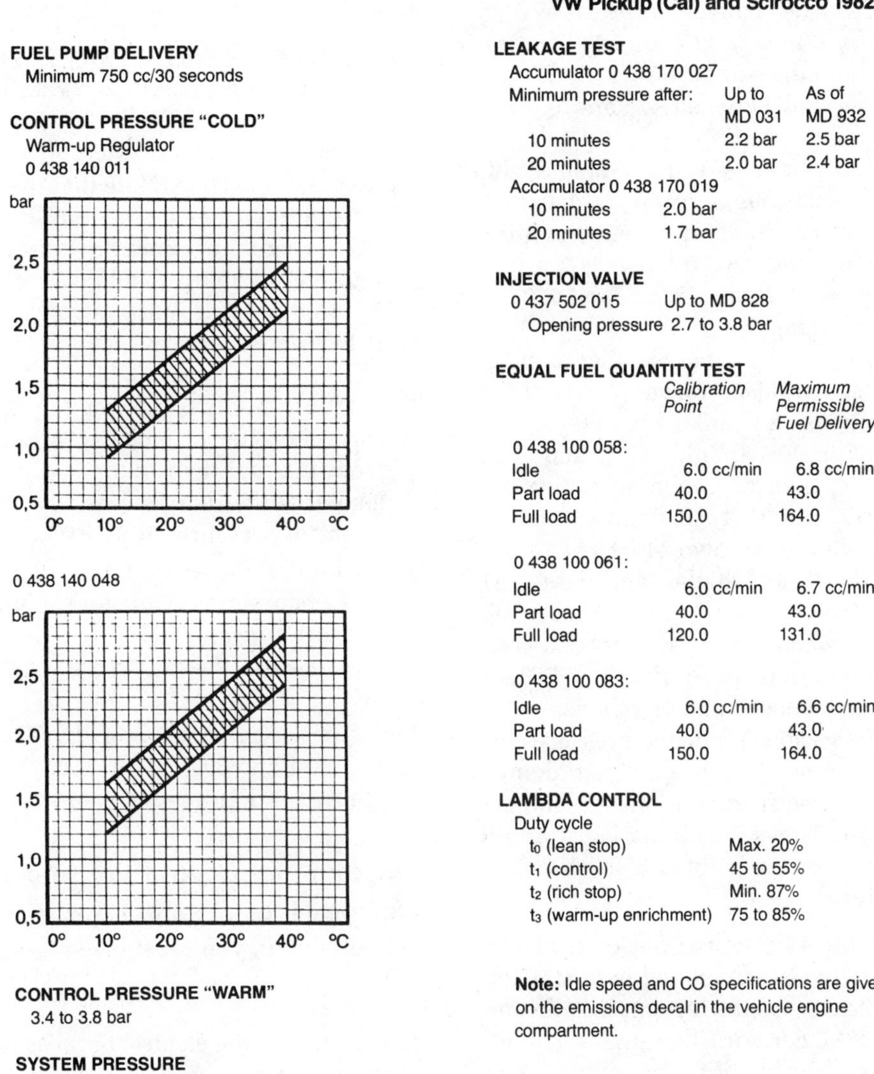

FUEL PUMP DELIVERY
Minimum 750 cc/30 seconds

CONTROL PRESSURE "COLD"
Warm-up Regulator
0 438 140 011

0 438 140 048

CONTROL PRESSURE "WARM"
3.4 to 3.8 bar

SYSTEM PRESSURE

Checking Value	4.7 to 5.4 bar
Setting Value	4.9 to 5.1 bar

LEAKAGE TEST
Accumulator 0 438 170 027

Minimum pressure after:	Up to MD 031	As of MD 932
10 minutes	2.2 bar	2.5 bar
20 minutes	2.0 bar	2.4 bar

Accumulator 0 438 170 019

10 minutes	2.0 bar
20 minutes	1.7 bar

INJECTION VALVE
0 437 502 015 Up to MD 828
Opening pressure 2.7 to 3.8 bar

EQUAL FUEL QUANTITY TEST

	Calibration Point	*Maximum Permissible Fuel Delivery*
0 438 100 058:		
Idle	6.0 cc/min	6.8 cc/min
Part load	40.0	43.0
Full load	150.0	164.0
0 438 100 061:		
Idle	6.0 cc/min	6.7 cc/min
Part load	40.0	43.0
Full load	120.0	131.0
0 438 100 083:		
Idle	6.0 cc/min	6.6 cc/min
Part load	40.0	43.0
Full load	150.0	164.0

LAMBDA CONTROL
Duty cycle

t_0 (lean stop)	Max. 20%
t_1 (control)	45 to 55%
t_2 (rich stop)	Min. 87%
t_3 (warm-up enrichment)	75 to 85%

Note: Idle speed and CO specifications are given on the emissions decal in the vehicle engine compartment.

Figure 6–24. K System test specifications—Volkswagen vehicles. [Courtesy of Robert Bosch Corporation, Broadview, Illinois]

Also at 850 cc/30 seconds are the DeLorean and the Volvo 79-82 260, GLE Coupe, and Sedan.

At 930 cc/30 seconds are the Audi 5000 turbo, the Mercedes-Benz 280E, CE and SE models; the Saab 900 turbo produces 950 cc/30 seconds.

At 1000 cc/30 seconds is the Mercedes-Benz 380 and Porsche 911SC; at 1050 cc/30 seconds is the Porsche 924 turbo. At 1100 cc/30 seconds is the Ferrari 308 GTSI and the Mercedes-Benz 450 SL and 6.9L models.

All vehicle models operate at fuel system control pressures (warm) between 3.4 and 3.8 bar (49–55 psi) with the exception of the BMW 320i, which runs between 3.3 and 3.7 bar (48–53.7 psi). System pressure is between 4.5 and 5.2 bar (65.3–75.5 psi) on the Saab, BMW, Volvo, and Porsche vehicles, while the system pressure on VW, Audi 4000, and 4000+5, Quattro, Peugeot 505, and Mercedes-Benz 80–81 280E, CE, and SE vehicles averages between 4.7 and 5.4 bar (68–78.4 psi).

The 80–82 Mercedes-Benz 450 SEL, SL, and SLC system pressure was between 4.7 and 5.6 bar (68–81.3 psi). The Audi 5000 turbo fuel system operating pressure is between 5.2 and 5.8 bar (75.5–84.2 psi) as was the Mercedes-Benz 79 280E, CE, SE, and 450 SL, SEL, SLC, and 6.9L engine.

Injector Release Pressures are very similar on most makes of vehicle engines. Injectors that release at the same pressure of between 2.7 and 3.8 bar (39–55 psi) are VW, 79 Volvo 260, 80–82 Volvo GLE and GLE Coupe, Audi vehicles, Porsche vehicles using injector P/N 0 437 502 013, and BMW 320i units. Injector release pressure on the 79–81 Volvo 240, DL, GL, GT, and GLT engines; the Saab vehicles; and the 80–81 Porsche 911SC is between 2.5 and 3.6 bar (36–52 psi). Injectors releasing between 3.0 and 4.1 bar (43.5–59.5 psi) are the 81 Volvo DL and GLT turbo, the Peugeot 505, and the Mercedes-Benz range of vehicles.

As you can see, the various regulated pressures for different makes of engine/vehicles does not change too much; however, these specifications are subject to change. Therefore always check the latest applicable service information from Robert Bosch or the vehicle manufacturer.

Test No. 1—Cold Control Pressure Check. This test is done when the engine has exhibited hard-starting and warm-up complaints. As the test implies, the engine must be "cold." Therefore the engine should have preferably been left standing overnight or at least for several hours.

Before connecting up the Special Tool (Robert Bosch P/N KDJE-P100 or KDEP 1034) as shown in Figure 6–25, refer to the CAUTION dealing with relieving fuel system pressure above, under the Warm-

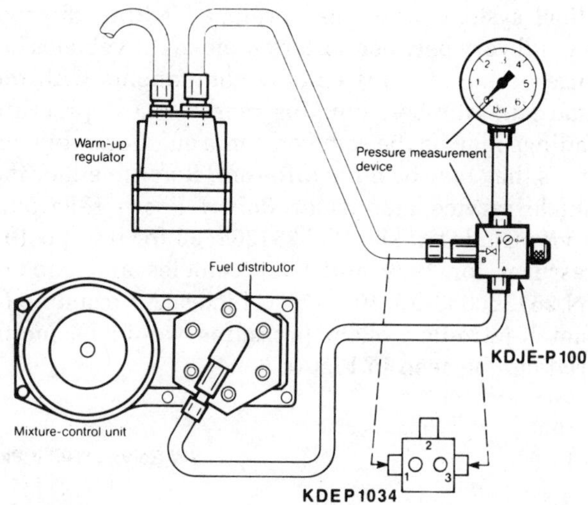

Figure 6–25. K System cold control fuel pressure check. (Courtesy of Robert Bosch Corporation, Broadview, Illinois)

Up Regulator system. Note that the gauge is teed into the fuel line between the fuel distributor and the warm-up regulator. Fuel flows from the fuel distributor to the warm-up regulator!

The fuel pressure test valve/gauge has three possible lever positions:

1. Position 2 Older Model (or open both hollow screws on newer models) = Fuel under pressure from the fuel distributor is directed to the valve/gauge, which is open, and it will therefore read "control pressure" in the fuel system.
2. Position 3 Older Model (or close hollow screw 1 and open screw 3 on newer models) = System primary pressure.
3. Position 1 = Engine off—rest or holding pressure.

Two types of gauge adapters have been used with the KDEP-1034 and KDJE-P100 fuel pressure tester; they are shown in Figures 6–26 and 6–27.

Procedure for Control Pressure Test—Cold

With the pressure gauge connected into the fuel system as shown, you must bleed the pressure tester as follows:

1. Disconnect the electric harness plugs at the warm-up regulator, the cold start valve, and the auxiliary air device; then bridge the safety circuit so that the electric fuel pump will operate when the ignition key is switched on. This action is explained under the heading "Air-Flow Sensor Check."

2. Gently let the gauge and its support hose hang lower than the fuel lines (helps to vent air from system).
3. Switch on the ignition key (electric fuel pump runs with safety circuit bridged).
4. If the former (earlier) model type is being used, simply move the control lever backward and forward about 5 or 6 times between positions 2 and 3; pause at each position for about 10 seconds.
5. If the newer model is being used, simply pump the directional control valve as you simultaneously open and close the hollow screw number 1; this effectively bleeds air from the fuel system.
6. Lift up the tester gauge and support or hang it in a safe position where it will not be damaged and so you can read it clearly during the test.
7. Place the gauge valve at position 2 on the older models, or open both hollow screws on the newer models.
8. Start and run the engine at an idle rpm for 30 seconds to a minute.
9. Note the pressure reading on the gauge and compare it to the manufacturer's specifications; if the cold control pressure is not correct, then the warm-up regulator may be at fault.

Test No. 2—Control Pressure Warm. This test is conducted in the same manner as that for the "cold pressure" check in No. 1. Generally if the engine only had a hard starting problem when cold and you conducted test No. 1 above, this test would not necessarily be required. However, if hard starting is experienced when the engine is warm, then proceed as follows.

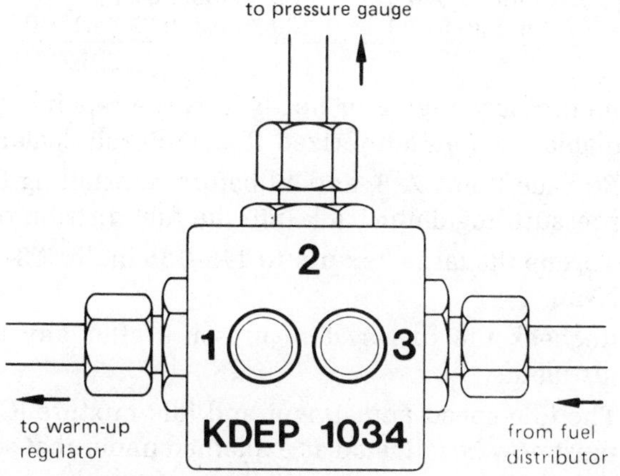

Figure 6-26. K System fuel pressure gauge adapters KDEP 1034. [Courtesy of Robert Bosch Corporation, Broadview, Illinois]

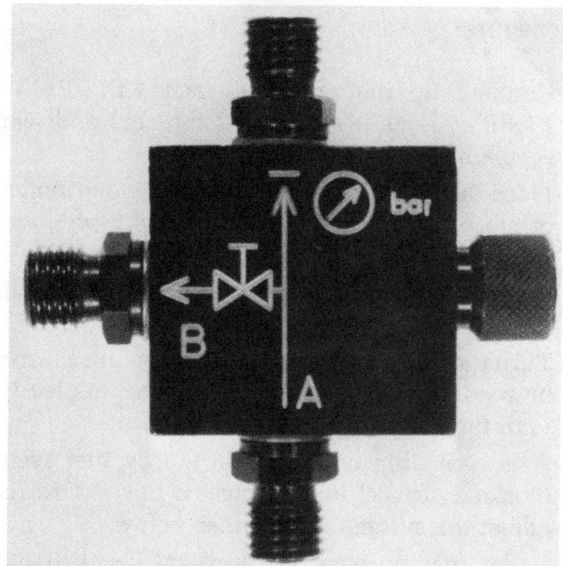

Figure 6-27. K System fuel pressure gauge adapters KDJE-P100. [Courtesy of Robert Bosch Corporation, Broadview, Illinois]

Procedure

1. Place the control valve at position 2 (older models), or open both screws (later model testers).
2. Reconnect the warm-up regulator electric cable plug.
3. Turn the ignition key on (safety circuit bridged) to allow the electric fuel pump to raise the system pressure.
4. After a few seconds, read the pressure gauge.
5. If the pressure is low, start and idle the engine and check that suitable voltage (11.5 V minimum) is available at the warm-up regulator plug contacts.
6. If full voltage is available at the plug, then the warm-up regulator is faulty and must be replaced.

Test No. 3—Primary Pressure Check. The fuel system primary pressure is established by the action of the electric fuel pump and a spring loaded pressure valve located in the side of the fuel distributor unit (see system illustration in Figure 6-6). All 1978 and later vehicles with K fuel systems employ a push valve that helps to maintain system pressure at the warm-up regulator. (See system description.)

Before conducting the primary system pressure check, make sure that the electric fuel pump is delivering its correct fuel quantity as described earlier and that the fuel filter is not plugged. Replace the fuel filter if in doubt.

Procedure

1. Connect up the pressure tester KDJE-P100 or KDEP 1034 as described for the cold and warm regulator tests.
2. Place the tester lever to position 3 (older models), or close hollow screw 1 and open screw 3.
3. Disconnect the cold start valve electric connection and bridge the safety circuit in order to allow the electric fuel pump to operate.
4. Turn the ignition key *on* until system pressure can be read on the test gauge; compare this reading with the manufacturer's specs.
5. A low reading could be caused by fuel system blockage or fuel line leakage; if this is OK, then adjust the system as explained below.
6. If the fuel pressure is too high, check that no blockage exists in the fuel return line to the tank. If no blockage is evident, then the fuel system pressure would also require adjustment.

Isolating Possible Problem Area. A fuel pump problem or primary regulator problem (Figure 6–6) can be checked quickly by plugging the fuel return line to the tank from the fuel distributor. Place the gauge so that you can see it, and turn the ignition key *on* just long enough to read the gauge pressure; a good fuel pump would be confirmed by the fact that the fuel pressure would rise quickly to 100 psi (6.9 bar) or higher. If the fuel pump checks out OK, then the problem of low pressure can be traced to the primary regulator.

Fuel System Pressure Adjustment. To actually adjust the system pressure, the fuel distributor may have to be removed from the air-flow sensor plate assembly as shown in Figure 6–28. Figure 6–29 illustrates the system pressure regulator in disassembled form, and Figure 6–30 illustrates the pressure regulator push valve in assembled schematic form.

NOTE: Bleed down the system fuel pressure as explained earlier under Air-Flow Sensor Check and clean all fuel line connections before loosening them off in order to keep dirt out of the system.

1. Remove the fuel distributor by taking out the three retaining screws on the top.
2. Exercise care when removing the fuel distributor top so as not to let the fuel control piston fall out.
3. Remove the large hex plug (Item 6) on the side of the fuel distributor and withdraw the components shown in Figure 6–29.

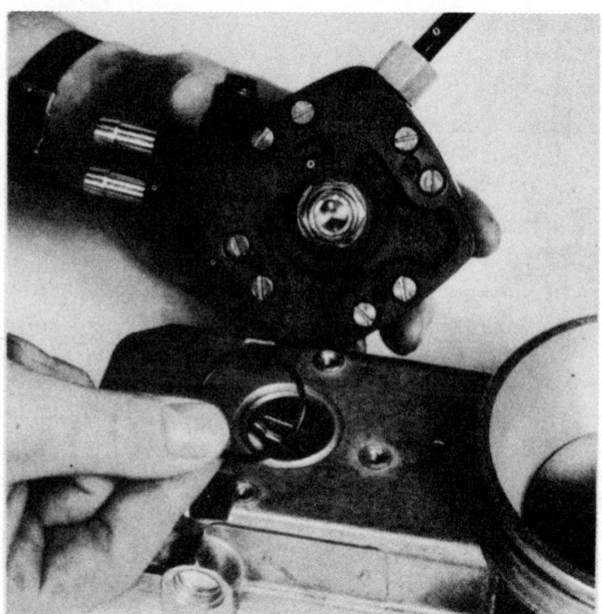

Figure 6–28. Removing K system fuel distributor from the air-flow sensor plate. (Courtesy of Robert Bosch Corporation, Broadview, Illinois)

4. Item 9 in Figure 6–29 and Item 8 in Figure 6–30 are shims that control the fuel system pressure; measure this shim or shims for thickness. To increase fuel system pressure, you have to increase the thickness of the shim(s); to reduce pressure, you have to decrease the thickness of the shim(s).

NOTE: Fuel pressure can be changed as follows:

Shim Thickness	*Pressure Change*
0.1 mm (0.004″)	2.175 psi (0.15 bar)
0.15 mm (0.006″)	3.335 psi (0.23 bar)
0.30 mm (0.012″)	6.525 psi (0.45 bar)
0.40 mm (0.016″)	8.7 psi (0.6 bar)
0.50 mm (0.020″)	10.875 psi (0.75 bar)

If the pressure regulator needs repair, a repair kit is available from an authorized Robert Bosch dealer.

5. Replace Items 7, 8, and 13 before reinstalling the pressure regulator back into the fuel distributor.
6. Torque the large hex nut to 115–135 in. lb. (13–15 N•m).
7. Recheck the fuel system pressure after any adjustments.
8. The idle speed adjustment and idle mixture (CO) must now be adjusted as explained under that section.

Test No. 4—At Rest Fuel Pressure. When the engine has been switched off, the fuel pressure will

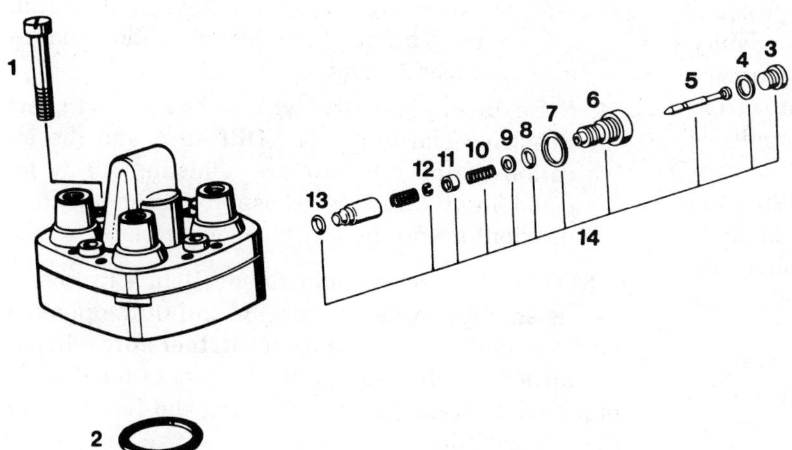

1 **Fillister Head Screw**
2 **Seal**
3 **Allen Head Screw Plug**
4 **Washer**
5 **Push Valve**
6 **Hex Head Screw Plug**
7 **Washer**
8 **O-ring**
9 **Shim (.1mm, .15mm, .3mm, .4mm, .5mm)**
10 **Spring**
11 **Spacer**
12 **Snap Ring**
13 **O-ring**
14 **Repair Kit including 3-13 above**

Figure 6–29. Component parts of a K fuel system distributor. (Courtesy of Robert Bosch Corporation, Broadview, Illinois)

drop after a period of time. However, the action of the spring loaded fuel accumulator between the electric fuel pump and the fuel line filter will maintain system pressure to facilitate ease of start-up especially when the engine is warm. If there is a loss of system pressure, hard starting when warm will result because a vapor lock can occur within the injector fuel lines.

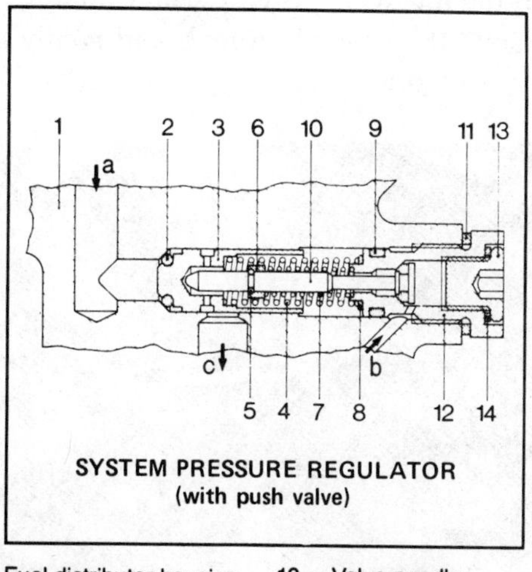

SYSTEM PRESSURE REGULATOR
(with push valve)

1 = Fuel distributor housing
2 = O-ring
3 = Control piston
4 = Spring
5 = Retainer
6 = Retaining ring
7 = Spring
8 = Shims
9 = Seal
10 = Valve needle
11 = Seal ring
12 = Screw plug
13 = Screw plug
14 = Flat seal ring

a = from fuel distributor
b = from warm-up regulator
c = fuel return

Figure 6–30. K Fuel system pressure regulator. (Courtesy of Robert Bosch Corporation, Broadview, Illinois)

Conditions for checking the "at rest or stopped" fuel pressure are the same as that for the cold and warm regulator tests conducted earlier. With gauge fixture KDJE-P100 or KDEP-1034 in position and the engine running proceed as follows.

Procedure

1. Run the engine until it is at normal operating temperature.
2. Shut off the engine and note the pressure maintained in the system.
3. If the at rest pressure is low, restart the engine until the fuel pressure builds back up and then stop the engine.
4. Withdraw the cold start injector from the intake manifold.
5. Since the engine is warm, the cold start injector should not be in operation; therefore if fuel is dripping from the cold start valve, replace it because it is faulty (excess raw fuel will cause hard starting on a warm engine).
6. Recheck the at rest pressure.
7. If the cold start valve is OK, then install it again.
8. Restart the engine until fuel pressure builds up again.
9. Stop the engine.
10. Close off the three-way valve (KDEP-1034) or KDJE-P100.

If the fuel pressure is still low, replace the warm-up compensator. If the fuel pressure drops rapidly once you close the three-way valve, then the problem is at the inlet side of the fuel system such as a faulty fuel pump, fuel accumulator, or primary pressure regulator. The fuel pump can be checked as explained earlier by conducting a flow check (delivery volume).

The fuel accumulator is good if, once you stop the engine (watch the pressure gauge closely), the pressure drops and then stops followed by a slight increase in gauge pressure as the spring loaded accumulator comes into action.

A faulty accumulator will not allow this slight pressure increase. Also, if the accumulator is faulty, it will usually cause the fuel pump to emit unusual noises because of fuel pressure fluctuations. If low at rest pressure still exists, then the primary pressure regulator at the side of the fuel distributor is faulty.

Idle RPM and Idle Mixture Settings

The K (CIS) Jetronic system employs both an idle rpm adjustment screw and an idle mixture (CO) adjustment screw. The location of these two adjustment screws can be seen by referring to the K-Jetronic operational diagram (Figure 6–1 and 6–31). The engine must be at operating temperature when performing these two adjustments.

The idle adjustment screw is actually a screw that is designed to restrict or increase air flow past the throttle butterfly valve within the intake manifold. Screwing it in will result in a decrease in engine idle speed, while screwing or turning it out will result in an increase in engine idle rpm by allowing more air to flow past it at idle. This screw is shown in Figure 6–3(a) to the immediate right of the start valve.

The adjustment of these two screws must be done together (a trial and error adjustment) to ensure that both the air/fuel mixture is correct as well as the idle rpm. In the K system, when you adjust the air/fuel (CO) mixture, this will affect the complete speed range of the engine, whereas in a carburetor application this is not the case.

Procedure

1. Make sure that the throttle butterfly valve is correctly adjusted to place it against the idle stop; if not, adjust the linkage until it is against the idle stop.
2. Depending on the particular make of vehicle, various auxiliary items may have to be disconnected before setting the idle speed.
3. Adjust the idle speed with an accurate tachometer to that range specified by the vehicle manufacturer (idle speeds can range from as low as 600 rpm on some units to as high as 1050 rpm on others). To adjust the air/fuel ratio mixture (CO), proceed to Step 4.
4. Refer to Figure 6–31, which shows the Robert Bosch special tool P/N KDEP-1035 and the location of the idle mixture adjusting screw located in the air-flow sensor between the fuel distributor and the air-flow sensor plate.

NOTE: This screw adjustment (Item 4 in Figure 6–1) is actually threaded into the end of the pivoted air-flow sensor plate balance lever; therefore when it is adjusted, it will react on the bottom of the distributor fuel plunger thereby affecting the fuel flow at the plunger slits.

5. To accurately adjust the idle mixture screw, you have to use a CO meter.
6. Remove the access plug at the air-flow sensor to allow insertion of special tool KDEP-1035 into the adjustment screw.
7. Start and idle the engine, read the CO meter and compare the readings to manufacturer's published specifications for either the Federal or California emissions regulations.

NOTE: All vehicles have a sticker plate located usually under the hood that states the allowable CO for that engine.

8. Turning the special wrench KDEP-1035 to the right will create a richer mixture, while turning it to the left will create a leaner mixture.
9. Remove the special wrench and gently accelerate the engine.

Figure 6–31. Location of K system idle mixture adjustment screw. [Courtesy of Robert Bosch Corporation, Broadview, Illinois]

10. Let the engine drop back to an idle speed and recheck the CO content on the meter. To obtain a correct reading, you may have to alternately work between the idle speed adjustment screw and the mixture control screw.
11. Replace the plug in the air-flow sensor housing when you have finished your adjustments.

If an accurate CO mixture reading cannot be obtained, check the air system for vacuum leaks (Figure 6–14) as explained earlier. If no vacuum leaks can be found, then it's possible that the fuel distributor is faulty.

Gasoline Fuel Injector Service

The fuel injectors used with the Bosch K-Jetronic (CIS) system continuously inject fuel while the engine is running. There is one injector per engine cylinder. The most common problems with fuel injectors are usually associated with dirt or water in the system. The injectors are mounted in a holder screwed into the intake manifold so as to insulate them from engine heat as much as possible.

The quantity of fuel delivered to the injectors is established at the fuel distributor. The injectors are designed basically as a fuel delivery valve, and they will open (spray fuel) at a fuel pressure usually within the range of 47–54 psi (3.2–3.7 bar), although some Mercedes-Benz units can go as high as 59 psi (4.1 bar). Each particular engine injectors are calibrated to open within a specific range of pressure. This must be checked as per the manufacturer's specifications during injector testing.

The injector valve has a nozzle through which the gasoline is forced (injected): Due to the design of the injector, which uses a spring loaded needle valve internally, the fuel pressure will cause this valve to open and close many times. This opening and closing action is similar to that found in diesel fuel injectors and is commonly referred to as injector "chatter." This noise is caused by the needle valve leaving its seat and then being forced back onto its seat.

To effectively service/test injectors requires the use of Robert Bosch special tool KDJE-P400, which is illustrated in Figure 6–69 and also shown in Figure 6–32. Gasoline is not and should never be used when testing injectors because of the safety considerations such as flammability, fumes, etc; therefore a special calibrating fluid is used instead. This calibrating fluid can be obtained from a variety of sources and suppliers. Robert Bosch recommends that either test oil VS14 492-CH or Shell Mineral Spirits 135 be used

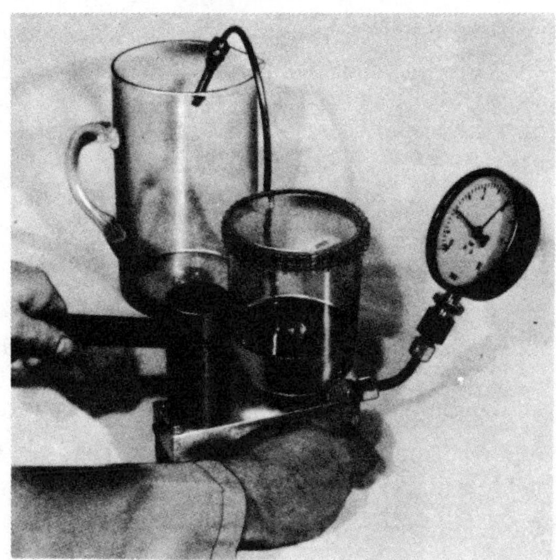

Figure 6–32. K Injector installed in KDJE-P400 pop tester to check spray pattern. (Courtesy of Robert Bosch Corporation, Broadview, Illinois)

where available or a fluid that conforms to the following: a viscosity rating of 1.8 to 2.8 cSt (centistokes), a density of 0.77–0.81 g/cm^3, a boiling point of 140–200°C (284–392°F), and a self-ignition temperature of at least 245°C (473°F).

The following tests and procedures are used to test the fuel injectors.

Contamination Test.

1. Loosely attach the injector to the fuel line of the pop tester.
2. Manipulate the tester handle (up and down) until clear fluid runs out from the loosely connected fuel line at the injector, which will bleed air from the unit.
3. Tighten up the fuel line to injector nut.
4. Open the tester control valve and pump the handle slowly without causing the gauge to exceed 1.5 bar (21.75 psi).
5. The injector is faulty if it leaks or pressure fails to build up.

NOTE: Before performing the remaining tests, the pop tester and injector should be bled as follows: close the tester valve and pump the handle until clear, air-free fluid leaves the nozzle tip.

Injector Opening Pressure Test.

1. Open the pop tester control valve.
2. Pump up the handle until the injector sprays fuel from its tip.

3. Carefully note the pressure at which the injector sprays fluid; compare the opening pressure with that in the specifications of either the vehicle service manual or Robert Bosch technical information bulletins.

Injector Leakage Test.

1. Pump the tester handle until the gauge pressure reads 0.5 bar (7.25 psi) below the opening pressure that was achieved during the above test. This pressure is not to be less than 2.3 bar (33.4 psi).
2. Maintain this pressure for about 15 seconds and closely watch the injector tip for signs of fuel leakage or drips. If drips appear, discard the injector.

Injector Spray Pattern Test. The injector spray pattern test indicates whether or not the injector is in fact atomizing the injected fuel properly. If it is not, then poor engine operation will result (raw fuel).

1. With the injector installed onto the pop tester fuel line, close the tester gauge valve to protect it.
2. Pump the test handle at a rate of about a stroke per second.
3. Carefully note the type of spray pattern that comes out of the injector nozzle tip and compare it with the spray patterns shown in Figures 6–33 and 6–34.
4. Listen for an audible chatter from the injector as you pump the handle because this indicates that all internal parts are moving freely and that the injector is free from carbon formation or dirt.
5. No drops should form on the injector tip during the spray pattern test.
6. Use new seal rings when replacing the injectors back into their holder at the intake manifold.

Unlike a diesel engine where the injectors should always be replaced in the same cylinder from which they were removed, the gasoline injectors can be replaced in any intake manifold position.

Equal Fuel Quantity Test

The purpose of this test is twofold: one test is used to check that the fuel distributor is in fact delivering equal fuel at each one of its outlets, and the second test confirms that each individual fuel injector is operating correctly.

This test requires the use of Robert Bosch special tool P/N KDJE-P200, which is illustrated in Figure 6–67.

Figure 6–33. K System—correct injector fuel spray patterns. (Courtesy of Robert Bosch Corporation, Broadview, Illinois)

If the vehicle is equipped with steel fuel delivery lines to the fuel injectors rather than polymide tubing, you will require test lines P/N KDJE-P200/25, Figure 6–68. The use of this special test tool allows an on-vehicle test without having to remove the fuel distributor and with the injectors inserted between the fuel distributor and the test unit as shown in Figure 6–35.

Figure 6–35 illustrates the tester connected to the fuel system for the test sequence.

The special tool KDJE-P200 allows you to check the fuel delivery and injector condition at an idle speed, part and full-load conditions. If one or more fuel distributor outlets are in excess of the maximum flow specification, repeat the flow test to be certain that this is the case. An example of the specifications for an equal fuel quantity test can be seen in Figure 6–24.

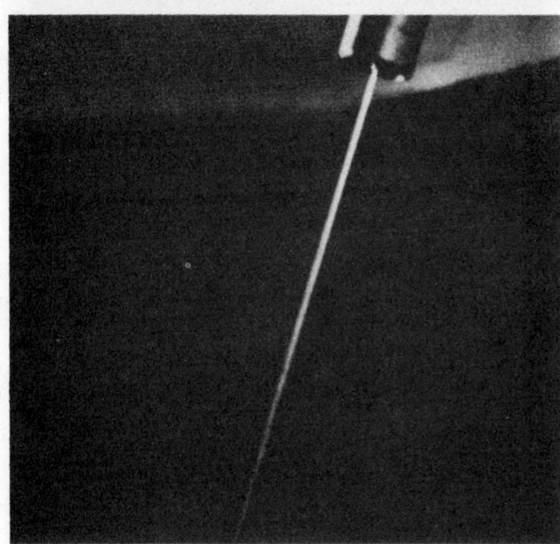

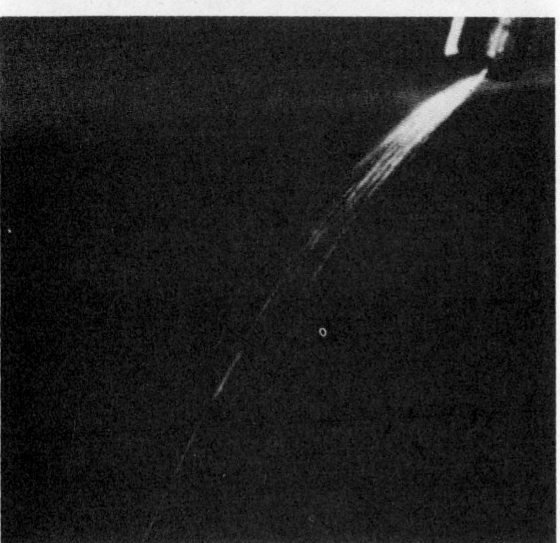

Figure 6–34. K System—incorrect injector spray patterns (faulty injectors). (Courtesy of Robert Bosch Corporation, Broadview, Illinois)

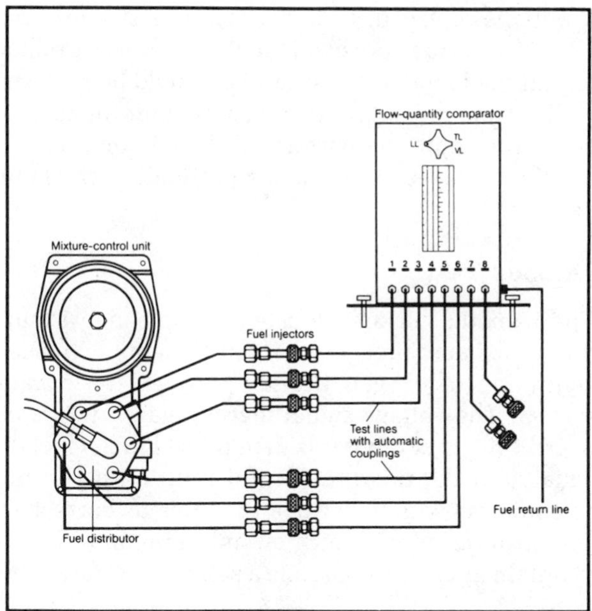

Figure 6–35. K Fuel system special tool KDJE-P200 connected into position (formerly tool KDJE-7451). (Courtesy of Robert Bosch Corporation, Broadview, Illinois)

The test tool is arranged so that the idle flow test is read on the left-hand side of the test tool while the right-hand side registers both the part and full-load flow rates.

The safety circuit must be bridged as shown in Figures 6–15 and 6–16 to allow the electric fuel pump to operate.

Located along the bottom of the tester is a row of buttons that is designed to handle readings from up to eight injectors. When the safety circuit has been bridged, each button is pressed to allow that particular injector circuit to be monitored while gently applying pressure manually to the air-flow sensor plate to duplicate the idle calibration point shown on the test spec sheet (example shown in Figure 6–24).

Once the air-flow sensor plate position has been obtained, use a feeler gauge between the plate and the throat bore to hold it in a fixed position so that the other fuel injector circuits can also be monitored by activating the test tool selector button for that specific circuit. Establish what injector has the lowest delivery and set this one to the exact calibration point given in the vehicle test spec sheet. Repeat the test for all injectors.

To isolate whether the problem lies in the fuel injector or in the fuel distributor assembly, switch the fuel injector with one that indicated a lower fuel delivery and conduct the flow test again. If the fuel delivery is now within the published specifications, this indicates that the injector was in fact faulty and should be replaced. If, however, the distributor outlet still delivers too much fuel, there is a problem within the fuel distributor and it should be replaced.

The same procedure that was used for the idle test is used for both the part and full-load conditions as per the test spec sheet for the particular vehicle under test.

Lambda Test

The Lambda sensor is located in the exhaust manifold or exhaust pipe and is used to monitor the concentration of oxygen flowing through the exhaust system. The voltage signal that is relayed to the onboard vehicle computer is then passed to the fuel distributor in the mixture control unit by means of a solenoid valve/timing valve in order to correct the composition of the air/fuel mixture and thereby maintain as close as possible a stoichiometric air/fuel ratio.

Tester KDJE-P600 shown in Figure 6–70 is used to check the operation of this unit.

When the engine is cold and or the exhaust temperature is lower than 300°C (572°F) approximately, the Lambda sensor signal is ignored by the electronic computer and the fuel system operates in what is commonly called the open-loop condition. When the exhaust temperature is hot enough, the Lambda sensor sends out a voltage signal to the computer, which now uses this signal to vary the pulse width of the injectors. Under such a condition the engine is said to be in a closed-loop condition.

The Lambda sensor is a replacement item, and experience has shown that its life expectancy is about 30,000 miles, or 48279 km.

Troubleshooting the K-Jetronic Fuel System

Figure 6–36 illustrates a Robert Bosch troubleshooting chart that can be used to effectively and quickly isolate a problem with the fuel injection system.

Robert Bosch D-, L-Jetronic, and Motronic Systems

Both the D- and L-Jetronic gasoline fuel injection systems have undergone a number of product improvements since they were first designed and released into mass production in the early 1970s. Minor differences exist between these two types of systems; however, both systems operate on the principle of intermittent low pressure injection rather than the CIS (continuous injection system) design that is used with the Bosch K-Jetronic unit.

The D system employs an air pressure sensor in the intake manifold along with a set of trigger contacts in the ignition distributor; both feed signals to an ECU (electronic control unit). The ECU sends the necessary signals to the fuel injectors to regulate the amount of fuel injected.

The L system relies on an air-flow sensor in older models and on a hot-wire air-mass meter within the intake manifold on newer models to generate a voltage signal proportional to the volume of air drawn into the intake manifold. As with the D system, the L system uses an ECU that receives information on the quantity of air drawn into the engine, as well as engine coolant and cylinder head temperature, the position of the throttle valve, the starting process, and engine speed as well as the start of injection to control engine operation.

The D-Jetronic fuel system has been superseded by the L and a newer system known as Motronic, which uses an on-board vehicle computer to control both the fuel injection and ignition systems.

The L system is available in three basic models:

1. LE = L-Jetronic, 2nd generation, Europe
2. LU = L-Jetronic, 2nd generation, USA
3. LH = L-Jetronic with hot-wire air-mass meter

The Motronic and Japanese equipped L-Jetronic fuel systems operate basically the same with a few minor variations. The Motronic fuel system is an L-Jetronic design with a crankshaft sensor and a different ECU; however, the testing procedures follow that of the L-Jetronic.

D-Jetronic System Operation

The D injection system, although not widely used due to the initial reluctance of vehicle manufacturers to invest the extra money required to convert from the cheaper and less efficient conventional carburetor system, was the first mass-produced gasoline injection system to be made available for production passenger car use and was therefore a leader in this field.

The D fuel system derives its name from the fact that the German word for pressure is *druck*. This system employs a pressure sensor in the intake manifold along with a set of trigger contacts in the ignition distributor, which are designed to transmit electrical pulses to the electronic control unit to establish the beginning of injection and also for information on engine speed.

Although the D system has been superseded by the more advanced and reliable L system and Motronic

Trouble-shooting Guide for K-Jetronic® fuel injection

This guide is designed to be used in conjunction with the Bosch Service Guide for K-Jetronic® (CIS). The page numbers refer to the appropriate pages in the Service Guide.

Engine does not start/starts hard cold	Engine does not start/starts hard warm	Irregular idle during warm-up	Irregular idle with warm engine	Engine will not accelerate, backfires	Engine backfires under load	Poor performance	Engine runs on	Fuel consumption too high	Driving performance unsatisfactory	CO concentration at idle too high	CO concentration at idle too low	Idle speed cannot be adjusted (too high)	Engine starts but immediately dies	CAUSE	REMEDY	PAGE
	●	●	●	●		●			●		●			Leaks in air intake system (false air)	Check air system for leaks	10
●	●		●	●	●	●	●		●	●	●			Sensor plate and/or plunger not moving freely	Check for free movement	11
	●						●							Air flow sensor plate stop incorrectly set	Check and reset	12
●		●												Auxiliary air valve does not open	Check valve for correct function	13
												●		Auxiliary air valve does not close	Check valve for correction function	13
●	●				●								●	Electric fuel pump not operating	Check pump fuse, pump relay and pump	13
●														Cold start system defective	Check cold start system	14
		●	●				●	●		●				Cold start valve leaking	Check cold start valve	14
●														Thermo-time switch defective	Test thermo-time switch	14
				●					●				●	Excessive fuel delivery to control circuit	Check fuel delivery	17
●		●												"Cold" control pressure outside tolerance	Test pressure	18
	●		●	●	●	●			●				●	"Warm" control pressure too high	Test pressure	19
			●	●		●		●	●	●			●	"Warm" control pressure too low	Test pressure	19
					●	●			●				●	Primary system pressure out of tolerance	Test pressure and adjust with shims	20
	●													Fuel system leakage	Inspect fuel system for leaks	21
●	●	●	●		●		●							Injector leaking, opening pressure low	Check injectors on tester	24
●	●	●	●			●			●					Injected fuel quantities unequal	Check for equal fuel quantities	27
●	●	●	●	●			●	●	●	●	●			Basic idle setting incorrect	Check and adjust CO level	32
						●								Throttle butterfly does not open completely	Check butterfly and stops in throttle venturi	32
		●	●	●		●		●	●	●	●			Lambda control system not functioning correctly	Check Lambda control system	29

Figure 6-36. K Fuel system troubleshooting chart. [Courtesy of Robert Bosch Corporation, Broadview, Illinois]

system, its operation is described here for two main reasons:

1. The basic operation of the D system is not too much different from the current L and Motronic systems.
2. There are a number of vehicles still in operation that are equipped with the D system.

The layout of the D fuel system is illustrated in Figure 6–37. You will note from Figure 6–37 (and Figure 6–43) that both the D and L fuel systems employ many similar components to that already discussed on the K (CIS) Jetronic system. Such components are:

1. Thermo-time switch
2. Start valve
3. Electric fuel pump
4. Fuel filter
5. Fuel pressure regulator (similar to K accumulator)
6. Auxiliary air device

Since these components were discussed in detail as to their function and operation under the K-Jetronic system, we will not describe them again in this discussion.

The ECU is designed to electronically monitor the following:

1. Engine throttle position
2. Engine coolant temperature or cylinder head temperature
3. Inlet air temperature
4. Exhaust gas content (Lambda sensor voltage signal)
5. Ignition timing
6. Start of fuel injection
7. Engine starting sequence
8. Engine speed

The ECU receives voltage signals from various engine sensors. After monitoring these signals, the ECU controls the time that the fuel injectors are open as well as controlling the air/fuel mixture as close to stoichiometric as possible, usually within 0.02%. Minor differences do exist between various makes of vehicles that employ the same fuel system. For example, you may find that some L-Jetronic equipped

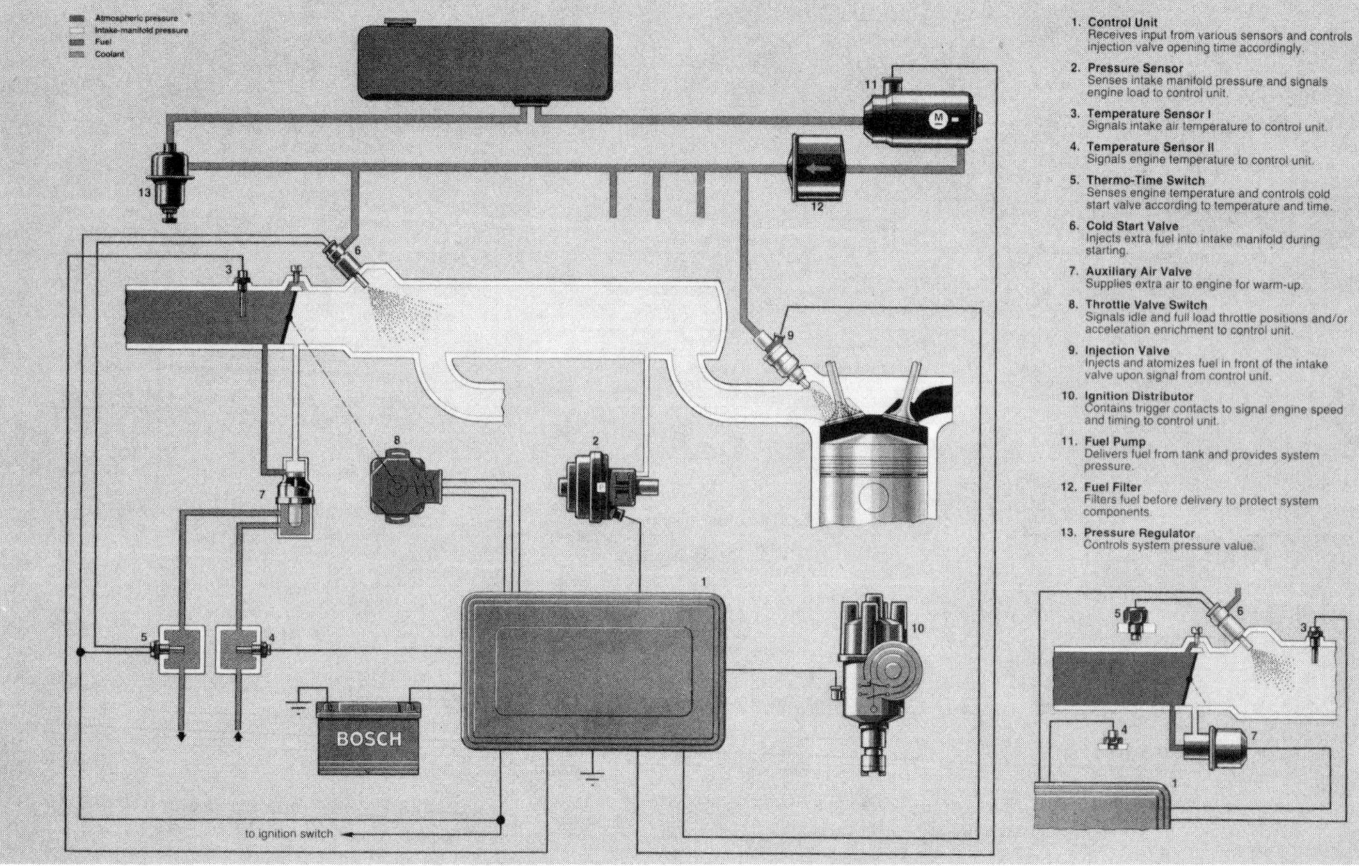

Atmospheric pressure
Intake-manifold pressure
Fuel
Coolant
BOSCH
to ignition switch
1. Control Unit
Receives input from various sensors and controls injection valve opening time accordingly.
2. Pressure Sensor
Senses intake manifold pressure and signals engine load to control unit.
3. Temperature Sensor I
Signals intake air temperature to control unit.
4. Temperature Sensor II
Signals engine temperature to control unit.
5. Thermo-Time Switch
Senses engine temperature and controls cold start valve according to temperature and time.
6. Cold Start Valve
Injects extra fuel into intake manifold during starting.
7. Auxiliary Air Valve
Supplies extra air to engine for warm-up.
8. Throttle Valve Switch
Signals idle and full load throttle positions and/or acceleration enrichment to control unit.
9. Injection Valve
Injects and atomizes fuel in front of the intake valve upon signal from control unit.
10. Ignition Distributor
Contains trigger contacts to signal engine speed and timing to control unit.
11. Fuel Pump
Delivers fuel from tank and provides system pressure.
12. Fuel Filter
Filters fuel before delivery to protect system components.
13. Pressure Regulator
Controls system pressure value.

Figure 6–37. Component layout of EFI-D (Electronic Fuel Injection) D-Jetronic fuel system. (Courtesy of Robert Bosch Corporation, Broadview, Illinois)

European vehicles may not use a fuel damper in the system while both Datsun and Nissan do employ this fuel damper to reduce fuel pump pulsations. Another change is that of the Datsun 200SX car, which does not use a cold start valve but uses instead a Datsun/Nissan designed cold start arrangement. An operational description of both the D and L fuel systems follows.

An example of the basic layout of the components of the D series electronic fuel injection system for a 6-cylinder Volvo 164E engine is shown in Figure 6–38. The basic flow and operation of the system shown in Figure 6–38 is as follows: Fuel from the tank is supplied to the system by an electric fuel pump (2). Pump pressure is controlled by the use of a pressure regulator (28.5 psi) after passing through the fuel filter (3) and to the injectors (8). The injectors are solenoid operated via a signal received from the electronic control unit of the vehicle. Item 11 is a start valve that sprays additional fuel into the intake manifold during starting at low ambient temperatures. The pressure regulator (4) bypasses fuel back to the tank (1), in order to control the system fuel pressure. A check valve installed in the fuel pump prevents an immediate pressure drop in the fuel line when the engine is switched off.

Figure 6–39 illustrates the components and electronic controls used with the D-Jetronic fuel injection system.

Operation of the D-Jetronic Fuel System

The electronic control unit (ECU) (Item 20 in Figure 6–39) is located under the passenger seat on the model 164E Volvo and is accessible by removing the seat cushion. Battery voltage powers the ECU through relay 19 when the ignition is turned on. To allow fuel pressure build-up immediately after the ignition switch is turned on, a time switch installed in the ECU lets the pump run for one second to ensure adequate fuel pressure prior to start up. The fuel pump is controlled through relay 18, which will only allow pump operation when the starter motor is energized through the terminal (50), or when the engine

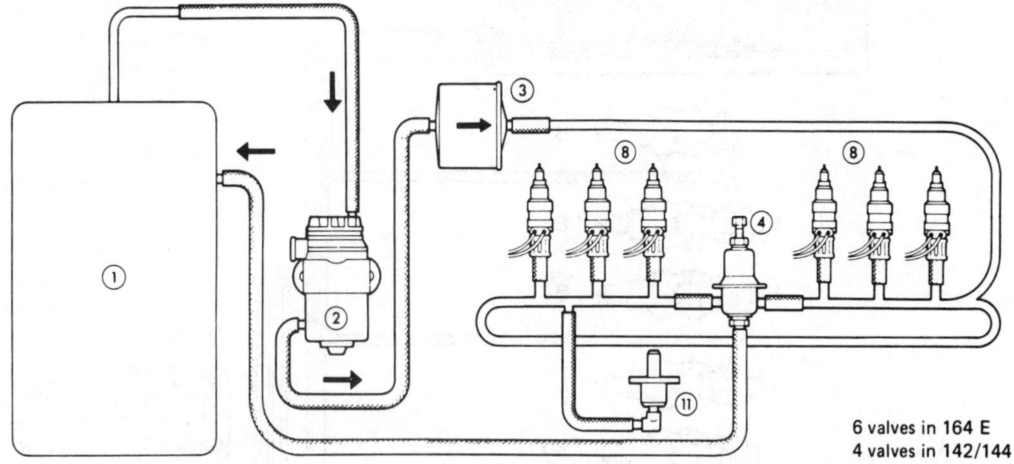

Figure 6-38. Basic D-Jetronic fuel system operational diagram—Volvo. (Courtesy of Robert Bosch Corporation, Broadview, Illinois)

speed is in excess of 200 rpm. This feature is necessary in order to prevent flooding if an injection nozzle valve becomes defective.

The amount of fuel injected is directly proportional to the time that the injection valve remains energized (or open) because the electric fuel pump and pressure regulator maintain a steady system pressure of 2 to 2.2 kgf/cm^2 or 2 to 2.2 bar (28 to 31.9 psi), with later models using the higher setting. How long the fuel injection valves remain open is controlled by the signal from the ECU, which computes this time based on the signals received from the following engine mounted sensors:

1. Intake manifold pressure sensor.
2. Temperature sensor (engine temperature) located in the cooling system or cylinder head on some Japanese vehicles (Datsun/Nissan).
3. Throttle valve switch signals idle and full-load throttle positions and/or acceleration enrichment.
4. Ignition distributor contains trigger contacts to signal engine speed and timing.
5. A second temperature sensor that monitors intake manifold air temperature (hot-wire sensor).

The trigger contacts located under the centrifugal advance device in the distributor shown in Figure 6–40 control the actual point or moment of injection according to the position of the camshaft on the distributor shaft. Engine speed is relayed to the ECU from the trigger contacts based on the interval of time between the trigger pulses. In addition, the pressure sensor in the intake manifold (suction side), which is basically related to throttle position, converts this pressure into an electrical impulse and relays it to the ECU, thereby effectively analyzing both engine speed and load conditions. For example, high vacuum = closed throttle/low speed and load; low vacuum = open throttle/high speed and load.

The signals received by the ECU from the pressure sensors and trigger contacts of the distributor are processed by the ECU, which then directs an appropriate signal to the solenoid of the injectors to open for a longer (open throttle) or shorter (closed throttle) time period. Obviously, since the fuel system pressure is constant (controlled by the fuel pressure regulator—Item 13 in Figure 6–37), the volume of injected fuel depends on the time that the injectors are open. To ensure smooth start-up and warm-up periods in cold weather, it is necessary to inject an additional amount of fuel to the engine during this time period (similar to a choke feature in a carburetor) and therefore the start valve (Item II in Figures 6–38 and 6–39) will inject additional fuel into the intake manifold.

The start valve is controlled by the thermo-time switch (Item 24 in Figure 6–39 or Item 5 in Figure 6–37), which is installed into the engine cooling system. This time switch electrically opens or closes the circuit to the start valve based on the engine temperature. For example, the start valve activation period can range from 5 to 20 seconds at a coolant temperature of −20°C, depending on the particular engine make that it is fitted to.

At higher coolant temperatures, this time period will be shortened, and at temperatures between +20°C and 40°C, it reaches zero time because extra

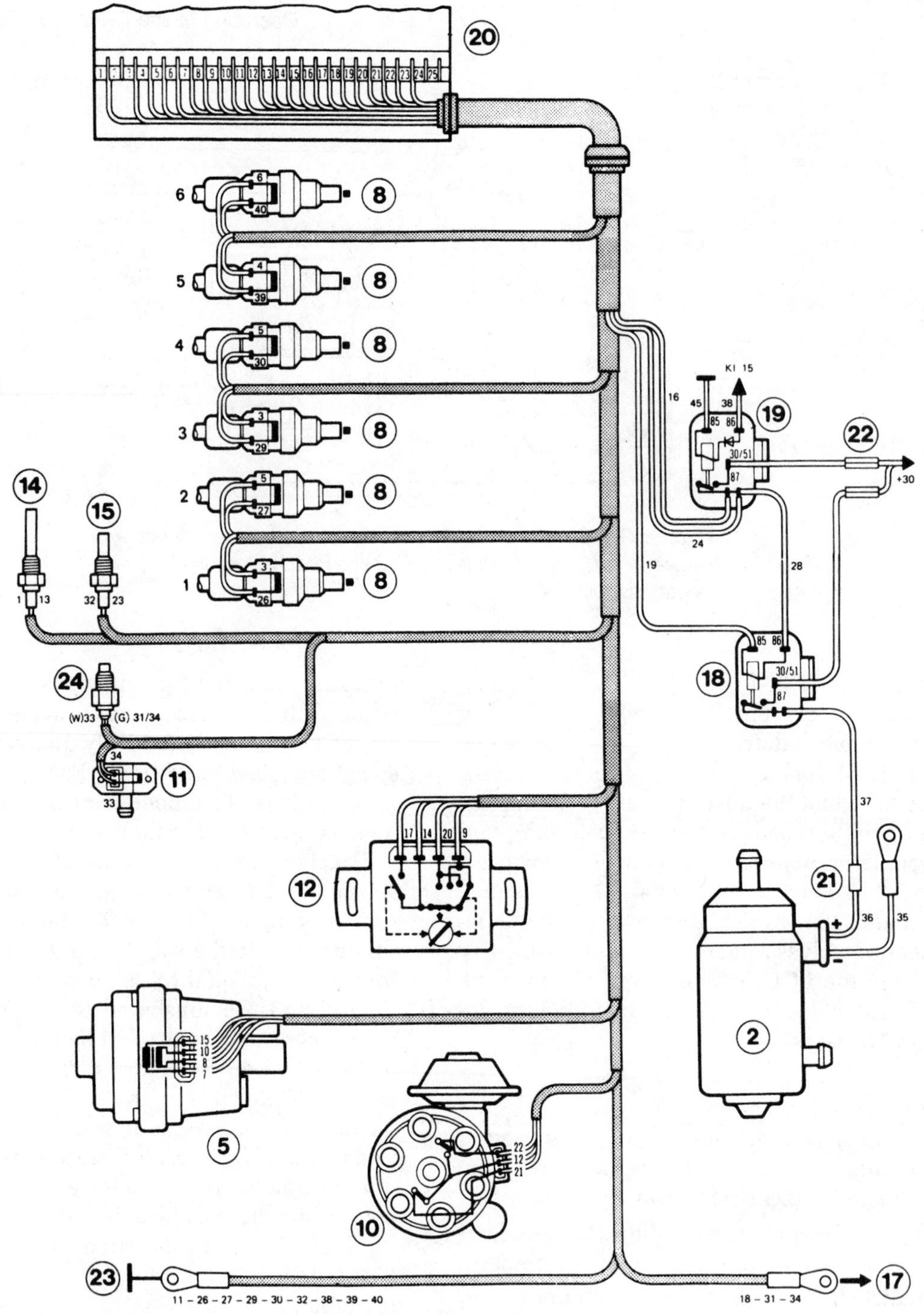

② Electrically-operated fuel pump
⑤ Pressure sensor
⑧ Solenoid-operated injection valves
⑩ Ignition distributor with trigger contacts
⑪ Start valve
⑫ Throttle valve switch
⑭ Temperature sensor I (intake air)
⑮ Temperature sensor II (coolant)
⑰ Starting motor terminal 50
⑱ Pump relay
⑲ Main relay
⑳ Electronic control unit (ECU)
㉑ Connector
㉒ Twin fuse box
㉓ Ground
㉔ Thermo-time switch

Figure 6-39. D-Jetronic fuel system electronic controls and components. (Courtesy of Robert Bosch Corporation, Broadview, Illinois)

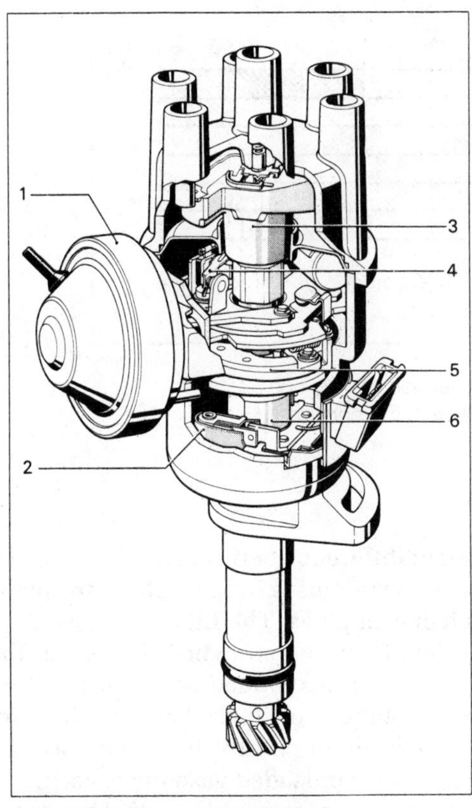

Figure 6–40. Ignition distributor with trigger contacts for a 6-cylinder engine using contact breaker point ignition. (Courtesy of Robert Bosch Corporation, Broadview, Illinois)

starting fuel is not necessary. The ECU receives a signal from the temperature sensor (15) shown in Figure 6–39 (coolant) plus a signal from the temperature sensor (14, intake air) to correct the injected quantity of fuel. The throttle valve switch (12) signals "overrun" or "foot off the throttle" (engine braking condition) so that the ECU will signal the injectors to cut off all fuel. It should be noted, however, that this overrun condition will only exist if the engine rpm is allowed to reach 1700 rpm. At a decreasing engine speed of 1000 rpm, the fuel supply is again switched back on to ensure a smooth changeover to an idle condition.

Both of the rpm's quoted are increased by approximately 300 rpm when the engine is cold because of the greater frictional resistance to engine rotation.

NOTE: The fuel cutoff condition described applied to earlier model D-Jetronic systems. In late 1971 systems, the fuel supply was not cut off during overrun; therefore, the throttle valve switch (12) gives the ECU the information to idle instead during an overrun.

Another function of the throttle switch is to supply the ECU with the necessary information for increased fuel when the throttle pedal or accelerator is depressed. A unique feature of the injection system is that the electrical signals from the ECU to the injectors causes them to open in two groups: namely, group 1, which controls engine cylinders 1-5-3, and group 2, which controls cylinders 6-2-4. This is basically the firing order sequence of the engine—1-5-3-6-2-4. This feature ensures that three injection valves of any one group are open at the same time. Fuel is actually injected (fine spray) onto the closed intake valve from the injector, which is located in the intake manifold. It is temporarily stored there until the valve is opened by the engine camshaft and rocker arm mechanism, at which time it will flow into the engine cylinder with the flow of intake air, thereby premixing the air/fuel ratio.

Figure 6–41 illustrates the sequence of events involved in the opening of the two groups of injection valves through 600 crankshaft degrees for the Volvo 164E vehicle engine.

To appreciate the extremely short time during which both the injection valve and cylinder intake valves would remain open, consider the following situation. With the engine running at an idle rpm of 600, the intake valves are open for a total duration of 217.5 degrees of crankshaft rotation, which is equivalent to 1/993 of a minute, or 0.0604 second. The injection valve would be injecting for approximately 22.5 degrees of crankshaft rotation based on the example related to 600 rpm, and injection time would therefore be 1/9600 of a minute, or 0.00625 second.

A power stroke would occur every 120 crankshaft degrees since the engine is a 4-stroke cycle unit, meaning that two full rotations of the crankshaft are necessary to produce one power stroke. With 360° in a full rotation, the crankshaft will rotate 2 × 360° = 720° divided by the number of cylinders, which is 6. This gives us 120° between firing impulses.

No possibility can therefore exist of fuel puddling around the intake valve with such a short period between injections. In addition, you have to realize that with the pressure of fuel leaving the injection valve, it will be in a reasonably atomized state. Also, when the engine is warm, this assists the gasoline in vaporizing more readily than it tends to do when cold.

The time period for injection is of course lengthened (open throttle) or shortened (closed throttle) by the signal from the ECU to vary the quantity of fuel injected as described earlier.

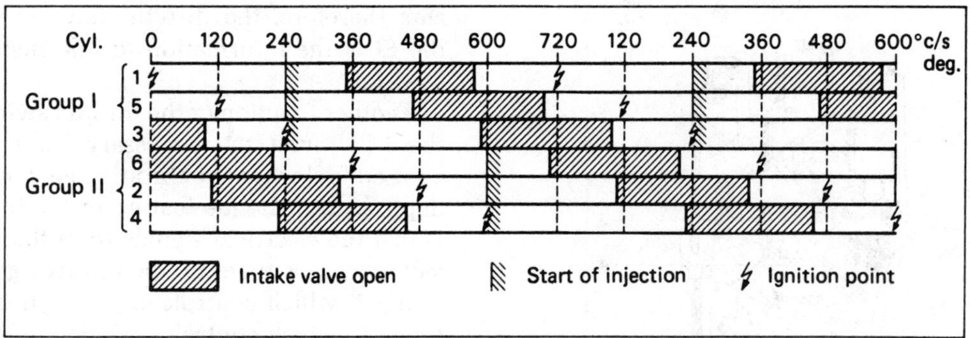

Figure 6–41. Volvo 164E D-Jetronic injection valve group opening period. (Courtesy of Robert Bosch Corporation, Broadview, Illinois)

L-Jetronic System Operation

The L system derives its name from the fact that the German word for air is *luft*, therefore the letter L. The system relies on an air-flow sensor that generates a voltage signal proportional to the volume of air actually drawn into the engine intake manifold. The EFI L-Jetronic system is a more advanced system than the D series, and it therefore provides a more effective system to meet exhaust emissions regulations. Since the air-flow sensor on the L-Jetronic system is different from that on the K, refer to Figures 6–42 and 6–43 for a more detailed explanation of its operation.

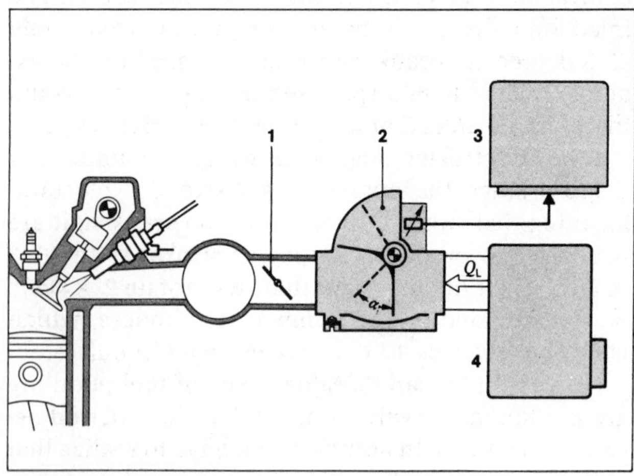

1 Throttle valve
2 Air-flow sensor
3 Control unit
4 Air filter
Q_L Amount of air drawn in

Figure 6–42. L-Jetronic air-flow sensor. (Courtesy of Robert Bosch Corporation, Broadview, Illinois)

The main difference between the L/LE and the LH fuel injection systems is in the type of air-flow sensor that each one employs. The LE system [see Figure 6–44(a)] differs from the straight L system in that the LE system has an expanded scope of functions for precise adaptation of fuel delivery. In addition, Lambda closed-loop control is possible as an additional function if unleaded gasoline is used. The system is then known as an LU, with U standing for unleaded fuel.

Specific differences lie mainly in the improvements over the original L model system as follows:

1. Improved functions, reduced costs, and optimized circuitry of the electronic control unit for a reduced power loss.
2. Injectors with a high internal resistance (without series resistor).
3. An air-flow sensor with an integrated intake-air temperature compensation.
4. A reduced number of connectors.
5. Expanded scope of functions for increased economy in the metering of fuel to suit the respective operating condition.
6. Control of the fuel pump through an electronic control relay; therefore, elimination of the pump contact in the air-flow sensor.

The LE system can be divided into three main functional areas which are:

1. Fuel supply.
2. Measurement of operating data.
3. Fuel metering.

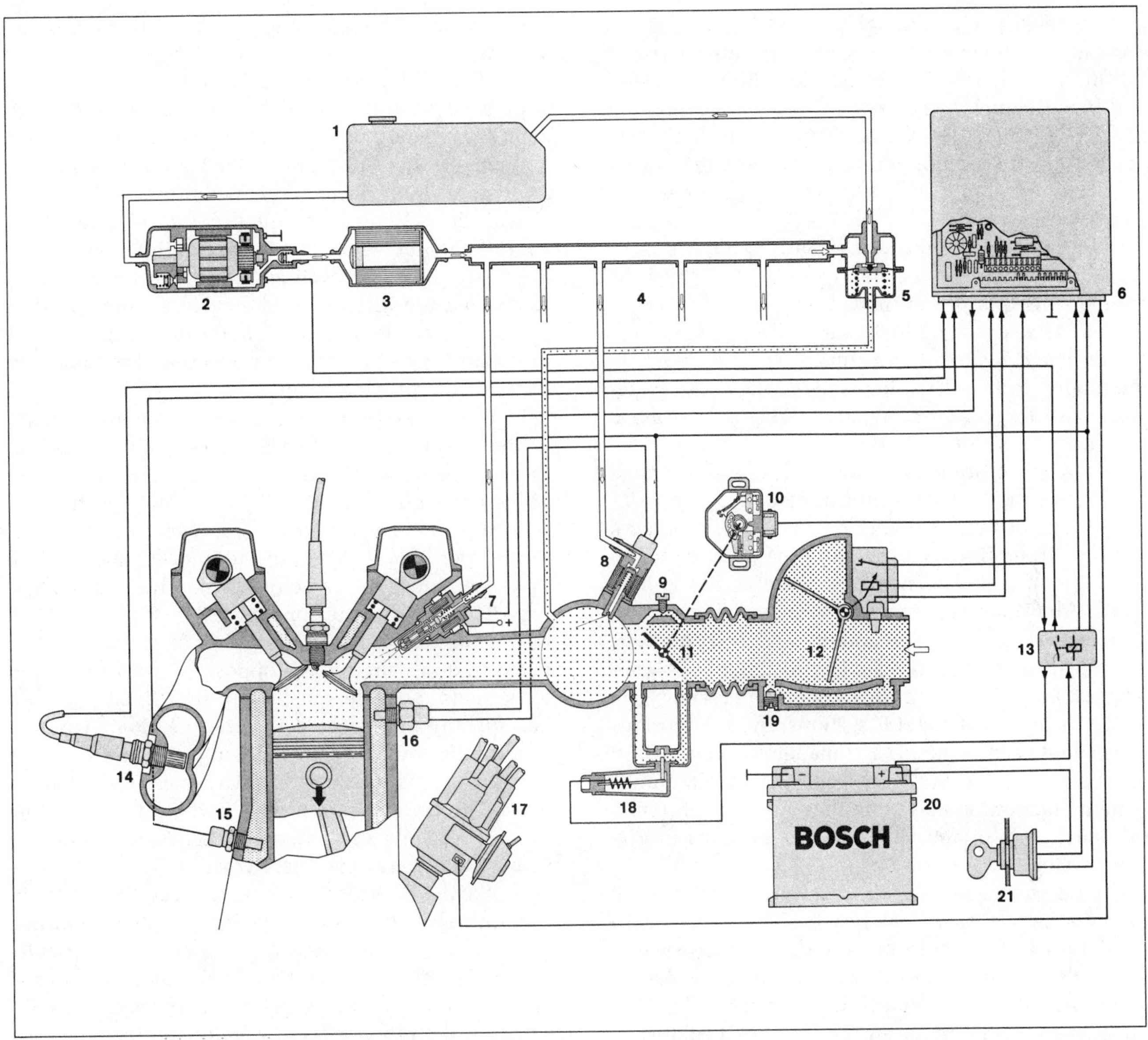

1 Fuel tank, 2 Electric fuel pump, 3 Fuel filter, 4 Distributor pipe, 5 Pressure regulator, 6 Control unit, 7 Injection valve, 8 Start valve, 9 Idle-speed adjusting screw, 10 Throttle-valve switch, 11 Throttle valve, 12 Air-flow sensor, 13 Relay combination, 14 Lambda sensor (only for certain countries), 15 Engine temperature sensor, 16 Thermo-time switch, 17 Ignition distributor, 18 Auxiliary-air device, 19 Idle-mixture adjusting screw, 20 Battery, 21 Ignition-starter switch

Figure 6–43. L-Jetronic fuel system component layout. [Courtesy of Robert Bosch Corporation, Broadview, Illinois]

The LH system [see Figure 6–44(b)] on the other hand can be divided into five main functional areas which are:

1. Fuel supply.
2. Measurement of operating data such as air mass, engine speed, load condition (idle/full load), and engine temperature.
3. Metering of basic fuel quantity and adaptation to cold starting, warmup, idle, full load, acceleration, and overrun.
4. Idle speed control.
5. Emission control with Lambda closed-loop control.

NOTE: Closed loop means that the engine exhaust

oxygen sensor is hot enough to create a reaction. As a result, a voltage signal is sent from this sensor to the ECU, which will influence the duration (pulse width) of the fuel injectors and therefore the quantity of fuel injected. By this method, the air/fuel ratio control can be maintained at nearly stoichiometric, or ideal, conditions of 14:1 under all operating conditions.

In an open-loop mode, the exhaust oxygen sensor does not influence the signal from the ECU and the engine runs in a preset mode.

The LH system employs a hot-wire mass air-flow meter that measures the air mass being drawn into the engine rather than the vane-type air-flow meter used with the L and LE models. This hot-wire air-flow meter takes into account the possibility of temperature and altitude variations. This hot-wire mass air-flow sensor is shown in Chapter 7 in Figure 7–35(a), 7–35(b), and 7–35(c) and is used by General Motors on the Buick 3.8L V6 and the Chevrolet Corvette engines, just to name two commonly known North American passenger performance cars.

Of Robert Bosch design and manufacture, this system of air-flow measurement is a first of its kind. Rather than using an air-flow sensor flap (vane-type), which is used in the L/LE systems, the hot-wire arrangement used in the LH system employs a very thin platinum wire (70 μm thick) set into the air intake system and contained within the bore area of item 5 in Figure 7–35(b) and shown in heated form in Figure 7–35(c).

Air flowing over the wire draws heat from it so that the wire's electrical resistance changes with temperature. An electronic amplifier in the system responds instantly to any electrical resistance change and regulates the current flow of from 500 to 1200 mA (depending upon the air-flow rate) to the hot wire in order to maintain it at a virtually constant temperature. The current flow required to maintain a stable temperature at the wire becomes in effect a measure of the air mass flowing over the wire with the output signal being a voltage.

Major advantages of the hot-wire system of measurement are its very small dimensions and the complete lack of any moving parts.

Another unique feature of the hot-wire system of air-flow measurement is that because impurities may accumulate on the heated wire's surface, the voltage output signal could change. Therefore the system includes a provision for automatically increasing the wire's temperature for one second each time the engine is switched off to allow the wire to burn off impurities or corrosion. The "burn-off" command comes from the electronic control unit of the LH system.

Another difference between the L and the LH systems is that the L control unit is available in either analog or digital versions while the LH control unit is digital with a microcomputer governing its adaptation to engine parameters. In addition, both the LE/U and the LH can be adapted for deceleration fuel shutoff such as when in overrun (going down a hill) or in city traffic, as well as for engine speed limitation when the maximum allowable engine speed is reached. In the latter situation, the electronic control unit suppresses the injection pulses and protects the engine from over-revving.

Figure 6–44a illustrates the actual component parts of the LE-Jetronic system as they would appear on the vehicle while Figure 6–44b illustrates the newer LH system components. This visual identification will help you to associate the numbered items shown in the operating diagram in Figure 6–43 and to readily locate parts on the vehicle when tracing the system. The air-flow sensor flap (2) in Figure 6–42 is moved in relation to the volume of air flow being taken into the engine. An opposing restoring spring force maintains the sensor flap in a fixed position, i.e., for any pre-selected throttle position. A potentiometer is connected to the flap position and sends a voltage signal back to the ECU (electronic control unit). The compensation flap used with the air-flow sensor prevents oscillations of the sensor flap because it has the same effective area.

Unlike the D-Jetronic system, the L system does not require trigger contacts in the ignition distributor because all injection valves are connected electrically in parallel and will simultaneously inject half of the required amount of fuel twice during each camshaft rotation; therefore, a fixed relationship between the cam angle and the start of ignition is no longer necessary. The injection pulses are controlled from the ignition distributor contact points or reluctor pickup coil in electronic ignition systems.

In a 6-cylinder engine, the distributor must relay/time a high-tension spark six times in one operating cycle of the engine, namely intake, compression, power, and exhaust (720° or two complete revolutions). Therefore, the frequency of injection must be divided in half in the control unit (ECU) for each engine revolution, which, for the 6-cylinder engine, would be divided by a factor of 3. In a 4-cylinder, the distributor delivers a high-tension spark four times during each operating cycle or 720°, and since fuel is injected only twice for each engine operating cycle, the frequency has to be divided in half at the ECU.

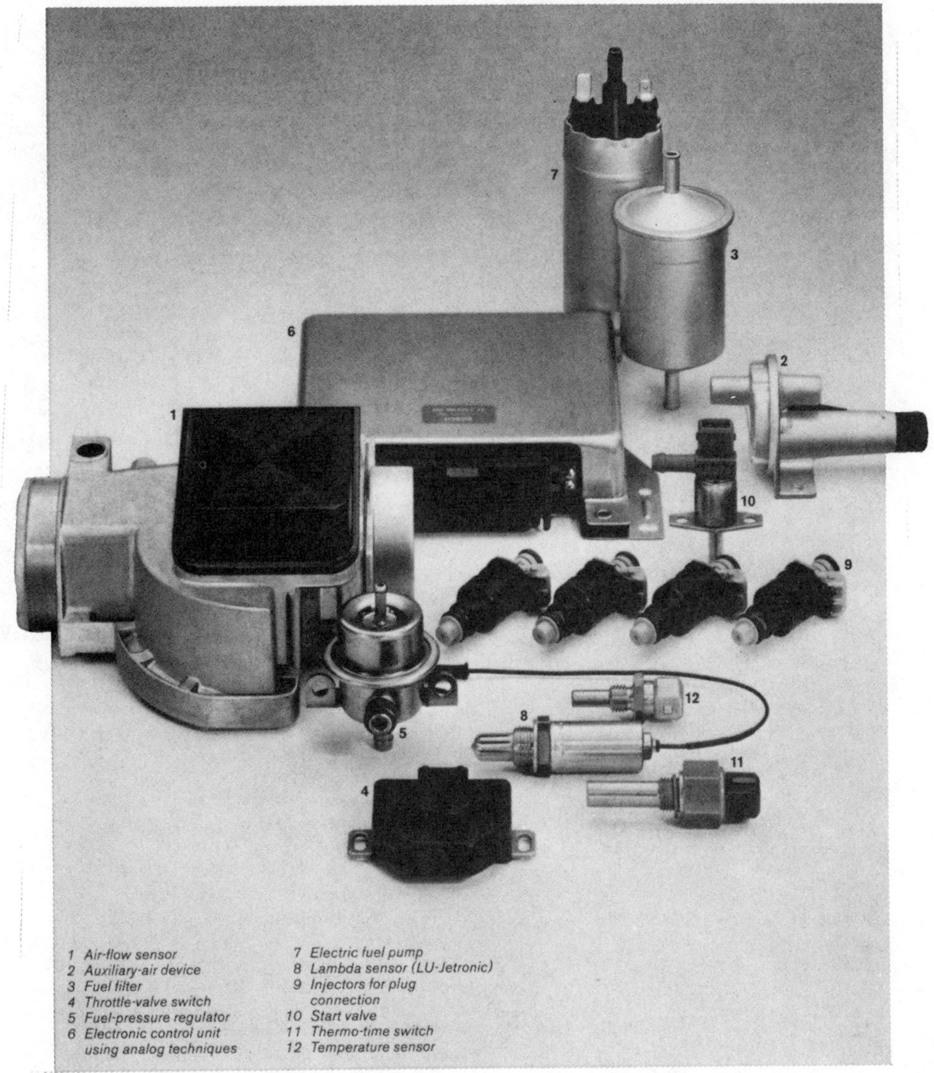

Figure 6–44(a). Components of the LE-Jetronic fuel system. [Courtesy of Robert Bosch Corporation, Broadview, Illinois]

With no trigger contacts in the L system, as there were with the D system, the pulses from the ignition distributor are converted to rectangular pulses in a pulse shaper, such as shown in Figure 6–45, within the throttle valve switch.

An example of how the pulse shaper controls the injection system is best explained by reference to Figure 6–46.

NOTE: In the breakerless (solid state) high energy-type ignition distributor, pulses of the ignition voltage would cause the same basic action to occur as explained in Figure 6–46.

Because fuel injection will take place twice for every camshaft rotation related to four ignition distributor pulses, the frequency of the system must be divided in half by the frequency divider shown in the block diagram in Figure 6–47. A capacitor is charged by these rectangular pulses; therefore, each injection pulse is initiated by the discharge of the capacitor. Air flap position determines the duration of injection (wider throttle position, longer fuel injection period, and vice versa). The other input signals shown in Figure 6–47 also contribute to the actual length of the injection period.

An example of an L-Jetronic fuel injection system for a 6-cylinder BMW is shown in Figure 6–48 and 6–49.

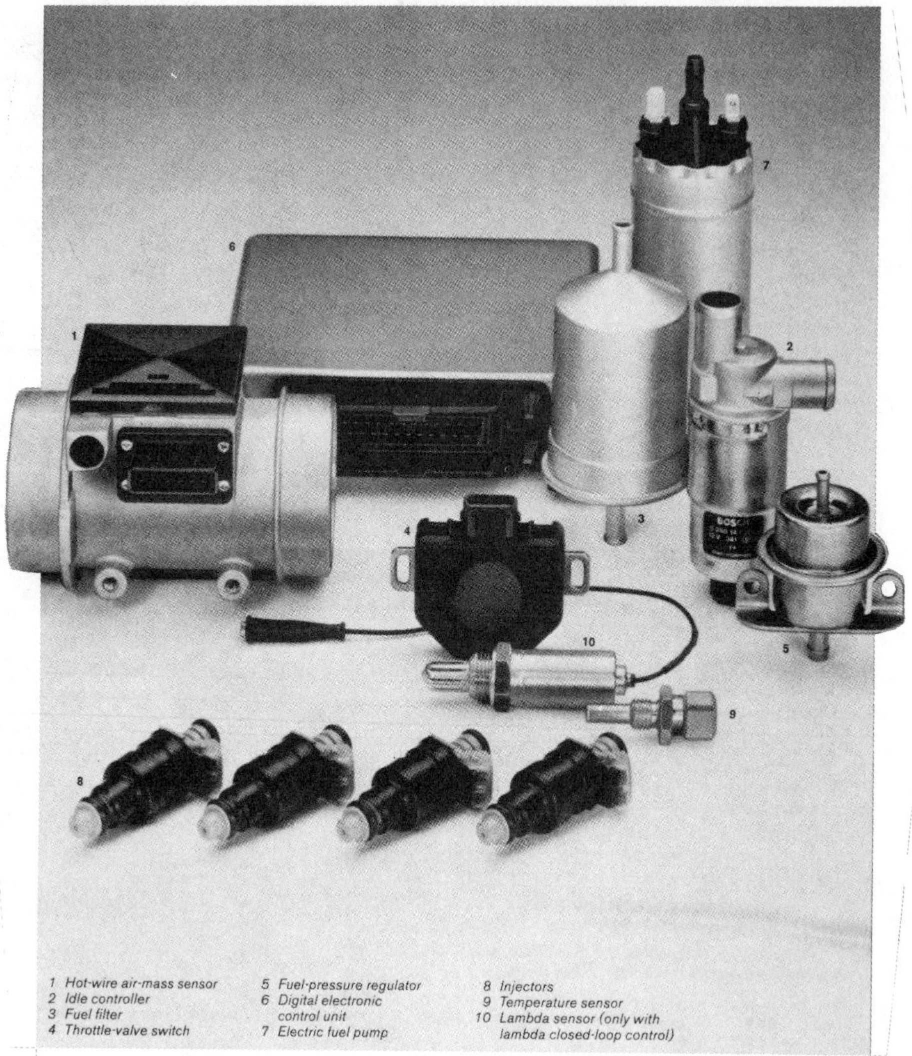

Figure 6-44(b). Components of the LH-Jetronic Fuel System. (Courtesy of Robert Bosch Corporation, Broadview, Illinois)

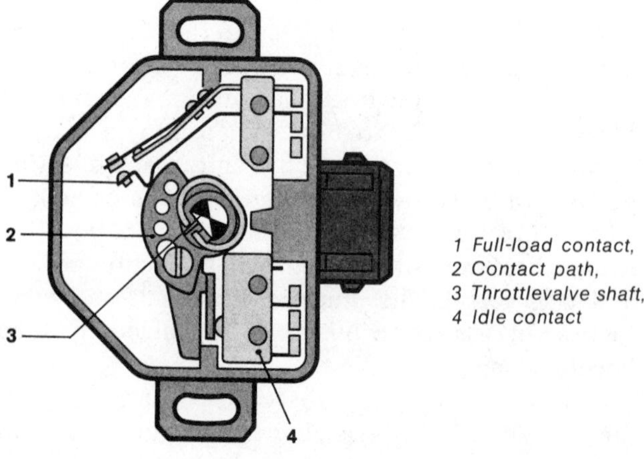

Figure 6-45. L Fuel system throttle valve switch. (Courtesy of Robert Bosch Corporation, Broadview, Illinois)

In Figure 6–48, when the ignition is switched on, battery voltage is relayed through the relay set (9) and (10) to the ECU (1) and injection valves (5). When the engine starts, the relay set switches on the electric fuel pump, the start valve, the thermo-time switch (8) and the auxiliary air device. If the engine fires and runs, the power supply for the pump and the auxiliary air device is maintained through a contact in the air-flow sensor. If the engine fails to start, the relay set (9) and (10) interrupt the electric circuit to the fuel pump to prevent a flooding condition through continued pump operation with the engine being cranked continually.

Fuel injection pressure in the L system is controlled by the fuel pressure regulator to approximately 36 psi (2.5 bar). A relief valve in the pump is

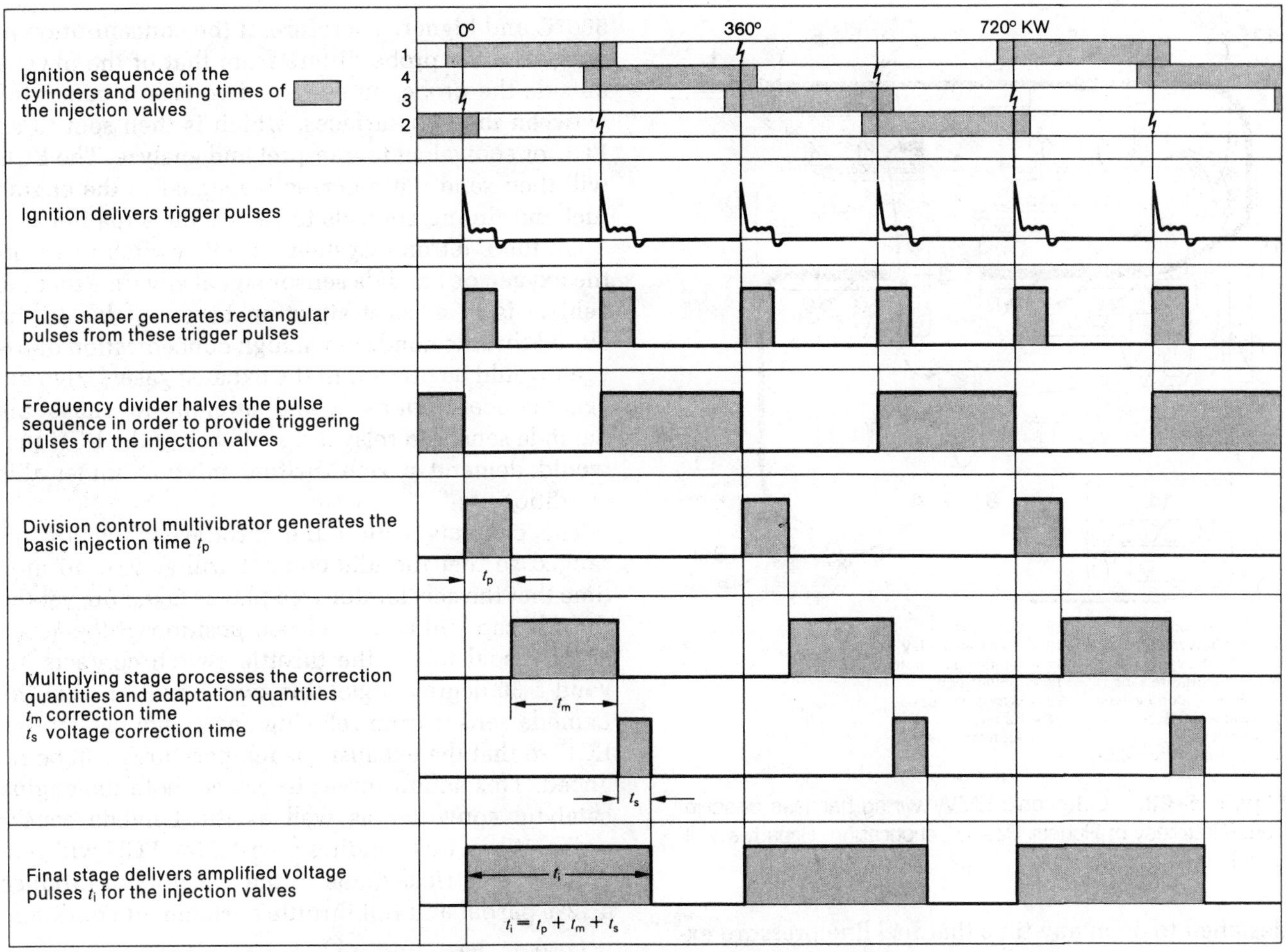

0°
360°
720° KW
Ignition sequence of the cylinders and opening times of the injection valves
1
4
3
2
Ignition delivers trigger pulses
Pulse shaper generates rectangular pulses from these trigger pulses
Frequency divider halves the pulse sequence in order to provide triggering pulses for the injection valves
Division control multivibrator generates the basic injection time t_p
t_p
Multiplying stage processes the correction quantities and adaptation quantities
t_m correction time
t_s voltage correction time
t_m
t_s
Final stage delivers amplified voltage pulses t_i for the injection valves
t_i
$t_i = t_p + t_m + t_s$

Figure 6-46. Generation of the injection pulses for a 4-cylinder L-Jetronic fuel system. (Courtesy of Robert Bosch Corporation, Broadview, Illinois)

Speed
Load range
shaper
divider
Final stage
½ n
t_i
Injection valves
Division control multivibrator
t_p
Multiplying stage
t_i Injection pulses corrected
t_p Basic injection time
n Speed
Air quantity
Engine tem-perature
Air tem-perature
Vehicle voltage

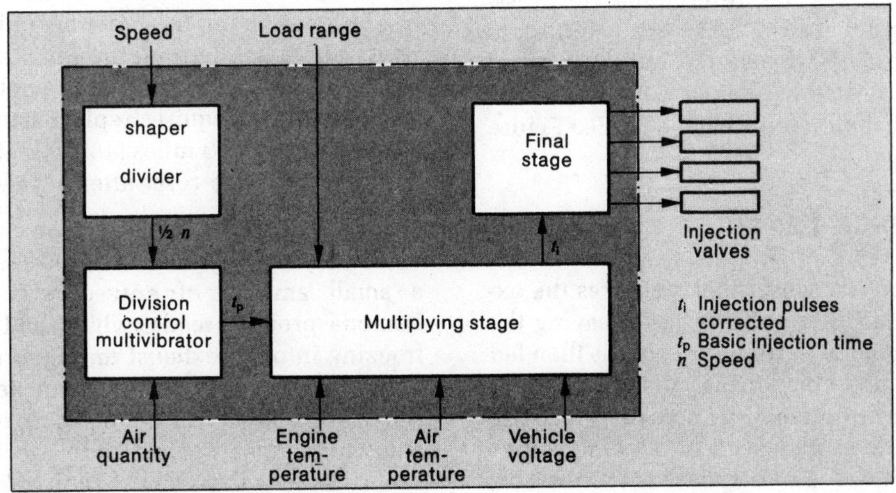

Figure 6-47. Block diagram of the L-Jetronic system control unit. (Courtesy of Robert Bosch Corporation, Broadview, Illinois)

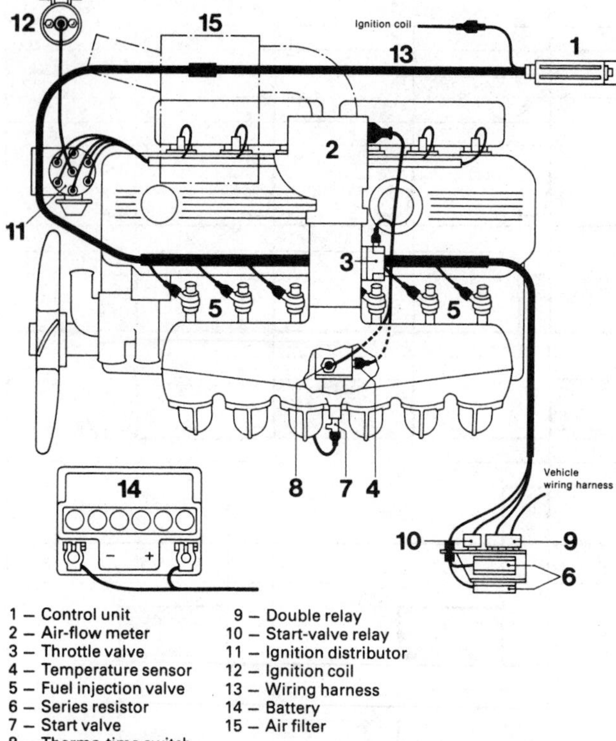

Figure 6–48. L-Jetronic BMW wiring harness description. (Courtesy of Robert Bosch Corporation, Broadview, Illinois)

designed to open any time that fuel line pressure exceeds 43–64 psi (2.96–4.41 bar).

Injection Valves

The injection valves used in both the EFI D series and L series Jetronic systems differ in that the cross-sectional opening of the valves in the L units are smaller than in the D because the fuel is injected twice per camshaft rotation in the L, but it is injected only once per camshaft rotation in the D. (See Figure 6–50.)

Lambda Sensor Probe

The Lambda probe is a sensor that measures the oxygen concentration in the exhaust gases leaving the engine. A signal generated from the probe is then fed to the ECU to regulate the air/fuel mixture in order to reduce the concentration of nitric oxides emanating from the engine as a by-product of combustion.

Figure 6–51 shows the Lambda probe sensor, which operates on the principle of conduction. The ceramic material and platinum electrodes used become conductive for oxygen ions at temperatures of 300°C and higher. Therefore, if the concentration of oxygen at the probe differs from that of the oxygen outside the probe, an electrical voltage is generated between the two surfaces, which is then sent to an ECU or equivalent to interpret and analyze. The ECU will then send out a corrective signal to the engine fuel and timing controls to rectify the situation.

On the L-Jetronic system, an idle switch turns off the oxygen or Lambda sensor signal any time that the vehicle is in a coast situation because, due to the closed throttle condition, a high concentration of oxygen would be present in the exhaust gases. High oxygen concentrations would attempt to cause the Lambda sensor to relay a signal back to the ECU that would demand a rich air/fuel mixture under this condition.

The contacts within the throttle switch are arranged so that the idle contact will switch off anytime that the accelerator free-play is taken up, yet the throttle can still be in a closed position. If the accelerator pedal forces the throttle switch contacts beyond a 30 degree angle, this switch will prevent the Lambda sensor from relaying any signal back to the ECU so that the exhaust gas temperatures will be reduced. This action serves to protect both the engine catalytic converter as well as the Lambda sensor probe. When this condition exists, the ECU will generate an electrical signal to the injectors to produce a 12% partial and full throttle enrichment condition.

NOTE: The use of leaded fuel will contaminate the Lambda sensor and render it inoperative on LU systems. After 10 to 12 tanks of leaded gasoline, HC (hydrocarbon) emissions are 4 to 5 times the baseline levels, CO (carbon monoxide) emissions 2 to 3 times, and NO_x (nitric oxide) emissions 1.5 to 2 times. The LU designation implies that only unleaded (U) fuel be used. Current L-Jetronic systems employ a service light that will light up on the vehicle dash when the Lambda sensor requires replacement. This occurs at about every 30,000 miles (48279 km), and the service indicator must be reset after a new Lambda sensor is installed.

When installing a new Lambda sensor unit, apply a small amount of anti-seize compound to the Lambda probe threads, which will prevent it from freezing into the exhaust manifold once it heats up. This will allow ease of removal next time around. Care should be exercised to ensure that the anti-seize compound does not get onto the Lambda sensing probe because this material will bake onto the probe surface causing an insulation effect that will create a false interpretation of how much oxygen is actually flowing through the exhaust system.

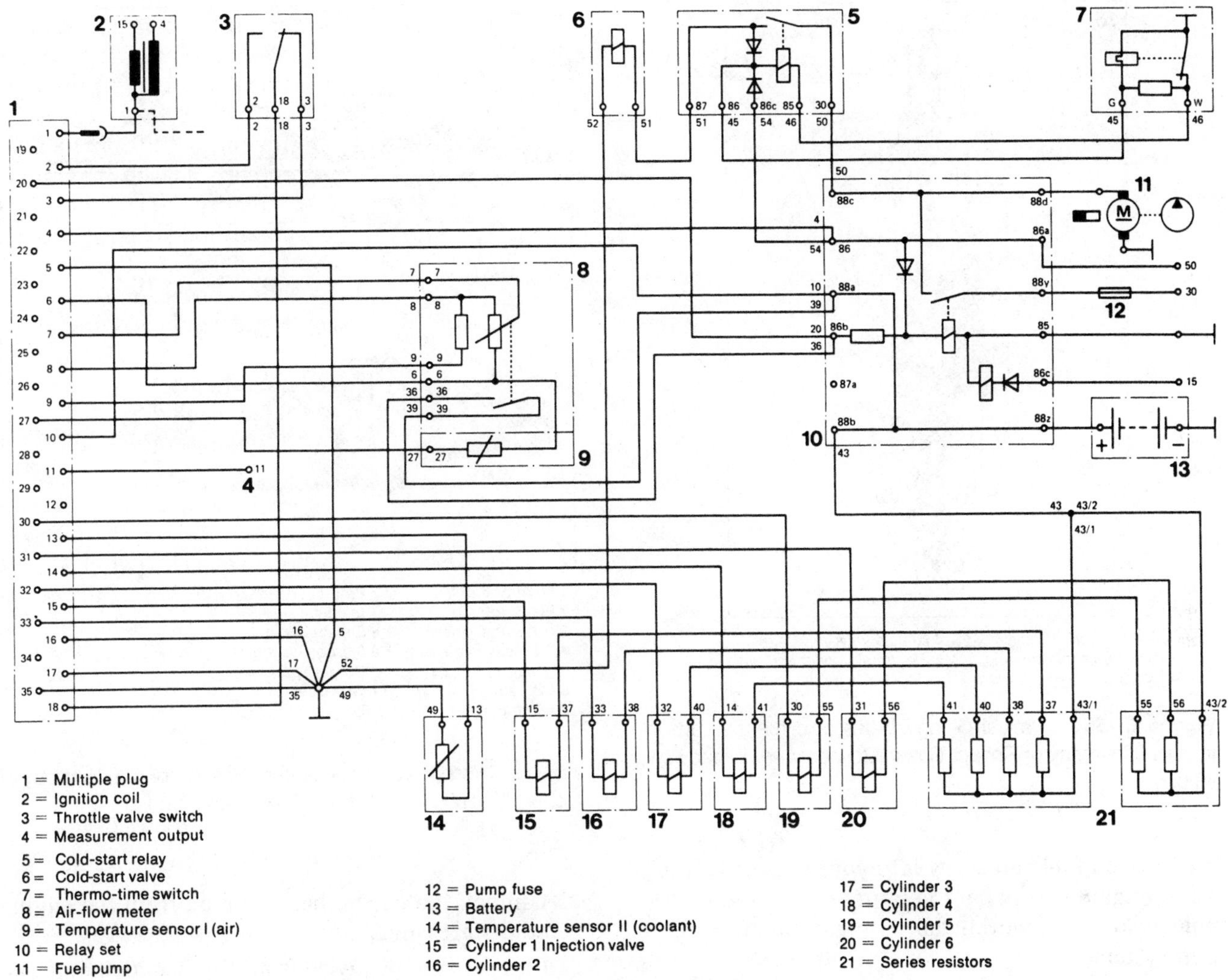

Figure 6-49. L-Jetronic BMW wiring diagram. (Courtesy of Robert Bosch Corporation, Broadview, Illinois)

D and L Checks and Adjustments

Before undertaking any checks or adjustments to these fuel injection systems, a number of safety recommendations should be followed; otherwise, damage to the system can occur. These recommendations are as follows.

Safety Recommendations

1. Never start the engine unless the battery cables are securely tightened.
2. Never boost/jump start the battery in order to start the engine.
3. Always disconnect the battery cables before charging.
4. Never disconnect either battery cable while the engine is running.
5. Never remove or attach the wiring harness plug to the ECU (electronic control unit) while the ignition is on.
6. When checking engine cylinder compression, unplug the red cable from the battery to the relays.
7. Before testing the L-Jetronic system, ensure that the ignition timing, dwell, and spark plugs check out OK and are all within vehicle manufacturer's specifications.

CAUTION: All Robert Bosch fuel injection systems are designed to maintain fuel pressure in the system after engine shutdown for a certain time (accumulator/regulator function). Therefore *do not* attempt to loosen off any fuel fittings or lines because

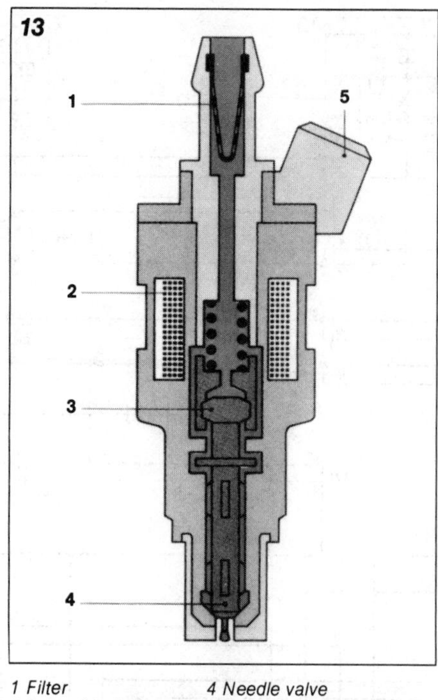

1 Filter
2 Solenoid winding
3 Solenoid armature
4 Needle valve
5 Electrical connection

Figure 6–50. Typical D- and L-Jetronic fuel system injector. (Courtesy of Robert Bosch Corporation, Broadview, Illinois)

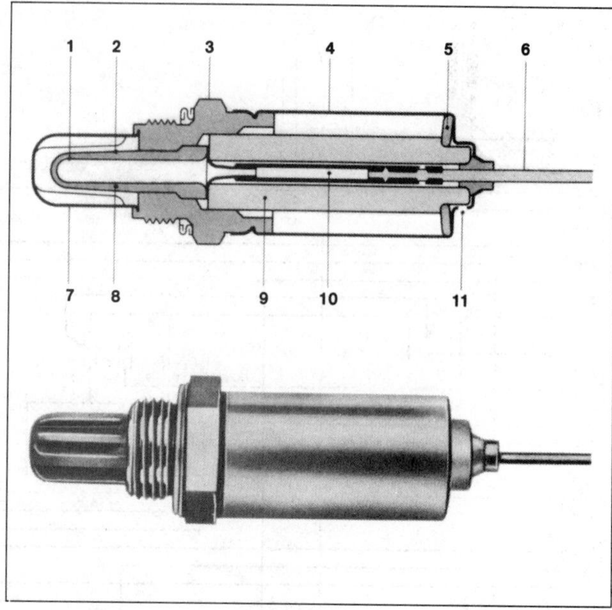

1 Electrode (+)
2 Electrode (−)
3 Housing (−)
4 Protective sleeve (air side)
5 Disc spring
6 Electrical connection
7 Protective tube (exhaust side)
8 Sensor ceramic
9 Support ceramic
10 Contact part
11 Vent opening

Figure 6–51. Lambda sensor construction. (Courtesy of Robert Bosch Corporation, Broadview, Illinois)

pressurized fuel can spray into your eyes or land on a hot engine component creating a fire hazard. *Do not* smoke when working around gasoline fuel injection systems.

To Relieve Fuel System Pressure

1. Disconnect the vacuum hose from the fuel pressure regulator.
2. Connect a hand vacuum pump (Kent-Moore P/N J-23738 or equivalent) to the regulator and pump the handle until a reading of 20 inches (50.8 cm) is registered on the vacuum pump gauge, which will draw (vent) system pressure back to the fuel tank.

Alternate Method to Relieve Fuel System Pressure

This alternate method can be used on any Bosch gasoline fuel injection system (previously explained under the K operation).

1. Disconnect battery ground cable.
2. Unhook the wiring harness plug from the engine cold start valve.
3. Run one jumper wire from the negative and one from the positive battery terminals to each contact of the cold start valve for 5 to 8 seconds to lower the system fuel pressure.

Troubleshooting D- and L-Jetronic Fuel Injection Systems

Before undertaking any specific checks or tests on either of these Robert Bosch fuel injection systems, refer to the troubleshooting charts in this section (Figures 6–52 and 6–53), which will assist you in determining where the actual problem area might be. If you are already familiar with these fuel systems, you may quickly recognize a problem from your own past experiences. However, sometimes a symptom that you detect as being related to a given problem may not always be cured by a repair to this unit. In this case, it often helps to be able to review the troubleshooting chart in order to evaluate what other possible areas of the fuel system may be contributing to this complaint.

Trouble-shooting guide for D-Jetronic fuel injection

This guide is designed to be used in conjunction with the Robert Bosch Jetronic Service Manual.

Engine cranks but does not start	Engine starts but then dies	Rough or unstable idle	Idle speed incorrect	Erratic running	Engine misses when driving	Fuel consumption too high	No maximum power	CAUSE	REMEDY
●	●	●	●	●	●	●	●	Defect in ignition system	Check battery, distributor, plugs, coil and timing
●	●	●	●	●	●	●	●	Mechanical defect in engine	Check compression, valve adj. and oil pressure
●								Fuel pump not operating	Check pump fuse, pump relay and pump
●	●							Relay defective; wire to injector open	Test relay, check wiring harness
●	●	●					●	Blockage in fuel system	Check fuel tank, filter and lines for free flow
	●	●	●	●	●		●	Leaks in air intake system	Check all hoses and connections; eliminate leaks
		●		●	●	●	●	Fuel system pressure incorrect	Test and adjust at pressure regulator
	●				●			Trigger contacts in distributor defective	Replace trigger contacts
●		●				●		Cold start valve defect	Check for spray or leakage
●								Thermo-time switch defective	Test thermo-time switch for correct function
	●		●					Auxiliary air valve not operating correctly	Must be open with cold engine; closed with warm
●	●					●		Temperature sensor II defective	Test for 2-3 k Ω at 68° F.
●	●	●		●		●	●	Pressure sensor defective	Test with ohmmeter
		●	●				●	Throttle butterfly does not completely close or open	Readjust throttle stops
		●	●	●			●	Throttle valve switch incorrectly adjusted or defective	Adjust as necessary or replace
	●	●	●					Idle speed incorrectly adjusted	Adjust idle speed with bypass screw
		●	●	●	●		●	Defective injection valve	Check valves individually for spray
●	●	●		●	●			Loose connection in wiring harness or system ground	Check and clean all connections
●	●			●	●		●	Control unit defective	Use known good unit to confirm defect

(SYMPTOM columns at left.)

Figure 6-52. D-Jetronic troubleshooting chart. (Courtesy of Robert Bosch Corporation, Broadview, Illinois)

Trouble-shooting guide for L-Jetronic fuel injection

This guide is designed to be used in conjunction with the Robert Bosch Jetronic Service Manual.

Engine cranks but does not start	Engine starts but then dies	Rough or unstable idle	Idle speed incorrect	CO value incorrect	Erratic running	Engine misses when driving	Fuel consumption too high	No maximum power	CAUSE	REMEDY
●	●	●	●		●	●	●	●	Defect in ignition system	Check battery, distributor, plugs, coil and timing
●	●	●	●		●	●	●	●	Mechanical defect in engine	Check compression, valve adj. and oil pressure
●	●	●	●	●	●	●		●	Leaks in air intake system (false air)	Check all hoses and connections; eliminate leaks
●	●	●						●	Blockage in fuel system	Check fuel tank, filter and lines for free flow
●									Relay defective; wire to injector open	Test relay; check wiring harness
●									Fuel pump not operating	Check pump fuse, pump relay and pump
●	●	●		●	●		●	●	Fuel system pressure incorrect	Check pressure regulator
●									Cold start valve not operating	Test for spray, check wiring and thermo-time switch
●	●	●	●	●	●		●		Cold start valve leaking	Check valve for leakage
●									Thermo-time switch defective	Test for resistance readings vs. temperature
●	●	●	●						Auxiliary air valve not operating correctly	Must be open with cold engine; closed with warm
●	●	●		●			●		Temperature sensor defective	Test for 2-3 kΩ at 68° F.
●	●	●	●	●	●		●	●	Air flow meter defective	Check pump contacts; test flap for free movement
		●	●					●	Throttle butterfly does not completely close or open	Readjust throttle stops
		●						●	Throttle valve switch defective	Check with ohmmeter and adjust
	●	●	●	●					Idle speed incorrectly adjusted	Adjust idle speed with bypass screw
		●		●	●	●		●	Defective injection valve	Check valves individually for spray
	●	●	●	●	●		●		CO concentration incorrectly set	Readjust CO with screw on air flow meter
●	●	●			●	●			Loose connection in wiring harness or system ground	Check and clean all connections
●	●	●				●		●	Control unit defective	Use known good unit to confirm defect

(SYMPTOM columns at left.)

Figure 6-53. L-Jetronic troubleshooting chart. (Courtesy of Robert Bosch Corporation, Broadview, Illinois)

Vacuum Leaks Check

Any air leaks on the suction side of the air system will adversely affect the operation of the fuel control system similar to that discussed for the K-Jetronic system earlier (Figure 6–14). Since these systems rely on air flow, any leaks will cause air/fuel ratio problems.

Procedure To Check for Vacuum Leaks

1. Disconnect one hose from the auxiliary air valve, then blow compressed air through the hose while holding the throttle valve wide open.
2. Spray soapy water around all joints (spray bottle) and look for leaks, or spray non-flammable solvent around potential leakage areas. If the idle speed smooths out, the solvent has found a leak.

Don't discount possible air leaks at places such as the engine oil dipstick tube and oil filler cap because these areas can also be the source of air in the engine—which will cause a poor running engine.

Fuel Pump Power Check

The fuel pump is usually located under the vehicle and in front of or very close to the fuel tank because access to the electric pump is required in order to check the voltage supply to it when it is suspected that the pump may be at fault.

If you suspect that the fuel pump is not receiving power, you should first of all remove the wiring harness plug from the pump and, with the ignition on, check the system voltage supply, which should be a minimum of 12 volts.

Also check the fuel pump fuse or fusible link located close to the battery. If the fusible link is burned, it will feel spongy when you squeeze the wire insulation. If the fusible link is burned out, check the electrical wiring back to the pump. A faulty pump (one that sticks) that may have been jammed by freezing water could cause this situation. Ensure that you find the reason for the burned out fusible link before accepting it as just one of those things.

Fuel Pressure Check

An accurate check of the fuel system pressure requires the use of a gauge (such as that shown in Figure 6–64). This gauge is available from Robert Bosch under P/N KDEP-1034 (earlier models) and KDJE-P100 (later models).

If this gauge is not available, select and use a pressure gauge that is graduated in increments of 0.5 psi (3.44 kPa) that will read to at least 50 psi (3.44 bar or 345 kPa).

Procedure

1. Disconnect the hose from the start valve at the intake manifold.
2. Connect up pressure gauge KDEP-1034/KDJE-P100 or equivalent in series with the disconnected fuel line by using a "Y" connection and extra piece of rubber hose.
3. Remove the fuel pressure regulator vacuum hose from the intake manifold.
4. Start and idle the engine.
5. Note and compare the fuel pressure gauge reading with the vehicle manufacturer's specifications.

   NOTE: Since the L-Jetronic system employs intermittent fuel injection, this will be reflected by a gauge needle that fluctuates when the engine is running; therefore, take the pressure as that between the lowest and highest needle positions. Pressures will vary but usually will run between 2.3 and 3.2 bar (33.3 to 46.4 psi).
6. Reconnect the fuel pressure regulator vacuum hose to the intake manifold.
7. The fuel pressure should now drop to about 2 bar (29 psi) at idle.
8. Accelerate the engine slightly and notice if the pressure rises; if it does not, recheck the vacuum hose for leaks. If fuel pressure is still low, check the fuel filter for possible plugging or kinked or collapsed fuel lines. If these are OK, then check for pump problems.
9. A high fuel pressure reading is generally indicative of a faulty fuel pressure regulator.
10. Stop the engine and carefully watch the pressure gauge; fuel pressure should drop and hold to about 1.17 to 1.4 bar (17 to 20 psi).

If the fuel system pressure drops lower than this or to zero, it is a positive indication of a fuel pressure leak somewhere in the system. If no external leaks are visible, check the following components for internal leakage:

1. Electric fuel pump (outlet check valve—one way valve)
2. Fuel pressure regulator
3. Cold start valve
4. Fuel injectors

Troubleshooting Sequence (Gauge Still Connected)

1. Take a pair of vise-grips or the equivalent and squeeze closed the fuel pressure regulator line to the fuel tank.
2. Turn the ignition key on/off several times to cause the electric pump to operate.
3. Check the pressure gauge to see if system pressure is now maintained; if so the regulator is at fault.
4. If fuel pressure still doesn't hold, have someone perform Step 2 above while you place another pair of vise-grips over the hose from the fuel pump to the fuel rail.
5. The fuel pump one-way check valve is faulty if the pressure under this condition holds steady (no leak back because of vise-grips on fuel rail line).

NOTE: If the fuel pump check valve is at fault, this will also be confirmed by the fact that the engine will be hard to start when warm because the start valve doesn't operate under this condition and the fuel pressure will be low due to the leaking pump one-way check valve. Vapor lock can occur because of low pressure (leak back of fuel to the tank).

Fuel Pump Delivery Test

In order to clarify that a problem may exist in the electric fuel pump, you should take what is known as a "volume test" of the pump. This test simply confirms whether or not the electric fuel pump is in fact delivering the correct quantity of fuel under pressure to the system.

To check the pump's delivery, simply disconnect the fuel pump delivery line at a convenient point in the system. Energize the electric pump and measure the amount of fuel spilled into a container in a 30-second time period. There should be at least 1 liter of fuel delivered—otherwise the pump is faulty, unless of course there is a plugged fuel filter or restriction in the fuel line to or from the pump.

Testing Auxiliary Air Valve (AAV)

To test the operation of the auxiliary air valve, simply remove both hoses from the valve body. If you cannot see directly into the valve, use a mirror and light to reflect the position of the internal valve.

If the auxiliary air valve is operating correctly, it should be partially open when the engine is cold and fully closed when the engine is warm (at operating temperature).

Specific Check for Auxiliary Air Valve (AAV)

1. Tests should be conducted with the AAV assembly at a temperature of approximately 68°F (20°C).
2. Disconnect the AAV hoses and wiring.
3. Place an ohmmeter across the AAV terminals and note the reading; 29–49 ohms is considered average.
4. Now place a voltmeter across the AAV connector terminals and crank the engine over on the starter; a voltage reading should be apparent.
5. No voltage reading while cranking the engine over indicates an obvious electrical fault somewhere in the system.
6. Reconnect the AAV hoses and electrical plug.
7. Start and run the engine at an idle speed.
8. To check the operation of AAV, simply squeeze shut the AAV supply hose; if the engine coolant temperature is lower than 140°F (60°C), there should be a drop in engine rpm of approximately 100 to 125 rpm; once the engine reaches operating temperature, the rpm drop under the same test should not exceed 50 rpm.

Testing Thermo-Time Switch (TTS)

The thermo-time switch (TTS) performs the same basic function in all Robert Bosch fuel injection systems. This check was discussed under the K-Jetronic system earlier.

A detailed description of the thermo-time switch operation was discussed earlier; however, its main function is to control or energize the cold start valve operation for ease of start-up at lower engine coolant temperatures.

Since the wiring harness is connected in series between the thermo-time switch and the cold start valve, when the engine is started, electricity flows to both units. Within the TTS is a bimetallic switch and heating coil to control how long the cold start valve will be energized.

On the L-Jetronic system, test the thermo-time switch in the vehicle for correct resistance values at specified engine temperatures as per the manufacturer's test procedures.

Typical Test Procedure

1. Disconnect the electric plug from the thermo-time switch.
2. Refer to Figure 6–54, which illustrates the condition of the thermo-time switch terminals when

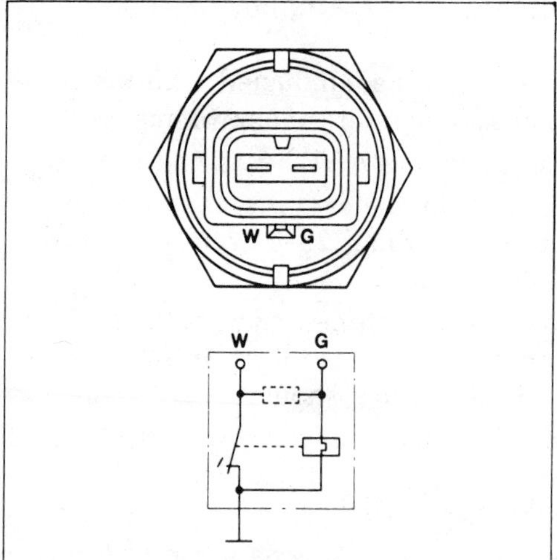

Figure 6–54. Thermo-time switch terminal identification. [Courtesy of Robert Bosch Corporation, Broadview, Illinois]

the engine coolant temperature is lower than 50°F (10°C).

3. Using an ohmmeter, measure the resistance between terminal G on the thermo-time switch and ground.
4. The resistance value should be between 40 and 70 ohms at any engine temperature.
5. A reading higher or lower than the above means that the thermo-time switch is defective.
6. Measure the resistance between terminal W and ground, which should be ZERO ohms continuity if the engine coolant temperature is below 50°F (10°C), or infinite resistance if the engine temperature is above 68°F (20°C).
7. If the resistance values in Step 6 are outside the limits, then the thermo-time switch is defective.
8. Measure the resistance between terminals G and W.
9. If the engine temperature is above 68°F (10°C), an infinite resistance should be read; if the engine coolant is lower than 50°F (10°C), then a resistance of between 40 to 70 ohms should be obtained; the switch is defective if readings other than those stated are obtained.

Optional Method (Thermo-Time Switch) TTS Check. Each TTS has stamped on its hex nut flats a given time period and temperature for which it should operate. This time in seconds at a given temperature should be your guide when testing the TTS.

1. With the engine cold, remove the TTS from the coolant passage and temporarily install a blanking plug in its place.
2. Submerge the TTS probe in cold water and connect the necessary test wiring to the TTS connector points.
3. Connect up one lead of a 12-volt test lamp to the hot side (+) of the battery while the other lead should be connected to the black wire of the TTS.
4. Run a test lead from the negative side of the battery to the body of the TTS, which should cause the test lamp to glow (light) if in fact the TTS is operating properly.
5. Repeat the same procedure described in Steps 2 through 4, only this time the water in the container should be heated and agitated to ensure even heat rise. Do not allow the TTS to sit on the bottom of the container because a false reading will occur and the TTS may be damaged (similar to testing a thermostat).
6. As the water temperature increases, the test lamp should go out somewhere between 88° and 102°F (31° and 39°C); if the test lamp fails to go out at these temperatures, then it must be replaced because excess fuel will be delivered to the engine when the engine is warm.

Air-Flow Meter Check

A check of the air-flow meter is required to ensure that it is functioning properly. Remember that the function of the air-flow meter (air valve) sensor is to generate a voltage signal proportional to the amount of air drawn into the engine. To check the air-flow meter requires the use of an ohmmeter to measure a resistance. Figure 6–55 illustrates a typical air-flow meter assembly with its electrical contacts shown on the right-hand side near the top.

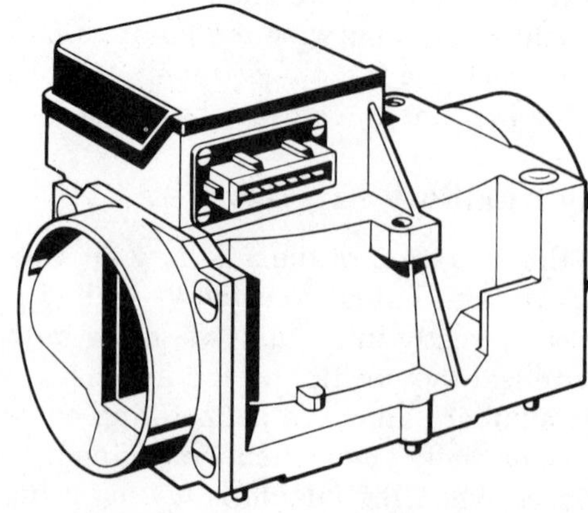

Figure 6–55. Air-flow meter and electrical contacts. [Courtesy of Robert Bosch Corporation, Broadview, Illinois]

Also contained within the air-flow meter housing is an air-flow temperature sensor (see Figures 6–56 and 6–57) that sends an electrical signal to the ECU. Failure of this air temperature sensor within the air-flow meter will result in having to replace the complete air-flow meter unit. If this sensor or the water temperature (cylinder head sensor in later models) sensor fails, you will never be able to tune the engine properly.

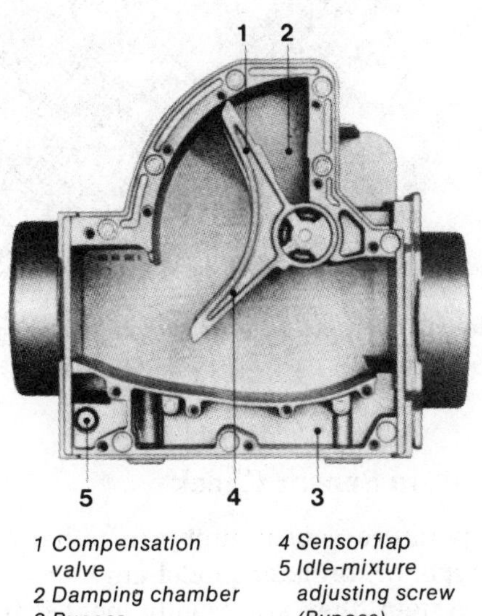

Figure 6–56. Air-flow sensor (air side). (Courtesy of Robert Bosch Corporation, Broadview, Illinois)

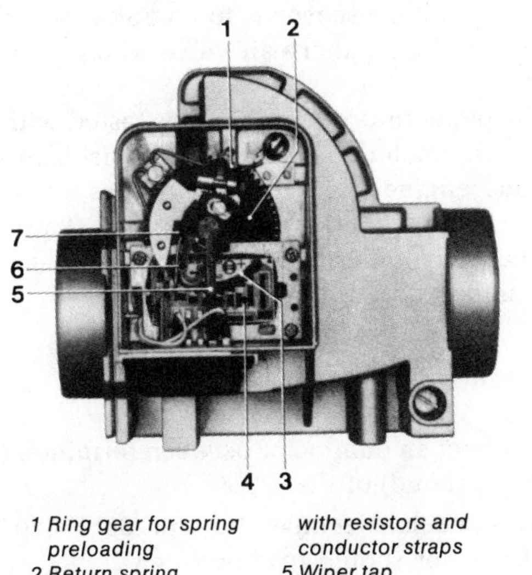

Figure 6–57. Air-flow sensor (connection side). (Courtesy of Robert Bosch Corporation, Broadview, Illinois)

Procedure Check

1. Remove the upper section of the air filter.
2. Loosen off the hose clamp and pull off the large hose from the air filter to the inlet side of the air-flow meter.
3. Connect up an ohmmeter between terminals 36 and 39 on the air-flow meter, which should read infinity ohms with the air flap or inlet valve closed.
4. Put your hand inside the air-flow meter inlet and gently deflect the air valve slightly; the ohmmeter should read zero ohms. If there is no variation from an infinity to a zero ohms reading, replace the air-flow meter assembly.
5. Connect up a CO meter to the exhaust pipe.
6. Using the bypass (mixture adjustment) air screw located on the side of the air-flow meter housing, adjust it to obtain the correct idle mixture setting to the CO stated on the vehicle EPA or ECE sticker while the engine is running at its normal operating temperature. Turning the idle mixture screw clockwise will increase CO; turning it counterclockwise will reduce CO.
7. Now adjust the engine idle speed to that stated on the vehicle EPA or ECE sticker, which is usually located under the hood.

Throttle Valve Switch Adjustment Check

The throttle valve switch illustrated in Figure 6-45 shows the idle and full-load electrical contacts. When the operator depresses the throttle pedal, linkage will cause the throttle valve switch to come into play at idle and maximum speeds. This causes an electrical output signal to be processed by the ECU when determining the duration of injection.

The throttle valve switch is located on the side of the engine throttle body or venturi unit. To test the throttle valve switch, an ohmmeter is required to obtain and compare resistance readings.

Procedural Check

1. Place the accelerator in the idle position.
2. Connect the ohmmeter across terminals 2 and 18 of the ECU plug.
3. The ohmmeter should register zero.
4. If the ohmmeter reads infinite resistance, then the throttle valve switch requires adjustment. Adjustment of the throttle valve switch may vary slightly between makes of vehicles; however,

1 – Fastening screws holding the throttle valve switch
2 – Hose leading to auxiliary-air valve
3 – Central ground terminal in L-Jetronic
4 – Air-flow meter
5 – Common intake manifold

6 – Throttle stop screw (idle speed)
7 – Stop bracket
8 – Full-load stop
9 – Pressure regulator connection on intake manifold
10 – Screw for adjustment of idle speed

Figure 6-58. Throttle valve switch adjusting screws. [Courtesy of Robert Bosch Corporation, Broadview, Illinois]

Figure 6-59. Throttle valve switch adjusting screws. [Courtesy of Robert Bosch Corporation, Broadview, Illinois]

Figures 6–58 and 6–59 illustrate a typical example (BMW) of how this should be done.

5. Loosen screws (1) in Figure 6–58 and connect the ohmmeter across terminals (2) and (18) of the switch.
6. Manually open the throttle valve until the stop bracket (7) in Figure 6–59 is about 3 mm (0.060" or 1/16") off the stop screw (6) in Figure 6–59.
7. Carefully adjust the position of the throttle valve switch until the ohmmeter needle just swings from an infinite reading to a zero reading.
8. Now place the throttle in the wide open position.
9. Connect an ohmmeter across terminals (3) and (18) of the ECU plug and note the reading; it should be zero ohms.
10. If infinite resistance is noted, then the throttle valve switch is incorrectly adjusted, or it is possible that a short circuit exists in the wiring harness cables.
11. Pull the harness plug off the throttle valve switch.
12. If the ohmmeter still registers infinite ohms, then the cable is defective.
13. If the ohmmeter doesn't register infinite ohms, adjust or replace the throttle valve switch. The resistance reading obtained across the throttle valve switch terminals should be infinite ohms at any position between idle and wide open throttle.

Temperature Sensor Check

The temperature sensor unit can either be of the coolant type, or, on later model engines with the L-Jetronic system, they are usually located in the cylinder head. The function of this sensor is to send a signal to the ECU, which in turn will control how long the injector remains open in relation to engine operating temperature. Another function of the coolant temperature sensor is to complete the electrical circuit to the auxiliary air valve when the engine is cold.

The temperature sensor unit is tested with the use of an ohmmeter to establish its resistance value at various engine coolant temperatures since a thermistor is used within the sensing unit. The electrical resistance of a thermistor is such that it decreases as the temperature of the unit increases.

Procedure

1. Connect an ohmmeter between terminals (13) and (17) (ground) of the ECU.
2. At a coolant temperature of 68°F (20°C), the ohmmeter should read between 2–3K.
3. At a coolant temperature of 14°F, the ohmmeter should read 7–12K.
4. At a coolant temperature of 176°F (80°C), the ohmmeter should read 250–400 ohms.

5. If the meter reads infinite ohms, check the temperature sensor at its terminals with the ohmmeter; if the resistance readings are satisfactory here, replace the harness cable to the sensor.
6. If the ohmmeter reads zero ohms when placed across the sensor terminals, replace the sensor unit.

NOTE: If the ohmmeter reads zero ohms, pull the plug from the sensor; if the meter when placed across the ECU plug then reads infinite ohms, replace the sensor. If not, then replace the cables.

Cold Start Valve Check

The cold start valve can be checked in two ways as follows.

Method 1 Procedure

1. Run the engine at an idle speed with the electrical plug disconnected from the start valve.
2. Measure the voltage supply at the cold start valve plug across terminals (51) and (52); between 11 and 12.5 volts should be obtained with the engine temperature below 50°F (10°C).
3. With the engine coolant temperature above 68°F (20°C), there should be no voltage reading to the cold start valve plug terminals because it is controlled from the thermo-time switch.
4. Any readings other than the above would require a check of the thermo-time switch (described above).
5. Also check cables (51) and (52) for continuity with an ohmmeter.
6. Connect the ohmmeter between terminals (51) and (52) of the cold start valve.
7. A reading of approximately 4 ohms should be observed.
8. If a reading of infinite ohms is obtained, then the cold start valve windings are open.
9. If a reading of zero ohms is obtained, then the cold start valve windings are shorted.
10. If either condition in Step 8 or 9 is observed, replace the cold start valve.

Method 2 Procedure

1. Remove the retaining screws from the cold start valve (CSV) at the intake manifold, but do not disconnect either the fuel line or electrical plug from the CSV.
2. Obtain either a glass or plastic jar that you can place the CSV into so that the fuel flow from the CSV can be observed.
3. Generally, when the engine coolant temperature is lower than 35°C (95°F), the CSV should spray fuel for between one to twelve seconds when the starter motor is activated.
4. If the CSV continues to spray fuel after this time period or raw fuel drips from its tip, replace it.
5. If the CSV does not operate at temperatures below 35°C (95°F), replace it.

L-Jetronic Injector Tests

The individual cylinder injectors are opened and closed by an electrical control signal from the ECU (electronic control unit). As you know from the earlier description of the L system, the injectors are pulsed on and off by these signals in groups of two (parallel) so that they do not continuously inject fuel as is the case with the Robert Bosch K-Jetronic (CIS) continuous injection system. The time that the injectors remain open (deliver fuel) is controlled by the duration of the pulse, which is determined or computed within the ECU.

In the earlier description of the L injectors, it was explained that they operate on the electric solenoid valve principle as shown in Figure 6–50. When the injector solenoid is energized, the plunger is pulled into the solenoid and fuel is injected into the intake manifold towards the intake valve. The L injector needle valve is lifted approximately 0.004″ (0.1 mm) and remains open between 1 and 1.5 ms (milliseconds), which is time calculated in thousandths of a second. Unlike the injectors used in the K-Jetronic system, which are mechanically operated and can be pop tested to check their release pressure and spray pattern, the L injectors cannot be placed on the pop tester because these injectors are pulsed on/off by electrical impulses from the ECU of the vehicle.

Problems with injectors cause the same symptoms in all fuel injection systems. These symptoms are generally as follows:

1. Poor start-up and idling conditions
2. Acceleration hesitation
3. Lack of power
4. Poor restart when the engine is warm

Poor injector spray formation can cause conditions 1, 2, 3, and 4, but condition 4 is usually caused by raw fuel dripping from injectors when the engine is stopped. An excess fuel mixture collects in the cylinder or manifold area making restart hard when the engine is hot.

L Injector Test Procedure

1. Check each injector electrical connection to ensure that it is clean and tight.
2. Dry around each injector so that any signs of leakage (especially at the plastic seam around the injector body when cold) can be detected since the leak will sometimes disappear when the engine warms up.
3. A quick method to pinpoint a faulty injector is to use a mechanic's stethoscope (a screwdriver can be used if no stethoscope is available) to listen to the individual injectors while the engine is running. Run the engine at an idle rpm and place the stethoscope against the injector body. You should hear a regular pulsing or clicking sound as the electrical solenoid within the injector is switched on/off by the signal from the ECU.
4. You can gently accelerate the engine, and the clicking sound at the injector should increase as it cycles on and off more frequently (which you can hear with the stethoscope).
5. If any injector does not have the same sound as the others, then it is not operating correctly. A poor electrical connection at the injector might be the cause.
6. Disconnect the battery ground cable and remove the electrical leads from each injector. Take an ohmmeter and measure the resistance between the terminals on each injector, which should be between 2 to 3 ohms; if the injector winding is open (infinite resistance) or shorted (zero ohms), replace the injector.
7. You should also check the resistance of the injector series resistors, which should be 5 to 7 ohms each; an example of the location of the relays, which are mounted together with the injector valve series resistors, is shown in Figure 6–60 where the relays are mounted on the driver's side firewall.
8. Turn the ignition key on, and using a voltmeter from terminal (43/1) and (43/2) of the injector series resistors to ground, check the voltage; there must be between 11 and 12.5 volts.
9. To check the condition of the relay unit, refer to Figure 6–61, which illustrates the numbered contacts of the relay wiring harness.
10. Using a voltmeter with the ignition turned on, place the voltmeter probe leads first across terminal 88z and ground and then across terminal 88b and ground. If full system voltage is not read on the voltmeter scale, then the relay is defective.

NOTE: Each injector uses a seal ring between its body/adapter and the intake manifold. If these seals are leaking, a vacuum leak condition problem similar to any other intake system leak will cause poor starting, rough idle, and poor performance. A quick check can be made of the injector seal ring by running the engine at an idle speed, then spraying the area of each seal ring with some carburetor and choke cleaner while using a vacuum gauge hooked up to the intake system. Carefully notice any change to the vacuum gauge reading.

1 – Series resistor group 1 (4 resistors)
2 – Series resistor group 2 (2 resistors)
3 – Plug connector for series resistor groups
4 – Double relay
5 – Post-start relay

Figure 6–60. Location of injector series resistors and relays. (Courtesy of Robert Bosch Corporation, Broadview, Illinois)

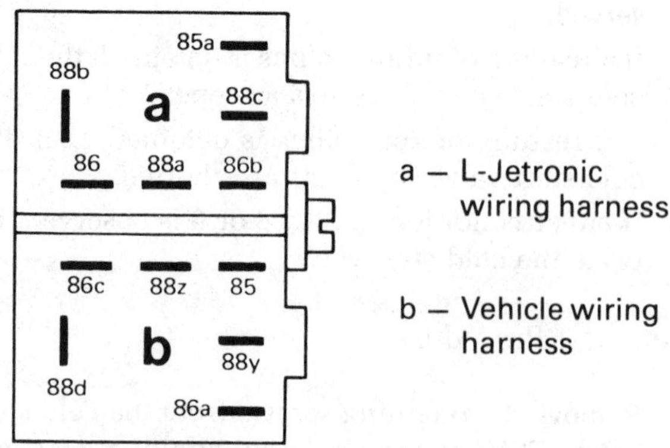

Figure 6–61. Relay wiring plug terminal numbers. (Courtesy of Robert Bosch Corporation, Broadview, Illinois)

Additional Injector Check

After having performed the above checks/tests, you can actually remove the complete fuel rail with the injectors attached. By switching the ignition on/off, you can establish if any of the injectors are dribbling or leaking raw fuel. Robert Bosch indicates that no more than two drops per minute should drip from each injector tip per minute, otherwise replace them. Any more than this will cause hard starting on a warm/hot engine.

Motronic Fuel Injection System

The Robert Bosch Motronic gasoline fuel injection system uses an L-Jetronic system coupled with an electronic ignition system, with both systems being controlled by a common digital microcomputer. The same sensors can be used for both the fuel injection and the ignition, thus more is achieved at a lower cost than with two separate systems. The operation of the sensors depends upon a reference voltage that is generated from the vehicle battery and fed through the ECU circuit at about 5 volts to each sensor.

The various sensors are constructed in such a way that they generate a voltage based upon either a temperature rise/fall or a pressure change. A pre-programmed system is contained within the microcomputer and is used as a base reference for engine operation. Voltage signals received at the microcomputer from the sensors are compared with the actual values stored in the computer. Any deviation from the base reference voltage or optimum operating conditions causes the microprocessor to send out corrected voltage signals to both the injection and ignition control circuits.

The Motronic system therefore offers all of the advantages of the L-type systems plus increased fuel savings by adapting the fuel quantity and ignition advance to all operating conditions by means of the Lambda sensor map and the ignition-advance map.

The ignition angle can be modified according to engine coolant and intake air temperature as well as throttle position and other parameters. Rather than using the conventional mechanically operated centrifugal and vacuum advance mechanisms within the distributor, Motronic has a spark-advance characteristic map stored in its control unit.

The microcomputer calculates the ignition angle between every two spark impulses from the information it receives about engine load and speed as well as temperature and throttle position; thus it is able to adjust quickly to every operating condition and give optimum performance, fuel consumption, and emissions control.

The intermittent, electronically controlled fuel injection is based upon the L system with the important difference being found in the signal processing—which is digital with the Motronic. The Motronic fuel system is shown in Figure 6–62 with its various sensors and engine controls, both for the ignition and the fuel injection systems.

Since the fuel injection portion of the Motronic system operates on the L-Jetronic principle, we will not repeat this information here (refer to the explanation earlier for the L operation). The important difference is that the single ECU (electronic control unit) with digital microcomputer, on the basis of the characteristic map stored in it, determines both the ignition and fuel settings once the vehicle ignition key is turned on and the engine is cranked. Figure 6–63 illustrates a block diagram of the Motronic system that shows the microcomputer system contained within the ECU. A 35-pin connector is used to connect the ECU to the battery, the sensors, and the controlling elements.

Motronic Operation

When the ignition is turned on and the engine is stationary, no fuel is delivered. A power transistor in the ECU controls an external pump relay shown as item 28 in Figure 6–62. When the ignition key is rotated to the start position, relay 50 shown in Figure 6–63 is connected to the battery positive, the electric fuel pump is energized, and the system is primed and ready to run at a fuel pressure of about 2.5 bar (36 psi). The ignition system is energized and it is ready to operate. During the primary current flow time, the coil is connected to ground through an output transistor in the ECU.

As the engine is cranked, fuel will be delivered via the cold start valve and from the individual injectors, and the engine fires and runs at an idle speed. For a detailed description, see the L system explanation. With the engine running, the ECU processes the sensor's input signals and calculates from them the fuel injection duration (quantity) and the optimum dwell and ignition angles. The microcomputer's output signals are too weak for the system's correcting elements and must therefore be amplified at their output stages before they are suitable for controlling an ignition coil or fuel injection valve.

The time for current to flow in the ignition coil is specified by the microcomputer as a function of the battery voltage and the engine speed. This stage also contains a control circuit for limiting the primary current of the ignition coil.

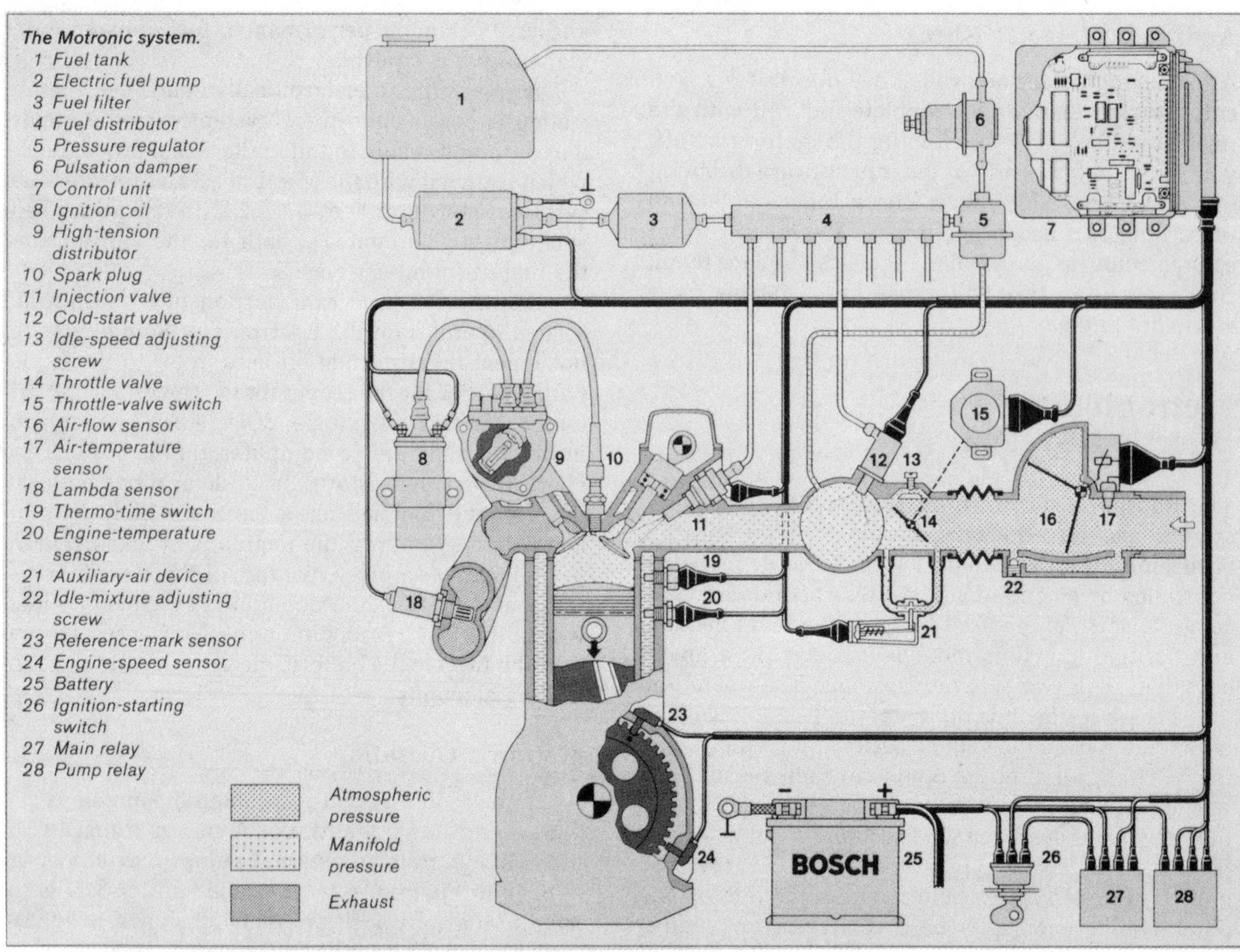

Figure 6-62. Operational diagram—Motronic fuel injection system. [Courtesy of Robert Bosch Corporation, Broadview, Illinois]

The engine speed is sensed by item 24 in Figure 6–62, which is an induction-type pulse generator (magnetic pickup) located over the rotating flywheel ring gear teeth. This unit provides pulses for the ECU. The engine crankshaft angle is determined by item 23 in Figure 6–62, which is an induction-type reference mark pulse generator on the flywheel ring gear. Pulses picked up from the rotating reference mark are sent to the ECU.

The air quantity flowing into the engine through the air-flow sensor is determined via the potentiometer mounted on the air flap as its angle of opening/closing is established by the throttle pedal. A voltage signal from here is also sent to the ECU.

The air-flow sensor flap measures the entire air quantity inducted by the engine, thereby serving, in addition to engine speed, as the main control quantity for determining the load signal and basic injection quantity. Current to the injection valves is regulated or pulsed on and off thereby energizing and de-energizing the solenoid contained within the fuel injector body. The longer the solenoid is energized, the greater the flow of fuel (quantity) into the cylinder.

The ECU controls this pulse width time at the injectors based on the input signals from the sensors. Maximum switch-on current to the individual injectors is approximately 7.5 A in a 6-cylinder engine and falls back at the end of injection to a low holding current of about 3 A. The injectors when energized only remain open from between 1 and 1.5 milliseconds.

Continuous monitoring of the engine sensors up to 400 times a second allows rapid correction to both ignition dwell and fuel injection pulse width.

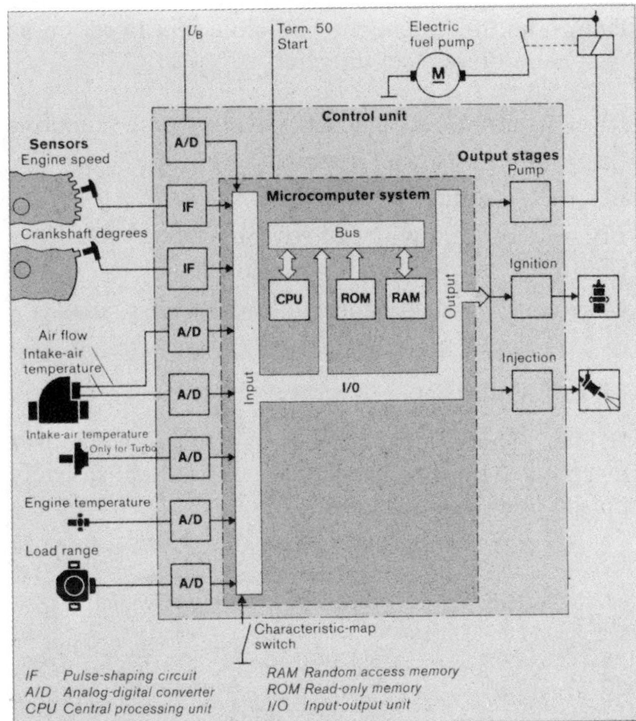

Figure 6–63. Motronic fuel system operational block diagram. (Courtesy of Robert Bosch Corporation, Broadview, Illinois)

Limiting Engine Speed

The maximum speed of the engine is controlled by suppressing the voltage signal to energize the injectors when the engine speed sensor relays a signal to the ECU. This is set to occur at 80 rpm ± the maximum speed that has been programmed into the ECU.

Stop/Start Operation

The Motronic system can be programmed to save fuel by stopping and starting the engine. If the vehicle is stationary at traffic lights or in heavy traffic jams, the ECU will interrupt the fuel injection signal if the ECU has determined that fuel will be saved, since additional fuel is required to restart the engine.

The engine will be shut off if the clutch pedal is depressed and the vehicle speed is less than 2 km/hr. (1 mph). To restart the engine, the driver must keep the clutch pedal depressed and simply steps on the accelerator within the first ⅓ of its travel.

Cylinder Cutout

Another feature of the Motronic system, which is still in the experimental stage, is the option of having several cylinders cut off during heavy city traffic so that the others will work more efficiently because of poor thermal efficiency at low loads and speed.

The Motronic cylinder cutout system allows only the working cylinders to be filled with air/fuel mixture and a minimum amount of throttling. Hot exhaust gases circulate through the inactive cylinders to maintain them at operating temperature.

The electronic circuitry in the ECU recognizes from the air-flow sensor's signal when cylinders can be switched.

Although this system is not yet in full production, Cadillac employed a similar system on their V8 engine in 1980–81 whereby the engine would run on either 4, 6, or 8 cylinders depending on engine /vehicle load demands.

Motronic Servicing

Service and maintenance of the Motronic fuel injection system is similar to that for the L-Jetronic system since the Motronic system employs all of the same components and design. Refer to the L section earlier for this information.

Mono-Jetronic System

The Mono-Jetronic system is a compact single-point injection system where fuel is metered at a centralized point within a throttle body from one solenoid operated injection valve. This Robert Bosch designed system is in wide use on a number of passenger cars worldwide.

In the chapters in this book dealing with GM, Ford, and Chrysler throttle body injection systems, a detailed description of operation can be found that is similar to that for the Mono-Jetronic system. Service and maintenance of these systems is also given in their respective chapters.

Robert Bosch Gasoline Fuel Injected Vehicles

The following information lists current vehicles that are equipped with Robert Bosch gasoline fuel injection systems. With experience in the manufacture of gasoline fuel injection systems dating from 1951, the Robert Bosch Corporation is undoubtedly the leader in this field.

There are now over 100 models of passenger cars using Robert Bosch gasoline fuel injection systems including General Motors, Ford, and Chrysler in North America who have adapted Bosch designs to suit their vehicles.

Many European vehicles employ either the suffix "e," "i," or "s" after a vehicle model. For example, BMW use all three of these suffixes after many of their vehicle models. The "e" is from the Greek word "eta," which translated basically means high efficiency. The letter "s" comes from the German word for gasoline which starts with an "s"; the "i" simply indicates that the vehicle is fuel injected.

Robert Bosch, through extensive study, has proved that a gasoline fuel injected vehicle can save up to 200 liters (53 US gallons or 44 Imperial gallons) of fuel per year over 20,000 km (12,428 miles); this equates out to approximately an 11–16% saving over its gasoline carbureted equivalent. As you can see from the extensive listing of passenger cars from "A" to "Z," Robert Bosch is truly a leader in the manufacture and technology of gasoline fuel injection as well as in diesel fuel injection.

KE = K-Jetronic with additional electronic functions
LE = L-Jetronic, 2nd generation, Europe
LU = L-Jetronic, 2nd generation, USA
LH = L-Jetronic with hot-wire air-mass meter
Motronic I = 1st generation Motronic
Motronic II = 2nd generation Motronic

List of Vehicles

NOTE: The K-Jetronic fuel injection system uses mechanical injectors; the others all use electronically pulsed open fuel injectors.

Vehicle	*Engine cc/cu.in.*	*CR*	*Power*	*Jetronic System*
Alfa Romeo				
Spider Veloce, 2.0L (USA)	1962/120	9.0:1	115/5500	L
Alfetta 2.0L	1962/120			Motronic II
GTV6–2.5L	2492/152	9.0:1	154/5500	L
AMC				
Alliance, Encore (USA) (Also equipped with Bendix TBI)	1397/85	8.8:1	64/4500	LU
Audi				
4000S	1780/109	9.0:1	88/5500	K, KE
4000S Quattro	2226/136	8.5:1	115/5500	KE
Coupe (5E) GT	2144/131	8.2:1	100/5100	K
Quattro Turbo	2144/131	7.0:1	160/5500	K
5000S	2144/131	8.2:1	100/5100	K
5000S Turbo	2144/131	8.0:1	140/5500	K
Bentley				
Camargue, Corniche, T 2	6750/412	8.0:1	260/4000	K
Bertone				
X1/9	1498/91	8.5:1	75/5500	L
BMW				
318i (4-cylinder)	1766/108	8.8:1	101/5800	K, LU
320i/323i				K, LE
325e (6-cylinder)	2693/164	9.0:1	121/4250	Motronic
518i				K
520i				K, LE
525i/528i				L, LE
528e (USA)	2693/164	9.0:1	121/4250	Motronic
533i	3210/196	8.8:1	181/6000	Motronic
M535i				L
628CSi				L, LE
633CSi	3210/196	8.8:1	181/6000	Motronic
635CSi				L, Motronic, 1, 11
725/728i				L, LE
732/732i				L, Motronic 1, 11
733i	3210/196	8.8:1	181/6000	Motronic
745i Turbo				L
3.0Si				D, L
3.3Li				L

(Continued)

Vehicle	Engine cc/cu.in.	CR	Power	Jetronic System
Buick				
Century 3.8T-Type	3791/231	8.0:1	125/4400	
(Bosch-GM MFI = Multi-Port Fuel Injection)				
Riviera 3.8 Turbo T-Type (Bosch-GM SFI = Sequential Port Fuel Injection)				
Citroen				
CX 2400 GTI				L
Chrysler				
Laser/Daytona 2.2/2.2 Turbo	2198/135	8.0:1	142/5600	
[Bosch-Chrysler Electronic TBI or MPFI (Multi-Port Fuel Injection)]				
DeLorean				
DMC 12				K
Ferrari				
308GTBi/GTSi Quattrovalvole	2926/179	8.8:1	230/6800	K
Mondial Quattrovalvole	"	"	"	K
BBi 512				K
400 i, GTi				K
Fiat				
Argenta 2000 ie, 1.e				L, LE
Brava/Spider 2000 (USA)				L
Strada Iniezione/X1/9 Iniezione (USA)				L
132-2000 Iniezione				L
Ford/Mercury				
Escort/Lynx 1.6				
EFT/Turbo	1599/98	9.5:1	86/5400	
(Bosch-Ford Electronic)				
Escort XR 3i/RS 1600 i				K
EXP 1.6, 1.6 EFI, 1.6 Turbo (Bosch-Ford Electronic Injection)				
Mustang SVO	2300/140	8.0:1	175/4500	
(Bosch-Ford Electronic Injection)				
Thunderbird 2.3L Turbo	2304/140	8.0:1	142/5000	L
Capri V6 2.8L/Granada 2.8i				K
Sierra XR 4i				K
Banksia (Australia)				LE
Isuzu				
Impulse	1949/119	9.2:1	90/5000	L
(Bosch-Isuzu L-Jetronic)				
Jaguar				
XJ6, Vanden Plas	4235/258	8.0:1	176/4750	L
3.6, XJ-SC 3.6	3590/219	9.0:1	200/5000	L
(Bosch-Lucas L-Jetronic)				
XJ 12 and early XJ-S				D
XJ-S	5343/326	11.5:1	262/5000	
(Bosch-Lucas L-Jetronic)				
Lancia				
Beta Trevi 2000 IE				L, LE
Gamma 2500 IE				L
Lamborghini				
Jalpa	3485/213	9.0:1	250/7000	
Countach LP500S	4754/290	9.2:1	350/6500	
(Both use KE-Jetronic)				
Mazda				
RX-7, S, GS, GSL, GSL-SE	1308/80	9.4:1	135/6000	L

(Continued)

Vehicle	Engine cc/cu.in.	CR	Power	Jetronic System
Mercedes-Benz				
190E 2.3	2299/140	8.0:1	113/5000	KE
230E, CE, 280E, CE, SE, SEL, SL, SLC				K
280E, CE, SE, SEL, SL, SLC				K
350 SE, SEC, SEL, SL, SLC				K
300SD, 380SE, 380SL	3839/234	8.3:1	155/4750	K
450SE, SEL, SL, SLC, 450SLC, 5.0L, 450 SEL, 6.9L				K
500SEL/SEC	4973/303	8.0:1	184/4500	K
Nissan/Datsun				
Stanza	1974/120	8.5:1	97/5200	L
200SX 2.0L; 1.8L Turbo*	*1809/111	8.0:1	120/5200	L
Maxima	2393/146	8.9:1	120/5200	L
280ZX Coupe/2+2	2753/168	8.3:1	145/5200	L
280ZX Turbo	2753/168	7.4:1	180/5600	L
300ZX	2960/181	9.0:1	160/5200	L
300ZX Turbo	2960/181	7.8:1	200/5200	L
Opel				
Ascona B 2.0E, Ascona C 1.8L, Ascona 400 Sport				L, LE, L
Commodore 2.5E, Kadett C GT/E - 1.9/2.0L				L
Manta B 1.9E				L
Manta B 2.0E, GT/E, Monza 2.5E, 3.0E				L, LE
Rekord 2.0E, Senator 2.5E, 3.0E				L, LE
Kadett, D 1.8L				LE
Peugeot				
504 V6, TI, 505 TI, STI, *505 GL 2.0L	*1971/120	8.4:1	97/5000	K
604 TI				K
Pininfarina				
Spider	1995/122	8.2:1	102/5500	L
Pontiac				
2000 Sunbird 1.8 ohc, 2.0 *1.8 Turbo	1836/112	8.0:1	150/5200	L
Porsche				
911 SC, 911 Turbo				K
911 Carrera (Motronic)	3164/193	9.5:1	200/5900	M
924, 924 Carrera GT, 924 Turbo				K
928, 928S	4644/285	9.3:1	234/5250	K, L
930 Turbo				K
944 (Motronic 11 or L)	2479/151	9.5:1	143/5500	L
Renault				
R5 Alpine Turbo				K
R17 (USA)				D
R18/Sportwagon 18i	2165/132	8.7:1	110/5500	L
Fuego Coupe 4	1647/100	8.6:1	82/5500	L
Fuego 2.2; *1.6 Turbo	*1565/95	8.0:1	107/5500	L
R30 TX				K
Rolls-Royce				
Corniche, Camargue	6750/412	8.0:1	260/4000	K
Silver Shadow II, Silver Wraith II				K
Silver Spirit, Silver Spur	6750/412	8.0:1	260/4000	K
Rover				
SD 1 (USA)				L
Saab				
99 EMS, GLE, Turbo				K
900 EMS, GLE, Turbo, Sedan				K
900, 900S	1985/121	9.3:1	110/5250	L
900 Turbo	1985/121	8.5:1	135/4800	L

(Continued)

Vehicle	Engine cc/cu.in.	CR	Power	Jetronic System
Subaru				
Turbo-Traction GL	1781/109	7.7:1	95/4800	L
Toyota				
Camry 2.0	1995/122	8.7:1	92/4200	L
Celica ST, GT, GT-S	2366/144	9.0:1	105/4800	L
Supra	2759/168	9.2:1	160/5600	L
Cressida	2759/168	8.8:1	143/5200	L
Van	1998/122	8.8:1	90/4400	L
TR 7 (USA)				L
TR 8 (USA)				L
TVR				
Tasmin	2792/170	9.2:1	145/5700	K
Vauxhall				
Royale 1, 3.0L				L
Cavalier 1.8 Sri				LE
Volkswagen				
Beetle				L
Rabbit 1.7L FI	1715/105	8.2:1	74/5000	K
Rabbit GTI (Golf)	1780/109	8.5:1	90/5500	K
Golf Cabrio GLI				K
Jetta 1.8 GLI	1780/109	8.5:1	90/5500	K
Passat GLI (Dasher)				K
Scirocco GTI, GLI	1780/109	8.5:1	90/5500	K
Rabbit Pickup				K
Quantum 2.1	2144/131	8.2:1	100/5100	K
Type 2, Type 25 Bus				L
Vanagon (Bosch Digital Injection)	1915/117	8.6:1	82/4800	
Volvo				
240 Series, DL, GL, GLE, Turbo				K
240 Series, 2.1L (USA)				K, LH
GTL	2127/130	7.5:1	162/5100	K
GL 2.3L	2316/141	9.5:1	111/5400	L
260 Series, GL, GLE, TE				K
360 GLT				LE
760 Series, GLE 2.8L V6	2849/174	8.8:1	134/5500	K
760 Series, GLE 2.3L (USA)				LH
760 Series, GLE 2.3L				Motronic II

Special Tools

There are a number of special tools that are required or are very useful to have when working on fuel injection systems. They are shown in Figures 6–64 through 6–75.

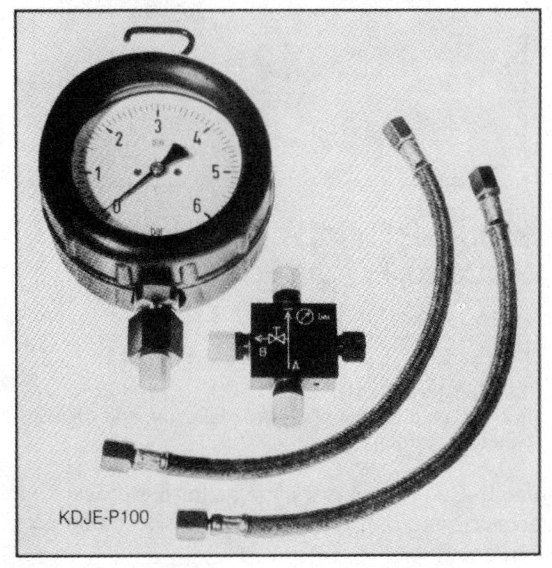

For all pressure measurements and leak tests on vehicles equipped with K, D and L-Jetronic systems; consists of control valve, 0-6 bar gauge, two hoses.

Figure 6–64. KDJE-P100 fuel pressure tester. (Courtesy of Robert Bosch Corporation, Broadview, Illinois)

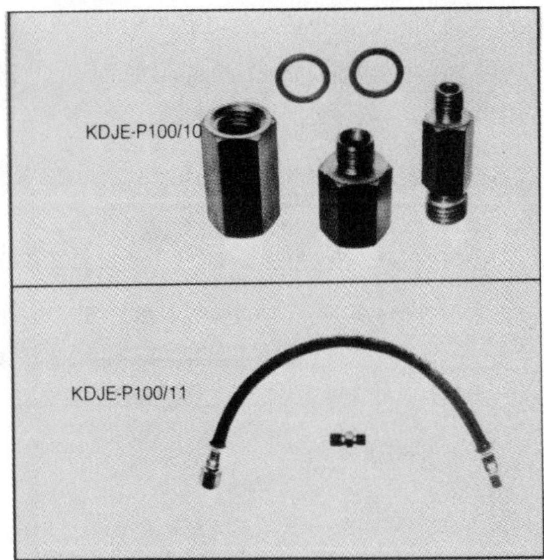

KDJE-P100/10
for Volvo, Porsche, Peugeot

KDJE-P100/11
for Mercedes-Benz

Figure 6-65. Accessory fittings—KDJE-P100/10 for Volvo, Porsche, Peugeot, and 11 for Mercedes-Benz. [Courtesy of Robert Bosch Corporation, Broadview, Illinois]

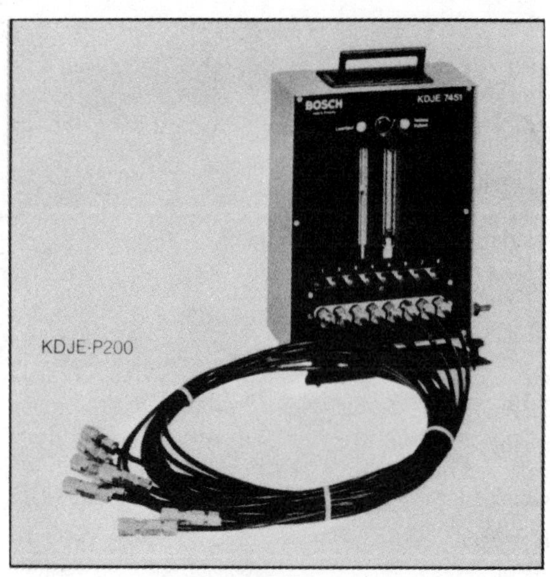

For testing of fuel distributor and the injection valves on the engine without removing the fuel distributor by taking comparison measurements on the individual injection nozzles

Figure 6-67. KDJE-P200 differential flow tester. [Courtesy of Robert Bosch Corporation, Broadview, Illinois]

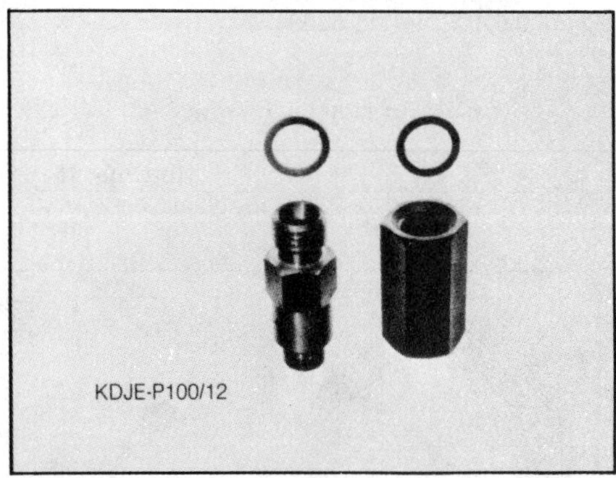

KDJE-P100/12
for Volvo, Ferrari, Porsche

KDJE-P100/13 (not shown)
Three Way Hose
For on-vehicle fuel pressure measurement of L-Jetronic systems

Figure 6-66. Accessory fittings for fuel pressure check, L-Jetronic system. [Courtesy of Robert Bosch Corporation, Broadview, Illinois]

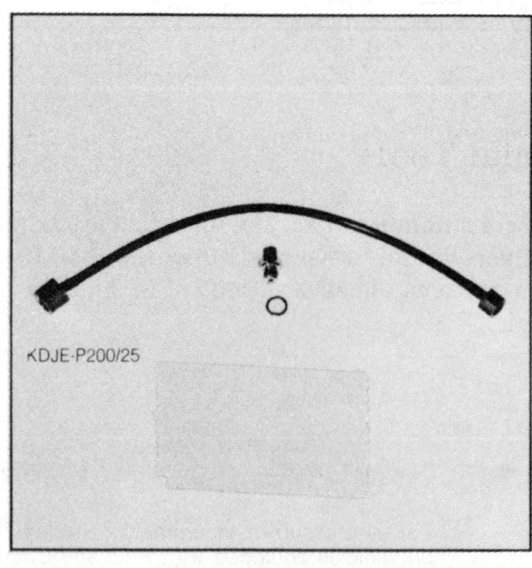

For use with KDJE-P200 when used on vehicles with steel fuel lines (one required per cylinder)

Figure 6-68. KDJE-P200/25 accessory adapter. [Courtesy of Robert Bosch Corporation, Broadview, Illinois]

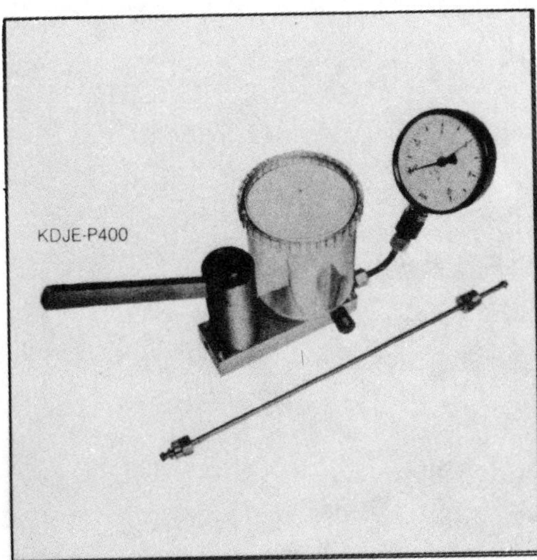

For checking opening pressure, tightness against leaks, spray and chatter characteristics of K-Jetronic injection valves when removed from engine. Use test oil VS14 492-CH or Shell mineral spirits 135

Figure 6-69. KDJE-P400 injection valve tester. [Courtesy of Robert Bosch Corporation, Broadview, Illinois]

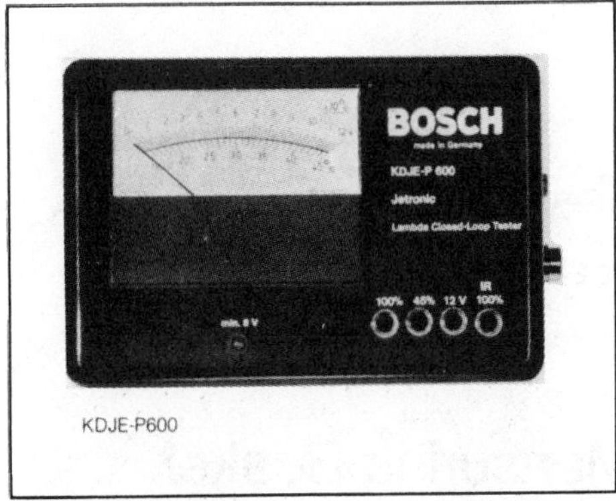

KDJE-P600
Lambda Tester

For use with all Lambda closed loop systems to check:

- Electrical function of the Bosch idle speed control on all Jetronic systems
- Output signals of control unit
- On-off ratio of frequency valve
- On-off ratio of idle speed regulator
- L-Jetronic integrator voltage

Accessory Cables
(Order separately)

- KDJE-P600/51 Universal application
- KDJE-P600/52 for Mercedes-Benz
- KDJE-P600/54 Extension for all cables
- KDJE-P600/55 for Porsche

Figure 6-70. KDJE-P600 Lambda tester; order accessory cables separately as shown. [Courtesy of Robert Bosch Corporation, Broadview, Illinois]

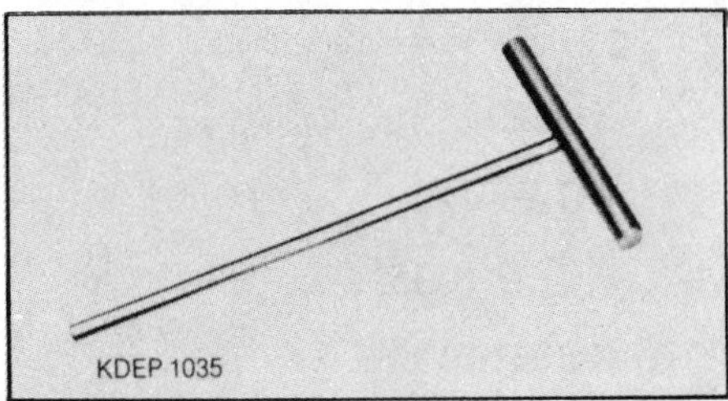

For adjusting idle mixture screw in the mixture control unit

Figure 6-71. KDEP 1035 idle mixture screw adjusting wrench. [Courtesy of Robert Bosch Corporation, Broadview, Illinois]

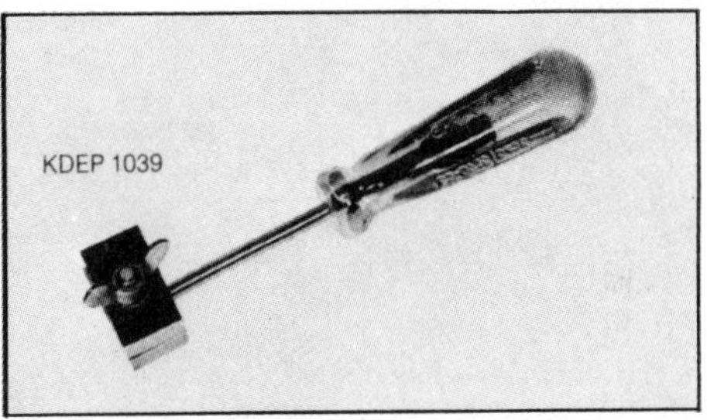

For pressing Polymide tubing on to fittings

Figure 6-72. KDEP 1039 polymide tubing assembly tool. [Courtesy of Robert Bosch Corporation, Broadview, Illinois]

For centering the air flow sensor plate in the air funnel of the air flow sensor. Four guide plates included with each guide ring.

KDEP 1040/10	**80mm dia.**
KDEP 1040/11	**60mm dia.**
KDEP 1040/12	**76mm dia.**
KDEP 1040/13	**85mm dia.**
KDEP 1040/14	**110mm dia.**
KDEP 1040/15	**105mm dia.**

Figure 6-73. Air-flow sensor plate centering guide rings—6 sizes as shown. [Courtesy of Robert Bosch Corporation, Broadview, Illinois]

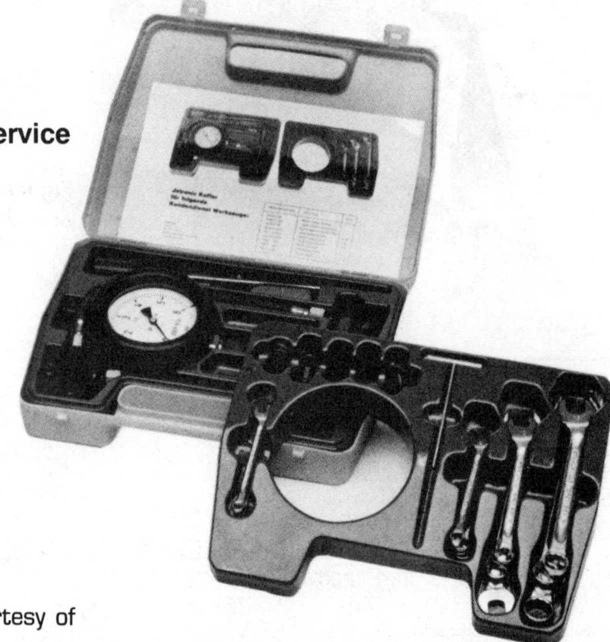

Plastic case containing the following:
KDJE - P100
KDJE - P100/10, /11, /12
KDEP 1035
KDEP 1039
KDJE 7461—15 common tools needed for K-Jetronic service

Figure 6-74. K-Jetronic service tool set. [Courtesy of Robert Bosch Corporation, Broadview, Illinois]

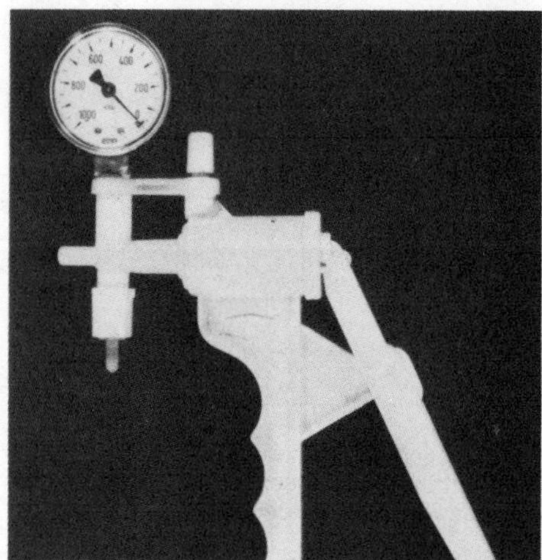

Figure 6-75. Hand-operated vacuum pump. [Courtesy of Robert Bosch Corporation, Broadview, Illinois]

Test Cable

KDEP 7450/70
For direct electrical connection to components for testing

Measuring Beaker

Ca. 1.5 liter capacity for testing fuel pump delivery

Tach-Dwell Meter

For measuring engine idle speed

CO Meter

For adjusting CO at idle

7

General Motors Throttle Body Injection Systems

The throttle body injection system was created as a means of meeting the EPA and ECE exhaust emissions regulations as well as improving the corporate average fuel economy (CAFE) and performance of the engine. CAFE is the average fuel economy in miles per gallon that a corporation's vehicles (across the model range) have to meet as per EPA specifications.

In North America, the throttle body injection (TBI) system is employed on vehicles produced by the four major manufacturers: General Motors Corporation, Ford Motor Company, Chrysler Corporation, and American Motors. In addition, certain vehicle models produced by the three major automobile manufacturers in North America also use "multiport" fuel injection systems whereby one injector is used for each engine cylinder.

American Motors, in concert with Renault in France, has produced the Alliance small car, which uses TBI rather than carburetion or one of the Robert Bosch fuel injection systems. The Renault/American Motors Alliance vehicle presently employs a TBI system that is produced by the Bendix Company; it operates on the same basic principle as other TBI systems.

Although carburetion is still offered in certain product lines, North American manufacturers now employ fuel injection systems for use on most of their vehicles; these systems are similar to the Robert Bosch K, L, and Motronic systems. They have until now chosen to use TBI on the majority of their standard production engines because TBI is cheaper to produce than multi-port fuel injection and supplies the results that they desire. However, they are now offering multi-point fuel injection systems on a greater number of vehicles, especially on the in-line four and V6 powerplants that are equipped with turbochargers. These fuel systems are basically a derivative of the Robert Bosch gasoline injection systems with modifications to suit the particular engine produced by that manufacturer.

In the late 1950s and early 1960s, General Motors did employ an injection system with individual injectors on some of their Corvettes where high performance and racing results were desired. However, this system did not find its way onto production vehicles in place of the then popular and widely used four-barrel carburetor because of price and serviceability.

Various aftermarket fuel injection systems such as Hilborn and others were installed onto high performance engines by local enthusiasts, but there were few mass-produced cars available at that time with fuel injection as a factory option.

General Motors Cadillac Motor Car Division did produce an electronic fuel injection (EFI) system for use on their 1977–79 Seville models as standard equipment and on 1977–79 full-size Cadillacs as optional equipment. This EFI port system employed individual fuel injectors at each cylinder, which sprayed fuel towards the intake valve similar to the existing Robert Bosch systems.

Cadillac also offered on its 1980 Seville, and subsequently on all its larger models, a digital fuel injection (DFI) system. This system employs a carburetor-type body but uses two electronically controlled injectors, which are mounted in the throat of the throttle body and which spray fuel into the intake manifold.

Ford Motor Company and Chrysler employ a similar system to that now being used by General Motors on their vehicles.

The type of TBI system being employed by each one of these manufacturers is described in detail in their respective chapters in this book. The following information describes both the TBI and MPFI (multi-port fuel injection) systems used on General Motors vehicles.

GMC Vehicle Listing—Fuel Injection

As mentioned in the introductory comments to this chapter, General Motors Corporation has been involved with gasoline fuel injection in one way or another for 25 years now. The 1984 line of General Motors vehicles is available with both TBI and port injection systems. These fuel systems are not available on all product lines; however, the following list indicates systems that are in use on this company's various vehicle models.

NOTE: Engines produced by one GMC Division can be installed in other General Motors cars/pickups manufactured by another Division of the Corporation. For this reason, the following list will show models of cars identified under another Division name. Although the car may be known by its model designation as belonging to one division, its engine has been manufactured by another division.

1. GMC "F" Body Camaro and Firebird
   a. 82–84 4-151 cu.in. (2.5L Pontiac Engine) TBI V.I.N. Code-2
   b. 82–84 V8-305 cu.in. (5L Chevrolet Engine) TBI V.I.N. Code-7, S
2. GMC "J" Body Cavalier, Cimarron, Firenza, Pontiac 2000, and 82–84 Buick Skyhawk
   a. 82–84 4-112 cu.in. (1837 cc) TBI, V.I.N. Code-0
   b. 83–84 4-122 cu.in. 2L TBI, V.I.N. Code-B
   c. 84 Pontiac 2000 SE equipped with modified Bosch L-Jetronic Injection and Turbocharged
3. BUICK/GMC
   a. 1984 3.8L (232 cu.in.) 90 degree V6 available in the Buick Century and Oldsmobile Cutlass Ciera with MFI (multi-port fuel injection) using Robert Bosch fuel injectors
   b. 1984 3.8L (232 cu.in.) V6 available with SFI (sequential port fuel injection) standard equipment on Buick Regal T Type Turbo, and Riviera
4. CADILLAC/GMC
   a. 77–79 V8 425 cu.in. (6970 cc/7L) EFI V.I.N. Code-T
   b. 77–79 V8 350 cu.in. (5.7L) EFI V.I.N. - R, B
   c. 1980 V8 368 cu.in. (6035 cc/6L) DFI used in the Seville and Eldorado models
   d. 81–82 V8 368 cu.in. (6035 cc/6L) DFI with Modulated Displacement Engine in the Seville and Eldorado models, V.I.N. - 9
   e. 81–84 V8 250 cu.in. (4.1L) DFI in the Seville/Eldorado VIN - 8
5. CORVETTE/GMC
   82–84 V8 350 cu.in. (5.7L) TBI V.I.N. - 8
   85 V8 350 cu. in. (5.7L) MPFI
6. PONTIAC/GMC
   a. A and X Body Front-Wheel Drive
   b. Celebrity, Century, Cutlass Ciera, 6000, Citation, Omega, Phoenix, and Skylark
   c. 82–84 4-151 cu.in. (2.5L/2476 cc) TBI V.I.N. Code - R
   d. 84 Fiero 4-151 cu.in. (2.5L/2476 cc) TBI

Specific vehicles or models not shown that are equipped with either TBI or port injection operate as described below.

General Motors TBI (Throttle Body Injection)

Current General Motors vehicles equipped with TBI—DFI (digital fuel injection) on Cadillac vehicles—are used in conjunction with the on-board vehicle computer commonly known as an ECM (electronic control module).

The ECM, which is located under the instrument panel, constantly receives voltage signals from a variety of engine/vehicle sensors and therefore is the control mechanism for the fuel injection system. These TBI units can employ either a single-bore-type throat such as all 1982 and later cars equipped with the Pontiac 4-cylinder 1.8L (112 cu.in.), 2.0L (122 cu.in.), and 2.5L (151 cu.in.) engines, or on Vee type engines, two injectors (one per throat).

General Motors 1.8L and 2.5L engines employ a Model 300 TBI system, and the 2.0L engines use a Model 500 TBI system. The 5.0L Chevrolet Camaro engine uses a Model 400 TBI unit with what is commonly referred to as cross-fire injection.

All TBI systems operate on the same principle and differ only in injector calibration and several other minor items. Therefore, the operational description for one TBI system can be considered common to all GMC engines.

Identification of a particular TBI unit can be made by the ID number that is stamped on the low mounting flange located on the TPS (throttle position switch) side of the throttle body shown in Figure 7–1.

The TBI injectors are located in the top of the TBI housing as illustrated in Figures 7–2 and 7–3, which show both a Model 300 and Model 500 TBI unit respectively. The injector mounted on the throttle body is electronically actuated and therefore meters the

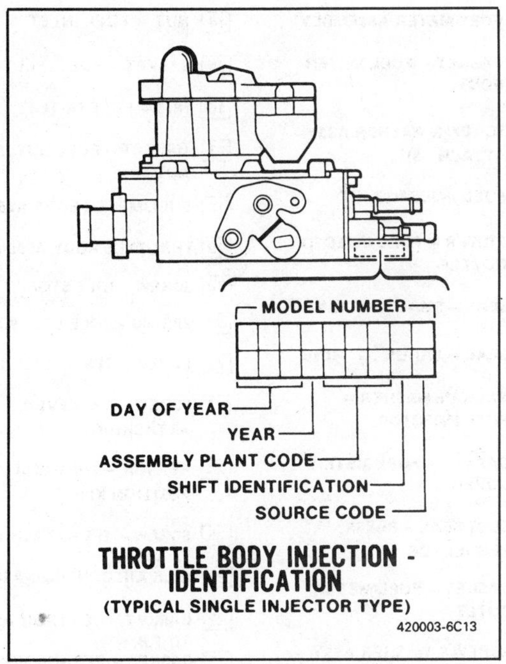

Figure 7–1. Throttle body injection (TBI) identification. (Courtesy of Pontiac Motor Division, GMC)

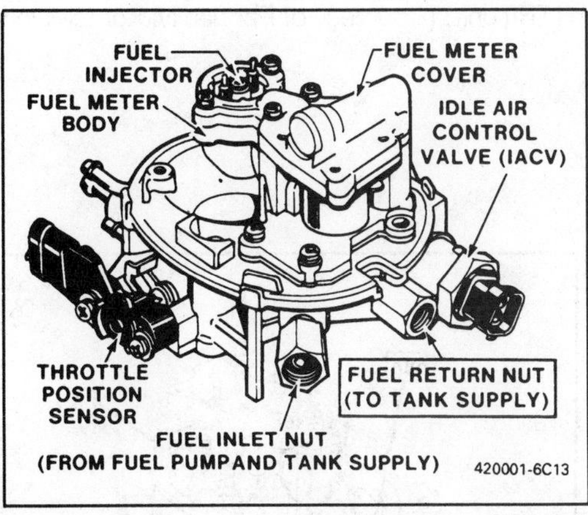

Figure 7–2. Model 300 throttle body injection (TBI) unit. (Courtesy of Pontiac Motor Division, GMC)

fuel into the throttle body above the throttle blades. An exploded view of a model 300 TBI assembly is shown in Figure 7–4, which is also typical of the components that would be found on both a model 400 and 500 TBI unit. No fuel can flow through the injector until the solenoid coil is electrically energized (similar to Robert Bosch L injectors).

When energized, the throttle body injector raises the plunger to allow the spring to push the check ball valve away from its seat; fuel will be injected into the intake manifold above the throttle blades. An example of a fuel injector can be seen in Figure 7–5 and vacuum port locations in Figure 7–6.

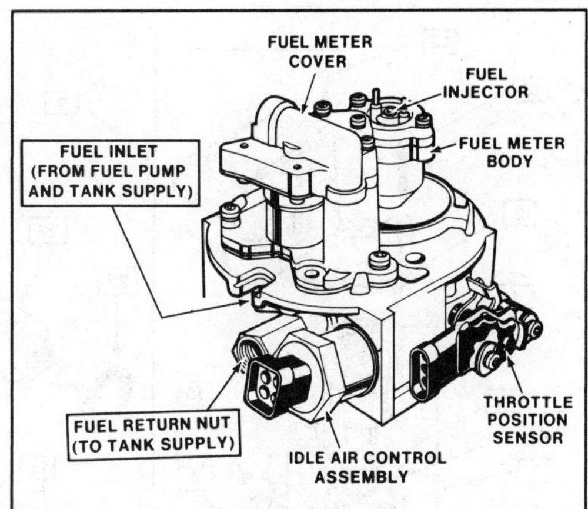

Figure 7–3. Model 500 throttle body injection (TBI) unit. (Courtesy of Pontiac Motor Division, GMC)

Closed-Loop versus Open-Loop Operation

All current fuel injection systems employ a Lambda (oxygen) sensor in the exhaust system to monitor the percentage of oxygen concentration passing through the system. A detailed explanation of the construction and arrangement of this Lambda sensor is given under Chapter 6, dealing with Robert Bosch fuel injection systems (both the K and L types).

The function of the Lambda (oxygen) sensor is to generate a signal related to the degree of oxygen ions flowing through the exhaust stream. This signal is relayed to the ECM, which then analyzes and sends out a suitable electrical signal to the throttle body injectors to pulse them open and closed. In this way, the air/fuel mixture is varied by just how long the injectors remain open. Through the Lambda sensor, the air/fuel mixture is maintained as close to ideal (stoichiometric) as possible under all operating conditions.

An open-loop condition exists when the oxygen sensor (Lambda) is not monitoring the exhaust condition. A closed-loop condition exists when the oxygen (Lambda) sensor is monitoring the exhaust gases and it sends a voltage signal to the ECM, which then monitors this information and sends out a corrected signal to the fuel system to control the air/fuel ratio.

On a normal operating engine, the dwell meter needle at both idle and part throttle will be between 10 and 50 degrees dwell and varying, which is termed the closed-loop cycle. In the open-loop cycle, the dwell reading will not vary with the engine cold and the oxygen sensor below 360°C (680°F) or at wide open throttle.

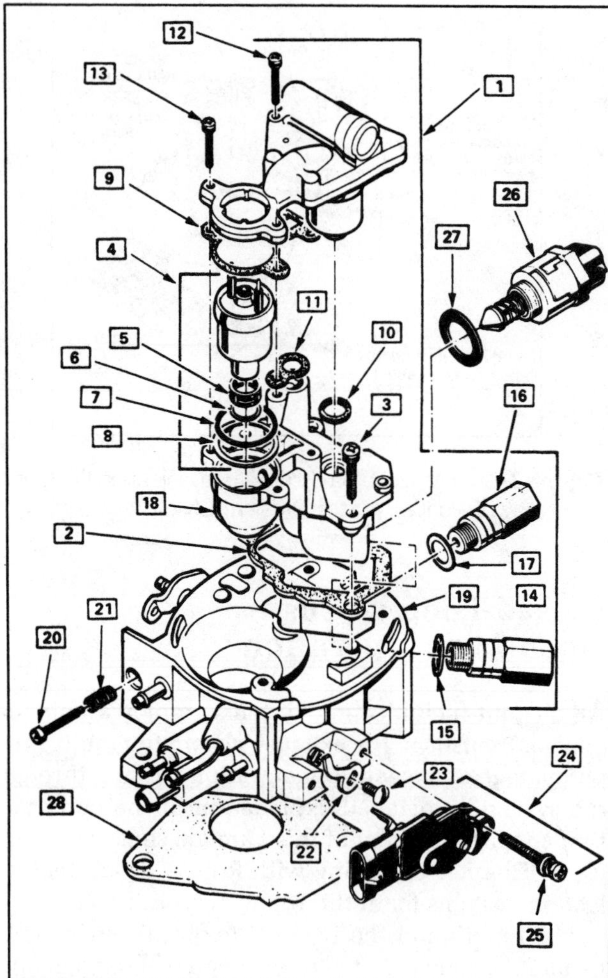

1 FUEL METER ASSEMBLY
2 GASKET – FUEL METER BODY
3 SCREW & WASHER ASSY– ATTACH. (3)
4 FUEL INJECTOR KIT
5 FILTER – FUEL INJECTOR NOZZLE
6 SEAL – SMALL "O" RING
7 SEAL – LARGE "O" RING
8 BACK-UP WASHER – FUEL INJECTOR
9 GASKET – FUEL METER COVER
10 DUST SEAL – PRESS. REGULATOR
11 GASKET – FUEL METER OUTLET
12 SCREW & WASHER ASSY – LONG (3)
13 SCREW & WASHER ASSY – SHORT (2)
14 NUT – FUEL INLET
15 GASKET – FUEL INLET NUT
16 NUT – FUEL OUTLET
17 GASKET – FUEL OUTLET NUT
18 FUEL METER BODY ASSEMBLY
19 THROTTLE BODY ASSEMBLY
20 SCREW – IDLE STOP
21 SPRING – IDLE STOP SCREW
22 LEVER – TPS
23 SCREW – TPS LEVER ATTACHING
24 SENSOR – THROTTLE POSITION KIT
25 SCREW – TPS ATTACHING (2)
26 IDLE AIR CONTROL ASSY-
27 GASKET – CONTROL ASSY- TO T.B.
28 GASKET – FLANGE MOUNTING

Figure 7-4. Exploded view model 300 throttle body injection [TBI] unit. [Courtesy of Pontiac Motor Division, GMC]

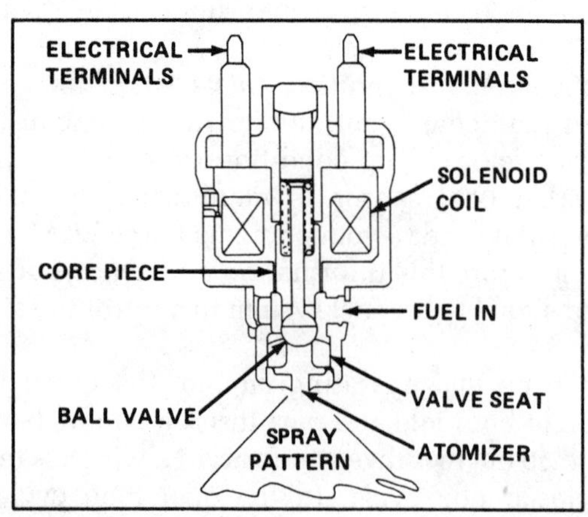

Figure 7-5. Fuel injector—digital fuel injection [DFI] system. [Courtesy of Cadillac Motor Car Division, GMC]

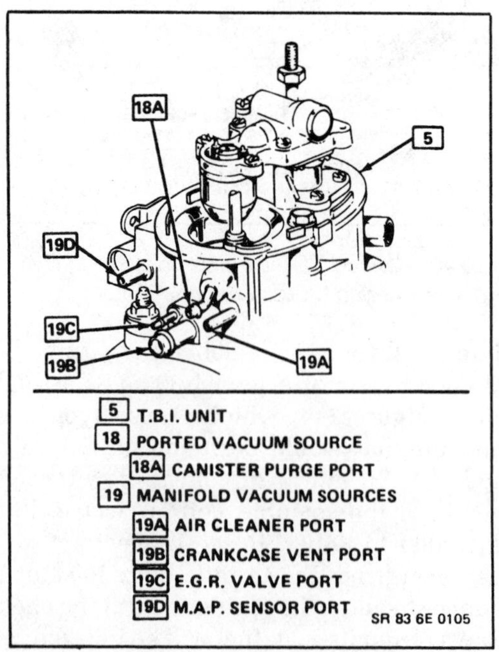

Figure 7-6. TBI vacuum ports location. [Courtesy of GMC Service Research]

On some engines, the oxygen sensor will cool off after the engine has been idling for a few minutes, which will place the system into an open-loop condition. In order to restore the system to a closed-loop condition for testing purposes when diagnosing the system at any time, the engine has to be run at part throttle to initiate enough exhaust gas flow to restore the system back into the closed-loop mode. An example of the fuel system components that form the closed-loop system is illustrated in Figure 7–7.

The quantity of fuel injected by the TBI system is controlled by the length of time that the valve is held open (as in the Bosch L system). The ECM (electronic control module) controls the timing and the amount of fuel injected into the engine by energizing the injectors in the throttle body (open) or de-energizing them (closed). The ECM receives and monitors signals from various sensors such as the engine coolant temperature, throttle position, vehicle speed, intake manifold absolute pressure (MAP), intake manifold air temperature (MAT), and a barometric pressure sensor (BARO) as well as from the oxygen, or Lambda, sensor. Figure 7–8 shows the typical operating conditions that are sensed by the ECM and what systems are controlled by a voltage signal from the ECM.

The sensing elements continually update the information sent to the ECM every 1/10 of a second for general information situations. For situations such as emissions and vehicle driveability, this information is delivered every 12.5 milliseconds. Since the ECM controls both the engine idle rpm and the mixture (air/fuel ratio), no adjustments are required to these two areas.

Current GMC TBI systems employ an idle speed air control motor that is designed to vary the position of a tapered-type valve in order to allow additional air to bypass the throttle when the engine is cold, therefore providing a higher engine speed for warm-up. This same motor will act to increase engine idle speed when accessory loads such as an air conditioner are switched on. The idle speed control motor is controlled from the ECM.

TBI System Operation

A simplified diagram of the flow through the basic TBI system is shown in Figure 7–9 from the in-tank mounted electric fuel pump up to the TBI unit fuel injector. When the ignition key is turned on, the ECM energizes the fuel pump relay for approximately 2 seconds to ensure that adequate fuel pressure exists at the fuel injector for initial start-up.

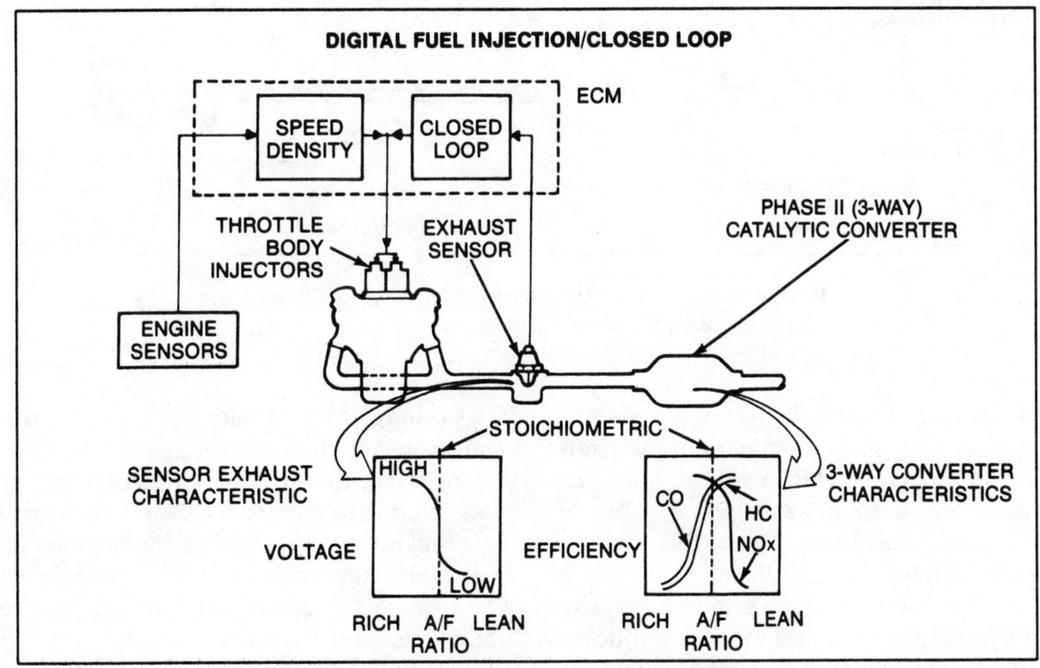

Figure 7–7. Closed-loop DFI operation. (Courtesy of Cadillac Motor Car Division, GMC)

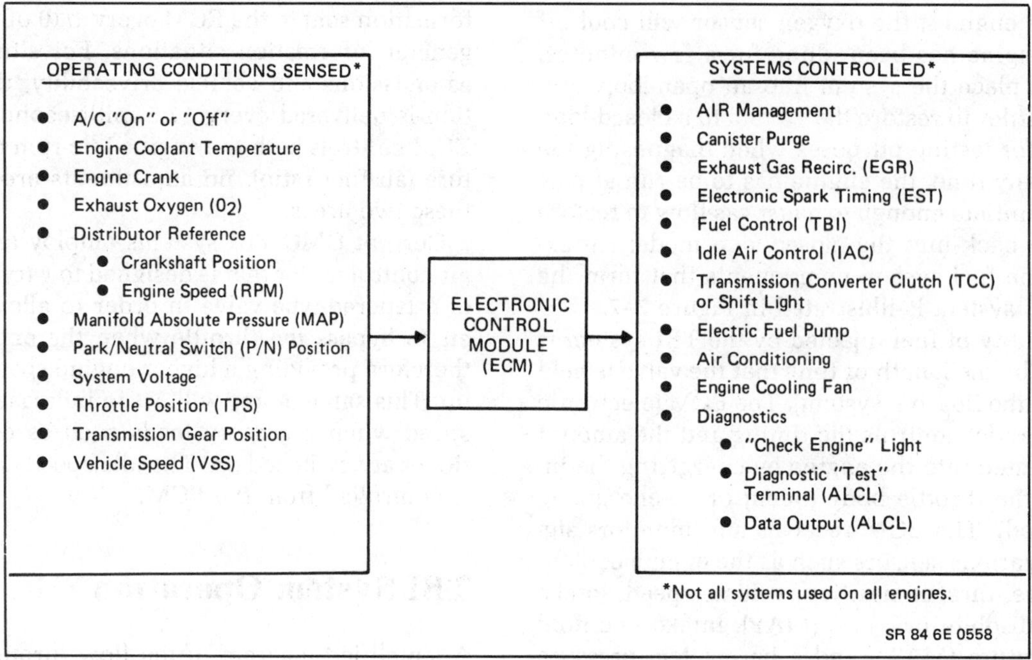

Figure 7–8. Electronic control module (ECM) operating conditions sensed and systems controlled chart. (Courtesy of GMC Service Research)

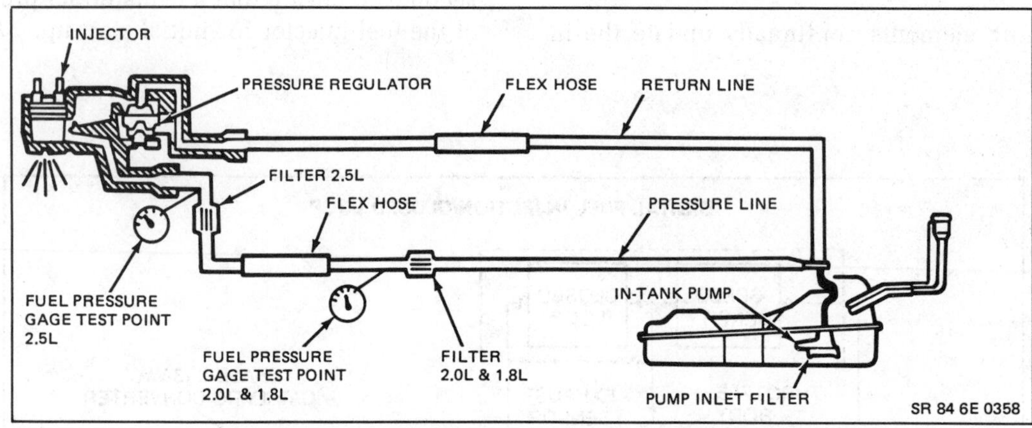

Figure 7–9. Basic TBI fuel system flow—2.0L engine. (Courtesy of GMC Service Research)

The air/fuel ratio at start-up will vary greatly, depending upon the ambient air temperature. At low temperatures of −36°C (−33°F), a very rich air/fuel ratio is required, and this can be as high as 1.5:1; in hot weather when the engine is warm, this mixture will climb to 14.7:1 at 94°C (201°F).

If the engine floods on a start-up attempt, this can be corrected by pushing the accelerator pedal down all the way; the ECM will then pulse (energize) the fuel injector from on to off very rapidly to supply an air/fuel ratio under this condition of as high as 20:1 for as long as the accelerator is held in the wide open position and the engine speed remains below 600 rpm. When the throttle is released to a position corresponding to less than 80%, the ECM will return the air/fuel ratio to the starting mode based upon the ambient air temperature conditions. Fuel pressure in the TBI system is controlled through the use of a pressure regulator that is an integral part of the throttle body. In Figure 7–10, fuel from the electric motor driven twin-turbine fuel pump located integrally within the fuel tank float unit is delivered through a

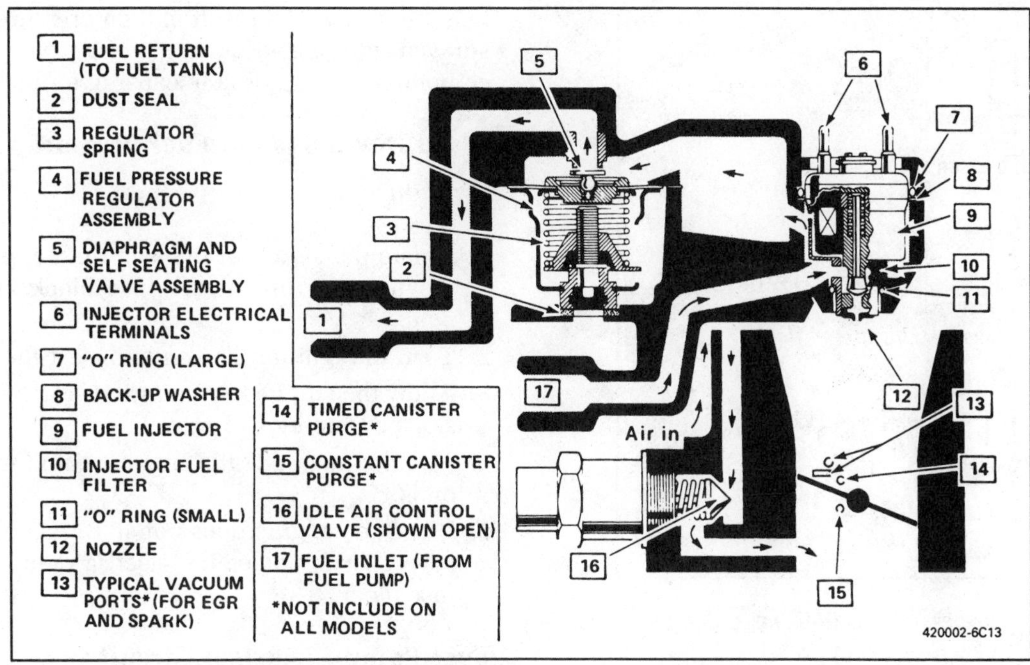

Figure 7–10. Fuel metering system—TBI. (Courtesy of Pontiac Motor Division, GMC)

filter to the throttle body, which contains the pressure regulator. The electric fuel pump is capable of delivering fuel at pressures as high as 18 psi (125 kPa). As you will note, fuel is directed to the injector assembly as well as to the regulator assembly simultaneously.

Fuel under pressure acts upon the top of the diaphragm, which is spring loaded. When fuel pressure reaches approximately 9–13 psi (62–90 kPa), the self-seating valve assembly is opened to allow fuel to bypass back to the fuel tank via a return line. Therefore, the pressure regulator maintains a constant pressure drop across the throttle body injectors.

If the fuel pressure is lower than 9 psi (62 kPa), either through a faulty fuel pump or a faulty regulator, then the engine will operate poorly, while too high a pressure can result in detonation and a strong smell of gasoline will be noticeable.

On dual injector TBI systems, a fuel pressure compensator is used on the second TBI unit assembly to compensate for a momentary fuel pressure drop between the two units. The constant flow of fuel through the throttle body prevents component overheating and vapor lock. If an injector fails to open (electrical problem), the engine will fail to start because no fuel can be injected into the throttle body. If the injector sticks in the open position, raw fuel will dribble into the throttle body and inlet manifold causing poor performance and a strong smell of gasoline. When the engine is turned off, two things will happen: one is that the engine may continue to diesel or run for a short time, and, two, pressure in the fuel system will bleed off making it very hard to start the engine after it has been shut off for any length of time.

Idle Air Control Valve

Figure 7–11 illustrates the various IAC (idle air control) valves used on the various models of engines. Two different types of IAC pintle valves are used on the Model 300 TBI systems employed on the 1.8L and 2.5L engines.

The Model 500 TBI system is used on all 2.0L engines and uses a dual taper valve when an automatic transmission is employed and a blunt pintle when a manual transmission is used.

These valves are designed to be used with a specific model of engine as shown in Figure 7–11, and if they are intermixed in different engines, poor performance and rough idle will result.

The location of the idle control valve can be seen in Figure 7–2 in the exploded view of the Model 300 TBI unit. This valve controls bypass air around the throttle plate and is similar to the idle air adjusting screw and auxiliary air bypass valve used on Robert Bosch gasoline fuel injection systems.

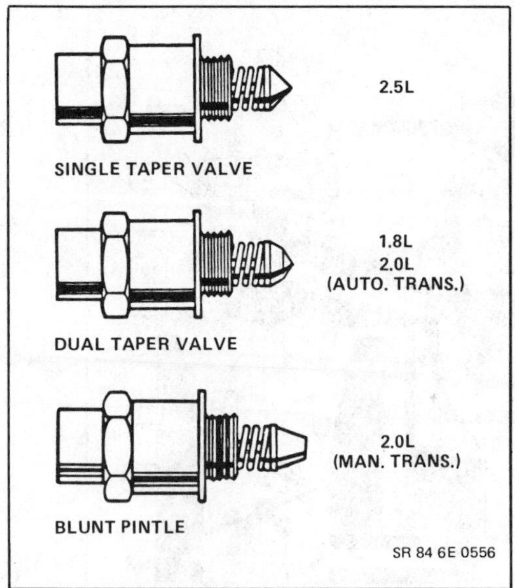

Figure 7–11. Idle air control (IAC) valve designs. (Courtesy of GMC Service Research)

The IAC valve receives voltage signals from the ECM, based upon engine coolant temperature, speed, and load as well as battery voltage. Anytime that the throttle plate is closed, the ECM will control the IAC valve position based upon the barometric pressure sensor (MAP) or manifold absolute pressure sensor.

The ignition should always be turned off anytime that the IAC valve is to be disconnected or connected to/from the system. Setting the valve too far open allows more bypass air, and the idle speed will be too high. If it is too far in or stuck, insufficient air will flow and the idle speed will be too low or rough and the engine could stall continually.

TBI System Tests and Adjustments

In order to perform the necessary tests and adjustments on the TBI systems employed by GMC, a number of special tools are required, which can be seen in Figure 7–56, located at the end of this chapter.

SPECIAL NOTE: On any gasoline fuel injection system, the system will remain under pressure once the engine is stopped through the use of a pressure regulator on GMC systems or by an accumulator on Robert Bosch fuel injection systems. Before attempting any service adjustments or repairs to the fuel system that would require you to loosen a pressure filled line, the fuel system pressure must be bled down to relieve this pressure. Failure to relieve the fuel system pressure could result in high pressure fuel being sprayed into your eyes, or spraying onto hot engine components and possibly causing a fire.

Bleed Down System Fuel Pressure

Procedure

1. Isolate the action of the electric fuel pump by removing the fuse at the fuse block marked *fuel pump*.
2. Turn the ignition key *on* and start the engine.
3. Allow the engine to run until it stumbles and then stops due to lack of fuel.
4. Attempt to restart the engine again for 3 to 5 seconds.
5. When the engine does not start, turn the ignition key off and replace the electric pump fuse at the fuse block.

Fuel Pump Electrical Circuit

The fuel pump is powered through a relay that is energized through the ECM for 2 seconds when the ignition key is first turned on in order to prime the injector(s). The ECM will send a voltage signal to the fuel pump relay to turn the fuel pump off after this time unless the ECM receives voltage signal reference pulses from the HEI (high energy ignition) distributor.

When the engine starts, the ECM feeds battery current to the relay. An oil pressure sender, which closes a set of contacts at approximately 4 psi (27.6 kPa), will supply power to the fuel pump and act as a backup power feed to the pump should the fuel pump relay malfunction. The in-tank fuel pump is illustrated in Figure 7–12 and is attached to the fuel

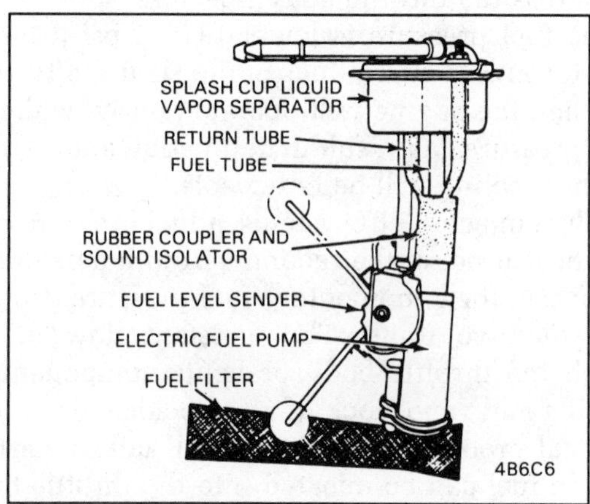

Figure 7–12. In-tank fuel pump arrangement. (Courtesy of Buick Motor Car Division, GMC)

sending unit. A woven plastic filter is located on the lower end of the fuel pickup pipe in the tank. Under normal conditions, the filter is self-cleaning and requires no maintenance unless a large amount of water or sediment gathers in the fuel tank—which would necessitate cleaning of the fuel tank and filter.

The fuel pump relay location will vary between different models of GMC vehicles. An example of three such locations are shown in Figures 7–13, 7–14, and 7–15 for Code E, P, and R engines.

Figure 7–13. Fuel pump relay location—engine code "E." [Courtesy of Buick Motor Car Division, GMC]

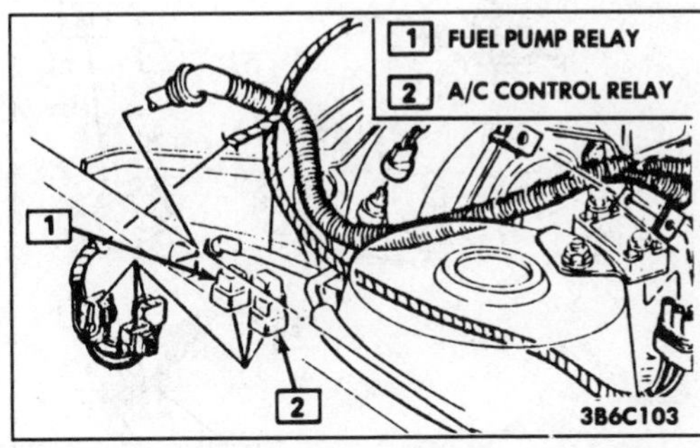

Figure 7–14. Fuel pump relay code location—engine code "P." [Courtesy of Buick Motor Car Division, GMC]

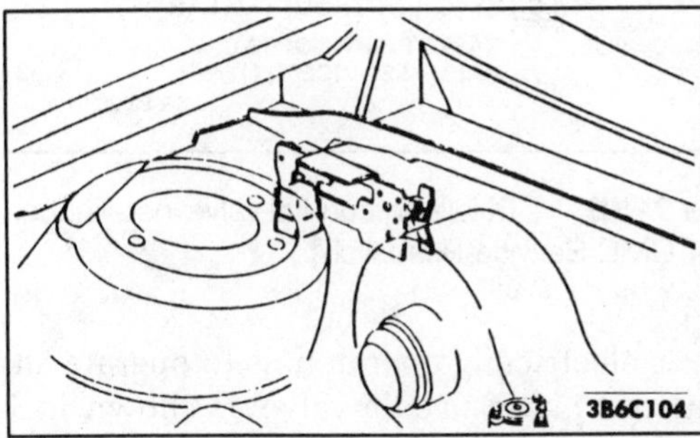

Figure 7–15. Fuel pump relay code location—engine code "R." [Courtesy of Buick Motor Car Division, GMC]

Fuel Filters

All TBI systems employ a fuel filter, located in the fuel feed line to the TBI unit. This filter can be replaced in the conventional manner after bleeding down fuel system pressure as described above. Figure 7–16 illustrates the fuel filter. The fuel injector(s) also contain a fine screen filter.

Fuel System Pressure Test

Before performing this check, relieve fuel system pressure as described above.

Procedure

1. Remove the air cleaner assembly.
2. Temporarily plug the thermal (heat) vacuum port on the throttle body.
3. Using a "TEE" fitting, install a fuel pressure gauge into the fuel line between the throttle body and the fuel filter.

   An example of a typical fuel gauge for a dual TBI test is shown in Figure 7–56, "Special tools." This gauge arrangement is available from any Kent-Moore tool dealer, P/N J-29658.

   NOTE: The gauge should be capable of registering at least 15 psi.

4. Start the engine (it may take a few seconds for the fuel pressure to come up to allow the engine to start because you relieved system pressure before installing the pressure gauge).
5. Once the engine starts, allow it to run for a minute or so and note the reading on the pressure gauge.
6. The fuel pressure should register between 9–13 psi (62–89.6 kPa).

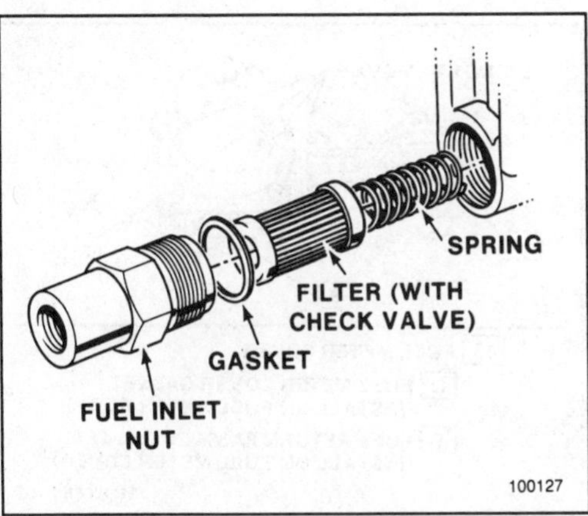

Figure 7–16. TBI unit fuel filter location. [Courtesy of Buick Motor Car Division, GMC]

SPECIAL NOTE: Relieve system pressure before disconnecting the fuel gauge. Follow the same sequence explained above to do this.

7. If the fuel system pressure is low, check the condition of the fuel filter, fuel lines for restriction/kinks, pump, and fuel pressure regulator.

Fuel Pressure Regulator

If it becomes necessary to check/replace the fuel pressure regulator/compensator, proceed as described below.

CAUTION: When removing the fuel pressure regulator, do not remove the four screws that hold the regulator to the fuel meter cover (see Figures 7–17 and 7–18). The regulator is non-adjustable and is also under spring pressure; therefore, removal of these screws is not required for any reason.

Procedure

1. Remove the air cleaner assembly.
2. Unhook the wiring harness plus connector to the injector(s).
3. Loosen/remove the five screws that hold the fuel meter cover to the body.
4. NOTE: If you are not replacing the fuel pressure regulator, do not submerge the cover/regulator assembly in cleaning solvents because damage to the regulator diaphragm and gaskets can result.
5. To reinstall the fuel pressure regulator, simply reverse the order of removal.

Idle Air Control Motor

The idle air control valve (IACV) is located as shown in Figures 7–19 and 7–20. The function of this valve,

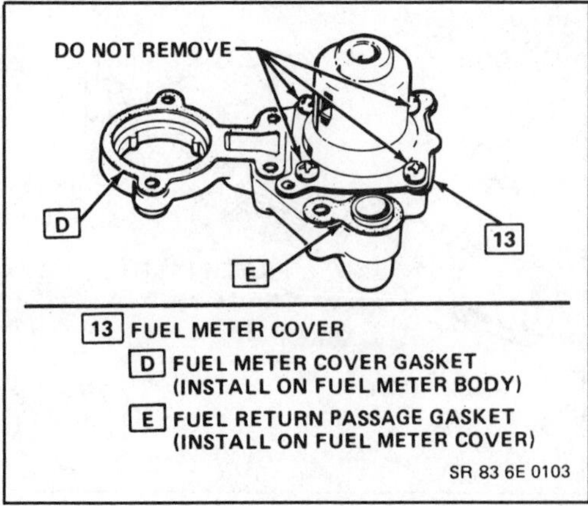

Figure 7–17. Installing fuel meter cover—TBI system. (Courtesy of GMC Service Research)

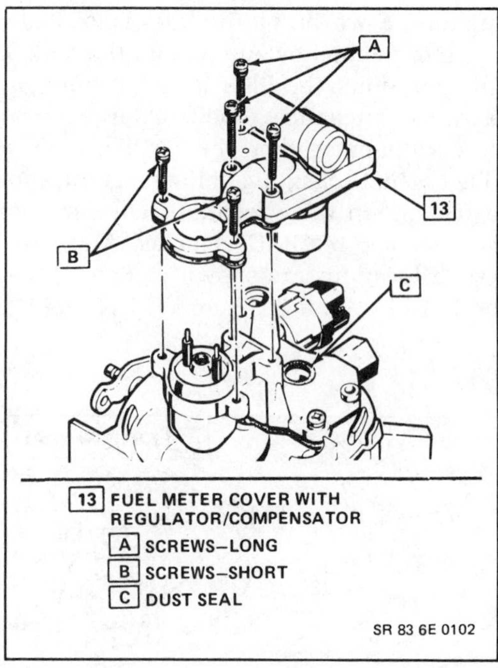

Figure 7–18. Fuel meter cover removal—TBI system. (Courtesy of GMC Service Research)

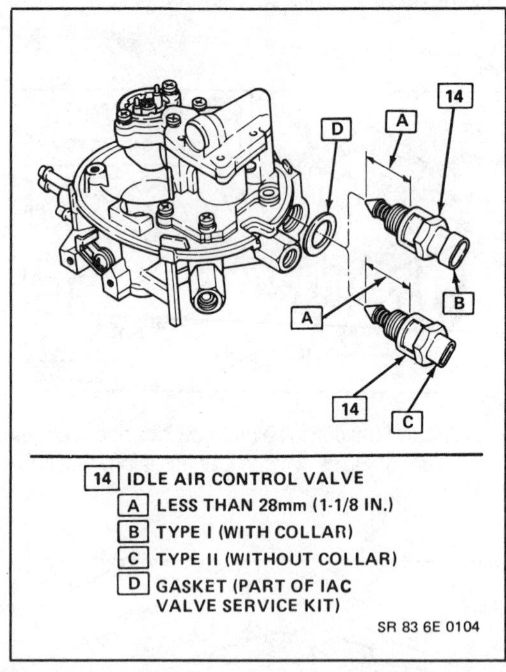

Figure 7–19. TBI Idle air control valve installation. (Courtesy of GMC Service Research)

which is electrically operated, is to operate an internal tapered pintle needle valve as shown in Figure 7–10. This IACV receives a signal from the ECM when the engine is cold so that the pintle valve will remain open. While open, the valve will allow air to

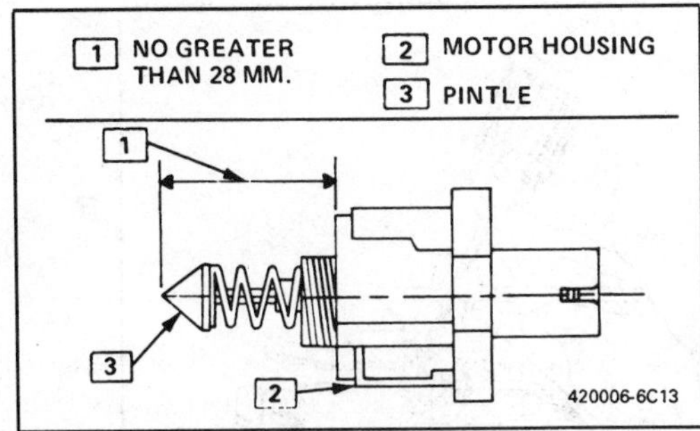

Figure 7–20. TBI Idle air control valve installation check. (Courtesy of Pontiac Motor Division, GMC)

bypass the throttle so that a fast idle speed can be maintained. The IACV also operates when accessory loads are applied to the engine when it is running at an idle speed so that an adequate idle rpm is maintained.

If it becomes necessary to replace the IACV, careful inspection of the dimension between the point of the pintle valve and the IACV body as shown in Figure 7–20 is required.

Removal/Installation of the IACV

1. Remove the air cleaner assembly.
2. Unhook and remove the wiring harness plug from the IACV.
3. Loosen off the IACV with a 1¼" (32-mm) wrench and remove it from the throttle body.
4. Before installing a new unit (IACV), measure the dimension shown in Figure 7–20 from the tip of the pintle to the IACV motor housing. This dimension must not be greater than 1.125" (28 mm); otherwise, IACV motor damage can result.
5. If adjustment of the IACV is required, determine if it is a Type 1 or a Type 2; Type 1 has a collar at the electrical terminal end, while Type 2 has no collar, which can be seen in Figure 7–19.

   Adjust Type 1 by exerting firm pressure against the conical valve while also using a side-to-side movement to retract it.

   Adjust Type 2 by compressing its retaining spring away from the conical valve while you rotate the valve with clockwise motion. Ensure that the return spring seats with its straight end aligned with the flat surface of the valve.

   NOTE: Tighten the IACV into the throttle body to 13 lb.ft. (18 N·m).

6. To check the successful operation of the IACV, start and run the engine until normal operating temperature is reached. If the engine fails to return to its normal idle speed and maintains a high idle speed, operate the vehicle at a road speed of 40 mph (64 kph) in order to allow the ECM to send out a voltage signal to the IAC valve; correct idle regulation will then occur.
7. On vehicles equipped with a manual transmission, the normal idle rpm is attained only after normal operating temperature is reached.
8. On vehicles equipped with an automatic transmission, place the gear selector lever into *D*, or *drive*, in order to allow the ECM to control the idle rpm.
9. Check the idle speed and if necessary, adjust the minimum air rate (2.5L engines) or the curb idle air rate (5.0L engines) with crossfire injection.

Minimum Idle Speed Adjustment

This particular adjustment is not performed unless components of the TBI system have been replaced or the throttle position sensor (TPS) has been adjusted. In all cases, adjustments to the TBI system should always be done with the engine at its normal operating temperature.

Procedure

1. Remove the air cleaner assembly.
2. Plug the vacuum port on the TBI body for THERMAC (thermostatic air cleaner) systems.

   SPECIAL NOTE: If the TBI unit employs a tamper-resistant plug, the TBI unit can either be removed to gain access to this plug, otherwise refer to Figure 7–21, which can be used to gain access to the adjustment screw while the TBI unit is still mounted to the intake manifold.

3. To gain access to the minimum air adjustment screw, remove the throttle valve (T.V.) cable from its bracket.
4. Connect a tachometer to the engine and disconnect the IACV wiring harness.
5. Start the engine (if not at operating temperature, run until it is).
6. Place gear selector in *neutral* (standard transmission) or in *park* on an automatic.
7. Install Kent-Moore tool P/N J-33047 (plug), see Figure 7–56, into the idle air passage of the throttle body unit; the tool *must* seat fully so that no air leaks exist (see Figure 7–22).
8. Adjust the minimum air screw using a No. 20 Torx bit screwdriver to obtain an engine idle speed of:

   a. 500 ± 25 rpm in *park* or *neutral* on a 2.5L engine connected to an automatic transaxle.

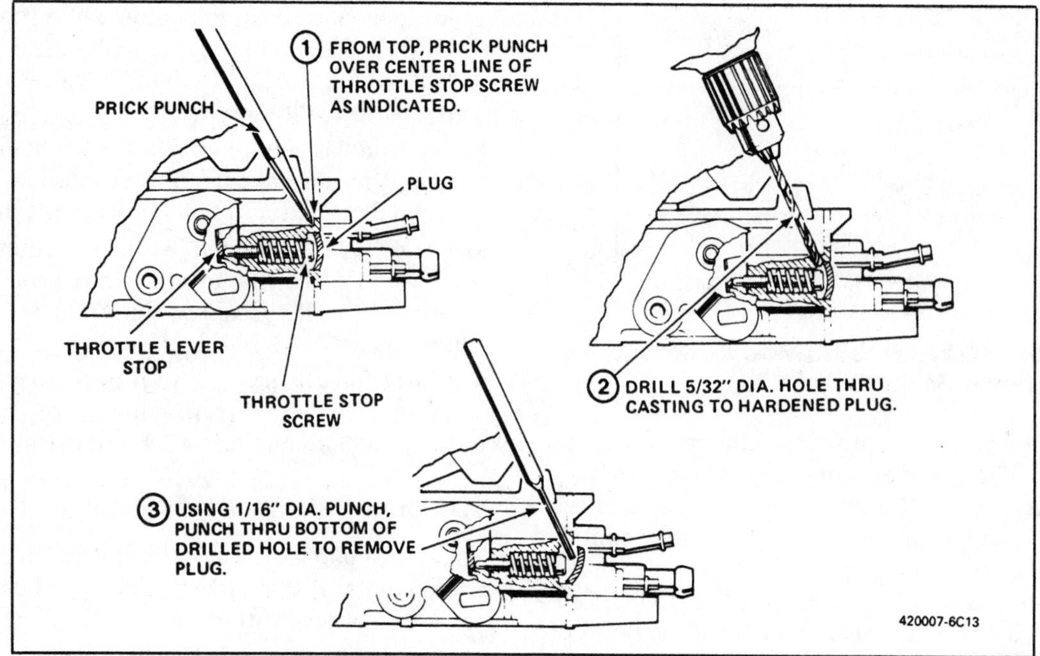

Figure 7–21. TBI Throttle stop screw plug removal. [Courtesy of Pontiac Motor Division, GMC]

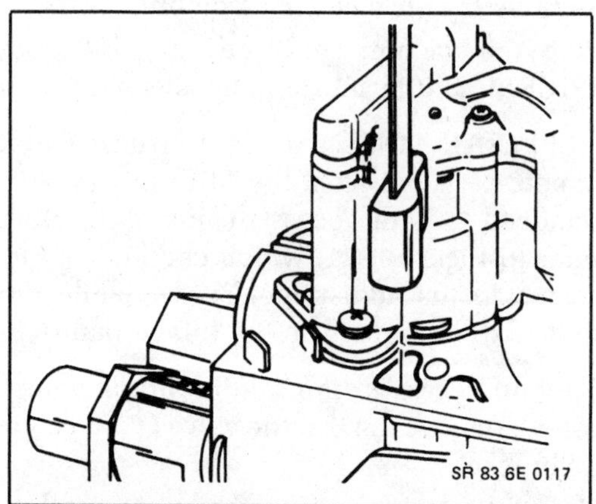

Figure 7–22. Installing tool J-33047 into throttle body. [Courtesy of GMC Service Research]

b. 700 ± 25 rpm for 1.8L engines with an automatic transaxle in *park* or *neutral*.
c. 800 ± 25 rpm for the 1.8L in *neutral* with a manual transaxle.
d. 775 ± 25 rpm for the 2.5L engine in *neutral* with a manual transaxle.
e. 650 ± 25 rpm for the 2.0L engine with gear selector in *D* on an automatic transaxle.

9. Stop the engine and remove tool J-33047 (plug).
10. Reconnect the T.V. cable.
11. Apply silicone sealant over the minimum air adjustment screw.
12. Reconnect IAC wire harness.
13. Reinstall the air cleaner assembly.

5.0L (305 cu.in.) Engine Curb Idle Air Rate Adjustment

This adjustment is only required when a suspected problem exists with the TBI throttle plates, or if new TBI units are installed or overhauled. This check/adjustment involves ensuring that both throttle body plates are in the exact same position within the throttle body bores. Failure to adjust these plates to the exact position will result in a greater air flow past one plate than the other with the result that the engine will run rough. The throttle plates are in correct adjustment if a reading of 0.45" Hg is obtained with a vacuum gauge at both the front and rear TBI units.

1 psi = 27.7" H_2O (Water) Displacement

1 psi = 2.036" Hg (Mercury) Displacement

Procedure

1. Remove the air cleaner assembly.
2. Start the engine and bring it to normal operating

temperature; if it is already warm, allow engine speed to stabilize.

3. Install Kent-Moore blanking plugs P/N J-33047 or equivalent into the idle air passages of each throttle body as per minimum idle adjustment routine for the 1.8L, 2.0L, and 2.5L engines described above.
4. With these plugs in place, the engine speed should decrease to its curb idle air rate.
5. If the engine speed does not drop, check for vacuum leaks at the blanking plugs or other areas.
6. Remove the cap from the rear TBI unit at the ported tube and connect up a vacuum gauge.
7. With the engine running at idle, the gauge needle should register approximately 0.45" Hg (6.122" H_2O).
8. If the vacuum gauge does not register 0.45" Hg (mercury), adjustment is necessary.
9. On some units it will be necessary to remove the tamper-resistant screw that covers the minimum air adjustment screw (Figure 7–21).
10. Adjust the minimum air adjustment screw until 0.45" Hg is achieved.
11. Perform Steps 6 through 8 on the front TBI unit.
12. Locate the "split lever" screw on the throttle linkage; if the screw happens to be welded, break the weld and install a new threaded screw with Loctite or equivalent.
13. Adjust the "split lever" screw until 0.45" Hg is obtained.
14. Stop the engine and disconnect the vacuum gauge and remove the idle passage plugs (J-33047).
15. Check the voltage at the throttle position sensor (TPS) and adjust as per TPS directions herein.
16. Reconnect the vacuum line to the TBI unit and also install the air cleaner assembly.

Throttle Position Sensor (TPS) Check/Adjustment

The TPS (throttle position sensor) is a variable potentiometer that is fitted to the side of the TBI unit as shown in Figures 7–2 and 7–3. The TPS is extremely important to the operation of the engine fuel system because it is designed to monitor the exact position of the air throttle valves (blades) inside the throat of the TBI unit.

Rotation of the throttle by the driver operating the accelerator pedal causes the TPS switch to vary its resistance relative to throttle blade angle. One end of the TPS is connected to a 5-volt feed from the ECM; the other end is connected to ground.

A third wire leads from the TPS to the ECM. The voltage signal from the TPS will vary from about 0.5 volt at an idle rpm up to 4.5 volts on a wide open throttle position and is sent through this third wire.

The TPS unit is installed and pre-adjusted at the factory, and the retaining screws are generally spot welded so that the TPS will not move. It is a sealed electrical unit and as such should not be immersed or saturated in a cleaning solvent.

If TPS replacement is required, the TBI unit must be removed and a 5/16" drill used to remove the spot welds on the retaining screws. New screws are supplied in the TPS service kit. Figure 7–23 shows an example of drilling out the spot welded TPS retaining screws.

Check/Adjustment Procedure

1. After a minimum idle air adjustment service check, the TPS should be checked.
2. Remove the air cleaner assembly and disconnect the TPS wiring harness plug.
3. Refer to Figure 7–24 and install three jumper wires between the TPS wiring harness plug and the TPS at the throttle body unit.
4. Switch the ignition key *on* (engine stopped) and measure the voltage between terminals B and C at the TPS as shown in Figure 7–24.
5. The acceptable voltage reading should be:
   a. 0.525 ± 0.075 volt on the 500 Model TBI unit used on the 2.0L engines.
   b. 0.450 to 1.250 volts at closed throttle on the 300 Model TBI unit used on the 1.8L and 2.5L engines.

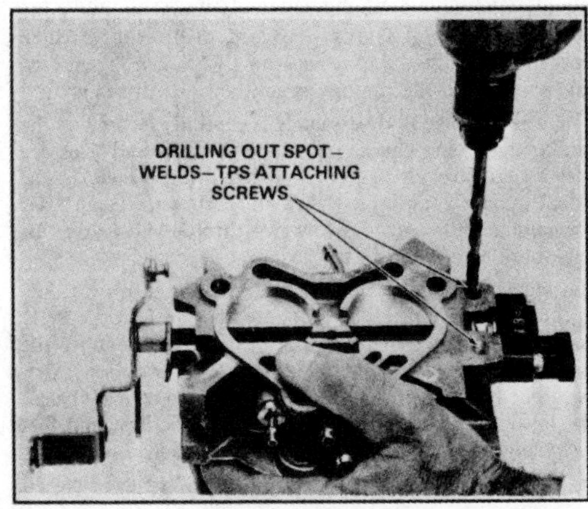

Figure 7–23. Throttle stop screw spot weld removal. (Courtesy of Cadillac Motor Car Division, GMC)

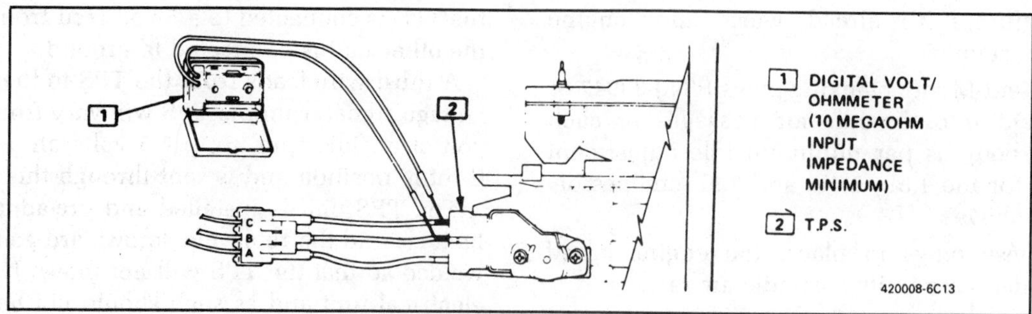

Figure 7-24. Throttle position switch voltage check. (Courtesy of Pontiac Motor Division, GMC)

NOTE: The TPS used on the 1.8L and 2.5L engines is not adjustable because the ECM uses the value obtained at idle (closed throttle) as its zero reading.

6. If these voltage readings are not obtained on the 300 Model TBI/TPS used with the 1.8/2.5L engines, check the condition of the wiring and the TPS switch.
7. On 2.0L engines, loosen the two TPS attaching screws and rotate the TPS until the voltage readings in Step 5 are obtained.
8. Install the air cleaner assembly.

SPECIAL NOTE: If the TPS is out of adjustment, open, shorted, or loose, engine performance will be poor. Any of these problems will cause the voltage signal that is normally fed to the ECM to be affected. The ECM will substitute a given value for the TPS voltage so that the vehicle can still be driven. However, a trouble code will be set and the check engine light will illuminate on the instrument panel to alert the driver of a problem situation.

If the sensor is open, the ECM will not receive a voltage signal. The ECM will therefore assume that the throttle is closed (idle), and fuel delivery to the engine from the electrically controlled injectors will not be adequate to throttle position—resulting in poor performance. Such a condition will trigger a Code 22 into the memory bank of the ECM when the TPS voltage signal is:

1. Less than 0.2 volt for 2 seconds on the 1.8L and 2.5L engines and engine speed is less than 1600 rpm.
2. Less than 0.2 volt for 5 seconds and engine speed is less than 1200 rpm on the 2.0L engine.

A Code 21 will set in the ECM when the TPS is shorted because the ECM now assumes that the throttle is wide open and the voltage signal is:

1. Greater than 2.5 volts for 2 seconds on the 1.8L and 2.5L engines with an engine speed less than 1600 rpm as well as the MAP (manifold absolute pressure) being less than 17.8" Hg or 60 kPa under a no-load condition.
2. Greater than 2.5 volts for 5 seconds on the 2.0L engine with an engine speed less than 1200 rpm as well as the MAP pressure being less than 8.7" Hg or 30 kPa under a no-load condition.

Detailed analysis and interpretation of these problems can best be referenced in the respective service manual for the make and model of vehicle under service. Refer to the driveability and emissions section of the service manual.

Problems that can cause the ECM to flash a Code 21 or 22 can be caused by such items as opens or shorts, which may require repairs to the wiring harness or replacement of the TPS or even the ECM in some instances.

Fuel Injector Removal/Installation

The fuel injector(s) are cycled on/off by an electrical signal from the ECM. One injector is used on smaller GMC engines; two can be found on the larger Vee-type engines.

If an on-car test indicates that the injector(s) dribble fuel with the electrical connections disconnected, then the injector(s) should be replaced. A rough idle speed will exist if the injector O-rings are cut, distorted, or damaged in any way. In addition, the steel backup washer must be installed below the large upper O-ring. Flooding will occur as a result of these conditions. Engine hesitation, on the other hand, can often be traced to plugged or damaged injector fuel filters. Hard starting can also be caused by either of the above conditions.

Injector Removal

The fuel injector(s) can be removed from the TBI unit as described below (see Figures 7–25 and 7–26).

Procedure

1. Bleed/relieve fuel system pressure (see above).
2. Remove the air cleaner assembly.
3. Disconnect the wiring harness plug to the injector(s).
4. Remove the TBI unit fuel meter cover discussed above under fuel pressure regulator removal.
5. Using a screwdriver as shown in Figure 7–26 or a pair of small pliers, place a protective cover around the center of the injector, very gently grasp the center collar of the injector between the electrical terminals, and gently remove the injector with a slight, combined lifting and twisting motion.

   NOTE: Since the fuel injectors are electrically operated, never submerge them in gasoline, solvent, or any liquid cleaner, although they can be cleaned externally. Faulty injectors must be replaced as an assembly because no components are available for repair.

6. Rotate the injector fuel filter back and forth in order to remove it from the base of the injector assembly.
7. Remove the large O-ring and steel backup washer, then the small O-ring at the bottom as shown in Figure 7–25.

Injector Installation

Figure 7–27 illustrates the injector being installed back into the TBI bore. Fuel injector installation should be performed as follows.

Procedure

1. Clean or replace damaged injector fuel filters; the fuel filter is cone shaped with the large end of the filter facing up towards the electrical connections. Place the larger end of the filter facing the injector so that the filter covers the raised rib at the base of the injector; use a twisting motion to position the filter against the base of the injector.
2. Always use new O-rings when reinstalling the injectors, whether they are used injectors or new ones.
3. To install the injector(s), the O-rings should be lightly lubricated with a light oil (Dexron 11 transmission fluid can be used).

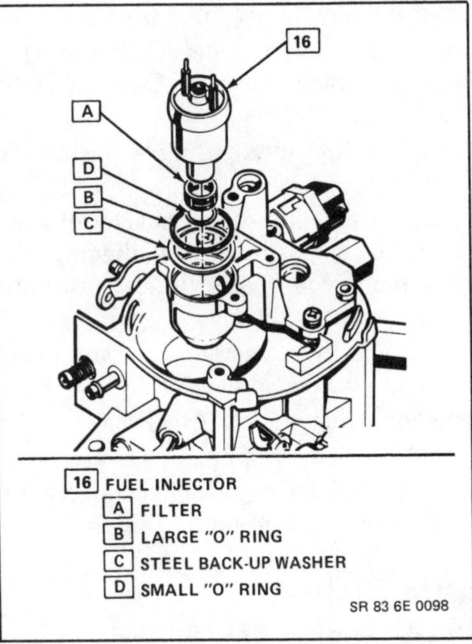

Figure 7–25. TBI Fuel injector components. [Courtesy of GMC Service Research]

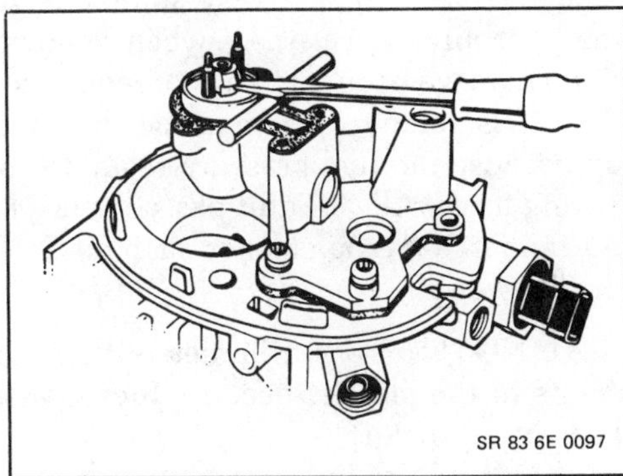

Figure 7–26. Removing TBI system fuel injector. [Courtesy of GMC Service Research]

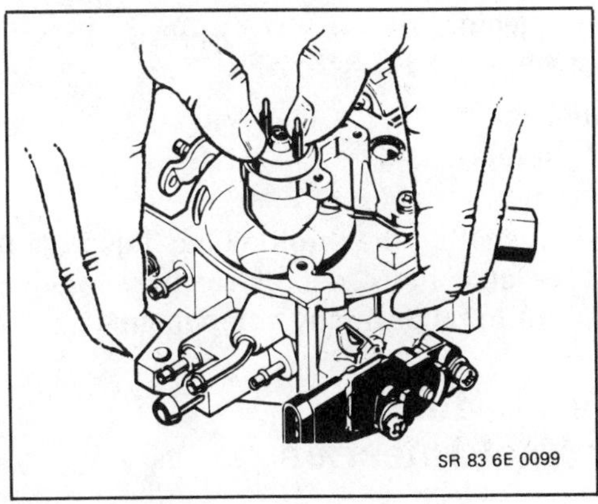

Figure 7–27. Installing TBI fuel injector. [Courtesy of GMC Service Research]

4. Install the new O-ring over the injector nozzle.
5. Position the large upper O-ring steel backup washer and new lubricated O-ring into the injector housing bore.
6. Install the injector(s) using a pushing-twisting action.
7. Align the injector electrical terminals in an across-the-car position or by aligning the raised lug on the injector base with the cast-in notch in the fuel meter body cavity.
8. Install the fuel meter cover using Loctite or equivalent on the screws.
9. Install the injector electrical leads and the air cleaner.
10. Start and run the engine, checking for leaks and correct injector operation.

Throttle Body Removal/Installation

Relieve fuel system pressure as explained above! Disconnect the battery before removing the TBI unit(s). Removing the TBI unit(s) is very similar to that of removing a carburetor. Take care when removing the unit that no damage occurs to any components. Once the TBI unit is removed, it must be disassembled carefully because the fuel pressure regulator unit is under spring force. All TBI unit parts can be cleaned in a cold immersion-type cleaner such as Carbon-X (X-55) or equivalent.

SPECIAL NOTE: Do not immerse the following components in the cleaner because these parts will swell, harden, or distort:

1. Idle speed control
2. Throttle position sensor
3. Fuel meter cover with pressure regulator
4. Fuel injectors
5. Fuel filter
6. Rubber parts
7. Diaphragms

When installing the throttle body injection unit(s), simply reverse the order of removal—again being careful not to damage any components.

General Motors Port Fuel Injection

GMC offers two types of gasoline fuel injection systems on their line of vehicles: the TBI (throttle body injection) system described above, plus what is commonly called "port injection."

In the TBI system, either one or two electronically controlled fuel injectors are mounted into the top of a throttle body casting, which looks similar to a carburetor body.

The port fuel injection system employs a fuel injector for each engine cylinder and is available in two basic models:

1. The MFI (multi-port fuel injection) system
2. The SFI (sequential port fuel injection) system

Control for the system uses the patented GM electronic control module (ECM), which receives voltage inputs from the following sensor units:

1. Exhaust gas oxygen (Lambda) sensor
2. Coolant temperature sensor
3. Detonation sensor
4. Hot film mass air flow sensor
5. Throttle position sensor
6. Engine speed sensor
7. Vehicle speed sensor
8. Transmission gear sensor
9. Power steering mode sensor
10. Air conditioning mode sensor

Both systems operate on voltage signals received from the ECM unit similar to that used with the TBI system. The main difference is that the MFI system, which is used on engine Code 3 vehicles, energizes the fuel injectors every crankshaft revolution. General Motors technically refers to this fuel system as a "simultaneously double-fire" design because all injectors fire once in each crankshaft revolution. This results in two injections of fuel being mixed with the incoming air to produce an excellent air/fuel charge for each combustion cycle.

The SFI system, which is found on engine Code 9 vehicles, utilizes a turbocharger, and the injectors turn on independent of each other—or once every two crankshaft revolutions. The SFI system will provide longer fuel injector pulse widths (time energized, or open) at low engine speeds, which will therefore provide better idle stability as well as low speed operation, such as when driving in the city.

SFI was offered initially on the 1984 Turbo Buick Regal T Type, Riviera T Type, and was optional on the Riviera convertible.

Multiple and sequential port fuel injection is found on the GMC 1.8L turbo LA5 (J Series) VIN Code "J" and also on the 3.8L LN3 (A Series) VIN Code "3" and 3.8L turbo LM9 (E and G Series) VIN Code "9" vehicles.

The major difference between TBI and MFI/SFI systems is shown in Figure 7–28, which illustrates the position of the individual fuel injector in relation to the intake valve of the cylinder. The basic system arrangement for the 1.8L 4-cylinder and 3.8L V6 engine fuel systems is shown in Figure 7–29, which is a flow diagram chart of the basic system arrangement.

The port fuel injection system offers better performance than does the TBI system because each cylinder receives a near stoichiometric mix every time the intake valve is opened regardless of engine speed/load.

Consider, for example, that the 1985 Corvette, using a Bosch hot-wire air meter and tuned port fuel injection system rather than the dual TBI system that was in use prior to the 1985 models, allows its engine to deliver 240 bhp (brake horsepower) rather than the 205 bhp that was available using TBI.

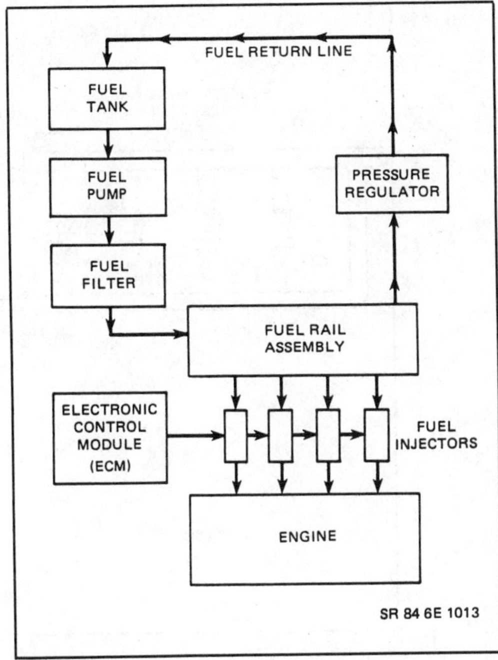

Figure 7–29. Operational block diagram of port-type fuel injection system. (Courtesy of GMC Service Research)

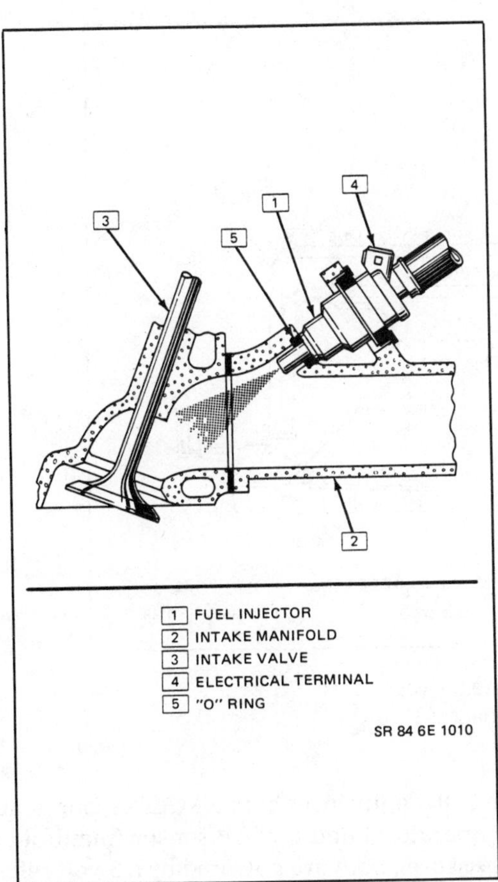

Figure 7–28. Port-type fuel injector location. (Courtesy of GMC Service Research)

System Operation

The major fuel rail components used with the 1.8L and 3.8L engines are shown in Figures 7–30 and 7–31. System components used with the MFI/SFI system are similar to that used with the TBI system except that a fuel rail is required to allow fuel circulation to all injectors while the engine is running. This removes air and vapors and assists in keeping the fuel cool in hot weather situations.

The electric fuel pump is an in-tank unit similar to that shown for the TBI system and pumps fuel through an in-line type fuel filter.

When the ignition key is turned to the on position, the starting sequence for the port fuel injection system follows the same pattern as that described for the TBI system.

Electric fuel pumps used with the MFI/SFI port fuel systems employ a higher operating pressure than that used with the TBI systems. TBI fuel systems operate at between 9–13 psi (62–90 kPa) with the pressure regulator set at about 18 psi (124 kPa). The fuel operating pressure on the port system runs between 28–36 psi (193–248 kPa) on engine Code 3 vehicles, while on the engine Code 9 vehicles it runs between 28–50 psi (193–345 kPa).

Port injection systems do not require a thermac, an EFE, barosensor, A.I.R. (air injection reaction) system, or dual bed exhaust converter.

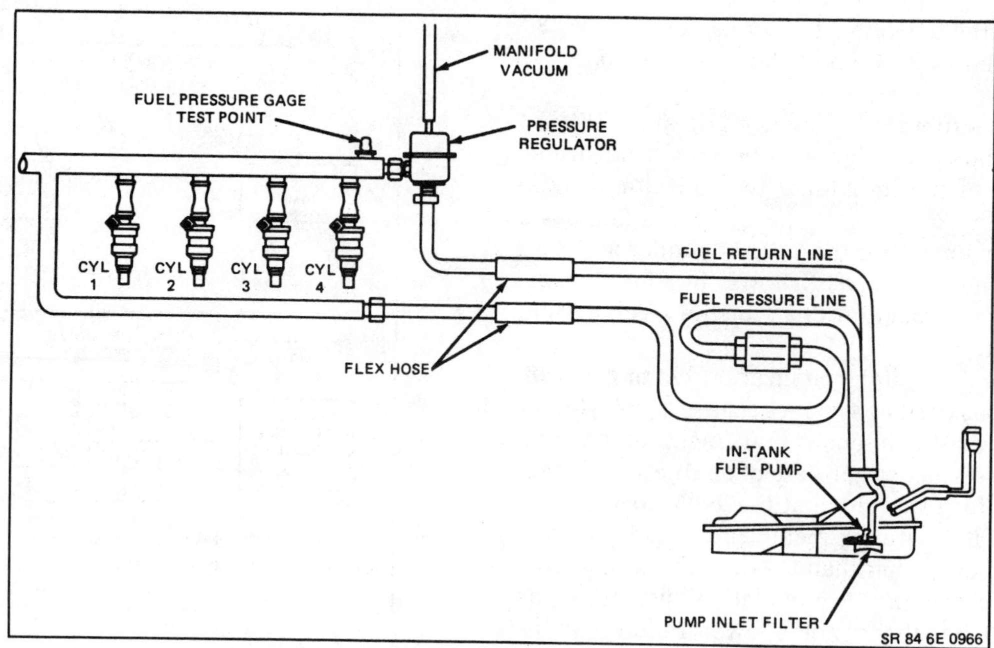

Figure 7–30. Port fuel injection system layout—1.8L engine. (Courtesy of GMC Service Research)

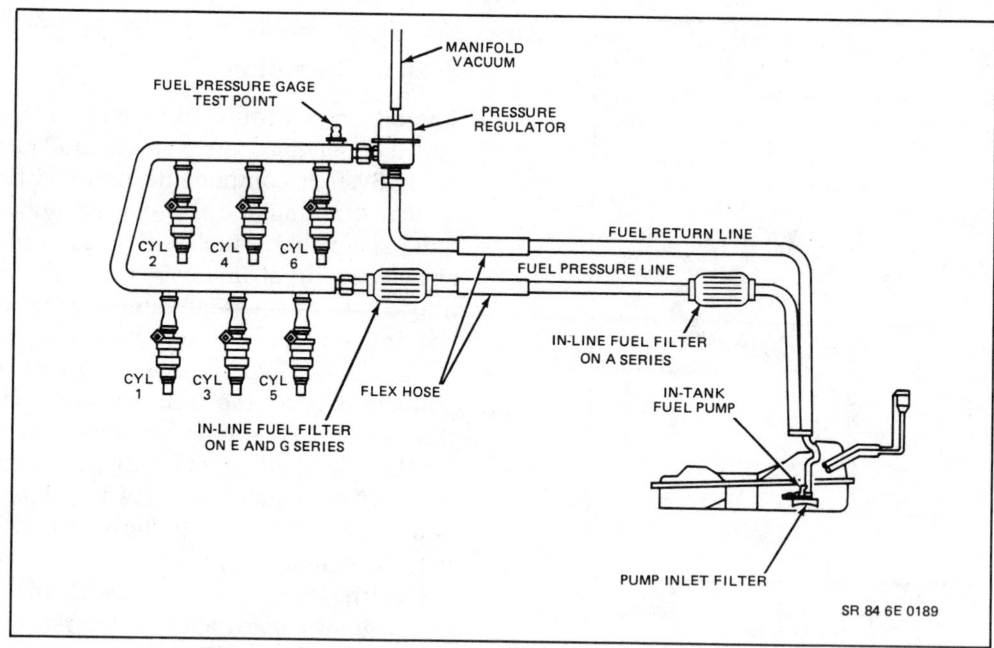

Figure 7–31. Port fuel injection system layout—3.8L V6 engine. (Courtesy of GMC Service Research)

The port system offers much better cold driveability, less exhaust emissions, and a better throttle response.

The intake manifold used with the MFI/SFI systems is a tuned arrangement that offers between 8–10% vehicle performance improvement.

The 1.8L engine uses both a MAT sensor (manifold air temperature) and a MAP sensor (manifold absolute pressure); both are powered by a 5-volt reference signal from the ECM. The MAT sensor is a thermistor mounted in the intake manifold as shown in Figure 7–32, which is designed to produce a high-resist-

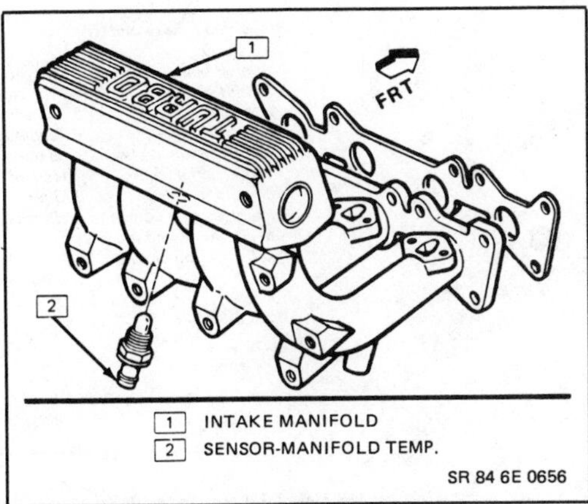

Figure 7–32. Manifold air temperature [MAT] sensor location—1.8L engine. [Courtesy of Buick Motor Car Division]

ance value (100,000 ohms at −40°C/−40°F) or a low-resistance value (70 ohms at 130°C/266°F). The MAP sensor shown in Figure 7–33, on the other hand, measures intake manifold pressure and, under certain conditions, will also measure barometric conditions. Its voltage signal along with that from the MAT sensor is relayed to the ECM to determine how long the fuel injectors will be energized (held open) in order to spray fuel into the combustion chamber.

The ECM uses the MAP sensor signal to establish both fuel delivery characteristics and also ignition timing.

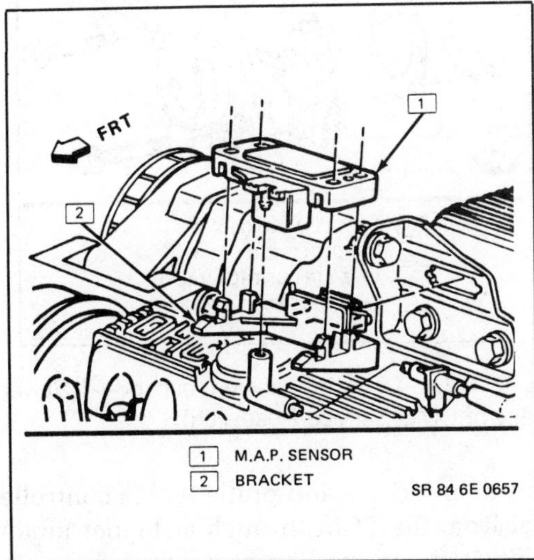

Figure 7–33. Manifold absolute pressure [MAP] sensor location—1.8L engine. [Courtesy of GMC Service Research]

The 3.8L V6 engines do not use either a MAT/MAP sensor but instead use a MAF sensor (mass air-flow sensor) in the intake manifold, as shown in Figure 7–34, to measure the volume of air flowing into the engine. The voltage signal produced by the MAF sensor is fed to the ECM for control of injector pulse width (energized time—open).

Mass Air-Flow (MAF) Sensor

In order to measure the volume of air (mass) flowing into the engine as accurately as possible, this system uses another Robert Bosch concept, which is a heated film that is maintained at a temperature of 75°F (24°C). The air flow is calculated electronically by determining the energy required to keep this heated film at a steady temperature of 75°F (24°C) above that of the incoming air. A resistor is used to measure the inlet air temperature.

The warmer the incoming air, the lower the energy requirements to keep the wire (heated film) at the predetermined temperature. This lowering of the air mass (air expands when it is heated) will cause a resistance change through the sensor wire, and a voltage signal corresponding to this air temperature increase (air mass decrease) will be fed to the ECM, which will send out a signal to the injectors to alter the time period at which they remain open. In this way, the air/fuel ratio is maintained as close as possible to ideal (stoichiometric—14.7:1) under varying throttle, engine load, and speed conditions. This system is also used on the 1985 Corvette engine. Figures 7–35(a), (b) and (c) illustrate the mass air-flow sensor that is located between the air filter and the throttle body assemblies. See page 90 for a more detailed explanation of the hot-wire mass air-flow sensor wire.

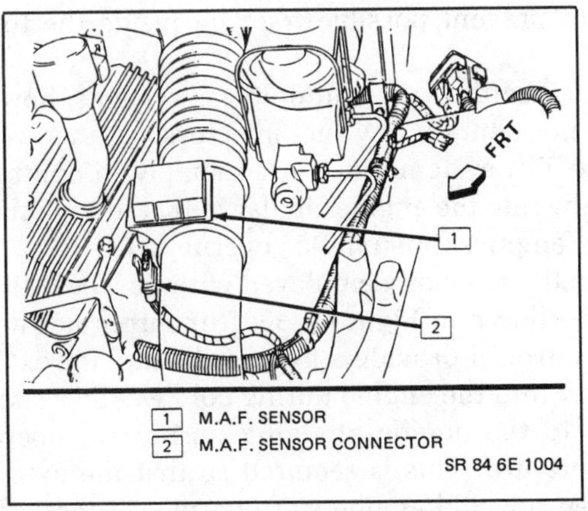

Figure 7–34. Mass air-flow sensor location 3.8L V6 turbo engine. [Courtesy of GMC Service Research]

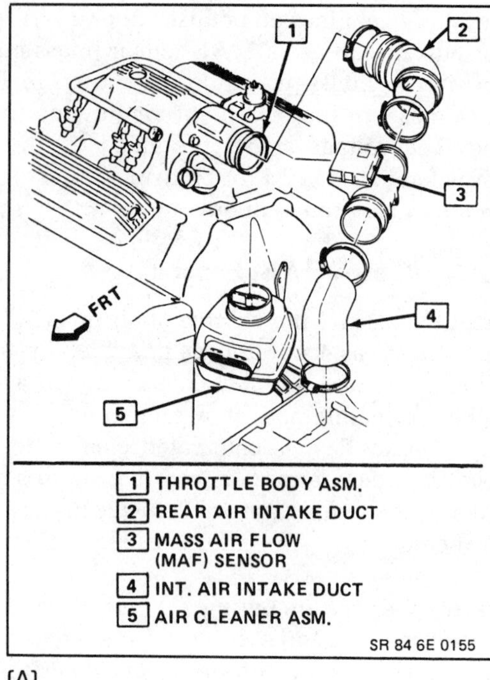

[A]

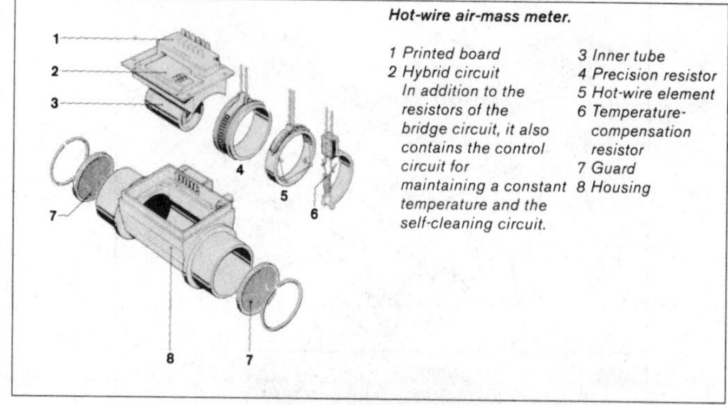

[B]

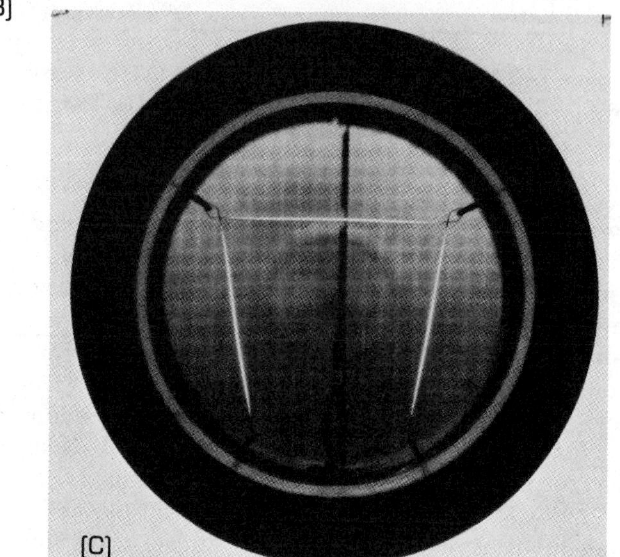

[C]

Figure 7-35. Mass air-flow sensor. a. Mass air-flow [MAF] sensor and surrounding components—V6. [Courtesy of GMC Service Research] b. Cutaway view of Buick mass air-flow sensor—V6 engine. [Courtesy of Robert Bosch Corp.; Broadview, Illinois] c. Mass air flow sensor—platinum wire inside air flow measuring tube. [Courtesy of Robert Bosch Corp.; Broadview, Illinois]

MFI Throttle Body

The throttle body used with the ported fuel injection systems has a 68-mm (2.677") single bore throat with two engine coolant lines arranged to supply engine coolant to and through the throttle body for rapid warm-up characteristics especially in cold weather and to prevent possible freezing inside the throttle bore.

The throttle body unit on these systems, however, does not contain any fuel injectors such as we find in the TBI system but is used simply to control the air flow into the engine similar to that found on some diesel engines (pneumatic governors).

The throttle body employed with the GM/MFI system is shown in Figure 7-36. It incorporates an idle air control to provide a bypass channel for extra air to flow into the engine during cold weather starts or prior to the engine attaining its normal operating temperature. This is required so that the extra fuel needed for cold engine start/warm-up is assisted by sufficient bypass air flow.

The idle air control unit used with the MFI system consists of an orifice and pintle that is controlled by a signal from the ECM through a stepper motor. As the engine temperature increases, the bypass air passage is gradually reduced. All information relating to the IAC (idle air control) valve can be found under the IAC valve described in the TBI system.

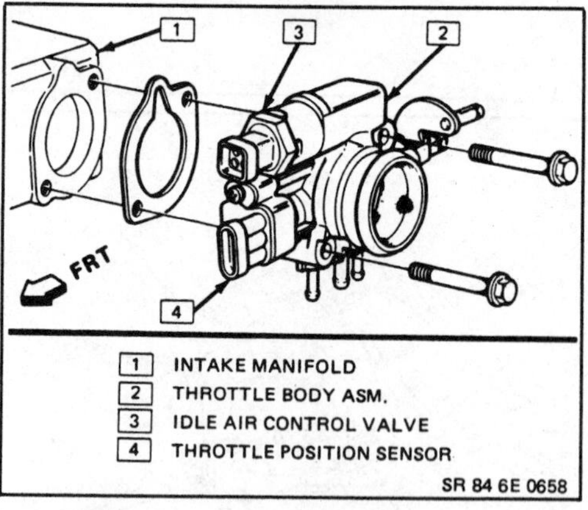

Figure 7-36. Throttle position sensor—1.8L engine. [Courtesy of GMC Service Research]

MFI Exhaust Gas Recirculation

On GM engines prior to the 1984 model year, EGR was basically an on/off system; however in the 1984 models, EGR is controlled by an electronic vacuum regulator valve that functions to provide both a variable rate and timing arrangement of EGR flow.

All GMC carbureted, TBI, or port fuel injected engines employ an oxygen sensor sometimes referred to as a "Lambda sensor" to monitor the oxygen content of the exhaust gases. A detailed explanation of an oxygen sensor and its makeup can be found in Chapter 6, dealing with Robert Bosch fuel injection systems.

The oxygen sensor relays a voltage signal of between 0.1 volt (when a high oxygen content due to a lean air/fuel mixture is present) and 0.9 volt (when a low oxygen content due to a rich air/fuel mixture is present). This voltage signal is fed to the ECM in the closed-loop position to allow constant monitoring of the air/fuel ratio. The open-loop versus closed-loop condition is described in detail under the TBI section above.

Location of the oxygen sensor for both the 1.8L and 3.8L turbo V6 engines is illustrated in Figures 7–37 and 7–38.

A TPS (throttle position sensor) is used with the port fuel injection system as it is with the TBI system described above. The TPS switch is designed to act as a variable resistor or potentiometer with a low voltage output (0.4 volt) at idle and between 4.5–5.0 volts at wide open throttle. The function of the TPS is to tell the ECM where the throttle blades are so that an accurate fuel pulse signal can be sent to the fuel injectors. Details of the TPS and its adjustment are similar to that given for the TBI system.

An example of the location of the TPS on the 1.8L and 3.8L turbocharged engines is shown in Figures 7–36 and 7–39.

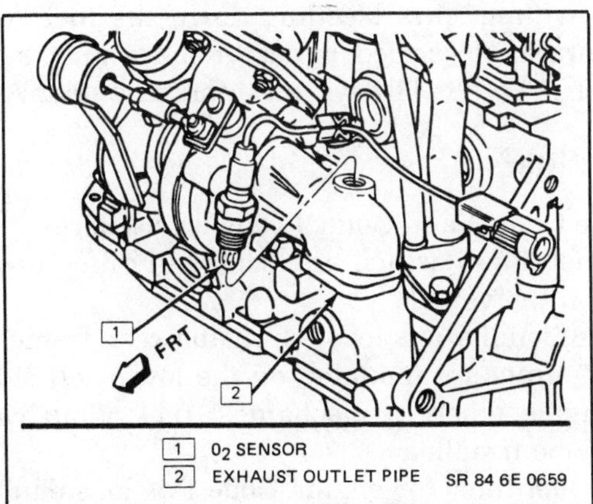

Figure 7–37. Oxygen sensor location—1.8L engine. (Courtesy of GMC Service Research)

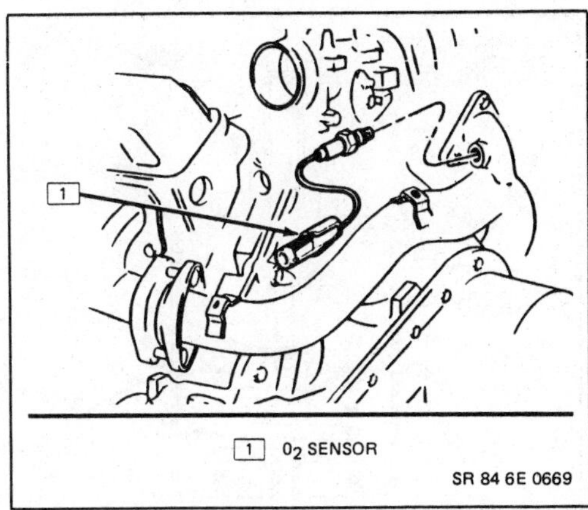

Figure 7–38. Oxygen sensor location—3.8L V6 turbo engine. (Courtesy of GMC Service Research)

Pressure Regulator Unit

A spring-loaded fuel pressure regulator assembly shown in Figure 7–40 is used to maintain a set pressure within the fuel rail of the injection system while the engine is running and also to maintain that pressure when the engine is shut off for long periods of time.

If the engine is not started for a considerable time period and the fuel pressure is lost, then when the ignition key is turned to the on position, the fuel pump relay will be turned on for two seconds by a voltage signal from the ECM to allow pressure buildup for starting purposes.

The pressure regulator is mounted on the fuel rail as shown in Figures 7–41 and 7–42. If fuel system

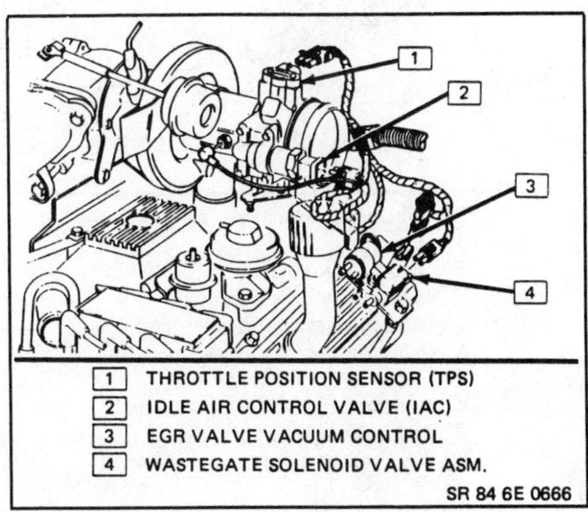

Figure 7–39. Throttle position sensor—3.8L V6 turbo engine. (Courtesy of GMC Service Research)

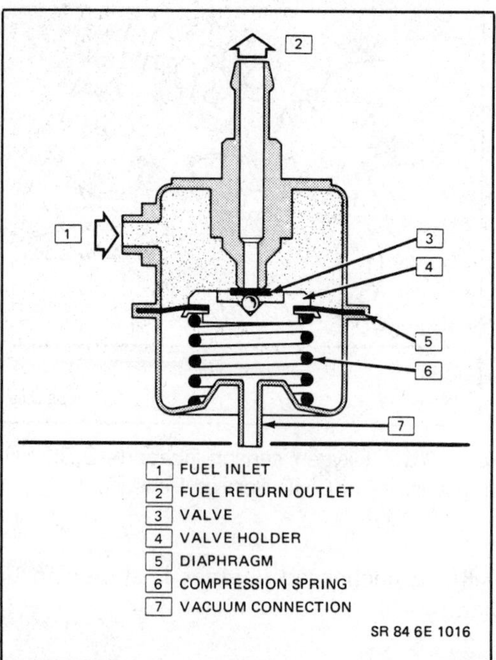

Figure 7-40. Fuel pressure regulator operational diagram. [Courtesy of GMC Service Research]

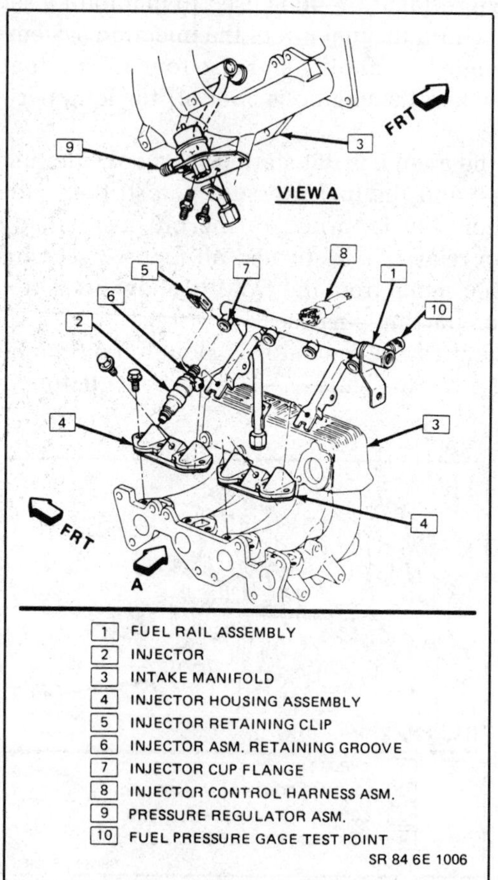

Figure 7-41. Fuel rail and injectors—1.8L port fuel injected engine. [Courtesy of GMC Service Research]

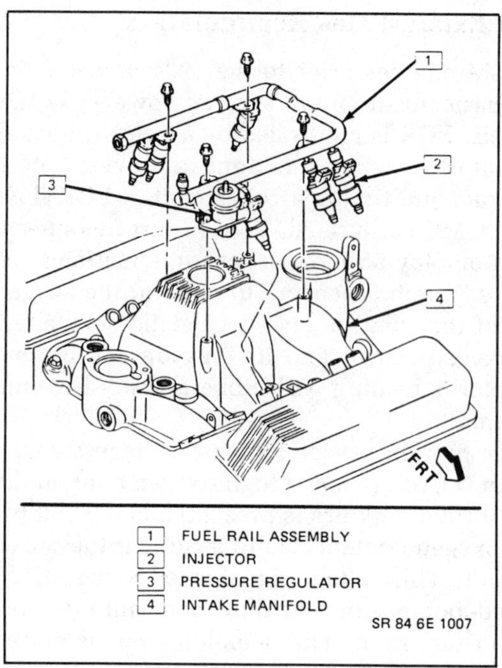

Figure 7-42. Fuel rail and injectors—3.8L port fuel injected V6 turbo engine. [Courtesy of GMC Service Research]

pressure is too low, engine performance will suffer due to lack of fuel atomization at the injectors and low delivery characteristics. Too high a fuel pressure on the other hand can cause detonation and excessive fuel delivery leading to a strong smell of gasoline.

Engine Code 3 vehicles also utilize an accumulator that is located in the fuel feed line near the cowl area. The accumulator is used to dampen out pressure pulsations that will exist as a consequence of the injector firing sequence (pulses open and closed).

NOTE: Never remove any fuel system components without first bleeding down the fuel system pressure as described under "Relieve Fuel System Pressure" in the TBI system information above.

Fuel Filters

Engine Code 3 and Code 9 vehicles use large in-line type fuel filters, which are located in different positions on the vehicle.

Code 3 units are located on the rear frame rail; Code 9 units are mounted on the lower left side of the engine where a mechanical fuel pump would usually be installed.

The fuel filter for engine Code P is located at the rear of the fuel tank mounted on the body bar. Engine Codes O and R have the filter attached to the engine on the left fender side; engine Code 8 vehicles have an in-line fuel filter located on the right side of the frame in the engine compartment. Figures 7–43 and 7–44 illustrate the fuel filters for the Code 3 and Code 9 engines.

Fuel Pump Flow Test

When it is suspected that the electric in-tank fuel pump is at fault, a fuel flow test should be performed as follows.

Procedure

1. Disconnect the fuel feed line after bleeding or relieving fuel system pressure as described above.
2. Hold the fuel feed line in a suitable container, preferably a clean, graduated one.
3. On engine Code E vehicles, turn the ignition switch to the *on* position.

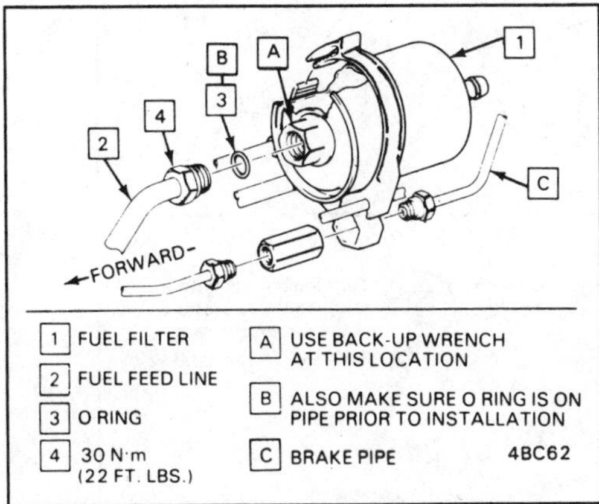

Figure 7–43. Fuel filter components—engine code 3. [Courtesy of Buick Motor Car Division, GMC]

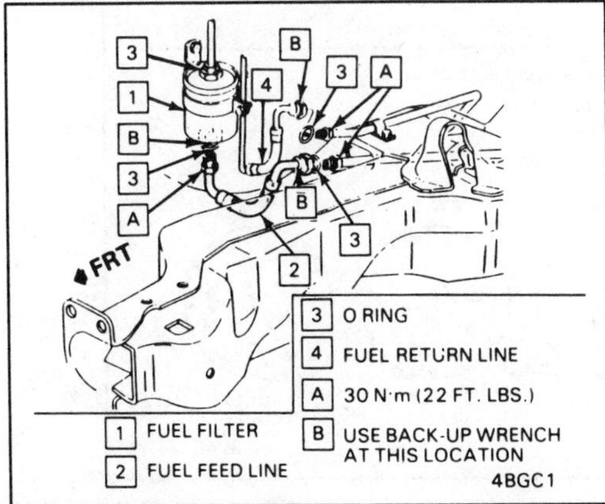

Figure 7–44. Fuel filter location—engine code 9 [G Series]. [Courtesy of Buick Motor Car Division, GMC]

4. On engines with multi-port injection systems, apply battery voltage to the fuel pump test terminal G of the assembly line communication link (as shown in Figure 7–49 under the ECM Checkout procedure at the end of this chapter).
5. The fuel pump should deliver at least ½ pint or more within a 15-second time span.
6. If the fuel flow is less than in Step 5, check the system for possible fuel line restriction or a faulty pump.

Fuel Injectors

The fuel injectors are manufactured by Robert Bosch, see Figure 6–50, who has been the leader in gasoline injection technology for many years; however, the fuel injectors are assembled in the U.S.

The injectors used with the MFI system are pulsed open/closed by a signal from the ECM and therefore operate in the same manner as Bosch L-Jetronic units. The quantity of fuel delivered is proportional to the length of time that they are open.

An electric solenoid (within the injector body) when energized pulls open the armature and lifts a pintle valve from its seat to allow the pressurized fuel to be sprayed into the intake manifold towards the inlet valve as shown in Figure 7–28.

Fuel injector problems with the MFI/SFI systems will tend to exhibit the same characteristic problems as with the TBI system. If the system has one faulty injector, one cylinder will possibly receive no fuel because in the MFI/SFI system each cylinder is fed from its own injector. In the TBI system, if only one injector is used and it fouls up, then every cylinder will suffer.

An injector that doesn't open will cause hard starts on port-type fuel injection systems. An injector stuck in the partially open position will cause a loss of fuel system pressure (especially when the engine is stopped) and flooding due to raw fuel dribbling into the engine. With multiple fuel injectors, the engine may start after cranking but will run rough.

Dieseling would also tend to occur with an injector stuck open, due to fuel dribble when the ignition key is turned off.

Injector Balance Test

Before removing the fuel injectors, check them while still in the engine with the use of a special tester unit. This tool is manufactured by the Kent-Moore Tool Division and is used to cycle the individual fuel injectors open and closed. This tool is used in conjunction with a fuel pressure gauge that can be inserted into the fuel rail pressure tap point shown in Figure 7–41 for the 1.8L engine and in Figure 7–45 for the 3.8L V6 engine.

CAUTION: If the engine is warm, be careful that you do not spill gasoline onto hot engine parts during this test. Wrap a rag/cloth around the pressure gauge adapter on the fuel rail to minimize fuel spillage onto the engine.

Before beginning the test, disconnect all the wiring connectors at all injectors. Connect the tester J-

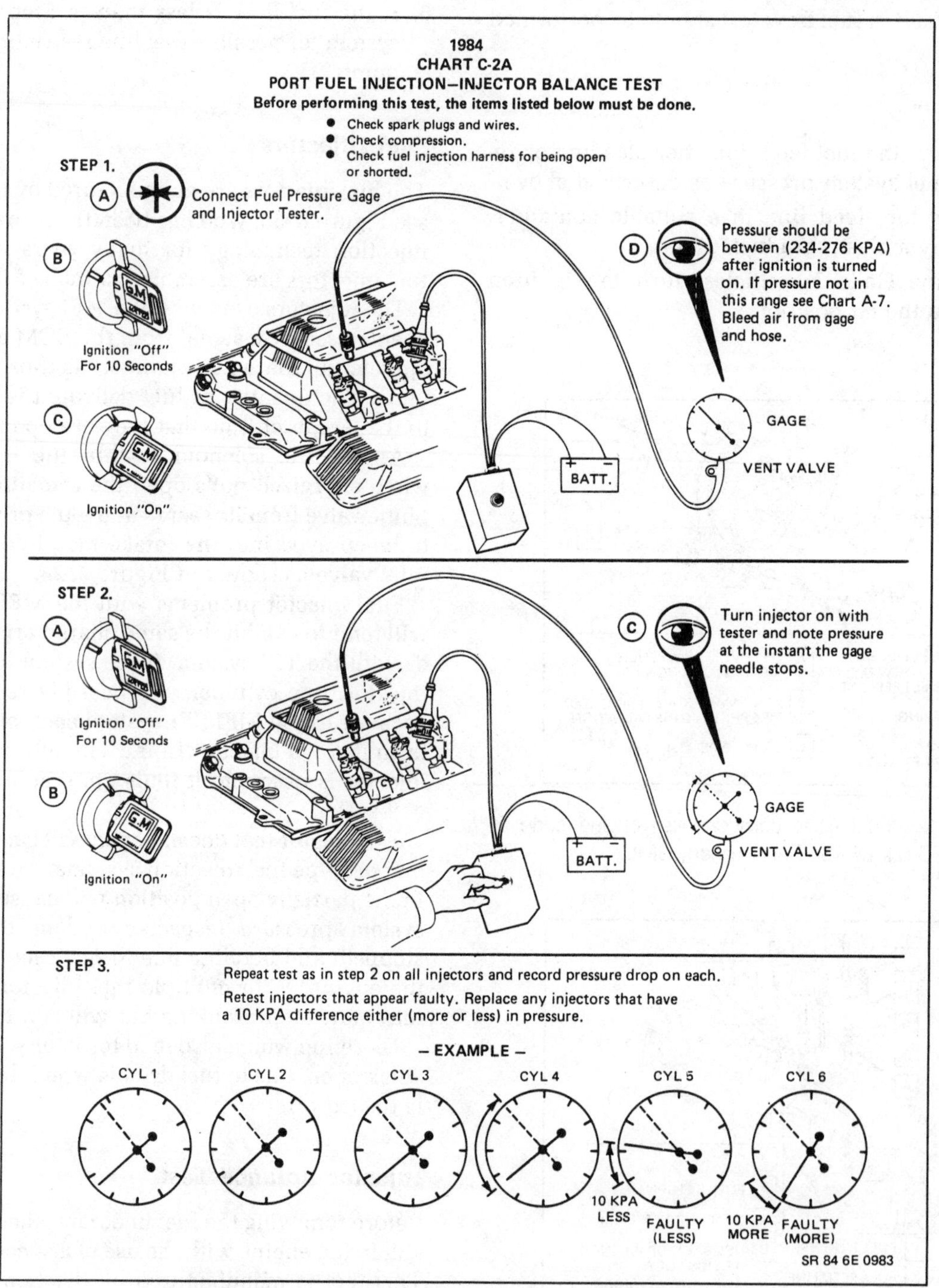

Figure 7–45. Port fuel injected system—injector balance test for both the 1.8L 4-cylinder and 3.8L V6 engines. [Courtesy of GMC Service Research]

34730-3 as shown in the diagram to one injector. On turbocharged engines, if you are unable to gain access to all of the injectors, connect up the adapter in the test kit that is included for this purpose.

Diagnosis. If after completing Step 1 of the injector balance test the fuel gauge pressure is not within 34–40 psi (234–276 kPa), refer to Chart A7 (Buick service manual). This chart is found in all GMC vehicle service manuals and is found in the driveability and emissions section.

The following information deals with the possible problem areas when low fuel pressure is found.

With the ignition turned off, the problem may be in the pressure regulator assembly. This can be confirmed by running the engine at an idle rpm with the pressure gauge connected to the fuel rail. Refer to Figure 7–40, which illustrates the design and operating mode of the fuel pressure regulator assembly. When the engine is running at an idle rpm, fuel pressure will always be lower (between 28–32 psi or 193–220 kPa) because the low pressure applied to the pressure regulator diaphragm will offset the spring pressure.

On a turbocharged engine, when the engine speed is increased and it is placed under load, the higher boost pressure within the intake manifold will also act upon the fuel pressure regulator diaphragm spring. It will therefore increase the fuel system pressure by approximately 1 psi or 6.895 kPa for every pound of turbocharger boost increase until a maximum fuel pressure of between 46–50 psi (317–345 kPa) exists.

To avoid having to run the engine on a chassis dynamometer or road test it with gauges connected back to the passenger compartment, use a small hand pump or clean filtered shop air, running it through a regulator valve.

Falling fuel pressure on the test gauge after the engine and ignition have been turned off can be caused by any of the following:

1. In-tank electric fuel pump check valve leaking.
2. Electric pump coupling hose leaking.
3. Fuel pressure regulator valve leaking.
4. Injector sticking open.

If the fuel pressure during each injector test remains fairly constant but it is less than the recommended 34–40 psi, the engine will run, but it will most likely exhibit hard starting when cold, and general overall poor acceleration and power. Pressures less than 24 psi (165 kPa) can result in stalling, and if the vehicle can be driven, it will tend to surge and be unable to sustain steady operation.

Low fuel pressure can be a direct result of restricted fuel flow and can be checked by performing a fuel pump flow test as described previously under the heading, "Fuel Pump Flow Test."

The fuel return line can also be plugged off and battery voltage applied to the fuel pump test terminal (terminal G of the ALCL—assembly line communication link) Figure 7–49. With the pump running and no return flow possible, the fuel system pressure should quickly rise to 75 psi (517 kPa) or higher. What we are determining here is whether the problem is in the fuel pressure regulator or due to a restricted fuel return line.

Injector Removal. Before attempting to remove the fuel injectors, relieve the fuel system pressure as described under the TBI fuel system by removing the fuel pump fuse, cranking the engine, and letting it run until it stalls because of no fuel. Crank the engine for 3 to 5 seconds longer, then proceed to remove the injectors. Leave the fuel pump fuse disconnected until you are ready to start the engine again.

Procedure

Refer to Figures 7–41 and 7–42, which illustrate the fuel rail and injector arrangement for both the 1.8L and 3.8L V6 turbocharged engines.

1. Disconnect the battery ground terminal.
2. Remove any accessories in your way to gain access to the fuel rail and injectors.
3. Remove the injector electrical connections.
4. Remove the fuel rail.
5. Remove the fuel injectors.

Distributorless Ignition System

Another interesting feature of the 1984 Buick turbocharged 3.8L V6 engines that employ SFI systems is the introduction of a "distributorless" ignition system. The coil of the ignition system is replaced by an electronic coil module that incorporates three separate ignition coils, each of which fires two spark plugs simultaneously (one from the positive end and the other from the negative end of the coil). This ignition system is designed to fire one spark plug at the beginning of the power stroke, while the other one fires at the end of the exhaust/beginning of the intake stroke.

This system is termed CCCI or C3I (computer controlled coil ignition) because it contains two Hall-effect sensors whereby one sensor monitors camshaft position and the other senses crankshaft position (see Figures 7–46(a), 7–46(b), and 7–47).

The camshaft sensor Hall-effect switch feeds the ECM with a voltage signal when No. 1 piston is on the compression stroke in order to allow the ECM to energize the injector and time the high tension spark. These signals from the sensors are fed to the on-board electronic computer, which controls the actual firing of the plugs.

The crankshaft Hall switch sensor triggers the ignition module to fire each one of the three coils at the correct time.

Hall-Effect System. A "Hall-effect" system is used extensively in high energy-type or breakerless ignition distributors now in wide use on many of the world's automotive engines. A Hall-effect ignition distributor is shown in Figure 7–48 whereby a rotating metallic-chopper spins between a magnet and a Hall-effect chip, which effectively causes the magnetic field to be subjected to a make-and-break effect commonly known as chopping.

NOTE: Figure 7–48 is shown only for clarification of the Hall-effect system; this distributor is *not* used with the CCCI distributorless system!

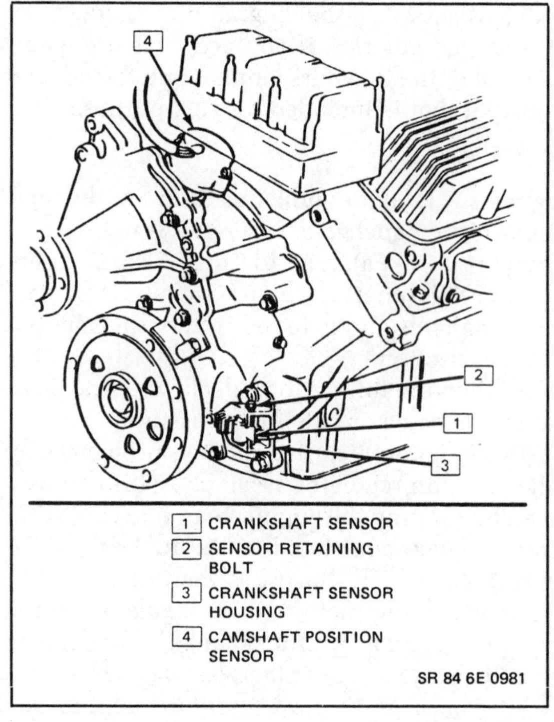

Figure 7–47. CCCI crankshaft sensor location. (Courtesy of GMC Service Research)

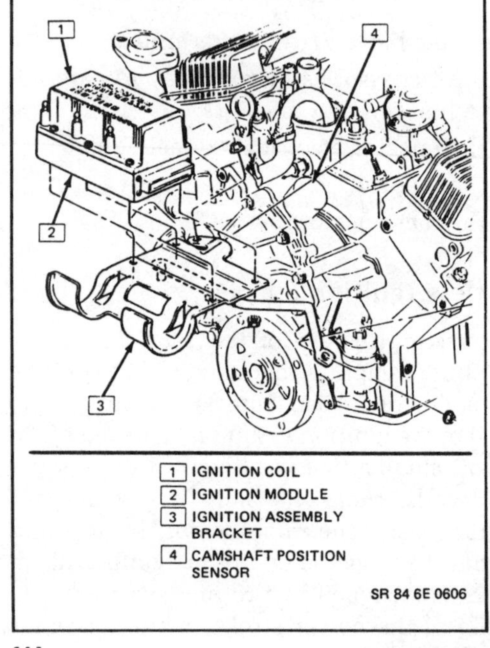

(A)

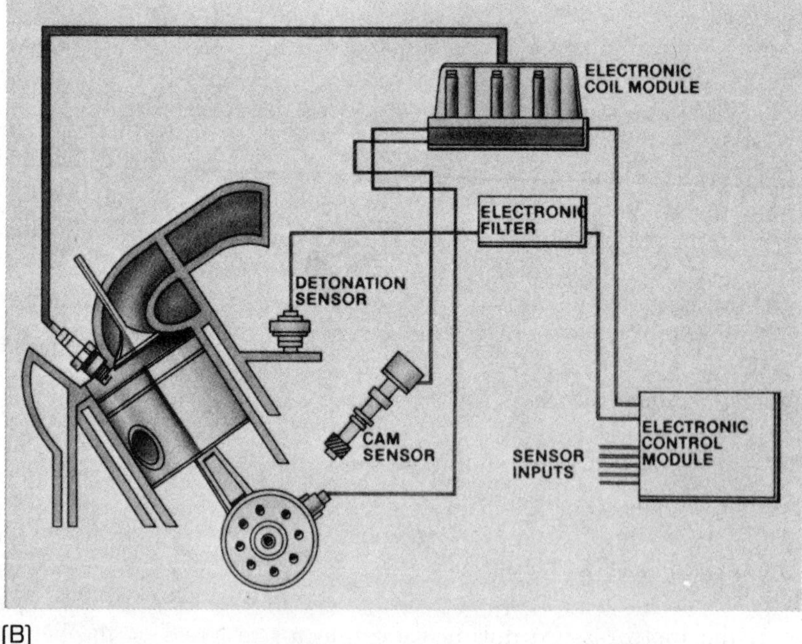

(B)

Figure 7–46. Computer controlled coil ignition (CCCI or distributorless). a. Coil location on engine. (Courtesy of GMC Service Research) b. Arrangement of CCCI system components. (Reprinted with permission © 1984 Society of Automotive Engineers, Inc.)

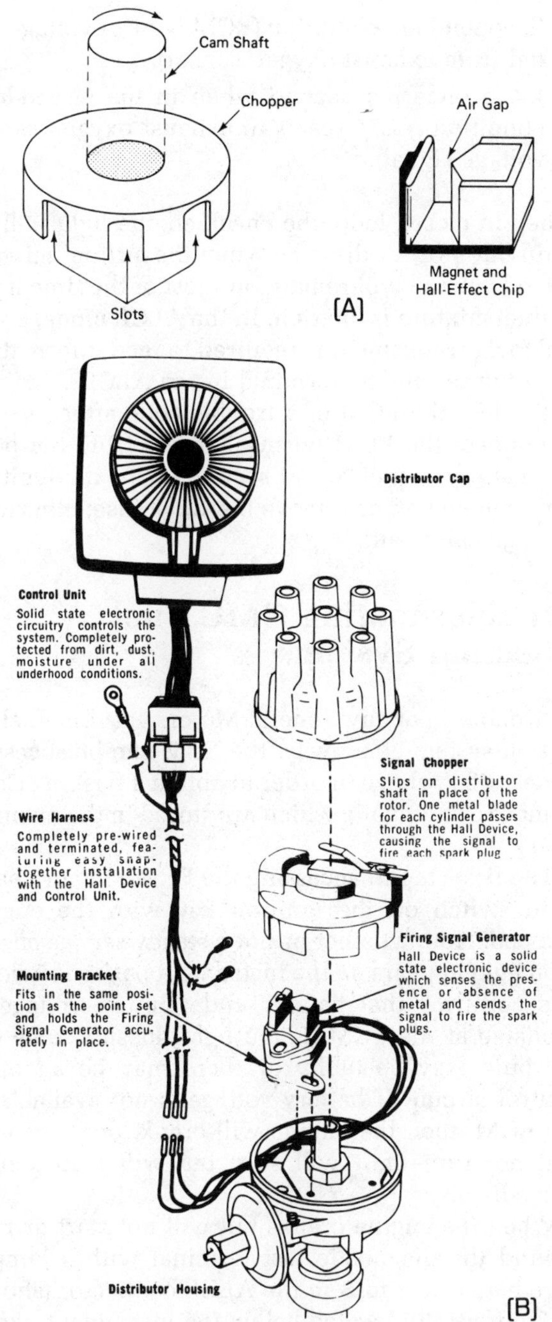

Figure 7–48. a. Hall-effect concentric chopper unit. b. Hall-effect ignition system distributor. (Courtesy of Reston Publishing Co., Inc.)

For detailed information on electrical ignition systems, refer to *Electric and Electronic Systems for Automobiles and Trucks* by Robert N. Brady, published by Reston Publishing Co., Inc., a Prentice-Hall Company, Reston, Va. 22090.

Both the camshaft and crankshaft Hall-effect sensors used with the CCCI system operate on the same principle as shown in Figure 7–48.

A plate with three vanes mounted to the harmonic balancer at the front of the engine allows the vanes to pass through slots in the crankshaft sensor thereby triggering a voltage signal to the ignition module and from there to the ECM. As the engine speed increases and decreases, the voltage signal will vary for automatic control of ignition timing.

Limp-Home Capability. The distributorless ignition system features "limp-home" capability in case the computer malfunctions. No regular engine ignition timing is required with this system, and except for regular spark plug service, no other ignition system maintenance is necessary.

Plans are under way to eliminate two of the existing three ignition coils and develop an electronics package that will be simpler still.

The system employs what GMC calls a "waste spark" method whereby each cylinder is paired with its opposite unit such as 1-4, 5-2, and 3-6 so that the high-tension spark occurs in both the cylinder coming up on exhaust and also coming up on compression. This results in maximum voltage occurring within the compressing cylinder with a small amount being used in the exhausting cylinder. This process will be reversed when the cylinder stroke roles are reversed. (For detailed information on the CCCI system, refer to the Buick Service Manual from GMC, Section 6E—Driveability and Emissions, Fuel Injection (Port).)

General Motors ECM Checkout

All General Motors vehicles are equipped with an ECM (electronic control module) that stores information on problem areas that exist with the engine and transmission. A *check engine light* will illuminate on the instrument panel as a general warning to the operator that a problem has occurred and that the vehicle should be checked out as soon as is reasonably possible. In addition, the service technician can access stored information within the ECM to isolate and repair a problem area.

The ECM receives its information from a number of engine, transmission, and vehicle speed sensors. Since the ECM is basically a miniature computer, it has memory capability and will therefore compare voltage readings from the various sensors. If this voltage reading is not what the memory bank has on file, then the ECM will activate the *check engine light* on the instrument panel and will also store a trouble code in the memory bank.

A device known as a CALPAC is used to allow fuel delivery if other parts of the ECM are damaged. This unit is accessible through an access door in the ECM.

In order to access the memory bank of the vehicle ECM, the service technician uses what is known as the ALCL (assembly line communication link) connector. The ALCL connector is located inside the vehicle, as shown in Figure 7–49, underneath the dash on the driver's side.

The ECM/ALCL operates in either a diagnostic, field service, or ALCL mode.

In the diagnostic mode, the TEST terminal B in Figure 7–49 must be grounded with the ignition key *on* and the engine stopped. The ECM flashes the check engine light along with a Code 12 to show that the system is in fact operational. This Code 12 consists of one flash followed by a short pause and then two quick flashes. The code will repeat three times, and if no other code is stored in the memory bank, Code 12 continues to flash until the TEST terminal is ungrounded. The trouble codes that are used by General Motors and interpreted by a number are shown in the Trouble Code Identification Chart, Figure 7–50.

In the field service mode, if the test terminal of the ALCL is grounded with the engine running, the check engine light will:

1. Flash two and one-half times per second when in the open-loop condition (ECM ignores voltage signal from exhaust oxygen sensor).
2. Flash once per second when in the closed-loop condition (ECM reacts to exhaust oxygen sensor voltage signal).

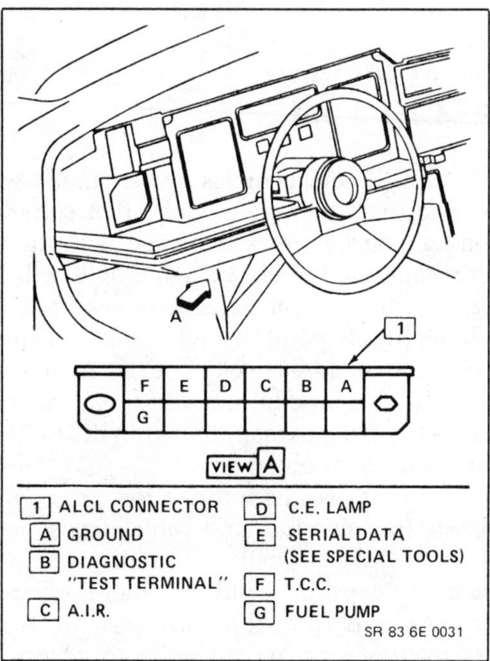

Figure 7–49. Assembly line communication link (ALCL) connector location. (Courtesy of Pontiac Motor Division, GMC)

When in closed loop, the check engine light will remain out most of the time when the air/fuel mixture is too lean and will remain on most of the time if the air/fuel mixture is too rich. In the ALCL mode, a special tool arrangement is required to read information from the serial data terminal in the ALCL.

To clear the ECM of a trouble code after repairs, disconnect the ECM wiring harness from the positive battery pigtail for 10 seconds with the ignition key switched *off* or remove the ECM fuse; otherwise damage can result.

Troubleshooting GMC Fuel Injection Systems

To troubleshoot any General Motors gasoline fuel injection system effectively, the ECM can be accessed as described above in order to obtain a flashing Code number or numbers, which are stored in the memory bank.

The first step in accessing the ECM memory bank is to switch on the ignition key with the engine stopped. Note whether or not a steady *service engine soon* light appears on the instrument panel. If it does, this confirms that battery and ignition voltage is available at the ECM. If the light doesn't come on, the bulb may be faulty or there may be an open control circuit. If battery voltage is not available to the ECM, then the engine will crank, but it usually will not run—or it will start but will stop almost immediately.

When the engine cranks but will not start or run, ground the diagnostic test terminal with a jumper wire between A to B in the ALCL connector (shown in Figure 7–49) located below the instrument panel. The ECM will immediately cause the *service engine soon* light to flash a Code 12 three times, which will be followed by any other trouble codes stored in the memory bank. Once any additional codes have each flashed three times, the system repeats itself starting with Code 12 again. When no other codes are stored in the memory bank of the ECM, Code 12 will continue to flash until the jumper wire is disconnected from the ALCL.

For codes other than Code 12 that appear, refer to Figure 7–50, Trouble code identification chart and proceed to check and correct the possible problem area.

TROUBLE CODE IDENTIFICATION

The "CHECK ENGINE" light will only be "ON" if the malfunction exists under the conditions listed below. It takes up to five seconds minimum for the light to come on when a problem occurs. If the malfunction clears, the light will go out and a trouble code will be set in the ECM. Code 12 does not store in memory. If the light comes "on" intermittently, but no code is stored, go to the "Driver Comments" section. Any codes stored will be erased if no problem reoccurs within 50 engine starts. A specific engine may not use all available codes.

The trouble codes indicate problems as follows:

TROUBLE CODE 12	No distributor reference pulses to the ECM. This code is not stored in memory and will only flash while the fault is present. Normal code with ignition "on," engine not running.
TROUBLE CODE 13	Oxygen Sensor Circuit — The engine must run up to four minutes at part throttle, under road load, before this code will set.
TROUBLE CODE 14	Shorted coolant sensor circuit — The engine must run two minutes before this code will set.
TROUBLE CODE 15	Open coolant sensor circuit — The engine must run five minutes before this code will set.
TROUBLE CODE 21	Throttle Position Sensor (TPS) circuit voltage high (open circuit or misadjusted TPS). The engine must run 10 seconds, at specified curb idle speed, before this code will set.
TROUBLE CODE 22	Throttle Position Sensor (TPS) circuit voltage low (grounded circuit or misadjusted TPS). Engine must run 20 seconds at specified curb idle speed, to set code.
TROUBLE CODE 23	M/C solenoid circuit open or grounded.
TROUBLE CODE 24	Vehicle speed sensor (VSS) circuit — The vehicle must operate up to two minutes, at road speed, before this code will set.
TROUBLE CODE 32	Barometric pressure sensor (BARO) circuit low.
TROUBLE CODE 34	Vacuum sensor or Manifold Absolute Pressure (MAP) circuit — The engine must run up to two minutes, at specified curb idle, before this code will set.
TROUBLE CODE 35	Idle speed control (ISC) switch circuit shorted. (Up to 70% TPS for over 5 seconds.)
TROUBLE CODE 41	No distributor reference pulses to the ECM at specified engine vacuum. This code will store in memory.
TROUBLE CODE 42	Electronic spark timing (EST) bypass circuit or EST circuit grounded or open.
TROUBLE CODE 43	Electronic Spark Control (ESC) retard signal for too long a time; causes retard in EST signal.
TROUBLE CODE 44	Lean exhaust indication — The engine must run two minutes, in closed loop and at part throttle, before this code will set.
TROUBLE CODE 45	Rich exhaust indication — The engine must run two minutes, in closed loop and at part throttle, before this code will set.
TROUBLE CODE 51	Faulty or improperly installed calibration unit (PROM). It takes up to 30 seconds before this code will set.
TROUBLE CODE 53	Exhaust Gas Recirculation (EGR) valve vacuum sensor has seen improper EGR vacuum.
TROUBLE CODE 54	Shorted M/C solenoid circuit and/or faulty ECM.
TROUBLE CODE 55	Grounded Vref (terminal "21"), high voltage on oxygen sensor circuit or ECM.

5-3-83Z
SR 84 6E 0340

Figure 7-50. Engine trouble code identification chart. [Courtesy of GMC Service Research]

Engine Cranks But Will Not Run

If the engine cranks but does not run, start by checking the conditions stated above under accessing the ECM memory bank.

Start by checking if fuel is available at the injectors. If fuel is available at the injector(s), it is possible that the engine has been flooded. To correct this condition, push the accelerator pedal all the way down, which allows the ECM to pulse the fuel injector(s) at a lean air/fuel ratio condition of approximately 20:1 for as long as the pedal remains down and the engine speed is below 600 rpm. When the throttle pedal is released to a position less than 80%, the ECM returns to the starting mode.

No fuel spray from the injector is indicative of a faulty injector or no ECM control of the injector(s). To check the injector, disconnect the wire harness at the injector(s), and crank the engine. If any fuel appears or drips at the injector tip, replace it. If fuel is available at each injector when its wiring harness is connected, the fuel control system is operating correctly.

Check the ignition system for a suitable spark at the spark plugs using an ST-125 spark plug gap tool. No spark would indicate a HEI (high energy ignition) problem. If a good spark is available at the plugs, then check the following.

Procedure

1. Check the throttle position sensor to ensure that it is not sticking or binding in a wide open throttle position because this will cause a voltage signal to the ECM. The ECM will place it into the "clear flood" condition and will cause an air/fuel ratio of between 18 and 20:1, which in a cold engine cranking situation can be too lean to effectively start a cold engine.

   NOTE: Remember that fuel-injected engines do not employ a choke as a carburetor system does; therefore, the TPS becomes important to the initial start-up routine.

2. A faulty coolant sensor especially in an open-loop condition with the ignition off will cause a voltage signal to the ECM that duplicates a cold engine condition (−40°F). The ECM will send out a signal to the fuel injector(s) that will pulse them open for longer periods leading to a very rich air/fuel ratio condition and possibly causing flooding of the engine. On a warm engine, very hard starting will result. The ECM cannot recognize an open coolant circuit condition until the engine has been running for at least 1 minute or longer.
3. In very cold ambient temperatures, do not rule out the possibility of gas line freeze-up. Water in the fuel can cause a no-start condition when it freezes. If the engine starts in warmer weather but experiences hard starting in cold weather, this may be the cause.
4. Check the EGR circuit for a stuck-open condition because this can also cause a high air/fuel ratio during cranking.
5. An open crank signal can also cause a no-start condition in very cold weather.
6. Check each injector circuit on CKT 439 for 12 volts. There should be a test light illumination on only one injector connector terminal. Reconnect the injector connector and using a 12-volt test light that is attached to a good ground, check for a light when you probe the white ECM connector terminal because the injector control circuit may be open.
7. No blinking light indicates that the ECM has no control to the injector(s). Using a voltmeter in the AC position and the scale at the 2-volt position, voltage should be higher than 0.7 volt; otherwise there is an open or a short-to-ground in the HEI reference circuit number 430.

With the engine running and the diagnostic terminal grounded, the ECM will respond to the oxygen sensor voltage signal and will use the *service engine soon* light to indicate this information in one of four ways:

1. A closed-loop condition (normal condition) indicates that the ECM is in fact receiving and using this signal to control the air/fuel ratio to the engine.
2. An open-loop condition that occurs on initial start-up and remains in this condition until the oxygen sensor attains a temperature of 360°C or 680°F. In the open-loop condition, the ECM ignores the voltage signal from the oxygen sensor. This open-loop condition usually exists for between 30 seconds to 2 minutes, depending on the engine temperature. If the system fails to enter the closed-loop cycle after this time, refer to the Trouble code identification chart, Figure 7–50.
3. *Service engine soon* light out indicates that the exhaust is in a lean condition.
4. *Service engine soon* light on indicates a rich exhaust gas flow condition.

NOTE: The *service engine soon* light may stay on too long, especially during steady state heavy acceleration modes. The light may be off too long due to long periods of deceleration, which will produce a lean mixture or fuel cutoff situation. Also the light may stay on too long with an engine idle speed less than 1200 rpm because of the oxygen sensor cooling down. Running the engine at increased speeds for several minutes can trigger the system back into a closed-loop cycle.

Cadillac EFI System

In the 1977–79 Cadillac Seville models as standard equipment and on 1977–79 full-size Cadillacs as an option, General Motors offered an EFI (electronic fuel injection) system. A microprocessor DFI (digital fuel injection) system was introduced in 1980 on Seville models. This DFI system is now standard on all 1981 and later 6.0L (368 cu.in.) V8 engines, and 1982 and later 4.1L (250 cu.in.) engines.

The DFI system is the same basic system now used on all similarly equipped TBI (throttle body injection) General Motors products. Since the TBI system is explained in detail in other areas of this book, this section deals with only the EFI system.

Cadillac first used this system (as mentioned) in 1977; since that time, this system has largely been displaced by the TBI system on most of General Motors' vehicles produced by each Division. However, the 1984 3.8L (232 cu.in.) 90 degree V6 available in the Buick Century and Oldsmobile Cutlass Ciera with MFI (multi-port fuel injection) using Robert Bosch fuel injectors is similar to the 1977 Cadillac EFI system in that each cylinder receives its fuel supply via an individual injector.

The 1984 3.8L (232 cu.in.) V6 on the Buick Regal T Type Turbo, Riviera T Type Turbo, and (optional on) the Riviera convertible uses SFI (sequential port fuel injection) as standard equipment. SFI is also of the individual cylinder injector type. Both of these systems, as well as the Cadillac EFI system, employ fuel injectors that are located in the intake manifold. The injectors are placed so that they spray fuel towards the intake valve of each cylinder.

The 1984 MFI and SFI systems make greater use of electronic sensors and use a different ECM than did the 1977 Cadillac unit.

EFI System Description

The EFI system is made up of four basic systems:

1. The fuel delivery system
2. The air induction system
3. Engine sensors
4. Vehicle electronic control unit (ECM)

An illustration of the EFI system used by Cadillac on these models is shown in Figure 7–51. This system was designed by Bendix and operates similarly to the Robert Bosch L-Jetronic fuel injection system in that the individual injectors (Figure 7–52) are pulsed open and closed by an electrical signal from an ECM (electronic control module).

The ECM receives a signal from an intake manifold pressure sensor as well as engine speed and equates and delivers a corrective signal to the individual injector units. The injectors are split into two separate groups with Cadillac employing cylinder injectors 1-2-7 and 8 into one group, while injectors for cylinders 3-4-5 and 6 form the other group.

This arrangement allows each group of four injectors to be pulsed on and off simultaneously, similar to the Robert Bosch L-Jetronic control system. This setup thereby produces four injectors opening and closing at the same time once every engine camshaft revolution.

Cadillac models that used the EFI system employed two electrically operated fuel pumps wired in parallel with the ECM (electronic control module). One of these pumps was an in-tank unit that functioned to deliver fuel to the system main pump, usually located outside of the fuel tank and chassis mounted. Therefore, the in-tank pump is generally referred to as a "boost pump" that will prevent any possibility of a vapor lock occurring on the suction side of the main pump. The electric fuel pumps are energized for a maximum of about 1 second when the ignition key is turned on to provide sufficient fuel pressure to the system. Otherwise, the electric fuel pumps will not operate unless the ignition switch is in the cranking position or the engine is running.

The fuel pressure regulator used with the EFI system is shown in Figure 7–53. Note that the fuel pressure regulator is designed so that the air and fuel chambers are separated by a spring loaded diaphragm.

A hose connected from the fuel pressure regulator to the intake manifold ensures that the pressure within the air chamber of the regulator will always be the same as that in the intake manifold.

In addition, fuel pressure acts against the diaphragm and spring as illustrated. Throttle position will change the vacuum within the intake manifold; therefore, this pressure change will affect the position of the diaphragm valve within the regulator, causing a small pressure change.

If the throttle is closed (high vacuum), the diaphragm will move to the left in Figure 7–53 because of fuel pressure on the one side with intake manifold vacuum on the other to move the valve away from

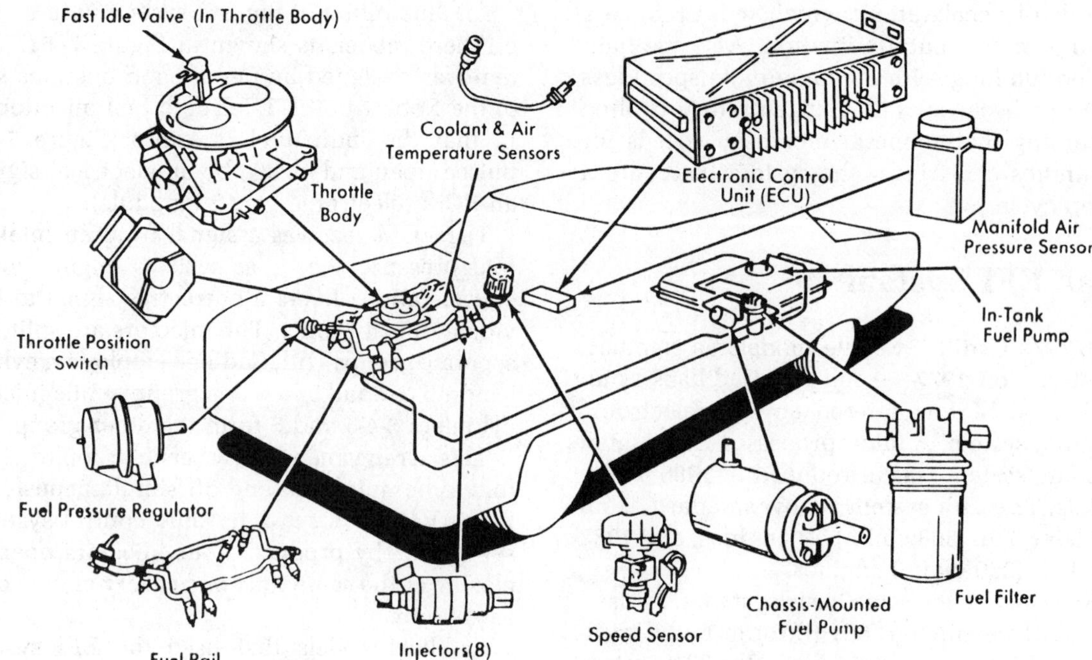

Figure 7-51. Cadillac/Bendix EFI [electronic fuel injection] system arrangement. [Courtesy of General Motors Corporation]

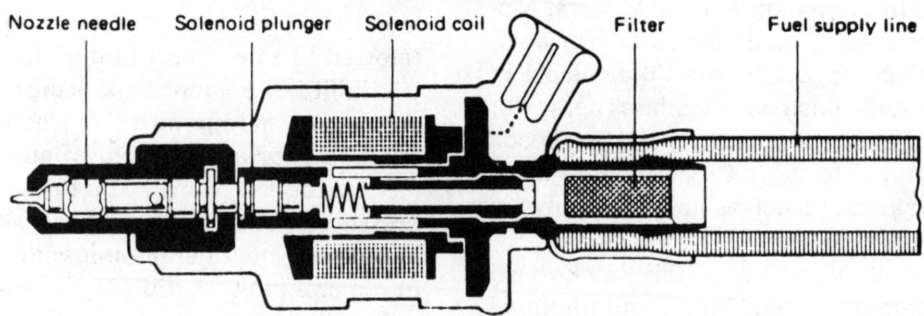

Figure 7-52. The Cadillac/Bendix EFI fuel injector. [Courtesy of General Motors Corporation]

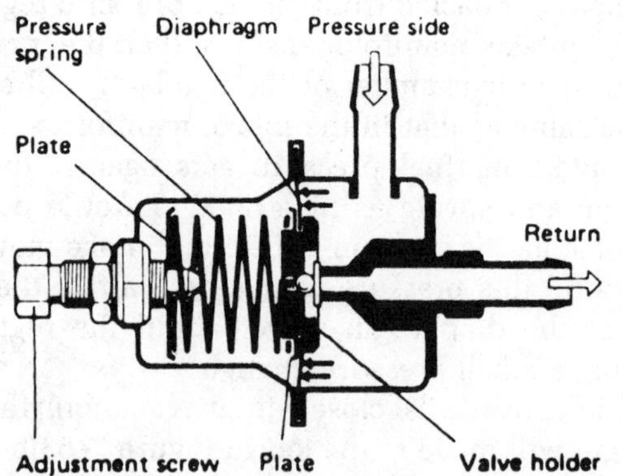

Figure 7-53. The Cadillac/Bendix EFI fuel pressure regulator. [Courtesy of General Motors Corporation]

the orifice. On the other hand, if the throttle is open (low vacuum), the diaphragm spring will overcome the fuel pressure and move the valve to the right against the orifice. Therefore, due to changing throttle positions, the valve orifice will be continually opened and closed. This action ensures that a fuel pressure between 38–40 psi (262–276 kPa) is maintained to the fuel injectors at all times.

The return fuel line carries excess fuel back to the fuel tank. The throttle body used with these systems is basically similar to that used with TBI systems, but it doesn't have any injectors in the body. The throttle body on the EFI system controls air flow through two bores, or throats, with a throttle valve in each one that is controlled by direct linkage to the accelerator pedal. Figure 7–54 illustrates the throttle body system used with this design. Note that an idle bypass passage adjustment screw is supplied to provide suitable idle speed control.

When the engine is cold, a signal from the ECM activates the fast idle valve (FIV) on top of the throttle body to allow additional air to flow into the engine in order to compensate for the fuel delivery from the injectors. This arrangement is not unlike that used with the TBI fuel systems in that a tapered pintle valve receives a signal from the ECM to perform this same basic function. In the EFI system, a thermostatic element allows extra air passage into the engine by keeping the plunger valve open. As the engine heats up, this valve is closed off by the expansion of the element.

Engine Sensors

Sensors used in the engine to feed signals to the ECM are:

1. The engine speed sensor
2. The throttle position switch (TPS)
3. The intake manifold absolute pressure sensor (MAP)
4. The use of two temperature sensors (one for coolant and one for air)

All of the sensors are designed to produce an electrical signal that is fed back to the ECM.

Both the air and coolant sensors, for example, are of the same basic design and are in fact interchangeable. The electrical resistance through these sensor wires will change in relation to the change in temperature so that a low temperature will produce a low resistance, while a high temperature will produce a high resistance. Depending upon the resistance created in the sensor wire, a varying voltage signal will be sent to the ECM.

The MAP sensor operates by sensing any change in intake manifold depression (vacuum) related to throttle position and either barometric pressure or altitude. Electrical signals therefore produced in the MAP sensor are sent to the ECM.

The throttle position switch connected to the throttle simply senses throttle position as the shaft is rotated (accelerator pedal position) and feeds a varying voltage signal to the ECM.

The engine speed sensor is commonly part of the ignition distributor unit. It is basically a trigger switch unit that functions to do the following:

1. Direct a signal to the ECU in order to time the opening/closing of an injector group in relation to intake valve timing.
2. Signal engine speed (rpm) to the ECU.

The trigger switch encompasses two small reed switches and a set of magnets with a rotor turning at distributor shaft speed. This combination causes the reed switches to continually open/close and therefore create a continuously varying voltage signal to the ECM.

A graphic representation of the ECU and the signals that it receives are illustrated in Figure 7–55. The ECU is calibrated for a particular vehicle model and cannot be interchanged between various Division models at random.

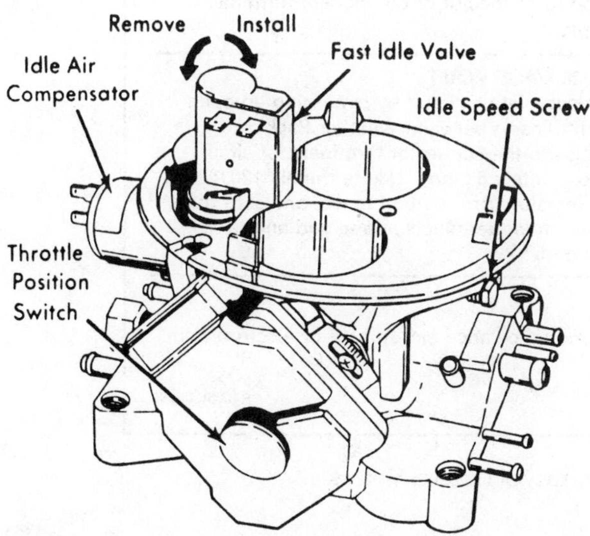

Figure 7–54. Cadillac/Bendix EFI throttle body arrangement. (Courtesy of General Motors Corporation)

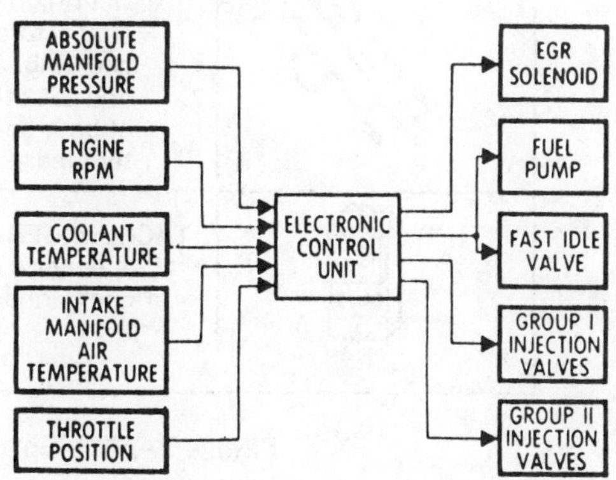

Figure 7–55. Operational block diagram of ECU (electronic control unit). (Courtesy of General Motors Corporation)

Special Tools

These tools are shown in Figure 7–56.

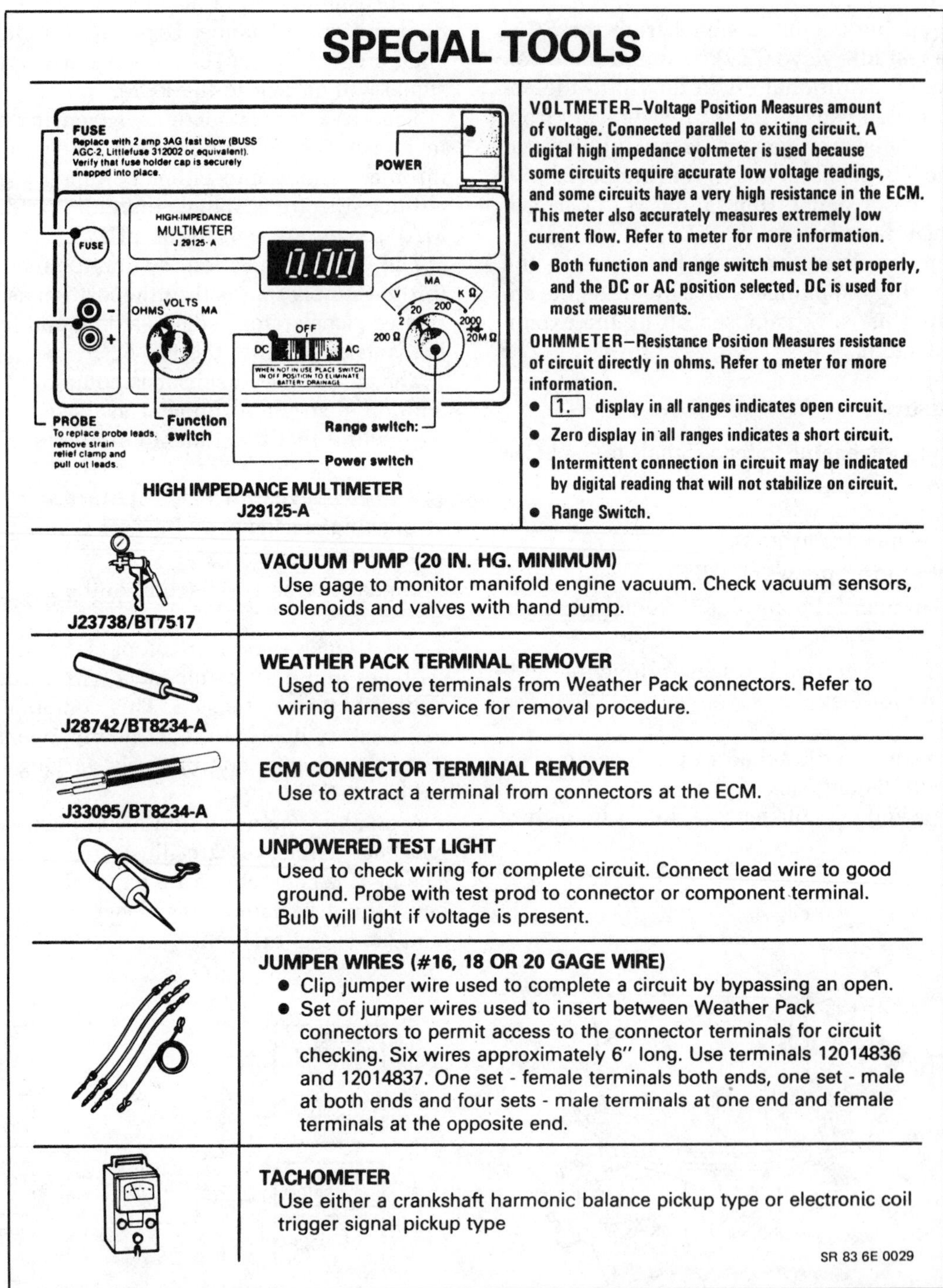

Figure 7–56. Special tools. [Courtesy of General Motors Corporation]

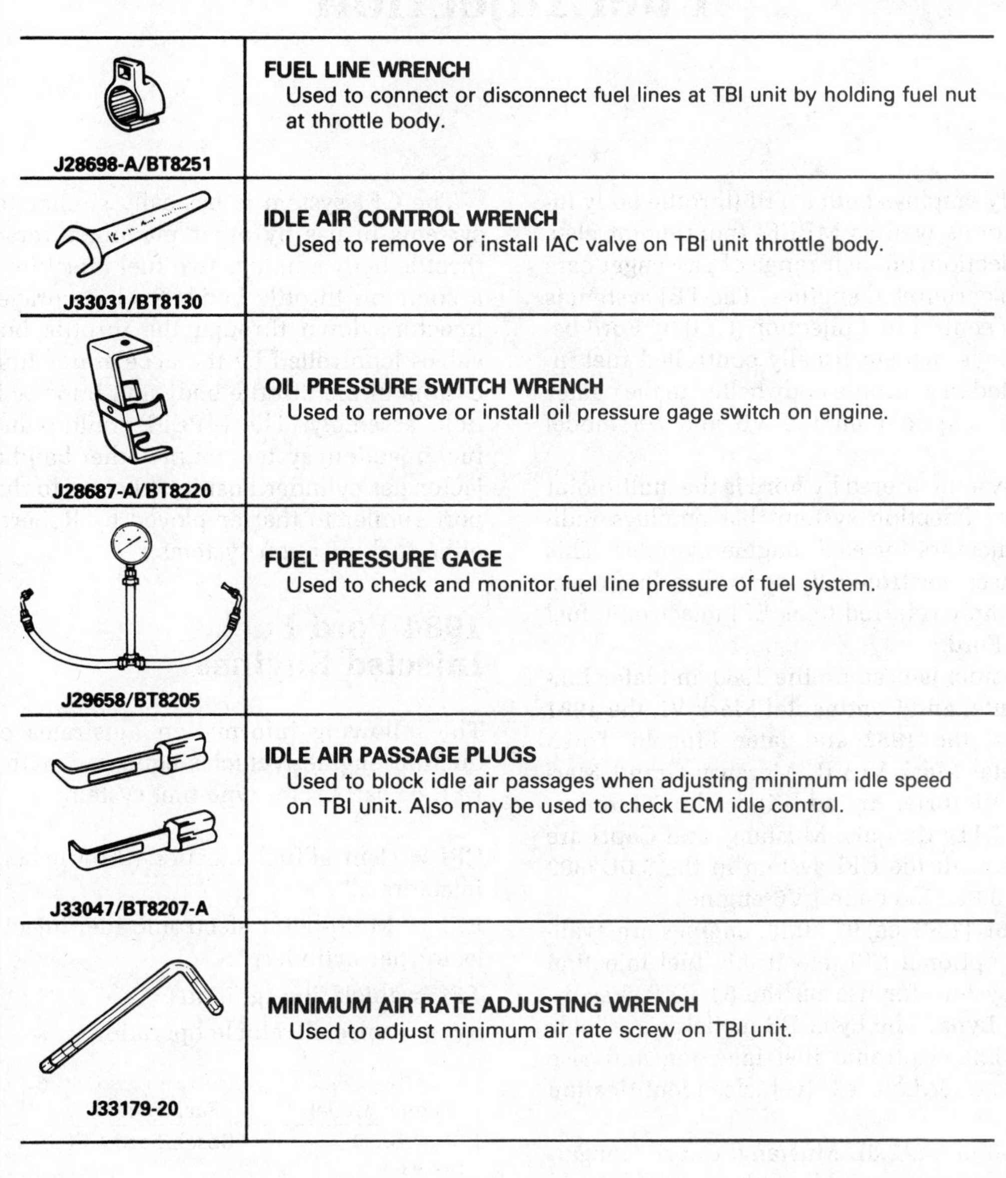

SPECIAL TOOLS
J28698-A/BT8251
FUEL LINE WRENCH
Used to connect or disconnect fuel lines at TBI unit by holding fuel nut at throttle body.
J33031/BT8130
IDLE AIR CONTROL WRENCH
Used to remove or install IAC valve on TBI unit throttle body.
J28687-A/BT8220
OIL PRESSURE SWITCH WRENCH
Used to remove or install oil pressure gage switch on engine.
J29658/BT8205
FUEL PRESSURE GAGE
Used to check and monitor fuel line pressure of fuel system.
J33047/BT8207-A
IDLE AIR PASSAGE PLUGS
Used to block idle air passages when adjusting minimum idle speed on TBI unit. Also may be used to check ECM idle control.
J33179-20
MINIMUM AIR RATE ADJUSTING WRENCH
Used to adjust minimum air rate screw on TBI unit.
SR 83 6E 0030

Figure 7-56. [*Continued*]

8

Ford Electronic Gasoline Fuel Injection

Ford currently employs both a TBI (throttle body injection) system as well as MPEFI (multi-point electronic fuel injection) on their range of passenger cars in addition to carbureted engines. The TBI system is referred to as central fuel injection (CFI) by Ford because it employs two electrically controlled fuel injectors mounted in a throttle body bolted to the center of the intake manifold on the V6 and V8 model engines.

The other system offered by Ford is the multi-point electronic fuel injection system that employs individual fuel injectors for each engine cylinder. This system also uses electronically controlled fuel injectors and is simply referred to as EFI (electronic fuel injection) by Ford.

The CFI system is used on the 1980 and later Lincoln-Continental and Continental Mark VI, the 1981 Ford/Mercury, the 1982 and later Lincoln Town Car/Continental Mark VI/VII, Mercury Grand Marquis/Crown Victoria. In addition, the Thunderbird/Cougar, LTD/Marquis, Mustang, and Capri are also available with the CFI system in the 5.0L (302 cu.in.) V8 or 3.8L (230 cu.in.) V6 engines.

1983–84 1.6L (1597 cc) 97 cu.in. engines are available with an optional EFI (electronic fuel injection multi-point) system for use on the 84 EXP, Escort, and Mercury Lynx. The Lynx RS model is available with multi-point electronic fuel injection and also a turbocharger. V.I.N. #2 (vehicle identification number).

1983–84 140 cu.in. 2.3L Mustang, Capri, Cougar, and Thunderbird are available with an optional turbocharger or a 5.0L (302 cu.in.) V8 engine. The 2.3L engine is equipped with multi-point EFI (individual injectors per cylinder). The 5.0L V8 is equipped with central point or CFI.

These injection systems are a Ford/Robert Bosch design derived from the Bosch L-Jetronic fuel system with Ford modifications.

The CFI system is basically similar to other TBI systems in use by other manufacturers in that the throttle body employs two fuel injectors mounted in a common throttle body. Fuel is sprayed from the injectors down through the throttle butterfly type valves (controlled by the accelerator linkage) at the bottom of the throttle body and into the intake manifold assembly. The MPEFI (multi-point electronic fuel injection) system on the other hand uses one injector per cylinder positioned close to the inlet valve port similar to that employed by Robert Bosch gasoline fuel injection systems.

1984 Ford Fuel Injected Engines

The following information illustrates current production engines/vehicles equipped with either CFI, EFI, or carburetor-type fuel systems.

CFI = Central fuel injection (throttle body type) two injectors

EFI = Multi-point electronic fuel injection (one injector per cylinder)

TFI = thick film ignition

SVO = special vehicle operation

Engine Model	*Fuel System*	*Ignition*
1.6L U.S. 50 States	Bosch/Ford EFI	TFI-1V
1.6L Turbo 49 States (Manual Transmission) SVO, EXP, Escort/ Lynx 84-1/4	Bosch/Ford EFI	TFI-1V
2.3L Turbo SVO 50 States A/T and M/T	Bosch/Ford EFI	TFI-1V

Engine Model	*Fuel System*	*Ignition*
Capri, Thunderbird/Cougar, Mustang		
3.8L (230 cu.in.) V6	Ford CFI	TFI-1V
50 States, Thunderbird/Cougar		
LTD/Marquis, Mustang/Capri		
5.0L (302 cu.in.)	Ford CFI	TFI-1V
Mustang/Capri AT/AOD H.O.		
LTD AT/AOD H.O.		
50 States Lincoln/Mark VII, Crown Victoria/Grand Marquis, Continental, Thunderbird/Cougar		

NOTE: Although a number of Ford vehicles are equipped with either CFI or EFI fuel systems, some of these engines are also available as carbureted units as follows

Vehicles with Carburetors
1.6L 50 States Escort/Lynx, EXP - Holley 740-2V Carburetor
2.3L 50 States Mustang/Capri, LTD/Marquis, Carter YFA-1V, FB Carburetor
2.3L HSC 50 States, Holley 6149-1V, FB Carburetor
2.3L HSC Tempo/Topaz, Holley 1949-1V Carburetor
3.8L (230 cu.in.) V6 Canada, Mustang/Capri, Thunderbird/Cougar, LTD/Marquis, Ford 2150-2V Carburetor
5.0L (302 cu.in.) V8, Mustang/Capri H.O. T50D M/T, Holley 4180C-4V Carburetor
5.0L Canada, Ford Mercury, Ford 2150-2V Carburetor
5.8L (351 cu.in.) H.O. 50 States/Canada Ford Mercury Police Only, Ford 7200-VV

All 1984 Ford Trucks use either Holley, Carter, or Ford manufactured carbs.

Central Fuel Injection (CFI) Systems

Figure 8–1(a) and (b) illustrate the Ford CFI system components.

CFI System Control Mechanism

As with other similar systems now in use, the CFI system injectors are controlled by an EEC (electronic engine control) system that receives voltage inputs from eight sensors. These sensors are:

1. An inlet manifold pressure sensor.
2. A barometric sensor mounted on the firewall.
3. An engine coolant temperature sensor threaded into the cooling system.
4. An inlet air temperature sensor threaded into the intake manifold.
5. A crankshaft position sensor located in the timing chain cover.
6. A throttle position sensor mounted on the side of the throttle body housing.
7. An EGR valve position sensor located at the base of the throttle body housing.
8. An exhaust gas oxygen sensor threaded into the exhaust manifold.

The basic identification and location of these individual sensors are shown in Figures 8–2 through 8–4 for Ford's 3.8L V6 (230 cu.in.) engine and the 5.0L (302 cu.in.) V8 engine.

All sensors operate on the principle of resistance through a wire. As the temperature increases, this resistance changes, and therefore a voltage signal from each sensor is constantly being monitored by the EEC unit. The EEC unit then sends out a signal to the CFI unit injectors to pulse them open or closed. The length of time that the injectors remain open establishes how much fuel is injected. With the signal from the sensors constantly being monitored, the engine air/fuel ratio is maintained as close to stoichiometric as possible under all operating conditions.

The Ford CFI system consists of six basic systems that are designed to maintain the air/fuel mixture as close to stoichiometric as possible in order that the exhaust emissions level, the fuel economy, and engine performance comply with the EPA regulations.

The six basic systems are:

1. The air control—two butterfly valves in the throttle body.
2. The fuel injector nozzles—one per bore in the top of the TB unit.
3. The fuel pressure regulator—mounted on the throttle body; maintains pressure to the injectors at 39 psi (269 kPa) and also keeps pressure in the system when the engine is shut off.
4. The fuel pressure diagnostic valve—located at the top of the throttle body and fitted with a Schrader-type valve; functions to allow bleeding air from the system as well as allowing a pressure tap point to monitor system pressure or to bleed down (relieve) fuel pressure before performing a service/repair to the system.
5. The cold engine speed control—attached to the throttle body unit; functions to raise engine idle speed when cold; a signal from the EEC unit kicks down the fast idle cam as the engine warms up.

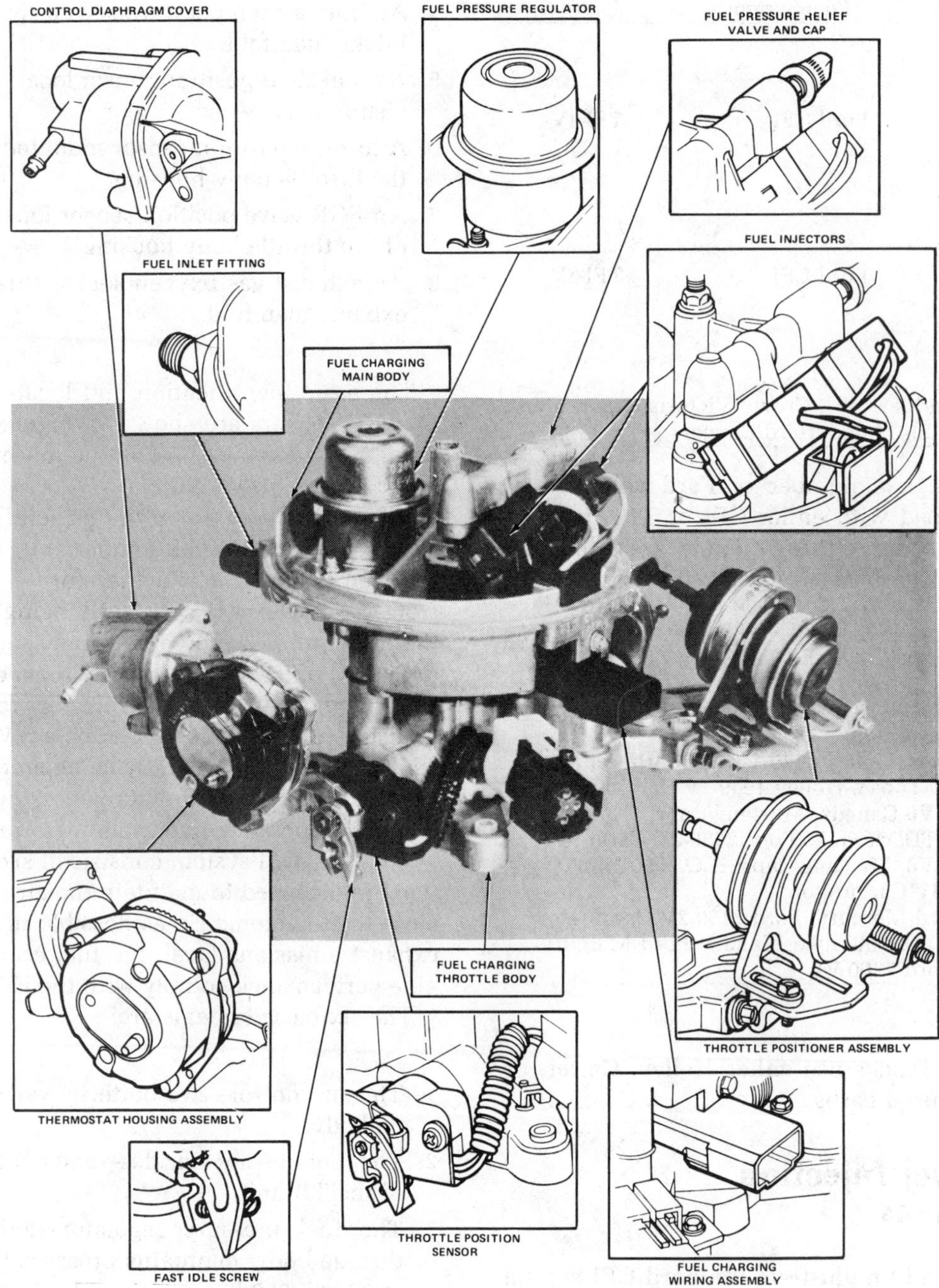

Figure 8-1(a). Central fuel injection (CFI) fuel charging assembly. 5.0L engine, right front view, (Courtesy of Ford Motor Company) Continued.

6. The throttle position sensor—attached to the throttle body in order to monitor the position of the throttle butterfly plates; this sensor sends a voltage signal to the EEC unit, which then controls how long the injectors will remain energized (open); EEC unit will also control spark timing and EGR action.

Open-Loop Versus Closed-Loop Operation

The fuel system can be operated in either an open- or a closed-loop condition similar to other current fuel injection systems and electronically controlled carburetor systems. In the open-loop mode, the engine air/fuel mixture is not controlled by the vehicle

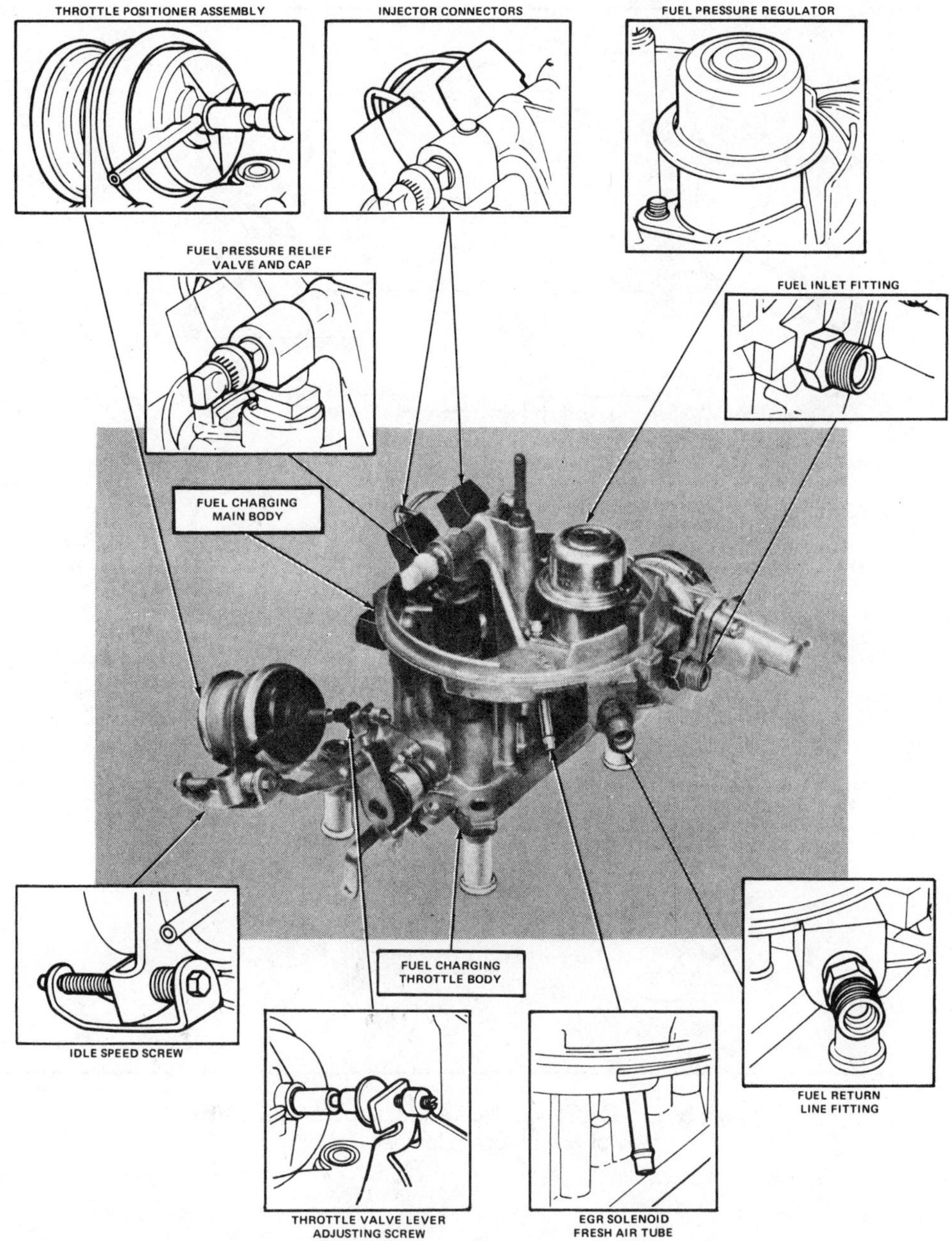

Figure 8–1(b). Central fuel injection (CFI) fuel charging assembly. 5.0L engine, left rear view. (Courtesy of Ford Motor Company) Continued.

engine EEC/MCU (electronic engine control/microprocessor control unit). The air/fuel ratio is controlled in the same manner as it would be in an engine that doesn't use sensor devices.

In the closed-loop mode, the engine air/fuel mixture is controlled by the EEC/MCU, which receives voltage signals from all the engine sensor units as well as the exhaust gas oxygen sensor (Lambda-type unit).

Electric Fuel Pump

An electrically operated in-tank fuel pump is used to force fuel through a fuel filter on its way up to the throttle body unit. At the TB unit, a regulator assembly maintains a fuel pressure of approximately 39 psi (269 kPa). Excess fuel not used by the injectors is rerouted back to the fuel tank.

The electric in-tank fuel pump used with the Lin-

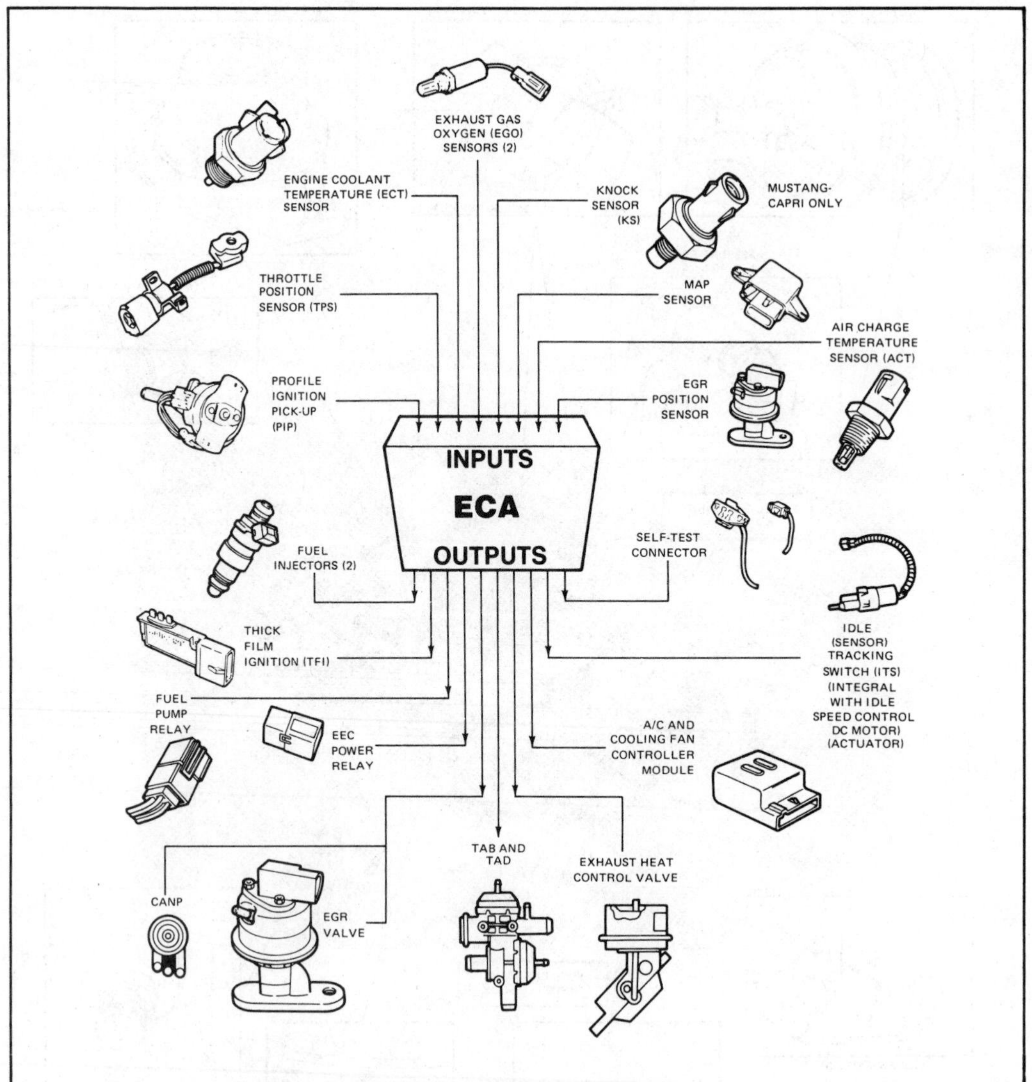

Figure 8–2. CFI Sensor inputs and outputs—3.8L V6 engine. (Courtesy of Ford Motor Company)

coln Town Car, Ford Crown Victoria, and Mercury Marquis is a high-pressure pump capable of delivering 100.0L (26.4 US gallons or 22 Imperial gallons) of fuel per hour at a steady pressure of 39 psi (269 kPa). An internal safety relief valve is used with this pump in case of a plugged filter, restriction, etc., that will bypass fuel when 120 psi (827 kPa) is reached. Figure 8–5 illustrates the in-tank electric fuel pump used with these vehicle models.

Electrical power to the in-tank fuel pump is controlled through three relays: the starter, fuel pump, and EEC (electronic engine control) power relays. As soon as the ignition key is turned on, the electric pump is energized through the EEC relay and its timing device. This allows reduced voltage to the pump through a resistance wire and the fuel pump relay. The EEC timing device will de-energize power to the fuel pump after 1 second if the ignition key is not turned to the crank position.

During engine cranking, the starter motor relay provides full power to both the starter circuit and the electric fuel pump, bypassing the resistance wire and the fuel pump relay. Once the engine starts, power to the fuel pump is once again supplied through the resistance wire and fuel pump relay. When engine speed drops below 120 rpm or is turned off, the EEC module, which is monitoring engine speed through a sensor pickup at the crankshaft pulley, shuts off the fuel pump by opening the ground circuit to the fuel pump relay.

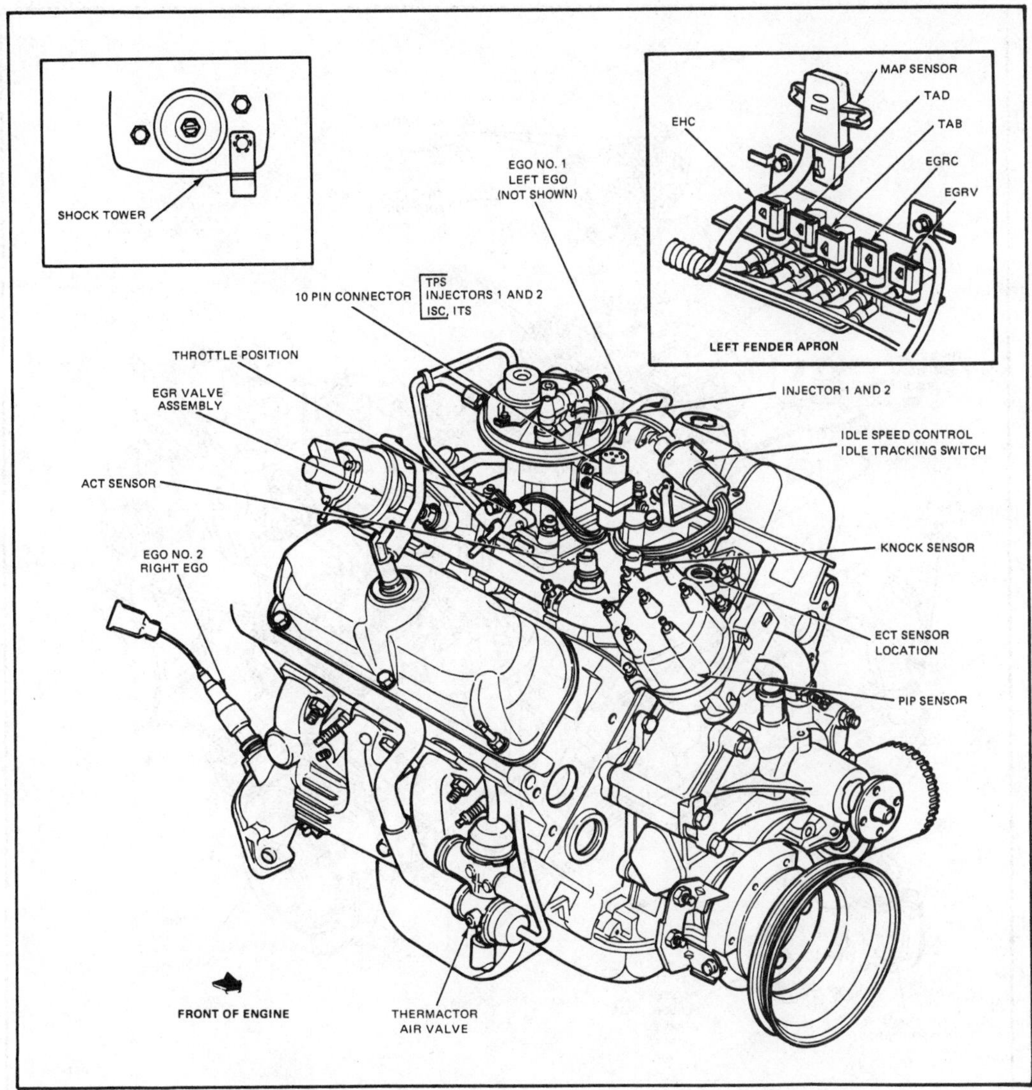

Figure 8–3. Sensor locations on 3.8L V6 engine equipped with CFI. (Courtesy of Ford Motor Company)

The electrical wiring circuit for the electric fuel pumps used with the different models of cars is shown in Figures 8–6, 8–7, and 8–8.

A different arrangement of electric fuel pump from that described above is used on the Mark VII/Continental, Thunderbird/Cougar, LTD/Marquis, Mustang, and Capri. This electric fuel pump is a low pressure in-tank unit used in conjunction with an externally mounted, high-pressure in-line electric fuel pump. The in-tank pump supplies fuel under low pressure to the inlet side of the high pressure unit. A nylon filter is used on the inlet side of the low pressure in-tank pump to prevent dirt from entering the system.

An external resistor is used to maintain a maximum of 11 volts to the in-tank low-pressure fuel pump. Delivery capability of the externally mounted high-pressure pump is 100L (26.4 US gallons or 22 Imperial gallons) per hour on Mustang/Capri turbocharged vehicles, and 60L (15.9 US gallons or 13.2 Imperial gallons) per hour on naturally aspirated vehicles at a delivery pressure of 39 psi (269 kPa). An internal pressure relief valve in the pump opens at 138 psi (950 kPa) if the filter becomes plugged or a restriction occurs. The electrical control system for the low- and high-pressure pump combination is similar to that explained for the in-tank high-pressure circuit.

Figures 8–9 and 8–10 illustrate the low pressure fuel pump arrangement used on the Mustang/Capri and Thunderbird/Cougar vehicles respectively.

An example of the externally mounted in-line high

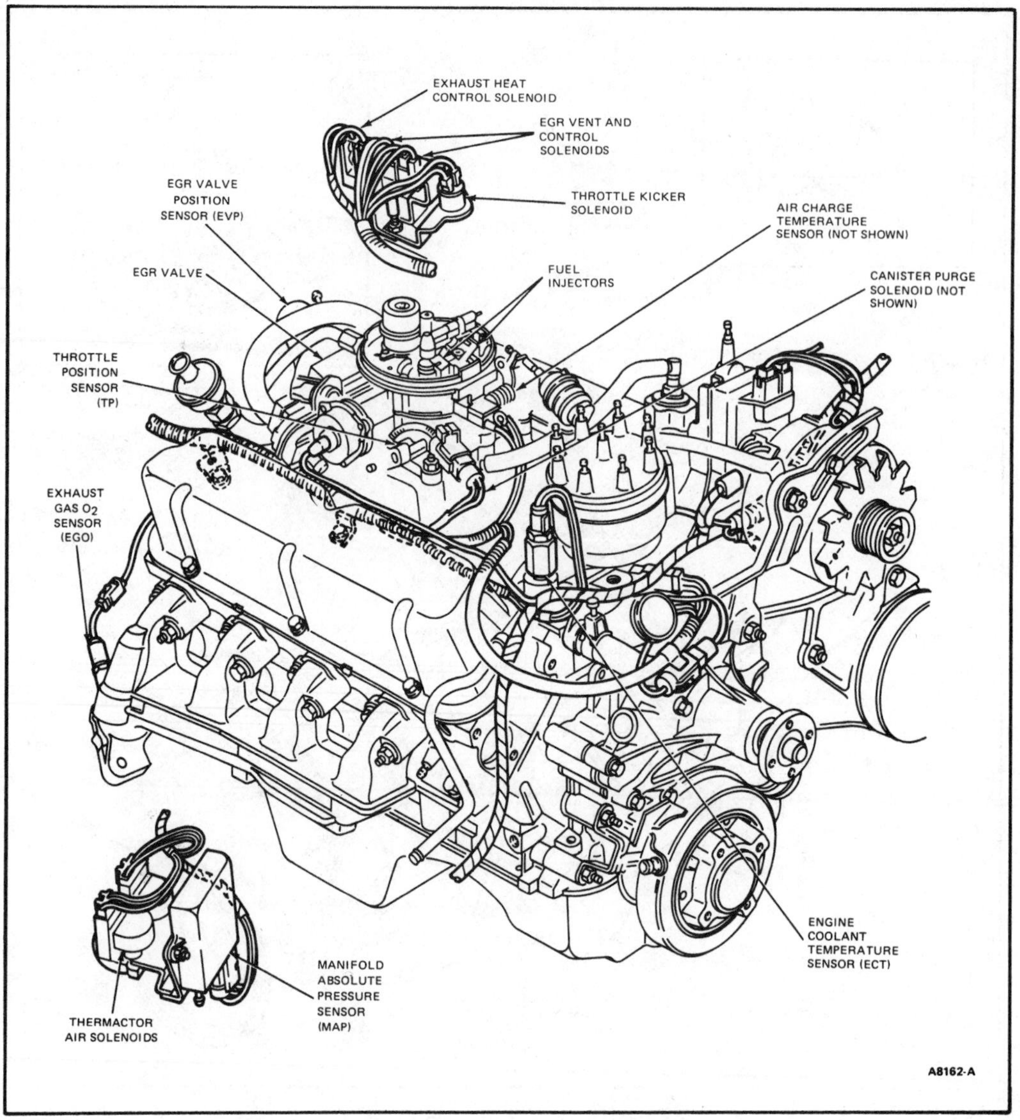

Figure 8–4. Sensor locations on 5.0L V8 engine fitted with CFI. (Courtesy of Ford Motor Company)

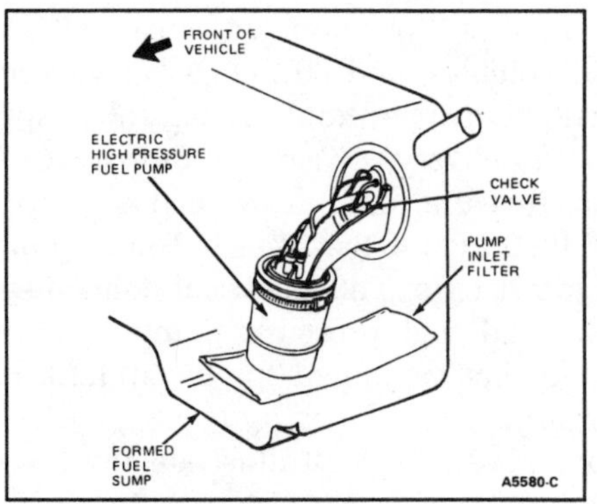

Figure 8–5. In-tank high pressure electric fuel pump. (Courtesy of Ford Motor Company)

pressure fuel pump used along with the low pressure in-tank pump assembly is illustrated in Figure 8–11.

On the Escort/Lynx and EXP vehicles, an externally mounted high-pressure electric fuel pump capable of supplying 80L (21.1 US gallons or 17.6 Imperial gallons) of fuel per hour at a pressure of 39 psi (269 kPa) is used. An internal safety relief valve is used that relieves fuel pressure due to plugged filters or a restriction when 123 psi (850 kPa) is reached.

SPECIAL NOTE/CAUTION: See heading "Relieving Fuel System Pressure" before performing any service work on either the CFI or EFI fuel injection systems.

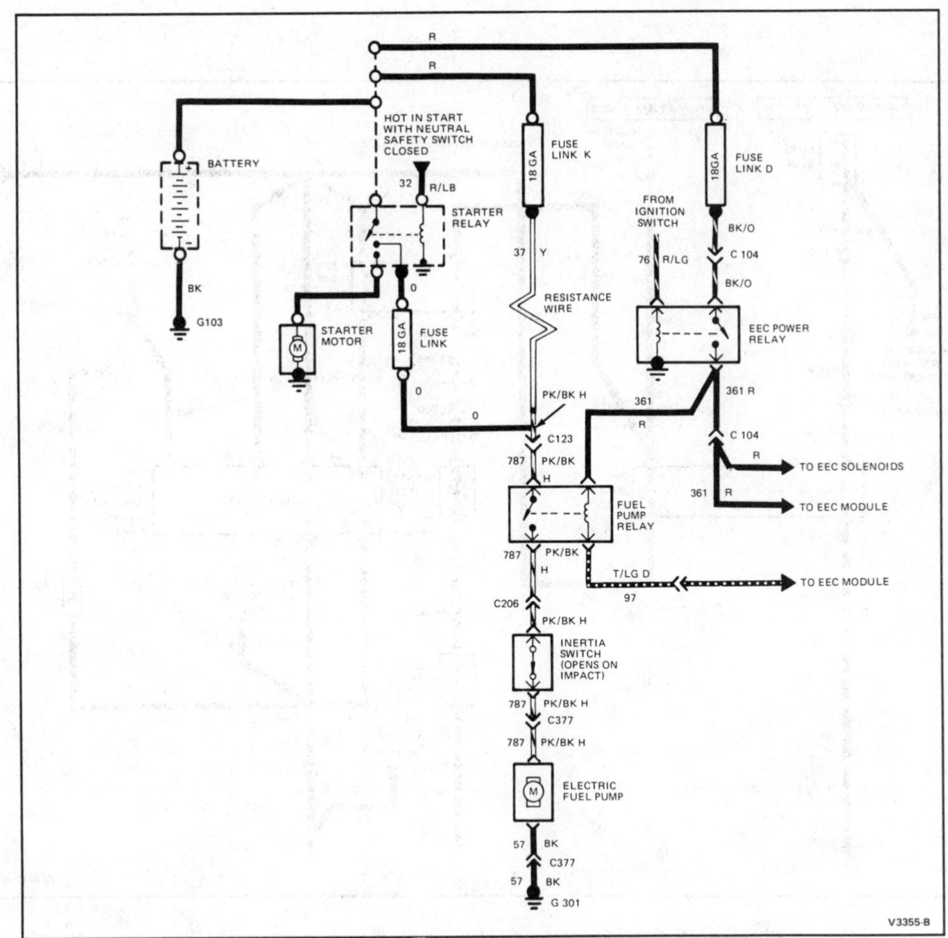

Figure 8–6. Electric fuel pump wiring diagram—Lincoln Town Car, Ford Crown Victoria/Mercury Grand Marquis. (Courtesy of Ford Motor Company)

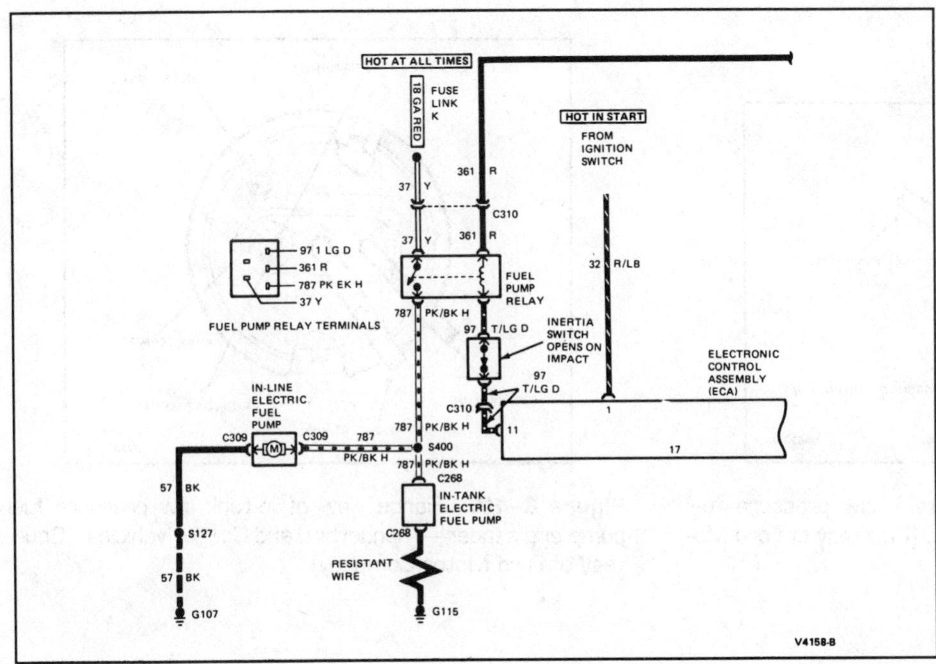

Figure 8–7. Electric fuel pump wiring diagram—Mark VII Continental, Thunderbird/Cougar, LTD/Marquis, Mustang, and Capri. (Courtesy of Ford Motor Company)

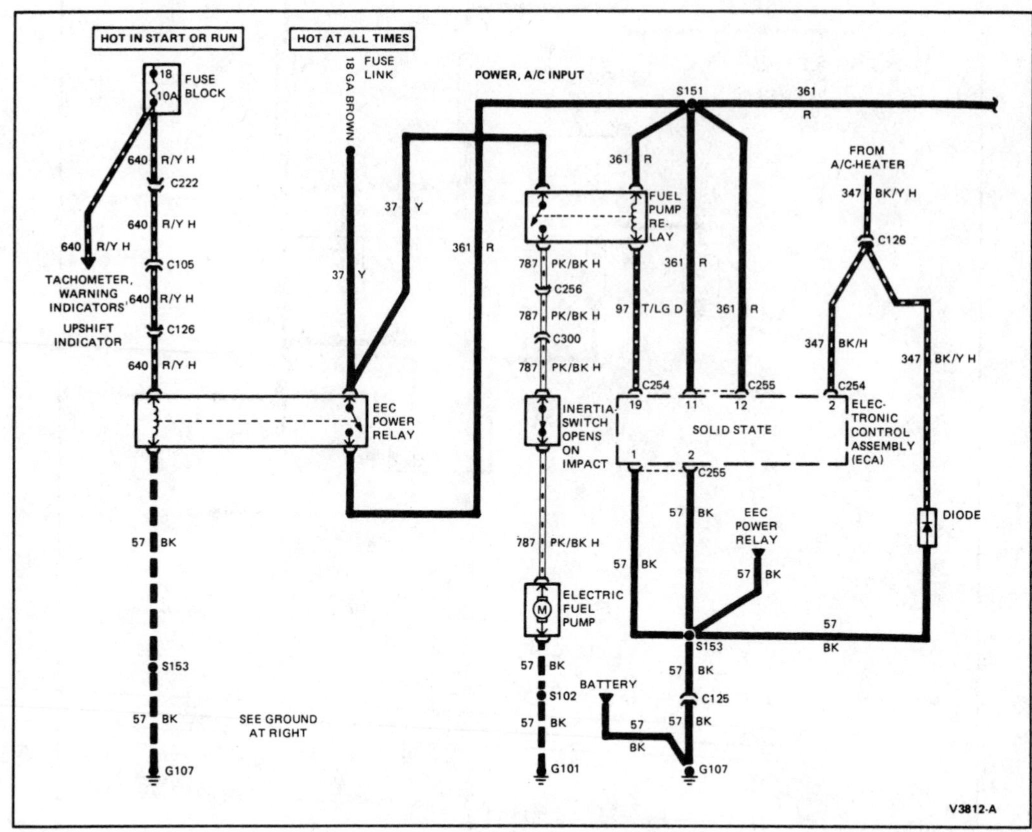

Figure 8-8. Electric fuel pump wiring diagram—Escort/Lynx and EXP vehicles with electronic fuel injection systems. (Courtesy of Ford Motor Company)

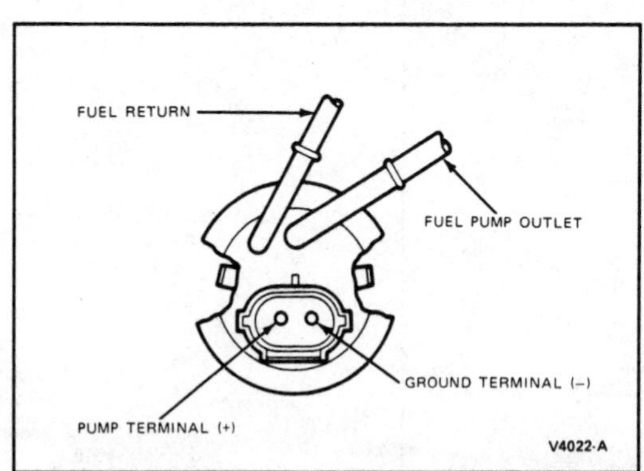

Figure 8-9. Flange view of in-tank low pressure fuel pump—Mustang and Capri vehicles. (Courtesy of Ford Motor Company)

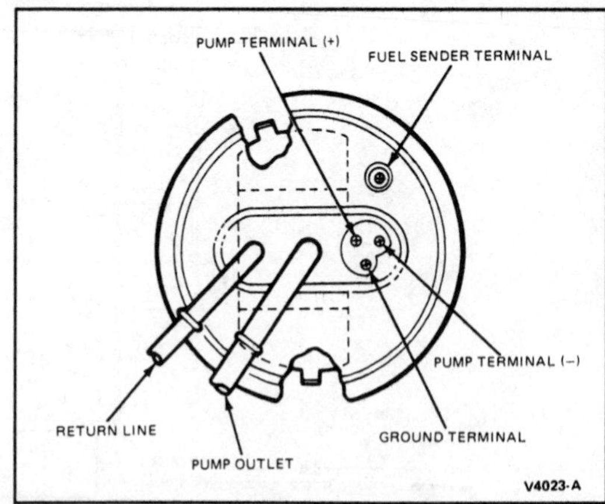

Figure 8-10. Flange view of in-tank low pressure fuel pump and sender—Thunderbird and Cougar vehicles. (Courtesy of Ford Motor Company)

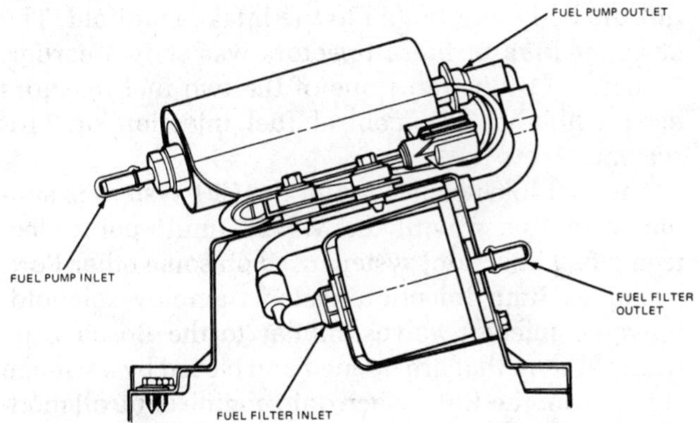

Figure 8–11. Externally mounted in-line high pressure fuel pump. (Courtesy of Ford Motor Company)

Fuel Pump Check. To check the fuel pump operating condition, proceed as follows.

Procedure

1. Make sure that no external leaks are evident.
2. Fuel tank should be at least half-full.
3. Check the pump or pumps for electrical continuity by disconnecting the electrical connector to the pump or pumps.
4. With a voltmeter connected to the fuel pump wiring harness, turn the ignition key on and note the voltage reading—which should read battery voltage for only about 1 second because of the action of the EEC timer relay.
5. Turn the ignition key momentarily to the *start* position; with the voltmeter connected across the high pressure in-tank unit wiring harness connections, voltage should be about 8 volts while cranking; if less than 8 volts—check electrical system.
6. On the two pump setup, if battery voltage is not 12 volts in Step 4 (no voltage), check the luggage compartment located (or other location) safety inertia switch (Figure 8–30 and 8–31) and reset it if necessary; if this fails, check the electrical system for a break.
7. On a CFI system, disconnect the fuel return line at the throttle body; on MPEFI systems, disconnect the return line at the fuel rail.
8. Use a calibrated container of at least 1 quart capacity for this check.
9. Connect pressure gauge P/N T80L 9974-A (Figure 8–28) to the fuel diagnostic valve (item 15 in Figure 8–32) on the fuel charging assembly of CFI systems or to the fuel rail on MPEFI systems. (Figure 8–26)
10. If not already done, disconnect the electrical connector to the fuel pump.
11. Connect an auxiliary wiring harness to the fuel pump connector and a fully charged 12-volt battery to activate the pump for at least 10 seconds, and note the fuel pressure on the test gauge hooked up in Step 9 above.
12. If no fuel pressure is apparent, check for adequate voltage at the fuel pump connector.
13. If the fuel pump does operate, let it run for 10 seconds and measure the quantity spilled into the container; the fuel pump is operating correctly if:
    a. Fuel pressure reaches between 34–45 psi (241–310 kPa).
    b. Minimum flow rate is about 10 ounces or 1/3 US quart (280 ml) after 10 seconds (7.5 ounces on the Escort/Lynx and EXP vehicles).
    c. Fuel pressure gauge reads a minimum of 30 psi after disconnecting the electric hookup.
14. Pump is satisfactory if all three conditions are met.
15. Good fuel pressure but low flow rate is usually indicative of plugged fuel filters or restricted lines.
16. Low fuel pressure after disconnecting the test wiring indicates a problem at the fuel pressure regulator or injectors; if these two items are OK, replace the high pressure fuel pump.
17. On the two-pump system, after replacement of the high pressure pump, lower the fuel tank and replace the filter on the low pressure pump.
18. If hot fuel problems exist or the system is noisy on the two-pump arrangement, remove the inlet push fitting and high pressure pump line and raise the end of the fitting above the fuel tank level to prevent any siphoning effect.
19. Turn the ignition key on and measure the amount of fuel into a container, which should be a minimum of 1.5 ounces (45 ml) for 1 second of operation.

Fuel Filters. Fuel is filtered at three locations on all CFI and EFI fuel injected applications. These filters are located in the following areas:

1. On the Lincoln Town Car, Crown Victoria, and Mercury Grand Marquis, a high-pressure in-tank fuel pump is used. This system employs a pump

inlet filter within the tank as shown earlier in Figure 8–5.

2. On the Mark VII/Continental, Thunderbird/Cougar, LTD/Marquis, and Mustang/Capri, an in-tank low-pressure fuel pump is used. A small nylon filter is used on the inlet side to prevent dirt entering the system.
3. An in-line fuel filter illustrated in Figure 8–12 located downstream of the electric fuel pump and mounted on the underbody forward of the right rear wheel well kick-up is used on the vehicles listed in 1 above. Vehicles listed in 2 above have a fuel filter incorporated into the in-line high-pressure fuel pump assembly shown earlier in Figure 8–5.

In addition to the filters listed above, each injector assembly (Figure 8–24) contains a filter located at the top of each injector. This filter is not serviceable; therefore if the injector screen/filter becomes plugged, the complete injector assembly must be replaced. Refer to the injector section for this information.

CFI System Fuel Injectors

There are two fuel injectors employed with the 3.8L V6 and 5.0L V8 engines that are mounted in the throttle body unit bolted to the intake manifold. The basic location of these injectors was shown earlier. Figure 8–13 illustrates one of the two fuel injectors used with the CFI (central fuel injection or TBI) system.

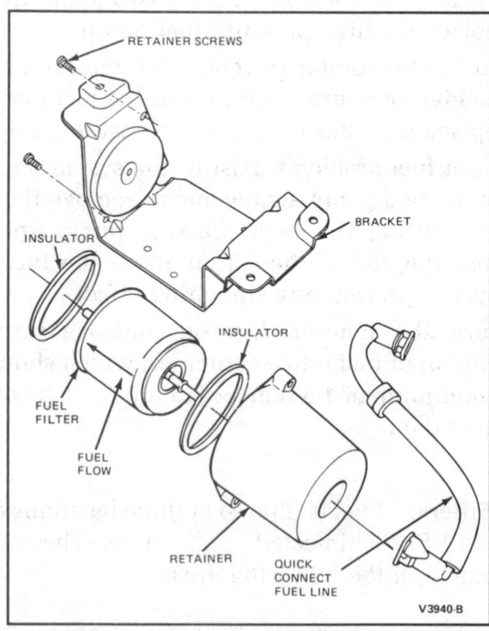

Figure 8–12. Fuel filter—in-line used with the Lincoln Town Car, Crown Victoria, and Mercury Grand Marquis vehicles. (Courtesy of Ford Motor Company)

The fuel injector used with the CFI system is similar to that used with the MPEFI (multi-point electronic fuel injection) system used on some other Ford products. Both injection systems employ solenoid-operated injector valves similar to the Bosch L jetronic system that are opened and closed by a voltage signal from the EEC (electronic engine control) module. The faster the engine speed, the faster the injector will open and close. The amount of fuel injected by the injector remains constant each time it opens; however, the length of time that it remains open establishes the quantity of fuel that will be injected. Therefore, the longer the voltage signal is supplied to keep the solenoid energized, the greater will be the amount of fuel injected.

When the injector is energized, the needle valve and armature are pulled back into the coil area of the body to allow pressurized fuel to leave the pintle area of the injector tip.

Coil resistance should be between 2 to 3 ohms when the injector is checked with its wiring connection disconnected and an ohmmeter is placed across the two pins on the injector body. Do not, however, attempt to apply 12 volts across the injector because the internal solenoid will be damaged almost immediately by the high battery current surge through it.

Fast Idle Control System Check/Repair (FICS)

To check/repair a possible fault in the FICS proceed as follows.

Procedure

1. Apply the parking brake, block the wheels, and remove the air cleaner.
2. Plug the vacuum hoses leading to the air cleaner.
3. Ensure that all vacuum hoses and electrical wires are in good condition.
4. Check and ensure that all charging assembly linkage is free from bind.
5. Connect an ammeter in series with the thermostat housing terminal and the electrical lead to the thermostat housing.
6. Start the engine; the ammeter should show an initial high reading followed by a drop to 2 amps or less after 20 seconds.

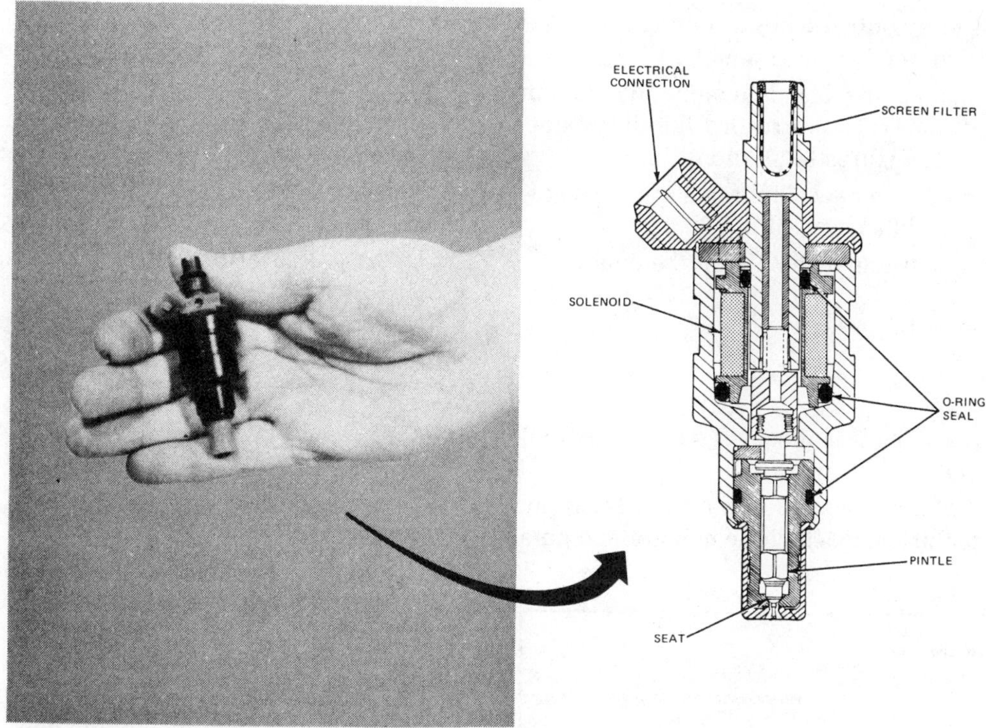

Figure 8–13. CFI System fuel injector unit. (Courtesy of Ford Motor Company)

7. If the ammeter registers a current draw in excess of 1 amp after 60 seconds, replace the thermostat housing.
8. If no current draw registers on the ammeter, connect a jumper wire from the ammeter lead to the center tap of the alternator. If the current draw is now to specs, replace the main feed wire; however, if the current draw is not to specs, replace the thermostat housing.

Fast Idle Pulldown Motor Vacuum Circuit Test.

Procedure

1. Remove the vacuum hose to the pulldown vacuum motor cover.
2. Using a vacuum source such as the Rotunda Pump 21-0014 or equivalent (see Figure 8–28, Special tools), apply an external vacuum source to the vacuum motor.
3. If the vacuum motor operates when the vacuum is applied, then proceed to Step 4; if the motor does not operate, proceed to Step 7.
4. To verify that the vacuum hose is serviceable, remove the hose connecting the fast idle pulldown motor to the vacuum source.
5. Plug one end of the hose and, using the Rotunda Pump 21-0014 or equivalent at the other end, apply a vacuum and see if it holds; if it doesn't, replace the hose.
6. Reconnect the vacuum hose to the vacuum source and pulldown motor assembly.
7. Check the fast idle pulldown diaphragm cover vacuum circuit by removing the fuel charging assembly.
8. Remove the fast idle pulldown diaphragm cover and spring.
9. Apply an external vacuum source with the Rotunda Pump 21-0014 or equivalent to the hose connection on the pulldown diaphragm cover; vacuum should not hold:
   a. If vacuum doesn't hold, proceed to section on "Fast Idle Cam Setting."
   b. If vacuum holds, replace the diaphragm cover.
   c. If replacement of the pulldown diaphragm cover is required, complete the fast idle cam setting check, then install a new plug in the fast idle cam adjusting screw access hole.
10. Reinstall the pulldown diaphragm cover and spring and recheck the system as explained in Steps 1 through 3 above.

11. If the fast idle control will not operate, replace the pulldown diaphragm assembly.
12. When replacing the fast idle pulldown motor diaphragm, measure and record the dimension "A" shown in Figures 8–14 and 8–15.
13. After reinstalling a new diaphragm assemply, recheck the fast idle cam setting.
14. Install fuel charging assembly on the engine.

Fast Idle Cam Setting.

Procedure

1. Manually place the fast idle lever on the first (top) step of the cam.
2. Using the Rotunda vacuum pump 21-0014, apply a vacuum to the fast idle pulldown motor and note the action of the fast idle cam, which should move and allow the fast idle lever to drop from the top step to the second step on the fast idle cam.
3. If the fast idle lever does not drop, or it drops to the third step, remove the fuel charging assembly.
4. Remove the plug that covers the fast idle cam adjusting screw.
5. Using a 3/32" Allen wrench, adjust the cam setting screw to set the fast idle cam on the second step and recheck the operation as per Step 2.
6. Install a new plug in the fast idle cam adjusting screw access hole.
7. Install the fuel charging assembly and reconnect all the hoses and electrical connectors.

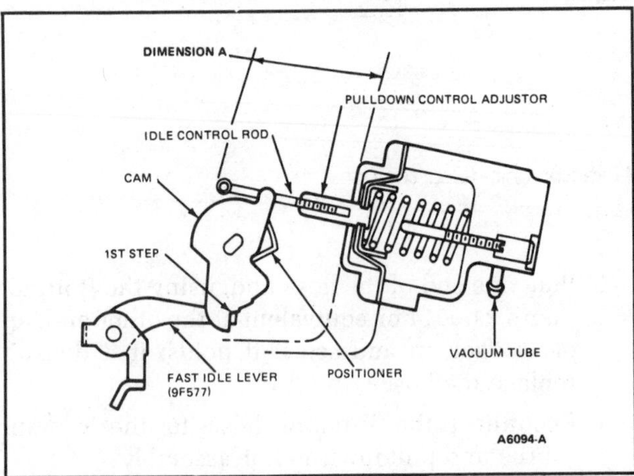

Figure 8–14. Pulldown motor adjustment—no vacuum. [Courtesy of Ford Motor Company]

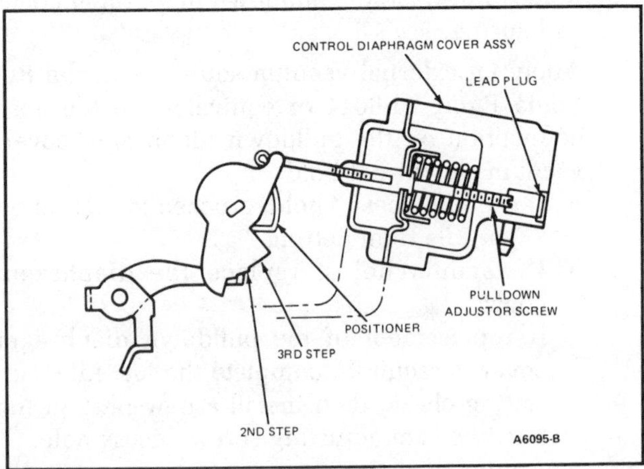

Figure 8–15. Pulldown motor adjustment—with vacuum. [Courtesy of Ford Motor Company]

Engine Cranks but Doesn't Start

Repeated attempts to start an engine that will not fire can result in excess amounts of raw fuel entering the engine and being expelled into the exhaust system. This condition can load up the exhaust system to the point that this raw fuel can severely damage the catalytic converter once the engine does start.

In order to avoid such a situation, you should disconnect the engine thermactor air supply plus the TBI injectors (wires) and run the engine until the excess fuel has been burned off before reconnecting the thermactor and injector wires.

The thermactor air injection system normally injects fresh air into the hot exhaust gas stream leaving the exhaust ports in order to promote further oxidation of both the hydrocarbons and carbon monoxide and convert them into harmless carbon dioxide and water.

Thermactor air flow to the catalytic converter is only possible when the thermactor air diverter solenoid is in the normal (operated) position. During engine warm-up situations, thermactor air is diverted to the exhaust manifold (thermactor air diverter solenoid does not operate). No air flows to either the catalytic converter or the exhaust manifold when the thermactor air bypass solenoid valve is operated. Instead, the air is immediately bypassed to the atmosphere.

Multi-Point Electronic Fuel Injection (MPEFI)

Ford uses a MPEFI (multi-point electronic fuel injection) system on some of its vehicles that differs in its operation from the CFI system in that it employs individual fuel injectors for each engine cylinder. The injectors are mounted in a tuned intake manifold so that each one will spray fuel towards the individual cylinder intake valve. Engines available with EFI are the 1.6L and 2.3L units used and offered in EXP, Escort/Lynx, Mustang/Capri, Thunderbird, and Cougar.

This system is similar to the CFI system in that an on-board electronic engine control (EEC) computer is fed voltage inputs from a variety of engine sensors. The EEC computer then analyzes these various inputs and controls the time that the electronic injectors remain open. The longer the injectors remain open, the greater the fuel delivery. This injector system is therefore similar in its operation to the Robert Bosch L-Jetronic system that also uses this arrangement to control the injector operation.

An illustration of the component location of a typical Bosch/Ford MPEFI system used on the 1.6L 4-cylinder engines is shown in Figure 8–16. Figures 8–17 and 8–18 show the arrangement used on the 2.3L EFI turbocharged 4-cylinder engines.

The EFI system receives voltage inputs to the EEC processor unit similar to that used with the CFI system discussed earlier. The actual sensor inputs are shown in Figure 8–19 for the 1.6L engine and in Figure 8–20 for the 2.3L turbocharged engine.

System Operation

The fuel system employs an electric fuel pump that is generally chassis mounted. The pump pulls fuel from the tank and then pushes it through a 20-mi-

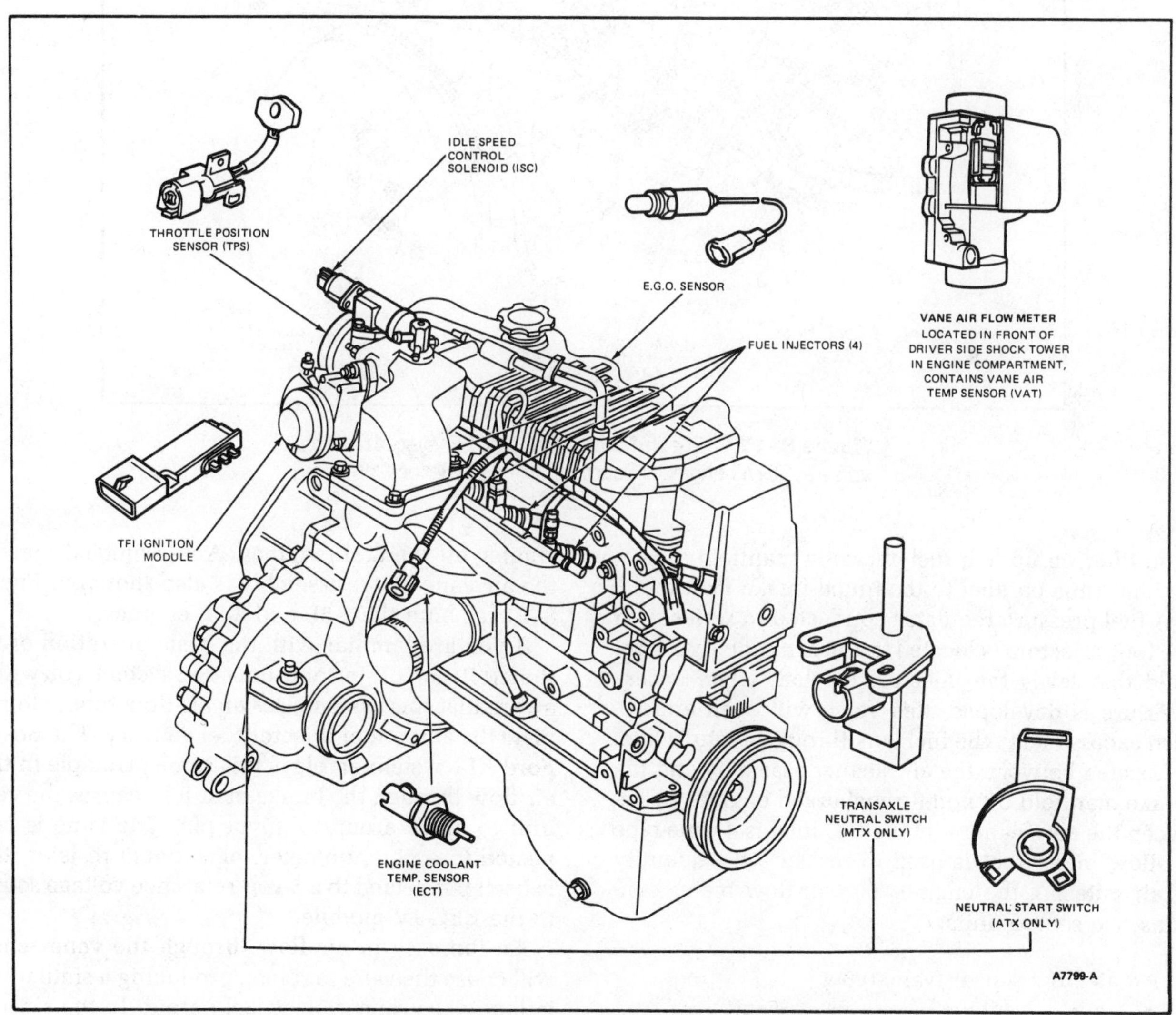

Figure 8–16. Bosch/Ford multi-point EFI system illustration—1.6L engines. (Courtesy of Ford Motor Company)

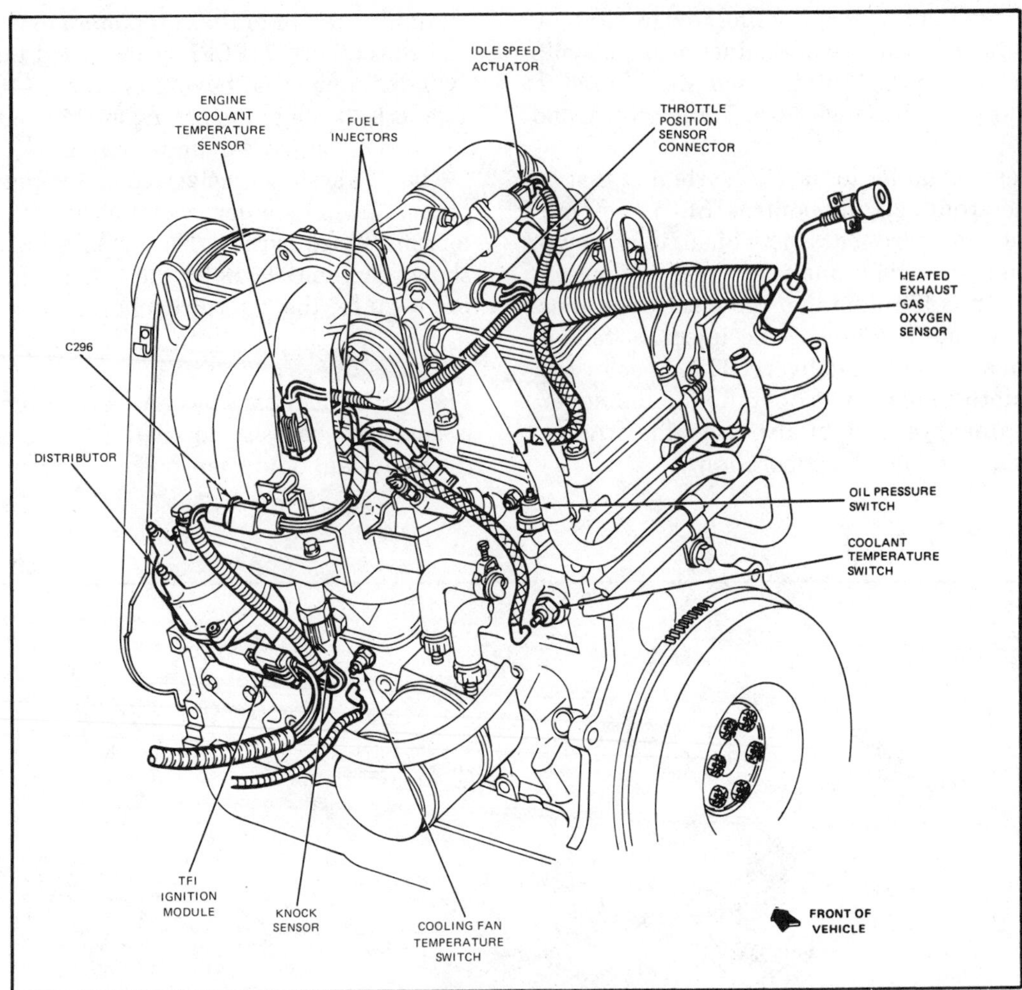

Figure 8–17. Bosch/Ford multi-point EFI system illustration—2.3LT/C engine. (Courtesy of Ford Motor Company)

cron filter on up to a fuel charging manifold assembly that runs parallel to the tuned intake manifold.

A fuel pressure regulator connected in series with the four injectors is located in the common fuel manifold that feeds the injectors so that, when excess pressure is developed, this valve will open and return excess fuel to the fuel tank through a return line.

Located between the air cleaner assembly and the intake manifold on non-turbocharged engines or between the air cleaner and the T/C inlet is a vane-type air-flow meter that is used to measure the quantity of air entering the engine. The air-flow meter contains two sensor units:

1. An air flow sensor (vane-type)
2. An air temperature sensor (vane-type)

Figure 8–16 (top right) illustrates the air vane meter assembly used with the 1.6L engines, which is similar on the 2.3L engines. An assembled view of the air vane meter assembly is also shown in Figure 8–21 for both the 1.6L and 2.3L engines.

If you are familiar with the basic operation of the Robert Bosch LE-Jetronic injection system, you will be aware that this system uses an air-flow sensor to regulate the amount of injector fuel delivery. The Bosch/Ford EFI system employs this same principle in that air flow through the intake manifold causes the vane unit to move around a pivot pin. The vane is connected to a potentiometer, or variable resistor, that is itself connected to a 5-volt reference voltage source to the EEC-1V module.

An increase in air flow through the vane sensor will cause the vane to rotate, producing a similar variation in the sensor electrical output. In the air temperature sensor, any change in inlet air temperature will cause a similar change in the resistance through the sensor wire. The EEC computer is designed to

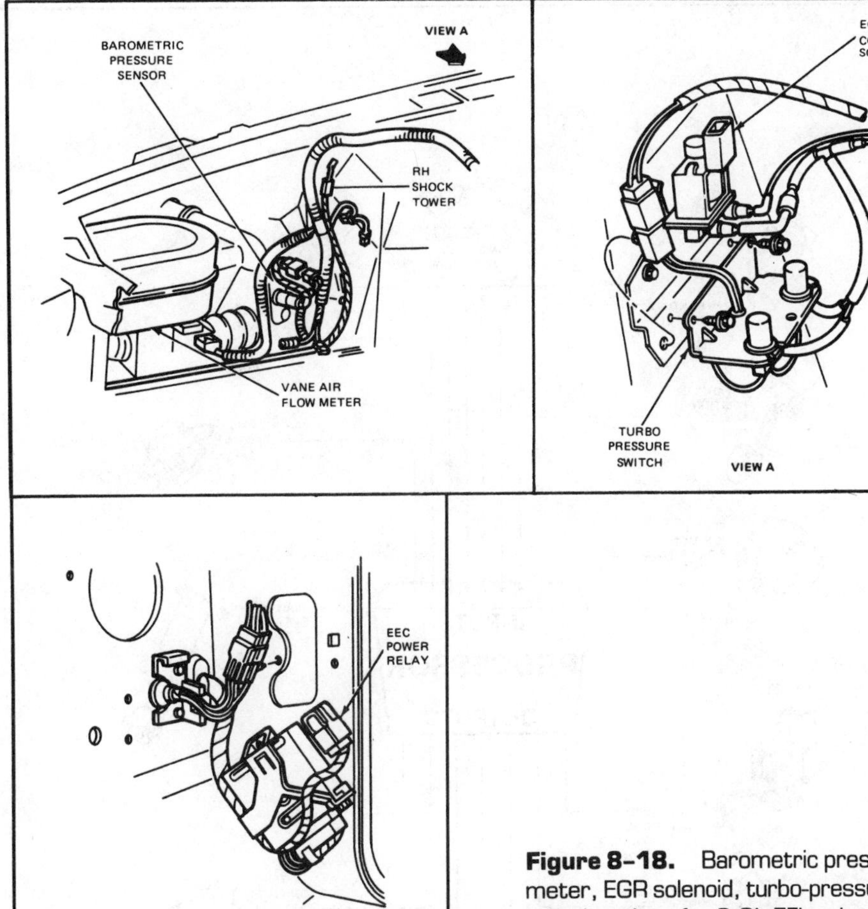

Figure 8–18. Barometric pressure sensor, vane air-flow meter, EGR solenoid, turbo-pressure switch, and EEC power relay locations for 2.3L EFI turbocharged engine used in the Mustang/Capri, Thunderbird and Cougar. (Courtesy of Ford Motor Company)

compute the voltage signals from both of these sensors and in turn relay a voltage signal to the individual injectors. All injectors are energized simultaneously, once every crankshaft revolution. The time that they remain energized controls the fuel delivery rate. The greater the air flow into the engine, the longer the injectors will remain open in order to deliver a greater volume of fuel so that the air/fuel ratio can be maintained as close to ideal as possible under varying loads and speeds. Figure 8–22 illustrates the components of the EFI system in exploded view form.

Air Throttle Body Assembly

This assembly is not unlike a single bore carburetor in that it contains a single butterfly-type valve that is controlled by the operator through mechanical linkage connected to the accelerator. As with most fuel injection systems, there must be a bypass channel that is used to control cold and warm engine idle air flow. Items 15 and 22 in Figure 8–22 illustrate the air intake charge throttle body and the throttle air bypass valve assembly respectively. Item 9 in the same figure shows the actual butterfly-type valve that is supported on a shaft and pivots within the bore of Item 15.

When the engine is cold, more air is required in order to offset the greater amount of fuel delivered during initial starting and engine warm-up. This extra air flow is controlled by an air bypass valve assembly (Item 22), which is directly mounted to the throttle body. The EEC computer controls the air bypass valve position through a linear actuator that in turn changes the position of a variable area metering valve. Also included on the air throttle body assembly are:

1. A throttle body position sensor that monitors the position of the butterfly valve in the bore.
2. An adjustment screw that is used to adjust the

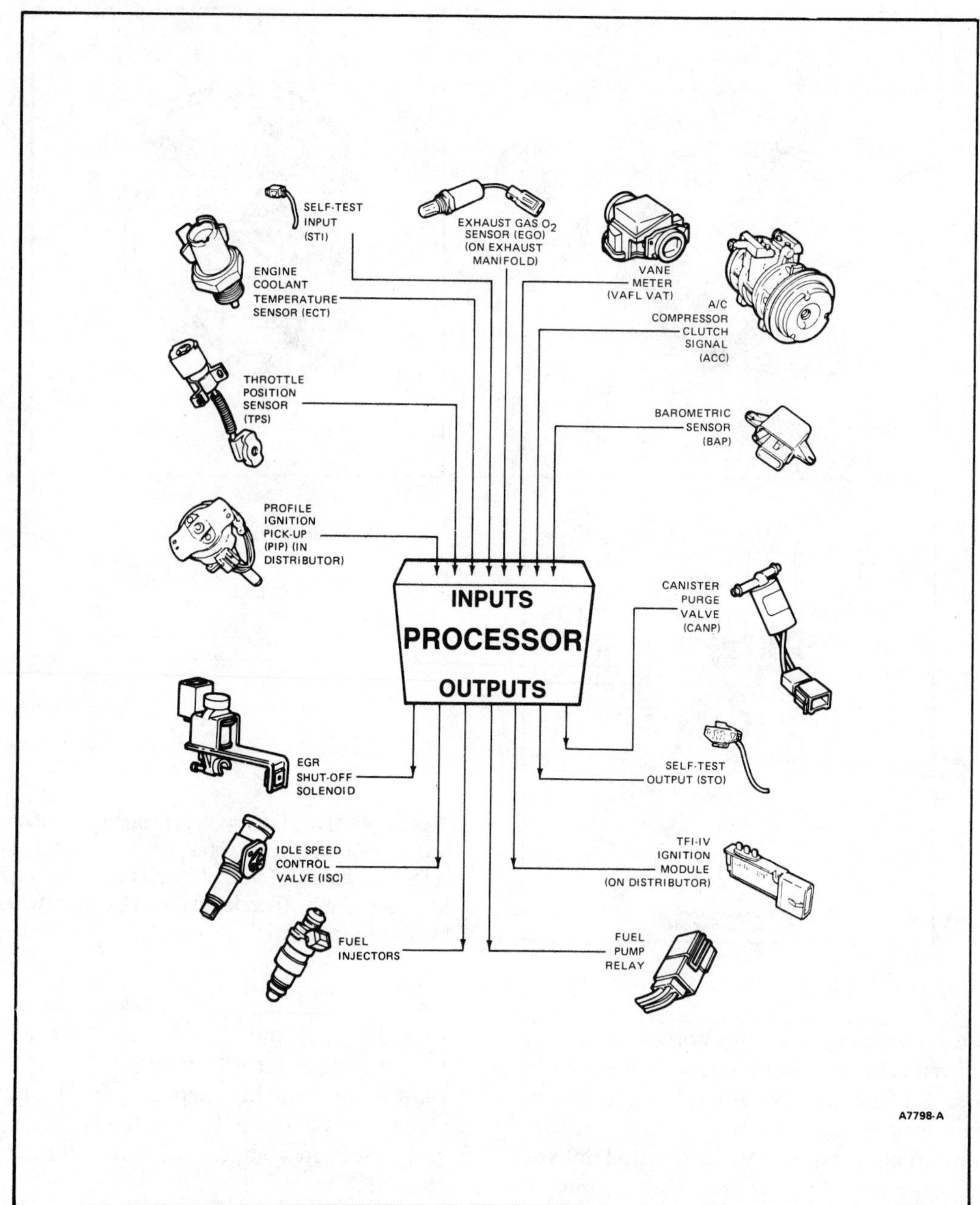

Figure 8–19. 1.6L Engine EFI sensor circuit components. (Courtesy of Ford Motor Company)

minimum air-flow position of the throttle plate (butterfly valve) for idle rpm.

3. An adjustment screw to limit the WOT (wide open throttle) position.
4. A PCV fresh air source that is located upstream from the throttle plate (butterfly valve).
5. Vacuum tap ports for the PCV (positive crankcase ventilation) and the EVAP (evaporative emission) controls.

Fuel Pressure Regulator

This unit is connected (bolted) to the fuel supply manifold and can be seen in Figure 8–22 as Item 9 (lower view). Its purpose is to maintain/control the fuel system pressure. Within the regulator assembly is a spring loaded diaphragm that is subjected to intake manifold pressure on one side and fuel pressure on the other. Basic fuel pressure is controlled by the action of balancing one side of the diaphragm with

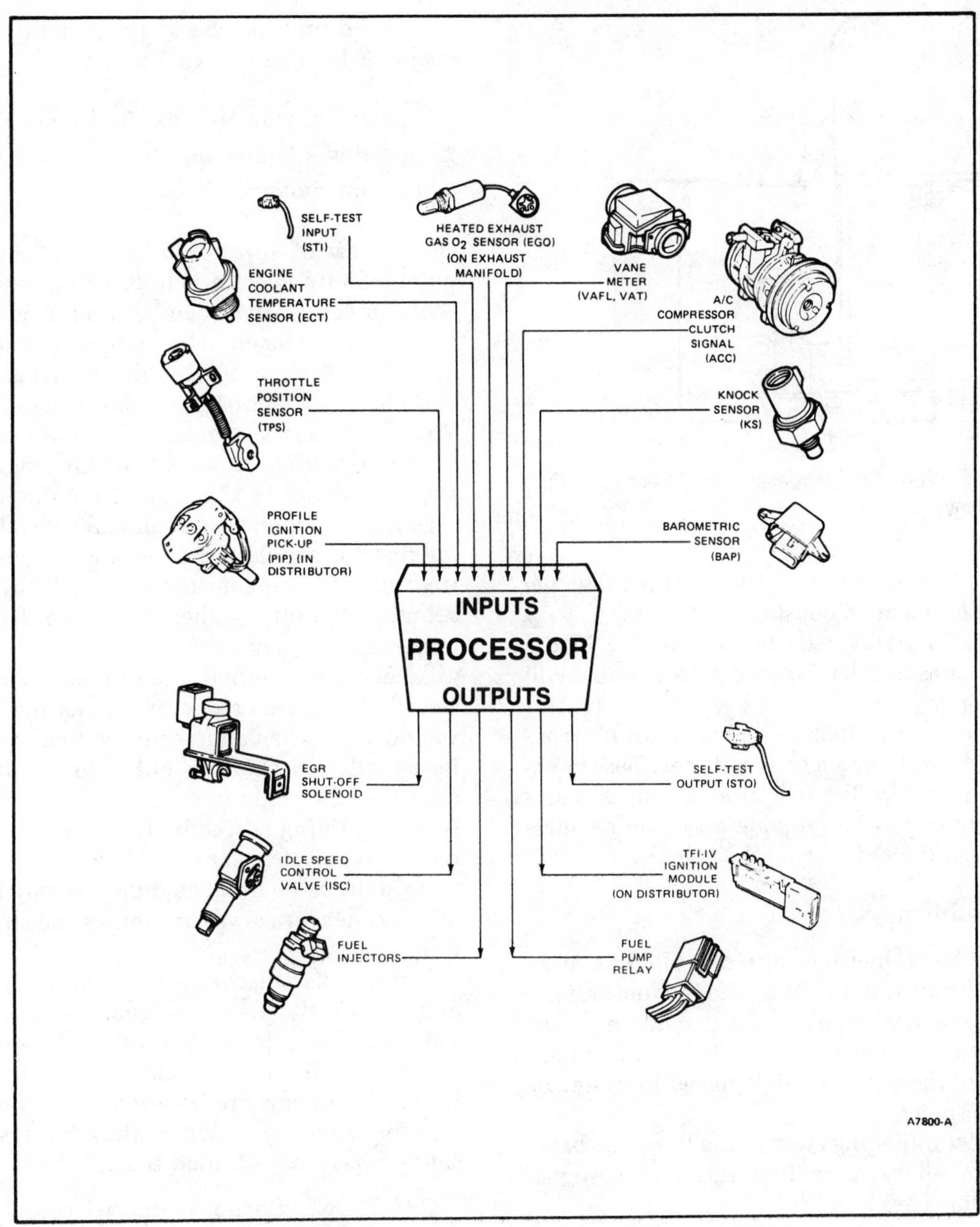

Figure 8–20. 2.3L Engine EFI sensor circuit components. (Courtesy of Ford Motor Company)

intake manifold pressure. This is achieved by the use of a spring on one side of the diaphragm as shown in the cutaway view of the regulator shown in Figure 8–23. Excess fuel is bypassed through the regulator to the fuel tank via a return line.

Fuel Injectors

The fuel injectors are similar to the Bosch L-Jetronic units in that they are cycled open and closed by a voltage signal from the EEC computer once per crankshaft revolution. The longer they are open, the greater the volume of fuel delivered. Figure 8–24 illustrates the injector used with the Ford MPEFI system.

The injector consists of an electrically operated solenoid that actuates a pintle and needle valve assembly. When the injector is energized by an electrical signal from the EEC processor unit, the needle/pintle will be pulled into the injector body thereby allowing fuel to flow into the intake manifold towards the in-

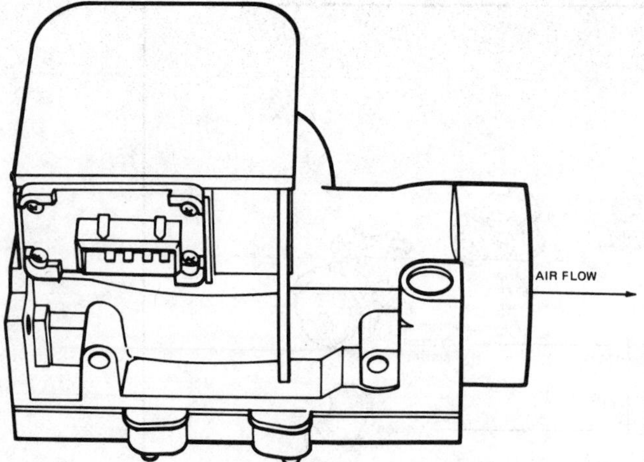

Figure 8–21. Vane air-flow meter unit. (Courtesy of Ford Motor Company)

take valves at each cylinder. The distance that the injector pintle opens is constant; therefore, to vary the amount of fuel delivered from the injector tip, the EEC computer establishes how long the solenoid will remain energized.

This action ensures that the pressure drop across the injector tip will remain constant regardless of engine speed; therefore the fuel atomization is maintained at a given level to provide adequate penetration of the fuel towards the intake valve.

Injector Problems

Problems with the fuel injectors can often be associated with dirt or water getting into the fuel system. Also, since the fuel injectors used with the Ford multi-point EFI system are connected to a common fuel manifold, they can all be damaged by water or dirt in the system.

Problems with fuel injectors cause the same basic symptoms in all gasoline fuel injection systems. These symptoms are:

1. Poor startup and idling conditions
2. Acceleration hesitation
3. Lack of power
4. Poor restart when the engine is warm

Poor injector spray formation can cause conditions 1, 2, 3, and 4, while condition 4 is usually caused by raw fuel dripping from injectors when the engine is stopped. An excess fuel mixture collects in the cylinder or manifold area making restart hard when the engine is warm/hot. Injectors should be checked for loose or dirty electrical connections and for correct voltage supply from the EEC unit.

Detailed injector checks are given under the subheadings later in the text entitled:

1. Injector Wiring Harness Resistance Check
2. Isolating a Faulty Injector
3. Injector Balance Check

Quick Test of Injectors. A quick method to pinpoint a faulty injector is to use a mechanic's stethoscope (a screwdriver can be used if no stethoscope is available) to listen to the individual injectors while the engine is running. Run the engine at an idle rpm and place the stethoscope against the injector body. You should hear a regular pulsing or clicking sound as the electrical solenoid within the injector is switched on/off by the signal from the EEC unit.

Gently accelerate the engine; the clicking sound at the injector should increase as it cycles on/off more frequently. If an injector does not have the same sound as the others, then it is more than likely not operating correctly.

An alternate method is to disconnect each injector electrical connector one at a time; the engine rpm should drop when an injector is isolated in this fashion (similar to shorting out a spark plug). If little change to the engine sound occurs, then the injector is not operating correctly. Each injector as it is disconnected electrically should cause the engine rpm to drop the same basic amount—usually in the range of 150 rpm average when running the engine at 2000 rpm.

You can also disconnect the battery ground cable and remove the electrical leads from each injector. Take an ohmmeter and measure the resistance between the terminals on each injector; the resistance is usually somewhere between 2 to 3.5 ohms. If the injector winding is open (infinite resistance) or shorted (zero ohms), then replace the injector.

CAUTION: Do not connect leads directly from the battery across the two injector terminals because the large surge of battery current will immediately burn out the internal solenoid.

To ensure that no vacuum leaks exist at the fuel injector-to-intake manifold, run the engine at an idle speed and spray the area around the individual injectors with some carburetor or choke cleaner while using a vacuum gauge hooked up to the intake system. Carefully notice any change to the vacuum gauge reading.

One other check that can be made, if you are still in doubt, is to remove all of the fuel injectors from the intake manifold and attach them to the common fuel manifold assembly. Place the injector tips into a

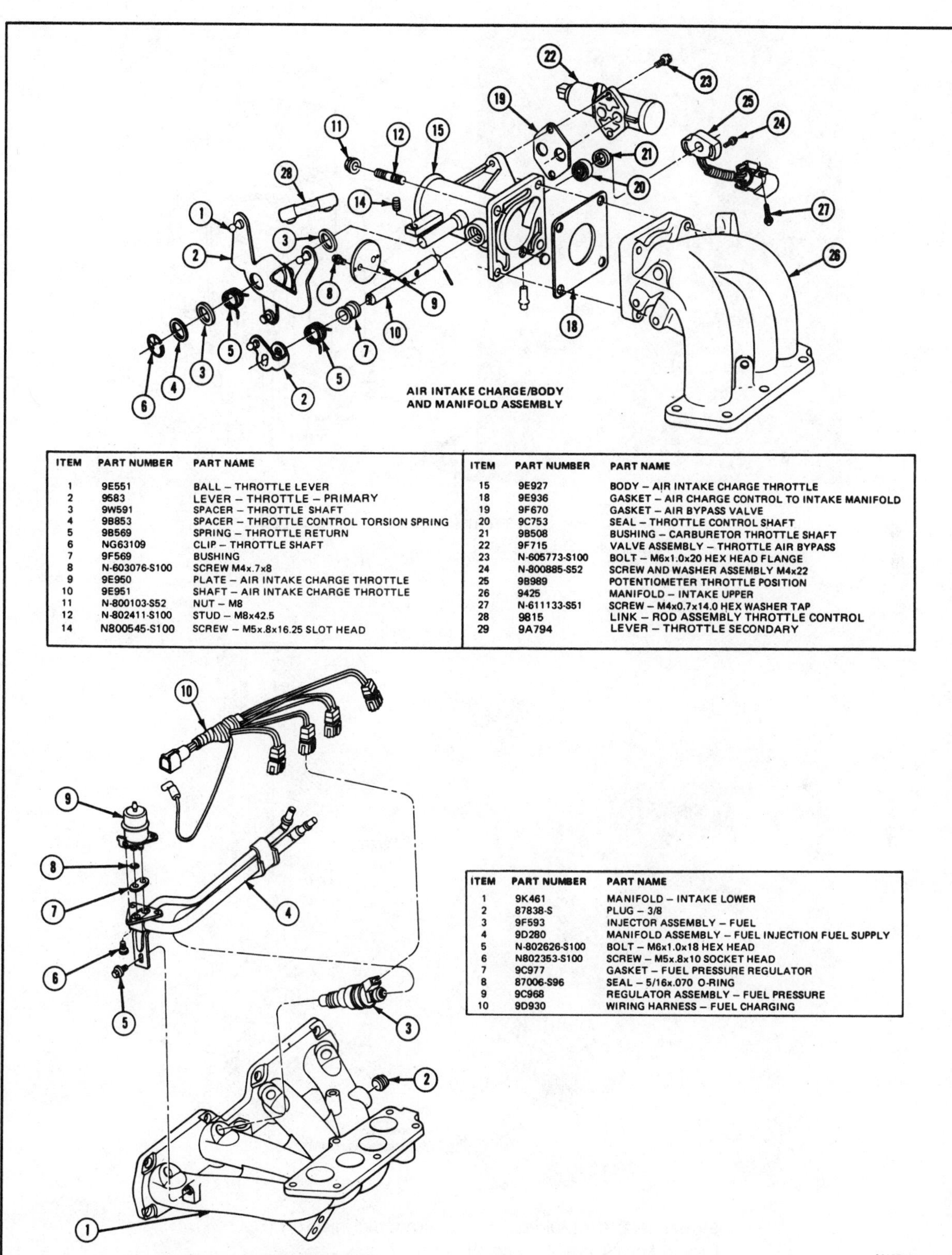

ITEM	PART NUMBER	PART NAME
1	9E551	BALL – THROTTLE LEVER
2	9583	LEVER – THROTTLE – PRIMARY
3	9W591	SPACER – THROTTLE SHAFT
4	9B853	SPACER – THROTTLE CONTROL TORSION SPRING
5	9B569	SPRING – THROTTLE RETURN
6	NG63109	CLIP – THROTTLE SHAFT
7	9F569	BUSHING
8	N-603076-S100	SCREW M4x.7x8
9	9E950	PLATE – AIR INTAKE CHARGE THROTTLE
10	9E951	SHAFT – AIR INTAKE CHARGE THROTTLE
11	N-800103-S52	NUT – M8
12	N-802411-S100	STUD – M8x42.5
14	N800545-S100	SCREW – M5x.8x16.25 SLOT HEAD

ITEM	PART NUMBER	PART NAME
15	9E927	BODY – AIR INTAKE CHARGE THROTTLE
18	9E936	GASKET – AIR CHARGE CONTROL TO INTAKE MANIFOLD
19	9F670	GASKET – AIR BYPASS VALVE
20	9C753	SEAL – THROTTLE CONTROL SHAFT
21	9B508	BUSHING – CARBURETOR THROTTLE SHAFT
22	9F715	VALVE ASSEMBLY – THROTTLE AIR BYPASS
23	N-605773-S100	BOLT – M6x1.0x20 HEX HEAD FLANGE
24	N-800885-S52	SCREW AND WASHER ASSEMBLY M4x22
25	9B989	POTENTIOMETER THROTTLE POSITION
26	9425	MANIFOLD – INTAKE UPPER
27	N-611133-S51	SCREW – M4x0.7x14.0 HEX WASHER TAP
28	9815	LINK – ROD ASSEMBLY THROTTLE CONTROL
29	9A794	LEVER – THROTTLE SECONDARY

ITEM	PART NUMBER	PART NAME
1	9K461	MANIFOLD – INTAKE LOWER
2	87838-S	PLUG – 3/8
3	9F593	INJECTOR ASSEMBLY – FUEL
4	9D280	MANIFOLD ASSEMBLY – FUEL INJECTION FUEL SUPPLY
5	N-802626-S100	BOLT – M6x1.0x18 HEX HEAD
6	N802353-S100	SCREW – M5x.8x10 SOCKET HEAD
7	9C977	GASKET – FUEL PRESSURE REGULATOR
8	87006-S96	SEAL – 5/16x.070 O-RING
9	9C968	REGULATOR ASSEMBLY – FUEL PRESSURE
10	9D930	WIRING HARNESS – FUEL CHARGING

Figure 8-22. EFI System exploded view, Thunderbird/Cougar, Mustang, and Capri 2.3L engine. (Courtesy of Ford Motor Company)

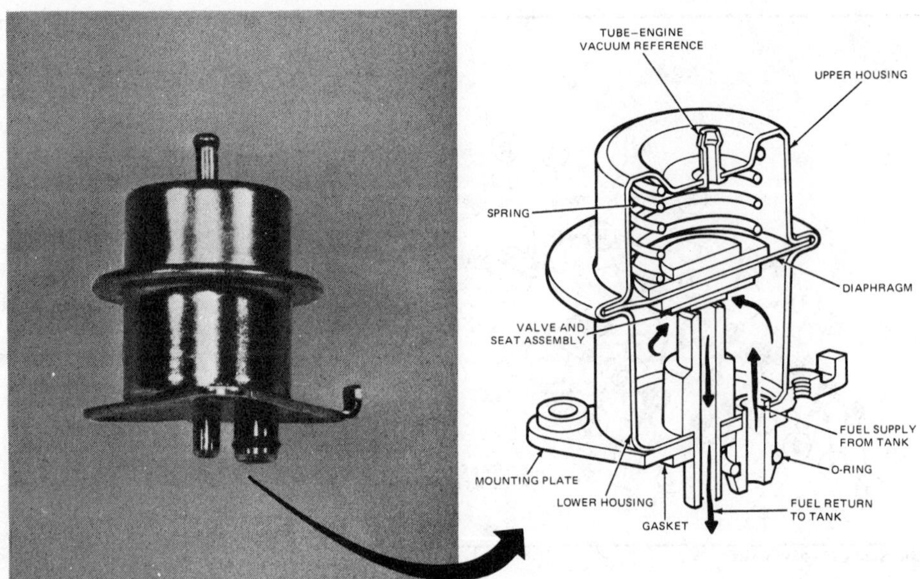

Figure 8-23. EFI System fuel pressure regulator. (Courtesy of Ford Motor Company)

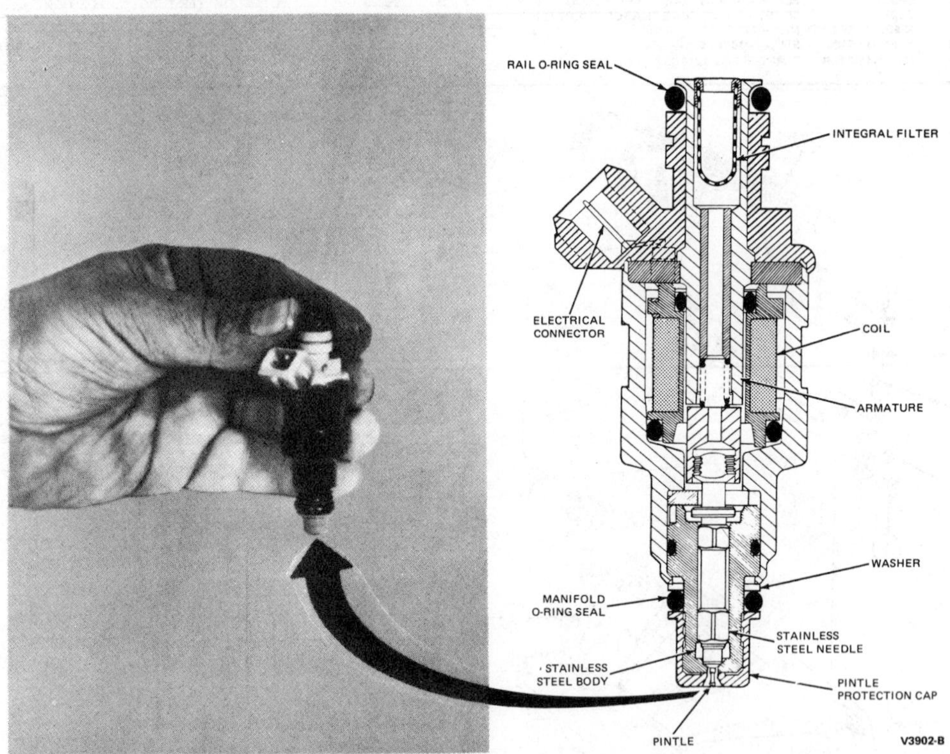

Figure 8-24. Multi-point EFI system fuel injector. (Courtesy of Ford Motor Company)

plastic or glass container and switch the ignition key on/off to establish whether or not any of the injectors are dribbling or leaking raw fuel. A slight sweat is acceptable and even a drop a minute is usually considered acceptable; however, any more than that will cause a hard starting condition on a warm engine. Therefore, injectors should be replaced if they exceed this.

Injector Wiring Harness Resistance Check

Sometimes an injector problem will originate simply from a poor electrical connection either at the injector itself or in the main wiring harness. Before removing an injector or proceeding to the additional checks/tests below, check out the wiring harness as follows:

Procedure

1. Turn the ignition key off and wait 10 seconds for the processor to attain a neutral state.
2. Disconnect the processor pin connector and closely inspect the connections for signs of possible damage or corrosion.
3. See Special tools, Figure 8–28 and connect the breakout box Rotunda P/N T83L-50-EEC-1V or equivalent to the disconnected wiring harness with the processor still disconnected.
4. Using a DVOM (digital volt ohmmeter), place it on the 200-ohm scale.
5. Make sure that the ignition key is *off* and disconnect the electrical connector from injectors No. 2, 3 and 4.
6. With the DVOM leads, measure the resistance of injector No. 1 between pins 37 and 58 as shown in the wiring circuit diagram in Figure 8–25 through the wiring harness connections and record the reading.
7. Disconnect injector No. 1 and reconnect injector No. 2; repeat the same test as in Step 6.
8. Disconnect injector No. 2 and reconnect No. 3; repeat Step 6, only place the leads of the DVOM across pins 37 and 59.
9. Disconnect injector No. 3 and reconnect injector No. 4 and repeat Step 8; compare all of the injector resistance readings.
10. If all resistance readings are within 2.0 to 3.5 ohms, proceed to "Injector Drive Signal Check" below.
11. If the resistance readings are not within 2.0 to 3.5 ohms, the main wiring harness is at fault; check for shorts or opens and if OK, then replace the faulty injector(s) and repeat the resistance check.

Injector Drive Signal Check

If Step 10 in the previous check was "yes," proceed as follows.

Procedure

1. Obtain a 12-volt non-powered test light.
2. Make sure that the ignition key is *off* and connect the breakout box as per Step 3 in the previous check.
3. Connect the processor to the breakout box.
4. Connect the test light leads between the breakout box pin 37 and 58 as shown in the wiring diagram in Figure 8–25 (previous check).
5. Crank or start the engine.

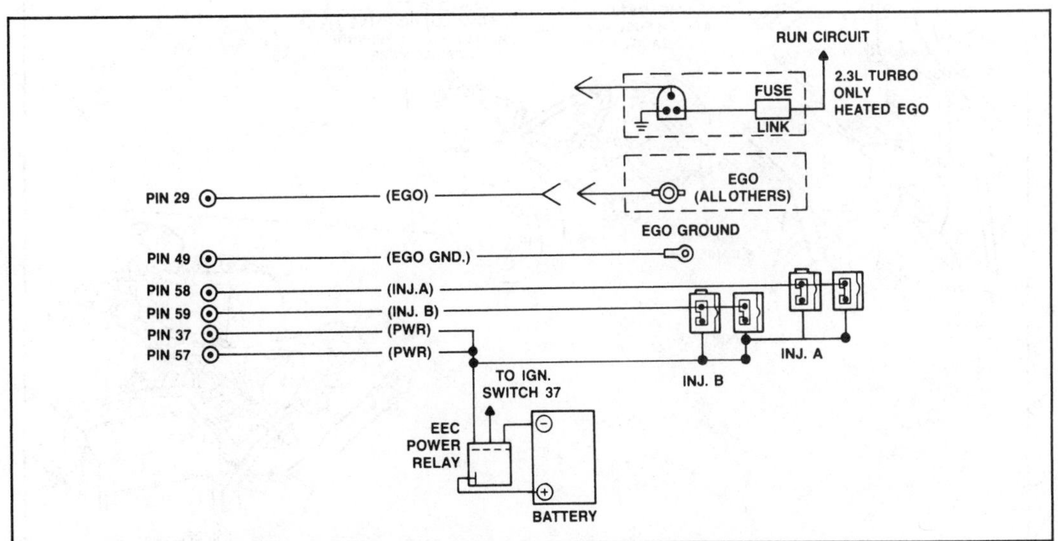

Figure 8–25. EFI injector wiring harness resistance check points. (Courtesy of Ford Motor Company)

6. Repeat the above test again, only place the 12-volt test light leads across pins 37 and 59.
7. If the 12-volt test light glows dimly on both Step 4 and 6, proceed to the "Injector Balance Test."
8. If the 12-volt test light does not glow on Step 4 or 6, check for 12-volt battery power at pins 37 and 57.
9. If the test light glows brightly on Step 4, Step 6, or both, check circuits 95 and 96 for shorts to ground; if this is OK, replace the processor and repeat the test again.

Isolating A Faulty Injector

To isolate a faulty injector or faulty EEC processor on the EFI system, proceed as follows:

Procedure

1. Turn the ignition key off.
2. Install Fuel Pressure Gauge Assembly Rotunda T80L-9974-A, shown in the Special tools, Figure 8–28 to the vehicle, as illustrated in Figure 8–26.
3. Remove all injector electrical connectors.
4. Pressurize the fuel system by turning the ignition key to Run for 1 second, then turn the key *off.* Wait for 10 seconds and repeat 5 times.
5. Electrically connect No. 1 injector and crank the engine for 5 seconds while closely watching the pressure gauge reading.
6. Disconnect the No. 1 injector and repeat Step 4.
7. Repeat Steps 5 and 6 for each of the injectors (repeat Step 4 after disconnecting each additional injector).
8. If all injector readings are not within 4 psi (27.5 kPa) of each other, replace those injectors that are not within 4 psi of one another.
9. If all injectors are within 4 psi (27.5 kPa) of each other, then disconnect the processor and inspect the pins for damage or corrosion and retest the injectors. If the problem still exists, replace the processor and repeat the injector test.

Injector Balance Check

To further clarify if an injector is not operating correctly proceed as follows:

Procedure

1. Run the engine until it reaches normal operating temperature.
2. Using an accurate tachometer, run the engine up to 2000 rpm by hand or run-in the curb idle set screw as shown in Figure 8–27 until 2000 rpm is obtained.
3. Pull off one injector electrical lead at a time, which will prevent the injector from opening (similar to shorting out a spark plug), and carefully record the engine rpm drop.

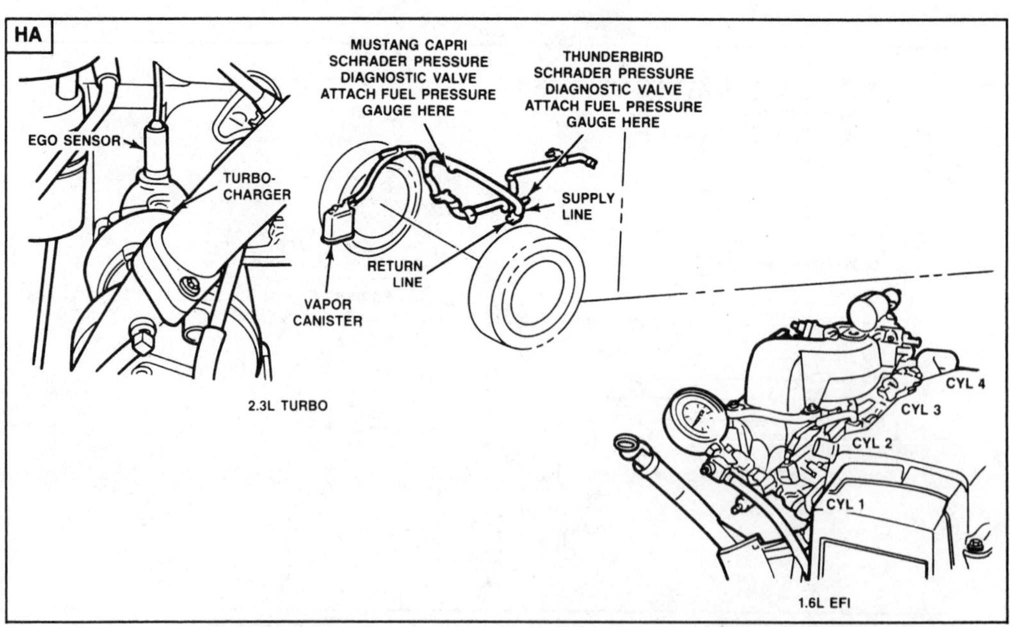

Figure 8–26. 1.6L/2.3L Engine EFI fuel diagnostic valve location. (Courtesy of Ford Motor Company)

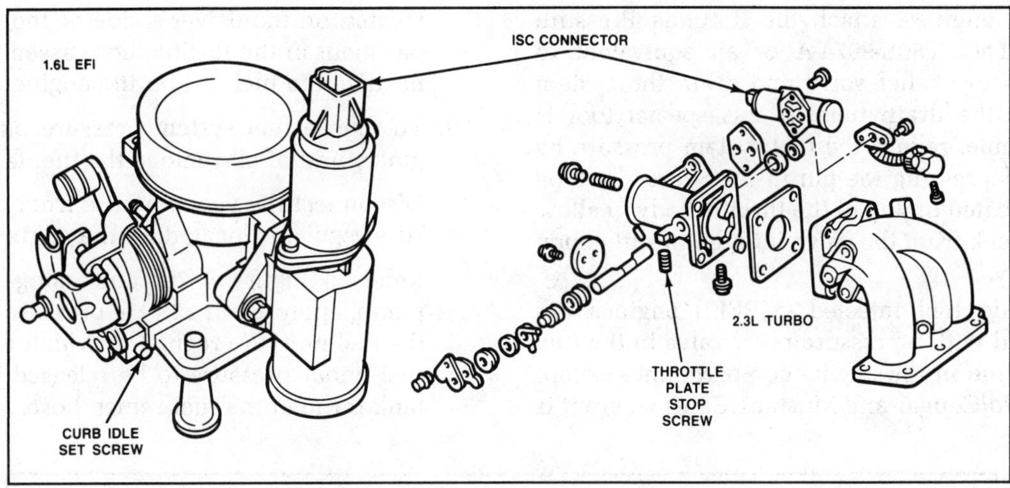

Figure 8–27. 1.6L EFI Curb idle set screw location. (Courtesy of Ford Motor Company)

4. Repeat Step 3 for each injector.
5. If each injector that is disconnected causes at least a 150 rpm drop in engine speed, then the injectors are operating correctly; the cause of the problem is more than likely an air/vacuum leak, contaminated fuel (dirt/water), or the EGR system.
6. If, however, there is less than a 150 rpm drop when a particular injector is isolated, replace that injector and repeat the test again.
7. Reset the curb idle as per "Curb Idle Adjustment" procedure herein.

Engine Runs Rich/Lean

When the engine runs rich or lean, this can usually be traced back to either the EGO (exhaust gas oxygen) sensor or a faulty processor.

Relieving Fuel System Pressure

All fuel injected systems employ a high-pressure electric fuel pump to deliver sufficient pressure to the injectors to ensure atomization of the injected fuel. This fuel pressure is maintained when the engine is shut off for long time periods through the use of a fuel pressure regulator or accumulator on some systems. Failure to relieve fuel system pressure before performing service work on the fuel system can lead to fuel being sprayed all over the engine compartment—which could cause a hazardous condition such as a fire or personal injury from fuel being sprayed into your eyes.

SPECIAL NOTE: Before attempting to perform any service work on the MPEFI system, ensure that the fuel pressure is bled off or relieved. This can be accomplished by attaching special tool P/N T80L-9974-A or an equivalent unit to the fuel manifold pressure relief valve as shown in Figure 8–26 earlier under the heading "Isolating a Faulty Injector." (This tool is shown in "Special tools," Figure 8–28.) Always wear safety goggles when relieving fuel system pressure. Also, remove the fuel tank cap to relieve fuel tank pressure during this procedure. A detailed explanation of this procedure entitled "Relieving Fuel System Pressure" is given under the CFI system earlier in this chapter.

Procedure

1. Turn off the ignition key.
2. Remove air cleaner assembly.
3. Special Tool T80L-9974-A is required for use with the CFI (central point fuel injection) or throttle body injection system; on the MPEFI system, a vacuum pump such as the Mityvac type unit is required (both of these tools can be seen under Special tools, Figure 8–28.
4. 5.0L (302 cu.in.) V8 engines and 3.8L (232 cu.in.) V6 engines with CFI are equipped with a pressure relief valve on the throttle body that is accessible after removing the air cleaner assembly [Figures 8–1(a) and (b)].
5. On 2.3L turbocharged EFI engines, the pressure relief valve is located in the flexible fuel supply tube about 12 inches back from where it connects to the engine fuel rail. (Figure 8–26)
6. On 1.6L EFI engines, the relief valve is connected to the fuel injection manifold/fuel rail. (Figure 8–26)

7. On CFI engines, attach the Rotunda Pressure Gauge Tool T80L-9974-A or an equivalent to the pressure relief valve and drain the system through the drain tube. If this special tool is unavailable, relieve the fuel system pressure by gently depressing the pin in the Schraeder-type valve located on top of the throttle body to allow fuel to leak from the valve into the throttle body assembly.
8. Electronic fuel injected (MPEFI) engines are equipped with a pressure relief valve in the fuel supply line in the engine compartment—except on T-Bird/Cougar and Mustang/Capri where it is located on the driver's side of the engine compartment in the flexible hose assembly that connects to the fuel rail on the engine.
9. To relieve fuel system pressure on MPEFI engines, first of all remove the fuel tank cap.
10. Disconnect the vacuum hose from the fuel pressure regulator located on the engine fuel rail.
11. Refer to Figure 8–29, and using the vacuum pump, apply about 25" Hg pressure (84.2 kPa) to the fuel pressure regulator, which will cause the fuel under pressure to be released into the fuel tank through the fuel return hose.

EQUIPMENT REQUIRED:

- Rotunda Self-Test Automatic Readout (STAR), No. 07-0004 with cable assembly No. 07-0010. Refer to STAR operation.
- Analog volt-ohmmeter (VOM), 0 to 20v DC, (alternate to STAR). Refer to Appendix A.

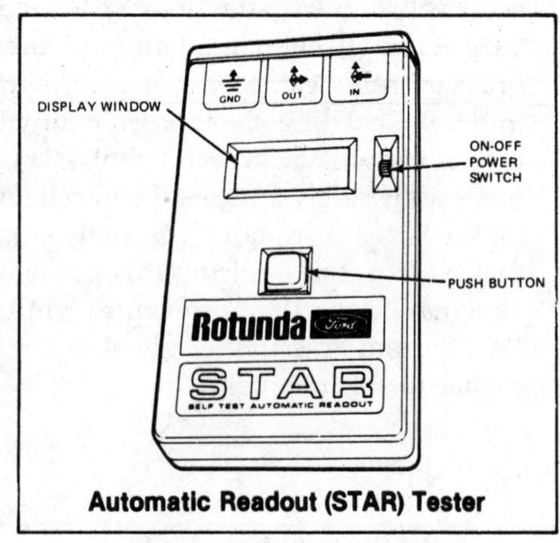

Automatic Readout (STAR) Tester

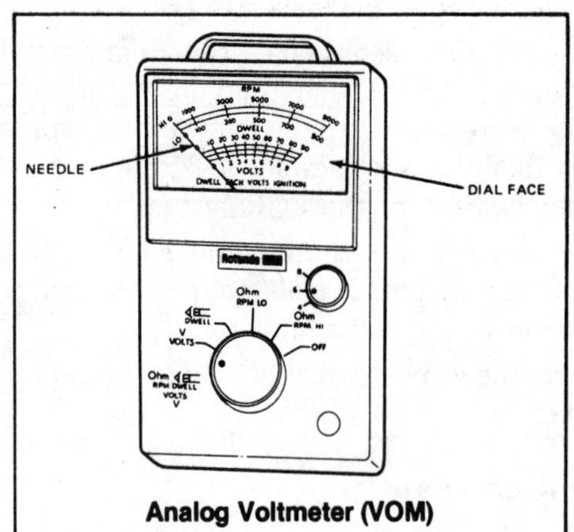

Analog Voltmeter (VOM)

- Jumper wire.
- Vacuum gauge, Rotunda 59-0008 or equivalent. Range 0-30 in. Hg. Resolution 1 in. Hg.

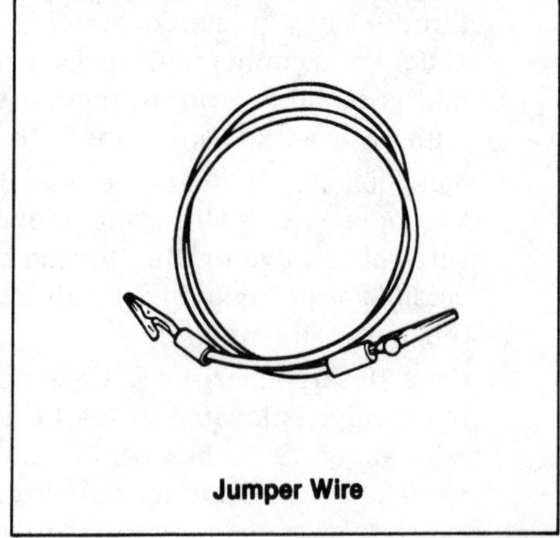

Jumper Wire

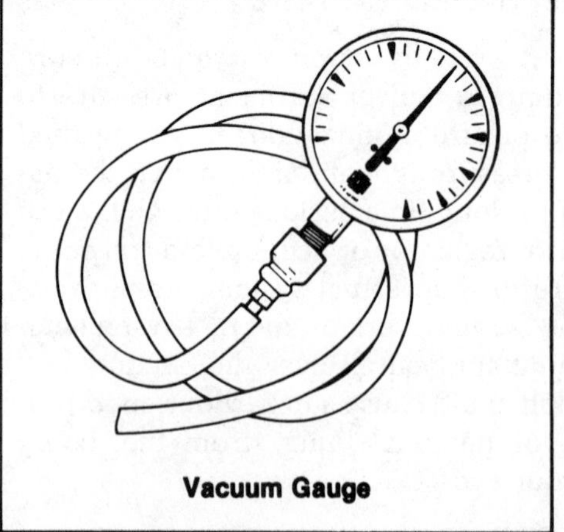

Vacuum Gauge

Figure 8–28. EFI System test equipment and special tools. (Courtesy of Ford Motor Company)

- **Tachometer, Rotunda No. 59-0010 or equivalent. Range 0-6,000 rpm. Accuracy ± 40 rpm. Resolution 20 rpm.**
- **Breakout Box, Rotunda T83L-50 EEC-IV or equivalent.**

Tachometer

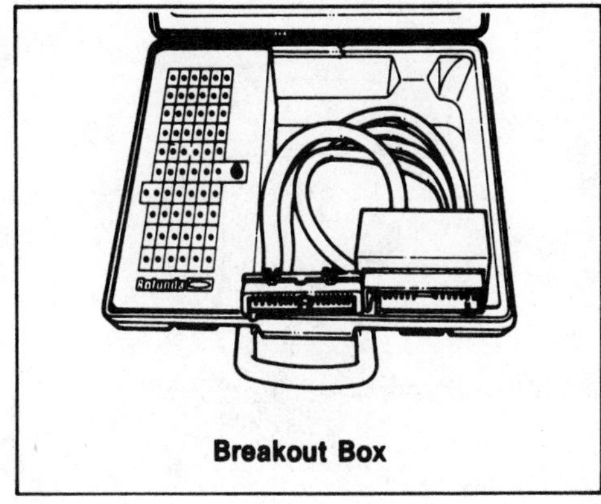

Breakout Box

- Vacuum pump, Rotunda No. 21-0014 or equivalent. Range 0-25 in. Hg.
- Digital volt-ohmmeter (DVOM), Rotunda No. 15-0031 or equivalent. Input impedance 10 Megaohm minimum.

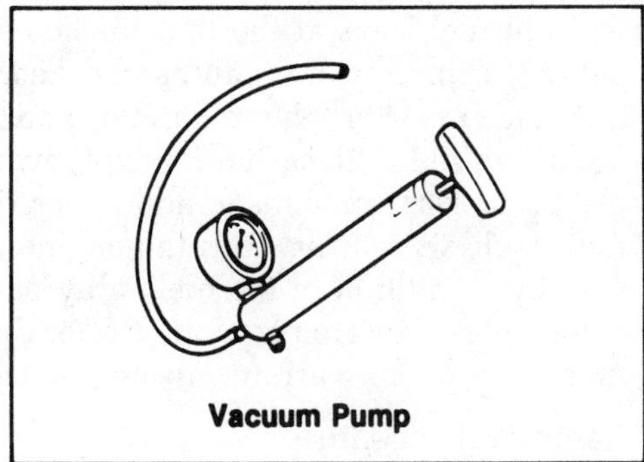

Vacuum Pump

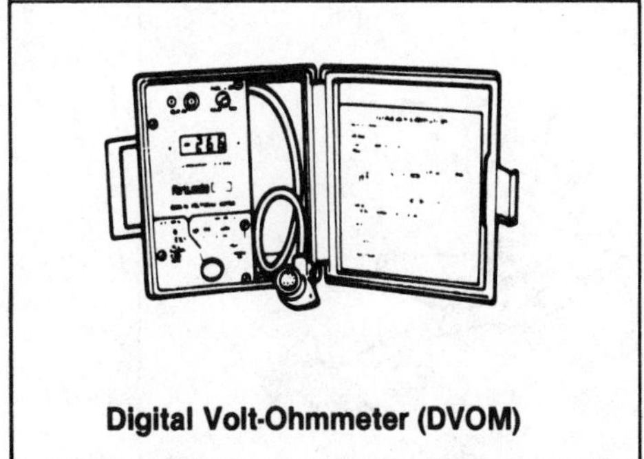

Digital Volt-Ohmmeter (DVOM)

- Electronic Fuel Injection Pressure Gauge EFI/CFI only, Rotunda No. T80L-9974-A or equivalent.
- Spark tester Tool D81P-6666-A or equivalent.

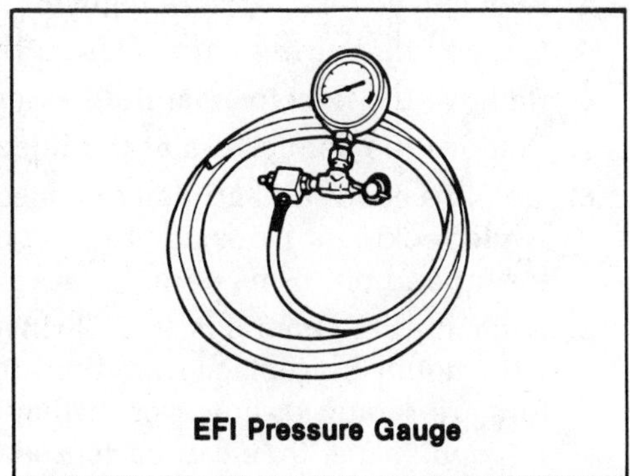

EFI Pressure Gauge

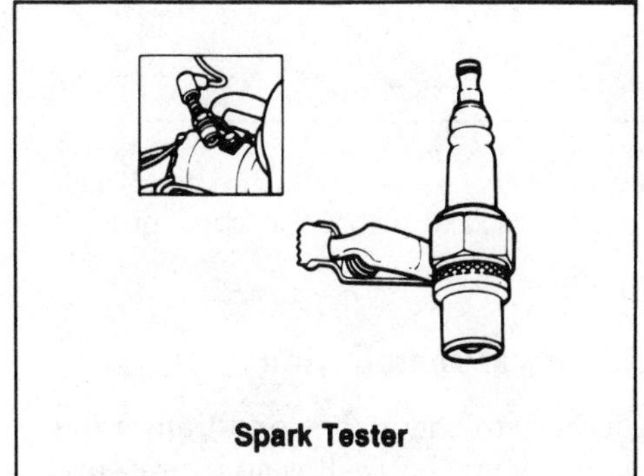

Spark Tester

Figure 8-28. *[Continued]*

- Timing light, Rotunda Tool No. 27-0001 or equivalent.
- Non-powered Test lamp.

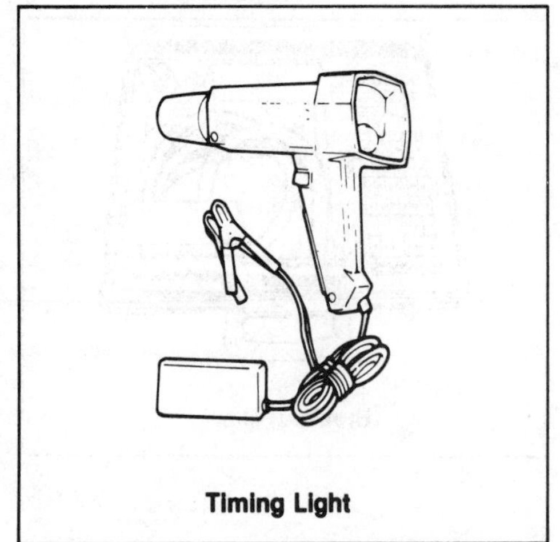

Timing Light

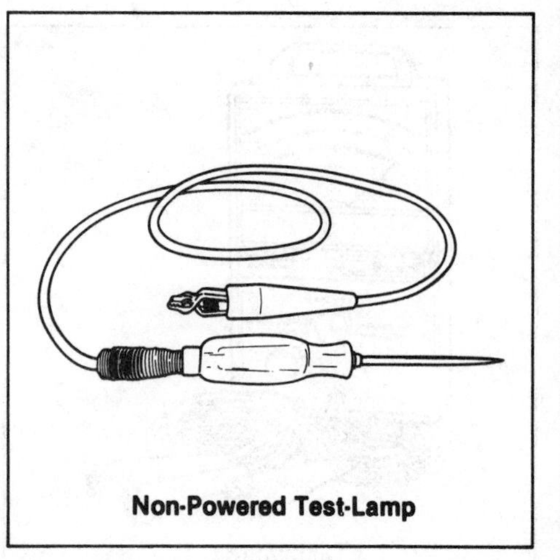

Non-Powered Test-Lamp

Figure 8-28. [*Continued*]

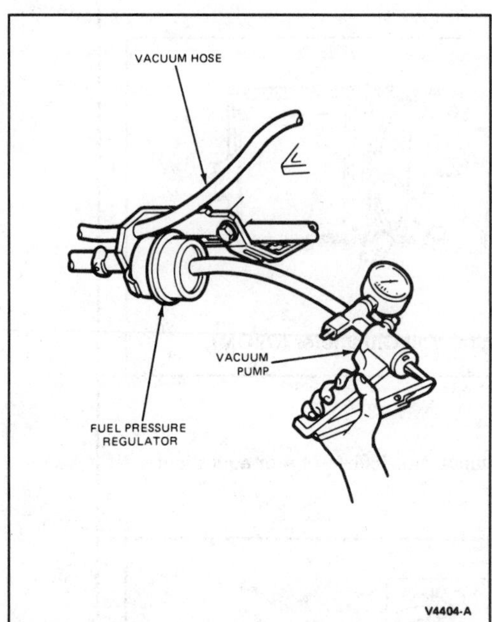

Figure 8-29. Relieving fuel system pressure on MPEFI-equipped engines with a vacuum pump. [Courtesy of Ford Motor Company]

Fuel Injector Removal/Installation

If it becomes necessary to check the condition of the injectors or replace them, the fuel system pressure must be relieved first as explained above. Before removing/condemning an injector, check that the voltage supply to each injector is the same. Check for air leaks at the injector-to-intake manifold as well as signs of fuel leaks at the injector bodies. The spray pattern of the injector can be checked by removing the injector from the manifold and plugging its mounting hole. Place the injector over/into a clear plastic or glass receptacle and cover the top of the receptacle with a rag. Run the engine and carefully note the condition of the fuel spray pattern into the receptacle. Poor atomization or after dribble are sufficient causes to warrant injector replacement.

Removal Procedure

1. Depressurize the system.
2. Disconnect the battery.
3. Disconnect both the fuel supply and return lines.
4. Remove the fuel pressure regulator vacuum line.
5. Unhook the wiring harness from the injectors.
6. Remove the injector manifold assembly.
7. Remove the connectors at the injector(s).
8. Obtain a good solid grip on the injector body and while rocking it from side to side, pull the injector up and out of its bore.
9. Each injector contains two O-rings that generally should be replaced anytime that the injectors are removed; however, if they appear to be in good shape, they can be reused.
10. Carefully inspect the injector plastic hat that covers the pintle at the injector tip as well as the washer; again these two items generally require

replacement, but if they are in good condition they can be reused. Figure 8–24 illustrates the injector O-ring seals and the injector plastic hat (protection cap).

NOTE: If a plastic hat is missing from the injector tip area, it will more than likely be found inside the intake manifold; make sure that you retrieve it before reinstalling the injector.

Injector Installation Procedure

1. When installing the injector(s), ensure that you always lubricate the injector O-rings with a light oil such as Ford ESF-M6C2-A or equivalent.

   SPECIAL NOTE: Never lubricate the injectors with a silicone grease because this material will tend to clog the injectors.

2. Install the injectors with a light, twisting/pushing motion.
3. Gently place the fuel injector manifold assembly back into position over the injectors, then bolt it into place and torque the bolts to 15–22 lb.ft. (20 to 30 N·m).

Special Tools—EFI and CFI Systems

A number of special tools are readily available through the Ford service dealerships or often through local tool/equipment supply houses. In some cases, the special tool can be substituted by using an equivalent make, and in some cases, regular shop tools may be open to use that will complete the same job.

The illustrations in Figure 8–28 show each of the tools that are required to service the CFI and EFI (electronic fuel injection) systems.

Fuel System Inertia Switch

Ford employs a fuel system inertia switch that is similar in operation to that employed on Robert Bosch fuel injection systems for safety reasons in case of a vehicle accident.

This inertia switch will cut off power to the electric fuel pump by inertia (tendency of a body to want to keep moving in the same direction) when the vehicle is suddenly brought to a stop—such as when involved in an accident or when the vehicle turns upside down. In the event of a collision, the electrical contacts in the inertia switch open and the fuel pump automatically shuts off.

Once an engine has stopped because of this action, it cannot be restarted again until the inertia switch has been reset manually. To reset the inertia switch, locate it in the luggage compartment of the left hinge support on Lincoln Town Car, Crown Victoria/Mercury Grand Marquis (Sedan only), T-Bird, Cougar, and Mustang/Capri 3-door models (shown in Figure 8–30). The Mustang/Capri switch is located near the left-hand corner of the spare tire well on the rear body panel on three-door models. The switch is located in the left side storage compartment on Ford Crown Victoria/Mercury Grand Marquis station wagons.

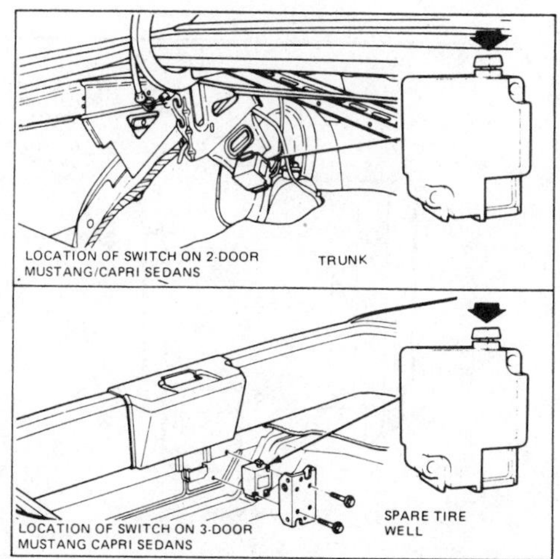

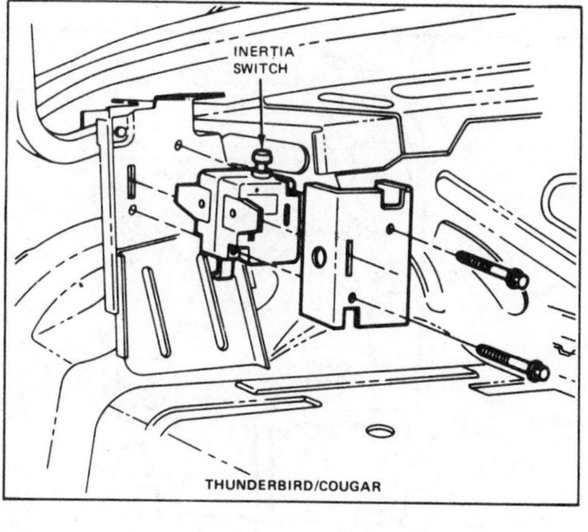

Figure 8–30. Fuel inertia switch locations. (Courtesy of Ford Motor Company)

On Escort/Lynx models, the inertia switch is located on the left-hand side of the vehicle behind the trim panel that is behind the rearmost seat. On EXP models, the switch is located on the left side of the vehicle inside the rear quarter panel shock tower door.

To reset the inertia switch, depress the button on the switch. Earlier inertia switches sometimes had two buttons.

Inertia Switch Operation

Figure 8–31 illustrates a cutaway view of an inertia switch unit used on all vehicles equipped with either CFI or EFI systems. The inertia switch contains a steel ball held in place by a magnet that is strong enough to hold the ball in place under normal driving conditions. If, however, the vehicle is involved in an accident whereby it is brought to a sudden stop by an impact, the steel ball will break loose from the magnet and roll up a conical ramp. The force of the steel ball hitting a target plate causes the inertia switch electrical contacts to open, thereby cutting off power to the electric fuel pump.

Test/Reset Inertia Switch

If a condition arises whereby no electrical power is available to the electric fuel pump(s), check the inertia switch assembly first.

Procedure

1. Always make sure that the ignition key is turned off.
2. CAUTION: If there is any sign of a leak or smell of gasoline, *do not* attempt to reset the inertia switch until you check the following: make sure that there is no leaking fuel in the engine compartment, at the fuel lines, or at the fuel tank(s) before you attempt to reset the inertia switch—otherwise an engine compartment or vehicle fire could result.
3. If no leaks are evident, push the reset button on the top of the switch.
4. Turn on the ignition key for a few seconds, then turn it off again and double-check for any signs of fuel leaks.
5. If the inertia switch doesn't seem to work when the reset button is pushed, use the DVOM (digital volt meter) with the LOS (limited operational strategy) button on and measure the voltage across both terminals of the switch. If the reading is greater than 0.3 volt, replace the inertia switch assembly.

CFI Throttle Body Assembly

The throttle body assembly used with the 5.0L (302 cu.in.) V8 is illustrated in Figure 8–32. The throttle body assembly used with the 3.8L (232 cu.in.) V6 is similar in layout to that for the 5.0L engine.

Removing the throttle body assembly is similar to that for removing a carburetor; however, *always* release the fuel system pressure as stated under "Relieve Fuel System Pressure" before performing any service work on the fuel system.

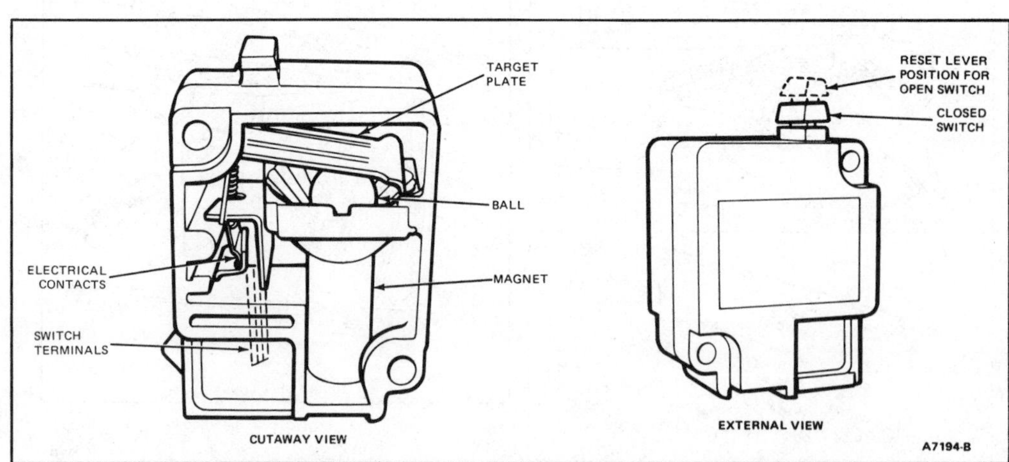

Figure 8–31. Cutaway view of CFI/EFI inertia switch. (Courtesy of Ford Motor Company)

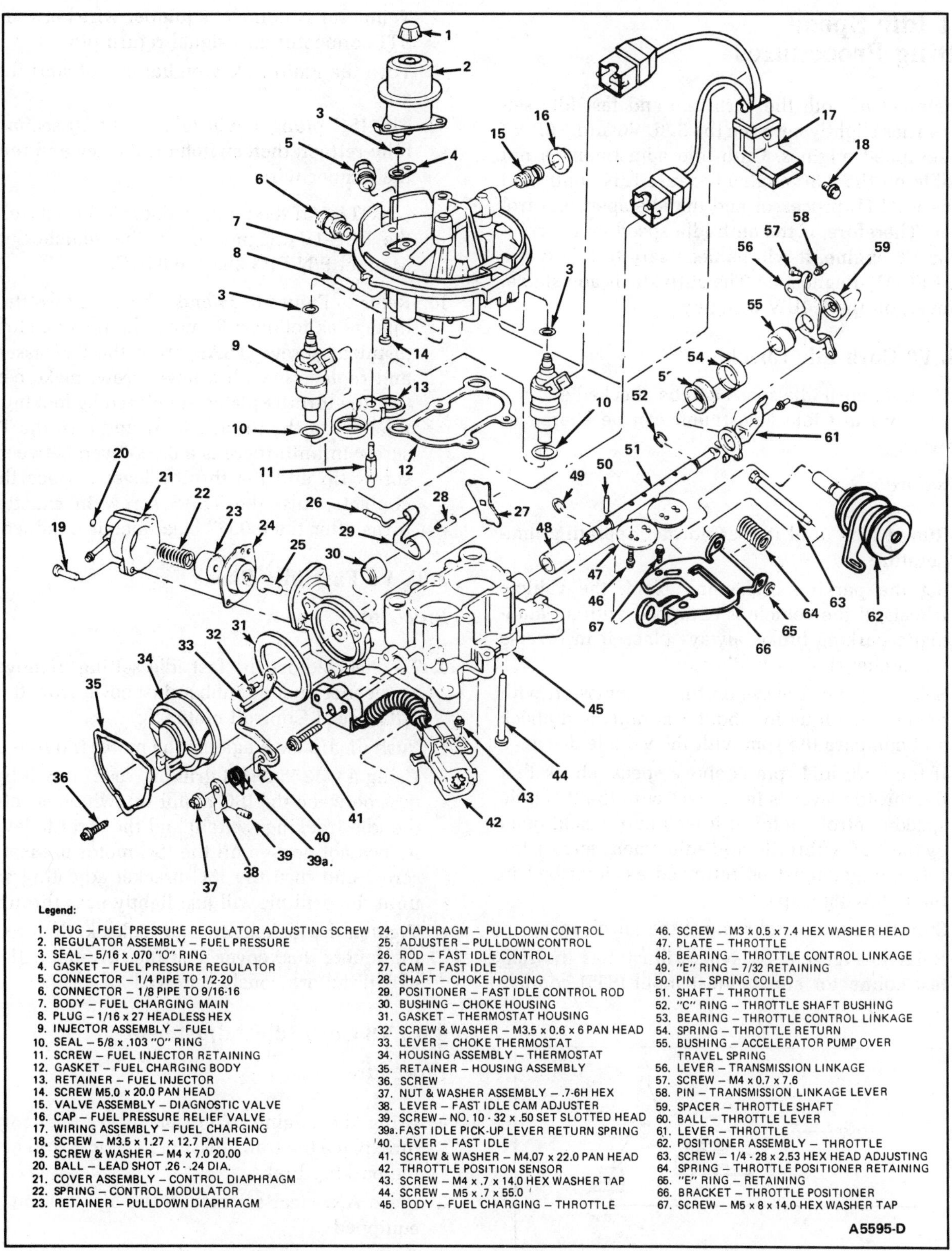

Legend:
1. PLUG – FUEL PRESSURE REGULATOR ADJUSTING SCREW
2. REGULATOR ASSEMBLY – FUEL PRESSURE
3. SEAL – 5/16 x .070 "O" RING
4. GASKET – FUEL PRESSURE REGULATOR
5. CONNECTOR – 1/4 PIPE TO 1/2-20
6. CONNECTOR – 1/8 PIPE TO 9/16-16
7. BODY – FUEL CHARGING MAIN
8. PLUG – 1/16 x 27 HEADLESS HEX
9. INJECTOR ASSEMBLY – FUEL
10. SEAL – 5/8 x .103 "O" RING
11. SCREW – FUEL INJECTOR RETAINING
12. GASKET – FUEL CHARGING BODY
13. RETAINER – FUEL INJECTOR
14. SCREW M5.0 x 20.0 PAN HEAD
15. VALVE ASSEMBLY – DIAGNOSTIC VALVE
16. CAP – FUEL PRESSURE RELIEF VALVE
17. WIRING ASSEMBLY – FUEL CHARGING
18. SCREW – M3.5 x 1.27 x 12.7 PAN HEAD
19. SCREW & WASHER – M4 x 7.0 20.00
20. BALL – LEAD SHOT .26 - .24 DIA.
21. COVER ASSEMBLY – CONTROL DIAPHRAGM
22. SPRING – CONTROL MODULATOR
23. RETAINER – PULLDOWN DIAPHRAGM
24. DIAPHRAGM – PULLDOWN CONTROL
25. ADJUSTER – PULLDOWN CONTROL
26. ROD – FAST IDLE CONTROL
27. CAM – FAST IDLE
28. SHAFT – CHOKE HOUSING
29. POSITIONER – FAST IDLE CONTROL ROD
30. BUSHING – CHOKE HOUSING
31. GASKET – THERMOSTAT HOUSING
32. SCREW & WASHER – M3.5 x 0.6 x 6 PAN HEAD
33. LEVER – CHOKE THERMOSTAT
34. HOUSING ASSEMBLY – THERMOSTAT
35. RETAINER – HOUSING ASSEMBLY
36. SCREW
37. NUT & WASHER ASSEMBLY – .7-6H HEX
38. LEVER – FAST IDLE CAM ADJUSTER
39. SCREW – NO. 10 - 32 x .50 SET SLOTTED HEAD
39a. FAST IDLE PICK-UP LEVER RETURN SPRING
40. LEVER – FAST IDLE
41. SCREW & WASHER – M4.07 x 22.0 PAN HEAD
42. THROTTLE POSITION SENSOR
43. SCREW – M4 x .7 x 14.0 HEX WASHER TAP
44. SCREW – M5 x .7 x 55.0
45. BODY – FUEL CHARGING – THROTTLE
46. SCREW – M3 x 0.5 x 7.4 HEX WASHER HEAD
47. PLATE – THROTTLE
48. BEARING – THROTTLE CONTROL LINKAGE
49. "E" RING – 7/32 RETAINING
50. PIN – SPRING COILED
51. SHAFT – THROTTLE
52. "C" RING – THROTTLE SHAFT BUSHING
53. BEARING – THROTTLE CONTROL LINKAGE
54. SPRING – THROTTLE RETURN
55. BUSHING – ACCELERATOR PUMP OVER TRAVEL SPRING
56. LEVER – TRANSMISSION LINKAGE
57. SCREW – M4 x 0.7 x 7.6
58. PIN – TRANSMISSION LINKAGE LEVER
59. SPACER – THROTTLE SHAFT
60. BALL – THROTTLE LEVER
61. LEVER – THROTTLE
62. POSITIONER ASSEMBLY – THROTTLE
63. SCREW – 1/4 - 28 x 2.53 HEX HEAD ADJUSTING
64. SPRING – THROTTLE POSITIONER RETAINING
65. "E" RING – RETAINING
66. BRACKET – THROTTLE POSITIONER
67. SCREW – M5 x 8 x 14.0 HEX WASHER TAP
A5595-D

Figure 8-32. CFI Throttle body exploded view—5.0L V8 engine. (Courtesy of Ford Motor Company)

CFI Idle Speed Setting Procedures

Adjustment of both the curb idle and fast idle settings varies slightly between the 3.8L V6 and 5.0L V8 CFI-equipped engines. Curb idle adjustment is not possible on the 3.8L engine because it is controlled by the EEC-1V processor and the idle speed control motor. Therefore, if the curb idle speed is incorrect on the V6 engine, it will be necessary to follow the 3.8L EEC-1V diagnostics. The curb idle is adjustable, however, on the 5.0L V 8 engine.

3.8L V6 Curb Idle Check

The 3.8L V6 curb idle speed can be checked and the throttle stop fast idle adjustment can be altered as follows:

Procedure

1. Run engine until it is at normal operating temperature.
2. Set the parking brake and block the vehicle wheels; if the vehicle is equipped with an automatic parking brake, always place it in *reverse* when checking curb idle rpm.
3. Select *drive* or *reverse* on the transmission with the engine idling for about 1 minute, and check and compare the rpm with the vehicle decal.
4. If the curb idle rpm is above specs, check that the throttle lever is in contact with the ISC (idle speed control) motor; if it isn't and is held open by the TSAS (throttle stop adjustment screw), the ISC plunger must be retracted as described in the following steps.
5. Stop the engine and remove the air cleaner.
6. Refer to Figure 8–33, which illustrates the self-test connector and self-test input (STI) connector.

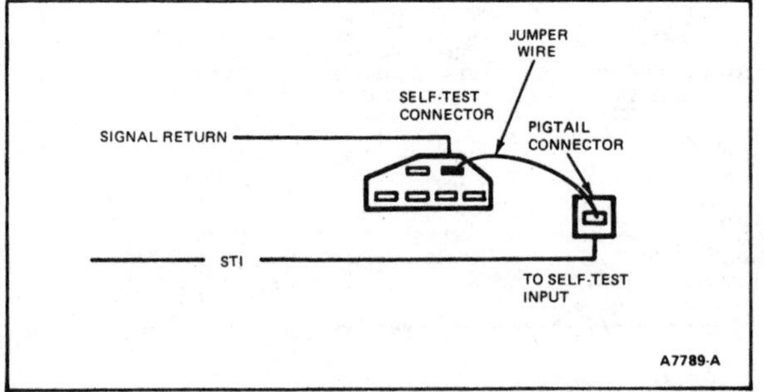

Figure 8–33. Jumper wire installation at STI and signal return pin for 3.8L V6 engine. (Courtesy of Ford Motor Company)

7. Temporarily connect a jumper wire between the STI connector and signal return pin.
8. Turn the ignition key *on* but *do not* start the engine.
9. The ISC plunger will take about 10 seconds to fully retract; then switch off the key and remove the jumper wire.

   NOTE: If the plunger doesn't retract, refer to the EEC-1V diagnostics in the vehicle Service Manual P/N FPS 365-126-84HC.

10. Refer to Figure 8–34 and adjust the throttle stop linkage as follows: Remove the throttle stop adjustment screw (TSAS) from the CFI assembly and replace it with a new screw; make certain that the throttle plates are closed by looking into the bore of the throttle body and turn the TSAS screw in until there is a 0.005" gap between the screw tip and the throttle lever surface that it contacts; turn the TSAS screw in exactly 1.5 turns after the 0.005" clearance is obtained.

3.8L V6 Fast Idle Check

Procedure

1. To check/adjust the fast idle setting, remove the air cleaner and the rubber dust cover from the ISC (Idle Speed Control) motor tip.
2. Push the tip back against the motor (no free play).
3. Using a 9/32" (.281") drill bit, see if it will lightly pass between the ISC motor tip (which is actually the idle tracking switch) and the throttle lever; if it does not, loosen off the ISC motor bracket lock screw and turn the ISC bracket adjusting screw until the drill bit will just lightly pass through.
4. Tighten the lock screw and recheck, then replace the rubber dust cover over the ISC motor tip and install the previously removed air cleaner.

5.0L V8 Curb Idle Adjustment

Procedure

1. Bring the engine to normal operating temperature, place the transmission in *N* or *P*, and engage the parking brake.
2. Place A/C Heat selector in the *off* position, if so equipped.
3. Stop the engine; restart the engine and run it for 1 minute at 2000 rpm in N.
4. Let the engine return to idle speed for 30 seconds, then run it at 2000 rpm for more than 10 seconds in *N*; return to idle for about 10 seconds.

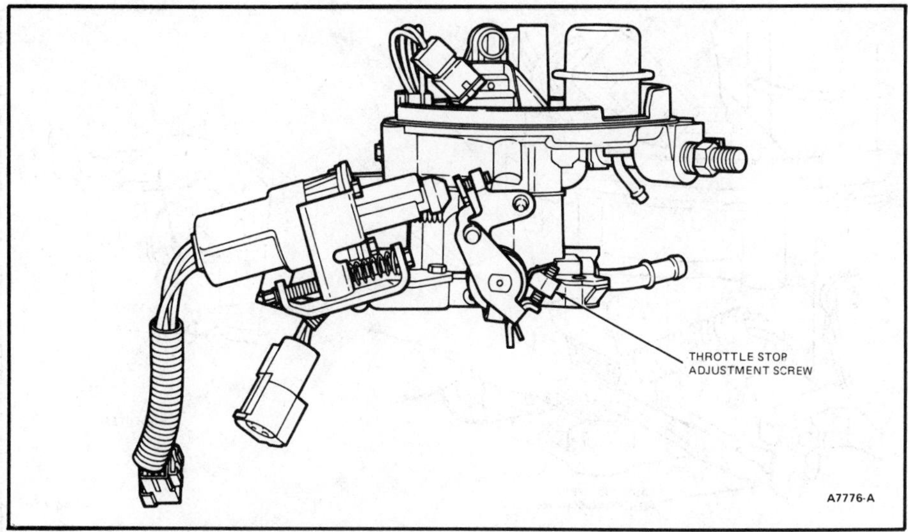

Figure 8–34. 3.8L V6 Engine TSAS on CFI throttle body (TSAS—throttle stop adjustment screw). (Courtesy of Ford Motor Company)

5. Place the transmission selector to *R* (reverse).

   SPECIAL NOTE: Idle speed should be adjusted with the transmission in *R* position only because if it is placed in *D* (drive), the parking brake will automatically be released. Do not press down on the brake pedal when setting curb idle speed because this causes the power brake booster (vacuum) to vary the idle rpm.

6. If idle adjustment is required, refer to Figure 8–35 and loosen off the saddle bracket locking screw to allow adjustment. Clockwise increases speed, counterclockwise decreases speed.

5.0L V8 Fast Idle Adjustment

Refer to Figure 8–36 for adjustment screw location on throttle body.

Procedure

1. Have engine at normal operating temperature; transmission in *N* or *P*.
2. Disconnect EGR valve vacuum hose and plug it.
3. Disconnect and plug vacuum hose at fast idle pulldown motor.
4. Disconnect and plug canister purge line.

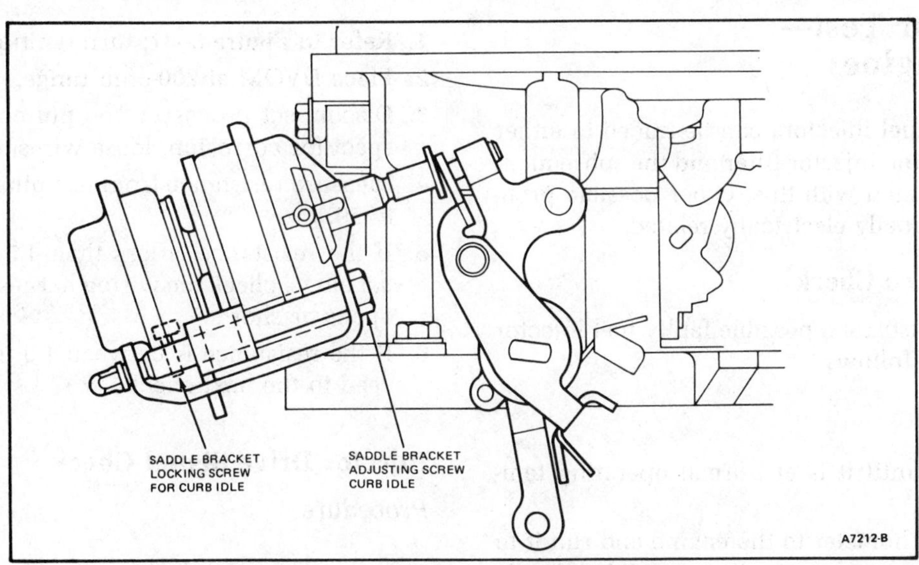

Figure 8–35. 5.0L V8 Engine curb idle adjustment screw—CFI unit. (Courtesy of Ford Motor Company)

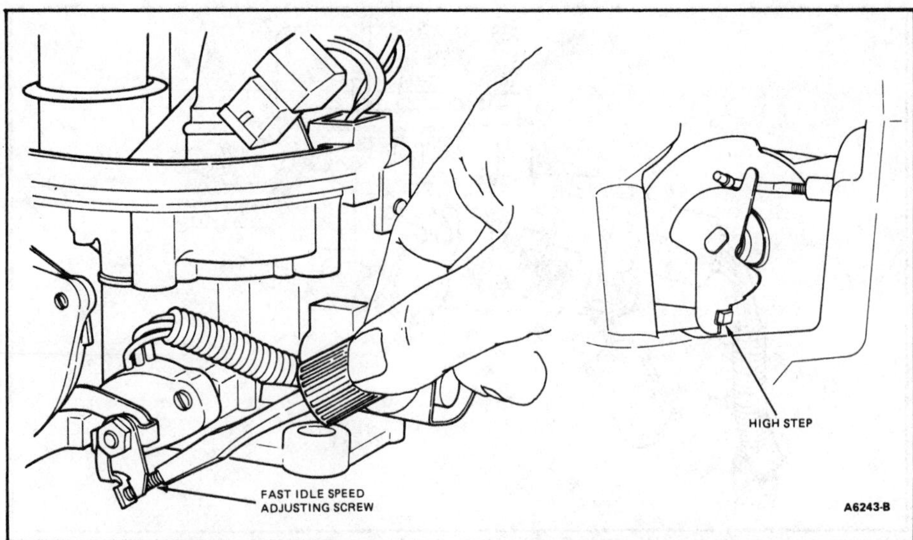

Figure 8–36. 5.0L V8 CFI Fast idle adjustment screw location. (Courtesy of Ford Motor Company)

5. In order to "initialize on-board computer," turn ignition off, then restart the engine.
6. Refer to Figure 8–36 and place the fast idle lever on the specified step of the fast idle cam.
7. Check/adjust the fast idle rpm.

   SPECIAL NOTE: On all vehicles except Mustang/Capri, the fast idle rpm must be set within 20–60 seconds after engine restart. If time is exceeded, repeat Steps 5, 6, and 7.

8. After adjustment, reconnect all disconnected vacuum hoses, etc.

CFI Injector Test— 5.0L V8 Engine

Problems with fuel injectors can be traced to either dirt or water in the injector filter and the subsequent problems associated with this. Other possible problem areas are usually electrically related.

Injector Balance Check

A quick way to isolate a possible faulty fuel injector is to proceed as follows.

Procedure

1. Run engine until it is at normal operating temperature.
2. Connect a tachometer to the engine and run it to 2000 rpm; you may have to disconnect the throttle kicker solenoid and use the throttle body stop screw to set the engine speed.
3. Isolate each injector one at a time by removing the electrical connector.
4. Carefully note the loss/drop in engine rpm when this is done—which should be at least 100 rpm, otherwise replace that injector.
5. Reset the curb idle and verify that the throttle kicker connection is satisfactory at the end of the test.

5.0L Engine—Injector Wiring Harness and Processor Connection Resistance Check

Procedure

1. Refer to Figure 8–37; turn ignition key *off*.
2. Place DVOM at 200-ohm range.
3. Disconnect processor #60 pin connector and inspect for corrosion, loose wires, or damage.
4. Measure resistance from test pins 57 to 58 and 57 to 59.
5. If the resistance is less than 1.5 or greater than 3.5 ohms, check the wiring harness connectors for opens or shorts.
6. If the resistance is between 1.5 to 3.5 ohms, proceed to the next check.

Injector Drive Signal Check

Procedure

1. You will require the "breakout box," Rotunda P/N T83L-50 EEC-1V or equivalent for this test (see Special tools, Figure 8–28).

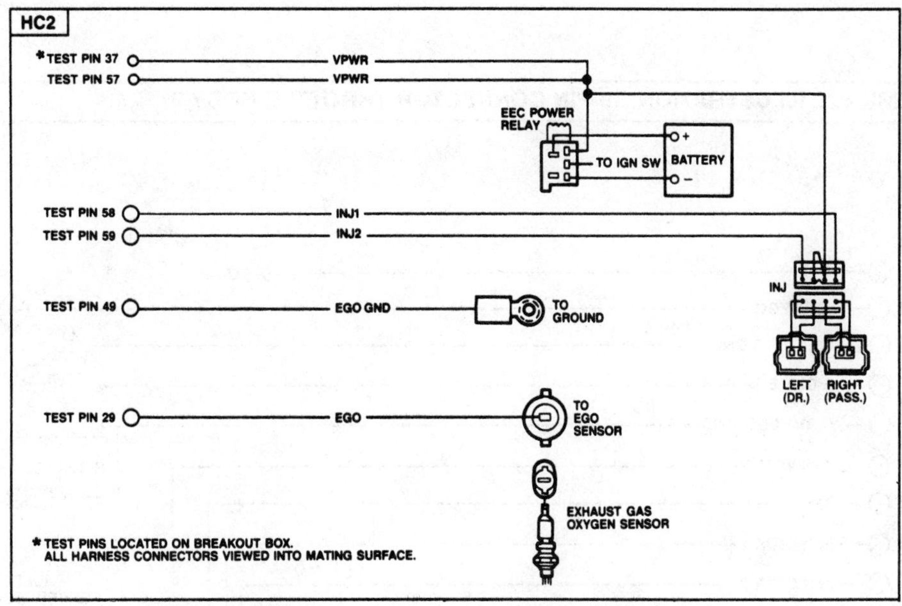

Figure 8–37. 5.0L/302 cu.in. V8 Engine injector/processor wiring harness arrangement. (Courtesy of Ford Motor Company)

2. You will also require a standard non-powered 12-volt test light.
3. Turn the ignition key off and install the breakout box; reconnect the processor.
4. Connect the 12-volt test light from test pin 57 to test pin 59 and note how brightly the test lamp glows.
5. Crank or start the engine and repeat Step 4 above.
6. If the test lamp has a dim glow on both tests (Steps 4 and 5), perform an injector balance test as explained above.
7. If the test lamp fails to light on one or both tests (Steps 4 and 5), make sure that there is 12 volts at pin 57; if voltage is present, then replace the processor unit.
8. If the test lamp glows brightly on one or both tests, check both injector circuits for shorts to ground; if OK, replace the processor unit.

CFI Injector Test—3.8L V6 Engine

To perform a quick check on the injector operation, proceed as follows.

Injector Balance Check Procedure

1. Bring engine to normal operating temperature.
2. Connect a tachometer to the engine.
3. Disconnect and plug the hose to the EGR valve.
4. Run the engine at 2000 rpm; you may have to disconnect the idle speed control and use the throttle body stop screw to set engine speed.
5. Isolate the left injector by unhooking its electrical connector.
6. The engine speed should drop at least 150 rpm; otherwise replace the fuel injector.
7. Repeat the same test for the right fuel injector.
8. If the engine speed does drop at least 150 rpm when each injector is disconnected, the problem is somewhere else in the fuel system.

NOTE: To clarify if the problem with an injector might be in the wiring harness, try switching the connectors from one to the other so that you can isolate each injector as in the above steps.

3.8L (232 cu.in.) V6 CFI Injector Wiring Harness Checks

1. Refer to Figures 8–38 and 8–39, which illustrate the two types of injector/processor wiring harness arrangements used with the 3.8L CFI fuel system.
2. Special tools required for the following electrical tests are the Rotunda DVOM P/N 15-0031 and Rotunda Breakout Box P/N T83L-50 EEC-1V or equivalent shown in Figure 8–28.
3. Short to battery at injector connectors—turn the ignition key off and wait 10 seconds.

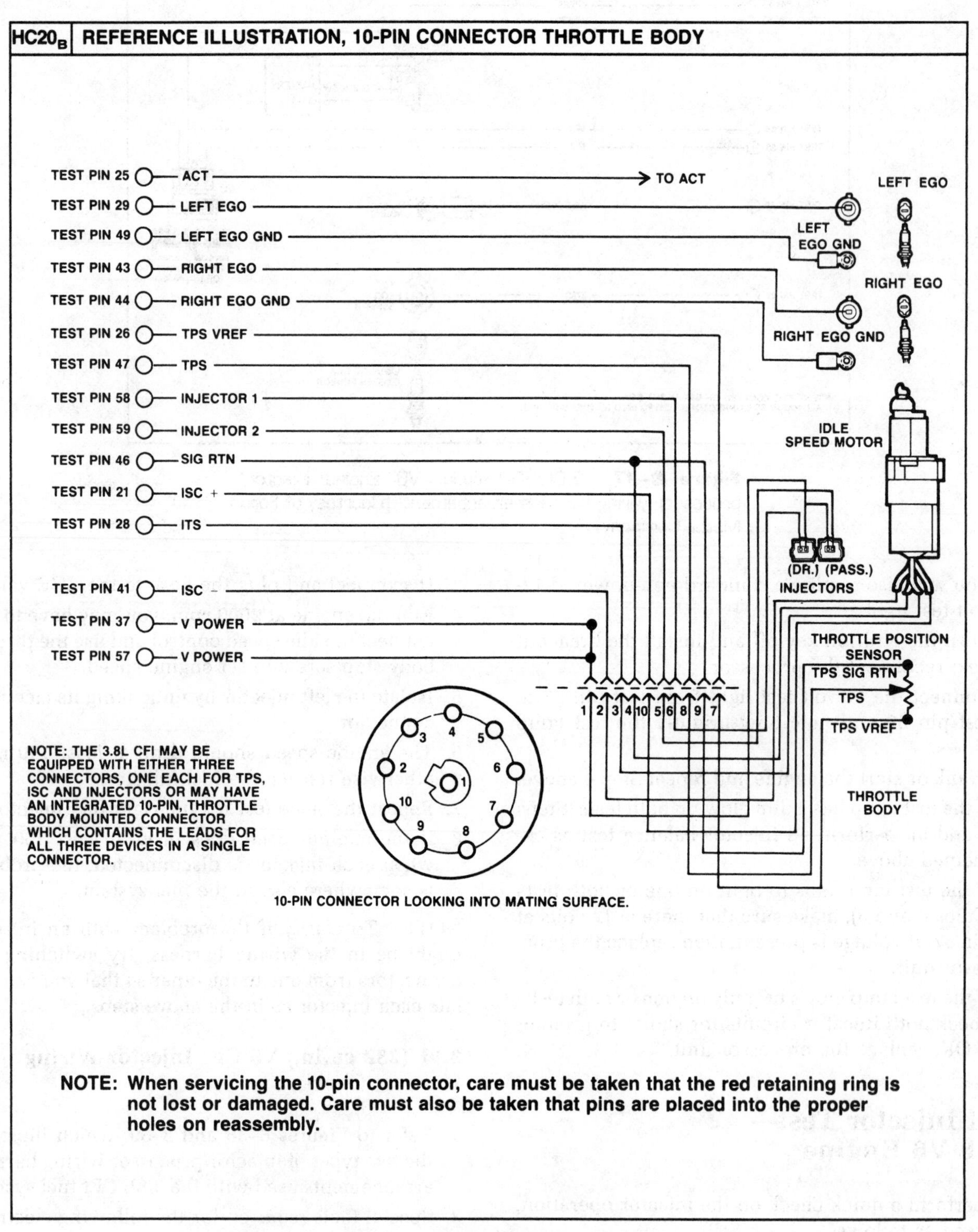
HC20B REFERENCE ILLUSTRATION, 10-PIN CONNECTOR THROTTLE BODY
TEST PIN 25 ACT
TO ACT
TEST PIN 29 LEFT EGO
TEST PIN 49 LEFT EGO GND
TEST PIN 43 RIGHT EGO
TEST PIN 44 RIGHT EGO GND
TEST PIN 26 TPS VREF
TEST PIN 47 TPS
TEST PIN 58 INJECTOR 1
TEST PIN 59 INJECTOR 2
TEST PIN 46 SIG RTN
TEST PIN 21 ISC +
TEST PIN 28 ITS
TEST PIN 41 ISC −
TEST PIN 37 V POWER
TEST PIN 57 V POWER
LEFT EGO
LEFT EGO GND
RIGHT EGO
RIGHT EGO GND
IDLE SPEED MOTOR
(DR.) (PASS.) INJECTORS
THROTTLE POSITION SENSOR
TPS SIG RTN
TPS
TPS VREF
THROTTLE BODY
NOTE: THE 3.8L CFI MAY BE EQUIPPED WITH EITHER THREE CONNECTORS, ONE EACH FOR TPS, ISC AND INJECTORS OR MAY HAVE AN INTEGRATED 10-PIN, THROTTLE BODY MOUNTED CONNECTOR WHICH CONTAINS THE LEADS FOR ALL THREE DEVICES IN A SINGLE CONNECTOR.
10-PIN CONNECTOR LOOKING INTO MATING SURFACE.
NOTE: When servicing the 10-pin connector, care must be taken that the red retaining ring is not lost or damaged. Care must also be taken that pins are placed into the proper holes on reassembly.

Figure 8–38. 3.8L CFI Engine 10 pin wiring harness connector. (Courtesy of Ford Motor Company)

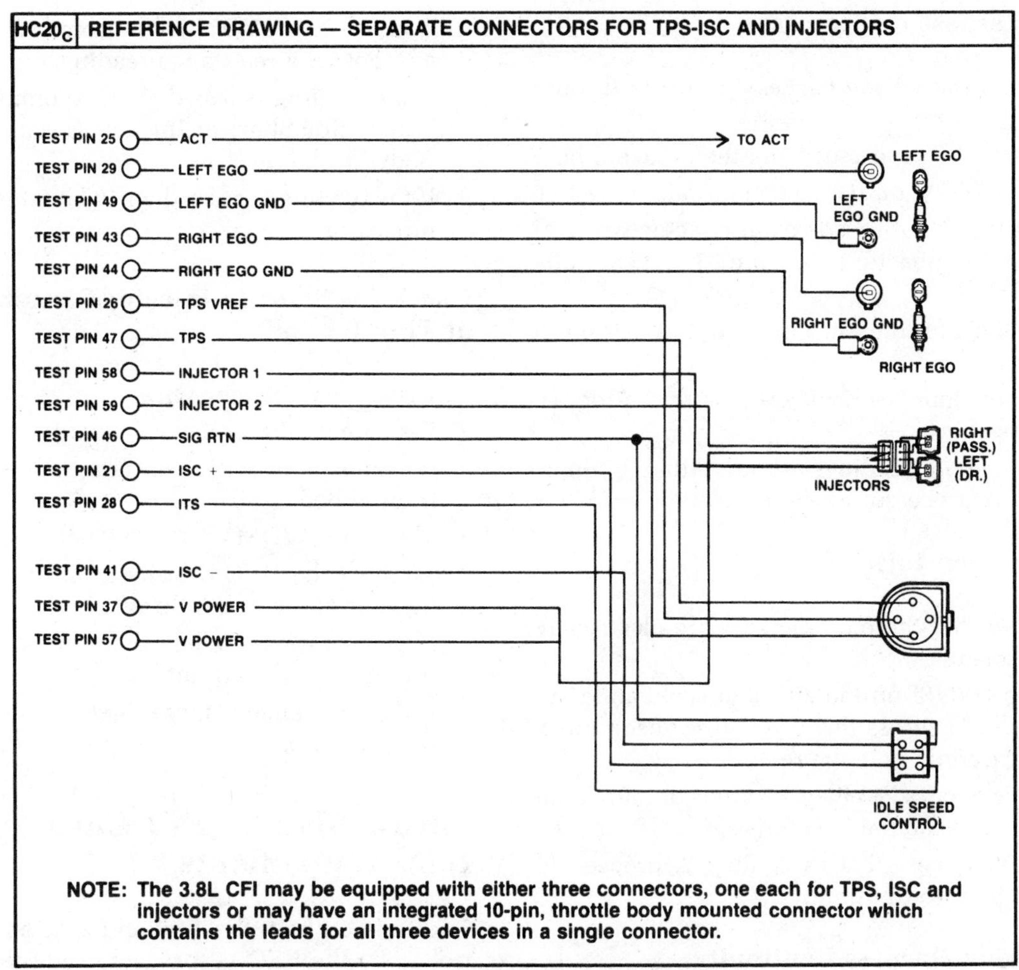

Figure 8-39. 3.8L CFI Engine separate wiring harness connectors. (Courtesy of Ford Motor Company)

4. Disconnect the injector wire connectors.
5. Place the DVOM scale at 20V and connect it between the injector connector and ground; the DVOM should read less than 1 volt; otherwise there is a short to ground somewhere (repeat on other injector).

Check Both Injectors for a Short to Ground

1. Turn ignition key off and wait 10 seconds; make sure that the injector wires are properly connected.
2. Install the breakout box.
3. Place the DVOM at the 200-ohm range and connect the DVOM leads between test pin 60 and pin 58 for the left injector and between 60 and 59 for the right injector test.
4. If the DVOM reads less than 5 ohms, proceed to "Isolate Short to Ground Check" below; if DVOM reads more than 5 ohms, proceed to "Check for Open Injectors" test below.

Isolate Short to Ground

1. Disconnect the wiring harness connector at the throttle body.
2. Place the DVOM on the 200-ohm scale.
3. Remove the wiring harness from each injector.
4. Connect the DVOM from No. 1 injector of the throttle body connector to ground and carefully note the resistance reading on the DVOM.
5. Repeat Step 4 for the No. 2 injector.
6. If the resistance reading is less than 5 ohms on either or both injectors, correct the short or replace the injector(s).
7. If the DVOM reading is less than 5 ohms, proceed to "Short in Harness or Processor" test.

Short in Harness or Processor

1. Disconnect the wiring harness from the throttle body.
2. Disconnect the processor from the breakout box.
3. Place the DVOM on the 200-ohm scale.
4. Connect the DVOM leads from injector No. 1 of the harness connector to test pin 58 and carefully note the reading.
5. Repeat Step 4 for the No. 2 injector, but to test pin 59.
6. If either or both readings are less than 5 ohms, repair the short.
7. If either or both readings are more than 5 ohms, replace the processor assembly.

Check for Open Injectors

1. Make sure that the fuel injectors are electrically disconnected.
2. Place the DVOM on the 200-ohm scale and connect the leads across the terminals of each injector and carefully note the readings.
3. If the resistance is less than 1.5 ohms or more than 2.5 ohms, replace the injector(s).
4. If the resistance is not as in Step 3, proceed to "Check Harness Continuity."

Check Wiring Harness Continuity

1. Make sure that the throttle body wiring harness is disconnected.
2. Place the DVOM scale at 200 ohms and connect the DVOM leads from No. 1 injector to test pin 58 and note the resistance reading.
3. Repeat Step 2 for No. 2 injector, but to pin 59.
4. If the resistance is less than 2 ohms, reconnect the wiring harness to the throttle body assembly, reconnect the processor to the breakout box, and proceed to "Check for Short Between Injector Inputs at the Processor" test.
5. If the resistance is more than 2 ohms, repair the wiring harness.

Check for Short Between Injector Inputs at Processor

1. Connect processor to the breakout box and make sure that the throttle body harness is connected.
2. Place the DVOM to the 2000-ohm scale.
3. Make sure that the injectors are connected to the wiring harness.
4. Connect the DVOM leads from test pin 58 to 59 and note the resistance reading.
5. If the reading is less than 10 kilohms, proceed to "Check for Short to Injector Circuits at Throttle Body" test.
6. If the reading is more than 10 kilohms, replace the processor.

Check for Short to Injector Circuits at Throttle Body

1. Disconnect injector wires.
2. Disconnect the processor from the breakout box.
3. Disconnect the wiring harness connector from the throttle body.
4. Place the DVOM on the 2000-ohm scale.
5. Connect the DVOM leads from injector No. 1 to injector No. 2 at the throttle body connector.
6. If the resistance is less than 10 kilohms, correct the short in the circuit.
7. If the resistance is more than 10 kilohms, replace the processor.

Ford 1.6L/2.3L EFI Curb Idle Adjustments

Curb idle on both the 1.6L and 2.3L engines is controlled by the EEC-1V processor and the idle speed control air bypass valve assembly. If the engine curb idle rpm is not within specs after performing the curb idle adjustment, it will be necessary to perform the appropriate EEC-1V diagnostics procedures as per the Ford Service Manual P/N FPS 365-126-84HC entitled "Emission Diagnosis/Engine Electronics."

1.6L Engine Procedure

1. The engine must be at its normal operating temperature before you can conduct this check.
2. Place the transmission in *N* or *P* on an automatic.
3. Disconnect the vacuum hose connector at the EGR solenoid and plug both lines.
4. Disconnect the idle speed control (ISC) electrical power lead.
5. The electric cooling fan *must* be on during the idle speed check/set procedure; otherwise a low idle speed will occur if you set it to specs and the fan cuts in later.
6. Start and run the engine at 2000 rpm for 1 minute.

7. Make sure that the hand brake is on and that the gear lever is in neutral on a manual transmission and in DRIVE for an automatic transmission (block wheels).
8. The curb idle speed must be checked and adjusted within a maximum of 2 minutes (120 seconds), otherwise the EEC processor will switch to a different operating mode; if the adjustment is not done within 2 minutes, switch the ignition key off and restart the engine.
9. Refer to Figure 8–40 and adjust the idle speed by turning the throttle plate screw clockwise to increase and counterclockwise to decrease the speed.
10. If the vehicle is equipped with an automatic transmission and the initial engine speed adjustment causes an increase or decrease of more than 50 rpm, refer to the A/T linkage adjustment procedure in Chapter 17 of the following:
    a. 1984 Ford Car Shop Manual, Powertrain, Lubrication, Maintenance All Models, P/N FPS 365-126-84D.
    b. 1984 Ford Car Shop Manual, Powertrain, Lubrication, Maintenance, Tempo/Topaz, Escort/Lynx, EXP P/N FPS 365-126-84E.
11. Switch the engine off, remove plugs from EGR vacuum lines, and reconnect.
12. Reconnect idle speed control (ISC) power lead.

2.3L Turbo Idle Adjustment Procedure

1. Bring engine to operating temperature and place the transmission in neutral; turn the A/C Heat selector off.

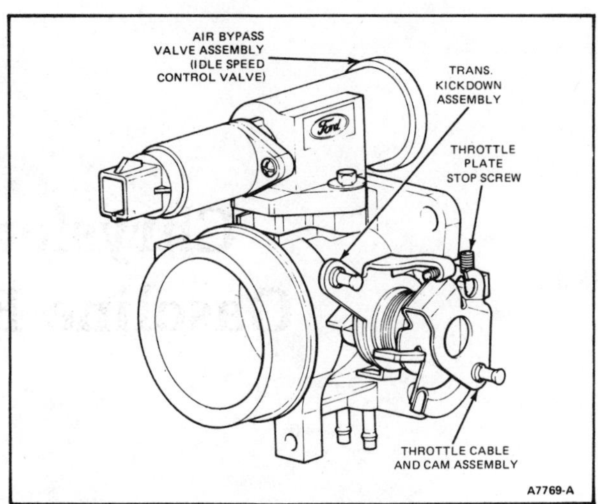

Figure 8–40. 1.6L EFI Engine—air intake charge throttle body assembly. [Courtesy of Ford Motor Company]

2. Disconnect the ISC air bypass valve power lead.
3. Start and run the engine at 2000 rpm for 2 minutes.
4. If the electric cooling fan comes on during idle speed set procedure, disconnect the electrical lead to the fan and proceed.
5. Let the engine idle and check the rpm, which should be 700–800 rpm.
6. Refer to Figure 8–41 and adjust the throttle plate stop screw to obtain the correct speed as per the vehicle EPA/manufacturer decal.
7. Reconnect disconnected components and start the engine; manually operate the throttle to make sure that the throttle plate is not stuck and that the engine idle speed is still correct.

NOTE: If excessive engine idle speeds are experienced when driving the vehicle on the idle system, turn the ignition switch to Off position and restart.

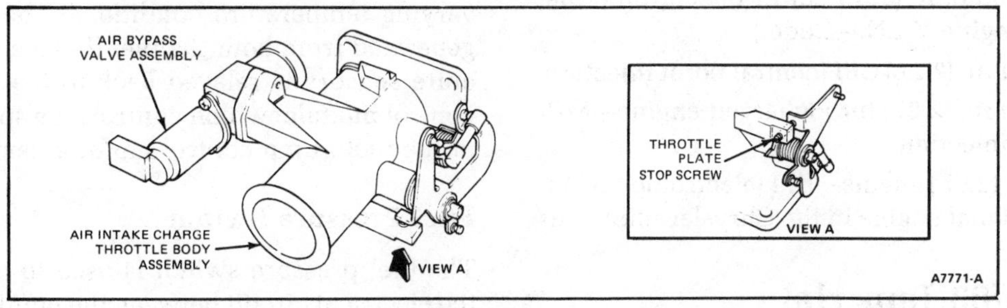

Figure 8–41. 2.3L EFI Turbo engine idle adjustment screw location. [Courtesy of Ford Motor Company]

9

Chrysler Electronic Gasoline Fuel Injection

Three basic types of EFI (electronic fuel injection) systems are employed by Chrysler Corporation on their vehicles. One system is of the TBI (throttle body injection) design and operates similarly to those employed by other vehicle manufacturers. It is used on their 4-cylinder engines with either a single injector, or in the case of the 1.6L turbocharged Dodge Colt, two throttle body mounted injectors are used. The second system is called multi-point EFI and is used on the 2.2L turbocharged engines found in the Laser/Daytona models that employ one injector for each cylinder. The third system used is not unlike the concept of the TBI design; however, it doesn't use individual electronically controlled fuel injectors such as we find in the TBI system. This third system was common to the 1981–83 Chrysler Imperial model.

The following information indicates what models of Chrysler products use which fuel system:

1979 V8 360 cu.in. (5904 cc) engine—EFM (electronic fuel metering), Engine V.I.N. Code—J and L

1981–84 V8 318 cu.in. (5215 cc) EFI (electronic fuel injection), Engine V.I.N.—Code J

1983–85 134 cu.in. (2.2L) CPI (central point injection)

1983–85 134 cu.in. (2.2L) turbocharged engine—MPI (multi-point injection)

1984 156 cu.in. (2.6L) engine—EFI (electronic fuel injection), Optional engine in the Chrysler mini-vans

EFI—1981/83 Imperial

This fuel system differs from others employed by Chrysler; it is shown in Figure 9–1. With this system, an in-tank electric fuel pump pushes fuel through a pair of parallel fuel filters to a control pump assembly, which is a positive displacement unit driven by a variable-speed electric motor. This control pump creates fuel system pressure between 24–60 psi (165–414 kPa), depending on engine speed/load demands.

This fuel under pressure flows through a fuel flowmeter and temperature sensor then on to a fuel pressure switch and into the fuel injection body assembly.

Fuel Flowmeter

The flowmeter assembly consists of a free-turning vaned wheel that is subjected to fuel flowing through it at all times. The greater the fuel flow through the flowmeter, the faster the speed of rotation of the vaned wheel. A light-emitting diode (LED) and a phototransistor are used with this arrangement so that as the vanes of the flowmeter rotate they will make/break the light path of the LED. The speed at which the light path is interrupted creates an electrical pulse condition that is monitored at the fuel flowmeter module assembly.

A fuel temperature sensor is used in conjunction with the flowmeter so that an accurate measurement of fuel rate (volume flow) can be recorded under varying temperature conditions. The voltage signal generated from both the fuel flowmeter and temperature sensors is relayed back to the electronic fuel control module, which controls the speed of rotation of the fuel pump control motor assembly.

Fuel Pressure Switch

The fuel pressure switch is used to open and close the electrical circuit between the ignition key and the fuel control pump motor. Normally, when fuel pressure is high enough to start and run the engine, this switch is in an open position. If however, the fuel system pressure was low due to the engine having been shut off for a long period, this switch would complete a bypass circuit any time that the ignition

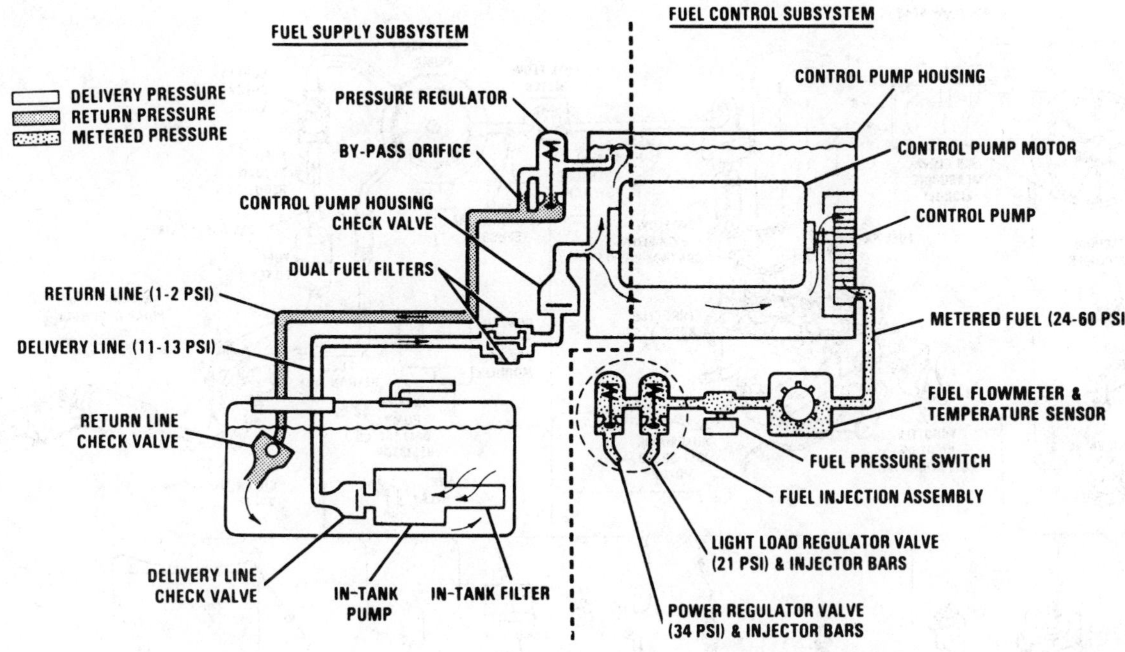

Figure 9–1. Chrysler EFI hydraulic system 81–83 Imperial with 318 cu.in. engine. (Courtesy of Chrysler Corporation)

key is turned to the start position. In this condition, the fuel switch would be closed, allowing the fuel control pump motor to be driven at its maximum rpm within 1–2 seconds in order to supply maximum system pressure for start-up. Fuel vapor lock is prevented through this arrangement.

Fuel Injection Assembly

The fuel injection assembly consists of two pressure-regulated valves that supply fuel under pressure to their own individual U-shaped fuel injection bars, which are located over the throttle body assembly.

The two pressure regulator valves are designed to split the fuel delivery system into a low and a high demand load arrangement. One of the valves operates to supply fuel to the throttle body "light load" injector bar when the pressure reaches or exceeds 21 psi (145 kPa). This light load circuit actually supplies not only all engine fuel requirements as long as the system fuel pressure is maintained between 21 and 34 psi (145–234 kPa) but also some of the demands beyond these fuel pressure ranges.

During initial engine starting, or when engine loads are high, the fuel control pump motor will rotate at its maximum speed in order to supply the greater fuel flow requirements and maintain sufficient delivery pressure. Under such a condition, the fuel pressure will always be in excess of 34 psi (234 kPa), therefore the second of the two regulator valves, known as the "power regulator valve," is forced open to deliver a greater supply of fuel to the engine.

Four air foil shaped holes (nozzles) located on the lower side of the U-shaped injector bar promote atomization of the fuel to ensure good mixing of the fuel and air.

Fuel Return Pressure Regulator

A fuel pressure regulator located within the fuel return line opens any time that the system fuel pressure exceeds a preset maximum; the regulator vents fuel back to the tank. Also within the pressure regulator is a bypass orifice that functions to vent fuel vapors from the control pump housing when the engine is shut off.

Although this system does not use an inertia switch to disengage the electric fuel pump (such as Ford and Robert Bosch Systems) in case of an accident, it does have a fuel return line check valve that will prevent a fuel loss if the vehicle rolls over in an accident.

EFI System Control Mechanism

The EFI system shown in Figures 9–1 and 9–2 is controlled through sensor inputs that are monitored/analyzed by the CCC (computer controlled combustion). Input voltages from the various engine sensors

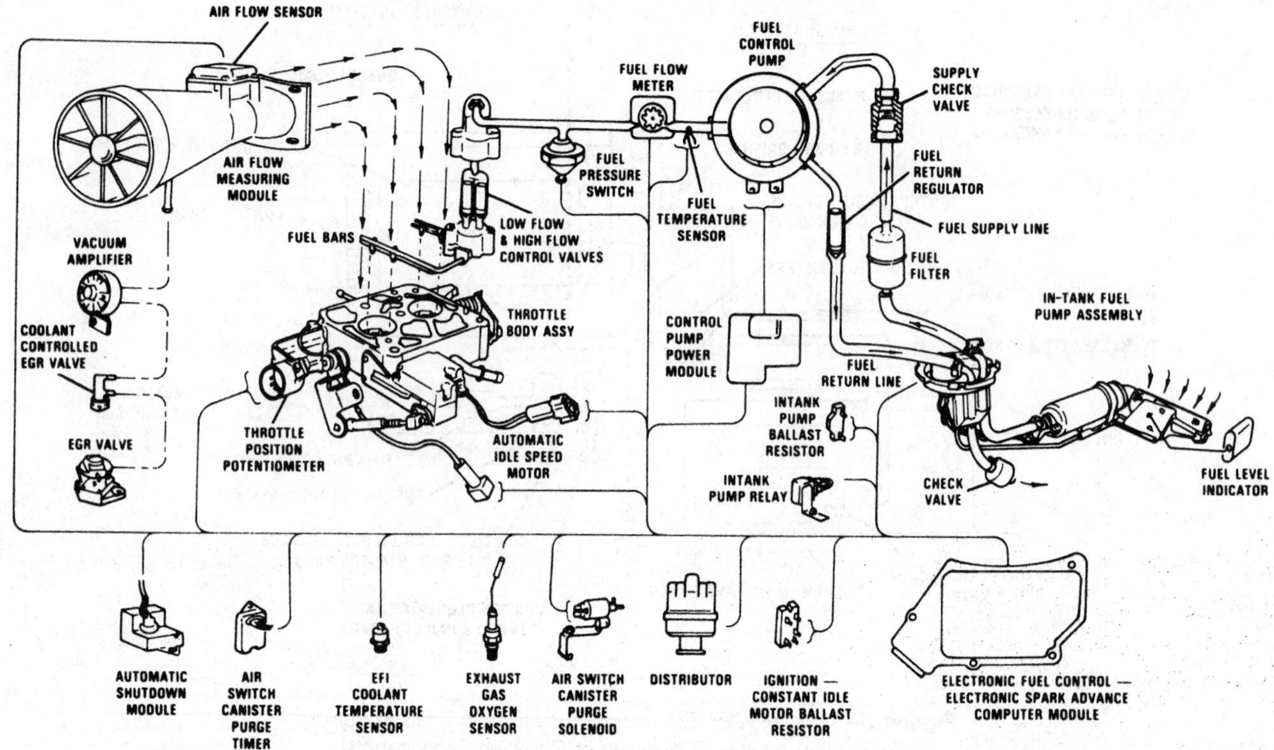

Figure 9-2. EFI air/fuel/ignition CCC system. (Courtesy of Chrysler Corporation)

to the CCC are then used to control the fuel flow and the ignition timing. The CCC operates in conjunction with a vehicle power module and a shutdown module.

In order to comply with EPA exhaust emissions, an oxygen (Lambda) sensor in the exhaust system monitors the concentration of oxygen ions flowing through the system and relays a voltage signal to the CCC. This system functions the same as all current oxygen sensors in that there is both an open- and closed-loop cycle of operation.

Basically, when the engine is started up or has failed to attain its normal operating temperature, the CCC is programmed to disregard the signal from the oxygen sensor; therefore, the CCC operates to control a preprogrammed air/fuel mixture to the engine. This is known as open-loop operation. Once the engine attains normal operating temperature, however, the oxygen sensor voltage signal causes the CCC to go into a closed-loop mode, and the air/fuel mixture is now influenced by this signal. For a detailed analysis of the operation/construction of an oxygen (Lambda) sensor refer to Chapter 6, dealing with Robert Bosch fuel systems.

The EFI system throttle body is similar to a carburetor system in that it contains a throttle plate (butterfly valve) mechanically connected to the accelerator pedal. In order to monitor the actual position of the throttle plate, a throttle position switch (variable resistance potentiometer) is mounted on the side of the throttle body housing. There is also a closed throttle switch and an automatic idle speed motor.

All of these units are designed to relay a position (voltage) signal to the CCC, which will alter the air/fuel ratio as required. Anytime that the accelerator pedal is released, the closed throttle switch energizes the automatic idle speed circuit and the CCC returns the ignition timing to its minimum position. If this system malfunctions, then the brake signal circuit acts as a backup because any time that the brakes are applied, the throttle is released simultaneously by the driver under normal driving conditions.

EFI Basic Troubleshooting

Basic pinpoint checks to perform to this system involve removing the air cleaner assembly as well as grounding the distributor cap secondary wire after removing it.

CAUTION: Do not smoke during this check.

Use a remote starter switch or have someone crank the engine while you look into the throttle body as-

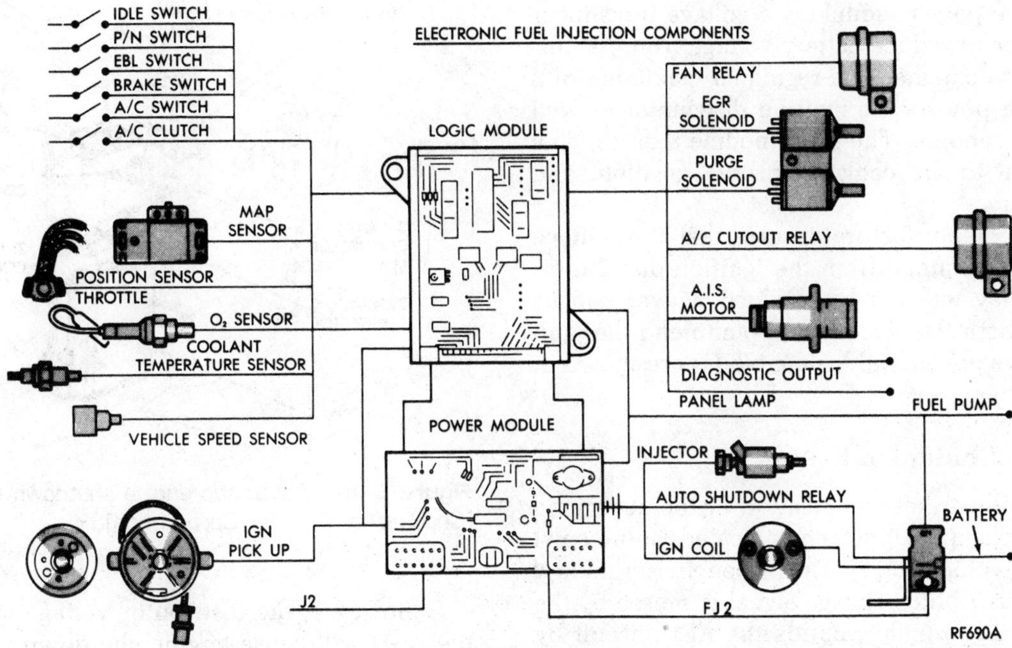

Figure 9–3. Electronic fuel injection component arrangement for the 2.2L engine—single point injection system. (Courtesy of Chrysler Corporation)

sembly to determine if fuel is in fact being sprayed from the injector bars. Sufficient fuel flow is generally indicative that the fuel system is operational; insufficient fuel flow would generally indicate that the fuel system requires service. However, if the amount of fuel flow seems to be excessive, check for possible flooding conditions.

If the fuel system appears to be operating, systematically check out the ignition system for problems.

2.2L Single Point Fuel Injection

The fuel injection system used with the standard non-turbocharged engines is of the TBI (throttle body injection) design. One CPI (central point injector) is utilized with these systems, which makes their operation common to other single point or throttle body systems now in use by other vehicle manufacturers.

NOTE: The 1.6L Dodge Colt turbocharged engine employs two injectors mounted in an injection mixer body (throttle body) arrangement.

The 2.2L fuel system is electronically controlled by a preprogrammed logic module unit that functions to regulate the air/fuel ratio, emission control devices, idle speed, and ignition timing. Typical sensor devices used with the 2.2L fuel system are illustrated in Figure 9–3, which depicts the basic layout of the electronic controls.

The sensors operate on the principle of varying resistance caused by a temperature/pressure increase through their pickups, which is transferred through a wire harness to the logic module. These voltage signals from the logic module are then relayed to the vehicle power module, which sends out the necessary corrected signals to change either the air/fuel ratio or, if necessary, the ignition timing.

If fault exists in the system, the logic module will actually store this information. When a mechanic/technician checks the system, the fault will be flashed by a light-emitting diode (LED) when the mechanic/technician hooks up a diagnostic read-out unit to access this information. The problem display code can then be referenced in the service information manual. This system is now used by many of the major vehicle manufacturers in one form or another so that the troubleshooting actions of the mechanic/technician are greatly simplified.

Power Module

This module is designed to provide the power source for the operation of the ignition coil as well as the fuel injector unit. In addition, the power module functions to actuate the automatic shutdown relay (ASD), which, when energized, will activate the fuel pump, ignition coil, and power module itself.

Within the power module is a voltage transducer that operates to reduce battery voltage from its normal 12 volts or higher to a regulated maximum of 8 volts, which powers the ignition distributor as well as the logic module. The logic module then sends a 5-volt signal to the coolant, throttle position, and MAP sensors.

A built-in safety feature ensures that if a voltage signal is not obtained from the ignition distributor, the ASD relay will not be actuated. Power will be shut off to both the electric fuel pump and the ignition coil. The power and logic modules are shown in Figures 9–4 and 9–5.

Automatic Shutdown Relay

Figure 9–6 illustrates the location of the automatic shutdown relay (ASD), which is located on the cowl side of the vehicle body. During engine cranking/starting, a distributor voltage signal is sensed by the power module, which grounds the ASD circuit by closing its contacts to complete the electric fuel pump, power module, and ignition coil circuits.

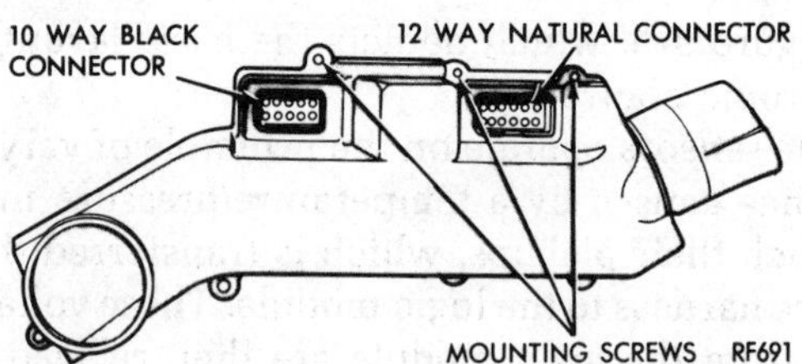

Figure 9–4. Power module unit. [Courtesy of Chrysler Corporation]

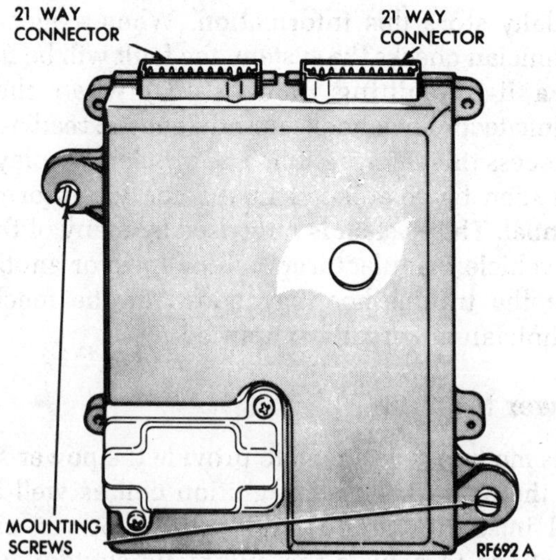

Figure 9–5. Logic module unit. [Courtesy of Chrysler Corporation]

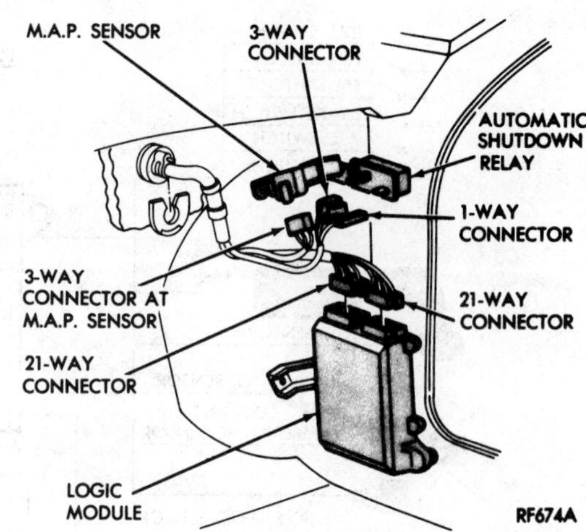

Figure 9–6. Automatic engine shutdown relay location. [Courtesy of Chrysler Corporation]

If, however, the distributor voltage signal is lost, the ASD will cause engine shutdown within 1 second by cutting off both the fuel and ignition electrical circuits.

To remove the ASD relay, it is necessary to first remove the glove box assembly in order to gain access to the relay.

Sensor Operation

As illustrated earlier in the system operation diagram, various sensors are used to feed voltage signals to the logic module while the engine is running. These signals are converted by the logic module and then sent to the power module, which will then monitor these signals with the preprogrammed inputs. If any correction to the engine fuel injector or ignition timing is required, then the power module will act accordingly.

Sensors used with the system are:

1. MAP (manifold absolute pressure)
2. Oxygen (O_2)
3. Coolant temperature
4. Throttle position
5. Vehicle speed

MAP Sensor. The MAP sensor shown in Figure 9–7 transmits intake manifold vacuum conditions and barometric pressure to the logic module. Engine load and throttle position will be sensed by the MAP sensor. This information is used with data from other sensors to determine the correct air/fuel mixture under all conditions of operation. The MAP sensor is usually located in the right side passenger compartment and is connected to a vacuum nipple on the

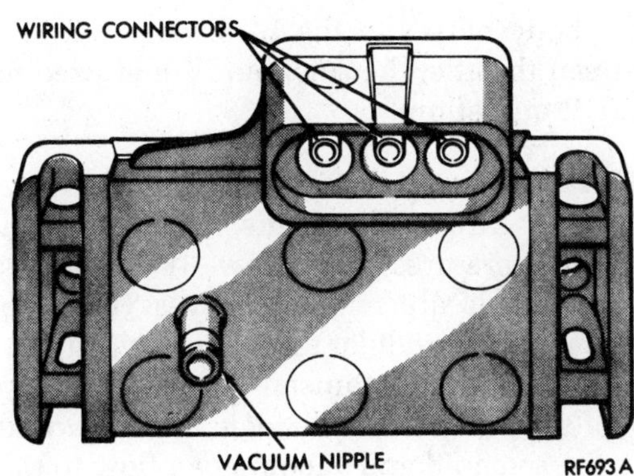

Figure 9–7. Manifold absolute pressure (MAP) sensor. (Courtesy of Chrysler Corporation)

throttle body unit; it is electrically connected to the logic module.

Oxygen Sensor. The oxygen sensor is often referred to as a Lambda sensor; its job is to sense/monitor the percentage of oxygen flowing through the exhaust system from the engine cylinders.

The body of the oxygen sensor is inserted into the exhaust manifold so that it is exposed at all times to exhaust gas flow while the other part of the body is exposed to the surrounding (ambient) air temperature. The body is made from a ceramic material (zirconium dioxide) and is fitted with electrodes composing a thin, gas-permeable platinum layer. In addition, a porous ceramic layer is provided at the end exposed to the exhaust gas flow to guard against contamination caused by combustion by-products (residues). The use of leaded fuel will render the oxygen sensor inoperable after a short time because of the residues contained in the leaded gasoline.

The operation of the oxygen sensor is based upon the fact that the ceramic material is oxygen-ion conductive once it reaches approximately 300°C (572°F); therefore, no voltage signal will radiate from the sensor until exhaust temperatures obtain this level.

Because of the differences in oxygen content that exist between the gas flow end and the surrounding air end, a voltage will be generated between the two interfaces. The generated voltage is representative of the difference between the oxygen in the exhaust manifold versus that of the surrounding ambient air.

The sensor signal is sent to the logic module and then on to the power module, which will alter the air/fuel ratio (injector pulse time) to a preset value under all conditions to ensure that the best fuel economy and engine performance are obtained.

When a large amount of oxygen is present in the exhaust gas flow (lean mixture), a low voltage is produced at the sensor; a small amount of exhaust gas oxygen indicates a rich mixture, and the sensor will produce a higher voltage signal.

To remove the oxygen sensor (refer to Special tools, Figure 9–25) use tool C–4589 to unscrew it from the exhaust manifold (see Figure 9–8). If the same sensor is being installed again, be sure to coat the threads with an anti-seize compound such as Loctite 771–64 or equivalent, otherwise the sensor will be very hard to remove the next time around. Take care not to get any of the compound on the sensor body that is exposed to the exhaust gas flow because a false reading can be obtained when the compound burns onto the ceramic coating. New sensors are already precoated with anti-seize compound on the threads. Torque the sensor to 20 lb.ft. (27 N·m).

Coolant Temperature Sensor. Located in the thermostat housing, the coolant temperature sensor is designed to send a signal to the logic module based on coolant operating temperature. When the engine is cold, a richer mixture is required, as well as a higher idle rpm, until the engine warms up and attains a preset temperature. At this time, the internal expansion of the liquid causes a different voltage signal to be sent to the logic module, which will then lean out the air/fuel ratio. Figure 9–9 illustrates the coolant temperature sensor.

Vehicle Speed Sensor. The vehicle speed sensor is located between the speedometer cable and the speedometer head, as shown in Figure 9–10. The speed sensor feeds a signal to the logic module when the vehicle is moving.

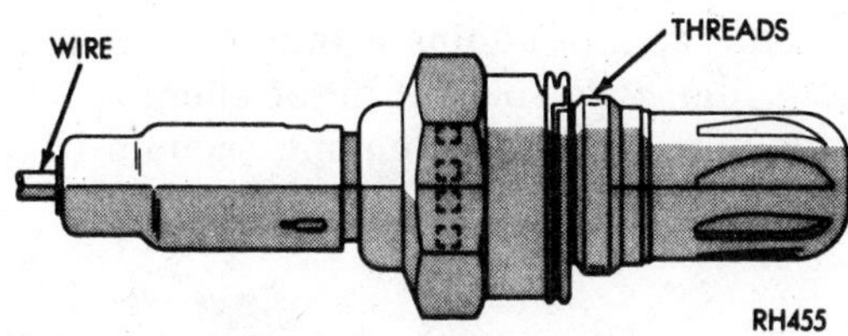

Figure 9–8. Oxygen sensor (Courtesy of Chrysler Corporation)

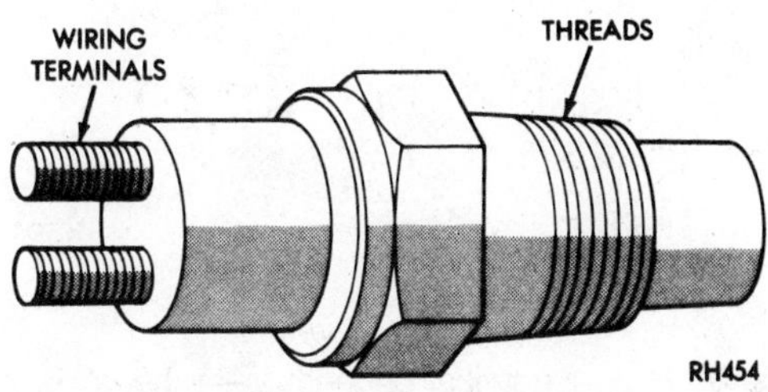

Figure 9–9. Coolant sensor. (Courtesy of Chrysler Corporation)

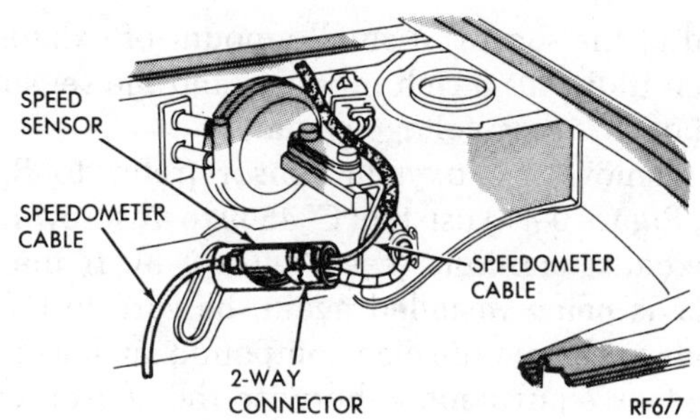

Figure 9-10. Vehicle speed sensor location. [Courtesy of Chrysler Corporation]

Throttle Position Sensor. The throttle position sensor is located on the side of the throttle body as shown in Figures 9–12 and 9–18. Basically, this sensor is a variable electric resistor-type switch that is activated by the rotation of the throttle shaft by the driver of the vehicle. As the throttle is opened and closed, a variable voltage signal will be sent to the logic module.

Other Switches

In addition to the sensors discussed above, various other switches are used with the system and affect the signals to/from the logic module. The following switches are used with the system.

EGR (Exhaust Gas Recirculation) Switch/Solenoid. This switch is shown in Figure 9–11 along with the purge solenoids. Engine temperatures below 21°C (70°F) permit the logic module to energize the EGR solenoid by grounding it, which prevents ported engine vacuum from reaching the EGR valve.

Once the engine attains a preset operating temperature, the EGR solenoid is de-energized by the logic module, and ported engine vacuum from the throttle body passes to the EGR valve. At idle and wide open throttle, the solenoid is energized to prevent EGR operation.

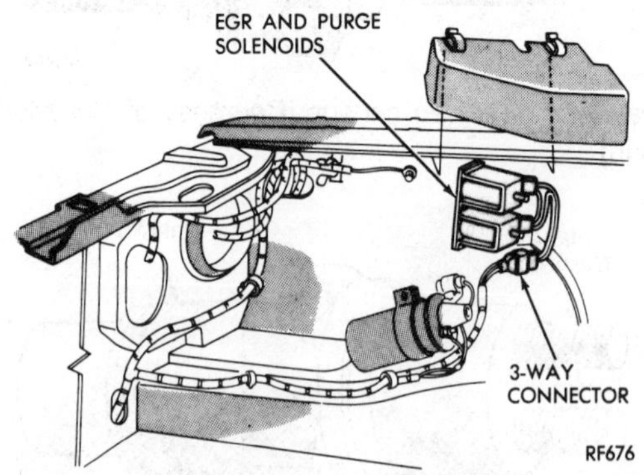

Figure 9-11. Electrical connector—exhaust gas recirculation [EGR] and purge solenoids location. [Courtesy of Chrysler Corporation]

The Purge Solenoids. These solenoids work similarly to the EGR solenoid discussed above. When engine temperatures are below 61°C (145°F), the purge solenoids are energized by a voltage signal from the logic module to prevent vacuum from reaching the charcoal canister valve. Once the preset temperature is reached, the logic module will de-energize the solenoid and vacuum can flow to the canister purge valve to vent vapors through the throttle body. The purge valve is shown in Figure 9–11 along with the EGR solenoids.

Various Other Switches. These include the idle, neutral safety, electric backlight, air conditioning, air conditioning clutch, and brake light switches. When in the *on* position, they cause the logic module to signal the automatic idle speed to a preset rpm.

The air conditioner cut-out relay is in the normally closed (NC) position (electrically on) during engine operation and is in series with the cycling clutch switch and low pressure cut-out switch. If a wide open throttle (WOT) position occurs, the throttle position sensor switch causes the logic module to energize the air conditioner relay. The relay then opens its contacts, which will prevent air conditioning clutch engagement at high engine rpm's to prevent damage.

If the logic module ever receives an incorrect signal or no signal at all from any of the engine sensors, a power-loss lamp on the instrument panel will illuminate to indicate that the logic module has entered the *limp-in mode.* During such a phase, the logic module substitutes information from other sources in order to let the vehicle limp home. Service should be performed immediately to find and correct the problem.

Another function of the power-loss lamp is that it can be used to display fault codes by cycling the ignition switch on/off/on/off/on within 5 seconds, which will allow any fault code stored in the logic module to be displayed. If, however, the fault is repaired or it ceases to exist, then the logic module cancels the fault code after 30 ignition key on/off cycles. The power loss lamp should glow for a few seconds anytime that the engine is started to indicate that the lamp bulb is in fact operational.

Automatic Idle Speed Motor (AIS)

Illustrated in Figure 9–12 is the AIS motor assembly that is operated from the logic module. Inputs from the various sensor units and switches cause the logic module to adjust the engine idle speed for various operating conditions. The AIS motor does this by varying the amount of air bypass on the back of the throttle body as a function of engine loads or ambient temperature conditions. Engine stall after a high-speed condition and a rapidly closed throttle condition are prevented by increasing engine idle from the logic module and the action of the AIS motor.

Electric Fuel Pump

An electric fuel pump assembly (Figure 9–13) is contained within the fuel tank as illustrated in Figure 9–14. A fuel filter is used with the fuel pump to prevent dirt entering the fuel pump system. A low pressure area is created around the fuel cap by the action of the return fuel. This allows main tank fuel to flow into the reservoir even when the fuel tank level is below the reservoir walls (Figure 9–15).

The electric fuel pump with this system maintains a working fuel pressure of approximately 36 psi (248 kPa) through the use of a fuel pressure regulator assembly. The pump is energized by a voltage feed from the automatic shutdown relay (ASD).

The fuel pump is an in-tank unit and is equipped with a sock-type filter on its inlet side as shown in Figure 9–13. The fuel pump is a positive displacement roller vane immersible design with a permanent magnet electric motor. The fuel is drawn in through the sock filter and leaves under pressure at the electric motor outlet. Two check valves are used with this pump. One valve regulates maximum output pressure, and the other one acts to prevent fuel flow in either direction when the engine is stopped.

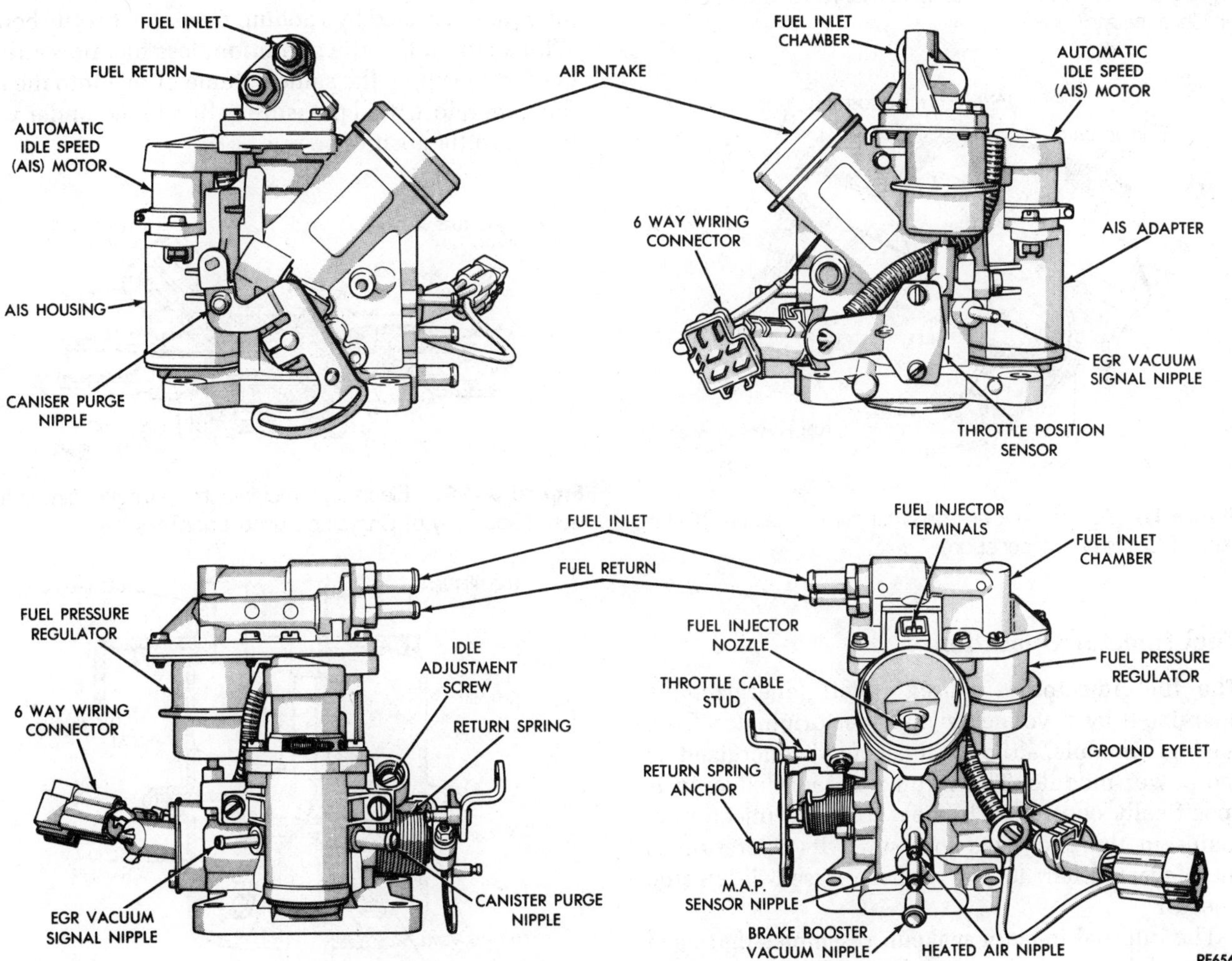

Figure 9–12. Single point throttle body injection unit and component location (four position view). (Courtesy of Chrysler Corporation)

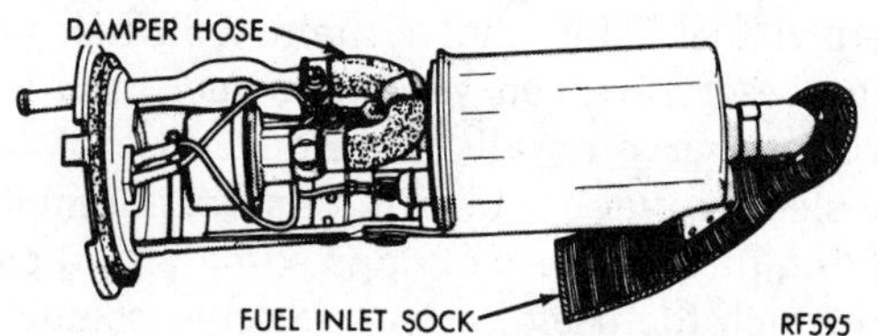

Figure 9–13. Electric fuel pump assembly. (Courtesy of Chrysler Corporation)

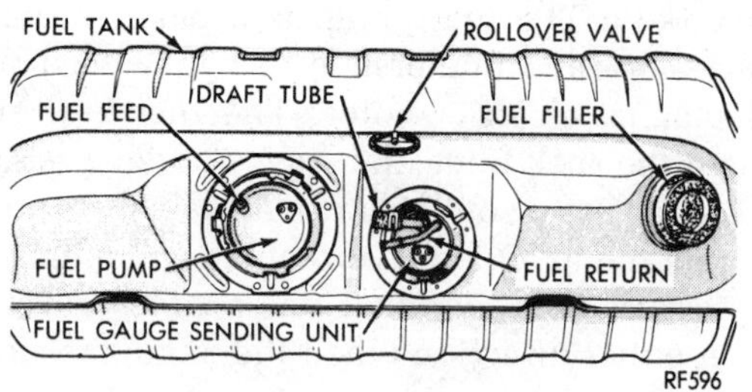

Figure 9–14. Fuel tank components. (Courtesy of Chrysler Corporation)

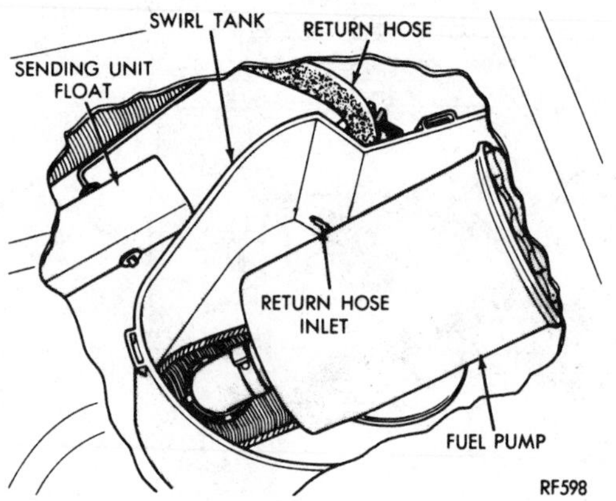

Figure 9–15. In-tank electric fuel pump location. (Courtesy of Chrysler Corporation)

Fuel Injector Operation

The fuel injector is pulsed on/off (energized/de-energized) by a voltage signal that originates from the logic module, although it is actually energized by the power module. The volume of fuel delivered is specifically controlled by how long the injector remains in the energized (on) state. Its operation is therefore similar to the Robert Bosch L-Jetronic injector.

The internal injector makeup resembles that used by Bosch in that an electric solenoid armature is connected to a pintle valve that is pulled into the injector body (when it is energized) against the force of a small spring. This action allows the pressurized fuel to be sprayed into the throttle body above the throttle plate (butterfly valve). A throttle position sensor switch monitors its position and feeds a voltage signal back to the logic module along with the other sensor signals to control the "opening time" of the injector.

The arrangement of the throttle body and the injector is illustrated in Figures 9–12 and 9–16.

Fuel Pressure Regulator

The fuel pressure regulator is located on the throttle body as shown in Figure 9–12. Steady fuel pressure originating at the electric fuel pump is maintained at 36 psi (248 kPa) by the regulator through the action of a spring loaded rubber diaphragm, which will uncover a fuel return port and relieve excess pressure back to the fuel tank.

Figure 9–17 illustrates a cross-sectional view of the fuel pressure regulator. The spring loaded diaphragm is assisted by vacuum from the throttle body. With a closed throttle condition, less fuel pressure is needed to supply the same volume of fuel into the air flow; therefore, fuel pressure is fine-tuned under various throttle positions.

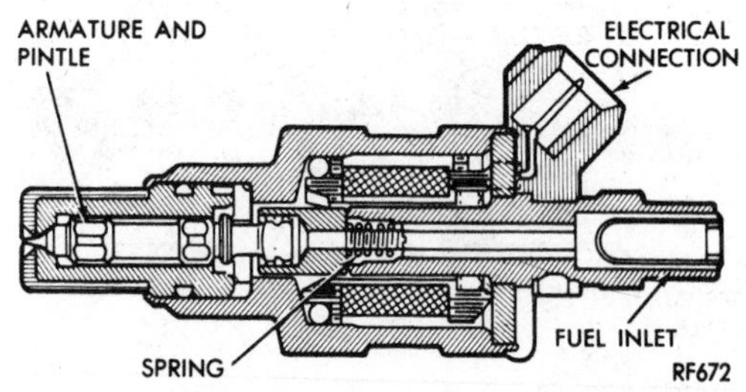

Figure 9–16. Electronic fuel injector—single point injection. (Courtesy of Chrysler Corporation)

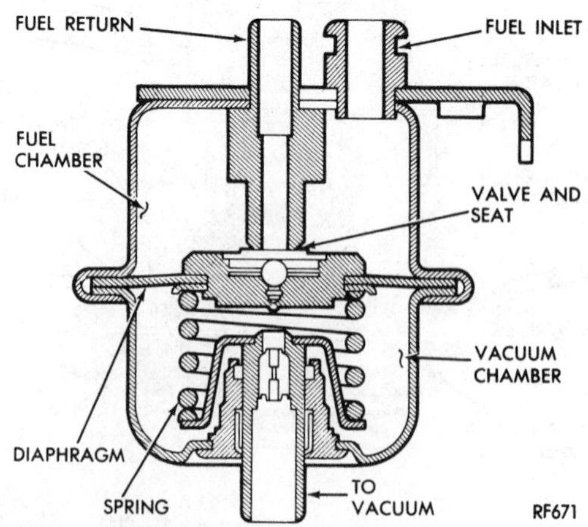

Figure 9–17. Fuel pressure regulator unit. (Courtesy of Chrysler Corporation)

Throttle Body Unit

The throttle body unit is shown in Figure 9–12 in a four-position assembled view. Figure 9–18 illustrates the throttle body in exploded view form with the location of all major components.

Single Point Injection Service

All current fuel injection systems are under pressure when the engine is stopped; pressure regulators or an accumulator arrangement are used.

SPECIAL NOTE: Before attempting to service or

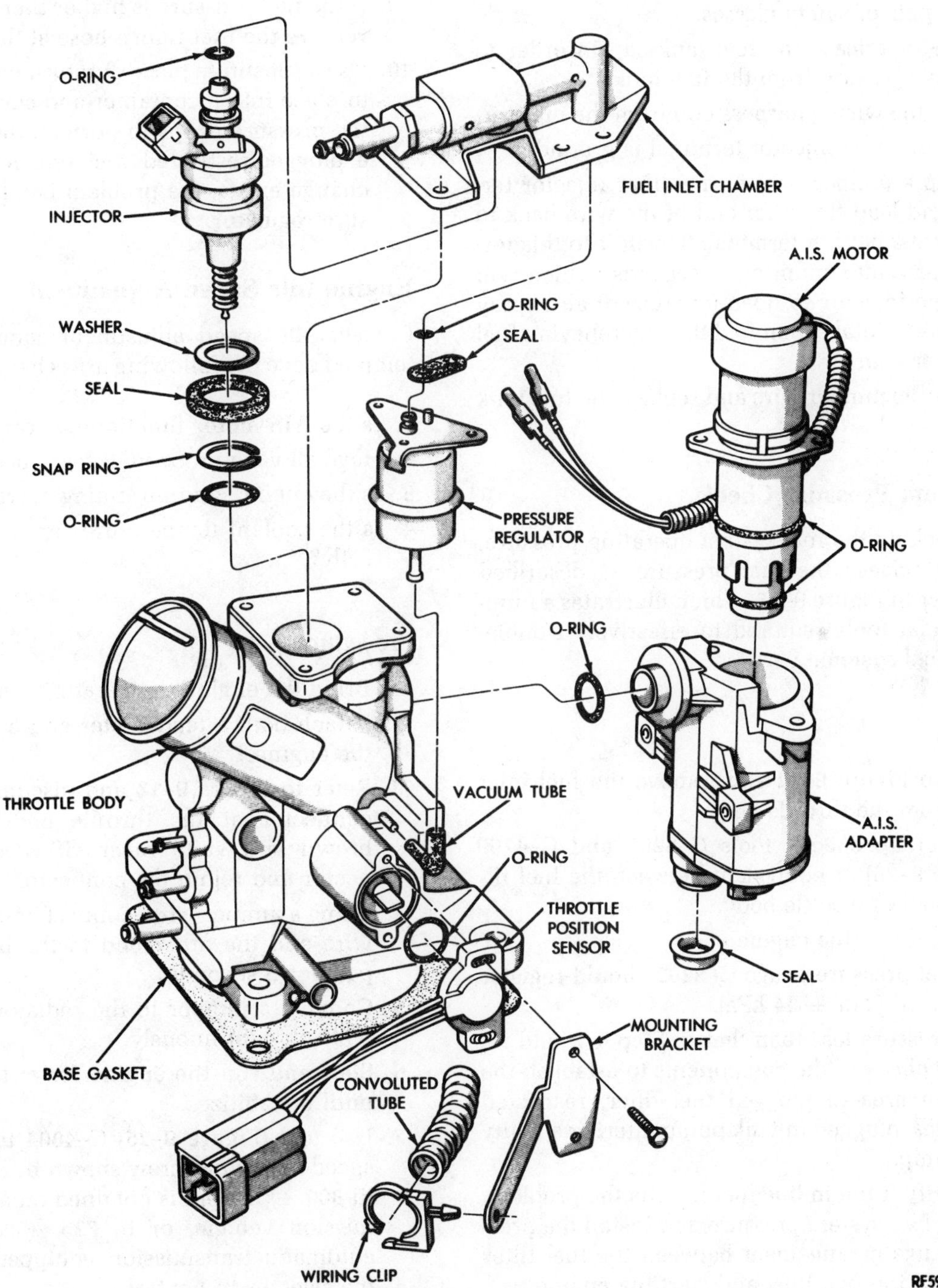

Figure 9–18. Throttle body injection unit exploded view. [Courtesy of Chrysler Corporation]

disconnect any fuel lines or components from the fuel system, the fuel pressure must be bled down or relieved to prevent personal injury or possible danger from fires should the fuel spray onto hot engine components or exhaust manifolds.

Procedure to Relieve Fuel System Pressure

1. *Do not smoke.*
2. Wear a pair of safety glasses.
3. Remove or release the fuel tank cap in order to vent any pressure from the fuel tank.
4. Remove the wiring harness clip from the injector.
5. Ensure that one injector terminal is grounded.
6. Hook up a jumper wire to the other injector terminal and lead the other end of the wire back to the positive battery terminal. Touch it to this terminal for a maximum of 10 seconds, which will cause the injector to open (energized) and spray fuel into the intake manifold thereby relieving fuel system pressure.
7. Remove the jumper wire and replace the fuel tank cap.

Fuel System Pressure Check

Before checking the fuel system operating pressure, be sure to release the fuel pressure as described above. Refer to Figure 9–25, which illustrates a number of special tools required to effectively troubleshoot the fuel system.

Procedure

1. Refer to Figure 9–12 and remove the fuel inlet hose from the throttle body.
2. Connect up special tools C–3292 and C–4799 (Figure 9–25) or equivalent between the fuel filter hose and throttle body.
3. Start and idle the engine.
4. The fuel pressure gauge C–3292 should register 36 ± 2 psi (248 ± 14 kPa).
5. Fuel pressure less than that in Step 4 would require a check of the components to establish the problem area—a plugged fuel filter, restricted fuel lines, plugged in-tank pump filter, or a faulty fuel pump.
6. To clarify if the in-line fuel filter is the problem, relieve fuel system pressure and install the pressure gauge arrangement between the fuel filter hose and the fuel line, and start the engine.
7. If the fuel pressure is correct, then replace the fuel filter. If, however, pressure is still low, gently squeeze the fuel return hose while watching the pressure gauge C–3292. If the pressure increases, the problem is a faulty regulator assembly.
8. No pressure increase in Step 7 after squeezing the fuel return hose indicates that the problem is the in-tank fuel filter or a faulty in-tank pump unit.
9. If the fuel pressure is higher than specifications, remove the fuel return hose at the throttle body.
10. Use a substitute piece of return hose long enough to place into a container and start the engine; if the pressure drops to normal, the problem is in a plugged/restricted fuel return hose, or if no change exists, the problem is a faulty fuel pressure regulator.

Engine Idle Speed Adjustment

Correct idle speed adjustment should only be attempted once the following areas have been checked:

1. Is the AIS motor functioning correctly?
2. Have all vacuum or EGR leaks been corrected?
3. Is the engine ignition timing correct?
4. Is the coolant temperature sensor operating correctly?

Procedure

1. Bring the engine to operating temperature.
2. Attach both a tachometer and a timing light to the engine.
3. Refer to Figure 9–12 and disconnect the 6-way connector at the throttle body. Remove the brown with white tracer AIS wire from the connector and rejoin the connector.
4. Using a jumper wire, connect one end to the AIS wire and the other end to the battery positive post for 5 seconds.
5. Connect a jumper to the radiator fan so that it will run continuously.
6. Start and run the engine for at least 3 minutes until it stabilizes.
7. Use tool (Figure 9–25) C–4804 to turn the idle speed adjusting screw shown in Figure 9–12 until 800 ± 10 rpm is obtained on a manual transmission vehicle, or to 725 ± 10 rpm on an automatic transmission equipped vehicle with the selector in neutral.
8. If the idle will not adjust correctly, check for binding linkage, etc.

9. Check engine timing which should be 18 ± 2 degrees BTDC (manual) or 12 ± 2 degrees BTDC on an automatic transmission unit.
10. If ignition timing is off, adjust it until it is as specified at the engine idle speeds in Step 7.
11. Stop engine; remove tach and timing light; remove jumper wire and reinstall AIS wire and 6-way connector.

Fuel Injector Removal/Installation

Release the fuel system pressure as described above! Refer to Figure 9–18 as a guide to component location and identification. The injector can be removed without having to remove the complete throttle body assembly because the injector is mounted into the fuel inlet chamber of the throttle body.

Procedure

1. Isolate the electrical system by disconnecting the battery ground cable (negative).
2. Loosen and remove the four TORX screws that hold the fuel inlet chamber to the throttle body assembly.
3. Remove the vacuum tube that runs from the throttle body to the fuel pressure regulator assembly.
4. Place a rag around the fuel inlet chamber to soak up any fuel as you lift this assembly along with the injector.
5. Pull the injector free from the fuel inlet chamber.
6. Remove the upper and lower injector O-rings as well as the snap-ring that retains the seal and injector washer.
7. New seal rings should always be installed any time that the injector has been removed; lightly lubricate the seal rings with a very light oil before installing the injector.
8. Once the injector and fuel inlet manifold have been installed, carefully torque the four TORX screws at 35 lb.in. (4 N·m).

NOTE: It is not necessary to relieve fuel system pressure when removing the following items:

1. Throttle position sensor (TPS)
2. Idle speed motor (AIS)
3. Automatic idle speed motor assembly

Fuel Filter Replacement

Replacement of the fuel filter at the throttle body unit requires that the fuel pressure be released first as described above. Other than the above precaution, filter replacement requires no special considerations.

Fuel Pump Replacement/Service

If the in-tank fuel filter, electric in-tank fuel pump, or fuel sending unit requires service or replacement, the fuel system pressure should be released first as described earlier.

Once this has been done, disconnect the battery ground cable in order to isolate the electrical system. The fuel tank must be removed to service the fuel pump, its filter, or sending unit. The tank must be drained of fuel by siphoning it out with the vehicle on a hoist.

To remove the fuel pump, refer to its location shown earlier in Figure 9–14. Using a hammer and non-metallic punch, tap the lock ring counterclockwise to release the pump from the tank. Be sure to replace the O-ring seal when replacing the fuel pump.

Fuel Tank Safety Valves

All fuel injected vehicles manufactured by Chrysler Corporation employ both a fuel tank pressure/vacuum relief valve cap and a safety rollover valve. The pressure/vacuum relief valve acts like a radiator pressure cap in that it opens to relieve pressure build-up in the fuel tank, while the vacuum part of the valve opens to prevent fuel tank and hose collapse when the engine is shut down.

The rollover valve (Figure 9–40) is used to prevent fuel leakage if the vehicle is involved in an accident and rolls over. This valve is shown under the discussion dealing with the 1.6L Dodge Colt ECI fuel system in this chapter.

2.2L Turbocharged Engine Fuel System

The fuel injection system used with the Laser/Daytona models that employ the 2.2L turbocharged/EFI engine is designed for a horsepower output (1984) of 142 (106 kW) versus the standard engine rating of 99 (74 kW).

The system operation follows the same basic pattern as that explained above for the 2.2L single point injection system.

To achieve this performance increase, the fuel system employs a Chrysler throttle body and fuel rail along with Robert Bosch fuel injectors (4) and fuel regulator that are mounted directly to the fuel rail and intake manifold. The four injectors are mounted to aim directly into the intake ports of each cylinder. This system is known as multi-point electronic fuel injection and is similar to that used by Ford on their 4-cylinder 1.6L and 2.3L engines.

The electric fuel pump is located in a swirl reservoir inside the fuel tank similar to that shown earlier (Figure 9–15) for the 2.2L single point injection system and pushes fuel through a filter to the fuel rail where a constant pressure of 53 psi (365 kPa) is maintained—compared to the standard engine's fuel pressure of 36 psi (248 kPa) using a single throttle body injector. The fuel pressure regulator (Figure 9–17) operates in the same manner as that for the non-turbocharged single point 2.2L engine; the difference is that it is set to relieve at a higher pressure.

The fuel injectors used with the turbocharged Laser/Daytona units are designed to be pulsed open and closed similar to Bosch L-Jetronic injectors once for each piston intake stroke and in the same way as the injectors used by Ford on their 1.6L and 2.3L multi-point injection systems. The injectors will remain open (energized) for about two milliseconds at idle to as long as 20 milliseconds at wide open throttle. In addition, these injectors fire in pairs, with cylinders No. 1 and No. 2 injectors firing together followed by cylinders No. 3 and No. 4 so that optimum air/fuel ratio can be obtained.

Figure 9–19 illustrates the 2.2L turbocharged engine. Its basic fuel system concept is shown in Figure 9–20.

The turbocharged 2.2L engine employs a knock sensor that is not found in the standard engine. This knock sensor is mounted on the intake manifold between cylinders 2 and 3 to detect any combustion detonation. When this occurs, a voltage signal is relayed to the electronic computer, which will retard the ignition timing until the knock disappears. The knock sensor and air charge temperature sensor are located side by side on the intake manifold as shown in Figure 9–21.

Sensors and injector operation for the turbocharged engine are similar to that explained earlier

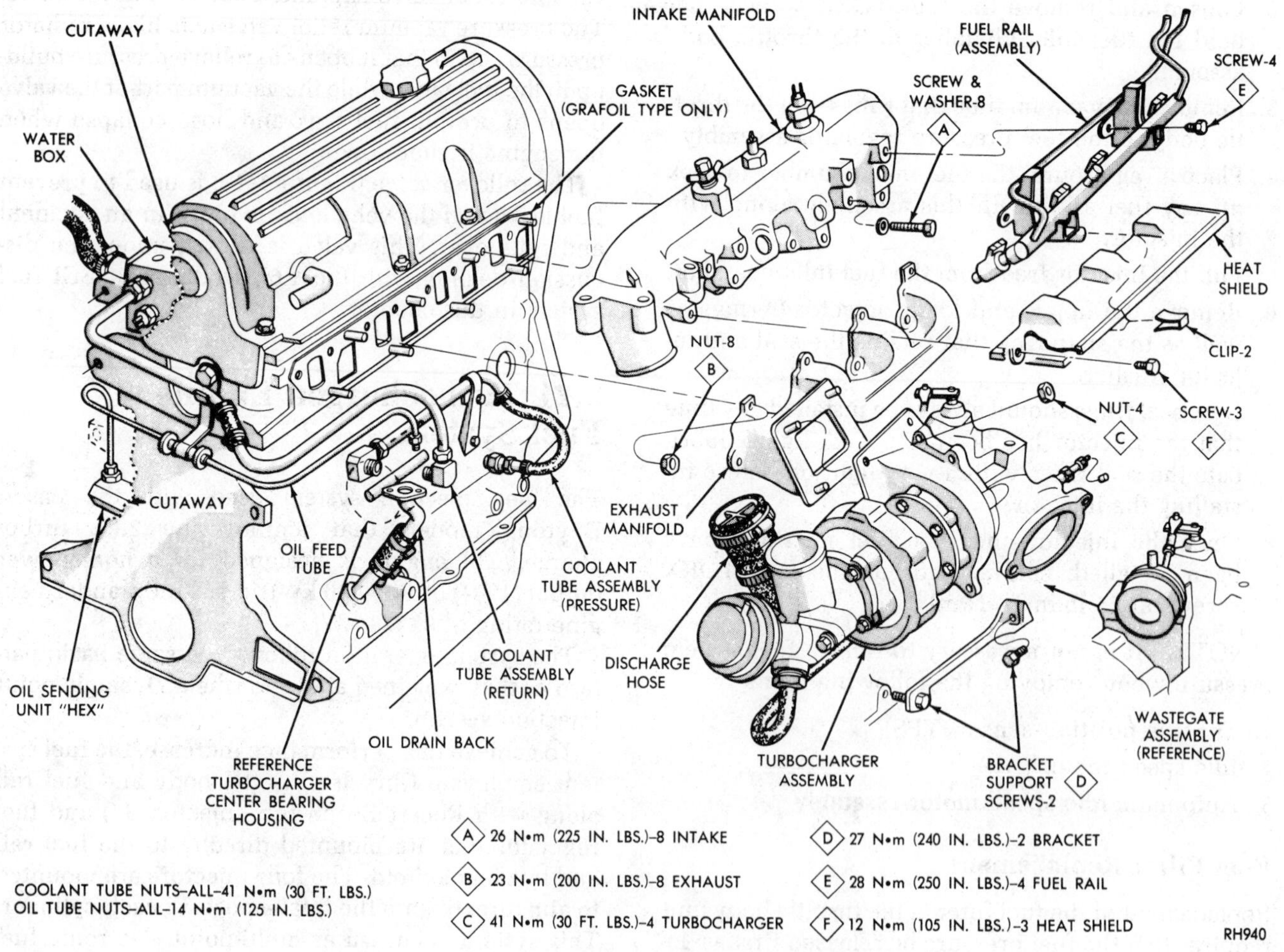

Figure 9–19. 2.2L Turbocharged Laser/Daytona engine. (Courtesy of Chrysler Corporation)

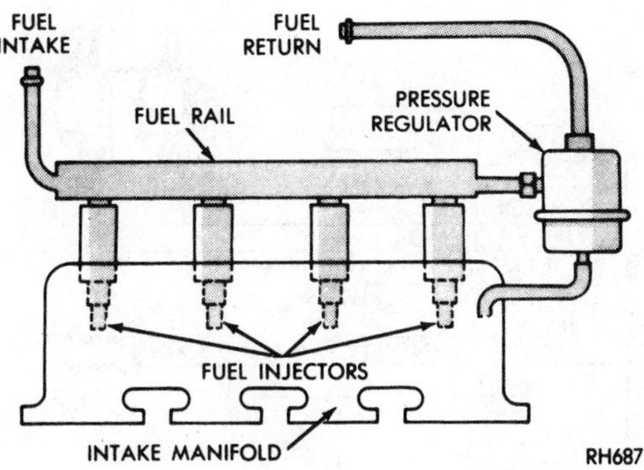

Figure 9–20. 2.2L Turbocharged Laser/Daytona fuel system arrangement. (Courtesy of Chrysler Corporation)

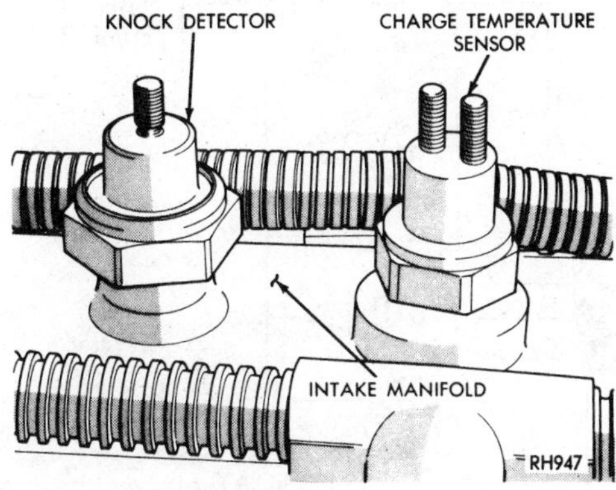

Figure 9–21. Turbocharger charge air temperature and knock/detonation sensor locations. (Courtesy of Chrysler Corporation)

for the single point 2.2L engine. The components of the 2.2L turbocharged engine fuel control system are illustrated in Figure 9–22.

Throttle Body

The multi-point electronic injection system employs a throttle body unit that contains a throttle position sensor. This sensor is basically a potentiometer (variable resistor) that feeds a varying voltage signal to the logic module where it is used along with other sensor inputs to vary the injectors pulse time (opening time).

A closed throttle results in a voltage signal that will pulse the injectors open for a short time period, while a wide open throttle results in a longer injector pulse time and therefore a greater fuel delivery. Figure 9–23 illustrates the throttle body unit used with the 2.2L turbocharged multi-point injection system.

Troubleshooting—Visual Inspection

Many problems that occur with the multi-point electronic fuel injection system can be caused by poor wiring connections or loose hose connections. Always inspect these two areas before conducting major checks/tests on the system (Figure 9–24).

Fuel Pressure Check

Follow the detailed procedure given under the 2.2L single point injection system above. The only difference is that the fuel pressure should be between 53 ± 2 psi (365 ± 14 kPa).

2.2L EFI Fault Code Description

To effectively troubleshoot a fuel system equipped with either the single point or multi-point injection systems, the on-board logic module has been programmed to monitor several different circuits of the fuel injection system. The mechanic/technician can very quickly access this information and therefore pinpoint the problem area without having to go through a long trial and error test procedure.

Any fault that occurs in the system often enough will actually trigger a memory storage bank in the logic module where it is stored until either the technician accesses it or the ignition key has been cycled on/off 30 times, after which time the logic module will cancel the memory of that particular problem.

Obtaining Fault Codes

In order to access the fault codes, special tool C–4805 shown in Figure 9–25 is required.

Procedure

1. Connect Diagnostic Readout Box Tool C–4805 to the vehicle diagnostic connector located in the engine compartment near the passenger side strut tower.
2. Start the engine if possible.
3. Cycle the transmission selector lever and the A/C switch if applicable, then shut off the engine.
4. Turn the ignition key to on/off/on/off/on, and within 5 seconds, record all of the diagnostic codes registering on the Diagnostic Box.
5. Make sure that the power loss lamp on the instrument panel does in fact illuminate for 2 seconds before it goes out, which will indicate that the bulb is OK.

Fault Code Description

Fault codes can appear either by the power loss lamp flashing or by hooking up the diagnostic tool C–4805

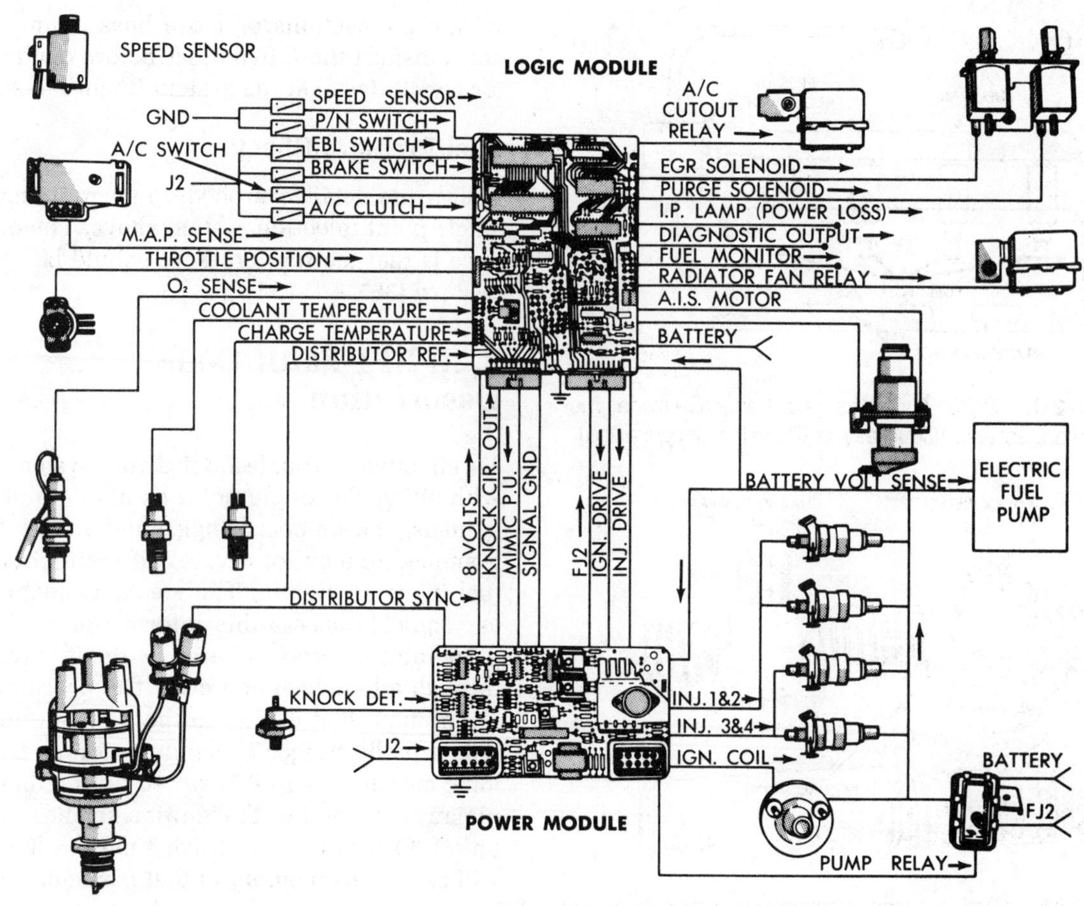

(1) Charge Temperature Sensor
(2) Coolant Temperature Sensor
(3) Throttle Position Sensor
(4) MAP Sensor
(5) Distributor pick ups

(E) Automatic Idle Speed Motor activated by FJ2 from Logic Module

(F) Logic sends information to Power Module:
(1) Ignition timing (anti-dwell)
(2) Injector turn on
(3) Injector turn off

(G) Power Module requires:
(1) Distributor signal within 1/2 second of cranking
(2) Distributor and injector pulses greater than 60 RPM

(H) Power Module then:
(1) Commands injector to supply fuel
(2) Triggers the ignition coil (−)

With the ignition key in the RUN position, the following information must be available to the Power Module and Logic Module or the engine will stop.

(A) Logic Module 21 way red connector	OUTPUT
(1) Number 2 pin—sensor ground	INPUT
(2) Number 3 pin—8 volts	OUTPUT
(3) Number 4 pin—5 volts	OUTPUT
(4) Number 5 pin—injector off	OUTPUT
(5) Number 7 pin—anti-dwell	OUTPUT
(6) Number 8 pin—injector on	OUTPUT
(7) Number 11 pin—distributor reference signal	INPUT
(8) Number 19 pin—distributor synchronization signal	OUTPUT
(B) Power Module 10 way black connector	
(1) Number 1 pin—coil (−)	OUTPUT
(2) Number 2 pin—J2	INPUT
(3) Number 3 pin—fused J2	OUTPUT
(4) Number 4 pin—injector driver number 2	OUTPUT
(5) Number 5 pin—injector driver number 1	OUTPUT
(6) Number 6 pin—ASD relay ground	OUTPUT
(7) Number 8 pin—battery supply	INPUT

Figure 9-22. 2.2L Turbocharged engine multi-point injection system components. (Courtesy of Chrysler Corporation)

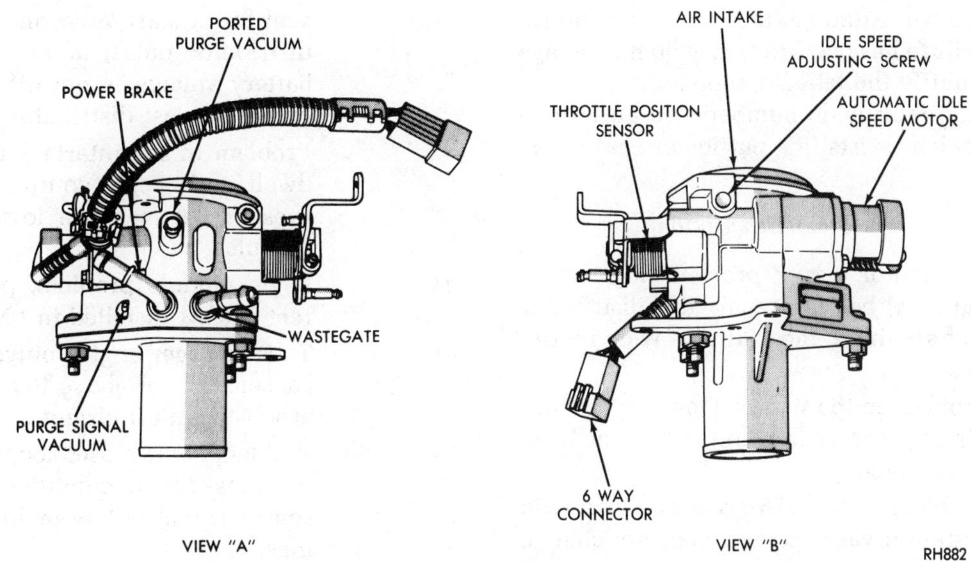

Figure 9-23. 2.2L Turbocharged engine throttle body system. [Courtesy of Chrysler Corporation]

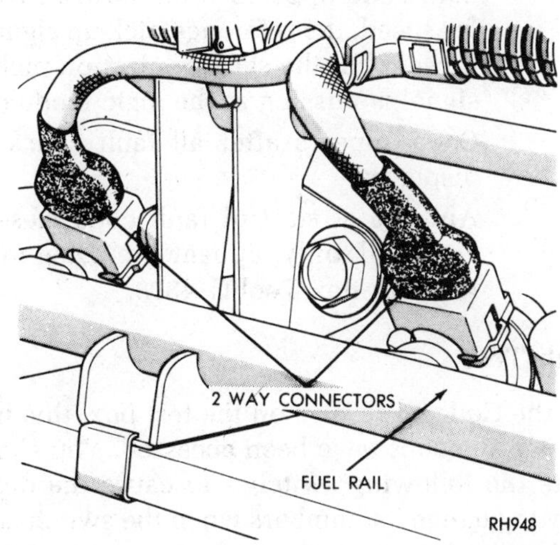

Figure 9-24. Fuel injector connections. [Courtesy of Chrysler Corporation]

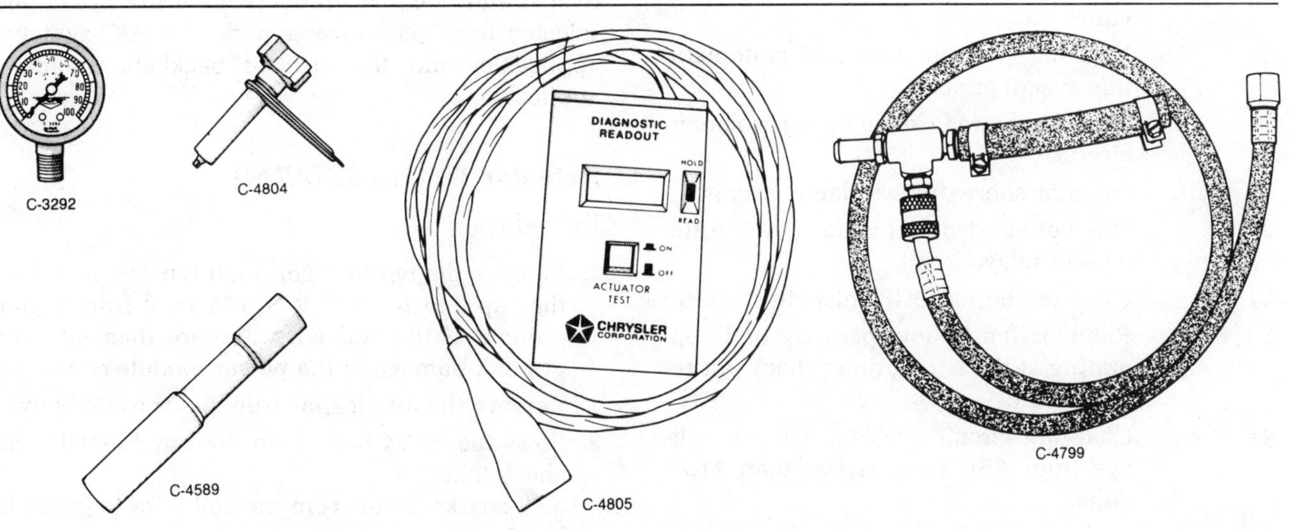

Figure 9-25. Special service tools. [Courtesy of Chrysler Corporation]

as described above. Although the fault codes do indicate that a failure has occurred, they do not always specifically identify the failed component.

The following fault code numbers are an indication that a problem exists in a particular area:

Fault Code	*Fault Description*
11	Distributor circuit problem; logic module (LM) has not received a distributor signal since the battery was reconnected.
12	Problem in the stand-by memory circuit; direct memory feed to the LM has been interrupted.
13	Problem in the MAP sensor pneumatic system if vacuum level does not change between start and start/run transfer speed (500–600 rpm).
14	MAP system electrical problem (voltage below 0.02 volt or above 4.9 volts).
15	Vehicle speed sensor problem; code will appear if engine is at idle and vehicle speed is less than 2 mph; code only valid if sensed while vehicle is moving.
21	Oxygen sensor feedback problem; code appears if engine temperature is above 77°C (170°F), engine speed is above 1500 rpm, and there has been no oxygen signal for more than 5 seconds.
22	Coolant temperature circuit.
23	Charge temperature problem in turbocharged engine only.
24	Throttle position sensor problem (sensor voltage is below 0.16 volt or above 4.7 volts).
25	Improper voltage from AIS (automatic idle speed) circuit.
31	Open or shorted canister purge solenoid circuit.
32	Open or shorted power lamp circuit.
33	Open or shorted A/C wide open throttle cut-out relay circuit.
34	Open or shorted EGR solenoid circuit.
35	Radiator fan is not operating or is operating at the wrong time; check fan relay circuit.
41	Charging circuit problem; battery voltage from ASD relay is less than 11.75 volts.
42	Code appears if during cranking, battery voltage from the ADS relay is not present for at least ⅓ second after the first distributor pulse; or after engine stall, battery voltage is not off within 3 seconds after last distributor pulse.
43	Problem in the interface circuit if anti-dwell or injector control signal is not present between the logic and power modules.
44	Logic module problem; possibly incorrect PROM installed in LM
45	Turbocharger engine only; MAP electrical sensor rises above 10 psi (69 kPa); overboost shut-off circuit.
51	Problem in the closed-loop system if during closed-loop conditions the oxygen sensor signal is low or high for 2 minutes.
52 and 53	Logic module internal failure.
54	Problem in synchronization pickup circuit; code appears if at start/run transfer speed, the reference pick-up signal is present but the synchronization pick-up signal is missing at the logic module.
55	Code appears after all fault codes are displayed.
88	Appears at start of fault code message and will only appear on Diagnostic Readout Box Tool C–4805.

Switch Test

When the Code 55 flashes on the test box, this indicates that all codes have been accessed. You should activate the following switches to cause the digital display to change its numbers when the switch is activated and released. Activate the brake pedal, gear selector from park-reverse-park, the A/C switch if applicable, and the electric backlight switch if applicable.

Actuator Test Mode (ATM)

Procedure

1. Remove the ignition coil high tension wire from the cap and place it ¼" (6.35 mm) from a good ground. If the coil wire is more than ¼" from ground, damage to the power module can result.
2. Remove the air cleaner from the throttle body.
3. Press the ATM button on the box (C–4805) and check that:
   a. 3 sparks occur from the coil wire to ground.
   b. Listen carefully for 2 AIS motor movements (1 open and 1 closed).

4. With the ATM button still depressed, connect a jumper wire between pins 2 and 3 of the gray distributor synchronizer connector. Listen for a click, which will indicate that one injector or one set of injectors has been activated. When you remove the jumper wire, listen for the other set of injectors being activated. Reconnect the distributor connector.
5. The ATM capability will only operate for 5 minutes after which time it will be cancelled. To reinstate this capability, cycle the ignition switch on/off three times ending in the *on* position.
6. When the ATM button is pressed, fault code 42 is generated because the ASD relay is bypassed. Do not use this code for diagnostics after ATM operation.
7. The ATM test confirms that the following areas are operational. When coil fires 3 times:
   a. Coil is operational.
   b. Logic and power modules are operational.
   c. Power/logic module interface is operational.
   d. AIS is operational.
   e. Injector fuel pulse into throttle body confirms that the fuel injector, fuel pump, and fuel lines are OK.

Fuel Usage Recommendations

For successful engine/vehicle performance, it is imperative that the correct octane level of gasoline be used. All high performance (Shelby) or turbocharged 2.2L engines will operate successfully on gasolines with an octane rating of 87, (R+M)/2, but for optimum performance and fuel economy, the use of Premium or Super Unleaded gasolines having a minimum octane rating of 91, (R+M)/2 is recommended.

Standard engines will operate satisfactorily on 87 octane, (R+M)/2.

Unleaded gasoline *must* be used in vehicles equipped with catalyst emission control systems. Using leaded gasolines in these engines will result in catalytic converter damage and poor engine performance.

Gasohol is a mixture of 10% ethanol (grain alcohol) and 90% unleaded gasoline and where available can be used where unleaded gasoline is normally recommended. However, if vehicle driveability problems occur as a result of using gasohol, revert to using 100% unleaded gasoline.

Fuel system cleaning agents should not be introduced into the fuel system because some of these contain highly active solvents that can affect gasket and diaphragm materials used in fuel system components.

Special Tools

The special tools shown in Figure 9–25 are required when servicing the Chrysler electronic fuel injection systems.

Dodge Colt 1.6L Turbocharged Engine Fuel System

The Dodge Colt 1.6L (97.4 cu.in.) in-line 4-cylinder engine is designated the G32B with a turbocharger (T/C). It has a compression ratio of 7.6:1 versus the non-turbocharged engine's 8.5:1. The engine firing order is 1–3–4–2, and initial ignition timing is 8 degrees BTDC versus the non T/C engine's 5 degrees BTDC.

The fuel system used with the turbocharged engine is of the throttle body design; however, this engine uses two single point electronically controlled fuel injectors rather than the one such as is found on the 2.2L design.

The throttle body is located downstream of the turbocharger compressor outlet as can be seen in Figure 9–26, which is a schematic of the air/fuel system used with the 1.6L T/C engine.

ECI (Electronically Controlled Injection) Control System

The 1.6L T/C engine fuel control system employs an ECU (electronic control unit) that receives various sensor inputs similar to that found on the 2.2L engines. The sensor inputs (parameters sensed) and output signals (parameters controlled) to and from the ECU are shown in Figure 9–27, where there are 13 input signals and 8 output signals. Table 9–1 shows the diagnostic items and their code numbers. The following system components are discussed in detail individually as to their function and operation in the overall system.

Table 9–1. *1.6L Engine code numbers and diagnosis items*

Code no.	*Diagnosis item*
1	Oxygen sensor and computer
2	Ignition pulse
3	Air-flow sensor
4	Pressure sensor
5	Throttle position sensor
6	ISC motor position switch
7	Coolant temperature sensor
8	Car speed

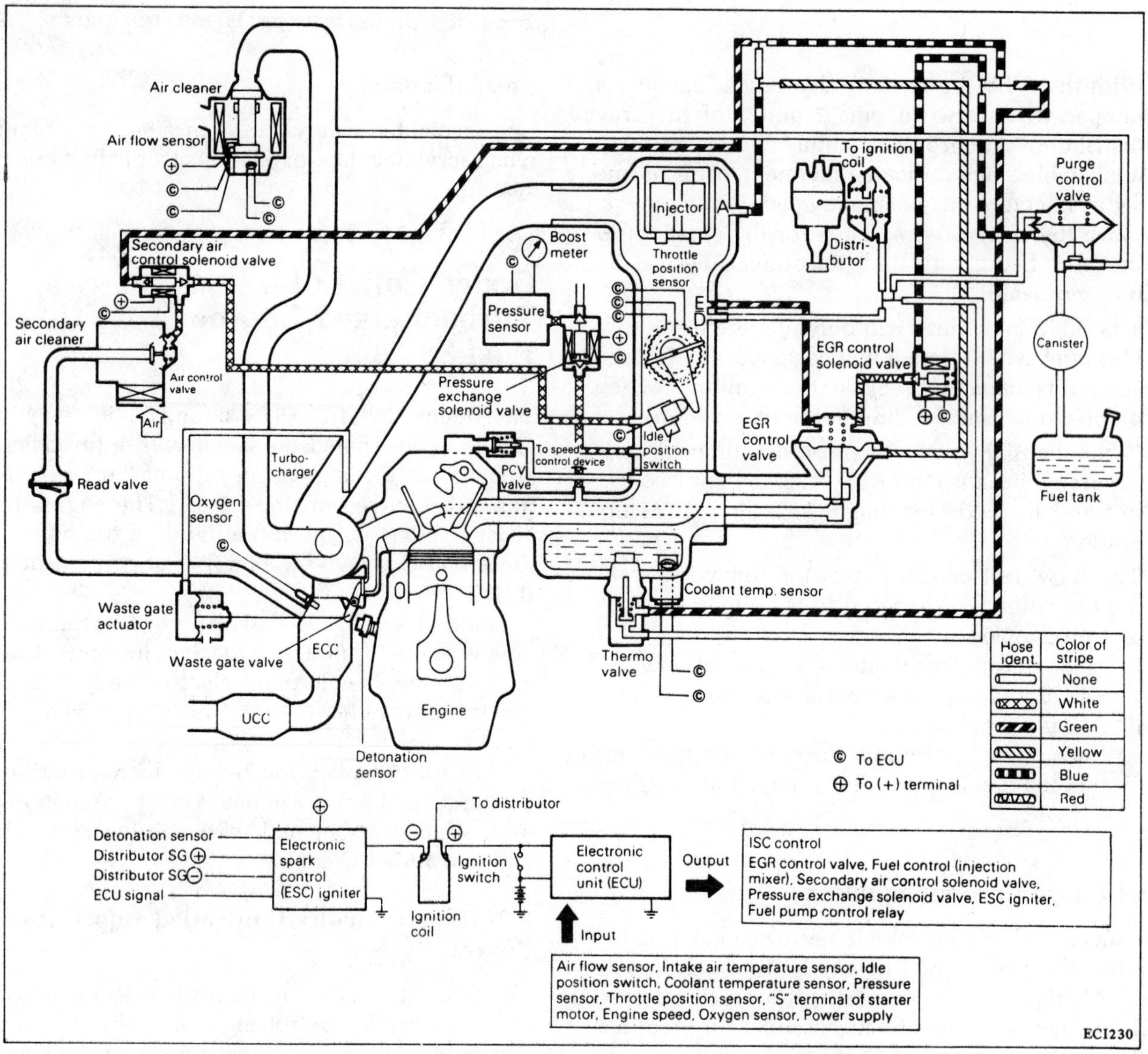

Hose ident.	Color of stripe
	None
	White
	Green
	Yellow
	Blue
	Red

Figure 9-26. Dodge Colt 1.6L T/C engine fuel system schematic. (Courtesy of Chrysler Corporation)

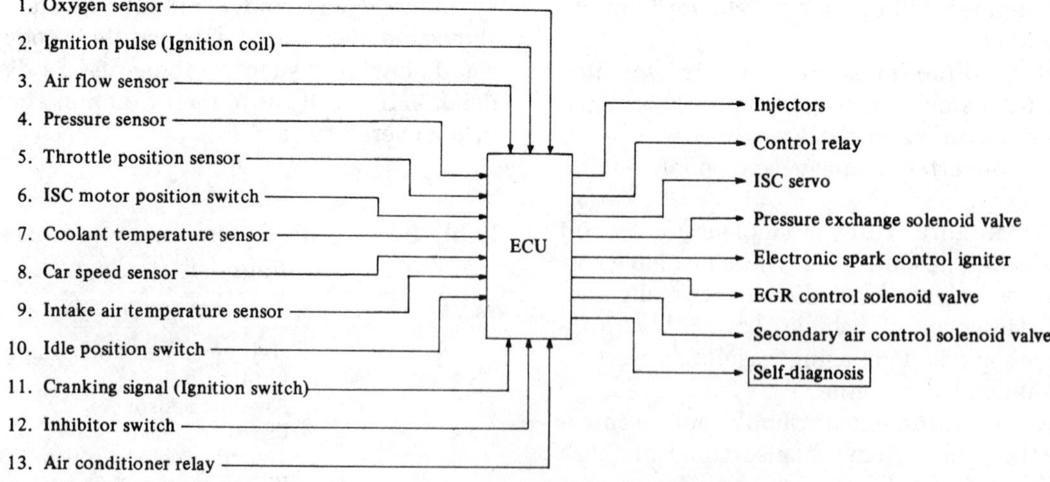

Figure 9-27. 1.6L Engine ECI/ECU sensor inputs and outputs. (Courtesy of Chrysler Corporation)

Air-FLow Sensor. The amount of fuel delivered to the engine under all operating conditions is a function of the volume/weight of air delivered to the engine cylinders. The air-flow sensor used with the 1.6L T/C engine is installed within the air cleaner assembly (Figure 9–28) and consists of a device known as a KARMAN Vortex unit (Figure 9–29).

The air-flow sensor operates on the basis of what is called the Karman Vortex phenomenon in the form of electric pulses that are converted by the modulator. Air entering the throat of the air-flow sensor is deflected around a small triangular casting that causes a velocity increase to occur. This action induces an air flow that creates a pulse condition in the form of ultrasonic waves with a constant frequency through the throat vortex in proportion to the air flow from the turbocharger.

The receiver detects these modulated waves (pressure and flow), and the modulator converts them into electric pulses that are relayed to the ECU. The ECU monitors these voltage signals and alters the pulse time of both fuel injectors as well as controlling secondary air management.

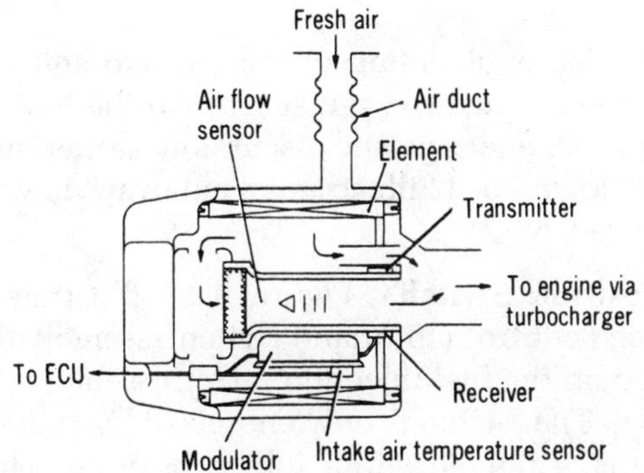

Figure 9–28. Air cleaner/air-flow sensor location. [Courtesy of Chrysler Corporation]

The longer the injectors are pulsed open (energized), the greater will be the rate of fuel delivery to the engine cylinders.

Air Temperature Sensor. An air temperature sensor is located on the modulator within the air cleaner assembly and functions to measure the temperature of the air flowing through the air cleaner.

A change in air temperature affects its density; therefore, the engine's volumetric efficiency will change. The generated sensor voltage signal is monitored by the ECU, which will then alter the fuel delivery characteristics by changing the pulse (opening) time of the two injectors.

Air Pressure Sensor. The air pressure sensor is located on the engine compartment bulkhead, or toe board area, and senses both ambient barometric pressure and absolute pressure within the intake manifold. Figure 9–30 illustrates the pressure sensor location and its connection to the ECU.

Voltage signals generated at the air pressure sensor are fed to the ECU based upon the weather and altitude conditions. Immediately after engine startup, the ambient barometric pressure is sensed by the air pressure sensor, which energizes the solenoid valve shown in Figure 9–30 via the ECU. After engine start-up, the solenoid valve is de-energized and the sensor detects absolute pressure in the manifold from the turbocharger and feeds a voltage signal to the ECU, which will control air/fuel mixture and idle speed.

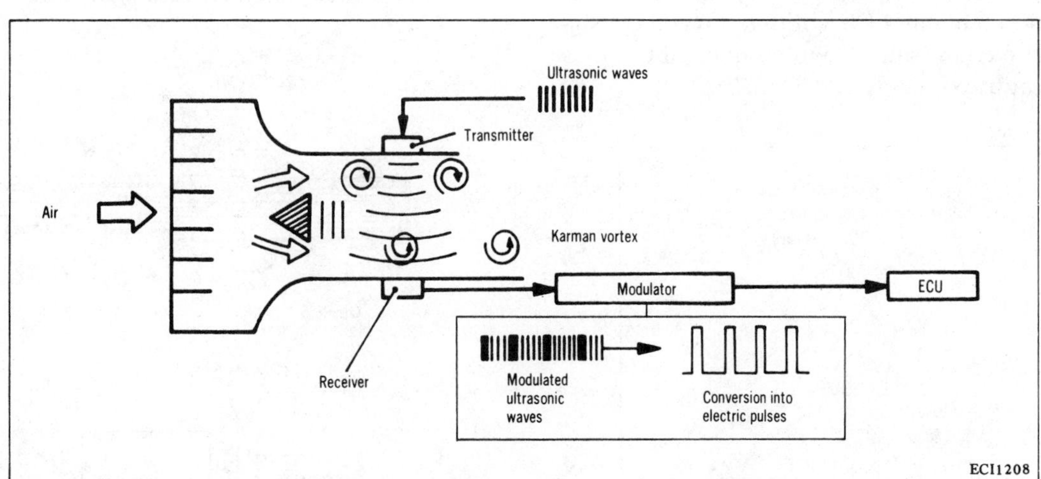

Figure 9–29. Karman air-flow sensor. [Courtesy of Chrysler Corporation]

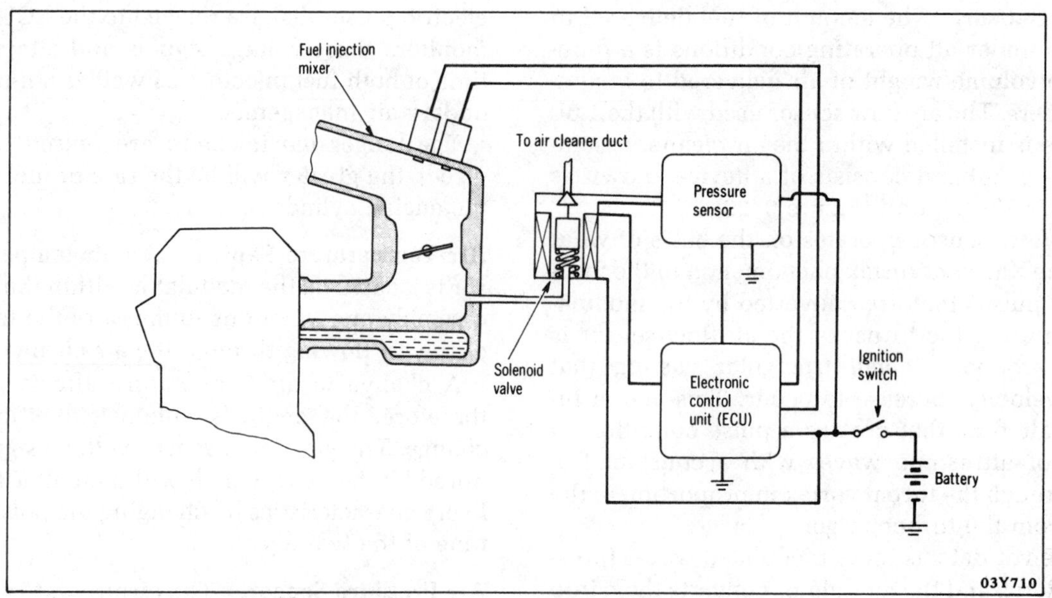

Figure 9–30. Pressure sensor/ECU arrangement. [Courtesy of Chrysler Corporation]

Temperature Sensor. This sensor is installed into the water-cooled intake manifold and operates on the thermistor principle (Figure 9–31). The generated voltage signal is fed to the ECU to vary fuel delivery time, EGR, secondary air management, and idle speed.

Exhaust Oxygen Sensor. The exhaust oxygen sensor is located between the turbocharger and the three-way catalyst. The voltage signal is sent to the ECU to control closed-loop operation of the fuel delivery time. Closed loop is the operation of the engine air/fuel ratio based upon the voltage signals from the ECI system and the oxygen sensor.

Open loop is the condition whereby the air/fuel ratio is determined by the ECU without the voltage signal from the oxygen sensor, which does not emit a signal until approximately 300°C (572°F).

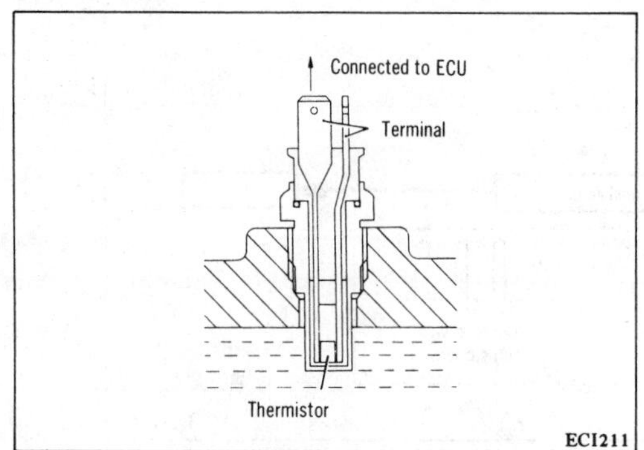

Figure 9–31. Coolant temperature sensor. [Courtesy of Chrysler Corporation]

A detailed explanation of the makeup and operation of the exhaust oxygen sensor can be found under the 2.2L fuel system discussion earlier in this chapter. Figure 9–32 illustrates a cutaway view of the oxygen sensor.

Idle Position Switch. Figure 9–33 illustrates the idle speed control motor and switch assembly that is installed on the fuel injection mixer assembly (throttle body). The switch is only energized (on) when the throttle is at the closed, or idling, position. The position of this switch provides a voltage signal to the ECU to control fuel delivery pulse time of the injectors—especially during vehicle deceleration, idle speed, and secondary air management.

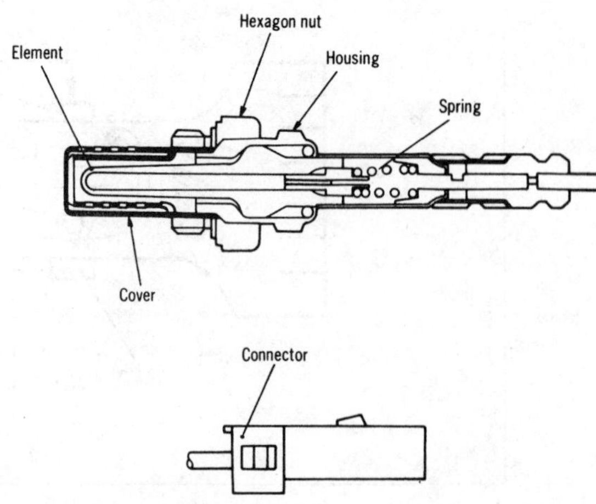

Figure 9–32. Exhaust oxygen sensor. [Courtesy of Chrysler Corporation]

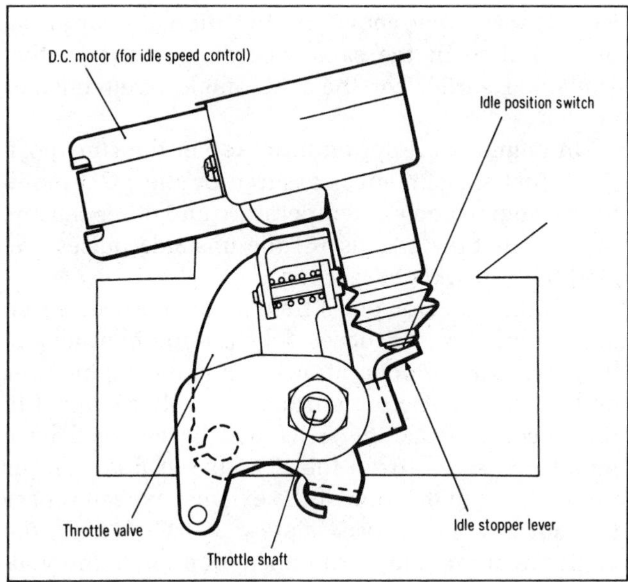

Figure 9–33. Idle speed control system. (Courtesy of Chrysler Corporation)

A throttle position sensor is also located on the throttle body shaft connected to the throttle blades so that when these are opened or closed a rotary potentiometer (variable resistor) feeds a voltage signal to the ECU to control fuel injector pulse time (opening time). The electronic components for the ISC system are integral with the ECU.

EGR Control. The EGR control valve is activated by an electric solenoid that is ECU (electronic control unit) monitored, based upon engine coolant temperature and speed.

Secondary Air Control. Pressure applied to the secondary air control valve is switched by the solenoid valve from the intake manifold pressure to the turbocharger pressure or vice versa. Figure 9–34 illustrates the secondary air system used with the 1.6L turbocharged engine.

The reed valve in the system is actuated by exhaust vacuum being generated from pulsation in the exhaust manifold. Extra air is supplied into the exhaust manifold through the secondary air control valve, which is opened by intake manifold pressure when the solenoid valve is energized by an ECU voltage signal. The reed valve supplies secondary air into the front catalytic converter to promote oxidation of exhaust gases during engine warm-up, engine hot start, and vehicle deceleration.

Vehicle Speed Sensor. A vehicle speed sensor employs a reed switch to pick up speedometer cable rpm and send these pulses to the ECU to control idle speed.

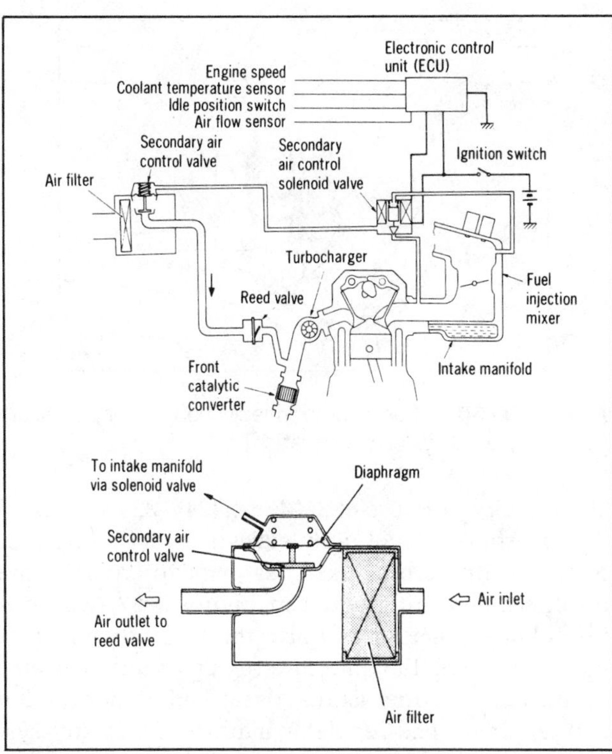

Figure 9–34. 1.6L Turbocharged engine secondary air supply system. (Courtesy of Chrysler Corporation)

1.6L Turbocharged Engine Fuel System Operation

An electric motor driven roller vane-type fuel pump is located underneath the vehicle rear floor area as shown in Figure 9–35. The fuel pump is capable of delivering fuel to the system at between 64 and 85 psi (441–588 kPa) at zero discharge flow rate. This pressure can be checked at the location shown in Figure 9–36.

The pump can deliver between 90–120 liters (24–32 US gallons) or 20–26.5 Imperial gallons an hour

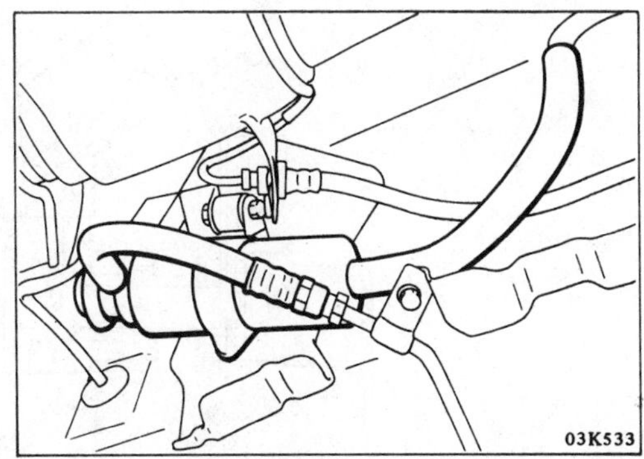

Figure 9–35. Electric fuel pump location. (Courtesy of Chrysler Corporation)

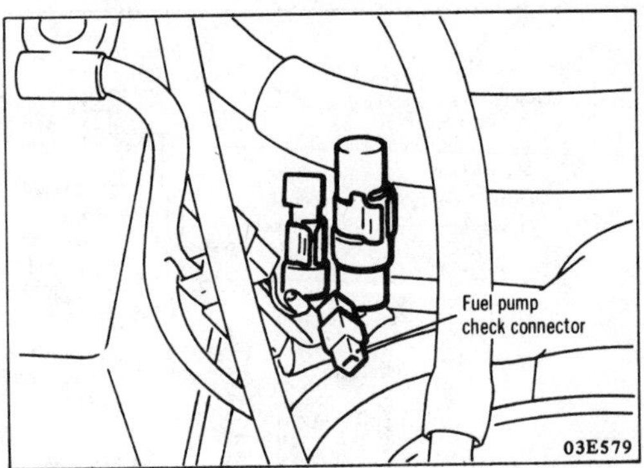

Figure 9-36. Fuel pump check connector location. [Courtesy of Chrysler Corporation]

at a delivery pressure of 35–47 psi (245–324 kPa). Fuel drawn from the fuel tank leaves the fuel pump and is passed through a fuel filter prior to entering the throttle body unit, shown in Figure 9–37, where it feeds both injectors and also the fuel pressure regulator unit. The fuel pressure regulator unit operates in the same manner as that described earlier for the 2.2L engines. The regulator maintains a steady system pressure of 35.6 psi (245 kPa) by returning excess fuel/pressure back to the fuel tank.

The fuel injectors are alternately energized by a voltage signal from the ECU. The longer the pulse time signal from the ECU to the injectors, the greater the volume of fuel delivered to the engine. The injectors spray fuel above the throttle valve in an atomized state in the same operating mode as that explained earlier for the 2.2L single point injector system.

On engine start-up (ignition key in the *start* position), fuel enrichment is assured by the ECU monitoring engine coolant temperature and other sensors but not air flow because of the unstable pulses created by the turbocharger lag.

A longer injector pulse time (open) ensures a rich air/fuel mixture for successful start-up under all ambient temperature conditions. When the engine fires and runs, this starting enrichment voltage signal to the injectors from the ECU will be changed by a monitored signal from the ECU based upon sensor inputs. During this time if the exhaust oxygen sensor has not attained approximately 300°C (572°F), the engine will operate in the open-loop mode (no voltage signal from the oxygen sensor).

During engine warm-up, the ECU processes information from the coolant sensor to provide fuel enrichment during the open-loop operation and part of the closed-loop operation (voltage signal from the oxygen sensor). This continues until the coolant temperature sensor reaches its preset cutoff temperature when the ECU will then switch (pulse) the injectors to a normal air/fuel ratio level based upon the voltage signals from the other input sensors. The engine will now be in a closed-loop condition.

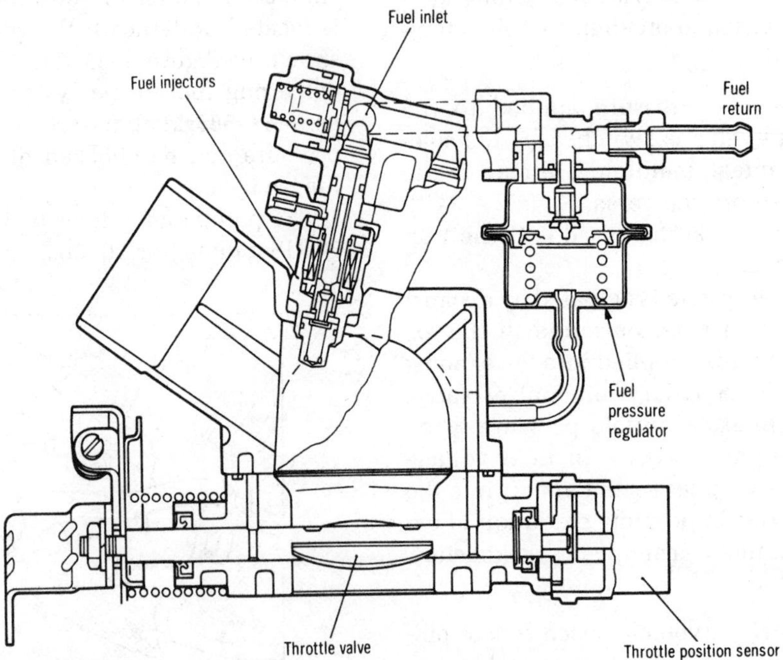

Figure 9-37. Fuel injection mixer assembly/throttle body unit for the 1.6L turbocharged Dodge Colt engine. [Courtesy of Chrysler Corporation]

When the vehicle is accelerated in either the open-loop (cold) or closed-loop (warm) condition, smooth driveability is ensured by a signal from the ECU, which processes information based upon the throttle position sensor, which is a rotary potentiometer (variable resistance).

The fuel mixer or throttle body assembly is shown in Figure 9–37 in assembled form, while Figure 9–38 illustrates the components of the assembly in a disassembled form for clarity.

Fuel Safety Valves

The 1.6L engine is equipped with both an overfill limiter and a fuel check valve as shown in Figures 9–39 and 9–40. The overfill limiter valve is designed to operate like a radiator pressure cap in that it relieves fuel pressure from the fuel tank to maintain a safe condition and also to open through its built-up vacuum valve when the engine is shut down to prevent fuel tank collapse.

The fuel check valve is designed to prevent fuel leaks from the system in case the vehicle rolls over in an accident. The valve is located in the fuel vapor line between the canister and overfill limiter (Figure 9–41) and is mounted as shown in Figure 9–42 on the vehicle engine compartment bulkhead.

Within the valve are two ball bearings that are normally in the position shown. However, if a vehicle rollover occurs, they would close off the fuel passage thereby preventing dangerous fuel leakage.

1.6L Turbocharged Engine Fuel System Adjustments/Checks

Before removing any components from the fuel system, you *must* release the fuel system pressure; otherwise, a danger exists of a fire or personal injury.

CAUTION: To relieve fuel system pressure, follow the same procedure as given earlier for the 2.2L single point injector system.

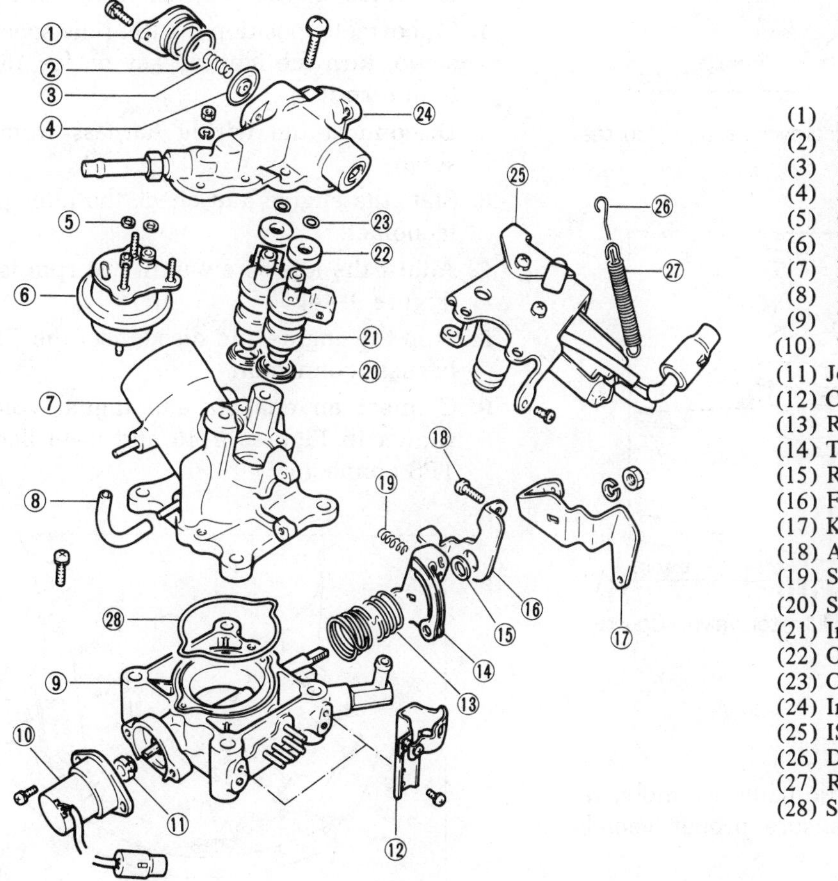

Figure 9–38. Disassembled view/components of the 1.6L turbocharged Dodge Colt fuel mixer/throttle body. [Courtesy of Chrysler Corporation]

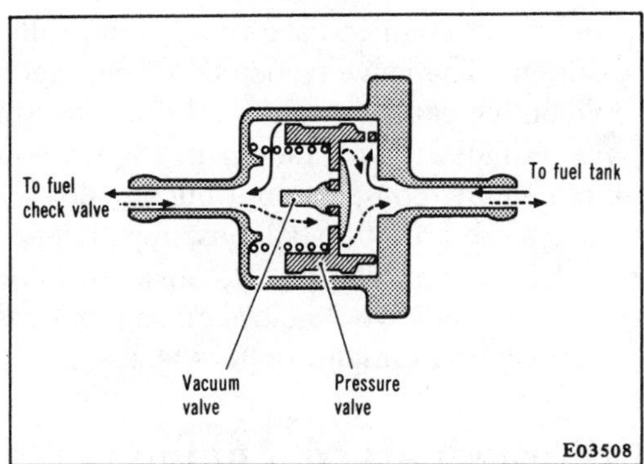

Figure 9–39. Fuel overfill limiter valve. [Courtesy of Chrysler Corporation]

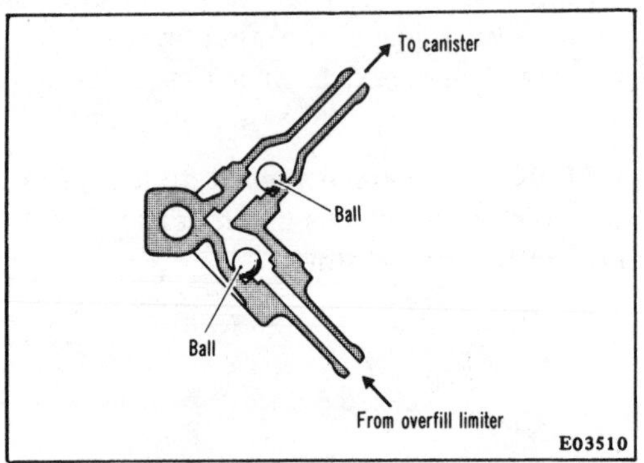

Figure 9–40. Fuel check valve/rollover safety. [Courtesy of Chrysler Corporation]

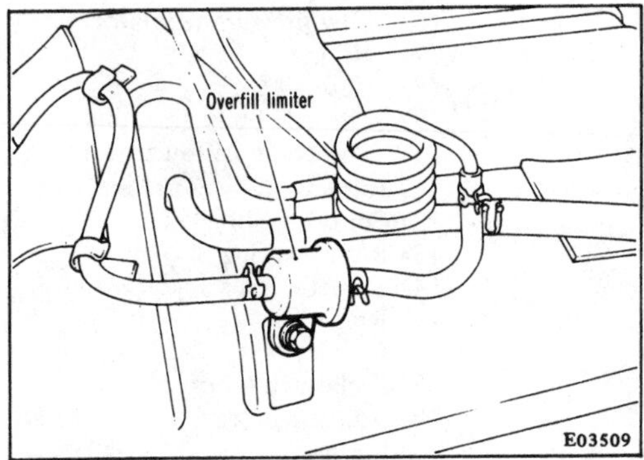

Figure 9–41. Location of overfill limiter valve. [Courtesy of Chrysler Corporation]

Injection Mixer Adjustments

The injection mixer, or throttle body assembly, requires two adjustments to ensure proper vehicle driveability:

1. Idle speed
2. Throttle position sensor

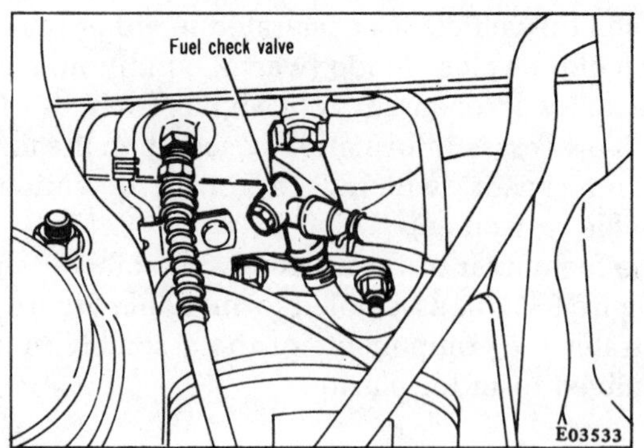

Figure 9–42. Location of fuel check/rollover valve. [Courtesy of Chrysler Corporation]

Procedure

1. Engine must be at normal operating temperature of 185–205°F (85–90°C).
2. Stop the engine and disconnect the accelerator cable from the throttle lever.
3. Refer to Figure 9–43; loosen the two retaining screws on the throttle position sensor (TPS), rotate it clockwise, and tighten the screws.
4. To correctly position the ISC (idle speed control) servo, turn the ignition key *on* for 15 seconds, then turn it *off.*
5. Disconnect the wiring harness from the ISC servo.
6. Start the engine and check the idle rpm with a tachometer.
7. Adjust the idle screw until 600 rpm is obtained (Figure 9–44).
8. Stop the engine and disconnect the TPS wiring harness connector.
9. Connect an adapter and digital voltmeter as shown in Figures 9–45 and 9–46 between the TPS connector.

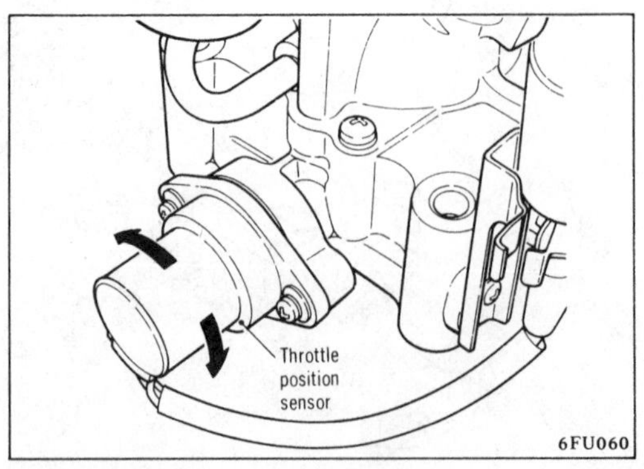

Figure 9–43. Adjusting throttle position sensor. [Courtesy of Chrysler Corporation]

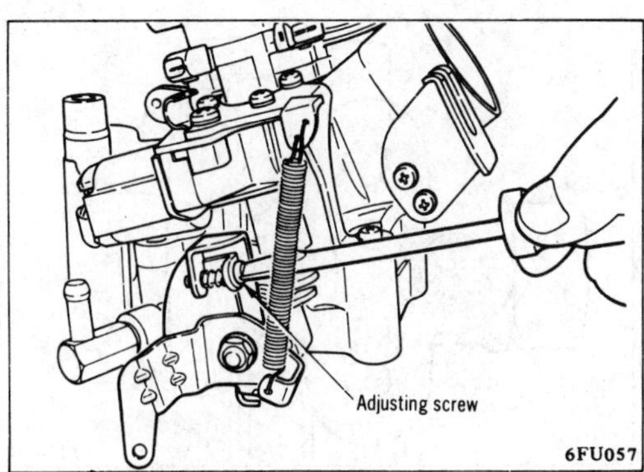

Figure 9–44. Adjusting idle speed screw. [Courtesy of Chrysler Corporation]

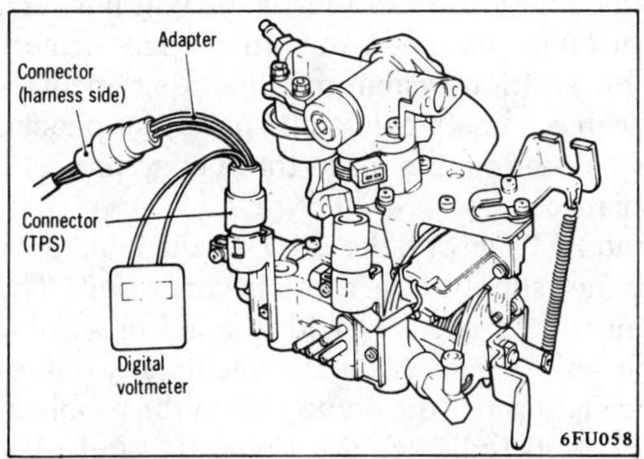

Figure 9–45. Digital voltmeter connections [1]. [Courtesy of Chrysler Corporation]

10. Do not start the engine but turn the ignition key *on*.
11. Read the TPS output voltage.
12. Loosen the TPS screws to obtain a voltage reading of 0.48 ± 0.03 volt by turning the TPS clockwise or counterclockwise; apply sealant to the screws and tighten them.
13. Manually open the throttle lever to wide open throttle once, and when it returns to its idle position, recheck the voltage reading as per Step 12.
14. When the voltmeter has been removed, reconnect the ISC harness; start the engine and check that the idle speed is 600 rpm; then stop the engine.
15. Turn the ignition key *on* for 15 seconds and then switch it *off* to position the ISC servo.
16. Reconnect the accelerator cable and adjust if necessary.

Injector Service

Faulty injectors can be checked by disconnecting one electrical lead at a time from each one while the engine is running to see if in fact they are operating—or by switching leads around. The injectors can be removed after releasing fuel system pressure and after taking off the retaining screws holding Item 24 (injector holder) as shown in the exploded view of the throttle body unit in Figure 9–38.

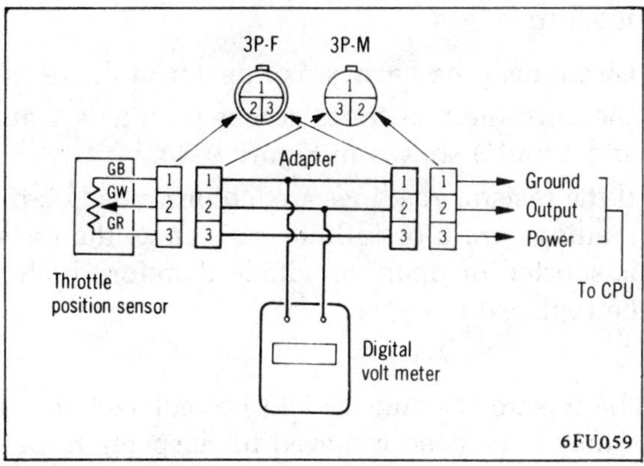

Figure 9–46. Digital voltmeter connections [2]. [Courtesy of Chrysler Corporation]

Do not immerse injectors or other parts such as the TPS and ISC servo in solvent because the insulated parts will be damaged.

The injectors can be checked for electrical resistance with an ohmmeter placed across the terminals and should read 2 ohms. Also check the resistor mounted near the ignition coil shown in Figures 9–47 and 9–48.

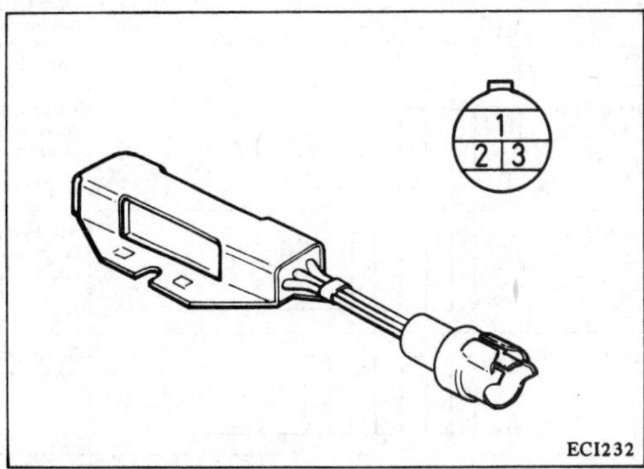

Figure 9–47. Resistor assembly. [Courtesy of Chrysler Corporation]

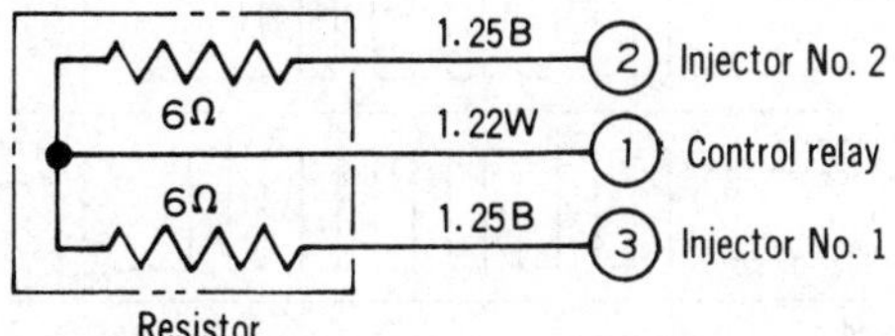

Figure 9–48. Resistor check readings. [Courtesy of Chrysler Corporation]

Procedure

1. Disconnect the harness connector of the resistor.
2. Measure the resistance across terminals 1 and 2 and 1 and 3 shown in Figure 9–48.
3. If the resistor readings are 6 ohms, it is OK: if the readings are 0 or infinite resistance, the resistor is shorted or open circuited—therefore it should be replaced.

The injector O-rings should be replaced any time that they have been removed for inspection. Before reinstalling item 24 (injector holder), inspect the injector filters as shown in Figure 9–49 for any dirt or damage.

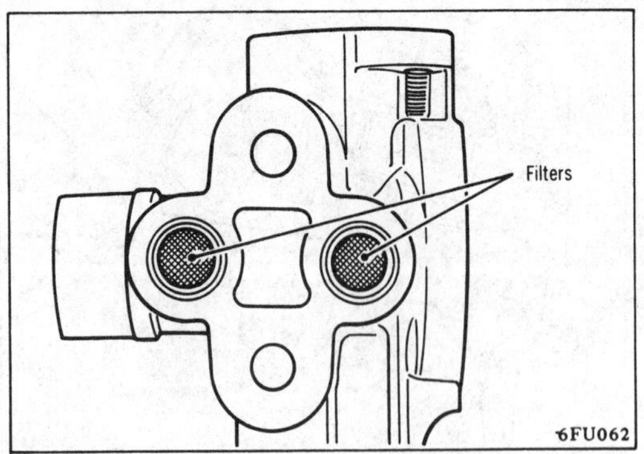

Figure 9–49. Injector filter screens location. (Courtesy of Chrysler Corporation)

Troubleshooting 1.6L Engine ECI System

CAUTION: Never disconnect the battery terminals unless the ignition switch is turned *off*. Removing a battery cable or terminal while the engine is running or the ignition key is *on* can cause a malfunction of the ECU or damage to semiconductors within the logic or power modules.

If the engine is hard to start or will not start, if it has unstable idle speed, or exhibits poor driveability, look for ignition system malfunctions, incorrect engine valve adjustment, etc. Therefore, check these areas before moving on to the fuel system as a possible cause.

If the ECI sensors, the ECU, or the injectors are at fault, fuel supply interruption can result. The ECI system can be checked with the aid of a voltmeter, which will deflect its meter needle anywhere from one to eight times depending upon the problem area. Figure 9–50 indicates the Code No. and diagnosis

Code No.	Diagnosis item	Voltage waveform (abnormal code)	Contents of diagnosis	Check item
1	Oxygen sensor	12V / 0V	In city driving mode, Oxygen sensor signal does not change for 20 seconds or more or engine stalls. When engine stalls, turn ignition switch from OFF position to ON position. If abnormal code is indicated, computer is normal. If no indication is made, computer is not in normal condition.	• Wire harness and connector • Oxygen sensor • Computer
2	Ignition signal	12V / 0V	While cranking the engine, input of ignition signal is not applied to computer for 3 seconds or more.	• Wire harness and connector • Igniter • Computer
3	Air flow sensor	12V / 0V	Air flow sensor output is 10 Hz or less while engine is idling, or it is 100 Hz or more when engine stalls.	• Wire harness and connector • Air flow sensor • Computer
4	Pressure sensor	12V / 0V	Boost sensor output is 1,460 mm Hg (4.5 V) or more, or it is 65 mm Hg (0.2 V) or less.	• Wire harness and connector • Boost sensor • Computer
5	Throttle position sensor	12V / 0V	Throttle sensor output is 0.2 V or less, or it is 4 V or more while engine is idling (idle switch is ON).	• Wire harness and connector • Throttle sensor • Computer
6	ISC motor position sensor	12V / 0V	Throttle sensor output is 0.39 V with L switch OFF.	• Wire harness and connector • ISC servo • Computer
7	Water temperature sensor	12V / 0V	Water temperature sensor output is 4.5 V or more, or it is 0.1 V or less.	• Wire harness and connector • Water temperature sensor • Computer
8	Car speed signal	12V / 0V	Air flow sensor output is 500 Hz or more and car speed is 2.5 km/h or less.	• Wire harness and connector • Car speed sensor • Computer

ECI241

Figure 9–50. Diagnosis chart 1.6L ECI fuel system. (Courtesy of Chrysler Corporation)

item to be checked based upon the voltmeter needle deflections.

In order to access these possible problems individually, it is necessary to follow the service procedures below:

1. Turn the ignition switch *off* and connect a voltmeter to the connector, which is located underneath the battery tray.
2. Turn the ignition switch *on*, and the ECU memory contents will cause a voltage signal to be sent to the voltmeter connections.
3. A deflection of the voltmeter to 12 volts is an indication that the system is normal.
4. Any abnormality stored in the ECU memory will cause the voltmeter pointer to deflect anywhere from 1 to 8 times.
5. Turn the ignition key *off*.
6. Check the diagnosis chart and check out the symptom related to the number of voltmeter needle deflections.
7. Once the defective parts have been repaired, it is necessary to disconnect the battery ground cable for 15 or more seconds, then reconnect it to ensure that the abnormal code has been erased.

ECU Tests

If it appears that the ECU is defective, perform all of the voltmeter tests and individual sensor tests; and if these check out OK, replace the ECU.

To check out all of the sensors and injectors, refer to Figure 9–51, which shows the identification of the ECU connector terminals, and with the aid of a voltmeter, check out each pin connector terminal as specified in Figure 9–52.

ELECTRONIC CONTROL UNIT (ECU)

If ECU appears defective, perform all voltmeter tests and individual sensor tests. If everything is in order, replace ECU.

A1	Throttle position sensor output
A2	Sensor power supply (5V)
A3	Coolant temperature sensor (+)
A4	Intake air temperature sensor (+)
A5	Idle position switch (+)
A6	Oxygen sensor
A7	Air flow sensor output
A8	Ignition coil (–) terminal
A9	Sensor ground
A10	Secondary air control solenoid valve
A11	–
A12	–
A13	Ignition switch (ST) terminal (cranking signal)
A14	MPS1
A15	Car speed (0 km/h)
A16	Diagnosis (output)
A17	Pressure sensor output
B1	Battery
B2	Ground
B3	Ground
B4	EGR control solenoid valve (–)
B5	Fuel pump relay
B6	ISC servo motor (Extension) (+)
B7	Battery
B8	Pressure sensor solenoid valve
B9	Injector No. 1 (–)
B10	Injector No. 2 (–)
B11	ISC servo motor (Retraction) (+)
B12	A/C relay
B13	Battery backup

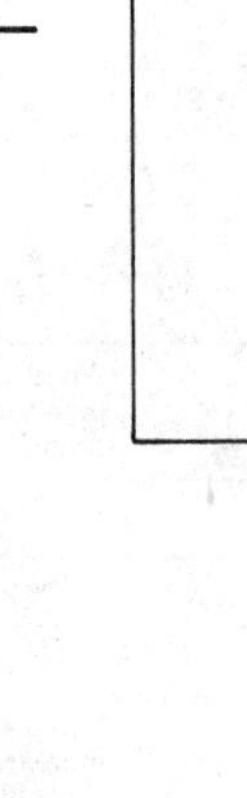

Figure 9–51. 1.6L Dodge Colt ECU connector terminals. (Courtesy of Chrysler Corporation)

Output Signals from Tester

Check item	Condition		Check meter reading when normal	Terminal location of computer
Power supply	Ignition switch OFF → ON		11–13V	B–1
Secondary air control solenoid valve	Ignition switch OFF → ST after warming up the engine		0.2–1V ↓ 30 seconds 13–15V	A–10
Throttle position switch	Ignition switch OFF → ON	Accelerator closed	0.4–1.5V	A–1
		Accelerator wide opened	4.5–5.0V	
Coolant temperature sensor	Ignition switch OFF → ON	0°C (32°F)	3.5V	A–3
		20°C (68°F)	2.6V	
		40°C (104°F)	1.8V	
		80°C (176°F)	0.5V	
Intake air temperature sensor	Ignition switch OFF → ON	0°C (32°F)	3.5V	A–4
		20°C (68°F)	2.6V	
		40°C (104°F)	1.8V	
		80°C (176°F)	0.6V	
Idle position switch	Ignition switch OFF → ON	Accelerator closed	0–0.4V	A–5
		Accelerator wide opened	11–13V	
ISC motor position switch	Ignition switch OFF → ON		11–13V *1	A–14
EGR control solenoid valve	Ignition switch OFF → ON	Return to 0 volt after 0.1 sec.	0–0.5V after 1 second	B–4
ISC motor for extension	Ignition switch OFF → ON		Pointer C swings a moment after 15 sec.	B–6
*2 A/C relay	Ignition switch OFF → ON	A/C switch OFF → ON	11–13V	B–12
Reed switch for car speed	Start engine, transmission in first and operate car slowly		0.2–1V 4–5V	A–15
ISC motor for retraction	Ignition switch OFF → ON		Pointer swings a moment after 15 sec.	B–11
Cranking signal	Ignition switch OFF → ST		Over 8V	A–13
Control relay	Idling		0–1V	B–5
ISC motor position switch	Idling		11–13V	A–14
Ignition pulse	Idling		12V–15V	A–8
	3,000 rpm	When engine speed is increased voltage drops slightly	11–13V	
Air flow sensor	Idling		2.7–3.2V	A–7
	3,000 rpm			
Injector No. 1	Idling		13–15V	B–9
	3,000 rpm	When engine speed is increased voltage drops slightly	12–13V	
Injector No. 2	Idling		13–15V	B–10
	3,000 rpm	When engine speed is increased voltage drops slightly	12–13V	
Oxygen sensor	Keep 1,300 rpm after warming up the engine		0–1V ↕ Flashing 2.7V	A–6
EGR control solenoid valve	Keep idling after warming up the engine Over 3,000 rpm, drops to zero volts		13–15V	B–4
Pressure sensor	Ignition switch: OFF → ON		1.5–2.6V	A–17 Switches back to 2.0V every two min.
	Idling		0.2–1.2V	
ISC motor for extension	Idling	A/C switch: OFF → ON	Over 5V *3	B–6
ISC motor for retraction	Idling	A/C switch: ON → OFF	Over 6V *3	B–11

Note *1: If ignition switch is turned on for 15 seconds or more, the reading drops to 1V or less momentarily and returns to 6 to 13V.
*2: A/C = Air conditioner
*3: It indiçates 6V a moment (very short time). If it is difficult to read the indication, repeat to turn the air conditioner switch off and on or on and off several times. If voltmeter swings, the components are in good condition.

Figure 9-52. Diagnosis chart—output signals from tester. (Courtesy of Chrysler Corporation)

10

History and Types of Diesel Fuel Injection Systems

History of Diesel Fuel Injection

Through the years, many companies and individuals have worked hard to develop and improve fuel injection systems capable of withstanding the requirements placed upon them. However, let us step back in time and consider some of the problems that faced Rudolf Diesel. In 1892 a patent was issued to him whereby he proposed an engine in which air would be compressed to such an extent that the resulting temperature would exceed by far the ignition temperature of the fuel. He calculated and anticipated that the fuel injection from top dead center (TDC) on would take place so gradually that combustion due to the descending piston and the expansion of the gas would occur without material pressure or temperature rise. Further expansion of the gas would continue after fuel injection ceased.

Diesel obtained the financial backing of Baron von Krupp and the giant Machinenfabrik Augsburg Nurnberg Company and set out to build an engine that would burn coal dust, then a useless by-product, mountains of which were piled up in the Ruhr valley. The actual commercial development of self-ignition of fuel in an internal combustion engine was carried out by Dr. Lauster and the engineers of the M.A.N. Company in Germany in collaboration with Diesel from 1893 to 1898. The first experimental engine that was built in 1893 used coal dust and high-pressure air to blast this fuel into the combustion chamber. This engine is said to have exploded, and further experiments with coal dust as fuel were unsuccessful. A successful compression ignition engine, however, was developed; it used oil as fuel, and a number of manufacturers were licensed to build similar engines.

The original oil burning engine experiments used mechanical injection. However, the results obtained were unsatisfactory, owing for the most part to the crude injection equipment with large dead volume.

Diesel again resorted to using air blast injection, which served the dual purpose of providing atomization of the fuel and turbulence of the mixture. His tests with air injection proved so successful that this became the accepted method of injection for many years. Failure of his first engines were due to his attempt to compress air to a pressure of 1500 psi and the lack of any provision for cylinder cooling. Diesel attempted to utilize so much of the heat of combustion by a long power stroke that no further cooling would be required.

His third engine, built in 1895, was a success. It was a 4-stroke cycle with a compression pressure of 450 psi, which is comparable to many present-day engines. It was water cooled, and fuel was injected by a blast of high-pressure air. It developed 24% brake thermal efficiency, which was actually 35% indicated, and was a great improvement over all previous engines. Much of the subsequent improvement and progress in diesel engine development has been largely dependent upon improvements in fuel injection technology.

Robert Bosch was most probably the one individual who contributed most to the success of the diesel engine as we know it today by manufacturing the first mass-produced fuel injection pumps for use as early as 1927. Robert Bosch spent some of his early years outside Germany, specifically with Edison in the United States and then with Siemens in Great Britain. He subsequently designed and produced a magneto and established the Bosch Magneto Company in 1906. This company started production in a factory in Springfield, Massachusetts, in 1909–1910, and by 1914 the U.S. output was higher than that in Germany. By 1914 there was also a branch in Japan.

In 1922, Bosch decided to embark upon the development of a fuel injection system for diesel engines. By 1923, a dozen pumps were tested, and by the end of 1924, a pump that could meet the requirements was ready. In 1925 Bosch signed an agreement with a developer, Franz Lang, to use a diesel engine system including its fuel injection equipment developed by Lang. This particular pump did not meet the particular needs; therefore, further development led to the design of an acceptable unit by 1927, and in August of that year, 1000 pumps had been produced; by March 1934, 100,000 pumps had left the factory.

As early as 1927, however, a diesel engine was installed in a passenger car, which was a 2.1-liter Acro Engine; it operated until 1929 and accumulated 35,000 kilometers. Then in 1932, a small truck diesel was put into a car for testing.

Bosch applied for a patent for the field of diesel injection equipment in 1926 and along the way acquired licenses or protective rights to the developments of Acro, REF-Apparatebau (established by the L'Orange company), Deutz, Schnurle, and Atlas Diesel, which all contributed to injection development in some part. Since then, over 1000 patents have been granted for fuel injection designs.

With the increasing demand for fuel injection equipment. Bosch in Germany was unable to keep up with the demand. Therefore, other agreements were reached with companies that would subsequently produce injection equipment. In France, the Societé des Ateliers de Construction Lavalette SA, which Bosch had already been a part owner of since 1928, was chosen as a partner. A manufacturing license was signed on October 10, 1931.

In England, a subsidiary of Joseph Lucas Ltd. had been producing injection pumps under license from REF-Apparatebau in limited quantity. Negotiations between Lucas and Bosch led in 1931 to a contract whereby Bosch would be a 49% participant in a Lucas subsidiary, C. A. Vandervell (bearings). One of the stipulations was that this company take on the name of CAV-Bosch, which would be supported by Bosch in the creation of production facilities for diesel equipment. This was dated from October 21, 1931.

In the United States, production facilities were established in Springfield, Massachusetts, under the name American Bosch, now "United Technologies—Diesel Systems." This contract was signed on January 1, 1934.

On August 10, 1938, a licensing agreement was signed with a group of Japanese, and the Diesel Kiki Company was founded, with which Bosch is still associated.

Today, licenses have been granted to manufacturing companies in Argentina, Australia, Brazil, India, Japan, Rumania, Spain, Turkey, and the United States; therefore, there are very few manufacturers of diesel fuel injection equipment today who do not use some form of Bosch designs. When diesel injection equipment is talked about in any circle, the name Robert Bosch stands at the forefront of any discussion. Figure 10–1 shows current Robert Bosch Corporation diesel injection equipment manufacturing facilities.

Types of Diesel Injection Pumps

In automotive passenger car and light pickup truck applications, fuel injection pumps are either one of two types:

1. Multiple plunger pump commonly referred to as an in-line type of injection pump
2. Distributor/rotary injection pump

In system number 1, individual pumps are used for each engine cylinder. These pumps are all contained within the one common injection pump housing and operate on what is commonly called a "jerk pump" concept. Figure 10–2(a) illustrates the basic concept of a multiple pump injection system while Figure 10–2(b) shows the concept of operation of the jerk pump arrangement.

In Figure 10–3, the pumping plunger used in the in-line pump can be moved up and down by the ac-

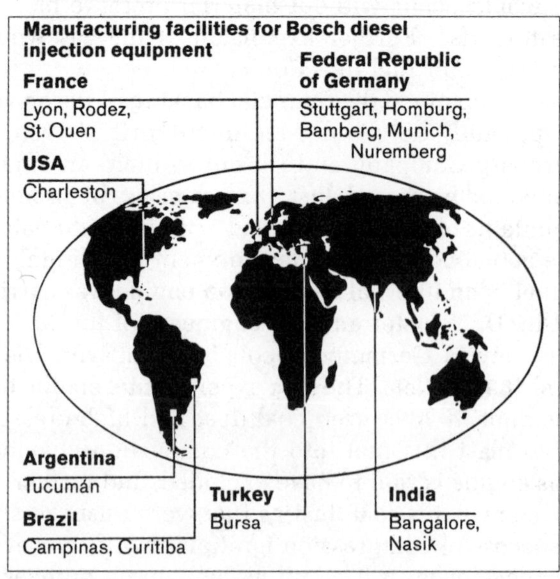

Figure 10–1. Manufacturing facilities worldwide—Robert Bosch Corporation. (Courtesy of Robert Bosch Corp.)

tion of a rotating camshaft that is driven from the engine gear train. Injection pump speed would of course be half that of engine rpm on a 4-stroke cycle engine.

A plunger return spring is used to maintain positive plunger movement and to stop it from bouncing on the pump camshaft as well as returning it to the cam base circle position at the end of injection. Fuel that is placed under high pressure by the action of the rotating pump plungers is delivered through a high pressure steel backed line to the individual cylinder fuel injectors in engine firing order sequence. Correct fuel delivery is obtained by timing the injection pump camshaft gear to the engine gear train. The number of pumping plungers will always be the same as the number of cylinders in the engine.

All of the individual pumping plungers are connected together by a toothed rod known as a fuel control rack in order to allow rotary motion of each plunger in addition to the up-and-down movement obtained from the pump camshaft rotation. A detailed description of the fuel control rack and the actual operation of the fuel injection pump is described in detail within this chapter.

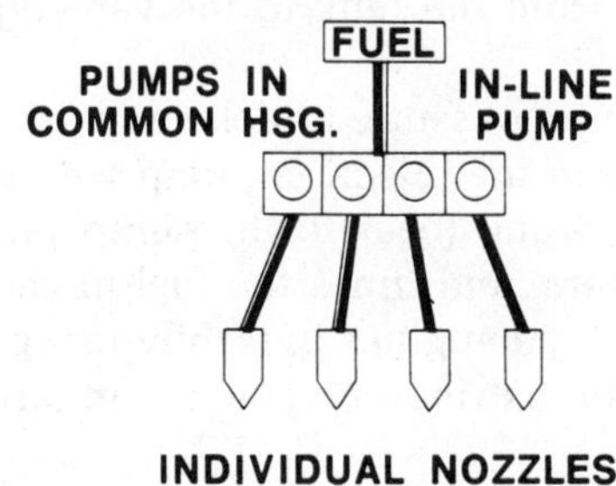

Figure 10-2(a). Basic arrangement of a high pressure in-line type fuel injection pump. (Courtesy of Deere and Company)

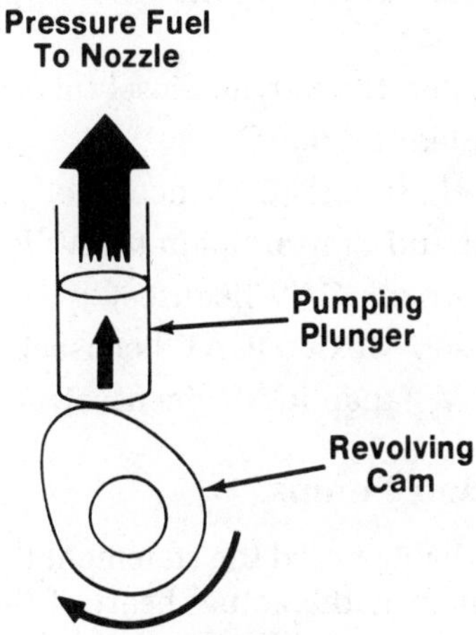

Figure 10-2(b). Basic operation of a jerk pump system. (Courtesy of Deere and Company)

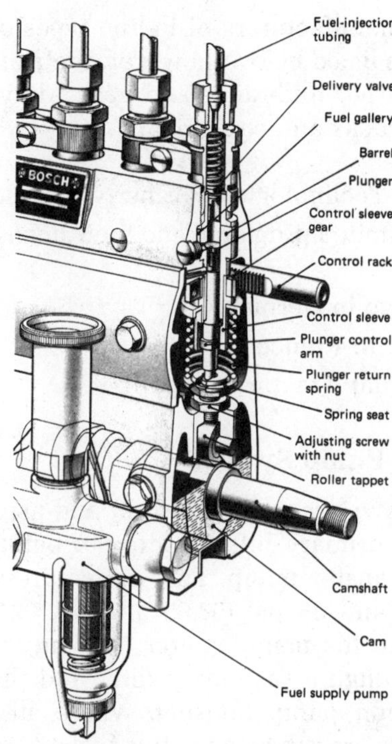

Figure 10-3. Typical in-line high pressure jerk-type fuel injection pump. (Courtesy of Robert Bosch Corporation)

The camshaft shape used with multiple pump (jerk-type) fuel injection systems basically controls the following:

1. The lift of the pump plunger and quantity of fuel delivered (in conjunction with the helix shape cut into the upper part of the plunger)
2. The basic timing of the plunger
3. Rate of fuel delivery

Fuel is supplied to the injection pump housing by a transfer or lift-type pump that is normally driven directly from the injection pump camshaft. Vehicles using the in-line type injection pump and described in this book are:

1. Mercedes-Benz, discussed in Chapter 13, Figure 13–3.
2. Datsun-Nissan SD25 pickup truck, discussed in Chapter 18, Figure 18–3.
3. Toyota Landcruiser, discussed in Chapter 19, Figure 19–3.

The fuel injection pumps used on the above vehicles are manufactured by:

1. Robert Bosch
2. Diesel Kiki (Robert Bosch licensee)
3. Nippondenso (Robert Bosch licensee)

Other manufacturers of in-line types of injection pumps are listed below; however, their products are used on larger mid-range and heavy-duty diesel engines in trucks and equipment.

1. United Technologies (Formerly American Bosch)
2. CAV Limited (England) and their licensees worldwide
3. L'Orange in Germany
4. SIGMA in France
5. Caterpillar Tractor Company, USA

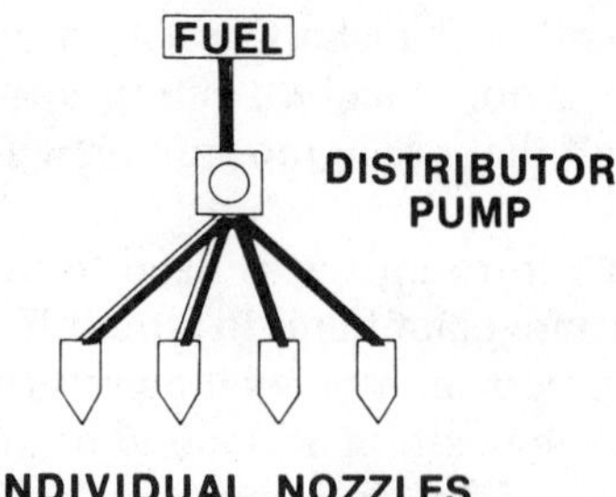

Figure 10–4. Basic arrangement of a rotary/distributor fuel injection pump system. (Courtesy of Deere and Company)

Multiple Pump System

Fuel is drawn from the tank first and usually passes through a primary filter or strainer before entering the fuel transfer pump. Fuel under pressure at between 10 and 35 psi (68.95 and 241.325 kPa), depending on the manufacturer, leaving the pump is forced through a secondary filter and then on into the injection pump housing, where all individual pump units are subjected to this fuel oil. Many of the newer pumps do not use any lube oil in their base to lubricate the rotating camshaft, but have the internal area of the injection pump filled with fuel oil at all times. Older pumps and some current ones have a separate lube oil compartment for the injection pump camshaft. Fuel at each individual jerk pump within the common housing is timed, metered, and pressurized and then delivered through a high pressure steel-backed line to each injector nozzle in firing order sequence at pressures ranging anywhere from a low of 100 atm (1470 psi) to as high as 300 atm (4410 psi). Some nozzles have a leak-off connection line on them to allow some fuel that is used for internal lubrication of the nozzle to return to the fuel tank.

Distributor-Type Injection Pumps

The basic concept of operation of the distributor-type pump is very similar to that of an ignition distributor used in a gasoline engine. The main difference is that instead of a high tension spark being sent to individual spark plugs, the fuel injection pump distributes high pressure fuel to individual fuel injectors for each engine cylinder. In both cases a rotor is used to time the spark to each cylinder or in the diesel, the rotor distributes fuel. A simplified line diagram of the distributor-type injection pump is shown in Figure 10–4.

Instead of a distributor cap, the injection pump employs what is commonly called a hydraulic head or rotor head. Attached to this head are fuel delivery lines to each engine cylinder. A detailed description of the operation of this type of pump can be found in both the Robert Bosch chapter (12), VE pump section, and also under the CAV DPA pump chapter (28).

The distributor-type pump contains a separate vane-type transfer pump in the drive end of the pump that delivers fuel to the injection pump hydraulic head. This vane-type pump pressure varies from about 35 psi (241 kPa) at engine idle to as high as 130 psi (896 kPa) at wide open throttle. Some vehicle applications use a mechanical or electrical lift pump to supply fuel from the tank to the vane-type transfer pump.

Vane pump pressure is delivered to a charging passage within the injection pump where a rotating and reciprocating (back/forth) pump plunger pressurizes, meters, and times the fuel to each injector. Automotive-type engines presently using the Robert Bosch rotary/distributor type pump are listed in Chapter 12.

Manufacturers of distributor type pumps for automotive applications are:

1. Robert Bosch
2. Diesel Kiki and Nippondenso, Japan (Robert Bosch licensee)
3. Roosa Master (Stanadyne diesel fuel systems)
4. ÇAV Limited (England)
5. Roto-Diesel, France (CAV licensee)
6. ConDiesel and Simsa, Spain (CAV licensee)
7. Lucas do Brazil (CAV licensee)
8. Inyec Diesel, Mexico (CAV licensee)
9. Nihon-CAV, Japan (CAV licensee)

Fuel Injection Pumps

You have probably heard the statement that "the fuel-injection system is the actual heart of the diesel engine." When you consider that indeed a high speed diesel could not be developed until an adequate fuel injection system was designed and produced, and that even Rudolf Diesel ran into problems basically associated with lack of a good injection system, then this statement takes on a much broader and stronger meaning.

From our previous discussion related to combustion systems in Chapter 3, you will recollect that efficient combustion is dependent upon the fuel being injected at the proper time and rate. In addition, the injection pressure must be high enough for adequate atomization and penetration. Involved in this is the compressibility and dynamics of the fuel column between the pump and nozzle, plus the mechanical characteristics of the pump, discharge tubing, and nozzle of the conventional jerk pump system.

Various methods of mechanical injection and metering control have been discarded; others have been improved. There have been many important developments in pumps and nozzles for high speed diesel engines.

Before delving into the specifics of individual injection pumps, let us consider what the actual demands and functions of a good injection system are.

Functions of a Fuel Injection System

The requirements of a fuel injection system can be summarized as follows:

1. In order to receive equal power from all cylinders, the amount of fuel injected must remain constant from cycle to cycle, and obviously from cylinder to cylinder. A smooth-running engine depends on even fuel distribution to each cylinder throughout the speed range; otherwise, cylinder balance will be upset and some cylinders will be working harder than others. Overloading and overheating would result. This function is commonly referred to as *metering*.
2. As engine load and speed vary owing to application and operating conditions, the point of actual injection for a given load and speed will vary with this condition. Therefore, the injection system has to adjust the timing or point of injection to the fluctuating demands of engine operation. In summary, the injection system must *inject fuel* at the correct point in the cycle regardless of the engine *speed* and *load*.
3. In (direct-injection) open combustion chamber engines and especially in modern high speed engines, a slow start or ending of the injection period affects both the initial and final portions of the injected fuel. In other words, the fuel will not be broken down or atomized into as fine a fuel droplet as it would be, for example, with a rapid start and rapid cutoff at the end of the injection period. Thus *injection must begin and end very quickly*.
4. Since fuel is compressible, there is a time lag between the actual beginning of delivery by the pump and the actual beginning of discharge from the nozzle. Also, the rate of delivery from the pump is not identical with the rate of discharge from the nozzle. Therefore, by controlling the rate at which fuel is injected, the performance of many engines can be improved. One of the most important characteristics is the spray duration, particularly at full load, since it directly affects engine power, fuel consumption, and exhaust smoke. In some engines, a small amount of fuel is injected 8 to 10 degrees ahead of the main injection charge so that it is already burning when the main injection process occurs. This produces smoother combustion and a relatively slow rate of pressure rise in the cylinder. The type of nozzle used can to some extent control the actual rate of injection. In summation, the injection system must inject fuel at a rate necessary to control combustion and the rate of pressure rise *during combustion*.
5. Good combustion is related to the degree of fuel atomization; therefore, the type of combustion chamber and engine speed affect these requirements. The type and size of nozzle plus the injection pressure will control the degree of atomization. The injection system must then *atomize the fuel charge* as required by the particular type of combustion chamber in use.
6. The volumetric efficiency of an engine generally decreases with an increase in speed, because of increasing resistance to air flow and inertia of air in the actual intake system. Therefore, the *power* developed by the engine and the completeness of *combustion* are really dependent on air flow and the uniformity of fuel distribution throughout the air charge within the combustion chamber. In direct-injection or open-chamber-type engines, the fuel must penetrate into the air mass in all directions within the combustion chamber. In smaller high speed engines, adequate penetration is also required; however, it is very undesirable to allow the fuel spray to strike either the piston crown or cylinder wall. Burning of the crown from direct flame impingement or cylinder wall wash, causing lube oil dilution, crankcase oil dilution, possible piston to liner scuffing, and eventual seizure,

could result. Incomplete combustion would be a further result, creating carbon deposits and ring sticking. In summation, then, the injection system must *distribute fuel evenly throughout the air mass in the combustion chamber.*

High Pressure In-Line Fuel Injection Pumps

Plunger Operation

In Figure 10–5(a), with the plunger at the bottom of its stroke, both ports are uncovered by the top of the plunger and the pump reservoir is full of fuel that will enter through both ports. The fuel delivery valve is closed at this time.

In Figure 10–5(b), as the plunger is starting its upward travel (by rotation of the injection pump camshaft), fuel will spill back out both ports into the reservoir until they are covered by the top land area of the plunger. This is termed *port closing* and is the basic start to injection. Note that the delivery valve is still on its seat.

In Figure 10–5(c), the fuel pressure will continue to rise until it is high enough (approximately 100 psi) to lift or force the discharge valve off its seat, therefore allowing the displaced fuel to pass through the fuel line to the nozzle and on into the cylinder, with the delivery valve open. Injection will continue until the lower helical land uncovers the control port. This port is uncovered slightly ahead of the actual ending of the upward-moving plunger, which will displace the remaining fuel back to the reservoir.

In Figure 10–5(d), the displaced fuel is allowed to escape down the relief area of the plunger and out the control port to the reservoir. The discharge valve closes, and the plunger completes its stroke and is positively returned to the next intake stroke [Figure 10–5(a)] by the plunger return spring.

At the end of injection when the control port is uncovered, the high pressure in the pump chamber flashes back into the reservoir; therefore, to prevent eventual erosion of the pump housing, this pressure is deflected by the hardened end of the barrel locating screw. Two ports will cut deflection pressures in half, while the conical port will disperse the spill deflection.

Metering Principle

The amount or volume of the fuel charge is regulated by rotating the plunger in the barrel as shown in Figure 10–6 to effectively alter the relationship of the control port and the control helix on the plunger. This is done by means of a rack and a control collar or control sleeve.

The *rack* is basically a rod with teeth on one side, which is supported and operates in bores in the housing. The rack is in turn connected to a governor. The geared segment or control collar is clamped to the top of the control sleeve with teeth that engage

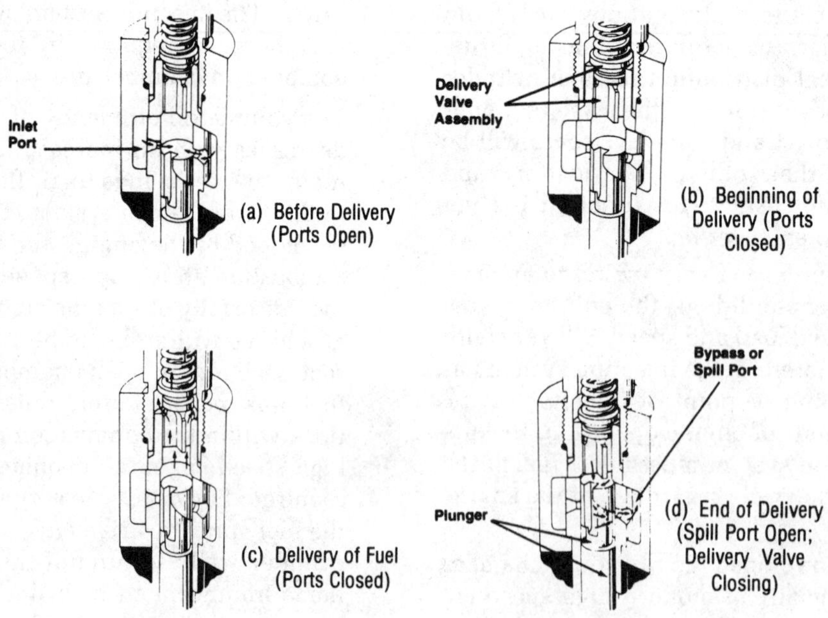

Figure 10–5. In-line pump plunger operation. [Courtesy of Robert Bosch Corporation]

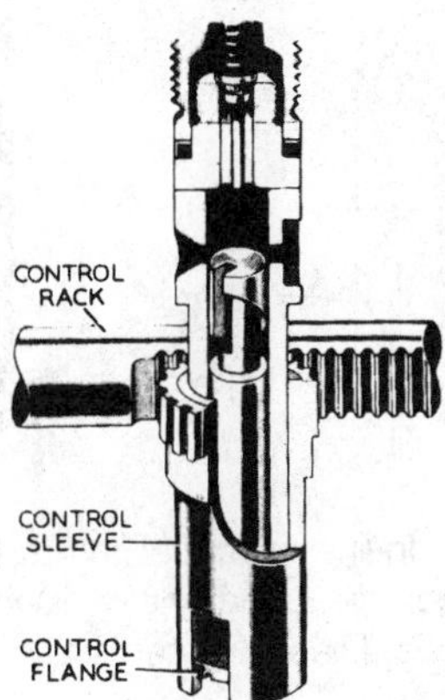

Figure 10–6. Regulating fuel charge by manipulation of fuel control rack (in-line pump). (Courtesy of United Technologies, Diesel Systems Group)

the rack. The control sleeve is a loose fit over the barrel and is slotted at the bottom to engage the flanges on the plunger so that as the rack is moved it will cause rotation of the collar, sleeve, and plunger.

The operation of Robert Bosch in-line pumps is basically the same as that of CAV and United Technologies Diesel Systems in-line pumps; however, let us quickly review the pumping plunger's operation and excess fuel device so that we thoroughly understand the principle.

The plunger within the barrel is moved up and down by the action of the rotating camshaft within the injection pump housing. It can also be rotated by the movement of the fuel control rack connected to the throttle and governor linkage. Anytime that the stop control is moved to the engine shutdown position, the plunger is rotated as shown in Figure 10–7, whereby the vertical slot machined in the plunger will always be in alignment with the supply or control port. Therefore, regardless of the plunger's vertical position within the barrel, fuel pressure can never exceed that delivered by the fuel transfer pump. This pressure will never be able to overcome the force of the delivery valve spring, so no fuel can be sent to the injectors.

Figure 10–8 shows the location of the *excess fuel delivery* and *retard* notch somewhat enlarged. Excess fuel is possible only during starting, since while the engine is stopped the speed control lever is moved to the *slow idle* position, thereby moving the fuel rack to place the plunger in such a position that excess fuel can be delivered. The instant the engine starts, however, the governor will move the fuel rack to a position corresponding to the position of the throttle lever. The retard notch, also in alignment with the control port, delays port closing and therefore retards timing during starting.

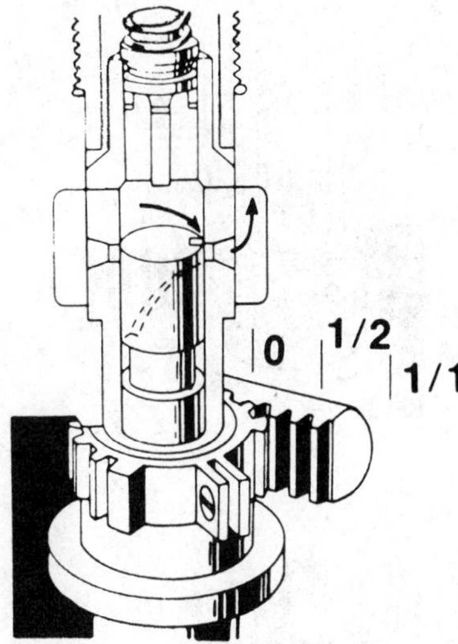

Figure 10–7. Plunger in no-fuel delivery position (in-line pump). (Courtesy of Robert Bosch Corporation)

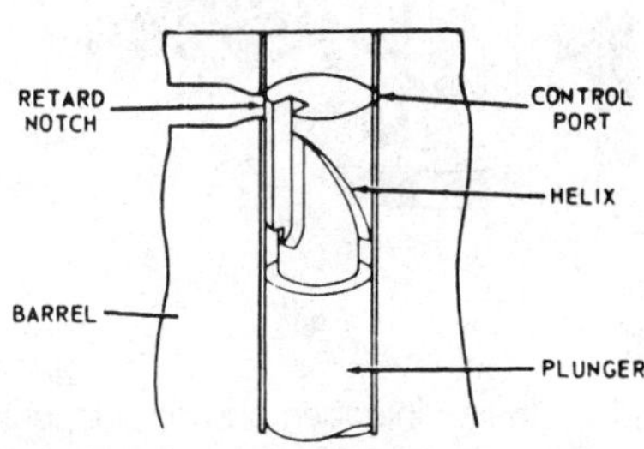

Figure 10–8. In-line injection pump, excess fuel delivery and retard notch machined into plunger. (Courtesy of Robert Bosch Corporation)

During any partial fuel delivery situation, the amount of fuel supplied to the injector will be in proportion to the *effective stroke* of the plunger, which simply means that the instant the supply port is covered by the upward-moving plunger, fuel will start to flow to the injector. This will continue as long as the control port is covered; however, as soon as the upward-moving plunger helix uncovers this port, fuel pressure to the injector is lost and injection ceases. Therefore, we only effectively deliver fuel to the injector as long as the control port is covered; this is shown in Figure 10–9(a) for any partial throttle position. This will vary in proportion to the throttle and rack position from idle to maximum fuel.

When the operator or driver moves the throttle to its maximum limit of travel, the effective stroke of the plunger, due to the rotation of the plunger helix, will allow greater fuel delivery because of the longer period that the control port is closed during the up-

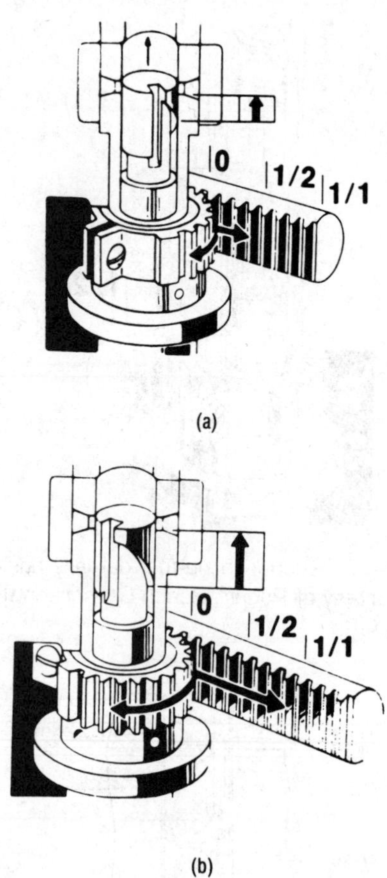

Figure 10–9. In-line injection pump, plunger fuel delivery positions. a. Partial. b. Maximum. (Courtesy of Robert Bosch Corporation)

ward movement of the plunger by the pump camshaft. This is shown in Figure 10–9(b).

Helix Shapes and Delivery Valve

Plungers are manufactured with metering lands having lower or upper helixes (see Figure 10–10) or both to give constant port closing with a variable ending, variable port closing with a constant ending, or both a variable beginning and ending. With ported pumps, good control of injection characteristics is possible due to the minimum fuel volume that is under compression. However, a disadvantage of conventional port control pumps is the rising delivery characteristics as speed increases. This is caused by the fuel throttling process through the ports resulting in less fuel being bypassed before port closing and after port opening as the speed of the pump increases.

When the plunger is rotated so that the vertical slot on the plunger is in line with the control port (locating screw side), all the fuel will be bypassed; therefore, there will be no injection. With the rack in the full-fuel position, the plunger is able to complete almost its entire stroke before the helix will uncover the control port. Remember, as the plunger is rotated, it will uncover the port earlier or later in the stroke (see Figure 10–11).

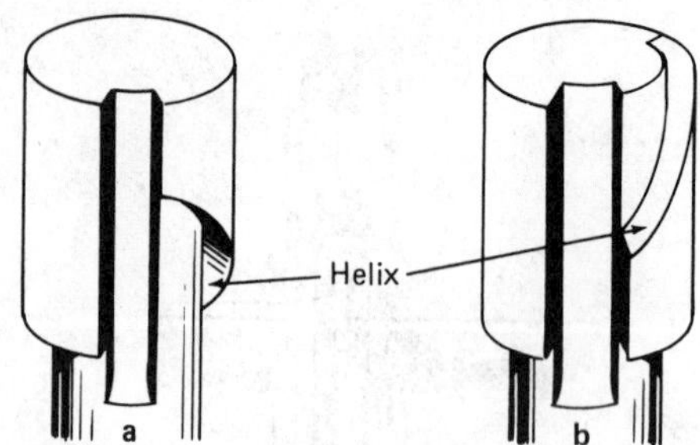

Figure 10–10. In-line injection pump, plunger types. a. Lower helix plunger. b. Upper helix plunger. (Courtesy of United Technologies, Diesel Systems Group)

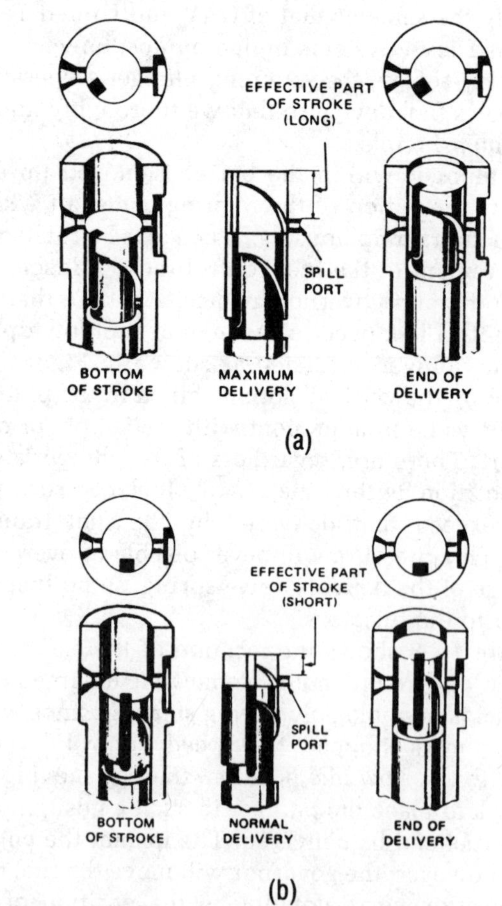

Figure 10–11. In-line injection pump, plunger in position for fuel delivery. a. Maximum. b. Normal. (Courtesy of United Technologies, Diesel Systems Group)

The standard basic type in use by United Technologies is the lower-right-hand helix where the start of injection is constant with regard to timing; however, the ending is variable. In some applications it is advantageous to advance timing as the fuel rate is increased. This is achieved by the use of an upper helix, which gives a variable beginning and a constant ending. The helix may be cut on the left- or the right-hand side of the plunger. It does not alter the injection characteristic except that the rack must be moved in opposite directions to increase or decrease fuel. There are other special adaptions, such as a short, shallow helix on top to give a slight retarding effect to the injection timing on engines that operate in the idle range for extended periods, and a double helix used by some manufacturers to provide rapid response with minimum rack movement.

Figure 10–12(b) shows a lower helix design; the beginning of delivery is constant and the ending of delivery is variable. The reason for the helix being on opposite sides is that the one on the left would be employed when the governor is on the left or when the fuel rack is in front of the plunger. Figure 10–12(a) shows an upper helix design; the delivery has a variable beginning but a constant ending. Figure 10–12(c) shows plungers with both upper and lower helixes; both the beginning and ending of delivery are variable. Other features of helix are shown in Figure 10–13.

The main function of any delivery valve in the injection pump is twofold:

1. At the end of the plungers' upward fuel delivery stroke, the delivery valve prevents a reverse flow of fuel from the injection line.
2. Figure 10–14 illustrates the sucking action that occurs at the delivery valve piston portion which controls the residual pressure in the injection line so as to effectively improve the injected spray pat-

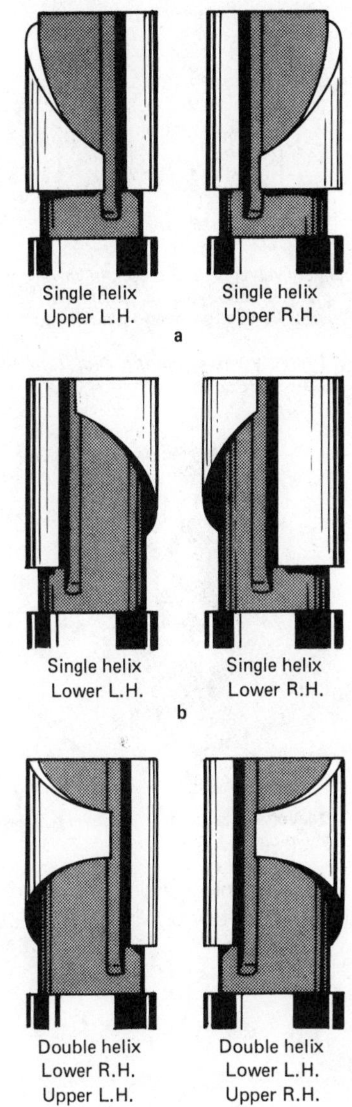

Figure 10–12. In-line injection pump plunger helix designs. (Courtesy of United Technologies, Diesel Systems Group)

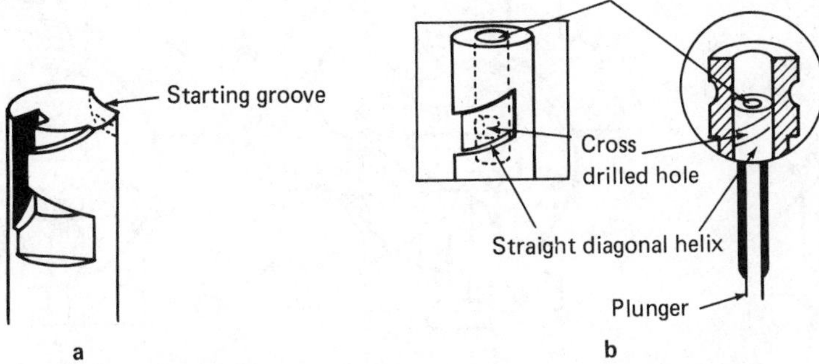

Figure 10–13. In-line injection pump, starting groove machined onto plunger helix. (Courtesy of Robert Bosch Corporation and C.A.V. Ltd.)

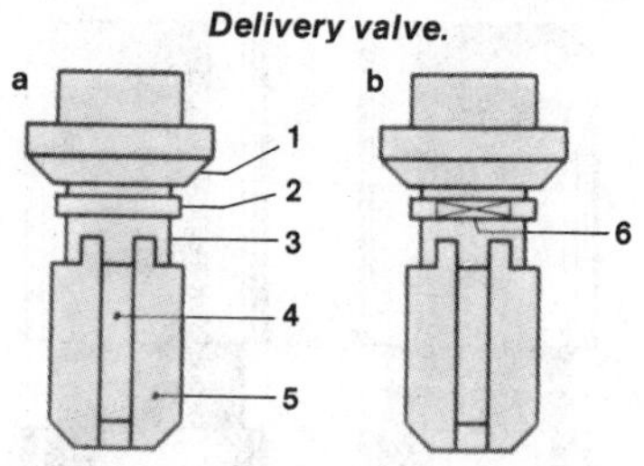

a Delivery valve, b Torque-control valve; 1 Valve cone, 2 Retraction piston, 3 Annular groove, 4 Stem, 5 Vertical groove, 6 Machined section.

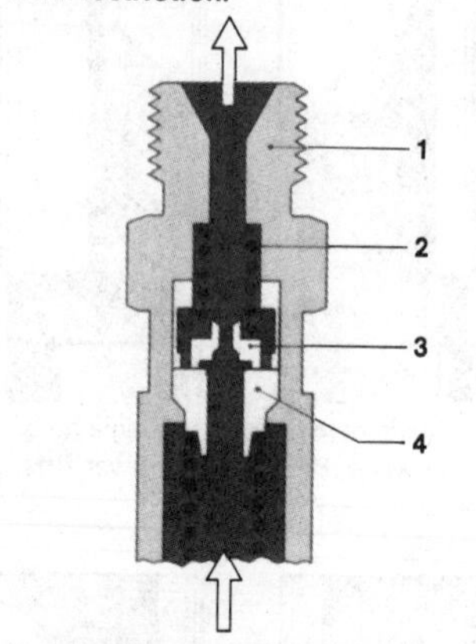

1 Delivery-valve holder, 2 Delivery-valve spring, 3 Valve plate, 4 Valve holder.

Figure 10–14. Pump plunger delivery valve constant and variable positions. [Courtesy of Nissan Motor Co., Ltd.]

tern of the fuel without fuel dribble and possible secondary injection. The sucking action that does occur at the delivery valve therefore effectively reduces the fuel pressure in the injection tube at the end of injection. (See Figure 10–14, Delivery Valve Closing Action.)

The delivery valve (see Figures 10–14 and 10–15), or what is sometimes referred to as a discharge valve, is specially designed to assist in providing a clean, positive end to injection. Below the valve face is a collar that is a precision fit in the valve bore. When pressure is created in the pump above the plunger by the closing of the ports, the valve must be raised far enough off its seat for the collar to clear the bore.

At the end of injection when pressure in the pump chamber is relieved by the opening of the control port, the valve drops down on its seat assisted by spring pressure. A volume of fuel equal to the displacement volume of the valve is added to the line and nozzle, reducing this pressure and allowing the nozzle valve to snap shut without the cushioning effect of pressure retained in the line and nozzle, such as with the closing of an ordinary valve. This is commonly called *line retraction,* which lessens the possibility of secondary injection or after-dribble at the spray nozzle. It is accomplished by an antidribble collar (accurately fitted relief or displacement piston) located at the upper end of the valve stem just below the seat.

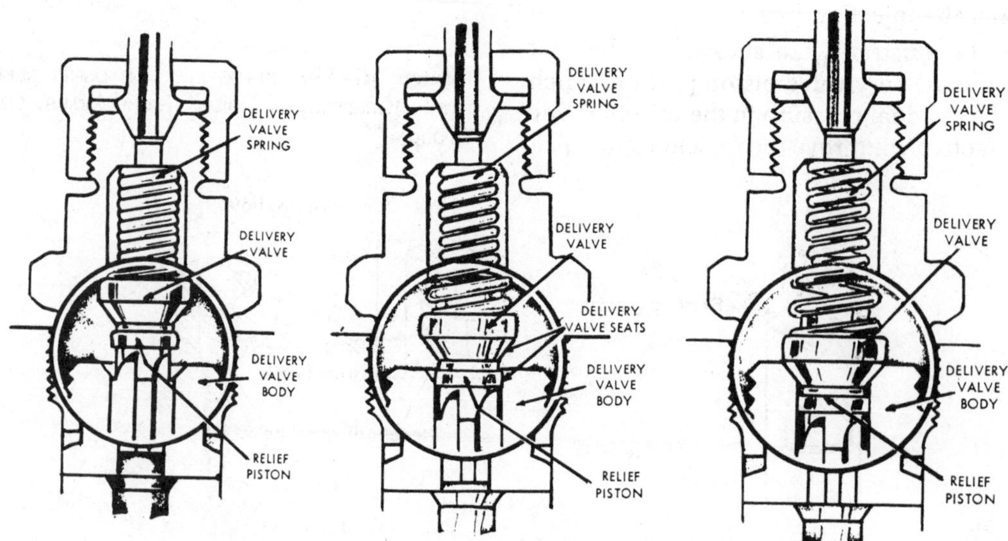

Figure 10–15. Pump plunger delivery valve cycle of operation. [Courtesy of United Technologies, Diesel Systems Group]

11

Basic Governor Operation

Gasoline engines are self-governing due to the fact that the air entering the engine is throttled by the use of a butterfly valve located in either the carburetor or fuel injected engine's throttle body. Through manipulation of the throttle pedal, the driver can change the engine speed to suit the demands placed upon the engine.

In contact-point distributor-type ignition systems, the engine's maximum speed was also controlled by the fact that at a given speed, point "bounce" would occur along with ignition coil saturation (insufficient build-up time).

Through these methods, the gasoline engine can and does operate without the need for a governor assembly to control engine speed. However, some gasoline engines are fitted with governor mechanisms. These are generally for the sole purpose of limiting the maximum road speed of the vehicle and the engine rpm to avoid poor fuel economy and engine abuse.

The diesel engine, on the other hand, operates with an excess amount of unthrottled air throughout its speed range, see Chapter 3. The fuel injection system is separately controlled from the air. When the diesel engine is started, the air is unthrottled and it remains in an unthrottled state as the operator opens the fuel control mechanism or throttle linkage to allow the injection system to inject more fuel into the engine cylinders. Without some form of fuel control mechanism, the diesel engine is capable of accelerating very rapidly and would self-destruct if the throttle was placed into the full-fuel position and left there with no means to regulate the fuel input.

Therefore a governor is required to regulate the injected fuel to prevent the engine from stalling at the low speed end and from overspeeding at the high speed end.

Governor Terms/Definitions

Speed Droop. This term is used to express the change in engine rpm from its maximum full-load speed to its maximum no-load rpm. This term is generally expressed as a percentage change rather than an rpm change. Governors used on automotive applications are "speed sensing" devices because they are driven from the engine.

The faster a governor can respond to a change in engine speed due to a change in engine load, the more sensitive it is said to be. The amount (percentage) of droop in any governor depends on the engine application—such as automotive, industrial, or marine.

The degree of droop used on most applications is usually between 5 and 10%. However, on special applications such as diesel engines driving power generators, a much closer degree of speed control is required, and droop characteristics of 0.5 to 1.5% are common. However, on power generators that are used in parallel, or load-sharing applications, zero droop governors are used.

Droop can be expressed in equation form as follows:

$$\text{Speed Droop} = \frac{\text{NNL} - \text{NFL}}{\text{NFL}} \times 100$$

where NNL = Normal No-Load Speed

NFL = Normal Full-Load Speed

Isochronous. This term is used to express a governor that is capable of maintaining the same speed at no-load and full-load—often known as a Zero Droop governor.

Sensitivity. Refers to how quickly the governor will

respond to a change in speed caused by a load increase or decrease. Sensitivity can vary from 0.25% for the best mechanical governors down to 0.01% for hydraulic servo-governors and solid state electronic governors.

Stability. This is the ability of the governor to return the engine to a new state of balance after a load correction without any tendency for the engine to drift up or down before settling at its new position.

Deadband. This is a very narrow speed range during which no measurable correction is made by the governor.

Hunting or Surging. The tendency of the engine to float from its normal steady idle speed due to the governor response being out of phase with the feedback signal. It can generally be adjusted with the use of a buffer or bumper screw on mechanical governors.

Work Capacity. This is the actual available work output expressed in foot-pounds that the governor is capable of overcoming at its output shaft. This work output depends on the size and speed of the weights and the distance through which their centrifugal force is transmitted. In hydraulic governors, this is the displacement of its power piston times the pressure of its working oil pressure.

Torque Control. This feature will take various forms in different governors; however, it depends on the engine's combustion characteristics as well as the fuel delivery and speed characteristics of the injection system. Generally the peak torque point will occur at about two-thirds of full-load throttle (speed), with the amount of torque increase (torque rise) being about 10–15% for most passenger car and light truck applications. However, current high-speed heavy-duty truck diesel engines now on the market can develop torque increases from full-load rpm down to the peak torque point of as high as 56%.

Forms and Types of Governors

Various forms and types of governors are used on diesel engines. They range from a straight mechanical-type to the newer electronic-type units, which employ a magnetic pickup located on the engine. The pickup sends a voltage signal back to an electronic module; the module then sends out a corrected signal to an actuator to vary the position of the fuel control rack mechanism. Hydraulic assisted governors are also used on a number of diesel engines to improve the response of the governor unit.

The concept of operation of most mechanical, hydraulic, and electronic governor assemblies is to monitor the speed of the engine crankshaft/flywheel; they are commonly known as "speed sensing" devices.

Most governors on automotive applications are designed to control the idle speed of the engine to prevent stalling and the maximum speed of the engine to prevent overspeeding. This type of a governor is commonly referred to as either a "limiting speed" or "minimum/maximum" type of unit. However, certain applications may use what is commonly referred to as a variable speed governor, which is designed to regulate the fuel input to the engine at any throttle position. This type of governor is sometimes known as an "all range" governor.

The mechanical and hydraulic assisted units depend upon two things to control the speed of the engine:

1. A set of flyweights driven from the engine that are designed to generate a given centrifugal force. The force of these weights is always attempting to pull the fuel control mechanism to a reduced fuel position.
2. A governor spring pack assembly, which can employ either one or two springs (one for idle and one for maximum speed); a limiting speed type of governor would employ usually both springs while the variable speed governor would employ only one spring. The force of the spring(s) is always attempting to increase the fuel control linkage position, and it is opposed by the centrifugal force generated by the rotating flyweights discussed in 1 above.

Operation of the Governor

Engine Stopped

When the engine is stopped, the weights are collapsed and the compressive force of the governor spring(s) pushes the fuel control linkage towards the full-fuel position for start-up purposes. However, on some engines, especially turbocharged units, an adjustable control mechanism is used to prevent a full-fuel position until the engine has actually started in order to prevent black exhaust smoke, which would result because of the turbocharger's lag on initial start-up (no exhaust flow to drive the turbocharger).

An example of a simplified governor control

mechanism is shown in Figure 11–1; it can take different forms in various engine makes. In Figure 11–1, with the engine stopped, the force of the spring will push down on one end of the lever while raising the opposite end. This action allows maximum fuel to flow into the engine cylinders for start-up purposes.

The same thing would occur in the diagrams shown in Figures 11–2 and 11–3. In Figure 11–2 the action is apparent; however, in Figure 11–3, the tension spring "F" would force items "C" and "D" to the left, thereby pulling the fuel control rack "E" to the full-fuel position.

Engine Starting

When the engine is cranked over and fires, the weight force will increase. The centrifugal force of the rotating flyweights, which are usually driven at twice engine speed for better control/response, would force the weights out and oppose the force of the governor spring, which can be preset by an adjusting screw. The greater the spring force, the faster the weights will have to rotate in order to balance out the spring force.

As the weights fly outwards, they will pull the fuel control linkage (rack) to a decreased fuel position un-

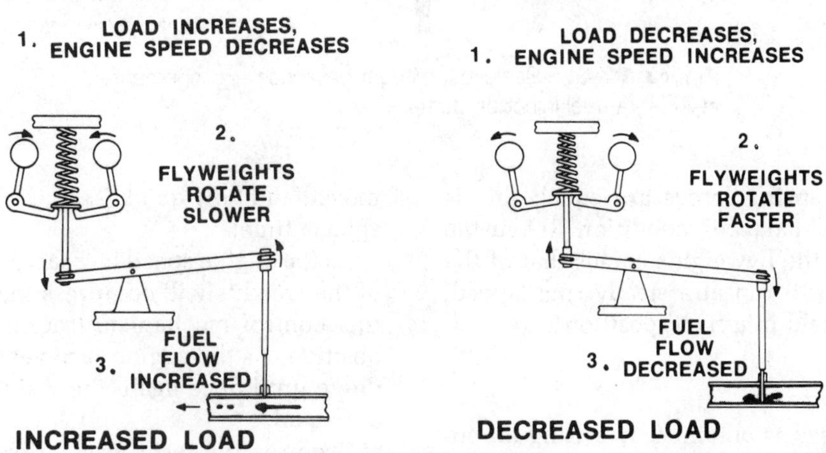

Figure 11–1. Mechanical centrifugal governor basic operation. (Courtesy of Deere and Company)

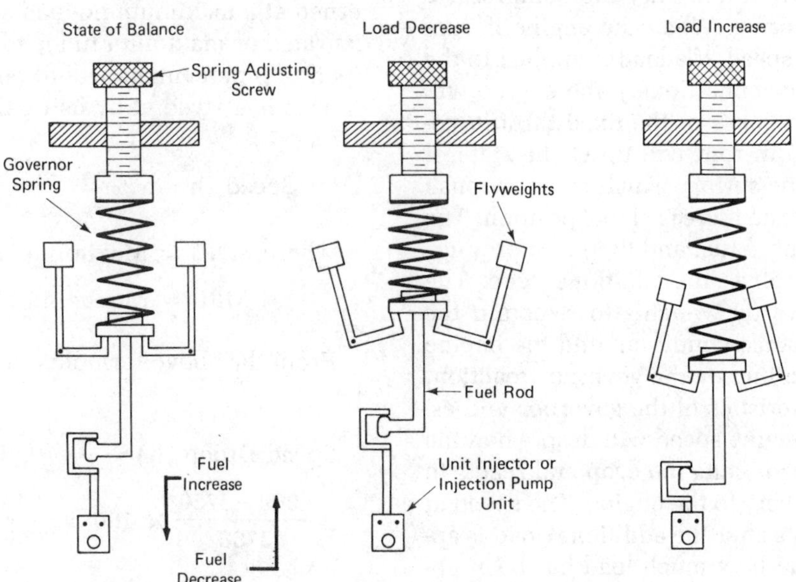

Figure 11–2. Basic governor weights and spring arrangement.

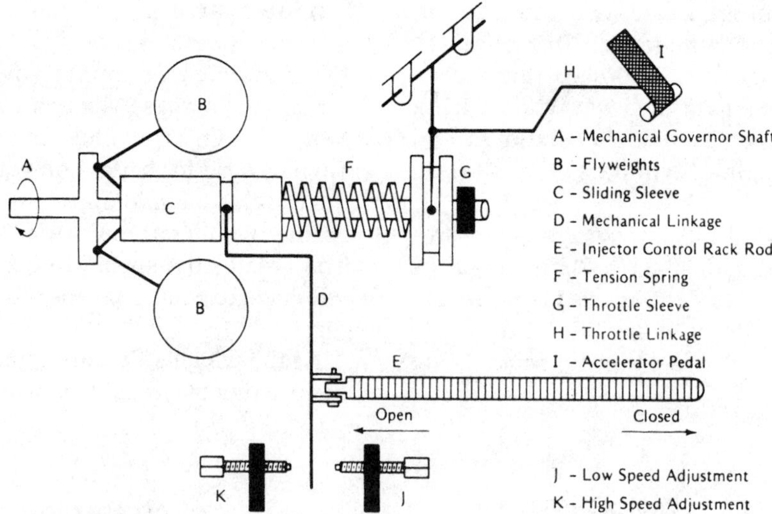

Figure 11–3. Basic centrifugal governor arrangement, in-line-type fuel injection pump.

til the weight and spring forces are equal; this is known as a "state-of-balance" condition. When the centrifugal force of the flyweights equals that of the spring, the engine will run at a steady rpm (speed) with the fuel rack held in a fixed position.

Engine Idling

The engine idle speed is obtained by setting an adjustment screw that varies the compressive force of the low-speed spring on a two-spring or limiting speed governor; on a variable speed governor, the idle adjustment screw would vary the compressive force of the single spring. When the engine is running at a steady idle speed, if a load is applied to the engine (such as an air conditioner), the engine will tend to slow down because of the fixed throttle position. This decrease in rpm will upset the state of balance in favor of the spring, which can now push the fuel racks towards an increased fuel position. The engine will receive more fuel and therefore produce more horsepower to offset the additional load. The additional fuel allows the weights to re-obtain the previous state of balance condition, and the engine is prevented from stalling by the governor reaction.

The droop characteristics of the governor will establish how far the engine speed will drop when the load is applied before it halts the drop and reacts to increase the fuel delivery to the engine. The speed at which the engine idles after the additional load is applied also depends on how much load has been applied and how responsive the governor is. If a load is removed from the engine at idle, the engine speed will increase in relation to the degree of load removed and how quick (sensitive) the governor is (response time).

As the engine speed increases, the centrifugal force of the weights will compress the spring and pull the fuel control mechanism (racks) to a decreased fuel position. As the engine receives less fuel, it will slow down until once again the weight and spring forces are equal.

Droop is the term used to express how sensitive the governor is to a change in engine speed. This sensitivity is usually expressed in terms of rated speed of the engine. For example, if an engine was governed at a maximum no-load speed of 4000 rpm and its rated or maximum full-load speed was 3750, then its droop percentage would be 6.66%.

This is arrived at by using the formula:

$$\text{Speed Droop \%} = \frac{\text{MNL} - \text{MFL}}{\text{MFL}} \times 100$$

where MNL = Maximum No-Load Speed

MFL = Maximum Full-Load Speed

From the above formula our engine figures would be:

Speed Droop (%) =

$$\frac{4000-3750}{3750} \times 100 = \frac{250}{3750} \times 100 = 6.66\%$$

In other words, the droop is the difference between the maximum no-load and the maximum full-load speed of the engine.

In order to effectively adjust a governor to ensure that the engine will run at its maximum rated full-load speed, we have to allow for its droop by adjusting the engine higher at no-load speed. The amount of additional rpm is determined by calculating the droop percentage as shown above.

Engine Accelerating

When the operator pushes the throttle pedal down, he/she overrides the governor temporarily. In a limiting speed governor (idle and maximum speed control only) that uses two springs (it can use either one or two sets of weights), once the idle spring has been compressed by the action of a set of rotating flyweights, the governor enters what is commonly called an "intermediate speed range." This term is used because no governor action is present and the operator now controls the engine speed by manipulation of the throttle linkage.

This is achieved by one set of weights or, in the two-weight set, by a larger set of weights known as the low-speed weights. In either case, when the weights have compressed the idle spring, no governor action can take place. In the single weight governor unit, a governor gap (free-play) allows the weights to continue to rotate with the engine without having any affect upon the governor mechanism. When they are rotating fast enough to produce enough centrifugal force, they will then act upon the governor high-speed spring to limit the maximum speed of the engine.

On the two-weight set governor, a lighter set of weights rides upon a shoulder of the larger low speed weights. When the larger low-speed weights rotate, they carry the smaller high-speed weights with them. The smaller/lighter high-speed weights have a leg extension that bears against the governor springs so that at a slow speed the combined action of both the low- and the high-speed weights oppose the idle spring force.

When the engine is accelerated, the larger low-speed weights will bottom on a shoulder of the weight carrier. Between this point and the point that the lighter high-speed weights are turning fast enough to leave their seat, no governor action occurs. As the engine is accelerated towards its high speed end, these smaller/lighter high-speed weights are thrown outwards by centrifugal force and they now compress the high-speed governor spring thereby limiting the maximum speed of the engine.

In a variable speed governor that uses one spring and one set of weights, the operator through throttle/governor linkage actually compresses the spring any time that the throttle is pushed down. This action can be seen in Figure 11–3. By increasing the force of the governor spring, the previous state-of-balance condition is disturbed in favor of the spring, which will force the fuel rack towards increased fuel. As the engine accelerates, the centrifugal force of the flyweights will increase until they are equal to the force of the spring. When this occurs, the governor is in a new state of balance and runs at a steady speed in the high-speed range.

With the variable speed governor, the operator can place the throttle at any position and, when the weight force balances the spring force, the engine will run at a steady speed.

NOTE: When a variable speed governor is used (all range), any load placed on the engine or taken off the engine will cause a decrease or increase respectively of engine speed throughout the speed range. The governor will respond to increase or decrease fuel input to the engine any time that the state of balance between the weights and the spring is disturbed as explained under the idle speed governor reaction above.

On a limiting speed governor however, there is no governor reaction in the intermediate speed range between idle and maximum. Therefore, any load applied or removed from the engine that causes a change in engine speed must be compensated for by the operator moving the throttle to either a lower or higher speed setting.

Maximum No-Load Speed

When the engine is accelerated towards the high speed end under no-load, the centrifugal force of the flyweights will be greater than it would be if the engine was loaded. This is because the engine speed will be higher with no-load on it than when it is fully loaded.

Under such a condition, the higher force of the weights will transfer this motion to the governor spring thereby compressing it further than it would under a full-load condition. This greater weight force will pull the fuel control linkage (rack) to a lower fuel position at no-load than it would at full-load, thereby limiting the maximum rpm of the engine.

Maximum Full-Load Speed

When a load is applied to the engine, its speed will decrease because it now has to work harder to overcome the resistance to its motion. As the applied load reduces the engine speed, it will also reduce the centrifugal force of the rotating flyweights, which will upset the state of balance that previously existed between the weights and spring.

The spring now expands and forces the fuel linkage towards more fuel to attempt to offset the additional load. This greater amount of injected fuel causes the engine to produce more horsepower at a lower speed. The speed that the engine will level off at a new state of balance will depend on the amount of load applied. If the load is less than that required at the engines full-load rpm, then the engine will run somewhere between its maximum no-load and full-load speeds.

If the load placed on the engine is greater than the available horsepower that the engine can produce at its maximum full-load rpm, then the speed will continue to decrease. This reduction in engine speed will decrease the horsepower, but it will cause an increase in engine torque. If the engine torque is high enough to keep the crankshaft turning, then the engine may eventually pick up speed again and re-attain a new state of balance.

If, however, the engine speed continues to drop, then the operator will have to downshift the transmission; or in an automatic, it will downshift on its own. Failure of the operator to downshift a standard transmission will result in the engine speed dropping to a dangerously low point with an engine stall being the next result.

The governor can respond to the changing load characteristics of the engine only within its range of operation. Once the governor reponds, the engine must have enough reserve horsepower to handle the new load; otherwise the engine speed will drop.

In a variable speed (all range) governor, the governor will respond throughout the complete speed range; however, in a limiting speed governor, no governor response will occur below the maximum full-load speed range unless it drops low enough to allow low speed regulation. Therefore, in the limiting speed governor, the operator must manipulate the throttle and transmission shift points when the engine speed enters the intermediate speed range because the governor has already pushed the fuel control linkage to its maximum position at the full-load rpm point.

Some diesel engines used in trucks today are designed, however, to increase the fuel as the speed drops below the maximum full-load rpm. This action allows a high rate of torque rise with a fairly constant horsepower curve.

To review the governor action, let's consider a vehicle running along a highway at 55 mph (88 km/hr) with the throttle in such a position that the governor is holding the fuel racks somewhere between the maximum no-load (MNL) and the maximum full-load (MFL) speeds. In this condition, the engine is not producing its maximum horsepower because we are not in the MFL fuel position.

If the vehicle now encounters an incline or hill and the driver does not increase the throttle position, the engine speed will start to decrease because of the increased resistance to forward motion. A drop in engine speed will upset the state-of-balance condition that existed between the weights and the high-speed spring in favor of the spring. The spring will therefore force the fuel control linkage towards an increased fuel position in an attempt to maintain a state of balance.

If the engine can produce enough additional reserve horsepower to offset the demands placed upon the forward motion by the hill, then the vehicle may climb the hill with a small drop in road speed. If however, the hill is a long, steep incline, then the road speed and engine speed will drop. If the engine torque is high enough to keep the crankshaft turning, then the vehicle may be able to negotiate the hill without a transmission downshift having to be made, but it will do so at a much lower road speed. If the vehicle road speed drops to the peak torque curve point of the engine and continues to decline, then the driver is forced into selecting a lower gear, or in an automatic transmission, the downshift is done automatically.

If the vehicle was moving along the highway in a condition similar to the above, and now encountered a downhill situation, this would represent a decrease in the engine load for a fixed throttle condition. The tendency would be for the engine speed to increase and the centrifugal force of the governor flyweights would increase thereby upsetting the previous state of balance and compressing the governor spring. As the spring is compressed, the fuel racks would be pulled to a decreased fuel position for a decrease in engine load.

If the operator were to push the throttle to the maximum position (floored) while the vehicle was going downhill and the road speed were to increase to a point whereby the road wheels were to become the driving member, engine overspeed could result. During this situation, however, the faster the engine rotates, then the greater would be the centrifugal force of the governor flyweights. The fuel rack would therefore be forced to a decreased fuel position to limit the speed of the engine. The operator would, however, have to check/slow the speed of the vehicle by the use of the brakes. In such a condition, if the engine overspeeds and a valve floats and pounds into a piston crown, the governor has done its job—which was to decrease the fuel to the engine. The engine overspeed was caused by poor driver habits.

Governor Adjustments

Under normal operation, the only adjustments that would be undertaken to the governor while the injection pump is on the engine would be the low or idle speed adjustment and the high or maximum no-load speed adjustment.

Certain engines employ additional screw adjustments to prevent hunt or surge at idle speed. These are usually referred to as buffer or bumper screws.

Governor adjustments other than throttle linkage settings are normally not done unless the governor/injection pump is placed on a test stand.

Each diesel fuel system described in this book covers the low and high speed adjustment procedures.

12

Robert Bosch Diesel Fuel Systems

Introduction

More than any other company, the Robert Bosch Company of Stuttgart, Germany, and Charleston, South Carolina, was responsible for the accelerated acceptance of the diesel engine by manufacturing precision fuel injection equipment as early as 1922, although mass production didn't start until August 1927.

The American Bosch Company, Ambac Industries, Inc., now known as United Technologies–Diesel Systems, was an offshoot of Robert Bosch, and many companies today manufacture precision fuel injection equipment under license from Robert Bosch on a worldwide basis. However, there is no longer any association between Robert Bosch Corporation and United Technologies.

Table 12–1 lists those companies presently using Robert Bosch fuel injection equipment but does not take into account additional engine manufacturers using Robert Bosch equipment built under license. A good example of this is the Diesel Kiki Company Ltd. and Nippondenso of Japan, who install Bosch pumps and nozzles on a wide variety of engines.

Injection pumps manufactured by Robert Bosch are very similar in design and operation to those of United Technologies and CAV in-line units, with the injection nozzles also working on the same principle.

Robert Bosch Pump Types

Robert Bosch Corporation manufacture a wide variety of injection pumps and nozzles that are used on many different applications including passenger cars and light trucks, medium and heavy duty trucks, industrial engines, pleasure craft and work boats, deepsea freighters and luxury liners in marine applications, and locomotives and off-highway equipment.

In the automotive and light truck range, Bosch offer both in-line and distributor-type pumps. Four models of in-line injection pumps (which are identified by size) are the "M," the "A," the "MW," and the "P," with the "M" pump being the smallest of the four and the "P" being the largest. Since the only models of in-line pumps that are used on passenger cars and light trucks are the "M," the "A," and the "MW," we will not provide details on the model "P" injection pump.

Detailed information on all Robert Bosch injection pumps and governors can be found in the textbook entitled *Diesel Fuel Systems* (by Robert N. Brady and published by Reston Publishing Co., Inc., Reston, Virginia 22090). In addition, Robert Bosch Corporation offers a programmed training course entitled "Pre-Tech Diesel Service Training," which is used in conjunction with either a 35-mm sound slide program or in an Audiscan tape format.

Figure 12–5, however, illustrates the "P" model pump since it is constructed and designed very similarly to the "MW" model pump, which is shown in this chapter and Chapter 13. The adjustments shown in this chapter for the "P" pump can also be applied to the "MW" pump discussed in this chapter and Chapter 13 for the Mercedes-Benz passenger car.

The "M" pump is illustrated in Figure 12–1(a) as it applies to some models of Mercedes-Benz passenger cars, although later models of Mercedes-Benz cars are equipped with the "MW" model pump, which is discussed in more detail in Chapter 13.

The "M" and "A" model pumps are very similar in arrangement and operation, while the "MW" and the "P" pumps are similar to one another in both style and operation. Figure 12–1(b) illustrates both the "A" and "P" model injection pumps in an end-section view.

A description of operation of the in-line pumps can

Table 12–1. *Robert Bosch Applications*

Customer	*Pump Type*	*Governor Type*
Domestic		
A. M. General (M.A.N.)	P	RQ
Allis-Chalmers	P	RQV . . . K
Fiat-Allis	P	RQV . . . K
(Fiat)	A	RQV
J. I. Case	A	RSV
(David Brown)	A	RSV
(Unimog)	A	RSV
(Unimog)	M	RSV
(Unimog)	A	EP/M
(Scania)	P	RSV
John Deere	A	RSV
	P	RSV
International Harvester	A	RSV, RQV (RQ)
	P	RSV, RQV (RQ), RQV . . . K
	MW	RSV
(Neuss)	VA	
Mack	P	RQV . . . K
(Scania)	P	RQV (RQ)
Massey-Ferguson (Hanomog)	A	RSV
Murphy Diesel(MWM)	A	RSV
	P	RSUV
New Idea (GMC)	A	RSV
Thermo-King Mevosa (M-B)	A	RSV
Transicold/Carrier Mevosa (M-B)	A	RSV
Waukesha	P	RSV
	Z	
	ZWM	
(Scania)	P	RSV
	P	RQV
Imported		
Bomag (Hatz)	PFR	
DAF	A	RSV
Deutz (KHD)	A	RSV (RS)
	A	RQV (RQ)
	EP/VA	
Lamborghini	A	RSV
Lombardini	PFE, PFR	
Mercedes-Benz (car)	M	EP/MN
	MW	RW
Mercedes-Benz (bus)	A	RQV
Mercedes-Benz (truck)	A	RQV (RQ)
	A	RSV
	P	RQV
	P	EP/M
Mercedes-Benz (DO Brazil)	A	RQV
Peugeot (car)	EP/VA	
Peugeot (tractor)	PFR	
V. M. Motori	PFR	
	A	RSV
Volvo (truck)	P	RQV
	MW	RWV
Volvo-Penta (marine)	PFR	
	EP/VA	
	VE	
	P	RSV
	MW	RWV
Volkswagen (car)	VE	

Note: (   ) indicates imported engine.

be found in Chapter 10. Additional information on in-line pumps can be found in Chapters 13, 18, and 19. The Nissan/Datsun engine (Chapter 18) uses a Diesel Kiki in-line-type pump that is manufactured under an agreement with Robert Bosch Corporation, while the Nippondenso pump used on the Toyota Landcruiser in Chapter 19 operates on the same principle as the Bosch/Kiki units.

The letters PES 4M (for example) on an in-line pump, such as shown in Figure 12–1(a), simply indicate the following:

P = Pump

E = Enclosed camshaft within the base of the pump housing

S = Flange mounted pump

4M = 4-cylinder M size pump with a pneumatic governor

"M" pumps use a 7-mm (0.275") pump plunger diameter; the "A" uses an 8-mm (0.314") pump plunger diameter; the "MW" uses either an 8-mm or 10-mm (0.394") pump plunger diameter; while the "P" pump uses a 10-mm, 11-mm (0.433"), or 12-mm (0.472") diameter pump plunger.

The operation, adjustment, and maintenance of the Bosch distributor-type pumps VE/VA can be found later in this chapter with additional information available in Chapters 14, 15, 16, 17, and 20.

Also contained in this chapter is information dealing with the Diesel Kiki model VM distributor-type pump.

Mercedes-Benz (Robert Bosch Injection Pump Model PES 4M)

The operation of the injection pump (Figure 12–1 (a) and (b)) is explained under "Types of Injection Systems," in Chapter 10. Therefore we will not delve into the operation. However, let us consider the basic adjustments that would be typical on the in-line injection pump pneumatic governor used on some models of Mercedes-Benz cars.

The type of adjustment that can be undertaken

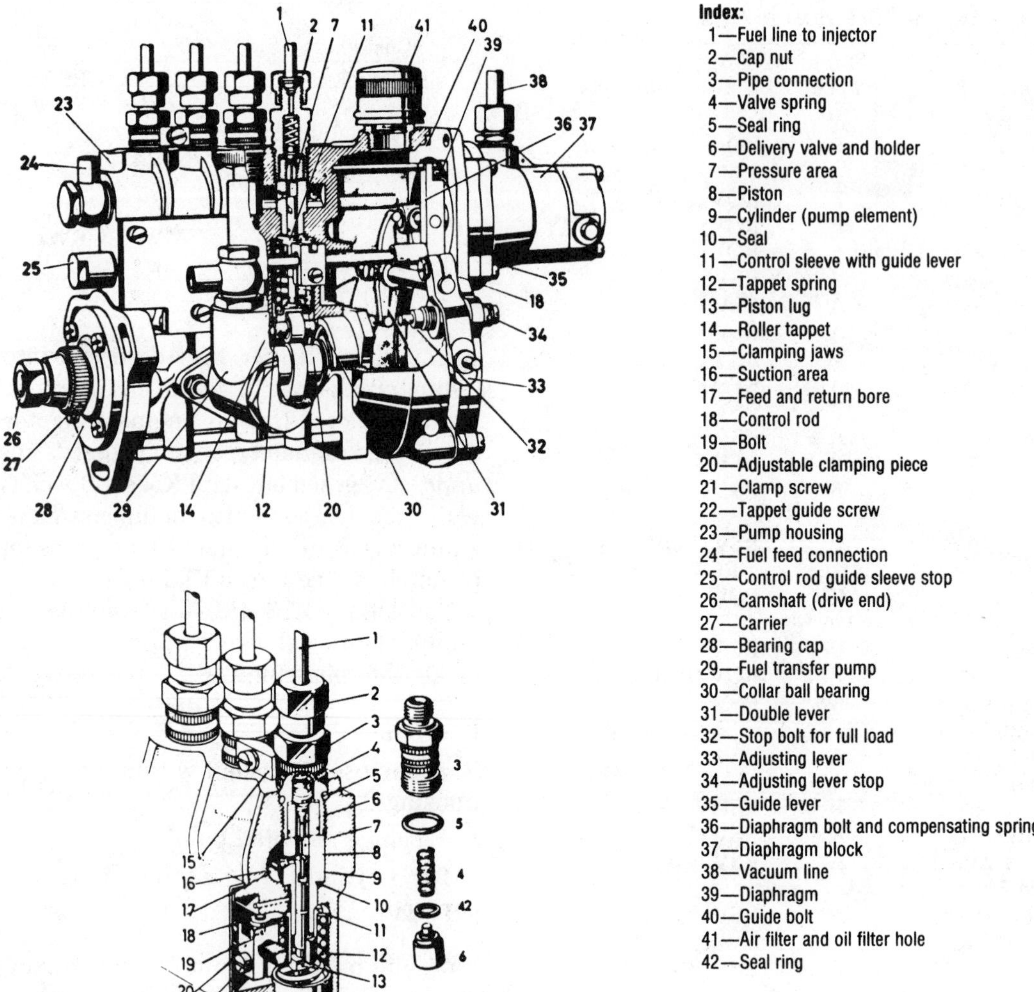

Index:
1—Fuel line to injector
2—Cap nut
3—Pipe connection
4—Valve spring
5—Seal ring
6—Delivery valve and holder
7—Pressure area
8—Piston
9—Cylinder (pump element)
10—Seal
11—Control sleeve with guide lever
12—Tappet spring
13—Piston lug
14—Roller tappet
15—Clamping jaws
16—Suction area
17—Feed and return bore
18—Control rod
19—Bolt
20—Adjustable clamping piece
21—Clamp screw
22—Tappet guide screw
23—Pump housing
24—Fuel feed connection
25—Control rod guide sleeve stop
26—Camshaft (drive end)
27—Carrier
28—Bearing cap
29—Fuel transfer pump
30—Collar ball bearing
31—Double lever
32—Stop bolt for full load
33—Adjusting lever
34—Adjusting lever stop
35—Guide lever
36—Diaphragm bolt and compensating spring
37—Diaphragm block
38—Vacuum line
39—Diaphragm
40—Guide bolt
41—Air filter and oil filter hole
42—Seal ring

Figure 12–1(a). Typical Robert Bosch in-line-type fuel injection pump, Model PES–4M. (Courtesy of Robert Bosch Corporation)

without removing the injection pump and governor from the engine would be related to complaints such as the following:

1. Lack of horsepower (kilowatts) complaint
2. Black smoke from the exhaust pipe
3. Rough running
4. Stalling and/or hunting (surging) during idling
5. Increase in maximum governed engine rpm

If these adjustments, when completed, fail to correct the problem, then the injection pump and governor assembly would have to be removed for further checks and tests on a fuel injection pump test stand. Figure 12–2 shows a cross-sectional view of the pneumatic-type governor used.

Testing the Diaphragm Vacuum Housing for Leaks

Any leaks on the vacuum housing side of the pneumatic governor will cause poor governor reaction because the force of the spring within the housing will be trying to push the fuel control rod toward an increased fuel position.

At the end of the injection pump opposite the governor, remove the protective cap from the pump fuel control rod (25 in Figure 12–1(a)).

1. Disconnect the vacuum line or hose from the connection (5 in Figure 12–2).
2. Hold the adjusting lever (33 in Figure 12–1(a)) in the *stop* position.

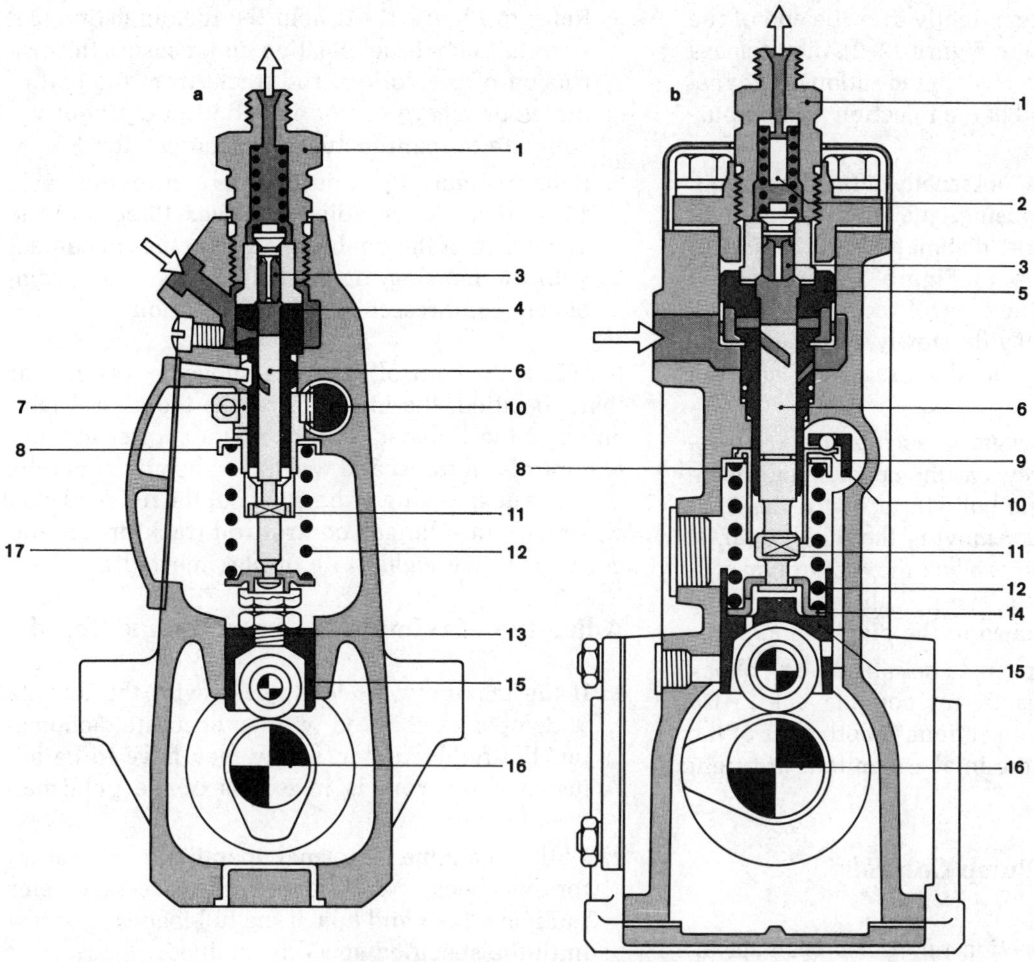

Figure 12-1(b). Comparison of A and P model injection pumps. (Courtesy of Robert Bosch Corporation)

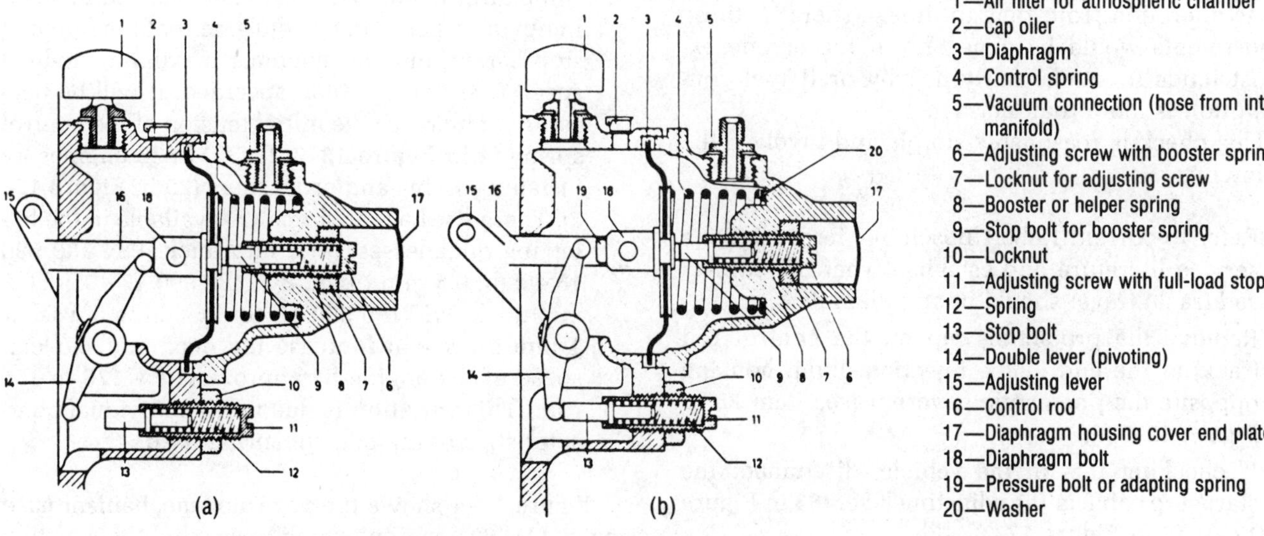

Figure 12-2. Cross-sectional view of a pneumatic-type governor (in-line injection pump). (Courtesy of Diamler-Benz, FRG.)

3. Place a finger or thumb tightly over the end of the vacuum connecter (5 in Figure 12–2), then release the adjusting lever and watch the action of the fuel control rod at the end of the injection pump housing.
4. If there are "no leaks" internally at the diaphragm and the vacuum housing, the control rod will move out only a short distance by the action of the control spring (4 in Figure 12–2) and stay there. If, however, the control rod moves out all the way, which it may do slowly, there is a vacuum leak at either the diaphragm or housing cover.
5. Access to the diaphragm is accomplished by removing the four screws at the cover, then disengaging the diaphragm bolt (18 in Figure 12–2) at the control rod, and removing the diaphragm. It is suggested that you replace the diaphragm if it is brittle or damaged, along with a new spring (control) that is matched to the diaphragm.
6. If the diaphragm appears to be satisfactory, check the cover and fittings for any possible leaks. Also make sure that the connections at both ends of the vacuum hose from the intake manifold are tight and leak free.

Checking Injection Pump Control Rod (Rack) Travel

Normally, once the injection pump has been set up on a test stand, further adjustment of the control rod should not be necessary. However, the full-load stop screw of the throttle butterfly and/or the adjusting screw of the feed volume full-load stop may require some alteration from time to time. Generally, these adjustments would be done only if the engine exhaust tends to smoke fairly heavily or if fuel consumption is more than normal.

This check is reasonably simple and involves the following:

1. Refer to current Robert Bosch or Mercedes-Benz service literature and establish what the control rod (rack) travel should be in millimeters.
2. Remove the protecting cap on the control rod (rack) at the end of the injection pump housing opposite the pneumatic governor (see item 25 in Figure 12–1(a)).
3. If checking this in the vehicle, disconnect the start/stop cable at the adjusting lever (33 in Figure 12–1; 15 in Figure 12–2).
4. Check that the control rod (rack) moves freely with no tight spots.
5. Refer to Figure 12–1; hold the fuel adjusting (33) lever in the full-fuel position and measure the protrusion of the control rod (rack) from the end of the guide sleeve on the injection pump housing; compare to manufacturer's specifications.
6. Refer to Figure 12–2; push in the control rod (rack) 16 by turning the adjusting lever 15 so that the lower arm of the double lever 14 comes in contact with the housing, or the helper spring 8 is completely compressed to its stop position.

NOTE: If the control rod (rack) movement is greater than specified, the injection rate is too high. Tightening of the full-load stop causes a shorter amount of control rod travel and will *lower* the injection rate, whereas unscrewing or backing out the full-load stop will result in a longer control rod (rack) travel and consequently a *higher* rate of fuel injected.

Adjusting Maximum No-Load Engine Speed

1. If the maximum no-load speed with the throttle wide open is checked with an accurate tachometer, the full-load stop screw may have to be adjusted if the rpm is in excess of the published specification.
2. With the engine at normal operating temperature for this check, if an adjustment is necessary, refer to Figure 12–3 and adjust the full-load stop screw until the specified speed is attained.
3. On Mercedes-Benz engine models OM 636, make sure that the adjusting screw (6 on Figure 12–2) with the booster spring is adjusted properly.
4. If the throttle butterfly valve has reached its maximum open position by adjustment of the full-load stop screw and the engine's maximum no-load speed is still lower than specified, it will be necessary to increase the initial tension of the control spring (4 in Figure 12–2) for OM 636 engines by installing shims and/or washers (20 in Figure 12–2). These washers or shims are available in the following thicknesses: (a) 1 mm thick, P/N 180 990 18 40; (b) 0.5 mm thick, P/N 180 990 17 40; (c) 0.2 mm thick, P/N 180 990 16 40. Installing a washer 1 mm thick will increase the maximum no-load speed of the engine by approximately 120 to 150 rpm. This variation is due to the individual characteristics of the control spring used.

Figure 12–4 shows the governor mechanism used on an OM 621 engine, which varies slightly from that on the OM 636. In Step 4 above, the same sequence would apply to the OM 621 engine, with the only ex-

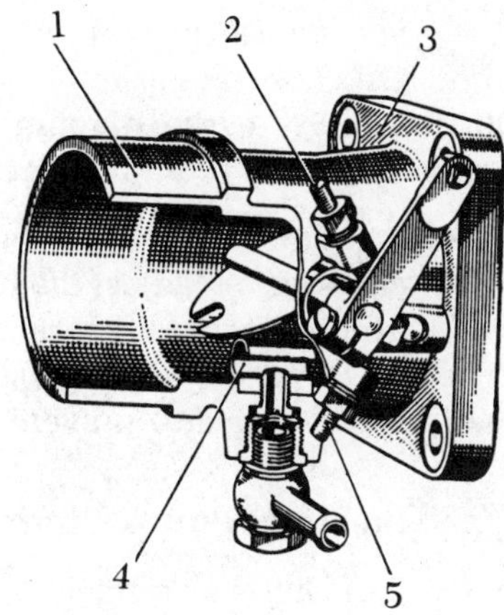

Index:
1—Air intake (venturi)
2—Idling stop screw
3—Mounting flange
4—Auxiliary venturi pipe
5—Full-load stop screw

Figure 12-3. Adjusting full-load stop screw. [Courtesy of Diamler-Benz, Aktiengesellschaft]

ception being that the spacer rings (14) rather than washers are used to alter the control spring tension. They have the same part number as that for the washers.

After any adjustment to the maximum no-load speed, the vehicle should be road tested for acceleration and signs of black smoke emanating from the exhaust pipe. If black smoke is visible, the maximum fuel discharge is too high. To correct this condition, proceed as follows:

1. Model OM 636 engines: refer to Figure 12–2 and turn in the adjusting screw (11) until there is no sign of smoke at the exhaust pipe.
2. Model OM 621 engines: refer to Figure 12–4 and turn in the adjusting screw (23) until the smoke disappears.

The smoke test should always be performed under *full-load* conditions; and a smoke tester/meter should be used, such as a Robert Bosch Model EFAW 78, consisting of a suction pump with accessories EFAW 65 and a photoelectric evaluator EFAW 68. This can be hooked up during a road test, engine dynamometer, or chassis dynamometer test. Refer to Chapter 30 for smoke meter information.

Adjusting Idling Speed

Before adjusting the idling speed, the engine must be at normal operating temperature (cooling water temperature at least 60°C or 140°F). This adjustment is done at the throttle linkage by the butterfly valve in the air intake, *not* at the injection pump.

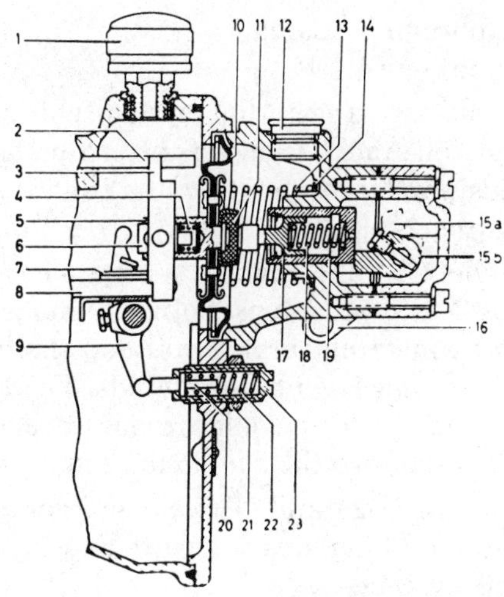

Index:

1—Air filter	13—Control spring
2—Guide rod	14—Spacer ring
3—Guide lever	15a—Engaging cam full-load position
4—Adaptation spring	15b—Engaging cam idling position
5—Diaphragm pin	16—Automatic advance control lever
6—Pressure pin for 4	17—Stop pin
7—Starting quantity stop	18—Additional spring
8—Control rod (rack)	19—Sliding spring capsule
9—Double lever	20—Full-load stop pin
10—Diaphragm	21—Fixing nut
11—Rubber buffer	22—Spring
12—Vacuum connection	23—Full-load stop adjusting screw

Figure 12-4. Governor mechanism used on OM 621 engines. [Courtesy of Diamler-Benz, Aktiengesellschaft]

Procedure

1. On vehicles with a Bowden cable idling control setup, make sure that you turn the knob on the instrument panel (dash) fully clockwise.
2. In this position, check that the throttle lever linkage is clear and not touching the adjustment ring of the Bowden cable. If it is, loosen the setting ring.
3. Refer to Figure 12–3 and adjust the idling screw until the engine speed is between 550 and 600 rpm on model OM 636 engines, and between 700 and 800 rpm on model OM 621 engines.
4. If the idle speed will not reduce to these speeds, proceed to check for an air leak at the following places:
    a. Vacuum line from the intake manifold to the governor

b. Diaphragm housing
c. Diaphragm itself

Any leaks existing in this system will cause the control rod (rack) to move toward an increased fuel position by the action of the control spring (4 in Figure 12–2).

5. On model OM 621 engines, a distance of approximately 1 mm must exist between the ball socket of the connecting throttle rod and the ball head on the angular lever at the cylinder head cover to ensure adequate throttle freeness. Adjust the length of the throttle connection rod if necessary.
6. On both OM 621 and OM 636 engines, after setting up the idling speed, ensure that the Bowden cable is correct.
   a. Turn the knob on the instrument panel (dash) fully clockwise.
   b. Clamp the setting ring on the cable so that a distance of between 0.1 and 0.2 mm exists between the setting ring and throttle lever or angular lever. Make sure that the cable is free to move without undue force being applied.

Adjustment to Correct Stalling or Hunting at Idle

Hunting, surging, or frequent stalling at an engine idle speed can be caused by vacuum variations in the line from the intake manifold to the diaphragm chamber of the pneumatic governor. This will cause small movements or oscillations of the diaphragm, which in turn, being connected to the fuel control rod (rack), will vary the amount of fuel being delivered to the cylinders and therefore cause the engine to hunt, surge, or stall.

To correct such a condition, the stop screw (6) must be adjusted as follows:

1. Bring the engine up to normal operating temperature.
2. Ensure that, with the knob on the instrument panel turned fully clockwise, the idling screw is against the stop at the throttle duct housing, or adjust if required.
3. Make sure that the engine is running at its proper idling rpm.
4. Refer to Figure 12–2 and screw in the adjusting screw (6) with Robert Bosch special wrench EFEP 95 with the engine running at an idle speed until all signs of hunting or surging disappear. Manually disturb the throttle; let the engine return to idle and make sure that it idles smoothly.

   NOTE: On MZ-type governors (newer versions), remove the end plate (17 in Figure 12–2) for this adjustment. On MZ governor versions EP/MZ60A39 d and 51 d, the diaphragm housing cover must be removed to adjust screw (6).
5. After this adjustment, recheck the maximum no-load speed. If necessary, readjust the full-load stop screw at the throttle duct. Also install a new end plate (17) in the governor housing. The governor should then be *lead sealed* to prevent tampering.

In-Line Type Injection Pump Timing

Since the majority of Diesel Kiki injection pumps are manufactured under a technical license agreement with Robert Bosch Company, the operation and therefore the maintenance of pumps produced by both companies is the same. The newer models of Bosch-type pumps, such as the MW pump used on the Mercedes-Benz 300D/SD engine, are discussed in Chapter 13. All Bosch inline pumps operate along the same lines; therefore, we shall limit this discussion to the basic adjustments required on the pump. The P pump is the largest in-line pump manufactured by Bosch and therefore is not used on either passenger car or light truck applications. It is used on heavy-duty trucks and equipment. Since the P and MW pumps are very similar, adjustments for one can be considered common to the other. Figure 12–5 shows a typical Robert Bosch P size in-line injection pump assembly with both one-piece and split-timing shims to effectively change "port closure" on each pumping unit.

The pump must be attached to a suitable test stand at major service intervals in order to correctly carry out these adjustments. With the pump mounted on the test stand, proceed as follows:

Adjusting Injection Timing. This requires adjusting the plunger prestroke.

1. Each pump must be checked out in firing order sequence. Begin by removing the delivery valve holder, spring, valve and gasket, and delivery valve stop from cylinder 1.
2. Refer to Figure 12–6. Install the plunger stroke measuring tool to the flange sleeve by turning it as far as it will go (tool 5782–419 Diesel Kiki; not required on other pump models which are simply spill timed by use of a test stand degree wheel).
3. Place the fuel control rod in the full fuel position. Manually rotate the pump camshaft until, with the use of the installed dial gauge, bottom dead center for pump 1 is established. At this point, zero in the dial indicator needle.

1. Delivery Valve Holder
2. Fill Piece
3. Delivery Valve Spring
4. Delivery Valve
5. Delivery Valve Gasket
6. Timing Shims
7. Spacer
8. O-rings
9. Delivery Valve Body
10. Flange Bushing
11. Barrel
12. Baffle Ring
13. Plunger
14. O-rings
15. Control Rack
16. Upper Spring Seat
17. Control Sleeve
18. Plunger Vane
19. Plunger Spring
20. Lower Spring Seat
21. Plunger Foot
22. Roller Tappet
23. Camshaft
24. Bearing End Plate
25. End Play Shim
26. O-ring
27. Bearing

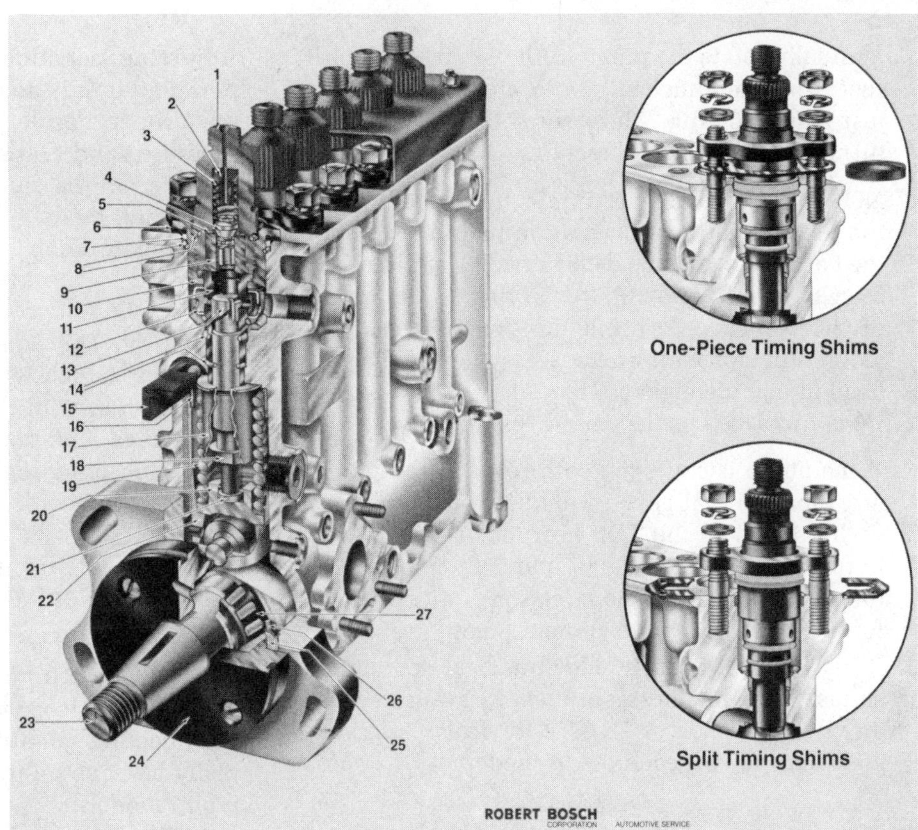

Figure 12-5. Robert Bosch "P" size in-line-type fuel injection pump. [Courtesy of Robert Bosch Corporation]

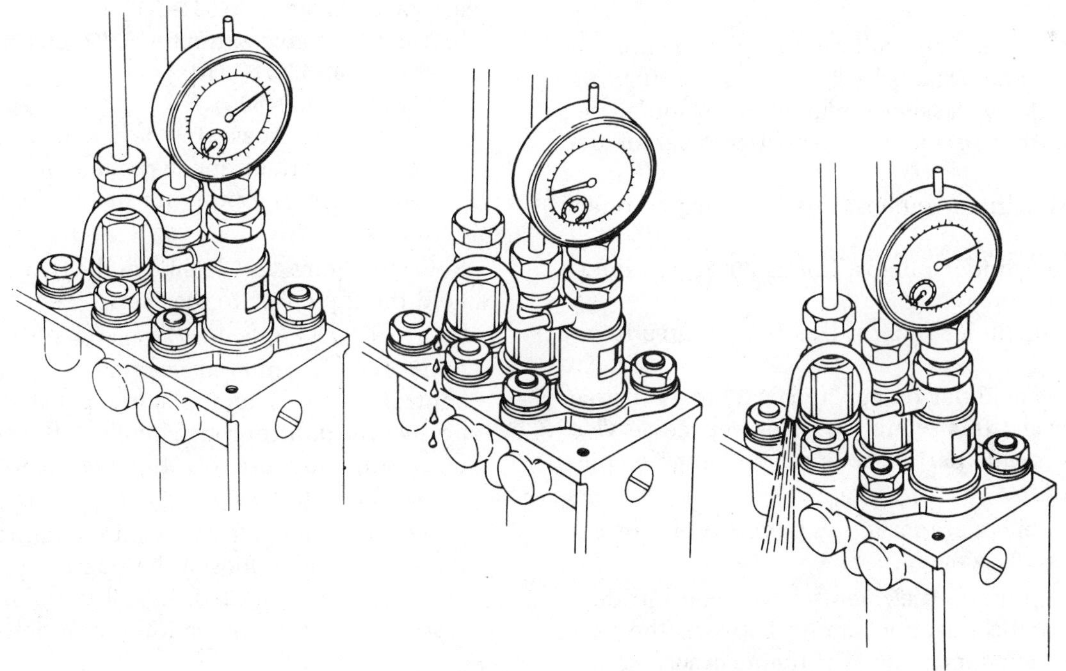

Figure 12-6. Pump plunger stroke measuring tool installed on No. 1 pump plunger. [Courtesy of Diesel Kiki USA Company Ltd., Irving, Texas]

4. Manually rotate the pump from its drive end until fuel from the attached nozzle pipe (Figure 12–6) just stops flowing. Make sure that the pump is turned in the proper direction.

   NOTE: The actual start of injection occurs when the plunger top just covers the fuel inlet hole in the barrel (can also be determined by spill-timing, see Chapter 18 and 19). By varying the thickness of the shims underneath the flange sleeve, the amount of camshaft lift necessary for the plunger head to completely close the inlet port in the barrel is increased or decreased respectively.

5. If the pump is equipped with reverse lead plungers, then the dial gauge would have to be zeroed with pump 1 in the TDC position. Then proceed to turn the pump camshaft manually in a reverse direction and repeat the same shim adjustment as for a normal plunger-equipped pump.
6. If fuel delivery is early, add shims; if late, remove shims. Shim thicknesses are readily available from any Robert Bosch or Diesel Kiki dealer in graduated sizes to suit each pump model.

Adjusting Injection Interval. Once pump 1 has been adjusted, set the pointer on the pump test bench to the desired position for further measurement checks.

1. Turn the pump camshaft manually in the normal direction and establish that fuel stops flowing from each successive pump in firing order sequence, and also at the correct number of degrees.
   a. For 4-cylinder pumps, this is 90 degree intervals.
   b. For 6-cylinder pumps, this is 60 degree intervals.
   c. For 8-cylinder pumps, this is 45 degree intervals.
   d. For 10-cylinder pumps, this is 27 to 45 degree intervals; this would be given on the service data for the particular pump and engine combination.

   Allowable tolerance for all pumps is plus or minus (±) 0.5 degree.

2. When all pumping elements have been set, double-check the injection timing between the first and last cylinders (pumps). If timing is correct, the mark on the bearing cover should align with the drive coupling, spline gear, or timing advance device flyweight holder mark. If not, the timing mark should be erased and a new one scribed in place.

Adjusting Injection Quantity. To check that each pumping unit is delivering the same amount of fuel, the proper nozzle, nozzle holder, nozzle opening pressure, and transfer pump pressure must be as specified in the calibration test sheet for the particular pump model (obtainable at any local fuel injection repair shop). Changing the amount of fuel delivered by the MW and P pumps simply requires turning the flange sleeve that rotates the plunger in the barrel. The flange sleeve is held in position by two nuts, which when loosened allow one to move the sleeve around the plunger up to a maximum of 10 degrees. The pump is then run on the test stand until each pumping unit has been adjusted to manufacturer's specifications.

Model PE-PD..A Plunger Block Assembly Replacement

If a pumping assembly becomes damaged during engine operation in the field, the plunger block assembly can be replaced with a new one without adjusting the injection quantity—an operation that must normally be done on the fuel pump test stand for other pump models.

Removal.

1. A variety of special tools is required for this purpose:
   a. Socket wrench 57914–050
   b. Delivery valve extractor 5792–004
   c. Stop bolt 57976–310
   d. Gauge 57990–520
   e. Cam stroke measuring device 5782–424, plus a screwdriver, torque wrench, and wrench to loosen the fuel pipe nut at the delivery valve holder
2. Refer to Figure 12–7. Remove the two screws that hold the top cover to the pump.
3. The fuel rack control rod must be placed into the mid-travel or centered position by removing the control rod cap and placing gauge 57990–520 against the control rod. Then lock the control rod in position by use of stop bolt 57976–310, as shown in Figure 12–7.
4. Since the flange sleeve is under plunger spring force, alternately loosen the retaining nuts; then, using extractor 5792–004, pull the plunger block assembly from the injection pump housing.

Installation. Remove stop bolt 57976–310, and mount the normal control rod stop. With the replacement plunger block assembly mounted into position on the pump housing, refer to Figure 12–8 and line

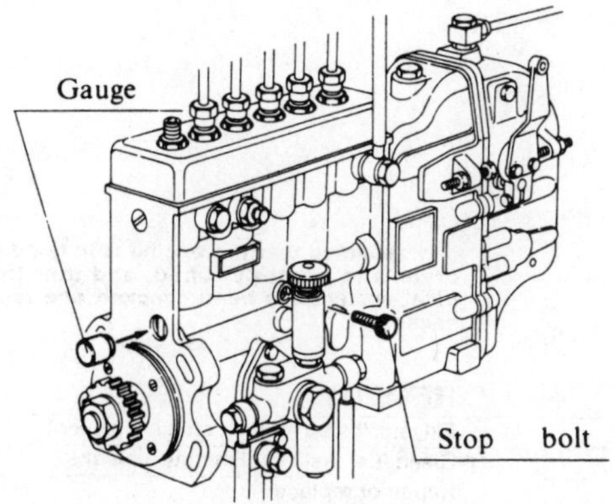

Figure 12-7. Placing the fuel rack into its centered position. (Courtesy of Diesel Kiki USA Company Ltd., Irving, Texas)

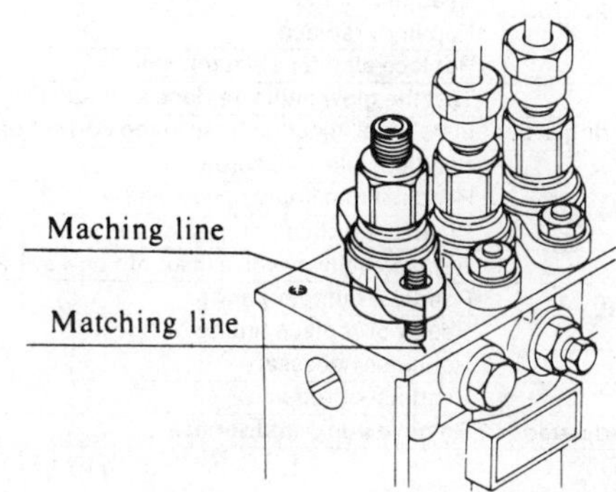

Figure 12-8. Aligning the plunger block assembly to the injection pump housing. (Courtesy of Diesel Kiki USA Company Ltd., Irving, Texas)

up the matching lines as shown. Remove the delivery valve assembly with tool 57914–050, and attach the cam lift measuring tool 5782–424 as shown in Figure 12–6.

By use of the transfer pump, prime the system and manually turn the injection pump camshaft until fuel stops flowing; read the dial gauge for lift. If necessary, change the shims until the reading is correct. Reassemble the delivery valve assembly, mount the top cover on the injection pump, attach all lines, and bleed the fuel system. Figure 12–9 indicates the fuel flow from the tank to all points in the system, and back to the tank (overflow fuel).

In-Line Injection Pump Troubleshooting Guide

Figure 12–10 provides a detailed chart to aid in troubleshooting Robert Bosch in-line fuel pumps.

Robert Bosch Governors

Governor Types

Robert Bosch governors used with in-line pumps "M," "A," "MW," and "P" can either be straight mechanical units or can be a combination mechanical/pneumatic such as shown earlier in Figure 12–4.

These governors can either be of the minimum/maximum (limiting speed) type whereby the engine

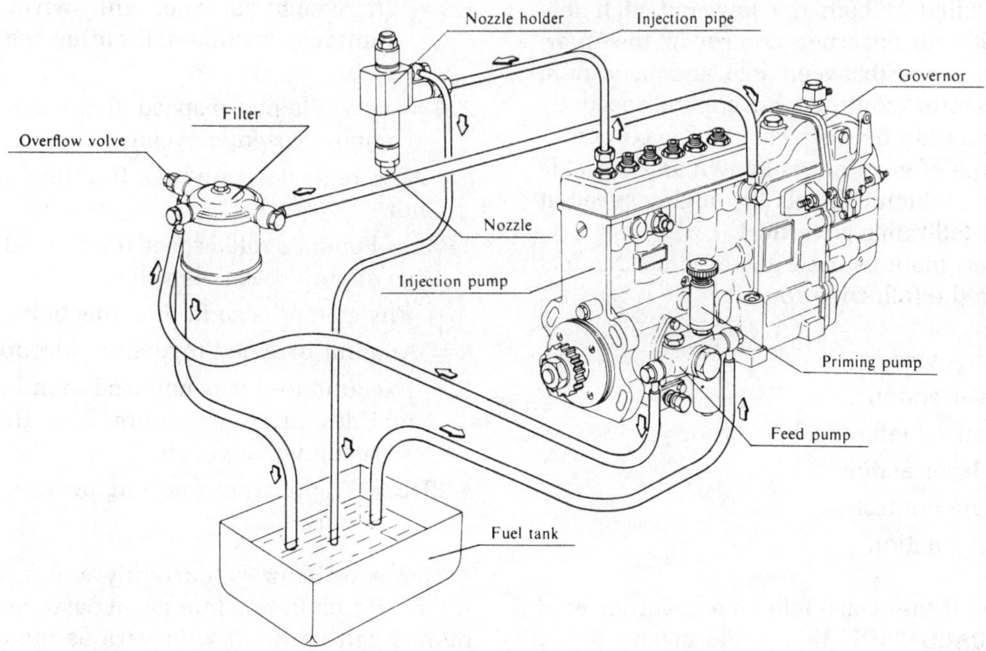

Figure 12-9. Typical in-line fuel injection pump system flow. (Courtesy of Diesel Kiki USA Company Ltd., Irving, Texas)

It is assumed that the engine is in good working order and properly tuned, and that the electrical system has been checked and repaired if necessary.

Starting Problem	Engine surges at idle	Rough idle when engine is warm	Engine misses under load	Low Power	Excessive Fuel Consumption	Engine cannot be shut off	Poor performance or black smoke or low power	Fog-like exhaust in full-load range (white or blue)	Incorrect idle or maximum speed	Engine does not rev up	Injection pump runs hot	CAUSE	REMEDY
●	●		●	●				●		●		Tank empty or tank vent blocked	Fill tank/bleed system, check tank vent
●	●	●	●				●	●		●		Air in the fuel system	Bleed fuel system, eliminate air leaks
●						●						Shut off/start device defective	Repair or replace
●			●	●				●		●		Fuel filter blocked	Replace fuel filter
●			●	●			●	●				Injection lines blocked/restricted	Drill to nominal I.D. or replace
●			●	●						●		Fuel-supply lines blocked/restricted	Test all fuel supply lines—flush or replace
●			●	●	●							Loose connections, injection lines leak or broken	Tighten the connection, eliminate the leak
●			●					●		●		Paraffin deposit in fuel filter	Replace filter, use winter fuel
●			●	●	●		●	●		●		Pump-to-engine timing incorrect	Readjust timing
		●	●	●	●		●			●		Injection nozzle defective	Repair or replace
				●	●		●					Engine air filter blocked	Replace air filter element
●												Pre-heating system defective	Test the glow plugs, replace as necessary
●		●		●	●		●	●				Injection sequence does not correspond to firing order	Install fuel injection lines in the correct order
	●				●				●			Low idle misadjusted	Readjust idle stop screw
				●					●			Maximum speed misadjusted	Readjust maximum speed screw
			●	●							●	Overflow valve defective or blocked	Clean the orifice or replace fitting
		●										Delivery valve leakage	Replace delivery valve (max. of 1 on 4 cyl., 2 on 6 cyl.)
	●								●			Bumper spring misadjusted (RS . . . governors)	Readjust bumper spring
				●	●		●	●		●		Timing device defective	Repair or replace timing device
●		●		●			●			●		Low or uneven engine compression	Repair as necessary
	●			●	●	●	●		●	●		Governor misadjusted or defective	Readjust or repair
●	●	●	●	●	●	●	●	●				Fuel injection pump defective or cannot be adjusted	Remove pump and service

Figure 12–10. Troubleshooting guide for diesel fuel injection systems with Robert Bosch in-line fuel injection pump. (Courtesy of Robert Bosch Corporation)

speed is controlled at both the low and high idle speeds but with no governor control in the intermediate speed range (between idle and maximum rpm). The operator controls the engine speed by throttle manipulation between idle and maximum.

The other type of governor is known as a variable speed governor, which controls the engine speed at all rpm ranges (all-range governor).

The letter designations used for these mechanical governors take the following forms:

R = Flyweight governor
S = Swivel lever action
V = Variable speed (all-range) governor
Q = Fulcrum lever action
K = Torque cam control
W = Leaf spring action

For example, if the nameplate on a governor read EP/RS275/1400A0B478DL, this would mean:

EP = Found on older governors, no longer used

RS = R/flyweight governor with swivel lever action, minimum/maximum (limiting speed) type governor

275 = Low idle pump speed (this would be 550 rpm engine speed, 4-cycle)

/ = Also indicates min/max (limiting speed) governor

1400 = Full-load rated speed (this would be 2800 rpm engine speed, 4-cycle)

A = Fits on "A" size in-line injection pump

0 = Amount of speed regulation (droop percentage)

B = Execution—this is not used to indicate the original design on governors. A = first change; B = second change, etc.

478DL = Application and engineering information only

Mercedes-Benz is currently the major user of Robert Bosch in-line pumps on passenger cars; these pumps can be fitted with various models of governors depending upon which model of pump is used. However, a commonly used governor is the RSV (variable speed) unit, which can be found on the "M," "A," "MW," and "P" pumps. The RSV governor is discussed in Figures 12–11, 12–12, and 12–15.

Other governors used by Mercedes-Benz are the RQ/RQV and the RW, which is shown in Figures 13–7 and 13–8 where its adjustments are discussed.

Governor Operation

Although governor terms were discussed under general governors in Chapter 11, it would be helpful to briefly review the basic operation of a governor mechanism and to ensure that we understand the

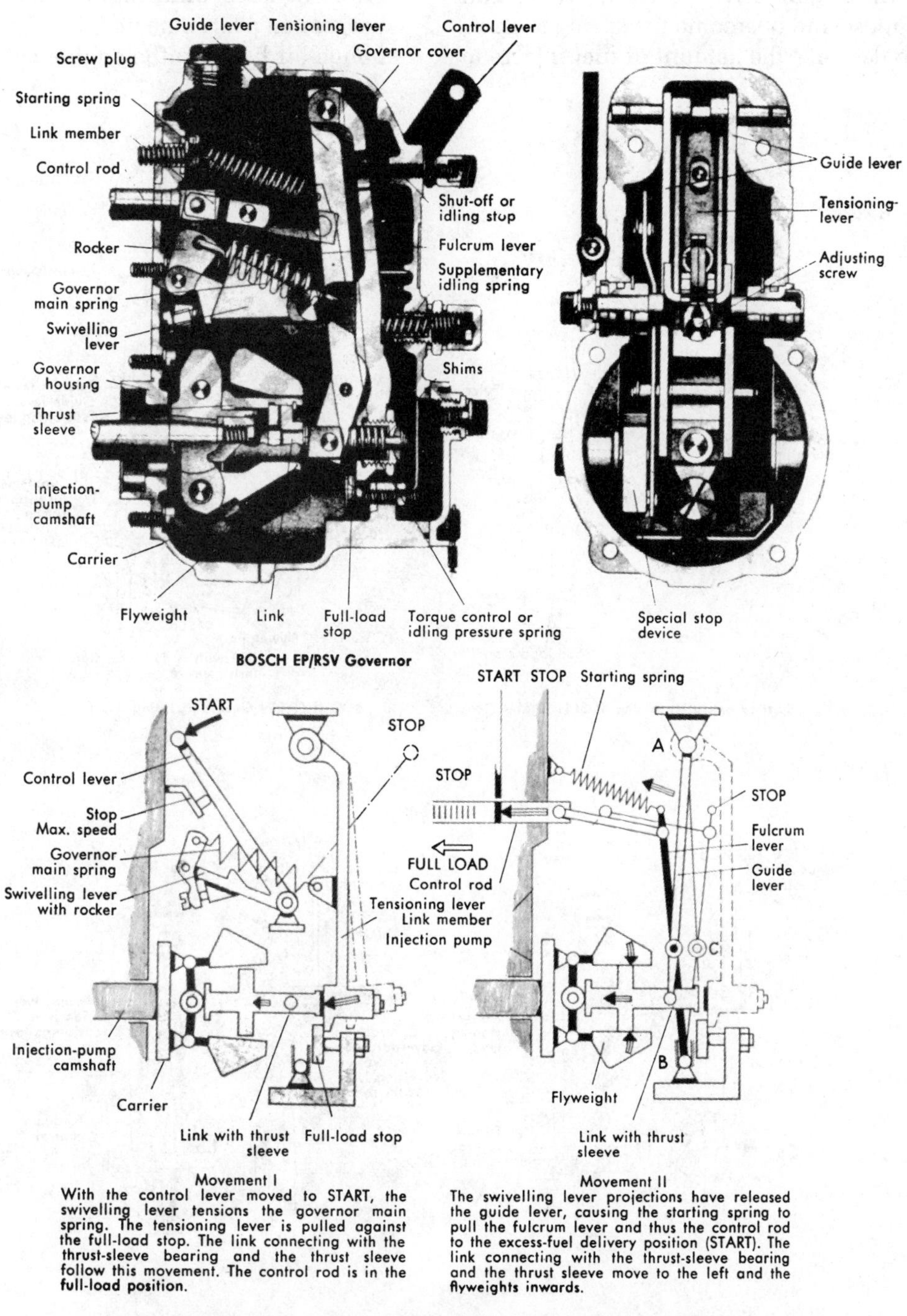

Figure 12–11. Method of operation of Bosch EP/RSV variable speed governor. (Courtesy of Robert Bosch Corporation) *Continued.*

meaning of those terms used to express various reactions related to the governor.

Refer to Figures 11–1; and 11–3. In any governor used on diesel engines, the basic principle is one of *weight* force against *spring* force. In other words, the tension of the governor spring assembly is always trying to *increase* the fuel delivered to the engine cylinders, and therefore the rpm; whereas the centrifugal force of the engine driven governor flyweights is trying to oppose and overcome the spring tension, and therefore *decrease* the amount of fuel injected.

Aneroid/Boost Compensator Control

On those engines using Robert Bosch injection pumps with a turbocharged engine, an aneroid/boost compensator control is used to prevent overfueling of the engine and hence black smoke during acceleration. This device controls the amount of fuel that can be injected until the exhaust gas driven turbocharger can overcome its initial speed lag and supply enough air boost to the engine cylinders. Such a de-

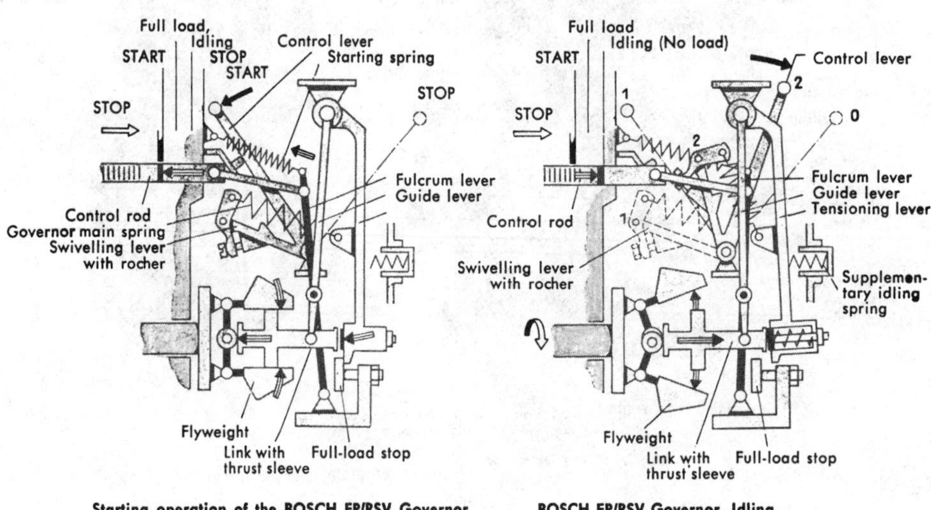

Starting operation of the BOSCH EP/RSV Governor

BOSCH EP/RSV Governor, Idling

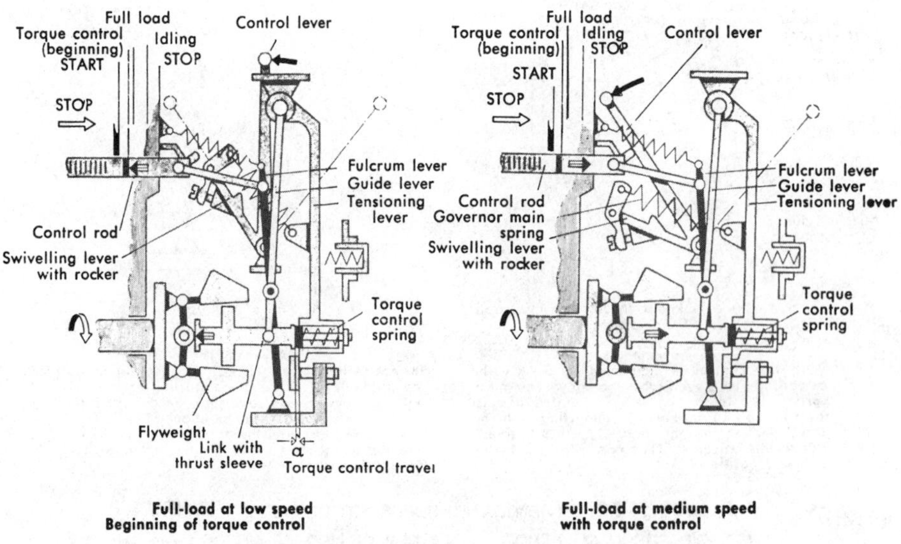

Full-load at low speed
Beginning of torque control

Full-load at medium speed
with torque control

Figure 12–11. *(Continued)*

vice is used extensively by all 4-stroke-cycle engine manufacturers today to comply with Federal Environmental Protection Agency smoke emission standards.

The aneroid is mounted on either the end or the top of the injection pump governor housing, with its linkage connected to the fuel control mechanism and a supply line running from the pressure side of the intake manifold (turbocharger outlet) to the top of the aneroid housing. Such a device is shown in Figure 12–13.

Figure 12–14(a) shows the position of the aneroid control linkage when the engine *stop* lever is actuated, which moves the aneroid fuel control link out of contact with the arm on the fuel injection pump control rack. Figure 12–14(b) shows the aneroid linkage position when the throttle control lever is moved to the *slow idle* position. This causes the starter spring to move the fuel control rack to the *excess fuel position.* During the cranking period *only* is excess fuel supplied to the engine. This is because the instant the engine starts, we have the centrifugal force

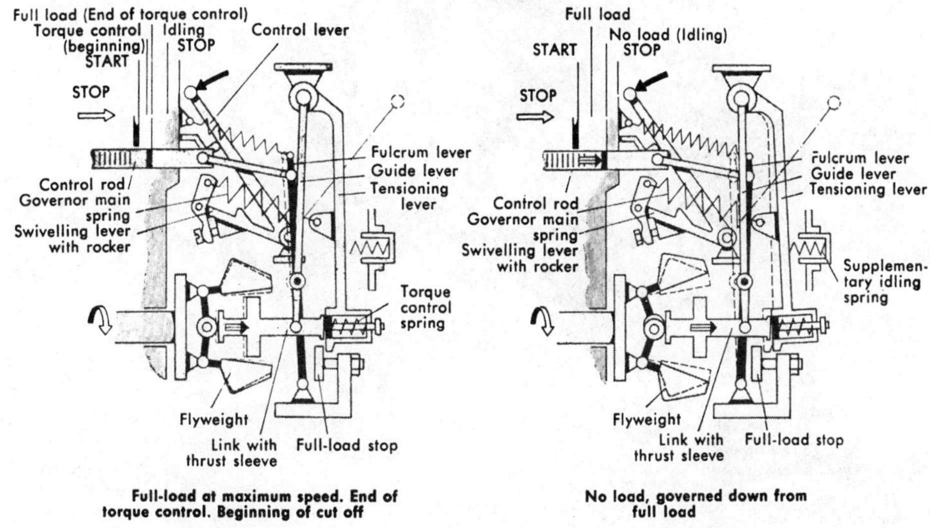

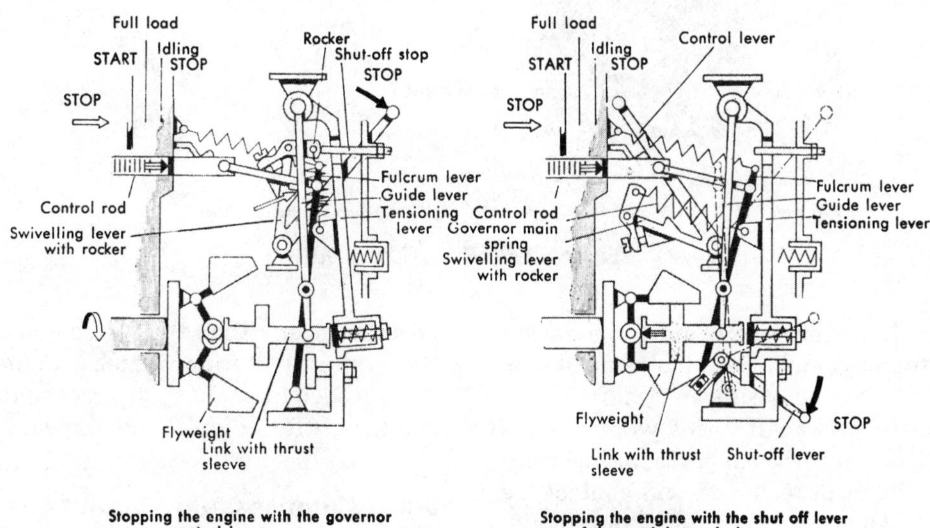

Figure 12–12. Additional governor positions, EP/RSV model. (Courtesy of Robert Bosch Corporation) *Continued.*

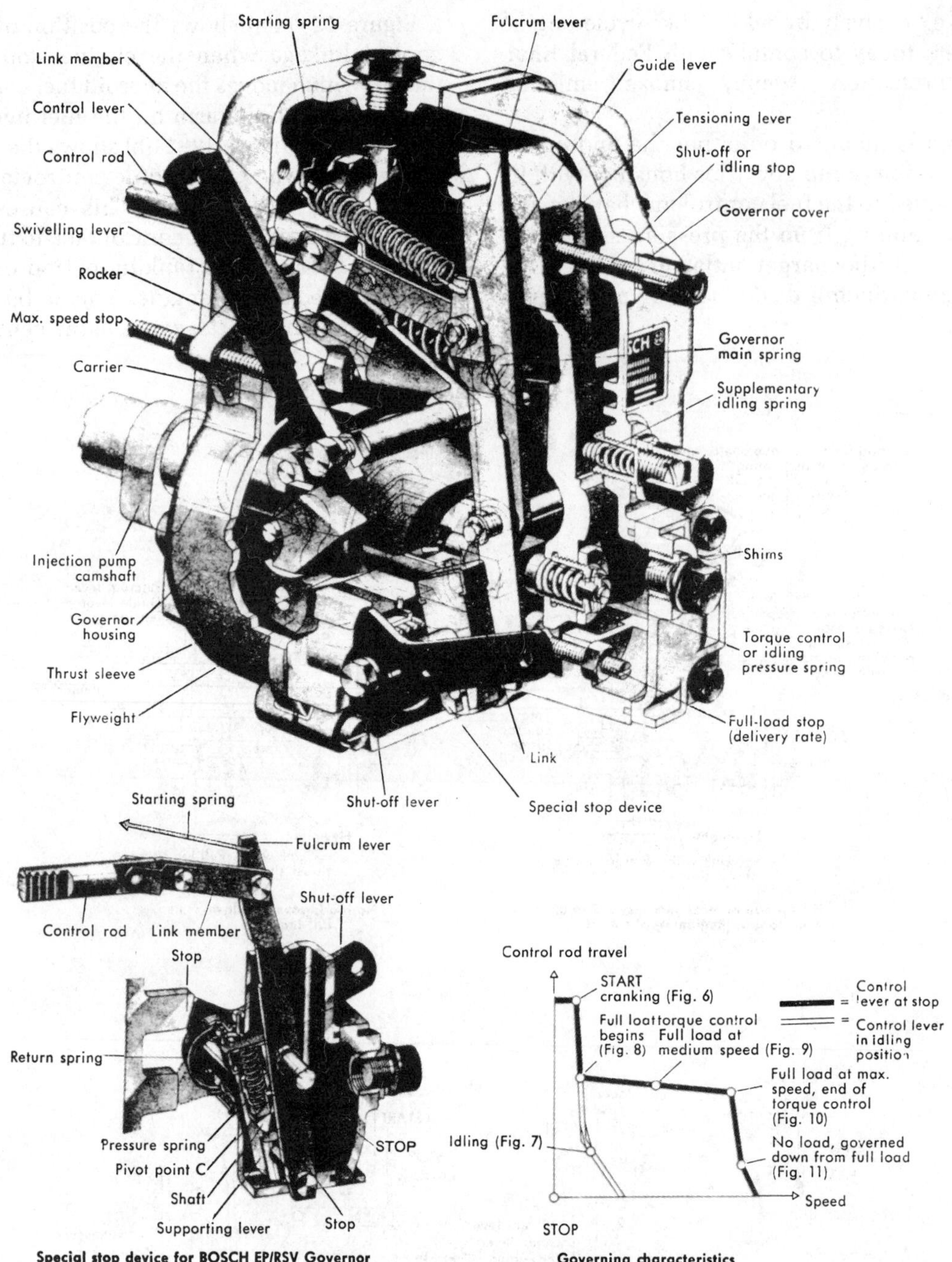

Figure 12–12. [*Continued*]

of the governor flyweights overcoming the starter spring tension, thereby moving the fuel control rack to a decreased fuel position. As this is occurring, the aneroid fuel control lever shaft spring will move the control link back into its original position. In Figure 12–14(b), the fuel control rack arm will contact the aneroid fuel control link, thereby limiting the amount of fuel that can be injected to approximately half-throttle and preventing excessive black smoke upon initial starting. The same lever will control the rack position at any time that the engine is accelerated, preventing any further increase in fuel delivery until the turbocharger has also accelerated to supply enough air for complete combustion.

Boost Compensator/Aneroid Operation

Basically, the boost compensator ensures that the amount of injected fuel is in direct proportion to the quantity of air within the engine cylinder to sustain

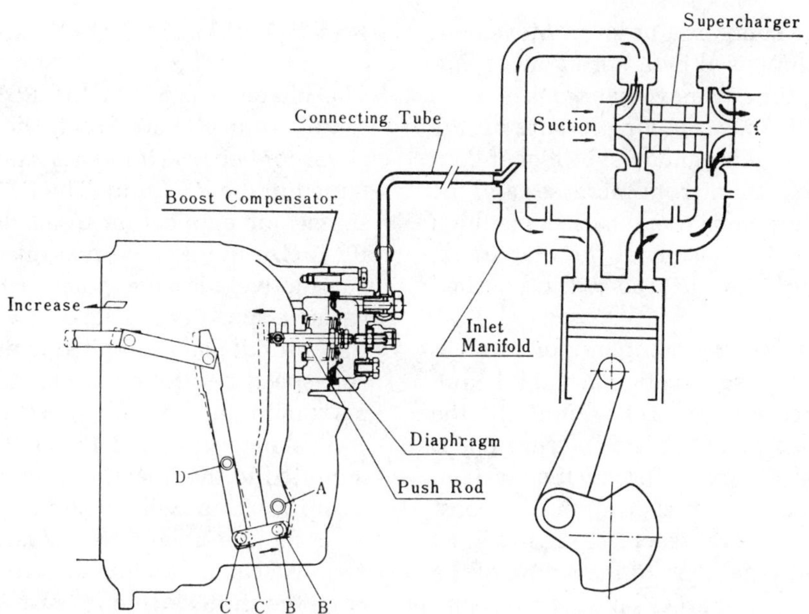

Figure 12-13. Turbocharged engine boost compensator arrangement. (Courtesy of Diesel Kiki USA Company Ltd., Irving, Texas)

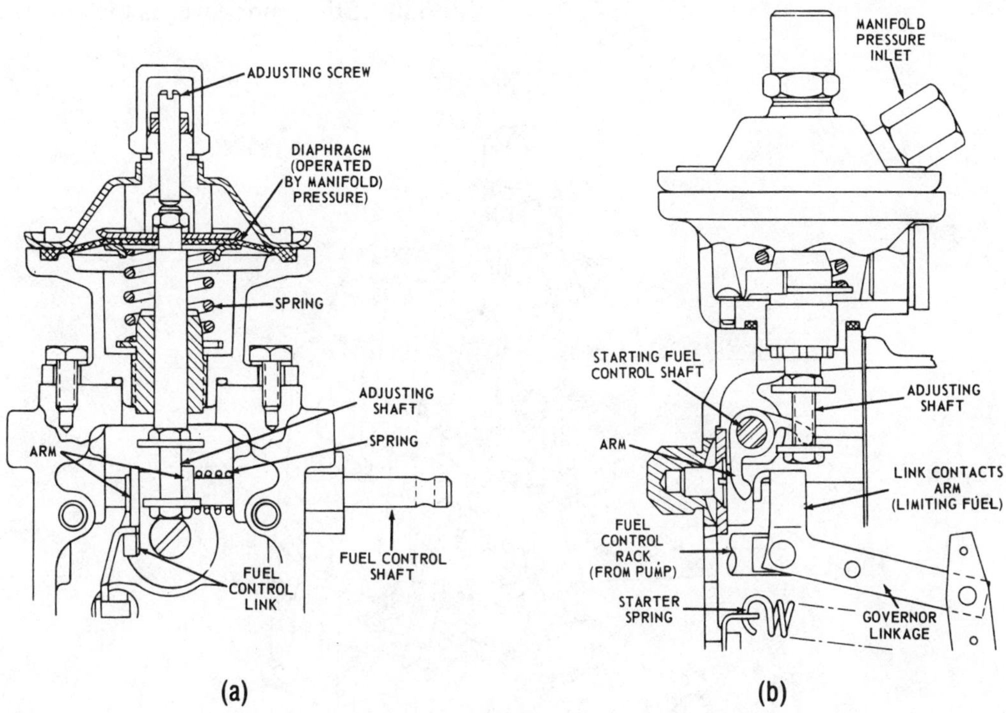

Figure 12-14. Governor aneroid control linkage—in-line-type pump. (Courtesy of Robert Bosch Corporation)

correct combustion of the fuel and therefore increase the horsepower of the engine.

The boost compensator unit is contained, as shown in Figure 12-13 in the upper portion of the governor case. Very simply, it operates as follows: With the engine running, pressurized air from the cold end of the turbocharger passes through the connecting tube from the engine air inlet manifold to the boost compensator chamber. Inside this chamber is a diaphragm, which is connected to a pushrod, which is in turn coupled to the compensator lever. Movement of the diaphragm is opposed by a spring at its left-hand side; therefore, for any movement to take place at the compensator linkage, the air pressure on the right-hand side of the diaphragm must be higher than spring tension. As the engine rpm increases and the air pressure within the connecting tube becomes high enough to overcome the tension of the diaphragm spring, the diaphragm and pushrod will be pushed to the left-hand side.

This movement causes the compensator lever to pivot in a counterclockwise direction around the fulcrum pin A shown in Figure 12–13. Similarly, the floating lever will also pivot around fulcrum pin D in a counterclockwise direction, forcing the fuel control rack toward an increased fuel position. The boost compensator will therefore react to engine inlet manifold air pressure regardless of the action of the governor. When the turbocharger boost air pressure reaches its maximum, the quantity of additional fuel injected will be equal to the stroke of the boost compensator in addition to the normal *full-load* injection amount that is determined by the governor *full-load stop bolt*.

RSVD Mechanical Governor

The characteristics of the RSVD governor are the same as the Robert Bosch/Diesel Kiki RSV type of governor because it incorporates all the features contained in the RSV unit. The RSVD was originally designed for application to all Robert Bosch PE-A or PE-B size in-line-type fuel injection pumps used on automotive diesel engines. However, since it has variable-speed characteristics similar to those of the RSV unit, it can be applied to engines requiring variable-speed control or, if necessary, both minimum–maximum and variable-speed governor control.

As shown in Figure 12–15, the flyweight carrier is mounted to the injection pump camshaft; therefore, pump rotation will cause the flyweights to transfer their motion axially to a sleeve and shifter mechanism via a ball-bearing thrust unit. A guide lever supports the shifter, which is suspended on the governor cover pin. As the engine is operated, the control lever shaft moves the lower end of the governor lever via the supporting lever, which is attached to the control lever shaft. Attached to a shaft located at the middle of the guide lever is the governor lever. Also

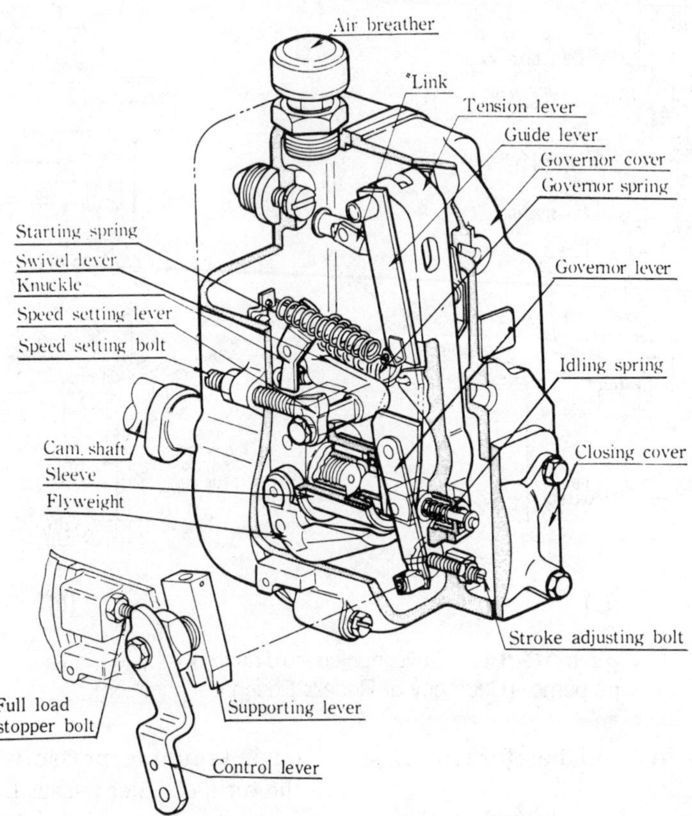

Figure 12–15. RSVD in-line pump mechanical governor arrangement. (Courtesy of Diesel Kiki USA Company Ltd., Irving, Texas)

at the lower end of the governor lever is a fixed pin, which acts as a pivot for the supporting lever at its lower end. At the top end of the governor lever is a link to which the fuel injection pump rack is connected. Hooked to the top end of the guide lever is the starting spring, which is coupled at its opposite end to a spring eye on the governor housing. A pin passing through the governor cover also supports both the tension and guide levers. In addition, a swivel lever shaft is fitted into the governor cover bushing.

The main governor spring is hinged between both the tension and swivel levers; the speed setting lever is also fitted to the swivel lever shaft (located in the governor cover). Governor main spring tension is of a predetermined value specified by the engine manufacturer. At the lower end of the tension lever is the idling spring. The speed of the engine is controlled by the movement of the control lever via the governor lever to the fuel-injection pump rack.

RSV Compared to RSVD

As previously mentioned, these two governors are identical in basic construction, the exception being that the speed-control lever on the RSV is used as a speed setting lever on the RSVD. In addition, the governor lever on the RSVD is shaped in a crank-type fashion around the middle of the guide lever, which has the effect of placing the control lever face toward the front of the injection pump, as shown in Figure 12–15.

Operation of the control lever on the RSV requires greater force than on the RSVD because the RSV, being of the variable-speed type, is opposed by spring tension or reaction, whereas on the RSVD minimum–maximum speed governor, there is no reaction force to overcome.

If it is decided to convert the RSVD to a variable-speed governor, a supplementary idling spring must be added; also, the speed setting lever must be used as a control lever, with the other lever used as a stopping lever.

As far as performance of the RSV versus RSVD is concerned, the following holds true:

1. RSV: the regulated speed of the engine is set by manipulation of the control lever angle (remember that the control lever does not move the fuel control rack directly on this unit).
2. RSVD: the control rack (fuel) is moved directly by manipulation of the control lever.
3. RSVD is a limiting speed type of governor; in other words, there is no governor control between the minimum and maximum speed ranges. Engine speed is controlled by movement of the control lever within these two ranges. This feature is of course a highly desirable quality on automotive applications because the operator will have more rapid response to throttle movement owing to the fact that he does not have to overcome control lever reaction force such as exists within the RSV unit.

RQV All-Speed Governor

The RQV governor is a widely used Robert Bosch unit that is designed to accomplish the following:

1. Maintain idle speed.
2. Control maximum speed.
3. Control speed throughout the operating range within the limits of its regulation.

This governor operates on the same basic setup as the RSV described earlier.

RQ Minimum–Maximum Governor

The RQ governor is used on highway truck applications and is therefore of the limiting speed type, since it will do the following:

1. Control idle speed.
2. Control maximum engine rpm.

Between idle and maximum speed there is no governor action, and the engine speed is controlled by the operator.

RQV/RQ Combination Governor

Identified by Robert Bosch Co. as the RQV 300–800/1500 governor, this unit provides all-speed (RQV) governor control from 300 to 800 rpm (600 to 1600 engine rpm) and min–max (RQ) governor control from 800 to 1500 rpm (1600 to 3000 engine rpm). The primary function of the RQV/RQ governor is to provide an adequate range of all-speed-type governor control to handle the majority of PTO (power take-off) applications such as required on garbage and refuse trucks, tankers, and some concrete mixer truck applications, while also providing the degree of min–max type governor control necessary for Allison-type transmission applications and limiting maximum speed.

The all-speed portion of this governor control is limited to the speed range as noted above owing to the transmission's shifting characteristics. The remainder of the governor control (2000 to 3000 rpm) is exactly like that of the RQ governor in that it is *load sensitive.* In PTO applications with speed requirements below approximately 2000 rpm, the all-speed portion of RQV/RQ combination governor reacts similarly to the standard RQV all-speed governor.

Robert Bosch Electronic Diesel Control

Throughout this book, several examples of electronic diesel fuel injection control systems are shown—such as the Bosch in-line pump in Figure 12–16, the Bosch VE injection pump used on the Ford/BMW 6-cylinder 2.4L turbocharged engine in Figure 27–6 as well as electronic diesel control systems used by Stanadyne on their model DB2 pump in Figure 21–3 and 21–4.

The major reasons for the adoption of electronic controls to diesel fuel injection pumps are to lower fuel consumption, provide a cleaner exhaust, and achieve smoother overall engine performance.

The injection pump mechanical governor still controls the fuel rack position with the governor receiving its primary signal from the accelerator pedal. In addition to this arrangement, however, the digital electronic control unit of Bosch's new system processes input signals from a number of engine sensors into electrical output signals. These output signals are then translated into mechanical actions that can then be made to regulate and control the position of the fuel injection pump rack or throttle position, depending on whether the injection pump is of the multi-plunger in-line-type or the distributor-type unit, which is more common to passenger car applications.

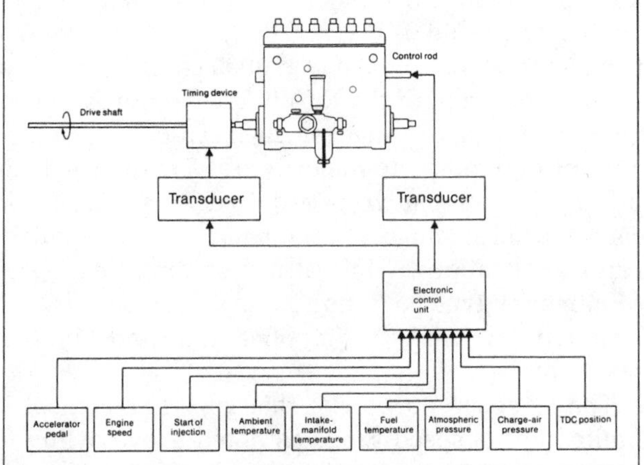

Figure 12–16. In line pump electronic control system. (Courtesy of Robert Bosch Corporation)

Major advantages of the diesel electronic control system are as follows:

1. Key start/stop function
2. Idle speed control
3. Part-load mixture regulation
4. Full-load limiter
5. Programmed driveability characteristics
6. Steep breakaway curve (speed droop = 0)
7. Temperature-dependent start-up fuel quantity
8. Cold start injection advance
9. Top speed limiter
10. Cruise control
11. Accelerator pedal programming
12. Accelerator pedal delay
13. Engine hunting suppression
14. Timing of the start of fuel injection
15. Temperature corrected fuel injection quantity
16. Fuel flow mass measurement

The adoption of these electronic control devices will become much more prominent throughout the 1980s and beyond.

Robert Bosch Model VE Injection Pump

The model VE fuel injection pump, Figure 12–17, takes its name from the German word "Verteiler," which means distributor pump, although it is also commonly referred to as a rotary-type design that operates upon the same basic principle as that of a gasoline engine ignition distributor. Rather than employing high tension pickup points inside a distributor cap, we have high pressure fuel outlet lines to carry diesel fuel to each individual cylinder. On a gasoline engine, the ignition distributor feeds the high tension spark via wire leads to each spark plug in engine firing order sequence. In the diesel engine, the fuel injection pump delivers high pressure fuel through steel-backed fuel lines to each cylinder's injector in firing order sequence.

The "E" designation in the pump model refers to the particular model of rotary injection pump produced by Robert Bosch Corporation. The pump is available in 2-, 3-, 4-, 5-, 6-, and 8-cylinder engine configurations to suit a variety of engines and applications. The VE injection pump is widely used on both passenger car and light truck applications worldwide. The pump, although of Robert Bosch design and manufacture, is also manufactured under license by both Diesel Kiki and Nippondenso in Japan.

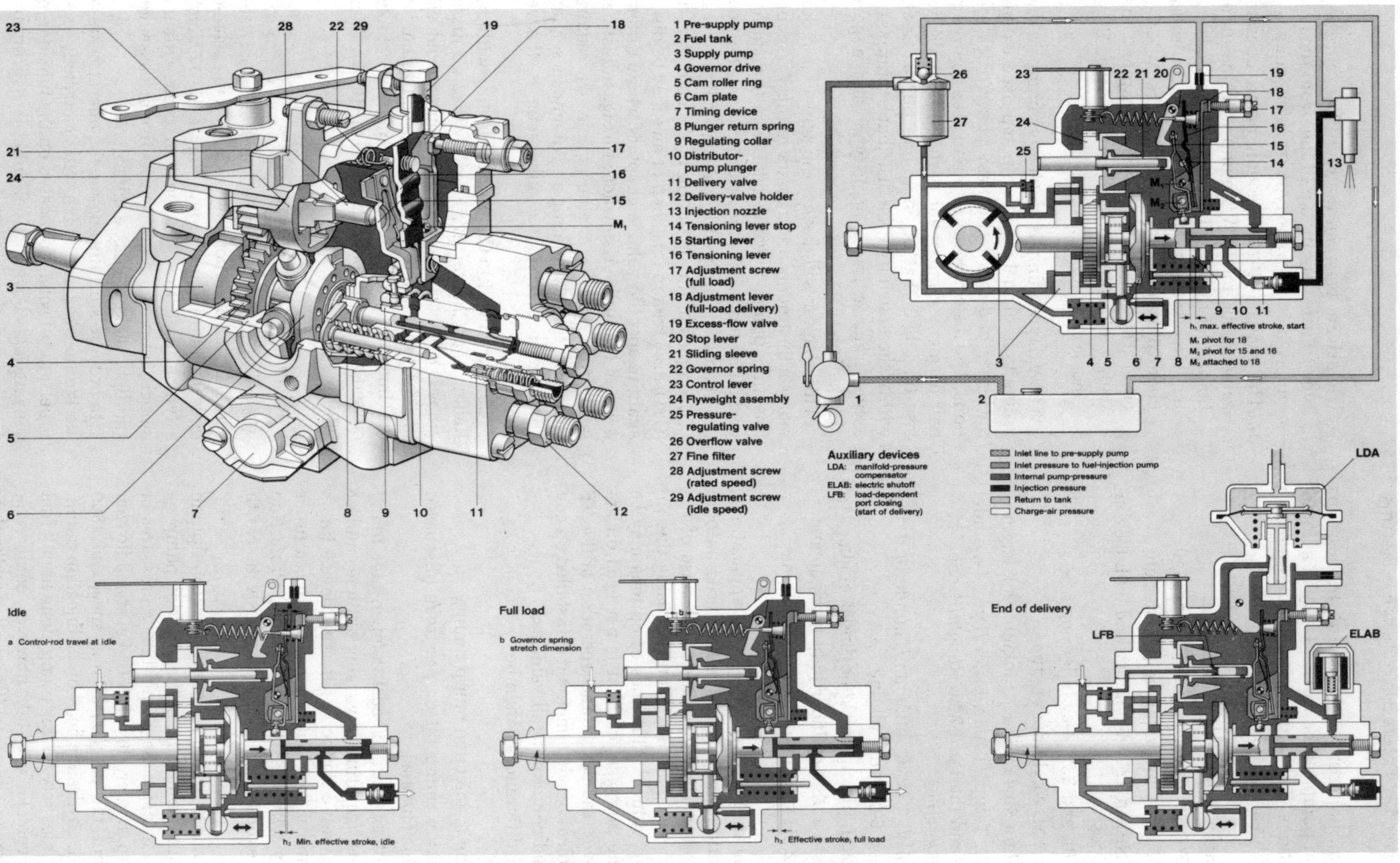
1 Pre-supply pump
2 Fuel tank
3 Supply pump
4 Governor drive
5 Cam roller ring
6 Cam plate
7 Timing device
8 Plunger return spring
9 Regulating collar
10 Distributor-pump plunger
11 Delivery valve
12 Delivery-valve holder
13 Injection nozzle
14 Tensioning lever stop
15 Starting lever
16 Tensioning lever
17 Adjustment screw (full load)
18 Adjustment lever (full-load delivery)
19 Excess-flow valve
20 Stop lever
21 Sliding sleeve
22 Governor spring
23 Control lever
24 Flyweight assembly
25 Pressure-regulating valve
26 Overflow valve
27 Fine filter
28 Adjustment screw (rated speed)
29 Adjustment screw (idle speed)
Auxiliary devices
LDA: manifold-pressure compensator
ELAB: electric shutoff
LFB: load-dependent port closing (start of delivery)
Inlet line to pre-supply pump
Inlet pressure to fuel-injection pump
Internal pump-pressure
Injection pressure
Return to tank
Charge-air pressure
h_1 max. effective stroke, start
M_1 pivot for 18
M_2 pivot for 15 and 16
M_2 attached to 18
Idle
a Control-rod travel at idle
h_2 Min. effective stroke, idle
Full load
b Governor spring stretch dimension
h_3 Effective stroke, full load
End of delivery
LDA
LFB
ELAB

Figure 12–17. Bosch distributor-type injection pump VE—F operational diagram. [Courtesy of Robert Bosch Corporation]

It is one of the most widely used distributor-type fuel injection pump on the market today in automotive applications. This pump can be found on such vehicle/engines as the following:

1. Volkswagen Rabbit 4-cylinder engine
2. Volvo D24 6-cylinder engine (VW design)
3. Audi 5000 5-cylinder engine (VW design)
4. Datsun/Nissan 1.7L 4-cylinder and 2.8L 6-cylinder engines
5. Isuzu 1.8L and 2.2L 4-cylinder engines
6. Peugeot 2.3L 4-cylinder engine
7. Dodge Ram 50 4-cylinder pickup truck engine
8. Ford 2.0L Tempo/Topaz, Escort/Lynx 4-cylinder engines and the Ford 2.2L Ranger/Bronco 11 4-cylinder engines
9. Ford/BMW 2.4L Continental/Thunderbird turbocharged 6-cylinder engine
10. Mazda 2.2L pickup truck 4-cylinder engine
11. Mitsubishi 2.3L 4-cylinder light truck engine
12. Toyota 2.2L 4-cylinder light truck engine

In addition to the above automotive applications, the VE injection pump is also widely used on both industrial and marine applications. Because of the various engine/vehicle manufacturers using this injection pump, minor differences or options may be found on one pump/engine that is not used on another; however, the design and operation of the VE pump regardless of what engine it is installed on can be considered common to all vehicles.

The major differences would be:

1. The engine to injection pump timing.
2. The injection pump lift or prestroke (discussed later in this chapter).
3. The use of an altitude/boost compensator found on engines operating in varying altitudes and or equipped with a turbocharger. This device limits the amount of fuel that can be injected in order to comply with EPA (Environmental Protection Agency) exhaust emissions standards in North America.
4. All VE pumps contain a vane-type transfer pump built within the housing of the injection pump assembly to transfer diesel fuel under pressures of from approximately 36 psi (250 kPa) at an idle rpm up to about 116 psi (800 kPa) at speeds of 4500 rpm into the hydraulic head of the injection pump. Some vehicles rely on this pump alone to pull fuel from the vehicle fuel tank; however, some vehicles employ an additional lift pump, usually electric-driven, between the fuel filter and the vane-type transfer pump to pull fuel from the tank and supply it to the vane-type transfer pump.
5. The model VP-20 injection pump operates the same as the model VE pump with the difference being that the VP-20 pump is controlled electronically while the model VE pump is controlled mechanically. An explanation of the VP-20 pump control system can be found in Chapter 27 dealing with the Ford/BMW 2.4L turbocharged diesel engine. (See Figure 27–6.)

The VE-type pump is much more compact than the in-line type of fuel injection pump used extensively on larger mid-range and heavy duty truck applications. The distributor-type pump uses approximately half as many component parts and usually weighs less than half that of an in-line pump. Contained within the housing of the distributor pump are both a fuel transfer pump (vane-type) and a governor mechanism.

This chapter will discuss the actual operation and major adjustments and repairs to the VE model fuel injection pump. Specific adjustments relating to a particular model of engine/vehicle can be found in the chapter dealing with that car or light truck.

Examples of the operation of the cold start device (CSD) are given in this chapter and may be discussed again in specific vehicle chapters where differences exist on that unit. Also where linkage adjustment relating to a specific engine/vehicle are required or where the troubleshooting and diagnosis sequence is common to that vehicle only, refer to the vehicle chapter.

Contained within Chapter 14, dealing with the VW Rabbit and Chapter 27, Ford/BMW 2.4L diesel engines is a Robert Bosch supplied pump troubleshooting chart that can be effectively used to troubleshoot the injection pump operation on any vehicle discussed within this book that uses the Bosch VE model pump.

Fuel Flow—Operation

Although minor differences may exist between the actual layout and fuel flow path from the vehicle fuel tank to the injection pump, Figure 12–17 illustrates a typical fuel flow arrangement used with the VE model pump as it applies to its use in passenger car and light pickup truck engines.

In order to start the engine, the operator must turn the ignition key *on*, which will electrically energize a fuel shutoff solenoid located on the injection pump housing just above the fuel outlet lines from the hydraulic head of the injection pump.

This solenoid is shown Figures 12–17 and 12–18 and when energized is designed to allow fuel under pressure from the vane-type transfer pump to pass into the injection pump plunger pumping chamber.

When the ignition key is turned *off*, the fuel solenoid is de-energized and fuel can no longer be supplied to the plunger pumping chamber; therefore, the engine will starve for fuel and stop immediately.

Some vehicles use only the vane-type transfer pump, which is contained within the injection pump housing to draw fuel from the tank to the injection pump, while others may employ either a mechanical diaphragm-type or an electrically operated lift pump to draw fuel from the tank and deliver it to the vane-type transfer pump.

Also, most vehicles today employ a fuel filter/water separator plus a secondary fuel filter in the system between the fuel tank and the vane-type transfer pump.

The vane-type transfer pump is shown as Item 3 in Figure 12–17. This pump is capable of producing fuel delivery pressures of about 36 psi (248 kPa) at an engine idle speed up to as high as 120 psi (827 kPa), although maximum pressures are generally maintained at around 100 psi (689.5 kPa). This fuel under pressure is then delivered through internal injection pump drillings to the distributor pump plunger shown as Item 10 in Figure 12–17. All internal parts of the fuel injection pump are lubricated by this fuel under pressure; there is no separate lube oil reservoir.

Maximum fuel pressure created by the vane-type transfer pump, which is located within the injection pump body, is controlled by an adjustable fuel pressure regulator screw.

On four-stroke engines, the injection pump is driven at one-half engine speed and is capable of delivering up to 2800 psi (approximately 200 bar) to the injection nozzles; however, the adjusted release pressure of the nozzle establishes at what specific pressure the nozzle will open.

An overflow line from the top of the injection pump housing allows excess fuel that is used for cooling and lubrication purposes to return to the fuel tank through a restricted bolt readily identifiable by the word OUT stamped on the top of it.

Since the vane-type transfer pump is capable of either left-hand or right-hand rotation, take care when servicing this unit that you assemble it correctly. Take careful note of holes 1, 2 and 3 shown in Figure 12–19. Hole 1 in the eccentric ring is farthest from its inner wall compared to hole 2. When looking at the eccentric ring, this hole must be in position 1 for right-hand rotation pumps and to the left for left-hand rotation fuel injection pumps. Hole 3 should be on the governor side when the transfer pump is installed. Also the pump vanes should always be fitted with the circular or crowned ends contacting the walls of the eccentric ring.

Fuel under pressure from the vane-type transfer pump is then delivered to the pumping plunger shown in Figure 12–17 and also in Figure 12–20, where it is then sent to the fuel injectors (nozzles). Let us study the action of the plunger more closely,

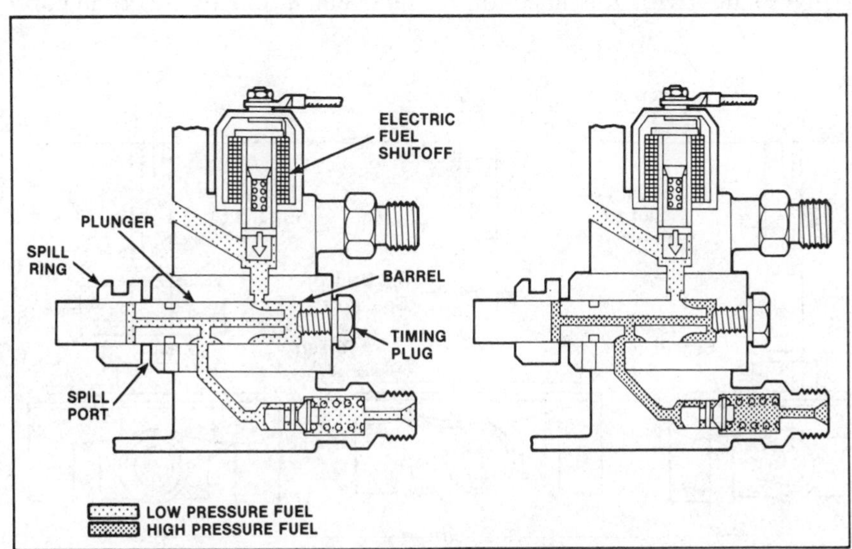

Figure 12–18. VE Injection pump plunger and barrel showing electric fuel shutoff. (Courtesy of Ford Motor Company)

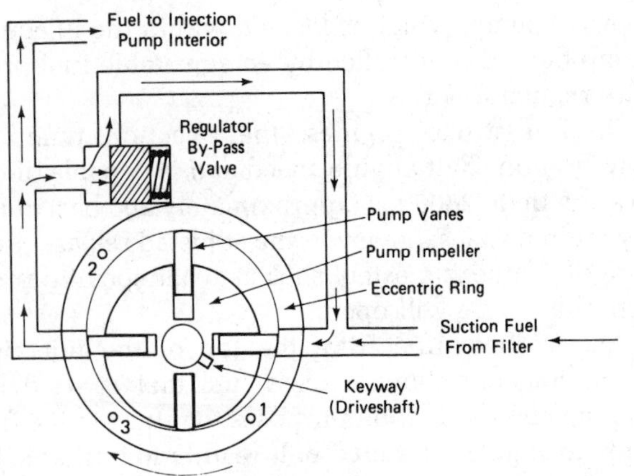

Figure 12-19. Four-cylinder VE vane-type transfer pump.

since it is this unit that is responsible for the distribution of the high-pressure fuel within the system. Figure 12–21 shows the actual connection between the cam rollers and the pump plunger, which is also visible in Figures 12–17 and 12–20.

Notice that the plunger is capable of two motions: (1) circular or rotational (driven from the drive shaft), and (2) reciprocating (back and forth by cam plate and roller action).

Reference to Figures 12–17 and 12–21 shows that the cam plate is designed with as many lobes or projections on it as there are engine cylinders. Unlike CAV and Roosa Master distributor-type injection pumps, the rollers on the VE pump are not actuated by an internal cam ring with lobes on it, but instead the cam ring is circular and attached to a round cam plate. As the cam ring rotates with the injection pump driveshaft and plunger, the rollers (which are fixed), cause the cam lobe to lift every 90 degrees (for example) in a 4-cylinder engine, or every 60 degrees in a 6-cylinder engine.

In other words, the rollers do not lift on the cam as in a conventional system, but it is the cam ring that is solidly attached to the rotating plunger that actually lifts as each lobe comes into contact with each positioned roller spaced apart in relation to the number of engine cylinders. With such a system then, the plunger stroke will remain constant regardless of engine rpm. At the end of each plunger stroke, a spring ensures a return of the cam ring to its former position as shown in Figure 12–17 (Item 8). Therefore the back-and-forth motion of the single pumping plunger is positive.

Anytime that the roller is at its lowest point on the rotating cam ring lobe, the pumping plunger will be at a position commonly known as BDC (bottom dead center); and with the rotating cam ring lobe in contact with the roller, the pumping plunger will be at TDC (top dead center) position as shown in Figure 12–21.

Distribution of fuel to the injector nozzles is via plunger rotation, and metering (quantity) is controlled by the metering sleeve position, which varies the effective stroke of the plunger.

If we consider the plunger movement, that is, *stroke* and *rotation*, Figure 12–22 depicts the action in a 90 degree movement such as would be found on a 4-cylinder 4-cycle engine pump. Even though there is a period of dwell at the start and end of one 90 degree rotation (one cylinder firing), the plunger movement during this time continues.

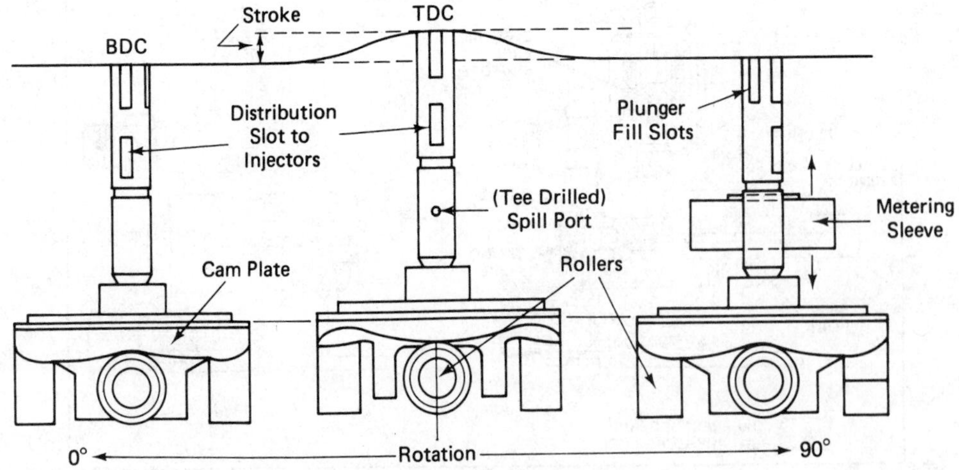

Figure 12-20. VE Injection pump distributor plunger location and design. (Courtesy of Volvo of America Corporation)

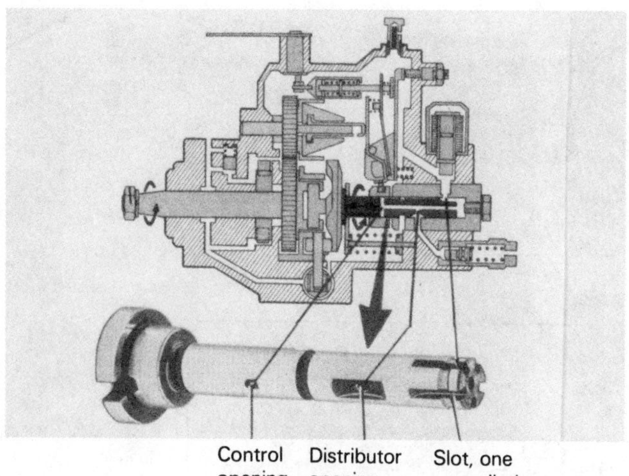

Figure 12–21. Arrangement and connection between VE injection pump cam rollers and plunger.

The sequence of events shown in Figure 12–22 is as follows:

1. The fill slot of the rotating plunger is aligned with the fill port, which is receiving fuel at transfer pump pressure as high as 100 psi (7 bar approximately), one cylinder only.
2. The rotating plunger has reached the *port closing* position. The plunger rotates a control spool regulating collar (see Figures 12–22, Item 8 and 12–17, Item 9). The position of the regulating collar is controlled by the operator or driver though linkage connected to and through the governor spring and flyweights. Because the plunger rotates as well as moving back and forth, the plunger must lift for port closure to occur; then delivery will commence.

   Because the rotating plunger does stroke through the metering sleeve in the VE pump, this pump is classed as the *port closing* type. Therefore, even though the roller may be causing the cam–ring–plunger to lift, the position of the regulating collar determines the amount of travel of the plunger or *prestroke,* so the actual effective stroke of the plunger is determined at all times by the collar position.
3. At the point of plunger lift (start of effective stroke), fuel delivery to the hydraulic head and injector line will begin in the engine firing order sequence.
4. The effective stroke is always less than the total plunger stroke. As the plunger moves through the regulating collar, it uncovers a *spill port,* opening the high-pressure circuit and allowing the remaining fuel to spill into the interior of the injection pump housing. This then is *port opening* or spill, which ends the effective stroke of the plunger; however, the plunger stroke continues.
5. With the sudden decrease in fuel delivery pressure, the spring within the injector nozzle rapidly seats the needle valve, stopping injection and preventing after-dribble, unburnt fuel, and therefore engine exhaust smoke. At the same time, the delivery valve for that nozzle located in the hydraulic head is snapped back on its seat by spring pressure.

In a 4-cylinder 4-stroke engine, we would have four strokes within 360 degrees of pump plunger rotation, which is of course equal to 720 degrees of engine rotation. In summation, the volume of fuel delivered is controlled by the regulating collar position, which alters the (*effective stroke*) time that the ports are closed.

In Figure 12–28, there is an annulus or circular slot located on the rotating plunger that is also visible at the right-hand side of the plunger in Figure 12–21; this is the reason that the plunger must lift for port closure to occur. Only after the annulus lifts beyond the fill port do we have port closure. Port closing occurs only after a specified lift from BDC.

Delivery Valve Operation

Contained within the hydraulic head (outlets) of the injection pump where the high pressure fuel lines are connected to the injection pump are delivery valves (one per cylinder) (Figure 12–23), which are designed to open at a fixed pressure and deliver fuel to the injectors in firing order sequence.

These valves function to ensure that there will always be a pre-determined fuel pressure in the fuel lines leading to the fuel injectors. Another major function of these individual delivery valves is to ensure that at the end of the injection period for that cylinder there is no possibility of secondary injection and also that any pressure waves during the injection period will not be transferred back into the injection pump.

If secondary injection were to occur, the engine would tend to misfire and run rough. The delivery valves ensure a crisp cutoff to the end of injection when the fuel pressure drops off in the line and also maintains fuel in the injection line so that there is no possibility of air being trapped inside the line.

Fuel Return Line

All model VE pumps use a percentage of the fuel delivered to the injection pump housing to cool and lu-

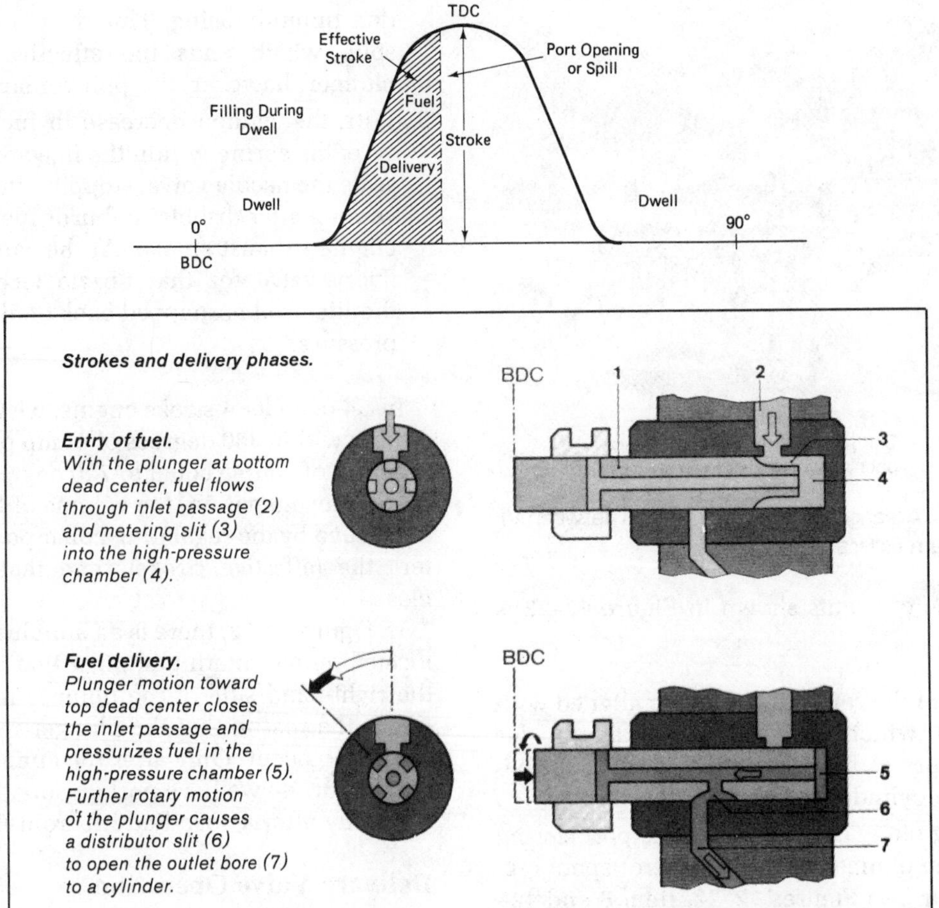

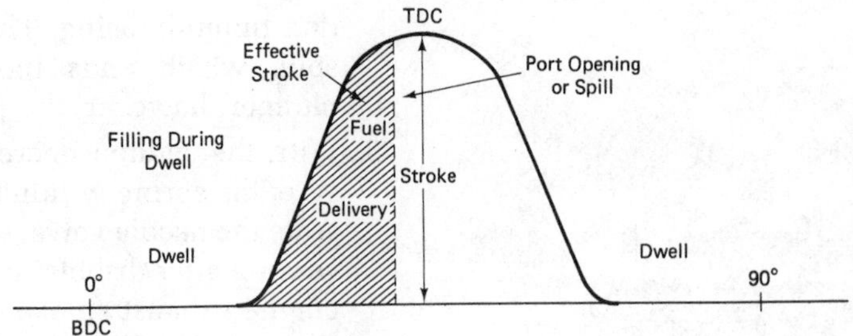

Strokes and delivery phases.

Entry of fuel.
With the plunger at bottom dead center, fuel flows through inlet passage (2) and metering slit (3) into the high-pressure chamber (4).

Fuel delivery.
Plunger motion toward top dead center closes the inlet passage and pressurizes fuel in the high-pressure chamber (5). Further rotary motion of the plunger causes a distributor slit (6) to open the outlet bore (7) to a cylinder.

End of delivery.
Fuel delivery ceases when the control spool (8) opens the cutoff bore (9).

Entry of fuel.
As the plunger returns to bottom dead center, its rotary-reciprocating motion closes the cutoff bore and the high-pressure chamber fills again.

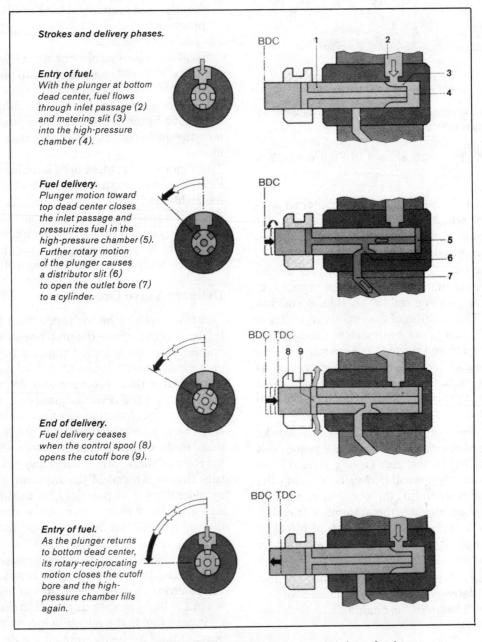

Figure 12–22. VE Pump plunger movement in a 4-cylinder, 4-cycle engine through 90° pump rotation. [Strokes/Delivery Phases, Courtesy of Robert Bosch Corporation]

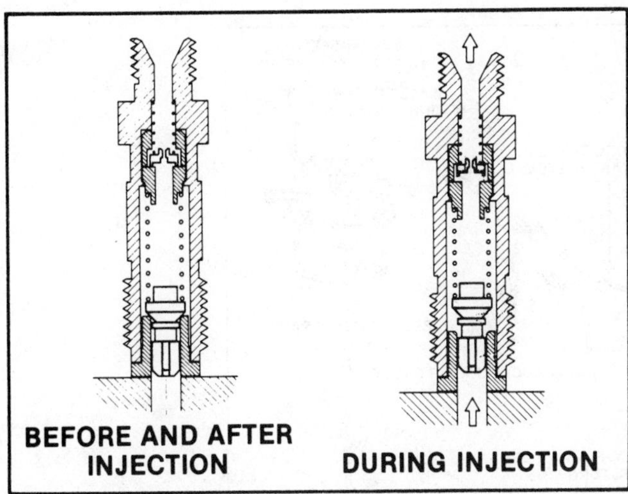

Figure 12-23. VE Injection pump delivery valve operation. [Courtesy of Ford Motor Company]

bricate the internal pump components. Since the diesel fuel will pick up some heat through this action, a bleed off or fuel return from the injection pump housing is achieved through the use of a hollow bolt with an orifice drilled into it as shown in Figure 12–24.

This bolt is readily identifiable by the word OUT stamped on the hex head, and if substituted with an ordinary bolt, no fuel will be able to return to the fuel tank from the injection pump.

Emergency Stop Lever

Should the fuel shutoff solenoid fail to operate when the ignition key is turned *off*, an emergency stop lever is connected to the injection pump housing and accessible underhood. This lever can be pulled to cut off fuel in the event of electric fuel solenoid failure. This lever is shown in Figure 12–17 as Item 20.

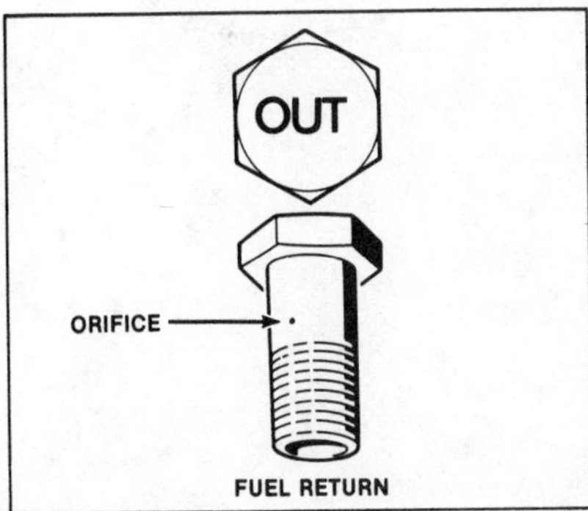

Figure 12-24. VE Injection pump fuel return bolt with restricted orifice. [Courtesy of Ford Motor Company]

Minimum/Maximum Speed Settings

The idle rpm and the maximum engine speed is controlled by adjusting two screws located on the top of the injection pump housing and shown as Items 28 and 29 in Figure 12–17. Both of these adjustments should always be done with the engine at normal operating temperature.

Turning the idle speed adjusting screw clockwise will increase idle rpm. Turning the high speed adjusting screw counterclockwise will increase the maximum speed setting of the engine. The minimum and maximum engine speed settings are listed on the Vehicle Emissions label/decal that is generally affixed under the hood in the engine compartment or at the front end of the engine compartment close to the radiator end.

Cold Start Device

All vehicles equipped with the VE fuel injection pump are equipped with either a manually operated or automatic cold start device (CSD). The year of vehicle manufacture and make establishes whether it has the former or the latter.

The main purpose of a cold start device is to provide easier engine starting and warm-up properties. When the cold start device is activated, the beginning of fuel injection is advanced through the movement of the injection pump cam roller ring in relation to the cam disc.

With a manually operated CSD such as shown in Figure 12–25, a control cable, which is mounted inside the vehicle, is pulled out by the operator and turned clockwise to lock it in place. This action causes a lever connected to a cam (Figure 12–26-a) to butt up against the injection pump advance piston and push it forward. Movement of the advance piston rotates the cam roller ring as shown in Figure 12–26(a) so that fuel injection will occur earlier in the cylinder BTDC. Another type of manual CSD uses a ball pin which is shown as Item 3 in Figure 12–26 (b) in order to rotate the roller ring (6).

The automatic CSD operates on the basis of engine coolant temperature in contact with a thermo-valve that contains a wax element similar to a thermostat. An example of this device is shown in Figure 12–27 with the linkage illustrated in both an engine-cold and engine-warm mode. Rotation of this linkage operates upon the timing control piston that will rotate the cam roller ring similar to the manually controlled system.

Various other automatic CSD systems can be seen in Chapters 25 and 27 dealing with the Ford 2.0L and 2.4L diesel engines. The degree of timing advance-

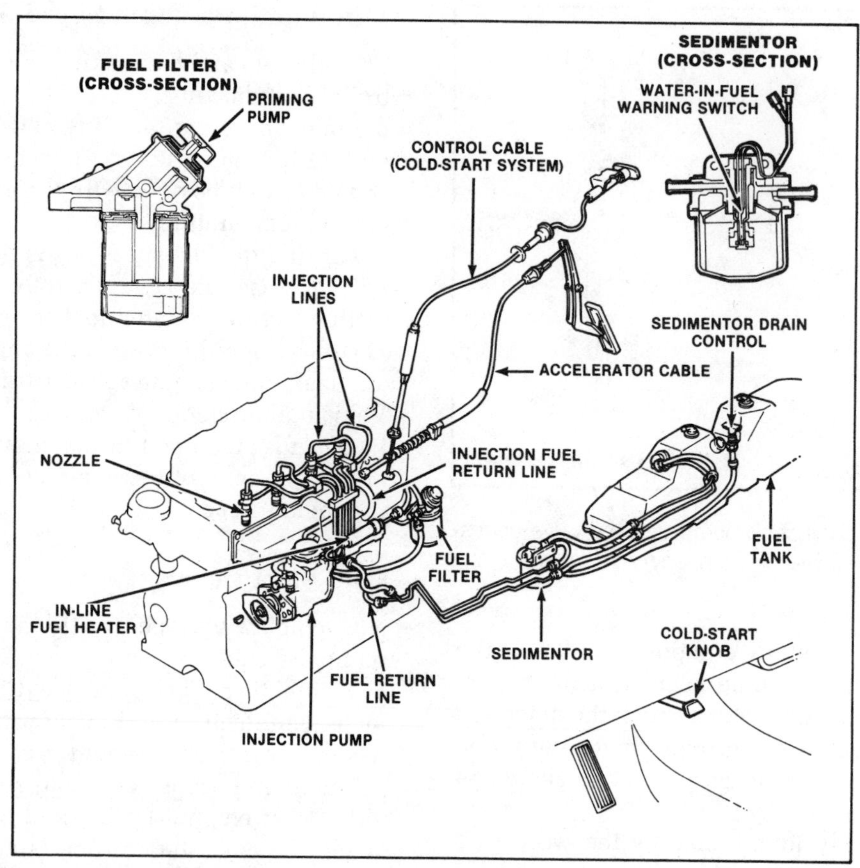

Figure 12-25. Manual cold start device components, VE injection pump. (Courtesy of Ford Motor Company)

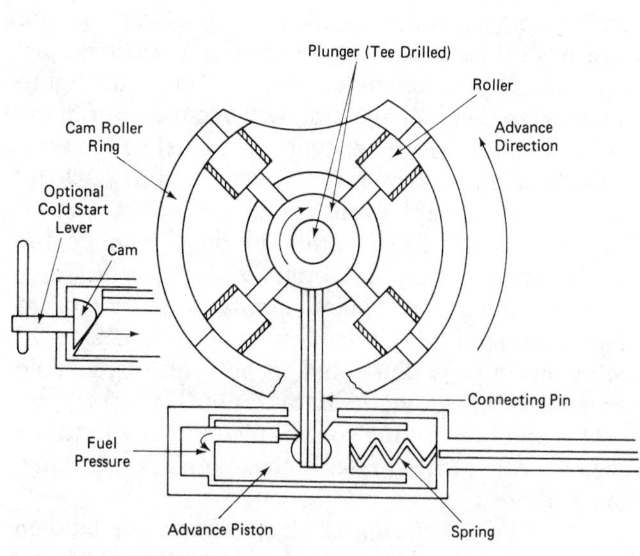

Figure 12-26(a). VE Injection pump automatic timing device arrangement.

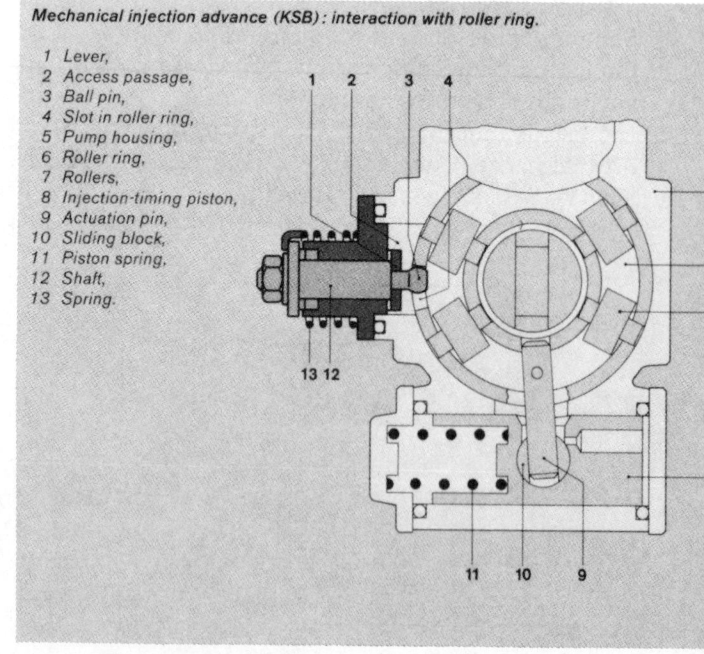

Mechanical injection advance (KSB): interaction with roller ring.
1 Lever,
2 Access passage,
3 Ball pin,
4 Slot in roller ring,
5 Pump housing,
6 Roller ring,
7 Rollers,
8 Injection-timing piston,
9 Actuation pin,
10 Sliding block,
11 Piston spring,
12 Shaft,
13 Spring.
1 2 3 4
13 12
11 10 9

Figure 26-26(b). KSB cold start mechanical injection advance linkage. (Courtesy of Robert Bosch Corporation)

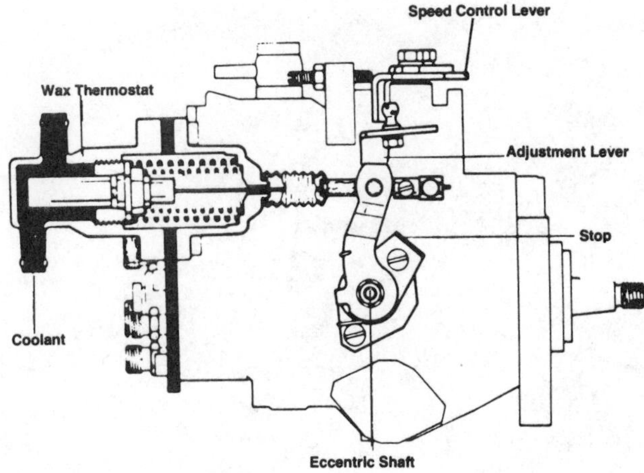

[A]

ENGINE WARM

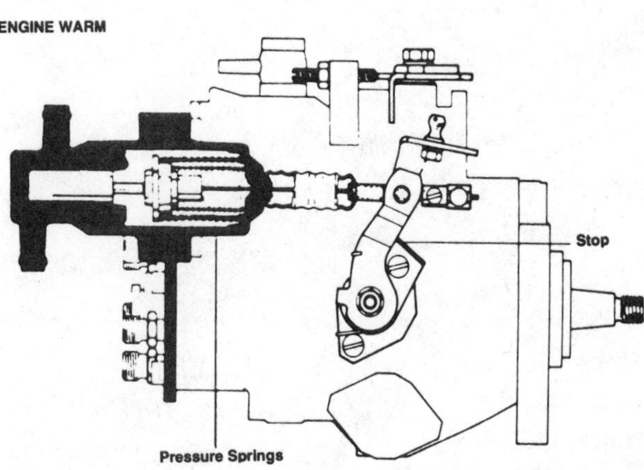

[B]

Figure 12–27. VE Pump cold start device linkage positions. (a) Engine cold; (b) Engine warm. (Courtesy of Robert Bosch Corporation)

ment will vary between makes of engines and is determined by the engine manufacturer.

Altitude/Boost Compensator

The operation of the altitude/boost compensator device is explained in detail under Chapter 27, Figure 27–7, related to the 2.4L Ford/BMW turbocharged engine used in the Lincoln Continental and Ford Thunderbird.

Governors/VE Pump

The Robert Bosch VE distributor/rotary injection pump is available with one of two mechanical governors to control the speed and response of the engine. These two types of governors and their functions are:

1. *A variable speed governor* that controls all engine speed ranges from idle up to maximum rated rpm. With this governor, when the throttle lever is placed at any position, the governor will maintain this speed within the droop characteristics of the governor. The variable speed governor and its operation are illustrated in Figures 12–28(a) and 12–28(b) with its actual location in relation to the other injection pump components being clearly shown in Figure 12–17.
2. *A limiting speed governor,* which is sometimes known as an idle and maximum speed governor since it is designed to control only the low and high idle speeds (maximum rpm) of the engine. When the throttle lever is placed into any position between idle and maximum, there is no governor control. Any change to the engine speed must be determined by the driver/operator moving the throttle pedal. This governor is shown in Figure 12–28(c).

The variable speed governor can be used on any application where all-range speed control is desired—such as on a stationary engine or on a vehicle that drives an auxiliary (power take off, i.e., PTO).

Operation of the Variable Speed Governor

If you are not already familiar with the basic operation of a mechanical governor, it may be advantageous to you to review the description of operation given in Chapter 11.

The thing to always remember is that the force of the governor spring is always attempting to increase the fuel delivery rate to the engine, while the centrifugal force of the governor flyweights is always attempting to decrease the fuel to the engine.

Anytime that the centrifugal force of the rotating governor flyweights and the governor spring forces are equal, the governor is said to be in a "state of balance" and the engine will run at a fixed/steady speed. You should also be familiar with the operation of the injection pump and how the "effective stroke" of the rotating pump plunger operates.

Engine Stopped. Refer to Figure 12–28(a). With the engine stopped there is no governor weight force and consequently the force of the idle spring (14) and the starting spring (6) force the governor linkage attached to the control spool (7) to a position whereby the "effective stroke" of the rotating pump plunger (9) will be at its maximum; therefore during engine cranking, maximum fuel will be delivered to the cylinders.

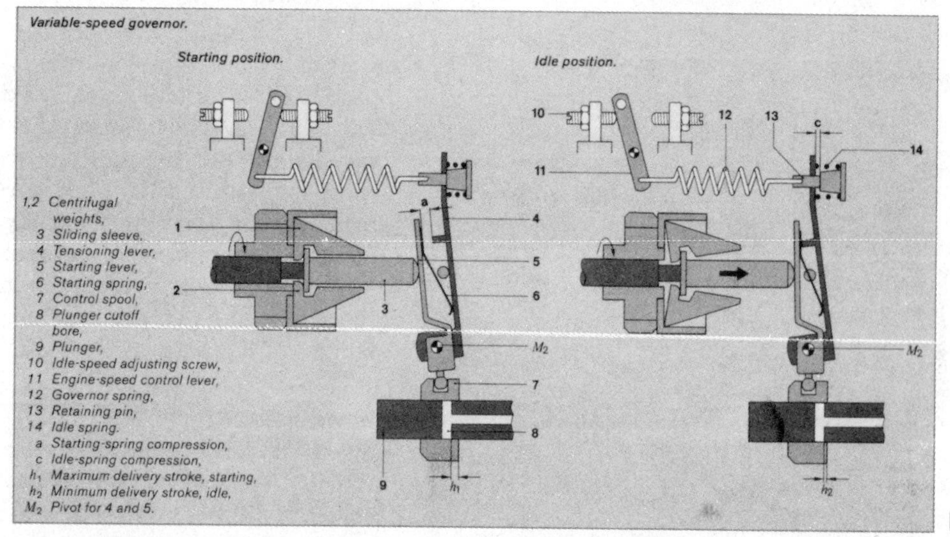

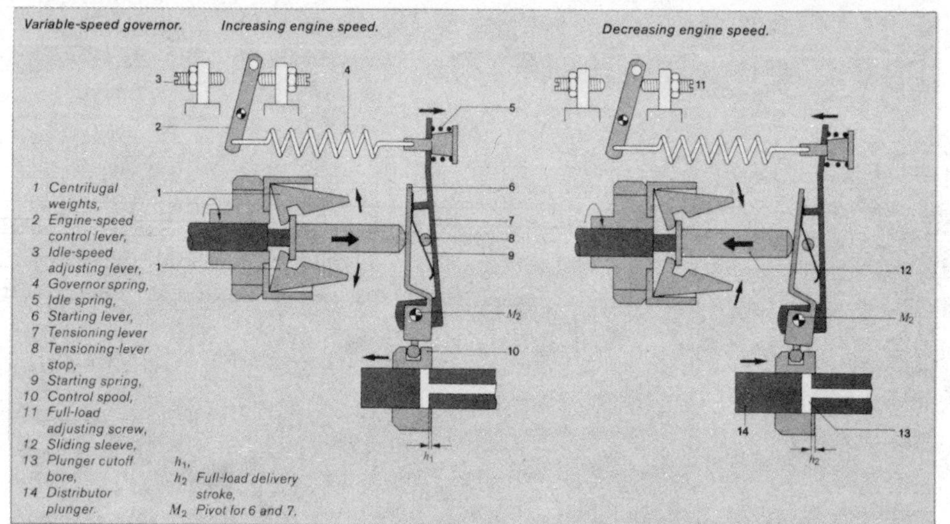

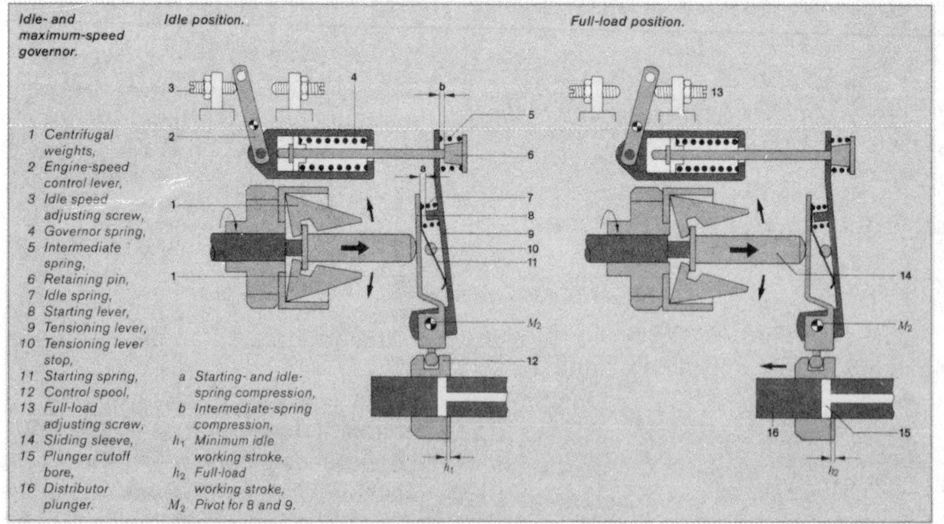

Figure 12-28. VE Injection pump mechanical governor component arrangement. [a] Variable speed governor—starting/idle position; [b] speed increase/decrease position; [c] idle/maximum speed governor, idle/full-load position. (Courtesy of Robert Bosch Corporation)

Engine Cranking/Starting. As the engine is cranked over, the centrifugal force developed by the rotating governor flyweights (1 and 2) will force the sliding sleeve (3) to the right in Figure 12–28(a) against the starting lever (5) and its spring (6). When the spring (6) is compressed, the lever (5) will butt up against a stop on the tensioning lever (4), which will now act directly against the force of the idle spring (14).

Movement of the tensioning lever (4), will pull the speed control lever on top of the governor back until it bottoms on the idle speed adjusting screw (10).

Once the centrifugal force of the flyweights equals the preset tension of the idle spring (14), the engine will run at a steady speed. A state-of-balance condition exists between the weights and the idle spring. If the throttle lever is moved above the idle speed, the spring will be collapsed by the distance "c" shown on the right-hand side of Figure 12–28(a).

Engine Acceleration. Refer to Figure 12–28(a). When the engine is accelerated beyond the idle rpm, the centrifugal force of the rotating governor flyweights will force the sliding sleeve (3) to the right, and with the starting lever (5) up against the tensioning lever (4), the idle spring (14) will be compressed. Additional engine speed and therefore weight force will now cause lever (4) to pull against the larger governor spring (12).

Refer to Figure 12–28(b). Movement of the throttle lever causes the engine speed control lever (2) to move away from the idle speed adjusting screw and towards the full-load adjusting screw (11).

The travel of the speed control lever is determined by the driver and just how fast he/she wants the engine to run. When the driver steps on the throttle, the previous state-of-balance condition that existed at idle is upset in favor of the governor spring (4). The control spool (10) is moved through lever (6) and (7) so that the "effective stroke" of the rotating pump plunger is lengthened by moving the control spool (10) initially to its right in Figure 12–28(b) under the heading "increasing engine speed."

As the engine receives more fuel and accelerates, the centrifugal force of the rotating flyweights (1) will push the sliding sleeve (12) to its right as shown in Figure 12–28(b) causing levers (6) and (7) to stretch the governor spring (4). When a state-of-balance condition exists once again between the rotating weights (1) and the spring (4), the engine will run at a steady speed with the throttle in a fixed position.

If the throttle is placed in full-fuel, then the speed control lever (2) will butt up against the full-load adjusting screw (11), which will limit the maximum speed of the engine. Weight force at this point is greater than spring force; therefore, the sliding sleeve (12) will cause the starting (6) and tensioning lever (7) to pivot around the support pin M2.

The control spool (10) will be moved to the left as shown in Figure 12–28(b) as shown under the heading "increasing engine speed," which will reduce the "effective stroke" of the rotating pump plunger. As a result, the engine will receive less fuel, thereby automatically limiting the maximum speed of the engine.

When the centrifugal force of the rotating governor flyweights (1) are equal to the governor spring force (4), the engine will run at a fixed rpm at maximum speed.

If the engine was started and accelerated to its maximum rpm with the vehicle in a stationary position, the action of the governor weights would limit the maximum amount of fuel that the engine could receive by moving the control spool to decrease the pump plunger's effective stroke. When the engine is running under such a condition (maximum no-load speed), it is not receiving full-fuel.

Decreasing Engine Speed. If the driver moves the throttle to a decreased speed position, then the engine speed control lever (2) will reverse the position of the control spool (10) through the levers (6) and (7). As the effective stroke of the pump plunger is reduced, the engine receives less fuel and therefore it will run at a lower rpm.

For a fixed throttle position at this lower speed, once the centrifugal force of the weights equals that of the governor spring (4), a new state of balance will occur and the engine will run at a steady speed.

Load Increase. Since this governor will control speed throughout the complete engine speed range, for a fixed throttle position, the engine will deliver a specific horsepower rating. As long as the engine is not overloaded at a given rpm position, the governor can control the speed within the confines of its droop characteristic.

NOTE: Droop is the difference between the maximum no-load rpm and the full-load rpm. Obviously, the engine speed will be lower under full load than it will be at no load. Similarly, when a load is applied to the engine for a given speed setting, it will tend to slow down since it now has to work harder to overcome the resistance to rotation.

A detailed explanation of droop can be found under the basic governor description in Chapter 11.

The reaction of the governor when a load is applied to the engine will be the same at any speed setting. A simplified description is as follows (Figure 12–28(b)):

1. Load applied at a given speed setting of the throttle, and engine slows down such as when going up a hill.
2. Upsets state of balance between weights (1) and spring (4) when above idle speed; if at idle, spring (5) in favor of the spring force.
3. Spring pressure is greater and therefore lever (6) and (7) acting through pivot point M2 moves the control spool (10) to its right to lengthen the effective stroke of the rotating pump plunger and supply the engine with more fuel to develop additional horsepower.
4. If the load on the engine continues to increase, the engine will receive more fuel to try to offset the load, but it will run at a slower rpm.
5. As long as the engine can produce enough additional horsepower, the governor will once again reach a state of balance between the weights and the spring, but at a slower speed than before the load was applied.
6. When the load was applied, the spring expanded (lengthened) to increase the fuel to the engine and in so doing lost some of its compression; therefore the weights do not have to increase their speed/force to what existed before in order to reestablish a new state of balance. The engine will produce more horsepower with more fuel, but will be running at a slower rpm.
7. Regardless of the governor's reaction to increase fuel to the engine, if the load requirements exceed the power capability of the engine, the rpm will continue to drop. In an automotive application, the only way that the speed can now be increased is for the driver to select a lower gear by downshifting.
8. If the engine was running at an idle rpm and an air conditioner pump was turned on, the engine would tend to slow down (load increase). The governor through the spring force/less weight force would increase the fuel to the engine to prevent it from stalling.

Load Decrease. When the load is decreased at a fixed throttle position, we have the following situation:

1. Engine speed increases; weights fly out with more force and they will cause the sliding sleeve (12) in Figure 12–28(b) to move levers (6) and (7) against the force of the spring (4).
2. The reaction is the same as shown under the heading "increasing engine speed" where the control spool (10) will move to its left to decrease (shorten) the effective stroke of the pump plunger and reduce fuel to the engine until a new corrected state-of-balance condition exists.
3. With less load on the engine, it requires less horsepower and therefore less fuel and as the engine slows down, so do the weights until the state of balance is reestablished.
4. If a vehicle goes down a hill, the load is reduced. If the driver does not check the speed of the vehicle with the brakes, it is possible for the driving wheels to run faster than the engine. If the drive wheels start to rotate the engine, the governor weights will also gain speed and in so doing they will reduce the effective stroke of the pump plunger and the engine's fuel will automatically be reduced.

Limiting Speed Governor Operation

The reaction in this governor is illustrated in Figure 12–28(c) and is the same as that described for the variable speed governor above with the exception that there is no governor control in the intermediate speed range—which is the speed range between idle and maximum rpm.

Engine Stopped. The engine will receive maximum fuel for startup since the force of the starting spring (11) and the idle spring (7) will move the control spool (12) to a position where the pump plunger will obtain its maximum effective stroke.

Cranking and Starting. As the engine is cranked, the centrifugal force of the governor weights (1) will force the sliding sleeve (14) to its right against the force of the starting spring (11) and the idle spring (7). As the starting levers (8) and (9) are moved to the right, the control spool (12) will be pulled back (left) to reduce the effective stroke of the pump plunger.

How far the spool (12) will be pulled back is established by the setting of the idle spring. When a state of balance exists between the weights (1) and the idle spring (7), the control spool (12) is held at a fixed position and the engine receives a fixed amount of fuel suitable for an idle rpm which is set by the adjusting screw (3).

Acceleration. When the throttle is moved initially beyond the idle range, the weights will compress the idle spring (7), and the weight force will now act upon the force of the intermediate spring (5) for a short time. This spring (5) allows a reasonably wide idle speed range, a large speed droop and a soft or gradual transition from the low idle speed range (governor control) to the point where the driver has complete control over the engine speed.

The intermediate spring (5) will be completely compressed (collapsed) shortly after the engine is accelerated from idle, and the throttle pedal now acts directly through the linkage to the sliding sleeve (14). There is not enough weight force to act upon the high speed spring (4) until the engine speed approaches the high end. Engine speed is now directly controlled by the driver.

High Speed Control. When the engine speed and therefore governor weight force is great enough, the centrifugal force of the weights will oppose the high speed spring (4) until a state of balance occurs.

When the weights and spring (4) come into play at the higher speed range, the maximum speed of the engine is limited by the fact that the weights as they fly out cause the sliding sleeve (14) to transfer motion through lever (8) and (9), which will compress the spring (4) and therefore move the control spool (12) to its left to shorten the effective stroke of the pump plunger. In this way, the engine receives less fuel and the maximum speed of the engine is therefore limited when the weights and spring (4) are in a state of balance.

As load is applied and released from the engine (up hill) and (down hill), the governor will react in the same way that it did for the variable speed governor described in detail earlier.

Automatic Timing Advance

The automatic advance mechanism employs the same principle of operation as does CAV and Roosa Master Stanadyne distributor-type injection pumps. Fuel pressure from the transfer pump is delivered to a timing piston whose movement is opposed by spring pressure. At low engine speeds, the relatively low supply pump pressure has little to no effect on the timing piston travel. As engine speed increases, the rising fuel pressure will force the timing piston to overcome the resistance of the spring at its opposite end. At the center of the piston, as shown in Figure 12–26(a), is a connecting pin extending up into the roller ring. The movement of the piston transmits this motion through the pin, which in turn rotates the roller ring in the opposite direction to drive shaft rotation, thereby advancing the timing of the cam plate lift from BDC to begin the plunger stroke. The timing piston travel should not be toyed with, but should be checked while the injection pump is mounted on a test bench.

Prestroke Compared to Non-Prestroke Pumps

Some VE injection pumps use a plunger whereby all the fill ports are interconnected by an annulus, or circular passage, running around the circumference of the plunger as shown in Figures 12–21 and 12–29. With this type of plunger containing the annulus, the unit is known as a *prestroke* pump. With this type, the fill ports cannot close by plunger rotation alone. The plunger must lift for port closure to occur. Only after the annulus lifts beyond the fill port do we have port closure. The plunger must be adjusted for a specific lift from BDC for port closure to happen. With this type, fuel pressure build-up within the Tee-drilled plunger takes a few degrees longer than for the zero prestroke type, which does not have the annulus and wherein port closure occurs by plunger rotation alone: the plunger lifts from BDC after rotation from port closure.

Overhaul of the Injection Pump

Repair and major overhaul of any injection pump should only be undertaken by personnel trained in the diversified and intricate work of fuel injection equipment. Since special tools and equipment are required, which are not always readily available to everyone, refer to the Robert Bosch publication 46, VDT-W-460/100 B, Edition 1, Repair of Distributor-type Fuel Injection Pump 04604-VE-F. This is obtainable through your local Robert Bosch dealer, or from one of the Robert Bosch licensees.

Bleeding the Fuel System

Anytime that fuel lines have been opened/loosened or the fuel system has been serviced, it will be necessary to vent all air from the fuel system in order to start the engine.

Procedure

1. On engines equipped with an electric fuel lift pump, this procedure is relatively easy. However, if the fuel system does not have a separate lift pump, then it will take a little longer because the vane-type transfer pump inside the drive end of

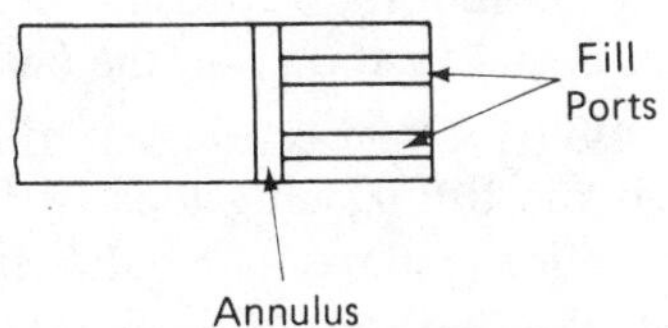

Figure 12–29. VE Injection pump plunger annulus interconnection slot.

the injection pump will have to pull the fuel from the tank to the pump on its own.

2. If the engine is equipped with an electric lift pump, make sure that all filter and injection pump vent screws are tight.
3. Turn the ignition key switch to the *on* position to energize the fuel cut-off solenoid and allow the electric lift pump to operate for 1 to 2 minutes.
4. Crank the engine over, and if it starts and runs correctly without misfire or stumble, then the system is properly bled of all air.
5. If the engine doesn't start, then loosen off the individual fuel line nuts (place a rag around the nut to absorb the spilled fuel) at the injectors and crank the engine over until air-free fuel appears at each line, then tighten them up.
6. If the engine is not equipped with an electric lift pump and only has a vane-type injection transfer pump, perform the sequence in Step 5 above while cranking the engine.
7. If the vehicle is equipped with a hand priming pump on the fuel filter/water separator, use this pump to bleed the filter first after opening the vent screw on top of the filter until air-free fuel appears. The inlet fuel stud on top of the injection pump housing can also be loosened off to vent air from the system right up to the injection pump. Place a drain tray underneath the fuel filter and pump to catch any leaking fuel. Step 5 can then be performed to bleed fuel up to the individual fuel injectors.
8. Once the engine starts and runs, if it is running rough, loosen off each injector fuel line nut one at a time (engine idling) to bleed each unit with a rag placed around it then tighten the nut.
9. Wipe all spilled or bled fuel from the engine and compartment.

SPECIAL NOTE: On a fuel system that has been completely emptied by running the engine out of fuel, it may be necessary to perform additional bleeding of the system by cranking or attempting to run the engine as follows:

1. Loosen off the fuel return fitting on the injection pump that is stamped OUT on the head of the hollow bolt (Figure 12–24).
2. Loosen off the timing plug located in the center of the injection pump distributor head. (12–30)
3. Loosen off the fuel shutoff solenoid.
4. Loosen off the injector pressure outlet valves.

Checking Injection Pump Static Timing

Contained within each engine chapter is a description of the various adjustments and timing checks for that particular engine. The following "static timing" check can be considered common to all model VE injection pumps with the major difference being in the dimension given by the manufacturer for a particular model engine. Several engines using the model VE pump will have the same setting while others will differ slightly.

Generally, a static timing check is only required when a new pump is being installed or when an engine has been rebuilt or the pump has been removed for one reason or another.

A dynamic timing check (engine running) is described under various chapters of engines using the VE pump with the use of special test equipment. Example Ford/BMW 2.4L Continental/Thunderbird 6-cylinder turbocharged diesel engine, Chapter 27. Figure 27–1 and Chapter 24 (Test Equipment).

Procedure

1. Manually rotate the engine over to place No. 1 piston at TDC on its compression stroke (both intake and exhaust valve closed). Align the timing mark on the crankshaft front pulley with the stationary pointer timing reference mark on the engine front cover.
2. Refer to Figure 12–30 and remove the center bolt from the injection pump hydraulic head along with its sealing washer. A dial indicator adapter is available for use with the particular engine that you are checking to allow the dial gauge to be held in position during the static timing check.

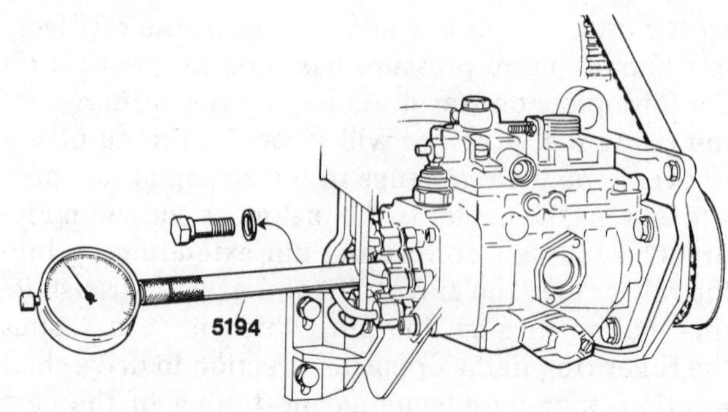

Figure 12–30. Checking VE injection pump plunger lift/timing with dial gauge installed into hydraulic head of pump. (Courtesy of Volvo of America Corporation)

One example of the timing gauge adapter is shown in Figure 12–31 for the Ford 2.0L 4-cylinder Tempo/Topaz, Escort/Lynx vehicles.

3. The adapter and dial gauge are installed onto the injection pump so that the plunger portion of the adapter projects into the injection pump. This will allow the dial gauge plunger to be in contact with the fuel injection pump plunger when installed. To do this correctly, ensure that the dial gauge shows at least 0.100" (2.54 mm) of preload on its face.

   Note, however, that VW recommends a preload of only 0.040" (1 mm), while Volvo on their D24 engine recommend 0.080" (2 mm) of gauge preload. The key here is that adequate preload be applied to the dial gauge to ensure that the pump plunger movement as you rotate the engine over during the static timing check will be felt/registered by the dial gauge plunger—otherwise a false reading will be obtained.

4. Manually rotate the engine in its normal direction of rotation until the dial gauge registers its lowest reading, then set the dial gauge to ZERO by rotating the face bezel to place the needle at zero.

5. Continue to rotate the engine manually in its normal direction of rotation smoothly until No. 1 piston is at TDC on its compression stroke. Some engine manufacturers supply a TDC aligning pin that is installed through a hole in the block to index with a hole in the flywheel so that the engine cannot be moved during this timing check. (Example is the Ford/BMW 2.4L 6-cylinder turbocharged engine in the Continental and Thunderbird, Figure 27–21.)

   If such a device is not available, ensure that either the timing marks between the crankshaft pulley/damper and stationary timing pointer are in correct alignment or that the flywheel timing marks such as found on the VW and Volvo diesel engines are in alignment.

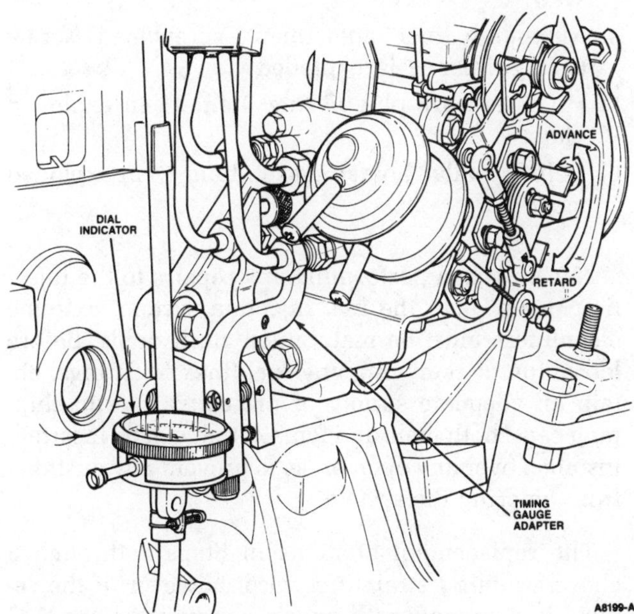

Figure 12–31. Dial gauge and support bracket installed into VE injection pump to check pump to engine timing. (Courtesy of Ford Motor Company)

6. The measurement on the dial gauge face should be noted and compared with the engine manufacturers specification. For example, if the static timing was given as 1 mm (0.03937"), then the gauge should register this specification. If it doesn't, then the injection pump-to-engine timing needs adjustment.

7. To change the injection pump-to-engine timing, loosen off the injection pump housing retaining bolts and move the pump towards the engine if the measurement on the gauge is too small (this will advance the timing); move the pump housing away from the engine if the gauge reading is too large (this will retard the timing).

   SPECIAL NOTE: What you are actually doing when you move the injection pump towards or away from the engine is adjusting the pump plunger lift from the BDC position to the point of port closure by turning the cam ring away from or towards the rollers.

   ADDITIONAL NOTE: Certain engine manufacturers supply a special adjusting bracket that can be bolted onto the injection pump housing to facilitate accurate adjustment of the timing. This allows the pump to be held in position as you tighten up the retaining bolts.

8. A specified tightening sequence is also given by various engine manufacturers to ensure proper seating of the pump-to-engine block.

9. Always rotate the engine over manually at least twice when you have completed your adjustment to double check that the setting is in fact correct. If the setting is incorrect, repeat Steps 1 to 8 above.

Robert Bosch VE Injection Pump Troubleshooting

Problems related to the VE injection pump are basically similar regardless of the type of engine and vehicle that it is installed on. The Volkswagen Rabbit and Ford/BMW 2.4L engine, Chapters 14 and 27, contain trouble-shooting charts that list the typical types of problems that might be encountered on that engine when using a VE injection pump.

When an engine exhibits heavy smoke after a cold start, the cold start device should be checked by monitoring the engine idle rpm. The cold start device used with the VE pump is controlled by a wax-type thermostat arrangement shown in Figure 12–27 that responds to engine coolant as it warms up. When the vehicle attains its normal operating temperature, the cold start device (CSD) does not operate.

Actual testing of the CSD can only be done properly with the injection pump mounted on a test bench (stand). However, a simple test of the CSD can be made on the engine as follows: Engine idle rpm should usually be about 200 rpm higher when the engine is cold compared to when it is at operating temperature. In addition, when the engine is at operating temperature, the cold start device lever should not contact the lever on the injection pump as shown in Figure 12–27(b). On vehicles equipped with an automatic transmission, an emergency stop lever is fitted to the side of the injection pump as shown in Figure 12–17. If the engine fails to shut off when the ignition key is turned *off*, there is a fault with the fuel solenoid located on the injection pump housing.

On a standard transmission equipped vehicle, the engine can be stopped by placing the transmission in gear with the engine idling and with your foot on the brakes, engaging the clutch to stall the engine.

On automatic transmission equipped vehicles, refer to Figure 12–17, Item 20 and pull the emergency stop lever. If the engine fails to start, the cause may well be the fuel solenoid on the injection pump as illustrated in Figure 12–17, Item "ELAB," and Figure 12–18. Check the fuel solenoid valve by placing a voltmeter across its terminal and ground. A voltage of less than 10 volts will fail to open (energize) this valve, while at least 8 volts is required to keep the valve in an open state while the starter motor is cranking the engine.

Injection Pump Service

Information following deals with those repair jobs that can be done to the injection pump while it is both on and off the engine.

Repairs other than O-ring seal replacement that are required to the pump when it is off of the engine require a number of special tools and the use of a test stand to check out the pumps delivery, etc.

On Vehicle Pump Adjustments

Before performing any seal ring replacement to the injection pump while it is on the vehicle, clean the external areas of the pump.

CAUTION: Do not apply steam directly to the injection pump housing because it is made from an aluminum alloy. This heat can cause the aluminum to expand at a different rate than the internal steel components of the pump causing serious damage, especially if the engine is running when this is done. Figure 12–32 illustrates the injection pump in an exploded view form. Use this figure as a guide when performing service procedures to the injection pump both on and off the vehicle.

Typical "on-vehicle" injection pump service items would include:

1. Removal and installation of the fuel return pipe washers/gaskets Item 137 in exploded view, Figure 12–32.
2. Removal and installation of the fuel inlet pipe washers/gaskets Item 245 in exploded view.
3. Tachometer plug O-ring Item 296 in exploded view.
4. Maximum fuel adjustment screw/load screw O-ring; Item 91 in exploded view.
5. High pressure plug O-ring; Item 15 in exploded view.
6. Delivery valve holder gasket, Item 55 in exploded view.

NOTE: When performing any repairs to the injection pump or to the fuel injection system, extreme cleanliness must be maintained at all times. Before loosening or removing any fuel lines or fittings, obtain an adequate supply of protective plastic shipping caps, both male and female, that can be inserted/installed over any open areas to prevent any dust/dirt from entering the system.

The replacement of items in Steps 1 through 3 above are fairly straightforward; however, if the vehicle is equipped with an air conditioner, the A/C drive belt and the A/C compressor will have to be removed to complete the jobs.

In job number 3, torque the tachometer plug to 8 lb.ft. (10 N·m) after replacing the O-ring.

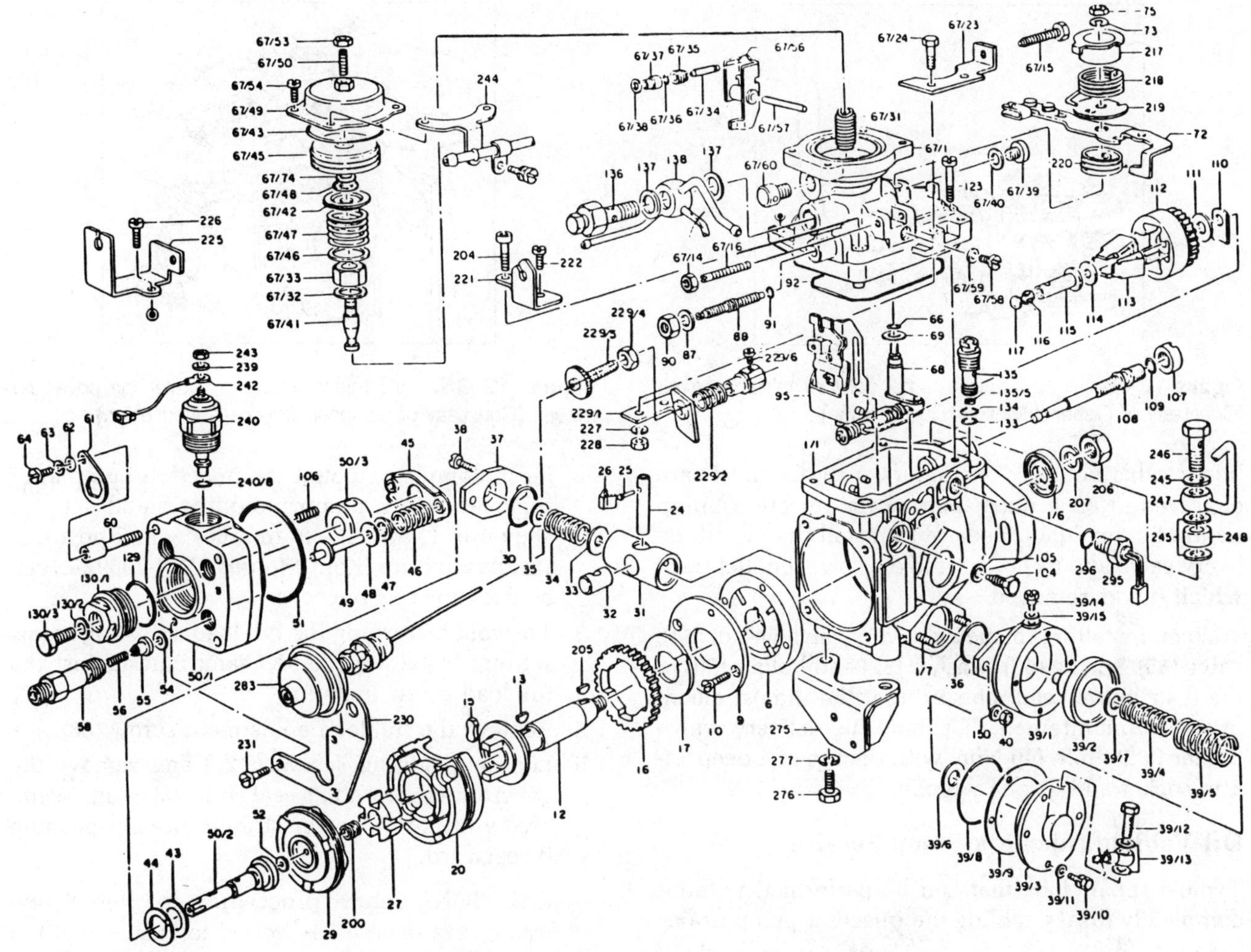

Figure 12–32. Exploded view of VE distributor-type injection pump. (Courtesy of General Motors Corporation)

In job 4, after removing the A/C drive belt and compressor, install a M8.0X1 jamb nut against the full-load adjustment screw lock nut to prevent disturbing this setting before removing the full-load adjusting screw. Remove and replace the O-ring. When installing the full-load adjusting screw, torque the jamb nut to 6 lb.ft. (8 N·m).

In job 5 above, after the A/C drive belt and A/C compressor have been removed, remove the high pressure plug using Kent-Moore special tool J-33309 shown in Figure 12–33. When the O-ring has been replaced, torque the high pressure plug to 50 lb.ft. (68 N·m).

If it becomes necessary to replace the delivery valve holder gasket at one or more outlets in the injection pump hydraulic head, refer to Figure 12–34 and after removing the A/C drive belt and A/C compressor, remove the fuel lines to the injectors as an assembly. Using a deep 14-mm socket, remove the leaking delivery valve from the injection pump hydraulic head.

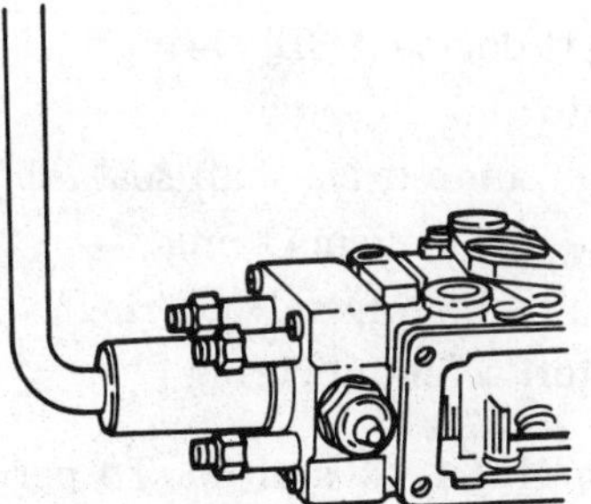

Figure 12–33. VE Injection pump, high pressure plug removal. (Courtesy of General Motors Corporation)

NOTE: The delivery valve and its seat assembly (55) are a matched set, therefore be careful that these two components are not intermixed with other de-

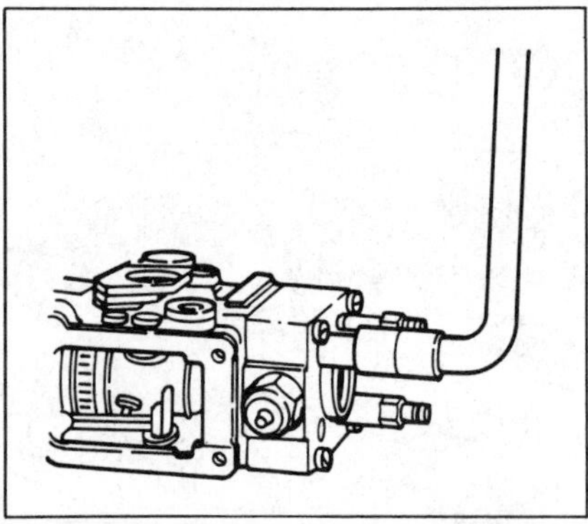

Figure 12–34. VE Injection pump, delivery valve removal. (Courtesy of General Motors Corporation)

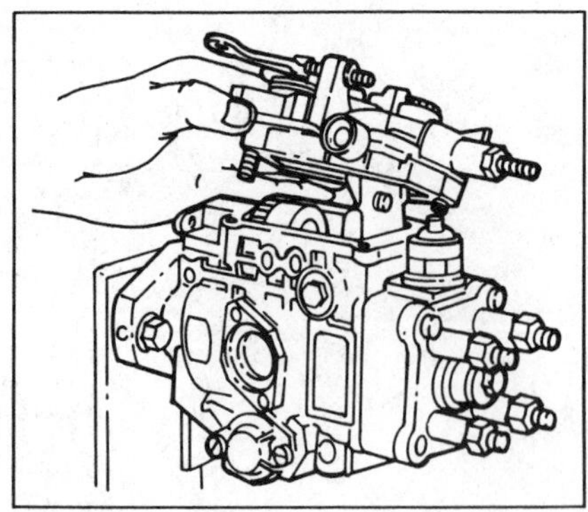

Figure 12–35. VE Injection pump, governor cover removal. (Courtesy of General Motors Corporation)

livery valve holders. The letters A, B, C, and D are engraved at each outlet port of the injection pump hydraulic head (50). Be sure to install the mated delivery valve and its holder into the correct outlet from which it was removed.

When installing the new delivery valve gasket (54), valve (55), spring (56), washer (57) and holder (58) to the distributor head, be sure that the utmost cleanliness is maintained. Tighten the delivery valve holder to 28 lb.ft. (40 N·m) with the use of a deep 14-mm socket and torque wrench.

Off Vehicle Injection Pump Service

Typical repair jobs that can be performed without completely disassembling the injection pump are as follows:

1. Throttle shaft O-ring replacement
2. Governor cover O-ring
3. Full-load adjusting screw O-ring
4. Cold start device (CSD) O-ring
5. Distributor head O-ring
6. Injection pump drive shaft seal
7. Advance mechanism O-ring
8. Pressure regulating valve O-ring
9. Fuel cutoff solenoid O-ring

Repair Job No. 1, 2, and 3. To perform replacement of the O-rings in 1, 2, and 3 in the above list, refer to the exploded pump view shown in Figure 12–32, which shows the numbered location of all injection pump components.

Procedure

1. Refer to Figure 12–35, which illustrates the removal of the governor cover in order to gain access to the throttle shaft O-ring.
2. To make these service procedures faster and easier to complete, mount the injection pump after removal onto a pump mounting stand such as shown in Figure 12–35 in a horizontal position. Holding fixture (Kent-Moore) P/N J-29692-A can be used for this purpose.
3. To avoid disturbing the full-load adjusting screw setting, install an M8.0X1 jamb nut against the full-load screw lock nut.
4. Remove the full-load adjustment screw (88).
5. Loosen retaining screws (123) and remove the governor cover and its seal ring (92) as an assembly by lifting the cover straight up and pushing it rearward.

   CAUTION: Before proceeding to Step 6, extreme care must be exercised to ensure that the control lever 67/5 does not come off of the throttle shaft until you have scribed a reference mark on the throttle shaft 67/2 and the control lever 67/3. Each serration on the throttle shaft that the lever is misaligned when reinstalled will cause a 15 degree change to its setting. This can cause a change to the full-load fuel delivery rate. The full-load delivery rate can only be adjusted with the injection pump on a test stand.

6. Remove the M10X6 nut (67/7) from the control lever 67/5 while holding onto the lever. Make a match mark on both the lever and the shaft to ensure that it can be installed back into the same position after replacing the seals.
7. Push the throttle shaft (67/2) out of the governor cover.
8. Remove the old O-ring.

The following steps deal with the installation of the above components:

9. Install the shim 67/4 back onto the throttle shaft.
10. Prior to inserting a new O-ring 67/3 onto the

throttle shaft, lubricate it with clean diesel fuel, synkut oil, or fuel pump test/calibrating fluid.

11. Gently hand press the throttle shaft back into the governor cover.
12. Install the throttle lever and spring back onto the throttle shaft making certain that the previously scribed marks are in alignment.
13. Torque the retaining nut to 5 lb.ft. (7 N·m).
14. Install a new governor cover O-ring.

    NOTE: Carefully lower the governor cover into position with the throttle lever in the idle position making certain that the control shaft engages with the tension lever (square cut into the slot in the tension lever). Also check that the damper spring wafer is on the distributor side of the tension lever.

15. Look through the fuel overflow pipe hole to confirm that the control shaft is fully bottomed in the tension lever slot.
16. Apply pressure to force the governor cover forward towards the injection pump drive shaft and down into position over the dowel pin.
17. Install the four cover retaining screws and torque to 6 lb.ft. (8 N·m).
18. Insert a new O-ring on the full-load adjusting screw.
19. Install the full-load adjusting screw taking care that you do not disturb the full-load setting; torque the jamb nut to the same setting as in Step 17 above.
20. Remove the previously installed safety backup jamb nut that was installed earlier.

Repair Job No. 4. Refer to Figures 12–36 and 12–37 which illustrate the location of the CSD (cold start device) components. To replace the CSD O-ring, proceed as follows:

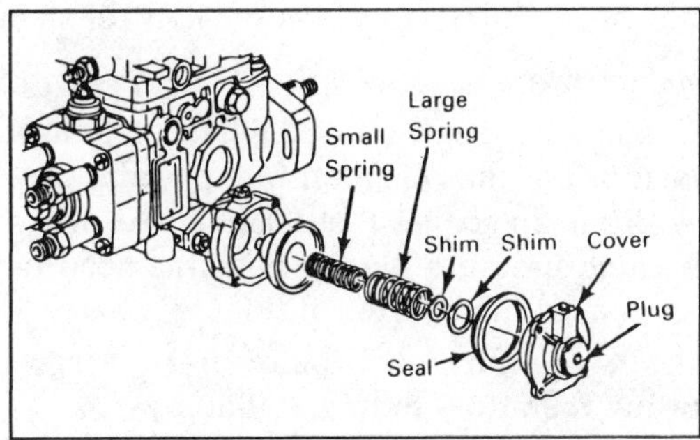

Figure 12-36. VE Injection pump, removal of CSD (cold start device) internal components. (Courtesy of General Motors Corporation)

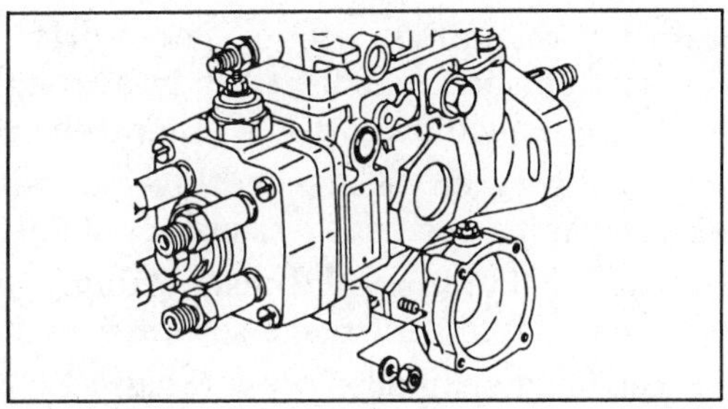

Figure 12-37. VE Pump, CSD housing removal. (Courtesy of General Motors Corporation)

Procedure

1. Refer to Figure 12–36 and using a 10-mm wrench, remove the center plug from the CSD cover *slowly* since the cover is under spring pressure.
2. Take careful note of the number and thickness of shims in behind the CSD cover.
3. Discard the O-ring and remove the small spring.
4. Remove the four bolts from the CSD retaining cover and also take note of the number and thickness of shims behind the cover.
5. Remove the timer spring, washer, and piston.
6. Remove and discard the cover seal.
7. Refer to Figure 12–37 and unscrew the two housing retaining nuts; pull off the housing and the O-ring.
8. Install a new O-ring (36).
9. Install the CSD housing and its two retaining nuts and torque these to 5 lb.ft. (7 N·m).
10. Carefully insert the large CSD spring, its shims, and piston as an assembly into the CSD housing with a new cover seal, and torque the retaining bolts to 5 lb.ft. (7 N·m).
11. Insert the smaller spring into the plug hole.
12. Make sure that you insert the correct number of shims that were removed previously from between the spring and the CSD cover plug; insert these shims into the plug recess after installing a new O-ring on the plug and tighten the plug.

Repair Job No. 5. To replace the injection pump distributor head O-ring, proceed as follows:

Procedure

1. Remove the governor cover (Figure 12–35) as described above.
2. Place the injection pump into a vertically mounted position on its holding fixture for ease of working access for this job.

3. Manually rotate the injection pump drive shaft to place the shaft woodruff key in a straight-up position perpendicular to the pump centerline.
4. If the delivery valves are tied together with retaining clamps, remove these clamps at this time.
5. Remove the four hydraulic head retaining screws (60) with a large wide blade screwdriver.

   Refer to the exploded view (Figure 12–32) of the injection pump, paying particular attention to the arrangement of the component parts at the left center of the diagram because these are the parts that you will be removing.

   CAUTION: Before removing the injection pump hydraulic distributor head, be sure to exercise care when performing this job. If possible, do not yank it out and disturb the injection pump plunger (50/2). If the head can be removed without pulling the pumping plunger out, the process will be easier.

6. Carefully pull the distributor head (50/1) from the injection pump body with its O-ring (51), two springs (106), guide pin (49), shim (48), and spring seat (47).
7. Should the plunger (50/2) come out with the removal of the hydraulic head, remove it along with its control sleeve (50/3), plunger spring (46), spring seat (45), shim (43), and washer (44).

Installation. Before reassembly, refer to Figure 12–32 and also to Figure 12–38, which illustrates the actual stack-up of the components at the pump plunger.

During assembly, lightly dip all mating injection pump components in test calibrating fluid or clean filtered diesel fuel, which will act as a lubricant.

8. Place the shim (43), washer (44) with the oil groove, and spring seat (45) to the plunger in that order, then mount the control sleeve on the plunger.

   SPECIAL NOTE: Mount the control sleeve (50/3, Figure 12–32) and also shown in Figure 12–38 so that the small hole at the bottom faces down towards the cam disc (29 in Figure 12–32). Ensure that the shim (52 in Figure 12–32) between the plunger and cam disc is inserted into the bottom of the plunger using light oil (not grease).

9. Install the plunger and shim (52) into the injection pump housing by carefully inserting the ball pin of lever assembly (95) into the control sleeve hole and then insert the groove at the base of the plunger into the knock pin of the cam disc (29).

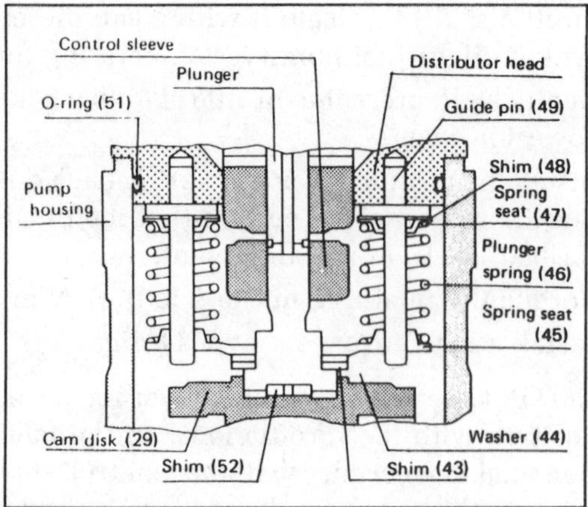

Figure 12–38. VE Injection pump, plunger shim location. [Courtesy of General Motors Corporation]

10. Refer to Figure 12–38 and mount the plunger spring (46) onto its spring seat (45). Place the large O-ring into the distributor head and assemble the guide pins (49), shims (48), and spring seat (47) into the distributor head.
11. Refer to Figure 12–32 and apply light grease to springs 106, and insert them into the hydraulic head.
12. Gently install the assembled hydraulic distributor head onto the injection pump housing while ensuring that the two springs (106) face the governor lever assembly.
13. Take care also that the end of the guide pin (49) is seated in the guide hole of the spring seat (45) and that the ball pin of the governor lever (95) is in the hole of the control sleeve and that shim (52) is inserted into the recess at the bottom of the flange of the plunger.

    SPECIAL NOTE: Make certain that the knock pin on the cam disc is in line with the drive shaft woodruff key and also that the plunger slides freely inside the distributor barrel in the distributor head (50/1) as well as the control sleeve.

14. Make certain that the large O-ring is in position.
15. To facilitate insertion of the hydraulic head into position on the injection pump body, use Kent-Moore deep socket P/N J-33309 over the end of the high pressure plug to push the head into position as shown earlier in Figure 12–33.
16. Install the four distributor head screws and torque them to 9 lb.ft. (13 N.m).

    CAUTION: After torquing the four retaining screws, operate the governor lever manually to see if there is any sign of resistance or a tight spot. By moving this lever, you are causing the control sleeve inside the injection pump to slide back and forward over the pumping plunger. Since this sleeve is a precision fit over the plunger, any misalignment of parts that occurred during assembly will reflect itself now by causing a tight action at the governor control lever. A smooth lever action indicates that the parts have been correctly assembled.

17. Finally, with the high pressure plug removed, look through the plug hole while rotating the pump drive shaft manually and check by the rotation of the plunger that the knock pin is correctly seated in the pumping plunger flange groove.

Repair Job No. 6. If it is necessary to replace the injection pump drive shaft seal, proceed as follows:

Procedure

1. Remove the woodruff key (205 in Figure 12–32, exploded view).
2. To remove the leaking seal (1/6 in Figure 12–32) from the shaft, use a seal remover tool or alternately a thin blade screwdriver or an awl. If the seal is extremely tight, in extreme cases drill two small holes into the seal casing and insert two self-tapping screws that can be used for pulling against the seal.
3. Clean up the seal bore and shaft area, then install a seal protector such as Kent-Moore P/N J-33308 onto the drive shaft.
4. To prevent the seal from being forced over a dry surface, lightly lubricate the seal protector tool with either test/calibrating fluid, vaseline, or diesel fuel.
5. Install a new drive shaft seal over the seal protector tool and using a piece of small round thick wall pipe or a deep socket, drive the seal into position until it is flush with the injection pump housing.

Repair Job No. 7. Refer to the exploded view (Figure 12–32) and locate the advance mechanism (items 31 through 38) at the left center of the diagram.

To replace the seal ring (30) proceed as follows:

Procedure

1. Remove the fuel overflow (return) pipe, eye bolt, fuel inlet pipe, and copper gaskets to allow access to the advance mechanism.
2. Before removing the advance cover (37) after taking out retaining screws (38), some tension will be exerted on the cover as you release the screws due to the internal spring (34).
3. Note that a shim(s) (item 33 and 35) should exist on either side of the advance spring 34.
4. Replace the O-ring seal with a new one and reinstall the components.

Repair Job No. 8. The fuel pressure regulating valve (item 135 in Figure 12–32) contains two O-rings that may require replacement. These can be replaced as follows:

Procedure

1. Unscrew the regulator valve and remove both O-rings.
2. Install two new O-rings onto the regulator valve and lubricate them with test/calibrating fluid or clean diesel fuel.
3. Install the fuel pressure regulating valve back into the injection pump housing and torque it to 6 lb.ft. (8 N·m).

Repair Job No. 9. The fuel cutoff solenoid is an electrically energized unit that allows diesel fuel to flow, and when it is de-energized, the fuel supply is cutoff thereby stopping the engine.

To replace the solenoid O-ring, refer to Figure 12–32, items 240 and 240/8 as well as to Figure 12–39. With a 24-mm wrench, loosen and remove the fuel cutoff solenoid. Be careful not to lose the small spring and solenoid armature as you remove it.

Remove the old O-ring and install a new one into position, then reinsert the solenoid and torque it to 31 lb.ft. (42 N·m).

This completes the possible seal ring replacements to the VE model fuel injection pump.

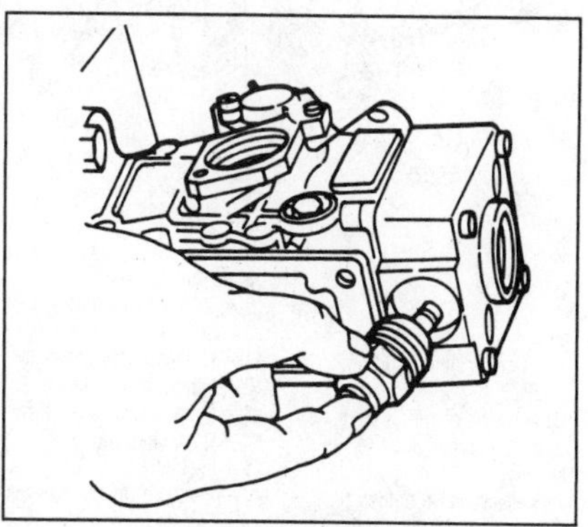

Figure 12–39. VE Pump, fuel cutoff solenoid removal/installation. (Courtesy of General Motors Corporation)

VA Model Pump

An earlier model of the Robert Bosch distributor pump, which is found on some International Harvester Company farm and light industrial equipment, Peugeot cars, Deutz (KHD) equipment, and Volvo-Penta marine applications, is the EP/VA . . . /H, commonly called the VA series distributor injection pump (Figure 12–40). It is very similar in operation to the VE model discussed herein; the major difference is that the main plunger is shaped (stepped) in such a fashion that it supplies fuel under moderate pressure for the governing circuit.

The fuel quantity to the injectors is governed by the *control piston,* which reciprocates (back and forth) in a certain phase relationship with the main plunger, which itself rotates as well as reciprocating.

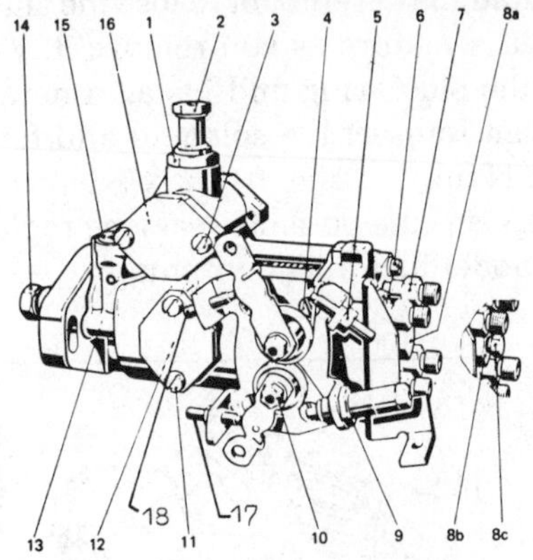

Index:
1—Overflow valve
2—Inlet union screw
3—Screw on cover plate for automatic timing device
4—Bolt for throttle and spill piston cover plate
5—Maximum speed stop screw
6—Slotted head screw
7—Injection nozzle delivery valve holder
8a/b—Screw plug with 8, 9, 10, 11, or 12 mm piston diameter
8c—Hex screw
9—Stop lever adjustment
10—Spill piston hex nut
11—Timing pointer cover screw
12—Timing pointer cover
13—Countersunk flat bolt for vane-type pump
14—Drive shaft nut
15—Fuel inlet
16—Pressure regulating valve
17—Full-load stop screw
18—Idling speed screw

Figure 12–40. VA Series distributor injection pump. (Courtesy of Robert Bosch Corporation)

The end of injection will occur earlier or later with respect to the travel of the main plunger, as engine load is changed with a given speed setting position (Figure 12–41).

Fuel delivery between full load and low idle is therefore achieved by this type of governing arrangement. The governor and automatic timing device are integral with the injection pump housing.

Figures 12–42 and 12–43 show the throttle and spill piston arrangements that are commonly found on the model VA pump.

With the injection pump installed on the test stand, the test sequence follows the same basic pattern as that for the VE model pump. The *notch* shown on the spill piston (5 in Figure 12–42) must point away from the throttle shaft in its initial position. With the pump running on the test stand at approximately idle rpm, refer to the test sheet, and adjust the throttle position until the correct quantity of fuel is delivered.

NOTE: If the throttle is rotated one half-turn from this position, no delivery will occur.

Major overhaul instruction for this injection pump can be found in Robert Bosch booklet 46, VDT–WJP 161/4 B. Complete test instructions can be found in Robert Bosch booklet VDT–WPP 161/4 B.

Diesel Kiki Model VM Distributor Pump

If you are familiar with the Robert Bosch VA and VE type distributor pumps, you will readily understand the operation of the VM (Figure 12–44). The injection pump is driven from the engine on 4-cycle engines at one-half engine speed. Also driven from the engine is the diaphragm-type fuel lift pump, which draws fuel from the tank, which may first pass through a primary filter or strainer. If water in the fuel is a problem, a fuel/water separator can be used. Fuel from the diaphragm feed pump is delivered to a secondary filter and then on into the vane-type transfer pump, which is driven from the injection pump drive shaft via a key and keyway.

Vane pump pressure can approach 100 psi (approximately 7 bars); it is controlled by the action of the main pressure regulator valve, which is adjustable. Fuel oil under pressure is directed through an external steel line to the fuel metering valve (14 in Figure 12–45) and to the piston of the automatic advance timing mechanism.

The cam disc, which is the same setup as the Bosch VE pump described earlier, has a number of cam lobe projections and rollers dependent on the number of

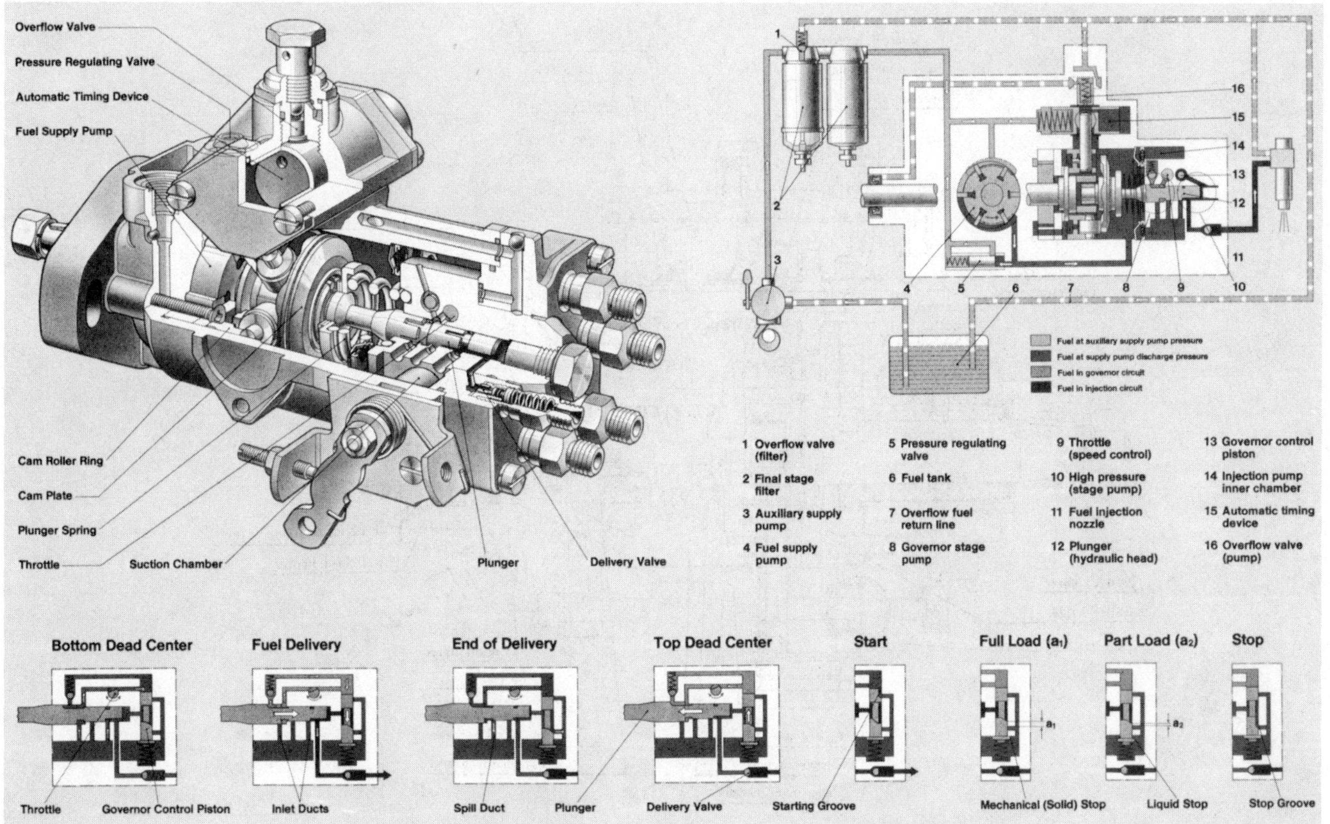

Figure 12-41. Bosch distributor-type injection pump, model EP/VA operational diagram. (Courtesy of Robert Bosch Corporation)

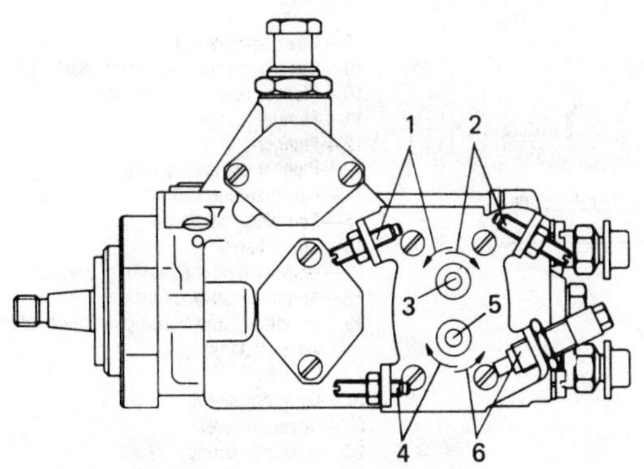

Figure 12-42. VA Pump, throttle, and spill piston arrangements. (Courtesy of Robert Bosch Corporation)

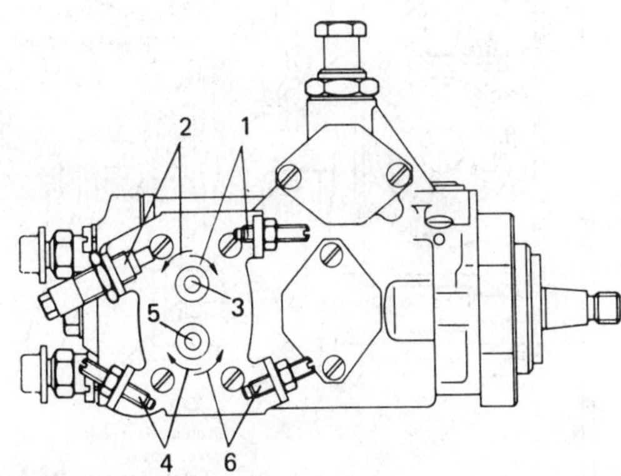

Figure 12-43. Throttle and spill piston arrangements. (Courtesy of Robert Bosch Corporation)

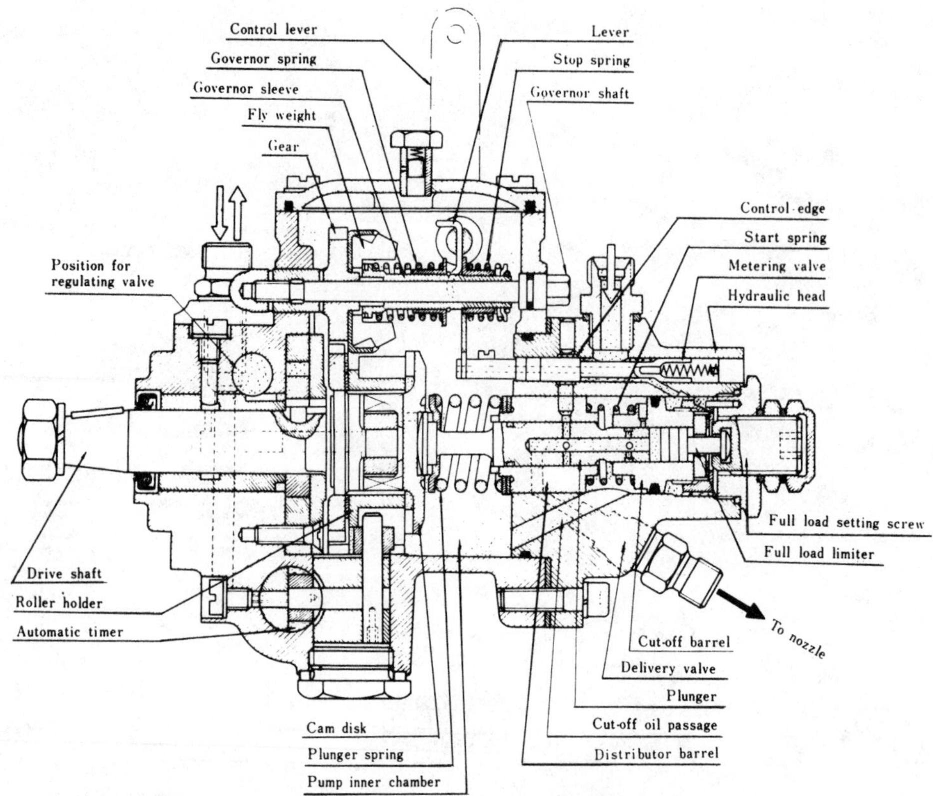

Figure 12-44. VM Injection pump, general construction. (Courtesy of Diesel Kiki USA Company Ltd., Irving, Texas)

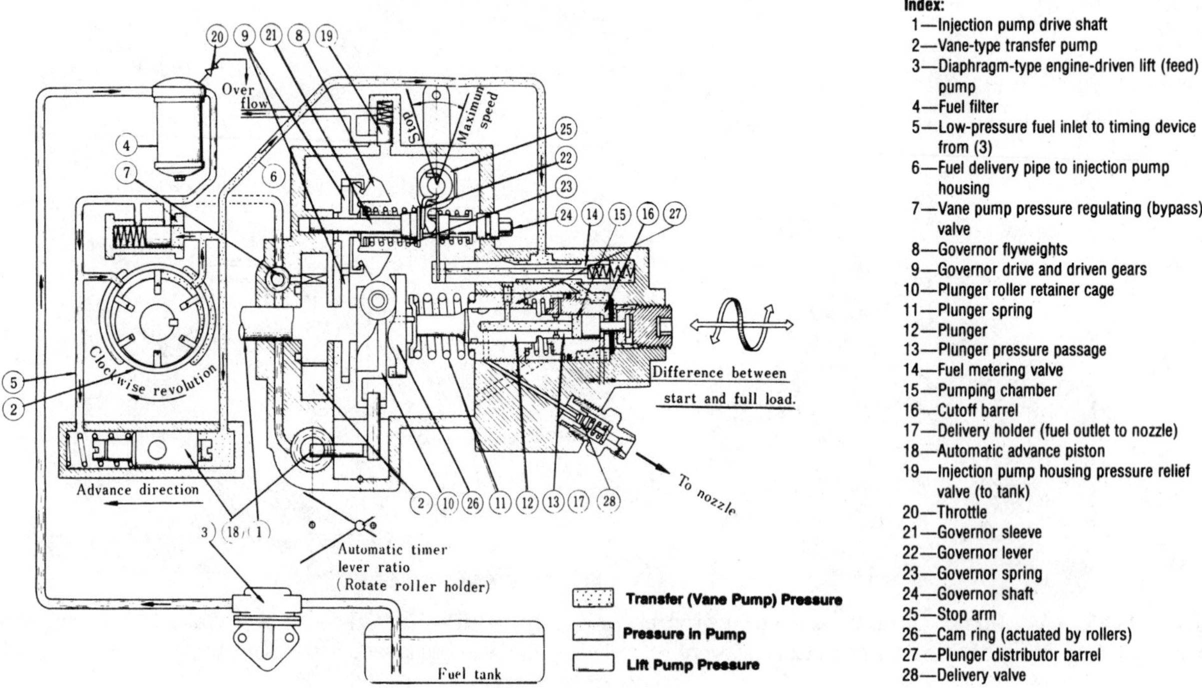

Figure 12-45. VM Pump, basic fuel system flow. (Courtesy of Diesel Kiki USA Company Ltd., Irving, Texas)

engine cylinders. As with the VE pump, the rollers do not lift, but stay stationary. The cam ring, which is driven directly by the injection pump drive shaft as it rotates, will cause the plunger attached to it to lift as each cam lobe projection on the cam ring comes into contact with the individual rollers. Therefore, the plunger operates exactly as in the VE model pump, that is with reciprocating and rotary motion (see Figure 12–20, 12–21 and 12–22).

Unlike the VE pump, which uses a metering sleeve through which the rotating plunger travels (strokes back and forth), the VM pump uses a separate metering valve (14) whose position is controlled directly by throttle and governor action, as shown in Figure 12–45.

Fuel from the vane transfer pump is directed into and around the metering valve and its bore. This fuel can flow from the metering valve bore into the rotating plunger inlet port when the plunger and barrel inlet ports are in alignment.

The charging or filling sequence is as follows: as the plunger continues to rotate, inlet port closure occurs and, through pump timing, the plunger lifts as the cam ring lobe runs over the individual roller for that cylinder. Plunger lift opens the outlet port (distributor slot), allowing fuel at increasing pressure to be distributed to one cylinder at a time in engine firing-order sequence through the corresponding delivery valve in the hydraulic head. Remember that the plunger may have as many holes or ports drilled in it as there are cylinders, but it only has one outlet port; therefore, delivery can occur to only one cylinder at a time. Injection will cease the instant the cutoff groove of the barrel aligns with the plunger's circumferential groove, allowing the spilled fuel to enter the injection pump housing.

The VM model injection pump is not at all dissimilar in its operation to other Robert Bosch distributor-type pumps. With the ever-growing popularity of diesel engines, the import and export of a variety of makes of vehicles and equipment, the Diesel Kiki (Robert Bosch licensee) VM pump will be found more and more on exported Japanese equipment.

Figure 12–45 shows the basic flow of fuel within the injection pump and is followed by a more detailed explanation of its operation.

Throttle position varies the opening of the inlet port between the metering valve and barrel and hence the volume of fuel that will be directed to the plunger fill ports. The position of the cutoff barrel spill port determines the maximum effective stroke of the plunger and, thus, the maximum amount of fuel that can be delivered.

Reference to Figure 12–46 shows that the larger diameter of the pumping plunger contains as many inlet ports as there are engine cylinders, while the smaller forward portion of the plunger contains the *cutoff groove,* which once in alignment with the barrel *cutoff port* will end injection.

Figure 12–47 shows the sequence of events once the fill port has rotated to the point that fuel supply to it has just stopped. As this is happening, simultaneously the rotating plunger cam plate or disc lobe is contacting the roller, which will start to pressurize the trapped fuel within the plunger. As the rotating plunger outlet port (one only) aligns with a delivery

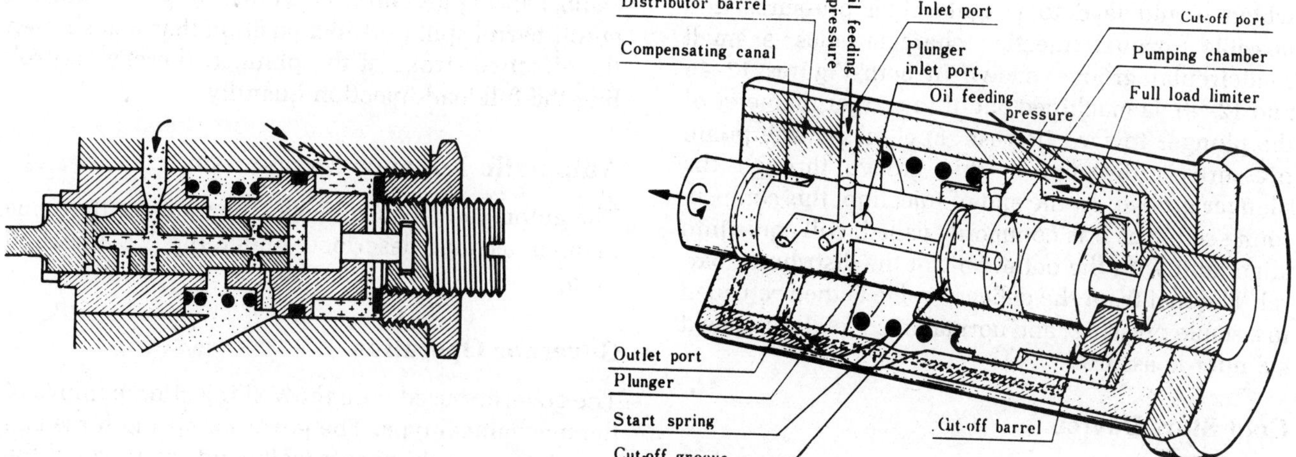

Figure 12–46. VM Pump, port filling or changing cycle. (Courtesy of Diesel Kiki USA Company Ltd., Irving, Texas)

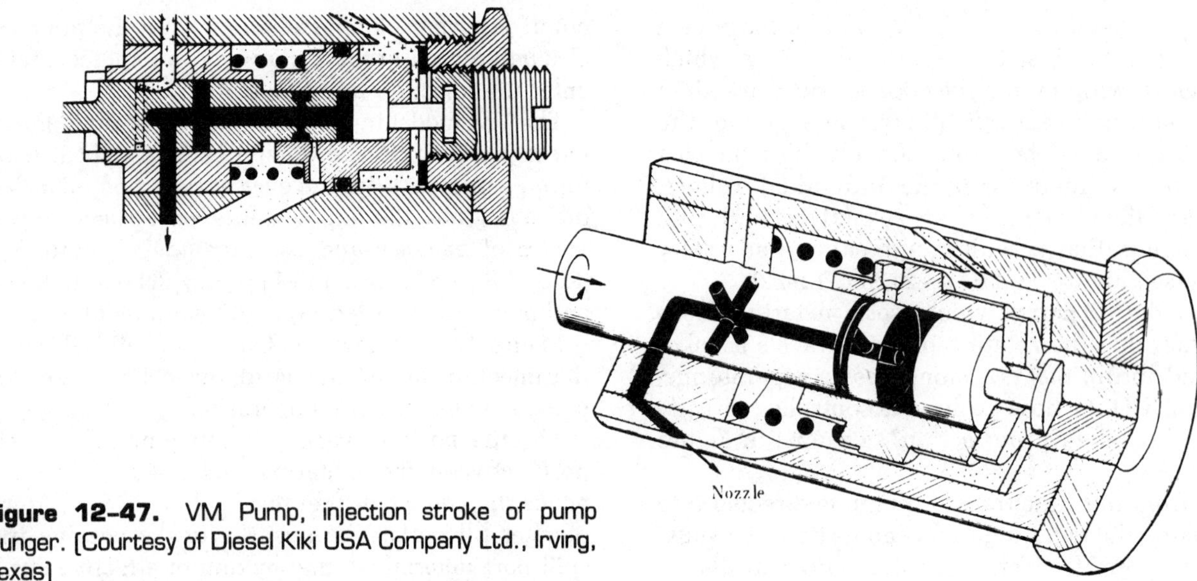

Figure 12–47. VM Pump, injection stroke of pump plunger. (Courtesy of Diesel Kiki USA Company Ltd., Irving, Texas)

port in the distributor barrel, fuel at high pressure will unseat the delivery valve within its holder, sending fuel to the injection nozzle, which overcomes the valve spring seating force, thus opening the nozzle and spraying fuel into the combustion chamber.

Fuel will continue to be injected as long as there is adequate pressure to the nozzle. However, as the plunger strokes through the barrel, the cutoff groove around the outer circumference of the smaller plunger diameter will come into alignment with the spill port or cutoff port in the barrel. The instant this occurs, fuel injection will cease because the fuel is vented through the excess fuel passage drilled diagonally within the hydraulic head. This is shown in Figure 12–48.

To prevent the possibility of pressure surges, which would lead to irregular fuel pressures and possible unequal injection characteristics, a small semicircular groove, shown in both Figures 12–45 and 12–49, is machined out of the larger diameter of the plunger toward its base. Fuel oil at vane pump pressure is directed to this groove through the plunger groove. At the end of injection, this *compensating canal,* as it is commonly called, will come into alignment with the outlet port of the distributor barrel. The fuel from the delivery valve is then returned to a stable pressure, and normal injection for the next cylinder is assured.

Cold Start Device

Unlike the small cold start cam on some VE model pumps, which is manually operated to advance the timing piston and so rotate the roller ring, the VM pump employs a starting spring that is sandwiched between the cutoff and distributor barrel. This spring can be adjusted by the use of shims, which has the net effect of forcing the cutoff barrel to the right when the engine is stopped. This retards the finish of injection by allowing a longer effective stroke of the plunger, which will therefore increase the quantity of fuel delivered during starting to the engine cylinders.

Figure 12–50 shows the action of the cold start device. When the operator or driver cranks the engine over on the starter, fuel pressure from the vane-type transfer pump is increasing, which is not only being sent to the metering valve, but also to the area surrounding the cutoff barrel. When the engine starts, this fuel pressure will force the cutoff barrel back against spring pressure. This effectively positions the cutoff barrel spill port in a position that will shorten the effective stroke of the plunger, thereby controlling the full-load injection quantity.

Automatic Advance Mechanism

The automatic advance device operates in the same manner as that described under the VE injection pump.

Governor Operation

The governor used with the VM injection pump is of the mechanical type. The governor spring force can be adjusted by shims internally and the travel of the throttle shaft lever through adjustment of the external screws. The governor operation is as follows:

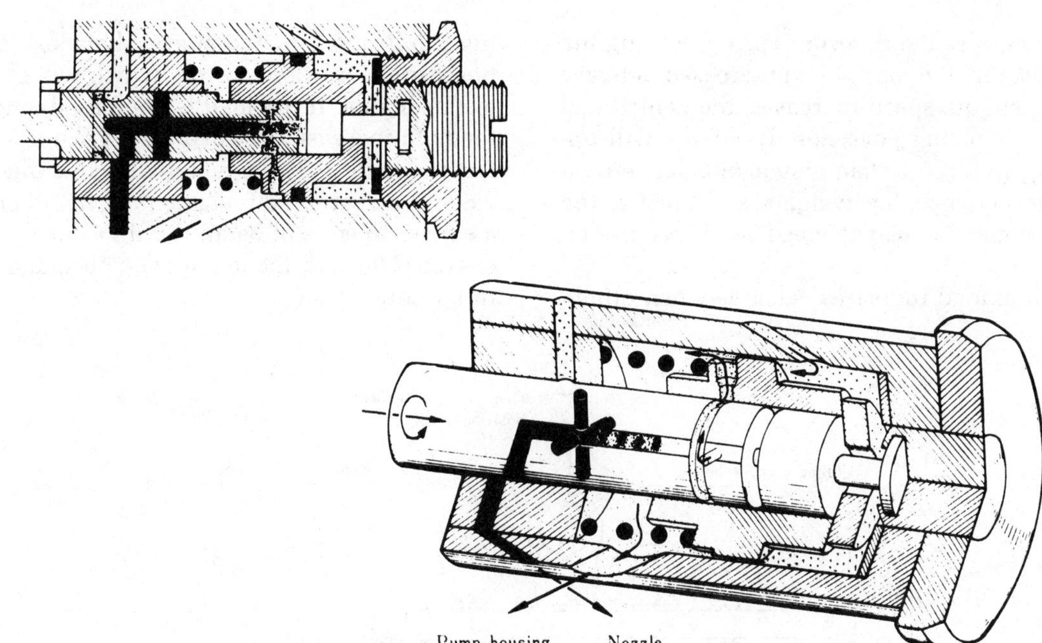

Figure 12-48. VM Pump, conclusion of injection. (Courtesy of Diesel Kiki USA Company Ltd., Irving, Texas)

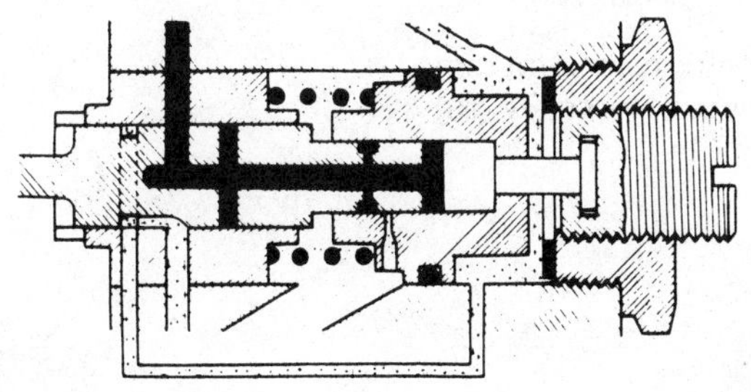

Figure 12-49. VM Pump compensating groove or canal. (Courtesy of Diesel Kiki USA Company Ltd., Irving, Texas)

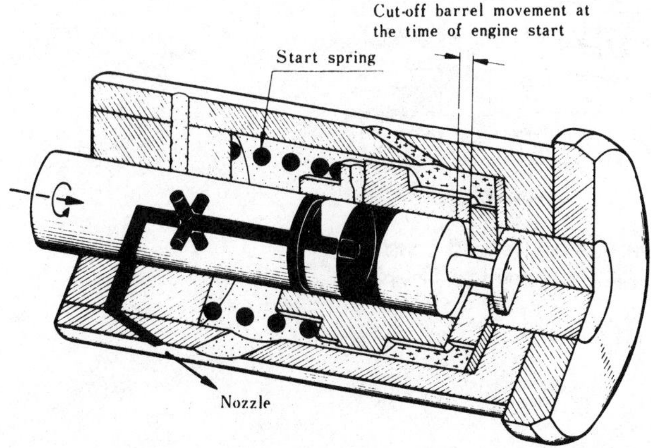

Figure 12-50. VM Pump cold start device. (Courtesy of Diesel Kiki USA Company Ltd., Irving, Texas)

Starting. Before the engine is started, the weights are in a stationary position, and the starting spring has placed the cutoff barrel into the retarded position as described earlier. The position of the throttle shaft will determine the actual position of the metering valve within its bore since this will manually be placed to a predetermined position by the adjustment of the idling screw stop bolt shown in Figure 12–51.

Idling. With the throttle control lever up against the idle stop bolt screw, the metering valve will be as shown in Figure 12–51. When the engine starts, fuel pressure forces the cutoff barrel back and the centrifugal force of the rotating governor flyweights will balance out the force of the established governor spring. If there is a change in engine speed, the opening and closing of the plunger inlet port will vary in proportion to the speed change, thereby controlling the engine rpm by throttle position alone.

High Speed Control. Normal speed regulation on this style of distributor pump can be controlled by throttle opening and therefore metering valve and inlet port opening time. However, at higher speeds, this function alone is not responsive enough to ensure accurate speed control under varying load conditions.

As the throttle lever position is changed by the operator or driver accelerating the engine, the throttle lever will compress the governor spring, forcing the fuel metering valve to move to an increased delivery position. As engine speed increases, the centrifugal force of the revolving governor flyweights will oppose the spring force. When a state-of-balance condition exists between the weights and spring, the engine will run at a constant speed, as shown in Figure 12–52.

If the engine load increases, such as when climbing a hill in a car, the state-of-balance condition is upset in favor of the spring, which will increase fuel to the engine. If the load is decreased, engine speed tends to increase; however, the greater force of the faster turning flyweights will move the metering valve to a decreased fuel position. In either situation, a corrected state of balance will occur as long as the engine is not overloaded or overspeed does not occur in too low a gear.

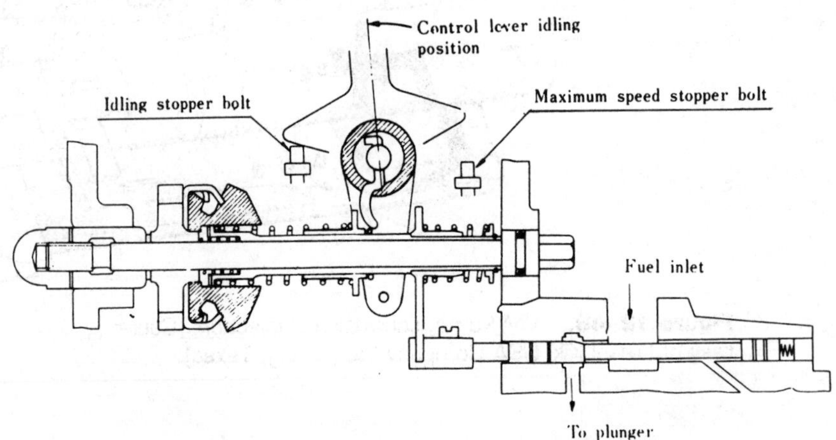

Figure 12–51. VM Pump governor idling control system. [Courtesy of Diesel Kiki USA Company Ltd., Irving, Texas]

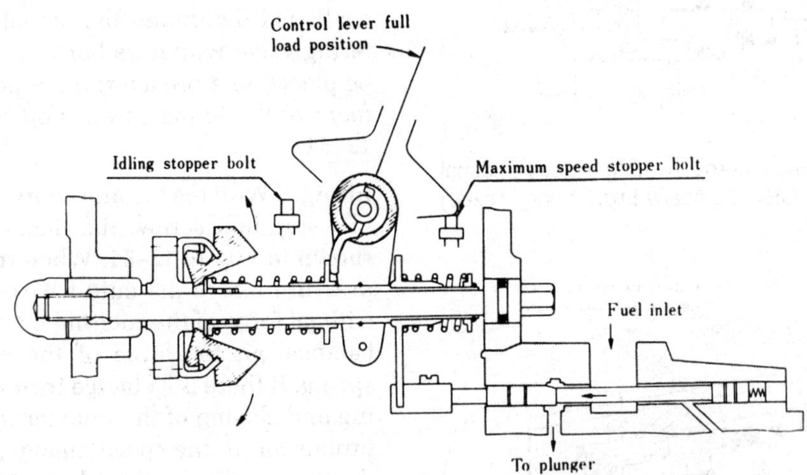

Figure 12–52. VM Pump governor high speed control circuit. [Courtesy of Diesel Kiki USA Company Ltd., Irving, Texas]

13

Mercedes-Benz Diesel Fuel Systems

Mercedes-Benz has been involved in the manufacture of diesel passenger cars for many years. In 1936, Diamler-Benz introduced its first diesel passenger car model, a 260D 4-cylinder engine of 2.6L (158 cu.in.). Mass production of Mercedes-Benz diesel cars was actually started in 1948 with the introduction of the 170D. This was followed up in 1958 with the OM 621 overhead camshaft 190D. Current 4-cylinder engines are based upon the OM 616 of 2.4L displacement. Mercedes-Benz introduced their 5-cylinder OM 617 A model (3005 CC/183 cu.in.) diesel engine in the 300D series sedans in 1974 followed by their turbocharged version in 1978.

An historical survey of the development of Mercedes-Benz passenger cars is illustrated in Figure 13–1.

TYP	ENGINE	1.PRODUCTION YEAR	DISPLACEMENT cm^3	STROKE mm	BORE mm	MAX.OUTPUT AT SPEED (DIN) kW	MAX.OUTPUT AT SPEED (DIN) 1/min	TORQUE AT SPEED Nm	TORQUE AT SPEED 1/min	NUMBER OF CYLINDERS	DRY ENGINE WEIGHT KG
 260D	OM138	1936	2545	100	90	33	3300			4	
 170D	OM636	1949	1697	100	73,5	28	3200	96	2000	4	
 180D	OM636	1953	1767	100	75	31	3500	101	2000	4	
190D	OM621	1958	1897	83,6	85	37	4000	108	2200	4	
 200D	OM621	1965	1988	83,6	87	40	4200	113	2400	4	
 200D	OM615	1967	1988	83,6	87	40	4200	113	2400	4	
220D	OM615	1967	2197	92,4	87	44	4200	126	2400	4	
240D	OM616	1973	2404	92,4	91	48	4200	137	2400	4	
300D	OM617	1974	3005	92,4	91	59	4000	172	2400	5	
 200D	OM615	1976									195
220D	OM615	1976									197
240D	OM616	1976			AS ABOVE						197
300D	OM617	1976									229
 300CD	OM617	1977									229
 300SD	OM 617 A	1978	2998	92,4	90,9	85	4200	235	2400	5 TURBO-CHARGED	244

Figure 13–1. Historical development of Mercedes-Benz diesel passenger cars. [Reprinted with permission © 1984, Society of Automotive Engineers, Inc.]

Mercedes-Benz also produces a large range of truck and industrial diesel engines that are used in many applications worldwide.

In 1977, 16 class and 3 world records were won using a prototype of the 5-cylinder turbocharged engine, which produced 190 horsepower (140 kW): for example, 10,000 miles at an average speed of 156 mph (251.798 km/h) installed into an advanced version of the experimental C 111 vehicle. Compared to the 5-cylinder naturally aspirated version, the turbocharged engine shows a 43% increase in performance with only 7% added weight. The turbocharged 5-cylinder and naturally aspirated versions both have a compression ratio of 21.5:1.

An example of the 5-cylinder diesel engine and its combustion chamber design can be seen in Figure 3–23, dealing with combustion systems.

Introduction

Engine Firing Order

The firing order for the 4-cylinder engines is 1-3-4-2, while the 5-cylinder engines is 1-2-4-5-3.

Compression Testing

Compression testing on the 4- or 5-cylinder engines can be done simply by removing a glow plug and installing a threaded gauge into an adapter screwed into the glow plug hole. Use a gauge calibrated to 500 psi because average pressure readings will be between 284 and 327 psi. Cylinder pressures should not vary more than 10% from the highest to the lowest reading.

Valve Adjustment

Valve adjustment is shown in Figure 13–2. Valve clearances should be as follows:

Year	*Model*	*Exhaust clearance*	*Inlet clearance*
78–84	240D	0.016″ (0.406 mm) Cold	0.004″ (0.1 mm) Cold
78–84	300D	0.012″ (0.3 mm) Cold	0.004″ (0.1 mm) Cold
78–81	300CD	Same	Same
79–84	300TD	Same	Same
78–84	300SD	0.014″ (0.355 mm) Cold	0.004″ (0.1 mm) Cold
81–84	300TD Turbo	Same	Same
82–84	300D Turbo	Same	Same
82–84	300CD	Same	Same

Procedure

1. Valve clearances should be adjusted when the engine is cold.
2. Refer to Figure 13–2 for the location of the adjustment screw.
3. Before setting the valve clearance, check the cylinder head and rocker arm bracket hold-down bolts for the correct torque limit.
4. Manually bar the engine over until the base circle of the cam is in the position shown in Figure 13–2. Check the clearance with a feeler gauge and compare it to the specifications given above.
5. To adjust the clearance, loosen off the hex nut (8) and turn the nut (7) to obtain the correct clearance. Retighten nut (8) after any adjustment.

Idle/Maximum Speed Settings

Idle and maximum speeds should always be adjusted with the engine at normal operating temperatures.

Also ensure that the instrument panel mounted idle speed knob is turned back all the way counterclockwise before setting the idle rpm.

All 4-cylinder engines should idle between 750–800 rpm, while all non-turbocharged 5-cylinder engines should idle at 700–800 rpm and the 5-cylinder turbo should idle at 650–850 rpm.

The idle adjustment is located at the injection pump by the throttle/speed control lever as is the maximum speed adjusting screw.

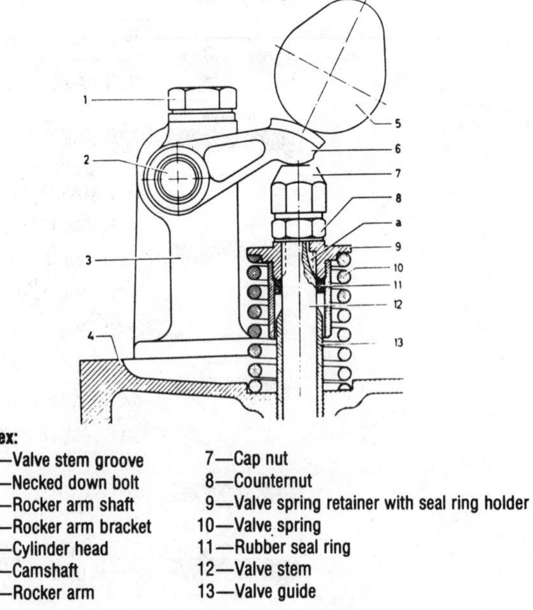

Figure 13–2. Mercedes-Benz engine valve adjustment. (Courtesy of Mercedes-Benz, Aktiengesellschaft)

Basic Injection Pump Design

The basic operation of the model MW in-line fuel injection pump (Figure 13–3(a) and (b)) used with the 300D/SD engine is discussed within this chapter. Additional information can be found under Chapter 10 dealing with History and Types of Fuel Injection Systems as well as in the Robert Bosch diesel fuel pump Chapter 12. Earlier models of Mercedes-Benz engines (OM 621) used the injection pump shown in Figure 12–1(a) and similar to that used on the Datsun/Nissan SD25 and Toyota Landcruiser vehicles shown in Chapters 18 and 19.

One feature that is used with the turbocharged 300SD diesel engine to prevent engine overboost protection is shown in Figure 13–3(a) top right, and explained in Figure 12–14(a) and (b). The turbocharger used with the 300SD engine is of conventional design and is manufactured by Garrett (see turbocharger chapter). The T/C uses the conventional wastegate system to bypass exhaust gases when the maximum boost pressure has been attained. A pressure sensing switch is employed as shown in Figure 13–4, which senses when maximum boost pressure has been obtained and energizes a three-way valve located in the connecting line to the aneroid of the fuel injection pump ALDA unit—which is the absolute pressure sensing system.

Maximum T/C boost pressure occurs after about 3 seconds when the vehicle is accelerated from a standing start under full load and reaches about 1.65 bar (24 psi).

The three-way valve vents the pressure line to atmosphere from the fuel injection pump, which effectively reduces the volume of fuel that can be delivered and thereby protects the engine from turbocharger overboost and excess fuel delivery.

Figure 13–3(a). Bosch MW in-line-type fuel injection pump with absolute pressure sensing system (ALDA).

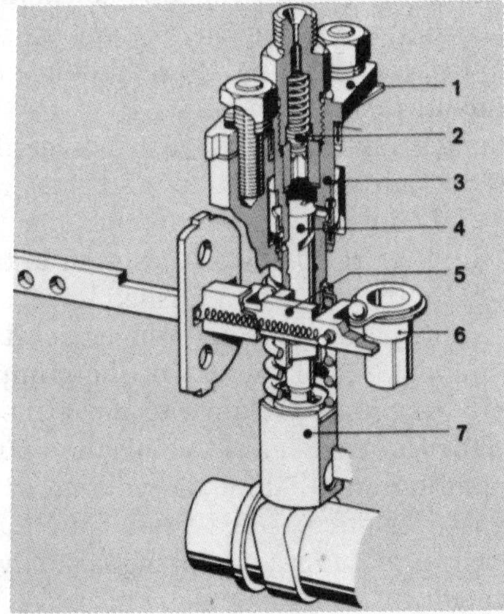

Index:
1 Fastening flange of plunger and barrel assembly, 2 Delivery valve, 3 Pump barrel, 4 Pump plunger, 5 Control rod, 6 Control sleeve, 7 Roller tappet

Figure 13–3(b). Partial section through a type MW injection pump. (Courtesy of Robert Bosch Corporation)

MW Model Fuel Injection Pump

The model MW pump is the latest addition to the broad range of Robert Bosch in-line style injection pumps. It is presently being used extensively on International Harvester equipment, Mercedes-Benz cars (including the 5-cylinder 300D and turbocharged version), Volvo trucks, and Volvo-Penta marine applications.

The basic operation of the pump follows the same

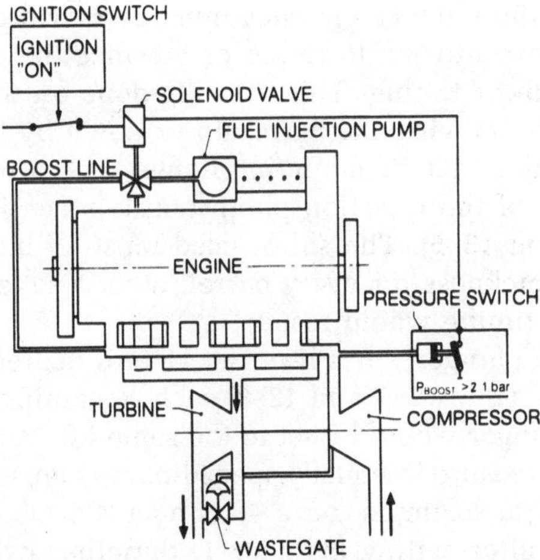

Figure 13–4. Component location of the turbocharger/fuel injection pump overboost/overfuel protection system. (Reprinted with permission © 1984, Society of Automotive Engineers, Inc.)

sequence of events as other in-line Bosch pumps (see Figures 10–5 and 12–1(b)) in that all pumping plungers are moved up and down by a camshaft contained within the base of the injection pump housing through roller-type lifters or followers. Rotary motion is by the accepted method of rack and gear, with the rack being controlled by a combination of operator or driver throttle position and by governor action. The MW pump is of similar construction to the P model pump shown in Figure 12–5; therefore, the description of the MW pump operation also applies to the P.

With the plunger at BDC within its barrel, fuel delivered to the injection pump housing by its transfer pump is free to flow into the charging port in the conventional Bosch manner. The rotation of the pump camshaft will raise the plunger within its barrel until it closes the supply or charging port (this is commonly referred to as *lift to port closure*). Delivery to the injector nozzle can begin only after *port closure* because prior to this, the fuel will only be at transfer pump pressure, which is in itself too low to unseat the delivery valve within the top portion of the pump housing. The rotation of the injection pump camshaft will cause an increase of the trapped fuel within the barrel area, which will unseat the delivery valve, (Figure 10–14 and 10–15) sending fuel at this high pressure to the injection nozzle and on into the combustion chamber.

The timing of this fuel to the injection nozzle depends on the point of port closure or amount of lift required by the camshaft to place the plunger into a position whereby the supply port at the lower end of the barrel is closed. To change the point of port closure on the model MW injection pump, the barrel in which the plunger operates must be raised or lowered, respectively, to retard or advance the start of injection or timing. This is easily done on the MW pump by moving the barrel up or down by the use of shims under its mounting flange, which rests on the top of the injection pump housing (see Figures 12–5 and 13–5). The shims used must all be of the same thickness for every barrel; otherwise, a variation in timing would result.

Once plunger 1 has been set at zero degrees, (see Figures 12–6, 12–7 and 12–8) each remaining barrel and plunger would be set to the same lift from BDC to port closure in equally spaced pump camshaft degrees. For example, on a 4-cylinder 4-cycle engine pump, after setting cylinder 1, the other cylinders would be shimmed in firing order sequence 90 degrees apart. This adjustment is commonly called phasing, or setting internal pump timing.

With the injection pump correctly timed, the following advantages will result:

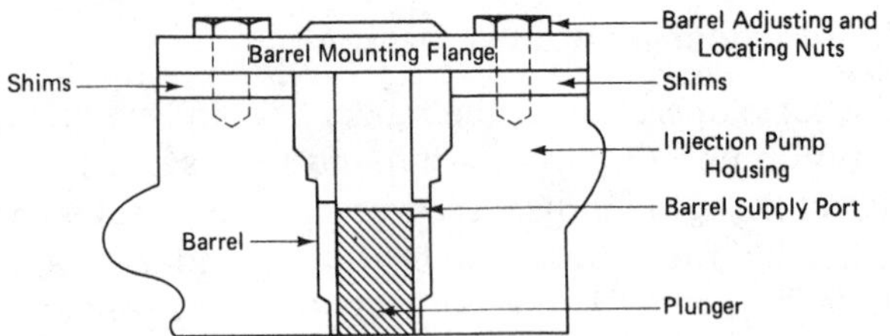

Figure 13–5. Use of shims to adjust the point of port closure on the Bosch "MW" model fuel injection pump.

1. Proper power developed
2. Good fuel economy
3. Ease of starting
4. Smoothness of operation
5. Proper exhaust smoke emissions

Remember that in any injection pump the quantity of fuel delivered is controlled by the plunger's *effective stroke*, which is the time between port closure and port opening. The start of the effective stroke is controlled by the shape of the helix cut on the plunger. The plunger rotation is controlled by injection pump rack movement coupled to the governor and throttle linkage, as shown in Figure 10–9.

When rack movement is checked by use of a dial gauge mounted on the end of the pump housing, we are in effect establishing the rotational movement of each and every plunger within its barrel, and thus the position of the plunger helix that controls the length of the effective stroke because the plunger's up and down movement remains constant with the pump camshaft and roller lift. For any specified rack position, then, each plunger will produce a particular volume of fuel delivery.

To change the volume of fuel delivered by each plunger, we can simply rotate the barrel within the housing of the injection pump, which effectively varies the supply port position in relation to that of the plunger helix. This is accomplished by loosening the retaining nuts passing through the barrel flange (Figure 12–5 and 13–5) at the top of the pump housing and gently tapping the flange to move the attached barrel to the front or rear of the injection pump housing, depending on whether the volume of fuel is to be increased or decreased. This is how the MW injection pump is calibrated for equal delivery from each plunger.

Checking Adjusting Injection Pump Timing

Both the current 4- and 5-cylinder engines are equipped with the Robert Bosch model PES/MW in-line-type fuel injection pump, which uses a mechanical centrifugal weight style governor. When a mechanical governor is used, no throttle valve is used in the intake manifold such as is found on some earlier engines that used the pneumatic-type governor assembly (Figure 12–1(a) and 12–2). The pumps are equipped with an altitude/boost compensator device, which is explained in detail in Chapter 12, Robert Bosch diesel fuel injection pump, and also in Chapter 18 (Nissan).

If the engine is non-turbocharged, then an altitude compensator regulates the position of the fuel rack to ensure that the engine will receive less fuel with an increase in altitude so the EPA exhaust emissions regulations can be met.

On the 5-cylinder turbocharged 300SD, the altitude compensator (see Figure 12–13 and 12–14) is also responsive to turbocharger intake manifold boost pressure and therefore responds not only to altitude variations but also to changes in turbo boost to meet EPA exhaust emissions regulations.

If it becomes necessary to check and adjust the engine to injection pump timing, the procedure is similar for all engines employing in-line-type injection pumps.

Contained in Chapter 18, Figures 18–35, 18–36 and 18–37, as well as in Chapter 19, Toyota Landcruiser, are directions for accurately spill timing an in-line-type of injection pump. Figures 19–11, 19–12, and 19–13 illustrate the three basic steps required when accurately setting/checking the engine/injection pump timing.

The following information deals with the general routine required to check pump timing. It is suggested that you read this over first before actually performing this check on the engine.

Mercedes-Benz Pump Timing Check Procedure

1. On model 616 engines, set the pre-glow switch located on the instrument panel to the driving position and remove the No. 1 and No. 2 fuel injection lines from the pump.
2. On later model 616 and 617 engines equipped with the Robert Bosch model MW or M/RSF injection pumps, remove the vaccum hose from the control box and the No. 1 fuel line from the injection pump.
3. Remove the No. 1 injection line delivery valve holder from the injection pump along with its spring, delivery valve, and sealing ring.
4. Reinstall the delivery valve holder back into the pump.
5. Attach a fuel spill pipe similar to that shown in Figure 18–36 and Figure 19–11 under the Toyota Landcruiser pump timing check.
6. Manually rotate the crankshaft in the normal direction of engine rotation (CW) to place the No. 1 piston on its compression stroke (both the intake and exhaust valves should be closed at this time).

   CAUTION: Rotate the engine steadily and slowly towards a point just short of 24 degrees BTDC on the harmonic balancer pulley/wheel.
7. Open the hollow vent/bleed screw on the top of the secondary fuel filter and manually operate the injection pump mounted transfer pump priming handle until fuel flows freely from the overflow pipe attached to the No. 1 pumping element (see Figure 18–36).

   SPECIAL NOTE: On models 617.910 MW, 616, 617 MW, and 616, 617 M/RSF engine injection pump combinations, be sure to push the fuel control/regulating lever on the injection pump to the full-load stop position before actually proceeding to Step 8.
8. Very gently rotate the crankshaft clockwise while watching the fuel flow from the No. 1 injection pump spill pipe. Stop turning the engine over when fuel flow from the overflow spill pipe decreases to approximately one drop per second (see Figure 18–37).
9. Check the position of the timing marks on the crankshaft harmonic balancer and the stationary pointer on the engine. If the reading is not 24 degrees BTDC, then the pump timing requires adjustment.
10. To change the injection pump timing, simply loosen off the injection pump retaining nuts on the pump flange and supporting brackets. It may also be necessary to loosen off the remaining fuel injection lines at the delivery valve holders.
11. Grasp the injection pump housing and push the pump towards the engine to advance the timing, and pull it away from the engine to retard the timing (see Figure 19–13). Move the pump while watching the fuel flow at the spill pipe on No. 1 pumping element until it reduces to one drop per second.

Automatic Timing Device (In-Line Injection Pumps)

The purpose of the automatic timing device used with in-line injection pumps is to advance the point of injection with a relative increase of the engine speed. Figure 13–6(a) shows the basic components of this unit. Also see Figures 19–18 and 19–19.

Index:
1—Injection pump drive gear
2—Segment plate
3—Contact surfaces for the centrifugal weights on the segment plate and flange
4—Centrifugal weights
5—Spring seat bolts
6—Tension spring
7—Stop bolt in the tension spring for limiting adjustment
8—Segment flange
9—Woodruff key groove

Figure 13-6(a). Typical components of the mechanical automatic timing device. [Courtesy of Robert Bosch Corporation]

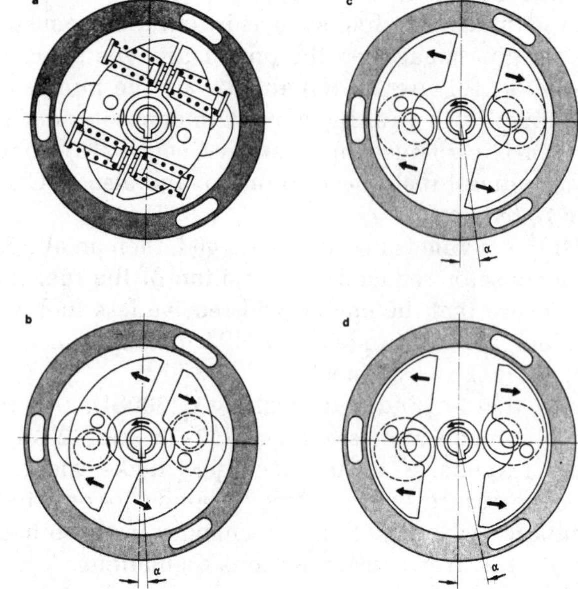

Figure 13-6(b). Eccentric automatic timing advance device. [Courtesy of Robert Bosch Corporation]

Operation. The operation of this timing device is basically that the centrifugal force created by the weights is opposed by the tension of the springs. As the engine accelerates, the motion transfer of the weights causes relative movement of the drive plate to the injection pump, drive gear and pump camshaft thereby advancing the point of initial injection to the engine.

NOTE: An eccentric timing device is also available which works on the same principle whereby the weights moving outward under centrifugal force cause two eccentric rollers to turn as shown in Figure 13–6(b).

RW Governor

The RW governor is being used by Robert Bosch on their model MW fuel injection pump, discussed in detail in this chapter. This governor is found on such engines as Mercedes-Benz cars, with the RWV (variable speed) governor being found on Volvo trucks and Volvo-Penta marine applications.

The governor operates on the same principle as other Bosch units (see Figure 12–28(c)) in that the centrifugal force created by a set of rotating flyweights is trying to reduce the position of the injection pump fuel rack, whereas the spring force within the governor is trying to increase the position of the fuel rack. As with any governor of this type, a state of balance can exist at an idle speed position or at the high end of the engine's rpm range.

NOTE: Refer to Chapter 11 for a detailed description of basic mechanical governor operation.

Therefore, this governor controls the engine speed during starting, normal idling, high idle or no-load rpm, rated rpm or full-load speed, and breakaway, which is simply the speed range between full-load and no-load rpm. At any speed above normal low idle, the operator or driver controls rack travel and therefore engine speed through the throttle linkage connected to the fuel control rack through the governor linkage.

Figure 13–7 shows the typical setup of the governor linkage and the actual component part stackup of the RW governor. Due to the fact that the RW governor contains linkage within the tensioning lever, Figure 13–7 depicts the governor linkage with the tensioning lever removed to show the parts hook up to the injection pump fuel control rack, with the lower view showing the tensioning lever in position with the various spring adjustments.

The initial force on the single leaf-type idle spring is established by the idle screw contacting up against it. In addition to the idle screw, a bumper spring adjustment is used whereby the idle regulation or stability is controlled by turning this bumper spring inward until it just touches the idle leaf spring with the engine running at its normal idle rpm.

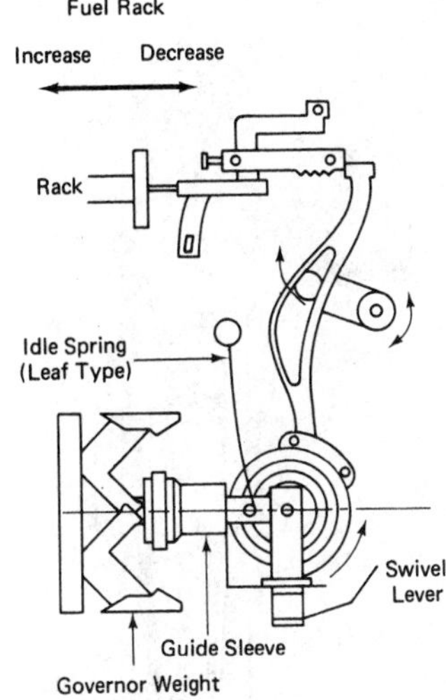

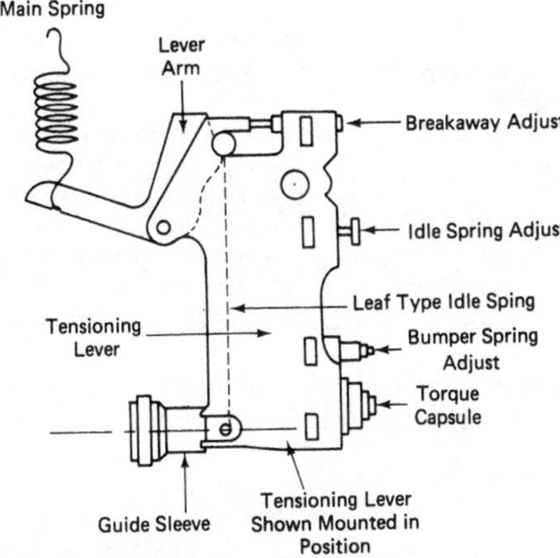

Figure 13-7. Typical RW mechanical governor linkage arrangement, Bosch in-line injection pump.

At engine start-up, the fuel rack will therefore be in the maximum fuel position; and once the engine starts, the centrifugal force of the rotating governor flyweights will force the guide sleeve to the right. This will transfer this horizontal motion through the swivel lever and linkage to actually move the fuel rack to the right, also thereby changing the position of the rack toward less fuel. When the force of the rotating weights equals that of the idle spring, a state-of-balance condition will allow the engine to run at a steady idle speed.

During any part-load situation above idle speed, the driver controls the rack travel through the use of the throttle linkage and governor. During any full-load situation, the governor controls the rack travel, as we shall see later. At a full-load condition, a state-of-balance condition will exist; however, neither the weight nor spring force is controlling the rack, but the full-load stop screw butting up against the throttle lever controls its maximum travel.

The governor also contains a torque control device to avoid a smoking condition. At full-load, the torque control allows the weights to pull the fuel rack to a decreased position, thereby reducing the full-load fuel limit.

As the rpm increases toward full-load, the weights oppose the internal torque capsule spring contained within the tensioning lever, which effectively reduces the full-load fuel delivery. The torque control capsule spring can expand, at lower engine speeds due to the reduced flyweight force and therefore push the fuel rack up to full load. With the engine under full power, the governor controls the rack for rated speed, at which time the force of the governor weights will cause the torque capsule to collapse. Turning of the collapsed torque capsule screw inward against the guide sleeve will increase the fuel delivery by increasing rack travel. This adjustment is commonly referred to as *torque backup.*

If at any time the engine were to rev past its rated rpm (called *breakaway*), the governor will pull the rack to a decreased fuel position because at rated speed the weights will overcome all spring force in the tensioning lever. Just prior to breakaway, the governor main spring continues to oppose the flyweight force working through the lever arm and tensioning lever.

Any adjustment of the breakaway screw inward causes the lever arm to stretch the governor main spring, which will therefore pull harder on the tensioning lever. This breakaway adjustment means higher rpm and a consequently higher weight force; therefore, after breakaway, the main spring determines rack cutoff.

Figure 13–8 shows the position of the various gov-

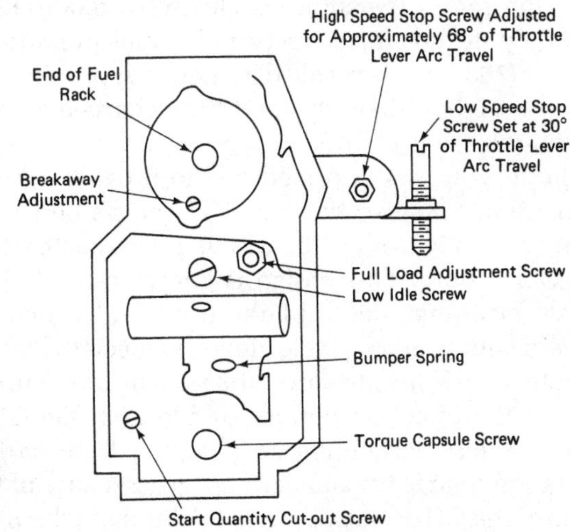

Figure 13-8. RW Mechanical governor adjustments.

ernor adjustments when viewed from end on with the cover removed.

The preceding adjustments should be made only when the injection pump and governor are mounted on a suitable fuel pump test stand; otherwise, serious problems can result if attempted while on the vehicle.

Anytime that adjustments are to be made to the governor while it is mounted on the test stand, they must be done in the following sequence:

1. Low idle
2. Loading
3. Unloading
4. Bumper spring
5. Rated speed
6. Torque control
7. Full load
8. Breakaway
9. High idle
10. Fuel cutoff

Injection Nozzles

A detailed description of the operation, maintenance, inspection, cleaning, and repair of injection nozzles can be found in Chapter 29, dealing with nozzles; therefore, refer to this chapter for service/repair information.

Nozzles used in Mercedes-Benz engines are Robert Bosch KAC-type, which are illustrated in Figures 29–4 and 13–9. Because of the higher combustion chamber pressures that occur on the 5-cylinder turbocharged engine versus its naturally aspirated version, the pintle injection nozzles were modified so that they do not inject fuel until a fuel pressure of 140 bar (2033 psi) is reached, which is approximately 200 psi higher than the non-turbocharged engine nozzle release pressure.

The nozzle was also modified to provide a longitudinal bore hole in the center of the pintle as can be seen in Figure 13–10, which is connected to a transverse bore. This design is known as a "CHIP" nozzle meaning "central hole in pintle," which ensures a much more stable flow of injected fuel especially at low needle valve lifts such as in the initial start to the injection period. In addition, the CHIP design is less susceptible to plugging from carbon formation than is the standard nozzle. Because of this feature, the CHIP nozzle tends to eliminate the pinging noise associated with a clogged-up annular slot

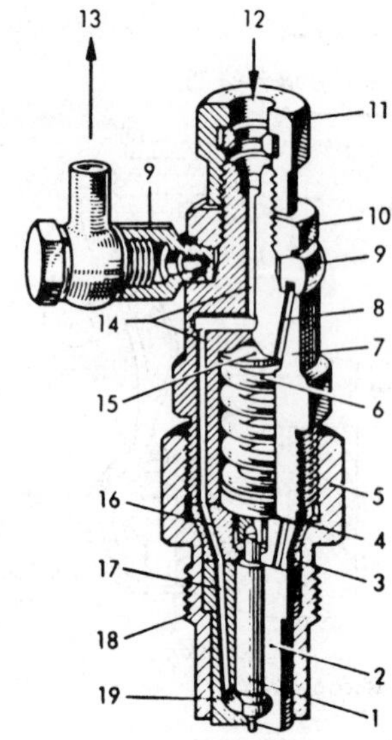

Index:
1—Nozzle needle valve
2—Nozzle head
3—Nozzle holder insert
4—Pressure bolt
5—Cap nut
6—Pressure spring
7—Nozzle holder
8—Leak-off passage
9—Leak-off adapter
10—Hex nut
11—Fuel-injection line cap nut
12—Fuel inlet
13—Leak-off line return to tank
14—High-pressure fuel passage
15—Spring adjusting washers
16—Fuel inlet holes in nozzle holder insert
17—Fuel passage in nozzle head
18—Mounting threads
19—Fuel pressure chamber in nozzle head

Figure 13–9. Bosch KAC-type injection nozzle and holder. [Courtesy of Robert Bosch Corporation]

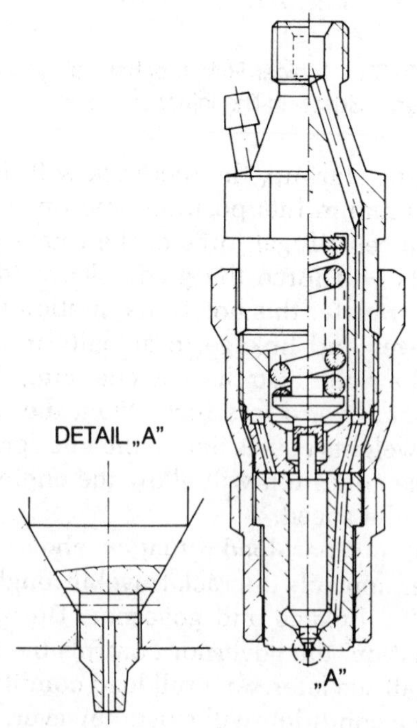

Figure 13–10. Bosch central hole in pintle [CHIP] type nozzle. [Courtesy of Society of Automotive Engineers]

in a standard nozzle especially after a cold start situation.

Maintenance and repair of the CHIP nozzle shown in Figure 13–10 follows the same basic routine as that for the standard nozzle. Refer to Chapter 29 dealing with Injection Nozzles—operation and repair.

Nozzle release pressures can be found in the injector overhaul chapter.

Nozzle Removal

To remove the nozzles, proceed as follows:

1. Remove any accessories that are in your way.
2. Refer to Figure 13–9 and loosen off the fuel injection nozzle line cap nut (item 11).

   NOTE: Always place plastic shipping-type protective caps over all open fuel lines and components to prevent dirt/dust from entering the fuel system.

3. Loosen the hex nut (10) that secures the leak-off adapter (9) to the nozzle body.
4. If the nozzle starts to turn in the cylinder head when loosening nut (10), prevent it from turning with a 24-mm wrench.
5. Using a deep 24-mm socket or box wrench, carefully unscrew the nozzle holder (injector) from the cylinder head. Remove the copper seal ring at the base of the bore once the injector assembly has been withdrawn if it is not already stuck to the bottom of the nozzle. Place a plastic cap over the nozzle bore to prevent dirt or any foreign objects from dropping into the cylinder.

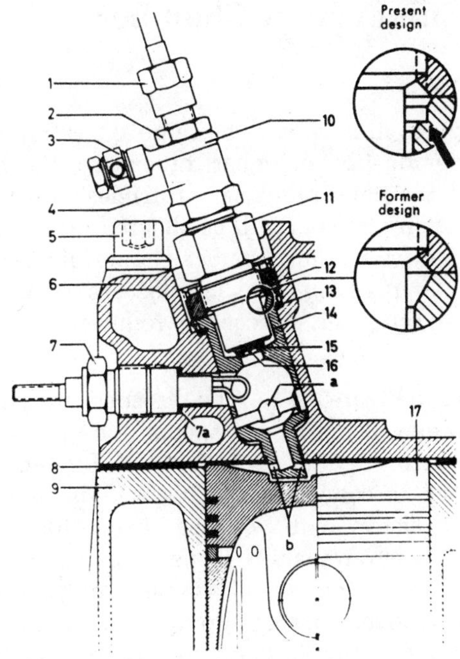

Index:

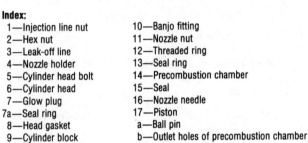

1—Injection line nut	10—Banjo fitting
2—Hex nut	11—Nozzle nut
3—Leak-off line	12—Threaded ring
4—Nozzle holder	13—Seal ring
5—Cylinder head bolt	14—Precombustion chamber
6—Cylinder head	15—Seal
7—Glow plug	16—Nozzle needle
7a—Seal ring	17—Piston
8—Head gasket	a—Ball pin
9—Cylinder block	b—Outlet holes of precombustion chamber

Figure 13–11. Mercedes-Benz diesel engine glow plug/nozzle location. [Courtesy of Diamler-Benz, Aktiengesellschaft]

Glow Plugs

All pre-combustion type diesel engines require the use of electrically operated glow plugs located within the prechamber to facilitate starting and initial warm-up. The glow plug is shown as Item 7 in Figure 13–11, with one glow plug being used per cylinder.

All 1980 and later models of engines are equipped with parallel hookup plugs that allow the others to operate even if one of the others is damaged. Pre-1980 glow plugs were of a spiral design while the 1980 and up engines are equipped with a pin-type glow plug that offers increased burn-off resistance over the spiral type.

NOTE: The old and new type glow plugs are not interchangeable. Quick pre-glow type plugs are readily identifiable by the fact that they use a brass hex nut on the electrode.

The glow plugs are operated through a relay that was connected to a coolant temperature sensor on pre-1981 models, while the 1981 and later engines use a pre-glow relay with a temperature sensor instead of the coolant temperature sensor.

The glow plug system is protected by a 50–80 amp fuse or fusible link located in the engine compartment as well as by the fact that the material in the control coil has a positive temperature coefficient that increases the resistance to current flow with a temperature increase. This condition reduces the pre-glow current and thereby prevents thermal overload. Testing of the glow plugs can be done in the same manner as that described for other engines in this book in Chapters 14, 15, 17, 18, 19, 20, 22, 24 and 26.

Glow plug removal simply involves disconnecting the wire from the glow plug then screwing it out of the cylinder head.

Pre-Combustion Chamber

Removal

Glow plugs must always be removed before pulling a pre-combustion chamber; otherwise, the heating wire of the glow plug that protrudes into the pre-combustion chamber will be sheared off.

A pre-combustion chamber should always be removed anytime that the nozzles are removed for checking purposes. Proceed as follows:

1. Refer to Figure 13–11. The threaded ring (12) must be removed by using a special tool (part no. 636 589 01 07 or P/N 636 589 02 07 for 9-mm-wide groove threaded ring). This tool is basically a keyed sleeve with two pegs extending from its base. It is circular in shape, with its size corresponding to the outside diameter of the threaded ring groove. Passing through the center of the sleeve is a spindle that is threaded at its base to screw into the inside threads of the threaded ring.
2. With the spindle screwed into the threaded ring and the keyed sleeve properly located as to the groove of the threaded ring, tighten the spindle to the sleeve by tightening the large threaded nut on the spindle up against the sleeve.
3. By turning the keyed sleeve, the threaded ring can now be removed.
4. To extract the pre-combustion chamber, obtain special tool P/N 636 589 01 33, which is similar to the tool used for pulling the threaded ring. The center spindle of the tool is screwed into the pre-combustion chamber as far as possible. The puller itself that the spindle threads through must be rotated until a cutout groove is aligned with a small groove in the cylinder head in line with the glow plug. With this done, the pre-combustion chamber can now be removed by tightening up a hex nut that is threaded to the spindle and up against the puller sleeve.

CAUTION: While pulling the pre-combustion chamber, make certain that the puller sleeve itself does not turn; otherwise, the nose of the pre-combustion chamber can be sheared off.

Installation

1. Install a new sealing ring (13 in Figure 13–11) into the pre-combustion bore of the cylinder head. *Be sure* that the proper thickness ring is used (these are available in different thicknesses) to properly establish the distance between the pre-combustion chamber (14) and the cylinder head.
2. Install the pre-combustion chamber into its bore by reversing the removal procedure.
3. Screw in the threaded ring and torque to specification.
4. Install the glow plug.
5. Install the nozzle.

Bleeding the Fuel System

Anytime that the fuel filters are changed, the injection or fuel lift pump is serviced, the injection nozzles are removed and reinstalled, or the fuel lines have been removed, it will be necessary to *bleed* the fuel system of all entrapped air. Otherwise, air in the fuel system can cause a heavy knocking sound from the engine cylinders, a reduction in engine performance, and hard starting characteristics.

1. On top of the main fuel filter is a bleeder screw (the one with the slotted screw and pin through it like a wing nut), which should be opened up between one and two turns.
2. Turn the control knob of the hand primer (fuel lift pump in Figure 13–12) counterclockwise until it is free from the body, and pump it up and down until fuel, free from any air bubbles, flows steadily out of the main filter bleeder screw; then close the screw. When bleeding the fuel system, place a suitable drain tray under the engine to retain dripping fuel.

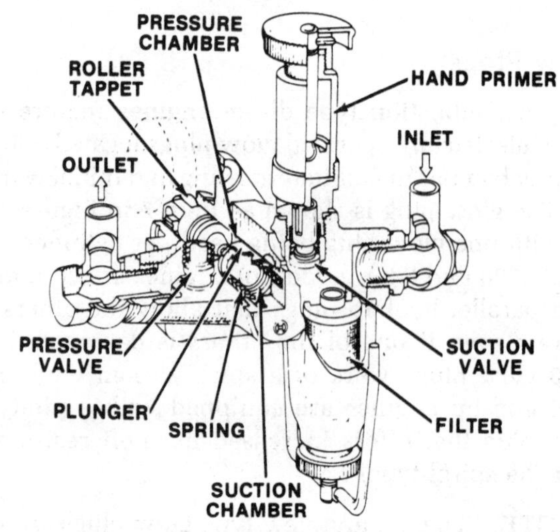

Figure 13–12. Bosch in-line-type injection pump fuel transfer pump assembly. (Courtesy of Robert Bosch Corporation)

3. The location of the bleeder screw or screws will vary between different models of Robert Bosch injection pumps; however, there are two common locations:
   a. Two bleeder screws located on the injection pump usually with slotted heads, positioned on the same side of the housing as the fuel lift pump, but located high up on the housing, with one at the front and the other at the rear (one screw each opposite the no. 1 and no. 4 pump fuel lines).
   b. On those injection pumps with just one bleeder screw, it is located on the top of the housing at the rear of the pump just beside the fuel line to the nozzle.

   In either case, loosen the bleeder screw or screws and pump the handle of the lift pump until a good flow of fuel, free from any air bubbles, appears; then tighten up the screws. Also at this time, push down the handle of the lift pump hand primer and turn it clockwise until it is tight on the barrel of the housing.

4. Bleeding of the fuel lines is usually necessary only if they have been disconnected and therefore drained of fuel. The quickest way is to loosen the fuel line nut at the nozzle, and crank the engine over until fuel appears at the loosened line; then tighten it up.
5. Once the engine fires and runs, you can quickly loosen each nozzle line one at a time to ensure that all air has in fact been removed; then tighten it again. Check all bleeder screws for signs of fuel leakage.

14

Volkswagen Rabbit/Jetta Diesel Engine

Volkswagen introduced its first 4-cylinder 4-stroke cycle diesel engine for passenger car use in the fall of 1976. The engine was of the precombustion swirl chamber design with a bore of 76.5 mm (3.012″) and a stroke of 80 mm (3.150″) for a displacement of 1471 cc (1.5L or 90 cu.in.), with a compression ratio of 23:1, which produces compression pressures of 483 psi (3330 kPa). In the fall of 1980, the engine displacement was increased to its current size of 1588 cc (1.6L or 97 cu.in.) by lengthening the piston stroke to 86.4 mm (3.40″).

The naturally aspirated diesel engine is rated at 52 SAE horsepower at 4800 rpm; the turbocharged version of this engine is rated at 70 SAE horsepower at 4500 rpm.

This engine design was one of the first high-speed diesel passenger cars to run at speeds in the 5000 rpm range. This higher operating range has removed the previous concept of a diesel engine being strictly a slow-speed, large engine. Volkswagen of America claim that their 1984 Rabbit Weltmeister diesel model returns about 65 mpg (4.3L/100 km) average, which is excellent fuel economy.

The engine is equipped with a Robert Bosch VE-type distributor injection pump and uses Robert Bosch injectors DNO SD 293. These injectors have a popping pressure of 155 bar (2249 psi) on the turbocharged engine, while the naturally aspirated engine injectors pop at 130 bar (1886 psi).

The same basic engine design used in the Rabbit/Jetta is also used in the 5-cylinder Audi 5000 and the VW 6-cylinder light truck diesel engine that is now in use by Volvo in their D24 car. All these engines use the same basic fuel system and combustion chamber design.

Turbochargers from Garrett-Airesearch and KKK (Kuhnle, Kopp and Kausch) are used on these engines. A wastegate control system is used to limit the maximum boost from the turbocharger (T/C). (See Chapter 31.)

Brake mean effective pressure in the 4-cylinder T/C engine is 155 psi and 161 psi in the Audi 5000 T/C engine.

The 4-cylinder diesel engine is available as a transverse mounted or longitudinal mounted powerplant for various models of vehicles. The engine block is very similar in design to that used on the gasoline version except for the cylinder head. Both the gasoline and diesel versions employ an overhead camshaft arrangement.

The turbocharged version of the 4-cylinder engine employs wastegate control to limit the maximum T/C boost pressure; however, if the wastegate becomes stuck, then an intake manifold mounted safety blow-off valve will open at approximately 11.6 psi (0.8 bar) to prevent engine and turbocharger damage.

Fuel Injection Pump

The fuel injection pump is a Robert Bosch VE model unit, which takes its designation from the German word for distributor—"Verteiler." Since this pump is a rotary-type distributor unit, it is aptly named. Detailed information on this injection pump will not be discussed in this chapter because this information is in the Robert Bosch diesel fuel systems, Chapter 12 (Figure 12–17).

Fuel flow through the VW system is similar to that of other vehicles using this same VE pump and is illustrated in the Robert Bosch diesel fuel systems, Chapter 12, VE pump section.

Adjustments below deal with external pump adjustments only, plus injection pump timing, removal, and replacement.

Idle and Maximum Speed Adjustment

To check/adjust the engine idle rpm, the engine must be at its normal operating temperature. If the idle rpm is checked on a cold engine, then the CSD (cold start device) will be in operation and the engine will usually be running about 200 rpm higher than it would when at operating temperature.

A detailed explanation of the operation of the CSD can be found under the Robert Bosch chapter VE pump section.

Use a photoelectric tachometer to check the engine idle speed or install adaptor P/N US 1324 or P/N VW 1324 which can be connected to the fuel injection pump, to pick up fuel pulsations and convert these into electrical signals that are interpreted by either a Robert Bosch EFAW 166C or Sun TDT-12 tachometer connected to the adapter.

The location of the idle and maximum speed adjusting screws can be seen in Figure 12–17, Items 28 and 29. If the idle rpm is not between 770–870, loosen off the lock nut and turn the adjusting screw clockwise to increase or counterclockwise to decrease the speed setting, then tighten the retaining nut.

To check/adjust the maximum rpm, ensure that the engine is at operating temperature, then place the speed control lever to its full-fuel position and quickly note the maximum rpm. Do not run the engine under a wide open throttle condition any longer than necessary to obtain the reading on the tachometer.

Earlier engines were designed for maximum full-load speeds of 5000 rpm; however, 1981 and later engines are designed for a maximum full-load speed of 4800 rpm. Therefore, on the pre-1981 engines, the maximum no-load rpm should be between 5400–5450 rpm, while the later engines should be between about 5175–5225 rpm. Clockwise adjustment of the high speed screw reduces rpm while counterclockwise rotation increases top rpm.

Throttle cable adjustment may be required in some cases to ensure that when the accelerator pedal is pushed all the way down full rpm is in fact being obtained.

With the engine stopped, place the accelerator pedal in its full down position by having someone hold it there, or jam it down with a suitable arrangement. By rotating these two nuts, the throttle lever can be moved towards or away from the maximum speed adjusting screw.

Current engines employ an automatic CSD (cold start device), which is discussed in Chapter 12, Robert Bosch VE pump section. Earlier engines used a cold starting knob on the vehicle dash (Figure 12–25 and 12–26(a)) that had to be pulled out to activate the CSD. In this way, the injection pump timing was advanced about 2.5 degrees to allow better cold start characteristics. Adjustment was simply a case of pushing the CSD knob and cable all the way in and ensuring that the other end of the cable was adjusted to keep the injection pump CSD lever in the non-engaged position. Check it by pulling the CSD knob out and making certain that the CSD lever does move to its fuel advanced position.

Checking/Adjusting Injection Timing

The injection pump static timing to the engine is at TDC with a pump prestroke of 0.032″ (0.83 mm). The timing check procedure follows the same process as that given in the Robert Bosch chapter, VE pump section in this book. (See Figures 12–30 and 12–31.) To check/adjust the timing, proceed as follows.

Procedure

1. Manually rotate the engine to place the No. 1 piston at TDC on its compression stroke; check that both the inlet and exhaust valves are closed.

   NOTE: Two methods can be used to set the No. 1 piston at TDC; these depend on whether the engine is in the vehicle or it has been removed. Figures 14–1 and 14–2 illustrate both of these methods.

2. Remove the plug from the center of the injection pump hydraulic head between the fuel delivery studs.
3. Install the dial indicator with its adapter (VW P/N 2066) into the plug hole and secure it in place, similar to Figure 12–30. This allows the dial gauge to measure the injection pump plunger lift/stroke when the No. 1 piston is at TDC/compression.
4. Follow the remaining steps listed under "static timing check" in the Robert Bosch Chapter 12, VE pump section.
5. Read the dial gauge with the TDC mark in alignment; it should read 0.032″ (0.83 mm).
6. If the gauge does not read as per Step 5, loosen off the injection pump mounting bolts slightly then grasp the pump with both hands and either push it towards the engine or pull it away from the engine until the correct reading is obtained on

Figure 14-1. VW Rabbit TDC timing mark on engine flywheel aligned with stationary pointer in inspection hole of bellhousing (engine in vehicle). (Courtesy of Volkswagen of America)

Figure 14-2. VW Rabbit TDC timing mark on engine flywheel aligned with special setting bar (engine out of vehicle). (Courtesy of Volkswagen of America)

the face of the dial gauge. What you are doing is actually adjusting the plunger lift (Figure 12–22) from the BDC position to the point of port closure by turning the cam ring away from or towards the rollers (Figure 12–26(a)) depending upon which way you turn the pump. In this way you can effectively advance or retard the start of injection.

7. Tighten the pump retaining bolts to 18 lb.ft. (25 N·m).
8. Perform the timing check after any adjustment is made to ensure that the pump is now correctly set.

Remove/Install Injection Pump

Removal and installation of the injection pump should always be done with cleanliness in mind. Any dirt allowed to enter the fuel lines or injection pump will cause serious damage to the internal components. Before removing an injection pump, procure a variety of plastic protective shipping caps both male and female that can be used to plug off all open fuel lines and openings as you remove them.

Pump Removal Procedure

1. Isolate the engine starting system by disconnecting the battery ground strap.
2. Refer to Figure 14–3 and remove the overhead camshaft drive belt, which is of cog design as described below.
3. Remove any accessories that prevent free access to the camshaft belt.
4. Remove the alternator drive belt after loosening off the retaining bolts.
5. Remove the crankshaft pulley after removing the four socket head bolts.
6. Remove the drive belt cover.
7. Remove the rocker cover.
8. Manually rotate the engine to place No. 1 piston at TDC/compression, which can be confirmed by ensuring that the flywheel mark is correctly aligned as well as both valves having clearance between the camshaft lobes.
9. Refer to Figure 14–4 and lock the camshaft sprocket (gear) in the No. 1 piston TDC position by inserting tool P/N 2065 (bar) across the rear of the cylinder head.
10. Loosen off the camshaft belt tensioner lock nut.
11. Place a wrench on the tensioner adjuster nut (larger nut) and rotate it counterclockwise to remove the tension from the camshaft drive belt, then remove the belt by working it towards the right-hand side of the vehicle.
12. Remove all the injection pump fuel lines and plug them and the open holes on the injection pump with protective plastic shipping caps.
13. Disconnect the pump fuel solenoid wire, throttle cable, and CSD cable (if used).
14. Loosen off the injection pump sprocket nut and install a puller (VW P/N 2036) as shown in Figure 14–5 to allow withdrawal of the sprocket from the pump tapered shaft.

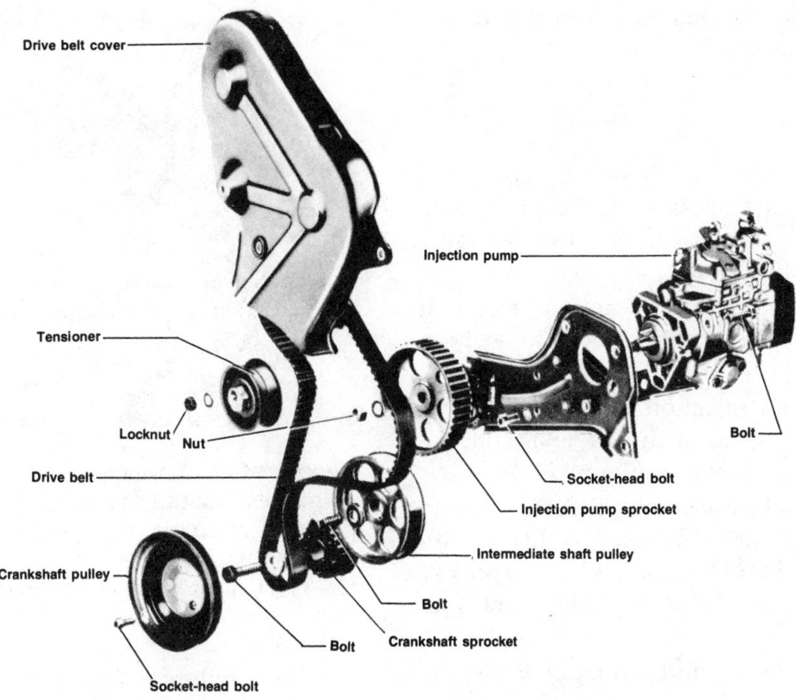

Figure 14-3. Components of overhead camshaft cog drive belt. (Courtesy of Volkswagen of America)

Figure 14-4. Engine camshaft (VW Rabbit) locked in position with special tool P/N 2065. Note also TDC marks in alignment (arrow). (Courtesy of Volkswagen of America)

Figure 14-5. Injection pump sprocket being loosened. Note spacer between the nut and the center screw of the puller to prevent thread damage to the pump drive shaft. (Courtesy of Volkswagen of America)

15. Remove the injection pump mounting plate retaining bolts to allow the injection pump to be removed from the engine.

Injection Pump Installation

Installation of the pump follows the reverse procedure to that of removal. The following sequence should be used.

NOTE: Be certain that the engine is still at TDC No. 1 piston/compression.

Procedure

1. Refer to Figure 14–6 and install the injection pump onto its mounting plate and support with the timing mark aligned; torque the retaining bolts to 18 lb.ft. (25 N·m).
2. Install the pump sprocket and nut and torque it to 32 lb.ft. (48 N·m) with pin P/N 2064 in position in the sprocket.
3. Manually rotate the injection pump sprocket until its mark is aligned with the mark on the mounting plate as shown in Figure 14–7.
4. With pin P/N 2064 in position, (Figure 14–7) install the camshaft drive belt and adjust it with the tensioner to obtain a reading of 12–13 on VW belt Tension Tester VW 210. Tighten the tensioner bolt.
5. Perform an injection pump timing check as described earlier.
6. When the alternator V-belt has been installed, pull the alternator away from the engine and adjust the belt so that it deflects between 3/8″ to 9/16″ midway between the crankshaft and alternator pulleys, then tighten the alternator bolts.
7. Install the camshaft belt cover and valve rocker cover.
8. Install all of the injection pump fuel lines and torque to 18 lb.ft. (24 N·m).

   SPECIAL NOTE: The injection pump fuel stud delivery valve holders are marked A-B-C and D.

Figure 14–6. Mark on pump aligned with mark on mounting plate (broken line). Fuel supply union bolt is at A. (Courtesy of Volkswagen of America)

Figure 14–7. Injection pump sprocket aligned with mark on mounting plate (black arrow). White arrow indicates installation of special timing pin P/N 2064 used during installation of the pump and engine camshaft cogged drive belt. (Courtesy of Volkswagen of America)

   No. 1 cylinder fuel injector line should be connected to A, No. 2 to D, No. 3 to B, and No. 4 to C. Also take care that the fuel return banjo union bolts are not intermixed because the return bolt (Figure 12–24) is a restricted fitting to ensure that fuel pressure is maintained within the injection pump housing. The return bolt is clearly marked OUT on the top.

9. Connect up the fuel solenoid wire, the throttle cable, and CSD cable if used.
10. Connect up the battery ground cable.
11. Bleed the fuel system as per the heading "Bleeding the Fuel System."
12. Start and run the engine and check for any fuel leaks.

Bleeding the Fuel System

Because of the vane-type pump used with the Robert Bosch VE model injection pump, it is seldom necessary to perform this routine. However, to avoid a long delay in starting the engine after a fuel injection pump and lines have been removed, it is advisable to bleed entrapped air from the fuel system. The sequence given under the Volvo D24 engine (Chapter 15) also applies to the VW engine because the Volvo D24 6-cylinder diesel engine is of VW manufacture.

The sequence for priming/bleeding the fuel system involves two basic methods:

1. Earlier vehicles used a hand-priming pump located on the fuel/water filter adapter that could be

unscrewed counterclockwise and then pumped in/out to deliver fuel to the pump and injectors. In the hand-pump system, a vent screw on the filter adapter could be loosened off to allow release of trapped air from the filter by operating the hand pump.

2. Newer vehicles may not have this arrangement; therefore, simply loosen off all injector fuel nuts and crank the engine until fuel that is free of air bubbles vents from the fuel lines.

Fuel Filter Service

Fuel filter service requires that water in the base be removed every 7500 miles (12,000 km) whether the instrument panel *water-in-fuel* light comes on or not. Should the *water-in-fuel* light come on, water should be drained immediately from the fuel/water filter separator.

The filter change period should be every 15,000 miles (24,000 km) unless the severity of service is such that a more frequent change period may be required.

To drain water from the filter, loosen off the vent screw on top of the filter or alternately loosen off the fuel return hose, then open the water drain plug at the base of the filter and drain the filter contents into a drain tray until all water has been removed and clear diesel fuel is evident. If the filter is fitted with a hand pump, operate the handle while you are draining the fluid from the filter.

Filter replacement can be done with the aid of a filter strap wrench in the usual manner. The filter is a spin-on type element and is therefore a throwaway item.

Fuel Injectors

The fuel injectors used with this engine are of Robert Bosch manufacture and are Model Numbers DNO SD 293. Figure 29–4 illustrates the injector used with the VW diesel engine, which is a throttling pintle-type nozzle mounted in a KC holder (injector body). As you can see, it is very similar in design to those used by Ford, Datsun/Nissan, Toyota, Volvo, Audi, Mercedes-Benz, Isuzu, Dodge Ram 50/Mitsubishi, Peugeot, and General Motors 6.2L in their diesel engines.

For detailed information relating to testing, disassembly, inspection, repair, and reassembly, refer to Chapter 29, which deals with all types of fuel injectors. Also shown in the injector overhaul chapter are exploded views of pintle-type nozzles similar to that used in the VW engine.

A quick and common method to determine if an injector is faulty is to run the engine at its idle rpm or slightly faster, then loosen the fuel line nut at the injector; this should cause the engine speed to decrease. If the speed and sound of the engine remain the same, then the injector is faulty and should be removed.

This check can be done at each injector to determine its condition. In addition, if a misfire only occurs at a particular speed range, then run the engine up to that speed range and loosen off the injector fuel line nut with a rag around it to prevent fuel spraying all over the engine compartment. Again, if there is no change to the engine speed and sound, then that injector is faulty.

Injectors are screwed into the cylinder head and are removed by using a special deep socket P/N US 2775 after removing the inlet and return fuel lines from the injector body.

Injector removal follows the same basic procedure as that used for the Volvo D24 6-cylinder diesel engine and other engines listed above. Always replace the heat shield underneath the injector, which sits at the bottom of the bore in the cylinder head. (See Figure 15–19)

Glow Plugs

The glow plugs used with the VW diesel engine operate in the same manner as that described for other diesel engines throughout this book; therefore, the description of operation will not be given in detail since Chapters 15, 17, 18, 19, 20, 22, 24, and 26 will serve to explain the characteristics of the glow plugs.

Testing and Replacing Glow Plugs. The test follows standard practice for other glow plugs. A voltage check of the glow plugs can be done by placing a 12-volt test light across the No. 4 cylinder glow plug and to ground. When the ignition key is turned to the *preheat* or *preglow* position, the test light should illuminate. If it doesn't, test the glow plug relay.

Each glow plug is connected to a bus bar, which serves to supply all of the glow plugs with power at the same time. If the previous check indicated that voltage was present, but the engine is hard to start, disconnect the bus bar to isolate each glow plug. Using a 12-volt test light, connect one end to the positive battery terminal and the other to the glow plug terminal (check each glow plug one at a time). If the test light fails to illuminate, then the glow plug is faulty and should be replaced.

Valve Adjustment

Valve adjustment on the Rabbit diesel is done with the use of discs that are removed in the same fashion as that explained under the Volvo D24 vehicle section (this is a 6-cylinder VW diesel engine) and shown in Figures 15–28 and 15–29.

The discs are available in 26 different thicknesses from 3.00 to 4.25 mm (0.1181 to 0.1673″). Experience has shown that the most commonly used discs fall within the range of 3.55–3.80 mm (0.1397–0.1496″). Each disc is stamped with its thickness in mm (millimeters) on one side to facilitate identification.

Valve adjustment should be performed in the same manner as that described for the Volve D24 6-cylinder diesel engine. They can be adjusted with the engine hot or cold—however, keep in mind that the clearances will be different. Valve clearances should be as follows:

1. Intake valve clearance Cold . . . 0.006–0.010″ (0.15–0.25 mm)
2. Intake valve clearance Hot . . . 0.008–0.012″ (0.20–0.30 mm)
3. Exhaust valve clearance Cold . . . 0.014–0.018″ (0.35–0.45 mm)
4. Exhaust valve clearance Hot . . . 0.016–0.020″ (0.40–0.50 mm)

A hot setting should be done with the engine coolant temperature at 95°F (35°C) minimum.

Valve clearance should be checked at each cylinder by manually rotating the engine until the overhead camshaft lobes for that cylinder are pointing upwards. In other words, the cam base circle should be over the valve follower so that both valves are completely closed and clearance exists between the cam lobe and the adjusting disc on top of the fol-

It is assumed that the engine is in good working order and properly tuned, and that the electrical system has been checked and repaired if necessary.

SYMPTOM: Starting Problem	Engine surges at idle	Rough idle when engine is warm	Engine misses under load	Low Power	Excessive Fuel Consumption	Engine cannot be shut off	Poor performance or black smoke or low power	Fog-like exhaust in full-load range (white or blue)	Incorrect idle or maximum speed	Engine does not rev up	Injection pump runs hot	CAUSE	REMEDY
●		●	●	●			●	●		●		Improper fuel (gasoline) in tank	Drain tank, flush system, fill with proper fuel
●	●		●	●				●		●		Tank empty or tank vent blocked	Fill tank/bleed system, check tank vent
●	●	●	●				●	●		●		Air in the fuel system	Bleed fuel system, eliminate air leaks
	●	●										Pump rear support bracket loose	Replace as necessary
●						●						Low voltage, no voltage or stop solenoid defective	Correct electrical faults/replace stop solenoid
●			●	●				●		●		Fuel filter blocked	Replace fuel filter
●			●	●			●	●				Injection lines blocked/restricted	Drill to nominal I.D. or replace
●			●	●				●		●		Fuel-supply lines blocked/restricted	Test all fuel supply lines – flush or replace
●			●	●								Loose connections, injection lines leak or broken	Tighten the connection, eliminate the leak
●			●					●		●		Paraffin deposit in fuel filter	Replace filter, use Diesel Fuel no. 1
●			●	●	●		●	●		●		Pump-to-engine timing incorrect	Readjust timing
		●	●	●	●		●					Injection nozzle defective	Repair or replace
				●	●		●					Engine air filter blocked	Replace air filter element
●												Pre-heating system defective	Test the glow plugs, replace as necessary
●		●		●	●		●					Injection sequence does not correspond to firing order	Install fuel injection lines in the correct order
					●				●			Low idle misadjusted	Readjust idle stop screw
				●					●			Maximum speed misadjusted	Readjust maximum speed screw
	●		●	●				●	●	●		Overflow fitting interchanged with inlet fitting	Install fittings in their proper positions
											●	Overflow blocked	Clean the orifice or replace fitting
●										●		Cold-start device not operating	Check bowden cable and lever movement
●		●		●			●			●		Low or uneven engine compression	Repair as necessary
●	●	●	●	●	●		●	●	●	●		Fuel injection pump defective or cannot be adjusted	Replace

Figure 14-8. Troubleshooting guide for diesel engine using a Robert Bosch VE model injection pump. (Courtesy of Robert Bosch Sales Corporation)

lower. Check the clearance with a feeler gauge and write it down because you will want to check all the clearances before performing any disc change. Refer to the Volvo D24 diesel engine section, which illustrates the method used to remove the discs and replace them.

CAUTION: The piston must not be at TDC when depressing the valve cam follower; otherwise, the valve will contact the piston, preventing you from removing the disc. Rotate the engine over until the piston is about 90 degrees past its TDC mark (power stroke).

The required new disc thickness can be arrived at by reading the disc thickness stamped on the one side, or by using a micrometer to measure it. Compare the thickness with the reading that you wrote down earlier when you checked the clearance and select the necessary new disc thickness to place the valve clearance within the recommended clearance range listed above.

When installing new discs, always install them with the numbered side down towards the cam follower.

Compression Check

A compression check also follows the same basic procedure as that for the Volvo D24. Compression pressures should be 483 psi (3330 kPa) with a minimum acceptable level of 398 psi (2744 kPa). Maximum allowable cylinder variation should not exceed 71 psi (490 kPa).

Troubleshooting Guide

A troubleshooting guide for the VW Rabbit diesel VE injection pump is given in Figure 14–8.

15

Volvo D24 Diesel Engine

The Volvo 6-cylinder D24 is a 2.4L (2383 cc) 145 cu.in. diesel engine which is of Volkswagen design and follows the same arrangement as the 4-cylinder VW 4-cylinder Rabbit engine and the 5-cylinder Audi 5000 diesel engines. The 6-cylinder Volvo engine is illustrated in Figure 15–2, which shows the location of the fuel injection pump, injectors, and main fuel filter assembly.

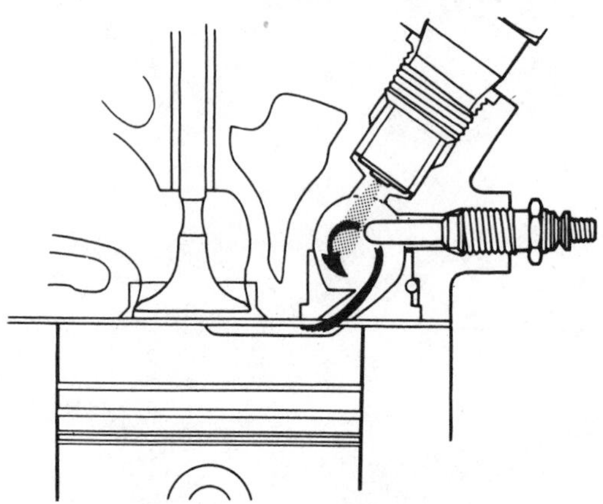

The engine has indirect fuel injection. The fuel is injected into the swirl chambers.
The swirl chambers are located in the cylinder head. They comprise approx. 50% of the total combustion volume.
During the compression stroke, air is forced into the swirl chamber and forced to rotate with high velocity. This air speed influences the combustion speed and therefore the engine can reach high engine speeds (= wide rpm range).
The fuel is injected just before the piston reaches top dead center. It mixes with the turbulent air within a short period of time.
During compression, the air/fuel mixture is heated to approx. 800° C = 1400° F which causes it to ignite. Combustion starts in the swirl chamber and spreads to the cylinder. The pressure increase is thus controlled resulting in a smoother running engine.

Figure 15–1. Volvo D24 6-cylinder diesel engine. (Courtesy of Volvo of America Corporation)

The engine is a 4-stroke unit and is mounted 20 degrees to the left of center for a reduced installation height. The engine is of Volkswagen manufacture and employs the conventional pre-combustion swirl chamber design, which is illustrated in Figure 15–1 and also discussed in Chapter 3.

Engine weight is 436 lb. (198 kg), which includes the engine mounts, starter motor, and alternator. The compression ratio is 23.5:1 in 1980–81 models, while 1982 and later models have a compression ratio of 23:1 from a bore of 76.5 mm (3.0118″) and a stroke of 86.4 mm (3.4016″). Compression pressures range from 400–485 psi (2758–3344 kPa).

Individual glow plugs are employed in each combustion chamber to facilitate starting and rapid engine warm-up.

A light alloy/aluminum cylinder head is employed along with an overhead camshaft which is driven from the front of the engine by a cogged belt. The rear of the camshaft drives the fuel injection pump also through a cogged belt drive (See Figure 15–4). The cogged belt drive allows for a quieter running engine. Naturally aspirated engine horsepower was 82 DIN at 4800 rpm in the 1980–81 engines, while 82 and later engines are rated at 76 horsepower SAE at 4800 rpm.

A conventional firing order of 1-5-3-6-2-4 is used; the engine rotation is clockwise from the front. The engine is designed for optimum performance on diesel fuel rated at 45 Cetane with a maximum sulphur content of 0.5%.

The injection pump is a Robert Bosch VE6-type unit mated to Robert Bosch DNO SD 193 nozzles in Bosch injector bodies KCA 30SD 27/4. The injectors are designed to open at between 1775–1920 psi (12239–13238 kPa).

Injection pump/engine timing is at TDC with a pump plunger reading of 0.028″ (0.70 mm).

Fuel System Operation

The fuel system operation follows the same basic arrangement as that used on other vehicles employing a Robert Bosch VE model injection pump.

NOTE: Detailed operation, maintenance, repair, and troubleshooting of the injection pump can be found under Chapter 12 dealing with the Robert Bosch VE model pump, which is a distributor-type pump.

Refer to Figure 12–17, which illustrates the injection pump components. Accompanying this figure is a description of the fuel flow under this same chapter from the fuel tank to the injection pump and nozzles.

When the ignition key is turned on, the fuel solenoid mounted on the injection pump hydraulic head (opposite drive end) is energized to allow fuel flow into the pump.

Fuel flow through the system is provided by a rotary vane-type transfer pump contained within the drive end of the injection pump housing. Fuel from the tank is drawn through a filter assembly that also acts as a fuel/water separator before it enters the injection pump. The vane-type transfer pump generally is capable of boosting the fuel pressure to about 100 psi (689.5 kPa), although this can vary between makes of vehicles.

This vane pump pressure is then directed into the injection pump to the pumping plunger area where the fuel is further pressurized to a high enough point that it can open the delivery valve at the hydraulic head of the injection pump. Fuel is distributed to each injector through its own delivery valve in engine firing order sequence, hence the term rotary or distributor-type pump.

The injection pump is lubricated by the fuel that flows through it, as are the fuel injectors. The volume of fuel delivered is always more than the injection pump and injectors can use; therefore, a cooling effect is assured at the injection pump at all times. Excess fuel is returned from both the injectors and the injection pump through a return line to the fuel tank through a restricted fitting.

Because of the large capacity of the vane pump and the recirculation of fuel, bleeding of trapped air in the fuel system is seldom required unless the system has been totally emptied of fuel. If it is necessary to bleed air from the fuel system, follow the instructions for bleeding the fuel system given below.

The inlet and return lines from the injectors and fuel pump are shown in Figures 15–2 while 15–3 illustrates the fuel/water separator drain cock and bleeder screw.

Figure 15–2. Engine fuel system component location. [Courtesy of Volvo of America Corporation]

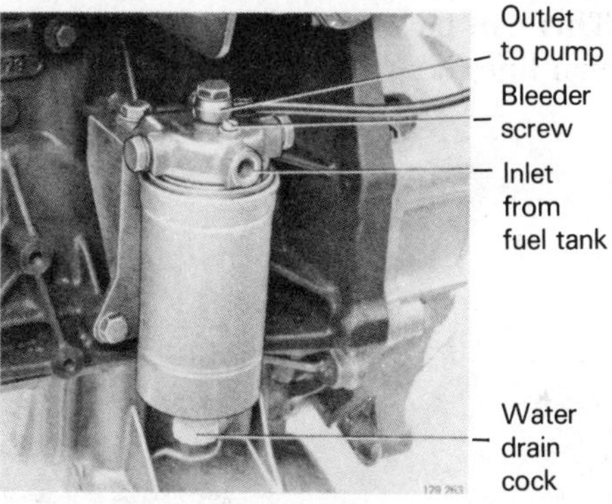

Figure 15–3. Fuel/water separator fittings. [Courtesy of Volvo of America Corporation]

Fuel/Water Filter Service

The fuel/water filter is designed to act as a water trap and to filter the fuel on its way up to the vane-type transfer pump inside the injection pump assembly.

The fuel/water filter contains a warning light that will illuminate a dash-mounted instrument panel lamp when an excess amount of water has collected in the filter. When this light comes on, the fuel/water filter should be drained immediately.

Normally, by way of proper service, the water should be drained from the filter assembly every 7500 miles (12069 km) by placing a drain tray underneath the filter and loosening the bleeder screw as shown in Figure 15–3 several turns as well as opening the drain valve until all the water has drained out and clear diesel fuel is seen to flow. Tighten the bleeder screw.

Fuel/Water Filter Replacement

The fuel/water filter should be replaced every 15,000 miles (24139 km) on an average. The severity of service and operating conditions may require a more frequent change interval.

Removal and installation of the fuel/water filter spin-on unit follows normal practice. Once the old filter has been removed, discard it and clean the mounting surface for the new unit. Apply a light coating of clean diesel fuel to the new filter seal (O-ring) and hand-tighten the filter until it makes contact with the filter body adapter. Once the filter contacts the adapter, hand-tighten it another ¼ turn.

Start and run the engine and check for signs of air leakage—which will be indicated by a rough running engine suggesting that the filter did not seal correctly. This can be confirmed by disconnecting the fuel return line, which will show air bubbles.

NOTE: Do not pour fuel directly into the fuel filter to prime it unless the fuel has been taken from a filtered source because some of the fuel will not pass through the filter but will be sucked up to the vane-type transfer pump within the injection pump and can cause damage if dirt or water is contained in it.

The bleeder screw can be removed to prime the filter if only non-filtered fuel is available.

Injection Pump General Checks/Adjustments

The following checks and adjustments are related strictly to external adjustments and removal/installation of the pump from the engine. Major injection pump repair requires the use of special tools and equipment plus experience in major overhaul/servicing procedures. For information related to injection pump repair, refer to Chapter 12 dealing with Robert Bosch VE model pumps.

Figure 15–4 illustrates the belt drive arrangement from the engine camshaft to the injection pump.

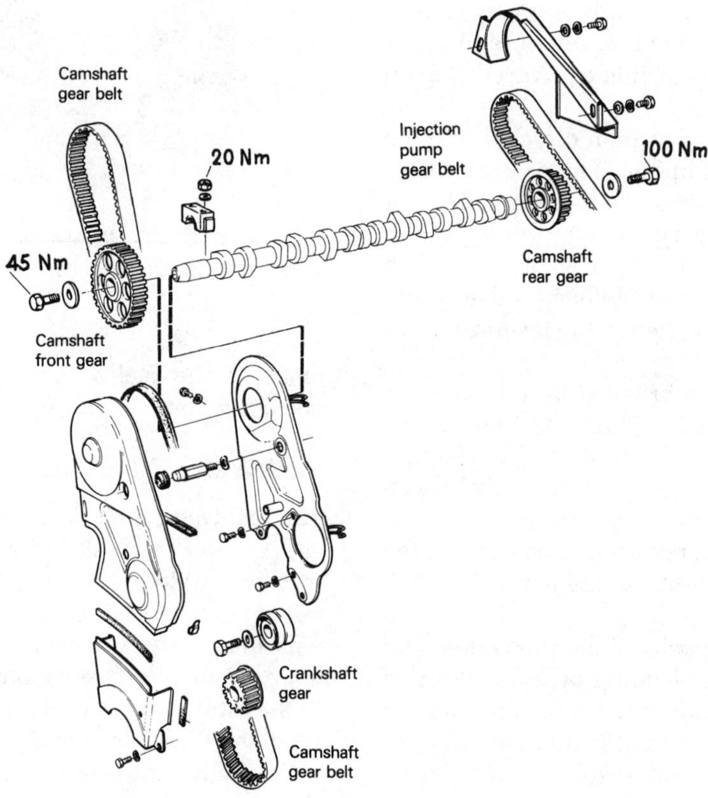

Figure 15–4. Injection pump drive belt location and arrangement. (Courtesy of Volvo of America Corporation)

Low and High Idle Speed Adjustment

Both the low and high idle speed adjustment screws are located on top of the injection pump housing and are illustrated in Figure 12–17 as Items 28 and 29.

To successfully perform both a low and high idle speed adjustment, proceed as follows.

Procedure

1. The engine should be at its normal operating temperature.
2. Start and run the engine while monitoring the speed with a photoelectric tachometer.
3. The idle rpm should be between 800 ± 50 rpm.
4. If the idle speed requires adjustment, refer to Figure 15–5 and loosen off the idle screw lock nut and turn the screw clockwise to increase the idle speed and counterclockwise to decrease the idle speed.
5. Paint and a tamper-resistant seal should be applied to the adjustment screw when you are finished.
6. To check the maximum speed of the engine (5200 rpm) gently accelerate the engine to wide open throttle by manually pushing the pump throttle lever, and carefully note the rpm with a photoelectric tachometer.
7. If the maximum speed does require setting, refer to Figure 15–6 and loosen off the high speed screw lock nut and turn the screw clockwise or counterclockwise to change the maximum rpm.

Setting Engine Speed Controls

When an adjustment is performed to the low/high idle speeds, the engine speed controls should be checked and adjusted if required.

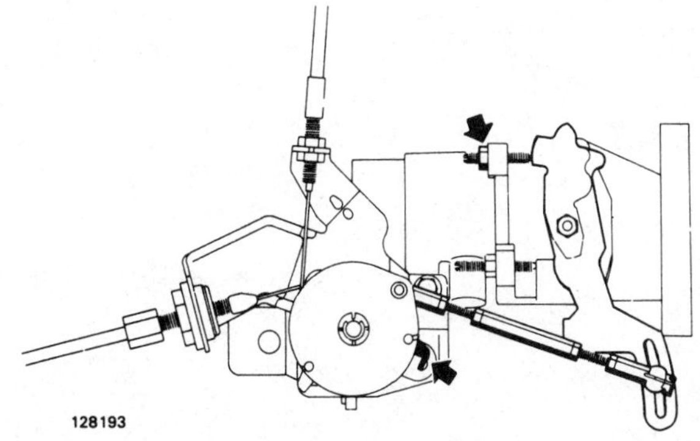

Figure 15–5. Idle speed adjustment screw location and pump link rod adjustment at an idle position. [Courtesy of Volvo of America Corporation]

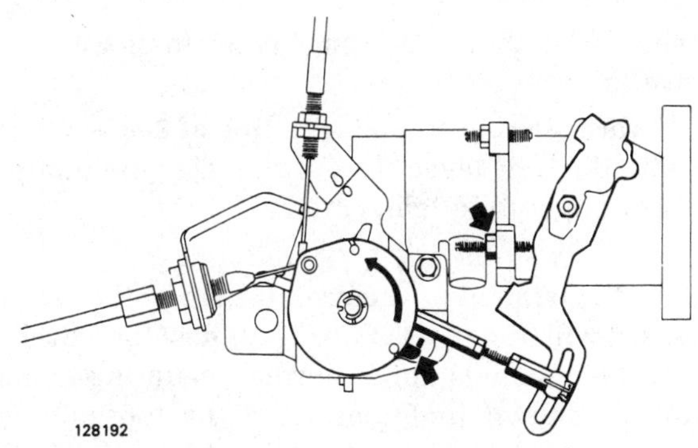

Figure 15–6. Maximum speed adjustment screw location and pump link rod adjustment in the maximum speed position. [Courtesy of Volvo of America Corporation]

This involves checks/adjustments to the following:

1. The cold start device (see location in Figure 15–2 above)
2. The accelerator cable
3. The automatic transmission kickdown cable
4. Injection pump link rod

Cold Start Device (CSD). The CSD is explained in detail in Chapter 12, Robert Bosch VE pump, and is illustrated in Figure 12–27. Basically speaking, this device is to allow a higher engine rpm when the engine is cold and to reduce the idle speed when the engine is warmed up.

This is achieved by routing coolant through the CSD, which operates on a wax element similar to a thermostat. Linkage connected to the CSD will extend the throttle lever towards an increased fuel position when cold and draw it back when the engine warms up as shown under the Bosch VE pump, Chapter 12.

CSD faults are generally indicated when any of the following are evident:

1. The engine is hard to start from cold.
2. Blue/white exhaust smoke is evident.
3. Engine will not start at all when temperatures fall to −10°C (14°F).

Accurate adjustment of the CSD can only be performed on a fuel pump test bench; however, a simplified check can be performed on the engine as follows:

1. Check the CSD with the engine both cold and at operating temperature.
2. Cold engine (below 20°C or 70°F) low idle rpm

should be about 200 rpm higher than when it is warm.

3. A warm engine should be idling at 800 ± 50 rpm with the CSD lever clear of the injection pump lever at operating temperature.

Other Adjustments. Before checking other adjustments, it will be necessary to disconnect the CSD linkage if the engine is not at normal operating temperature, as shown in Figure 15–7, by loosening off screw No. 1, pushing the lever forward, and turning the sleeve 90 degrees. Do not loosen off screw No. 2, otherwise the CSD will be put out of adjustment and the pump will have to be set on a test bench.

Accelerator Linkage Adjustment. Before checking/adjusting the accelerator linkage, disconnect the link rod from the lever on the injection pump as shown in Figure 15–8 (small arrow on the right-hand side).

Manually rotate the accelerator cable circular knob (sheath) located on top of the injection pump as shown in Figure 15–9 to place the cable in a tightly wound position. The pulley should come into contact with the *stop* as shown in Figure 15–9.

Manually place the accelerator into the maximum speed position and check to see if the pulley touches the full-speed stop as shown in Figure 15–9.

Check/Adjust Automatic Transmission Kickdown Cable. Refer to Figure 15–10 and push the throttle to the floor while checking that the cable does in fact move approximately 52 mm (2.05″) between its end positions. If not, adjust the cable. Refer to Figures 15–6 and 15–5, and adjust the injection pump throttle link rod in its maximum speed position ensuring that the speed control lever butts up against the high speed screw and the low speed screw respectively. To adjust the linkage at the idle screw, move the link rod ball joint in the slotted hole of the pump lever.

Maximum clearance allowed between the cable pulley and the maximum speed stop is 0.012″ (0.3 mm).

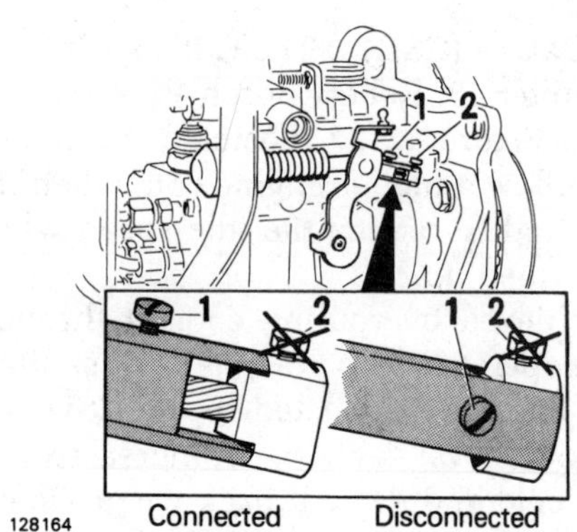

Figure 15–7. Disengaging cold start device linkage. (Courtesy of Volvo of America Corporation)

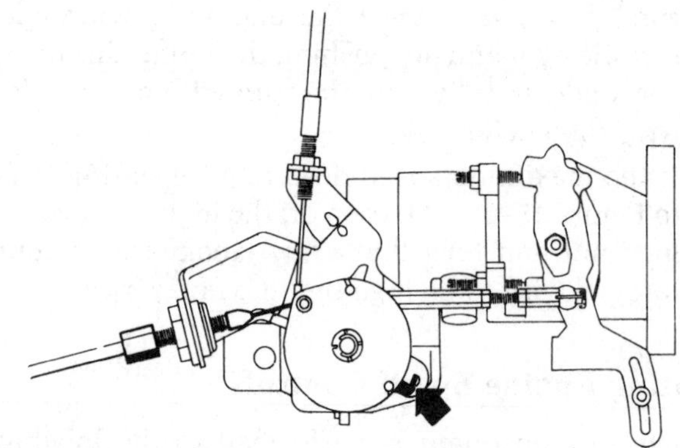

Figure 15–9. Checking accelerator linkage maximum speed position. (Courtesy of Volvo of America Corporation)

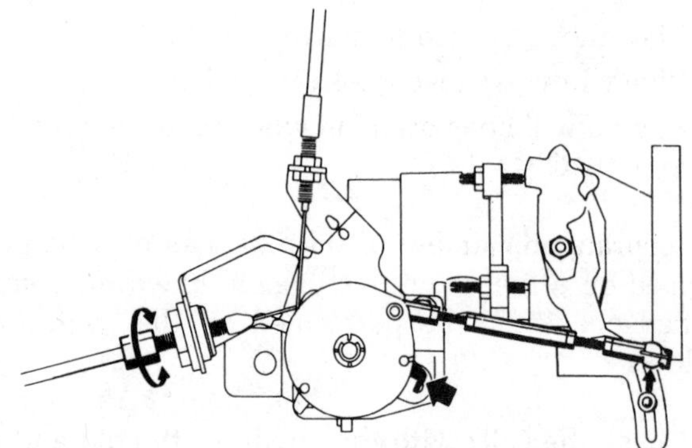

Figure 15–8. Disconnecting injection pump link rod. (Courtesy of Volvo of America Corporation)

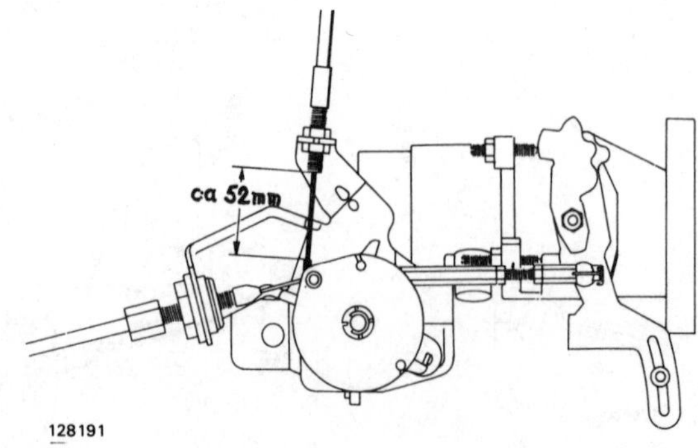

Figure 15–10. Automatic transmission kickdown cable adjustment specification. (Courtesy of Volvo of America Corporation)

Checking/Adjusting Injection Pump Timing

If it becomes necessary at any time to check/adjust the injection pump to engine timing, proceed as follows.

Procedure

1. Ensure that the engine is at operating temperature.
2. Refer to Figure 15–4 and remove the injection pump drive belt cover at the rear of the engine.
3. Disconnect the cold start device as stated under "Other Adjustments" above relating to "setting engine speed controls."
4. Follow the VE injection pump static timing procedure as described under the Robert Bosch, Chapter 12. (See Figure 12–30.)
5. Carefully read the dial gauge measurement with the engine flywheel timing mark aligned with its stationary pointer as shown in Figure 15–11; it should read between 0.0256–0.0287″ (0.65–0.73 mm), which is an acceptable tolerance. If the gauge does not read within these specifications, the injection pump should be adjusted to read 0.028″ (0.70 mm).
6. To adjust the injection pump /engine static timing loosen off the injection pump retaining bolts slightly as shown under "remove injection pump" below. (See Figure 15–12.)
7. If the timing gauge dimension is less than 0.028″ (0.70 mm), manually push the injection pump housing gently towards the engine block until this reading is obtained, then tighten up the pump retaining bolts.
8. If the timing gauge dimension is more than 0.028″ (0.70 mm), manually pull the injection pump housing away from the engine block until this specification is obtained, then tighten up the pump retaining bolts.
9. After any injection pump-to-engine static timing check/adjustment, the engine should be manually rotated at least two full crankshaft revolutions with the dial gauge and holder still in position and the dial gauge reading reconfirmed at 0.028″ (0.70 mm). (See Figure 12–30.)
10. Remove the dial gauge and adapter, install the hydraulic head plug and washer and torque it to 6.5 lb.ft. (9 N·m). Also install the injection pump drive belt cover and cold start device linkage.

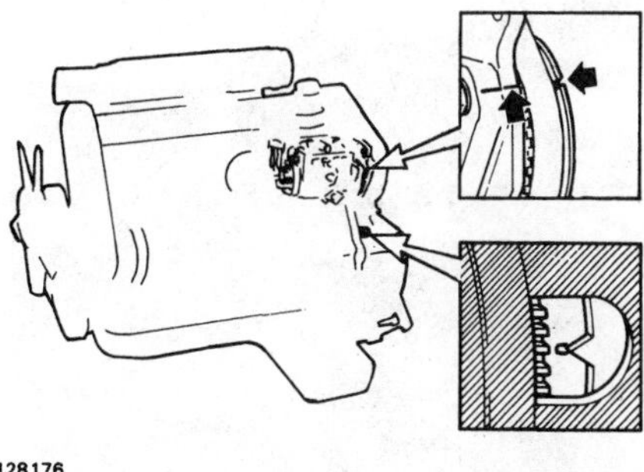

Figure 15–11. Injection pump and engine timing alignment mark locations. (Courtesy of Volvo of America Corporation)

Fuel Injection Pump Removal/Installation

If it becomes necessary to remove/install the injection pump, the following procedure can be followed. Figure 15–12 illustrates the injection pump and its related external components.

Removing Injection Pump

Procedure

1. Before disconnecting any components from the fuel system, obtain a drain tray to catch spilled fuel and also obtain a supply of various sized plastic male/female shipping protective caps that can be inserted into all open holes and fuel lines to prevent the entrance of dirt or foreign material into the fuel system.
2. Either drain the cooling system, or pinch shut the two water hoses connected to the cold start device at the hydraulic head end (opposite drive end) prior to disconnecting them.
3. Once the hoses in Step 2 have been disconnected, if water still remains in the coolant system, tie the hoses above the level of the cylinder head to stop water leakage.
4. Disconnect the accelerator and A/T kickdown cables and the stop cable.
5. Remove the injection pump rear timing belt cover as in the injection pump timing check.
6. Clean off all loose dirt around all fuel lines prior to removal.
7. Plug all open connections with plastic shipping caps as discussed in Step 1 above.
8. Remove the vacuum pump and plunger which is shown in Figure 15–2 in front of the injection pump.

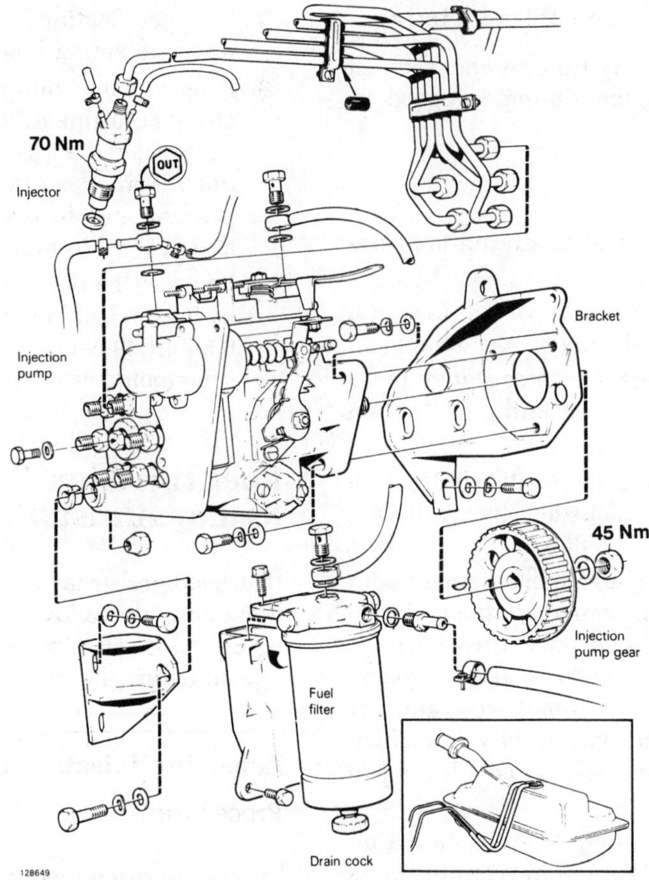

Figure 15–12. VE6 Injection pump and component arrangement. (Courtesy of Volvo of America Corporation)

9. Rotate the engine manually with a socket (27 mm or 1-1/16") placed on the crankshaft pulley/damper to place No. 1 piston at TDC on its compression stroke (check that both valves are closed).
10. Refer to Figure 15–13 and loosen off the retaining bolts on the injection pump bracket in order to remove the pump drive belt at the rear of the engine.
11. Using special tool 5199 or equivalent to hold the rear camshaft gear, loosen off the rear camshaft gear as shown in Figure 15–14 with wrench 5201 or equivalent just enough to allow the rear camshaft gear to rotate freely on the camshaft.

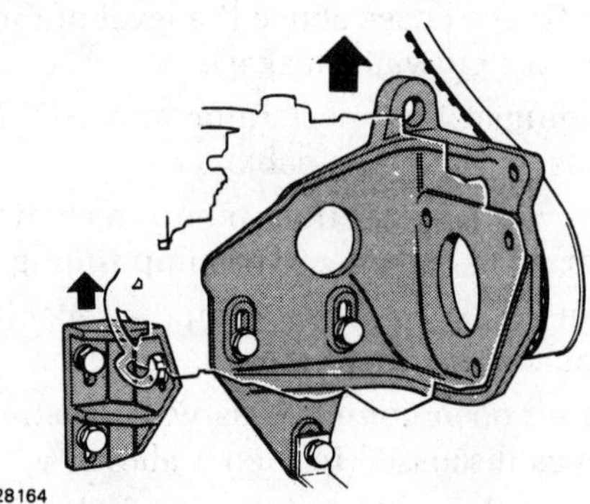

Figure 15–13. Loosening off injection pump support bracket bolts. (Courtesy of Volvo of America Corporation)

Figure 15–14. Loosening off rear camshaft gear retaining nut. (Courtesy of Volvo of America Corporation)

12. Using special tool 5193 to lock the injection pump gear in position, loosen off the gear retaining bolt with wrench 5201 and remove the nut as shown in Figure 15–15.
13. Once the injection pump gear retaining nut has been removed, leave tool 5193 in position to prevent gear rotation and install gear puller 5204 or equivalent in order to pull the gear from the injection pump drive shaft (Figure 15–16).
14. Remove the injection pump from the engine by loosening/removing all the bolts shown in Figure 15–17.

Figure 15–15. Removing injection pump gear nut. (Courtesy of Volvo of America Corporation)

Figure 15–16. Pulling injection pump gear from shaft. (Courtesy of Volvo of America Corporation)

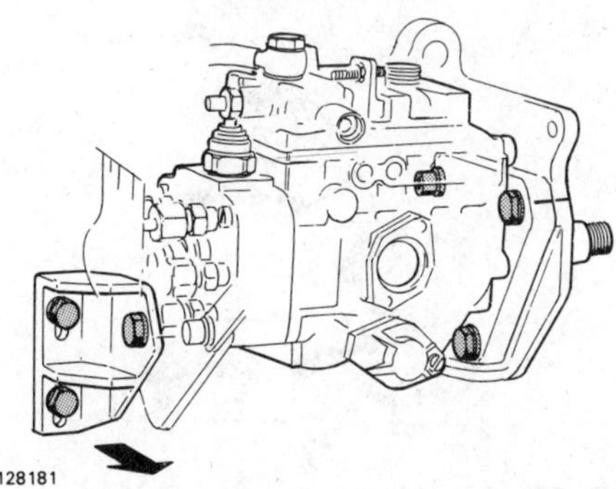

Figure 15–17. Removing injection pump retaining bolts. (Courtesy of Volvo of America Corporation)

Installing Injection Pump

When installing the injection pump, the procedure is basically a reversal of the removal process. Make certain that the engine is still at TDC for No. 1 piston on its compression stroke.

Procedure

1. Install the injection pump into position on the engine and tighten up the bolts finger tight that were removed in Step 14 under removal and shown in Figure 15–17.
2. Refer to Figure 15–17 (central right) and manually rotate the pump to align the scribe timing mark on the pump housing at its drive end with the mark on the engine cover.
3. Refer to Figure 15–15 and carefully insert the drive gear keyway over the shaft key; using the same special tools P/N 5193 and 5201, once the injection pump gear has been installed onto the pump drive shaft, install 5193 pin to prevent the injection pump from rotating while you torque the nut to 45 N·m (33 lb.ft.).
4. Check the injection pump to engine timing as described under "checking/adjusting injection pump/timing" earlier with the use of the dial gauge method.
5. If the injection pump has been rebuilt or if it is a new unit, remove the inlet pipe fitting from the top of the pump as shown in Figure 15–18 and pour clear/filtered fuel into the pump until it is full.
6. Install the rear timing belt and tighten the camshaft gear retaining nut with tool 5199 as shown earlier in Figure 15–14 to 45 N·m (33 lb.ft.).
7. Install the rear drive belt timing cover.
8. When connecting the injection pump fuel supply and return lines, make sure that you do not in-

Figure 15–18. Priming the fuel injection pump. (Courtesy of Volvo of America Corporation)

termix the supply and return screws, which look the same, otherwise the pump will starve for fuel and have low fuel pressure; the return screw is a restricted fitting type unit and is clearly marked OUT on the head of the screw (see Figure 15–12 and Figure 12–24).

9. Torque the inlet/return fuel screws to 25 N·m (18 lb.ft.).
10. Install all of the pump-to-injector fuel lines and torque them to 25 N·m (18 lb.ft.).
11. Connect the cold start device coolant hoses, accelerator cable, stop cable, and A/T kickdown cable.
12. Check and adjust the accelerator control cable as described earlier. (Figure 15–8 and 15–9.)
13. To start the engine, you may have to leave the fuel line nuts loose at the injector while you crank the engine until any entrapped air vents and clear fuel is flowing. Tighten the fuel lines one at a time when clear fuel is evident.
14. Torque the fuel lines 25 N·m (18 lb.ft.) and once the engine is running, check for any signs of fuel leaks.

NOTE: After replacing an injection pump on the engine, if the engine runs rough, emits black exhaust smoke at the exhaust pipe, and makes a heavy noise (clatters) like an engine bearing failure, the cause may be that an injector has jammed open; therefore, refer to Step 7 under "Injector Installation" following for correction procedures. Another cause could be that the injection pump may be incorrectly timed or assembled with the distributor plunger 180 degrees reversed; therefore, check the injection timing first. If OK, remove the pump and inspect or replace it with another one.

Fuel Injector Service

The fuel injectors used with the Volvo D24 diesel engine are Robert Bosch throttling pintle nozzles. The injector assemblies are Bosch model KCA 30SD 27/4 (Volvo P/N 1257144–4), while the nozzles are Bosch model DNO SD 193 mated to the Bosch VE6 distributor injection pump. Figure 29–4 under the fuel injector chapter clearly shows both the injector assembly and the type of nozzle used in this engine.

The injectors are designed to open at 1775–1920 psi (12239–13238 kPa) when new or being reset at overhaul, while an acceptable opening or popping pressure during checking them after removal from an engine is between 1700–1845 psi (11722–12721 kPa).

Injector Removal

Remove all the injectors as follows.

1. Remove all loose dirt and grime around the fuel lines and injector area of the cylinder head.
2. Obtain a variety of male/female plastic shipping caps that can be used to plug all fuel lines and open ports after they are removed.
3. CAUTION: *Never* place your hands under an injector when testing because the fuel can penetrate the skin and cause serious injury.
4. Remove the vacuum pump and its plunger from its position alongside the injection pump as shown in Figure 15–2 earlier.
5. Use a deep 1–1/16" (27-mm) socket to loosen off the injectors, then screw them out of the head.

Injector Installation

Use the following procedure.

1. Blow out the injector bore in the cylinder head with an air hose fitting, especially if there are any signs of loose carbon; sometimes you may have to scrape carbon from the cylinder head bore area.
2. Check that the injector body threads and cylinder head bore threads are burr free and not damaged.
3. Refer to Figure 15–19 and *always* install a new heat shield gasket whether the injector is an old or new one.
4. The injector should be torqued to 50 lb.ft. (70 N·m) with a deep socket once it has been run into position in its bore.
5. The injector fuel lines, once installed, should be left loose until the engine is cranked over and the trapped air can vent; when fuel pours from each individual line, tighten it up and then torque it to 18 lb.ft. (25 N·m).

Figure 15–19. Fuel injector nozzle body torque; fuel line torque and location of injector heat shield gasket. (Courtesy of Volvo of America Corporation)

6. Install the vacuum pump and plunger and install a new O-ring on the vacuum pump if it appears damaged or worn.
7. Once the engine has been started, look for any signs of leaks.

If after the engine is started, it sounds as though an engine bearing has failed (clatters excessively), it is possible that an injector may have jammed in the open position. This can be checked by loosening off each injector cap nut one at a time and looking for signs of bubbles. Any signs of bubbles would indicate that there is a cylinder compression leak; therefore, remove that particular injector.

Testing of the injectors should be done on a pop tester and they should be tested as described under fuel injector testing/repair, Chapter 29 as per the section dealing with Robert Bosch type pintle nozzles.

Glow Plug System

All current passenger car diesel engines employ individual glow plugs screwed into each pre-combustion chamber to facilitate cold weather start-up and rapid warm-up without stumble.

The glow plugs are energized electrically from the battery system through a switch/relay system. Most glow plug systems operate on the same principle although they will differ slightly in how they are wired into the electrical system. Figure 15–20 illustrates the system used by Volvo on their D24 diesel vehicles.

System Operation

Figure 15–21 illustrates the condition of the glow plug circuit when the engine coolant temperature is below 122°F (50°C) and the ignition key is turned to the *on* position. During this key position, the preheating time is indicated by a light that will come on in the instrument panel. The duration of the indicator lamp will depend upon just how cold the engine coolant is; however, the following are typical times for the lamp to illuminate:

4°F (−20°C)	45 seconds
32°F (0°C)	25 seconds
68°F (+20°C)	15 seconds
112°F (+50°C)	lamp will not come on

The glow plugs come on as soon as the ignition key is turned *on*, and they will remain on while the engine is cranking (see Figure 15–22).

The glow plugs are designed to remain *on* for about 10–25 seconds after the instrument panel light goes out to ensure smooth operation during cranking. Once the engine starts, the glow plugs will be de-energized, and the ignition key is returned to position 2. If the engine stalls, it will be necessary to turn the key back to position 1 to allow glow plug activation again.

The main purpose of the system *blocking relay* is to isolate the current flow between the control unit and the glow plug relay when the engine starts and the alternator starts to produce electrical flow to the battery. When the engine is running (Figure 15–23) and the alternator is not charging, the blocking relay "G" is connected to ground through the voltage regulator and alternator circuit.

Glow Plug Faults

Hard starting is often a direct result of one or more faulty glow plugs although the glow plug relay or temperature sending unit can also be the cause of the problem. Always check for loose or corroded connections especially at plug-in harnesses, which can cause a high circuit resistance and low voltage to the glow plugs.

The quickest way to check the glow plug circuit and to pinpoint the possible problem area is to refer to Figures 15–24 and 15–25.

NOTE: The engine coolant temperature must be 100°F (40°C) or less to effectively conduct this check. If the glow plug system turns off, turn the ignition key to the intermediate or position 1, then to the driving position 2 in order to allow the control unit to start another cut-in (energize) period.

Connect a 12-volt test light across one of the glow plug terminals and a good ground on the engine block as shown in Figure 15–24. Simultaneously check the condition of the test light and the glow plug preheat lamp on the instrument panel as shown in Figure 15–25.

The following conditions would indicate where the actual glow plug circuit problem is:

1. Test light and instrument panel light both *out* indicates that a failure has occurred at the glow plug control unit.
2. The test light does not glow but the instrument lamp does, is indicative of a glow plug relay failure.
3. The test light illuminates but the instrument panel lamp does not, is a sign that there is a problem at the temperature sender or the control unit.

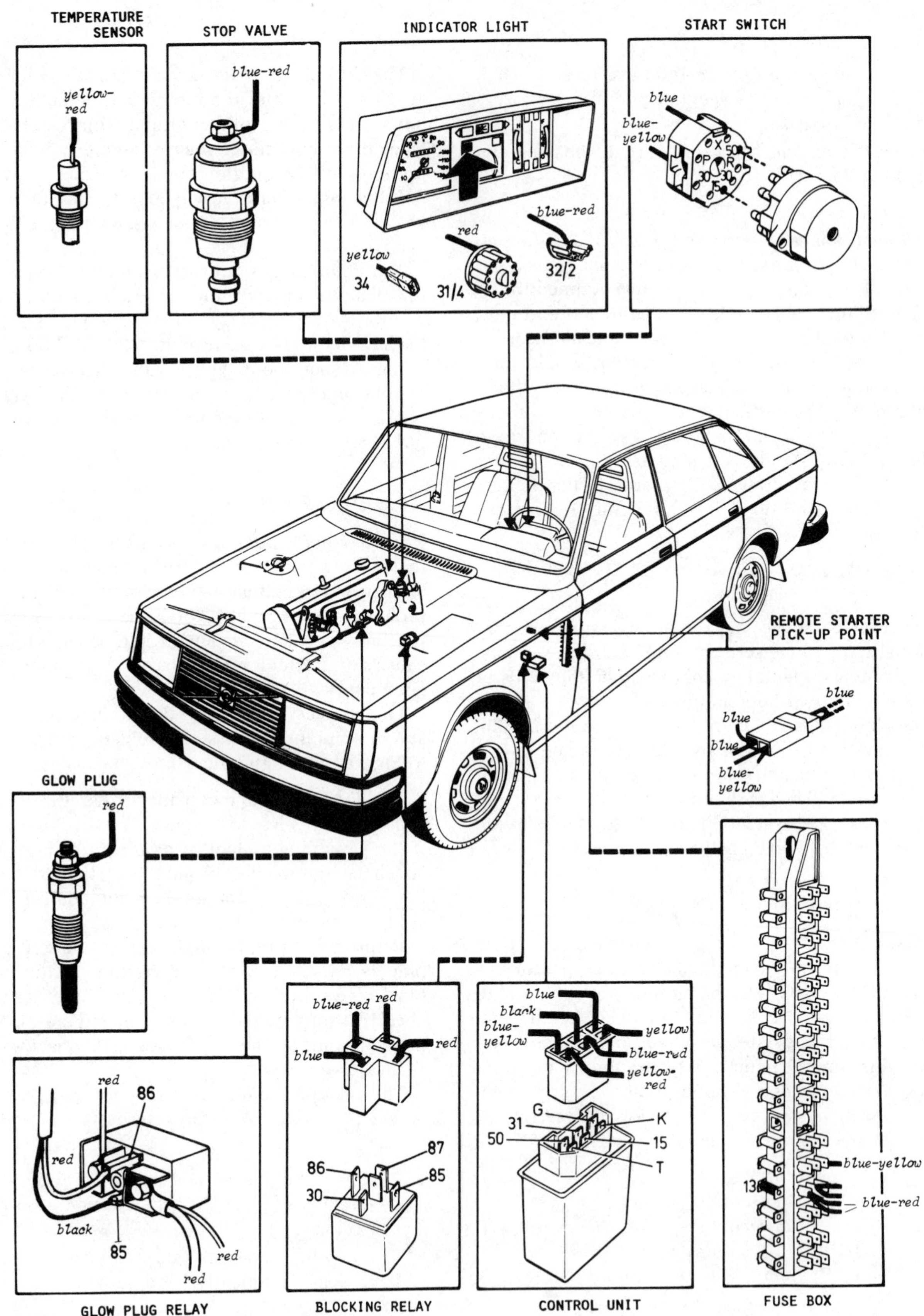

Figure 15-20. Volvo D24 glow plug system. (Courtesy of Volvo of America Corporation)

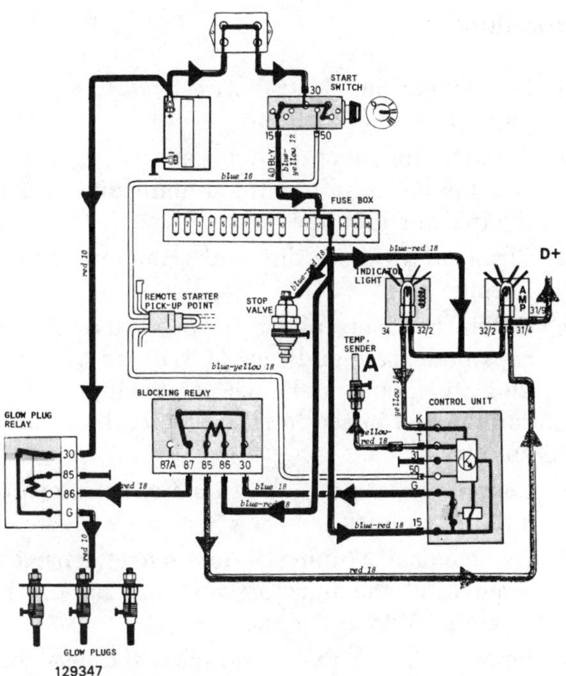

The blocking relay (G) is grounded through the voltage regulator and alternator circuit when the alternator is not charging.

Figure 15-21. Glow plug circuit—key *on* and coolant temperature below 122°F (50°C). (Courtesy of Volvo of America Corporation)

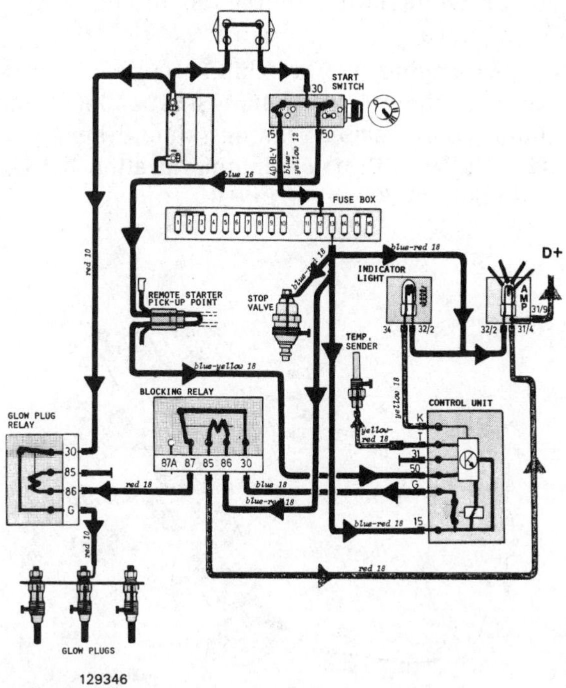

Figure 15-22. Glow plug circuit—starter motor operating. (Courtesy of Volvo of America Corporation)

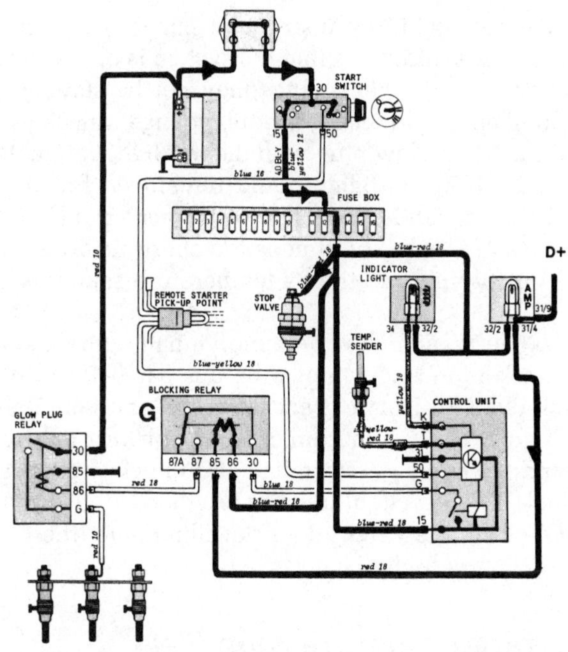

Figure 15-23. Glow plug system—engine running. (Courtesy of Volvo of America Corporation)

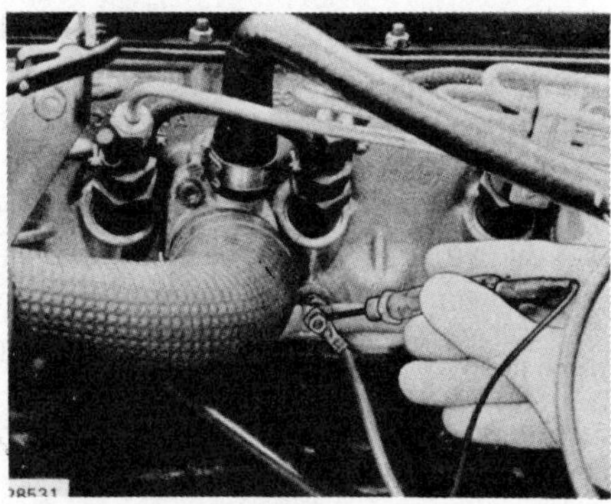

Figure 15-24. Test light connected across glow plug terminal and ground. (Courtesy of Volvo of America Corporation)

Figure 15-25. Checking test light and vehicle instrument panel glow plug lamp. (Courtesy of Volvo of America Corporation)

4. The test light and instrument panel light both illuminate, indicates that the system is operational; however, the glow plugs may not be staying on long enough or receiving full voltage. Check each individual glow plug with the test light connected to it. The test light should remain *on* for about 10–25 seconds after the instrument panel light goes out. If the *on* time is too short, insert a new control unit and then a temperature sender.

To check each glow plug individually, turn the ignition key *off* and remove the bus bar between the glow plugs to isolate them from one another. Refer to Figure 15–26 and connect a 12-volt test light between the battery + terminal and the glow plug terminal. If the test light illuminates, then the glow plug is OK; if the test light does not illuminate, then the glow plug is faulty.

Cylinder Compression Test

The fuel system and injectors are often blamed for a lack of performance, hard starting, poor fuel consumption, and exhaust smoke when in fact the problem can often be related to low compression in one or more cylinders.

Low compression can be a cause of either white or black exhaust smoke. White smoke is generally most noticeable when starting the engine from cold since combustion speed is lower than normal, which allows the fuel particles more time to condense due to the cooler air temperature and lower compression pressures.

To conduct a compression test, all injectors must be removed. Perform the compression test as follows.

Procedure

1. Disconnect the electric wire from the injection pump fuel solenoid.
2. Remove the vacuum pump and plunger alongside the injection pump and immediately below the injector fuel line group.
3. Clean off all loose dirt and grime around each injector.
4. Obtain a variety of plastic protective shipping caps both male and female that can be used to plug off all open fuel lines, cap the injectors, and cap the fuel injection pump hydraulic head fuel delivery studs.
5. Loosen and remove all of the fuel injector pipes and return lines.
6. Using a deep 27-mm (1-1/16″) socket, loosen and remove all the injectors and lift out the heat shield gaskets under the injectors.
7. Refer to Figure 15–27 and place the heat shield into the injector bore of the cylinder head, then screw in Volvo tool 9995191 and torque it to 50 lb.ft. (70 N·m).
8. Attach the compression tester to the nipple 5191.
9. Make sure that the air cleaner is not plugged and that there is no obstruction in the intake manifold.
10. Spin the engine over 5 or 6 times on the starter motor while noting the reading on the compression gauge.
11. A new engine should exhibit compression pressures in the region of 485 psi (3344 kPa); minimum compression readings should be 400 psi (2758 kPa) with a maximum variation between cylinders of 70 psi (483 kPa).

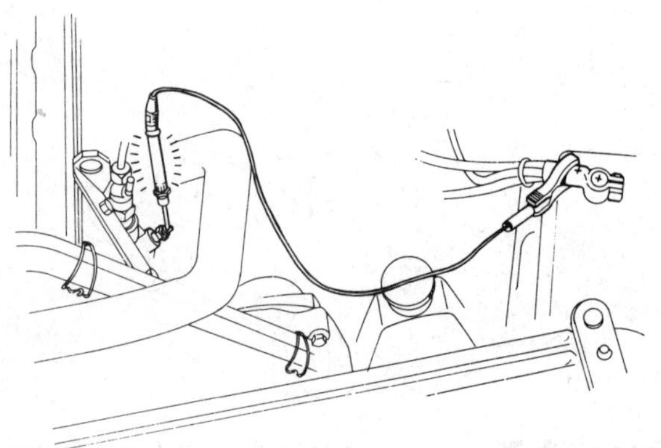

Figure 15–26. Testing glow plug. (Courtesy of Volvo of America Corporation)

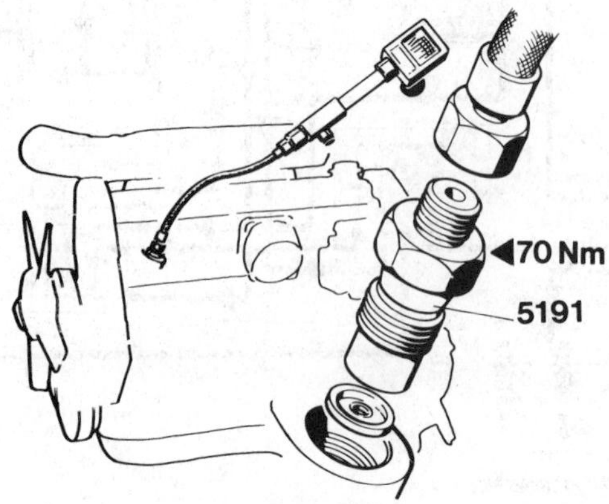

Figure 15–27. Installing compression gauge and nipple adapter. (Courtesy of Volvo of America Corporation)

12. Always install a new injector heat shield gasket when replacing the injectors back into the engine, and torque the injectors to 50 lb.ft. (70 N·m).

Valve Clearance Adjustment

The valve clearance adjustment on this engine requires the use of selected discs of different thicknesses. Since the engine is of the overhead camshaft design, no adjustment screw and lock nut is used.

The recommended valve clearances are as follows:

1. Intake valves cold setting . . . 0.008″ (0.20 mm)
   Intake valves warm setting . . . 0.010″ (0.25 mm)
   Acceptable intake valve clearance when checking cold . . . 0.006–0.010″ (0.15–0.25 mm)
   Acceptable intake valve clearance when checking warm . . . 0.008–0.012″ (0.20–0.30 mm)
2. Exhaust valves cold setting . . . 0.016″ (0.040 mm)
   Exhaust valves warm setting . . . 0.018″ (0.45 mm)
   Acceptable exhaust valve clearance when checking cold . . . 0.014–0.018″ (0.35–0.45 mm)
   Acceptable exhaust valve clearance when checking warm . . . 0.016–0.020″ (0.40–0.50 mm)
3. Available adjusting disc thicknesses in increments of 0.002″ (0.05 mm) start at 0.1299″ to 0.1673″ (3.30 to 4.25 mm)

To check the valve clearance, manually rotate the engine over with a 27-mm (1-1/16″) socket on the bolt of the crankshaft vibration damper pulley to place each individual cylinder at TDC as you check it. This can be confirmed by the fact that both cam lobes should face up and the cam base would be over each valve disc.

To alter the valve clearance, the engine should be rotated over with a 27-mm (1-1/16″) socket on the crankshaft vibration damper pulley to place the piston at about 90 degrees past TDC, otherwise the disc will not have enough clearance to be removed.

To remove a disc, refer to Figure 15–28, and using tool 5196, first of all line up the valve depressors so that their notches face as shown. The depressor grooves must be above the disc face in order to allow you to grab the disc with a pair of pliers as shown in Figure 15–29.

Once the disc has been removed from above the valve, check its thickness with a 0–1″ (0–25.4-mm) micrometer. All discs are identifiable with a number on one side as to its thickness. When installing new discs, lubricate them with engine oil and insert them on top of the valve retainer with the numbered side facing down.

Rotate the piston to TDC and check the valve to camshaft lobe clearance.

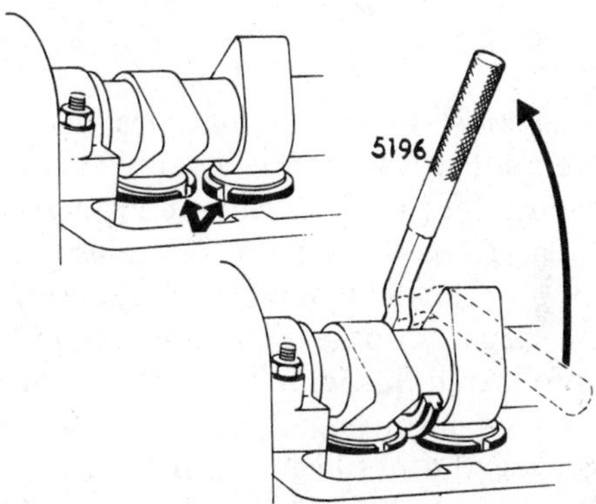

Figure 15–28. Aligning valve depressors and using tool P/N 5196 to depress them. (Courtesy of Volvo of America Corporation)

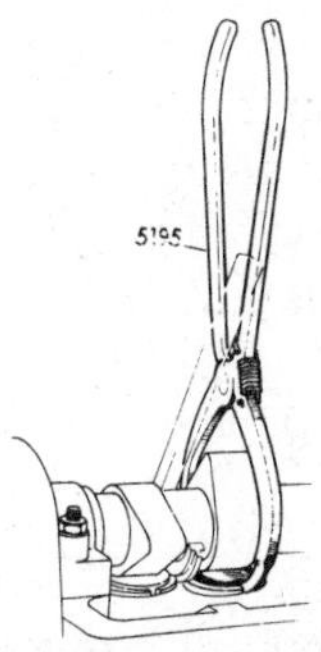

Figure 15–29. Removing valve adjustment disc with pliers P/N 5195. (Courtesy of Volvo of America Corporation)

16

Peugeot Diesel Fuel System

One of the most respected marques in the world when it comes to diesel cars is that of Peugeot, where 24% of their total car production in France is sold in that country. Their world diesel car sales account for about 20% of their total vehicle production.

Peugeot began producing industrial engines in 1928 and followed this with the introduction of their HL 50 passenger car engine in 1936. Mass production, however, did not begin until 1958 with the introduction of their XD4/85 diesel engine in the 403 model passenger car. This 4-cylinder diesel engine was subsequently installed in the 404 in 1963, the 504 in 1970, and the 604 in 1979. Figure 16–1 illustrates a history of Peugeot's production of diesel passenger cars.

In the North American market, Peugeot introduced in 1974 the 504D sedan, which was equipped with the XD4.90 diesel engine. This engine had a displacement of 2.1L (129 cu.in.). In 1976, the XD2 and XD2C engine with 2.3L (141 cu.in.) displacement was introduced. This engine is now offered as a turbocharged unit and is identified as the XD2S.

Year	*Vehicle*	*Engine*	*Displacement*	*Power Kw (HP)*	*Max. torque*
1958	403	XD4/85	1816 cc	37 (50) at 4000 rpm	70 lb. ft/2250 rpm
1963	404	XD4/88	1948 cc	41 (56) at 4500 rpm	77 lb.ft/2250 rpm
1968	204	XLD	1255 cc	29.5 (40) /5000 rpm	49 lb.ft/3000 rpm
1970	504	XD4/90	2112 cc	48 (65) at 4500 rpm	85 lb.ft/2500 rpm
1973	204	XL4D	1357 cc	32.5 (44) /5000 rpm	54 lb.ft/2500 rpm
1976	504	XD2	2304 cc	51.5 (70) /4500 rpm	95 lb.ft/2250 rpm
1977	304	XL4D	1357 cc	32.4 (44) /5000 rpm	54 lb.ft/2500 rpm
1979	305	X1D	1548 cc	37.5 (51) /5000 rpm	58 lb.ft/2500 rpm
1979	604	XD2S	2304 cc	61 (82.5) /4200 rpm	136 lb.ft/2000 rpm
77/84	504D	XD2			
	505D	XD2C	2304 cc	53 (71) at 4500 rpm	99 lb.ft/2500 rpm
81/84	505TD	XD2S Turbo	2304 cc	60 (80) at 4150 rpm	136 lb.ft/2000 rpm
82/84	604TD	XD2S Turbo	2304 cc	60 (80) at 4150 rpm	136 lb.ft/2000 rpm

Figure 16–1. History of Peugeot's production of diesel passenger cars. (Courtesy of Peugeot Motors)

The model numbering system used with the Peugeot engines indicates its bore in millimeters. For example, an XD4.90 indicates that it is a 4-cylinder engine with a bore of 90 mm (3.54"). The XD2/XD2C and XD2S engines have a bore and stroke of 3.70" × 3.26". The naturally aspirated diesel engines have a compression ratio of 22.5:1 versus the 21:1 used on the turbocharged 604 engine.

The combustion chamber design is of the Ricardo Comet V layout in all of their diesel engines; this layout is used in a number of pre-combustion chamber diesel passenger car engines such as the Volkswagen Rabbit/Jetta, the Audi 5000, and the Volvo D24.

The engine firing order is 1-3-4-2 for all engines with No. 1 cylinder being at the flywheel end of the engine and No. 4 cylinder being at the front of the engine.

Fuel System Arrangement

The fuel injection system used with the Peugeot diesel engines is supplied by several manufacturers. Robert Bosch distributor-type injection pumps are installed on a number of engines, while CAV (RotoDiesel France) are used on other models. The Bosch pumps used range from the Model EP/VA, the EP/VM, to the current VE model that is used on the turbocharged 604 engine.

The CAV injection pumps are manufactured under license in France by a company known as RotoDiesel. Whether it is a CAV or a RotoDiesel pump that is installed on the engine, both operate on the same principle since they are of the DPA (distributor pump assembly) design.

The 1980 and later engine models use a Robert Bosch EP/VE injection pump that replaces the CAV-RotoDiesel DPA-type injection pump.

The Robert Bosch injection pump models EP/VA and VE units are discussed in detail in Chapter 12 Robert Bosch; therefore, they will not be discussed here in detail. Refer to Chapter 12, Robert Bosch, VE pump section for detailed information relating to operation, maintenance, repair, and troubleshooting of the VE, VA, and VM injection pumps. Similarly, the CAV DPA injection pump is discussed in detail in the Lucas CAV chapter (Chapter 28).

The EP/VM distributor pump is manufactured by Diesel Kiki in Japan, which is a licensee of Robert Bosch in Germany. This injection pump is also used on certain models of Peugeot vehicles.

All three injection pump types employ an electrical fuel shutoff solenoid to stop the engine. In addition, the EP/VA pump is equipped with a deferred injection accumulator and a hydraulic governor, while the VE model pump incorporates a mechanical governor and a hydraulically activated timing advance mechanism. The VE pump uses a minimum/maximum speed governor that controls the low speed and maximum speed of the engine. Speeds in between this are controlled by the engine throttle manipulated by the driver of the vehicle.

To comply with U.S. EPA exhaust emissions, all pumps are equipped with an EGR (exhaust gas recirculation) control system. The XD2S turbocharged diesel engine uses two load sensors to regulate EGR according to engine load. The EGR device is commonly referred to as an "altitude/boost compensator," and its operation is described under the Robert Bosch (Chapter 12) VE pump section, and also in Chapter 18 (Nissan).

Injection Pump Timing Check

A description of the procedure required to check the static timing of the injection pump is listed in the Robert Bosch (Chapter 12) VE pump section, see Figures 12–30 and 12–31. Refer to this section when checking and setting the injection pump on the engine for additional information not contained in this chapter.

The following specifications are the current settings for Peugeot injection pumps.

Robert Bosch Pump Model	*Pump Lift*
EP/VM AR5 and AR7	0.38 mm (0.0149")
EP/VM AR8, AR 10, AR 12	0.55 mm (0.0216")
EP/VA	0.65 mm (0.0255")
EP/VE	0.50 mm (0.020")
80–83 XD2 Engine	1.35 mm (0.053")
80–83 XD2C Engine	0.97 mm (0.038")
81–83 XD2S Engine	0.40 mm (0.0157")

In addition to the above specifications, the 1978–84 XD2/XD2C engine models have an injection pump setting of 13 degrees BTDC, while the 1981–84 XD2S turbocharged engine pump setting is 0.016" (0.406 mm) BTDC; in Canada the setting is 0.031" (0.787 mm) BTDC.

NOTE: Always refer to the vehicle exhaust emissions decal/tune-up information sticker located under the hood for specific injection pump/engine timing settings.

SPECIAL NOTE: When checking the injection pump timing with a dial gauge installed onto the pump as shown in Figures 12–30 and 12–31, if the correct pump lift cannot be obtained on a gear-driven-type pump by rotating the pump towards or away from the engine because there appears to not be enough spacing in the pump mounting bolt hole slots, then the pump drive gear teeth are mistimed by at least one tooth. If this condition is encountered, remove the pump and realign the drive gears.

Anytime that the injection pump has been installed onto the engine and the dial gauge reading set to the specifications shown above, doublecheck the engine/pump timing as follows:

1. Make sure that the dial indicator mounted on top of the No. 4 exhaust valve (No. 4 cylinder is at the front of the engine) indicates that the engine is still at TDC (dial gauge should read ZERO ± 0.02 mm or 0.0007".
2. Manually rotate the crankshaft counterclockwise seven dial indicator revolutions and when the dial gauge pointer approaches the end of the seventh revolution, check to see that the injection pump mounted dial gauge is still at the BDC position.
3. Gently/slowly rotate the crankshaft clockwise and check that at TDC the pump lift corresponds to the readings listed in the above chart.

Injection Pump

Injection Pump Removal

Removal of the injection pump requires that the battery ground cable be disconnected first to isolate the starting system.

Procedure

1. Remove all fuel inlet and return lines from the pump and cover them with plastic shipping caps to prevent the entrance of dirt.
2. Disconnect the stop control cable (if used) or remove the fuel solenoid wire.

3. Disconnect the accelerator cable and fast idle cable from their connections on the injection pump.
4. On engines with a chain driven pump, remove the two pump-to-bearing support retaining screws. On engines with a gear driven pump, remove the two pump intermediate flange-to-timing gear housing retaining screws.
5. Move the injection pump backwards and pivot it towards the engine as you attempt to remove it.

Injection Pump Installation

Before installing the fuel injection pump onto the engine, the following procedure must be performed. Regardless of the type/model of injection pump used on a Peugeot diesel engine, accurate engine timing must first of all be established before installing the injection pump onto the engine.

If the engine gear train timing marks have not been disturbed, proceed as described below. If, however, the engine gear train timing has been disturbed during engine repair, with the front timing cover removed, align the timing marks as follows:

1. On chain-driven timing components, rotate the crankshaft to place the copper link in the chain opposite the timing mark on the crankshaft gear. The gear keyway should be at the 12 o'clock position with the copper link and gear timing mark at the 6 o'clock position. Also ensure that the chain links marked with a timing line are in fact opposite the camshaft and injection pump gear timing marks. The reference marks on the timing chain are located at link No. 1, 41 and 67 with link No. 1 always being identified by a white paint mark.

   NOTE: Once the timing marks are in alignment, if the engine is turned over for any reason, it requires 90 complete revolutions of the crankshaft to realign these marks.

2. On gear-timed engines, simply align all the dots on the gears to correctly time the engine. On these engines it will require 22 complete crankshaft revolutions in order to realign the timing marks if the engine is turned over for any reason after initial alignment of the marks on the gears.

The following procedure should be followed for all injection pumps before installation:

Procedure

1. Rotate the engine clockwise to place the No. 1 piston at TDC.
2. Using Peugeot special tool P/N 8.0105 or equivalent, slide the No. 4 exhaust valve rocker arm towards the rear of the engine with the rocker arm pallet (valve side) facing upwards.
3. Using the special tool in Step 2, depress and remove the No. 4 exhaust valve spring.
4. Check that the valve moves freely in its guide, then allow the valve to drop onto the piston.
5. Remove the plug connector bar and the No. 2 and No. 4 cylinder glow plugs to allow the engine to be turned over more easily.
6. Install a dial indicator gauge directly on top of the No. 4 exhaust valve stem and preload the dial gauge; rotate the gauge bezel to *ZERO* the dial gauge.

   NOTE: The purpose of placing the dial gauge on top of the No. 4 exhaust valve is to obtain an accurate TDC reference position mark.

7. Manually rotate the engine counterclockwise from the crankshaft pulley at least 4 full dial indicator revolutions.
8. Manually rotate the crankshaft clockwise from the pulley in a slow and steady fashion (don't jerk it), until a dial gauge reading of 1.40 mm (0.055") for the AR5, 7, or 8 injection pump is obtained; or until the gauge reads 1.46 mm (0.0574") for the AR10 or 12 injection pump is obtained.

   NOTE: Do not remove this dial gauge yet. Leave it in place because it will be referenced later once the injection pump has been installed onto the engine when it is used to double check that the pump-to-engine timing is in fact correctly set.

Pump Installation PreCheck

With the engine at TDC No. 1 compression stroke, follow the procedures below for the particular model of pump being used.

Procedure for Bosch/Kiki EP/VM

1. Refer to Figure 16–2 and install a Peugeot timing tool and dial gauge P/N 6.0168 or its equivalent into the top of the injection pump and bolt it into position.
2. Ensure that at least 20 mm (0.787") exists between the feeler arm of the tool and the dial indicator gauge mounting arm. Ensure that all components move freely.
3. If the pump is chain driven, manually rotate the pump in order to place the "double width" tooth of the pump drive gear in horizontal alignment

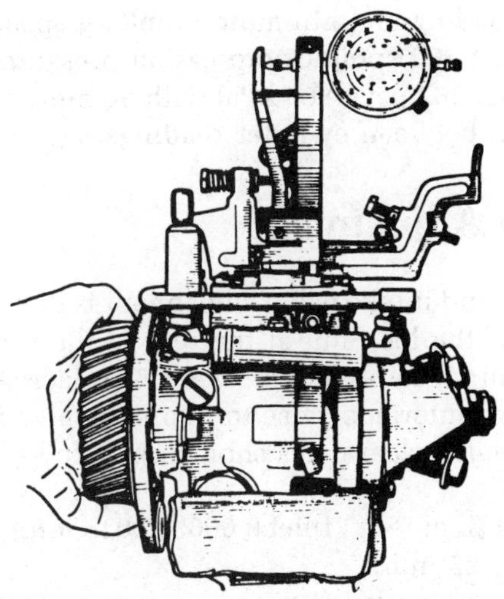

Figure 16–2. Checking EP/VM injection pump timing with a dial indicator. (Courtesy of Peugeot Motors)

with the delivery valve outlet fuel stud to No. 4 cylinder on the hydraulic head of the pump.

4. If the pump is gear driven, manually rotate the pump drive gear so that the tooth with the timing mark on it aligns with the "D" mark stamped to the left of the injector delivery valve fuel outlet stud when viewing the pump from the non-drive end.
5. Manually rotate the injection pump drive gear back and forward until the dial gauge registers at its lowest position and ZERO the gauge.
6. Manually rotate the pump drive gear gently/slowly in its normal rotation and stop the instant that the dial gauge just starts to move, which is the start of the pump plunger upstroke (approximately 0.02 mm or 0.0007").
7. The EP/VM injection pump can now be installed onto the engine.
8. Align the scribe timing mark on the pump housing with the stationary mark on the engine timing cover and snug up the retaining bolts, then proceed to check the pump lift with a dial gauge as described above.

EP/VA and EP/VE Pump Installation Precheck

Procedure

1. Remove the deferred injection accumulator immediately above the four injection fuel outlet studs on the pump hydraulic head; this accumulator must be removed in order to look into the pump through the hole at the center of the four outlet fuel studs on the end of the pump.
2. Look into the hole in the center of the pump (outlet end) and manually rotate the injection pump in order to align the timing groove with the outlet coupling fuel stud marked with a "B."
3. Using Peugeot dial indicator support tool P/N 8.0117F or equivalent, install a dial gauge into the center outlet of the injection pump as shown in the "Static Pump Timing Check" described under the Robert Bosch (Chapter 12), VE pump section, see Figures 12–30 and 12–31.
4. The injection pump can now be installed on the engine as per Steps 7 and 8 described above for the EP/VM pump.

Idle and Maximum Speed Adjustments

The location of the idle and maximum speed adjusting screws can be seen in Figure 12–17 of the Robert Bosch pump chapter.

Recommended idle and maximum speeds can be found on the vehicle exhaust emissions decal/sticker located under the engine hood. Typical idle rpm is between 780–830 on the XD2/XD2C engine or 830–860 when equipped with an air conditioning option. Idle speed for the XD2S turbo diesel is between 650–850 rpm and a minimum of 700–800 when equipped with an air conditioner option.

Fuel Injectors

The fuel injectors and nozzles used on Peugeot vehicles are of Robert Bosch manufacture and are of the pintle or throttling pintle type. A description of the operation of these nozzles can be found in Chapter 29, "Injection Nozzles." Engines using the CAV-RotoDiesel injection pump employ a Bosch nozzle with a side inlet for the fuel and a top outlet for the return fuel.

Engines using Bosch pumps employ an injector similar to that shown in Figure 29–4, with fuel entering the top of the injector and returning through two small leak-off lines just below the inlet. All information relating to removal, testing, and repair can be found in Chapter 29, diesel fuel injectors.

When an engine is suspected of having faulty nozzles, run the engine until it is at operating temperature, then slacken off the injector fuel line at the nozzle end with the engine running at an idle rpm. There should be a reduction in engine speed and a positive change to the operating sound of the engine.

If no reduction in rpm or change in engine operation occurs when you do this, then the nozzle in that cylinder is faulty and should be removed and tested.

New nozzle release pressures for the XD2/XD2C engines is 1740–1813 psi (11997–12500 kPa), while used nozzle release pressures are 1668–1813 psi (11500–12500 kPa). On units produced before 1983, the XD2S turbocharged engine nozzles should all release at 1813 psi (12500 kPa). The 1983 and later turbocharged engine nozzles should release at 2175 psi (14996 kPa).

Compression Check

Remove all injectors to perform this test or all of the glow plugs and install a compression gauge and adapter into the cylinder head. The engine should be at normal operating temperature. Place the fuel injection pump control in the fuel cutoff position. Crank the engine (minimum cranking speed should be at least 200 rpm). Compression pressures should be at least 262 psi (1806 kPa) with no more than 10% variation between cylinder readings.

Valve Adjustment

Exhaust and intake valve adjustment is done by loosening off the lock nut at the rear of the rocker arm and turning the screw clockwise to decrease valve lash or counterclockwise to increase valve lash.

Cold valve clearances should be as follows:

XD88/90 Engine . . . Inlet 0.006" (0.015 mm), Exhaust 0.010" (0.25 mm)

1978–84 XD2/XD2C Engine . . . Intake/Exhaust 0.010" (0.25 mm)

1981–84 XD2S Turbo . . . Intake 0.006" (0.015 mm), Exhaust 0.010" (0.25 mm)

17

Dodge Ram 50 Diesel Engine

The Dodge Ram 50 diesel engine is a 4-stroke cycle 4-cylinder engine designated the 4D55-TBD engine because it is a 55-horsepower turbocharged unit of 2.3L or 2346 cc (143.2 cu.in.) displacement.

The engine employs an overhead camshaft design with a bore of 3.59" and a stroke of 3.54" (91.1 and 90 mm) with a compression ratio of 21:1 for a minimum compression pressure of 384 psi (2648 kPa) at 250 rpm cranking speed. The firing order is 1–3–4–2 with the engine rotating clockwise from the front.

The engine is of the pre-combustion chamber swirl design and employs a glow plug in each combustion chamber to facilitate starting and warm-up. Initial injection pump to engine timing is 5 degrees ATDC at 1 mm (0.03934" or 0.040") pump plunger stroke.

The inlet valves open 20 degrees BTDC and close 48 degrees ABDC, while the exhaust valves open at 54 degrees BBDC and close at 22 degrees ATDC for a positive valve overlap condition of 42 degrees.

Inlet and exhaust valve clearances (hot) are 0.010" (0.25 mm).

Fuel System

The fuel system employs a distributor-type injection pump of the Robert Bosch/Diesel Kiki VE design and four Bosch throttling pintle-type injection nozzles. Detailed injection pump operation and service procedures can be found in Chapter 12 dealing with Robert Bosch VE injection pumps.

The letters VE originate from the German word "Verteiler," which means distributor, although many people refer to these pumps as rotary pumps. Diesel Kiki is a licensee of Robert Bosch; therefore both the Bosch and Diesel Kiki injection pumps operate in the same manner.

Simplified Fuel System Arrangement

A simplified layout of the basic fuel system is shown in Figures 17–1 and 17–2, which illustrates the flow from the fuel tank to a fuel/water filter separator now commonly used on all diesel automotive engine applications.

There is no separate fuel transfer pump used with the TBD engine because a vane-type pump (see Figure 12–17) located in the injection pump is used both to pull fuel through the system from the fuel tank and to place the fuel under pressure to feed the distributor injection pump operation. Fuel in the tank is first of all drawn through an in-tank fuel filter (item 18), which can be seen in Figure 17–1 showing the fuel tank components.

The fuel then passes to the fuel/water separator and then onto the vane feed pump contained in the injection pump housing where the fuel is placed under approximately 100 psi (689.5 kPa) pressure. Additional fuel pressure is created within the injection pump by the action of a set of rollers lifting a pumping plunger thereby increasing the fuel pressure to a high enough point that the injection pump delivery valve will open and supply fuel through a high pressure steel line to an injector.

The delivery valve plunger opens at approximately 306 psi (2108 kPa) on this injection pump to allow fuel to pass through to an injector. Fuel is delivered to each injector in engine firing order sequence by the action of the rotating/reciprocating injection pump plunger described in detail in the VE pump section (Chapter 12). The injection nozzles on the Dodge Ram 50 are adjusted to open at 1707 psi (11,768 kPa).

Fuel in the fuel injection pump and nozzles that is used for cooling and lubricating purposes is re-

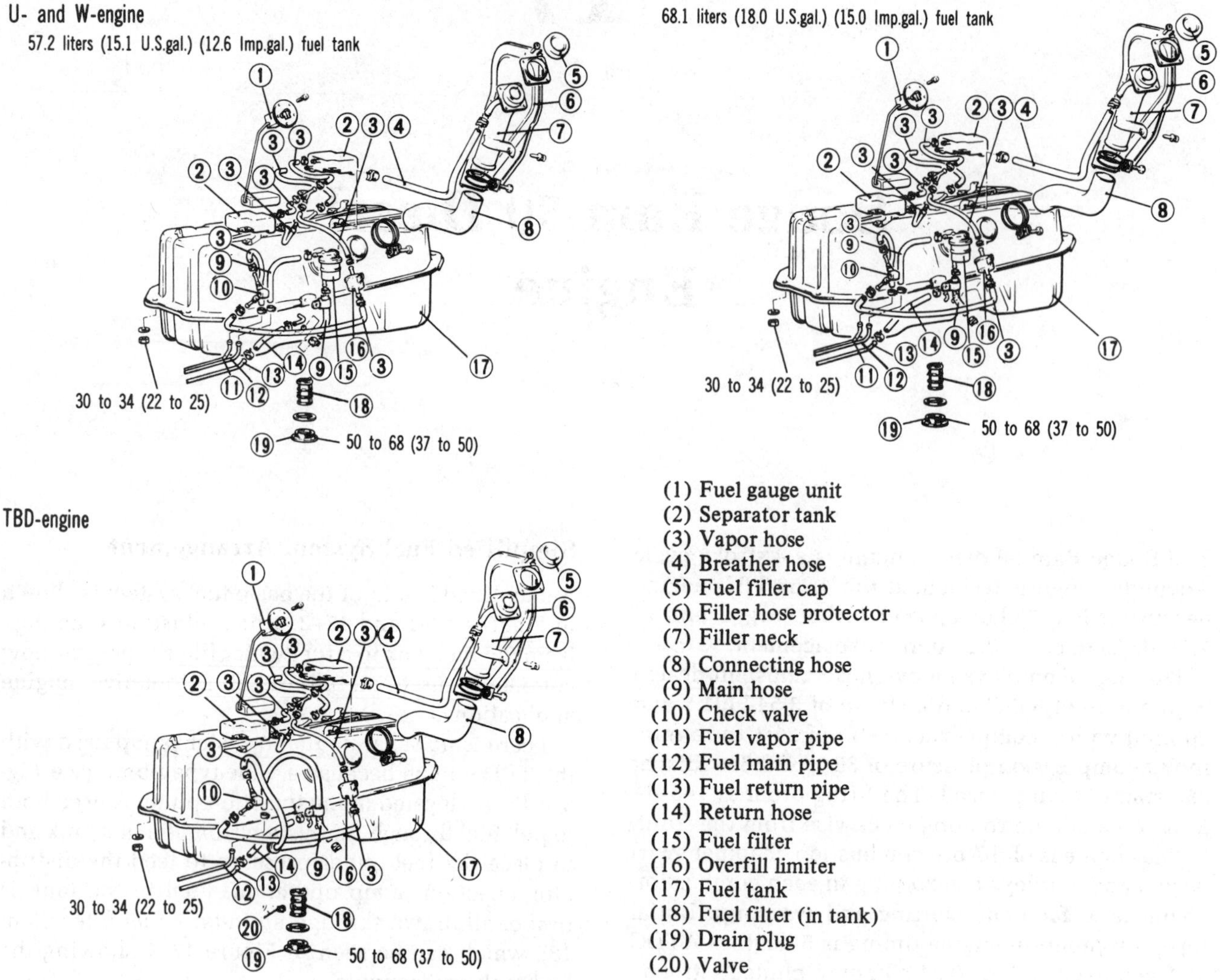

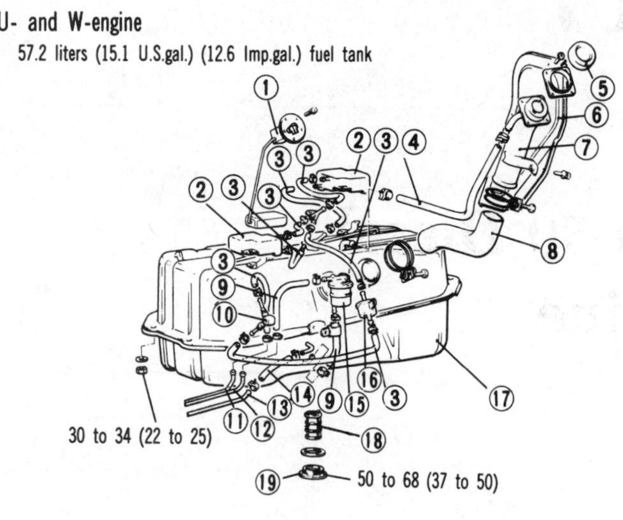

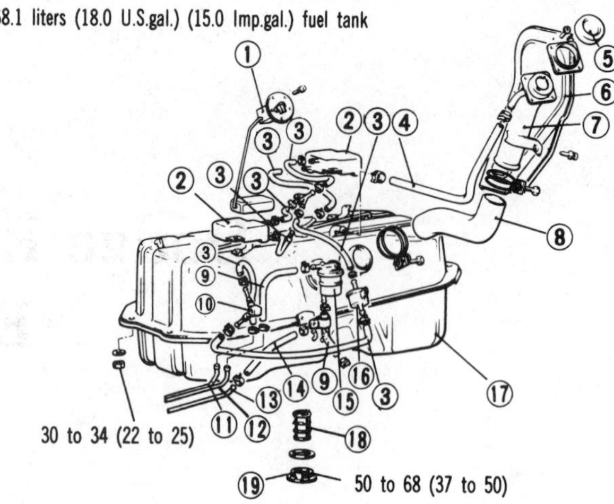

U- and W-engine
57.2 liters (15.1 U.S.gal.) (12.6 Imp.gal.) fuel tank
30 to 34 (22 to 25)
50 to 68 (37 to 50)
68.1 liters (18.0 U.S.gal.) (15.0 Imp.gal.) fuel tank
30 to 34 (22 to 25)
50 to 68 (37 to 50)
TBD-engine
30 to 34 (22 to 25)
50 to 68 (37 to 50)
(1) Fuel gauge unit
(2) Separator tank
(3) Vapor hose
(4) Breather hose
(5) Fuel filler cap
(6) Filler hose protector
(7) Filler neck
(8) Connecting hose
(9) Main hose
(10) Check valve
(11) Fuel vapor pipe
(12) Fuel main pipe
(13) Fuel return pipe
(14) Return hose
(15) Fuel filter
(16) Overfill limiter
(17) Fuel tank
(18) Fuel filter (in tank)
(19) Drain plug
(20) Valve

Figure 17–1. Ram 50 TBD fuel tank components. (Courtesy of Chrysler Corporation)

turned through a return fuel line to the fuel tank (as shown in Figure 12–16 and as item 13 in Figures 17–1 and 17–2). In this way the fuel is constantly recirculating back to the fuel tank where it can cool after picking up heat at both the injection pump and nozzles.

Fuel Lines and Hoses

The arrangement of the fuel lines and hoses to and from the fuel tank along with the fuel/water separator are shown in Figure 17–2.

Basic Fuel System Maintenance

Preventive maintenance of the fuel injection system generally consists of filter changes and idle and maximum rpm adjustments. Any other service requirements are discussed under "Injection Pump Checks/Adjustments" below.

Diesel fuel used with the engine can either be a No. 1D or a No. 2D grade, depending on the severity of service/ambient temperature conditions. For best fuel economy and performance at all ambient temperatures above −7°C (20°F), a No. 2D fuel should be used. At ambient temperatures below this, a No. 1D

(1) Main hose
(2) Return hose
(3) Fuel heater
(4) Hand pump
(5) Air plug
(6) Fuel temperature sensor
(7) Fuel filter pump body
(8) Protector
(9) Fuel filter cartridge
(10) Water level sensor
(11) Drain plug
(12) Fuel main line
(13) Fuel return line

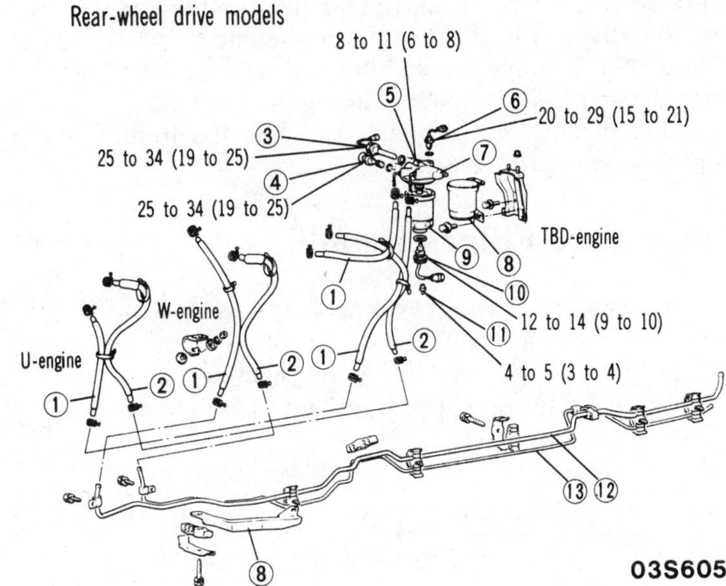

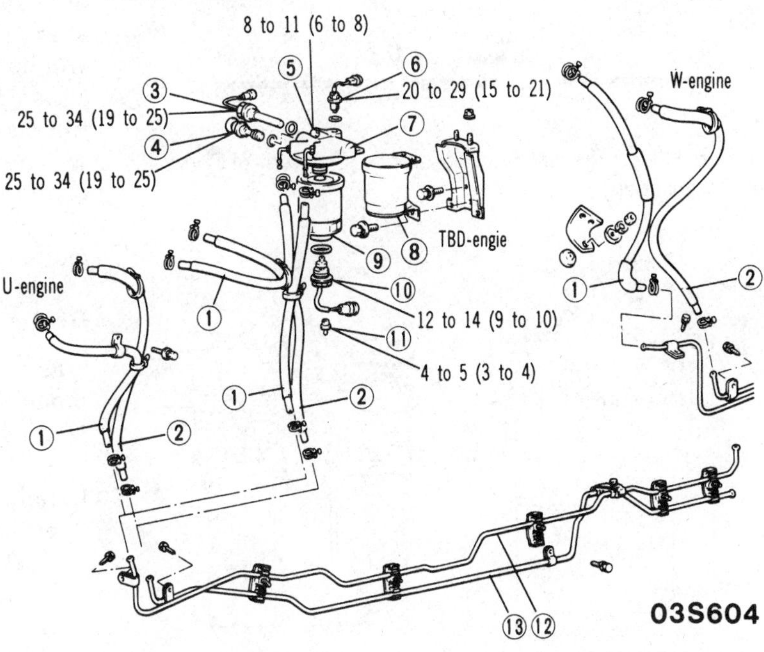

Tightening torque: Nm (ft-lbs.)

Figure 17-2. Vehicle fuel lines and filter components. [Courtesy of Chrysler Corporation]

or a winterized No. 2D fuel should be used to prevent wax formation at the filters that will plug them up and prevent fuel transfer to the pump, causing hard starting problems.

A detailed analysis of fuel oils can be found in Chapter 4 dealing with gasoline and diesel fuels.

In-Tank Fuel Filter

The replacement of the fuel tank filter is not a common preventive maintenance item. If it becomes necessary to change it at any time, the fuel tank has to be drained first.

Figure 17–3 illustrates the in-tank filter unit that is removed along with the tank drain plug, item 19 in Figure 17–1. Remove the old filter and slide the new filter element over the in-tank filler pipe. Then, after inserting the unit into the fuel tank, tighten the drain plug to 37–50 lb.ft. (50–68 N·m).

Fuel/Water Filter Assembly

The fuel/water filter assembly is actually a water separator as well as a fuel filter.

A sectional view of the fuel/water separator is illustrated in Figure 17–4. This filter unit also contains two main accessories that are important to the system operation:

1. A fuel heater element
2. A hand-priming pump

The fuel heater element is actuated by an electrical signal from a fuel heat controller at temperatures of 3 ± 3°C (37 ± 5°F) to ensure that wax crystals do not form in the filter assembly. The heater is capable of 110 watts output when the fuel temperature sensor senses a drop to the levels mentioned above; it simultaneously illuminates a fuel heater light on the instrument panel.

When the ignition switch is turned to the crank position, the fuel heater is turned off because of the large current draw to the starter motor. Once the engine starts however, as long as 9 volts or more is present at the "L" terminal of the alternator and the temperature of the fuel is between 4.5 ± 3°C (40 ± 5°F), the fuel heater is turned off. If the fuel temperature is between 3 ± 3°C (37 ± 5°F), the fuel heater will be turned on.

As with all fuel/water separators, the bottom of the filter is designed as a sediment trap for both dirt and water accumulations. When approximately 80 cc (4.9 cu.in.) of water gathers at the base of the filter, the water level sensor activates the instrument panel-mounted *water-in-fuel* light.

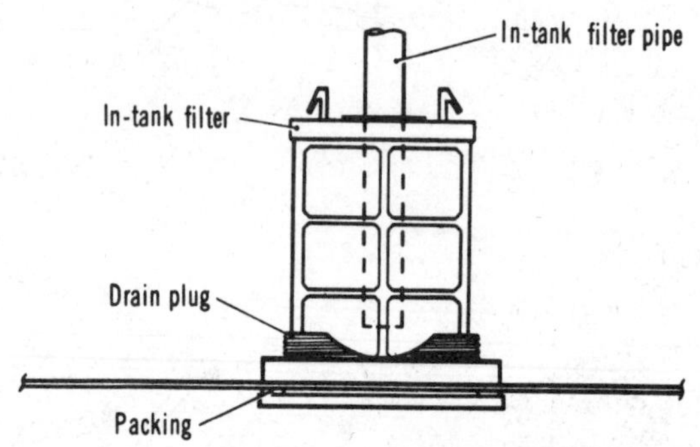

Figure 17–3. In-tank filter assembly. (Courtesy of Chrysler Corporation)

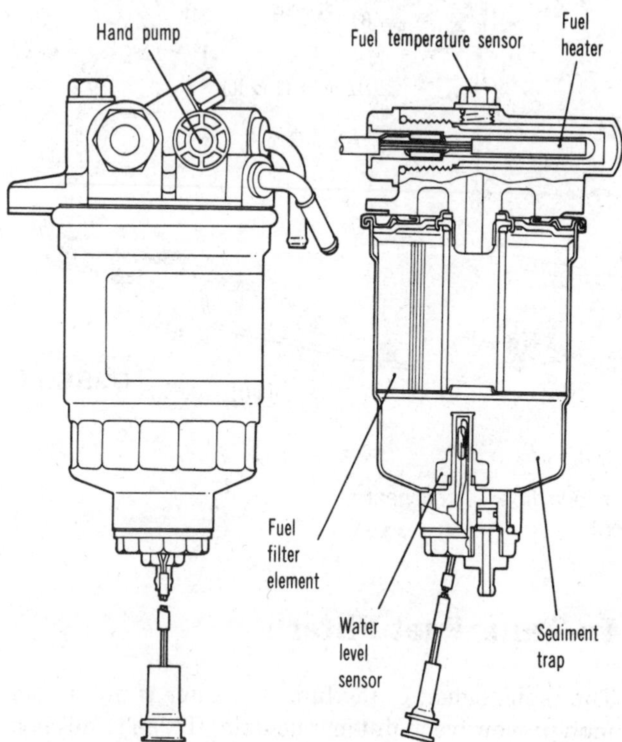

Figure 17–4. Sectional view of fuel filter. (Courtesy of Chrysler Corporation)

Draining Water From the Filter Assembly

When the *water-in-fuel* light illuminates, the water should be drained from the fuel/water assembly as soon as possible, as described below.

1. Refer to Figure 17–5 and loosen off the drain plug at the base of the filter assembly.

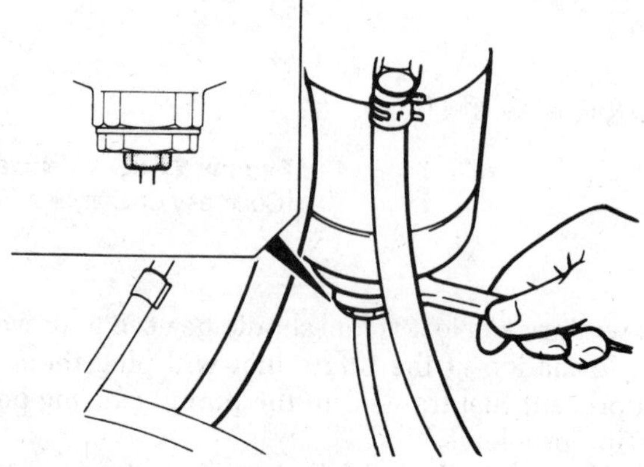

Figure 17–5. Loosening/tightening fuel/water filter drain plug. (Courtesy of Chrysler Corporation)

2. To ensure that the water drains completely, turn the hand pump knob on the upper part of the filter body counterclockwise until it can be pulled out.
3. Push/pull the priming pump handle until all water has been expelled from the filter assembly and water-free diesel fuel is seen to flow.
4. Tighten up the drain plug to 3–4 lb.ft. (4–5 N·m).

Fuel/Water Separator Removal

Recommended change period for the fuel filter on the Dodge Ram 50 diesel engine is every 30,000 miles or 48,000 km. This of course may be required sooner, depending upon engine operating conditions and severity of service.

Procedure

1. Remove the fuel/water filter assembly as shown in Figures 17–2 and 17–4 after disconnecting the water level sensor electrical connector, fuel heater electrical connector, fuel temperature sensor, and main hoses.
2. Remove the support bracket and take the complete filter assembly off the engine and place it on a work bench.
3. Refer to Figure 17–2 and screw the filter cartridge (9) out of the fuel filter pump body (7) by hand.
4. Remove the temperature sensor, fuel heater, hand pump, and air bleeder plug from the filter pump body.
5. Remove the water level sensor (Figure 17–4) and drain plug (Figure 17–5) by holding the base of the filter gently in a vise—preferably with soft jaw protectors; carefully inspect the water level sensor by connecting it to its wiring harness and making sure that the instrument panel warning light comes on when the ignition key is turned *on*.
6. If the warning light fails to illuminate, gently move the float up and down as shown in Figure 17–6 while checking to see if the warning light flashes on/off as you do this.
7. If the warning light doesn't flash on/off, replace the warning light bulb and recheck its proper operation; make sure that the warning light goes out when the engine is started.
8. Clean disassembled components in kerosene and replace necessary parts.
9. When assembling the fuel/water separator components, the following torques should be used:
   a. Fuel Heater and Hand Pump—19–25 lb.ft. (25–34 N·m)

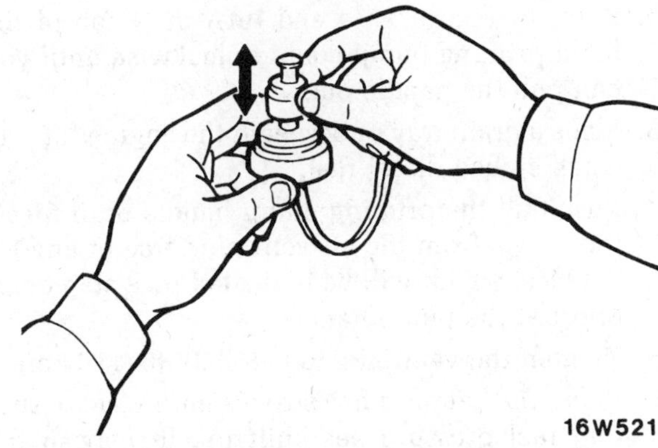

Figure 17-6. Water level sensor inspection check. (Courtesy of Chrysler Corporation)

   b. Fuel Temperature Sensor—15–21 lb.ft. (20–29 N·m)
   c. Air Vent Plug—6–8 lb.ft. (8–11 N·m)
   d. Water Level Sensor—9–10 lb.ft. (12–14 N·m)
   e. Drain Plug—3–4 lb.ft. (4–5 N·m)

Priming/Bleeding The Fuel System

When the fuel/water separator has been removed for service or if the fuel lines, injection pump, or fuel injector lines have been removed or loosened for service purposes, it will be necessary to vent all air from the fuel system in order to start the engine. The procedure to do this successfully is as follows:

Procedure

1. Refer to Figure 17–7 and loosen off the air vent plug on the top of the filter assembly.

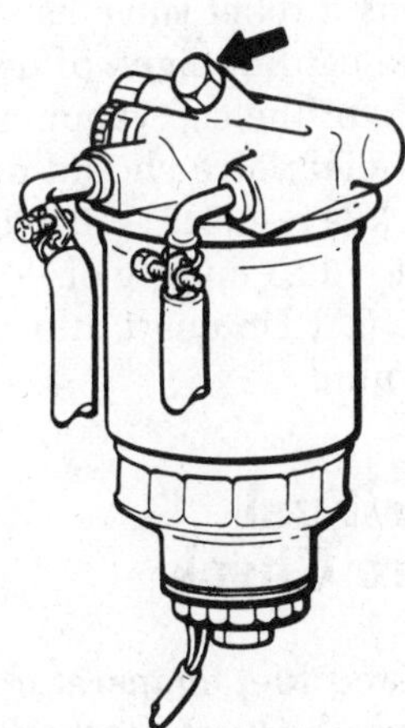

Figure 17-7. Loosening air vent plug. (Courtesy of Chrysler Corporation)

2. Refer to Figure 17–4 and turn the knob of the hand priming pump counterclockwise until you can pull the handle out.
3. Place a drain tray underneath the engine/filter to catch spilled diesel fuel.
4. Push/pull the priming pump handle until diesel fuel flows from the air vent plug free of any air bubbles, which will be indicated by a step-up in effort at the pump handle.
5. Tighten the vent plug to 6–8 lb.ft. (8–11 N·m).
6. Pump the handle a few strokes until you are sure that fuel pressure has built up, then push the handle in and turn the knob clockwise until it is tight.
7. If the injector fuel lines have been removed previously, leave the fuel line nut at the injector end loose, and prime the system by the hand pump until fuel flows from the end of each injector line.
8. You may have to crank the engine over once or twice to allow fuel to be bled from each injector fuel line.
9. Once fuel flows freely from each injector fuel line, tighten them up.
10. Dry off all spilled fuel.
11. Start the engine and check for any fuel leaks.
12. If the engine runs rough or stumbles, loosen off one injector line at the injector end and allow fuel to flow until it is free from any sign of air bubbles; repeat this procedure at each injector using a rag to prevent fuel from spraying all over.

Fuel Filler Cap

The fuel filler cap is often overlooked as a maintenance item when servicing the fuel system. The fuel filler cap contains a relief valve as shown in Figure 17–8. Inspect the cap for signs of deterioration and possible plugging of the cap vacuum relief valve.

The vacuum relief valve should open at between −4.413 ± 1.275 kPa (−0.640 ± 0.185 psi). The valve open flow rate at −33.1 mm Hg or −16.25 psi should be 1 liter/minute (1.1 US quart/min or 0.88 Imperial quart/min) minimum.

Fuel Heater/Fuel Temperature Check

The fuel heater and fuel temperature sensor are located in the upper body of the fuel/water filter assembly as shown earlier in Figure 17–4. The fuel heater/temperature sensor circuit is shown in Figure

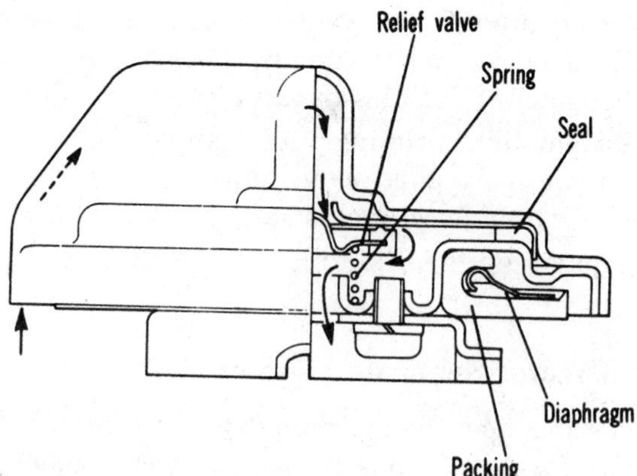

Figure 17–8. Fuel tank filler cap assembly. (Courtesy of Chrysler Corporation)

17–9 while the heater components can be clearly identified in Figure 17–10.

A fuel temperature sensor located in the top of the fuel/water filter body is designed to activate the fuel heater circuit through a controller when the fuel temperature drops to 37 ± 5°F (3 ± 3°C). At this temperature, the switch contacts in the fuel temperature sensor close and complete the circuit to the fuel heater controller to supply current to the fuel heater, which is rated at 110 ± 10 watts.

When the fuel heater is energized (on), an instrument panel indicator light comes on to signify that the heater is operational.

The fuel heater is set to go off when the fuel temperature obtains approximately 50°F (10°C), which will be signified by the indicator light going off on the instrument panel.

The fuel heater description and specifications are as follows:

Description	*Specifications*
Heater Rating	110 ± 10 watts
Thermo Switch ON	140 ± 9°F (60 ± 5°C)
OFF	104°F or less (40°C)
Fuse Rated Temperature	Approx. 246.2°F (119°C)

Fuel Heater Check

To check the operation of the fuel heater shown in Figures 17–4 and 17–9, proceed as follows:

1. Refer to the circuit wiring diagram in Figure 17–9.
2. Using an ohmmeter, measure the resistance between terminals 0.85G and 0.5BR; a reading of about 1 ohm indicates that the heater is OK.

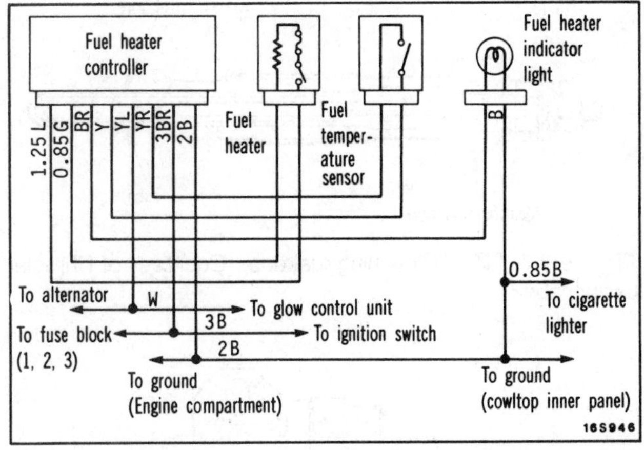

Figure 17–9. Fuel/heat sensor circuit. [Courtesy of Chrysler Corporation]

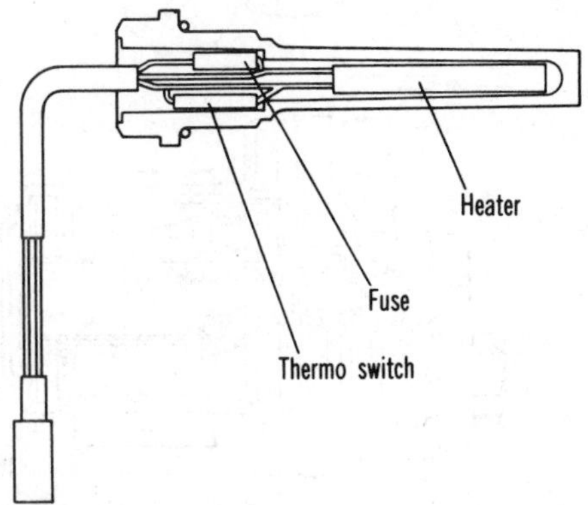

Figure 17–10. Fuel heater components. [Courtesy of Chrysler Corporation]

3. A low resistance reading in Step 2 indicates that the heater coil is shorted out, while a high resistance reading indicates that the coil has a break in it somewhere; in both cases, replace the heater element.
4. Using a jumper wire, apply 12 volts between the terminals in Step 2, and with the ohmmeter, check for continuity between the 0.85L terminal and the 0.85G terminal at the fuel heater temperatures indicated below:

Operating Temperature	*Continuity*
104°F (40°C) or less	Continuity should exist
140°F (60°C) or more	No continuity should exist

Fuel Temperature Sensor Check

To check the fuel temperature sensor correctly, a test arrangement similar to that for a thermostat or shutterstat sensor as shown in Figure 17–11 is required. Check the fuel temperature sensor as follows:

Test Water Temperature	*Resistance Readings*
50°F (10°C)	3.2–4.6 kilohms
68°F (20°C)	6.2–8.0 kilohms
86°F (30°C)	12–19 kilohms

Fuel Heater Controller Check

To check the fuel heater controller unit for correct operation, make sure that all wiring is connected and that the indicator lamp glows with a fuel temperature of 32°F (0°C) or less and goes out when the temperature reaches 50°F (10°C).

If the light does not go out, check for continuity with an ohmmeter between the 3BR and 1.25L terminal of the fuel heater controller shown in Figure 17–9 earlier. If, however, continuity does exist, then the controller is at fault and should be replaced.

Glow Plug Circuit

The TBD engine uses a glow plug threaded into each pre-combustion swirl chamber to facilitate cold weather start-up and rapid engine warm-up. The glow plugs are operated in a fashion similar to that explained in Chapters 18 and 19 for the Datsun/ Nissan and Toyota pickup trucks. Refer to the detailed explanation for either of these pickups and apply the information to the Dodge Ram 50 unit.

Figure 17–12 illustrates the wiring diagram for the glow plug circuit, which uses both a No. 1 and No. 2 glow plug relay that are designed to apply either full power or reduced power to the glow plugs, depending on the engine coolant temperature.

When the ignition key is turned on, the glow plug wait lamp will illuminate on the vehicle instrument panel. The engine should not be cranked until the light goes out or the driver is signalled to do so.

On cold days, the glow plugs receive full battery

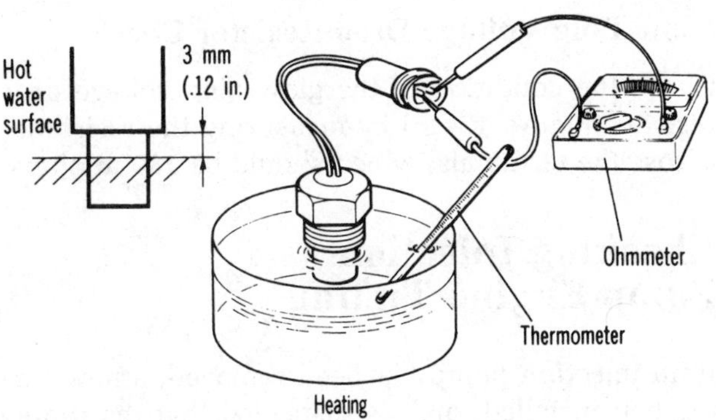

Figure 17–11. Fuel temperature sensor test arrangement. [Courtesy of Chrysler Corporation]

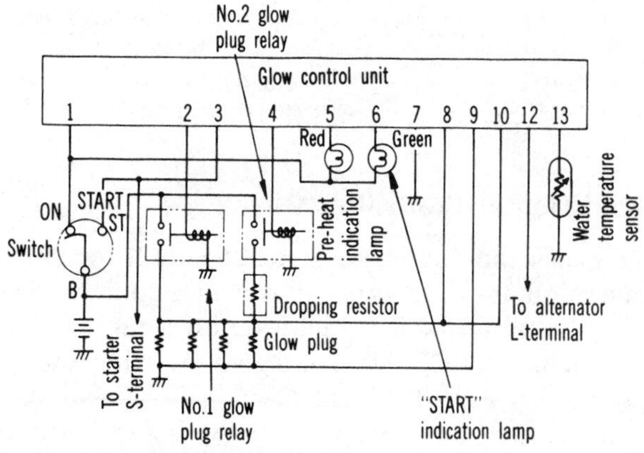

Figure 17–12. Glow plug wiring circuit. (Courtesy of Chrysler Corporation)

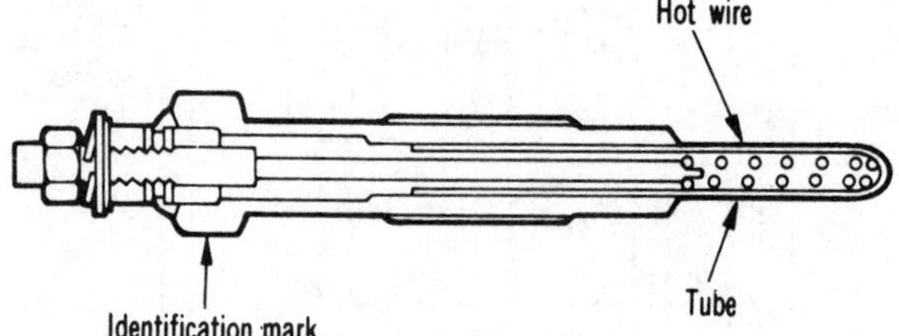

Figure 17–13. Glow plug makeup. (Courtesy of Chrysler Corporation)

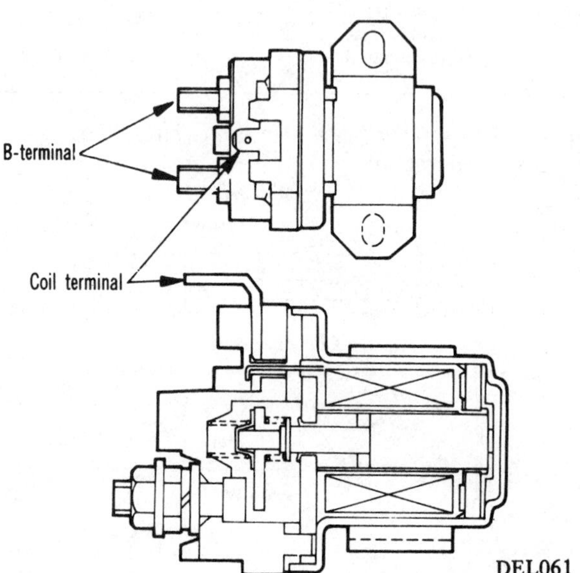

Figure 17–14. Glow plug relays. (Courtesy of Chrysler Corporation)

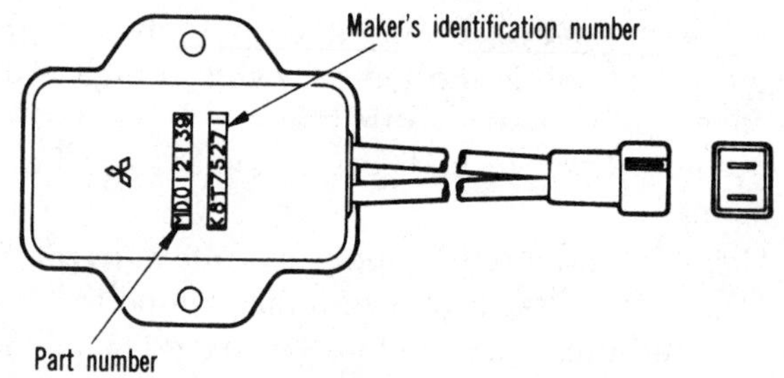

Figure 17–15. Glow plug voltage drop resistor. (Courtesy of Chrysler Corporation)

voltage; on warmer days or when the engine coolant temperature is warm, the water temperature sensor regulates the preheat and afterglow time of the glow plugs.

A voltage drop resistor is used to reduce the voltage applied to the glow plugs for warm-up after the engine starts.

Glow Plug Check

To check the glow plugs, use an ohmmeter with one lead across the terminal and the other lead grounded to the body. A good glow plug should exhibit a resistance value of 0.1 ohm at 68°F (20°C). If the glow plug has been removed, check it for any signs of tube damage (Figure 17–13).

Glow Plug Relay Check

The glow plug relays (No. 1 and No. 2) can be checked for continuity between the B terminals when 6 volts are applied between the glow plug relay and the coil terminal as shown in Figure 17–14.

Glow Plug Voltage Drop Resistor Check

Check the condition of the glow plug voltage drop resistor (Figure 17–15) by measuring the resistance across the terminals, which should be 130 m ohms.

Checking Injection Pump/Engine Timing

If an injection pump has been removed, a new timing belt installed, or it is suspected that the timing may not be right, a check of the injection timing can be done as follows: Injection timing should be 5 degrees ATDC with an injection pump plunger prestroke of 1 ± 0.03 mm (0.03934 ± 0.0011").

Procedure

1. Engine at operating temperature and stopped.
2. Remove the injection pump/engine timing belt upper front cover (Figure 17–21) to expose the

camshaft and injection pump sprocket (gear) timing marks which are shown in Figure 17–16.

3. Manually rotate the engine in its normal direction of rotation, which is clockwise from the front, by placing a socket on the crankshaft pulley bolt until the No. 1 piston is at TDC on its compression stroke as shown in Figure 17–16. If you are in doubt as to No. 1 piston being on its compression stroke, remove the valve rocker cover and check that both the inlet/exhaust valves are fully closed (clearance exists between camshaft lobe and valve mechanism).
4. In order to accurately check the injection pump timing, you must have Special Tool P/N MD998384, which is a *prestroke measuring adapter* and is shown in Figure 17–17.
5. Refer to Figure 17–17 and make certain that the inner push rod is protruding at least 10 mm (0.400"). If it is not, adjust it by disassembling the special tool and adjusting the inner nut.
6. Insert the prestroke adapter tool MD998384 along with a dial gauge into the center of the injection pump hydraulic head after removing the center plug as shown in Figure 17–18.
7. Manually rotate the crankshaft clockwise from

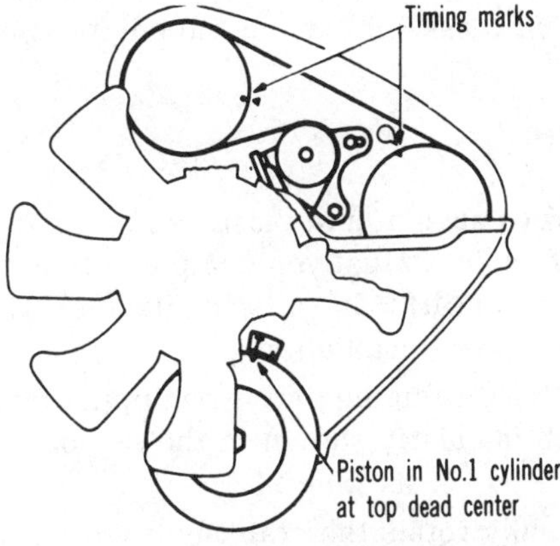

Figure 17–16. No. 1 Piston at TDC compression and camshaft and injection pump timing gear marks aligned. [Courtesy of Chrysler Corporation]

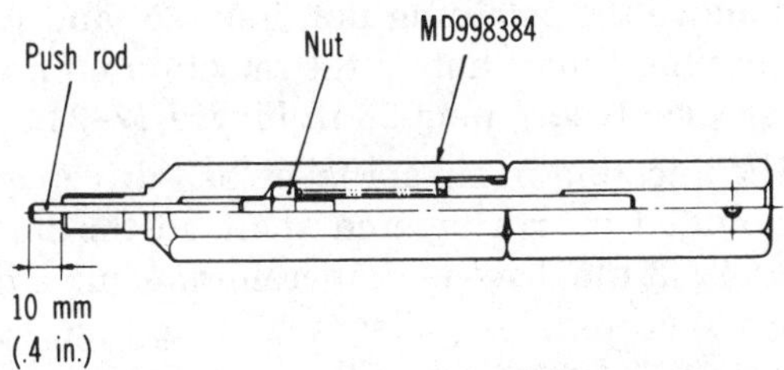

Figure 17–17. Prestroke measuring tool. [Courtesy of Chrysler Corporation]

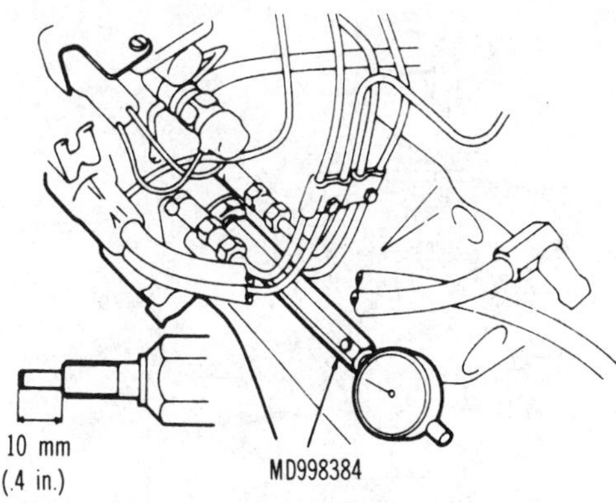

Figure 17–18. Attaching prestroke measuring adaptor. [Courtesy of Chrysler Corporation]

the front to place the timing notch on the pulley about 30 degrees BTDC on the compression stroke of No. 1 piston.

8. Lightly preload the dial gauge and set the pointer to zero.
9. Rotate the crankshaft pulley very slightly clockwise and then counterclockwise to make certain that the dial indicator needle does not move from its ZERO position; if it does, rotate the crankshaft to place the timing mark exactly at 30 degrees BTDC.
10. Gently but firmly rotate the crankshaft pulley to place the timing mark on the pulley in alignment with the stationary pointer at 5 degrees ATDC.
11. Read the dial indicator; it should be 1 ± 0.03 mm (0.03934 ± 0.0011") if the injection pump to engine static timing is correct.
12. If the dial gauge reading is not as specified in Step 11, then the timing must be adjusted by loosening off the injection pump retaining bolts/nuts as well as the injector fuel line nuts on the pump.
13. Move the pump towards the engine if the dial gauge reads less than 1 mm (0.03937") or away from the engine if the dial gauge reads more than 1 mm (0.03937").

Idle Speed Adjustment

The curb idle adjustment is done by adjusting Item No. 4 in Figure 17–19, which shows the location/identification of the injection pump external components. The engine should be at its normal operating temperature when this adjustment is made, which is between 185–205°F or (85–95°C). After the engine is at operating temperature, run it at between 2000 and 3000 rpm for about 5 seconds, then let it idle for about 2 minutes.

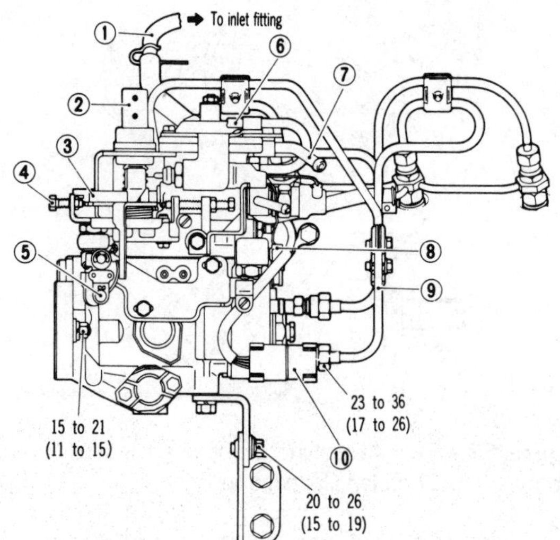

(1) Hose
(2) Vacuum regulating valve
(3) Accel lever
(4) Idle speed adjusting screw
(5) Fast idle lever
(6) Hose (to constant pressure valve)
(7) Fuel inlet nipple
(8) Connector (for timing control solenoid valve, fuel cut-off solenoid valve and accel switch)
(9) Fuel injection pipe assembly
(10) Connector

Figure 17–19. Fuel injection pump components. [Courtesy of Chrysler Corporation]

Check the idle speed with a tachometer and adjust it if necessary to 750 ± 50 rpm.

If the vehicle is equipped with an air conditioner, it may be necessary to adjust the idle-up adjustment as follows:

Procedure for Idle-Up Adjustment

1. Refer to Figure 17–20 for the adjustment screw location.
2. Engine should be at operating temperature, all lights and accessories should be turned off, the transmission must be in neutral, the parking brake applied; place the steering wheel in the straight-ahead position to remove any load from the power steering pump if so equipped.
3. Make sure that the curb idle is between 750 ± 50 rpm.
4. Refer to Figure 17–20 and loosen off nuts 1 and 2.
5. Turn the adjusting nut to place the vacuum actuator rod clear of the accelerator lever by 1 mm (0.03934") and retighten nuts 1 and 2.

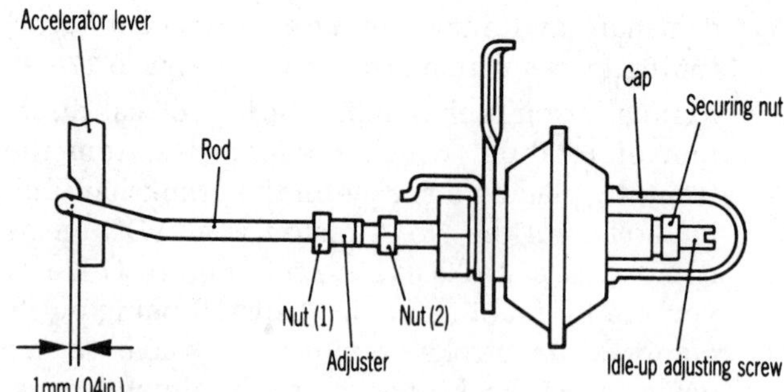

Figure 17–20. Idle-up adjustment. [Courtesy of Chrysler Corporation]

6. Turn the A/C control switch *on/off* a few times and then check that the clearance in Step 5 still exists.
7. Now loosen off the cap over the adjusting screw of the idle-up adjuster shown in Figure 17–20 and adjust the engine rpm to between 850 and 900 rpm and lock the nut.
8. Cycle the A/C switch on/off several times to check the vacuum actuator lever for proper up/down movement.

Injection Pump Removal

If it becomes necessary to remove or install an injection pump assembly from the engine, proceed as follows:

Procedure

1. Obtain a selection of plastic shipping caps (male and female) so that you can plug off all fuel lines to prevent dirt from entering the fuel system during removal/installation.
2. Remove the engine timing belt upper front cover, which is clearly shown in the exploded view in Figure 17–21 as Item 6.
3. Manually rotate the crankshaft with a socket on the front pulley bolt in a clockwise direction to place No. 1 piston at TDC on its compression stroke as per Figure 17–16 (engine timing check) earlier.
4. Remove the retaining nut, item 25, and washer from the center hub of the injection pump drive sprocket (gear), item 26 in Figure 17–21.
5. The injection pump sprocket (belt-driven gear) is mounted over a tapered shaft on the injection pump at the drive end; therefore, using a locally available suitable puller (similar to Figures 14–5 and 15–16) attached to the sprocket so as not to damage it, pull the sprocket free from the injection pump drive shaft.

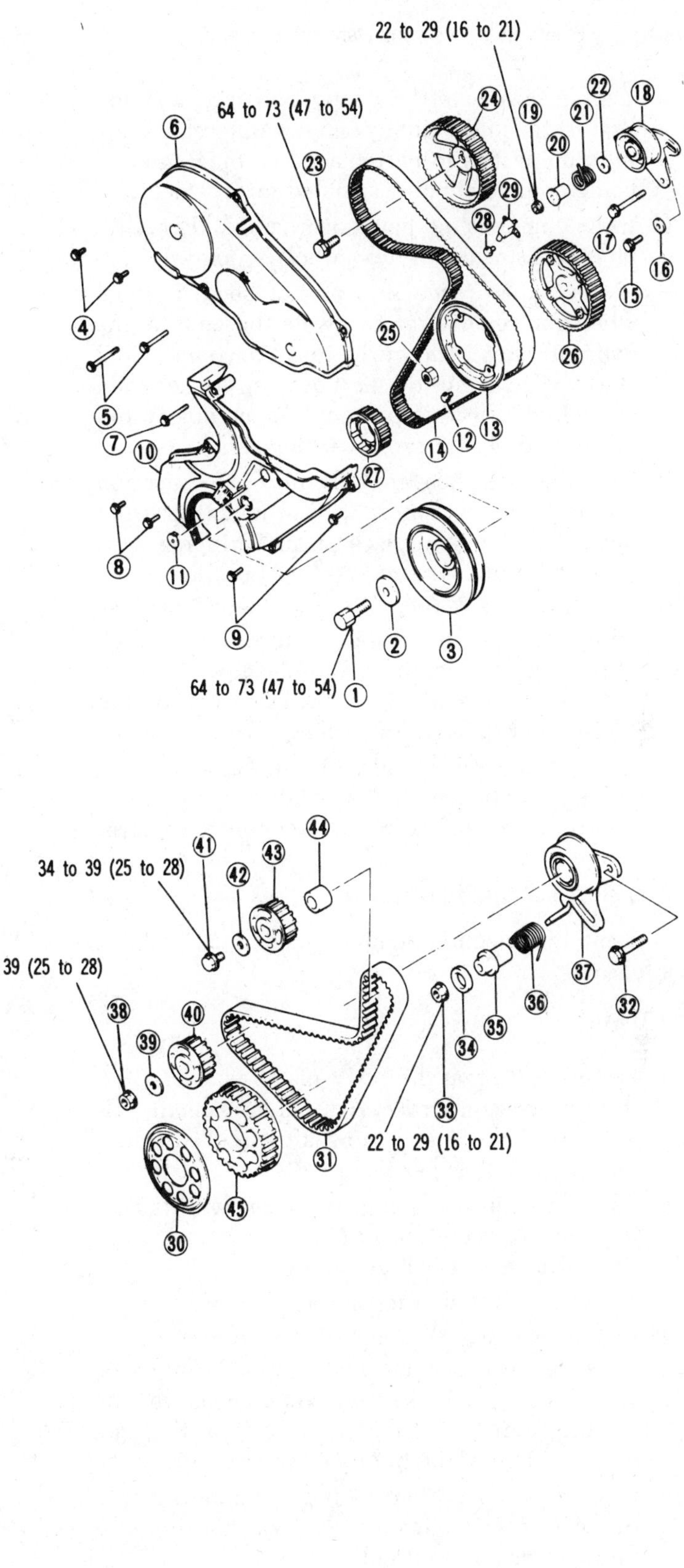

(1) Crank pulley bolt
(2) Special washer
(3) Dumper pulley
(4) Flange bolt (2)
(5) Flange bolt (2)
(6) Timing belt front upper cover
(7) Flange bolt
(8) Flange bolt (2)
(9) Flange bolt (3)
(10) Timing belt front lower cover
(11) Access cover
(12) Flange bolt (4)
(13) Flange
(14) Timing belt
(15) Flange bolt
(16) Washer
(17) Flange bolt
(18) Timing belt tensioner assembly
(19) Flange nut
(20) Tensioner spacer
(21) Tensioner spring
(22) Plain washer
(23) Flange bolt
(24) Camshaft sprocket
(25) Nut
(26) Injection pump sprocket
(27) Crankshaft sprocket
(28) Flange bolt (2)
(29) Switch assembly
(30) Flange
(31) Timing belt "B"
(32) Flange bolt
(33) Flange nut
(34) Gasket
(35) Tensioner spacer
(36) Tensioner spring "B"
(37) Timing belt tensioner assembly "B"
(38) Flange nut
(39) Washer
(40) Counterbalance shaft sprocket
(41) Flange bolt
(42) Washer
(43) Counterbalance shaft sprocket
(44) Spacer
(45) Crankshaft sprocket "B"

Figure 17–21. Timing belt train arrangement. (Courtesy of Chrysler Corporation)

6. The sprocket can be gently laid into position against the lower timing belt cover for safe keeping unless it is to be replaced or the engine is being completely disassembled for overhaul.
7. If the engine is not being overhauled, do not rotate the crankshaft after sprocket removal.
8. If the radiator coolant has not been drained, when you remove the two water hoses from the hydraulic head end of the injection pump (cold start mechanism), tie the hoses up higher than the cylinder head to prevent loss of coolant and install two plastic caps over the ends.
9. Disconnect the hose from the top of the injection pump boost compensator (turbocharger/inlet manifold hose) and place plastic shipping caps over both the hose end and injection pump.
10. Remove all the fuel injection lines from the pump, namely the inlet and return lines and all of the injector fuel lines, and place plastic shipping caps over all open fittings; remove the throttle linkage and fuel solenoid wire.
11. Remove the pump support bracket bolts and pump mounting nuts, then withdraw the injection pump assembly from the engine with care.

Injection Pump Installation

To install the injection pump to the engine, proceed as follows.

Procedure

1. Doublecheck that the No. 1 piston is at TDC on its compression stroke. Are the crankshaft pulley timing marks and camshaft sprocket timing marks aligned as per Figure 17–16?
2. Place the injection pump sprocket with its timing mark to about the 11 o'clock position while lifting the timing belt up with it.
3. Note the position of the sprocket keyway!
4. Rotate the injection pump drive shaft to place the drive key at a position approximately opposite where the sprocket keyway will be when the injection pump is inserted into position through the back side of the timing drive cover.
5. Insert the injection pump into position through the rear of the engine timing cover housing and through the sprocket hub.
6. Install two nuts and bolts to temporarily support the injection pump in position.
7. Install the injection pump drive shaft sprocket washer and nut over the drive shaft threads and tighten the nut to a standard torque figure.
8. Adjust the timing belt tension.
9. Perform an injection pump/engine timing check as described earlier.
10. Install all injection pipes and fuel lines, coolant hoses, and boost compensator hose, as well as the throttle linkage and the fuel solenoid wire.
11. Bleed the fuel system and start and run the engine.
12. Check for any fuel leaks.

Hand Throttle Cable

The hand throttle cable arrangement used with the diesel engine consists of a knob shown in Figure 17–22 connected to a cable that is in turn held by a retaining bolt to Item 5 in Figure 17–19. The hand throttle is located at the bottom of the instrument panel inside the cab on the driver's side.

If the hand throttle is sticky or hard to move, it should be removed for inspection. To remove the cable when damaged, remove the cable clamp and retaining bolt at the injection pump as well as the retaining screw inside the vehicle cab as shown in Figure 17–22 in order to withdraw the knob.

Remove the cable serrated/knurled retaining nut shown immediately in front of the knob and screw to allow you to pull the cable assembly through into the cab.

Fuel Injectors

Information contained in this chapter is basic and only relates to injector removal and installation because detailed information on the operation, maintenance, service, and repair of these nozzles can be found in Chapter 29, dealing with diesel fuel injectors.

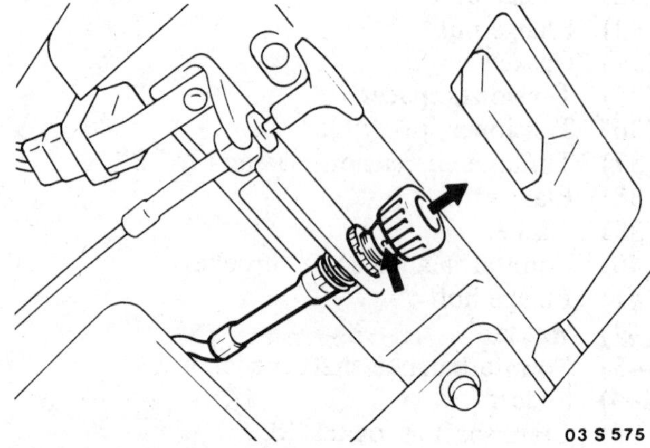

Figure 17–22. Hand throttle knob retaining screw. (Courtesy of Chrysler Corporation)

The fuel injectors used with this engine are Bosch throttling-type pintle nozzles designed for use with swirl-type pre-combustion chambers. These nozzles are similar in design and operation to those used by Ford, Nissan/Datsun, Toyota, Peugeot, Volkswagen, Volvo, Mercedes-Benz, Dodge Ram 50, and Mitsubishi vehicles and GMC in their Isuzu-powered diesel passenger cars.

The injectors are threaded into the cylinder head above each cylinder with both an inlet and return fuel line connected to them as shown in Figure 17–23.

The nozzles are adjustable for correct release pressure by the use of shims behind the internal needle valve (pintle) return spring. The nozzles are designed to release fuel at a popping pressure of 1707–1849 psi (11,768–12,749 kPa) when new, although the service limit when repairing used nozzles can be a minimum of 1565 psi (10,787 kPa).

Removal and installation of the injectors follows the standard pattern of removing the fuel supply and return lines, placing plastic shipping caps over all exposed fuel lines and openings, and using a deep socket as shown in Figure 17–24 to successfully loosen off the injector body from its bore in the cylinder head.

NOTE: When removing the injector fuel return lines, hold the nozzle body with a wrench to prevent it from loosening off, at the same time as you slacken off the return pipe nut.

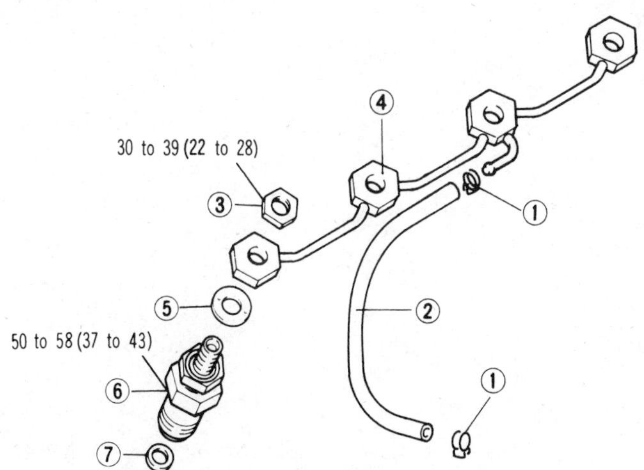

(1) Hose clip (2)
(2) Fuel hose
(3) Nut (4)
(4) Fuel return pipe
(5) Gasket (4)
(6) Nozzle (4)
(7) Nozzle tip gasket (4)

Figure 17–23. Dodge Ram 50 injection nozzle and return fuel lines. [Courtesy of Chrysler Corporation]

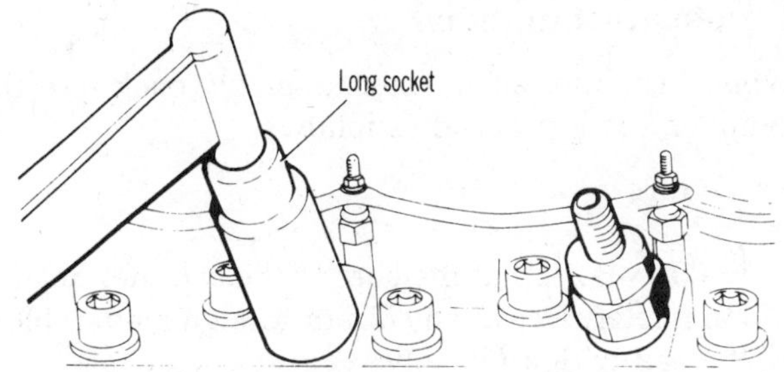

Figure 17–24. Removing injector with a deep socket. [Courtesy of Chrysler Corporation]

Nozzle Checks

Removal of injectors is often done as a matter of trial and error troubleshooting when in fact the nozzles may not be the cause of the problem. Nozzle problems are generally indicated when:

1. The engine is hard to start (Are the glow plugs operational?).
2. Lacks power (Is the governor/linkage correctly set?).
3. Exhibits black/gray smoke (Is the air filter plugged, or are you using the wrong grade of fuel?).
4. Exhibits white smoke (Is the correct grade of fuel being used or is the engine compression low?).

When it is suspected that the nozzles are at fault, perform a quick check by running the engine at an idle rpm while you loosen off one injector fuel supply line at a time, with a rag around the nut to prevent fuel spraying all over.

As you loosen off the injector fuel line nut, the engine rpm should drop, since you are preventing the injector from releasing fuel. This is similar to shorting out a spark plug on a gasoline engine. If the engine rpm doesn't drop, then the injector is not operating correctly. There should be a positive drop or reduction in engine rpm when the fuel line is loosened. Tighten the fuel line after this check.

Repeat this same check on each injector to determine their condition. If the engine exhibits a misfire at one particular rpm, accelerate it to this speed and loosen off each injector fuel line nut one at a time until you determine which cylinder is at fault.

When injectors are confirmed as being the problem area, they should be removed and tested for spray pattern and proper popping (release) pressure on an injector pop tester such as described in Chapter 29, fuel injectors.

Injector Installation

When installing an injector assembly back into the cylinder head, proceed as follows.

Procedure

1. Make certain that the bore of the cylinder head is completely free of any dirt or foreign matter; blow it clean with an air hose.
2. Always use a new nozzle seat gasket as shown in Figure 17–25 with a new or used nozzle assembly.
3. Make sure that the threads on both the injector and in the cylinder head are burr free and clean.
4. Insert the injector into the cylinder head and tighten it to spec with the same deep socket that was used for removal.

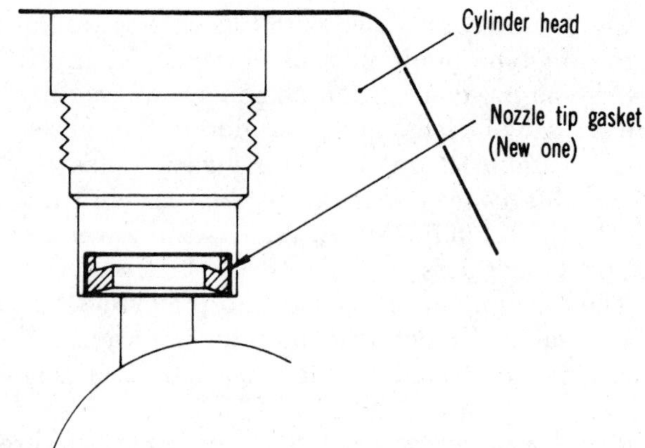

Figure 17–25. Location of nozzle seat gasket. (Courtesy of Chrysler Corporation)

18

Nissan (Datsun) SD 25 Diesel Pickup Truck

The Nissan diesel pickup uses a 4-cylinder 2488 cc (152 cu.in.) 4-stroke engine with a bore and stroke of 89 × 100 mm (3.504 × 3.937 in.) and a compression ratio of 21.4 to 1.

The combustion chamber is of the IDI (indirect-injection) swirl chamber design with glow plugs to facilitate starting when the engine is cold.

The fuel injection system is a Diesel Kiki (Robert Bosch Licensee) in-line-type injection pump feeding pintle-type nozzles. The injection pump is controlled via a DPC (digital processor control) unit, which is explained in detail later on in this chapter.

The engine firing order is 1–4–3–2. The engine serial number is stamped on the right side of the cylinder block just above the in-line-type injection pump.

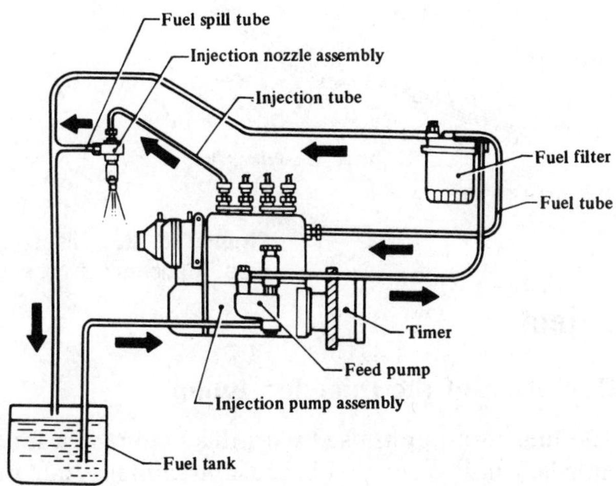

Figure 18–1. Basic fuel system layout. (Courtesy of Nissan Motor Co., Ltd.)

Fuel System

Fuel System Arrangement

The fuel system arrangement is illustrated in Figure 18–1 in the basic layout that it is found on the vehicle.

Diesel fuel is drawn from the tank by the action of a mechanically driven fuel transfer or feed lift pump that is bolted to and driven from the injection pump camshaft. Fuel under low pressure at the outlet port of the feed pump is delivered to the fuel filter assembly where excess fuel is returned to the fuel tank through an overflow valve on the top of the filter assembly.

Fuel from the filter is forced into the injection pump where the position of the throttle establishes the quantity of fuel that will be delivered through the individual pumping plungers to each nozzle via the delivery valve contained at the top of each plunger in the injection pump. Accumulated fuel in each injection nozzle that is used for cooling and lubricating purposes is returned to the fuel tank through the injector return lines.

Operation of the Fuel Feed Pump

The fuel feed or transfer pump for the fuel injection pump is similar in design and operation to that used by Toyota on their Landcruiser diesel engine. (This pump can be seen in Figures 19–7 and 19–8.) The fuel feed pump is bolted to the side of the injection pump and is driven from the injection pump camshaft through a roller plunger arrangement. The function of the fuel feed pump is to suck diesel fuel from the tank and deliver it to the filter assembly and then on to the injection pump.

Vehicle Fuel Tank and Lines

The fuel tank and fuel lines arrangement for the diesel pickup is shown in Figure 18–2.

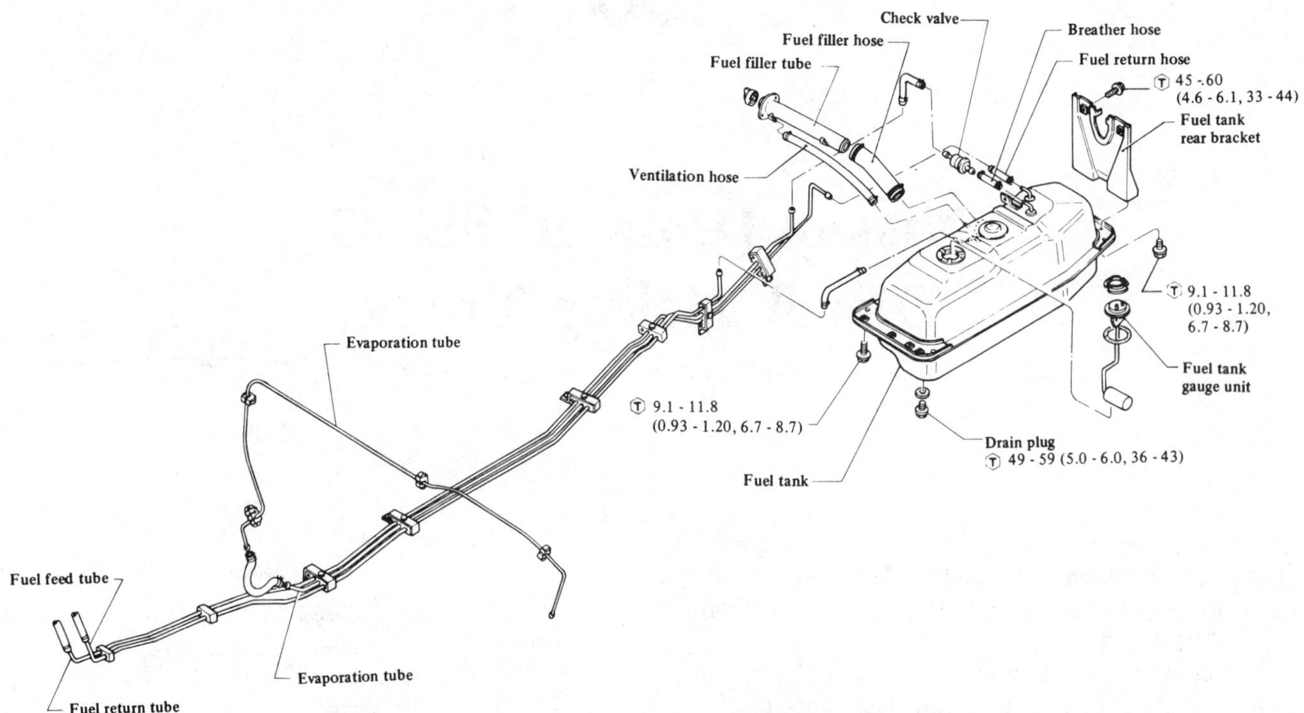

Figure 18–2. Nissan pickup SD25 engine fuel tank and lines. (Courtesy of Nissan Motor Co., Ltd.)

Injection Pump

Operation of the Injection Pump

The injection pump used with this engine is a multiple-jerk unit so named because it employs four individual pumping plungers within the aluminum body of the main injection pump housing, which is of an in-line design.

The concept of operation of the in-line injection pumps now in use on high speed automotive type diesel engines is explained in detail in Chapter 10.

In the in-line injection pump a gear driven pump camshaft (engine driven) raises and lowers a pumping plunger that rides within a barrel or bushing with very close tolerances or clearances. Figure 10–2(b) illustrates in simple form the principle of the jerk-type injection pump.

Fuel from a lift or transfer pump is directed to the one piece or common injection pump housing where it is distributed to the individual pumping plungers within the housing (the number of pumping plungers being the same as there are cylinders in the engine). Figure 10–2(a) shows the simplified concept of four pumps contained within a common housing.

Diesel Kiki (Robert Bosch) In-Line Injection Pump Operation

The operation of the in-line injection pump can be considered common to basically all makes of automotive engines now using such a unit. Certainly, specific differences do exist between pumps and also between similar makes of vehicles; however, the operating principle remains the same.

Detailed information on the design and operation of in-line fuel injection pumps can be found in Chapters 10 and 12; therefore, a detailed description of the operation of this fuel injection pump will not be given here.

Figure 18–3 illustrates in cutaway form the typical components and their arrangement for an in-line Diesel Kiki (Bosch design) injection pump.

In Figure 18–3, movement of the throttle will cause the injection pump control rack to move right or left. This movement will cause the rack to rotate the plunger through a gear that is attached to the plunger. The basic operation of the plunger is shown in Figure 18–4 during one complete stroke.

Injection Pump Control Mechanism

The Diesel Kiki (Robert Bosch design) in-line injection pump that is used with the 4-cylinder diesel engine model SD25 has a pneumatic governor assembly that operates in conjunction with an electronic control unit to control the start-up, running, load/speed control, and stopping of the engine.

A description of the function and operation of these control components follows.

Figure 18–5 is a schematic diagram of the current

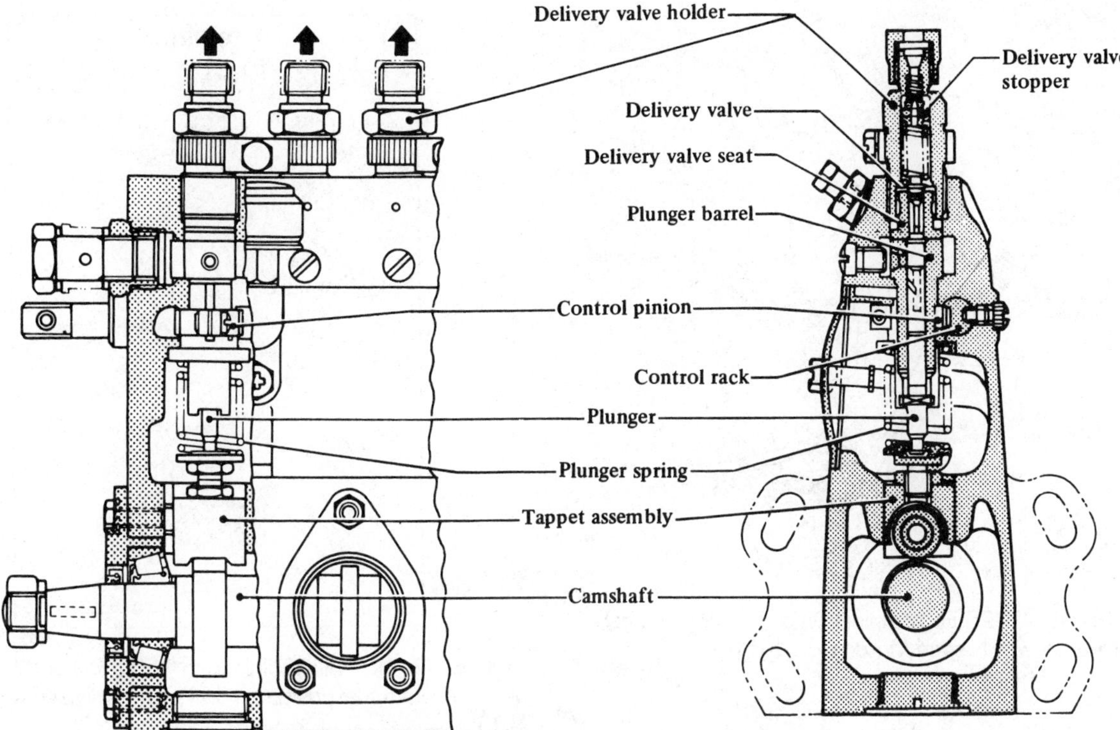

Figure 18-3. Diesel Kiki in-line injection pump. [Courtesy of Nissan Motor Co., Ltd.]

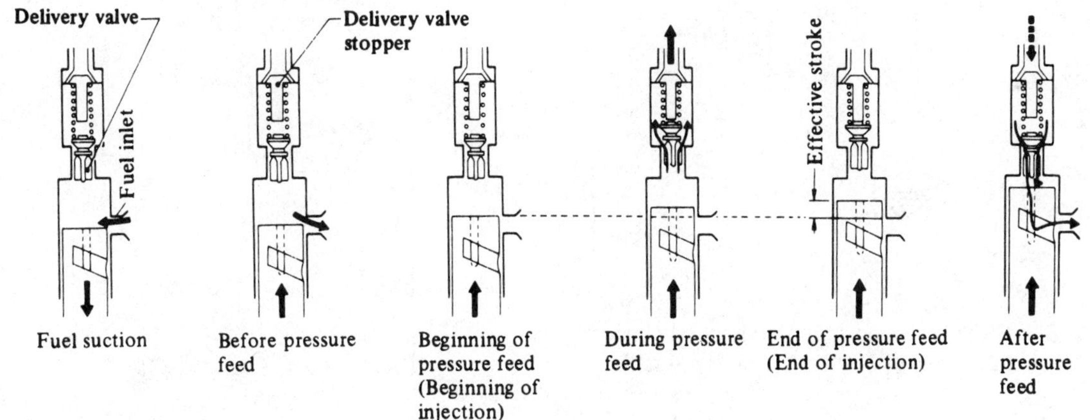

Figure 18-4. Basic plunger operation—in-line injection pump. [Courtesy of Nissan Motor Co., Ltd.]

Diesel Kiki injection pump that is found on the Nissan/Datsun pickup truck using the SD25 diesel engine. As shown in Figure 18–5, the injection pump controller is electrically energized from the ignition switch. When the controller is energized, it activates the injection pump control lever that is connected to the fuel rack inside the pump to place the plungers into an excess fuel position for starting. This action is achieved within the controller mechanism as shown in Figures 18–6 through 18–9.

In Figure 18–6, a miniature armature is contained within the controller mechanism. This armature is activated by an electrical signal received from the DPC (digital processor control) unit via the ignition switch. A drive shaft connected to the small armature has a screw thread on one end that engages with a "geneva gear." Anytime that the armature is energized, this shaft will rotate the geneva gear, which is free to contact any of the three terminals located inside the housing. The particular contact that the

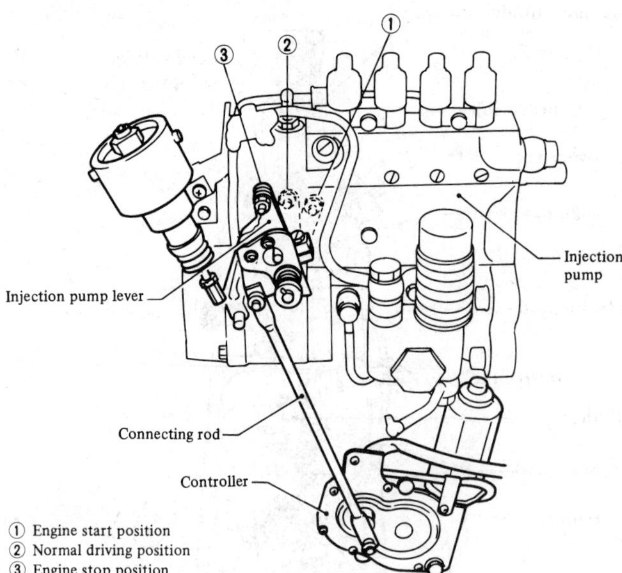

Figure 18-5. SD25 Engine injection pump control mechanism. (Courtesy of Nissan Motor Co., Ltd.)

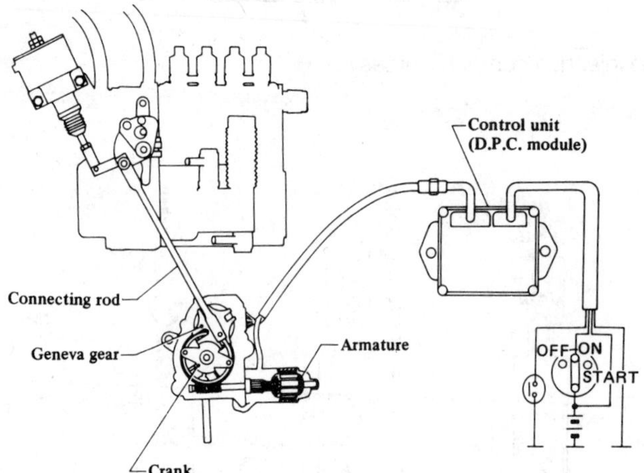

Figure 18-6. SD25 injection pump control mechanism electrical layout. (Courtesy of Nissan Motor Co., Ltd.)

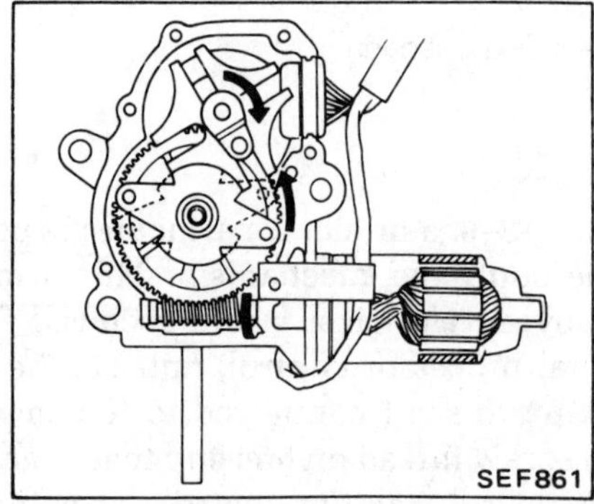

Figure 18-7. SD25 Injection pump controller geneva gear mechanism. (Courtesy of Nissan Motor Co., Ltd.)

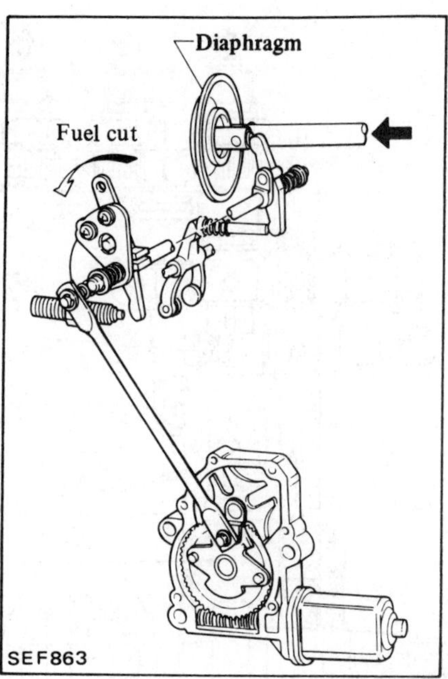

Figure 18-8. SD25 Injection pump controller mechanism—excess fuel position. (Courtesy of Nissan Motor Co., Ltd.)

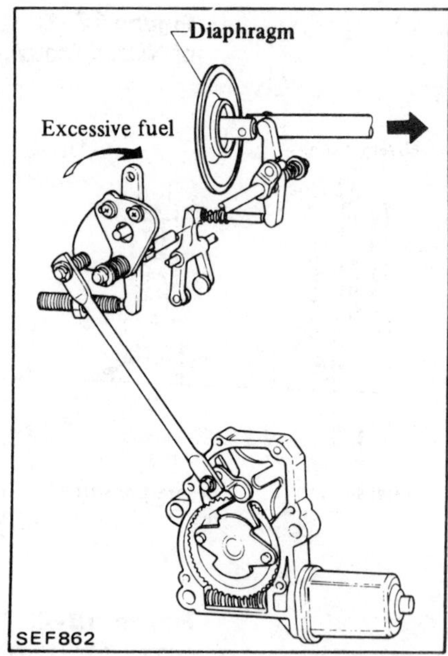

Figure 18-9. SD25 Injection pump controller mechanism—fuel cutoff position. (Courtesy of Nissan Motor Co., Ltd.)

geneva gear will stop at depends on the position of the ignition switch.

The three contact positions shown in Figure 18-5 are:

1. Engine start position
2. Normal driving position
3. Engine stop position

Engine Start-Up

During engine start-up, the injection pump control rack is moved to position number 1 by the connecting rod attached to the geneva gear to provide excess fuel for start-up. The actual mechanical connection to the fuel control rack is illustrated in Figure 18–8. The fuel control rack is connected to a diaphragm that is part of the pneumatic governor assembly, which is explained later.

The actual electrical sequence for engine start-up is shown in Figure 18–10. Rotating the ignition key switch to the *start* position allows current to flow from terminal "O" at the digital processor control (DPC) unit to rotor "A" of the injection pump controller and on to terminal "C" back at the DPC unit. This energizes the rotor, which turns counterclockwise to place the fuel rack in the injection pump at the excess fuel position for easy starting.

Rotation of the rotor "A" stops when it reaches its *start* position, which effectively cuts off current flow between terminal "O" and terminal "C" at the DPC unit.

Engine Idling

Once the engine starts, the governor mechanism will return the engine to its normal idle rpm, which places the injection pump control (throttle) lever in position number 2. The electrical flow during this time period is illustrated in Figure 18–11.

Because the ignition switch is spring loaded, once the engine starts, the operator will release the switch, which is pushed back to the *on* position from the previous *start* position. This action allows current flow now from terminal "D" of the DPC unit to and through rotor "A" and on to terminal "C" at the DPC unit. Current flowing through rotor "A" causes it to turn until it reaches its *drive* position when current flow between terminal "D" and "C" is broken. Note the position of rotor "A" between Figures 18–10 and 18–11.

Stopping the Engine

When the operator chooses to stop the engine, the ignition key switch is turned to the *off* position. The engine will also stop if at any time while it is running the oil pressure drops low enough to activate the oil pressure switch contacts.

Either of the above conditions will activate the fuel injection pump control unit allowing current to flow from terminal "S" through rotor "A" and back to terminal "C" of the DPC unit. This action also energizes the controller's motor, which will also rotate along with rotor "A" to its stop position thereby cut-

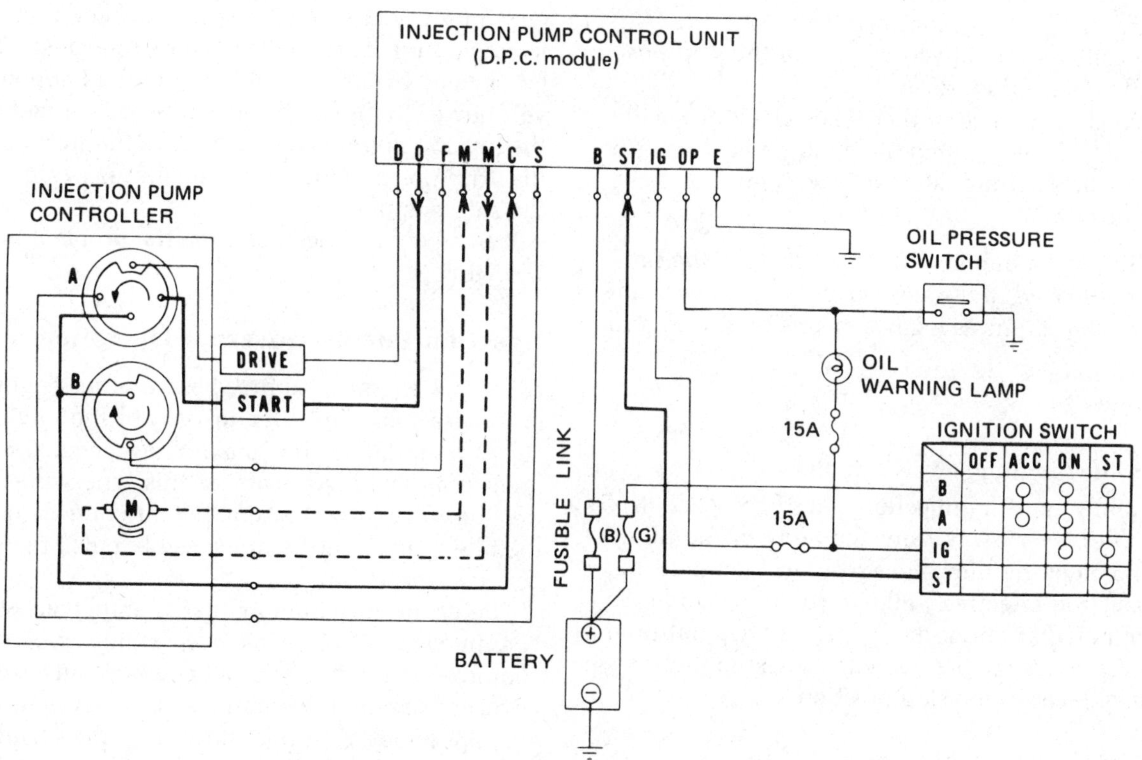

Figure 18–10. Electrical sequence for engine startup. [Courtesy of Nissan Motor Co., Ltd.]

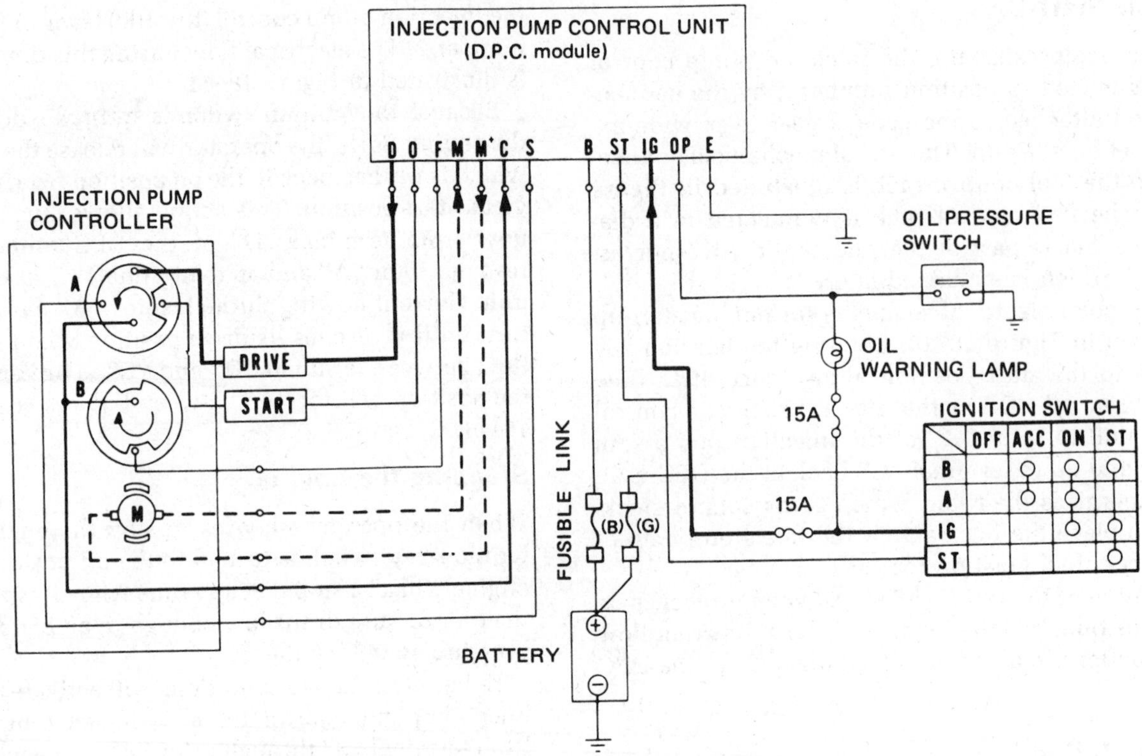

Figure 18–11. Electrical sequence during engine idle. (Courtesy of Nissan Motor Co., Ltd.)

ting current flow between terminals "S" and "C." The controller is effectively placed in the *stop* position at this point (Figure 18–12).

The mechanical action that takes place internally within the injection pump at this time is shown in Figure 18–9, whereby the fuel rack is moved to the no fuel position.

NOTE: If the engine fails to stop when the ignition key is moved to the *stop* position, you can manually stop the engine as follows by referring to Figure 18–13.

Procedure

1. Turn the ignition key switch *off*.
2. Manually pull the injection pump speed control (throttle) lever away from the body of the pump as indicated by the solid arrow in Figure 18–13. This action effectively allows you to override the lever pin that connects the injection pump and control levers so that you can move the fuel rack control lever to the *stop* position.

DPC Unit Troubleshooting

Although adapters can be fabricated and utilized to test out the operation of the digital processor control (DPC) unit, you should first of all check that the injection pump controller (armature) operates. This can be done by bypassing the DPC unit and applying 12-volt power to the controller to see if it in fact moves the injection pump control lever to the *start* position, the *drive* position (or normal operating position), and the *stop* position.

If this action is correct, then the problem is in the DPC unit.

Injection Pump—Governor Operation

All diesel engines require the use of an engine governor assembly in order to control the engine idle speed and the engine maximum speed. Some engines may use a governor of this type while others may use a governor assembly that controls not only the idle speed and the top speed but also the rpm in between these two.

The governor assembly that is employed with the Nissan SD25 4-cylinder diesel engine is of the mechanical/pneumatic type. What this means is that the governor assembly contains a set of flyweights that are opposed by a spring. The force of the spring tends to want to always increase the fuel rack position while the centrifugal force of the rotating flyweights is always attempting to decrease the position of the

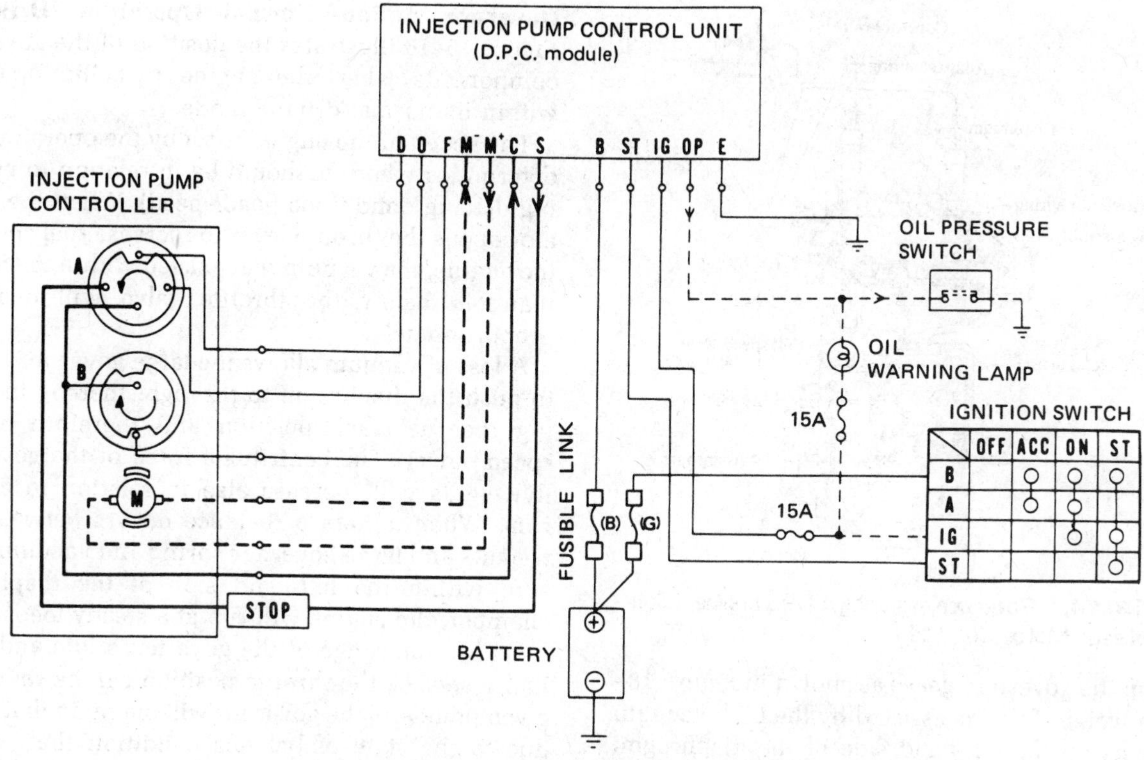

Figure 18-12. Engine stop operation—electrical sequence. [Courtesy of Nissan Motor Co., Ltd.]

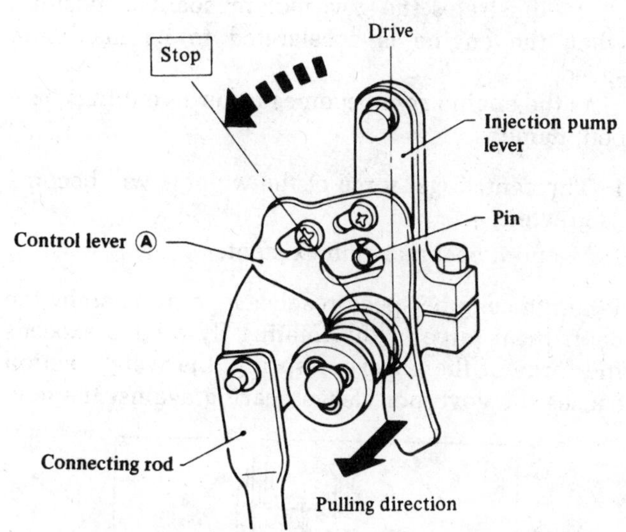

Figure 18-13. Manual engine stop operation. [Courtesy of Nissan Motor Co., Ltd.]

fuel rack. Under certain operating conditions, a state of balance will occur between the weight and spring forces as explained under Chapter 11 dealing with basic governor operation.

In addition to the weights and spring used in the governor assembly, this unit employs a pipe connected between the intake manifold and the governor housing. This arrangement is illustrated in Figure 18–14. Note that the diaphragm is exposed on one side to a pipe connection that is vented to atmosphere on the right-hand side while the left-hand side is connected through a pipe that runs back to just below the butterfly valve in the intake manifold. This side is therefore exposed to vacuum in relation to the throttle position.

Operation—Engine Stopped. When the engine is stopped, no vacuum exists, consequently the governor spring will force the diaphragm to the right, which places the injection pump control rack towards the full-fuel position. Remember that when the ignition key switch is turned *on*, the controller will force the rack to the excess fuel position for start-up.

The flyweights will be in the collapsed state; therefore, the spring bearing against the governor sleeve removes any free-play in the system.

Operation—Engine Start-up. Once the engine starts, the electronic controller returns the injection pump control lever to the normal driving position. When the engine starts, the governor flyweights will be forced outwards due to centrifugal force. This force is opposed by the governor weight spring act-

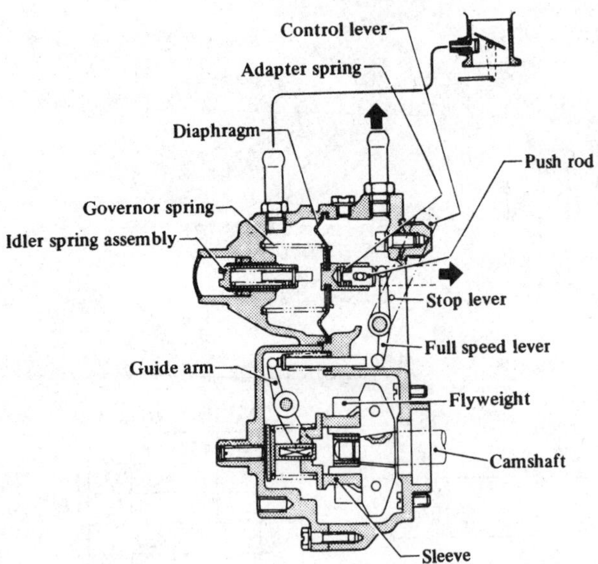

Figure 18–14. Governor assembly/SD25 engine. (Courtesy of Nissan Motor Co., Ltd.)

ing upon the governor sleeve as shown in Figure 18–15. The weight force is assisted by the high vacuum that exists on the left-hand side of the diaphragm. The action of the weights bearing against the governor sleeve pushes against the guide arm and plunger, which reverse this motion at the full-speed lever to effectively push the fuel rack back to a decreased fuel position.

The combined action of the governor weights and high vacuum created within the governor diaphragm (left-hand side) allows the governor spring (large one) to be placed under compression and the diaphragm to bear directly against the governor idle spring (small one). A state of balance (steady engine speed) will be maintained when the centrifugal force of the flyweights balances the force of the idle spring, which is also established by the amount of vacuum within the governor chamber. The force of the idle spring can also be adjusted.

Governor Action—Normal Operation (Driving). Figure 18–16 illustrates the position of the governor components when the engine is being operated within its normal driving mode.

The speed of the engine is set by the operator who determines where it should be in relation to changing driving conditions (load/speed). When the operator opens the throttle lever to increase fuel/speed of the engine, the vacuum that existed within the intake manifold below the throttle valve will decrease proportionately.

A loss of vacuum allows the large governor spring to push the diaphragm to the right thereby increasing the fuel rack position and therefore engine speed/power. The centrifugal force of the governor flyweights will increase also in relation to engine rpm. When a state of balance occurs between the weights and large governor spring and vacuum pressure within the left-hand side of the diaphragm chamber, the engine will run at a steady load/speed.

Within the range of the governor's idle and maximum speeds, the throttle position can be set at any given point and the governor will maintain that speed due to the state of balance condition that will be created.

Governor Operation—Maximum Speed. Figure 18–17 illustrates the governor mechanism position when the engine is accelerated to its maximum speed.

As the engine rpm becomes faster two things happen, namely:

1. The centrifugal force of the weights will become greater.
2. Vacuum becomes non-existent.

When the engine speed reaches a point whereby the centrifugal force of the rotating flyweights exceeds the force of the governor spring, the weight action causes the governor sleeve bearing against the gov-

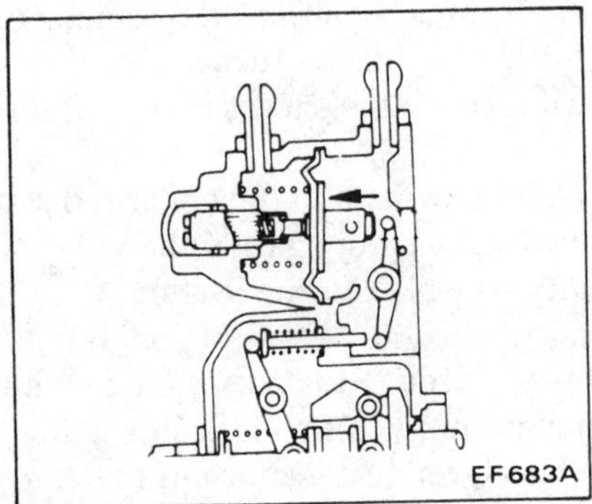

Figure 18–15. Governor action—engine started and idling. (Courtesy of Nissan Motor Co., Ltd.)

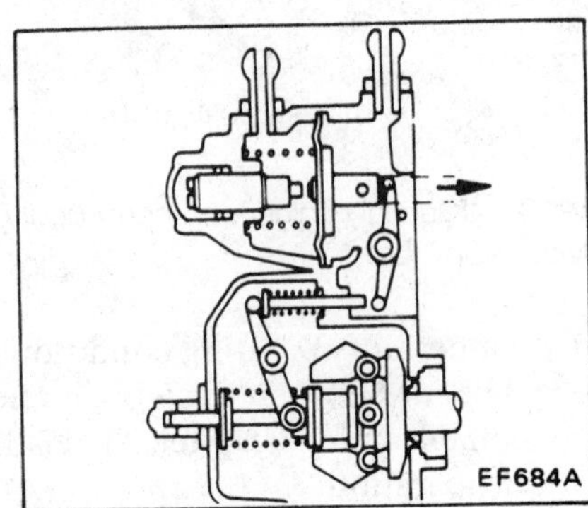

Figure 18–16. Governor action—normal operation (driving). (Courtesy of Nissan Motor Co., Ltd.)

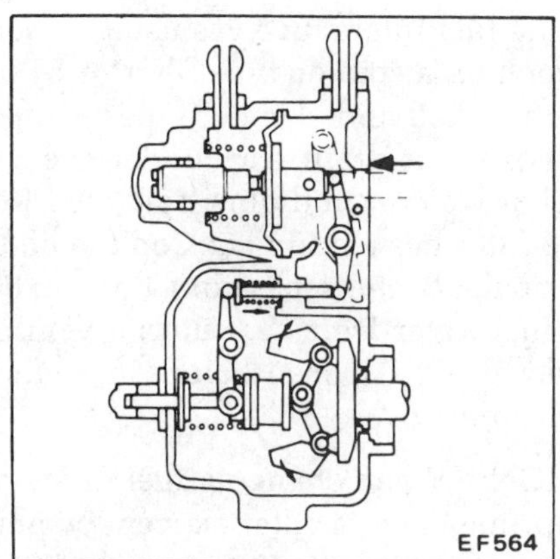

Figure 18–17. Governor action—maximum speed position. [Courtesy of Nissan Motor Co., Ltd.]

ernor guide arm to transfer this motion through the plunger to the full-speed lever. Due to the center pivot point of the full-speed lever, the top end moves back against the governor diaphragm, which compresses the high speed spring. This backward movement pulls the fuel control rack of the injection pump to a decreased fuel position thereby limiting the maximum speed that the engine can attain.

As the engine speed slows down, a state of balance will once again be established between the weights and the governor high speed spring.

Altitude Compensator

An altitude compensator must be used on the fuel injection pump for a number of reasons: due to the varying geographical areas that these pickup trucks are used in, in order to comply with EPA (Environmental Protection Agency) exhaust smoke emissions standards, and to maintain suitable engine performance. Figures 18–18 and 18–19 illustrate the location and the operation of this unit.

The altitude compensator is mounted on a bracket alongside the fuel injection pump. It operates as follows: A spring-loaded bellows is contained within the housing of the altitude compensator and is connected through an adjustable pushrod that extends out of the housing and bears against the injection pump control lever as shown in Figure 18–19.

At high altitudes there is a decrease in atmospheric pressure (14.7 psi at sea level)—the higher the altitude, the lower the atmospheric pressure.

The pressure inside the bellows will therefore be greater at high altitude than the surrounding atmospheric pressure. This results in the bellows expanding, which forces the pushrod to retract the injection pump control lever to a lower fuel delivery position. Without this altitude compensator, there would not

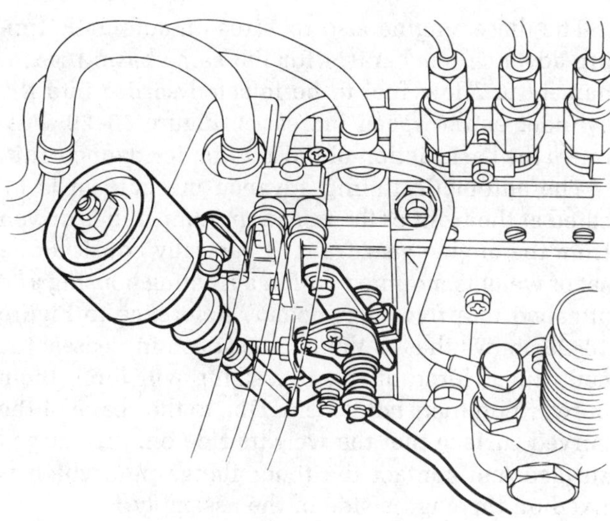

Figure 18–18. Location of altitude compensator. [Courtesy of Nissan Motor Co., Ltd.]

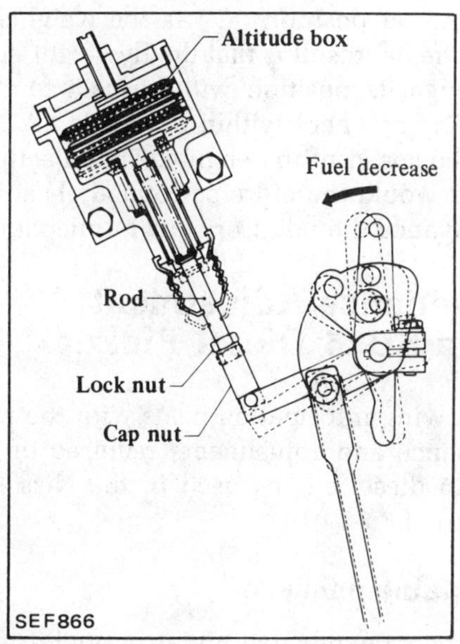

Figure 18–19. Arrangement of altitude compensator. [Courtesy of Nissan Motor Co., Ltd.]

be sufficient air to provide complete combustion of the injected fuel and the engine would not only lose power but would also emit high concentrations of black smoke.

Automatic Advance Unit

In a gasoline engine, the ignition timing is automatically advanced either by a vacuum advance at lower speeds or by a centrifugal advance mechanism at higher speeds. The advance mechanism allows the high tension distributor spark to fire the spark plug earlier in relation to piston position in the cylinder when the engine speed increases.

The diesel engine also requires an automatic timing advance mechanism for the same basic reason, namely to allow fuel to be injected earlier into the cylinder as the speed increases. Figure 18–20 illustrates the basic action of the automatic advance unit.

The automatic timing advance mechanism is located at the front of the injection pump and is driven from the engine gear train. It basically consists of a set of weights mounted within a rotating housing and attached to a flyweight holder. Reference to Figure 18–20 shows that as the engine speed increases, the centrifugal force of the flyweights will force them outward on their holder pin. Due to the shape of the curved surface that the weights ride on, this curved surface will contact the timer flange pin, which is fixed on the engine side of the assembly.

The timer flange pin will therefore act as a cam, causing the force of the flyweights to compress the timer spring, shown in Figure 18–20, from position "B" to that at position "C" as the weights fly outwards. The net result is that the flyweight holder pin will change its position with respect to the timer flange pin anywhere within the angle "A" depending on the speed of the engine. The injection pump camshaft would therefore be rotated ahead to effectively advance the point or start of injection.

Maintenance/Adjustment Nissan SD25 Diesel Pickup

The following information deals with the necessary maintenance and adjustments required on the Nissan SD25 diesel engine used in the Nissan/Datsun pickup truck.

Filter Maintenance

Nissan recommends that under normal driving conditions that the primary and main fuel filters be changed every 30,000 miles, 48,000 kilometers, or 24 months—whichever occurs first.

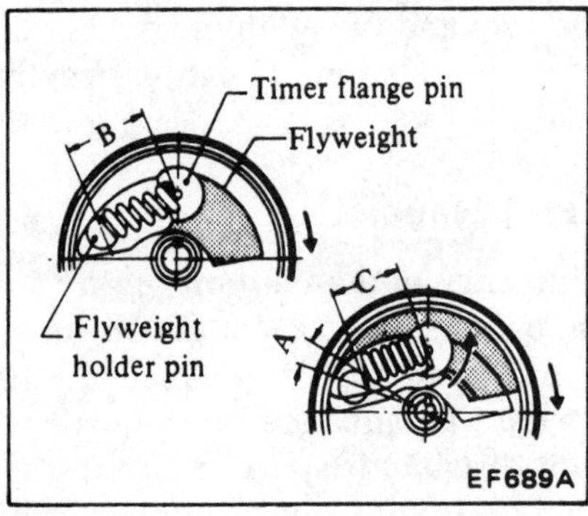

Figure 18-20. Injection pump—automatic timing advance unit. (Courtesy of Nissan Motor Co., Ltd.)

Changing fuel filters involves using either a strap-type wrench on a spin-on-type filter or by loosening the bolt on a shell and element type.

Always place a drain tray under the vehicle to catch fuel or water as it drains from the filter assembly. If the filter has a drain cock on the bottom such as found on fuel/water separators, this can be opened to drain fuel/water from the filter; a vent screw on top should also be opened before loosening off the used filter unit.

CAUTION: When changing fuel filters *do not* attempt to prime the new filter element by pouring old or used fuel from a shop container into it unless this fuel has been pre-filtered. Any addition of unfiltered fuel to a new filter assembly results in a portion of this fuel bypassing the filter element because, for example, fuel in the center of the filter will immediately be sucked up to the feed pump or pass straight to the injection pump.

Draining water from the fuel filter/water separator is described under Chapter 17 dealing with the Dodge Ram 50.

Testing Fuel Filter Overflow Valve

Fuel pressure in the system is established by the fuel feed pump and regulated between 16–21 psi (108–147 kPa or 1.1–1.5 kg/cm^2) on average by the action of the overflow valve at the main fuel filter bypassing fuel back to the fuel tank. Figure 18–21 shows the location of the overflow valve in the filter assembly, while Figure 18–22 shows the hookup requirements

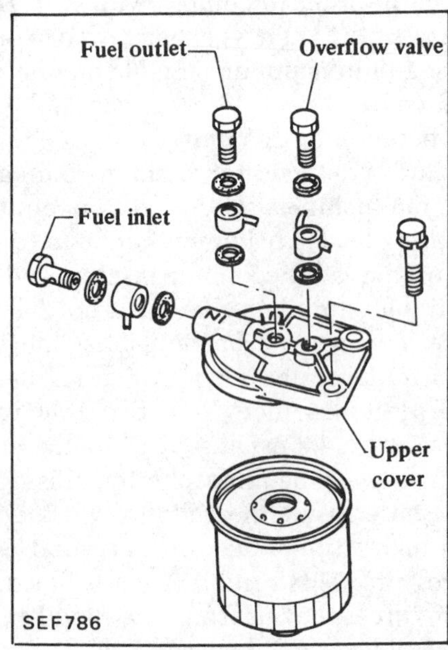

Figure 18-21. Fuel filter assembly. (Courtesy of Nissan Motor Co., Ltd.)

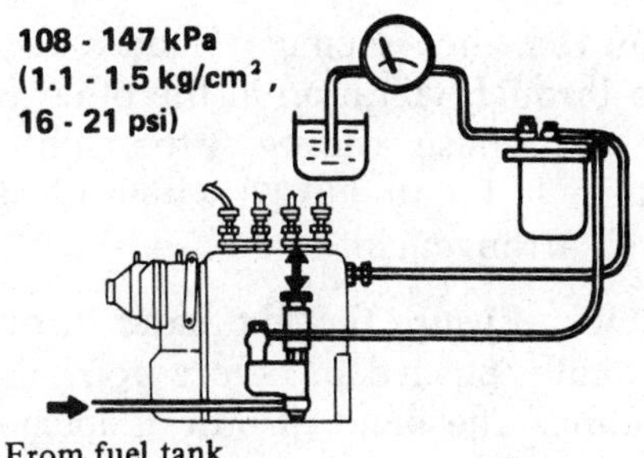

Figure 18-22. Overflow valve test hookup. (Courtesy of Nissan Motor Co., Ltd.)

to test the operation of the overflow valve. The overflow valve opening pressure can be checked by using the feed pump priming handle to raise system pressure. Should the overflow valve fail to open within the range specified, replace it.

Bleeding the Fuel System

Anytime that the fuel filters are changed or any fuel lines have been disconnected for service work to any fuel system components, it will be necessary to bleed the fuel system of any entrapped air. This can be done as follows:

Procedure

1. Refer to Figure 18–23.
2. Remove the plastic protective cap that covers the priming pump handle (Item 2 in Figure 18–23).
3. Unscrew the priming pump handle knob counterclockwise.
4. Loosen off the air vent screws (Item 1) on the injection pump (one at each corner) and place a drain tray underneath to catch the vented fuel.
5. Pull/push the priming pump handle up and down, which will allow fuel to pour from the open screw on the injection pump. Continue to pump the feed pump handle until fuel free of air bubbles flows from the vent screw holes.
6. Tighten both air vent screws.
7. Push down the feed pump handle until it bottoms, then tighten the handle knob clockwise and install the protective cap back over the handle.
8. Wipe/dry off any fuel around the injection pump and engine.

Checking and Adjusting Injection Pump/Engine Timing

If the injection pump has been removed from the engine during an engine overhaul or major pump repair, it is necessary to align the injection pump to engine timing marks when the pump is reinstalled onto the engine.

The two things that must be checked and aligned are:

1. The injection pump drive gear, which also includes the automatic timing advance unit, must be timed to the engine gear train. The correct timing alignment marks are shown in Figure 18–24.
2. The alignment mark on the injection pump housing and on the engine front cover (Figure 18–25). If the marks are not in alignment, loosen off the injection pump mounting bolts, grasp the injection pump with both hands, and rotate it until the marks are in proper alignment. Tighten the bolts to 20–25 N·m (14–18 lb.ft.) or 2–2.5 kg-m.

NOTE: If it is necessary to install a new front cover on the engine or an unmarked injection pump housing, it will be necessary to "spill time" the injection pump in order to establish an accurate alignment mark between the pump and engine front cover. Refer to the sub-heading dealing with "spill timing" in this chapter.

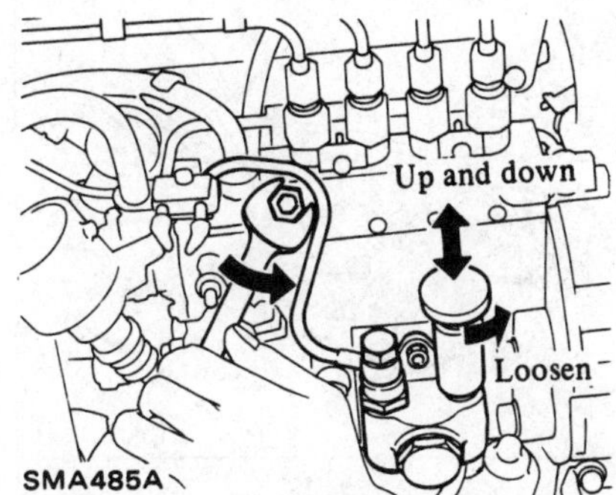

Figure 18-23. Bleeding the fuel system of air. (Courtesy of Nissan Motor Co., Ltd.)

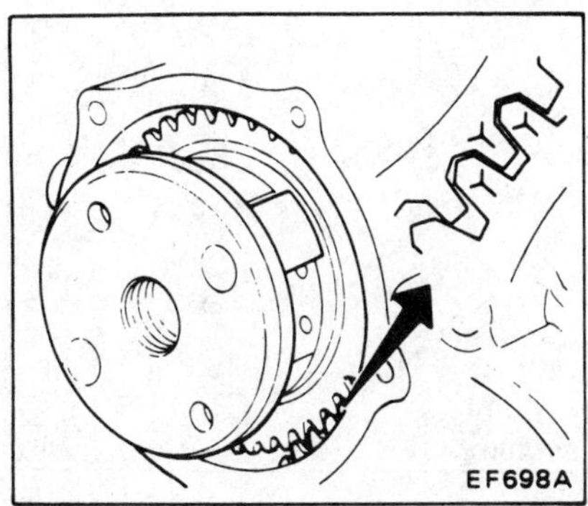

Figure 18-24. Checking injection pump/engine timing. (Courtesy of Nissan Motor Co., Ltd.)

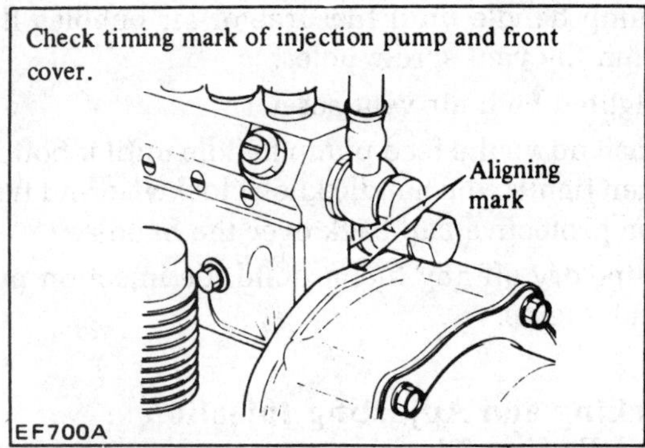

Figure 18–25. Alignment mark on injection pump housing and engine front cover. (Courtesy of Nissan Motor Co., Ltd.)

Injection Pump Diaphragm Oil Check. Refer to Figure 18–26 and remove the filler plug on the top of the injection pump housing. This plug is on the atmospheric side of the pneumatic governor diaphragm; the recommended service interval for the addition of oil to this unit is every 5000 miles (8000 km) or 24 months, whichever occurs first. The oil capacity for this unit is 4 ml or 0.14 fluid ounce.

Check/Adjust Idle Speed

Recommended idle speed for the Nissan SD25 diesel engine is anywhere between 650 and 800 rpm. Before checking the idle speed, ensure that the engine is at its normal operating temperature. Engine rpm can be checked with a variety of digital-type engine tachs now on the market.

Diesel engine tachometers are now available in clip-on fashion whereby an adapter clip is placed over the No. 1 cylinder fuel injection line in order to pick up the injection pulses through the line, which are then converted to an rpm reading on the tach.

For the SD25 diesel engine, a more accurate reading is obtained by removing the clamp on the No. 1 injection line. The reason for this is that if the clamp is not removed, then secondary fuel pulses can be picked up through vibration at the other fuel injection lines and a false reading at the tach can be reflected. Refer to Figure 18–27, which illustrates the type of tach arrangement to be used.

CAUTION: Ensure that the hand throttle inside the cab is fully pushed in before you take an idle speed reading. The hand throttle is located on the left-hand side of the steering column.

Procedure

1. Run the engine with the tach connected as shown, at approximately 2000 rpm for about 2 minutes to allow the pickup to sense and stabilize the pulses through the fuel line.
2. Run the engine at idle rpm for about 1 minute before reading the actual rpm on the tach.
3. If the idle rpm is not within specs, then proceed to adjust the idle screw as shown in Figure 18–28 by loosening off the lock nut and turning the adjusting screw clockwise to increase the idle speed, or counterclockwise to decrease the idle speed.

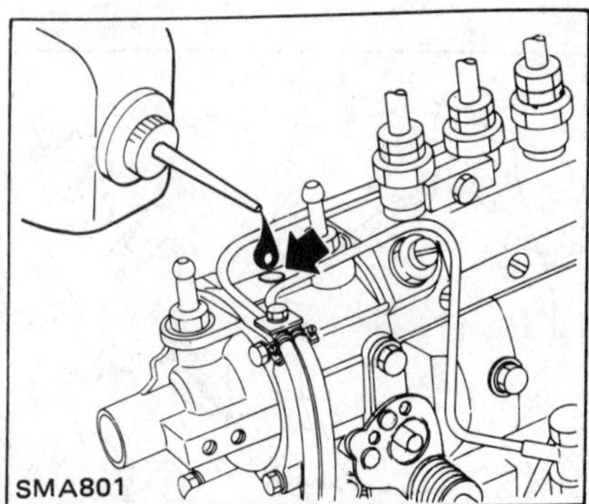

Figure 18–26. Adding oil to injection pump pneumatic governor. (Courtesy of Nissan Motor Co., Ltd.)

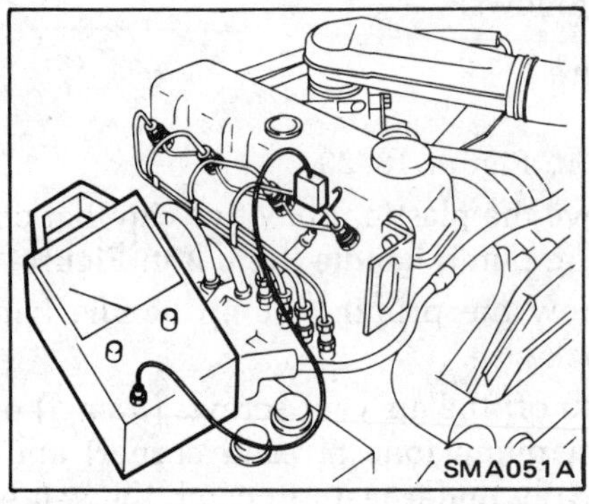

Figure 18–27. Tach hookup. (Courtesy of Nissan Motor Co., Ltd.)

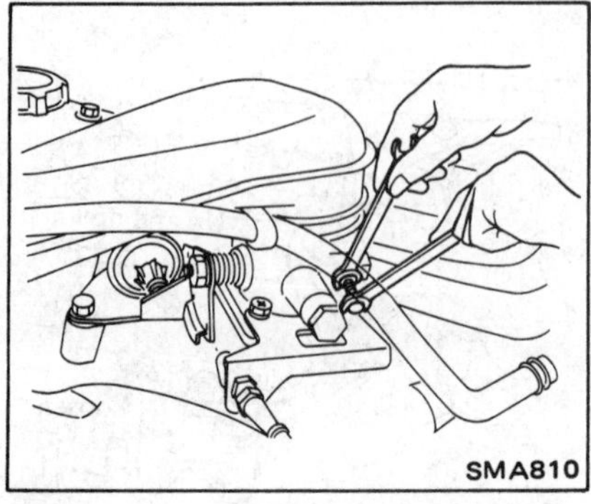

Figure 18–28. Idle adjusting screw. (Courtesy of Nissan Motor Co., Ltd.)

4. Carefully note the rpm registered on the tach, then manually rev up the engine let the engine settle back down to idle, and again read the tach.
5. If the rpm is correct, tighten the lock nut on the adjusting screw.

Once the idle speed has been set, the dashpot adjustment must be checked and set.

Procedure

1. Leave the tach connected as per the idle speed setting instructions.
2. Refer to Figure 18–29 and loosen off the dashpot lock nut.
3. Manually speed up the engine by manipulating the throttle lever at the air cleaner-to-intake manifold, to a speed between 1280–1350 rpm and adjust the dashpot until its plunger lightly contacts the throttle control lever, then lock it there.

Checking the Injection Nozzle

There is no specific mileage/kilometers at which the injection nozzles have to be serviced. The type of service to which the vehicle is subjected determines when service may be required or when the nozzles may require checking. Generally speaking, if hard starting or exhaust smoke becomes a problem, the nozzles probably are part of the problem, if not the main problem.

A quick check that will help you determine if the nozzles are the main problem is as follows.

Procedure

1. Run the engine until it is at operating temperature.
2. With the engine running at an idle rpm, loosen off the fuel line at the nozzle (place a rag around it).

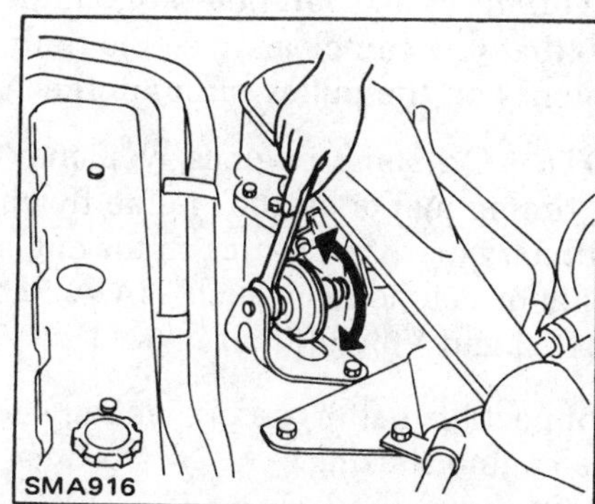

Figure 18–29. Loosening off the dashpot lock nut. (Courtesy of Nissan Motor Co., Ltd.)

There should be a reduction in engine rpm with the fuel line loosened off. If no rpm reduction occurs, then that particular injector is not operating properly.

NOTE: There is always the possibility that there is a problem within the injection pump plunger feeding that nozzle; however, this would involve pump removal and repair.

Nozzle Removal.

Procedure

1. Refer to Figures 18–30, 18–31, and 18–32 and remove both the injection nozzle fuel (spill) return lines as well as the injection lines between the injection pump and nozzles.

   NOTE: Place plastic shipping caps over the delivery valve fuel studs at the injection pump to prevent the entrance of dirt.

2. Remove the nozzle fuel spill tubes by using a socket and ratchet while holding onto the nozzle with an open-end wrench.

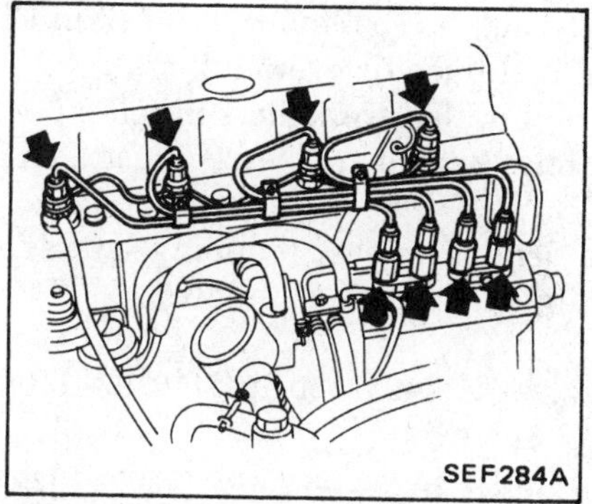

Figure 18–30. Injection nozzle fuel supply and (spill) return lines. (Courtesy of Nissan Motor Co., Ltd.)

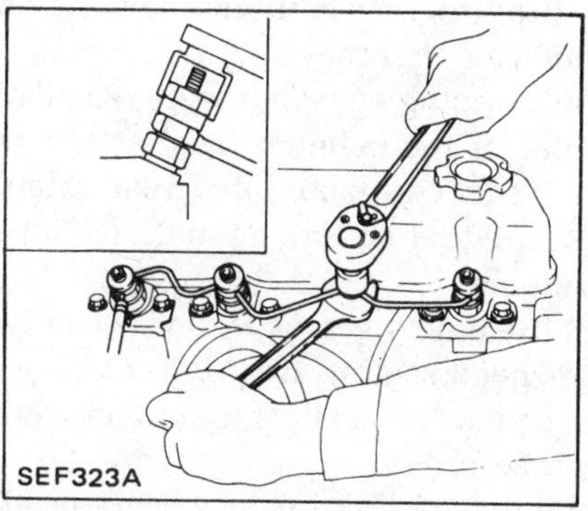

Figure 18–31. Removing nozzle fuel spill tubes. (Courtesy of Nissan Motor Co., Ltd.)

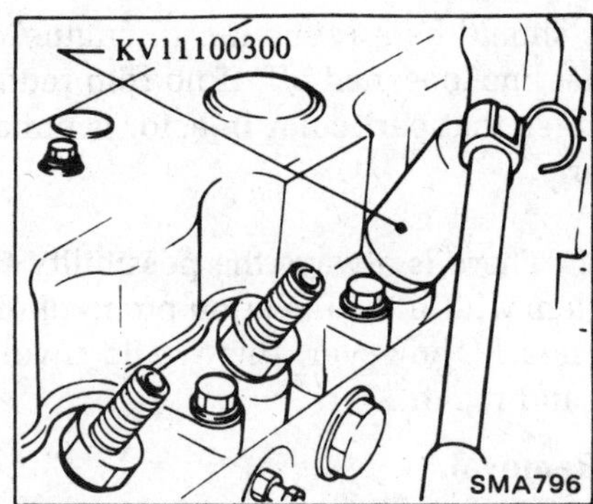

Figure 18–32. Loosening and removing nozzle from bore in cylinder head. (Courtesy of Nissan Motor Co., Ltd.)

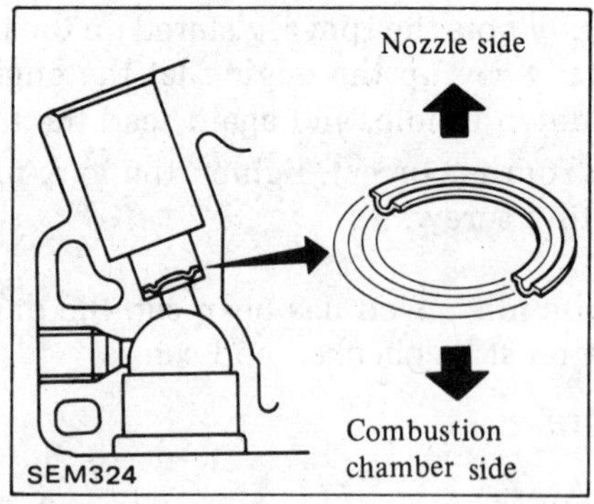

Figure 18–33. Nozzle gasket installation. (Courtesy of Nissan Motor Co., Ltd.)

3. Using Tool KV11100300 or equivalent, loosen the nozzle, then withdraw it from its bore in the cylinder head.
4. Place plastic shipping caps into the nozzle bores of the cylinder head to prevent the entrance of any foreign material.

Nozzle Testing

Testing, repair, maintenance, and overhaul of the injection nozzles can be found in detail in Chapter 29 on diesel fuel injector overhaul.

Injection nozzles are Robert Bosch-type throttling pintle nozzles (which are clearly illustrated in Figure 29–4).

Recommended release (popping) pressure for the SD25 engine nozzles is as follows:

New nozzles . . . 1493–1607 psi (10297–11082 kPa) or 105–113 kg/cm²

Used nozzles . . 1422–1493 psi (9807–10297 kPa) or 100–105 kg/cm²

Nozzle Installation—Engine

Installation of the nozzle should be done once it has been established that they operate correctly. Remove the plastic shipping caps that were installed into the nozzle holes of the cylinder head at removal. Blow the bores clean with shop air (wear safety glasses) and make sure that no foreign material has settled in the nozzle bore.

Always use a new gasket at the base of the nozzle-to-cylinder head seating area. Refer to Figure 18–33 to ensure that you install this new gasket the right way into the bore.

Once the injection nozzle has been installed along with its new gasket, torque it to 43–51 lb.ft. (59–69 N•m or 6.0–7.0 kg-m).

CAUTION: To prevent possible breakage of the nozzle spill (fuel return tubes), the spill tube nuts should be gradually tightened in sequence. Once the nozzles have been installed and all lines and fittings connected, bleed the fuel system as described herein.

Injection Pump Spill Timing

It will often become necessary to check the accurate timing of the injection pump to the engine. This type of check is normally required anytime that new components have been fitted to the engine such as a new engine front cover, timing gears, or injection pump components. This check may also be required during the life of the engine if and when it is suspected that the injection pump timing may be incorrect.

The sequence given herein is general in nature because minor differences will exist between makes of engines. For example, the static timing position or number of degrees BTDC that initial timing occurs will vary between engines.

Procedure for In-Line Injection Pump

1. Clean off the crankshaft pulley or damper assembly so that you can clearly see the timing marks or notches on the pulley circumference.

   NOTE: On some engines, you may have to align timing marks on the engine flywheel with a stationary mark or pointer on the engine clutch housing or bell housing such as VW Rabbit (Figure 14–2) and Volvo (Figure 15–11).

2. Disconnect the battery to prevent any possibility of the engine starting.
3. Manually rotate the engine over with a socket on the crankshaft pulley nut or at the flywheel ringgear to place No. 1 piston at TDC compression.

4. Align the manufacturer's specified timing mark on the crankshaft pulley or flywheel with the stationary timing fixture or mark similar to that shown in Figure 18–34.
5. Remove all the injection lines from the pump to the nozzles.
6. Disconnect any linkage or governor hoses.
7. Use plastic shipping caps over the ends of the nozzle fuel inlets and also over the ends of the injection pump delivery valve holders (fuel studs).
8. Refer to Figure 18–35 and using a deep socket, loosen off the No. 1 injection pump delivery valve holder (a lock plate may have to be removed first on some injection pumps), unscrew the delivery valve holder carefully, and remove the valve stop and spring.
9. Reinstall the delivery valve holder *without* the delivery valve spring and valve stop and tighten it to specs.
10. If the injection pump fuel feed hose from the transfer pump is not connected up, do it now.

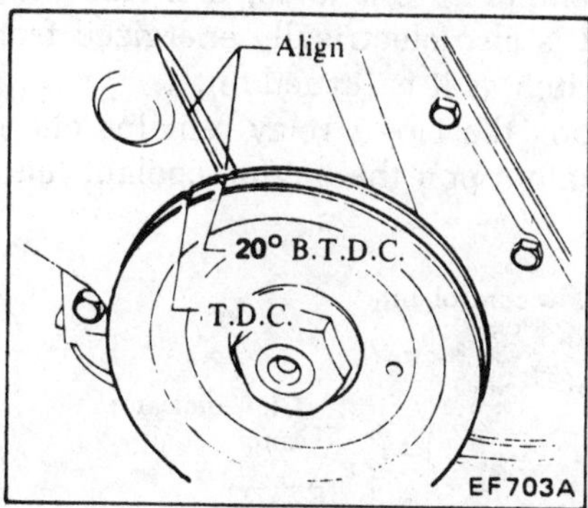

Figure 18–34. Aligning crankshaft pulley timing mark with stationary pointer. (Courtesy of Nissan Motor Co., Ltd.)

11. Loosen off the injection pump-to-engine mounting bolts/nuts and push the injection pump towards the engine block.
12. Refer to Figure 18–36 and connect a used fuel injection line to the No. 1 delivery valve holder. Operate the feed pump priming handle so that fuel flows freely from the bent goose neck injection line into a container.
13. While operating the priming pump handle, slowly pull the injection pump away from the engine block until the flow of fuel from the No. 1 delivery valve holder pipe stops as shown in Figure 18–37.
14. The point at which the fuel flow stops is the accurate injection pump-to-engine static timing position.
15. You may want to doublecheck this timing point by repeating the procedure. Once you are satisfied that this is correct, tighten up the injection pump to engine mounting bolts/nuts, as the case may be.
16. Check that the alignment mark on the injection pump housing to engine front cover is correct. If it is not, then using a small chisel, gently place a scibe mark on both the injection pump housing and engine front cover so that both marks are in proper alignment.
17. Remove the test tube from the No. 1 delivery valve holder.
18. Remove the No. 1 delivery valve holder and install the previously removed spring and valve stop along with the delivery valve holder.
19. Torque the delivery valve holder to the manufacturer's specifications.
20. Install all of the necessary injection pipes, etc., and bleed the fuel system.

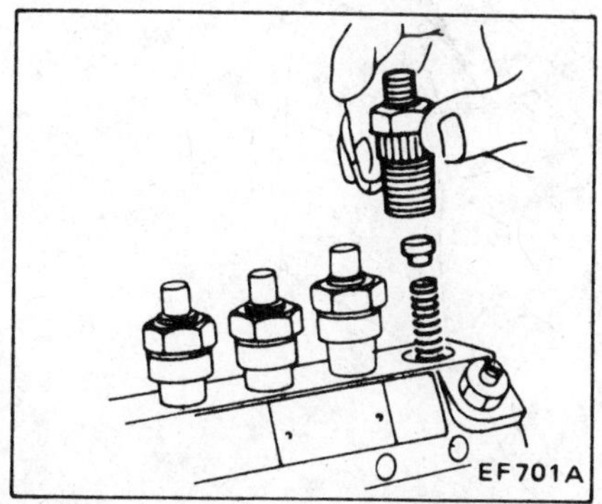

Figure 18–35. Delivery valve holder removal. (Courtesy of Nissan Motor Co., Ltd.)

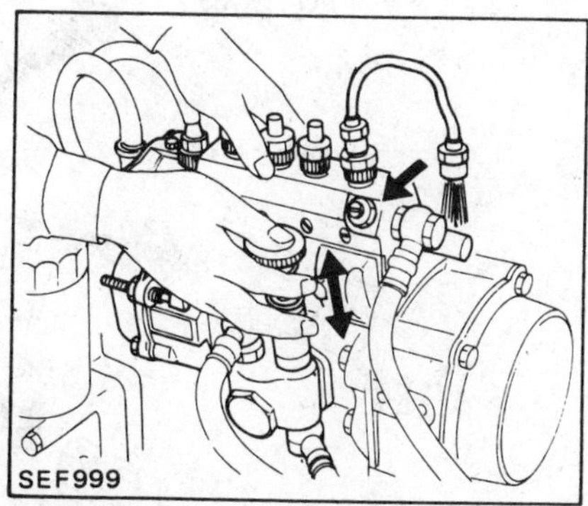

Figure 18–36. Fuel flows freely from spill timing tube. (Courtesy of Nissan Motor Co., Ltd.)

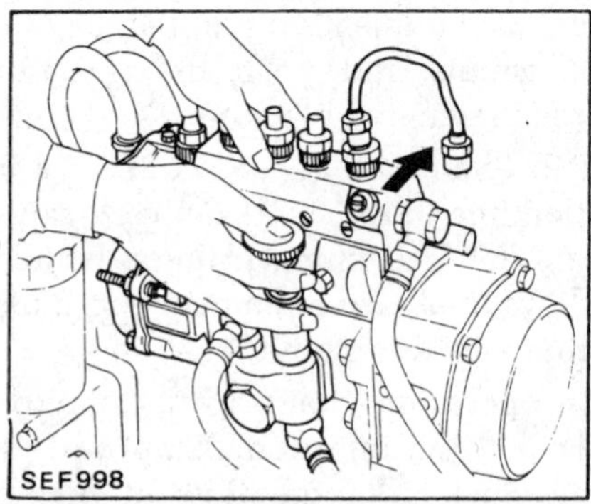

Figure 18-37. Fuel delivery stops—correct timing. [Courtesy of Nissan Motor Co., Ltd.]

Nissan SD25 Diesel Glow Plug Circuit

The SD25 diesel engine used in the Nissan/Datsun diesel pickup is a 4-cylinder IDI (indirect-injection) type engine and therefore requires the use of combustion chamber glow plugs to facilitate starting of the engine.

There is one glow plug for each cylinder; the glow plug is screwed into the cylinder head as shown in Figure 18–38.

Glow Plug Circuit Component Location. The actual glow plugs, as mentioned, are screwed into the cylinder head. However, the electrical controls for the glow plugs are located as shown in Figure 18–38.

Description of Operation—Glow Plug Circuit

The glow plug circuit is designed to provide preheat to the pre-combustion chamber prior to start-up and after start-up to ensure rapid engine warm-up and smooth low speed operation.

Two basic situations exist to control the glow plug preheat time, namely what is known as "pre-glow" and "after-glow." Figure 18–39 is a simplified electrical schematic of the glow plug wiring circuit.

When the ignition switch is first turned on, relay No. 1 shown in the wiring diagram is electrically energized and battery current flows through to the grounded glow plugs in the cylinder head. This high current flow allows rapid heating of the glow plug elements (1652°F/900°C) for a period of between 6 to 7 seconds after which time relay No. 1 is turned off (de-energized).

During cold weather operation or when the engine coolant temperature is lower than 122°F (50°C), relay No. 2 is also electrically energized from the ignition switch as it is turned on.

How long the No. 2 relay remains energized depends totally upon the engine coolant temperature.

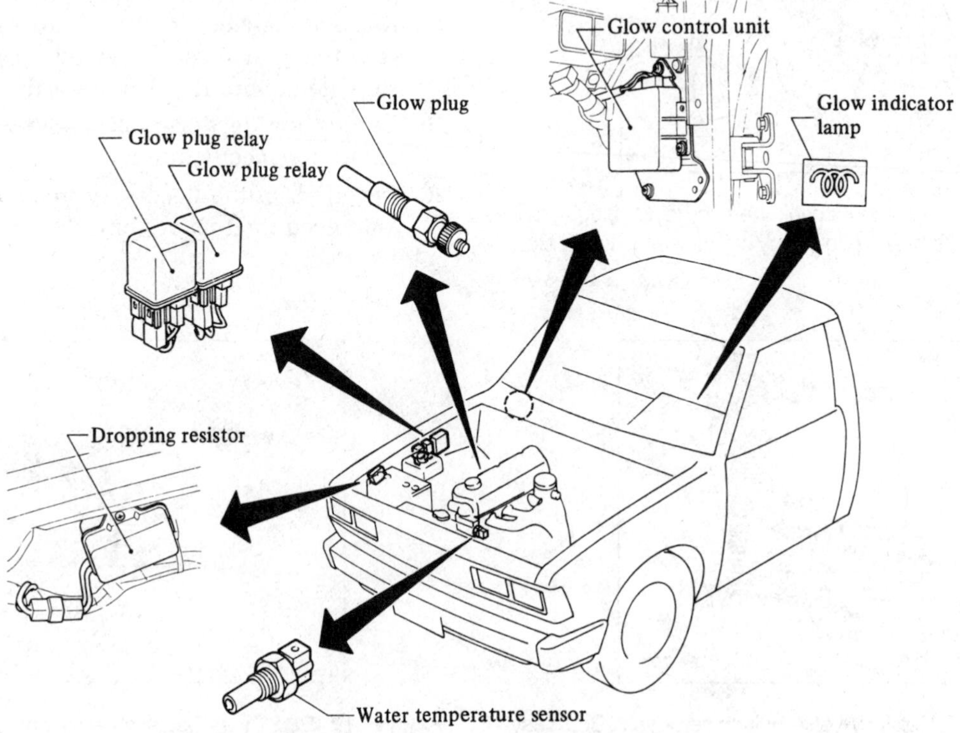

Figure 18-38. Location of all glow plugs and components. [Courtesy of Nissan Motor Co., Ltd.]

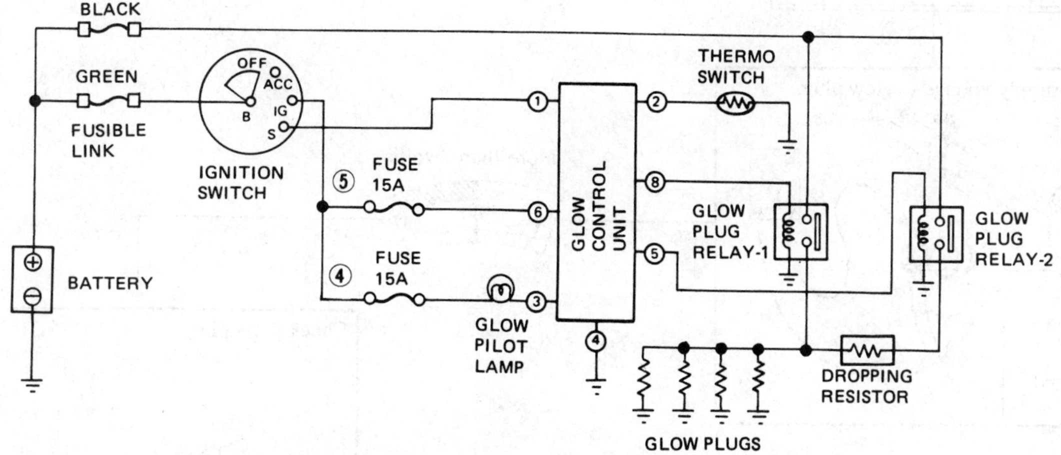

Figure 18-39. Electrical schematic—glow plug circuit. (Courtesy of Nissan Motor Co., Ltd.)

This relay can remain energized from 20 to 45 seconds during which time a lower current will flow to the glow plugs than that which would flow through the No. 1 relay because of the resistor placed into the circuit. The main function of the No. 2 relay is simply to improve the combustion efficiency after the engine has started, as long as the engine coolant is lower than 122°F (50°C). Regardless of the engine coolant temperature being higher than 122°F (50°C), the No. 2 relay will supply current to the glow plugs during engine cranking, then switch off. The actual wiring circuit as it would appear on the vehicle is shown in Figure 18–40 with the fusible links and wire colors illustrated.

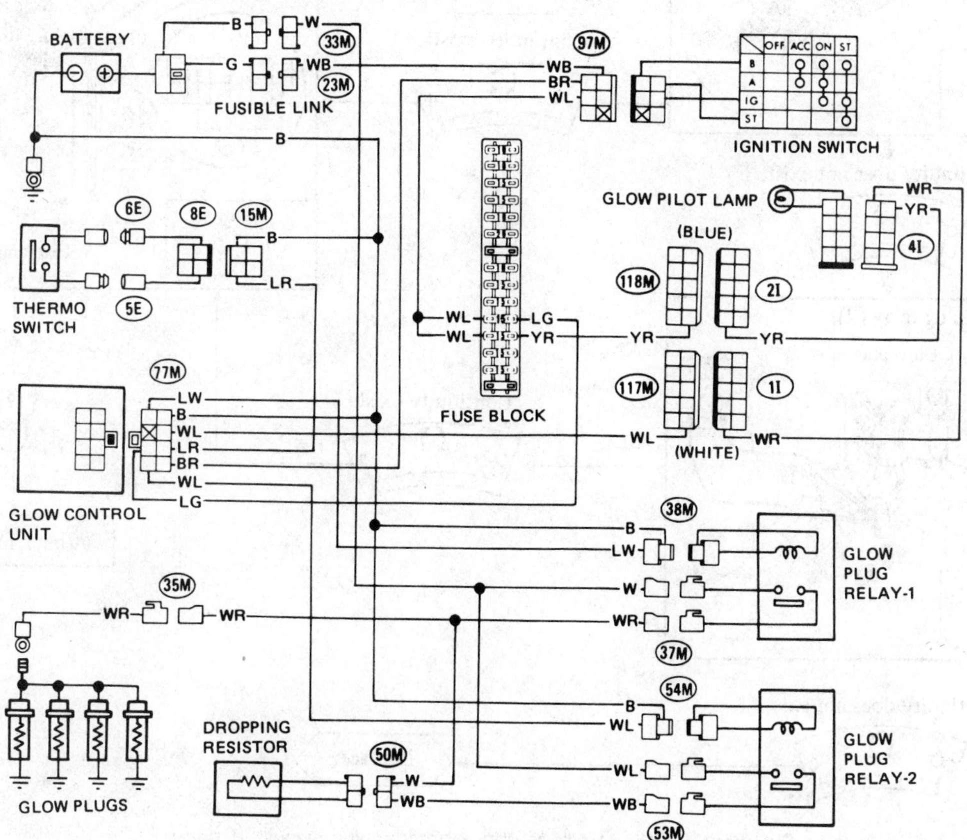

Figure 18-40. Nissan diesel pickup glow plug wiring circuit connections. (Courtesy of Nissan Motor Co., Ltd.)

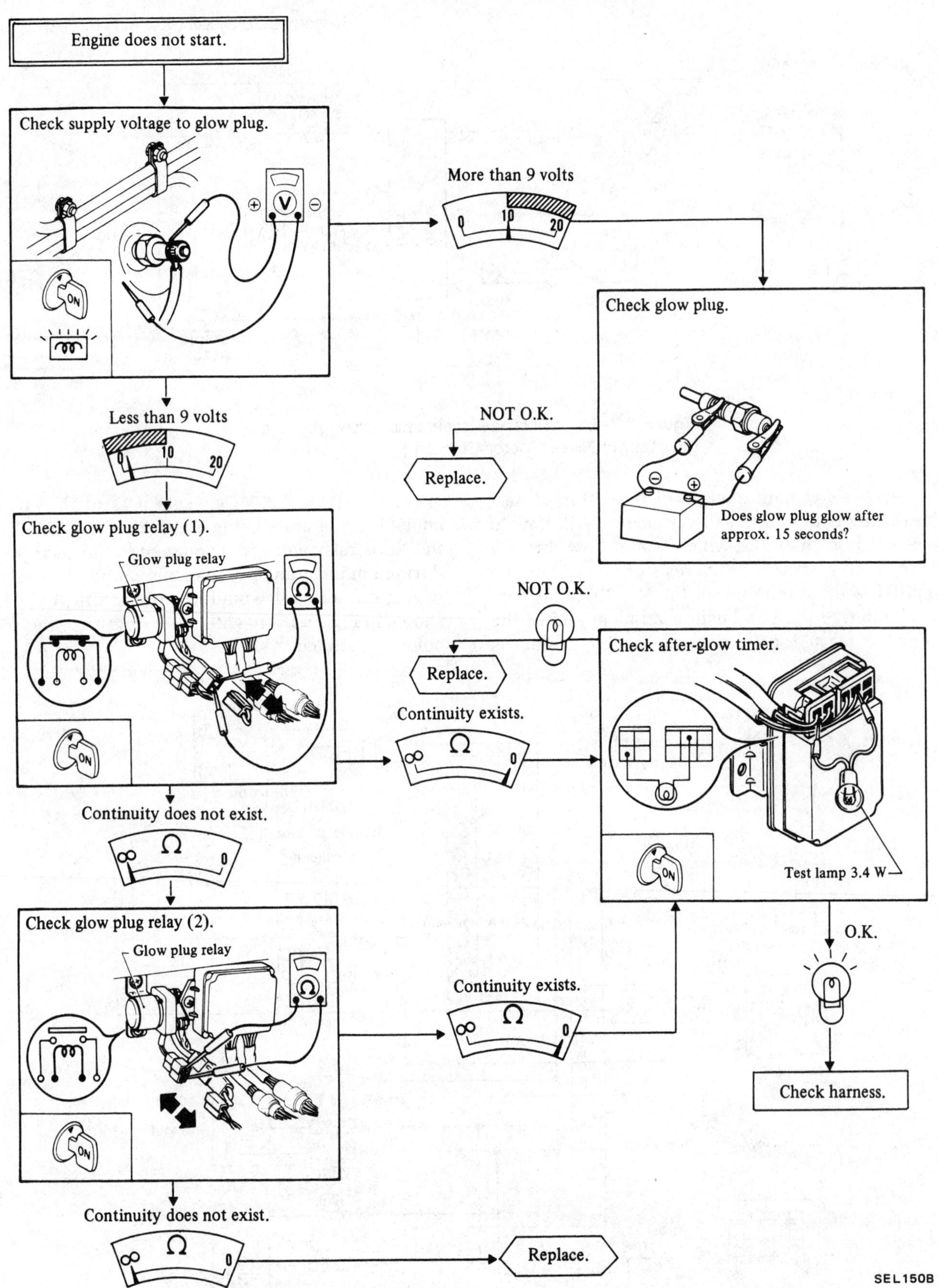

Figure 18-41. Voltage check of glow plug circuit. (Courtesy of Nissan Motor Co., Ltd.)

Troubleshooting the Glow Plug Circuit. Problems in the glow plug circuit are usually indicated when hard starting is experienced because without adequate preheat to the air coming into the combustion chamber, the injected diesel fuel cannot vaporize properly; therefore, hard starting and black smoke at the exhaust are typical results. In addition, once the engine does start, it will run rough because there is no "after-glow" to the glow plugs to assist warm-up of the combustion chamber area.

The first thing to check when the glow plug circuit is suspected of being faulty is each individual glow plug wiring connection at the cylinder head. In addition, check that the battery is in a good state of charge.

If the answer to each of the above questions is OK, then proceed as follows in order to isolate the problem to the glow plugs themselves or to the glow plug control circuit.

Procedure

1. Refer to Figures 18–41 and 18–42 and complete these checks in the sequence shown to determine if the problem is low voltage supply or a faulty glow plug. When checking the voltage supply to the glow plug, the ignition key switch should be on as well as the glow plug indicator lamp on the vehicle instrument panel.

   CAUTION: When applying direct battery voltage to the individual glow plugs, take care not to burn yourself and do not continue to apply voltage once the plug has glowed.

2. If the supply voltage to the glow plug is less than 9 volts, it usually indicates that the problem is in the relay or the after-glow timer.

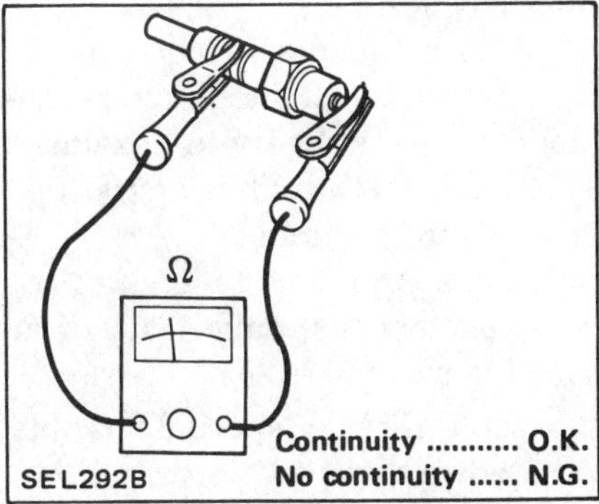

Figure 18–42. Continuity check of a glow plug. (Courtesy of Nissan Motor Co., Ltd.)

To check the glow plug relay, water temperature (thermal) sensor, and drop resistor connected to relay No. 2, proceed as follows:

1. Refer to Figure 18–43 and apply 12-volt battery power across terminals 3 and 4 while checking for continuity across terminals 1 and 2.
2. Refer to Figure 18–44 and suspend the water temperature (thermal) sensor in a beaker of water. Preheat the water and when the temperature reaches that shown in the illustration, check the resistance across the unit and compare the readings with that in the chart.

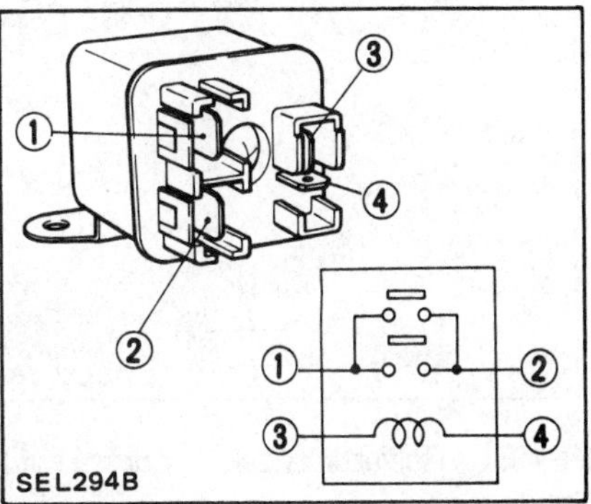

Figure 18–43. Glow plug relay check. (Courtesy of Nissan Motor Co., Ltd.)

Temperature °C (°F)	Resistance kΩ
−10 (14)	7.0 - 11.4
20 (68)	2.1 - 2.9
50 (122)	0.68 - 1.00

EF329A

Figure 18–44. Water temperature sensor check. (Courtesy of Nissan Motor Co., Ltd.)

3. To check the condition of the drop resistor used with relay No. 2, simply connect an ohmmeter across terminals 1 and 2 as shown in Figure 18–45.
4. To check the glow plug control unit refer to Figure 18–46 and connect the necessary test wires as shown. Connect the following resistors as indicated in Tables 18–1, 18–2, and 18–3 to terminal No. 2 in Figure 18–46 and ensure that lamp "A," "B," or "C" in Figure 18–46 goes out within the time specified in Tables 18–1, 18–2, and 18–3.

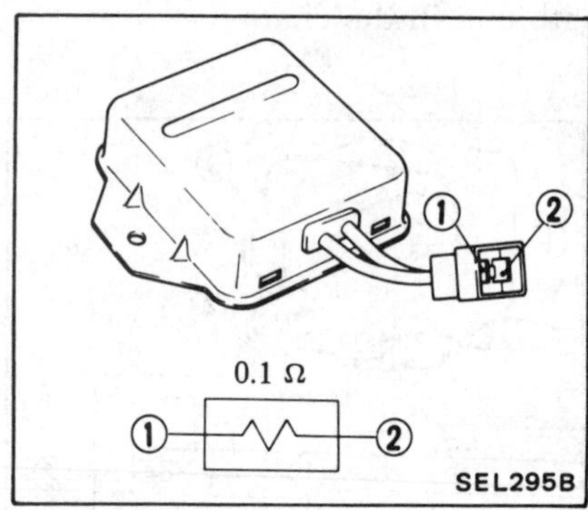

Figure 18–45. Resistor check. (Courtesy of Nissan Motor Co., Ltd.)

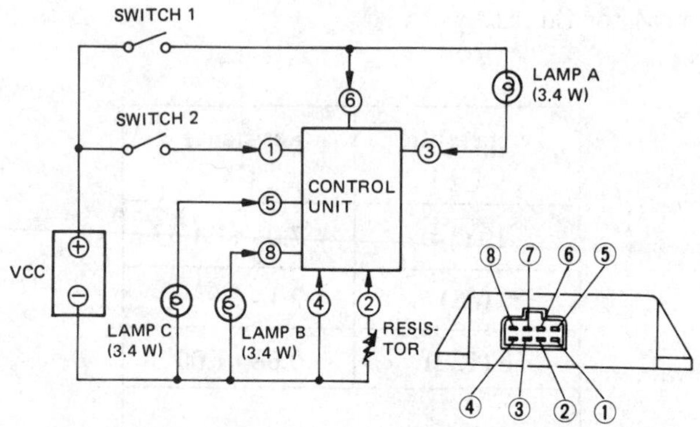

Figure 18–46. Glow plug control unit check. (Courtesy of Nissan Motor Co., Ltd.)

Table 18–1. *Glow Lamp Check Chart*

Vcc [V]	*Switch 1*	*Resistor [kΩ]*	*Time in which lamp goes out (Seconds)*
			Lamp A
10.5	ON	19.0	5–7
		11.5	4.5–6.3
		5.6	3.5–5.1
		3.7	2.9–4.3
		1.2	0.6–1.9

Table 18–2. *Pre-Glow Check Chart*

Vcc [V]	*Switch 1*	*Switch 2*	*Resistor [kΩ]*	*Time in which lamp goes out (Seconds)*
				Lamp B
10.5	ON	OFF	30	5.3–6.7
	ON	ON	20	5.5–7.5

Table 18–3. *After-Glow Check Chart*

Vcc [V]	*Switch 1*	*Switch 2*	*Resistor [KΩ]*	*Time in which lamp goes out (Seconds)*
				Lamp C
12.0	ON	ON first, then OFF	19.0	38–55
			11.5	33–49
			5.6	25–39
			3.7	20–31
			1.2	15–25

Nissan Diesel—Troubleshooting

Contained within the fuel system section dealing with the SD25 diesel engine used in the Nissan/Datsun pickup is information dealing with all aspects of operation and maintenance of the fuel system.

Effective troubleshooting is only possible if you have a thorough understanding of the function and operation of the various components within the fuel system. Typical problems that you may be called upon to deal with would be for example:

1. Engine fails to start or is difficult to start. Possible problem areas related to this situation would be:
   a. Battery condition or connections, slow cranking speed, low current flow to glow plugs, activation of injection pump controller (DPC) unit.
   b. Fuel system:- adequate fuel, fuel delivery, filters plugged (or cloud point reached in low ambients), fuel delivery to nozzles, faulty nozzles. Check DPC switch operating positions, injection pump timing.
   c. Mechanical problems = engine compression low, intake/exhaust valves out of adjustment, glow plug system.
2. Unstable engine idle speed. Possible problem areas related to this situation could be:
   a. Fuel system = injection pump timing incorrect, low idle screw adjustment required, fuel starvation/filters/lines/leaks, faulty nozzles.

b. Mechanical problems = air starvation, low compression, engine timing, injection pump timing, valve clearances.

3. Excessive white or black smoke. Possible problem areas related to this situation could be:
   a. White smoke: Fuel system = injection pump timing, air in system, low compression, feed pump problem, injection pump problem, grade of fuel.
   b. Black smoke: Air starvation (plugged air filter or restriction at intake manifold venturi butterfly valve), fuel injection pump timing, check compression, check injection nozzles, check diesel injection pump controller (excess fuel), injection pump problem, grade of fuel.
4. Engine lacks power. Possible problem areas related to this situation could be:
   a. Fuel system = grade of fuel used, fuel starvation, air in fuel system, butterfly valve operation in intake manifold, injection pump timing, injection nozzles, fuel injection pump controller.
   b. Mechanical system = air starvation, low compression, intake/exhaust valve adjustment.

Nissan Diesel Compression Check

A compression check on a diesel engine is similar to that on a gasoline engine with the exception that you cannot use a hand-held compression gauge such as you can use on some gasoline engines. The reason for this is that the cylinder pressures produced in a diesel engine are considerably higher than that produced within the gasoline engine. Attempting to use a hand-held compression gauge to effectively monitor the compression on a diesel engine can result in personal injury.

A compression check is normally undertaken when the engine is:

1. Hard to start
2. Emits exhaust smoke when cold or warm
3. Lacks power
4. Burns oil

If the fuel system and glow plug circuit have been checked out to satisfaction, a compression check can be undertaken as follows.

Procedure

1. Run the engine until it attains its normal operating temperature.
2. Remove the injection nozzle spill (return) fuel lines as well as the feed lines from the injection pump to the nozzles.
3. Install plastic shipping caps over the injection pump delivery valve holder fuel studs.
4. Install plastic shipping caps over the injector fuel inlet studs.
5. Remove all injection nozzle assemblies with their base gaskets.
6. Refer to Figure 18–47 and install compression gauge P/N ED19600000 into No. 1 injector hole in the cylinder head and torque the gauge adapter to 56–58 lb.ft. (76–78 N•m or 7.7–8.0 kg-m).
7. Disconnect the wiring harness connector between the injection pump control unit (DPC) and the pump controller.
8. Place the injection pump control lever into the stopped position and wire it securely in this position.
9. Check to see that the compression gauge release button is in the closed position.
10. Block the engine throttle (intake manifold) wide open and proceed to crank the engine over on the starter motor for about 4 to 6 revolutions.

The following compression readings should be obtained at a cranking speed of 200 rpm:

Normal readings—427 psi (2942 kPa or 30 kg/cm^2)
Minimum readings—356 psi (2452 kPa or 25 kg/cm^2)

Acceptable variation between cylinders should not exceed 43 psi (294 kPa or 3 kg/cm^2).

Push the compression gauge release button to vent the air from the cylinder and gauge and remove the gauge and adapter.

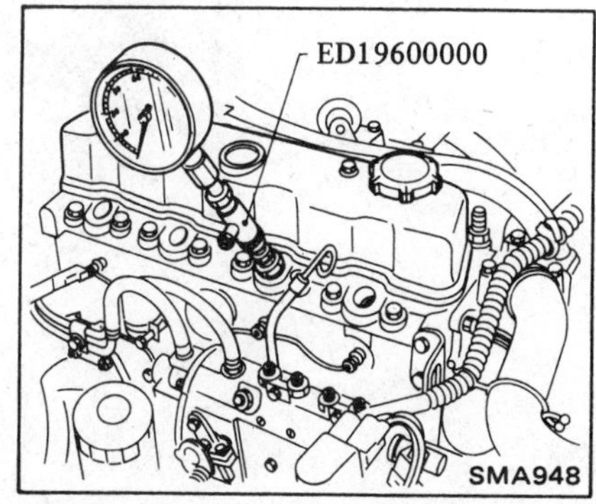

Figure 18-47. Compression gauge installed into No. 1 injector hole. (Courtesy of Nissan Motor Co., Ltd.)

Interpretation of Readings. Low compression is generally due to worn piston rings or faulty valves. Piston rings are generally the cause when the pressure on the gauge is low on the first crank but tends to build up on the following strokes; however, the pressure still remains lower than the minimum manufacturer's specifications.

Some diesel engine manufacturers are against adding engine oil to the cylinder to confirm whether the problem is in the rings or the valves because, with such a high compression ratio in the engine, serious damage can result because of the non-compressibility of the oil in the cylinder. Such problems can be blown head gaskets and possible connecting rod damage due to a hydrostatic lock occurring.

Nissan, however, believes the addition of oil to be acceptable; therefore if compression in one or more cylinders is low, squirt a few drops of clean engine oil into the cylinder and recheck the compression reading.

An increase in compression with oil added generally indicates that the piston rings are worn. If the pressure still remains low, then the valves are generally at fault (leaking).

It is possible, when two adjacent cylinders read low and the addition of oil does not improve the compression that the cylinder head gasket is leaking.

19

Toyota Landcruiser 2.4L Diesel

The Toyota Landcruiser diesel engine is a 4-cylinder unit of 4-stroke design displacing 2.4L (146 cu.in.) in the 1984 model.

Fuel System

The fuel system employs an in-line type fuel injection pump of Nippondenso manufacture (Robert Bosch licensee). This fuel injection pump design follows the same basic layout and arrangement as other in-line type pumps in that it follows the Bosch multiple plunger concept. In other words, four individual pumping plungers operate within barrels contained within the one injection pump housing (as shown in Figure 19–23).

If you are familiar with the in-line fuel injection pump system used by Robert Bosch (Chapter 12) and CAV, then the Nippondenso pump will present you with no unusual features. Examples of other vehicles using an in-line pump of this general design are Datsun/Nissan SD25 pickup (Chapter 18) (Diesel Kiki Pump), Isuzu 4- and 6-cylinder QD engines, Mercedes-Benz, and many others.

The general layout of the fuel system is very similar to that used on the SD25 Datsun/Nissan pickup truck illustrated in Figure 18–1. The fuel is drawn from the fuel tank where it then passes through a fuel filter on its way to the injection pump camshaft driven feed/transfer pump. Fuel is placed under pressure within this reciprocating piston-type pump and delivered to a secondary fuel filter from where it flows to the inlet fuel fitting attached to the fuel injection pump. Fuel then flows to a common manifold within the injection pump housing where it can continually supply the demands of the individual injector pumping plungers.

Fuel is metered and placed under pressure as explained in detail in Chapter 10 (see Figure 10–5 or Figure 18–4). It is then delivered to the individual fuel injectors in firing order sequence. Excess fuel that is used for cooling and lubrication of injector components leaks past internal parts and is returned to the suction side of the feed transfer pump or fuel tank.

Fuel Filter Maintenance

There are two fuel filters employed with this system: the primary fuel filter, located between the fuel tank and the feed transfer pump, and the secondary filter, located between the outlet (pressure) side of the feed transfer pump and the injection pump fuel inlet line. (These two filters can be seen in the basic fuel system layout in Figures 19–1 and 19–4.)

The primary fuel filter shown in Figure 19–1 is placed into the system between the fuel tank and the fuel lift pump (mounted on the injection pump) and is designed to remove most of the dirt and sediment from the fuel as well as act as a water trap. A *water-in-fuel* warning light (WFWL) is activated anytime

Figure 19–1. Primary fuel filter.

that enough water collects in the primary filter assembly. A float arrangement is used within a water level detection switch so that when a given volume of water collects in the base of the filter, this float completes an electrical circuit, which in turn energizes a *water-in-fuel* warning light on the vehicle instrument panel. The specific gravity of water is heavier than that of diesel fuel; therefore, any water in the system settles at the base of the primary filter assembly. Since water is not as good an insulator as diesel fuel, when sufficient water collects in the base of the filter the WIF (water-in-fuel) warning light will be activated. Figure 19–2 illustrates the wiring circuit used with the WFWL.

Primary Filter Replacement. Filter replacement depends on the severity of service in which the vehicle is used. However, an average filter change period can be anywhere from 10,000–30,000 miles (16,000–48,000 km) or every 6 months to 2 years. Nissan/Datsun, for example, recommends a filter change every 24 months or 30,000 miles (48,000 km), whichever occurs first.

Plugged or restricted fuel filters will cause a lack of power because the engine will starve for fuel.

A suitable filter change period should be established depending upon the type of average service the vehicle is subjected to.

When replacing the primary filter/water separator (Figure 19–1), you should also inspect the parts for physical damage. Always replace the filter seal with a new one. You should also check the operation of the water-in-fuel unit for continuity through the use of an ohmmeter across the switch connector as shown in Figure 19–3.

Secondary Filter Replacement. The secondary filter assembly can be either of the spin-on throwaway-type or the replaceable element type. Figure 19–4 illustrates the components of the spin-on-type filter, which can be removed using a Toyota special fuel filter wrench P/N SST 09228–34010 or a suitable strap wrench.

With the replaceable element-type fuel filter, a new retaining bolt gasket and large O-seal ring between the filter canister and body should be used each time the filter is serviced. With the replaceable element (non spin-on) type unit, remove the filter simply by

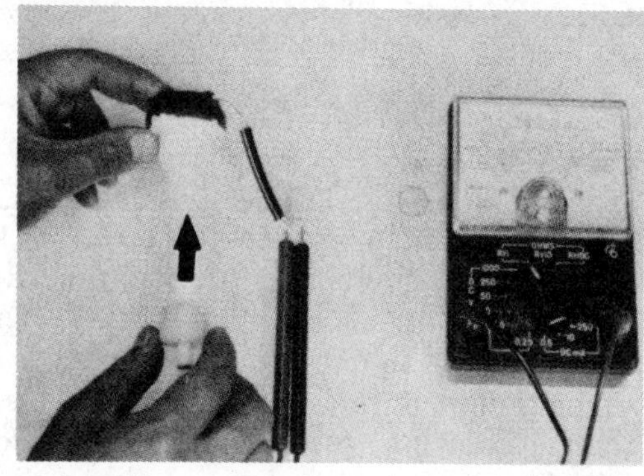

Figure 19–3. Water-in-fuel warning switch check.

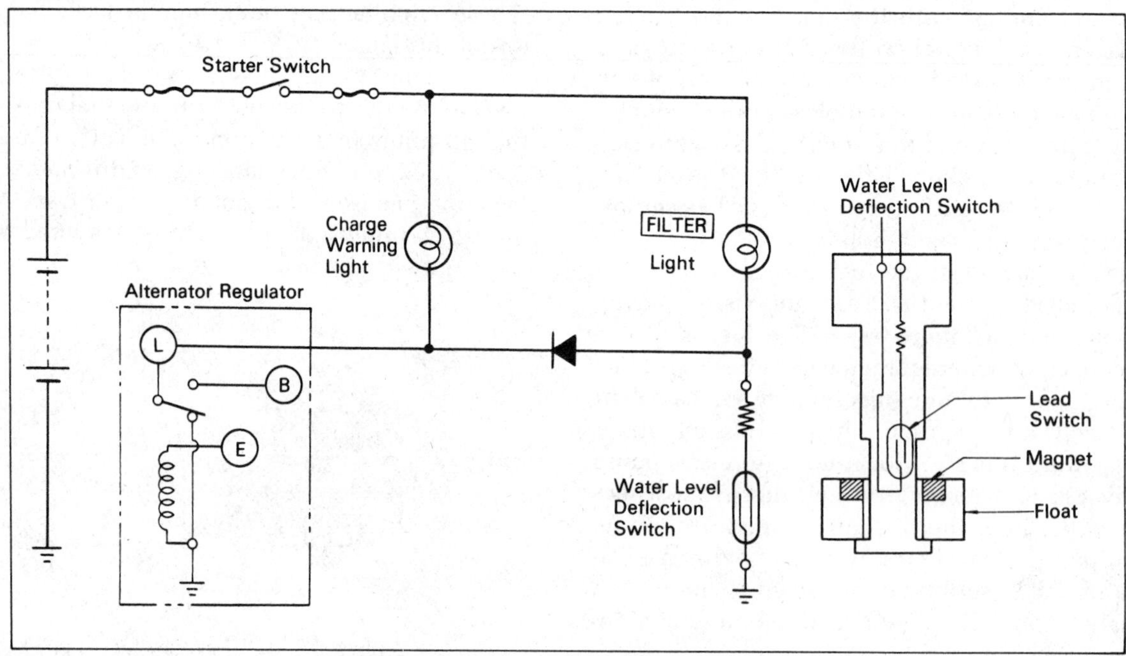

Figure 19–2. Water-in-fuel wiring circuit.

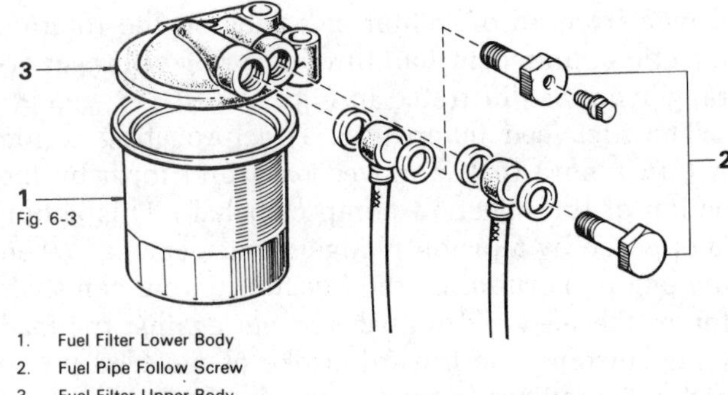

Figure 19–4. Spin-on fuel filter assembly.

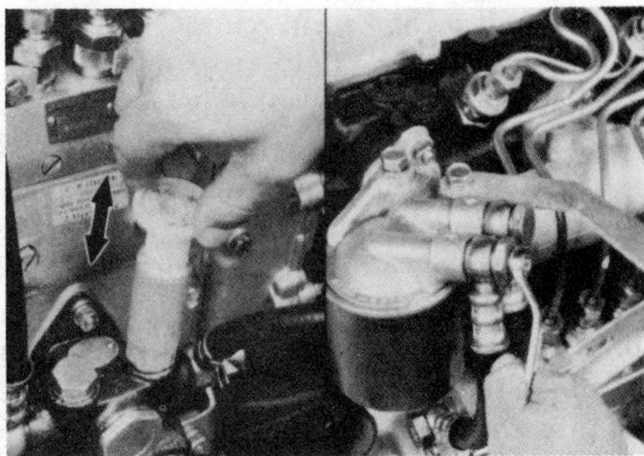

Figure 19–5. Loosening filter bleeder screw (right-hand side). Loosening feed pump priming handle (left-hand side).

loosening off the center bolt and withdrawing the shell along with the fuel filter element. Wash the components in clean solvent or diesel fuel and blow dry all parts or use a lint-free rag.

SPECIAL NOTE: Fuel filters should not be primed by filling the bowl with diesel fuel unless the fuel has already been filtered from a shop supply. Using unfiltered fuel to fill the filter or bowl can cause either unfiltered fuel or water to pass directly to the fuel injection pump. Any fuel that is contained within the inside bore area of the filter does not have to pass through the actual filter element. In addition, if the priming fuel contains any water, this water can corrode internal injection pump components and in extreme cases, damage injector nozzle tips.

The filter should be assembled dry if "filtered priming fuel" is unavailable. Apply a small amount of light oil to the O-ring seal or the captive seal on the top of the spin-on-type filter before they are tightened up.

Tighten the center bolt on the shell-type unit. With the spin-on-type filter assembly, tighten it by hand until it makes contact with the filter base and snug it up by hand until tight.

Bleeding Air From Fuel System

Anytime that the fuel filters are changed or the fuel injection pump lines have been removed, it will be necessary to bleed/vent all air from the fuel system. To successfully vent the air from the system, proceed as follows.

Procedure

1. Refer to Figure 19–5 and loosen off the small fuel filter bleeder plug (right-hand side of Figure 19–5).
2. Refer to the left-hand side of Figure 19–5 and turn the feed transfer pump knob to the left (counterclockwise) until you are able to pull this priming handle up.
3. Manually operate the feed pump priming handle down/up in order to draw fuel from the fuel tank into the feed pump. When you push the handle down, fuel is forced out to the filter assembly.
4. Continue pumping the feed pump handle until fuel that is free of air flows readily from the filter bleed plug, then tighten it.
5. In some cases it may be necessary to bleed the injection pump and injector nozzles if they have been disturbed or removed; in order to do this, proceed to Step 6.
6. Refer to Figure 19–23 and loosen off the injection pump bleeder screw, which is shown to the immediate right of the injection pump identification plate and just where the pump body and governor bolt together.
7. Continue to operate the feed pump handle until fuel flows freely from the injection pump bleeder screw, then tighten it.
8. If it is necessary to bleed fuel up to the injectors, this can be done in one of two ways:
   a. Refer to Figure 19–6 and loosen off all of the injector pipe nuts at the injectors between

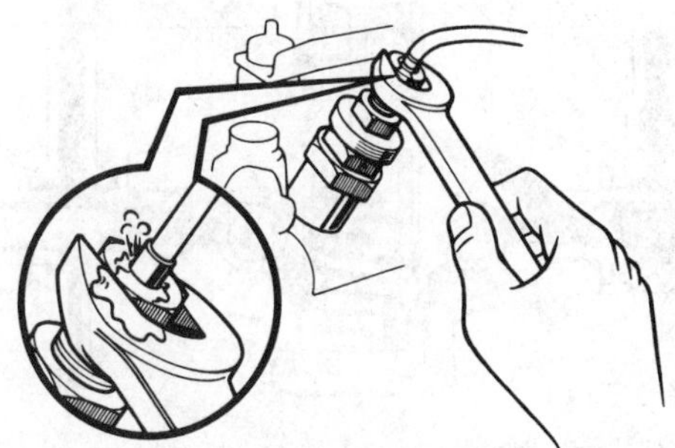

Figure 19–6. Bleeding fuel to the injectors.

one-half to one turn; operate the feed pump handle until fuel flows from the injector nuts, then tighten them. It may be necessary to manually rotate the engine over in order to allow fuel to flow from all injector pipes; be sure to push down the feed pump priming handle when finished and turn it to the right (clockwise) until it is tight.

b. After loosening the injector nuts, tighten down the feed pump priming handle, then depress the accelerator pedal and rotate the engine by engaging the starter motor until fuel flows freely from the injector lines, then tighten the nuts.

NOTE: Do not over- or undertighten the injector pipe nuts; they should be torqued to 15–21 lb.ft. (2.0–3.0 kg-m).

9. Start and run the engine and check all areas for signs of fuel leakage.
10. In some cases, it might be necessary to rebleed the system if the engine runs rough. This can be done by running the engine at an idle rpm, loosening off one injector nut at a time, and allowing any air to escape from the system. Be sure to place a rag around the injector nut while you do this.

Fuel Feed/Transfer Pump

The fuel feed pump assembly is bolted onto the side of the injection pump housing. This feed pump is driven from an off-center cam lobe on the rotating injection pump camshaft through a roller plunger arrangement as illustrated in Figure 19–7.

The fuel feed pump uses a reciprocating piston within a bore that is moved back and forth by the action of the injection pump camshaft. This action is opposed by a piston plunger return spring. When the engine is running, the injection pump camshaft forces the feed pump push rod out against the feed pump piston. The inward stroke of the feed pump piston is achieved through the force of the return spring, which causes the inlet check valve to be pulled open allowing fuel to be drawn into the cavity behind the piston while the fuel in front of the piston is placed under pressure and delivered to the fuel filter. Fuel feed pump delivery pressure averages between 26–31 psi (179–214 kPa or 1.8–2.2 kg-m) at 1200 engine rpm (600 pump speed) with a minimum delivery rate of 900 cc (0.237 US gallon or 0.197 Imperial gallon) per minute at 1000 rpm pump speed or 2000 rpm engine speed.

This type of feed pump is common to all Robert Bosch/Diesel Kiki and Nippondenso model in-line injection pumps.

The action of the priming pump handle/piston of the feed pump is shown also in Figure 19–7. When the handle is pulled up, the piston sucks open the inlet check valve, while allowing the outlet check valve to be closed by the spring above it. When the handle is pushed down, the displaced fuel closes the inlet valve and directs the fuel through an internal passage towards the outlet check valve. The fuel opens the outlet check valve against the force of its

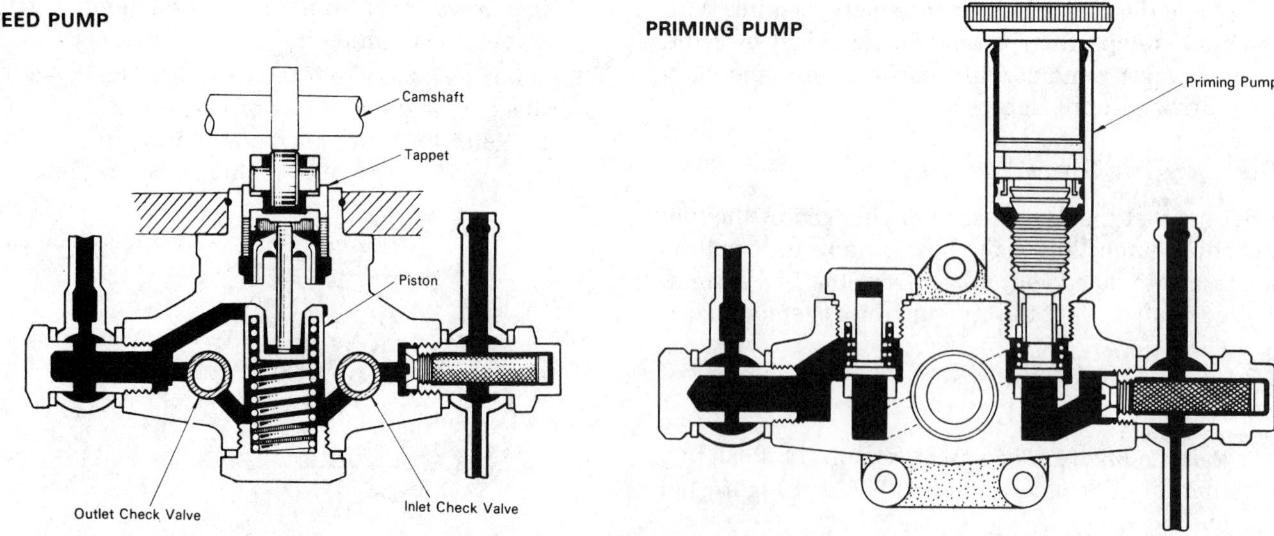

Figure 19–7. Fuel feed/transfer pump.

light spring, and the fuel flows out of the feed pump and over to the fuel filter assembly.

Feed Pump Service

Service to the fuel feed/transfer pump requires that occasionally the internal filter should be removed and washed in clean solvent or diesel fuel. Figure 19–8 shows the component parts that make up this feed pump assembly.

To gain access to the feed pump filter, refer to Figure 19–8 and loosen off the fuel inlet retaining nut. Unscrew it and remove the filter. After cleaning the filter, blow it dry with compressed air and reinstall it.

Procedure

1. Remove the fuel inlet and outlet lines from the feed pump body.
2. Loosen and remove the feed pump to injection pump housing retaining nuts and withdraw the pump assembly.
3. Refer to Figure 19–8 and inspect the feed pump components.

   NOTE: Do not remove the pump push rod unless it appears to be damaged since it is a precision fit. However, if it has to come out, take careful note of the way that it came out.

4. The allowable tolerances (wear limits) for the piston and push rod clearances in the feed pump body are:
   a. Piston—0.009–0.013 mm (0.0004–0.0005").
   b. Push rod—0.003–0.006 mm (0.0001–0.0002").
5. Refer to Figure 19–9 and check the condition of the check valves and their seat.
6. Closely inspect the condition and compression characteristics of the fuel feed pump piston and check valve springs with new ones.
7. To check the priming pump condition, refer to Figure 19–8 and by placing your finger over the bottom of the pump, pull the handle up in order to check its suction capability. Push the handle down to check that it is in fact capable of building up pressure.

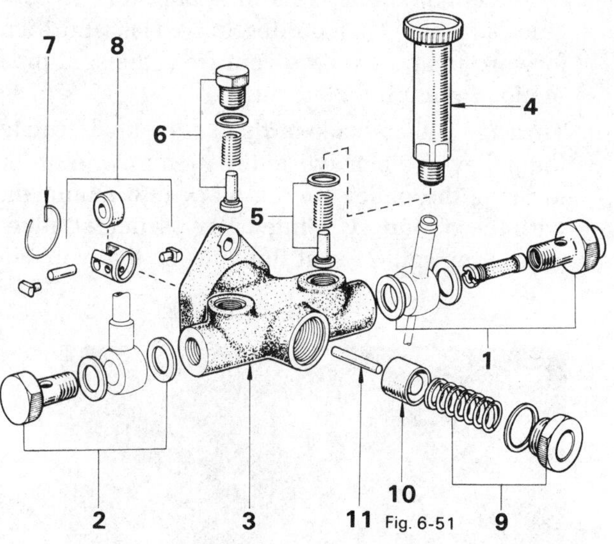

1. Inlet Hose
2. Outlet Hose
3. Feed Pump Assembly
4. Priming Pump
5. Inlet Check Valve & Spring
6. Outlet Check Valve & Spring
7. Snap Ring
8. Tappet
9. Piston Chamber Plug & Spring
10. Piston
11. Push Rod

Figure 19–8. Fuel feed/transfer pump components.

Lubrication—Injection Pump

The injection pump camshaft and bearings are pressure lubricated by engine oil through a delivery line (Item 6 in Figure 19–16). Oil is returned from the injection pump housing at its drive end by overflowing into the engine front timing cover.

This pump employs a pneumatic governor assembly that requires periodic lubrication by the addition of three or four drops of oil to the governor diaphragm. To do this, refer to Figure 19–10 and remove the small plug on top of the governor housing.

NOTE: Make sure that there is not an accumulation of engine oil or diesel fuel in the diaphragm chamber. If there is, it can be drained by removing the drain bolt on the housing. Check the reason for this oil or fuel accumulation.

Figure 19–9. Fuel feed pump check valves and seat.

Figure 19–10. Adding oil to the pneumatic governor diaphragm housing.

Injection Timing—Check/Adjust

It is often necessary after many miles on an engine to check and adjust the injection pump-to-engine timing. This is also required any time that the pump has been removed for service.

Procedure

1. Place a socket over the crankshaft pulley retaining nut as shown in Figure 19–11 and rotate the engine with a ratchet or bar until the pulley timing mark is positioned at 14 degrees BTDC or to whatever position your year of engine calls for (simply align the mark on the pulley that is before the TDC mark as you rotate the engine clockwise from the front). At this position, the pulley mark should be aligned with the engine front timing cover stationary pointer.

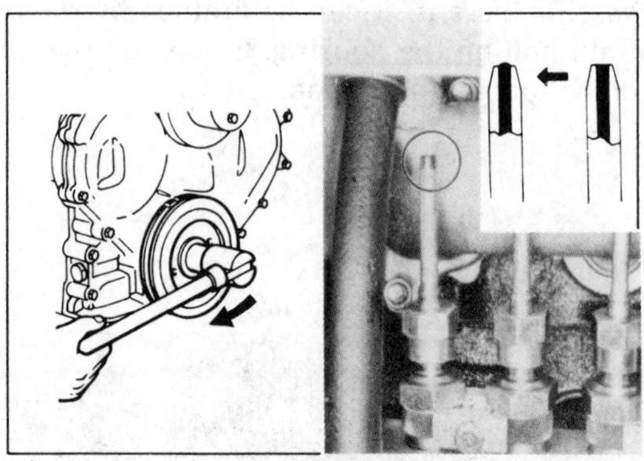

Figure 19–11. Placing engine to 14 degrees BTDC on the compression stroke (left-hand side of split figure). Fuel spill pipe fuel level check to establish actual timing (right-hand side of split figure).

2. At 14 degrees BTDC, both the intake and exhaust valve for No. 1 cylinder should be fully closed. If the intake and exhaust valves are partly open, the engine will have to be rotated another 360 degrees to place the No. 1 piston on its compression stroke.
3. Refer to Figure 19–12 and make sure that the scribe marks on both the injection pump housing and engine front cover are aligned.
4. With Steps No. 1, 2, and 3 completed as per instructions, the injection pump would appear to be correctly timed; however, due to gear backlash and general engine coupling/injection pump drive wear, the pump may not in fact be correctly timed. To ensure that it is in fact correct, proceed to Step 5.
5. Remove the fuel injection line from the No. 1 cylinder to injection pump.
6. Obtain or make up a small "spill pipe" similar to that shown in Figure 19–11 that can be installed over the No. 1 pumping element of the injection pump.
7. Refer to Figure 19–23 and remove item 22; reinstall the delivery valve holder (22) without its spring.
8. If necessary, bleed the fuel filter and injection pump as explained under "Bleed Fuel System."
9. Refer to Figure 19–11, and with a socket and bar on the engine pulley retaining bolt, turn the engine back slightly (counterclockwise), and then forward to allow fuel to vent from the spill pipe of No. 1 pumping element.
10. Turn the pulley backwards again; slowly rotate the pulley in a clockwise direction now in order to bring the pulley timing mark into alignment with the stationary pointer. Pay particular attention to the spill pipe at this time so that you can

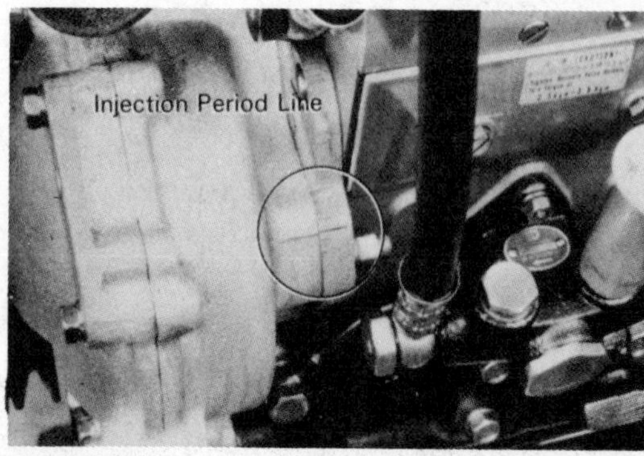

Figure 19–12. Checking injection pump to engine front cover alignment marks.

accurately establish exactly when the fuel level starts to rise in the spill pipe. It may be necessary to gently wipe fuel from the top of the spill pipe with your finger, back the engine up slightly, then bring it ahead to make sure that you have in fact spotted just when the fuel starts to rise in the spill pipe.

11. The spot at which the fuel rises in the spill pipe is an accurate injection pump-to-engine timing position and is commonly known as "spill timing."
12. With the engine in the position described in Step 9, check the following:
    a. That the pulley timing mark is in alignment with the stationary engine front cover pointer.
    b. That the injection pump to engine front cover scribe marks are aligned.

    NOTE: If the two conditions in Step 11 are met, then the injection pump is accurately timed; if, however, the pulley mark is before or after the stationary pointer, the injection pump mounting bolts would have to be loosened off in order to reposition the pump-to-engine cover scribe/timing marks.

13. In order to position the injection pump correctly, refer to Figure 19–13 and, with the injection pump mounting bolts loose, push the upper injection pump housing towards the engine to "advance" the timing, and pull the housing away from the engine in order to "retard" the timing.

    CAUTION: It may be necessary to loosen off all of the injection pump to nozzle lines in order to change the pump timing, otherwise these fuel lines may be bent or placed under considerable strain.

The pulley timing mark should be in alignment with the stationary pointer, and the fuel should just be rising in the spill pipe to ensure that the injection pump-to-engine timing is correct.

After any timing adjustment, repeat the check at least once more in order to positively assure that the timing is correct after the pump-to-engine timing bolts have been tightened.

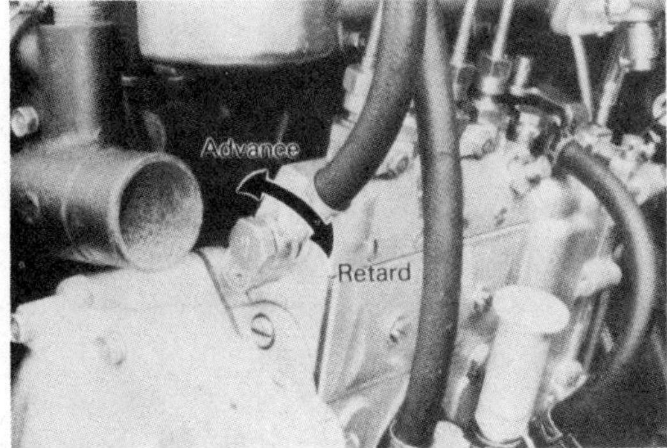

Figure 19–13. Adjusting injection pump timing "advance and retard."

Idle and Maximum Engine Speed Adjustments

After any timing adjustment to the injection pump, the idle and maximum speeds should be checked and adjusted if necessary. Both of these speeds *must* be checked with the engine at its normal operating temperature to ensure an accurate setting.

Procedure

1. Using an accurate digital tachometer, monitor the idle speed of the engine; the idle speed should be 650 rpm.
2. If the idle speed requires adjustment, refer to Figure 19–14 and loosen off the lock nut on the stop plate and turn the adjusting screw clockwise to increase the idle speed and counterclockwise to decrease the idle speed while watching the tachometer; or alternately, make your adjustment, then recheck the idle speed with the tachometer after tightening the lock nut.
3. To check the maximum no-load (high-idle) engine speed, check that the butterfly valve within the intake manifold is wide open when the throttle is pushed to the floor.
4. Aim an accurate digital photoelectric tach on the reflective tape that has been placed on the engine crankshaft pulley and carefully note the maximum engine speed, which should be between 4050–4100 rpm.

Figure 19–14. Idle speed adjustment screw.

5. If the butterfly valve within the intake manifold doesn't open fully, loosen off the throttle pedal adjustment bolt, which is located underneath the pedal itself inside the cab of the vehicle. Adjust it until the butterfly valve within the intake manifold is horizontal when the throttle pedal is pushed to the maximum position.
6. Once the throttle has been adjusted to the wide open position, recheck the maximum rpm; if the maximum speed is still not within 4050–4100 rpm, refer to Figure 19–15 and adjust the screw on the governor end of the injection pump.
7. To increase the engine maximum speed, turn the adjusting screw clockwise; and to decrease the speed, turn the screw counterclockwise then tighten the lock nut. This adjustment screw should be sealed with lockwire after any adjustment to prevent any movement of the screw and to deter anyone from making unwarranted adjustments.

Injection Pump Removal and Installation

Injection Pump Removal

If it becomes necessary to remove the injection pump from the engine for service/repair, proceed as follows.

Procedure

1. Disconnect the battery ground cable.
2. Manually rotate the engine with a socket/bar on the crankshaft front pulley bolt until the TDC mark on the pulley aligns with the stationary pointer on the engine front cover. Make sure that the No. 1 piston is at TDC on its compression stroke. If in doubt, remove the rocker cover and check that both the No. 1 inlet and exhaust valves are closed.

Figure 19–15. Maximum speed adjustment screw.

3. Note the alignment marks between the injection pump housing and the engine front cover (are they in alignment, or have they been changed?).
4. Refer to Figure 19–16 and remove the injection pump components in the numbered sequence 1 through 7.

SPECIAL NOTE: Always have enough plastic shipping/dust caps that can be placed over all open fuel lines and fittings to keep dirt out of the injection system.

Injection Pump Installation

To install the injection pump to the engine, proceed as follows.

Procedure

1. Rotate the engine manually with the aid of a socket/bar placed over the crankshaft pulley retaining bolt until No. 1 piston is at TDC on its compression stroke (both valves closed).
2. Refer to Figure 19–17 and align the match mark on the injection pump drive shaft with the mark on the backside of the engine timing cover gear hub.
3. Slide/engage the injection pump and drive shaft into the hub of the drive gear and over the captive mounting studs.

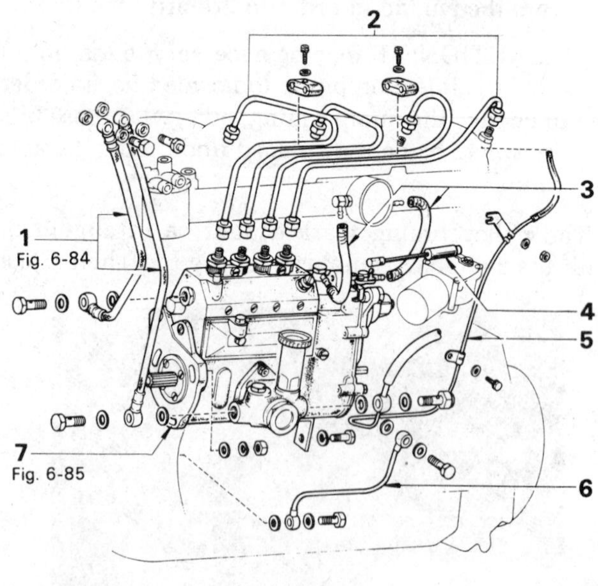

1. Fuel Hose
2. Injection Pipe
3. Vinyl Hose
4. Control Rod
5. Fuel Pipe
6. Oil Pipe
7. Injection Pump

Figure 19–16. Injection pump removal sequence.

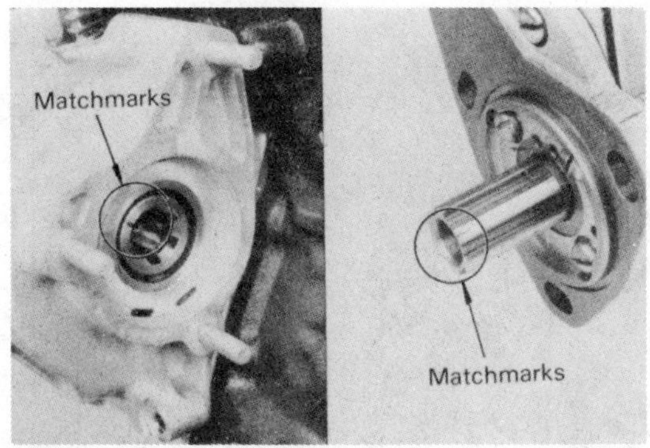

Figure 19–17. Timing mark alignment—injection pump drive shaft to gear drive hub.

4. Run the injection pump housing retaining nuts onto the studs until they are finger tight.
5. Refer to Figure 19–16 and install the rest of the components as per the numbered sequence 2 through 7.

SPECIAL NOTE: Leave the No. 1 injector delivery line off for now; grasp the injection pump body and move it until the scribe mark on the housing is aligned with the stationary scribe mark on the engine timing cover as shown in Figure 19–12. Check engine timing. Bleed the air from the fuel system and proceed to follow the detailed sequence explained earlier for "Checking Injection Pump Timing" by the "fuel spill" method.

Automatic Timing Advancement

All diesel engines employ some form of automatic fuel injection advancement mechanism so that fuel can be injected earlier in the compression stroke to obtain maximum power and fuel economy as well as complying with EPA exhaust emissions levels.

The system used by most manufacturers using inline type injection pumps is one whereby a set of weights acts between the timing gear and the hub that is splined to the injection pump driveshaft. The weights are so arranged that when the engine is running, centrifugal force causes the weights to move outwards. The weight action is then transferred through a set of pins to the injection pump drive shaft hub to rotate it forward (advance it) with an increase in engine rpm.

The weights are opposed by springs to ensure that tension is always maintained within the automatic timing unit to avoid any slack, which would affect the response of the timing unit.

An example of the timing device used with the Toyota diesel engine is shown in Figure 19–18 in exploded view form. In addition, Figure 19–19 illustrates the actual location of the assembled timing advance unit in relation to the engine gear train.

NOTE: The operation of the automatic timing advance unit cannot be checked unless it is installed with the injection pump onto an injection pump test

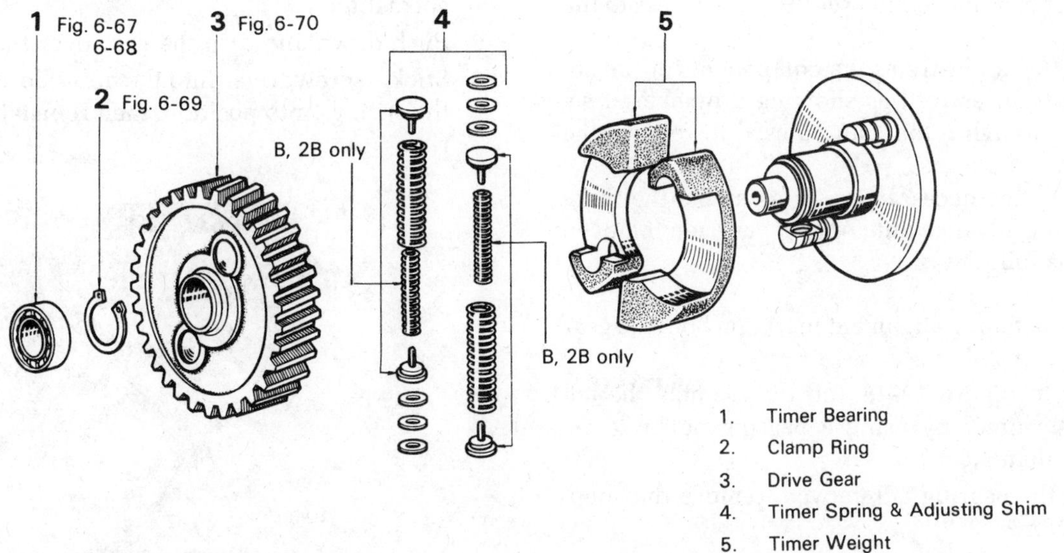

Figure 19–18. Exploded view, automatic timing advance unit.

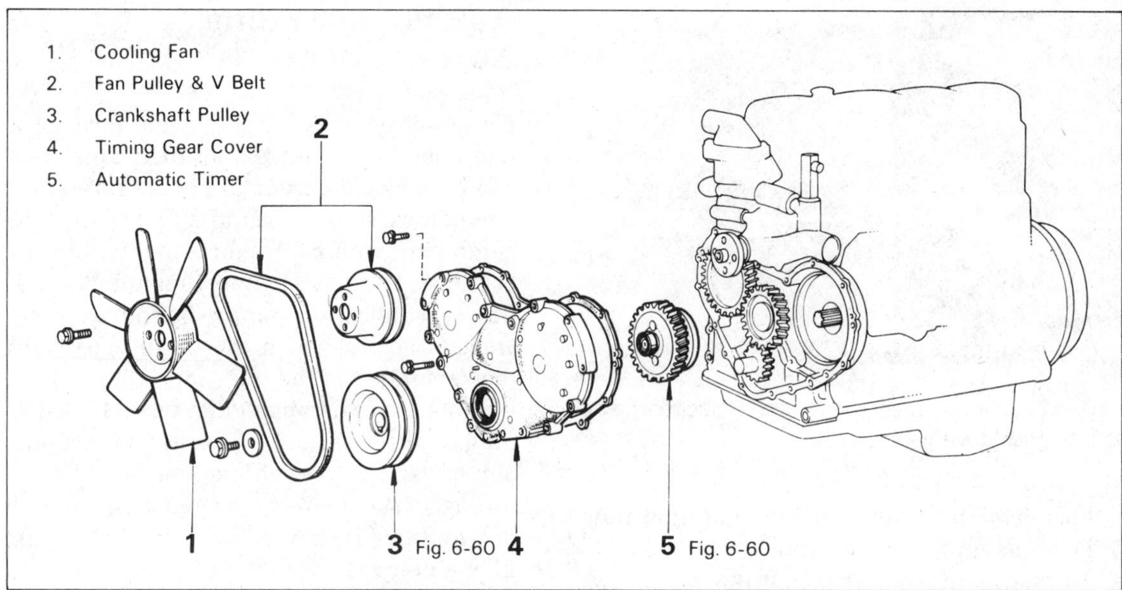

Figure 19–19. Engine gear train/automatic timing advance location.

stand. Timing can be changed by adding or removing shims from the inner and outer springs. If the timing unit is too far advanced, shims must be added to strengthen the spring force opposing the weights. If the timing unit is retarded, shims should be removed in order to decrease the force of the springs opposing the weight action.

Removal—Automatic Timing Advance Unit

If it becomes necessary to check the engine gear train to automatic advance unit timing, the engine front cover must be removed in order to gain access to the gear train.

Figure 19–19 illustrates the components of the engine gear train as well as showing a numbered sequence 1 through 5 that can be used to remove the components.

If it becomes necessary to disassemble the automatic timing advance unit, once it has been removed proceed as follows:

1. Place matching alignment marks on both the gear and inner hub.
2. Refer to Figure 19–18 and disassemble the automatic timer by using a bearing puller to remove item No. 1.
3. After the bearing is removed, remove the snapring No. 2.
4. Take care now, when you pull up on the drive gear to separate it from the hub, that the springs and shims No. 4 inside do not fly out.
5. Clean all the components in solvent or diesel fuel, then inspect them for wear or damage.
6. Reassembly is basically the reverse of disassembly; however, you will require special tool P/N SST 09280-76010 as shown in Figure 19–20 to hold the springs and shims in position as you assemble the unit.
7. With the special tool in position on the gear, align the match marks that you previously placed on the gear and hub.
8. While holding the top of the gear, remove the special tool.
9. Push down and turn the gear over the hub.
10. Stick a screwdriver into the hub side slots to pop the springs into position, then reinstall the snapring.

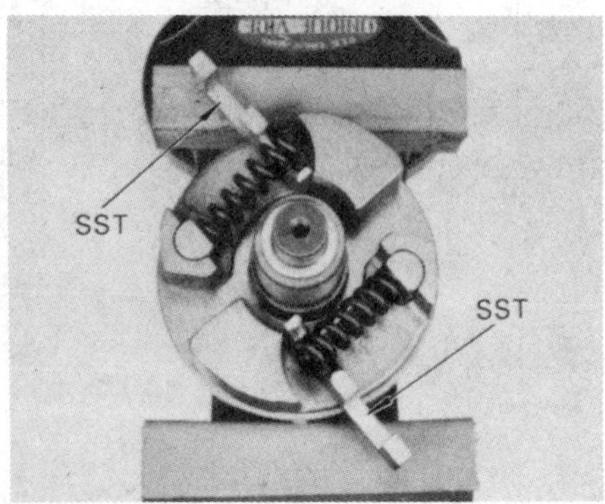

Figure 19–20. Special tool SST 09280-76010 for automatic timing device assembly.

11. Drive the bearing into position in the hub.
12. When reinstalling the automatic timing device onto the engine, lightly coat the splines and bearing with MP (multi-purpose) grease; make sure that you align the matchmarks on the timing gear hub with the one on the injection pump splined shaft as shown in Figure 19–21.
13. Recheck the alignment of the engine gear train timing marks; they should appear as shown in Figure 19–22.

SPECIAL NOTE: Preheat the engine timing cover to 140°F (60°C) in oil or water before installation!

Injection Pump Overhaul

Major overhaul of fuel injection pumps and injectors is not a practice that is carried out in the majority of automotive service departments due to the high cost of purchasing test equipment and special tools as well as having to train individuals to become specialists in this field.

Overhaul of an injection pump requires not only access to special tools and test stand equipment but also a commitment to excellence of repair techniques and extreme cleanliness when working with all injection system components.

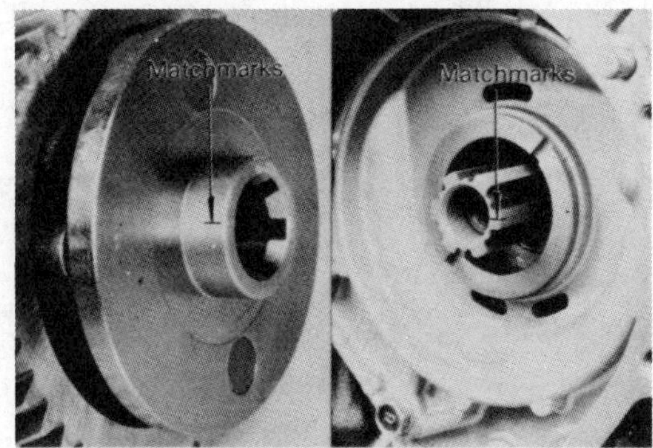

Figure 19–21. Automatic timing gear hub to shaft alignment marks.

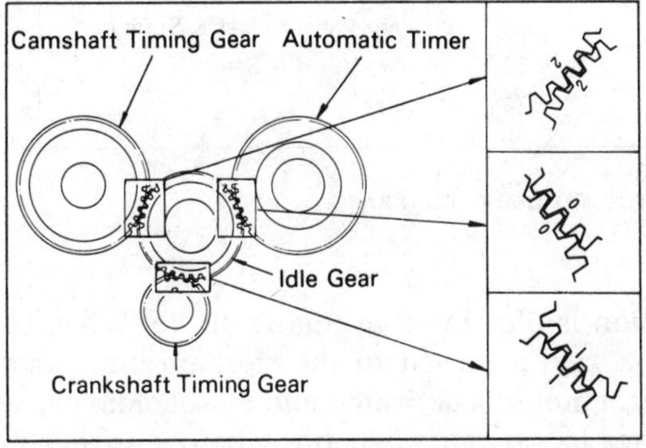

Figure 19–22. Engine gear train alignment marks.

Repair of injection equipment today requires considerable expense in both training of personnel and the purchase of necessary service repair literature, tools, and equipment. For this reason, most injection pump, nozzle, and governor repairs are left to specialists in this field.

Many major companies involved in service/repair work are members of the Association of Diesel Specialists, which is a worldwide affiliation of individuals and companies—service repair specialists who are dedicated to the highest caliber of service/repair of diesel injection equipment and assorted components.

Technician certification exams are conducted on a regular basis at various locations throughout the country.

Regular local meetings as well as an annual general meeting are held not only to give association members and manufacturers an opportunity to meet and exchange the latest technical information, but also to become familiar with new equipment and techniques in the industry. Membership application can be made to: Association of Diesel Specialists, 9140 Ward Parkway, Kansas City, Missouri, 64114 U.S.A.

The automotive or diesel mechanic will however be called upon to remove and install the injection pump on the engine as well as being capable of performing the necessary tests and adjustments required to time it to ensure maximum performance from the engine.

In addition, the ability to maintain and troubleshoot the fuel injection system is a must to prevent the unwarranted removal of an injection pump or nozzles.

The internal components and their location on the Toyota (Nippondenso) in-line-type fuel injection pump is shown in Figure 19–23.

Fuel Control Motor

An electronically activated fuel control motor is connected through linkage to the fuel injection pump to facilitate ease of engine starting, normal running, and engine shut-off. This motor is mounted on a bracket alongside the injection pump as shown in Figure 19–24.

SPECIAL NOTE: The fuel control motor used with this system is similar in its design and construc-

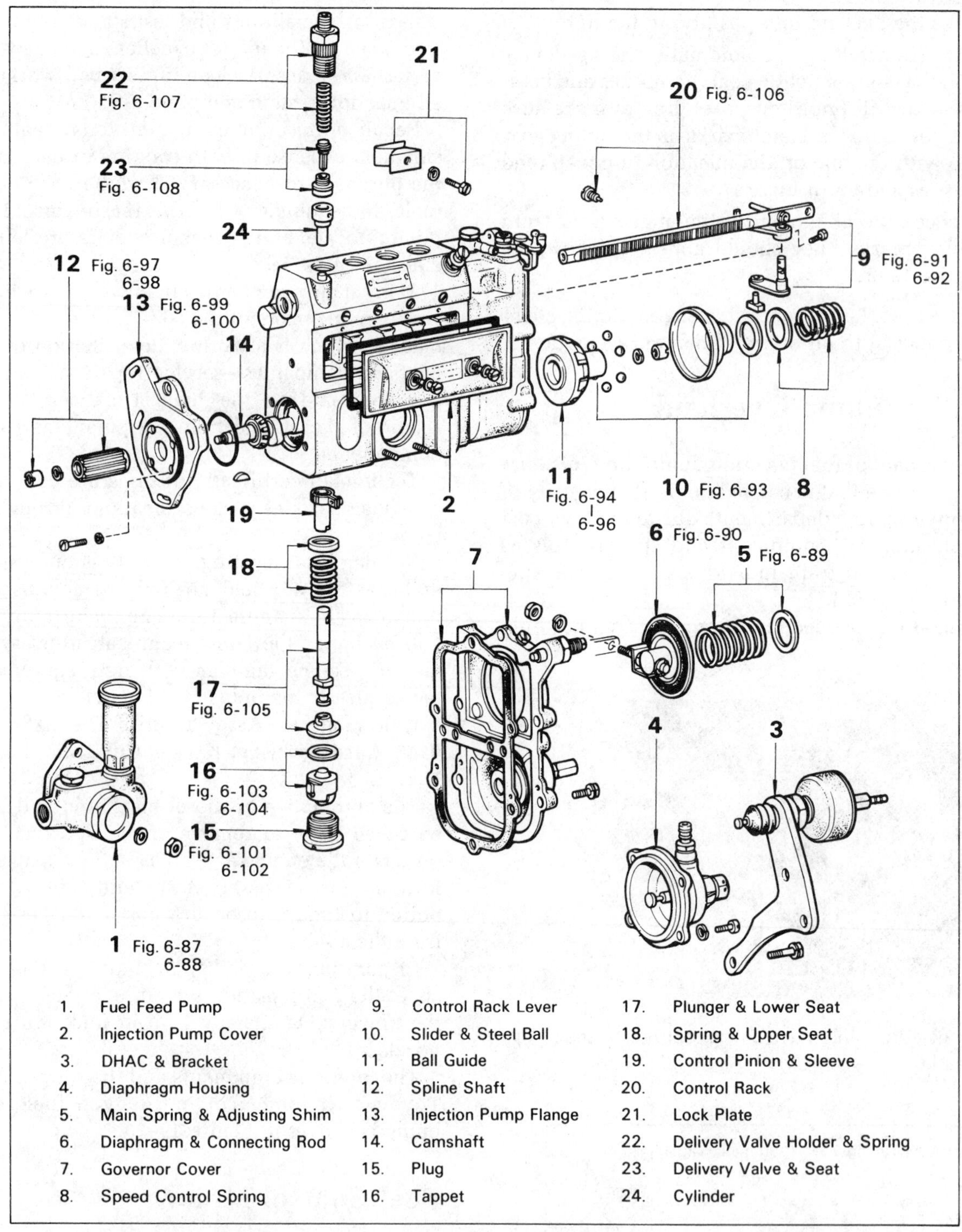

Figure 19–23. In-line type fuel injection pump components.

tion to the one used by Nissan/Datsun on their SD25 engine. For a detailed description of this motor, refer to Chapter 18, dealing with the Datsun engine under the heading "Injection Pump Control Mechanism."

The electrical system for the fuel control motor activation is illustrated in Figure 19–25. When the ignition key is turned to the *start* position, the fuel control motor is activated and its solenoid causes its linkage to move the injection pump governor mechanism to place the fuel rack into an excess fuel po-

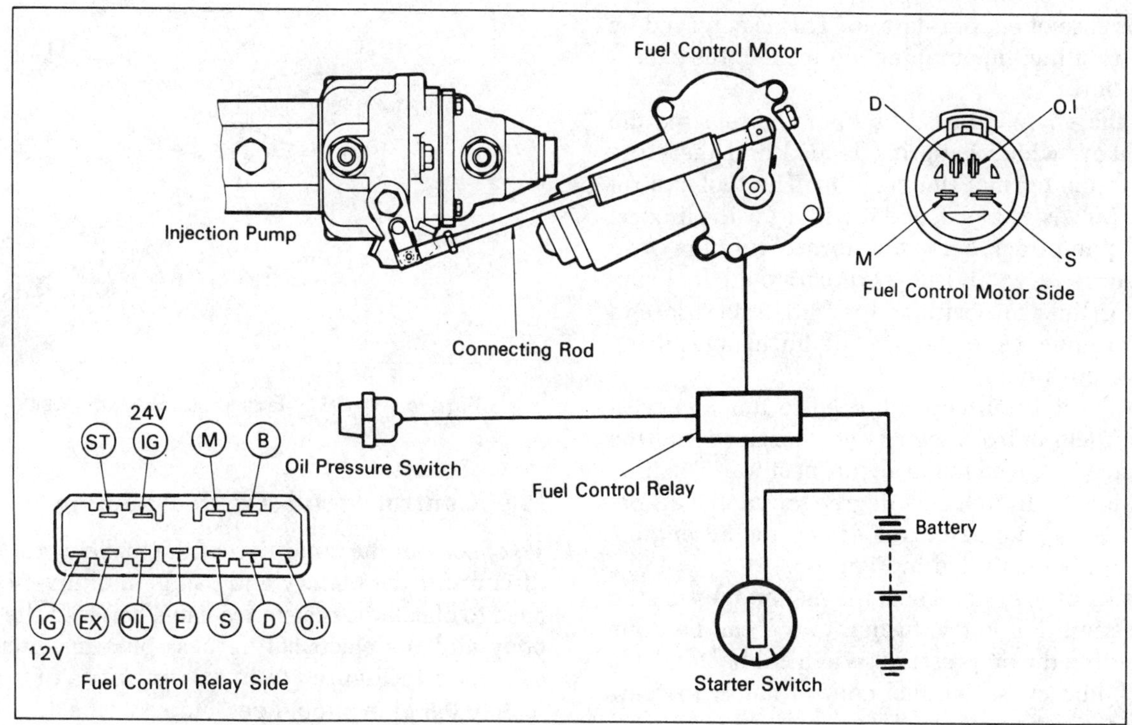

Figure 19-24. Electronic fuel control motor arrangement.

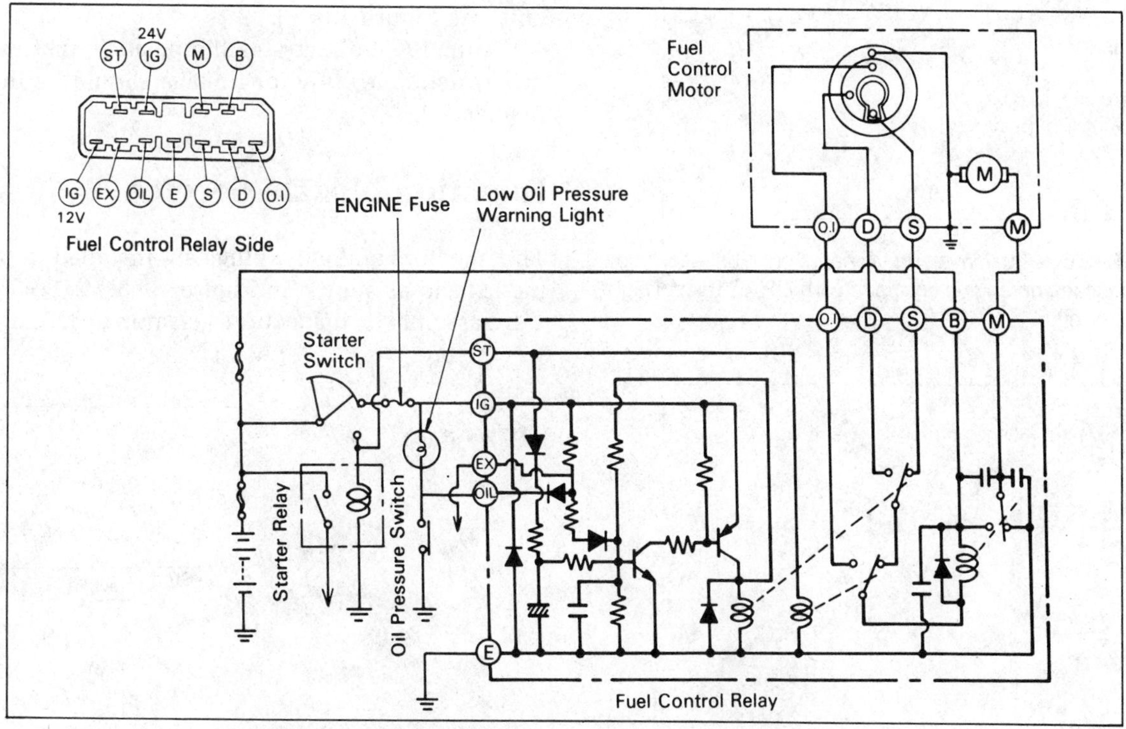

Figure 19-25. Toyota fuel control motor electrical circuit.

sition for ease of engine start-up. This is achieved by the rack rotating the pumping plungers to the excess fuel position.

Once the engine starts, the operator releases the ignition key, which is spring loaded and therefore returns to the normal *run* position. The fuel control motor supply is reduced, and it will draw the linkage and fuel pump rack back to a normal running position. Figures 19–26 and 19–27 illustrate the fuel control motor linkage positions for initial start-up and normal running. Note the travel of the linkage during these two positions.

When the ignition key is turned to the *stop* position, the fuel control motor is de-energized and the fuel linkage is moved to the *no fuel* position. The linkage will move through a 49 degree arc on the 2B engine and through a 52 degree arc on the 3B engine. This action is illustrated in Figure 19–28.

A check of the system, or if the engine fails to stop when the ignition key is turned to *off*, can be done by grounding the oil pressure switch connector. This action should cause the fuel control motor to draw the linkage to the fuel shutoff position.

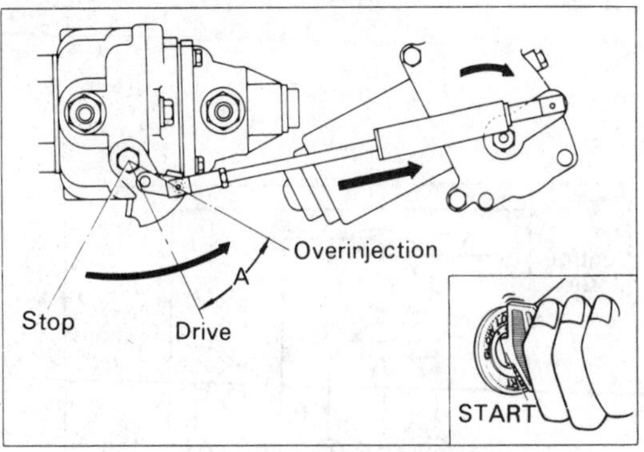

Figure 19–26. Fuel lever movement for initial start-up; A = 41 degrees on engine equipped with DHAC (diesel high altitude control).

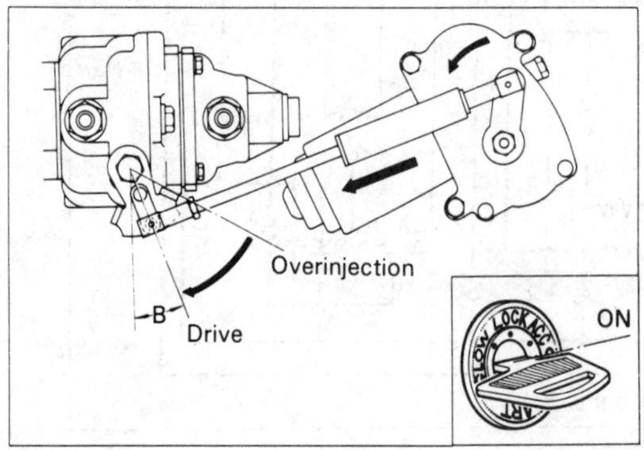

Figure 19–27. Fuel lever position after start-up; 2B engine = 19.5 degrees and the 3B engine = 22.5 degrees.

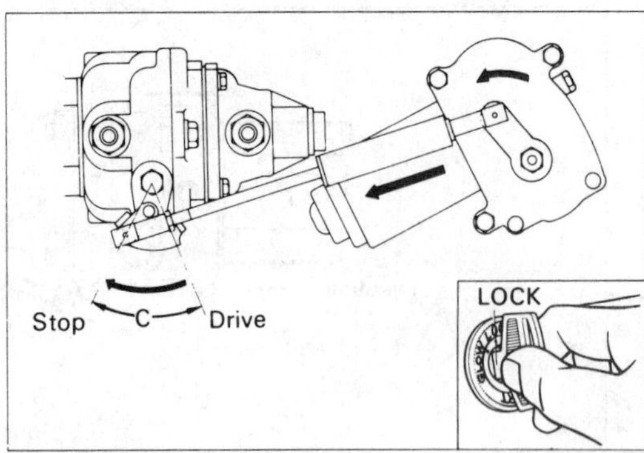

Figure 19–28. Engine stop/key off position.

Fuel Control Motor Troubleshooting

To check out the condition of the fuel control motor, disconnect the battery and using an ohmmeter, proceed to check the resistance value between the motor body and the electrical harness plug terminal "M" as shown in Figure 19–29. A resistance of approximately 0.8 ohm should be obtained on a 12-volt motor, or approximately 3.3 ohms on a 24-volt motor system. Also check for continuity between terminal D and 0.1, S and 0.1, and S and D of the wiring harness plug. Position D to 0.1 is the motor Stop Circuit, position S to 0.1 is the Normal Run Position Circuit, while position S to D is the Excess Fuel Position Circuit (see Figure 19–26).

Figure 19–30 illustrates the position that the governor/fuel control motor linkage should be in when installed.

Injection Nozzle Service

The injectors and glow plugs are installed in the cylinder head as shown in Figure 19–31.

Disassembly, inspection, cleaning, testing, and

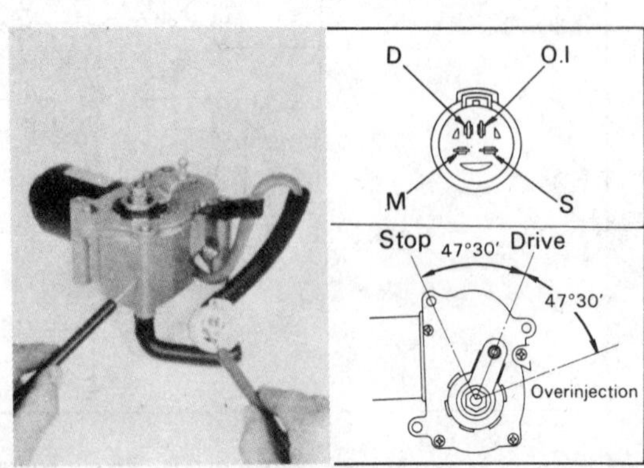

Figure 19–29. Fuel control motor electrical checks.

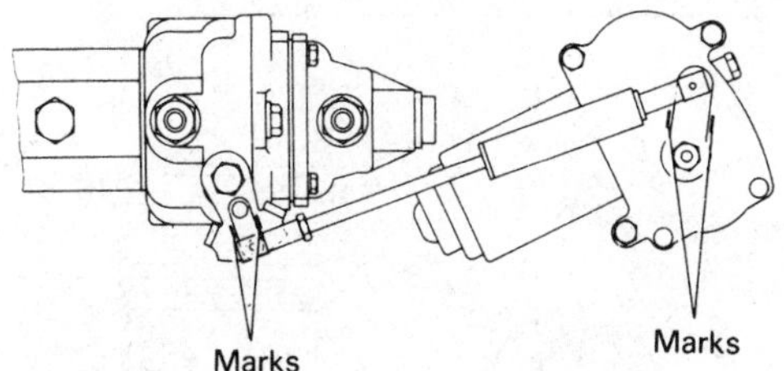

Figure 19-30. Position control linkage between marks shown on pump and motor bodies.

overhaul of injectors can be found in detail in Chapter 29, dealing with fuel injectors. The fuel injectors used with the Toyota engine are Bosch design pintle-type nozzles, which can be seen in Figure 19–31. The operation of these nozzles is also explained under the fuel injector chapter (Figure 29–4). Nozzle release pressures on the Toyota diesel engine are:

New Nozzles . . 1636–1778 psi (115–125 kg/cm²)
Used Nozzles . . 1493–1778 psi (105–125 kg/cm²)

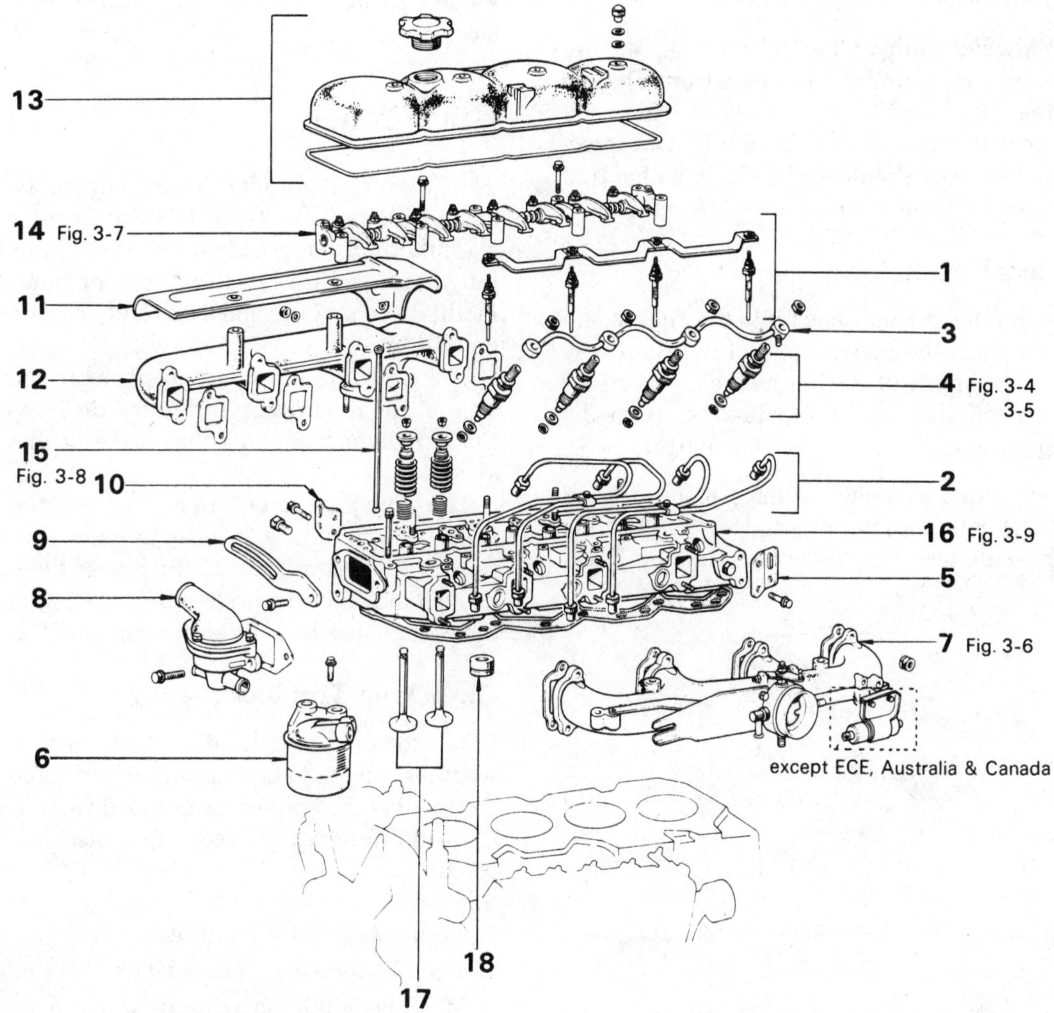

1. Glow Plug & Glow Plug Connector
2. Injection Pipe
3. Leakage Pipe
4. Injection Nozzle Holder & Gasket
5. Rear Engine Hanger
6. Fuel Filter
7. Intake Manifold with Overinjection Magnet
8. Water Outlet Housing
9. V Belt Adjusting Bar
10. Front Engine Hanger
11. Heat Insulator (for 3B)
12. Exhaust Manifold
13. Cylinder Head Cover & Gasket
14. Valve Rocker Shaft Assembly
15. Valve Push Rod
16. Cylinder Head
17. Valve & Compression Spring
18. Combustion Chamber

Figure 19-31. Injector and glow plug location.

Injector Removal

SPECIAL NOTE: Before disconnecting any fuel lines, make sure that you have an adequate supply of male/female plastic shipping caps to cover all exposed fuel lines and fittings in order to keep dust/dirt out of the system as well as plugging the injector cylinder head hole to prevent the entrance of any foreign material into the cylinder bore.

To remove the injector(s), refer to Figure 19–32 and disconnect the numbered components 1 through 5. Use special tool SST 09268–46011 or an equivalent deep socket to loosen the injector holder.

NOTE: Always number the injectors to ensure that you know what cylinder they came from. In case of an engine disassembly problem, you can accurately diagnose the reason for failure if it is an injector problem. Injectors should always be installed into the same cylinder from which they were removed.

Injector Installation

When installing the injector back into the engine, ensure that you place the injector gasket the right way up, as shown in Figure 19–33. Screw the injector into the cylinder head and torque it to 44–57 lb.ft. (6–8.0 kg-m) using special socket SST 09268–46011.

NOTE: The fuel system will have to be primed (bled of all air) before attempting to start the engine. Refer to the sub-heading "Bleeding air from Fuel System" in this chapter.

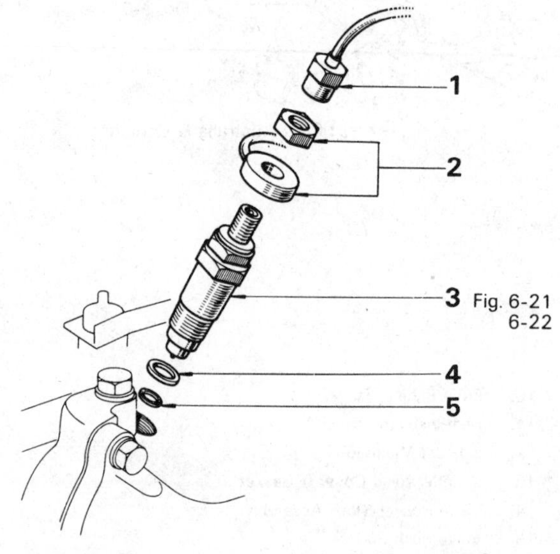

1. Injection Pipe
2. Leakage Pipe
3. Injection Nozzle Holder
4. Injection Nozzle Seat
5. Injection Nozzle Seat Gasket

Figure 19–32. Injector lines/parts arrangement.

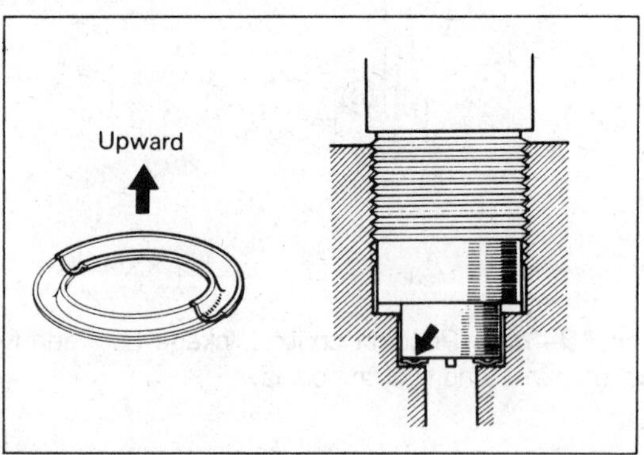

Figure 19–33. Correct installation of injector seat gasket.

Glow Plugs

The Toyota 4-cylinder diesel engine is a 4-stroke cycle unit that is of the IDI (indirect-injection) design; therefore, it requires the use of individual glow plugs screwed into each pre-combusion chamber to facilitate ease of starting and engine warm-up conditions.

The glow plugs are energized electrically as can be seen in the illustration in Figure 19–34 which shows the typical components that make up the glow plug circuit.

The glow plug circuit in Figure 19–34 is the system used on vehicles sold into Canada (BJ60 Designation), while Figure 19–35 illustrates the circuit used on vehicles sold in areas other than Canada which are designated by the term "Except BJ60."

Glow Plug Troubleshooting

When the engine is hard to start—especially in cold weather—the problem could be attributed to various causes. Before condemning the glow plugs, consider that the following areas can also affect engine starting.

1. Is the battery fully charged?
2. Are all electrical connections clean and tight?
3. What weight oil is being used in the engine (heavy oil will reduce the cranking speed of the engine and overload the starter)?
4. Does the starter appear to be cranking the engine at normal speed?
5. Is the fuel control motor moving the injection pump linkage to the excess fuel position for start-up?
6. Is the engine in a good state of mechanical tune (low compression)? White smoke?

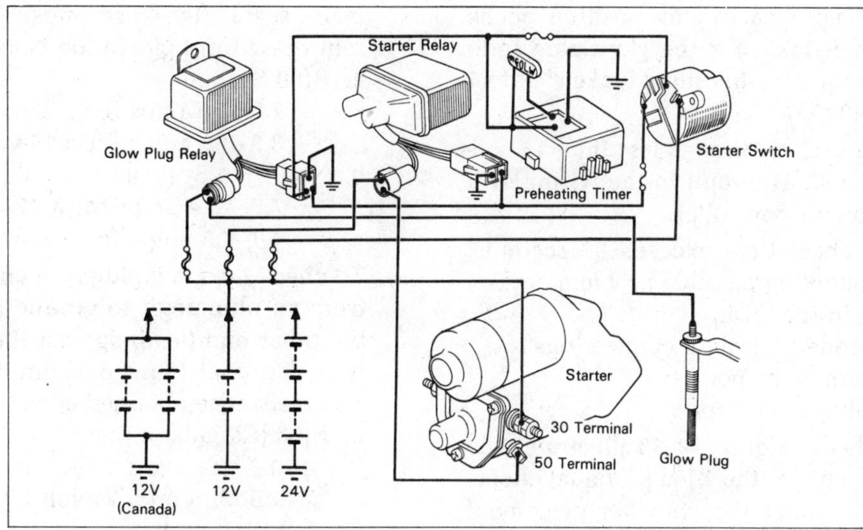

Figure 19–34. Typical glow plug circuit components BJ60.

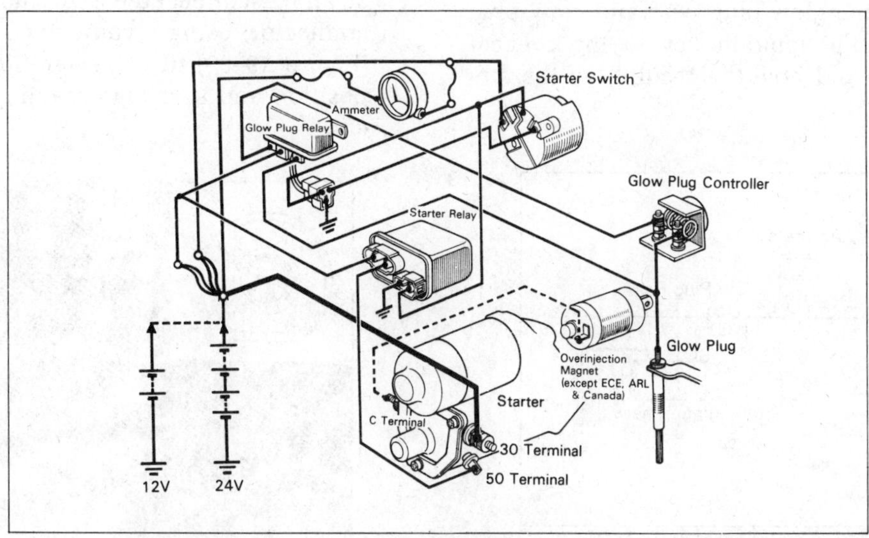

Figure 19–35. Typical non-BJ60 glow plug circuit components.

7. Are the injectors in good condition (color of exhaust smoke when the engine is running)? Black/gray smoke?
8. What type of fuel is being used? No. 1 or No. 2 grade?

If the glow plug circuit appears to be at fault, perform the following tests:

1. On BJ60 (Canada) models, turn the ignition key switch to the *preheat* position and note how long it takes the instrument panel glow plug light to go out; the glow plug light should remain illuminated for approximately 15 seconds at 0°C (32°F).
2. Once the engine is started, the glow plug timer controls the after-glow action of the glow plugs whereby they still receive current to maintain a smooth engine operation and warm-up condition. Check the glow plug timer by placing an ohmmeter across the 2nd and 3rd terminals of the timer as shown in Figure 19–36; no continuity should exist across these terminals once the engine has started.
3. On non-BJ60 models, proceed as follows. With the

ignition key in the preheat or glow position, check to see how long it takes for the glow plug indicator light to illuminate; this should take about 15 seconds at 0°C (32°F).

4. If the glow plug preheat time is less than 15 seconds, check for a short circuit in the wiring or a shorted glow plug or controller.
5. If glow plug preheat time exceeds 30 seconds, check out the following possible problem areas:
   a. Open circuit in the wiring
   b. Loose or corroded terminal connections
   c. Burnt switch or relay points
   d. Open glow plug or controller
6. Load current check: Figure 19–36 illustrates the ammeter placement for the BJ60 (Canada) check; Figure 19–37 illustrates the ammeter placement for the non-BJ60 circuit.
7. To check the current draw to each individual glow plug unit, remove the glow plug connector. With the ammeter connected as shown in the illustrations between the glow plug wire and glow plug terminal, turn the ignition key to the preheat (glow) position and note the reading on the ammeter scale after 40 seconds. The acceptable current draw for a glow plug is as follows:
   a. BJ60 System:
      7.2–9.8 amps for a 12-volt system
      3.8–5.2 amps for a 24-volt system
   b. Non-BJ60 System:
      7.6–10.4 amps for a 12-volt system
      4.1–5.5 amps for a 24-volt system
8. To check the glow plug resistance, place one lead from an ohmmeter to ground (engine block) and the other end (lead) against the actual glow plug wire terminal (wire disconnected for this test). The resistance obtained at 20°C (68°F) should be:
   a. BJ60 (Canada) units:
      0.2 ohm on a 12-volt circuit
      1.0 ohm on a 24-volt circuit
   b. Non-BJ60 units:
      0.2 ohm on a 12-volt circuit
      0.8 ohm on a 24-volt circuit
9. To check the glow plug relay, refer to Figure 19–38 and disconnect the EDIC motor wiring harness connector; using a voltmeter at the ST terminal, (Figure 19–39) turn the ignition key to the *start* position and a voltage reading should be apparent.

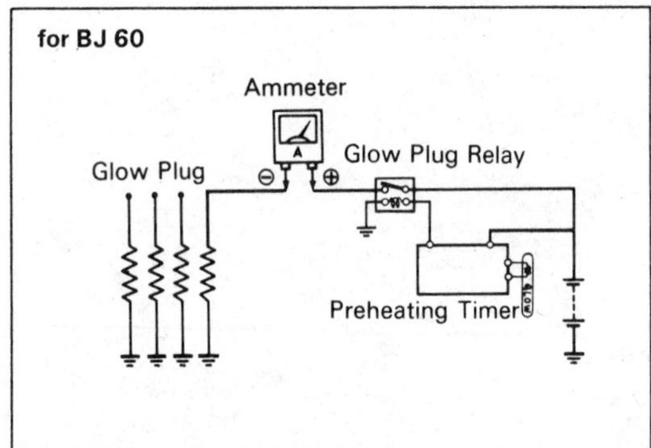

Figure 19–36. BJ60 Current draw check.

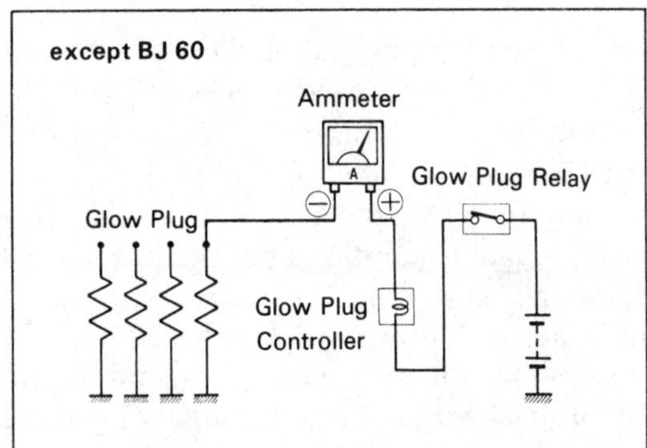

Figure 19–37. Non-BJ60 current draw check.

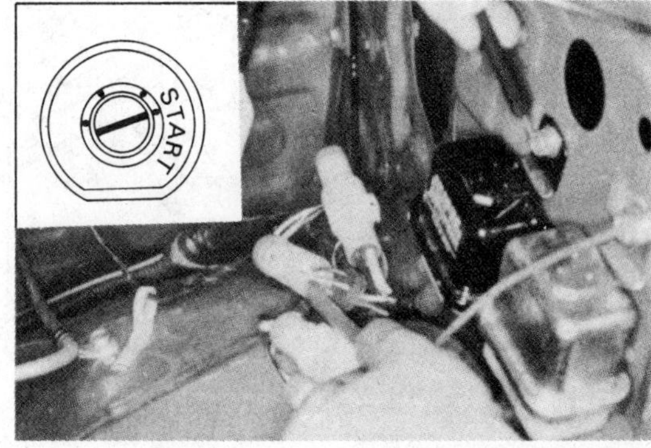

Figure 19–38. Disconnect EDIC motor connector.

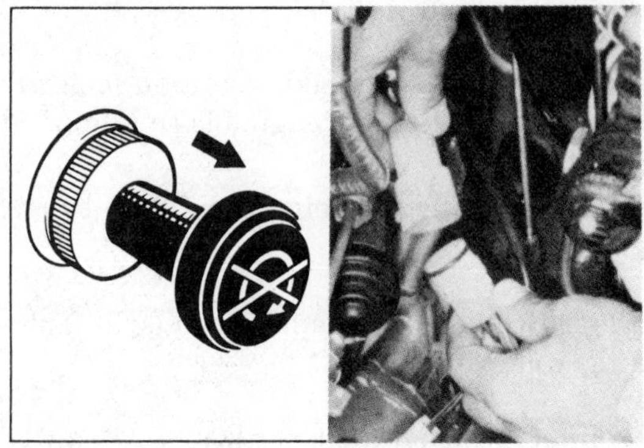

Figure 19–39. Placement of voltmeter leads to ST terminal.

Engine Compression Check

A cylinder compression check can be performed anytime that hard starting, lack of power, and unacceptable exhaust smoke are apparent after having established that the fuel injection system is not at fault. To perform a compression check, proceed as follows.

Procedure

1. Run the engine until it is at operating temperature.
2. Remove all glow plugs from the cylinder head.
3. Tie back the glow plug leads so that there is no possibility of them grounding out on the engine block or other metal surfaces.
4. Obtain special tool SST 09992–00021, which is an adapter that threads into the glow plug hole; a compression gauge is then attached to this adapter. Figure 19–40 illustrates this adapter installed in a glow plug hole.
5. On engines equipped with the EDIC (electrical diesel injection control) system, disconnect the EDIC motor wiring harness plug as shown in Figure 19–41 to prevent engine start-up during the test.
6. On units without the EDIC system, simply pull the stop button on the vehicle instrument panel all the way out to prevent any possibility of the engine starting during the compression test.

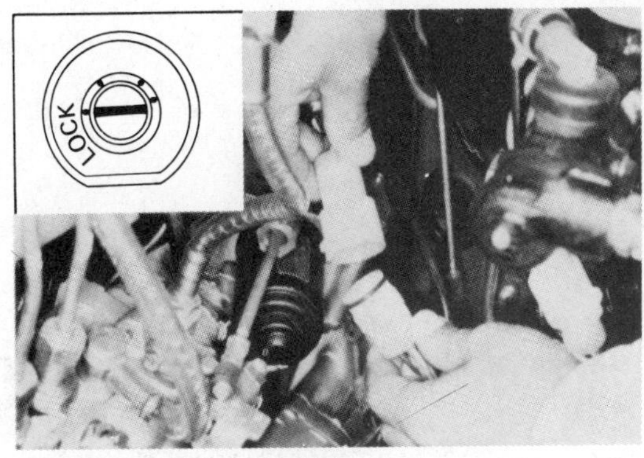

Figure 19–41. Disconnecting EDIC wiring harness plug.

7. Install the compression gauge onto the adapter in the glow plug hole as shown in Figure 19–42.
8. Jam the throttle wide open and spin the engine over on the starter for 4 to 6 revolutions and note the compression gauge reading.
9. For an accurate reading, the engine should be at normal operating temperature plus have a minimum cranking speed of 250 rpm.
10. Acceptable compression pressures should be:
    a. New or rebuilt engine: 427 psi (30.0 kg/cm^2)
    b. Minimum acceptable reading: 284 psi (20.0 kg/cm^2)

    NOTE: The maximum allowable variation between engine cylinders should not exceed 28 psi (2.0 kg/cm^2) regardless of the readings obtained between 284 and 427 psi!

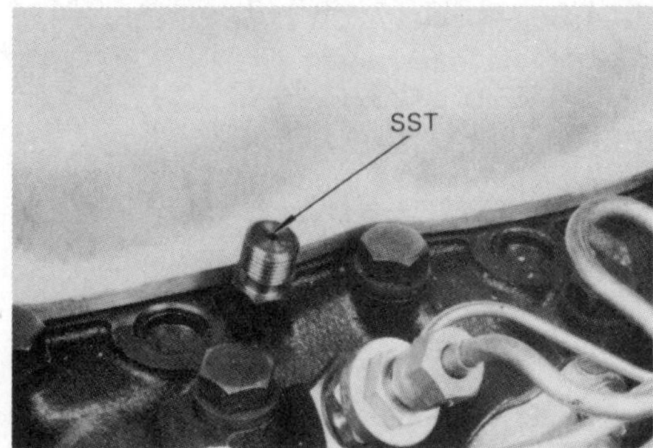

Figure 19–40. SST 09992-00021 Adapter in glow plug hole.

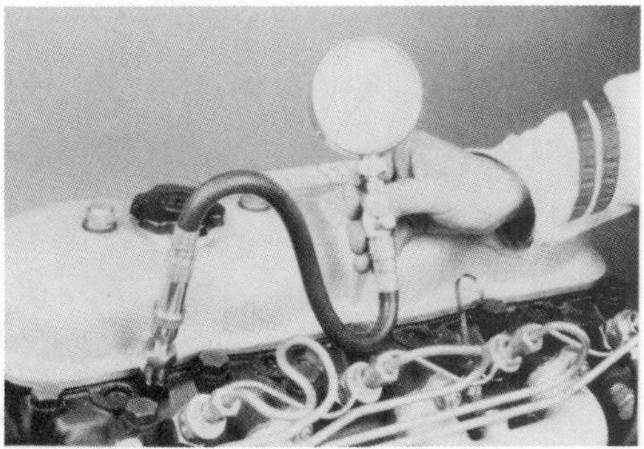

Figure 19–42. Compression gauge installed on adapter.

20

Isuzu/GMC Diesel Engines

Isuzu/GMC 1.8L/2.2L Diesel Engines

General Motors and Isuzu have been closely linked for a number of years now; General Motors actually own 39% of Isuzu. Isuzu has had many years of experience in the design and development of diesel engines. The company was started in 1916 and it produced its first diesel engines for trucks in the 1930s. Current production Isuzu diesel engines range from a 2-cylinder engine of 860 cc (52.4 cu.in.) displacement up to a 16.8L (1024 cu.in.) V12. GMC has chosen to offer a number of Isuzu diesels in a number of their cars and small pickup trucks as well as their mid-range diesel trucks starting with the 1984 model year.

Both the Pontiac 1000 and Chevette small cars have offered the 4-cylinder Isuzu 1.8L (111 cu.in.) diesel engine as an option since 1981. The Isuzu code name for the 1.8L diesel engine is the 4FBI. The basic engine is shown in Figure 20–1.

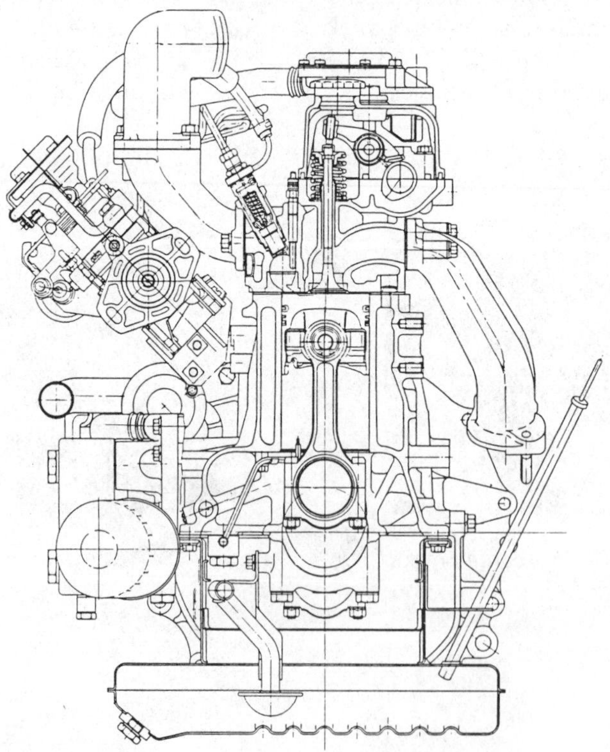

Figure 20–1. Cross section of 4FB1 engine. (Courtesy of General Motors Corporation)

In the 1984 model year for the first time, GMC's Truck and Bus Group is offering a 2.2L (134 cu.in.) 4-cylinder Isuzu diesel engine in the GMC "S" Series compact size pickup trucks furnished by Isuzu Motors, Ltd; in Japan, through American Isuzu Motors, Inc.

These engines produce 62 bhp at 4300 rpm and 96 lb.ft. of torque at 2200 rpm. The engine is an overhead cam design with a forged steel crankshaft and autothermatic pistons for low noise levels in low temperature conditions.

Expected EPA fuel economy of these S series trucks is expected to average 43 mpg (US gallon) or 51.5 mpg Imperial gallon in highway driving along with 33 mpg (US gallon) or 39.6 mpg Imperial gallon plus in city driving. These engines will be mated to either a four- or five-speed manual transmission.

Approximately 5% of all Isuzu cars produced are fitted with diesel engines. In 1980, Isuzu averaged a production capacity of 10,600 diesel engines per month, or 127,200 diesel passenger car engines per year; therefore they are one of the larger manufacturers of passenger car diesel engines. In 1980, the company produced a total of 402,000 diesel engines for a variety of applications.

Information on the 1.8L engine is basic because the 2.2L engine fuel system is covered in detail. Both engines employ the same basic fuel pump and nozzles as well as the same basic glow plug circuit.

A detailed description of the VE type injection pump can be found under Chapter 12, which deals with Robert Bosch in-line and distributor-type pumps. Since the Isuzu uses a VE distributor-type injection pump, refer to this section for an operational description as well as adjustment and repair information.

Isuzu 2.2L Diesel Engine Specifications

The 4-cylinder in-line diesel engine specifications for the Isuzu unit used in the GMC S-10 pickup truck are shown in Table 20–1.

Isuzu 2.2L Diesel Engine

The 2.2L diesel engine is an indirect-injection combustion chamber arrangement (swirl chamber), similar to the 1.8L engine. It uses electrical glow plugs to assist starting, especially in cold weather.

The fuel injection pump used with the 2.2L Isuzu engine is of Robert Bosch design but is manufactured under licence by Diesel Kiki in Japan.

Table 20–1. *Specifications for 4-cylinder in-line diesel engine.*

Items		
Engine type		Water-cooled, 4-cycle in-line, overhead valve
Combustion chamber type		Swirl chamber type
Cylinder liner type		Combined with cylinder block
Timing gear system		Belt drive
Number of piston rings		Compression ring: 2, oil ring: 1
Number of cylinders—bore × stroke	mm (in.)	4-88 × 92 (3.46 × 3.62)
Piston displacement	cc (cu.in.)	2238 (136.6)
Compression ratio		21.0 to 1
Engine dimensions:		
length × width × height	mm (in.)	687.5 × 560 × 683 (27.1 × 22 × 26.9)
Engine weight (dry)	kg (lb.)	190 (419)
Fuel injection order		1-3-4-2
Fuel injection timing (BTDC Static)		15°
Type of fuel used		Number 2-D diesel fuel (above −7°C [20°F]) Number 1-D diesel fuel (below −7°C [20°F])
Injection pump type		Bosch distributor VE type
Governor type		Mechanical, variable speed (half all speed) with aneroid compensator
Injection nozzle type		Throttle type
Fuel injection starting pressure	kg/cm^2 (psi)	105 (1493)
Compression pressure	kg/cm^2 (psi)	31 at 200 rpm (441 at 200 rpm)
Idle speed	rpm	800
Valve clearances	mm (in.)	Intake and exhaust, cold; 0.40 (0.0158)
Intake valves		
open at		16° (B.T.D.C.)
close at		54° (A.B.D.C.)
Exhaust valves		
open at		56° (B.B.D.C.)
close at		14° (A.T.D.C.)
Lubrication method		Pressurized circulation
Oil pump type		Trochoid type
Oil filter type		Paper element type (Full-

(Continued)

Table 20–1. *(cont.)*

Items		
		flow and by-pass flow)
Lubrication oil capacity	liters (gal.)	5.7 (1.15) with oil filter and oil cooler
Oil cooler type		Water cooled type
Cooling method		Pressurized circulation
Cooling water capacity—engine only	liters (gal.)	5 (1.32)
Water pump type		Impeller type
Thermostat type		Waxpellet type (with jiggle valve)
Air cleaner type		Viscous type paper element
Starter voltage-output	V-KW	12–2.0

The distributor/rotary injection pump is a Model VE, which is used extensively on a number of well-known vehicles. A list of these vehicles can be found under the Robert Bosch injection pump chapter, VE section, Chapter 12. Within this book you will find other engines equipped with the Model VE injection pump. A detailed description of the operation and adjustment of the VE pump can be found in Chapter 12.

Because of the minor differences that exist between engines using the same fuel injection pump, information in this chapter will deal mainly with external pump adjustments.

Idle Speed Adjustments

The engine should always be at its normal operating temperature before any speed adjustments are made. Two speed adjustments can be done—the low and the fast idle rpm settings. Figure 20–2 illustrates the location of both the low or base idle and the fast idle speed setting screw.

Both the low and fast idle adjustments follow the same first four steps, which are as follows:

Procedure

1. Apply the parking brake and ensure that the driving wheels are blocked securely.
2. Place the transmission selector lever in *neutral.*
3. Connect up an accurate tachometer or stick a piece of reflective magnetic tape on the crankshaft pulley and aim an electronic magnetic pickup tachometer at the rotating pulley.
4. Start and run the engine (if the engine is not at operating temperature, run it until it is).

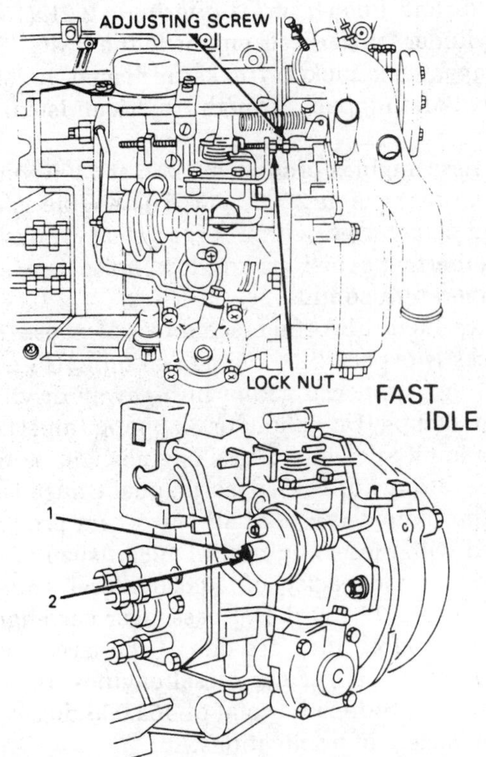

Figure 20–2. Base and fast idle speed setting screws. (Courtesy of General Motors Corporation)

5. To set the low idle speed, simply loosen the lock nut on the idle speed adjusting screw and rotate it clockwise to increase the speed and counterclockwise to decrease it; lock up the nut after the rpm specified on the vehicle emissions label has been obtained.
6. To set the fast idle speed, apply vacuum to the fast idle actuator with a hand pump.
7. Loosen the lock nut on the fast idle adjustment screw.
8. Adjust the knurled nut either clockwise or counterclockwise to obtain the fast idle rpm stated on the emissions label.
9. Tighten the lock nut.

Fuel Filter Replacement

Filter service follows the same basic procedure as that for other diesel engines discussed in this book. Figures 20–3 and 20–4 illustrate the actual fuel filter and its location on the S-10 pickup truck.

The fuel filter is designed to act as a water trap as well as removing impurities from the fuel. A thermostatically controlled fuel heater is used to prevent

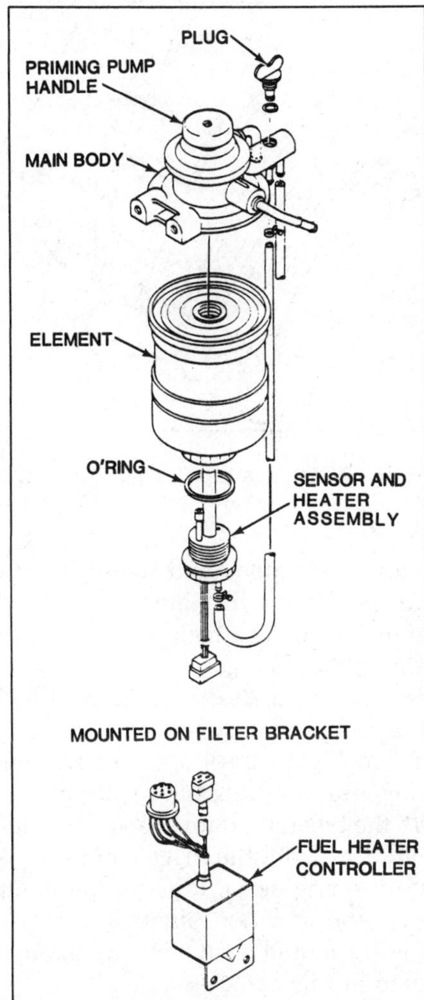

Figure 20-3. Fuel/water filter assembly. [Courtesy of General Motors Corporation]

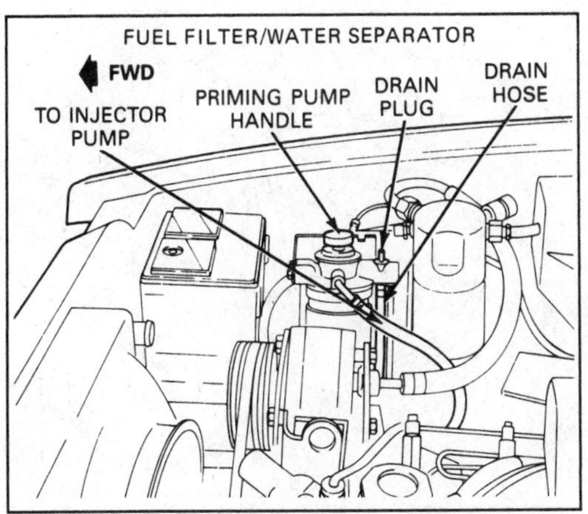

Figure 20-4. Fuel/water filter location. [Courtesy of General Motors Corporation]

fuel waxing at low ambient temperatures, which would prevent the engine from starting and running.

When changing the filter proceed as follows:

1. Disconnect the negative battery terminal to isolate the fuel heater and water sensor unit, which are electrically controlled.
2. Place a drain tray underneath the fuel filter to catch water and fuel.
3. After disconnecting the water sensor wiring harness, use a filter wrench, such as Kent-Moore P/N J-22700 or equivalent to loosen the filter.
4. Once the filter has been drained, remove the water/heater element from the base of the old filter since it will be installed onto the new unit.
5. Lightly lubricate the water sensor O-ring with diesel fuel before placing it onto the base of the new filter.
6. Apply a light film of diesel fuel to the new filter seal ring and rotate the new filter until its gasket contacts the sealing surface.
7. Turn the filter 2/3 of a turn after it contacts the filter base.
8. Connect up the water sensor wiring.
9. Temporarily disconnect the injection pump fuel return hose and place it into a drip container.
10. Prime the filter by operating the fuel transfer pump priming handle mounted on top of the filter assembly as shown in Figure 20–3.
11. When fuel flows from the injection pump return hose free of air, reconnect the hose to the injection pump.
12. Start the engine and check for signs of fuel leaks.

Drain Water From Separator

A water sensor located at the base of the fuel/water filter or separator as shown in Figure 20–3 will activate an instrument panel warning light when an excess amount of water collects in the base of the unit.

When this warning light comes on, the water should be drained as soon as possible by following the listed steps below:

1. Park the vehicle and shut off the engine.
2. Engage the parking brake and block the wheels if required.
3. Open the hood and place a drain tray or container under the drain hose that runs from the base of the separator.

4. Loosen the wing nut at the top of the filter and pull/push the filter priming pump until all signs of entrapped water have disappeared—which will be evident by clear diesel fuel flowing from the drain line.
5. Close the drain plug shown in Figure 20–4 and continue to manipulate the priming pump handle until a pressure build-up is felt (which primes the fuel system).
6. Start and run the engine to see if the *water-in-fuel* light has gone out; if it has not, it may be necessary to check the water sensor for a fault or there may be additional water in the system that requires draining and cleaning the fuel tank.

Injection Pump Removal

If it becomes necessary to remove the fuel injection pump from the engine for reasons of repair or during an engine overhaul situation, follow the necessary procedures given below. Before attempting to remove the fuel injection pump, obtain an adequate supply of protective plastic shipping caps, both male and female, that can be used to cover all open fuel lines and holes to prevent the entrance of dirt into the fuel system.

Before removing the actual injection pump, it will be necessary to:

1. Remove the timing belt and engine front covers.
2. Remove the injection pump drive gear.
3. Remove the injection pump bracket bolts and retaining nuts.

Special tools required for this procedure are as follows:

1. Kent-Moore gear puller P/N J-22888.
2. Floor jack or vehicle hoist.

Procedure:

1. Isolate the starter and fuel solenoid electrical system by disconnecting the battery negative cable.
2. Remove the injection pump timing drive belt as follows:
   a. Manually rotate the engine from the crankshaft pulley until the timing mark on the crank pulley is aligned with the stationary pointer as shown in Figure 20–5.
   b. Remove the crankshaft pulley (you will have to remove the radiator) by removing the larger retaining bolt at its center hub.

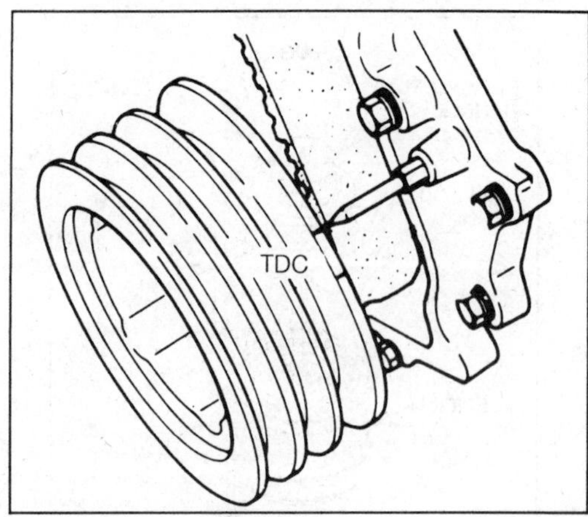

Figure 20–5. Crankshaft pulley at TDC. (Courtesy of General Motors Corporation)

   c. Remove the upper and lower engine timing covers as shown in Figure 20–6.
   d. Remove the bolts attaching the injection pump timing pulley flange.
   e. Refer to Figure 20–7 and remove the belt tension spring.
   f. Refer to Figure 20–8 and remove the tension pulley retaining nut and center hub.
   g. With the tension removed from the timing belt, lift it clear from the drive sprockets.
   h. With the engine at TDC on the compression stroke, the injection pump and camshaft timing gears should have their alignment marks as shown in Figure 20–9.
3. Refer to Figure 20–10 and remove the injection pump drive pulley using Kent-Moore puller P/N J-22888 after removing the retaining bolt from the center of the pulley.
4. Remove all the fuel lines from the injection pump

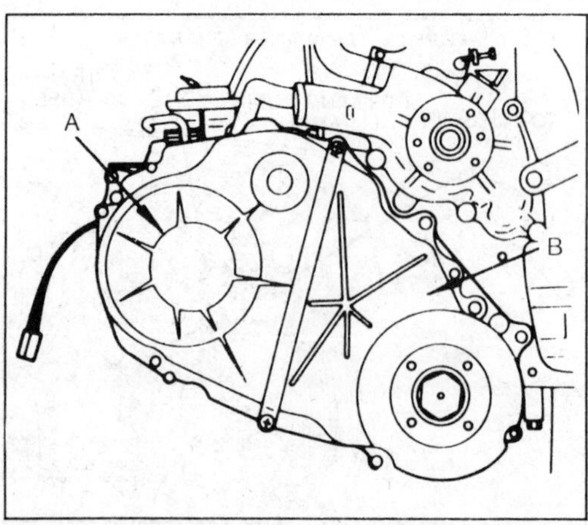

Figure 20–6. Engine timing covers. (Courtesy of General Motors Corporation)

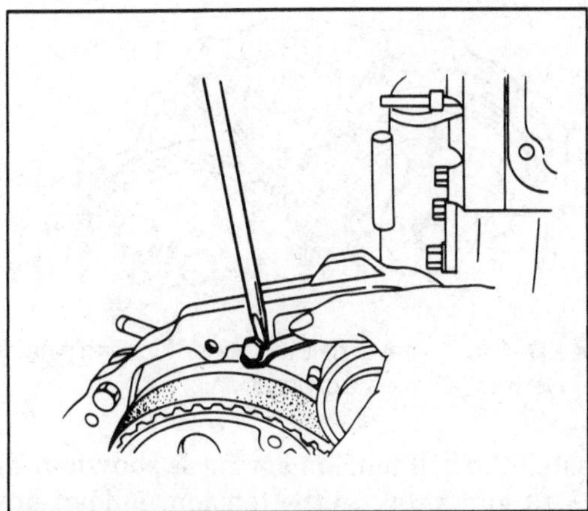

Figure 20-7. Removing tension spring. [Courtesy of General Motors Corporation]

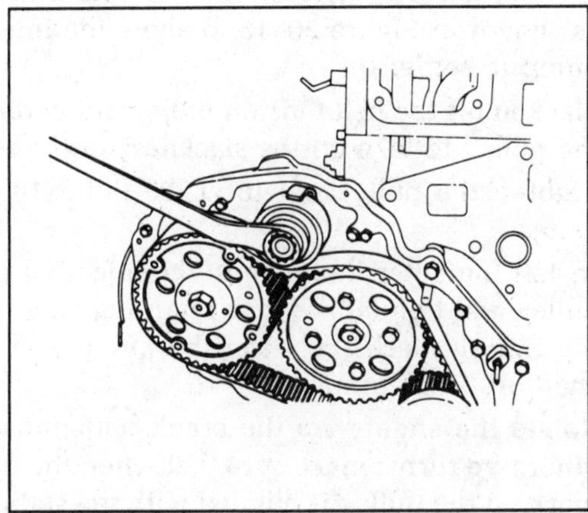

Figure 20-8. Belt tensioner removal. [Courtesy of General Motors Corporation]

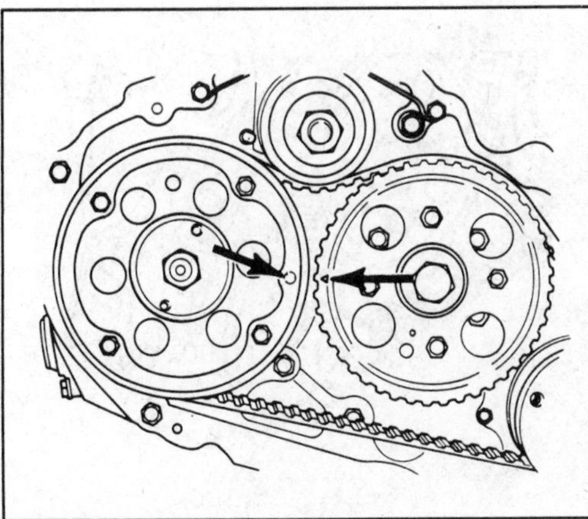

Figure 20-9. Timing marks aligned at injection pump and camshaft. [Courtesy of General Motors Corporation]

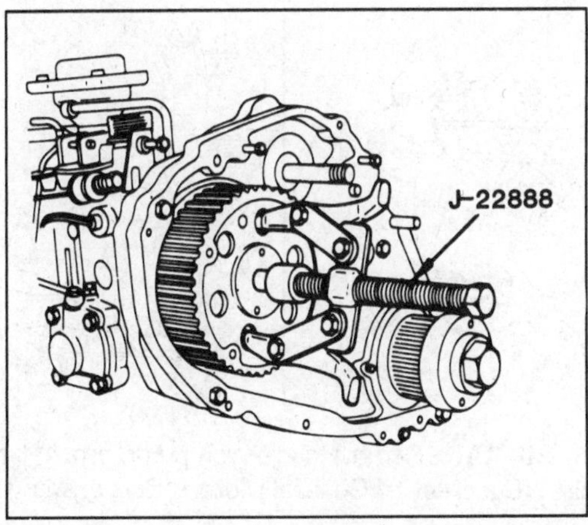

Figure 20-10. Removing injection pump pulley. [Courtesy of General Motors Corporation]

along with the fuel cutoff solenoid wire, the throttle cable, and return spring. Plug all open fuel lines and openings.

5. Jack up the vehicle or raise it on a hoist so that you can gain access to the injection pump bracket bolts shown in Figure 20-11.
6. Once the injection pump bracket bolts have been removed, remove the two pump mounting flange nuts around the circumference of the pump mounting flange.
7. Figure 20-12 illustrates the timing scribe mark that appears between the injection pump mounting flange and the engine timing cover.
8. If the timing scibe marks in Step 7 are not in alignment, make a match mark on the pump housing and timing cover to facilitate alignment upon reinstallation of the injection pump.
9. Pull the injection pump out of the housing from the backside of the front timing cover.

Injection Pump Installation

Installation of the injection pump is basically the reversal of removal and is further discussed under "Check Injection Pump Timing."

The timing belt must be properly adjusted as follows:

1. Place the engine at TDC No. 1 cylinder on the compression stroke as shown earlier.
2. Install the drive belt over the crankshaft pulley sprocket, the camshaft, and injection pump pulley sprocket.
3. Refer to Figure 20-13 and install the belt tension center hub and pulley.

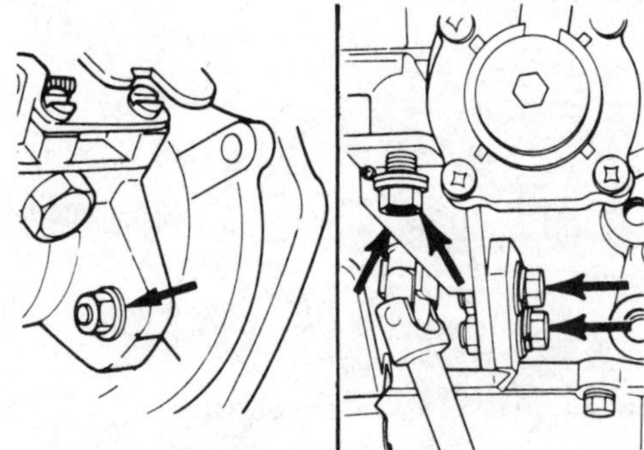

Figure 20–11. Removing injection pump bracket retaining bolts. (Courtesy of General Motors Corporation)

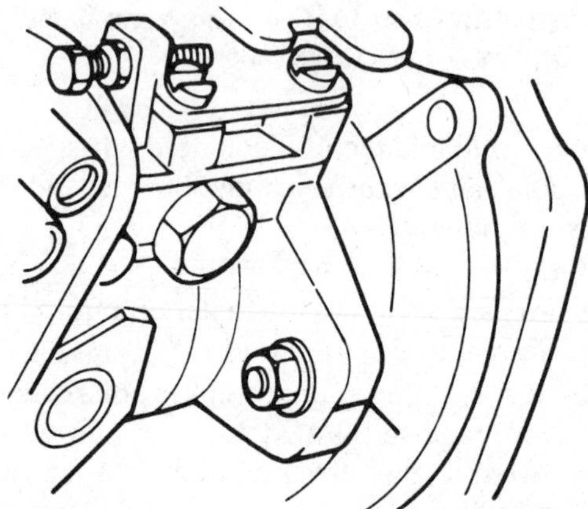

Figure 20–12. Injection pump mounting flange timing alignment marks. (Courtesy of General Motors Corporation)

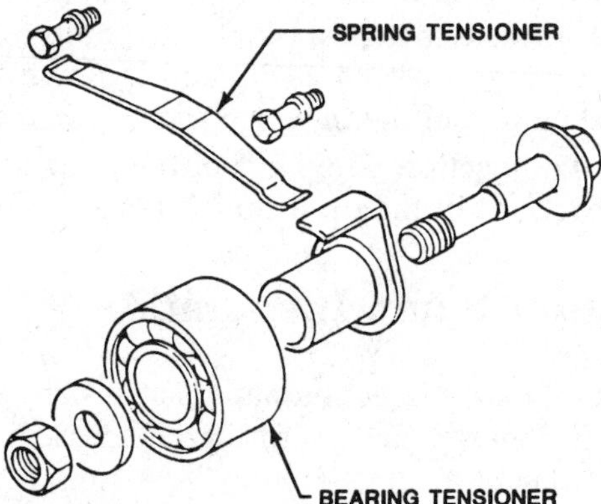

Figure 20–13. Timing belt tensioner installation. (Courtesy of General Motors Corporation)

4. Refer to Figure 20–14 and ensure that the end of the belt tension center hub is in proper contact with the two pins on the timing pulley housing.
5. Hand tighten the retaining nut so that the pulley can slide freely.

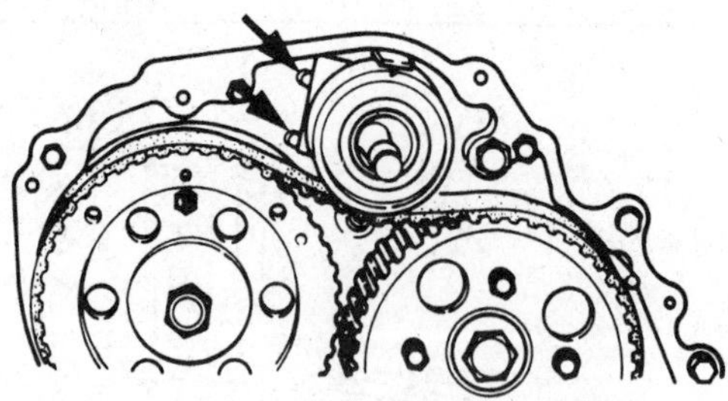

Figure 20–14. Timing belt tensioner locating pins. (Courtesy of General Motors Corporation)

6. Install the belt tension spring as shown in Figure 20–13 and snug up the tension pulley retaining nut.
7. Manually rotate the engine at least two complete turns to allow the belt to seat correctly on the pulley sprocket teeth.
8. Rotate the crankshaft another 90 degrees ATDC as shown in Figure 20–15 to allow the injection pump to settle.
9. Slacken off the belt tension pulley nut and allow the pulley to take up the slackness in the belt.
10. Tighten the pulley nut to 78–95 lb.ft. (106–129 N·m).
11. Install the outer flange onto the injection pump pulley and tighten the four retaining bolts when the timing marks between the pump and camshaft are in alignment.
12. Rotate the engine via the crankshaft pulley another two turns; make sure that when the timing mark on the pulley is aligned with the static timing pointer that the No. 1 piston is in fact on its compression stroke. If in doubt, remove the valve rocker cover and check that both the intake and exhaust valve on No. 1 cylinder have clearance; if they do not, then rotate the engine another 360

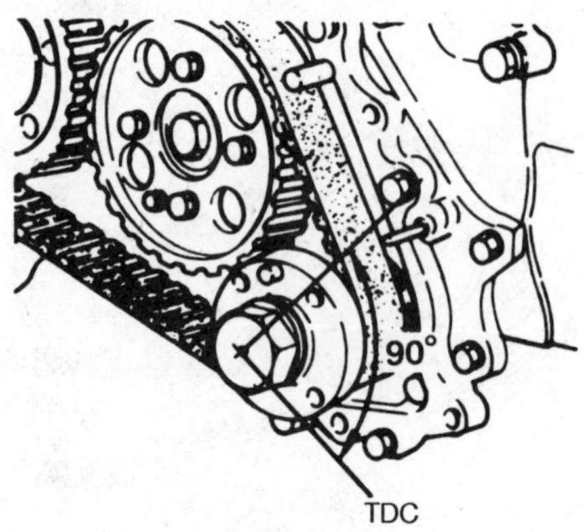

Figure 20–15. Placing crankshaft 90 degrees ATDC. (Courtesy of General Motors Corporation)

degrees since it is on the end of the exhaust stroke and the start of the intake stroke at this position.

13. Refer to Figure 20–16 and check the belt tension in the position illustrated. The tension should read between 213–356 psi.
14. Install the upper and lower timing covers and crankshaft pulley.

Injection Pump Timing Check

The injection pump timing follows the same basic pattern as that for other engines using the Robert Bosch or Diesel Kiki VE model fuel injection pump.

Incorrect injection pump timing generally results in poor engine performance, lack of power, and poor fuel economy as well as extra smoke at the exhaust. When it is suspected that the injection pump timing is incorrect, check the timing as per the detailed procedure under the heading "Static Timing Check" described in Chapter 12, Robert Bosch fuel injection pump, VE pump section. See Figures 12–30 and 12–31.

NOTE: The Isuzu specification for static pump timing is 15 degrees BTDC on the crankshaft pulley. At this position, the dial gauge installed into the injection pump hydraulic head as shown in Figures 12–30 and 12–31 should read 0.5 mm (0.020") if the injection pump to engine timing is correct. If it is necessary to adjust the injection pump timing, recheck that the scribe line on the pump flange is aligned with the engine timing cover mark. If it is not, remark the pump housing so that its mark is opposite the engine mark.

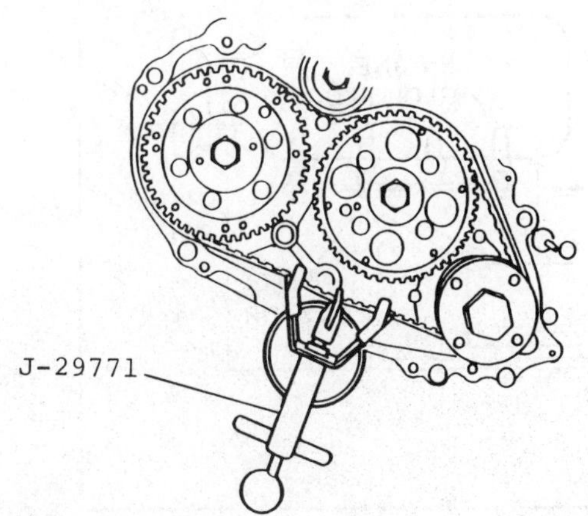

Figure 20–16. Checking timing belt tension. (Courtesy of General Motors Corporation)

Injection Nozzles

The injection nozzles used with the 2.2L Isuzu engine are Bosch/Kiki DN-type throttling pintle units, which are mounted in KC style holders and are set to open at 1493 psi (10,294 kPa) (Figure 29–4).

These nozzles are similar to those used in engines such as the Volkswagen Rabbit; the Audi 5000; the Volvo D24; Mercedes-Benz vehicles; Ford 2.0L, 2.2L, and 2.4L engines; and the Datsun-Nissan and Dodge Ram 50 pickup.

Removal procedures follow the same pattern as for those engines. A detailed description of the testing, disassembly, inspection and reassembly of the nozzles can be found in the information contained in Chapter 29 dealing with "Injection Nozzle Testing."

Typical nozzle patterns for these nozzles are as shown in Figure 29–38. Nozzle leakage should be checked at 284 psi (1958 kPa) lower than the popping (release) pressure. The nozzle tip should remain dry, although a slight sweat is acceptable after 10 seconds if no droplets are formed.

Glow Plug Circuit

The glow plug circuit used on most diesel engines in passenger cars operates on the same principle and often has the same basic electrical circuitry. A basic line diagram of the glow plug circuit is illustrated in Figure 20–17. The operation is as follows: When the engine coolant temperature is lower than 50°C (122°F), a thermo-time switch located at the thermostat housing is designed to provide a completed circuit to the glow plugs for maximum preheat.

Under these conditions, Relay No. 1 in Figure 20–18 is energized and the glow plug indicator lamp on the instrument panel will glow for 3.5 seconds or longer, depending upon how cold the engine is.

Glow plug preheat temperatures can approach 900°C (1652°F) when the engine is cold to facilitate rapid heating of the combustion chamber area and provide maximum temperature rise to the incoming ambient air.

When the engine temperature is higher than 50°C (122°F), Relay No. 1 is inoperative and thus Relay No. 2 is activated to provide a lower voltage through the dropping resistor to ensure stabilized glow plug heating during initial starting of the engine.

A sensing resistor (shunt wound) is used in series with the glow plugs to cause a small voltage drop, which is monitored by the controller. Two in-line fu-

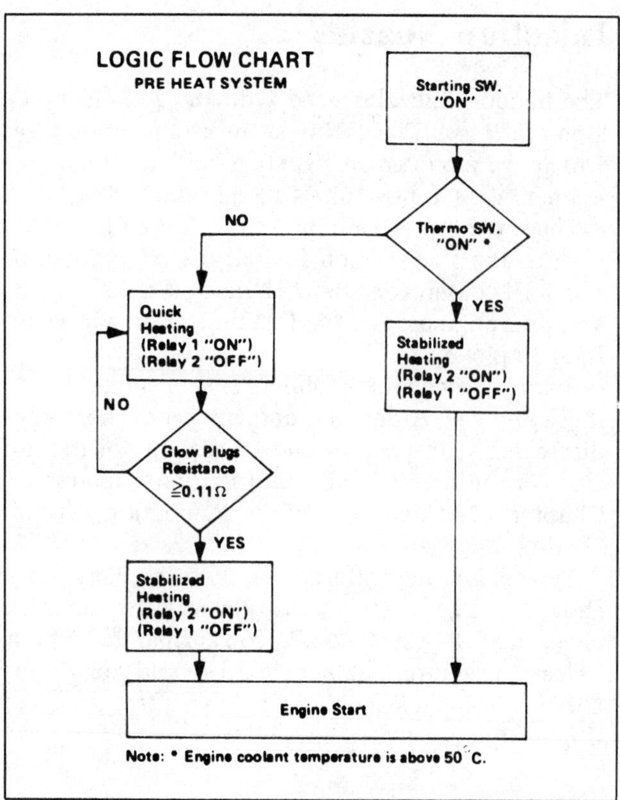

Figure 20–17. Glow plug operational flowchart. (Courtesy of General Motors Corporation)

sible links are used in the glow plug electrical wiring to protect the circuit.

When a problem occurs with the glow plug circuit, check for loose or corroded wiring first, then feel the fusible links for damage. Checks that can be made to the glow plug circuit are shown in Figures 20–19 through 20–24.

Isuzu/Bosch VE Pump Electronic Controls

Isuzu recently completed the development of a diesel engine control system employing electronics technology. They used the 1.8L 4FBI engine as the base engine; the electronic system was supplied by Robert Bosch, who also assisted in the manufacture of the prototype car. Once the prototype vehicle was produced, Diesel Kiki (a Robert Bosch licensee in Japan) then took responsibility for the manufacture of the prototypes and most of the components for volume production.

Robert Bosch refers to the system as an EDC (electronic diesel control) system because it uses a number of sensors, an electronic control unit (ECU), and actuators for controlling such variables as fuel delivery, start of injection, and exhaust gas recirculation (EGR).

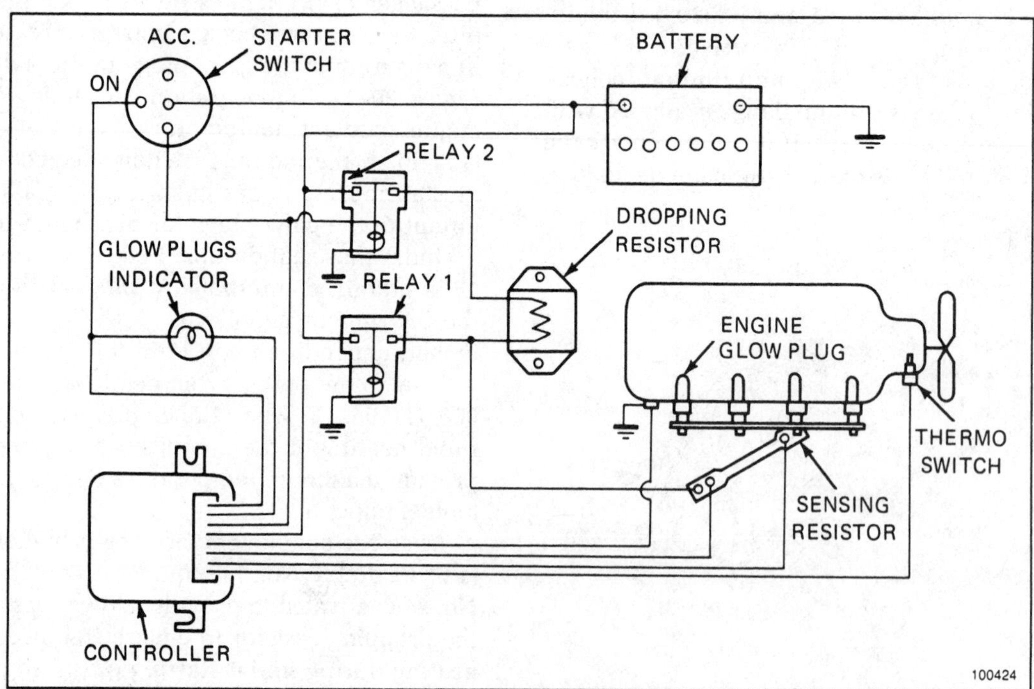

Figure 20–18. Glow plug electrical circuit diagram. (Courtesy of General Motors Corporation)

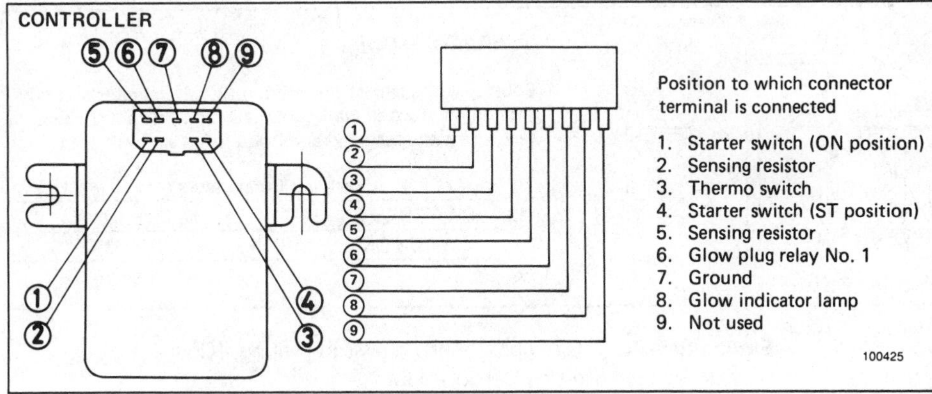

Figure 20-19. Glow plug controller connections. (Courtesy of General Motors Corporation)

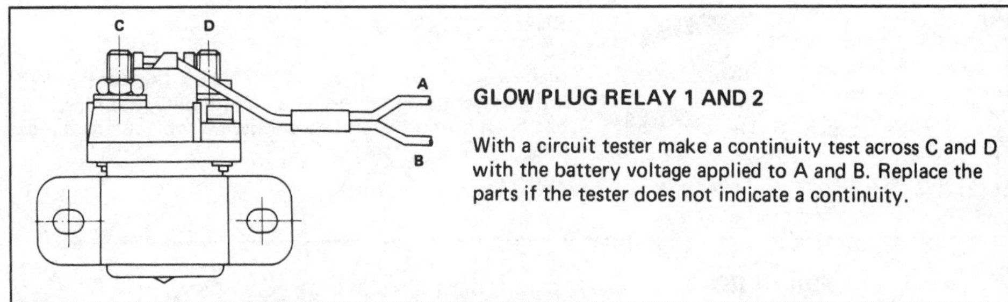

C
D
A
B
GLOW PLUG RELAY 1 AND 2
With a circuit tester make a continuity test across C and D with the battery voltage applied to A and B. Replace the parts if the tester does not indicate a continuity.

Figure 20-20. Glow plug relays (checkout). (Courtesy of General Motors Corporation)

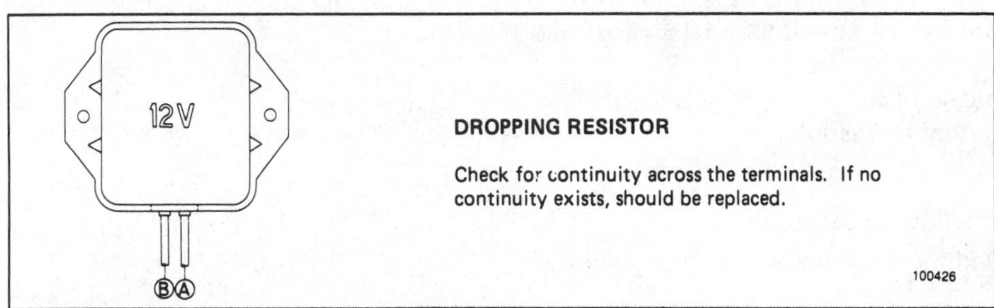

12V
DROPPING RESISTOR
Check for continuity across the terminals. If no continuity exists, should be replaced.
B A
100426

Figure 20-21. Glow plug dropping resistor check. (Courtesy of General Motors Corporation)

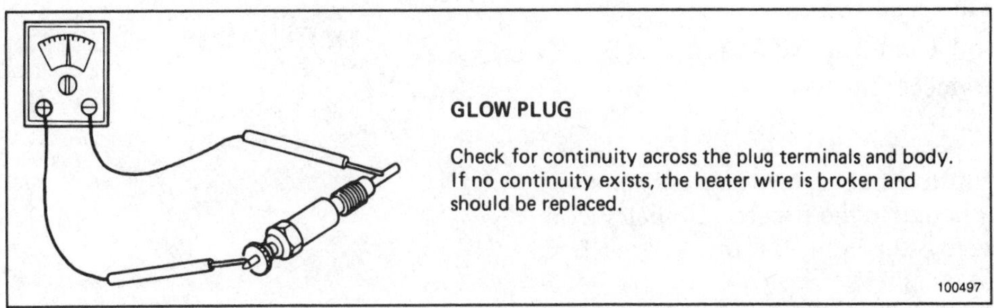

GLOW PLUG
Check for continuity across the plug terminals and body. If no continuity exists, the heater wire is broken and should be replaced.
100497

Figure 20-22. Glow plug testing. (Courtesy of General Motors Corporation)

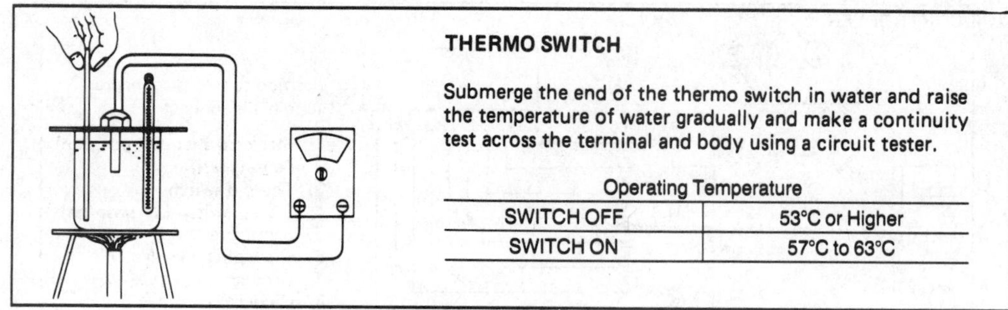

Figure 20–23. Glow plug thermo-switch testing. (Courtesy of General Motors Corporation)

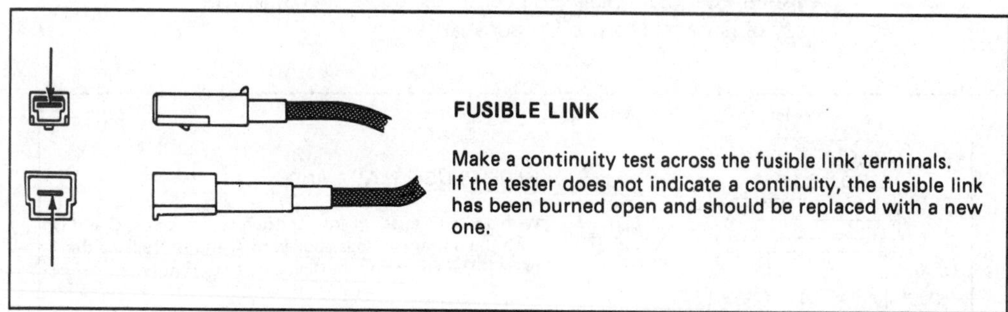

FUSIBLE LINK
Make a continuity test across the fusible link terminals.
If the tester does not indicate a continuity, the fusible link has been burned open and should be replaced with a new one.

Figure 20–24. Glow plug fusible link testing. (Courtesy of General Motors Corporation)

The finished product is known as the I-TEC-Diesel (Isuzu-Total Electronic Control-Diesel). This system has been available since late 1982 on Isuzu diesel-powered vehicles.

Major advantages of the electronically controlled fuel injection system are as follows:

1. Improved acceleration response
2. Improved maximum horsepower output performance
3. Better fuel economy
4. Reduced idling noise
5. Improved cold startability
6. Automatic idle speed control
7. Automatic cruise control feature
8. Improved serviceability

Some of these improvements were achieved through a minor design change to the Ricardo Comet V combustion chamber by changing the throat area from the conventional design.

The basic system is shown in Figure 20–25, which illustrates the required sensors. The air-flow sensor is similar to that employed by Robert Bosch in their L-Jetronic gasoline fuel injection system. The operation of this sensor is discussed in Chapter 6. This

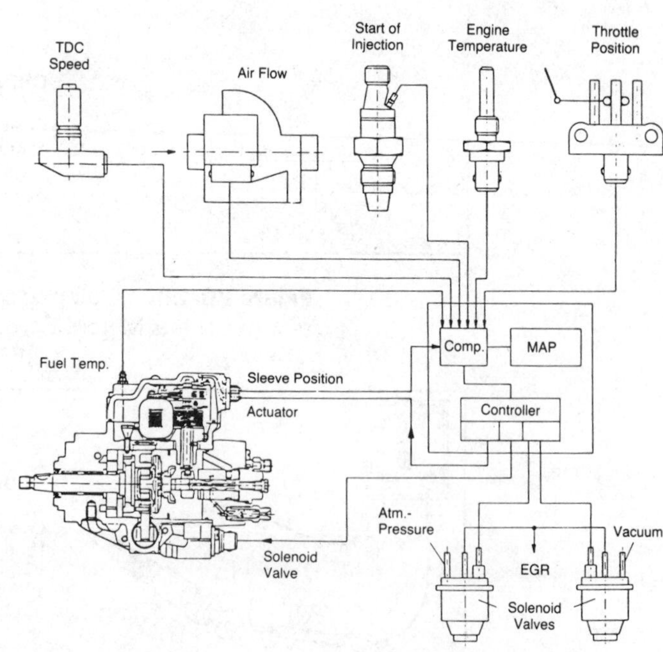

Figure 20–25. Diesel electronic control system components (Reprinted with permission © 1984 Society of Automotive Engineers, Inc.)

air-flow sensor determines the engine's consumption of air. Fuel quantity is evaluated uniformly from the control sleeve position inside the injection pump (Item 9 in Figure 12–17). Engine speed is established from a magnetic pickup from the crankshaft.

A Hall-effect sensor inside an injector (Figure 27–16) establishes nozzle needle movement to determine speed and fuel quantity. Operation of a Hall-effect sensor is discussed in Figure 7–48.

The electronic control system is designed to be "fail-safe" so that in the event of a malfunction the fail-safe system initiates a restrictive condition. This condition can take the form of either a restrictive engine operating condition or of a total engine shutoff, depending upon the seriousness of the problem. Figure 20–26 shows a block diagram of the fail-safe system, which involves both detection and evaluation of system failure.

Once a failure has been detected by one of the system sensors, the action taken is one shown in Figure 20–27. The actions listed in Figure 20–27 such as 11(a), 11(b), 11(c), 11(d), and 111 are discussed below.

Stage 1 would allow retention of an engine operating mode. Stage 11 would change the engine operating mode, while Stage 111 involves a shutoff of the engine. In Stage 11, the operating modes would be:

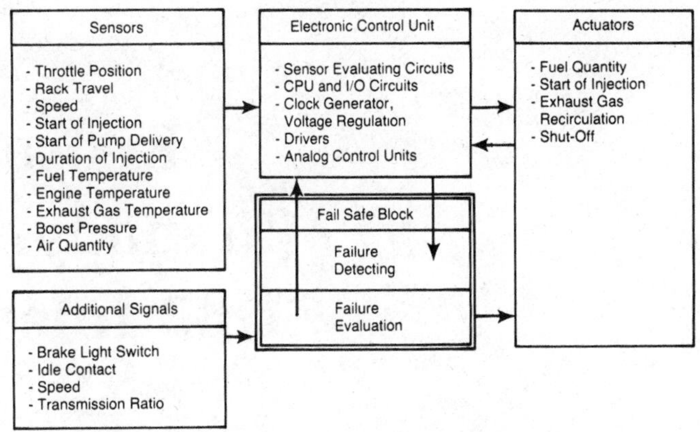

Figure 20–26. Fail safe detection/evaluation operational block. [Reprinted with permission © 1984 Society of Automotive Engineers, Inc.]

1. 11(a). A reduction in fuel quantity
2. 11(b). An increase in fuel quantity
3. 11(c). A reduction in engine speed
4. 11(d). Limp-home capability under controlled operation

This system is one of many electronic systems that will become much more prevalent in the successful use of diesel passenger cars and light trucks.

Components	Failure	Failure Detecting	Actions IIa	IIb	IIc	IId	III
Sensors and Evaluating Circuits	Fuel Temperature	Signal-Range-Check	x				
	Exhaust Gas Temperature	"	x				
	Engine Temperature	"		x			
	Air Temperature	"	x				
	Boost Pressure	"	x				
	Air Quantity	"	x				
	Start of Delivery	"				x	
	Start of Injection	"				x	
	Speed	"					x
	Rack Position	"				x	
	Throttle Position	"			x		
Actuators and Controllers	Fuel Quantity - Jammed	Steady State Control Deviation					x
	- Rack Controller Defective						
	- Ground Short Solenoid Coil	"					
	- Short Circuit Output Stage	"					
	- De-Energized	None					
	Start of Injection	Steady State Control Deviation	x				
	Exhaust Gas Recirculation	Reduced Air Flow	x				
Data Processing System	Software	Selfconfidence Check Watch Dog Program-Branch-Test				x	
Engine	Overspeed	Overspeed					x

Figure 20–27. Failure detection/corrective measures chart. [Reprinted with permission © 1984 Society of Automotive Engineers, Inc.]

21

Roosa Master (Stanadyne/Hartford) DB2 Injection Pump

Roosa Master fuel injection pumps take their name from Vernon Roosa who designed and manufactured the first mass-produced distributor-type diesel fuel injection pumps for use on high-speed engines. Today these injection pumps are manufactured by the parent company, which is Stanadyne's (Hartford Division) Diesel Systems.

These distributor pumps were manufactured in the early 1950s and were a radical new design in comparison to the then typical in-line injection pumps. These distributor, or rotary injection pumps, have been extremely successful, and to date, over 12 million have been sold worldwide in a variety of applications fitted to a number of well-known diesel engines.

These pumps have gained increased prominence since 1977 when the Oldsmobile Division of General Motors Corporation introduced their 350 cu.in. (5.7L) V8 diesel engine to the North American market.

Oldsmobile also offered a V8 260 cu.in. (4.3L) diesel through the 1982 model year, which has now been superseded by a new V6 263 cu.in. (4.3L) passenger car diesel.

Since the 1982 model year, the Chevrolet Division of GMC has offered a 6.2L (378 cu.in.) V8 diesel engine as an option in pickup trucks, vans, and light trucks such as motorhome chassis and recreational vehicles.

The 1984 Ford full size F-350 and E-350 vehicles are also offered with a 6.9L (420 cu.in.) V8 diesel engine that is manufactured by International-Harvester. This engine also employs a Roosa Master/Stanadyne Model DB2 fuel injection pump.

The GMC engines use the Roosa Master DB2 distributor injection pump with either the Roosa Master pencil-type injection nozzle; the CAV Microjector, which is now being manufactured under licence by GM-DED (Diesel Engine Division); or a Robert Bosch-type nozzle.

The Ford engine uses either CAV or United Technologies Diesel Systems Group fuel injectors. The Roosa Master pencil nozzle, which was first introduced in 1965 by the company, was the first major change in diesel engine nozzle design in over 30 years—therefore it was considered quite revolutionary.

Although Roosa Master does produce another model of distributor injection pump, namely the DM4, only the DB2 model is used on automotive passenger car and light pickup truck engines; therefore, only the DB2 will be discussed here. The DM4 operates in a similar manner to the DB2 and is of the same general design.

The Model DB2 Pump

The model DB2 pump incorporates a single pumping chamber, and is described as an opposed plunger, inlet metering, distributor-type pump. The major component parts are shown in Figure 21–1. The distributor rotor located in the pump's hydraulic head is driven by the drive shaft, and the pumping plungers are actuated toward each other by an internal cam ring through rollers and shoes. The internal workings of all Roosa Master pumps are basically the same. Figure 21–2 shows the main rotating parts of the pump.

Component Description for Figure 21–1

Figure 21–1 is a cutaway schematic of a typical Roosa Master fuel injection pump. The numbered components have the following function:

1 – The drive shaft is gear-driven from the front of the engine at the rear of the camshaft.

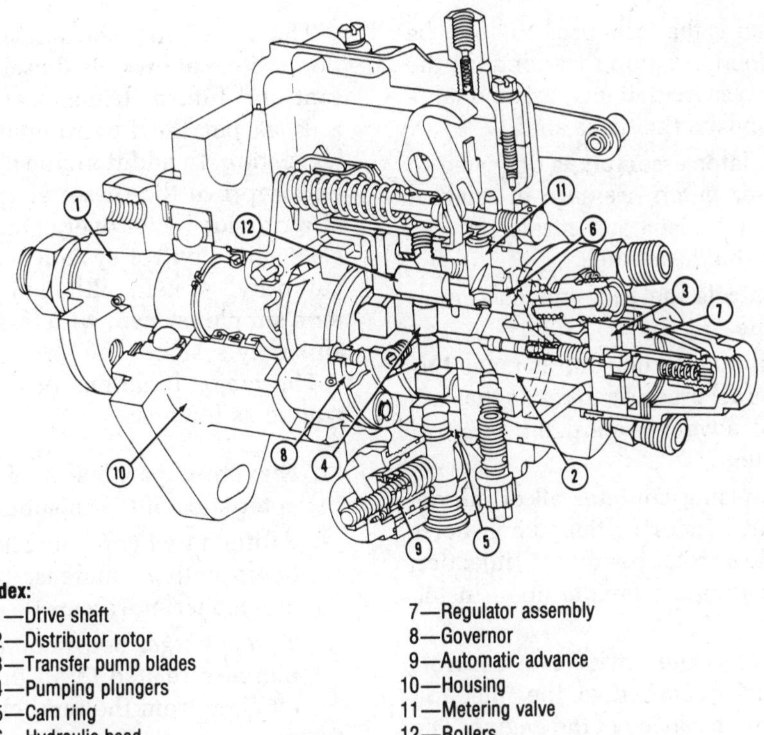

Figure 21-1. DB2 Distributor injection pump. (Courtesy of Stanadyne Diesel Systems)

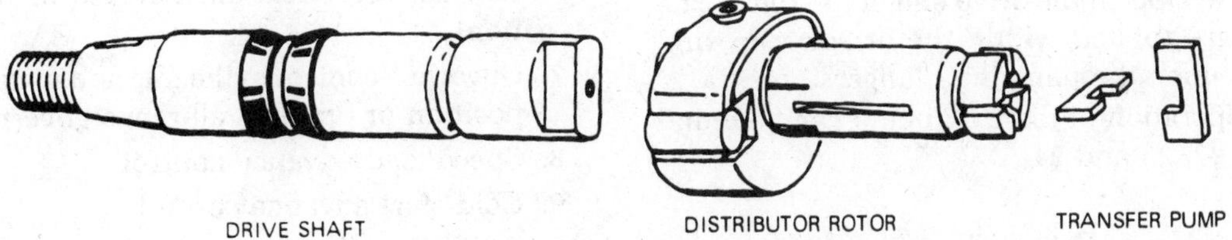

Figure 21-2. Main rotating parts of the DB2 model injection pump. (Courtesy of Stanadyne Diesel Systems)

2 - The injection pump distributor rotor acts like a rotor in an ignition distributor in that it delivers fuel at the correct point in the engine cycle to the individual cylinders as it rotates.

3 - The transfer pump is of the multiple vane type, and it is driven from the injection pump drive shaft. This pump delivers approximately 12 psi (83 kPa) at an idle rpm up to approximately 130 psi (896 kPa) at wide open throttle with an increase in fuel delivery volume.

4 - The pumping or charging plungers receive fuel at transfer pump pressure. They are forced in and out as they come in contact with the rotating injection pump cam ring lobes. This action causes the plungers to exert force on the fuel to a high enough pressure to open the delivery valve in the hydraulic head that leads to an injector.

5 - The internal cam ring is driven at injection pump drive shaft speed which is 1/2 engine speed (4-cycle engine). Within the inside diameter of the cam ring are lobe projections that provide the necessary lift for the fuel pumping plungers. The number of lobe projections in the cam ring are equal to the number of engine cylinders. The cam ring is also rotated by the injection advance mechanism to vary the injection timing.

6 – The hydraulic head is the term used to describe the area of the injection pump that contains the hydraulic fuel pressure/delivery components and is the end opposite the drive shaft.

7 – The pressure regulator assembly is designed to control the transfer pump pressure in relation to the engine speed. It also compensates for viscosity changes in the diesel fuel.

8 – The governor controls the idle and maximum speed of the engine.

9 – The automatic timing advance unit is moved by the transfer pump fuel pressure to rotate the cam ring slightly or to advance/retard the point of fuel injection timing.

10 – Injection pump housing contains all of the operating components. Its drive flange bolt holes are slotted to allow rotation of the injection pump in order to time the engine upon installation.

11 – The metering valve is the device that controls the amount of fuel delivered to the injection pump and therefore the speed of the engine. The metering valve has a helix, or scroll, cut into it that will vary the volume of fuel delivery. Throttle and governor linkage determine its position.

12 – The rollers are carried in roller shoes, which are located in slots in the drive end of the rotor. The rollers are rotated within the driven cam ring and act upon the pumping plungers.

Detailed operation of the DB2 pump can be found in Chapters 22, 23 and 24.

DB2 Injection Pump—Electronic Controls

To this time, control of the fuel delivery in the Stanadyne line of fuel injection pumps has been done through the use of a mechanical metering valve. The metering valve in turn was controlled directly by the foot throttle or pedal. The mechanical governor was also connected to this same metering valve, as can be seen in Figure 22–4.

For several years now, fuel injection pump manufacturers have been researching and developing the application of electronics to the diesel engine fuel injection system.

In 1981, Stanadyne and Motorola's Automotive Products Division entered into an agreement to develop an electronic engine control system. The success of this venture is reflected in the fact that limited production of the electronically controlled injection pump will begin in the 1985 model year with full production scheduled for the 1986 model year.

The use of an electronically controlled fuel injection system allows the diesel engine to meet the current and future stringent exhaust emissions standards as per the Environmental Protection Agency legislation. In addition, the use of electronic controls will improve the diesel engine's fuel economy and also provide better overall engine performance. This electronic control system will be known as a "full authority" system, although Stanadyne will also offer a simpler system with less capability than the full authority system.

The major functions of the full authority system will be as follows:

1. Maximum part-load fuel shaping (delivery characteristics of the injection pump)
2. Altitude fuel compensation (decrease in fuel delivery with an increase in altitude due to lower atmospheric pressure)
3. Turbocharger boost compensation (fuel delivery increase related to an increase in pressure and air flow from the turbocharger)
4. Excess fuel delivery control
5. Transient fuel trimming (during acceleration/deceleration)
6. Throttle progression tailoring (specific control of fuel delivery with an increase in throttle position)
7. Governor control in the idle and maximum speed position or optional all-range governor control
8. Speed/load advance control
9. Cold start advance control
10. Altitude timing compensation
11. Pump installation error
12. Drive shaft wear compensation
13. Timing map flexibility

The electronically controlled injection pump will be known as a PCF pump (pump cam follower). A schematic diagram of the arrangement of the electronic control system used with the full authority system is shown in Figure 21–3. Major components of the system include a PCF (pump cam follower) electronic fuel control pump, engine-mounted sensors, and a CPU (central processing unit), which is based on Motorola's 6805 8-bit microprocessor.

The sensor inputs that are fed into the CPU are shown in the Electronic Fuel Injection Control block diagram in Figure 21–4. Sensor signals can either be fed directly to the CPU or multiplexed into an analog-to-digital converter.

For the start of injection, a Hall-effect sensor, mounted in an injector nozzle, is used to relay a volt-

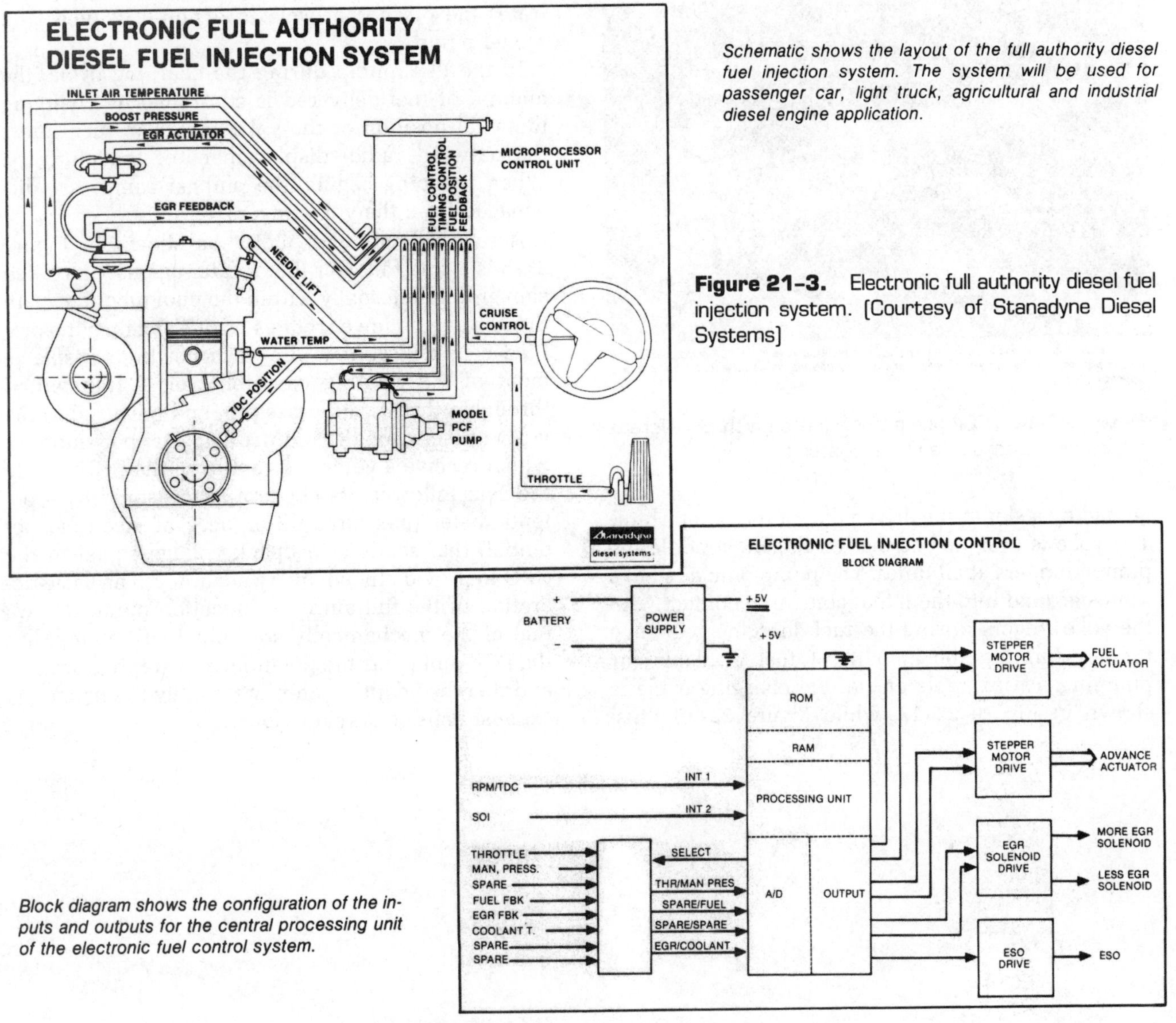

Schematic shows the layout of the full authority diesel fuel injection system. The system will be used for passenger car, light truck, agricultural and industrial diesel engine application.

Figure 21–3. Electronic full authority diesel fuel injection system. (Courtesy of Stanadyne Diesel Systems)

Block diagram shows the configuration of the inputs and outputs for the central processing unit of the electronic fuel control system.

Figure 21–4. EFI control block diagram. (Courtesy of Stanadyne Diesel Systems)

age signal back to the CPU. Operation of a Hall-effect sensor can be found by referring to Figure 7–48.

The CPU for the system is partitioned into a 3.8 K read-only memory, a 128-word random-access memory, an analog-to-digital converter, output rack, and processing unit. The CPU analyzes the various sensor inputs and calculates an error between the desired and actual fuel delivery. A voltage signal is then relayed to small stepper motors to wipe out the error signal. Small stepper motors are used to control the fuel and advance actuators as well as the EGR solenoid drive to increase or decrease the EGR setting and control the engine shutoff drive.

Electronic governing is used in place of the previous mechanical governor system on non-PCF pumps to provide a more compact and lighter assembly. Figure 21–5 illustrates the pump concept employed with the PCF (pump cam follower) system.

In the PCF system, a unique plunger control mechanism is used to control fuel delivery rather than using a mechanically controlled metering valve such as was used in earlier pumps (shown in Figure 22–4).

The PCF system is designed to limit the radial displacement of the pump plungers to effectively control fuel delivery. This is achieved by using a fingered yoke (shown in Figure 21–5) that is located and

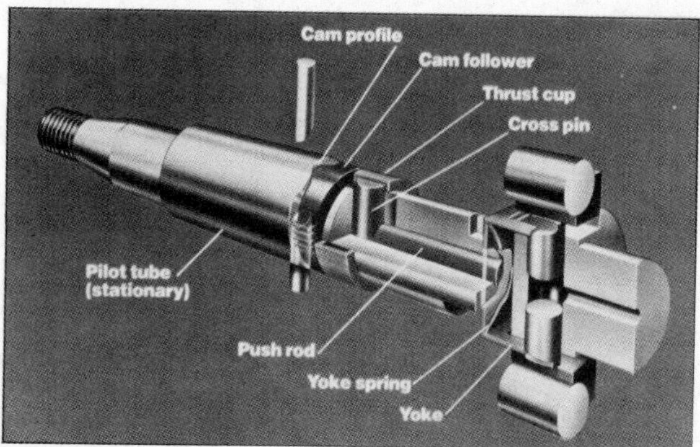

A novel plunger control mechanism is used in place of a metering valve in the PCF pump. Limiting the radial displacement of the plungers by means of a yoke controls the fuel available for the fuel charging sequence.

Figure 21-5. PCF pump plunger used with EFI system. (Courtesy of Stanadyne Diesel Systems)

guided by a slot at the driven end of the pump rotor. The yoke is designed so that its fingers straddle the pump plungers at all times. The pump plungers have ramps ground into them that come into contact with the yoke fingers during the fuel charging sequence thereby limiting the amount of fuel available for pumping. An example of the fuel charging cycle is shown in Figure 22–17, while Figure 22–18 illustrates the discharge cycle in a non-PCF type DB2 model pump.

In the PCF pump during the charging cycle, the amount of fuel delivered is controlled by changing the axial position of the yoke, which in turn limits the outward radial displacement of the plungers. When pumping begins, the plunger ramps are disengaged from the yoke fingers.

Assembled into the pilot tube shown in Figure 21–5 is a cam follower that rotates against a profiled cam ground internally within the pilot tube. The path that the cam follower rotates through against the profiled cam will control the yoke position. Displacement of the rotating cam follower is transmitted through a thrust cap, cross pin, and push rod to the yoke. Electronically controlled stepper motors, which receive a voltage signal from the CPU, rotate the cam follower; its position is measured by a potentiometer (measures differences of electrical potential) that sends a feedback voltage signal to the CPU to provide fuel delivery accuracy. The basic operation of the full authority injection pump follows that of the mechanically controlled DB2 pump, but the PCF pump has the potential to be reprogrammed and thereby adapt much more readily to changes in exhaust emissions regulations.

22

General Motors Corporation 4.3L V6 and 5.7L V8 Diesels

The 5.7L (350 cu.in.) Oldsmobile V8 diesel engine was designed as a possible alternative choice to the consumer and not as a replacement for the gasoline-powered 350 cu.in. engine that was used for many years.

To avoid lengthy delays and hundreds of millions of dollars required to design an automotive diesel engine option for their line of full-size vehicles, General Motors Corporation undertook, through the Oldsmobile Division, the conversion of an existing gasoline powerplant—the 350 cu.in. V8 engine.

Several reasons were considered for this approach: first, that Olds was experienced in the design and manufacture of large-bore rugged gasoline engines; second, corporate policy dictated that in order to meet the EPA CAFE (Corporate average fuel economy) requirements, GMC would have to be able to produce an engine that was capable of obtaining much better fuel economy than the existing large displacement gasoline V8's of that time, which were the powerplants of the full-size vehicles.

Third, GMC felt that the market was right for a diesel passenger car in North America that could be mass-produced fairly soon, that would meet the CAFE requirements, and that would still allow the consumer the choice of owning a full-size family sedan.

Extensive design and testing went into the final product, and the engine was ready for release in the 1977 model year cars and pickups. Since then, other GMC Divisions have offered the 5.7L diesel as an option in their product lines. Due to a variety of reasons attributable to the inherent different operating characteristics of a diesel engine, the 5.7L engine has had some troubled times. However, in its defense, many owners have received excellent performance results from the engine over many service miles.

At this time, the 5.7L diesel engine is still being offered in various GMC products with on-going improvements being a fact of life.

More recently, the Oldsmobile Division also offered a smaller diesel version—the 263 cu.in. 4.3L V6 unit, which shows the ability of gaining some improvement over its bigger brother, the 5.7L unit.

Another V8 diesel engine being offered by GMC is the 6.2L (378 cu.in.) engine that is manufactured by the Detroit Diesel Engine Division for use in pickup trucks, recreation vehicles, etc.: it is discussed in Chapter 23. This engine has been improved based upon the mistakes encountered with the 5.7L because reports on its performance seem to be very good.

The latest offering of the 5.7L engine shows it rated at 105 bhp at 3200 rpm versus its earlier rating of 135 bhp in the 1977 model year at 2800 rpm. The 4.3L V6 engine is rated at 85 bhp at 3600 rpm for the 1984 model year.

Although General Motors diesel car sales were off more than 50% in 1983 versus 1982, in 1984, diesel powerplants were offered in more of the company's car lines than in 1983 and also in more light trucks. General Motors obviously believes that there is potential in the domestic market for this type of engine, but more than likely, this will favor the pick up truck rather than the passenger car.

General Motors new C-body cars were designed to replace the regular Cadillac deVille and Fleetwood, Buick Electra, and Oldsmobile 98 models. These C-body cars offer front-wheel drive and a new 4-speed automatic transmission. GM also offers new front-wheel drive station wagons in the intermediate class, which are available with the diesel engine option and are being marketed by Chevrolet, Pontiac, Oldsmobile, and Buick Divisions. These vehicles will be available with the new 4.3L (260 cu.in.) V6 diesel powerplant. General Motors announced in late 1984 that it will drop the optional diesel engine in some 1986 mid-size, large and luxury car lines.

Both the 4.3L and 5.7L engines operate with the same basic fuel system layout, although minor differences do exist between them.

4.3L and 5.7L Diesel Improvements

To meet 1984 exhaust emissions regulations, a barometric pressure sensor was fitted to all GMC federal (meets all states except California emissions standards) and altitude automotive diesel engines. This unit will determine—when a vehicle is at altitudes above 4000 feet (1219 m)—at what point both the injection timing and the EGR calibration will be electronically adjusted to ensure optimum vehicle/engine driveability and continue to meet sea-level emissions standards.

The 1984, 4.3L V6 diesels sold into California will be equipped with an electronic emissions control system that uses a microprocessor to monitor throttle position, engine rpm, vehicle speed, and vacuum modulator output. This information will then be fed to the on-board computer system that will control the EGR and exhaust pressure regulator valves as well as the torque converter lockup clutch on automatic transmission equipped vehicles. The 5.7L diesel will not be available in California during the 1984 model year.

Any continued diesel engine use by General Motors in their vehicles will undoubtedly rely heavily upon electronic injection pump controls similar to those described in Chapter 21 and shown in Figure 21–3 and 21–4.

Another feature that has been added to improve the engine idle quality is that the opening (popping) pressure of the injector nozzles has been reduced. This allows a more even fuel flow and smoothness of operation. Product improvement has led to an injector design that is said to make injector nozzles self-cleaning as a result of a clearance increase between the injector needle valve and the body. This increased clearance allows a slight movement of the nozzle valve; in order to avoid a carbon build-up around the nozzle tip.

The glow plug setup for both the 4.3L and 5.7L engines has also been improved so that fewer components are required with the new system. The glow plugs are designed to shut off automatically if the engine is not started within a prescribed period of time; the earlier units would cycle on/off until the batteries were flat. In addition, a self-limiting feature also prevents the maximum temperature that the glow plugs can attain, thereby extending their service life.

General Information—4.3L/ 5.7L GMC Diesel Engines

The V6 4.3L (262 cu.in.) and V8 5.7L (350 cu.in.) diesel engines are of the same general design in that they are both 4-stroke cycle 90 degree V-type engines that also use the same basic fuel system.

Both the V6 and V8 engines employ the same bore and stroke, which are 103.05 mm (4.057") and 85.98 mm (3.385") respectively, making these engines oversquare in design. The compression ratio on both engines is 22.5:1.

On the V6 engine, the firing order is 1-6-5-4-3-2 with cylinders 2, 3, and 5 being on the left bank (front), while cylinders 2, 4, and 6 are on the right (rear) bank, in transversely mounted engines.

On the V8 engine, the firing order is 1-8-4-3-6-5-7-2 with cylinders 1, 3, 5, and 7 on the left bank and cylinders 2, 4, 6, and 8 on the right bank.

In both cases, the odd-numbered cylinders are on your right-hand side when you stand at the front of the engine. The exception to this is the gasoline V6 used in transverse mounted form in the newer front-wheel drive units.

Buick, Oldsmobile, and Pontiac V6 engines have cylinders 2-4-6 closest to the front of the vehicle, while Chevrolet has cylinders 1-3-5 closest to the front of the car. Compare this with the 2.8L gasoline engine firing order which is 1-2-3-4-5-6.

SPECIAL NOTE: The V6 diesel engine cylinder heads can be either cast iron or aluminum. Use a magnet to determine which type of material is used on a specific cylinder head. Immersing an aluminum cylinder head in a caustic cleaning solution at overhaul can damage the cylinder head.

Both the V6 and V8 engines are of the pre-combustion chamber design and therefore use glow plugs to facilitate starting. Current engines use 6-volt glow plugs, which are cycled on/off to receive 12-volt battery current. Glow plugs, when energized, glow cherry red somewhat like the cooking element on a kitchen stove.

General Motors recommends that all diesel engines in temperatures above 20°F (−7°C) should use No. 2 diesel fuel, while in temperatures below 20°F (−7°C), No. 1 diesel fuel is recommended. A blended fuel is available in certain areas of North America; this fuel combines both No. 1 and No. 2 diesel fuels and can be used where available.

No. 2 diesel fuel at temperatures below 5°F (−13°C) tends to have the paraffin wax crystals settle out in the fuel, which will restrict fuel flow and cause hard starting. For detailed information on diesel fuel characteristics, refer to Chapter 4, dealing with gasoline and diesel fuels.

The pre-combustion chambers in the cylinder head are made from stainless steel and are serviced separately from the head. These pre-combustion chambers are similar to those shown in Figure 23-3. The pre-combustion chamber inserts can be pushed out for servicing after cylinder head removal and after both the glow plugs and nozzles have been removed.

The basic difference between the V6 and V8 fuel systems can be seen in Figures 22–1 and 22–2.

The V6 engine uses an electric fuel lift pump mounted on the engine, while the V8 engine employs a diaphragm-type mechanical pump mounted on the right side of the engine and driven by an eccentric on the crankshaft (not the camshaft as on a gasoline GMC V8). Both of these pumps create approximately 5 psi (35 kPa) of fuel pressure.

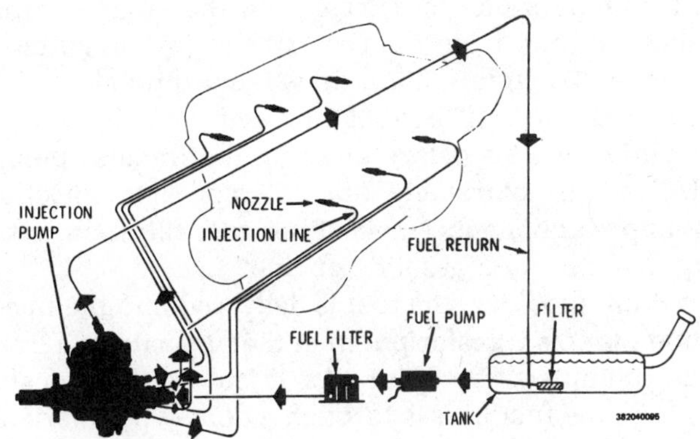

Figure 22–1. V6/4.3L Diesel engine basic fuel system circuit. (Courtesy of Oldsmobile Division, GMC)

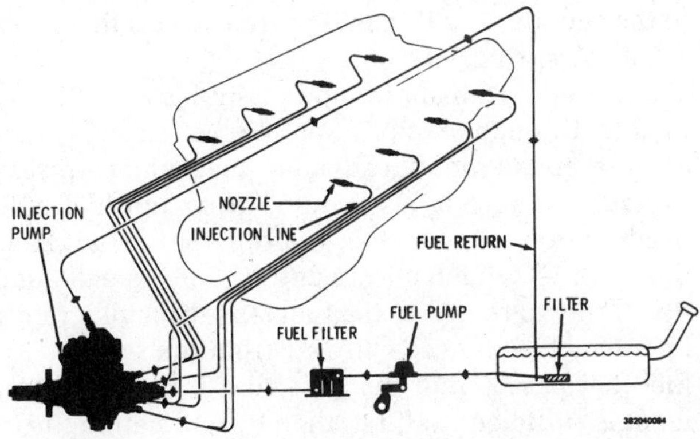

Figure 22–2. V8/5.7L Diesel engine basic fuel system circuit. (Courtesy of Oldsmobile Division, GMC)

The in-tank filter is a sock-type element (Figure 22–57) that contains a bypass valve arrangement that will open anytime that the filter sock is covered with fuel wax at low temperatures in order to allow fuel to flow to the fuel heater, described below.

A fuel filter is located on both engines between the fuel lift pump and the injection pump. Fuel that passes through the fuel filter is then directed into the suction side of the transfer pump, which is driven from the injection pump drive shaft. This pump is a vane-type unit that produces between 12–130 psi (83–896 kPa).

A fuel line heater is available as an option on engines that operate in cold weather situations because low ambient temperatures can cause fuel waxing to occur, which will prevent fuel flow through to the injection pump—and the engine will not run. Fuel waxing is discussed in detail under Chapter 4, dealing with fuel oils.

The optional fuel heater, shown in Figure 22–3, is an in-line heater that is a thermostatically controlled electrical resistance unit. This heater is placed into the fuel system so that all fuel must pass through it before it enters the engine fuel filter, otherwise plug-

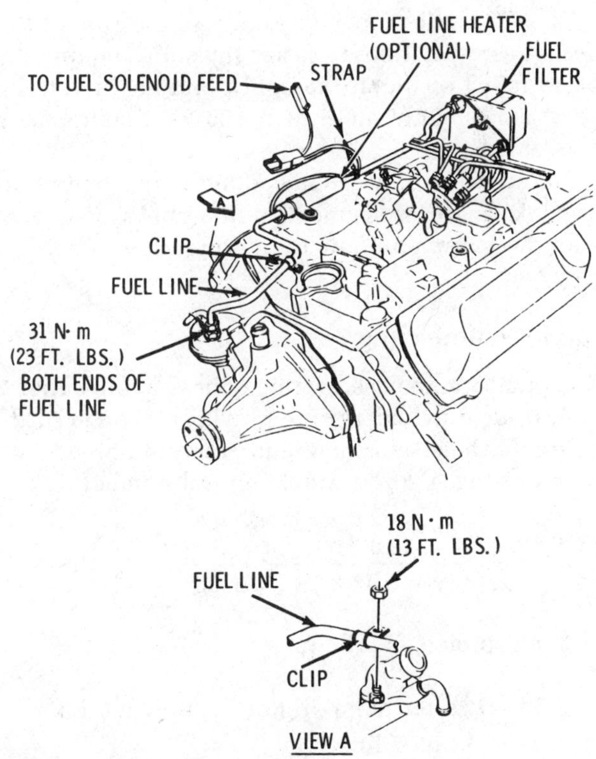

Figure 22–3. V8/5.7L Diesel engine fuel line arrangement. (Courtesy of Oldsmobile Division, GMC)

ging of the filter will occur in cold weather through waxing. A temperature sensor in the fuel system triggers (energizes) the fuel heater when the fuel temperature drops to −6°C (20°F) or lower.

Inside the heater is a spiral-wound strip wrapped around the fuel pipe controlled by a bimetallic (thermal) switch that senses fuel temperature. When the fuel temperature is approximately −5°C to +5°C (22°F to 41°F), the switch closes the electrical circuit when the ignition key is in the *run* position. The circuit to the fuel heater is automatically opened at 12–22°C (54–72°F) by the expansion of the bimetallic switch.

To check the operation of the fuel heater, always ensure that the fuel temperature is below −6°C (20°F), otherwise no current will flow to it.

Fuel Injection System

The fuel injection pump and pre-1980 engine injection nozzles were both manufactured by Roosa Master/Stanadyne.

Injection nozzles used since 1980 are the CAV Microjector, which is now manufactured under licence by the GM/DED (Diesel Equipment Division).

The DB2 injection pump follows the same basic design as other Roosa Master injection pumps. The basic injection pump has only four main rotating parts and slightly more than 100 total components altogether.

Before describing the injection pump's operation, let's take a look at the model numbering system used with the pump and how to interpret its symbols.

Model Numbering

The pump can be used with engines having either 2, 3, 4, 6, or 8 cylinders simply with a change to the hydraulic head assembly and some minor internal pump changes. For example, model number:

a	*b*	*c*	*3*	*e*
DB2	8	33	JN	3000

is interpreted as follows:

a: DB2—*D* series pump, *B*-Rotor, 2nd generation.

b: 8, number of cylinders.

c: 33, abbreviation of plunger diameter, e.g., 25, 0.250″ (6.35 mm); 27, 0.270″ (6.86 mm); 29, 0.290″ (7.37 mm); 31, 0.310″ (7.87 mm); *33*, 0.330″ (8.38 mm); 35, 0.350″ (8.89 mm).

d: JN, accesory code that relates to a variety of special pump options such as electrical shutoff, automatic advance, and variable speed droop adjustment.

e: 3000, specification number that determines the selection of parts and adjustments for a particular pump application.

Fuel System Operation

The model DB2 injection pump's operation can be likened to that of an ignition distributor. However, instead of the ignition rotor distributing the high-voltage spark to each cylinder in firing order sequence through the distributor cap, the DB2 pump distributes pressurized diesel fuel as two passages align during the rotation of the pump rotor, also in firing order sequence. Figure 22–4 shows the fuel system layout as it pertains to the Oldmobile 350 cu.in. (5.7L) diesel engine.

The basic fuel system flow is as follows: The fuel lift pump, which is driven from the engine crankshaft and not from the camshaft as on the gasoline engine 350 cu.in. (5.7L), draws fuel through a filter contained within the fuel tank and sends it on to an additional filter before entering the transfer pump. Due to the extremely fine tolerances of injection pump components, much finer fuel filtration is required than in a gasoline engine.

From the filter, the fuel is delivered into the injection pump's transfer pump at the opposite end from the pump's drive shaft. As it enters the transfer pump, the fuel passes through a cone-type filter and on into the injection pump's hydraulic head assembly.

Fuel under pressure is also directed against a pressure regulator assembly, where it is bypassed back to the suction side if the pressure exceeds that of the regulator spring.

Fuel under transfer pump pressure is also directed to and through a ball check valve assembly and against an automatic advance mechanism piston. Pressurized fuel is also routed from the hydraulic head to a vent passage leading to the governor linkage area, which allows any air and a small quantity of fuel to return to the fuel tank through a return line, which self-bleeds any air from the system. The fuel that passes into the governor linkage compartment is sufficient to fill it, thereby lubricating the internal parts.

Most of the fuel leaving the hydraulic head is di-

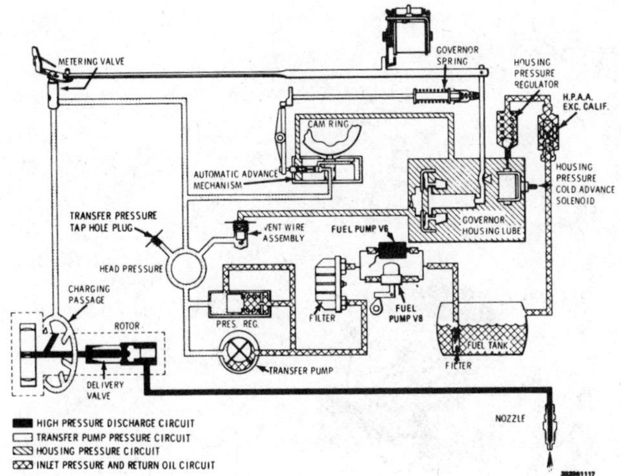

Figure 22–4. 1984 and Later fuel injection pump circuit. [Courtesy of Oldsmobile Division, GMC]

rected to a fuel metering valve, which is controlled by operator throttle position and governor action. This valve controls the amount of fuel that will be allowed to flow on into the charging ring and its ports. Rotation of the rotor by the pump's drive shaft aligns the two inlet passages of the rotor with the charging ports of the charging ring, thereby allowing fuel to flow on into the pumping chamber.

The pumping chamber consists of a circular cam ring, two rollers, and two plungers (see Figure 22–17 and 18). As the rotor continues to turn, the rotor's inlet passages will move away from the charging ports, allowing fuel to be discharged as the rotor registers with one of the hydraulic head outlets. With the discharge port open, both rollers come in contact with the cam ring lobes, which forces them toward one another, causing the plungers to pressurize the fuel between them and sending the fuel on up to the injection nozzle and into the combustion chamber.

The nozzle contains a spring loaded check valve that is held on its seat until fuel pressure is high enough to lift it from its seat against this spring pressure. When this occurs, fuel will be injected through holes in the end of the nozzle. When the fuel pressure drops off because of the rotor rotation past the cam ring lobes, the rollers and plungers move outward, effectively reducing this pressure. This reduction in pressure allows the spring inside the injection nozzle to force the valve back onto its seat, effectively cutting off any further injection of fuel.

Automatic advancement is accomplished in the pump by fuel pressure acting against a piston, which causes rotation of the cam ring, thereby aligning the fuel passages in the pump sooner. Its action can be compared to the vacuum and weight advance of the breaker plate inside an ignition distributor. See Figure 22–26 for more detailed information on this aspect. Maximum engine rpm is accomplished through a mechanical governor that affects the position of the fuel metering valve; this is covered later in the chapter.

The injection pump assembly shown in Figure 22–5 depicts the general layout of the relative parts. The drive shaft engages with and drives the distributor rotor in the hydraulic head. Also contained in the hydraulic head in which the rotor revolves are the metering valve bore, the charging ports, and the head discharge fittings, which have the high pressure steel injection lines connected to them.

Detailed Description of Components in Figure 22–5. The listed components have the following function:

1 – The minimum/maximum governor is basically a limiting speed governor assembly; it is designed to control the engine's low idle rpm and also control the maximum speed of the engine. In between these two speed ranges, the operator/driver selects a throttle position that corresponds to the demands placed upon the vehicle. A single-piece throttle shaft is used to actuate the assembly.

2 – The electric (12-volt) fuel shutoff solenoid is used to stop the flow of fuel and thereby shut down the engine by rotating a shutoff lever that will cause the fuel metering valve to block fuel flow from the vane-type transfer pump into the hydraulic head charging passages.

3 – During starting and warm-up periods, the injection pump housing pressure is reduced to provide a greater degree of timing advance by the use of the housing pressure cold advance solenoid.

4 – The fuel metering valve is designed with a helix, or scroll, on its body to control the volume of fuel delivery; a cantilevered spring is used to hold the metering valve against the guide stud and prevent vertical vibration of the valve in its bore.

5 – Some DB model pumps used a vent wire in the head area to allow any air in the fuel pump housing to vent back to the fuel tank; newer model pumps employ the air vent shown in the diagram, which is only accessible by removing the governor cover.

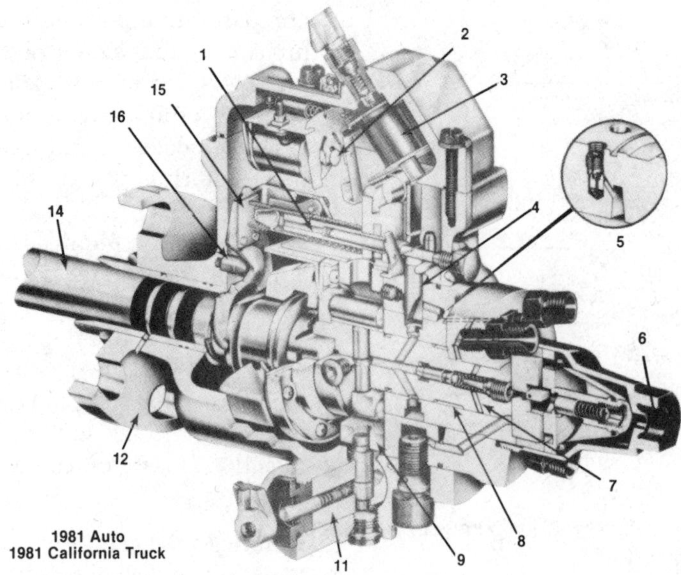

Figure 22-5. DB2 Injection pump assembly component identification. (Courtesy of Oldsmobile Division, GMC)

6 – The fuel transfer pump is a multi-vane-type unit, and the end cap employs a 7/16″ × 20 inverted male flare inlet. The regulating piston stop bushing replaces the seal in the pressure regulator on vehicle engines.

7 – A series of small vent ports (shown in Figure 22–6) incorporated in the rotor lead to the delivery valve spring chamber and simultaneously register with the head outlets after each injection period so as to balance the residual pressure variations between injection lines.

8 – A thermal relief groove is designed into the hydraulic head rotor unit as shown in Figure 22–7. A reduced diameter at the rotor area between the ports is used to ensure that sufficient clearance exists during cold, high speed acceleration situations. Head-to-rotor seizure can occur otherwise.

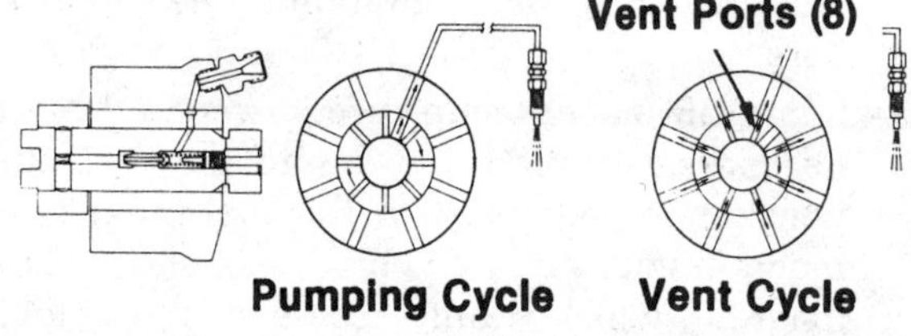

Figure 22-6. DB2 Pump residual pressure balancing rotor vent ports. (Courtesy of Stanadyne Diesel Systems)

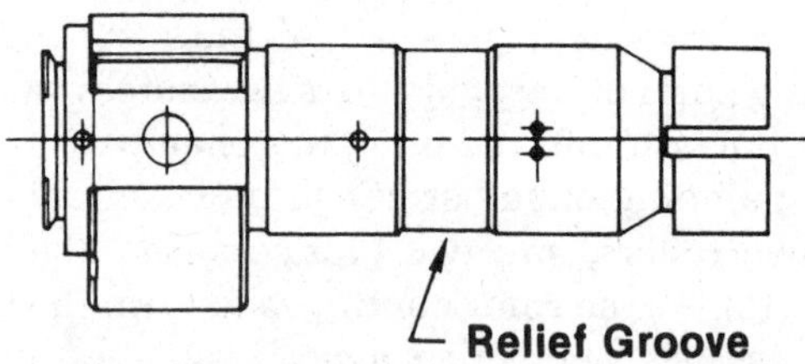

Figure 22-7. DB2 Hydraulic head rotor thermal relief groove. (Courtesy of Stanadyne Diesel Systems)

9 – The cam ring is made of high strength sintered metal. An axial ID groove is located on one face of the ring.

11 – To reduce exhaust emissions at low throttle openings, the injection timing is advanced by the use of the mechanical/light load advance unit.

12 – The injection pump housing mounting flange employs three cast-in bosses on the right-hand side for mounting a vacuum-operated transmission modulator (automatic transmission equipped vehicles). In addition, a dynamic timing mark appears on the flange that is used to align the pump with a stationary engine timing mark.

14 – 1981 and later model DB2 automotive pumps used an O-ring, which seats in a groove on the drive shaft.

CAUTION: Figure 22–8 illustrates the location of the O-ring seal. When pressure testing the injection pump assembly with air for signs of leaks, the drive shaft must be held in place mechanically to prevent possible serious personal injury.

15 – To ensure correct engine idle speed as the ambient temperature changes, a bimetallic strip is attached to the governor arm.

16 – 1981 and later DB2 automotive pumps use a ball-pivot governor arm cast into the housing rather than the earlier governor arm pivot shaft. These newer housings are identifiable by the lack of holes for a through pivot shaft.

DB2 Automotive Pump Minimum/Maximum Governor Operation

The DB2 injection pump is available for a variety of applications that may require either an all-range variable speed governor or the more commonly used limiting speed (minimum/maximum) governor found on automotive applications. The minimum/maximum speed automotive governor is designed to control the "low idle" speed of the engine and the "maximum" speed of the engine—which is also referred to as either "maximum engine speed cutoff" or "high idle" in some cases. Both terms mean the same.

In the speed range between low and high idle, the driver of the vehicle is the "governor." He/she manipulates the throttle to establish a given engine speed for a given condition. This is accomplished by the use of two different springs on the governor capsule; one spring controls low idle speed while the other spring controls high idle speed.

The linkage arrangement and its reaction under various operating conditions is shown in Figures 22–9 through 22–12.

Low Idle Speed Operation

The basic operation of a governor was explained in detail in Chapter 11, "Basic Governor Operation." In

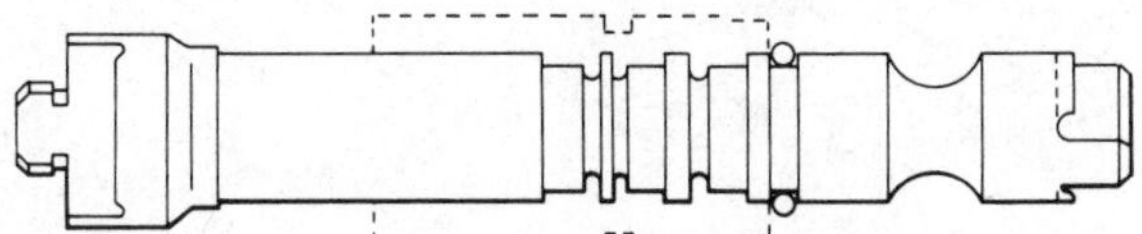

Figure 22–8. DB2 O-Ring retained injection pump drive shaft. (Courtesy of Stanadyne Diesel Systems)

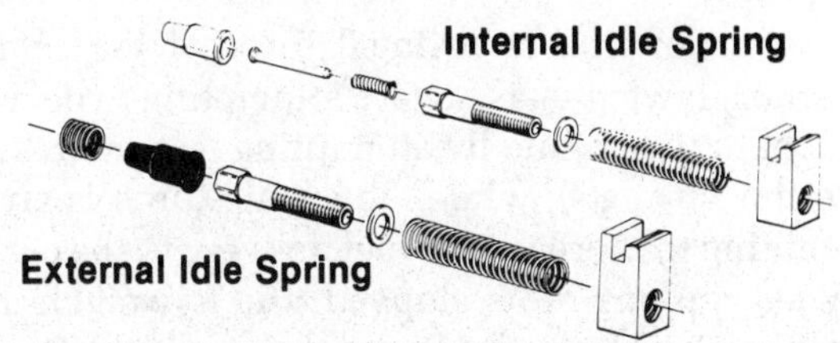

Figure 22–9. DB2 Low/high idle governor springs (two); available as either an internal or an external mounting. (Courtesy of Stanadyne Diesel Systems)

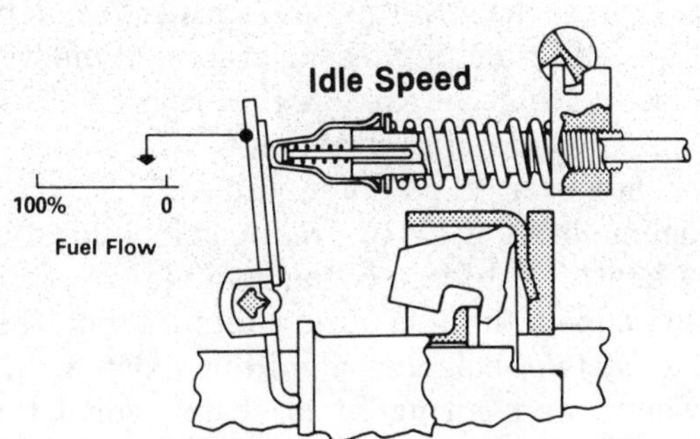

Figure 22–10. Governor reaction during normal idle speed. (Courtesy of Stanadyne Diesel Systems)

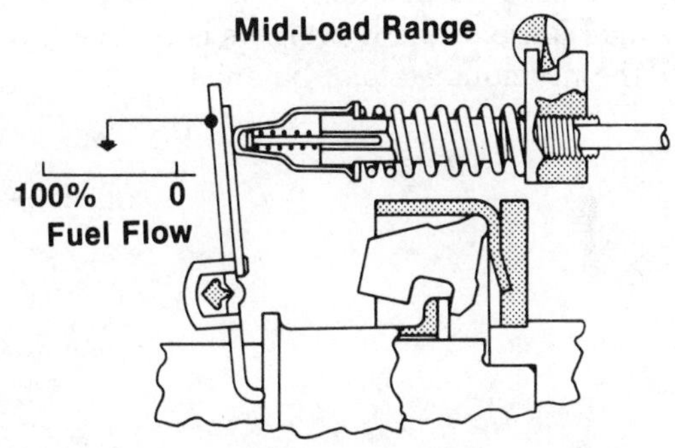

Figure 22–11. Governor mechanism position at half-throttle. (Courtesy of Stanadyne Diesel Systems)

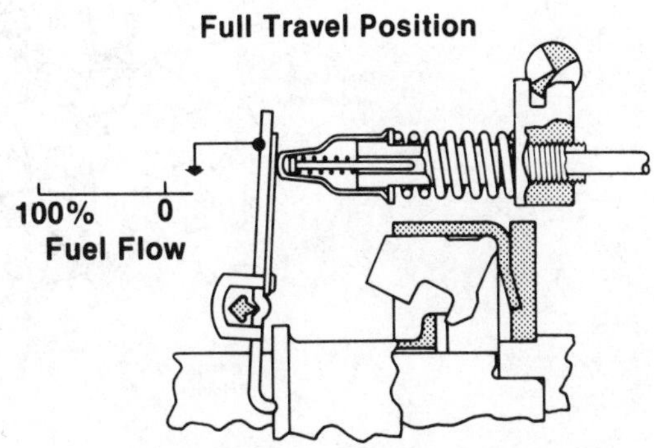

Figure 22–12. Governor mechanism positon at wide open throttle. (Courtesy of Stanadyne Diesel Systems)

any governor, the centrifugal force of the rotating governor flyweights is always attempting to decrease the fuel to the engine by attempting to compress the governor spring(s), while the spring force is always attempting to increase the fuel delivery to the engine.

When the engine is stopped, the flyweights have no force; therefore, the force of the idle spring will act upon the governor linkage to place the fuel control linkage towards a full-fuel position (see Figure 22-13). As the engine is cranked over on the starter (200–240 rpm), the weights are attempting to fly out and compress the governor idle spring. When the engine fires and runs, rapid acceleration of the weights will occur. This weight force working against the force of the idle spring attempts to compress the spring; however, in so doing, the weight force pulls the fuel metering valve to a reduced fuel position until the centrifugal force of the rotating governor flyweights equals the idle spring force. When this occurs, a "state-of-balance" condition exists with both the weight and spring forces being equal (Figure 22–10).

If, during the idle range, a load was applied to the engine—such as an air conditioner coming on—the tendency would be for the engine to slow down due to its fixed throttle position. When this occurs, the centrifugal force of the flyweights is reduced, which upsets the previous state of balance.

As the weights slow down, the force of the idle spring will be greater and it will push the fuel control linkage towards more fuel. As the spring expands, it loses some of its force; therefore, as the engine receives more fuel, the rotating flyweights will once again reach a state-of-balance condition and the engine is prevented from stalling.

If a load is taken off the engine at this time, the speed will increase for a fixed throttle position. The state of balance is upset again, only this time in favor of the weights because of the increase in speed. As the weights fly out, they compress the idle spring and also pull the fuel control linkage and metering valve to a decreased fuel position. With less fuel, the engine slows down until once again the weight and spring forces are equal. When the weights flew out and compressed the idle spring, they increased its compressive force; therefore, as the engine speed is reduced, this spring force balances out with the weights.

Intermediate Speed Range

When the engine is accelerated by the driver pushing the throttle pedal down, the centrifugal force of the flyweights will increase. As the weights move out, they fully compress the idle spring. In Figure 22–11, the engine is at about a half-throttle position.

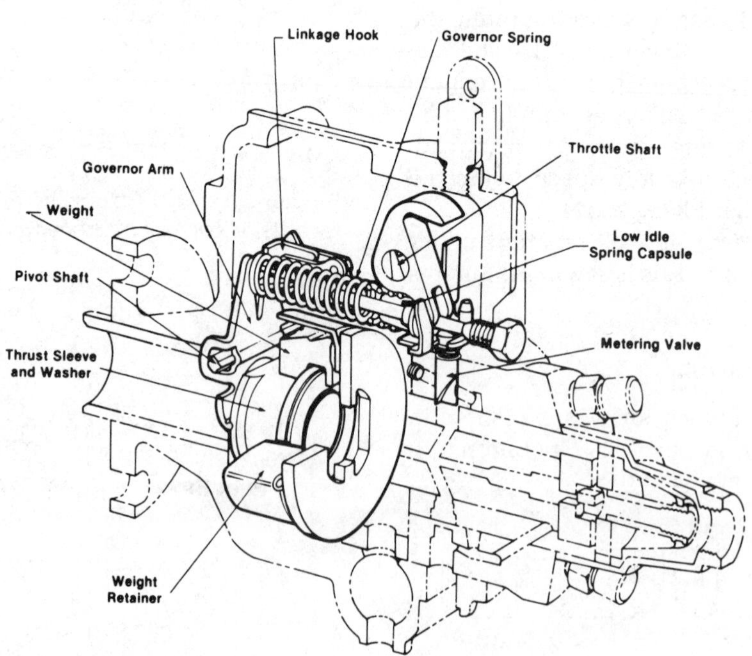

Figure 22–13. Governor linkage setup. (Courtesy of Stanadyne Diesel Systems)

The force of the governor high speed spring is too strong for the rotating weights to compress until the engine speed approaches the maximum setting; therefore, in the intermediate speed range such as half-throttle, the governor capsule assembly acts as a solid link against the governor arm. In this situation, the driver can control the position of the fuel metering valve throughout the speed range between low and high idle.

Governor Full-Speed Position

There are two maximum speeds in any governor, namely its no-load rpm and its full-load rpm, unless the governor is capable of a zero droop condition (isochronous). With no-load on the engine, the engine will run faster than it will when it is fully loaded because it doesn't have to work as hard and therefore requires less fuel at no-load speed.

When the throttle is opened to its maximum no-load position (Figure 22–12), the centrifugal force of the rotating governor flyweights will be greater at this higher speed than they would be at its full-load speed. The higher weight force is therefore capable of compressing the high speed spring. When this happens, the fuel control linkage of the governor will pull the fuel metering valve to a reduced fuel flow position thereby limiting the maximum no-load rpm of the engine to prevent damage and runaway.

When a load is applied to the engine, such as when climbing a hill, the state-of-balance condition between the weights and the high speed spring is disturbed in favor of the spring, which will increase the fuel position of the metering valve until a state of balance is re-established as explained for the low idle situation of load increase.

If the engine can develop enough additional horsepower to offset the increased load, then the vehicle will be able to climb the hill at a slightly slower speed without a downshift having to be made. If however, the additional fuel reaches a point where the engine is now at its full-load rpm, no further horsepower can be developed because no more fuel is available. The engine and vehicle speed will drop if this is the case, and a downshift will have to be made if the increasing engine torque caused by the reduction in engine speed is insufficient to keep the engine crankshaft turning at a high enough speed. If the torque increase is enough, the vehicle will negotiate the hill but at a reduced road speed.

If a downshift is made, either in an automatic or standard shift transmission, then the engine rpm is allowed to rise and the vehicle can move up the hill but at a slower road speed because of the lower overall gear ratio.

If the engine rpm is allowed to drop low enough, the weight force will be insufficient to oppose the high speed spring and the driver will have to manipulate the throttle and gearing to suit the operating conditions.

If a load is removed from the engine, such as when the vehicle goes down a hill, then the engine speed will increase for a fixed throttle position. This increased rpm will cause the weight force to become greater than the high speed spring force, and the fuel control linkage will be moved towards a decreased fuel position. The vehicle road speed will have to be controlled with the use of the vehicle brakes; otherwise, in a standard transmission or lock-up torque converter automatic transmission, a direct power path exists back up the driveline to the engine.

If the road wheels become the driving member, then engine overspeed could result, and engine damage would occur such as valve float—where the valves would strike the piston crown.

Components of Fuel System

Fuel Transfer Pump

The delivery capacity of the fuel transfer pump is capable of exceeding both the pressure and volume requirements of the engine, with both varying in proportion to engine speed. Excess fuel can, however, be recirculated back to the inlet or suction side of the pump by the regulator.

Figure 22–14 shows the basic component parts of the rotor and transfer pump, which is a positive displacement vane-style unit consisting of a stationary liner with spring-loaded blades that ride in slots of

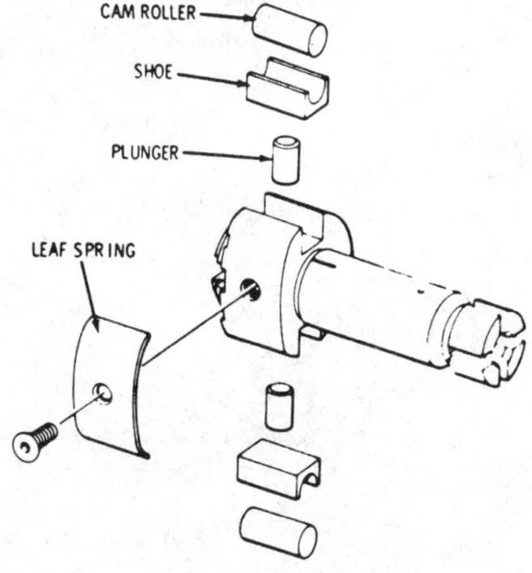

Figure 22–14. Component parts of the rotor and transfer pump. (Courtesy of Roosa Master, Stanadyne/Hartford Diesel Systems)

the rotor. The cam rollers move in or out with the rotor's rotation, causing the shoe to force the plungers inward when the lobe of the internal cam ring makes contact with them simultaneously. The leaf spring controls the maximum plunger travel and therefore the maximum fuel delivery, since it limits the travel of the roller shoes. Fuel pressure fed into the area between both plungers comes from the rotor to the charging ring, where it then enters the plunger chamber. Since fuel pressure and volume are proportional to engine speed, the plungers will reach their maximum outward travel only when the engine is running under full-load conditions.

Transfer pump pressure is adjusted with the pump on a test stand by increasing or decreasing the pressure regulator spring tension by clockwise or counterclockwise rotation of the spring plug. Maximum top end pressure should never exceed 130 psi (896.35 kPa).

The fuel pressure delivered to the plungers within the injection pump assembly comes from the vanes of the transfer pump riding in four slots at the opposite end from the plungers and rollers. The right-hand end of the rotor shown in Figure 22–14 shows these slots. Figure 22–15 depicts the action of the vanes as the rotor completes one full revolution of its cycle.

In Figure 22–15, as the rotor turns within the eccentric liner, the vanes riding in the ends of the rotor slots will be forced to move in and out. In position A of the diagram, vanes 1 and 2 are in a central position between the inlet port and are drawing fuel

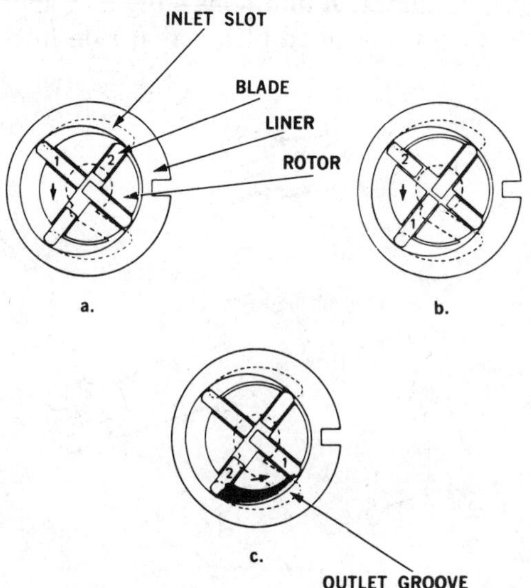

Figure 22–15. Vane action as rotor completes one full revolution. (Courtesy of Stanadyne Diesel Systems)

into the area formed between them. Moving to sequence B, we see that vanes 1 and 2 have now passed completely the inlet port of the liner, thereby trapping and carrying around with them the volume of fuel that was drawn in at position A.

As the rotor continues to turn, the vanes will be forced back into their slots because of the shape of the eccentric liner bore. Position C shows that owing to the ever-reducing space between vanes 1 and 2, the volume of fuel that was drawn in initially at point A has been forced into this smaller area, causing an increase in fuel pressure, which will leave the outlet groove in proportion to engine speed.

The pressurized fuel is directed through the groove of the pressure regulator assembly, past the rotor retainers, and into an area on the rotor leading to the hydraulic head passages. The fuel volume will continue to decrease at the vanes until blade 2 passes the outlet groove in the liner assembly.

Pressure Regulator Assembly

The action of the pressure regulator is the same as any pressure regulator, whether it be in a fuel, lube, or hydraulic system, that is, to limit the maximum pressure developed in the system by relieving or by-passing fuel back to the low-pressure or suction side of the system. Obviously, then, the only time that the regulating piston assembly will operate is when the engine is running. Remember from earlier discussion that the pump output will vary with engine speed.

Figure 22–16 shows the action of the regulator assembly. Fuel under pressure leaving the transfer pump is directed against the pressure regulator piston, which applies pressure to the spring in behind the piston. As the engine speed increases, so do pump volume and pressure, which will force the regulating piston back until it starts to uncover the pressure regulating slot, therefore limiting and controlling the maximum pump delivery pressure. Some regulators incorporate what is known as a high-pressure relief slot, which simply prevents excessively high transfer pump pressure if the engine is forced into an overspeed condition such as when on a vehicle.

Fuel Viscosity Compensation

Since different grades of diesel fuel may be used with the pump and temperature variations are bound to occur in different geographical locations, the pump employs a rather unique feature within the regulating system just described that offsets pressure changes due to these two factors. A thin plate con-

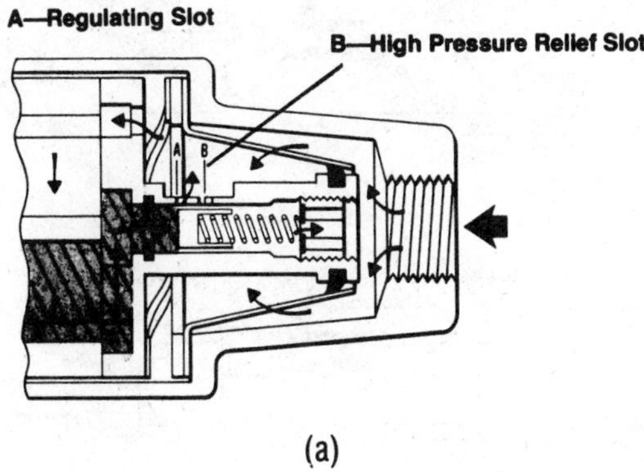

(a)

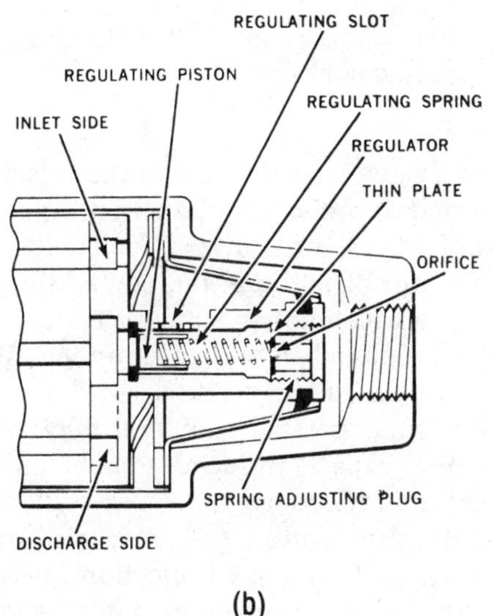

(b)

Figure 22-16. Action of fuel pressure regulator assembly. [Courtesy of Stanadyne Diesel Systems]

tained in the spring adjusting plug incorporates a sharp-edged orifice that allows fuel leakage that passes between the piston and its bore to return to the suction side of the transfer pump. The rate of fuel flow through any short orifice is so small with a change in fluid temperature that the pressure exerted against the back side of the regulating piston will be controlled by the fuel leakage past the piston and its bore and the resulting pressure drop through the sharp-edged orifice.

When the ambient air temperature is low, the fuel leakage past the regulating piston will be very small indeed; therefore, any pressure on the back side of the piston will be extremely low. However, if the ambient air temperature is high and the engine is running at its normal operating temperature, the viscosity of the fuel oil will be less, allowing a higher rate of fuel leakage past the piston. This will increase the fuel pressure in behind the regulating piston within the spring cavity and therefore fuel flow through the orifice.

The change in pressure behind the piston will therefore vary the piston's position as it reacts to the transfer pump's outlet pressure normally acting upon it at the front end. This variation in regulating piston position will compensate or offset the fuel leakage that would occur as the fuel thins out, and in this way, the fuel transfer pump pressure can be maintained over a broad range of fuel temperature changes.

Charging and Discharging Cycle

Charging Cycle. Rotation of the rotor allows both inlet passages drilled within it to register with the circular charging passage ports. The position of the fuel metering valve connected to the governor linkage controls the flow of fuel into the pumping chamber and therefore how far apart the two plungers will be. The maximum plunger travel is controlled by the single leaf spring, which contacts the edge of the roller shoes. Maximum outward movement of the plungers will therefore only occur under full-load conditions. Figure 22–17 shows the fuel flow during the charging cycle. Any time that the angled inlet fuel passages of the rotor are in alignment with the ports in the circular passage, the rotor discharge port is not in registry with a hydraulic head outlet and the rollers are also off the cam lobes.

Discharging Cycle. The actual start of injection will vary with engine speed since the cam ring is automatically advanced by fuel pressure acting through linkage against it. Therefore, as the rotor turns, the

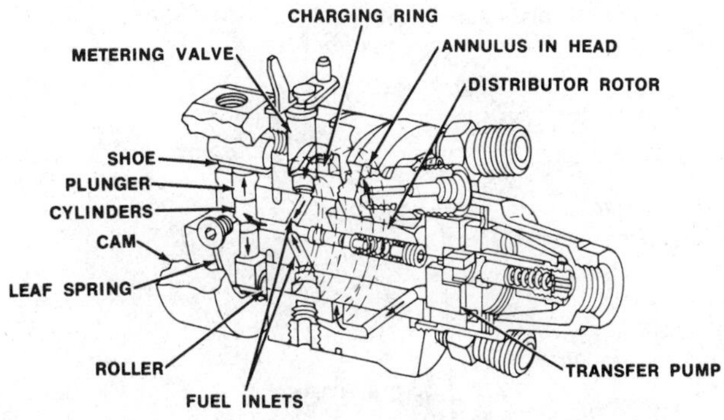

Figure 22-17. Fuel flow during charging cycle. [Courtesy of Stanadyne Diesel Systems]

angled inlet passages of the rotor move away from the charging ports. As this happens, the discharge port of the rotor opens to one of the hydraulic head outlets.

Also at this time, the rollers make contact with the lobes of the cam ring, forcing the shoes and plungers inward and thus creating high fuel pressure. The fuel flows through the axial passage of the rotor and discharge port to the injection line and injector. This fuel delivery will continue until the rollers pass the innermost point of the cam lobe, after which they start to move outward, thereby rapidly reducing the fuel pressure in the rotor's axial passage and simultaneously allowing spring pressure inside the injection nozzle to close the valve. The sequence of events is shown in Figure 22–18.

Delivery Valve Operation

To prevent after-dribble, and therefore unburnt fuel with some possible smoke at the exhaust, the end of injection, as with any high speed diesel, must occur crisply and rapidly. To ensure that the nozzle valve does in fact return to its seat as rapidly as possible, the delivery valve within the axial passage of the pump rotor will act to reduce injection line pressure after fuel injection to a value lower than that of the injector nozzle closing pressure.

From some of the views shown so far you will recollect that the delivery valve is located within the rotor's axial passageway. To understand its function more readily, refer to Figure 22–19. The delivery valve requires only a stop to control the amount that it can move within the rotor bore. No seals as such are required, owing to the close fit of the valve within its bore. With a distributor-type pump such as the DB2, each injector is supplied in firing order sequence from the axial passage of the rotor; therefore,

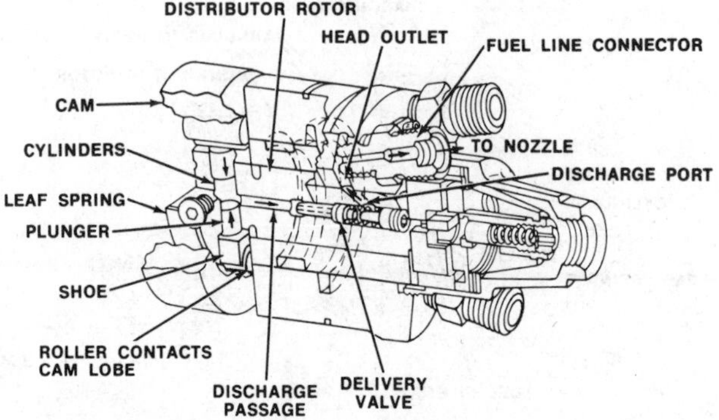

Figure 22–18. Fuel discharge cycle. (Courtesy of Stanadyne Diesel Systems)

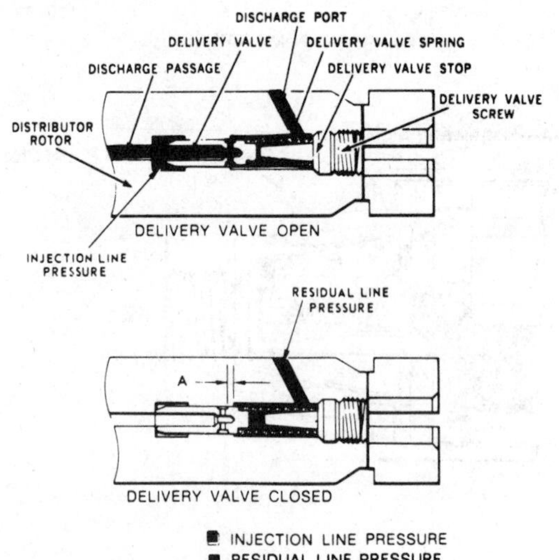

Figure 22–19. Delivery valve action. (Courtesy of Stanadyne Diesel Systems)

the delivery valve operates for all the injectors during the period approaching the end of injection.

In Figure 22–19, pressurized fuel will move the valve gently out of its bore, thereby adding the volume of its displacement as shown (section A) to the delivery valve chamber, which is under high pressure. As the cam rollers start to run down the lobe of the cam ring, pressure on the delivery valve's plunger side is rapidly reduced, thereby ending fuel injection at that cylinder.

Immediately thereafter, the rotor discharge port closes totally, and a residual injection line pressure is maintained. In summation, the delivery valve will seal only while the discharge port is open because the instant the port closes, residual line pressures are maintained by the seal existing between the close-fitting head and rotor.

Fuel Return Circuit

A small amount of fuel under pressure is vented into the governor linkage compartment. Flow into this area is controlled by a small vent wire that controls the volume of fuel returning to the fuel tank, thereby avoiding any undue fuel pressure loss. Figure 22–20 shows the location of the vent passage, which is behind the metering valve bore and leads to the governor compartment via a short vertical passage. The vent wire assembly is available in several sizes to control the amount of vented fuel being returned to the tank, its size being controlled by the pump's particular application. In normal operation, this vent wire should not be tampered with because it can be altered only by removal of the governor cover. The

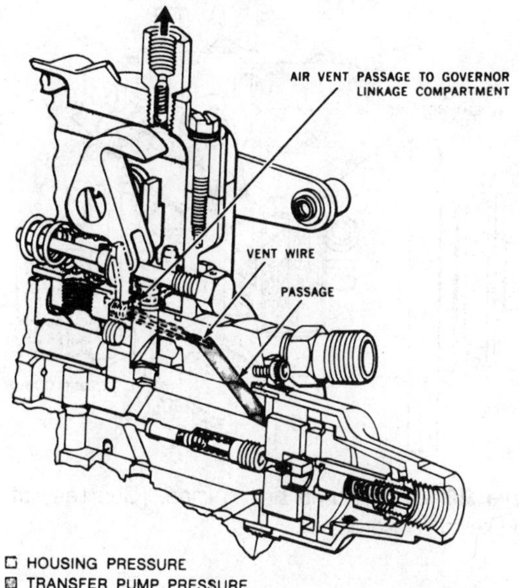

Figure 22-20. Location of air vent passage. (Courtesy of Stanadyne Diesel Systems)

correct wire size would be installed when the pump assembly is being flow tested on a pump calibration stand.

The vent wire passage, then, allows any air and a small amount of fuel to return to the fuel tank. Governor housing fuel pressure is maintained by a spring-loaded ballcheck return fitting in the governor cover of the pump.

Plunger Replacement

There are various roller shoe sizes for the DB2 pump. These are identified by the number on their end, from a minus 20 to a plus 10 in individual increments of 5.

It is imperative that all four shoes are of the same dimension for the particular application. This shoe size is determined by a part number only.

Always insert a cam roller by sliding it into the shoe from the end, not the top; otherwise, damage will result.

Another important point here is that if at any time plungers are to be replaced, note that etched on the head of the rotor assembly is a letter indicating the specific diameter of plunger to be used with the rotor.

For example, a plunger of 0.330″ (3.38-mm) diameter has a mated part number 11076, and the graded sizes, A through D, are identified by the part numbers 11077, 11078, 11079, and 11080, respectively. Therefore, the replacement plungers for a 0.330″ (8.38-mm)-diameter plunger with a pump rotor etched with the letter C would be P/N 11079. However, always refer to current specifications for part numbers.

You may come across some pumps with a −2 etched on the rotor following the letter grading, which indicates that the plunger bore is 0.002″ (0.05 mm) oversize.

Maximum Fuel Output Adjustment. Adjustment of maximum fuel delivery for all Rossa Master injection pumps must be done with the pump disassembled and mounted to the proper test fixture as shown in Figure 22–21. Of course the pump must then be reassembled, and with it mounted on a fuel injection pump test stand, its flow rate compared to specifications along with any other relevant adjustments.

For a detailed analysis and rebuild information, always refer to the manufacturer's operation and instruction manual for your particular model of injection pump.

Figure 22–21 shows a DB2 pump rotor and plunger assembly mounted to Roosa Master special test fixture no. 19969, which can be clamped into a vise on its flat sides.

The dial indicator can be mounted to its bracket, but slide it to its outer limit of travel for now (not in contact with the plunger roller).

In order to check the distance between each set of opposed rollers, they must be forced outward by the use of clean, dry shop air supplied by a line as shown. Maintain air pressure between 40 and 100 psi (276 and 689.5 kPa). Manually rotate the rotor until the

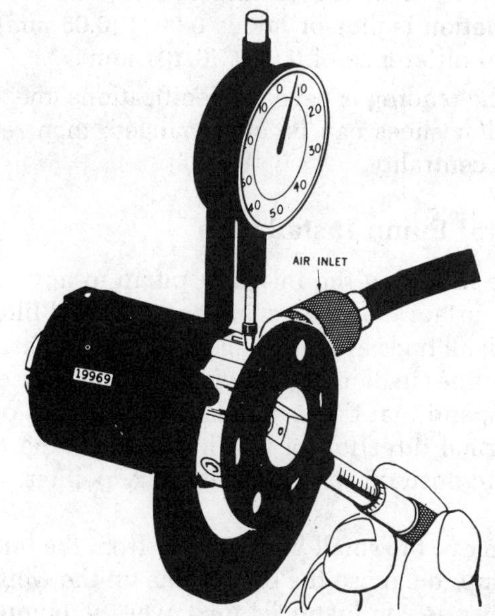

Figure 22-21. Checking the roller-to-roller clearance. (Courtesy of Stanadyne Diesel Systems)

rollers attain their maximum outward movement by butting up against the leaf spring.

Using a 1-" to 2-" (25.4- to 50.8-mm) micrometer, measure the distance between the crown of the opposed cam rollers and compare with specifications for your particular pump.

If the specification is incorrect, each leaf spring must be adjusted alternately by turning its adjusting screw inward (clockwise) to increase and outward (counterclockwise) to reduce the dimension.

It may be necessary to change out or invert the leaf springs in order to obtain the correct specification on each set of rollers.

The roller settings of each pair must be within 0.003" (0.076 mm).

To check that the rollers are concentric to each other, i.e., to make certain that each pair will start its pumping stroke at the same time, proceed as follows:

1. Manually turn the rotor around until any roller is directly under the dial indicator stem plunger. Loosen off the clamp on the bracket holding the dial indicator in position and gently slide the indicator down until the dial gauge is pre-loaded at least 0.010" (0.254 mm) on its face and lock the retaining screw of the bracket. Carefully rotate the knurled bezel or outer dial to zero in the dial indicator.
2. Manually rotate the rotor in either direction until the next roller comes into contact with the dial indicator plunger and note the reading on the dial face. Do this with all four rollers. The allowable variation is plus or minus 0.002" (0.05 mm) for a total difference of 0.004" (0.101 mm).
3. If the reading is beyond specifications, the rollers and/or shoes can be interchanged; then recheck the centrality.

General Pump Installation

Before installing the injection pump to any engine, make certain that the engine has new fuel filters installed; all lines are blown clean; mounting area and timing lines in damper, pulley, or flywheel are clearly visible; and that the engine has been turned over in its normal direction of rotation to place the timing mark in position with the stationery pointer.

1. Remove the small timing cover from the injection pump to expose the timing line on the weight retainer as you manually turn over the pump drive shaft. Make certain that the timing line is in alignment with the line on the cam ring outside diameter. See Figure 22–22.

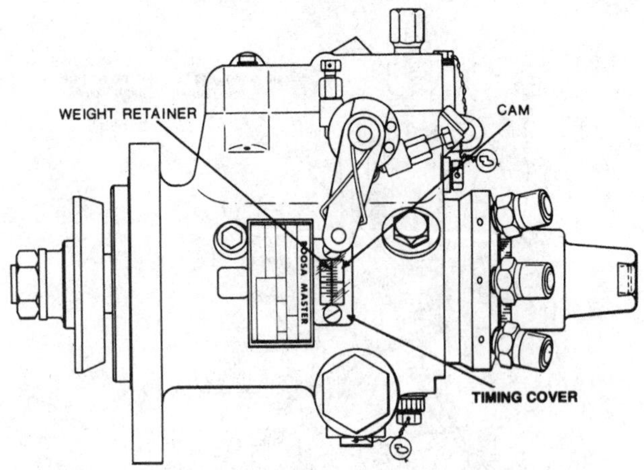

Figure 22–22. Pump timing lines. (Courtesy of Stanadyne Diesel Systems)

2. Slide the pump into its mated position on the engine; install the washers and nuts; and torque to specifications.
3. Depending on the engine, you may now have to install the pump drive gear to the hub or shaft.
4. To remove all gear backlash in the engine-to-pump drive train, bar the engine backward from its normal direction of rotation at least one half-turn; then slowly bring it forward until the timing lines appear opposite one another in the pump timing window. Check the marks on the engine damper, pulley, or flywheel. If incorrect, loosen the pump flange retaining nuts and turn the pump housing by firmly grasping it and twisting to advance or retard the timing. Which way it must be turned will depend on the rotation of the pump. However, after any correction, repeat the previous procedure to ensure that the pump is properly timed. Install the timing cover on the pump housing.
5. Remove all protective caps on all lines and connect their nuts to the correct location and torque to specifications. Leave them loose at the individual injectors for now. Doublecheck that the lines are connected to the proper injector as per the engine firing order. Connect all necessary linkages and fittings.
6. Open the secondary filter bleed screw if a hand priming pump is fitted before the filter itself, or prime the filters first if no hand priming pump is fitted, until a steady flow of fuel free of air bubbles appears; then tighten the transfer pump inlet line.
7. Crank over the engine until fuel drips from each injector line and tighten the nut of each in firing order sequence, after which time the engine

should fire and run. Let it idle for a few minutes while you check both lube and fuel pressures. If the engine runs rough, individually crack open the injector line nut to rebleed each one, and then tighten. If you have trouble starting the engine, you may have to rebleed the fuel system.

Housing Pressure Cold Advance Mechanism

Contained within the injection pump governor cover is a solenoid assembly that receives an electrical signal from a sensing unit mounted on one of the cylinder heads. This sensing unit is immersed in the engine coolant and will relay a voltage signal to the solenoid, depending upon engine coolant temperature. When the solenoid is energized, it provides more injection timing advance during cranking and engine warm-up situations.

Two basic conditions are shown in Figures 22–23 and 22–24.

In the cold advance position illustrated in Figure 22–23 the electrical solenoid (1) is energized by a voltage signal from the coolant sensor. This action causes the plunger (2) to move in the direction shown with the rod (3) contacting the return connector ball (4). With the ball lifted off its seat, the injection pump housing pressure is lowered because of the increased fuel flow through the connector. This lower fuel pressure allows the advance piston to move further towards the advance direction.

In Figure 22–24, the engine coolant temperature has achieved a predetermined level that causes the voltage signal to the solenoid to be cut off. With the solenoid now de-energized, the plunger (2) is returned to the position shown. The ballcheck valve is controlled by the return connector spring pressure, which seats the ball to increase the injection pump housing fuel pressure to normal. Normal injection advance operation is now maintained.

The injection pump automatic advance group components are shown in detail in Figure 22–25.

Automatic Timing Advance

Similar in function to the distributor breaker plate movement in a gasoline engine, the injection pump can also be equipped with a simple direct-acting advance mechanism, controlled by fuel pressure, that acts to rotate the cam ring and thereby vary engine pump timing (Figure 22–26). Fuel pressure from the transfer pump will vary in direct proportion to en-

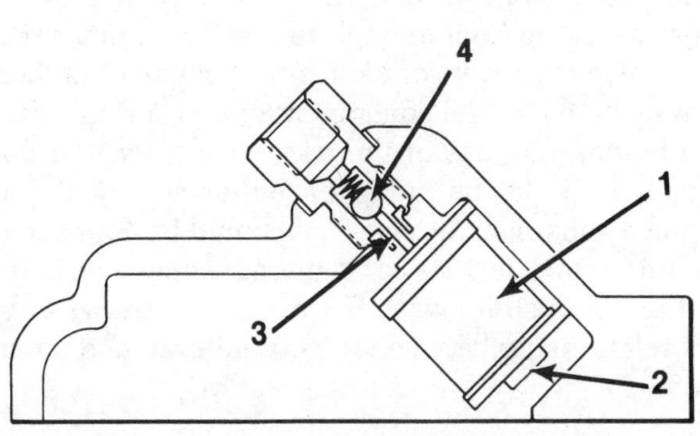

Figure 22-23. Housing pressure cold advance position. [Courtesy of Stanadyne Diesel Systems]

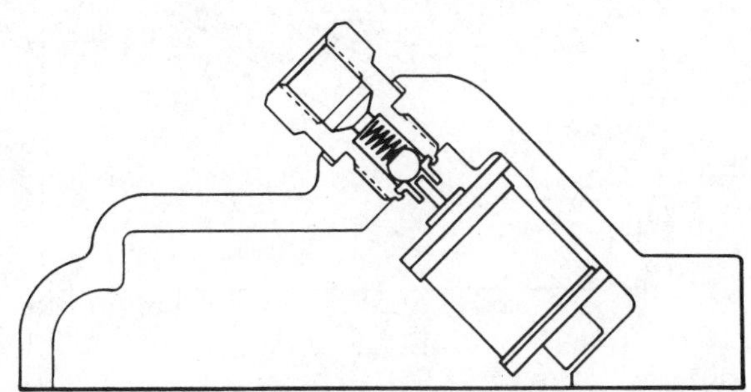

Figure 22-24. Housing pressure normal advance position. [Courtesy of Stanadyne Diesel Systems]

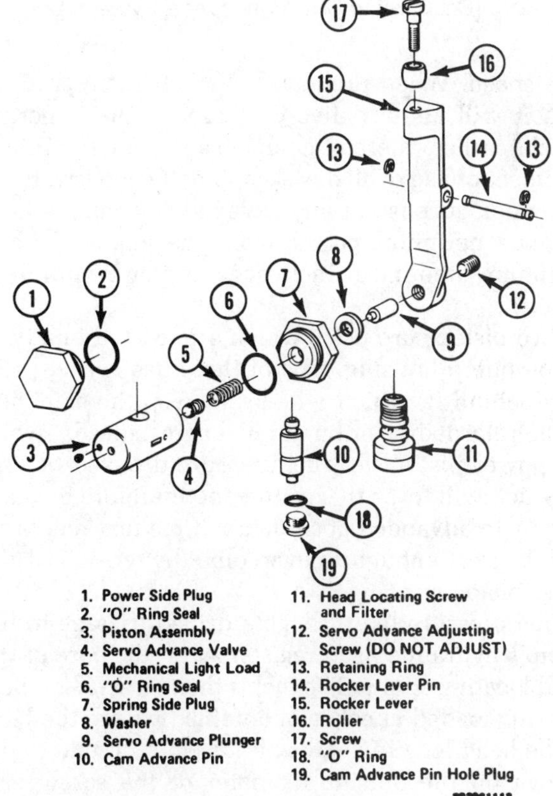

Figure 22-25. Injection pump automatic advance group. [Courtesy of Oldsmobile Division, GMC]

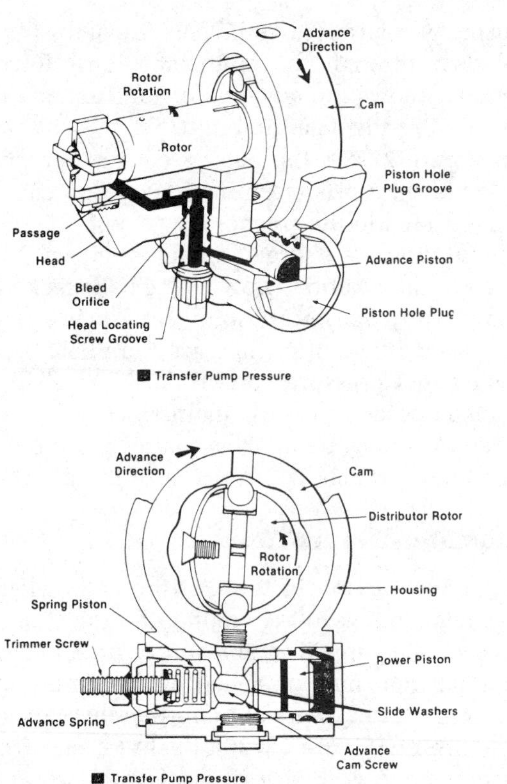

Figure 22-26. Automatic timing advance mechanism operation. (Courtesy of Stanadyne Diesel Systems)

gine speed. When the advance mechanism is in effect, it will allow delivery of fuel to the injection nozzle earlier, thereby compensating for the inherent injection lag of all diesel engines. Obviously, then, the major purpose of any advance mechanism is to produce optimum power from the engine with a minimum amount of fuel usage and also minimum smoke.

Two pistons are used, which are located in a bore in the pump housing. One of these has spring pressure behind it, which will act to keep the cam ring in the retarded position at low engine speeds, while the power piston has fuel pressure directed behind it, which will force this piston and therefore the cam ring to an advanced position any time that fuel pressure is great enough to overcome the tension of the other piston's spring.

Fuel pressure fed through a drilled passage in the pump's hydraulic head registers with the bore of the head locating screw. The fuel is then directed on past a spring-loaded checkball contained within the bore of the head locating screw, after which it enters the groove on the outside diameter of the screw and passes on up to the back side of the power piston of the automatic advance mechanism. In order for any timing advance to occur, the fuel pressure must be high enough to overcome not only the opposing piston's spring pressure but also the dynamic injection loading on the cam ring.

At low engine speeds, the start of injection timing will be controlled by the position that the injection pump was installed in and timed to the engine upon initial installation. The amount of actual advance will therefore be proportional to engine speed since fuel pressure is tied to this also. As the engine speed leaves the low range, fuel pressure will force the power piston to push against the advance pin, thereby causing relative movement of the cam ring toward the rotation of the moving rotor shaft. This simply means that injection will occur earlier in the stroke. The one-way check valve prevents the normal tendency of the cam ring to return to the retarded position during injection by trapping the fuel in the piston chamber.

Any reduction in engine speed will likewise cause a reduction in fuel pressure, allowing the other piston spring to return the cam ring to a retarded position in direct proportion to the speed and fuel reduction. Any fuel in the piston chamber can then bleed off through a control orifice in the piston hole plug (below the ballcheck valve in the head locating screw).

The actual start of cam ring movement will be determined by both piston spring pressure and the fuel pressure being high enough to overcome this spring resistance. The start of cam ring movement is therefore set by the initial adjustment of this spring, which can be done by use of the trimmer screw. For convenience, it can be found on either side of the advance mechanism; however, it should be adjusted on the fuel pump test stand. Total advancement is limited by the piston length. Figure 22–27 shows a typical relationship between injection pump and engine timing.

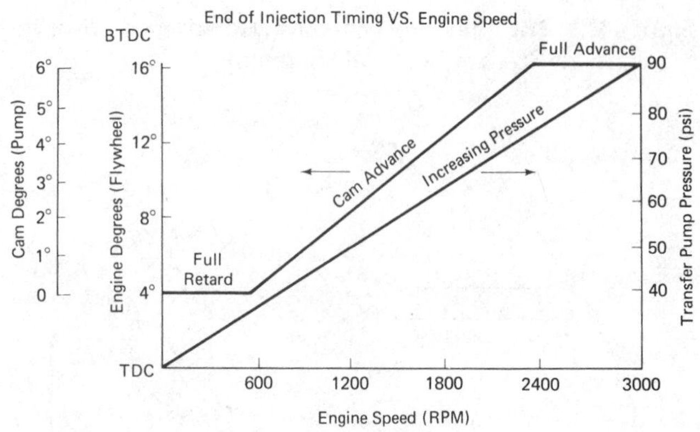

Figure 22-27. Typical relationship between injection pump and engine timing. (Courtesy of Oldsmobile Division, GMC)

Any adjustment of the speed and light-load advance device should be done only on a fuel injection pump test stand because engine damage could result if the engine is operated at excessive advanced timing.

Figures 22–58 and 22–59 show a cutaway view of the engine at the front, which lets you see the injection pump drive mechanism. The pump drive gear is driven off the camshaft by a bevel gear drive immediately behind the chain-driven camshaft gear. Lubrication for the injection pump driven gear is by engine oil pressure directed through a passage from the top of the front camshaft bearing. Oil then passes through an angled passage in the gear shaft to the rear driven gear bearing. The fuel pump eccentric cam on the crankshaft is lubricated from the right bank main oil gallery by oil sprayed from a small orifice.

On-Vehicle Injection Pump Repairs

There are many small repair jobs that can be done to the injection pump while it is in place on the engine. These repairs deal mainly with seal and gasket replacements on the injection pump.

No detailed description of these procedures are given for the 4.3L (262 cu.in.) V6 and 5.7L (350 cu.in.) V8 engines here because these engines use the Roosa Master DB2 injection pump. This same injection pump is also used on the 6.2L (378 cu.in.) V8 diesel engine produced by the Detroit Diesel Allison Divison of GMC and the 6.9L Ford V8 used in light truck product lines. Chapter 23 deals with the 6.2L GMC engine's fuel system and a detailed explanation of the sequence of repairs to the DB2 injection pump and its seals/gaskets. Refer to Chapter 23 for this repair information. Figure 22–28 illustrates the injection pump and its components in exploded view form, which shows the location of the various seals and gaskets.

Typical adjustments, removal, and replacements that the automotive mechanic will be expected to perform on this particular engine would be the following:

1. Engine idle speed setting
2. Throttle position switch adjustment
3. Vacuum regulator valve adjustment
4. Fuel injection line removal/replacement
5. Injector removal, test, and replacement
6. Glow plug testing, removal/replacement
7. Fuel filter service
8. Fuel heater service
9. Injection pump timing check

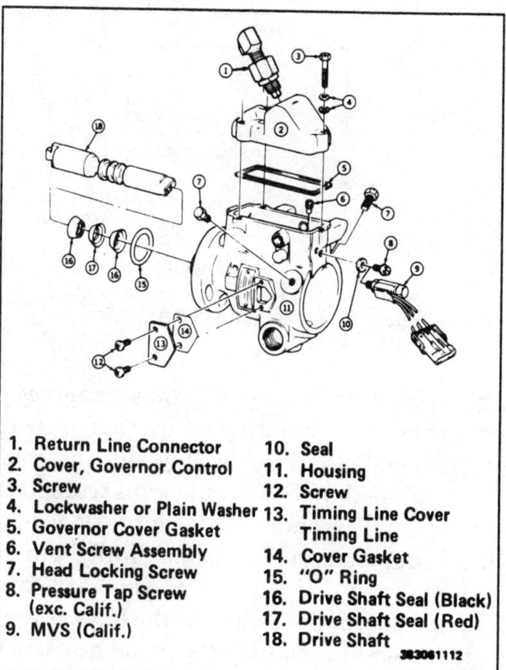

Figure 22–28. DB2 Injection pump housing and drive group. (Courtesy of Oldsmobile Division, GMC)

10. Injection pump removal/replacement
11. Injection pump cover seal and/or guide stud seal replacement
12. Injection pump throttle shaft seal replacement
13. Injection pump solenoid testing, removal, and replacement
14. Injection pump side cover gasket replacement
15. Fuel pump flow test

The above maintenance items can be done while the injection pump is on the engine. The following service items would require injection pump removal from the engine:

1. Advance pin hole plug seal replacement
2. Automatic advance seals replacement
3. Hydraulic head seal O-ring replacement
4. Injection pump drive shaft seal replacement
5. Injection pump pressure testing

Checking/Adjusting Injection Pump Timing on Engine

Generally, when it is suspected that the injection pump-to-engine timing is not correct, you would visually check the scribe line marks on both the injection pump adaptor and flange mounting surfaces as shown in Figure 22–29 as viewed from above the engine.

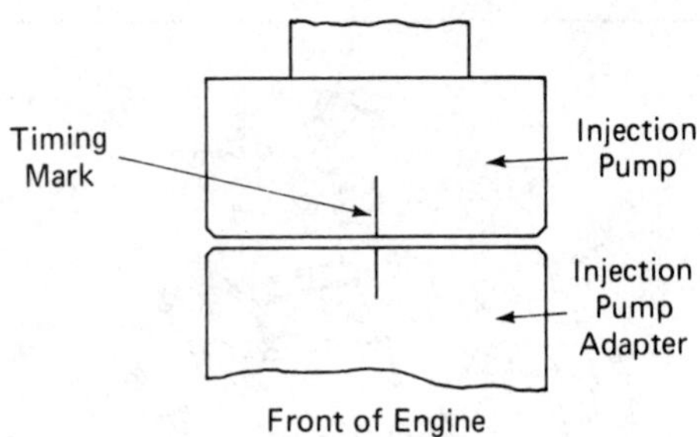

Figure 22-29. Alignment of scribe marks.

On early model 4.3L and 5.7L diesel engines, if the timing marks were not aligned as shown in Figure 22–29 then with the engine stopped, the injection pump mounting flange retaining bolts were loosened off with special wrench (Kent-Moore P/N J-26987), and the pump manually rotated until both marks were in alignment. If the pump was hard to rotate by hand, a 3/4″ open wrench on the boss at the front end of the injection pump (V8's) could be used, while a 1″ open end wrench (P/N J-25304) could be used on the V6 injection pump.

The three pump retaining nuts were then torqued to 35 lb.ft. (47.35 N·m).

Timing Method Using J-33075 Timing Meter. To improve the injection pump timing, a special Kent-Moore tool P/N J-33075 was introduced. This tool can be seen in "Special tools," Figure 22–55, which shows all tools required to effectively service the fuel injection system.

This timing meter incorporates a glow plug probe and a photoelectric cell pickup unit along with a tachometer pickup to monitor the actual start of the injection timing when the engine is running.

The tachometer pickup is installed into the same locating hole at the front of the engine above the crankshaft pulley as is the Mag-Tach engine rpm digital tachometer.

To effectively monitor the actual start of injection, the glow plug is removed from the No. 3 cylinder on a V8 engine and from the No. 1 cylinder on the V6 diesel engine.

SPECIAL NOTE: If timing readings are unsuccessful with the probe in the suggested glow plug holes of the V8 and V6, install the timing probe into cylinder 2 or 3 on the V8 engine and on 1 or 4 in the V6. If a difference in injection pump-to-engine timing exists between cylinders, use the trial-and-error approach until you establish which one of the two timing positions offers the best performance and fuel economy.

The timing probe is then threaded into the glow plug hole and torqued to 8 lb.ft. (11 N·m). *Do not* overtorque the probe, otherwise the pickup lens can be cracked. The timing meter probe that is installed into the Mag-Tach hole above the crankshaft pulley will monitor both engine speed and crankshaft position. The luminosity signal generated at the glow plug probe actually determines when combustion (injection timing) takes place.

Procedure

1. Bring the engine up to normal operating temperature and then shut it off.
2. Place the transmission in *park* on an automatic transmission; apply the parking brake (automatic and standard transmission); block the drive wheels.
3. Install protective cover J-26996-1 inlet manifold cover screens (shown in Figure 22–55 "Special tools") after removing the air cleaner assembly and disconnecting the EGR hose.
4. Make sure that the crankshaft pulley/balancer circumference is wiped clean as well as the probe hole above the pulley before you install the rpm probe holder.
5. It is very important that you clean both ends of the probe that fits into the glow plug hole as well as the lens on the photoelectric pickup. It may be necessary to scrape off any loose carbon that appears in the glow plug hole. Inaccurate readings can result if the probe is not clean.
6. Open up the timing meter case and set the selector switch to position "B" (99.5) for V8 engines and to position "A" (20) for V6 engines.
7. Connect up the timing meter battery leads (red to +) and (black to −).
8. Disconnect the alternator-to-lead connector.
9. Find out what speed the Vehicle Emission Control Information Label specifies for an engine idle rpm; adjust the engine to this specified speed.
10. With the engine running at this speed, carefully note the degrees that the timing meter pointer indicates. Wild fluctuations of the timing needle is indicative of a faulty firing cylinder; therefore, you will not obtain a satisfactory reading until this has been corrected.
11. Allow the engine to idle for 2 minutes and again note the reading on the timing meter face.
12. A stable reading should be taken over a 2-minute interval; the reading should then be compared to that specified on the Vehicle Emission Control Information Label.

NOTE: The timing reading when set to the specification is always after top dead center (ATDC); therefore, it is considered as a "negative reading."

Timing Adjustment. If the injection pump-to-engine timing is incorrect, the engine must be shut off to effect a change to the timing. Marks on the injection pump-to-adapter flange are usually aligned within 0.030" (0.762 mm) on V8 engines and within 0.050" (1.27 mm) on V6 engines. Carefully note the position of the timing marks as shown in Figure 22–29 and loosen off the three retaining nuts on the pump-to-adapter mounting flange.

Use either a 3/4" or 1" open-end wrench respectively on the V8 or V6 injection pump boss (front) to rotate the pump either clockwise (right) to retard the timing or counterclockwise (left) to advance the timing as the case may be to change the timing.

NOTE: One degree of timing change on the V8 engine is equal to about the width of the scribed timing mark, while on the V6, it is equal to about 2/3 of a degree.

Current retaining nut torques on the V6 pump flange adapter is 35 lb.ft. (47 N·m), and on the V8, they are 18 lb.ft. (24 N·m).

After any adjustment, always start the engine again and recheck the timing as described earlier.

Once the injection pump timing has been readjusted to specifications, remove the timing probe from the glow plug hole and reinstall the glow plug and torque it to 12 lb.ft. (16 N·m).

SPECIAL NOTE: On aluminum V6 cylinder heads, apply a light coat of lubricant (GM P/N 1052771 or its equivalent) to the threads of the glow plug only because damage to the heating element can occur if the lubricant gets onto it—it will bake onto the surface.

Reconnect the previously removed alternator-to-lead connector; then after removing the intake manifold screens, install the air cleaner assembly and EGR hose.

Throttle Linkage Adjustments. After any adjustment to the injection pump timing, it is usually necessary to check and adjust the throttle linkage as described below. Refer to Figure 22–30 and disconnect the transmission throttle valve or detent cable from the throttle assembly and also from the cruise control servo-rod if the vehicle is so equipped.

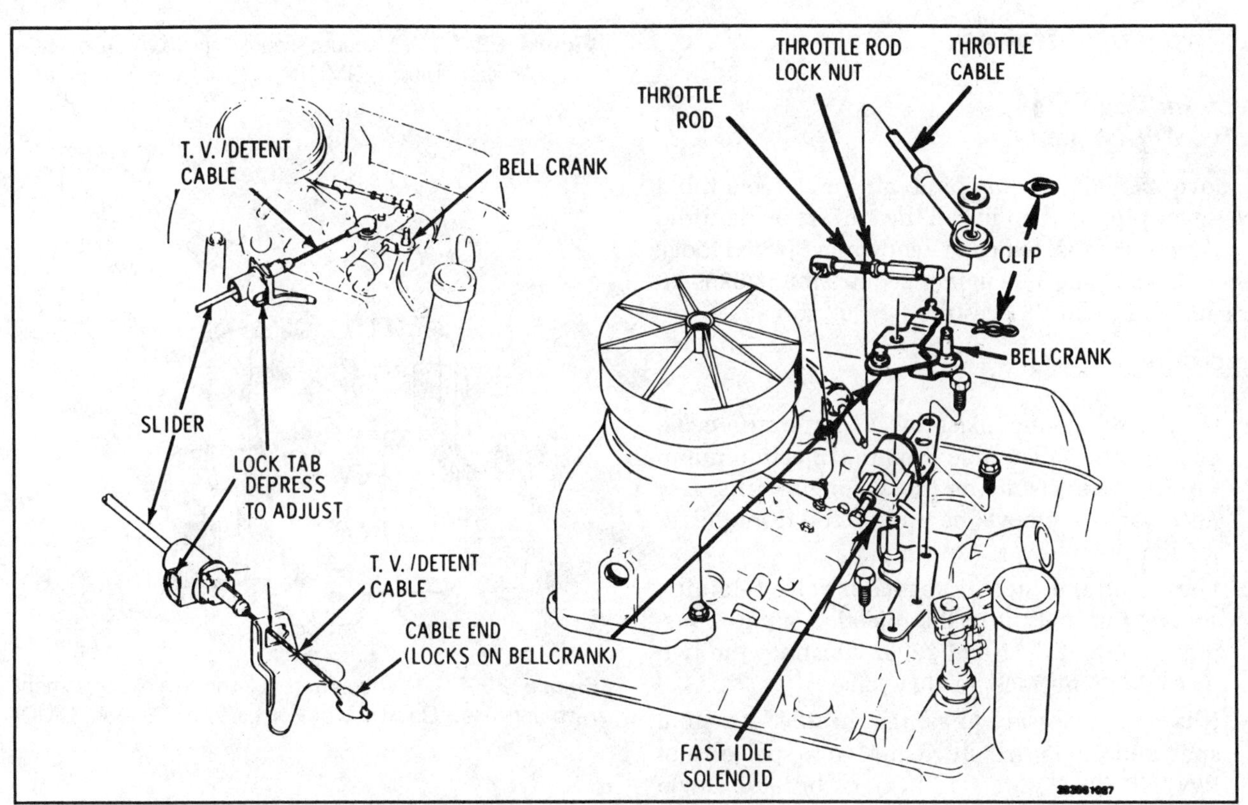

Figure 22–30. 5.7L (350 cu.in.) V8 Diesel throttle linkage. (Courtesy of Buick Motor Division, GMC)

Procedure

1. Manually rotate the bellcrank lever assembly and hold it in the full-fuel position; carefully note if the injection pump throttle lever actually makes contact with the full throttle stop.
2. If necessary, loosen the lock nut on the pump rod and lengthen the rod until the injection pump lever does make contact with the full throttle stop and tighten its lock nut.
3. Notice the metal lock tab in Figure 22–30; depress this tab on the throttle valve cable upper end and manually push the slider through the fitting away from the bellcrank until it stops against the metal fitting; release the metal lock tab.
4. If the vehicle is equipped with a cruise control unit, reconnect the cruise control servo-rod, then reconnect the transmission throttle valve cable.
5. Manually rotate the bellcrank lever to the full-fuel position and release it, checking to make sure that no bind exists in the linkage.

You should now proceed to check the vacuum regulator valve adjustment as well as the idle speed and cruise control servo-rod.

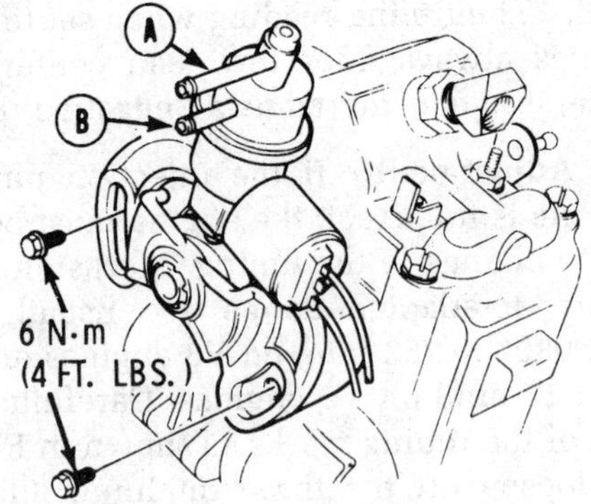

Figure 22–31. Vacuum regulator valve. (Courtesy of Buick Motor Division, GMC)

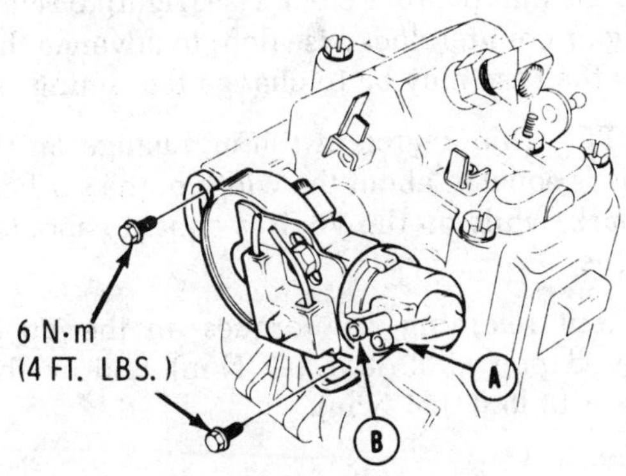

Figure 22–32. Vacuum regulator valve. (Courtesy of Buick Motor Division, GMC)

Other Adjustments

Vacuum Regulator Valve Adjustment

Remove the air cleaner and air intake manifold crossover piping, then install the air intake manifold protective screens, which are shown in Special tools, Figure 22–55. The V8 engines require the J-26996-10 screens while the V6 uses the J-29657 screens.

Procedure

1. On the V6 engine, disconnect the throttle cable and throttle valve cable from the injection pump throttle lever, while on the V8 engine it is only necessary to remove the throttle rod from the injection pump lever.
2. Loosen off the vacuum regulator valve retaining screws that hold it to the injection pump housing. Figures 22–31 and 22–32 illustrate the two possible mountings for this valve.
3. It is now necessary to set the throttle lever to a specified angle; install Kent-Moore special tool P/N J-26701-15 (or BT-7944) carburetor angle gauge adapter to the injection pump throttle lever by placing the angle gauge BT-7704 or J-26701 onto the adapter as shown in Figures 22–33 and 22–34.

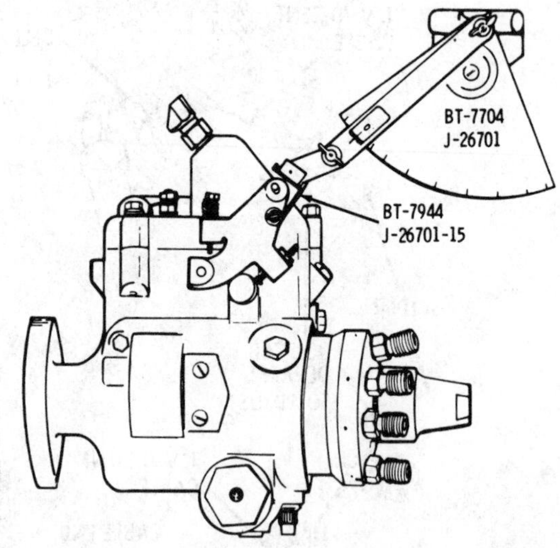

Figure 22–33. Vacuum regulator valve adjustment in zero degrees. (Courtesy of Buick Motor Division, GMC)

SPECIAL NOTE: Due to the thicker throttle lever used on the V6 diesel, you will have to lightly file the adapter tools to make them fit.

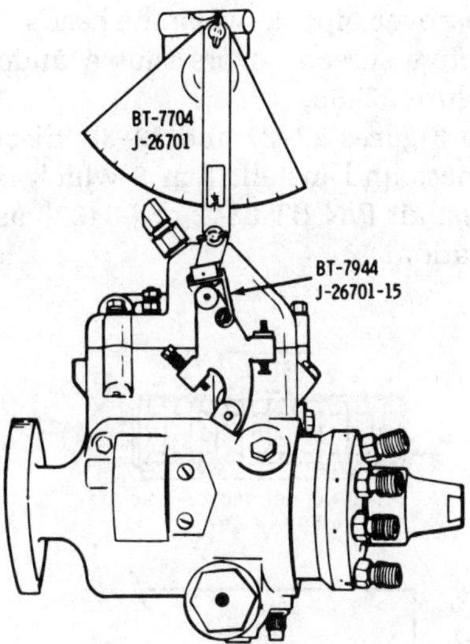

Figure 22–34. Vacuum regulator valve adjustment at specified angle. (Courtesy of Buick Motor Division, GMC)

4. With the gauges in position over the injection pump throttle lever, manually rotate the throttle lever to its full-fuel position and then set the angle gauge to *zero* degrees and set the bubble in the level to its centered position.
5. Refer to Figure 22–34 and set the angle gauge to 58 degrees on the V8 engines and to 49 degrees on the V6 units; rotate the throttle lever until the bubble is centered.
6. Refer to Figures 22–31 and 22–32, which show the two types of vacuum regulator valves used and apply a vacuum source to port "A" while a vacuum gauge is installed at port "B." Use a hand-operated vacuum pump or electric motor driven unit to create 75 kPa or 22″ Hg at port "A." Special tools P/N BT-7517 or J-23738 can be used as a vacuum source.
7. Automatic transmission equipped vehicles must have their vacuum valve adjusted by rotating the vacuum valve clockwise to produce 36 ± 1 kPa (10.6 ± 0.3″ Hg) of vacuum. Manual transmission equipped vehicles should have their vacuum regulator valve adjusted to produce 30 ± 1 kPa (8.9 ± 0.3″ Hg).
8. When the above readings have been attained, tighten the vacuum valve retaining bolts to 4 lb.ft. (6 N·m).
9. Reconnect the throttle linkage that was disconnected in Step 1 above.
10. Remove the intake manifold protective screens and install the air crossover pipe, followed by the air cleaner assembly.

Metering Valve Sensor Check/Adjustment

All 1984 and later 4.3L V6 diesel vehicles that are sold into the state of California are equipped with an electronic controlled Exhaust Gas Recirculation (EGR) emission system.

The vehicle ECM (electronic control module) receives three input signals that it uses to vary the on/off time of the EGR valve solenoid.

These three input voltage signals (sensors) are:

1. Engine rpm sensor
2. Vehicle speed sensor
3. Injection pump fuel metering valve position

EGR is necessary in order to reduce the combustion chamber temperatures; this in turn reduces the quantity of nitrogen oxides emitted into the atmosphere from the exhaust system. To monitor the ECM control of EGR, an intake manifold absolute pressure sensor is also used to measure the amount of absolute pressure in the EGR vacuum line.

It is therefore necessary to check/adjust the third sensor listed above, namely the "injection pump fuel metering valve position" unit, which is a variable resistor that sends a voltage signal back to the ECM. The sensor is connected to a voltage reference and is designed to have its highest resistance at a closed throttle position. At wide open throttle (full-fuel position), the resistance is at its lowest point and the voltage output from the sensor is approximately 5 volts.

A failure in the sensor will turn on the *check engine* light at engine speeds below 1500 rpm, while the *check engine* light will go off at speeds above 1500 rpm if there is a metering valve sensor (MVS) error. Figure 22–35 illustrates the MVS location on the injection pump assembly.

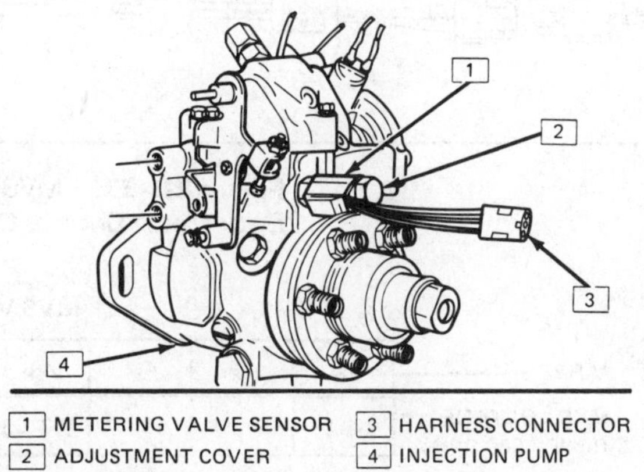

Figure 22–35. 4.3L V6 Diesel engine metering valve sensor location. (Courtesy of Buick Motor Division, GMC)

The wiring diagram used with the MVS sensor unit is shown in Figure 22–36; use this as a reference when checking the voltage readings as per the check sequence given below.

To check the MVS sensor, always run the engine until normal operating temperature is reached, then shut it off. Remove the air cleaner assembly and air intake crossover pipe between the heads, then install the protective screen covers shown under "Special tools," Figure 22–55.

Refer to Figures 22–37 and 22–38; disconnect the MVS harness and install Item 3 which is the MVS harness test kit P/N BT-8342 or J-34678 as shown in the illustration.

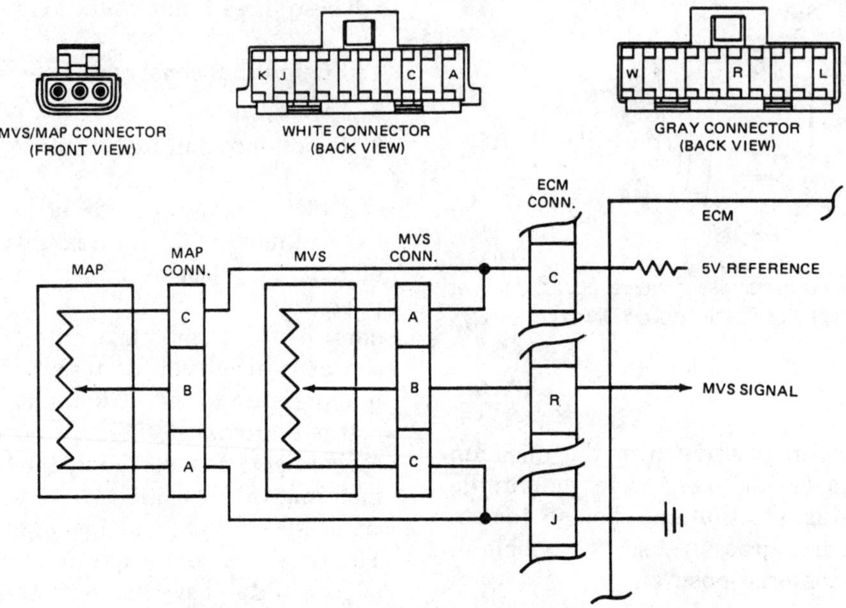

Figure 22–36. Metering valve sensor wiring diagram. (Courtesy of Buick Motor Division, GMC)

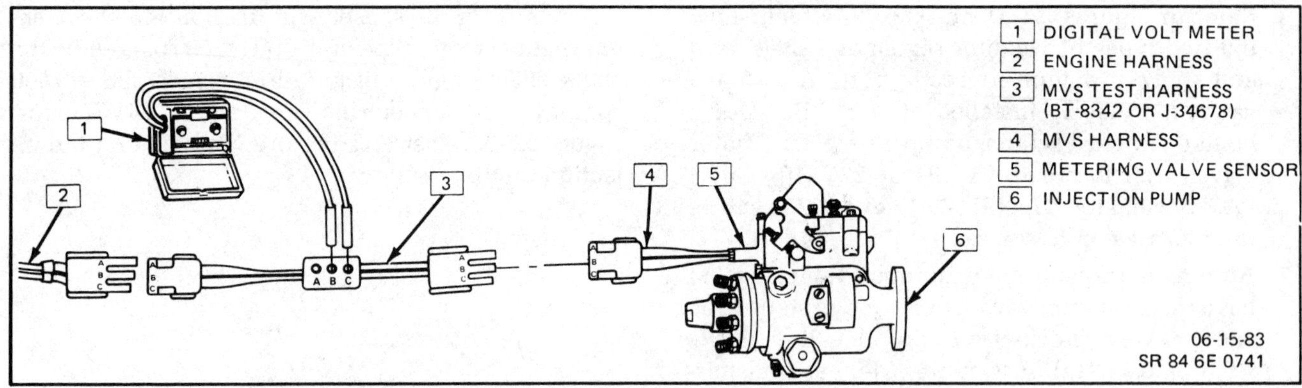

Figure 22–37. MVS Voltage hookup check. (Courtesy of Buick Motor Division, GMC)

MVS VOLTAGE TABLE

V-REF →	4.5	4.6	4.7	4.8	4.9	5.0	5.1	5.2	5.3	5.4	5.5
MVS VOLTAGE → (In "D" 650 RPM)	.53-.55	.54-.56	.55-.57	.57-.59	.58-.60	.59-.61	.60-.62	.61-.63	.63-.56	.64-.66	.65-.67

Figure 22–38. MVS Voltage reading chart. (Courtesy of Buick Motor Division, GMC)

Procedure

1. Install a Mag-Tach (tachometer) J-26925 with its probe into the probe hole immediately above the crankshaft vibration damper pulley (see Figures 22–39 and 22–40).
2. Make sure that the MVS unit is tight to the injection pump; check the retaining bolts which should be 30 lb.in. (3.5 N·m).
3. Start and accelerate the engine to 1500 rpm for 10–20 seconds in order to stabilize the fuel flow, then let it idle.
4. With the parking brake on and the drive wheels blocked, shift the transmission to *D* position and adjust the engine idle speed to 650 rpm.
5. Leave the transmission selector in the *D* position; refer to Figures 22–36 and 22–37 and measure the voltage between terminals "A" and "C," using J-29124 multi-meter or equivalent; this must be recorded because this voltage is in fact the reference voltage.
6. Shift the transmission selector to *P*.
7. Compare your measured voltage reading in Step 5 above with that shown in Figure 22–38.

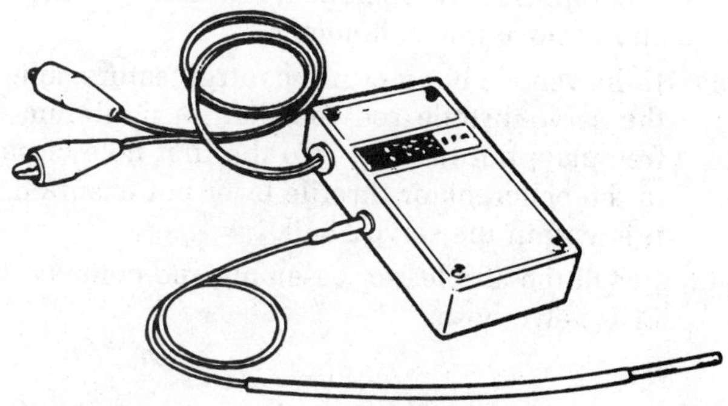

Figure 22-39. Using tool J-26925. (Courtesy of Buick Motor Division, GMC)

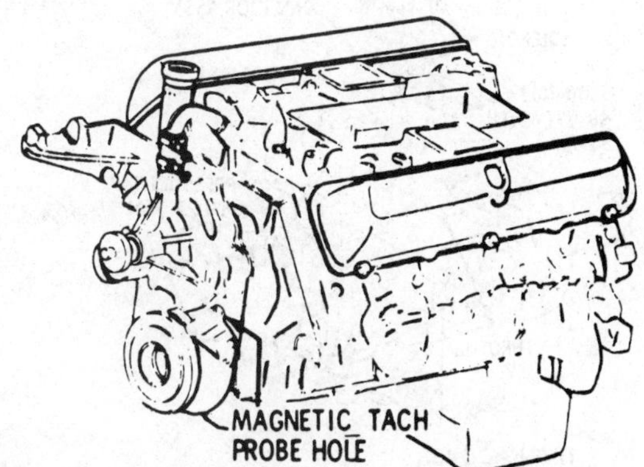

Figure 22-40. Location of Mag-Tach probe hole on engine.

8. With the transmission selector in *P*, measure the voltage across terminals "B" and "C" as you rapidly move the throttle from its idle position to wide open throttle (full-fuel) and back to idle speed. The recorded voltage on the multi-meter should go from somewhere less than 1 volt at idle to between 4 and 5 volts when the throttle is rapidly cracked open or flashed.
9. If the voltage reading in Step 8 is as per the voltage chart, Figure 22–38, then the sensor is operating correctly; otherwise proceed to the adjusting sequence below.

Adjustment—Metering Valve Sensor. To adjust the metering valve sensor (MVS), refer to Figure 22–35 and with the engine stopped, remove the MVS adjustment cover (Item 2) while taking care not to disturb the MVS sensor switch. Refer to the "Special tool" list (Figure 22–55) and obtain Item 4 (J-24182-2), which is the MVS adjuster tool.

Procedure

1. Insert the adjusting tool or equivalent into the MVS and by rotating the screw clockwise the voltage output will be increased, while counterclockwise rotation will decrease the voltage output. The screw has a total of 48 turn increments for fine adjustment purposes.
2. Reinstall the MVS hole plug finger tight and repeat Steps 3 through 7 as per the "checking procedure" listed above.
3. When you have adjusted the MVS voltage reading to specifications, make sure that you install the MVS hole plug with a new O-ring seal and while holding the MVS assembly, torque the hole plug to 30 lb.in. (3.5 N·m).
4. Reconnect the MVS connectors, and start and run the engine, and make sure that there are no fuel leaks at the hole plug. Reset the idle speed if it is not as per the Vehicle Emission Information Label.
5. Install the air intake manifold crossover pipe and the air cleaner assembly.

Idle Speed Adjustment

Refer to Figure 22–40 which shows the location of the magnetic tach probe hole at the front of the engine above the crankshaft pulley damper. This hole is used to install the probe unit of the Kent-Moore Mag-Tach digital tachometer P/N 26925 on all GMC cars and light pickup trucks and vans. The Mag-Tach unit is powered from the vehicle battery (12 volts) by placing the red lead to the battery positive terminal and the black lead to the negative terminal.

Before checking the engine's idle rpm, run the engine until it is at normal operating temperature, then switch it off, block the drive wheels, apply the parking brake, and place the transmission selector in *neutral*.

To ensure that the Mag-Tach will pick up an accurate count of the engine's rpm from the pulley/damper, clean off the circumference of the pulley.

Procedure

1. Make sure that you disconnect the two-lead connector from the rear of the alternator and turn off all electrical accessories; on the V6 engine, disconnect the A/C compressor clutch lead at the air compressor to prevent any possibility of this unit operating under load when you are attempting to set the idle speed.
2. Another caution that should be exercised is to ensure that no one rotates the steering wheel or applies the brake pedal during this idle check because both will cause a change to the idle setting speed.
3. Start the engine with the gear selector in *D* on an automatic transmission equipped vehicle and in *neutral* on a standard transmission.
4. Read the engines idle rpm from the Mag-Tach unit and compare it with that printed on the Vehicle Emission Information Label; reset the idle speed by adjusting the slow idle adjustment screw as shown in Figure 22–41.
5. To check the fast idle speed adjustment, it is necessary to first of all unplug the electrical connector from the EGR-TVS (thermal vacuum switch) (or HPCA and fast idle temperature switch on California certified engines only; HPCA = Housing Pressure Cold Advance).
6. Install a wire jumper between the connector terminals as shown in Figures 22–42 through 22–46.

   CAUTION: Do not let the jumper wire come into contact with a ground!

7. Check the fast idle speed of the engine as per that stamped on the Vehicle Emission Information label; if the speed is not as per the label, readjust the fast idle solenoid. The V6 engine solenoid is shown in Figure 22–47, while the V8 solenoid is shown in Figure 22–30, which shows the V8 throttle linkage arrangement.
8. Once the fast idle solenoid has been reset, remove the jumper wire and reconnect it to the EGR-TVS or HPCA temperature switch as shown in Figures 22–42 through 22–46.
9. Doublecheck that the idle speed has not changed; if it has, reset it.
10. Stop the engine and turn off the ignition key and reconnect the wire lead at the alternator and A/C compressor (if equipped) and also disconnect and remove the tachometer.
11. If the vehicle has a cruise control feature, adjust the servo-throttle rod to obtain a minimum of free play; put the clip into the first hole closest to the bellcrank or throttle lever but ensure that it is within the servo "bail."
12. Install the air cleaner assembly and connect the EGR valve hose.

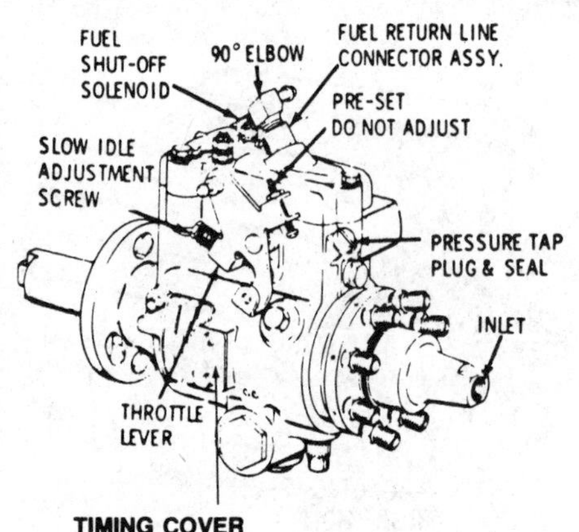

Figure 22–41. Injection pump idle speed adjustment screw location.

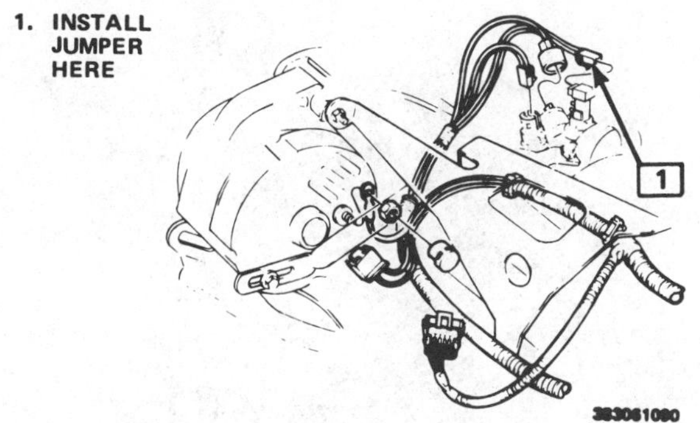

Figure 22-42. V8 EGR-TVS Switch jumper wire location. (Courtesy of Buick Motor Division, GMC)

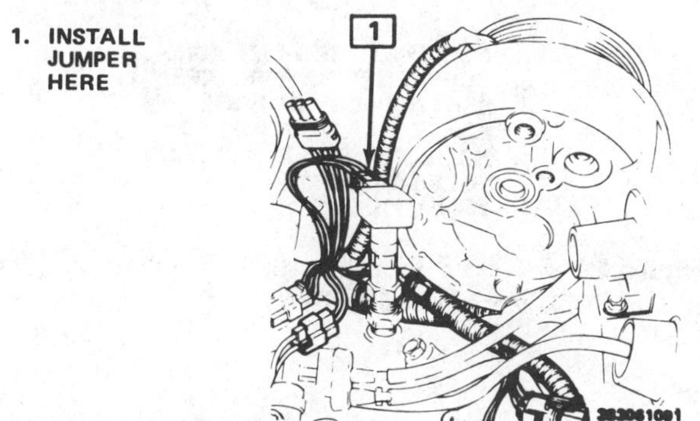

Figure 22-43. VIN V EGR-TVS switch jumper wire location, except California. (Courtesy of Buick Motor Division, GMC)

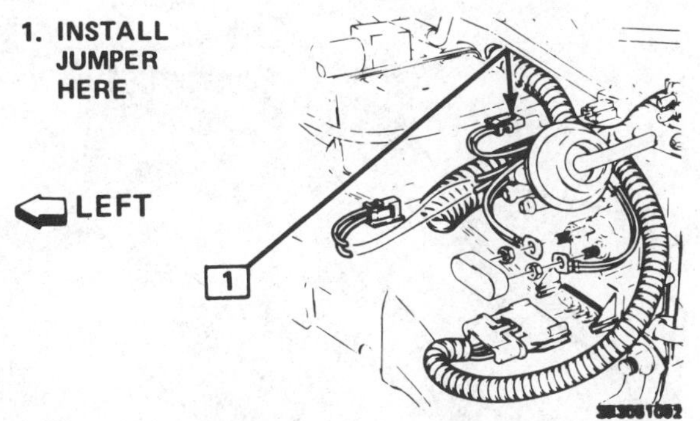

Figure 22-44. VIN T EGR-TVS Switch jumper wire location except California. (Courtesy of Buick Motor Division, GMC)

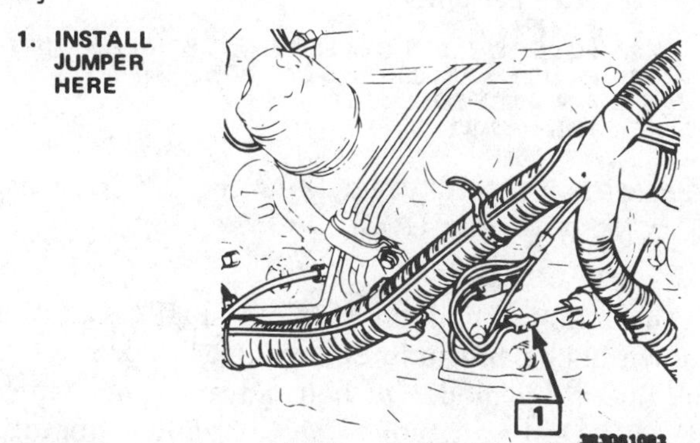

Figure 22-45. HPCA/Fast idle temperature switch VIN V California. (Courtesy of Buick Motor Division, GMC)

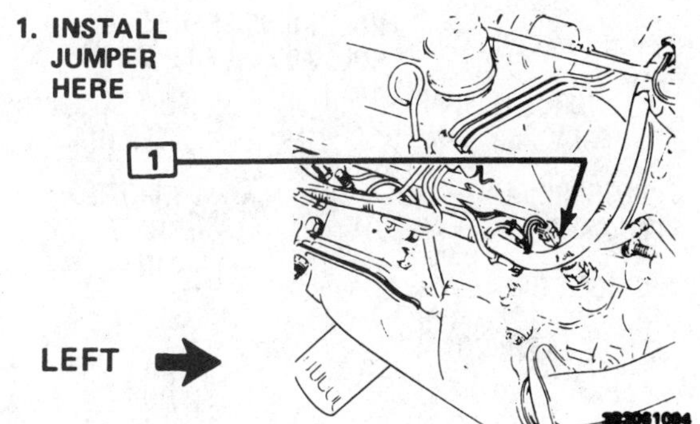

Figure 22-46. HPCA and fast idle temperature switch VIN T California. (Courtesy of Buick Motor Division, GMC)

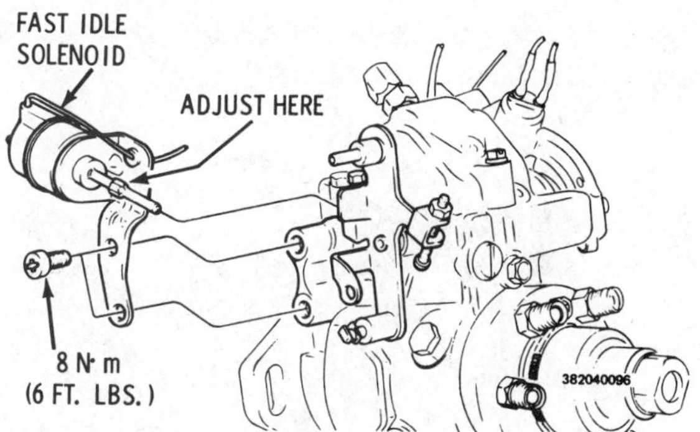

Figure 22-47. 4.3L V6 Diesel fast idle solenoid adjustment location. (Courtesy of Buick Motor Division, GMC)

Fuel Pressure

Checking Injection Pump Housing Fuel Pressure

Engine problems such as a lack of power or sluggishness, can be caused by low fuel injection pump housing pressure because this can affect the advance mechanism of the pump. Figures 22-48 and 22-49 illustrate the location of the injection pump housing pressure cold advance (HPCA) solenoid on the V8 engine and the location of the solenoids and connectors found on the V6 diesel engines, respectively.

The 1984 and later diesel engines, with the exception of the California certified V6 units, have injection pump timing controlled by two pressure regulators. One of these is the HPCA (housing pressure cold advance) unit, which is located as shown in Figures 22-23 and 22-24 in the injection pump housing. The second advance unit, known as the HPAA (housing pressure altitude advance), is a pressure regulator that is located in the fuel return line as shown in Figures 22-50 and 22-51, which illustrate both the V8 and V6 engine locations respectively.

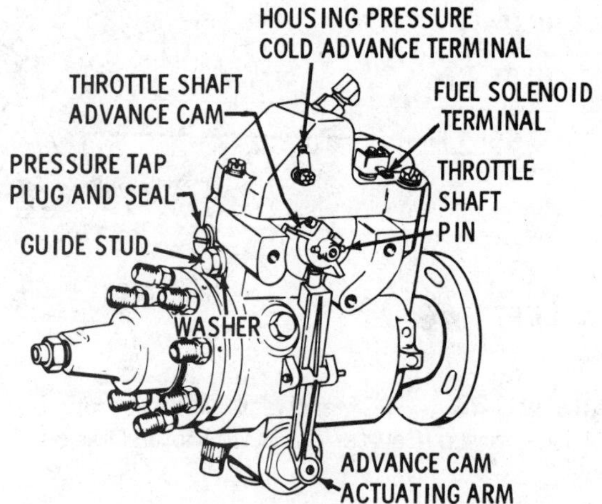

Figure 22-48. V8 Diesel HPCA solenoid location. [Courtesy of Buick Motor Division, GMC]

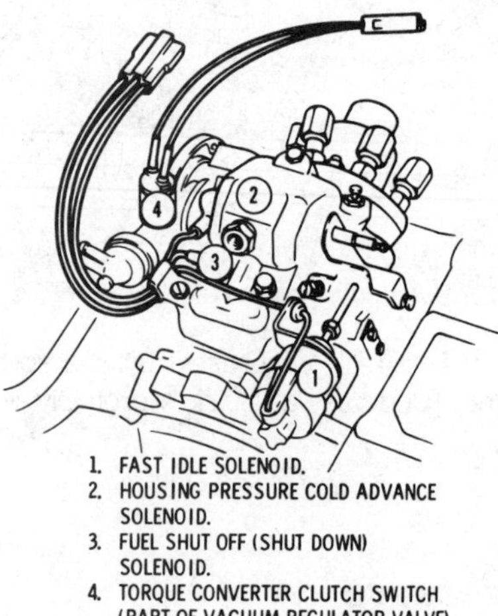

Figure 22-49. V6 Diesel solenoids and connectors. [Courtesy of Buick Motor Division, GMC]

An altitude control switch controls the HPCA, HPAA, and EGR trim; this switch is shown in Figures 22–52 and 22–53 for non-"A" series vehicles and also "A" series vehicles.

The HPAA electric solenoid regulates injection pump housing pressure in relation to altitude. When the solenoid is energized (on), the glass check ball is seated similar to that shown for the HPCA system in Figure 22–24 to regulate fuel pressure to its calibrated value. When the HPAA is de-energized (off), the check ball is off its seat, the fuel return line is opened, and there is no regulation of injection pump housing fuel pressure.

Because of the arrangement used for HPCA and HPAA, it is possible to have the HPCA (off) while the HPAA is (on) regulating fuel pressure within the injection pump housing.

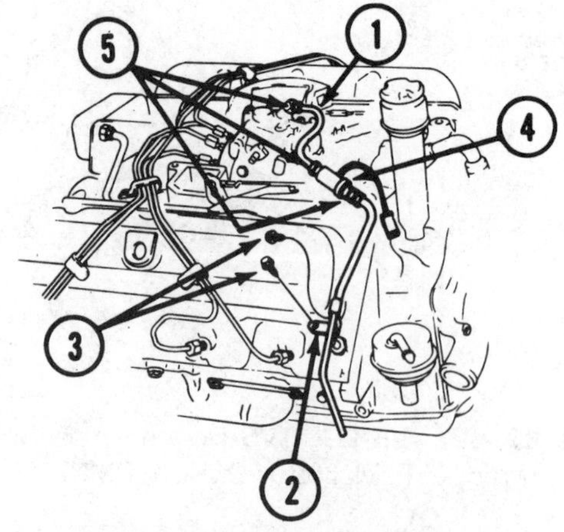

Figure 22-50. V8 Diesel engine HPAA location. [Courtesy of Buick Motor Division, GMC]

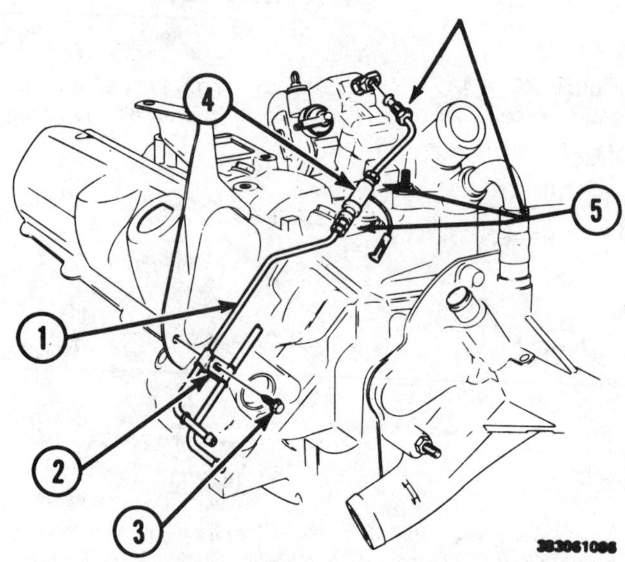

Figure 22-51. V6 (VIN V) HPAA Location. [Courtesy of Buick Motor Division, GMC]

The action of both the HPCA and HPAA units are shown in chart form in Figure 22–54 to demonstrate the operating modes of both advance mechanisms at different altitudes and engine operating temperatures.

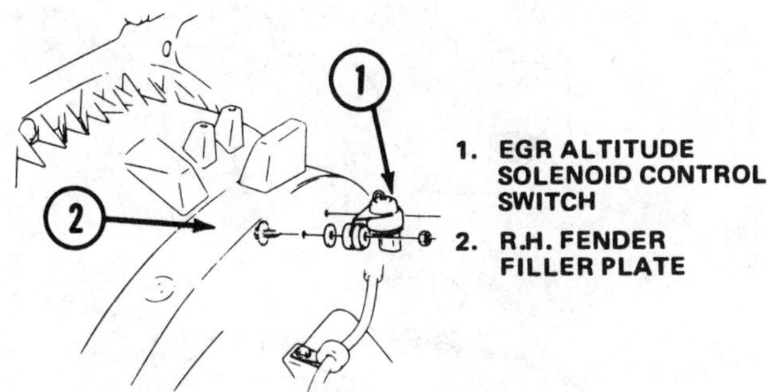

Figure 22-52. EGR/HPCA Altitude control switch—typical except "A" series. (Courtesy of Buick Motor Division, GMC)

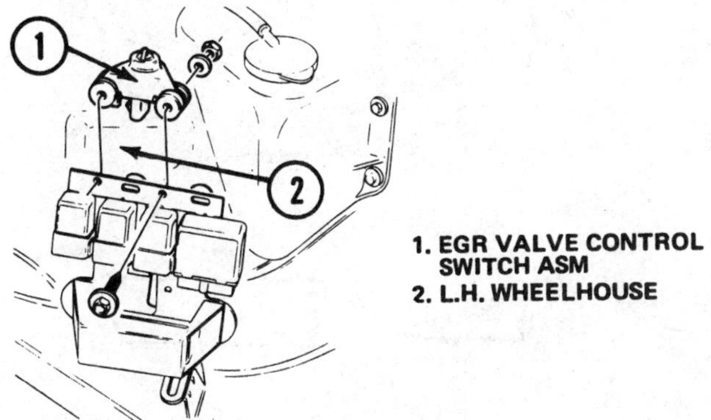

Figure 22-53. EGR/HPAA Altitude control switch—"A" series. (Courtesy of Buick Motor Division, GMC)

NOTE: All diesel engines are equipped with the HPCA unit, which is designed to advance injection pump timing by 4 degrees during cold engine operation. Activation of the HPCA unit is via the EGR-TVS switch on all engines except California certified units, which use a temperature switch on the left rear cylinder head bolt.

The HPCA switch is arranged so that it will open the circuit at approximately 41°C (105°F) on federally certified engines and at approximately 52°C (125°F) on California certified engines.

At any temperature below the switching point, injection pump housing pressure is decreased to zero psi from 9–11 psi (62–83 kPa) in order to allow timing advancement of about 4 degrees.

When the switch opens at the temperatures mentioned above, the electric solenoid is de-energized, which allows the injection pump housing fuel pressure to return to a fuel pressure between 8–12 psi (55–83 kPa). The switch will close to energize the solenoid once again when the temperature drops below 35°C (95°F).

It should be noted here that this same switch also energizes the fast idle solenoid.

Fuel Pressure Check—Injection Pump Housing

Poor engine throttle response and black smoke at the exhaust pipe can be caused by low injection pump housing pressure because this affects the advance mechanism operation. To check the fuel pressure of the injection pump, refer to Figure 22–48 and remove the pressure tap plug and seal on federally certified engines; while on California certified engines, remove the fuel metering valve sensor shown in Figure 22–35.

To gain access to these items, remove the air cleaner assembly and intake manifold crossover pipe between the cylinder heads and install the protective covers J-29657 (V6) or J-26996 (V8) shown under the "Special tools" list, Figure 22–55.

NOTE: The engine must be at normal operating temperature when performing this check!

Procedure

1. Refer to Figure 22–55 and, using Item 11, i.e., J-29382 injection pump pressure test adapter, screw it into the pressure tap hole in the injec-

	CONDITION		HPCA	HPAA	NOMINAL HOUSING PRESSURE kPa (psi)
	ALTITUDE	COOLANT			
FEDERAL PACKAGE	BELOW 1219m (4000 FT.)	COLD	ON	OFF	0
	BELOW 1219m (4000 FT.)	HOT	OFF	OFF	68.9 (10)
	ABOVE 1219m (4000 FT.)	COLD	ON	OFF	0
	ABOVE 1219m (4000 FT.)	HOT	ON	ON	48.3 (7)
ALTITUDE PACKAGE	ABOVE 1219m (4000 FT.)	COLD	ON	OFF	0
	ABOVE 1219m (4000 FT.)	HOT	OFF	OFF	68.9 (10)
	BELOW 1219m (4000 FT.)	COLD	ON	OFF	0
	BELOW 1219m (4000 FT.)	HOT	OFF	ON	89.6 (13)

HPCA = HOUSING PRESSURE COLD ADV. HPAA = HOUSING PRESSURE ALTITUDE ADV.
TIMING RETARDS WITH HIGHER HOUSING PRESSURE
383061100

Figure 22-54. HPCA and HPAA Injection pump timing control conditions. (Courtesy of Buick Motor Division, GMC)

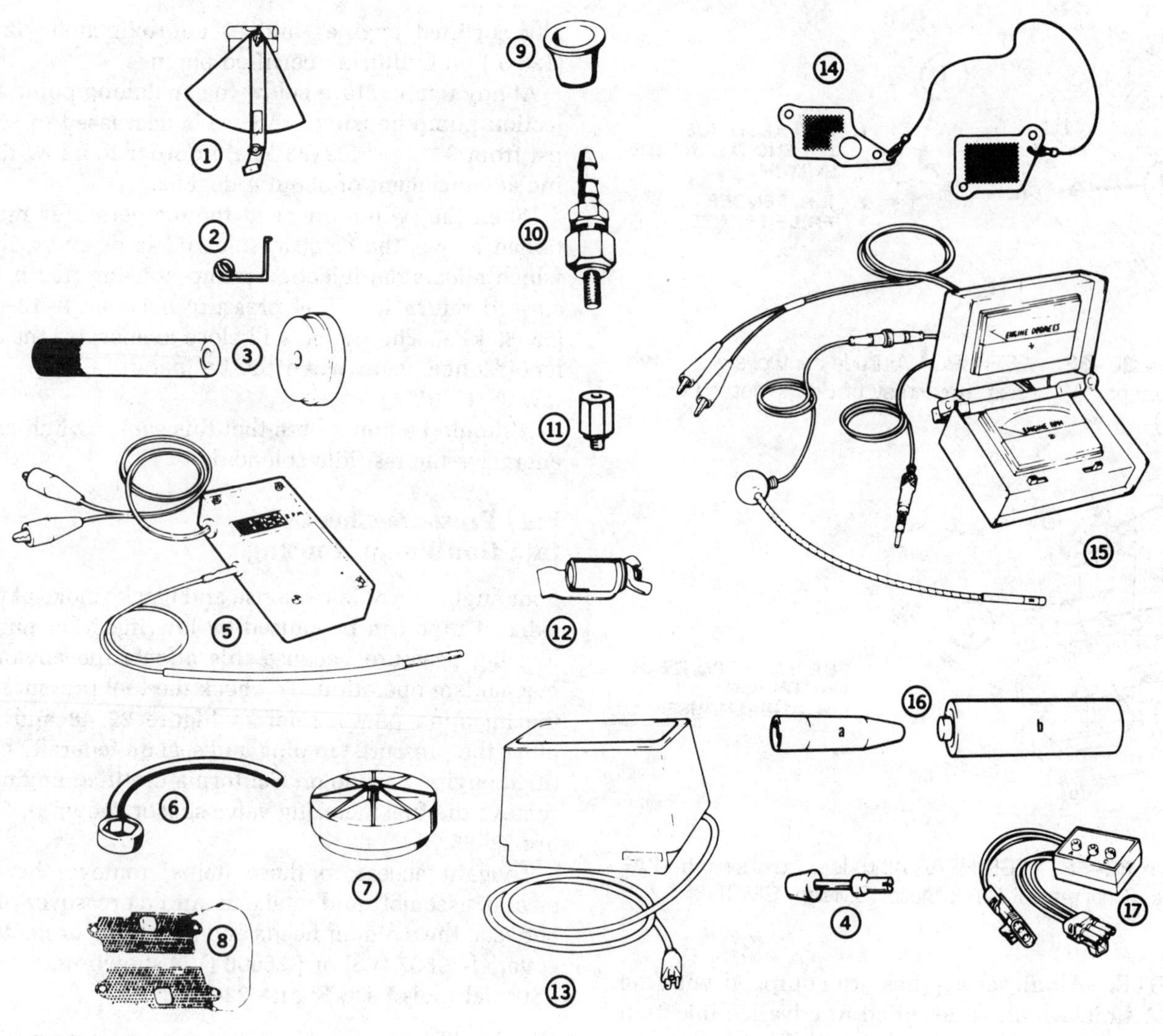

1. BT-7704 or J-26701 ANGLE GAGE
2. BT-7944 or J26701-15 ANGLE GAGE ADAPTER
3. BT-8018 NOZZLE ASSEMBLY TOOL
4. J-24182-2 METERING VALVE SENSOR ADJUSTOR
5. J-26925 TACHOMETER
6. J-26987 INJECTION PUMP WRENCH
7. J-26996-1 AIR CROSSOVER COVER
8. J-26996-10 MANIFOLD COVER SCREENS
9. J-28438 PLASTIC PLUGS - PUMP LINES & NOZZLES
10. J-28526 INJECTION PUMP PRESSURE TEST ADAPTER
11. J-29382 INJECTION PUMP PRESSURE TEST ADAPTER
12. J-29601 INJECTION PUMP THROTTLE SHAFT CAM TIMING ADAPTER
13. J-29653 ULTRA SONIC NOZZLE CLEANER
14. J-29657 MANIFOLD COVER SCREENS
15. J-33075 TIMING METER
16. a. J-29745-A DRIVE SHAFT SEAL PROTECTOR
    b. J33081 ADVANCE PLUNGER SEAL INSTALLER
17. J-34678 METERING SENSOR ADJUSTING HARNESS

Figure 22-55. Special tools. [Courtesy of Buick Motor Division, GMC]

tion pump housing using the seal from the pressure tap plug that you have removed.

2. Screw J-28526 adapter (Item 10 in Figure 22–55) into J-29382 in Step 1 above.
3. Adapt a low pressure fuel gauge (0–20 psi) 0–138 kPa range into the J-28526 adapter.
4. Install the Mag-Tach probe into the hole above the engine crankshaft pulley/damper adapter to monitor engine rpm.
5. Block the drive wheels and engage the parking brake; place the transmission selector in *neutral* in a manual transmission and in *park* in an automatic transmission.
6. Start and run the engine at 1000 rpm and carefully record the fuel pressure shown on the gauge; pressure should be between 9–11 psi (62–83 kPa) with no more than 1 psi (6.895 kPa) variation. Refer to Figure 22–54, which shows the fuel pressure that would apply to a specific vehicle depending on whether it is fitted with the altitude package or the federal package.
7. A *zero* pressure reading generally indicates a problem with the HPCA (housing pressure cold advance) unit, which can be checked as follows:
   a. Refer to Figure 22–48 or Figure 22–49 and remove the electrical lead from the HPCA terminal; with the engine still running at 1000 rpm, if the fuel pressure is still *zero*, then it will be necessary to remove the injection pump cover after stopping the engine and checking the advance solenoid for signs of binding. Replace the solenoid if a fault is found by removing the injection pump cover as shown in Figure 22–56 to gain access to the cold advance solenoid.
   b. If the fuel pressure comes up to normal when the electrical terminal is disconnected, check the operation of the temperature switch on the EGR-TVS (or cylinder head bolt).
8. If the fuel pressure remains low, replace the fuel return line connector assembly.
9. High fuel pressure is indicative of the HPAA or return fuel line being restricted.
10. If a new fuel return line connector is installed, the injection pump timing should be checked and adjusted as described herein.
11. If the fuel pressure is lower than previously when a new return line connector is installed, correct the restriction in the line; if it is still too high, replace the fuel return line connector assembly.
12. Fuel pressure still remaining too high would necessitate injection pump removal, disassembly, and repair.

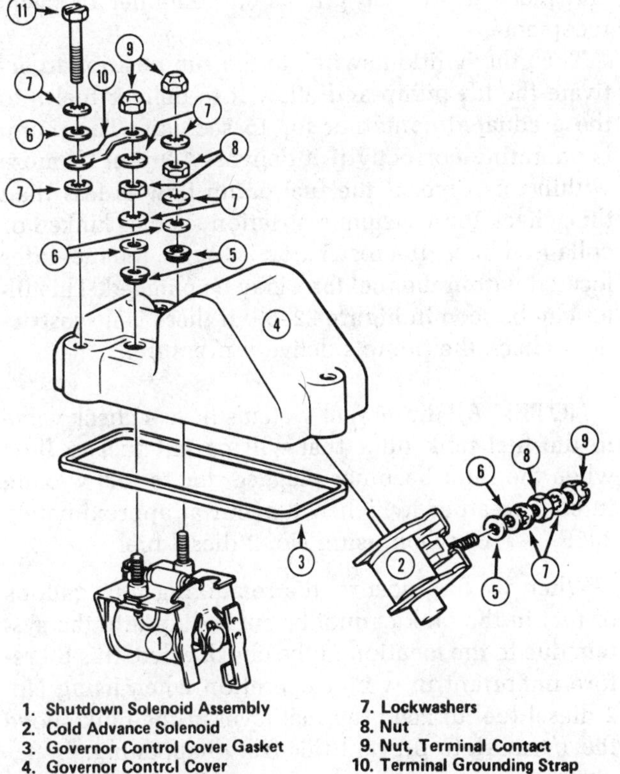

Figure 22-56. DB2 Injection pump solenoids. (Courtesy of Buick Motor Division, GMC)

13. Remove the Mag-Tach unit, the fuel pressure gauge, and the adapters.
14. Use a new pressure tap plug seal and, on California certified engines, install the MVS (metering valve sensor) using a new O-ring. Coat the threads with Loctite 242 or equivalent sealer, and torque the MVS to 30 lb.in. (3.5 N·m).
15. Check, and if necessary, adjust the MVS voltage output as described in this section.
16. Remove the intake manifold protective screens, install the air crossover pipe, and replace the air cleaner assembly and connect up the EGR hose.

Electric Fuel Pump

Flow Test

Current V6 4.3L engines employ an electric fuel lift pump that is mounted on the engine. To test the lift pump delivery, disconnect the fuel filter feed line and place it into the graduated container or clean receptacle.

Turn the ignition switch to the run position to activate the lift pump and allow it to deliver fuel into the graduated container for 15 seconds. The pump is operating correctly if it delivers 1/2 pint or more within this time. If the fuel pump flow is less than this, check for a system restriction such as kinked or collapsed lines. If a restriction is found, the fuel filter located within the fuel tank may be plugged. This filter can be seen in Figure 22–57. If there is no restriction, check the pump's delivery pressure.

NOTE: All diesel fuel systems have a check valve in the fuel tank filter that will permit fuel to flow when the filter becomes plugged due to fuel waxing in cold weather, which will occur at approximately 20°F (−7°C) when using No. 2 diesel fuel.

When waxing occurs, the remaining four gallons of fuel in the tank cannot be sucked up into the system due to the location of the check valve. It is therefore important in winter operation when using No. 2 diesel fuel to keep the fuel level in the tank above the 1/4 mark to prevent this from happening.

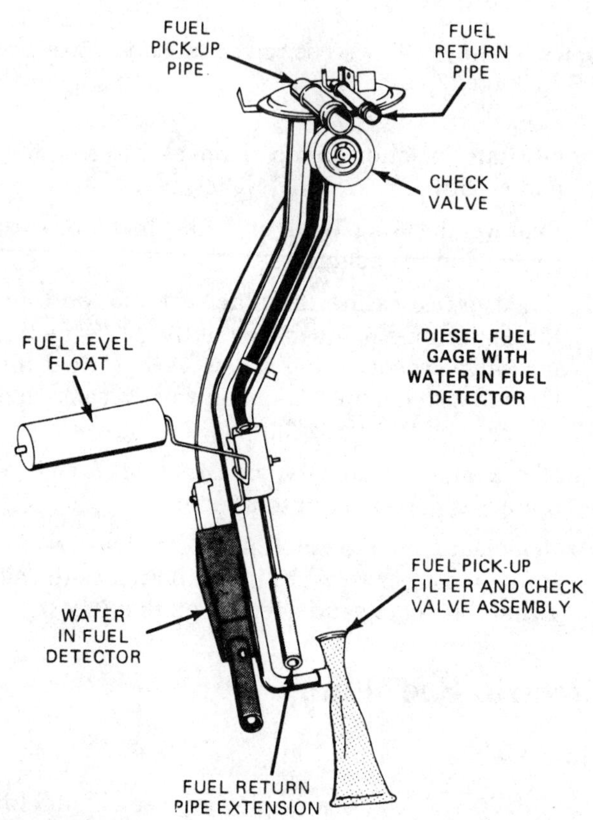

Figure 22–57. Typical water-in-fuel indicator. (Courtesy of Oldsmobile Division, GMC)

Fuel Pump Pressure/Restriction Check

If the fuel lift pump fails to deliver 1/2 pint or more of fuel within a 15-second time span as described above, proceed to check its operation by performing both a vacuum and pressure check.

Procedure

1. Disconnect the fuel inlet hose at the lift pump and connect a vacuum gauge.
2. Turn the ignition key to the run position and carefully note the vacuum gauge reading.
3. If the vacuum reading is less than 5.3" Hg or 17.9 kPa, the pump should be replaced—if no other restriction is found between the pump and the tank filter.
4. The fuel lines should be checked by disconnecting each section of line and connecting a vacuum gauge with the ignition turned to *run*; the reading on the gauge should be the same as it was in Step 3, otherwise replace the line or hose.
5. If the fuel pump and the lines are OK, then remove the tank unit, replace/clean the in-tank filter, and clean out the fuel tank.
6. Perform a fuel pump pressure test by installing a pressure gauge into the fuel line between the pump and filter; turn the ignition key to the *run* position and note the pressure.
7. A properly operating lift pump should create between 5.8–8.7 psi (40–60 kPa), otherwise replace the pump; if the pressure surges or is too high, the pump should also be replaced.

NOTE: The electric fuel lift pump receives battery current when the ignition key is turned to either the *run* or *start* positions. The pump ground circuit is through its mounting bracket. Maximum operating current of the pump is 3 amps.

A poor ground can cause the pump to operate poorly or even intermittently under certain conditions. Do not replace the pump before checking the state of the electrical connections and confirming that it is receiving 12 volts from the battery at a current draw of 3 amps.

If the lift pump operates according to all of the tests conducted above, but the engine seems to be starving for fuel and you have already checked the fuel filter for plugging, keep in mind the following point: The diesel fuel filter cannot be checked in the same manner as a gasoline fuel filter by simply blowing through it with your mouth because air will not pass easily through a wet filter. Use clean, dry regulated shop air pressure between 2–3 psi (13.8–20.7 kPa) to effectively determine if the filter is plugged.

Alternately, you can install a fuel pressure gauge between the filter and the injection pump and compare its reading to that obtained on its suction side (lift pump pressure). A pressure difference of between 2–3 psi is considered average. Pressure delivery to the injection pumps transfer pump in the hydraulic head should be at least 2 psi.

Injection Pump Removal

Removal of the fuel pump should be done in a systematic fashion, and care should be exercised when removing all items to avoid any possible damage. Start by removing the air cleaner, and proceed to remove all items that are either in the way or are necessary in order to remove the pump. Once all these items have been removed, loosen and remove the three pump-to-adapter nuts and withdraw the pump assembly.

A detailed injection pump removal procedure is given under the 6.2L GMC diesel engine in Chapter 23. Since the 4.3L, 5.7L, and 6.2L diesel engines all use the Roosa Master DB2 injection pump, the sequence for all engines is basically the same.

Injection Pump Installation

Before pump installation, you should check out the condition of the pump adapter. If there is any doubt about bushing wear or possible injection pump adapter drive shaft gear wear or damage, the engine front cover would have to be removed to get at the drive and driven gears. If one gear is replaced, then both must be replaced because of the matched angled drive setup used.

If the injection pump drive and driven gears are being replaced, refer to Figure 22–58 and make sure that the injection pump driven gear is within the 0.002 to 0.006″ (0.05 to 0.152 mm) stated by Oldsmobile for gear end play. If not, shim thrust washers are available in 0.003″ (0.076-mm) increments from 0.103 to 0.115″ (2.616 to 2.921 mm) in order to obtain the specified clearance.

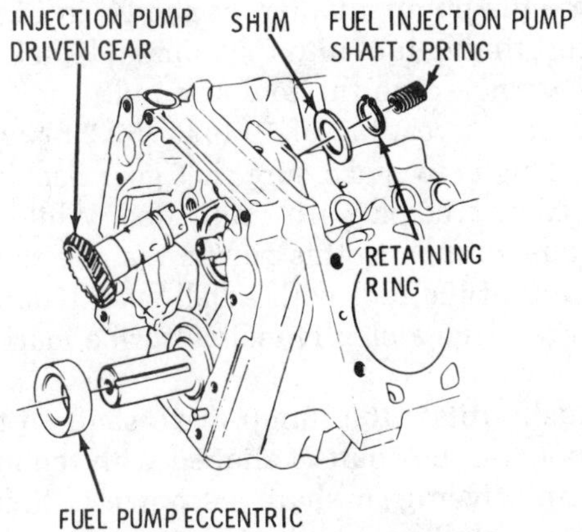

Figure 22–58. Shimming injection pump gear end play. [Courtesy of Oldsmobile Division, GMC]

Refer to Figures 22–58 and 22–59; position the camshaft dowel pin on the front end of the camshaft where the cam gear mounts, at the 3 o'clock position. Align the 0 marks up between the injection pump drive and driven gears, and with the camshaft gear and crankshaft gear timed to one another as shown in Figure 22–60, locate and slide the injection pump drive gear onto the camshaft with care; then slide the camshaft into its bore all the way. Install the bolt into the end of the camshaft, which holds the gear and chain in position, and tighten it to 65 lb.ft. (88.12 N·m) of torque.

CAUTION: Be sure to exercise care when installing the camshaft to avoid camshaft support bearing damage. Coat the camshaft and bearings with a good-quality lubricant or Oldsmobile recommended lubricant no. 1051396.

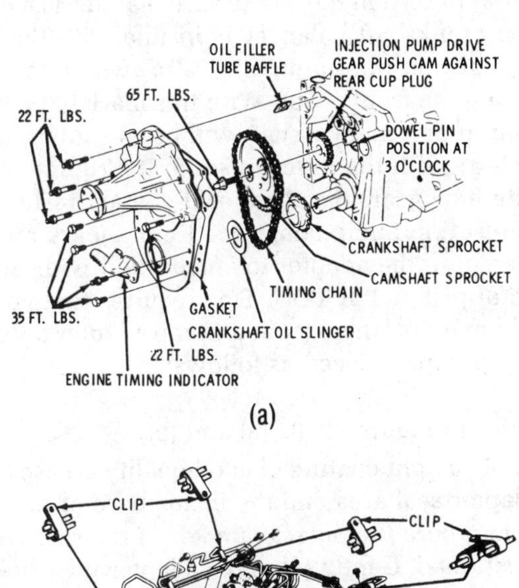

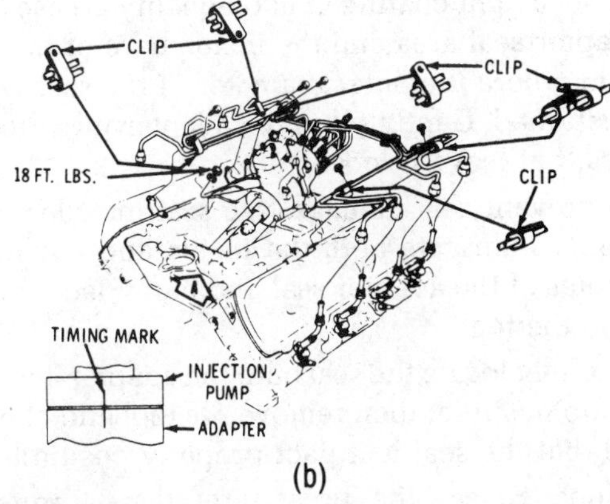

Figure 22–59. a. Engine front cover bolts. b. Timing marks and injection pump lines. [Courtesy of Oldsmobile Division, GMC]

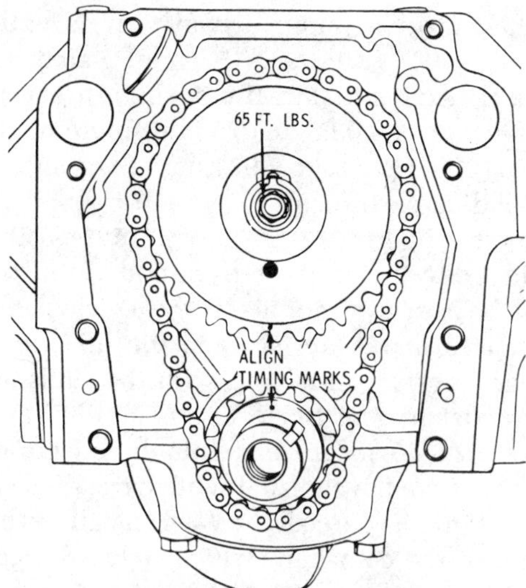

Figure 22-60. Checking alignment of engine timing marks. [Courtesy of Oldsmobile Division, GMC]

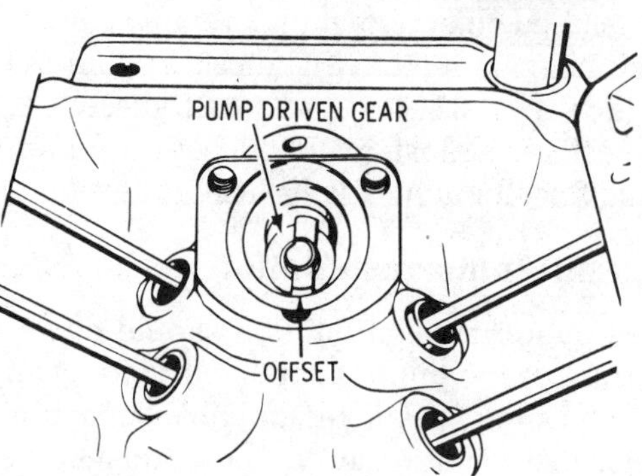

Figure 22-61. Index offset to right—Pump driven gear. [Courtesy of Oldsmobile Division, GMC]

Before installing the injection pump adapter (Figure 22–62), bar the engine over until cylinder 1 is at TDC. Cylinder 1 is on your right when standing at the front of the engine. Make sure that the TDC mark on the crankshaft balancer is in line with the 0, or zero, mark on the timing indicator, which is bolted to the engine front cover. With this mark in position, the pump driven gear shaft will be as shown in Figure 22–61 with the index *offset* to the right.

With new gears installed, or if you are installing a new injection pump adapter, a timing mark must be scribed onto the adapter to correspond with the position of piston 1 at TDC. This requires that you file the old mark off the injection pump adapter only, not off the pump. Proceed as follows:

1. Refer to Figure 22–62 (a) and (b).
2. Apply a light coating of good-quality grease to the adapter seal area and the inside bore of the manifold where the outer diameter of the seal will be positioned. Gently place the adapter into the bore area, but leave it loose.
3. To prevent seal damage, use seal installer tool J-28425. Lubricate both the inside and outside diameter of the adapter seal and tool. Place the seal onto the tool.
4. Carefully locate the seal onto the adapter and push it into position; then remove the tool and check to see that the seal is in fact properly positioned.
5. If not, repeat the procedure; then torque the adapter bolts to 25 lb.ft. (33.89 N·m).

Once the adapter and seal are in position, timing tool J-26896 must be used to place a new timing scribe line onto the pump adapter, as shown in Figure 22–63. Insert timing tool into the pump adapter. Place a torque wrench onto the tool and pull it toward cylinder 1 until a reading of 50 lb.ft. (67.79 N·m) is obtained. This is to make sure that gear backlash and timing chain free play are eliminated. While holding the torque wrench in this position, strike the button on the top of the tool with a ball peen hammer just enough to place a clearly visible scribe mark on the flange.

Manually rotate the pump drive shaft so that the offset tang on the shaft is aligned with the offset on the pump driven gear shaft, as shown in Figure 22–61; then carefully guide the injection pump into position and align the scribe mark on the injection pump housing with the scribe mark on the adapter.

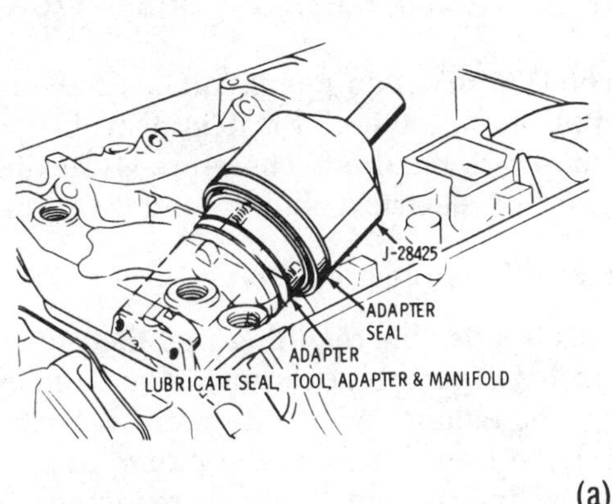

(a)

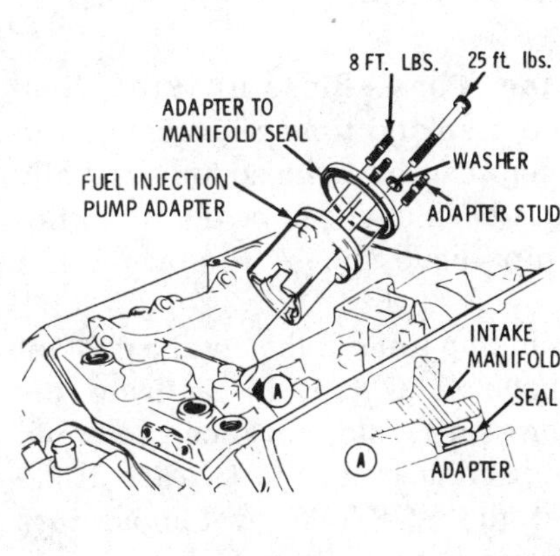

(b)

Figure 22-62. a. Using seal protector to install adapter to manifold seal. b. Exploded view of location of manifold seal. [Courtesy of Oldsmobile Division, GMC]

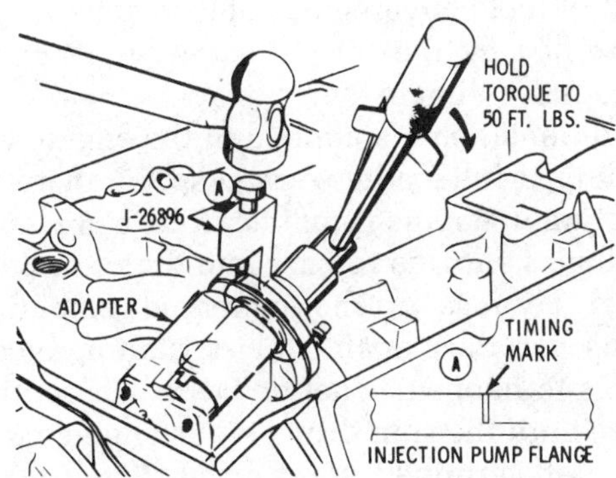

Figure 22-63. Use of timing tool to scribe new timing line. [Courtesy of Oldsmobile Division, GMC]

A ¾-in. (19.05-mm) socket or wrench on the boss at the front end of the injection pump may be required to turn the pump so that the timing lines are aligned. Tighten the pump to adapter retaining nuts to 35 lb.ft. (47.45 N·m). Remove all shipping caps from the pump and connect all fuel lines to nozzles, fuel filter bracket and filter, and all other necessary parts.

Bleeding the Fuel System

On vehicle applications, with all lines connected and the electric shutoff device wired up, place the injection pump throttle lever in its maximum speed position. All injection nozzle fuel lines should be left loose at the nozzle for now. When you turn on the ignition switch, the *wait* light will come on, indicating that the glow plugs are being supplied with electricity to preheat the pre-combustion chambers. Since the nozzle lines are loose at this time, no fuel will be delivered to them. Crank the engine over until fuel starts to bleed from each nozzle fuel line, and tighten them up individually as fuel flows freely from each one. With fuel having been bled to each injection nozzle, turn the ignition key to *wait*, and once the *wait* light on the dash goes out, the *start* light will come on.

Attempt to start the engine; if it does not start, loosen the injection nozzle lines as before and rebleed the system. Repeat the starting procedure again.

Once the engine starts, run it at a fast idle and check all fuel lines for fuel leakage. Inspect the pump for any signs of leakage. Warm up the engine to its normal operating temperature and do any adjustments required.

Reducing Cranking Time—Bleeding Fuel System. Generally the cranking time requirements to prime the fuel system can be reduced substantially once a fuel filter has been changed or the fuel lines or injection pump have been removed and reinstalled.

To reduce the cranking time, if the engine coolant temperature is above 41°C (105°F), activate the HPCA (housing pressure cold advance) solenoid (Figure 22–49) by disconnecting the two electrical lead connectors at the EGR-TVS or temperature switch at the cylinder head bolt on California engines. Bridge the connector with a jumper wire arrangement such as shown in Figure 22–42 through 22–44. Crank the engine until it fires and runs, then remove the jumper wire and reconnect the EGR-TVS or temperature switch wire leads.

Injection Pump Overhaul

Injection pump overhaul should only be done by those personnel experienced in the intricacies of such detailed work. Special equipment and test machines are required for this purpose. Therefore, if pump repair or testing is necessary, take it to a local fuel injection shop that is equipped to deal with such work.

If you are experienced in the overhaul and repair of fuel injection pumps and nozzles and have the necessary equipment readily available, obtain a copy of Roosa Master Operation and Instruction Manual for the Model DB2 Pump, Catalog Number 99009, along with the specifications adjustment sheet from your local fuel injection dealer, or write to Roosa Master, Stanadyne/Hartford Division, Box 1440, Hartford, Connecticut 06102, U.S.A.

Injection Pump Accessories

The DB2 injection pump can be used on a variety of applications; therefore, it is available with several options as required. These are (1) flexible governor drive, (2) electrical shutoff, and (3) torque screw.

Flexible Governor Drive

This is a flexible retaining ring that serves as a cushion between the governor weight retainer and the weight retainer hub. Any torsional vibrations that may be transmitted to the pump area are absorbed in the flexible ring, therefore reducing wear of pump parts and allowing more positive governor control.

Electrical Shutoff

This device is available as either an energized to run (ETR) or energized to shutoff (ETSO) model. In either case, it will control the run and stop functions of the engine by positively stopping fuel flow to the pump plungers, thereby preventing fuel injection.

Torque Screw

This device is used with Roosa Master pumps to allow a tailored maximum torque curve for the particular engine application. This feature is commonly referred to as *torque backup* because the engine torque will generally increase toward the preselected and adjusted point as the engine rpm decreases. The three factors that affect this torque are (1) the metering valve opening area, (2) the time allowed for fuel charging, and (3) the transfer pump pressure curve.

Do *not* attempt to adjust the torque curve on the engine at any time. This adjustment can be done only during a dynamometer test where fuel flow can be checked along with the measured engine torque curve or on a fuel pump test stand.

Turning in the torque adjustment screw moves the fuel metering valve toward its closed position. The amount of fuel delivered at full-load governed speed is controlled by the torque screw and not by the roller-to-roller dimension.

If additional load is applied to the engine while it is running at full-load governed speed, there will be a reduction in engine rpm; therefore, a greater quantity of fuel is allowed to pass into the rotor pumping chamber because of the increased time that the charging ports are open (slower running engine allows this to happen). Fuel delivery will continue to increase until the rpm drops to the engine manufacturer's predetermined point of maximum torque. It is at this point that the amount of fuel will be controlled by the setting of the pump roller-to-roller dimension (see Figure 22–21).

External Pump Adjustments

Several adjustments may be required from time to time on the injection pump, and these can be done quite readily. If minor external adjustments fail to solve the problem, do not attempt to disassemble the injection pump piece by piece on the engine. Remove the pump as an assembly and check it out on a test stand by yourself if you are experienced in fuel injection practices and procedures, or take it to a local fuel injection dealer for repair. Following are some of the typical checks and adjustments that you may have to do on the DB2 pump.

DB2 Injection Pump Testing

To effectively test the injection pump, it must be mounted onto a test stand, such as a Hartridge unit. Test directions for the particular model of pump are available from Roosa Master and the engine manufacturer; however, the following information can be considered as general routine to mount and test a DB2 injection pump on a test stand. Figure 22–64 illustrates the DB2 connections required for such a pump on the test stand.

Since the injection pump is driven at half engine speed on a 4-stroke cycle engine, the test stand should be capable of at least 2250 rpm. Most test machines

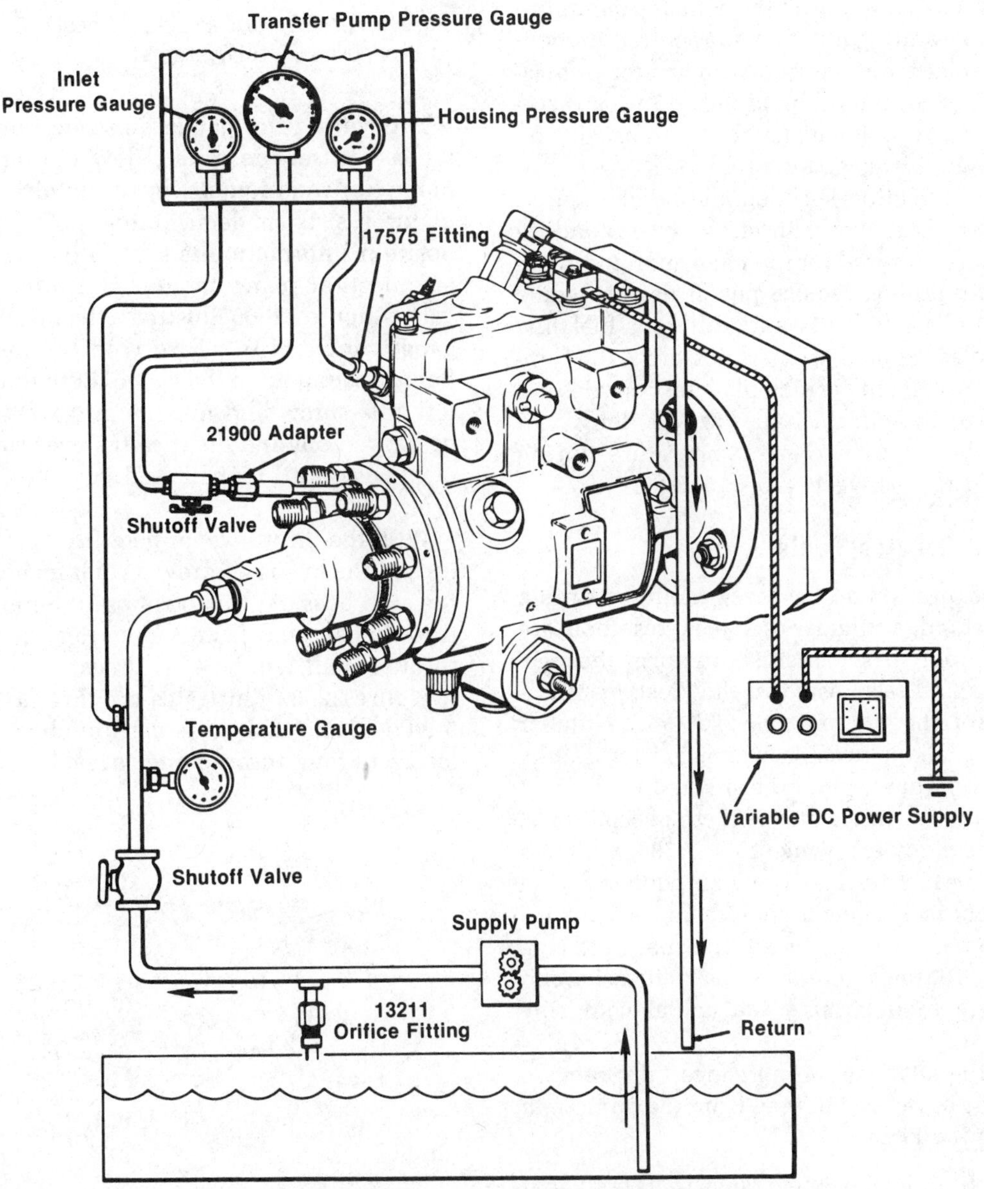

Figure 22–64. Roosa Master DB2 injection pump/test stand hookup. [Courtesy of Stanadyne Diesel Systems]

are currently equipped with both a zero backlash drive coupling as well as a digital tachometer. These items are required to ensure satisfactory test results.

A variable voltage source to duplicate the vehicle electrical system voltage is required to test the operation of the electric fuel shutoff solenoid. This variable voltage is also required to test the injection pump housing fuel pressure cold advance mechanism.

Required test gauges are common to most test machines and should present no problem; however, a 0–160 psi (0–1.130 mPa) is necessary for the vane transfer pump pressure check. A 0–30 psi (0–207 kPa) gauge is needed to check injection pump housing pressure.

A vacuum gauge calibrated from 0–30″ Hg or 0–0.101 mPa is needed to check the transfer pump lift.

A fuel temperature gauge is also required to monitor the fuel inlet temperature. A flowmeter to measure return fuel oil should be plumbed into a three-way valve so that the flowmeter is only in use during the return fuel oil leakage test.

The test specification sheet should be checked to establish what particular calibrating nozzles and injection lines are required for the particular pump under test. Some pumps require pencil nozzles while others require microjectors and different fuel line sizes. Example: Pencil nozzle lines are 0.072″ inside diameter, while the orifice plate test nozzles are 0.0931″—although both line types are 25″ long.

Normal calibrating fluid oil temperature during testing should run between 110–115°F (43–46°C).

Special Test Service Tools

Once the injection pump has been mounted onto a suitable test stand, a number of special test tools are required in order to successfully perform the necessary checks and tests listed in the "test specification" sheet for the pump. A test kit Part Number 22037 contains all the necessary tools to successfully test the injection pump on the test stand.

Figure 22–65 illustrates one of these special tools, namely the advance test gauge P/N 21734, which is designed to measure the injection pump internal cam ring movement in 1/4 degree increments. The gauge scale is designed to magnify the internal cam ring movement by 10 times in order to make it easier and more accurate to determine the actual cam ring movement.

To install the advance timing gauge unit, remove the timing line cover and its seal from the pump and install the timing kit.

NOTE: The timing kit has two plugs in its body that can be removed to allow installation of a housing pressure fuel gauge.

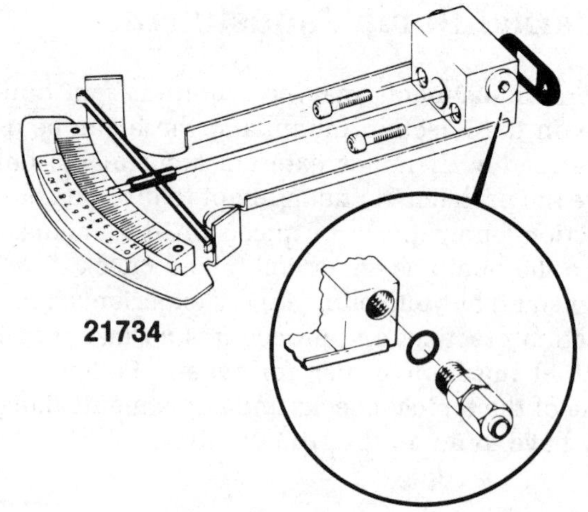

Figure 22–65. Advance test gauge 21734. (Courtesy of Stanadyne Diesel Systems)

A throttle lever gauge illustrated in Figure 22–66 is used to set the critical low idle angle. On automatic transmission equipped vehicles (vacuum modulator assists in determining shift points), throttle position information is sent to the transmission via an injection pump mounted throttle shaft external valve. Figure 22–66 illustrates the injection pump and gauge mounted in a vise with the pump mounting flange resting upon the gauge. Both the spring loaded low idle screw and maximum travel screws should be backed out and the throttle lever rotated counterclockwise until its ball stud just contacts the gauge arm.

With the throttle lever held firmly in this position, turn in the low idle screw until it makes contact with the stop boss. At this position, the movement of the throttle ball stud from the vertical centerline of the throttle shaft will be 41 degrees.

Figure 22–67 illustrates the throttle lever protractor P/N 22089, which is used to check both the part load and total throttle angle/travel.

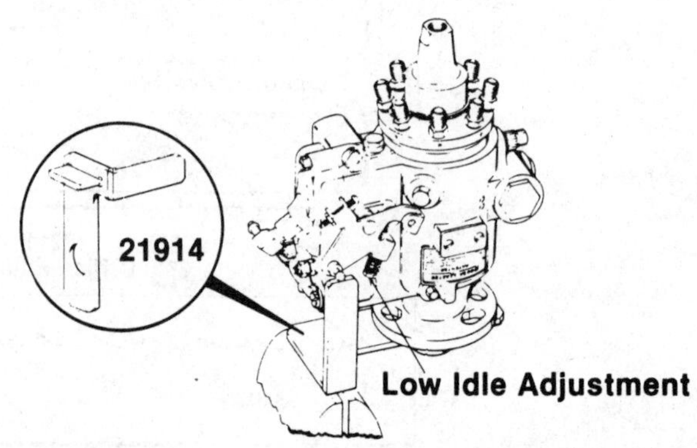

Figure 22–66. Throttle lever gauge 21914. (Courtesy of Stanadyne Diesel Systems)

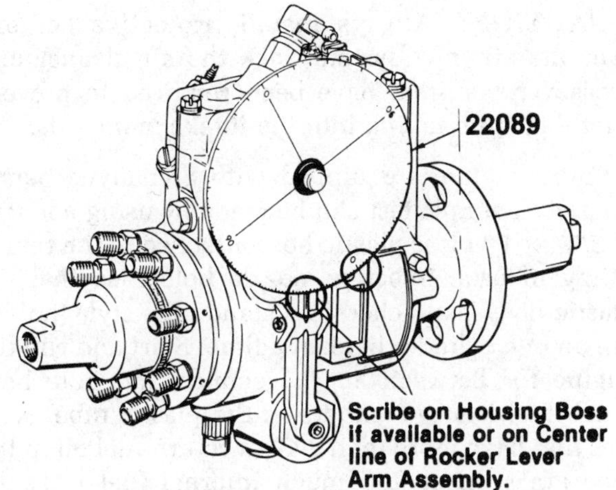

Figure 22-67. Throttle lever protractor. (Courtesy of Stanadyne Diesel Systems)

The air timing tool (Figure 22–68) is designed to allow an optical check of the position of the injection pump timing mark along with a scribe to re-mark the pump mounting flange if necessary. This check is done with applied air pressure of 60–100 psi (414–690 kPa) to the No. 1 discharge fitting of the pump, which is identified by the number "1" stamped on the housing.

The four socket head screws are then loosened off, and the tang locating fixture is rotated in the normal direction of pump rotation (see arrow on housing) until the cam rollers are felt to make contact with the internal cam ring, which is the actual beginning of injection.

If two contact points are felt, use the last one for timing purposes. Holding the drive shaft, rotate the scribe body to align the timing marks on the pump mounting flange and optical gauge scribe line. The actual number of degrees can be determined by counting the number of increments from the "0" mark on the locating fixture tang to the pointer on the scribe body.

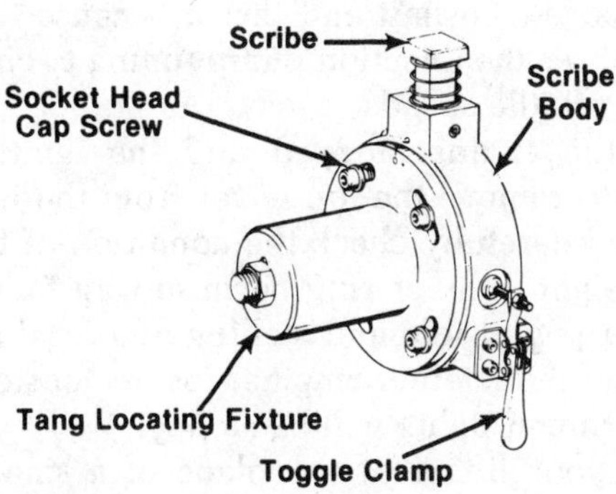

Figure 22-68. Air timing tool. (Courtesy of Stanadyne Diesel Systems)

Diesel Engine Precautions

Because of the design and operating characteristics of the diesel engine, several precautions should be considered any time that you are servicing or troubleshooting these engines.

1. Since the diesel engine operates with an unthrottled air supply at all times to the cylinders, no vacuum is created. This presents a problem on passenger cars and light trucks that depend upon a vacuum source to operate power brakes and automatic transmissions. The vacuum source on the Oldsmoble and Chevrolet diesel engines, namely the 4.3L (V6) and 5.7L and 6.2L V8's, is a camshaft-driven vacuum pump located at the rear of the engine in the drive hole where the ignition distributor is normally located on a gasoline engine. Running the engine without this vacuum pump in position will result in *no drive* to the engine oil pump. The engine will run but will receive no oil pressure.
2. *Never* adjust injection pump timing with the engine running.
3. Because of the heavier cranking load (higher compression) of the diesel engine, two 12-volt bateries connected in parallel are used. When jump starting the engine, always ensure that you *never* connect the boost batteries in "series" because this will produce 24 volts. Serious vehicle electrical system damage can result as well as damage to the glow plugs.
4. Always install screened covers over the air intake after removing the air cleaner or the air crossover pipe.
5. Anytime that any rocker arms have been loosened or removed for service work, the valve lifters must be bled down to prevent possible damage to the valve train because of the very small clearance that exists between the piston and the valves on high compression diesel engines.
6. Avoid applying 12 volts directly to a fast glow plug (6-volt unit) because the current must be pulsed on/off—otherwise the glow plug can burn out.

Troubleshooting GMC 263/350 Diesel Engines

Effective troubleshooting of the Oldsmobile 263 cu.in. (4.3L) V6 and 350 cu.in. (5.7L) V8 diesel engines parallels that for the newer 6.2L (378 cu.in.) V8 Chevrolet light truck diesel engine, as well as the Ford/IHC 6.9L, because all engines employ the same basic fuel system.

Initial problems that existed with the Oldsmobile diesel were generally associated with rough idle, exhaust smoke, and hard starting, as well as a commonly heard complaint of a lack of power.

The lack of power complaint was understandable because most people who initially purchased these vehicles had in the majority of cases been used to driving large gasoline V8's with considerably more horsepower and faster acceleration characteristics than its diesel counterpart. However, once personnel realized that the diesel is an inherently slower accelerating engine than the gasoline unit, they realized that their driving habits would have to change in order to obtain maximum fuel economy and performance from these engines.

Troubleshooting Technique

Troubleshooting techniques should always start with the simplest checks. Do not overlook what you may consider to be too simple a possibility because in most cases of effective troubleshooting, service personnel feel that if there is a problem with the engine, then it must be major.

This situation has been worsened by the fact that many automotive mechanics have not had the opportunity to receive training on the newer diesel engines now in use on cars and light trucks and therefore feel alienated from their usual effective troubleshooting techniques.

You should always start with the same basic routine that you would use on a gasoline engine when troubleshooting a diesel engine. Check the color of the exhaust smoke first and determine if it is unusual. Check for loose linkage, fuel leaks, oil leaks, air leaks, and other typical areas of possible basic trouble spots.

Rough Idle

A rough idle situation on a diesel can be caused by a variety of problems; therefore, before removing any items from the engine or fuel system, check for signs of fuel leakage at both the injection pump outlet lines and at the nozzle fuel lines as well as the nozzle-to-cylinder head seating area.

Checks for rough idle should be made with the engine at normal operating temperature.

Due to the location of the injection pump (which is between the cylinder heads and in the front of the block valley), any fuel leakage will tend to cause a fuel puddle under the pump. If signs of fuel leakage are evident, then remove the air cleaner and crossover manifold to allow access to the injection pump fuel lines.

CAUTION: Always install protective screens over the air inlet system once the air cleaner and crossover manifold have been removed to prevent anything from falling into the intake manifolds.

If the engine is equipped with Stanadyne pencil nozzles, a simple test can be made by using a length of 3/8″ × 6″ clear plastic hose attached to the return fitting of each injection nozzle holder so that the plastic hose is attached in a stand pipe style (hold it up) on one cylinder bank at a time. Start and run the engine for between 30–45 seconds and note how much fuel leakage gathers in the plastic tube. Nozzles that have more than 3/4″ of return fuel oil in the tube are suspect (too much internal fuel leakage); therefore, they should be removed and tested for proper operation.

If the engine is equipped with the newer CAV or GM/DED microjectors, which are a screw fit into the cylinder heads, cracked injector bodies can result from overtorqueing of the nozzle assembly when installed into the head. To check for cracked injector bodies, disconnect the high pressure tubing at the nozzle holder inlet (disconnect the wire lead at the electric shutoff solenoid at the injection pump to prevent the engine from starting), and crank the engine over on the starter.

Carefully observe the area around the injector body where it seats on the cylinder head. If signs of bubbles appear, then compression leakage exists. It may be helpful to squirt a small amount of fuel oil into the injector inlet as well as spraying a small amount around the injector seat area to help isolate a possible leak.

On pre-1981 diesel engines, pull the vacuum pump (located at the rear of the engine where the ignition distributor sits on a gasoline engine) and turn the shaft 120 degrees. An out-of-phase drive gear was known to cause a rough idle condition due to the fact that the camshaft and thrust was augmented by the vacuum pump cyclic loading characteristics.

Another source of possible rough idle was caused by excessive camshaft end thrust because this condition causes the injection pump timing to change at idle and off-idle speeds.

With the engine stopped and the ignition key turned *off*, remove the top cover from the injection pump and carefully check the condition of the governor weight retainer ring because ring failure can cause surging at idle and stalling and hard starting because, if the retainer ring has lost its tension, then the governor weights will be sloppy.

Using your finger or the blade of a small clean screwdriver, engage the retainer between the governor two weight sockets and force the retainer to move first one way and then the other. If the retainer doesn't spring back to within 1/8″ of its original position when released, the flex-ring should be replaced.

Another check that should be done at this time is to see if the pump incorporates the new hydraulic snubber kit and idle spring that was released by Stanadyne to cure a possible rough idle condition. If you are in doubt on this point, Stanadyne reference S.L 360A bulletin describes this kit, or check with your local GM dealer or local ADS (Association of Diesel Specialists) diesel repair shop.

If any nozzles are suspected of misfiring, it is good practice to remove all of them and check their popping/release pressures. Also check their spray patterns and look for signs of leakage. A glow plug resistance check can be used to effectively isolate possible faulty nozzles. This procedure can be found in detail under the "glow plug" section for these engines in their respective chapters.

When nozzles are removed from the engine, you should conduct a cylinder compression check, although this can also be done by removing the individual cylinder glow plugs and adapting a pressure gauge fitting.

Average compression pressures will run between 375–425 psi (2586–2930 kPa), although 300 psi (2068 kPa) is acceptable as a minimum. However, the lowest cylinder reading must be 80% or more of the highest cylinder reading.

Another problem area that has caused rough idle is varying inside diameters of high pressure steel injection tubing fuel lines, which has resulted in uneven delivery volumes to each injector. The individual cylinder fuel injection lines can be checked using a syringe or air flow gauge to determine if they are in fact delivering the same flow rates. Replace any fuel lines that do not balance.

Match up a set of injection line tubes as closely as is possible even if it means using some new lines. Do not mix tubings marked "B" with those marked "H," and be especially careful that tubings of the ferrule type are not intermixed with nozzles that require the flare or swaged type ends or vice versa.

Mid-range engine surges can be the result of a sticking or worn metering valve within the injection pump, or a standard delivery valve in an oversize bore; pump roller concentricity out of spec can also cause this complaint.

On 1980 models with lockup torque converter transmissions, a converter clutch switch can be responsible for a surge condition between 35–45 mph. If after resetting the switch, the surge still exists, change the pressure switch inside the transmission to a 42–45 psi type.

If after sudden release of the throttle, the engine idles rough or stalls, this could be an indication of camshaft or valve lifter problems. Check this with a dial indicator to determine the valve lift.

Figure 22–69 provides a guide for diagnosis of diesel engine idle roughness; Figure 22–70 for glow plug problems resulting in rough idle.

Hard Starting

Hard starting can be caused by a variety of reasons such as:

1. Waxed fuel in cold weather.
2. Low battery power (200–240 rpm required).
3. Engine oil too heavy (slow cranking speed).
4. Faulty glow plugs.
5. Poor atomization at injectors.
6. Low compression.
7. Plugged air filter.
8. Defective starter solenoid or starter drawing excessive amperage can result in failure to hold the electric shutoff solenoid in the open position.
9. Defective fuel shutoff solenoid.
10. Low fuel delivery at cranking speed especially noticeable during hot starts.
11. Plugged fuel filters.
12. Faulty fuel tank filler cap.
13. Faulty fuel check valve in fuel transfer pump allowing fuel to drain back to the tank can cause an intermittent no-start condition.
14. Air leakage into the suction side of the fuel system can cause hard starting, misfiring, excessive smoke, or low power (check this condition with the insertion of a clear plastic hose at the return line on the injection pump by looking for air bubbles in the return fuel).
15. Injection pump timing incorrect.

Lack of Power

A lack-of-power complaint can be caused by an air or fuel system problem (Figure 22–71) or by improper throttle linkage adjustment.

Injector problems, low cylinder compression and low fuel delivery can all cause low power problems as well as incorrect pump/engine timing. A plugged air filter can also cause air starvation and hence a low power complaint.

Low injection pump housing fuel pressure can cause low power. On 1978–79 vehicles, malfunction of the ballcheck connector assembly on the injection pump return line elbow affected the automatic advance function of the pump, leading to sluggish performance.

DIESEL ENGINE IDLE ROUGHNESS DIAGNOSIS PROCEDURE

CONDITION

IDLE ROUGHNESS

Idle roughness is defined as an uneven shaking of the engine in comparison to others with the same number of cylinders and in the same body style.

A rough idle condition may be caused by a difference in the output between cylinders on diesel engines. By selection of parts it is possible to alter the output between cylinders, and smooth out the idle quality.

CORRECTION

Follow the diesel engine idle roughness diagnosis procedure. Make all necessary adjustments and corrections. The idle roughness procedure must be followed step by step prior to performing the glow plug resistance check. The glow plug resistance check will only be successful after the idle roughness procedure is performed and corrections made.

CONDITION

COAST DOWN ROUGHNESS

A condition may exist where a roughness is observed on coast down at 50 mph or less with a closed throttle.

CORRECTION

Confirm that this condition is engine roughness rather than a tire waddle or a bent wheel by coasting down through the roughness period in neutral with the engine at 1500 to 2000 RPM. If roughness still exists, during the coast down, the condition is not caused by engine roughness. If the roughness condition is gone, follow the idle roughness diagnosis procedure. If not corrected prior to the glow plug resistance procedure, correct the roughness using the glow plug resistance procedures.

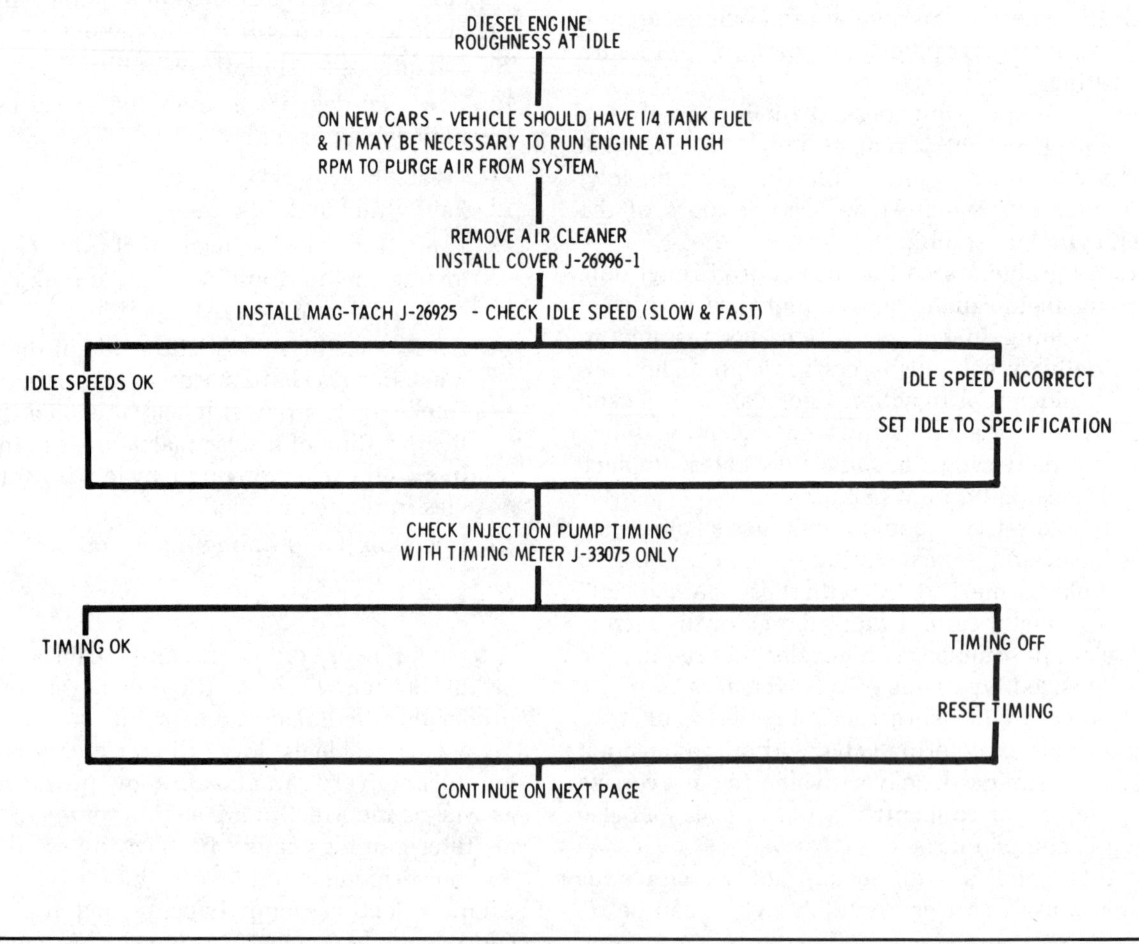

Figure 22-69. Guide for diagnosis of diesel engine idle roughness. (Courtesy of Oldsmobile Division, GMC) (Continued)

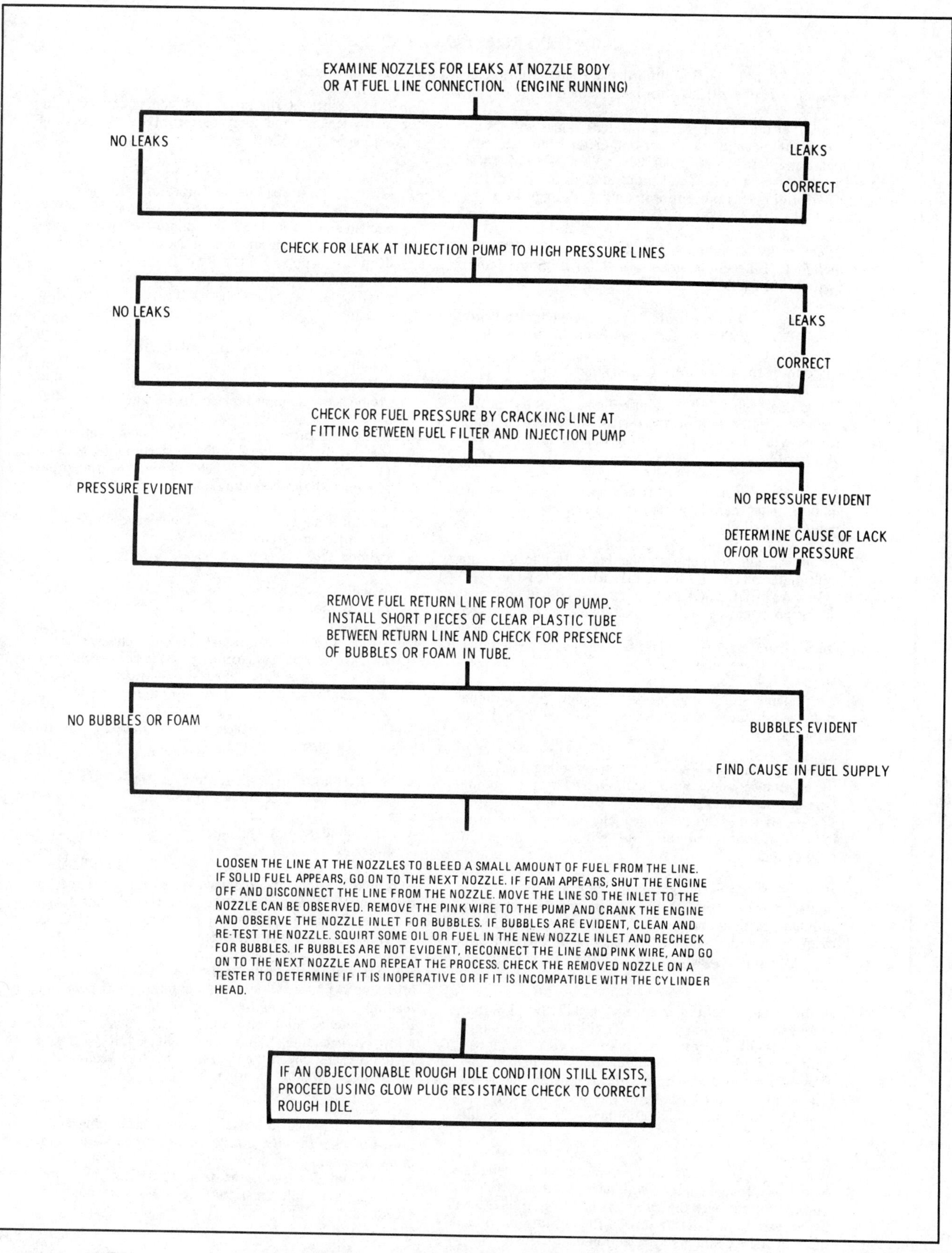

Figure 22-69. Guide for diagnosis of diesel engine idle roughness. (Courtesy of Oldsmobile Division, GMC)

GLOW PLUG RESISTANCE PROCEDURE

1. Use the Kent-Moore High Impedence Digital Multimeter (Essential Tool J-29125A) for measurements.

   If another ohmmeter is used different values will result. This does not mean that another ohmmeter is not accurate for measuring resistors or solenoids . Tool J-29125A was used in the development of this procedure. Its use is required when performing this procedure if similar readings are to be obtained.

2. Select scales as follows: LH Switch to "OHMS", RH Switch to full counterclockwise, "200Ω," Slide Center Switch to the left "DC.LO."

3. Start engine, turn on heater and allow engine to warm up. REMOVE all the feed wires from the glow plugs.

4. Disconnect the generator two lead connector.

5. Using Mag-Tach J-26925, adjust engine speed by turning the idle speed screw on the side of the injection pump to the worst engine idle roughness, but do not exceed 900 RPM.

6. Allow engine to run at worst idle speed for at least one minute. The thermostat must be open and the upper radiator hose hot.

7. Attach an alligator clip to the **black** test lead of the multimeter. THIS CLIP MUST BE GROUNDED TO THE FAST IDLE SOLENOID. It must remain grounded to this point until all tests are completed.

8. On a separate sheet of plain writing paper write down the engine firing order.

9. With engine still idling, probe each glow plug terminal and record the resistance values on each cylinder in firing sequence.

   If the car is equipped with an electric cooling fan, record the resistance values with the cooling fan not running. Do not disconnect the cooling fan electrically.
   The resistance values are dependent on the temperature in each cylinder, and therefore indicate the output of each cylinder.

10. If an ohm reading on any cylinder is about 1.3 or 1.4 ohms, check to see if there is an engine mechanical problem. Make a compression check of the low reading cylinder and the cylinders which fire before and after the low cylinder reading. Correct the cause of the low compression before proceeding to the fuel system.

11. If the engine misfires erratically, install the glow plug luminosity probe that is included with the Diesel Timing Meter (J-33075), into the cylinder with the lowest resistance reading. Observe the combustion light flashes. They will be erratic in time with the misfire that is felt. If it is not, move the probe to the next highest reading cylinder until the malfunctioning cylinder is found.

12. Examine the results of all cylinder glow plug resistance readings, looking for differences between cylinders. Normally, rough engines will have a difference of .4 ohms or more between cylinders in firing order. It will be necessary to raise or lower the reading on one or more of these cylinders by selection of nozzles.

13. Remove the nozzles from the cylinders in which you wish to raise or lower the ohm reading. Determine the pop off pressure of the nozzles as well as checking the nozzle for leakage and spray pattern. (Refer to Testing of Nozzles, Section 6C5.)

    A. Install nozzles with a higher pop off pressure to lower the ohm reading, and nozzles with lower pop off pressure to raise an ohm reading. Normally, a change of about 30 psi in pressure will change the reading by about .1 ohm. Nozzles normally will drop off in pop off pressure with miles. Use nozzles from parts stock or a new car. Use broken-in nozzles on a car with 1500 or more miles, if possible.

    B. Whenever a nozzle is cleaned or replaced, before installing the injection pipe, crank the engine and watch for air bubbles at the nozzle inlet. If bubbles are present, clean or replace the nozzle.

    C. Install the injection pipe, restart engine and check idle quality. If idle is still not acceptable, recheck glow plug resistance of each cylinder in firing order sequence. Record reading.

    D. Examine all glow plug resistance readings looking for differences of .4 ohms or more between cylinders.

       It will be necessary to raise or lower the reading on one or more of these cylinders as previously done.

    E. After making additional nozzle changes again check idle quality. Normally after completing two series of resistance checks and nozzle changes, idle quality can be restored to an acceptable level.

14. An injection pump change may be necessary if the following occurs:

    A. If the problem cylinder moves from cylinder to cylinder as changes in nozzles are made.

    B. If cylinder ohm readings do not change when nozzles are changed.

NOTE: It is important to always recheck the cylinders at the same RPM. Sometimes the cylinder readings do not indicate that an improvement has been made although the engine may in fact idle better.

A nozzle with a tip leak can allow more fuel than normal into the cylinder, which will raise the glow plug ohm reading. This will rob fuel from the next nozzle in the firing sequence and will result in that glow plug having a low ohm reading. If this is encountered, it is advisable to remove and check the nozzle with a high reading. If it is leaking, it could be causing the rough idle.

If you experience low readings on a glow plug and it does not change with nozzle change, then switch glow plugs between a good and bad cylinder. If the reading of each cylinder is not the same as before the switch, then the glow plug can not be used for rough idle diagnosis.

Figure 22-70. Glow plug resistance check for rough idle in diesel engine. (Courtesy of Oldsmobile Division, GMC)

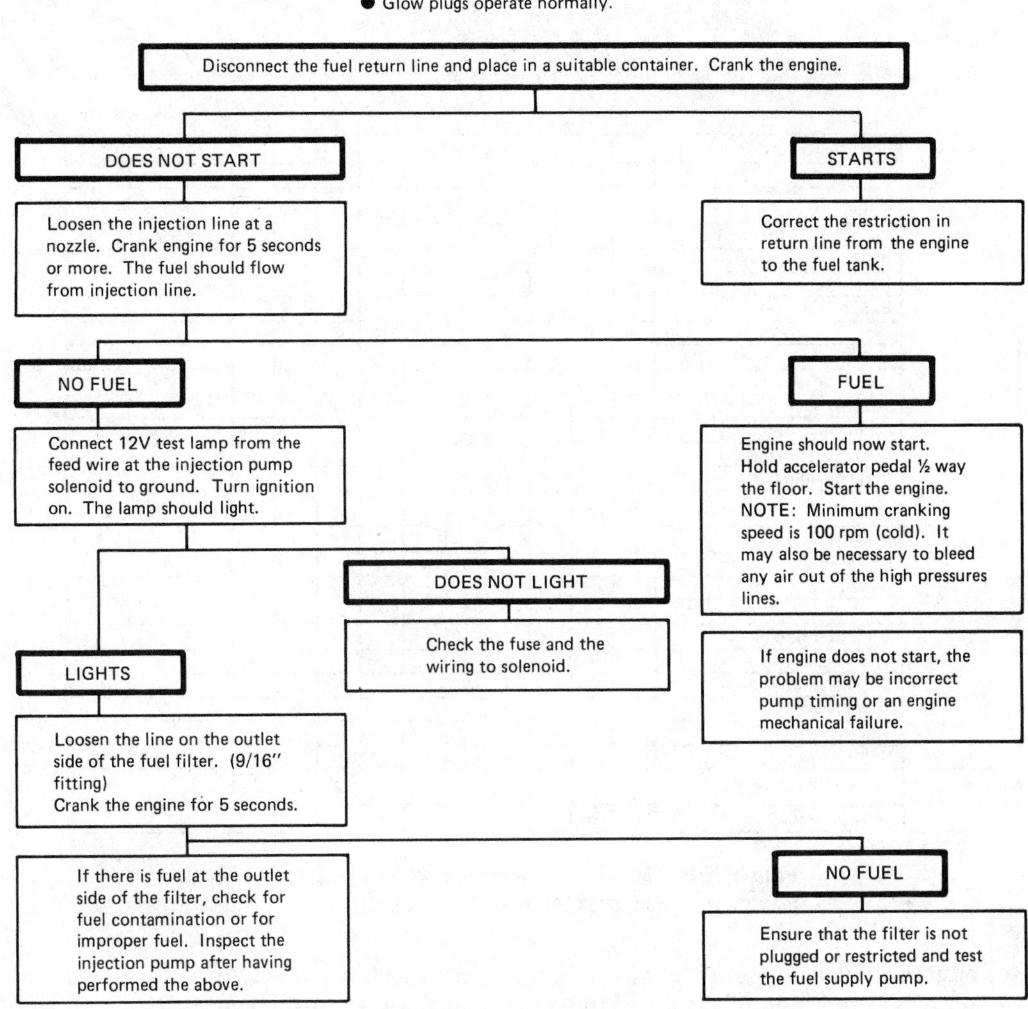

● Engine cranks normally but won't start
● Glow plugs operate normally.
Disconnect the fuel return line and place in a suitable container. Crank the engine.
DOES NOT START
STARTS
Loosen the injection line at a nozzle. Crank engine for 5 seconds or more. The fuel should flow from injection line.
Correct the restriction in return line from the engine to the fuel tank.
NO FUEL
FUEL
Connect 12V test lamp from the feed wire at the injection pump solenoid to ground. Turn ignition on. The lamp should light.
Engine should now start. Hold accelerator pedal ½ way the floor. Start the engine. NOTE: Minimum cranking speed is 100 rpm (cold). It may also be necessary to bleed any air out of the high pressures lines.
DOES NOT LIGHT
Check the fuse and the wiring to solenoid.
If engine does not start, the problem may be incorrect pump timing or an engine mechanical failure.
LIGHTS
Loosen the line on the outlet side of the fuel filter. (9/16" fitting) Crank the engine for 5 seconds.
If there is fuel at the outlet side of the filter, check for fuel contamination or for improper fuel. Inspect the injection pump after having performed the above.
NO FUEL
Ensure that the filter is not plugged or restricted and test the fuel supply pump.

Figure 22-71. Diesel engine fuel diagnosis chart. (Courtesy of General Motors Corporation)

Similarly, a sticky automatic advance device can cause the same complaint.

Troubleshooting/Diagnosis Charts for the 4.3L, 5.7L, and 6.2L Diesels

The following information deals with possible problem areas in the 4.3L, 5.7L, and 6.2L diesel engines produced by General Motors. These charts are not all-encompassing but will serve to cover the most probable causes that you might encounter on these engines. For specific information related to a given year and model of vehicle, refer to the manufacturer's service literature (shop manual).

The diesel engine diagnosis chart shown in Figure 22–72 is arranged so that a large dot and a small dot are shown on the same line. The larger dot indicates that this is more than likely the problem area; however, if this fails to rectify the problem, then sequentially follow the smaller dots until you can isolate the problem. The diagnosis sheets will serve to assist you in identifying the problem condition, its possible cause and suggested correction.

Injection Nozzles—4.3L and 5.7L

The operation of the injectors (nozzles) in the 4.3L (262 cu.in.) V6, 5.7L (350 cu.in.) V8, and 6.2L (378 cu.in.) V8 light truck diesel are very similar. Fuel injectors used in these engines will vary depending upon the year of manufacture.

Up until 1982, GMC also produced a V8 4.3L (260

	Condition	Will not start	Hard starting	Starts—then stops	Rough idle	Misses	Dilution of oil	Knocks	Low power	Black smoke at idle	Black smoke at load	White smoke	Excessive fuel consumption	No heat from heater
Air system	Restricted air intake		•	•					•	•	•		•	
	High exhaust back pressure			•					•		•		•	
Fuel system	Out of fuel	●		•										
	Restricted fuel return line			•	•	•			•			•	●	
	Air leaks in suction lines	•	•	•	•	•			•			•		
	Restricted fuel line or filter	•	•	•		•			•	•	•	•	•	
	External fuel leaks		•	•	•	•			•				•	
	Defective nozzles		•		●	●		●	●	•	●		●	
	Fast idle inoperative			•	•									
	Faulty fuel supply pump	•	•	●	•	•	•						•	
	Incorrect fuel (gasoline)	•	•			•		•	•		•		•	
	Paraffin deposit in filter	•	•	•					•					
	Idle speed too low			●	•									
	Injection pump defective	•	•	•	•	•	•	•	•		•	•	•	
Oil	Wrong grade for ambient	●	•											
Mechanical	Head gasket leaks								•	•			•	●
	Broken or worn piston rings				•		•	•	•	•			•	
	Valve leakage				•	•			•				•	
	Incorrect bearing clearance							•						
	Damaged bearings							•						
	Low compression	•	•		•	•					•		•	
	Loose timing chain				•	•			•			•	•	
	Timing advanced		•		•			•	•	●	●		•	
	Timing retarded		•		•				•			●	•	
	Camshaft worn		•		•	•			•		•		•	
Electrical	Batteries not charged	●												
	Glow plugs inoperative	●	●									•		

Figure 22-72. Diesel engine problem diagnosis chart. [Courtesy of General Motors Corporation]

cu.in.) diesel engine model; this engine was subsequently dropped from production in favor of the V6 4.3L (262 cu.in.) diesel.

Early model 350 cu.in. (5.7L) V8's and some 4.3L (260 cu.in.) V8's pre-1980 were equipped with the Roosa Master pencil nozzle, while later engines were equipped with Lucas CAV "Microjectors," which are also now being produced by the GM/DED (Diesel Equipment Division).

The 6.2L (378 cu.in.) V8 light truck engine uses Robert Bosch nozzles. The main difference between these injection nozzles is that the Roosa Master pencil nozzle is held in place by a clamp and a bolt while the other nozzles are all threaded into the cylinder head similar to a spark plug in a gasoline engine.

The 6.2L nozzles are discussed in detail under Chapter 23 dealing with that engine's fuel system and also in Chapter 29.

Injection Nozzle

Pre-1980 injection nozzles are located at each cylinder and held in place by a spring clamp and bolt screwed into the cylinder head. The nozzle is shown in Figure 22-73. Due to the compactness and thin shape of the injection nozzle, it is commonly referred to as a pencil nozzle with a closed end, since the actual nozzle valve does not project through an opening in the nozzle tip. The valve within the nozzle is held on its seat by spring pressure, and the current nozzle requires 1800 psi (12,411 kPa) to pop or open it, with an allowable tolerance of plus 100 psi (689.5 kPa) or minus 50 psi (344.75 kPa) of this figure.

The nozzle lift adjusting screw is generally set for five-eighths of a turn. An edge-type filter is retained in the inlet fitting as shown. Two seals are used with the nozzle body, a nylon seal immediately beneath the inlet fitting banjo, which prevents any compression leakage, and a Teflon carbon dam seal, which prevents the build up of carbon between the nozzle body and its bore.

Nozzle Fuel Flow

To prevent any variation in timing, the fuel lines to all nozzles are the same length. Fuel, which has been

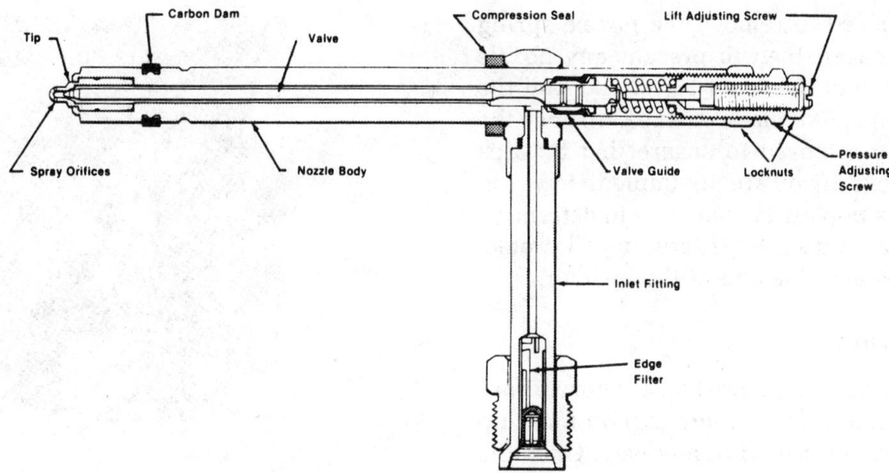

Figure 22-73. Pencil-type injection nozzle. (Courtesy of Stanadyne Diesel Systems)

metered within the injection pump by the position of the fuel metering valve, is pressurized by the inward moving pump plungers, which send the fuel on up to the injection nozzle.

The fuel flows through the edge filter and around the valve, thereby charging or filling the nozzle body. Fuel pressure must be high enough to overcome the spring tension within the nozzle to allow fuel injection into the pre-combustion chamber. This fuel pressure acts upon a differential area at the valve tip, which is the difference between the area of the major diameter of the nozzle valve (at the valve guide) and the area at the valve seat at the tip.

When the valve opens under the action of fuel pressure, it will lift up a distance that has been set by the adjustment of the lift adjusting screw. When fuel delivery from the injection pump ends, the nozzle spring will act to rapidly seat the valve, and sealing is maintained by an interference angle between the valve and its seat, as shown in Figure 22-74.

All internal nozzle parts are lubricated by the fuel oil, a small amount of which leaks between the clearance existing at the valve and its guide during injection. This fuel then flows through a leak-off boot at the top of the nozzle and back to the fuel tank.

For nozzle testing procedures, obtain a copy of Booklet 99002 from your local fuel injection dealer or from Roosa Master, Stanadyne/Hartford Division, Box 1440, Hartford, Connecticut, 06102, U.S.A.

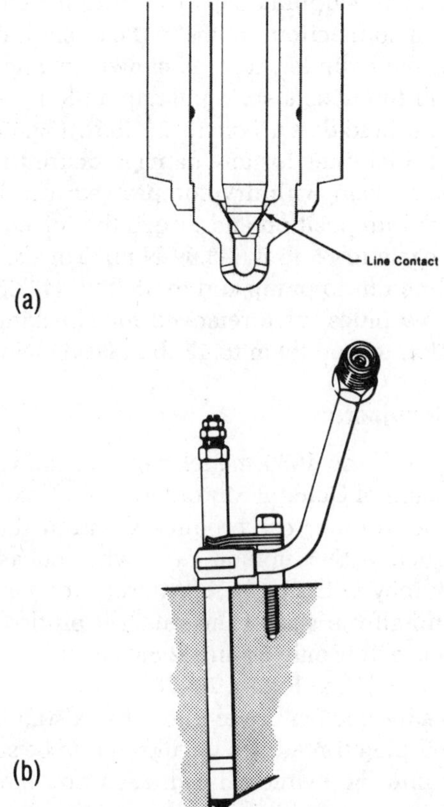

Figure 22-74. a. Nozzle to seat contact. b. Method of holding pencil nozzle in cylinder head. (Courtesy of Stanadyne Diesel Systems)

Injection Nozzle Removal

At any time when nozzle removal is required, take care that you do not bend or distort fuel lines or exert undue force on component parts of the nozzle. If only one nozzle is to be removed, then only the fuel supply and return line for that one needs to be disturbed; however, if all nozzles are being removed for service, then all fuel supply and return lines should be removed and plastic shipping caps placed over the nozle fuel inlet. Proceed to remove the nozzle spring clamp bolt and spacer; then to prevent any nozzle body damage, use tool number J-26952 and pull the nozzles one at a time. Exercise extreme care once the nozzles have been removed to ensure that the tips will not be damaged. If you are not immediately testing the nozzle on a pop testing machine to determine its release pressure and spray pattern, install protective shipping caps over the end of the injector.

Nozzle Installation

Before nozzle installation, ensure that the nozzle bore in the cylinder head is free of any carbon build-up and is not damaged or scored in any way. Once the nozzle has been removed, both a new compression and carbon stop seal, as shown in Figure 22–75, must be installed before the injector is reinstalled in the engine. With the seals installed, proceed to remove the protective shipping caps from the injector inlet and return connections, and slip the injector into position in the cylinder head as shown in Figure 22–75. Install the nozzle spring clamp and spacer, and torque the hold-down bolt to 25 lb.ft. (33.89 N·m). Check all fuel lines for any damage or fretting, and blow them clean with dry compressed air; then install them into position and torque the injection line nut to nozzle to 25 lb.ft. (33.89 N·m). Torque the injection line nut to pump end to 35 lb.ft. (47.35 N·m). If the glow plugs were removed for checking, after installation torque them to 12 lb.ft. (16.26 N·m).

The Microjector

Starting with the 1980 model engines, the Oldsmobile Division of General Motors, which offers its diesel engine to other car producers within the GMC line, signed a $65 million deal with Lucas CAV, whereby they will supply a "Microjector" which is less than half the size of the smallest pintle-type injector in use. It is only 65-mm (2.55″) long and weighs 52 g (0.114 lb). See Figure 22–76.

It has a unique feature in that it has a standard 14-mm spark plug thread, which allows it to be screwed directly into the cylinder head. Also no "fuel back-leak" pipe is needed bacause the nozzle valve guide is not required to seal against injection pressure. An outward opening valve creates a narrow spray evenly distributed into the pre-combustion chamber, with engine compression assisting a sharp cutoff. Both smoke and NOx emissions plus noise have been reduced in these engine models.

The Lucas CAV Microjector uses a nozzle with an outward opening, spring loaded poppet valve, in contrast with the inward opening valve of a conventional diesel fuel injector. Since both engine compression and combustion pressure forces on an outward opening valve are additive to that exerted by the nozzle spring, opening pressure settings of the Microjector are correspondingly lower than those of conventional injectors.

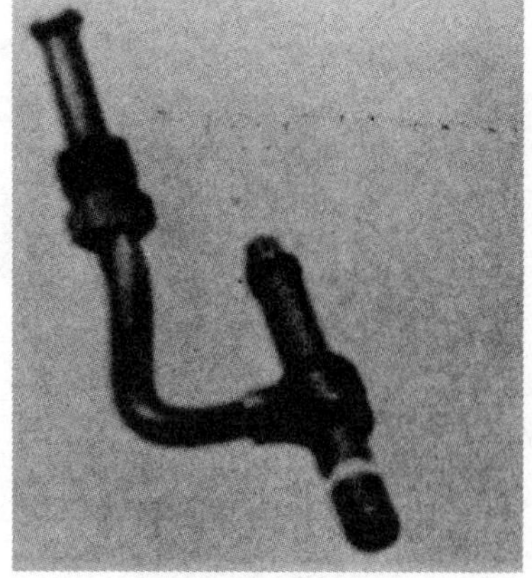

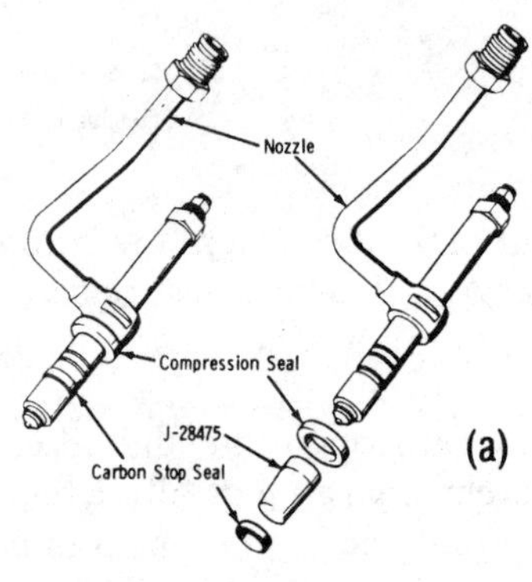

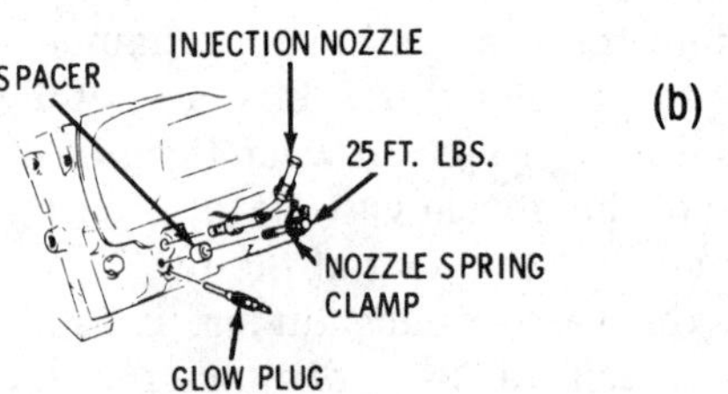

Figure 22–75. a. Installation of compression seal and carbon stop seal. b. Injection nozzle installation. (Courtesy of Oldsmobile Division, GMC)

During injection, a degree of swirl is imparted to the fuel before it actually emerges around the head of the nozzle valve, which forms a closely controlled annular orifice with the nozzle valve seat. The resultant high-velocity atomized fuel spray forms a narrow cone suitable for efficient burning of the fuel in the pre-combustion chamber of the engine.

Servicing the Microjector. The Microjector has been designed as basically a throwaway item. After

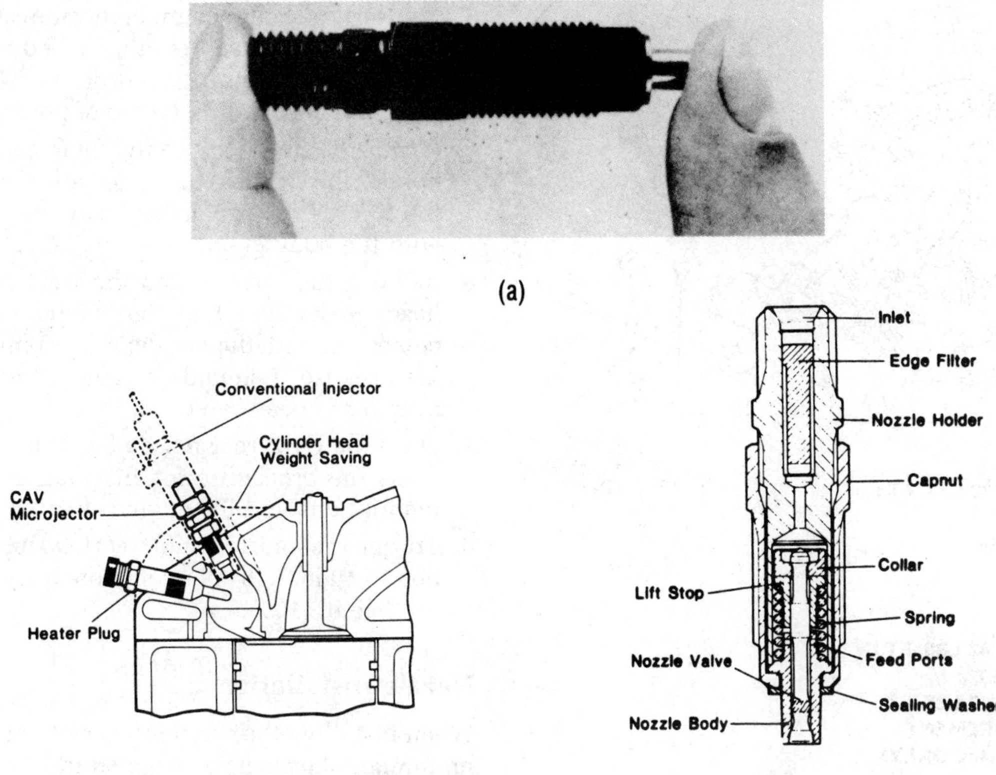

Figure 22–76. a. Microjector size. b. Location of injector in engine. c. Internal parts location of microjector. [Courtesy of Lucas CAV Ltd.]

a period of service, the functional performance may not meet the test specifications, as the nozzle is a matched and preset assembly. Lucas CAV recommends that after a service period of 50,000 mi (80,450 km), new Microjectors be fitted, The vehicle or engine manufacturers also recommend that the injectors be removed at specific service intervals for examination and testing. Refer to Chapter 29 for injector service/test information.

Microjector Nozzle Removal

Before removing any injectors from the engine, obtain a supply of plastic protective shipping caps that can be used to cover all exposed fuel lines after breaking them loose.

Always use two open-end wrenches on the nozzles when loosening or tightening the fuel lines—one wrench on the nozzle hex and the other one on the fuel line nut. Place the wrench on the largest hex of the nozzle to remove it.

The nozzles used on the V6 and V8 engines are identified in Figures 29–13 and 29–14.

SPECIAL NOTE: It is usually necessary when working on the right rear cylinder bank of VIN (vehicle identification number) "T" engine "A" series vehicles to jack the vehicle up in order to gain access to the injectors.

Procedure

1. Disconnect the intermediate steering shaft from the stub shaft by removing the stub shaft clamp bolt when it is in the up position.

   NOTE: A loss of steering control can occur if you fail to disconnect the intermediate shaft from the rack and pinion stub shaft.

2. Remove the engine support strut (Item 6) as shown in Figure 22–77 and place a floor jack under the front crossmember of the cradle; pump up the jack just until it starts to raise the vehicle.
3. Refer to Figure 22–78 and remove the two front body mount bolts No. 1 and No. 3 along with their lower cushions and retainers; remove the cushions from the bolts.

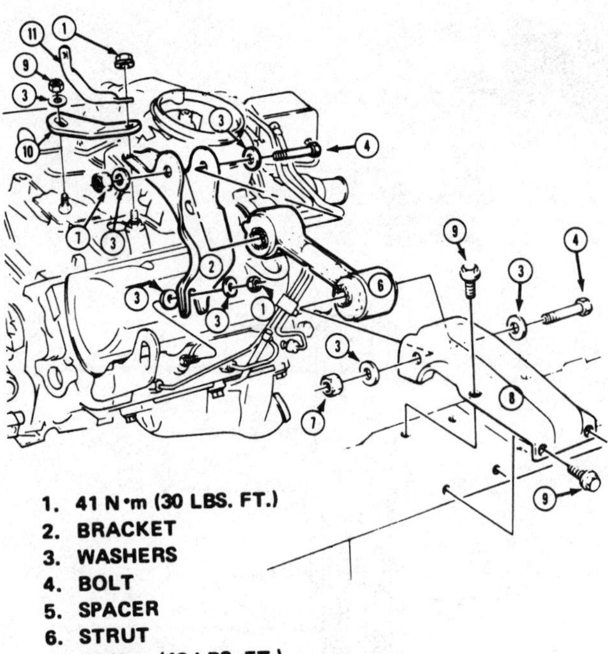

1. 41 N·m (30 LBS. FT.)
2. BRACKET
3. WASHERS
4. BOLT
5. SPACER
6. STRUT
7. 57 N·m (42 LBS. FT.)
8. BRACKET
9. 23 N·m (17 LBS. FT.)
10. REINFORCMENT
11. BRACE (A/C ONLY)

MOVE ENGINE FORWARD 10 mm (3/8")
AND HOLD WHILE TIGHTENING NO. 7

Figure 22-77. VIN "T" Engine "A" series mounting strut and brackets. (Courtesy of Buick Division, GMC)

4. Reinsert the body mount bolts minus the cushions but with their retainers into the body mount holes at least three full turns in order to make sure that the cradle doesn't move out of position.
5. Lower the floor jack gently (making sure that no hoses, lines, pipes, or cables are damaged) until the vehicle crossmember has come into contact with the body mount bolt retainers.
6. Remove the nozzle from the right rear cylinder head; make sure that the copper nozzle gasket comes out with the nozzle; if not, remove this gasket because it should be replaced each time the nozzle has been removed.
7. Once the right rear nozzle has been installed, reverse the procedure for installation of the body mount bolts and the engine strut.
8. Torque the intermediate steering stub shaft clamp bolt to 40 lb.ft. (54 N·m) and the body mount bolts to 77 lb.ft. (105 N·m).

Nozzle Installation

When installing the nozzles, check to make sure that no foreign objects have dropped into the bore in the cylinder head and that all possible signs of carbon have been removed from the bore seat and thread area. Always install new copper washers on the nozzles before screwing the injector into the cylinder

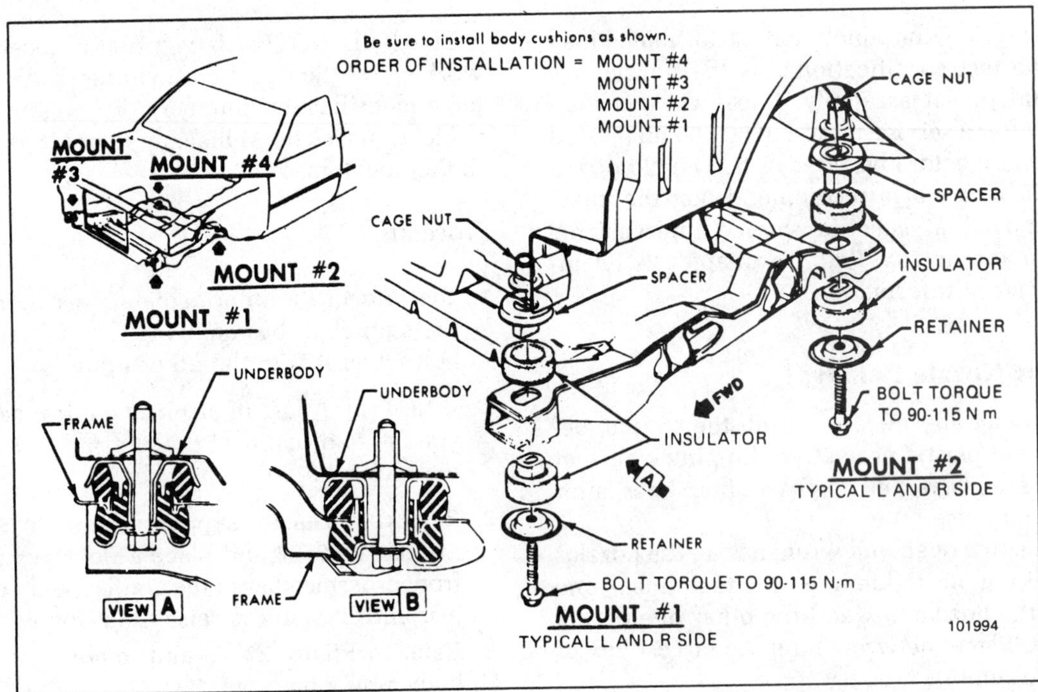

Figure 22-78. VIN "T" Engine "A" series body mount bolts (remove No. 1 and No. 3). (Courtesy of Chevrolet Division, GMC)

head. This usually requires the use of a pair of pliers to pull the old copper washer off as shown in Figure 22–79; however, exercise care not to damage the nozzle tip.

On aluminum cylinder head V6 engines, you *must* apply GMC lubricant P/N 1052771 or its equivalent to the threads of the nozzle before installing it into a VIN T engine.

Install the nozzle into the threaded bore of the head Figure 22–80 and torque it to 25 lb.ft. (34 N·m) applying force only to the large hex on the nozzle body.

Connect up all of the fuel lines and torque them to the same figure as that for the nozzle while using a back-up wrench on the upper injection nozzle hex.

Oldsmobile/GMC Glow Plug Circuits

There have been three basic types of glow plug circuits used on the Oldsmobile and Chevrolet Vee-Type diesel engines, and they have used both 12-volt and 6-volt glow plugs. The three types of glow plug circuits are:

1. Type One: 1978–79 Cars and 1978–79.5 Module Slow Start System, 12-volt system.

   CAUTION: While performing glow plug diagnosis, the system will shut down after 2 to 5 minutes. You must then cycle the ignition switch to re-energize the system.

2. Type Two: 6-volt System Module Fast Start (Truck and Oldsmobile and Buick 81 "A" 5-pin probe; Cars 80–81, Trucks 1979.5–1981).

   CAUTION: Ensure that the engine coolant temperature is below 120°F before attempting diagnosis. *Do not* apply a constant 12-volt power source to the glow plugs because they will burn out.

3. Type Three: 6-volt System Thermal Switch Fast Start (Passenger Car 6-Pin Thermal Switch); Cars 1980–81 Including some 1981 Oldsmobiles and Buick "A."

   CAUTION: Engine coolant must be below 120°F before performing diagnosis. *Do not* apply a constant 12-volt power supply to the glow plugs because they will burn out.

NOTE: All Fast Start Systems have a thermal probe or switch mounted beside the oil filler tube. Systems have either five or six wires connected at this point. Slow Start Systems have no thermal probe at the cooling system or switch at the oil filler tube.

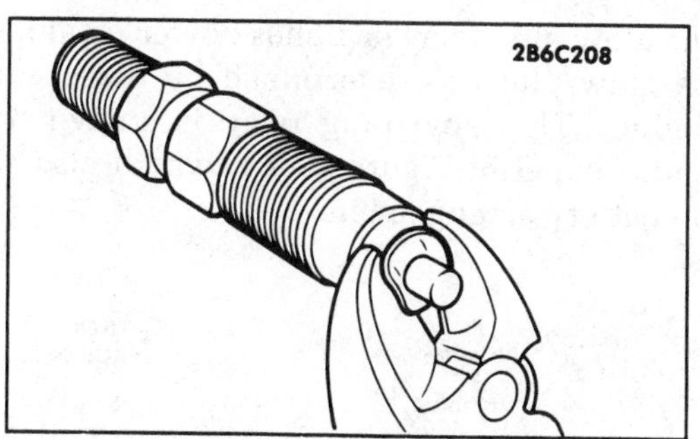

Figure 22–79. Removing old sealing washer from microjector. (Courtesy of Chevrolet Division, GMC)

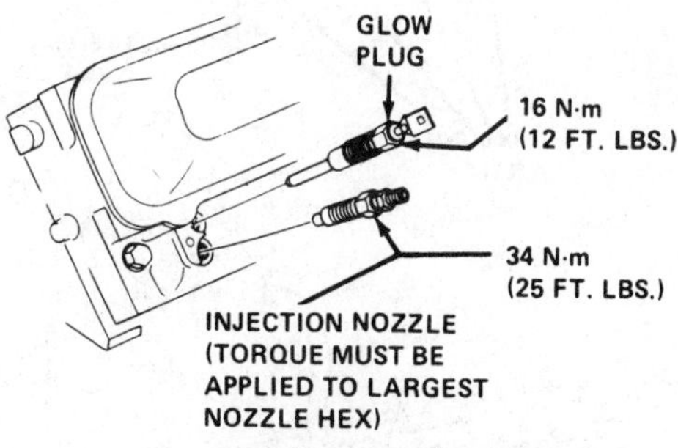

Figure 22–80. Installing microjector in cylinder head. (Courtesy of Oldsmobile Division of GMC)

GMC 4.3L V6 and 5.7L V8 Diesel Engine Glow Plug Circuit/Fuel Heater

The basic wiring of a vehicle equipped with a diesel engine is very similar to that of a unit with a gasoline engine. The major differences are in the following areas:

1. Because of the higher compression ratio of the diesel engine, two batteries are required to supply enough current to the starter motor as well as to the glow plugs for preheating of the combustion chamber especially in cold weather.

   Figure 22–81 shows the two 12-volt batteries connected in parallel.

2. A larger starter motor is employed with the diesel engine because of the greater energy required to crank the diesel engine. A cranking speed of at least 100 rpm is necessary to start the diesel engine.

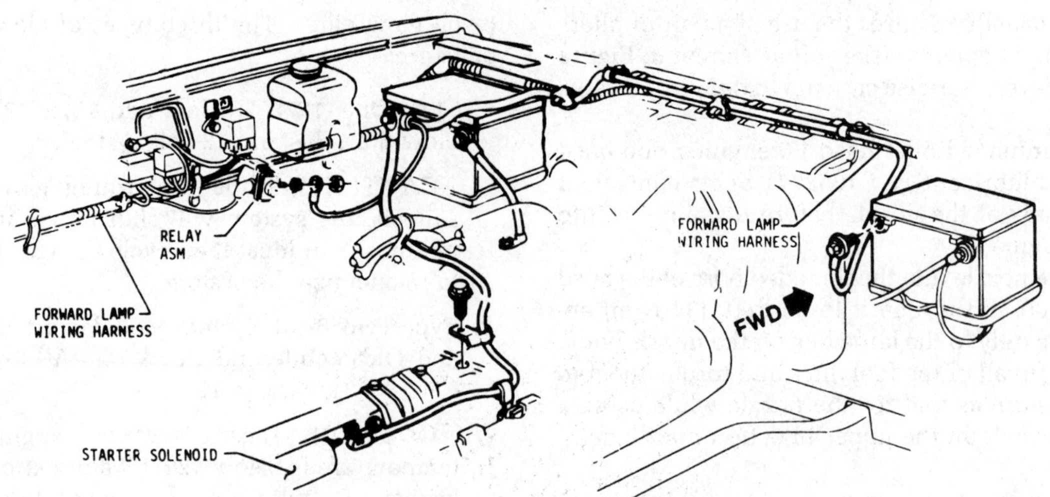

Figure 22-81. Vehicle battery wiring/glow plug relay assy. (Courtesy of Chevrolet Motor Division, GMC)

3. One glow plug per cylinder is employed to preheat the incoming air to the cylinder. The glow plugs are all 6-volt units; however, they are operated at normal system voltage or 12 volts in order to provide rapid heating. One of the main improvements that Oldsmobile made to the 1981 models was to improve the glow plugs so that they now operate to provide much faster preheat at freezing temperatures and below. Earlier engines using 12-volt glow plugs took up to 60 seconds before the start engine light would come on. Current engines fitted with 6-volt glow plugs, but operating on 12 volts have reduced this preheat time to 6 seconds.

   The glow plugs are screwed into each cylinder and are located as shown in Figure 22–82 just above the fuel injectors.

   It is imperative that former and current glow plugs not be intermixed in an engine because the earlier glow plugs were 12-volt units with steady current applied to them, whereas the current glow plugs are 6-volt units with controlled pulsing current applied to them. The former glow plugs were therefore known as slow glow plugs while the current ones are known as fast glow plugs. The former and current glow plugs can be identified by the size of the bayonet connection on the top of the glow plug as shown in Figure 22–83.

   To control both pre-glow and after-glow time periods, a controller located in a water passage (usually the front water passage) of the intake manifold senses coolant temperature and controls glow plug current. This controller also signals the cowl located lamp control relay to turn off the dash wait light, and when to start the engine.
4. The glow plug relay switches power on and off to the glow plugs as determined by the electronic module. The glow plug relay is shown on the fender panel in Figure 22–81, which also shows the battery arrangement.

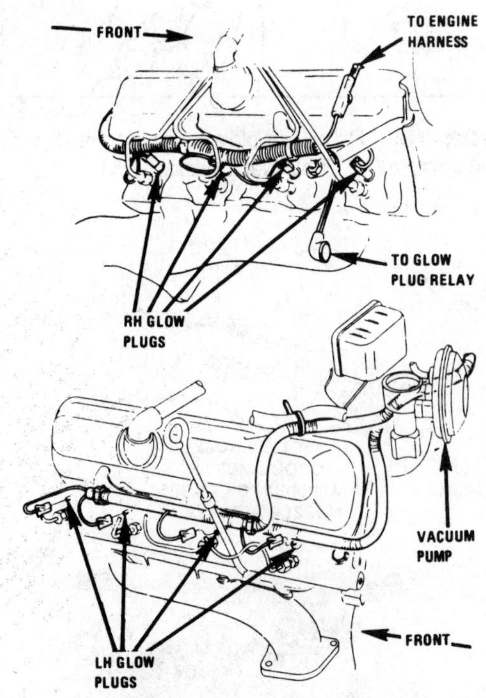

Figure 22-82. Glow plug location. (Courtesy of Oldsmobile Division, GMC)

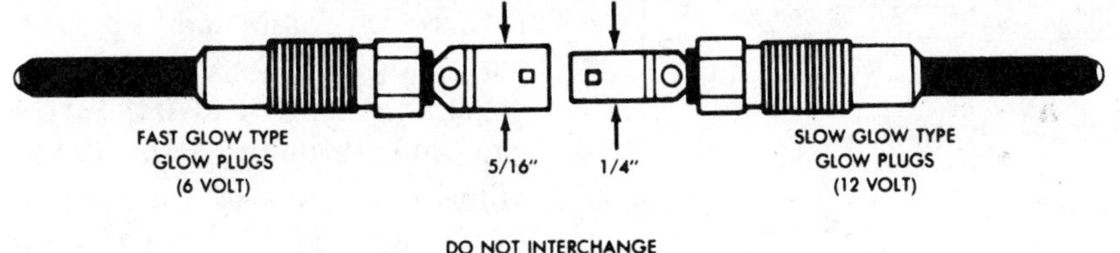

Figure 22-83. Former and current glow plugs (Courtesy of Cadillac Division, GMC)

CAUTION: Do no attempt to bypass the relay of the glow plugs by using a jumper cable, because the glow plugs are designed for intermittent operation and not continuous. Supplying constant current to the glow plugs through bypassing the relay will result in burning out the glow plugs' elements.

Figure 22–84 shows a control schematic for the glow plug circuit.

CAUTION: If the ignition switch is left *on* in early model vehicles (run position) without starting the engine, the glow plugs will continue to pulse on and off until the batteries are run down—which takes about four hours when the coolant switch is open. Also do not use more than a 2–3 candle power test light when making circuit checks.

The glow plug setup has been improved whereby fewer components are required with the new system. The glow plugs are designed to shut off automatically if the engine is not started within a prescribed period of time whereas the earlier units would cycle on/off until the batteries were flat. In addition, a self-limiting feature also prevents the maximum temperature that the glow plugs can attain, thereby extending their service life.

Glow Plug Circuit Description

A simplified wiring diagram for a 4.3L V6 or 5.7L V8 GMC diesel engine glow plug circuit is shown in Figure 22–85. The control module used with the circuit has the following functions:

1. Energizes the instrument panel *wait lamp* operation according to ambient temperature and/or system voltage.
2. Controls glow plug shutdown timing.

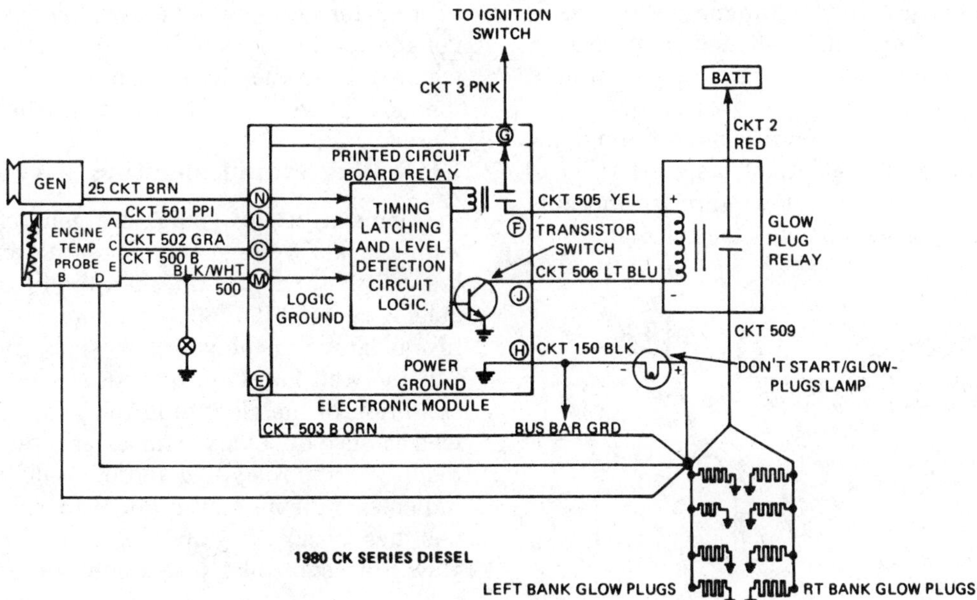

Figure 22-84. Glow plug control schematic. (Courtesy of Chevrolet Division, GMC)

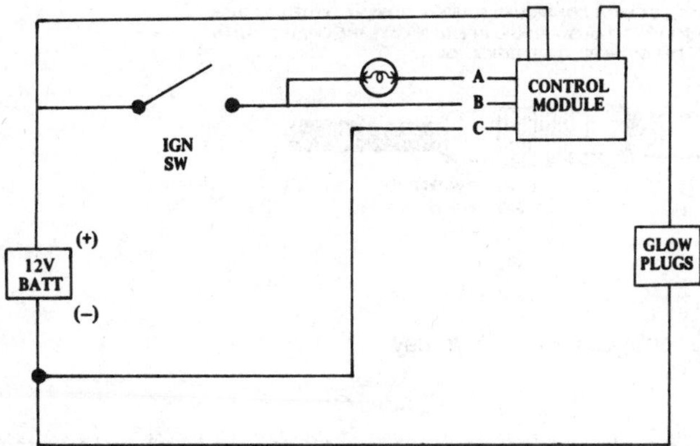

Figure 22-85. Glow plug wiring circuit diagram. (Courtesy of Oldsmobile Division, GMC)

3. Contains an over-voltage protection feature.
4. Disengages the glow plug battery current/voltage via a thermal cutout at module temperatures greater than 45°C (113°F).
5. The power relay that switches the voltage applied to the glow plugs.
6. A quick reset feature that lets the module recycle quickly after the initial shutdown "time-out."

The glow plug control module is illustrated in Figure 22-86.

System Operation

Battery voltage is applied at all times from a fusible link, and through the gauges' fuses and the ignition fuse when the key is turned to the *run* position. Current is allowed to flow to the wait lamp, the pre-glow switch, and the contacts of the reset relay to ground via the gauges' fuse.

Current through the ignition fuse flows through the thermal controls and contacts of the reset relay to ground, while current also flows through the contacts of the over-voltage protector, the coil of the glow

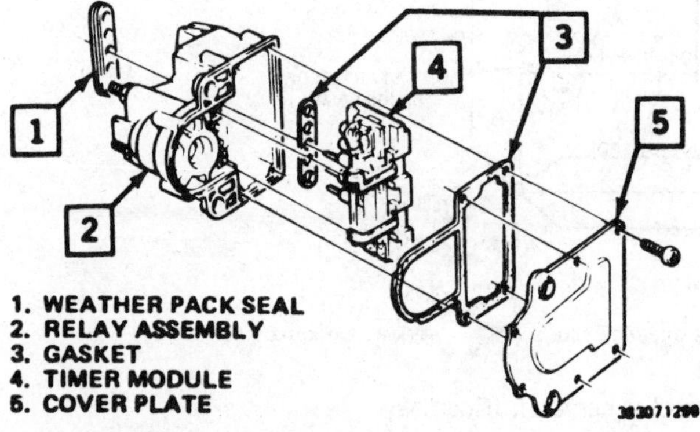

Figure 22-86. Glow plug control module. (Courtesy of Oldsmobile Division, GMC)

plug relay, the after-glow switch and the contacts of the reset relay to ground. The over-voltage protector is therefore connected in parallel with the coil of the glow plug relay.

With the ignition switch turned on, battery current can flow through to the individual cylinder glow plugs, which causes them to glow cherry red similar to the heating coil on a kitchen stove. As the glow plugs heat up, their resistance increases, which limits the flow of current through them. Engine temperature (thermal controls) establishes how long the pre-glow sequence will take; in colder weather the pre-glow will be on longer than it would in warm weather.

When the pre-glow switch heats up, the *wait* lamp will go off and the operator can now crank the engine to start it. With the pre-glow switch now in the *open* position, no current will flow through the gauges' fuse. Current can still continue to flow through the ignition fuse; therefore, the glow plug relay is energized and current still flows to the glow plugs—this is referred to as the after-glow period.

When the after-glow switch moves to its *hot* position, the glow plug relay has its path opened to ground and the power is removed from the glow plugs. The reset relay now energizes to lock out the thermal controls, and the glow plug circuit will be ready for the next attempt at engine start-up once the ignition switch has been turned *off* first and then back to *on*.

Anytime that the engine coolant temperature is above 140°F (60°C), the thermal controls will prevent current flow to the glow plug circuit. An over-voltage condition (14 volts or higher) causes the protector to open the circuit to the glow plug relay and the glow plugs will not receive any battery current.

Glow Plug Troubleshooting

Problems with glow plugs are generally associated with a lack of voltage and current flow. Generally, if the wait lamp comes on, then the fuses are serviceable; if the wait lamp doesn't come on, then either a fuse is blown or the wait lamp is burned out.

If the wait lamp and the fuses are OK, check the condition of the glow plug relay for voltage feed as well as the glow plugs themselves if there is a voltage reading at the relay. An audible *click* from the relay indicates suitable operation. If the glow plug relay operates, check the glow plug controller; replace the glow plug controller if in doubt, for an operational test.

To check the condition of the glow plugs and their wiring harness, turn the ignition key *off* and obtain a self-powered 12-volt test light. Disconnect all the glow plug harness connectors at the glow plugs and connect one lead of the test light between the green wire post of the glow plug relay and ground.

At each individual glow plug, touch the harness connector to the glow plug spade terminal; if the test light illuminates, the glow plug and harness lead are OK. Make sure that you disconnect the wiring harness connector after testing each glow plug.

If the test light fails to illuminate, touch the harness connector to a good ground surface. If the light now illuminates, then replace the glow plug. However, if the light still does not illuminate, replace the glow plug wire.

If all glow plugs fail to operate, the problem is most likely the glow plug controller unit.

Glow Plug Resistance Check

If a glow plug is suspected of being faulty, it can be removed from the vehicle and grounded on the body while battery voltage is applied to the connection terminal. Do not apply power for longer than 30 seconds to the glow plug in this manner; it can burn out (element damage). Compare the brilliance of the suspect plug with that of one known to be good or with a new one.

Glow plugs that fail to perform should be replaced. In addition to supplying preheat for starting purposes, a resistance check through the glow plugs can very effectively establish if one cylinder is producing more power than another, especially if a rough idle condition exists. Since the glow plugs are installed directly into the engine cylinder combustion chamber area along with the injector nozzle, they will be subjected to both the pressures and temperatures of combustion. A change in cylinder temperature will reflect a change in the resistance measurement through the glow plug when checked with an ohmmeter. This resistance method check is the recommended check by General Motors when a rough idle problem exists.

NOTE: The glow plug resistance check should only be undertaken once the idle roughness checks and adjustments have been done as per the systematic procedure found in all General Motors vehicle service manuals.

Once this has been done, the glow plug resistance check can be done as shown in the "Diesel Engine Idle Roughness Diagnosis" section within this chapter.

Water-in-Fuel Detector

Because of the extremely fine tolerances of the injection pump components and nozzles of the diesel engine, the presence of any water in the fuel system can do irreparable damage to these components. Therefore, a water-in-the-fuel detector is now used on current production vehicles fitted with the diesel engine.

Figure 22–57 shows the water-in-fuel detector which is located in the fuel tank, and Figure 22–87 shows how this unit should be tested if it is suspected of being faulty.

If the water-in-fuel detector indicates that there is in fact water in the fuel system, then the fuel tank must be drained, removed from the vehicle, and flushed clean.

In addition, all fuel lines should be blown clean with low pressure air towards the rear of the vehicle to prevent any dirt or water from being directed towards the fuel pump. The fuel filter should then be changed, and if not already done, the water-in-fuel sending unit should also be removed from the fuel tank and cleaned. The fuel system must then be bled in order to start the engine.

The wiring circuit for the water-in-fuel system does vary slightly between division vehicles of GMC. A typical system is shown in Figure 22–88. In addition to water in the fuel, if for any reason the fuel tank is filled with gasoline, the tank should be drained and refilled with fresh diesel fuel, all lines blown clean, the filter replaced, and the system bled.

Diesel Fuel Heater

In order to reduce the possibility of the fuel filter plugging through wax crystal formation at low ambient temperatures, an in-line 12-volt electrical heater Figure 22–89 is used along with an engine coolant heater. When the ambient temperature drops to about

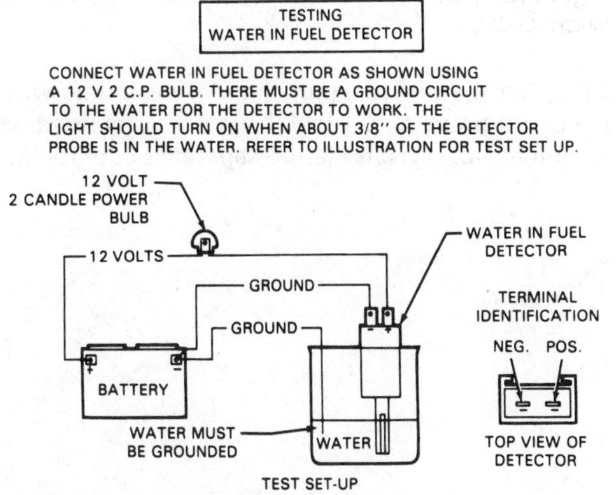

Figure 22–87. Testing water-in-fuel detector assembly. (Courtesy of Cadillac Division, GMC)

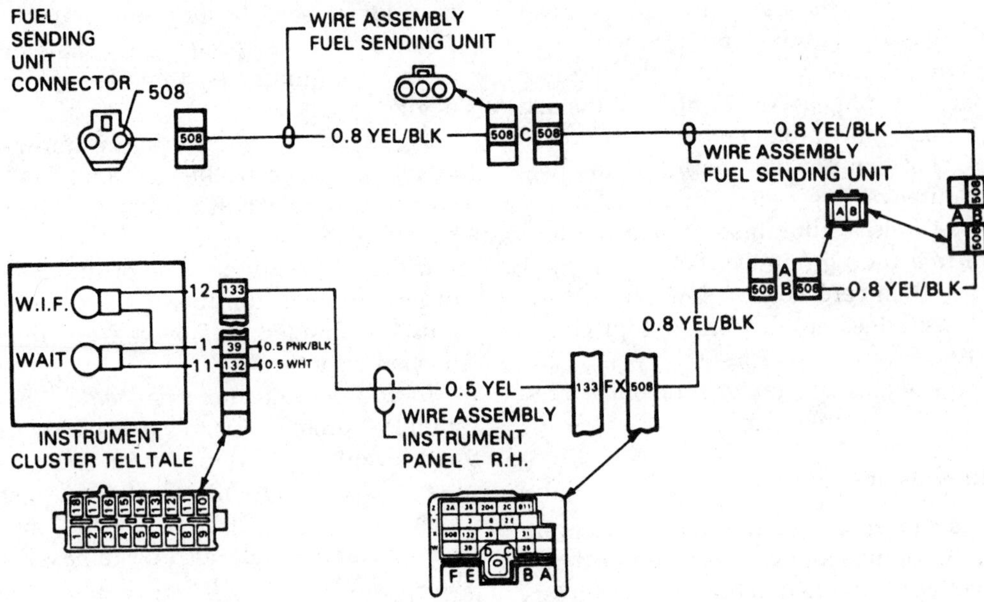

Figure 22-88. Typical wiring circuit for water-in-fuel system. [Courtesy of Cadillac Division, GMC]

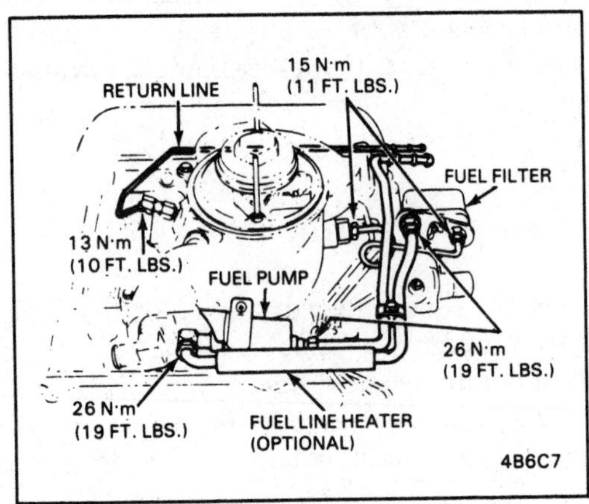

Figure 22-89. Diesel fuel heater. [Courtesy of Buick Division, GMC]

20°F, the diesel fuel approaches what is known as the cloud point of the fuel. This is the temperature at which wax crystals start to separate out in the fuel due to the fact that the fuel has a paraffin wax base. The diesel fuel heater is a thermostatically controlled electrical resistance-type heater that is designed to heat the fuel before it enters the fuel filter, and thereby prevent these wax crystals from settling out, which would subsequently plug the filter and prevent fuel flow.

When the ignition key is turned to the *run* position, battery voltage is applied to the heater, which consists of a spiral strip wound around the fuel inlet pipe.

To control when the fuel heater is used, a bimetallic thermal switch sensitive to ambient and fuel temperature closes the electrical circuit only when the fuel temperature drops to 20°F and opens it when the temperature reaches 50°F. Within the fuel tank filter sock, there is a bypass valve to allow fuel flow to the heater when wax crystals gather on the fuel tank filter. This wax will usually fall away after the pump has stopped.

23

General Motors Corporation 6.2L V8 Diesel

This engine is a 6.2L, 6217 cc (379.4 cu.in.) V8 diesel. It is similar in both its layout and design to the Oldsmobile V8 350 cu.in. (5.7L) unit; however the 6.2L (379.4 cu.in.) engine was designed as a diesel engine from its inception rather than a gasoline-derived base unit as in the case of the 5.7L Olds engine. The general engine arrangement is shown in Figure 23–1.

The U.S. Armed Forces recently took delivery of 53,000 Chevrolet Blazers and pickup trucks that are fitted with the 6.2L diesel engine.

The injection pump used with the 6.2L is the same basic unit as that used on the Oldsmobile/GMC passenger car and light pickup trucks as well as the Ford 6.9L V8 E and F series vehicles, namely the Roosa Master Stanadyne model DB2.

The injectors, however, are Robert Bosch supplied rather than the Stanadyne pencil nozzle, Lucas CAV Microjector, or GM/DED (Diesel Equipment Division) models, which are used in the 4.3L (262 cu.in.) V6 and 5.7L (350 cu.in.) GMC diesel engines.

The engine cylinders are numbered 1, 3, 5, and 7 on the left bank, and cylinders 2, 4, 6, and 8 on the right bank with a firing order of 1-8-7-2-6-5-4-3. The odd-numbered cylinders are on your right-hand side

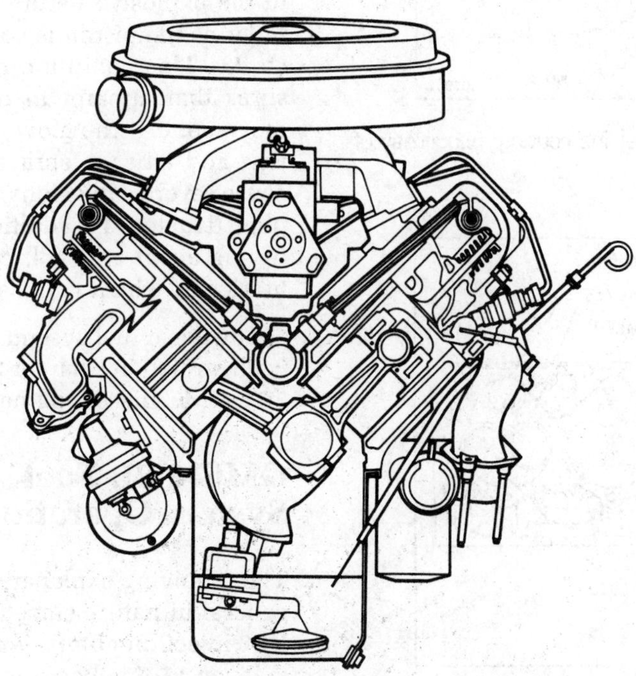

Figure 23–1. GMC 6.2L V8 Light truck diesel engine. (Courtesy of General Motors Corporation)

when viewing the engine from the front. Both the cylinder arrangement and the injection pump fuel line routing are illustrated in Figure 23–2.

The engine is an IDI (indirect-injection) type with special alloy steel pre-combustion chambers located within the cylinder head for each cylinder. Figure 23–3 illustrates the location of the pre-combustion chambers in the cylinder head relative to both the intake and exhaust valves. The high-swirl pre-combustion chamber along with a special cavity in the piston top as shown in Figure 23–4 ensures proper mixing of the injected diesel fuel with the high pressure-temperature air to provide an efficient fuel burn with low emissions. As with all IDI-type engines, this unit also employs a glow plug in each cylinder to facilitate engine starting. Both the glow plugs and the

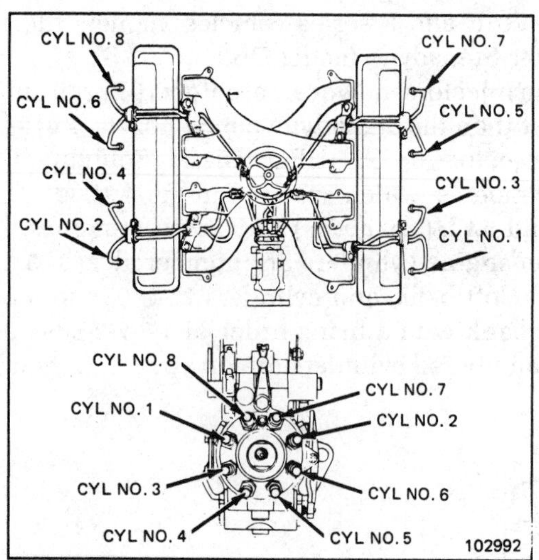

Figure 23–2. 6.2L Engine fuel line routing. (Courtesy of General Motors Corporation)

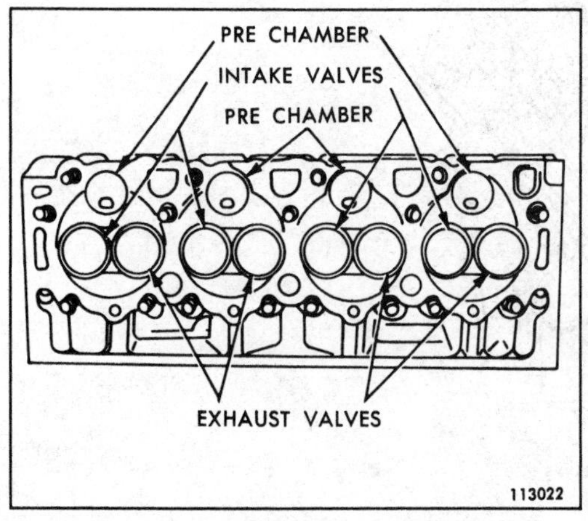

Figure 23–3. Pre-combustion chamber location in cylinder head. (Courtesy of General Motors Corporation)

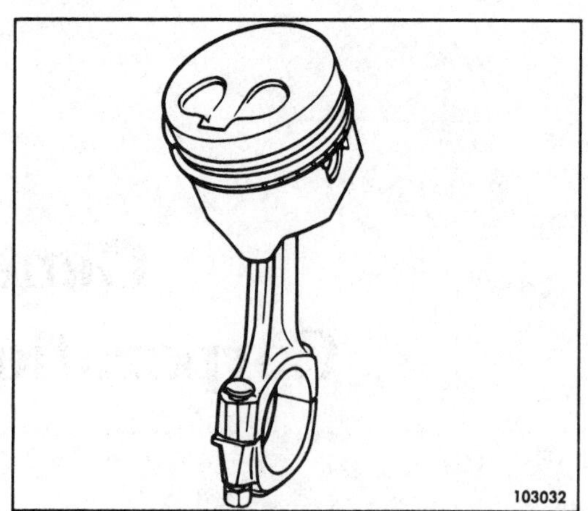

Figure 23–4. Piston crown shape—6.2L diesel engine. (Courtesy of General Motors Corporation)

injectors are threaded into the cylinder head with the injector nozzles being spring loaded and calibrated to open at a specified psi of fuel pressure.

The compression ratio of this engine is 21.5:1 and it is of oversquare design with a bore of 3.98" (101 mm) and a stroke of 3.8" (97 mm).

CAUTION: Due to the high compression ratio of this engine and the use of glow plugs, *do not* at any time attempt to use starting fluids to start the engine in cold weather. The use of such highly volatile liquids (ether) can result in serious engine damage due to the explosive nature of this fluid within the cylinder as the piston is coming up on its compression stroke. This condition creates extremely high pressures that attempt to oppose the piston's upward movement. If the glow plugs are in the heated position and ether is sprayed into the engine, this can cause severe explosions within the cylinder and glow plug damage. In addition, ether tends to act as a drying agent and will cause lack of upper cylinder lubricant, which can also cause engine damage.

The basic fuel system arrangement for this engine is illustrated in Chapter 22, dealing with the 4.3L and 5.7L GMC diesel engines.

GMC 6.2L Fuel System Operation

The following explanation of the fuel system flow is general in nature since a detailed explanation of the DB2 injection pump along with the system operation is given in this book under the 4.3L and 5.7L GMC diesel fuel system, Chapter 22.

The main components of the fuel system are:

1. The fuel tank, which includes a water sensor unit and a screen-type filter.
2. A primary filter with water draining provisions on the older models, or a fuel filter/water separator on the later engines.
3. A mechanical fuel pump.
4. A secondary fuel filter.
5. A fuel line heater.
6. The distributor-type injection pump.
7. High pressure fuel lines (injection pump to injectors).
8. Eight fuel injection nozzles.

The fuel transfer pump, which is located on the left-hand side of the engine when viewed from the front, is similar in operation to the fuel pump found on the GMC V8 gasoline engines. The pump is also driven from the engine in a like manner to that of the gasoline unit by an eccentric lobe on the camshaft through a push rod. When the engine is started, this pump will draw fuel from the tank; the fuel then passes through the primary fuel filter located on the front of the dash (bulkhead) on CK truck models shown in Figure 23–5, while on the G model van it is located on the vehicle chassis right-hand underbody cross sill ahead of the fuel tank as shown in Figure 23–6.

Since this is the suction side of the pump, any leaks here can result in air being drawn into the system, which will cause a reduction in fuel flow and a rough running engine.

Fuel is placed under pressure within the transfer pump body, and it is then forced to and through the secondary filter, which is generally mounted on the engine air inlet manifold. Later model vehicles are equipped with a fuel filter/water separator arrangement, illustrated in Figures 23–7 and 23–8. The fuel then proceeds to the injection pump, which is located in a similar position to that on the Oldsmobile 5.7L V8 engine at the front of the engine. The injection pump on the 6.2L engine, however, is mounted horizontally rather than at an inclined angle as is the

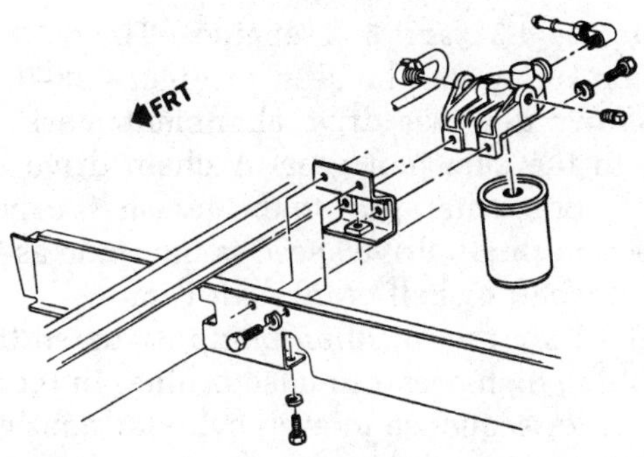

Figure 23-6. Primary filter G van. (Courtesy of General Motors Corporation)

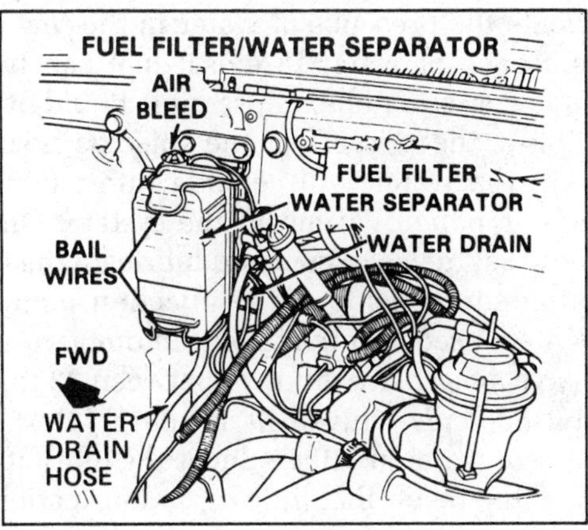

Figure 23-7. Fuel filter water separator—CK truck. (Courtesy of General Motors Corporation)

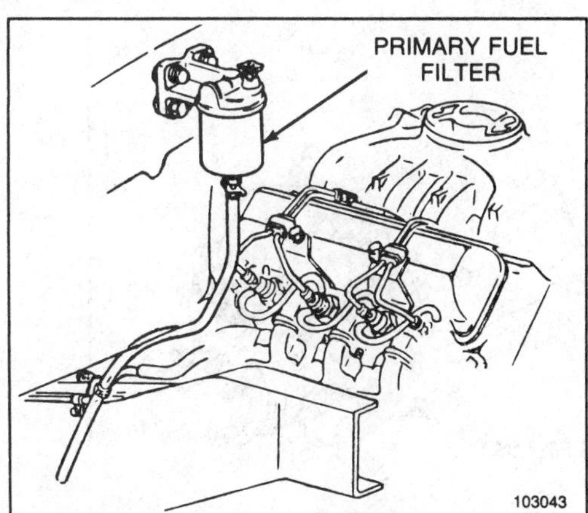

Figure 23-5. Primary filter—CK truck. (Courtesy of General Motors Corporation)

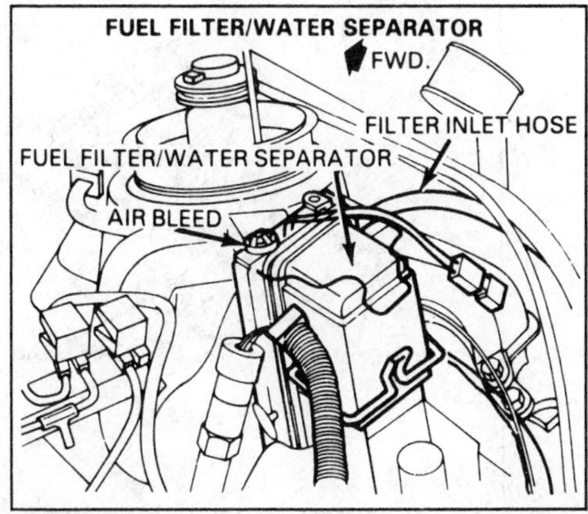

Figure 23-8. Fuel filter water separator—G van. (Courtesy of General Motors Corporation)

case on the 4.3L and 5.7L engines. The 6.2L engine injection pump can be seen in Figure 23-9, which also shows the gear drive alignment marks of the pump to the camshaft gear. A chain drive is used from the crankshaft gear to the camshaft drive gear.

Injection pump drive speed is the same as the engine camshaft or half crankshaft rpm.

Fuel within the injection pump is routed through each of the high-pressure injector lines in the engine firing order sequence to each cylinder nozzle.

Although each injector line is of a different shape, each one is the same length as the others because any differences in length would cause a change in the injection timing. A longer line would retard the timing, while a shorter line would advance the timing. The amount of fuel injected is controlled by the throttle position, which establishes how much fuel can flow past a helical cut metering valve.

Water-in-Fuel Protective System

The 6.2L engine is equipped with a warning system to indicate the presence of water in the fuel system. The presence of water in diesel fuel can be disastrous because as a non-compressible liquid, it can actually blow the tips off of the injector nozzles. In addition, the water will corrode internal injection system components if left in the system. One function of diesel fuel is to cool and lubricate the internal operating components of the injection pump. Most diesel fuels used in high-speed automotive applications have a viscosity rating of between 33 to 40 SSU (Saybolt Seconds Universal) at 100°F (37.7°C). The presence of water greatly reduces the lubricity of the fuel with the result that little or no lubrication exists between the moving parts, and seizure can result.

The water-in-fuel warning system used with the 6.2L engine is illustrated in Figure 23-10 for both the CK model light trucks and the G model vans.

Basically, the water-in-fuel detector (WIFD) employs a floating unit that will rise (or move upwards) when water is present within the fuel tank. Since the specific gravity of diesel fuel is less than that of water, the WIFD will not move upwards to close an electrical circuit when only the lighter specific gravity diesel fuel is in the fuel tank. If sufficient water is present in the tank however, the WIFD will float upwards to complete the circuit and a warning light flashes in the cab.

Figure 22-57 illustrates the fuel tank internal screen filter that screens out the water and lets it settle on the bottom of the tank due to its heavier specific gravity. Note from the figure that the fuel pump pickup line sits above the bottom of the tank.

If the WIFD light comes on, the water *must* be drained from the fuel tank immediately. To facilitate the removal of the water, a siphoning system connected to the tank and extending to the rear spring hanger on some models, and at the midway point of the right frame rail on other models, allows the service technician to attach a hose at the shutoff valve, as shown in Figure 23-10, and thereby siphon/remove the water from the fuel tank.

Drain Water from Filter

Generally, if the WIFD activates the dash warning light, it is more than likely that some water may have already passed through to the primary fuel filter assembly. To check this filter for water, simply open the petcock on the top of the housing to allow atmospheric pressure to force the water out of the filter when the petcock on the bottom of the filter housing is opened. Use a piece of rubber hose or

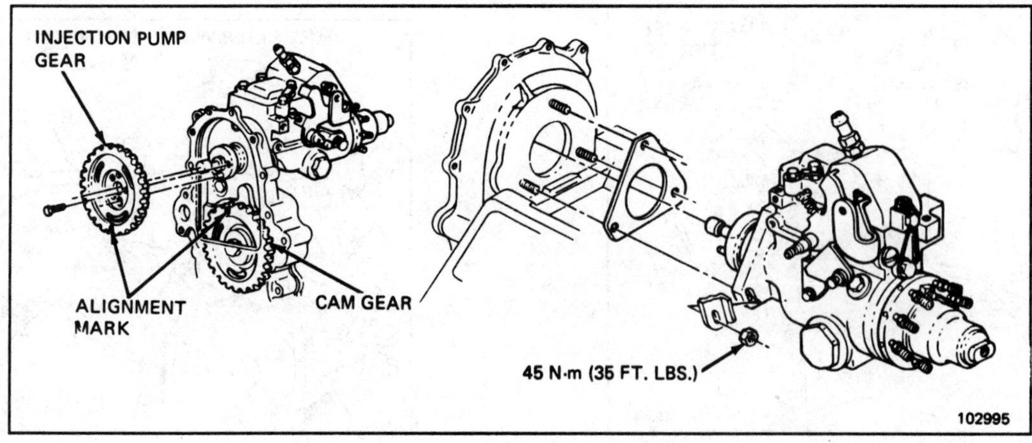

Figure 23-9. Injection pump drive gears and timing marks. (Courtesy of General Motors Corporation)

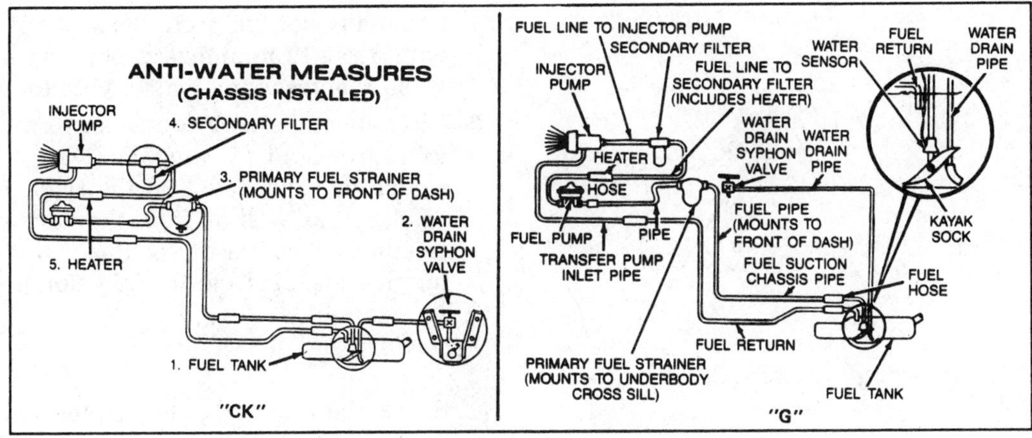

Figure 23-10. Vehicle water-in-fuel system layout. [Courtesy of General Motors Corporation]

equivalent attached to the lower petcock so that the drained fluid can be directed to a container as shown in Figures 23-7 and 23-8, which illustrate the primary filter location for both the CK truck and G van. Once the filter has been drained, close both the top and bottom petcocks tightly.

NOTE: If no water is visible when draining the primary filter, simply close the petcocks. If, however, most of the contents of the primary filter have been drained in order to get all of the water out, it will be necessary to remove the filter shell and refill it with clean diesel fuel. Otherwise, once the engine is started, the injection pump will draw fuel from the secondary filter creating a lack of fuel supply and an air lock in the fuel system. Since the primary filter is empty, it is possible to run out of fuel and the engine will stall.

On later 6.2L diesel engines equipped with fuel filter/water separators, a water drain line is connected to the base of the filter as shown in Figure 23-7 for the CK truck.

CAUTION: Any time that the 4.3L, 5.7L or 6.2L GMC automotive diesel engines are to be run, the mechanical vacuum pump at the rear of the engine must be in place—otherwise *no oil pressure* will be created because the vacuum pump acts as the drive medium for the oil pump. Figure 23-11 shows the location of the vacuum pump used with these engines, which is used to create vacuum for the power brake system.

After draining water from the primary filter, start and run the engine for several minutes to ensure that it does run correctly. The engine may run rough for a few minutes until it purges the fuel system of air; however, if the engine continues to run rough, check that both petcocks are tightly closed because any leaks from the fuel tank up to the suction side of the transfer pump will cause air to be drawn into the fuel system. (Suction side of system.)

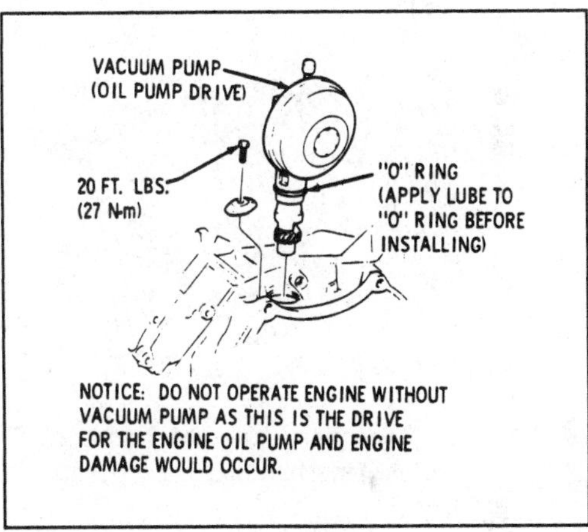

Figure 23-11. Typical gear driven vacuum pump. [Courtesy of General Motors Corporation]

Secondary Fuel Filter

The secondary fuel filter is mounted to the engine intake manifold, which is helpful because in cold weather operation the heat from the engine will radiate to the fuel within the filter to assist the fuel in staying above its cloud point (wax crystal formation in fuel) thereby preventing engine stalling from lack of fuel flow. Figures 23-12 and 23-13 illustrate the secondary filter location.

Priming the Secondary Fuel Filter. Removal of the secondary fuel filter differs slightly between the light truck and the van models. Both models require that the air cleaner assembly be taken off first to gain ac-

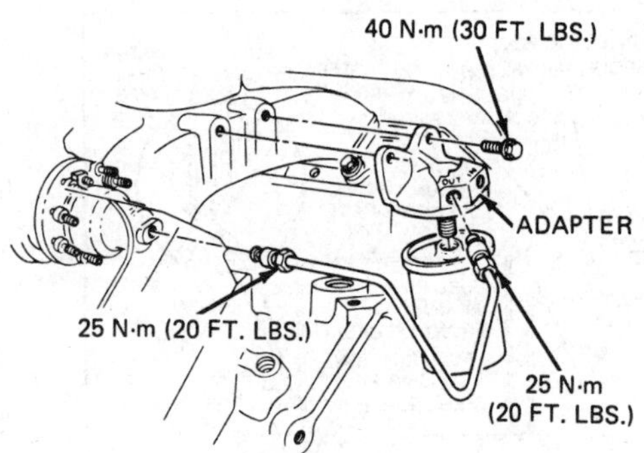

Figure 23–12. Secondary fuel filter—CK truck. (Courtesy of General Motors Corporation)

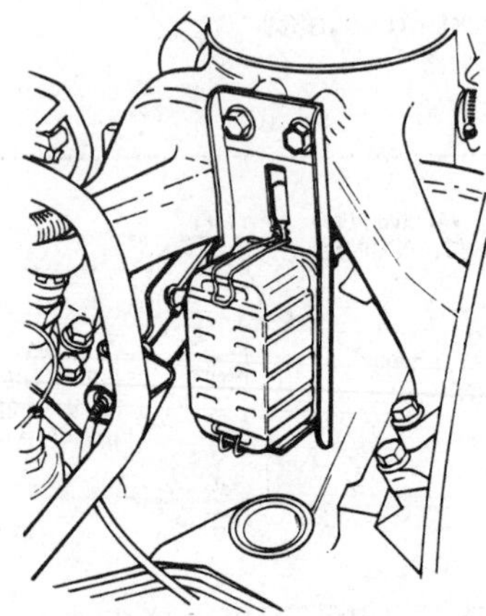

Figure 23–13. Secondary fuel filter location, G and P models. (Courtesy of General Motors Corporation)

cess to the filter. The fuel lines are then removed from the CK filter adapter followed by the fuel filter adapter itself, then the fuel filter.

On the van G and P models, once the air filter has been removed it is advisable to place a cloth under the filter because fuel will spill out once the filter is loosened off. Refer to Figure 23–13, unsnap the lower bail on the filter in order to relieve any fuel pressure within the filter assembly, then unhook the lower bail and remove the filter unit.

Filter Installation. CK Models.

CAUTION: Do not pour unfiltered fuel into the filter assembly as an aid to priming because not all of the fuel will have to pass through the filter. The fuel on the inside of the filter will immediately be drawn into the injection pump. Should any contaminants get into the injection pump, serious damage can result.

SPECIAL NOTE: If the filters are changed when the engine is hot (above 115°F, or 46°C), two conditions should be noted:

1. Catch any hot fuel from the secondary fuel filter with a rag to minimize vaporizing the spilt fuel when it comes into contact with the hot engine.
2. Disconnect the wire connector from the Housing Pressure Cold Advance (HPCA) temperature switch that is located on the right rear cylinder head bolt as well as from the jumper connector terminals. The reason for this action is to aid in purging any air from the injection pump.

Procedure

1. Screw the filter onto its adapter. Once the filter gasket contacts the adapter base, tighten it an additional 2/3 of a turn.
2. Bolt the filter and adapter to the intake manifold.
3. Connect up the filter inlet line only.
4. Disconnect the *pink* electrical wire at the fuel injection pump to de-energize the fuel solenoid and prevent the engine from starting while priming the fuel system.
5. Place a rag or cloth under the outlet of the fuel filter in order to catch fuel as you crank the engine over (10 seconds maximum).
6. If no fuel vents from the outlet line after 10 seconds, pause for 15 seconds then repeat Step 5 again.
7. Once fuel flows freely from the secondary fuel filter outlet, install the outlet line to the filter assembly, then reconnect the *pink* electrical wire to the injection pump solenoid.
8. Start the engine and let it idle for a few minutes to ensure that the engine runs smoothly and that no fuel leaks are apparent. If the engine runs rough after several minutes, then rebleed the system of air and restart.
9. Reinstall the air cleaner and filter assembly.

G and P Models (Secondary Filter Installation).

1. Ensure that both filter plate mounting fittings are dirt free.
2. Install the new filter into position and snap on only the upper bail clamp.
3. Repeat Steps 4, 5, and 6 as per the CK model above.
4. When fuel flows from the lower filter fitting, snap on the lower bail clamp.
5. Reconnect the *pink* wire to the injection pump solenoid.
6. Start and idle the engine for a few minutes to

check for fuel leaks and possible air entrapment (rough idle).

7. Install air cleaner and filter assembly.

Fuel Line Heater

All 6.2L diesel engine equipped GMC vehicles are fitted with a fuel line heater to aid in cold weather starting.

You may recollect from earlier discussions that since diesel fuel is basically extracted from a paraffin-base crude oil that as the ambient air temperature drops, wax crystals in the diesel fuel will begin to form (settling of these fuel particles). This condition is commonly known as the *cloud point* of the fuel, and when it occurs, fuel starvation will result because of the clogging up of the fuel filters by the formation of these wax crystals. Without the use of a fuel heater, problems of starting would be commonplace in cold weather operation.

An electrical fuel line heater is mounted on the engine block as shown in Figure 23–14 for both the CK and G van models. This heater is a 110-watt unit with a set of contacts that automatically closes at 4°C and contains a one-shot fuse assembly. The fuel heater is controlled from the ignition switch *on* and *crank* positions; however, if the ambient temperature is above 4°C, then the temperature controlled on/off switch within the fuel heater circuit will remain off.

The electrical circuit for the fuel heater is explained under the 4.3L and 5.7L GMC engine fuel system, Chapter 22.

6.2L Injection Pump Maintenance

If it becomes necessary to conduct tests, checks, or adjustments to the injection pump assembly, a limited amount of adjustment is available to the average service technician without having the skills obtained through certified service training programs.

In addition, if an on-engine adjustment fails to rectify the problem at hand, then few automotive service shops have the necessary injection pump test equipment to properly overhaul and test the injection pump. Most major injection pump servicing is being carried out by a Diesel Service Facility in your area. Most of these facilities specialize in diesel pump and nozzle repairs and are affiliated with the Association of Diesel Specialists. Their service personnel are highly skilled individuals who continually receive specialized training in the overhaul, testing,

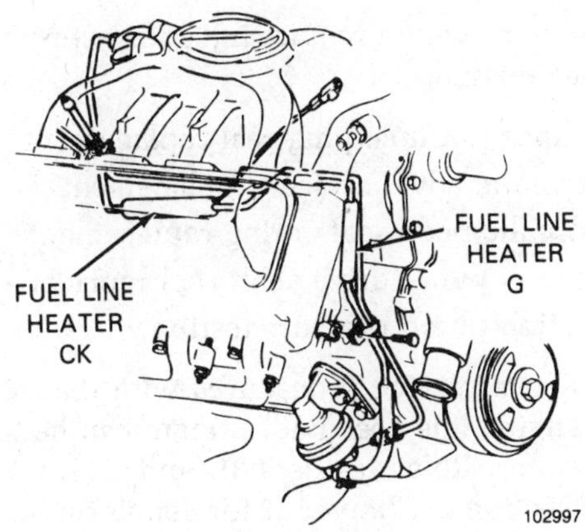

Figure 23–14. 6.2L GMC Diesel in-line fuel heater. (Courtesy of General Motors Corporation)

and calibrating procedures necessary for all types of fuel injection systems.

Major automotive diesel engine manufacturers are now setting up a training network for their own distributor/dealer personnel so that they will be capable of performing much of this specialized work.

With the number of diesel pickup trucks and vans along with cars gaining additional penetration of the marketplace, the automotive engine mechanic is required to be aware of the operating conditions, testing, and adjustment of diesel engines.

Typical adjustments, removal, and replacements that the automotive mechanic will be expected to perform on this particular engine would be as follows:

1. Engine idle speed setting
2. Throttle position switch adjustment
3. Vacuum regulator valve adjustment
4. Fuel injection line removal/replacement
5. Injector removal, test, and replacement
6. Glow plug testing, removal/replacement
7. Fuel filter service
8. Fuel heater service
9. Injection pump timing check
10. Injection pump removal/replacement
11. Injection pump cover seal and/or guide stud seal replacement
12. Injection pump throttle shaft seal replacement
13. Injection pump solenoid testing, removal, and replacement
14. Injection pump side cover gasket replacement

The above maintenance items can be done while the injection pump is on the engine. The following service items would require injection pump removal from the engine:

1. Advance pin hole plug seal replacement
2. Automatic advance seals replacement
3. Hydraulic head seal O-ring replacement
4. Injection pump drive shaft seal replacement
5. Injection pump pressure testing

The above service procedures with the exception of the engine idle speed adjustment can be found in this chapter. Refer to the 4.3L and 5.7L GMC fuel system section in Chapter 22 for details on setting the idle speed adjustment on the engine because this routine can be considered common to all of these diesel engines.

Throttle Position Switch Adjustment

The adjustment of the throttle position switch (TPS) is similar for both the LH6 engine and the LL4 and 700R4 transmission. Due to the minor differences in the actual electrical connections, Figures 23–15 and 23–16 describe the specific routine necessary to accurately adjust the TPS.

Vacuum Regulator Valve Adjustment (VRV)

Adjustment of the VRV is described in Figure 23–17 for the LL4 6.2L engine.

Checking/Adjusting Injection Pump Timing

The injection pump is timed to the engine gear train on installation and would not normally require any adjustment to this initial timing.

To check the pump to engine timing, refer to Figure 23–18, which illustrates the "timing marks." These timing marks are actually two scribe marks on all federally certified engine models, while all California certified engines employ two "half-circles" with one mark being on the pump mounting flange and the other mark being located on the top of the engine front cover. If the injection pump-to-engine timing is correct, then both the scribe marks or the half-circles will be in alignment.

SPECIAL NOTE: Refer to the 4.3L and 5.7L engine (Chapter 22) for information on the latest timing method using Kent-Moore P/N J-33075 "timing meter arrangement."

If the injection pump-to-engine timing is incorrect (marks not in alignment) on the 6.2L diesel engine, proceed as follows to correct this condition.

Procedure

1. Make sure that the engine is shut off.
2. Remove the air cleaner assembly to allow adequate working room.

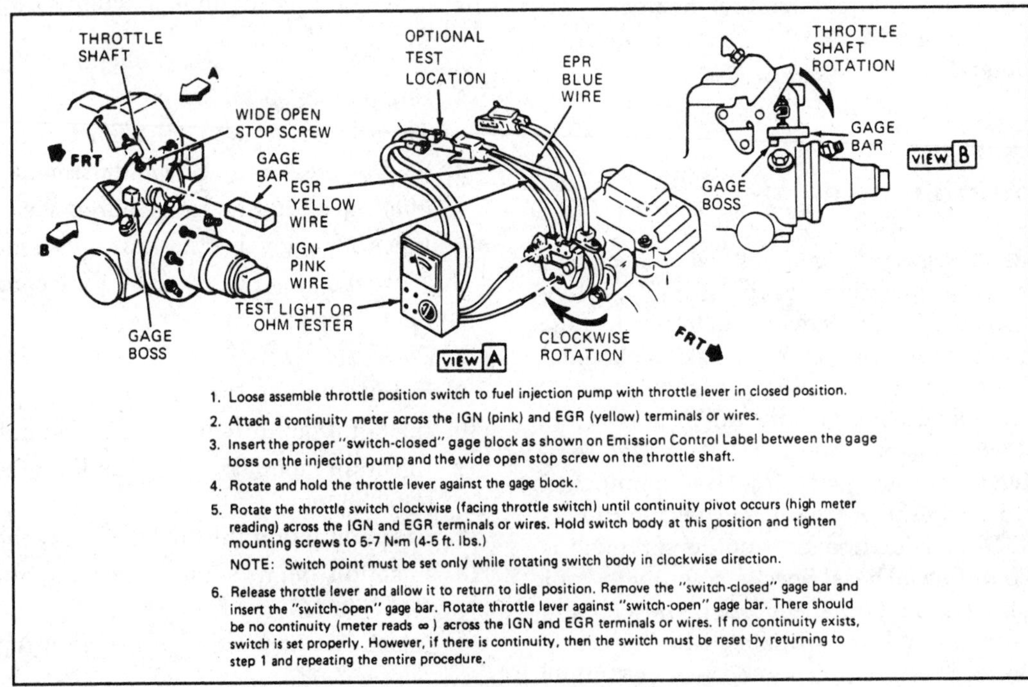
THROTTLE SHAFT
A
FRT
WIDE OPEN STOP SCREW
GAGE BAR
EGR YELLOW WIRE
IGN PINK WIRE
TEST LIGHT OR OHM TESTER
B
GAGE BOSS
OPTIONAL TEST LOCATION
EPR BLUE WIRE
VIEW A
CLOCKWISE ROTATION
FRT
THROTTLE SHAFT ROTATION
GAGE BAR
GAGE BOSS
VIEW B
1. Loose assemble throttle position switch to fuel injection pump with throttle lever in closed position.
2. Attach a continuity meter across the IGN (pink) and EGR (yellow) terminals or wires.
3. Insert the proper "switch-closed" gage block as shown on Emission Control Label between the gage boss on the injection pump and the wide open stop screw on the throttle shaft.
4. Rotate and hold the throttle lever against the gage block.
5. Rotate the throttle switch clockwise (facing throttle switch) until continuity pivot occurs (high meter reading) across the IGN and EGR terminals or wires. Hold switch body at this position and tighten mounting screws to 5-7 N·m (4-5 ft. lbs.)
NOTE: Switch point must be set only while rotating switch body in clockwise direction.
6. Release throttle lever and allow it to return to idle position. Remove the "switch-closed" gage bar and insert the "switch-open" gage bar. Rotate throttle lever against "switch-open" gage bar. There should be no continuity (meter reads ∞) across the IGN and EGR terminals or wires. If no continuity exists, switch is set properly. However, if there is continuity, then the switch must be reset by returning to step 1 and repeating the entire procedure.

Figure 23–15. T.P.S. Adjustment LH6 engine. (Courtesy of General Motors Corporation)

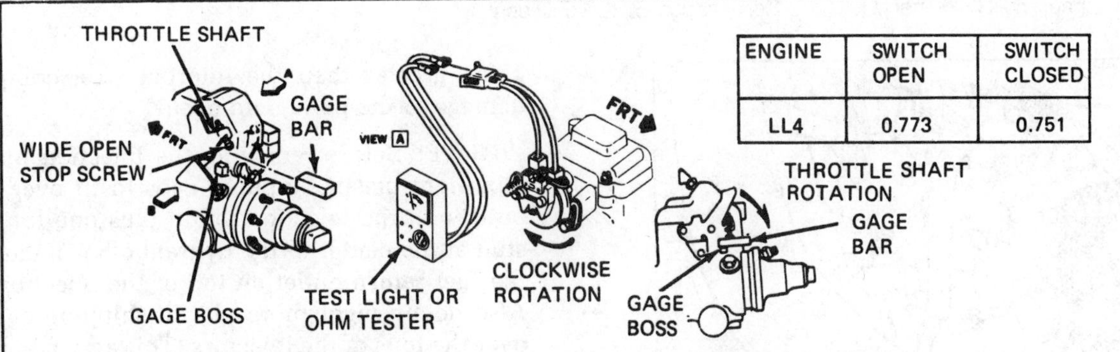

ENGINE	SWITCH OPEN	SWITCH CLOSED
LL4	0.773	0.751

1. Loose assemble throttle position switch to fuel injection pump with throttle lever is closed position.
2. Attach a continuity meter across terminals.
3. Insert the "switch-closed" gauge block between the gauge boss on the injection pump and the wide open stop screw on the throttle shaft.
4. Rotate and hold the throttle lever against the gauge block.
5. Rotate the throttle switch clockwise (facing throttle switch) until continuity just occurs (high meter reading) across the terminals. Hold switch body at this position and tighten mounting screws to 5.7 N•m (4-5 ft. lbs.)

   NOTE: Switch point must be set only while rotating switch body in clockwise direction.
6. Release throttle lever and allow it to return to idle position. Remove the "switch-closed" gauge bar and insert the "switch-open" gauge bar. Rotate throttle lever against "switch-open" gauge bar. There should be no continuity (meter reads ∞) across the terminals. If no continuity exists, switch is set properly. However, if there is continuity, then the switch must be reset by returning to step 1 and repeating the entire procedure.

105228

Figure 23-16. T.P.S. Adjustment LL4 and 700R4 transmission. [Courtesy of General Motors Corporation]

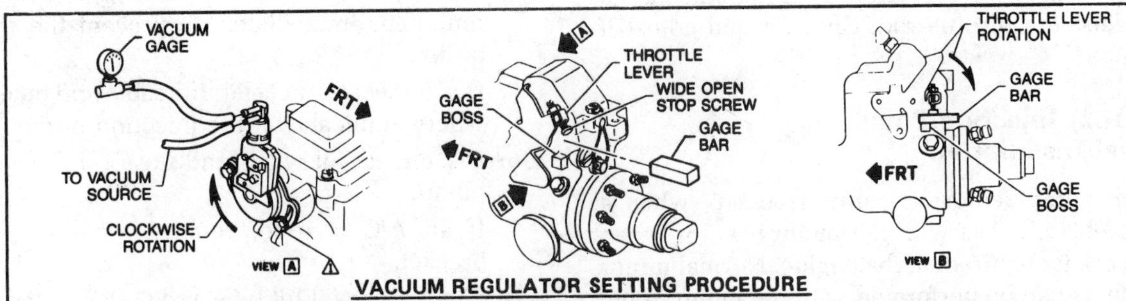

VACUUM REGULATOR SETTING PROCEDURE

1. ATTACH THE VACUUM REGULATOR VALVE SNUGLY, BUT LOOSELY TO THE FUEL INJECTION PUMP. THE VALVE BODY MUST BE FREE TO ROTATE ON THE PUMP.
2. ATTACH VACUUM SOURCE OF 67 ± 5 kpa TO BOTTOM VACUUM NIPPLE. ATTACH VACUUM GAGE TO TOP VACUUM NIPPLE.
3. INSERT VACUUM REGULATOR VALVE GAGE BAR (0.646) BETWEEN THE GAGE BOSS ON THE INJECTION PUMP AND THE WIDE OPEN STOP SCREW ON THE THROTTLE LEVER.
4. ROTATE AND HOLD THE THROTTLE SHAFT AGAINST THE GAGE BAR.
5. SLOWLY ROTATE THE VACUUM REGULATOR VALVE BODY CLOCKWISE (FACING VALVE) UNTIL VACUUM GAGE READS 27 ± 2 kpa. HOLD VALVE BODY AT THIS POSITION AND TIGHTEN MOUNTING SCREWS TO 5-7 N•m (4-5 ft. lbs.)

   NOTE: VALVE HAS BUILT IN HYSTERESIS AND MUST BE SET WHILE ROTATING VALVE BODY IN CLOCKWISE DIRECTION ONLY.
6. CHECK BY RELEASING THE THROTTLE SHAFT ALLOWING IT TO RETURN TO THE IDLE STOP POSITION. THEN ROTATE THROTTLE SHAFT BACK AGAINST THE GAGE BAR TO DETERMINE IF VACUUM GAGE READS WITHIN 27 ± 2 kpa. IF VACUUM IS OUTSIDE LIMITS, RESET VALVE.

103455

Figure 23-17. V.R.V. Adjustment LL4 engine. [Courtesy of General Motors Corporation]

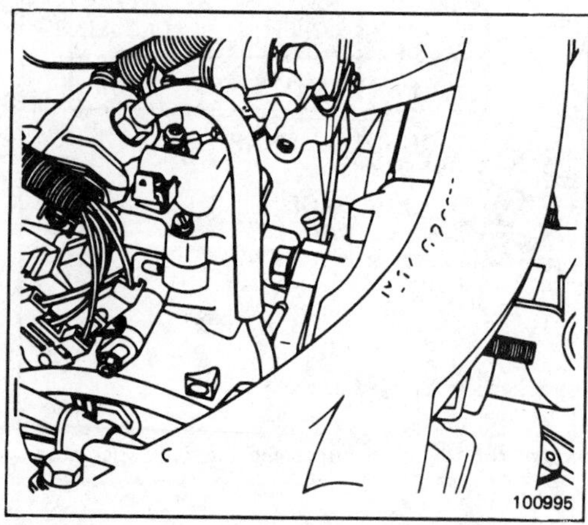

Figure 23–18. Injection pump-to-engine timing alignment marks. (Courtesy of General Motors Corporation)

3. Loosen off the three retaining nuts at the injection pump mounting flange.
4. Firmly grasp the injection pump body with both hands and rotate it either clockwise or counterclockwise until the timing mark on the injection pump is aligned with the stationary timing mark on the top of the engine front cover as shown in Figure 23–18.
5. Torque the three retaining nuts on the pump flange to 40 N·m or 30 lb.ft.
6. It is usually necessary after a timing adjustment to adjust the throttle rod. Check it and adjust if required.

6.2L (DB2) Injection Pump Removal/Installation

Injection pump removal is only necessary when a major problem exists with the pump itself or when major work is required on the engine. Normal minor adjustments can be performed with the pump in position on the engine.

Minor variations exist for injection pump removal between the CK model truck and the G model van. The procedure for injection pump removal is given below.

If it becomes necessary to remove the injection pump from the engine either due to repairs to the engine or to the fuel injection pump itself, always exercise care with the fuel injection lines and the injection pump so that you do not cause damage to these components.

CAUTION: If the engine is to be steam cleaned before injection pump removal *do not* direct the steam jet against the injection pump housing because it is manufactured from an aluminum alloy. Due to the rate of expansion of the pump housing being greater than the internal steel components, damage to the pump can occur.

NOTE: Before removing the injection pump, obtain eight plastic shipping caps to fit over the fuel outlets on the injection pump plus one for the fuel stud at the center of the hydraulic head and one for the fuel return outlet on top of the injection pump. Also obtain eight more plastic shipping caps to fit over the tops of the injectors. This action is required to ensure that no dirt or foreign material will enter the pump or injectors while the unit is in a state of disassembly.

Proceed to remove the injection pump on the model CK light truck as follows:

Procedure

1. Isolate the starting system and electrical system by disconnecting both batteries.
2. Remove the radiator fan shroud and fan assembly.
3. Remove the engine air cleaner/intake manifold assembly.
4. Remove all fuel lines (fuel feed line and return line from the injection pump) as well as all of the injection pump to injector high pressure lines and cap them with the protective plastic shipping caps immediately to prevent the entrance of dirt.
5. Disconnect the accelerator cable and detent wire where applicable at the injection pump.
6. Disconnect all wires and hoses at the injection pump.
7. If an A/C is used, remove the hose retainer bracket.
8. Remove the oil fill tube and C.D.R.V. (crankcase depression regulator valve) vent hose and grommet.
9. Check that the injection pump to timing cover (top) alignment marks are clearly visible (Figure 23–18). If they are not, then match mark by scribing or painting a mark on both.
10. Refer to Figure 23–19. It is now necessary to rotate the engine manually so that you can gain access as shown through the oil filler neck hole to the bolts on the injection pump drive gear.
11. Remove the three injection pump to front cover flange mounting nuts so that the injection pump can now be withdrawn.

Model "G" Van Injection Pump Removal. The removal of the injection pump from the van models dif-

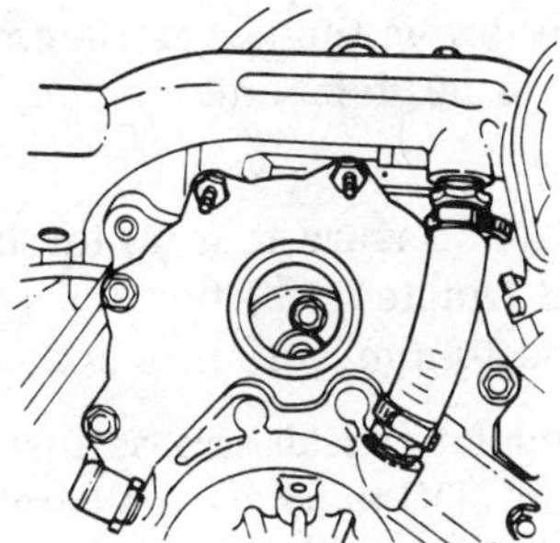

Figure 23–19. Pump drive gear bolts. [Courtesy of General Motors Corporation]

fers in that several other attachments and components require removal—such as the windshield washer bottle, fuel filter and bracket, oil pan dipstick bracket, and the TV cable on an automatic transmission.

The other major difference is that once the number 2, 4, 5, 6, 7, and 8 injection lines have been disconnected at the nozzles, the vehicle must be jacked up or raised so that number 1 and 3 injection lines can be removed from the nozzles.

Lower the vehicle and disconnect the injection lines from the hydraulic head outlets at the injection pump. Ensure that you tag the lines for correct installation later. As with the CK truck model, ensure that the pump to front cover timing marks are in alignment; if not, scribe a match mark or paint mark to assist in alignment when installing the pump.

Remove the three injection pump mounting flange nuts and withdraw the pump.

Injection Pump Installation

1. Before installing the injection pump, remove the old gasket between the pump mounting surface and the flange, and replace it with a new one.
2. Refer to Figures 23–9 and 23–20 and align the locating pin on the front of the injection pump hub with its mating slot on the back face of the injection pump drive gear, as well as aligning the pump to front cover timing marks.
3. Engage the injection pump into the front timing cover of the engine ensuring that the locating pump on the pump hub has engaged fully with the hole in the back of the drive gear.
4. Install the injection pump flange retaining nuts and draw them up snug. Ensure that the pump to cover alignment marks are perfectly aligned prior to torquing the retaining nuts to 40 N·m or 30 lb.ft.
5. From the front side of the timing cover on the engine, install the drive gear to injection pump bolts through the oil fill tube access hole and torque the bolts to 25 N·m or 20 lb.ft.

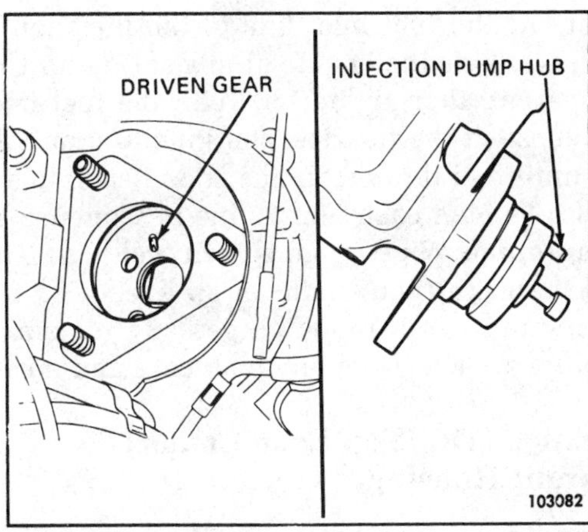

Figure 23–20. Injection pump locating pin. [Courtesy of General Motors Corporation]

6. Install all of the remaining components that were removed during pump removal.

CAUTION: Blow all of the injection lines clean through their bores with clean dry shop air prior to installation. Install the lines as per tagging procedure used during removal and route as per Figure 23–2 shown earlier.

NOTE: Be sure to tighten all of the fuel injection lines to the specified torque setting since undertightening can lead to fuel leaks and engine misfire.

Overtightening can cause the nipple on the end of the fuel injection lines to be deformed causing leakage as well as permanent line damage. Recommended torque settings are as follows:

1. Injection nut at the nozzle and the pump—25 N·m (20 lb.ft.)
2. Injection line at the bracket—20 N·m (15 lb.ft.)
3. Injection line at the intake manifold—40 N·m or 30 lb.ft.)
4. Injection line support clamps—3 N·m (26 lb.in.)

Priming/Bleeding Fuel System

If the fuel system has been drained completely because of a major repair to the engine or fuel injection pump, it will be necessary to bleed the system of entrapped air so that the engine can be started easily.

Refer to the earlier discussion/procedure dealing with fuel filter maintenance/change that indicates how to bleed the filters. It may also be necessary to bleed the fuel to and through the injection pump and up to the fuel injectors. To do this systematically, loosen off the fuel inlet line to the injection pump while cranking the engine until a steady flow of fuel is apparent, then tighten it. With the fuel injection lines loose at the nozzles, continue to crank the engine until fuel flows from the end of each line at the nozzle. Tighten each nozzle line in sequence as fuel spills from it. Wipe up all spilled fuel with a towel.

Start and run the engine at an idle speed for several minutes until it self-purges any remaining air from the system. Note any fuel leaks and correct.

Marking TDC (Top Dead Center) on Front Housing

If it becomes necessary to install a new engine front timing cover, it will have to be correctly marked as to true TDC. This procedure requires the use of a special Kent-Moore service tool, Part Number J-33042 which is illustrated in Figure 23–21.

Procedure

1. Manually turn (bar) the engine over until No. 1 piston is at TDC on the firing stroke (both valves closed). This can be confirmed by the fact that the injection pump locating pin hole in the driven gear should be at the vertical 6 o'clock position (See Figure 23–20). If it is not, then the engine *must* be turned another 360 degrees, which will place the engine camshaft and injection pump gear marks in alignment.
2. Install the special tool (timing fixture) J-33042 into the fuel injection pump mounting hold without the pump gasket.
3. Tighten the injection pump driven gear to tool J-33042 by the bolts accessible through the oil pump filler hole at the front of the timing cover (See Figure 23–19).
4. Insert a 10-mm nut to the upper stud of the timing cover housing finger tight to hold tool J-33042 flange.

Figure 23–21. Special tool J-33042. (Courtesy of Kent-Moore Tool Group, Sealed Power Corporation)

5. Proceed to torque the larger 18-mm bolt counterclockwise to 50 lb.ft. and tighten up the 10-mm nut.

   CAUTION: Ensure that (a) the crankshaft has not moved (rotated), (b) the tool J-33042 did not bind on the 10-mm nut.

6. Using a mallet, strike the scriber to positively place a true TDC mark on the front of the timing cover.
7. Remove special tool J-33042.
8. Install the fuel injection pump with its gasket as explained under injection pump installation herein.

On Vehicle Pump Repair

Repairs that can be attempted successfully to the fuel injection pump while it is on the engine are limited to basic adjustments and fuel leakage because any attempt to perform major repairs will meet with little or no success.

When major problems exist with a fuel injection pump, it should be removed and sent to a local fuel injection specialty shop where they have both the trained personnel and special test equipment to perform all repairs. The average automotive mechanic does not have access to this special equipment, and in many cases, they do not have the skills to perform these major repairs successfully. Personnel employed in the repair of fuel injection equipment are trained to perform all types of repairs to a wide variety of fuel injection pumps, injectors/nozzles, and governor assemblies. These same personnel are certified through tests and exams plus exposure to continuous upgrading programs so that they are "fuel injection repair specialists."

However, more and more large automotive dealerships are training some of their mechanics to be able to perform major fuel injection repairs with the necessary special equipment. In many cases, however, a dealership finds it faster and cheaper to have this major repair work conducted by a local diesel fuel injection shop that is capable of offering fast turnaround service plus an excellent warranty on their work since most of these local shops are independently owned and are members of the ADS (Association of Diesel Specialists).

Typical jobs that can be successfully performed by the automotive mechanic are as follows:

1. Injection pump cover seal and or guide seal replacement

2. Throttle shaft seal replacement
3. Solenoid replacement
4. Side cover gasket replacement

Pump Component Location

Before performing any repairs to the injection pump while it is on the engine, you should be familiar with the identification and location of the various components. Figures 23–22 and 22–56 illustrate clearly these components and their positions.

Injection Pump Governor Cover Gasket Replacement

This is a fairly straightforward operation and simply involves removal of all components that would otherwise prevent injection pump cover removal.

CAUTION: Always disconnect both negative battery cables before performing any job on the fuel injection pump.

Procedure

1. Remove the air cleaner and intake, and install protective screens (Kent-Moore P/N J-29664) in the cylinder heads, similar to item 14 shown in Figure 22–55.
2. Clean the area around the top of the injection pump to prevent the entrance of dirt or other foreign material when the cover is removed.
3. Disconnect the fuel return pipe and the wires from the fuel solenoid and the housing pressure cold advance (HPCA) unit in order to remove the fuel solenoid unit.

   NOTE: Diesel fuel will spill into the engine valley; therefore, use cloths to wipe this up.

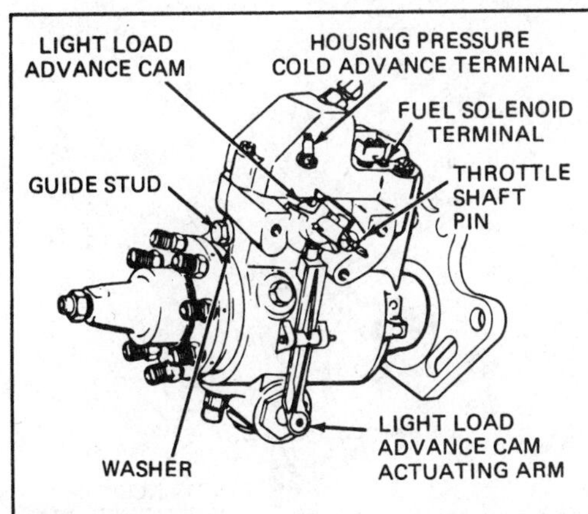

Figure 23–22. Injection pump housing right side view typical. (Courtesy of General Motors Corporation)

4. Remove screws from the cover and lift the cover away from the injection pump body (See Figure 22–56).
5. Refer to Figure 23–23, which illustrates the position of the metering valve spring.
6. Attention is drawn to the position of the metering valve spring located under the injection pump cover because, if it is necessary to replace the guide stud seal because of leakage, then extreme care must be taken to ensure that this spring is replaced in the exact same position upon reassembly.

   NOTE: If guide stud seal replacement is required, perform Steps 7 and 8; otherwise if only the injection pump cover seal is being replaced, move directly to Step 9.

7. Remove the guide stud along with its washer from the front center of the injection pump above the hydraulic head fuel injection line outlets taking careful note of the location of the metering valve spring as per Figure 23–24.

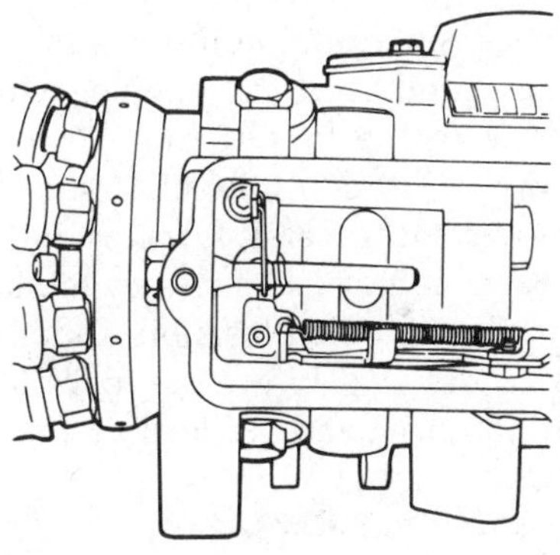

Figure 23–23. Metering valve spring position. (Courtesy of General Motors Corporation)

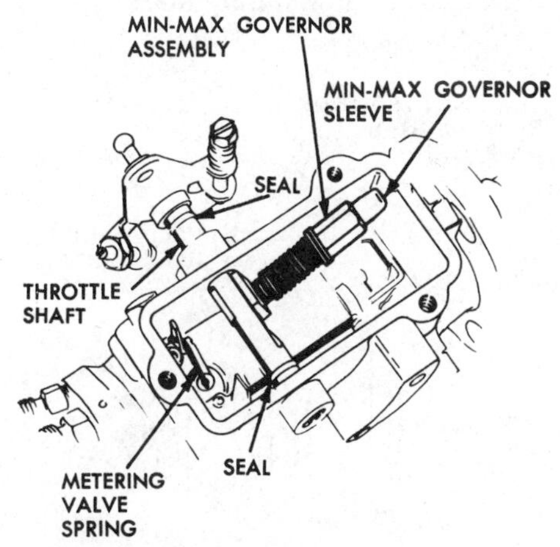

Figure 23–24. Injection pump cover removed. (Courtesy of General Motors Corporation)

8. Install a new washer over the guide stud after removal of the old one and insert the guide stud back into the housing, taking particular care that the upper extension of the metering valve spring rides on top of the guide stud. Torque the guide stud to 9.5 N·m (85 lb.in.).
9. To install the injection pump cover, refer to Figure 23–25. Hold the throttle in the idle position during the installation of the cover, and torque the retaining screws to 3.7 N·m (33 lb.in.).
10. With the negative battery cables reconnected, perform the following check: Ignition switch in run position, touch the pink wire to the fuel solenoid, which should be heard to click as you energize/de-energize the switch. If no click is heard, it is because the linkage inside the injection pump is jammed in the wide open position. Therefore remove the injection pump cover and ground the solenoid lead opposite the hot lead and connect the pink wire again. If the solenoid fails to move the linkage, replace the solenoid since minimum voltage across the solenoid terminals should be 12.0 volts.
11. Reconnect remaining components to the injection pump and start the engine, which will be rough due to the loss of fuel earlier. Should a rough operation continue after shutting the engine down for several minutes (air rises and self-purges), it may be necessary to bleed the fuel system as explained earlier under fuel filter maintenance. Check for fuel leaks at the injection pump cover and guide stud washer.

Throttle Shaft Seal Replacement

Fuel leakage at the throttle shaft involves the use of some special tooling as well as care during disassembly and reassembly of components.

Proceed as in injection pump governor cover gasket cover replacement (above) by disconnecting both battery negative leads to isolate the starting and electrical systems. With the air cleaner and crossover manifold removed, install protective screens (Kent-Moore P/N J-29664) in the cylinder head intake ports. Disconnect the fuel solenoid wire at the injection pump as well as the HPCA wire and the fuel return pipe.

Procedure

CAUTION: Match-mark components prior to disassembly to ensure that they will be reassembled in the same position.

1. Carefully mark the position of the throttle position switch or vacuum regulator valve. Remove both the throttle rod and return springs, and fast idle solenoid.
2. Remove the throttle bracket and install Kent-Moore tool P/N J-29601 as illustrated in Figure 23–26 over the throttle shaft advance cam and tighten the tool wing nut since this will now provide the correct alignment upon reassembly. Remove this special tool from the throttle shaft, then loosen the face cam screw.
3. Carefully drive the pin from the throttle shaft and remove the advance cam and fiber washer, and deburr the shaft in the pin area.
4. Clean around the injection pump cover and place some cloths to catch the spilled diesel fuel as the cover is removed.
5. Observe the position of the metering valve spring as per Figure 23–23.
6. Remove the guide stud and washer from the front center of the injection pump above the hydraulic head fuel stud outlets.

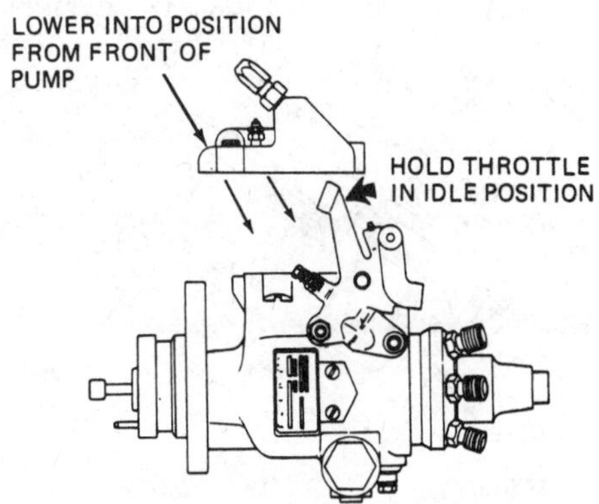

Figure 23–25. Installing injection pump cover. (Courtesy of General Motors Corporation)

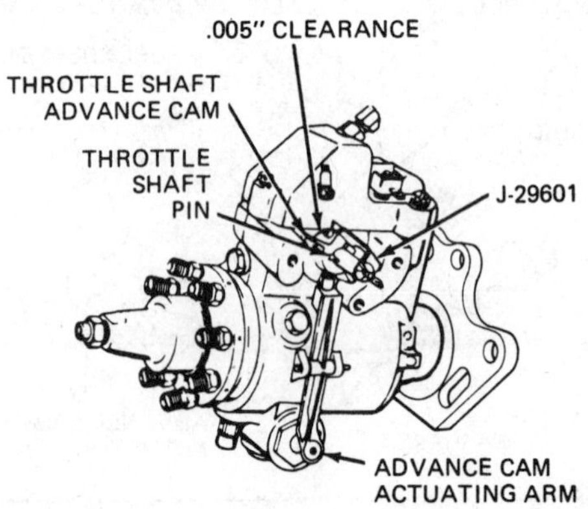

Figure 23–26. Injection pump with tool J29601 installed. (Courtesy of General Motors Corporation)

7. Refer to Figure 23–24 and rotate the min-max governor assembly up to ensure clearance and remove it from the throttle shaft.

   CAUTION: If the governor idle spring becomes disengaged from the throttle block, reinstall it with the tightly wound coils towards the throttle block.

8. Remove the throttle shaft and examine it and its bushings for wear or damage. Although the pump throttle shaft can be replaced, generally speaking, if this part is damaged, then the bushings will also be damaged and the injection pump should be removed and sent to the local Stanadyne or fuel injection dealer.
9. If the throttle shaft is serviceable, then remove the old leaking seals and replace them with new ones. Apply a light coating of Kent-Moore P/N J-33198 Synkut oil or clean chassis grease to the seals prior to installation on the throttle shaft, which has sharp edges—take care here not to cut the seals.
10. Refer to Figure 23–24 and install the throttle shaft back into the injection pump housing bore. Ensure that the throttle shaft does engage with the min-max governor assembly as you rotate it into position with the governor.
11. Install a new mylar washer, the throttle shaft advance cam (leave the screw loose), and a new throttle shaft drive pin.
12. Manually place the throttle shaft advance cam into a position whereby Kent-Moore tool P/N J-29601 (see Figure 23–26) can again be installed over the throttle shaft; pin the slots and the spring clip in position over the advance cam mechanism.
13. Refer to Figures 23–26 and 23–27 and insert a 0.005″ feeler gauge between the white mylar washer on the throttle shaft and the injection pump housing. Push against the throttle shaft and tighten the cam screw to 3.1 N·m (30 lb.in.); use Loctite 290 or equivalent to secure the cam screw, then remove the special tool J-29601.
14. Place a new washer on the guide stud and insert it into the injection pump housing.

    CAUTION: Ensure that the upper extension of the metering valve spring rides on top of the guide stud as shown in Figures 23–23 and 23–24. Torque the guide stud to 9.5 N·m (85 lb.in.).

15. The remaining procedures for fitting the pump cover, etc., are the same as that given in Steps 9 through 11 above dealing with pump governor gasket cover and/or guide stud seal replacement.

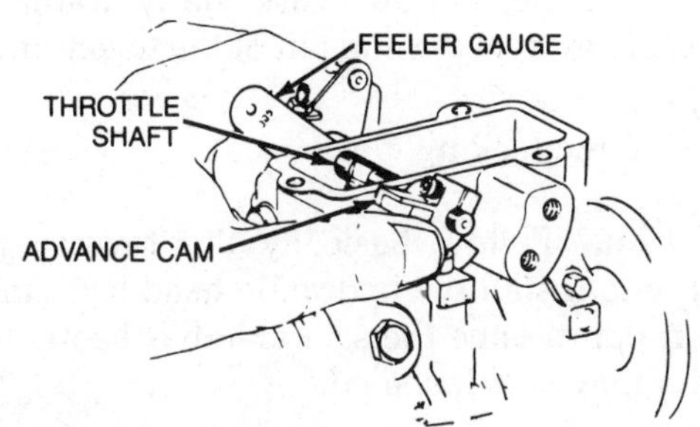

Figure 23-27. Feeler gauge installation. (Courtesy of General Motors Corporation)

Injection Pump Solenoids

Two electrical solenoids are used with the DB2 injection pump, which are:

1. The engine (injection pump) shutdown solenoid assembly.
2. The engine (injection pump) cold advance solenoid.

To remove the solenoids, remove the pump cover as explained earlier in order to obtain access to the solenoids (Figure 22–56).

Installation of Solenoids. Take the following precautions when installing solenoids:

1. On the shutoff solenoid, make certain that the linkage is free.
2. On the housing pressure cold advance solenoid, make certain that the plunger is centered so that it will contact the fitting check ball.
3. Ensure that all insulating washers are installed on the terminal studs and torque the nuts to 1 to 1.5 N·m (10 to 15 lb.in.).
4. Always check the operation of the solenoid before injection pump cover installation by using a 12-volt minimum DC power supply to test it. Also ensure that the pump shutoff linkage is free.
5. Reinstall the injection pump cover as stated earlier (above).

Injection Pump Side Cover Gasket

The removal and replacement of the injection pump housing side cover gasket simply involves the removal of the two screws, cover, and gasket as shown in Figure 22–28.

When replacing the gasket covers, be sure to torque the retaining screws to specifications.

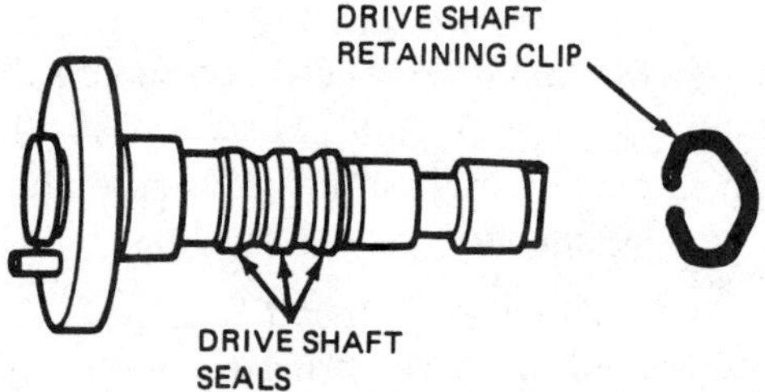

Figure 23-28. Drive shaft seal removal. (Courtesy of General Motors Corporation)

Injection Pump Service—Off Vehicle

When it is necessary to remove the injection pump from the engine for major repair, refer to the procedure herein dealing with this job. A number of repair jobs can be undertaken by the trained automotive mechanic to the injection pump when it is removed from the engine. These jobs are basically replacement of pump seals to prevent leakage.

The seals that would be replaced when the injection pump has been removed from the engine are:

1. The advance pin hole plug seal
2. The automatic advance mechanism seals
3. The hydraulic (injection pump) head seal O-ring
4. The injection pump drive shaft seal replacement

NOTE: Once any of the above seals have been replaced, the injection pump must be pressure tested to ensure that all leaks have in fact been corrected!

The procedure necessary to replace the injection pump seals is given below.

Advance Pin Hole Plug Seal Replacement

Refer to Figure 22–25 and tap Item 19 (advance pin hole plug) lightly with a small hammer to assist in loosening it. Loosen the plug and remove it along with its seal, discard the old seal, clean the seating area for the seal, lightly lubricate the new seal, and install it into position on the plug. Tighten the plug to 8.5 to 11 N·m (75 to 100 lb.in.).

Automatic Timing Advance Seal Replacement

Remove the advance pin hole plug (Item 19) as per the above explanation (see Figure 22–25). The two seals that will be replaced here are Items 2 and 6.

Procedure

1. Loosen and remove the spring side advance piston hold plug (Item 1) along with its piston, spring, and slide washer.
2. Loosen and remove the power side advance piston hole plug (Item 7), its piston and slide washer.
3. Disassemble both plugs and pistons.
4. Lightly lubricate both new seals, reinstall both plugs, and torque to 27 N·m (20 lb.ft.).
5. Install the advance screw hole plug with its new seal and torque to 8.5–11 N·m (75–100 lb.in.).

Hydraulic Head Seal—O-Ring Replacement

CAUTION: This job must be done with extreme care because the hydraulic head assembly is a major component of the injection pump. The hydraulic head should be match-marked to the injection pump body before removal. In addition, this job must be done in a dust-free area because the entrance of any dust or dirt into the hydraulic head area can result in severe damage to the injection pump assembly. A special injection pump holding fixture is available to mount the pump in during the performance of these jobs. If this holding fixture is not available (Kent-Moore P/N J-29692-B), then the pump body should be wrapped in soft cloths and lightly held in a soft-jaw vise. If this is not available, use small wooden blocks to protect the injection pump body from the vise jaws.

Procedure

1. Refer to "Throttle Shaft Seal Replacement" described previously in this chapter for information regarding removal of this assembly.
2. Refer to Figure 23–24 and remove the fuel metering valve.
3. Refer to Figure 22–28 and remove the injection pump housing vent screw (Item 6).
4. Refer to Figure 22–25 and remove the advance pin hole plug (Item 19) and the advance pin (Item 10).
5. With the pump mounted in its holding fixture so that the rear of the pump is sloping down, remove the hydraulic head retaining screws.
6. Firmly grasp the hydraulic head assembly and using a twisting motion, work the hydraulic head assembly away from and out of the injection pump body.
7. Remove the leaking O-ring seal.

Installation. Follow basically the reverse procedure here to install the hydraulic head back into the injection pump once the seal area has been cleaned off and a new seal installed.

1. Once the hydraulic head has been installed, install the two head locking screws finger tight; then turn the injection pump upside down.

2. Install a new seal on the head locating screw (Item 11 in Figure 22–25) and tighten it to 20–25 N·m (15–18 lb.ft.).
3. Torque the head locking screws (Item 7 Figure 22–28) to 20–25 N·m (15–18 lb.ft.).
4. Install the advance pin (Item 10) and its seal and plug (Items 18 and 19) in Figure 22–25.
5. Rotate the injection pump now so that the top cover of the injection pump is facing up, and install the metering valve.
6. Install the throttle shaft assembly and injection pump cover.

Injection Pump Drive Shaft Seal Replacement

Again extreme care must be exercised when removing the pump drive shaft so as to prevent the entrance of dirt or foreign material, which would cause serious damage to the pump assembly.

The following special Kent-Moore tools will be required to perform this job successfully:

1. P/N J-29692-B injection pump holding fixture
2. P/N J-9553-01 drive shaft retaining clip remover
3. P/N J-29745-A seal installer
4. P/N J-33198 Synkut seal oil

Procedure

1. Mount injection pump into tool J-29692-B.
2. Remove the pump cover and note the position of the metering valve spring, which should be over the top of the guide stud (See Figure 23–23). When reassembling, this same position must be duplicated.
3. Remove the guide stud and its washer as well as the min-max governor.
4. Manually rotate the pump drive shaft until you can see the hump on the drive shaft retaining clip (See Figure 23–28).
5. Using special Kent-Moore tool J-9553-01, remove the drive shaft retaining clip—which should always be renewed upon reassembly because it will stretch when removed.
6. Remove the drive shaft with the alignment pin located at the top.
7. Remove and replace the old seals using Kent-Moore special tool J-29745-A. Lubricate the seals with Kent-Moore Synkut oil P/N J-33198 or equivalent.

   CAUTION: There are three seals used on the injection pump drive shaft—the red seal is installed between the two black seals on the pump drive shaft.
8. Install a new drive shaft retaining clip taking care not to overstretch it during installation.
9. Gently install the drive shaft into the pump housing.

   NOTE: Make certain that the drill points on the drive shaft end are matched with the rotor drill points.
10. Install the min-max governor assembly.
11. When installing the guide stud be *sure* that it is under the metering valve spring (see Figure 23–23).
12. Install the pump cover and fast idle solenoid bracket.

Injection Pump Pressure Test

If at any time you suspect that the injection pump seals are leaking but you are unsure as to what ones, or if you have replaced any seals, it is good practice to pressure check the injection pump before reinstalling it onto the engine.

Procedure

1. If the pump has been removed for the pressure check or is already off of the engine, drain all diesel fuel from the pump.

   CAUTION: Air will be used to pressure check the pump; therefore, it is imperative that shop air be routed through a filter before hooking it up to the injection pump. The air *must* be clean and moisture free (dry).
2. Connect an air line to the injection pump inlet.
3. Seal off the injection pump return line fitting.
4. Immerse the injection pump assembly completely in a bath of clean test oil (preferably diesel calibrating fluid) that has been heated to normal test temperature [usually between 110–115°F (43–46°C)].
5. Allow the injection pump to sit in the test oil for at least 10 minutes, then apply air pressure to the pump at a maximum of 20 psi (137.9 kPa). Let the pump sit for another 10 minutes to allow any trapped air to escape.
6. If no leaks are apparent after this time period, reduce the air pressure to 2 psi (13.8 kPa) for 30 seconds; if no leaks exist, then increase the pressure to 20 psi (137.9 kPa) again. If no leaks are evident, then the pump is satisfactory.

GMC 6.2L Diesel Engine—Injection Nozzles

The injection nozzles used in the 6.2L engines differ between the model of vehicle that it is used in; therefore, the nozzles cannot be interchanged between the CK and G model vehicles produced by GMC. These internal differences are such that interchanging these nozzles will lead to poor engine performance and possible engine damage.

The type of nozzle used on the 6.2L engine is a Robert Bosch inward opening pintle unit. (For an explanation—see fuel injectors, Chapter 29.)

The injection pump-to-nozzle fuel pipes have a 2.5-mm bore with a 6.3-mm outside diameter and they are 616 mm (24.25") long. Figure 23–29 illustrates the pintle nozzle used with the 6.2L V8 diesel engine.

Nozzle Check—On Vehicle

A quick check of the nozzle's condition while it is still on the engine is performed as follows:

1. Run the engine until it is at operating temperature.
2. With the engine running at an idle rpm, loosen off the fuel line nut at the nozzle (place a rag around it). The engine should decrease in rpm (similar to shorting out a spark plug in a gasoline engine) if the nozzle is operating properly. If there is no change in engine rpm, then the nozzle is not operating correctly and should be removed.

Nozzle Removal

Before removing any fuel injection nozzle from the engine, keep in mind that "cleanliness" is of great importance when dealing with any fuel system components. Therefore, before proceeding with nozzle removal, obtain the necessary plastic shipping caps that will be required to cover each injection line and nozzle once disconnected and removed. Dirt and dust must be kept out of all fuel system areas.

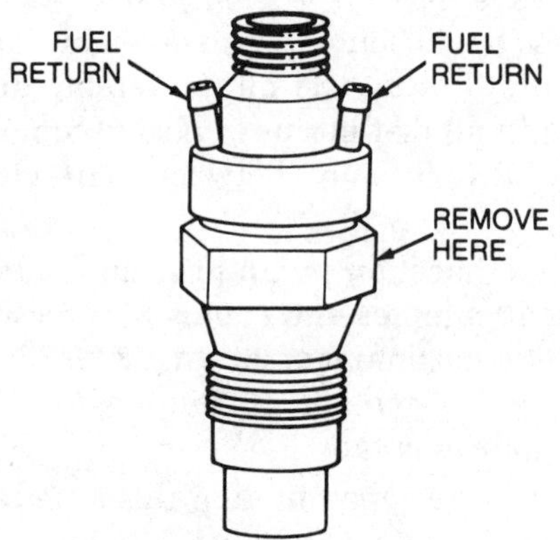

Figure 23–29. 6.2L GMC V8 Diesel pintle nozzle. [Courtesy of General Motors Corporation]

Procedure

1. Disconnect both battery negative cables.
2. Remove fuel line clip, fuel return hose, and fuel injection lines to the nozzles as discussed earlier in injection pump removal. Cap the nozzle and lines with shipping caps.
3. Special Kent-Moore tool J-29873 should be used if available to facilitate nozzle removal. Always remove the nozzle using the 30-mm hex nut on the body as shown in Figure 23–29.

Nozzle Installation

Installation is basically the reverse of removal. Refer to Figure 23–30.

Procedure

1. Install the nozzle and torque to 70 N·m (50 lb.ft.).
2. Remove the protective caps and connect the fuel injection lines; torque the line nuts to 25 N·m (20 lb.ft.).
3. Install the fuel return lines (hoses), fuel line clip, and reconnect the battery negative ground cables.

6.2L GMC Engine Glow Plug System

All automotive diesel engines are presently of the IDI (indirect-injection) type and require a form of combustion chamber preheater to facilitate rapid starting in cold weather due to the large radiated heat loss from the combustion chamber area. The most widely adopted starting aid is a number of glow plugs that are located (screwed) directly into the combustion chamber area of the engine at the cylinder head.

The glow plug is basically an electric heater that is energized by the operator through a switch that can be incorporated into the ignition switch, or in some cases it may be a separate switch. Once energized, electric current flows through the glow plug to cause it to glow or become red hot (1550–1650° F); therefore after a given time period when the starter motor is engaged, the ambient air that flows into the engine will be rapidly increased in temperature through the use of the hot glow plug within the combustion chamber.

Each cylinder in the 6.2L engine employs a glow

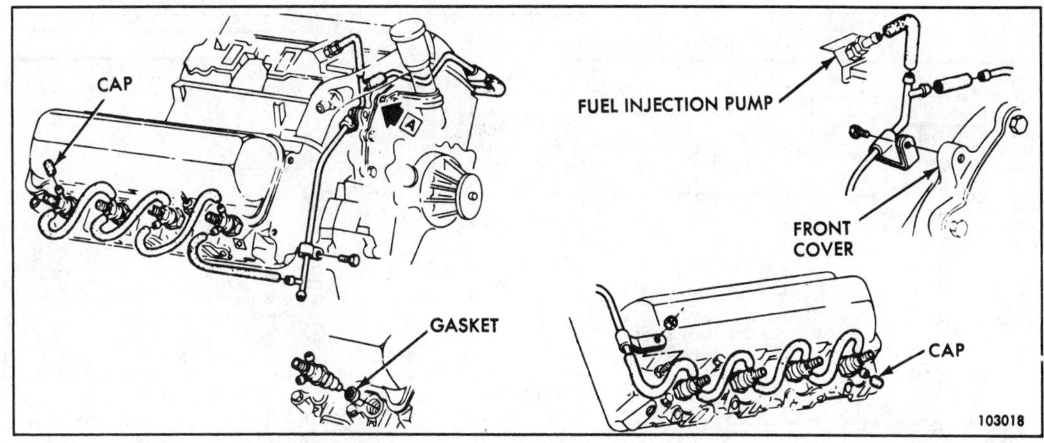

Figure 23–30. 6.2L Diesel engine nozzle installation. (Courtesy of General Motors Corporation)

plug that is actually a 6-volt unit operated from the 12-volt battery system when the ignition key is turned to the *run* position prior to engaging the starter motor. They remain pulsing for a short time after starting, then automatically turn off.

Within the instrument panel of the vehicle is a "glow plugs" light that will turn on immediately when the ignition switch is turned to the *run* position.

The major components of the glow plug system are described in the following paragraphs.

Glow Plug Controller. This unit is actually a thermal controller because it is located and mounted into the water passage at the rear of the engine so that the ground circuit to the glow plug can be opened or closed as necessary during engine start-up to control the glow plug preheat and after-glow cycles (engine running) by cycling the current to the glow plugs on and off.

Glow Plug Relay. This relay is mounted on the inner left fender panel with its main purpose being to relay current to the glow plugs. The action of the relay is controlled by the thermal controller, which will cause the relay to pulse on and off depending on the engine coolant temperature.

CAUTION: Do not attempt to bypass the glow plug relay with the aid of a jumper wire or by rewiring for the use of a manual glow plug bypass because this could result in the glow plugs being electrically energized for too long a time causing glow plug burnout and therefore very hard starting especially in cold weather.

Figure 23–31 shows the simplified diesel glow plug wiring circuit. During cold start-up, current will flow through both the thermal controller and glow plug relay coil to ground. The glow plugs will be electrically energized (heated) continuously on cold start-up for approximately 7.5 to 9 seconds at an ambient air temperature of 0°F in order to ensure that the pre-combustion chamber preheat is adequate for starting purposes. After this time period, the glow plugs will pulse on and off through the action of the thermal controller to maintain a suitable engine warm-up process. The controller is capable of varying glow plug operation for up to one minute upon a warm engine start-up where little or no heating would be required.

If the glow plugs thermal controller allows heat to the glow plugs for longer than 9 seconds, a circuit breaker in the thermal controller would open, cutting off current to the glow plugs and protecting the circuit.

Refer also to Chapter 22 on the 4.3L and 5.7L engine for additional information on the glow plug circuit operation.

6.2L Engine Troubleshooting

The troubleshooting techniques that can be applied to the 6.2L diesel engine are the same ones that are listed under the troubleshooting sections of Chapter 22 for both the 4.3L and 5.7L General Motors V6 and V8 diesel engines. In addition, the Ford 6.9L diesel V8 engine uses the same injection pump, namely the Roosa Master/Stanadyne model DB2 and has a detailed troubleshooting section in Chapter 24 that can also be followed to isolate possible problem areas with the engine fuel system.

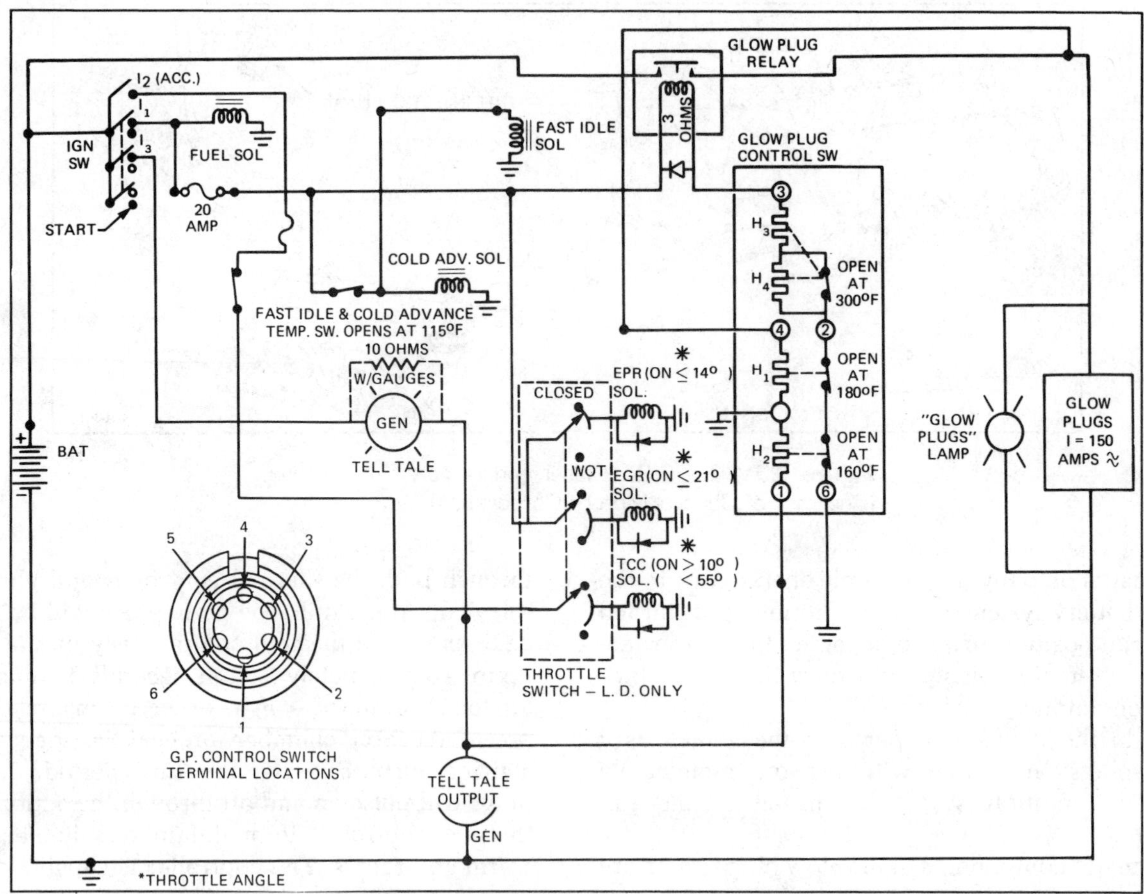

Figure 23-31. Glow plug wiring diagram 6.2L GMC engine. (Courtesy of General Motors Corporation)

In addition, Chapter 30, dealing with Diesel Engine Troubleshooting, can be referenced for general techniques and possible causes of poor engine performance.

The following information is typical of the areas of the fuel system that could be systematically checked to isolate a possible problem with the diesel engine fuel system. Detailed analysis and test procedures for these listed problem areas can be found under Chapter 24, dealing with the Ford 6.9L diesel engine fuel system.

Fuel System Checks

When the engine performance is poor, such as sluggish throttle response, low horsepower, and hard starting or stumble at idle, check the color of the exhaust smoke as a guide to where the problem might be.

Black exhaust smoke is generally indicative of either an air starvation situation or excess fuel delivery. Although the air cleaner element can be visually inspected for signs of plugging, this in itself is not conclusive proof that the element is in fact offering a high air inlet restriction. Use a water (H_2O) manometer or a vacuum gauge connected to the test port on the air cleaner to establish the amount of restriction to air flow. Run the engine at its maximum no-load rpm and note the reading—which should not exceed 25″ H_2O (maximum). Since 25″ is the maximum reading, anything close to this reading would be reason enough to change the filter element.

In the majority of cases, the air starvation problem is more common. When poor engine performance is evident with no black smoke or very little at the exhaust, this usually indicates a fuel starvation problem. Faulty injection pump timing and faulty injection nozzles can also be a cause of black exhaust smoke in addition to high exhaust back pressure.

To check the fuel system for fuel starvation, the following checks would be performed: The suction side of the system is between the fuel tank(s) and the inlet side of the lift pump. From the outlet side of the lift pump, the fuel is under low pressure until it enters the vane-type transfer pump in the injection pump. Fuel from the injection pump is at high pressure (2000–2150 psi or 13790–14824 kPa) and will spray fuel over the engine compartment if leaks exist.

1. Check fuel supply lines for kinks or damage (restriction) and also the return fuel line for restriction problems.
2. Check fuel inlet (suction) lines for loose fittings that would cause air to be drawn into the system.
3. Check the pressure side of the system for fuel leaks.
4. Check the condition of the fuel especially the Cetane rating with the test kit as well as for conditions such as dirt and water.
5. Check the fuel supply (lift) pump pressure as well as its delivery capacity.
6. Check the fuel filter outlet pressure.
7. Check for fuel tank restriction.
8. Check the fuel return pressure.
9. Check injection pump transfer pressure.

GMC 6.2L Engine Compression Check

A compression check can be performed at any time on the engine to establish the mechanical condition of the valves and rings. In order to successfully conduct a compression check on the engine, several special Kent-Moore tools will be required as follows:

1. Kent-Moore tool number J-26996-1 air crossover cover
2. Kent-Moore tool number J-29664-2 air crossover cover
3. Kent-Moore tool number J-26999-10 screw
4. Kent-Moore tool number J-26999 compression gauge

CAUTION: Because of the high compression ratio of this engine (21.5:1), do not attempt to add oil to any cylinder as a means of isolating the problem to valves or rings because a hydrostatic lock could occur that can result in extensive engine damage. In extreme cases, con-rod or piston damage can result.

NOTE: In order to receive an accurate compression reading, the batteries must be at or near a full state of charge in order to crank the engine over fast enough. A slow cranking rate will result in a low reading. The lowest reading obtained from any engine cylinder should not be less than 300 psi (2068 kPa).

In addition, the spread between cylinder pressure readings should be such that the lowest reading obtained is not less than 80% of the highest reading.

Procedure

1. Once the air cleaner is removed, install the safety air crossover cover J-26996-1 or J-29664-2 onto the air inlet manifold throat.
2. To prevent the engine from receiving fuel to the injectors, disconnect the *pink* wire from the fuel solenoid terminal at the injection pump.
3. Remove the glow plug wires, then remove all glow plugs because this is where the compression gauge will be fitted.
4. Attach (screw) tool J-26999-10 and compression gauge J-26999 into the glow plug hole of the cylinder to be tested.
5. The engine can be cranked on the vehicle starter switch or by using a suitable remote hand-held starter switch in one hand as you note the compression gauge reading. Usually about six puffs per cylinder is adequate to attain the maximum compression pressure in the cylinder.

Interpretation of Readings. If the problem is poor piston rings, this is generally evidenced by the fact that the compression reading on the gauge will be low on the first stroke but tends to increase on the remaining cranking strokes to a reading that is still less than the minimum of 300 psi (2068 kPa).

6.2L GMC Torque Specifications

The following information lists the recommended torque specifications for the various components of the fuel system.

Component	*Lb.Ft.*	*N·m*
1. Injection pump to gear bolts	13–20	18–27
2. Injection pump attaching nuts	25–37	34–50
3. Injection line nut to pump	15–24	20–32
4. Injection line nut to nozzle	15–24	20–32
5. Injection pump fuel filter inlet line	15–20	20–27
6. Injection pump fuel filter outlet line	15–20	20–27
7. Injection pump fuel inlet line	15–20	20–27
8. Injection nozzle to cylinder head	44–60	60–80
9. Glow plug	8–12	11–16
10. Controller	13–20	18–27
11. Fuel pump to block	20–30	27–40
12. Fuel pump adapter plate to block	4–7	6–10
13. Fuel filter to inlet manifold (G-Van)	25–37	34–50

6.2L Diesel Special Tools

A number of special tools are required for use with the 6.2L diesel engine. These tools are manufactured by the Kent-Moore Tool Division of General Motors Corporation. All tools start with the letter "J."

NOTE: Many of the tools required for the 6.2L diesel engine and those for the 4.3L and 5.7L GMC passenger car diesels are similar. Refer to Chapter 22, dealing with the 4.3L and 5.7L diesel engines, which illustrates many of these common special tools.

Tool Description	*P/N*
1. Compression gauge adapter	J–26999–10
2. Inlet manifold protective screens (cover)	J–29664–1
3. Nozzle socket	J–29783
4. Static timing gauge	J–33042
5. T.P.S. and vacuum regulator valve gauge block (0.646–0.668")	J–33043–2
6. T.P.S. gauge block (0.602–0.624")	J–33043–4
7. T.P.S. gauge block (0.751–0.773")	J–33043–5
8. Diesel tachometer	J–26925
9. Timing meter	J–33075
10. Injection pump wrench	J–26987
11. Plastic plugs—pump lines and nozzles	J–28438

24

Ford 6.9L V8 Diesel Engine

Ford 6.9L (420 cu.in.) V8 Diesel Engine

The 6.9L (420 cu.in.) V8 diesel engine is of the pre-combustion chamber design. The first year of manufacture for this unit was 1984 when Ford asked IHC (International Harvester Corporation) to build 72,000 of these engines for use in Ford's F350 pickup trucks and E350 Econoline vans. This same engine is also used by IHC in their own "S" Series trucks in the 1654 Models and 1753 school bus chassis.

One of the largest sectors of growth for this engine has been the heavy-duty pickup and van area from 6000 to 10,000 lb. GVW (gross vehicle weight).

The 6.9L (420 cu.in.) engine is available up to 170 horsepower (126 kW) and can be rated at 3000 or 3300 rpm. The engine weighs approximately 100 to 150 lb. more than its gasoline counterpart. Hydraulic roller cam followers for low friction and quiet operation are employed along with a matched-gear front drive to maintain exact valve and injection timing. An engine-mounted oil cooler is used to maintain oil temperatures at as low a level as possible for extended change periods.

This engine is certified in all 50 U.S. states.

Although not available in turbocharged form, an after-market kit is available through a company called Hypermax Engineering Inc., known for its line of International-Harvester diesel tractor pulling equipment. The addition of the turbocharger to the 6.9L engine is claimed to offer up to 40% horsepower improvement with no sacrifice in fuel economy as well as significantly reducing combustion noise and cold smoke on start-up.

Lucas CAV has introduced a 17-mm shank diameter low-spring-type injector for use in the 6.9L engine with a 7-mm diameter nozzle protrusion. This design allows increased cylinder head strength and cooling while providing flexibility in positioning of the nozzle and glow plug in the pre-combustion chamber. The injector needle valve features a special double angle seat.

The CAV injector will be used by IHC in conjunction with 17-mm nozzles from United Technologies Diesel Systems Group, who have been the major nozzle supplier to the 6.9L diesel engine program.

Figure 24–1 illustrates the 6.9L diesel engine, which uses in-board glow plugs and nozzles. The engine firing order is 1-2-7-3-4-5-6-8 with cylinders 1-3-5 and 7 being on the right bank. The bore and stroke of the engine is 4.00″ (101.6 mm) and 4.18″ (106.17 mm) respectively. Engine compression ratio is 20.7:1 with the piston using two compression and one oil ring. Figures 24–2 and 24–3 illustrate the location of the identification labels on the engine.

The injection pump used with the 6.9L is of Roosa Master/Stanadyne design and is the model DB2 unit. This injection pump is a distributor-type unit that is also used by GMC on their 4.3L, 5.7L, and 6.2L diesel engines. Figure 24–4 illustrates the location of the fuel injection pump, which is located between the cylinder heads in a recess at the front of the engine.

SPECIAL NOTE: Special tools for use in servicing and repairing the 6.9L diesel engine are featured at the end of this chapter (Figure 24–35). Refer to this information when a special tool is quoted within the service information throughout this chapter.

Basic Fuel System Flow

A detailed explanation of the operation/repair of the DB2 fuel injection pump will not be given in this chapter because this information can be found in

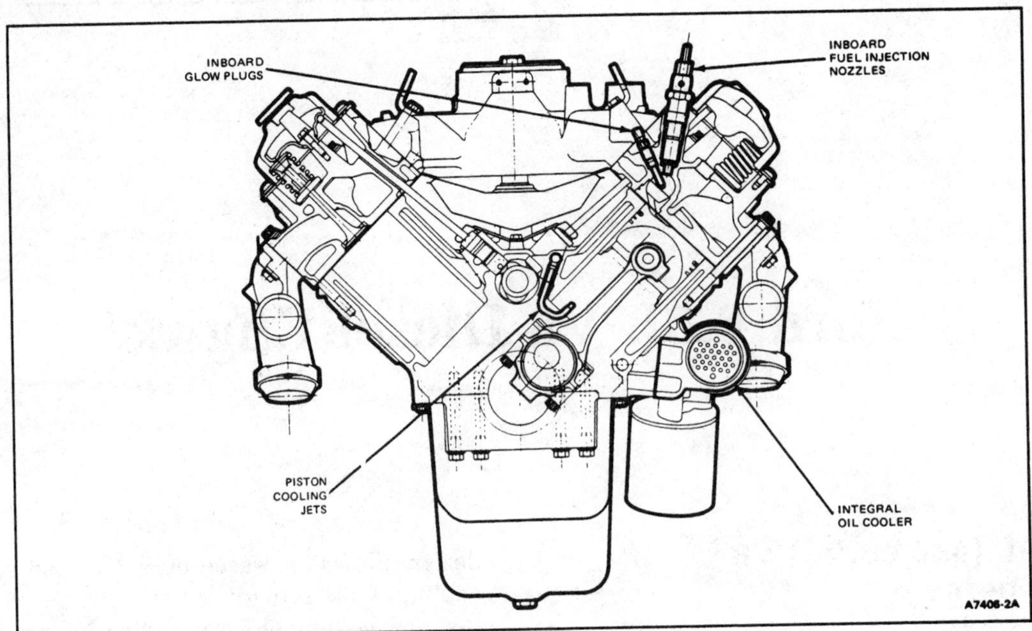

Figure 24–1. 6.9L (420 cu.in.) V8 Diesel engine cross section. (Courtesy of Ford Motor Company)

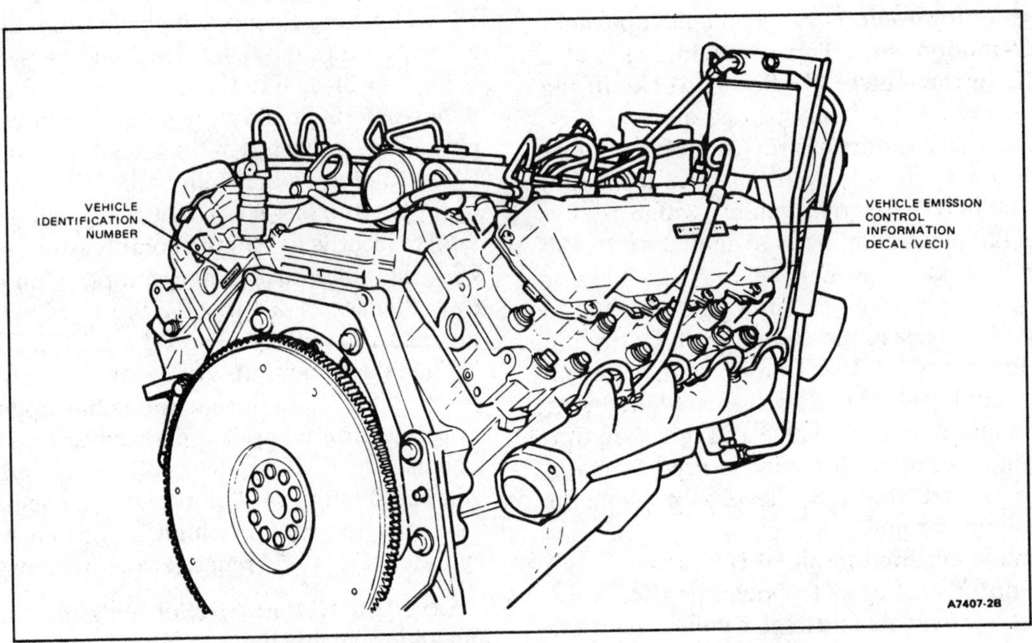

Figure 24–2. 6.9L Diesel engine—right side view. (Courtesy of Ford Motor Company)

Chapter 22 dealing with the 4.3L V6 and 5.7L V8 GMC diesel engines. Further information on the injection pump can also be found in Chapter 23 dealing with the GMC 6.2L V8 light truck diesel engine fuel system.

A schematic of the basic fuel system is shown in Figure 24–5, which illustrates the flow from the fuel tank up to the injection nozzles.

Fuel from the tank is drawn through a fuel/water separator up into a fuel (lift) pump that is driven from the engine. The lift pump is of conventional design similar to that found on gasoline engines.

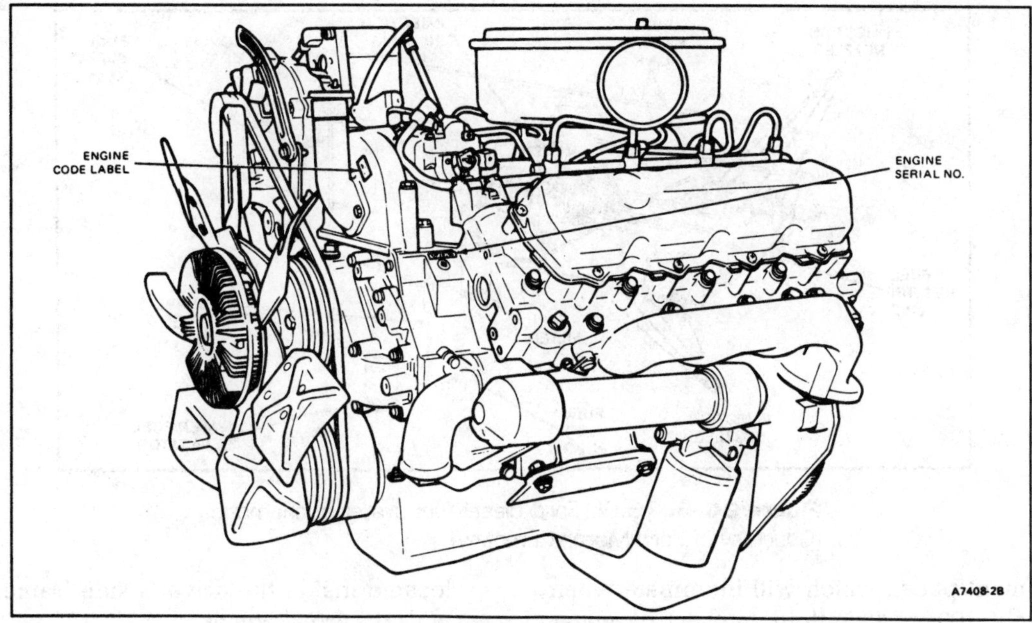

Figure 24-3. 6.9L Diesel engine—left side view. [Courtesy of Ford Motor Company]

To prevent waxing of fuel especially in cold weather, and also to allow use of the more economical No. 2 diesel fuel, a fuel heater is installed into the fuel line between the lift pump and the fuel filter (secondary).

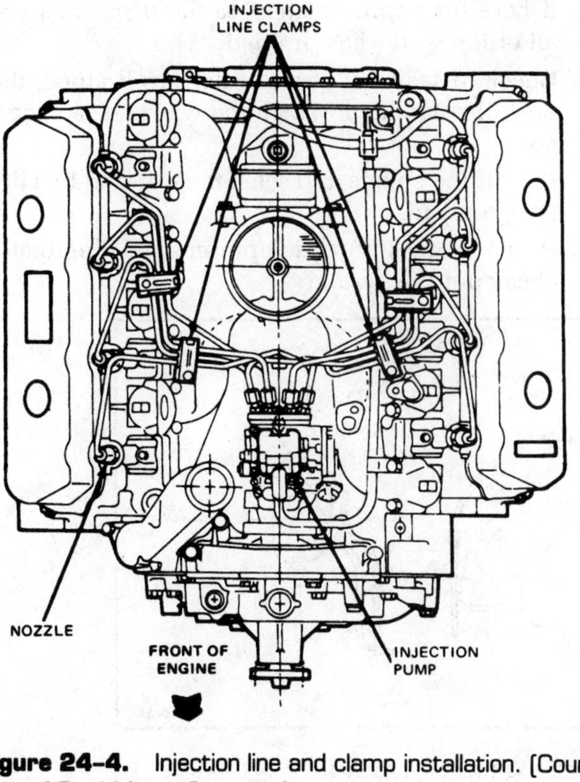

Figure 24-4. Injection line and clamp installation. [Courtesy of Ford Motor Company]

Filtered fuel, after passing through the secondary fuel filter, passes into the hydraulic head end of the injection pump, which contains a vane-type transfer pump running at injection pump speed. This pump is capable of delivering fuel to the injection pump plungers at pressures up to 130 psi (896 kPa).

An electric fuel shutoff solenoid that is energized from the ignition key switch in the *start* or *run* position allows fuel to enter the injection pump. When the ignition key is turned to *off*, the fuel shutoff solenoid is de-energized and no fuel can flow into the injection pump, thereby stopping the engine.

Fuel that is delivered into the injection pump is then increased in pressure by two plungers that deliver this fuel to the injection nozzles in firing order sequence. Normal opening pressure of the injection nozzles is between 2000–2150 psi (13790–14824 kPa).

Excess fuel that is delivered to the nozzles for cooling and lubrication purposes is then returned to the fuel tank through a fuel return line.

Fuel Filter/Water Separator Service

Both the fuel filter/water separator and the secondary fuel filter require service at various times on the engine. When a fuel filter/water separator is used, it should be drained at 5000-mile (8046-km) intervals. A *water-in-fuel* indicator lamp is provided on the ve-

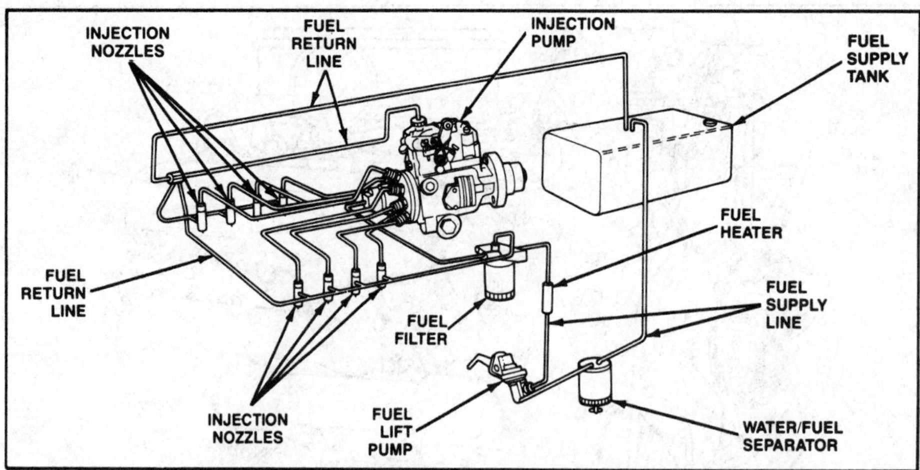

Figure 24–5. 6.9L Ford diesel fuel system schematic. (Courtesy of Ford Motor Company)

hicle instrument panel, which will illuminate when approximately 100 cc (3.5 fluid ounces) of water gathers in the separator bowl. This lamp will glow any time that the ignition switch is in the *start* position—which confirms that both the lamp and the water sensor are both functioning properly.

Although the normal drain period for the fuel water filter is listed above, if the water-in-fuel light comes on at any time while the engine is running, then the water must be drained immediately—otherwise there is the possibility of water entering the fuel system and causing serious component damage.

In order to drain water from the separator, refer to Figure 24–6 and proceed as follows.

Procedure for E (Econoline)-Series Vehicles

1. Shut off the engine and place a drain tray container underneath the water separator to catch drained water and fuel from the filter, which is located inside the driver's side frame rail in line with the front wheel.
2. Raise the plastic cover that is placed over the drain handle bracket on the floor of the cab to the right of the driver's seat.
3. Some plastic covers may have a conventional screwdriver head retaining screw on it; using a screwdriver or coin, back off the screw about 4 or 5 turns to loosen it off.
4. Pull up on the handle as far as it will go and hold it here for approximately 15 to 20 seconds until all of the water has drained.
5. Check to see if all of the water has drained, then push the handle down all the way and replace its cover.
6. Restart the engine and see if the water-in-fuel light has gone out.
7. If the water-in-fuel lamp remains illuminated, check out the system.

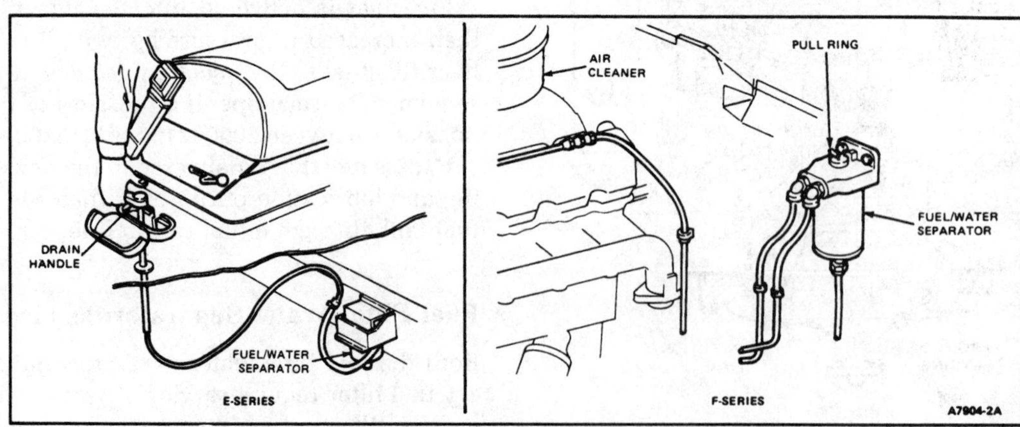

Figure 24–6. 6.9L Diesel engine fuel/water separator filter drain procedure. (Courtesy of Ford Motor Company)

Procedure for F-Series Vehicles

1. Refer to Figure 24–6 F-Series vehicles.
2. Repeat Step 1 as for the E-Series vehicles above.
3. Two configurations are used with this series of vehicles:
   a. Type "A" has a screw-type air vent at the top of the separator with a water drain plug at the bottom of the unit.
   b. Type "B" has a pull-ring as shown at the top of the separator as shown in Figure 24–6.
4. In Type "A," unscrew the air vent plug 2.5 to 3 turns; unscrew the water drain plug 1/2 to 2 turns to allow water to drain from the hose; close both drains when the water has vented.
5. In Type "B," pull up the pull-ring and hold it for about 15–20 seconds while the water drains; push the pull-ring back into position.
6. Start and run the engine, and if the water-in-fuel light remains on, check out the system for the cause.

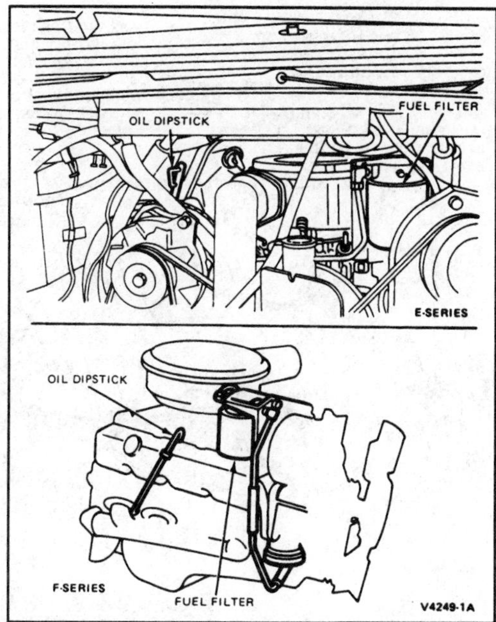

Figure 24–7. Fuel filter location 6.9L engine. (Courtesy of Ford Motor Company)

Fuel Filter Service

Fuel filter maintenance depends upon how many miles are accumulated on the engine. No specific time limit can apply to all engines/vehicles because operating conditions will vary greatly between applications.

Heavy-duty diesel engine applications, for example, generally check fuel filter condition at 250–300 hours (7500–9000 miles), although many current diesel manufacturers of light truck and automotive diesel engines recommend change periods of between 25,000 and 40,000 miles.

A filter change of once a year can be considered average, but some vehicles may require more frequent service intervals as mentioned above.

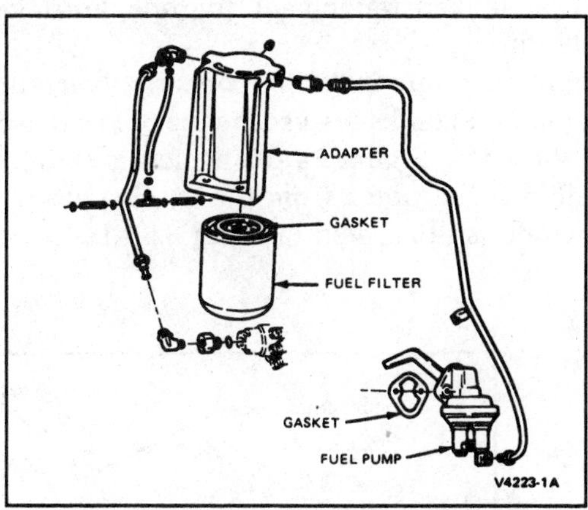

Figure 24–8. Fuel filter and adapter 6.9L engine. (Courtesy of Ford Motor Company)

Filter Service Procedure

1. Refer to Figures 24–7 and 24–8 and disconnect the battery ground cables from both batteries.
2. Unscrew the fuel filter from its adapter as shown in Figure 24–8.
3. Clean the gasket mounting surface of the adapter and lightly coat the new filter gasket with clean diesel fuel.

   CAUTION: Never fill the fuel filter with unfiltered diesel fuel before installation because any dirt or water in the fuel can damage the fuel injection components.

4. Once the filter has been screwed into place and contacted the gasket surface, tighten it another 1/2 to 3/4 turn.
5. Wipe up any spilled fuel; reconnect the battery ground cables; start and run the engine and check for any fuel leaks.

Fuel Tanks and Lines

A number of fuel tank configurations can be used with the 6.9L engine, depending upon the particular model of truck or van that it is used in. Since single tank vehicles do not differ in layout too much, let's look at the arrangement used by Ford when a "dual tank" installation is used. Figure 24–9 illustrates the

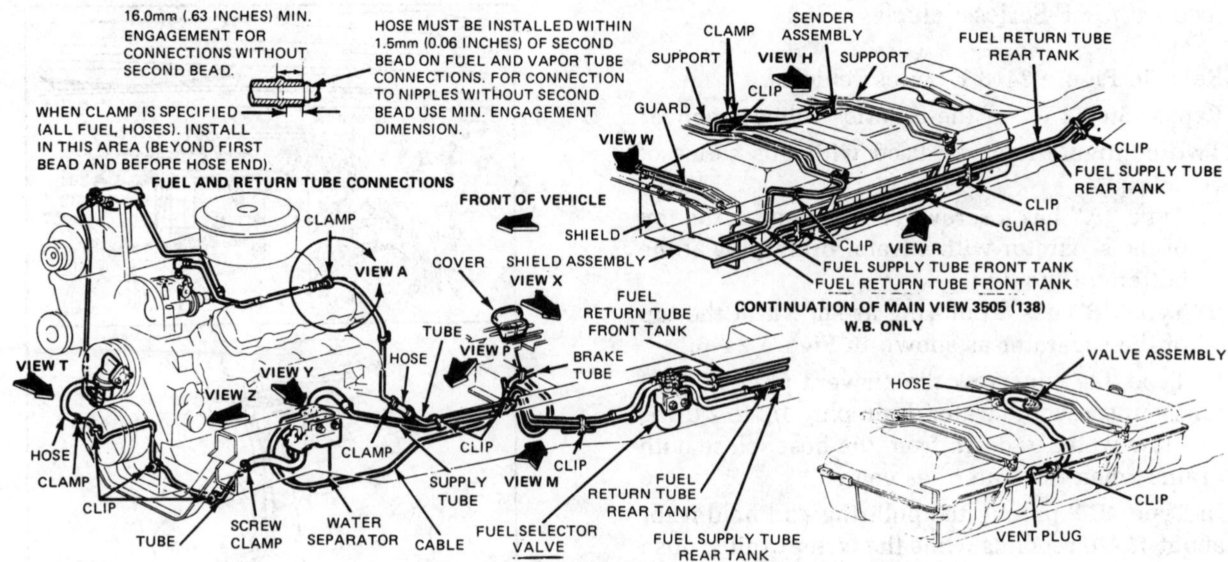

Figure 24–9. 6.9L Ford diesel engine dual fuel tank layout. (Courtesy of Ford Motor Company)

dual fuel tank arrangement for one Ford vehicle installation.

The main component to concern yourself with when dual fuel tanks are used is the 6-port diesel fuel valve shown in Figure 24–10, which is powered by a small electric motor from the vehicle electric circuit, which is illustrated in Figure 24–11.

The valve is operated as follows: When the ignition switch is *on*, battery power flows through the fuse panel and through the fuel tank selector switch to the motor of the fuel tank selector valve.

When 12 volts is applied to terminal No. 2 (wiring diagram) and terminal No. 1 is grounded, the valve will be shifted to the front position whereby the front

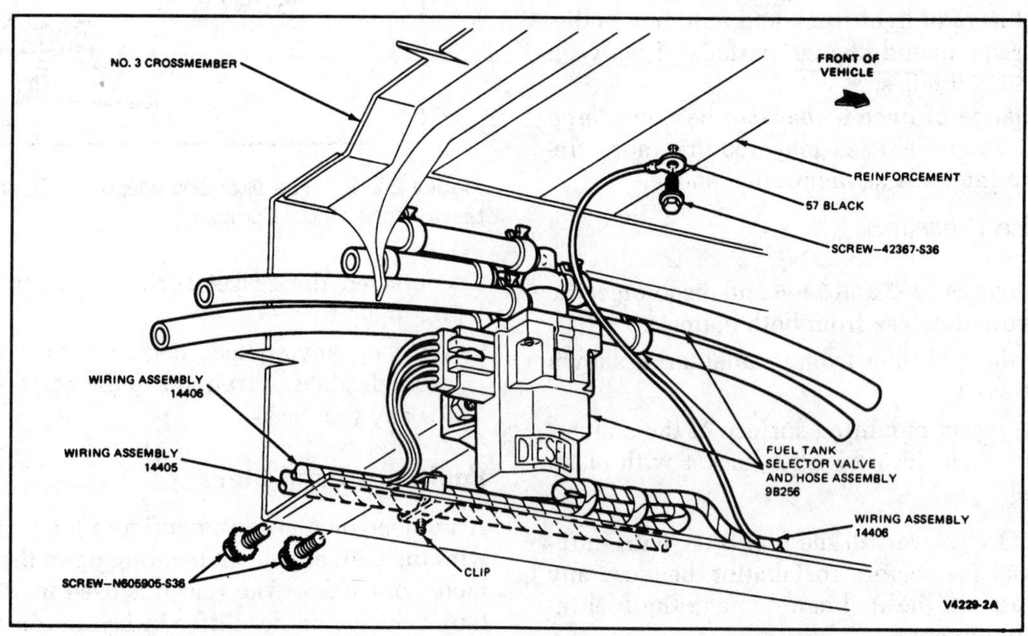

Figure 24–10. Dual fuel tank selector valve. (Courtesy of Ford Motor Company)

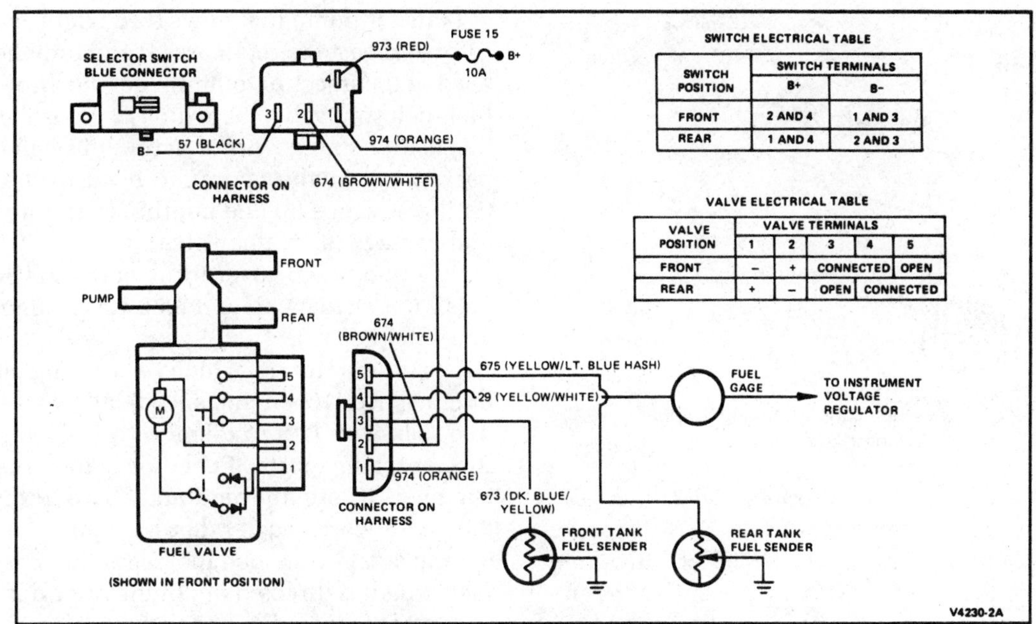

SWITCH ELECTRICAL TABLE

SWITCH POSITION	SWITCH TERMINALS	
	B+	B-
FRONT	2 AND 4	1 AND 3
REAR	1 AND 4	2 AND 3

VALVE ELECTRICAL TABLE

VALVE POSITION	VALVE TERMINALS				
	1	2	3	4	5
FRONT	–	+	CONNECTED		OPEN
REAR	+	–	OPEN	CONNECTED	

Figure 24-11. Dual fuel tank selector valve wiring diagram. (Courtesy of Ford Motor Company)

supply and return ports are open to the engine supply and return ports.

As the internal valve reaches the front operating position, an internal switch opens the circuit and stops the actuating motor. To switch the valve to the rear operating position, the power supply is simply reversed by applying battery power to terminal No. 1 and grounding terminal No. 2. Terminals No. 3 and No. 5 are connected to the individual fuel tank sending units (front and rear) while terminal No. 4 feeds the fuel gauge.

Fuel level in the front tank is registered on the fuel gauge through an internal switch that connects terminal No. 3 to terminal No. 4, while the rear tank fuel level is read when terminals No. 3 and No. 5 are connected through the switch.

When a problem exists with the fuel tank switching valve, if battery voltage is read at terminals No. 1 and No. 2 of the valve, always ensure that the ground wire is attached below the instrument panel.

Common problem areas to check when the fuel selector valve fails to switch tanks would be:

1. Fuse damage.
2. Check for voltage at selector valve in both the front and rear positions.
3. Check the ground circuit.
4. Check selector switch continuity.
5. Check fuel valve motor operation.
6. Check for fuel flow restrictions.

To check for fuel flow from both the front and rear tanks, refer to Figure 24-12 and perform a fuel system pumping capacity test on each fuel tank to determine where the problem exists.

Perform the fuel tank pumping capacity test as follows:

1. Turn the ignition to the *run* position and move the selector switch to the front position and turn the key *off*.
2. Pinch off the fuel supply line from the rear tank to the selector valve with a suitable tool such as vise-grips.
3. Remove the vacuum purge valve from the fuel filter adapter and install adapter P/N 3019 from the Rotunda Pressure Test Kit P/N 19-0002. (Figure 24-35)
4. Start the engine and let it idle, then open the clamp on the sample hose to allow fuel to flow into a bucket or container.
5. Check for a solid, continuous fuel flow.
6. Repeat the same procedure for the rear tank.

Fuel Injection Pump Flow

A detailed explanation of the injection pump operation can be found in Chapter 22 dealing with the 4.3L and 5.7L General Motors diesel engines, which also use this same model Roosa Master/Stanadyne pump.

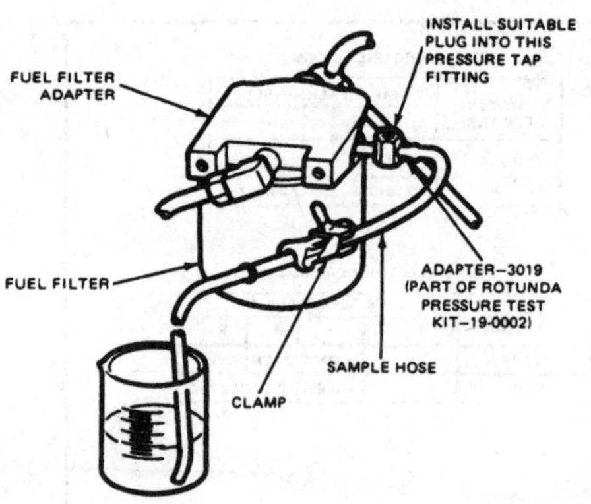

Figure 24–12. Dual fuel tank pumping capacity test. (Courtesy of Ford Motor Company)

Figure 24–13 illustrates the model DB2 injection pump and its major component parts. Flow through the injection pump is as follows: Fuel at lift pump pressure from the secondary fuel filter enters the injection pump at the hydraulic head end (injection line end). This fuel passes into the vane-type transfer pump (2) through a filter screen (1). To control maximum delivery pressure of the shaft driven transfer pump, a spring loaded pressure regulating valve will bypass fuel back to the inlet side of the transfer pump. This fuel pressure is set with the injection pump mounted on a fuel pump test stand and is usually limited to a maximum of 130 psi (896 kPa).

Transfer pump fuel flows through the center of the rotor and past the retainers (4) and into the hydraulic head of the injection pump. Fuel then flows up to the fuel metering valve (8), which is controlled by throttle position and governor action through connecting passage (5) in the hydraulic head to the automatic timing advance (6) and continues on through the radial passage (9) to this valve.

Injection pump governor operation is discussed in detail under the 4.3L and 5.7L GMC engine fuel system (Chapter 22).

The pump rotor, which is turning at injection pump speed (1/2 engine speed), allows the rotor fuel inlet passages (10) to align with the hydraulic head fuel charging ports. Fuel flows into the pumping chamber where the action of two rotor plungers moving towards each other by their rollers (11) contacting a cam ring lobe increases the pressure of the fuel, which is directed out of the rotor discharge port to one of the injection nozzle fuel delivery lines. This occurs in firing order sequence as the rotor revolves.

The purpose of the air vent passage (12) in the hydraulic head is to allow a percentage of fuel from the transfer pump to flow into the injection pump housing. This fuel is used to vent air from the system and also to cool and lubricate the internal pump components. This fuel flows back to the fuel tank via a return line.

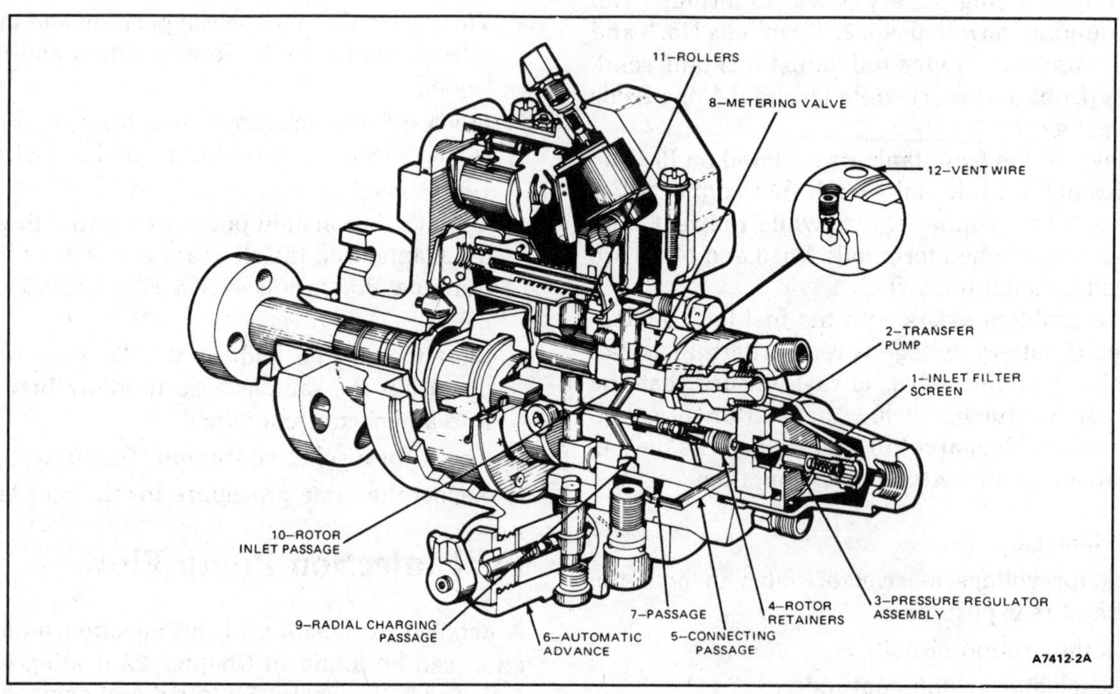

Figure 24–13. 6.9L Engine injection pump (Roosa Master Model DB2). (Courtesy of Ford Motor Company)

Injection Pump Throttle Linkage

Often a lack-of-power complaint can be caused by such things as a plugged fuel filter or a restriction in the fuel supply. An item that is often overlooked, however, when such a complaint is registered is that of incorrect throttle linkage adjustment.

Often, when other repair jobs are performed, the throttle linkage is disturbed and when the engine is returned to service, the lack-of-power complaint is often blamed on other things and often suspected to be tied in with the just-completed repair job.

Since a diesel engine relies on mechanical control of the fuel supply with an unthrottled air supply, incorrect throttle linkage adjustment can be more noticeable on a diesel engine than on a gasoline engine.

When the driver places the throttle pedal into the full-fuel position, check and ensure that the injection pump throttle lever has also gone all the way to its full-fuel position. This can be checked by disconnecting the throttle linkage at the pump, placing the throttle pedal to full-fuel, and seeing if the linkage can be connected to the pump lever without having to back off slightly on the injection pump throttle lever position. If you have to do this to engage the linkage, then the engine will not be receiving full-fuel when the throttle pedal is placed to full-fuel inside the cab.

The throttle linkage and its adjustment procedure for both the E-Series (Econoline) and F-Series vehicles is shown in Figures 24–14 and 24–15.

Injection Pump External Adjustments

The following information deals with external injection pump adjustments because detailed information on other necessary repairs can be found under the chapters dealing with the General Motors 4.3L, 5.7L (Chapter 22), and 6.2L (Chapter 23) diesel engine fuel systems.

Basic adjustments that the mechanic/technician should be able to perform on the 6.9L diesel engine are:

1. Idle speed adjustment
2. Fast idle speed adjustment
3. Static injection timing setting
4. Dynamic injection pump to engine timing setting

The above adjustments are explained below.

Idle Speed Adjustment Procedure. Figure 24–16 illustrates the location of the idle speed adjustment screw on the injection pump.

1. Idle speed adjustments should only be done when the engine is at normal operating temperature with the transmission in *neutral* on a standard shift transmission and in *drive* on an automatic unit.
2. Always check that the throttle linkage is adjusted as per the information herein dealing with throttle adjustments.
3. Check the vehicle emissions control information decal and establish what speed the engine should idle at.
4. Using a photoelectric tachometer such as the Rotunda 99-0001 or its equivalent (Figure 24–35), check the idle rpm and compare it to that on the vehicle decal.
5. If the idle is incorrect, adjust it by turning the curb idle screw on the injection pump clockwise to increase speed and counterclockwise to decrease speed. Manually rev up the engine after any adjustment and let it settle back down to idle and recheck the speed with the transmission in *neutral* or *park*, then place the transmission in *drive* on an automatic equipped vehicle and in *neutral* on a standard transmission vehicle.

Fast Idle Adjustment. When the engine is started from cold, a fast idle solenoid is used to increase the speed of the diesel engine by a voltage signal from a coolant sensor.

1. With the engine running at normal operating temperature and the transmission in *neutral* or *park*, disconnect the fast idle solenoid from the wiring harness as shown in Figure 22–49.
2. Using a jumper wire, apply battery voltage to the solenoid and momentarily rev up the engine to activate the solenoid; the engine idle speed should immediately increase to between 875 ± 25 rpm.
3. If the fast idle speed is not within specs, adjust it by turning the solenoid plunger clockwise to increase, and counterclockwise to decrease the setting.
4. Rev up the engine and let it settle; recheck the fast idle speed after an adjustment.
5. Disconnect the jumper wire and reconnect the solenoid wiring harness.

Static Injection Pump Timing. This check is usually done when the injection pump has been removed and reinstalled. When a power complaint is received, always employ the dynamic timing check to ensure that the injection pump timing is correct.

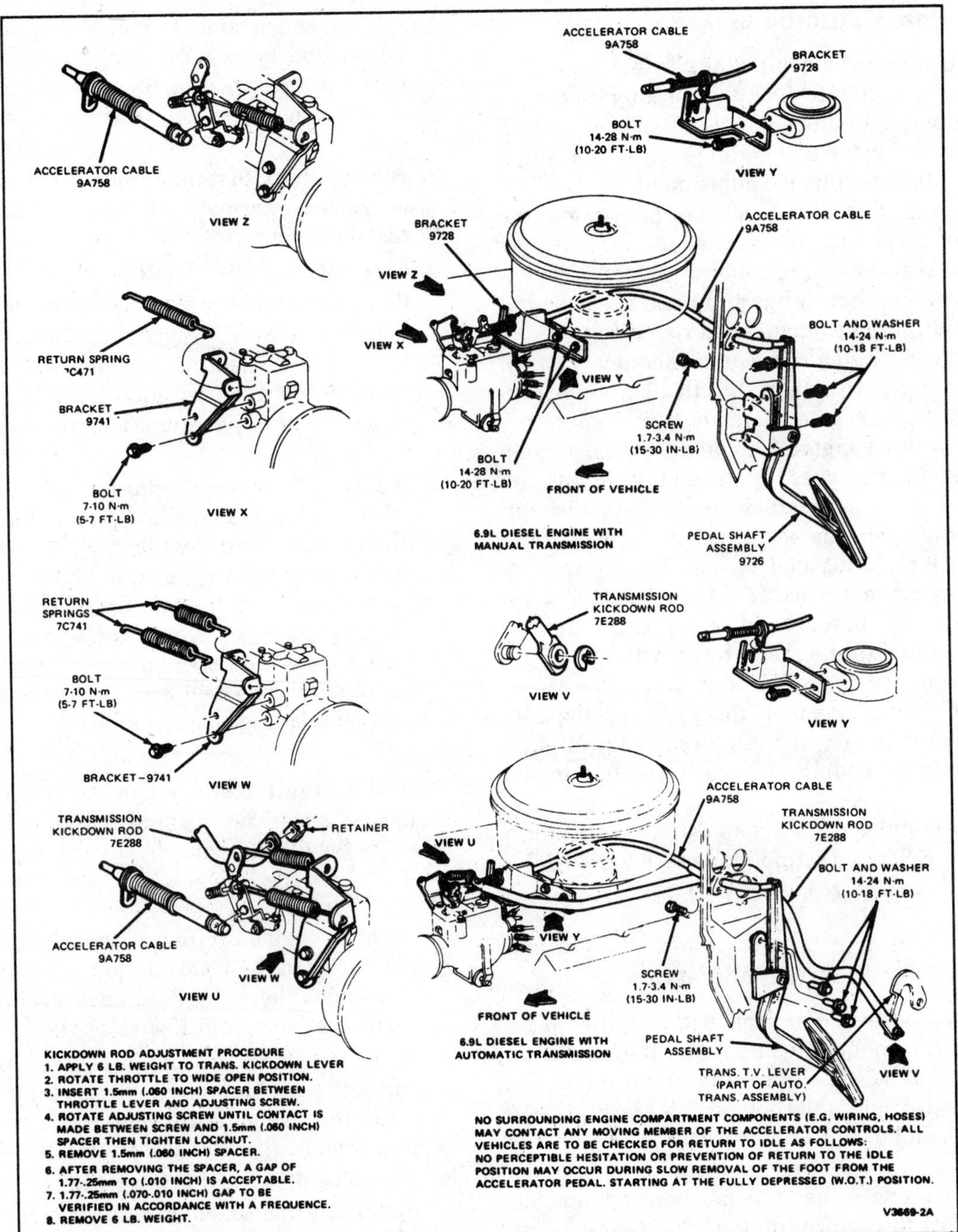

Figure 24–14. 6.9L Diesel engine throttle linkage adjustments F-250 and F-350 truck models. (Courtesy of Ford Motor Company)

Procedure

1. Remove the fast idle bracket and its solenoid from the injection pump.
2. Break loose the three injection pump retaining nuts slightly with the use of the special half-moon shaped box end wrench P/N T83T-9000-B. (Figure 24–35)
3. Refer to Figure 24–17 and either use special tool P/N Rotunda T83T-9000-C rotating tool or manually grasp the pump with both hands and move the injection pump so that the mark stamped on the injection pump housing aligns with the mark on the engine cover.

NOTE: The spec calls for a tolerance of 0.030″

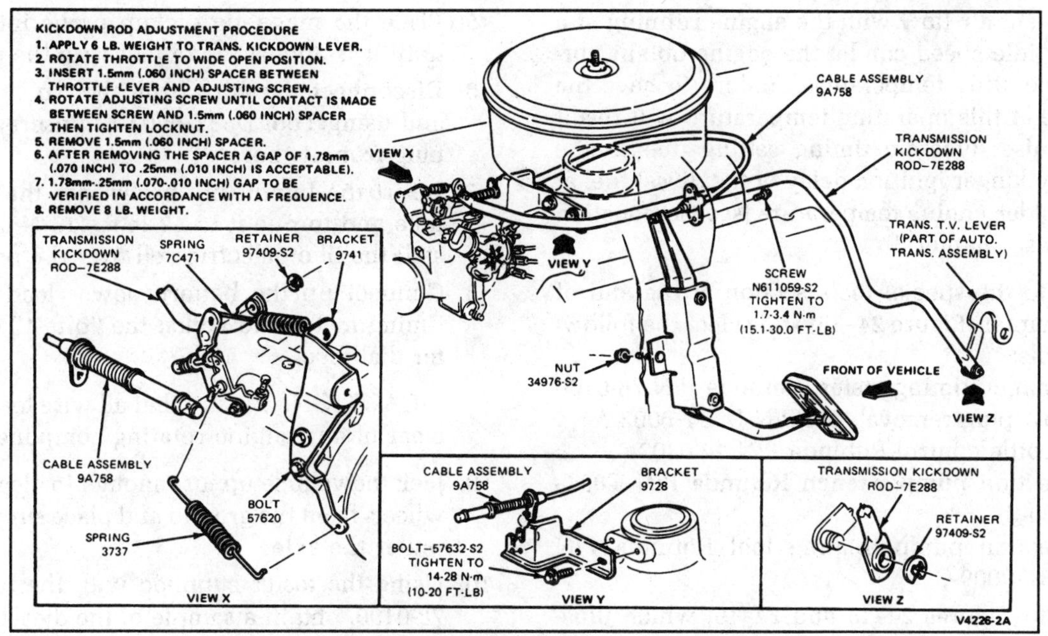
KICKDOWN ROD ADJUSTMENT PROCEDURE
1. APPLY 6 LB. WEIGHT TO TRANS. KICKDOWN LEVER.
2. ROTATE THROTTLE TO WIDE OPEN POSITION.
3. INSERT 1.5mm (.060 INCH) SPACER BETWEEN THROTTLE LEVER AND ADJUSTING SCREW.
4. ROTATE ADJUSTING SCREW UNTIL CONTACT IS MADE BETWEEN SCREW AND 1.5mm (.060 INCH) SPACER THEN TIGHTEN LOCKNUT.
5. REMOVE 1.5mm (.060 INCH) SPACER.
6. AFTER REMOVING THE SPACER A GAP OF 1.78mm (.070 INCH) TO .25mm (.010 INCH) IS ACCEPTABLE).
7. 1.78mm-.25mm (.070-.010 INCH) GAP TO BE VERIFIED IN ACCORDANCE WITH A FREQUENCE.
8. REMOVE 8 LB. WEIGHT.
VIEW X
TRANSMISSION KICKDOWN ROD–7E288
SPRING 7C471
RETAINER 97409-S2
BRACKET 9741
CABLE ASSEMBLY 9A758
BOLT 57620
SPRING 3737
VIEW X
CABLE ASSEMBLY 9A758
TRANSMISSION KICKDOWN ROD–7E288
TRANS. T.V. LEVER (PART OF AUTO. TRANS. ASSEMBLY)
VIEW Y
SCREW N611059-S2 TIGHTEN TO 1.7-3.4 N·m (15.1-30.0 FT-LB)
NUT 34976-S2
FRONT OF VEHICLE
VIEW Z
CABLE ASSEMBLY 9A758
BRACKET 9728
BOLT–57632-S2 TIGHTEN TO 14-28 N·m (10-20 FT-LB)
VIEW Y
TRANSMISSION KICKDOWN ROD–7E288
RETAINER 97409-S2
VIEW Z
V4226-2A

Figure 24-15. 6.9L Diesel engine throttle linkage adjustments E-250 and E-350 truck models. [Courtesy of Ford Motor Company]

(0.762 mm); however, attempt to place these lines as close as possible together since each 0.030" is equivalent to 2 degrees of timing.

4. Tighten up the injection pump retaining nuts to a torque of 14 lb.ft. (19 N·m) and recheck the timing alignment marks.
5. Reinstall the fast idle bracket and solenoid.

Dynamic Injection Pump Timing. This check should always be undertaken when a lack-of-power complaint or black exhaust smoke is noted.

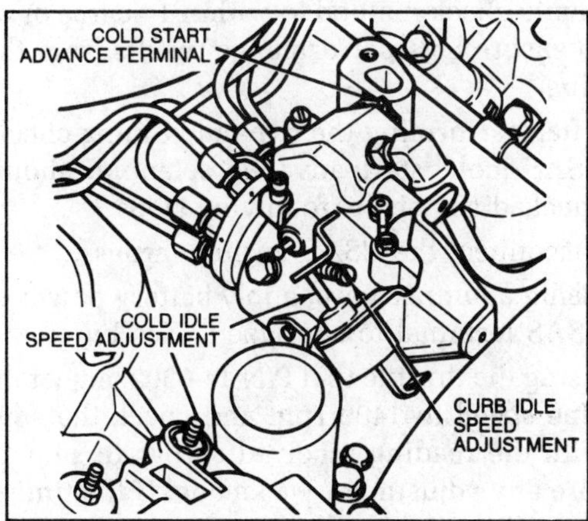
COLD START ADVANCE TERMINAL
COLD IDLE SPEED ADJUSTMENT
CURB IDLE SPEED ADJUSTMENT

Figure 24-16. 6.9L Diesel engine idle adjustment screw location. [Courtesy of Ford Motor Company]

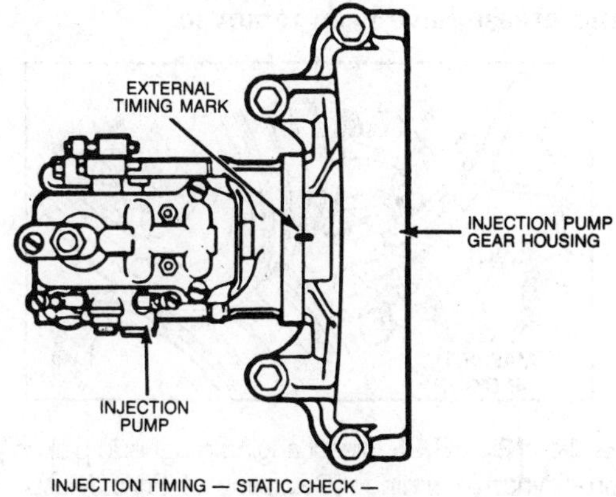
EXTERNAL TIMING MARK
INJECTION PUMP GEAR HOUSING
INJECTION PUMP
INJECTION TIMING — STATIC CHECK —

Figure 24-17. 6.9L Injection pump static timing marks location. [Courtesy of Ford Motor Company]

Procedure

1. A dynamic timing check can only be satisfactorily performed when the engine is at its normal operating temperature which should be between 192–212°F (89–100°C).

NOTE: To obtain this temperature, it may be necessary to road test the vehicle because running it at an idle rpm in a shop will result in a cold running engine due to the fact that the air is unthrottled and a minimum amount of fuel is being injected. Placing a cover over the radiator to prevent air flow with the engine running at a higher idle speed can let the engine obtain normal operating temperature. Failure to have the engine at this operating temperature will result in a false dynamic timing setting due to the slightly longer ignition delay of the diesel fuel at this colder engine temperature (see Combustion Systems, Chapter 3).

2. Refer to the special tools section at the end of this chapter (Figure 24–35) and select the following tools:
   a. Dynamic timing meter, Rotunda P/N 78-0100
   b. Glow plug removal tool P/N D83T-6002-A
   c. Throttle control Rotunda P/N 14-0302
   d. Injection pump wrench Rotunda P/N T83T-9000-B
   e. Injection pump rotating tool Rotunda P/N T83T-9000-C
3. Refer to Figures 24–18 and 24–19, which illustrate the magnetic pickup and luminosity probe locations.
4. With the engine stopped, clean off the vibration damper pulley to ensure an accurate reading of the engine rpm by sanding off any rust once dirt and grease have been removed.

Figure 24–18. 6.9L Diesel engine magnetic pickup probe location—dynamic timing. (Courtesy of Ford Motor Company)

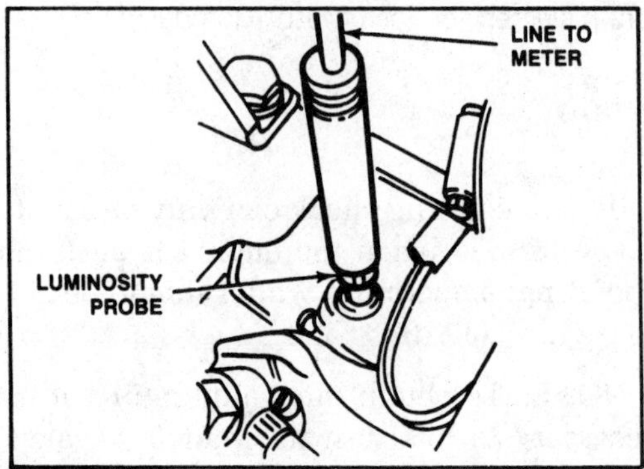

Figure 24–19. 6.9L Diesel engine dynamic timing luminosity probe mounted in glow plug hole. (Courtesy of Ford Motor Company)

5. Place the magnetic pickup probe into the hole until it is almost in contact with the pulley.
6. Disconnect the wire from the No. 1 glow plug and using Tool D83T-6002-A, unscrew the glow plug from its bore.
7. Insert the luminosity probe into the glow plug hole and torque it to 12 lb.ft. (16 N·m), then install the photoelectric cell over the probe.
8. Connect up the battery power leads from the timing meter and adjust the "offset" on the meter dial face.

CAUTION: Ensure that all wire leads are well clear of any engine rotating components.

9. Jack the vehicle up just enough to clear the drive wheels from the ground and place support stands under the axle.
10. Using the tester supplied with the tool kit P/N 78-0100, obtain a sample of the diesel fuel being used with the engine to check its Cetane rating value (see Figure 24–35).
11. With the transmission in *neutral*, start the engine and using the throttle control tool P/N 14-0302, raise the vehicle speed to 1400 rpm with no accessories in operation.
12. Refer to Figure 24–35 and having established what the Cetane rating of the diesel fuel is from Step 10 above, note and compare the injection timing on the dynamic timing meter, with Figure 24–20.
13. Timing is acceptable if it is within 2 degrees of the specification.
14. Adjust the timing if necessary after stopping the engine as per the instructions given for the "Static Timing Check" above.
15. Attempt to set the injection timing by the dynamic check method to within 1 degree or closer to ensure best performance results from the engine.
16. After performing the dynamic timing check, the CSAS (cold start advance solenoid) should be checked as per the following steps.
17. Disconnect the CSAS wiring harness.
18. Using a jumper wire, apply battery power to the CSAS terminal to energize it. See Figure 24–16.
19. Using the throttle tool P/N 14-0302, adjust the engine speed to 1400 rpm and check the advance with the reading received earlier in step 12 before any adjustment was made. If the timing advance is less than 2.5 degrees, it indicates a fuel injection pump problem and the pump should be replaced.

Fuel Cetane Value	Altitude	
	0-3000 Ft*	Above 3000 Ft*
38-42 43-46 47 or greater	6° ATDC 5° ATDC 4° ATDC	7° ATDC 6° ATDC 5° ATDC
*Installation or resetting tolerance for dynamic timing is ± 1°. Service limit is ± 2°.		

Figure 24–20. 6.9L Diesel engine dynamic timing specifications. (Courtesy of Ford Motor Company)

Injection Pump Removal

Injection pump removal should always be performed only after having cleaned the area around the pump. If steam cleaning is to be used, *never* direct the steam jet directly against the injection pump housing because the housing is manufactured from an aluminum alloy and the internal parts are manufactured from steel components. Since the rate of expansion of aluminum is approximately twice that of steel, serious damage can occur to the internal pump components if direct steam heat is applied to the pump housing whether the engine is running or not.

For reasons of safety, the engine should be turned off when steam cleaning the engine compartment.

If a steam cleaner or high-pressure washer is not available, then wash all dirt and grease from around the injection pump and nozzle areas with solvent or diesel fuel. The utmost cleanliness must be exercised anytime that service/repair work is to be performed on the diesel engine fuel system.

The injection pump is located at the front of the engine as shown in Figure 24–14. Before removing the injection pump, obtain a supply of plastic protective shipping caps such as P/N T83T-9395-A, which can be used to cover all open fuel lines as they are removed as well as the injectors and the hydraulic head fuel studs on the injection pump.

It is a good idea to place No. 1 piston at TDC on its compression stroke before removing the injection pump by manually rotating the engine over.

Removal Procedure

1. Isolate the starting system by disconnecting both battery negative ground cables; on the Econoline vehicles, remove the engine cover.
2. Remove the air cleaner assembly and install protective cover P/N T83T-9424-A or equivalent over the intake manifold.
3. Remove the engine oil filler neck as shown in Figure 24–21.
4. Refer to Figure 24–22 and remove the three bolts that attach the injection pump to its drive gear.

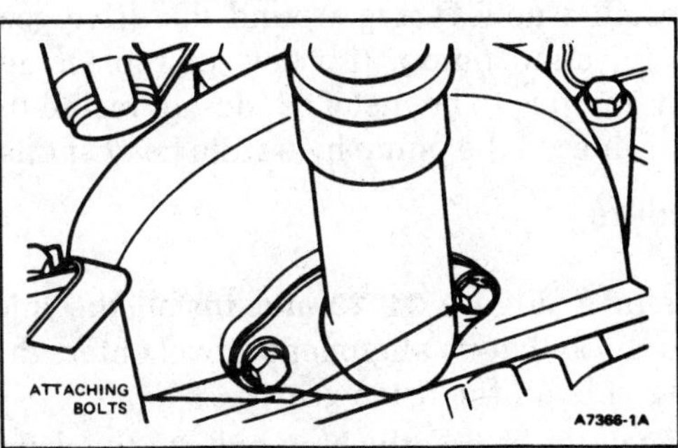

Figure 24–21. 6.9L Oil filler neck removal. (Courtesy of Ford Motor Company)

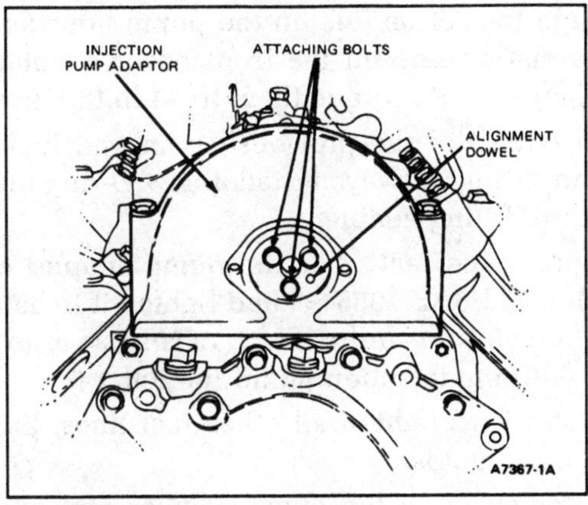

Figure 24–22. 6.9L Diesel injection pump drive gear attaching bolts. (Courtesy of Ford Motor Company)

5. Disconnect all electrical connections from the injection pump.
6. Disconnect the throttle linkage to the injection pump and the speed control cable, if used.
7. Remove all fuel inlet and return lines from the injection pump, which can be removed with the injection lines attached if the pump is not being replaced; be sure to cap all fuel lines and open fittings with the protective caps.
8. Remove the three injection pump housing-to-adapter retaining nuts with the use of the special tool P/N T83T-9000-B. (Back side of injection pump adaptor in Figure 24–22)

9. Lift the injection pump clear from the engine, supporting its weight by grabbing the pump housing and not by the attached fuel injection lines if they are still attached.

Injection Pump Installation

Check to see that the No. 1 piston is at TDC on its compression stroke; if not, manually rotate the engine to this position.

Clean the area around the pump mounting surface and install a new O-ring around the drive gear end of the injection pump. If a new or replacement injection pump is to be installed, do not install the fuel injector lines to the pump hydraulic head at this time.

Procedure

1. Refer to Figure 24–22 and install the injection pump so that its alignment dowel enters the mating hole on the drive gear.
2. Install the three attaching bolts to the drive gear and torque them to 25 lb.ft. (34 N·m).
3. Install the three pump housing retaining nuts and snug them up; before tightening them up, align the scribe line on the pump housing with its mating line on the front cover adapter, see Figure 24–17; torque them to 14 lb.ft. (19 N·m).
4. Install all of the previously removed fuel lines and fittings; always install a new O-ring into the pump fitting adapter.
5. Apply pipe sealant to the pump adapter elbow such as D8AZ-19554-A and tighten it to between 6 lb.ft. (8 N·m) and 10 lb.ft. (13 N·m) so as to align it with the injection pump fuel inlet line.
6. Install and tighten all other fuel lines, linkage, and solenoids.
7. After cleaning the pump adapter and oil filler neck sealing surfaces, apply a 1/8″ bead (3.2 mm) of RTV (room temperature vulcanizing) sealant P/N D6AZ-19562-A or equivalent in the adapter housing grooves and do not leave this sealant longer than 15 minutes, otherwise it will lose its sealing effectiveness.
8. Install the oil filler neck and tighten.
9. Connect up both battery ground cables.
10. Bleed air from the fuel system by leaving the injector fuel lines loose about 1/2 turn while cranking the engine; when air-free fuel appears at each individual line, tigthen it up, then torque it to 22 lb.ft. (30 N·m).
11. Start and run the engine and check for any oil or fuel leaks.
12. Install the dynamic timing tool kit components and check the engine timing as described above.

Engine/Injection Pump Gear Timing Marks

If it is necessary at any time to check the engine-to-injection pump drive gear timing marks, refer to Figure 24–23, which illustrates the alignment of these marks.

The injection pump drive gear meshes with the engine camshaft gear and rotates at camshaft speed, which is one-half engine speed. To check this timing, rotate the engine over by hand (remove glow plugs to facilitate rotation) to place No. 1 piston at TDC on its compression stroke. This can be confirmed as follows:

1. Position the injection pump drive gear dowel at the 4 o'clock position (see Figure 24–22), which should place the vibration damper at the TDC position.

   NOTE: The injection pump drive gear has one chamfered tooth that fits between the two chamfered teeth on the camshaft gear to facilitate visual inspection of the drive gear when the engine front cover is in position, see Figure 24–23.

Injection Nozzles

The injection nozzles used on the 6.9L engine are supplied by both United Technologies Diesel Fuel Systems and Lucas CAV. The nozzle is of the inward opening differential, hydraulically operated pintle type (see basic nozzle design, Chapter 29).

Fuel delivered to the nozzle from the injection pump must overcome the force of a needle valve return spring in order to open the nozzle and allow a metered quantity of fuel to be sprayed into the combustion chamber. Normal opening pressure of the nozzle is between 2000–2150 psi (13790–14824 kPa), although an acceptable service level on an engine

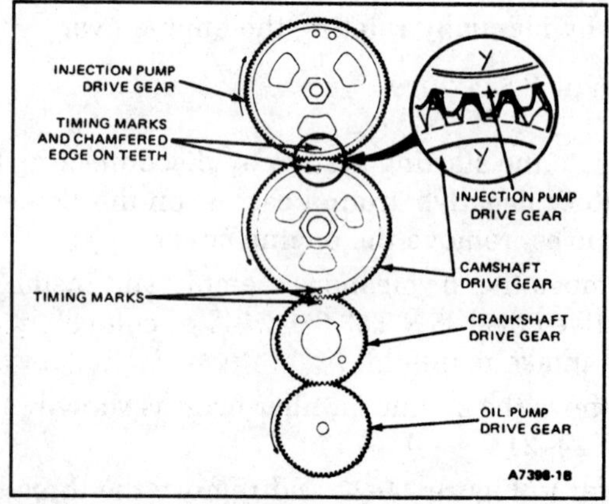

Figure 24–23. 6.9L Engine gear train timing mark alignment. (Courtesy of Ford Motor Company)

with a number of miles (kilometers) on it is 1600 psi (11032 kPa).

Nozzle opening pressure and spray pattern should be checked every 60,000 miles, or 96,500 km. Injection nozzle testing procedures can be found in this book under the chapter dealing with nozzle service/repair (Chapter 29).

Figure 24–24 illustrates the nozzle design used with the 6.9L diesel engine.

A small quantity of fuel inside the nozzle leaks past the internal parts to the injector spring cavity for lubrication purposes. This fuel is then returned back to the fuel tank through an injector leak-off line.

Nozzle Removal

Removal of the injection nozzles follows the same basic pattern as that for other diesel engines in that cleanliness must be exercised when removing or installing these units.

Obtain a supply of plastic protective shipping caps (Ford P/N T83T-9395-A) that can be used to cover the open ends of all fuel lines and nozzles to prevent the entrance of dirt.

CAUTION: If you are going to steam clean the engine before nozzle removal, make sure that you do not apply direct steam pressure against the injection pump housing. The reason for this is that the housing is manufactured from an aluminum alloy while the internal components are of steel. If the engine is running or is warm, the rate of expansion of these components is different with aluminum being about twice that of steel. Scoring of internal components can occur, and injection pump damage can result from this practice.

Clean around each injection nozzle with clean solvent or diesel fuel to remove any loose dirt that may fall into the cylinder head nozzle bore, and blow the area dry with compressed air.

Refer to Figure 24–4 and remove the nozzle fuel line retaining clamps and fuel lines along with their respective leak-off tees.

Remove the injection nozzles by using a deep socket after placing protective caps over the ends of the nozzle; once the nozzles have been removed, place plastic protective caps over the ends of the nozzle pintle as shown in Figure 24–25.

Be sure to remove the copper washer from the nozzle bore if it doesn't come out with the nozzle. Special tool P/N T71P-19703-C can be used for this purpose.

The nozzles should always be placed into a nozzle holding fixture as shown in Figure 24–25 in order that the nozzles can be placed back into the same cylinder from which they were removed. If a numbered holder is unavailable, label each nozzle so that it can be installed back into the same cylinder from which it was removed.

Nozzle Installation

Before nozzles are inserted back into the cylinder head bore from which they were removed, clean the seating area of the bore with tool P/N T83T-9527-A. If this tool is unavailable, remove any traces of carbon or other foreign material from the seating area and blow it clean with compressed air.

Remove the nozzle pintle plastic protective cap. New copper washers should always be installed over the nozzle pintle before installation. The washer can be retained in position on the nozzle with a small amount of DOAZ-19584-A grease or equivalent. A small amount of anti-seize compound should also be applied to the nozzle body threads during installa-

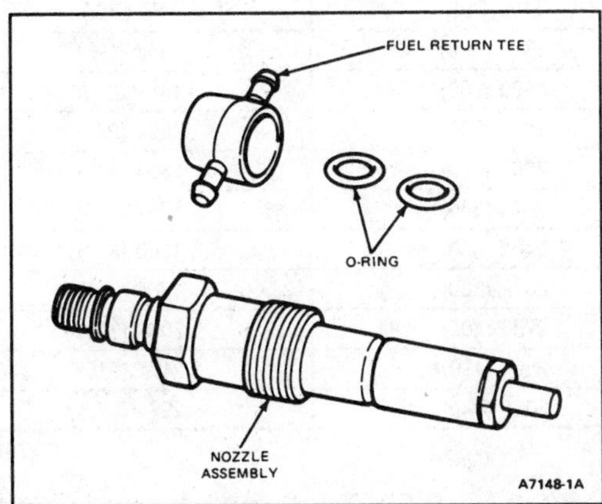

Figure 24–24. 6.9L Ford V8 diesel engine injection nozzle. (Courtesy of Ford Motor Company)

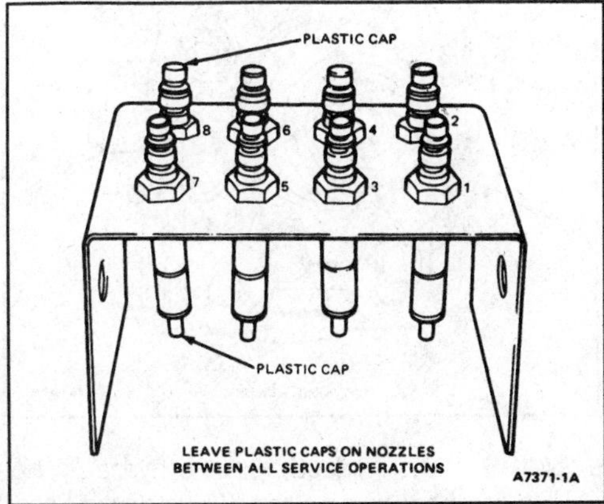

Figure 24–25. Nozzle holding fixture. (Courtesy of Ford Motor Company)

tion. Torque the nozzle body to 35 lb.ft. (47 N·m) in its bore.

Remove the protective caps from the nozzle threads when you are ready to install the fuel leak-off tees, which should have two new O-rings installed.

Connect the high pressure fuel line from the injection pump to the nozzle and torque it to 22 lb.ft. (30 N·m) with the aid of special tool P/N T83T-9396-A after bleeding the fuel system.

If the high-pressure fuel lines have been completely removed from the engine, refer to Figure 24–26, which shows the fuel line cylinder number location on the injection pump hydraulic head.

To bleed the fuel system, leave each injection nozzle fuel line loose by about one-half turn while you crank the engine. When fuel free from air bubbles appears at each line, tighten them up with a wrench, then torque them to specs. Wipe up spilled fuel with a rag.

Install fuel line retainer clamps as per Figure 24–4. Start and run the engine and check for any signs of compression leakage at each injector as well as for signs of fuel leakage at either the nozzle high-pressure line as well as at the nozzle fuel return tees.

Engine Compression Check

To perform a compression check of the engine, make sure that it is at operating temperature, then remove all of the glow plugs after stopping the engine. Remove the air cleaner and install protective intake opening cover P/N T83T-9424-A or suitable equivalent.

To prevent the engine from starting, disconnect the injection pump solenoid switch wiring harness.

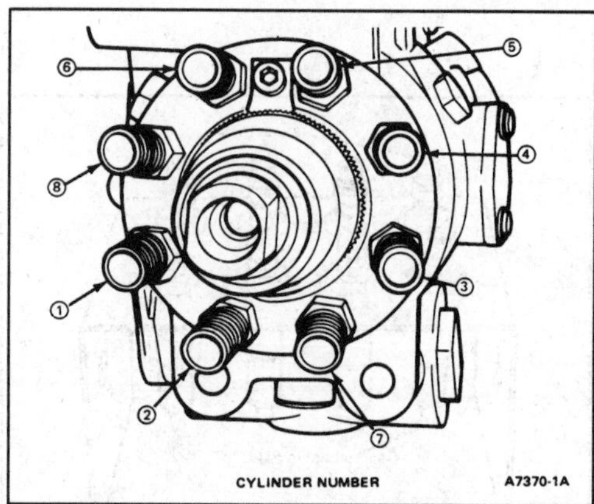

Figure 24–26. 6.9L Injection pump hydraulic head cylinder fuel line numbering sequence. [Courtesy of Ford Motor Company]

Make sure that the vehicle batteries are up to full charge in order to spin the engine fast enough, otherwise a low compression reading will result.

Install a suitable compression gauge with an adapter into each glow plug hole one at a time and crank the engine over on a remote starter switch hookup between 4 and 6 times per cylinder and note the cylinder pressure reading on the gauge. The lowest compression reading should not be any less than 75% of the highest reading, otherwise it is an indication of a piston ring or valve problem in that cylinder.

CAUTION: Due to the high compression ratio of the 6.9L diesel engine (20.7:1), *never* squirt engine oil into the cylinder in order to establish whether the problem of the low compression reading is attributable to the valves or the piston rings because a hydrostatic lock can occur and serious engine damage can occur.

Examples of cylinder compression pressure allowable variations are shown in Figure 24–27.

Ford 6.9L V8 Diesel Glow Plugs

The use of glow plugs is mandatory on all pre-combustion chamber diesel engines in order to facilitate ease of starting, especially in cold weather. Glow plugs are energized from the vehicle electric system when the ignition switch is turned to the *start* position.

Before the engine is cranked, a glow plug indicator lamp will illuminate on the vehicle instrument panel, which allows current to flow to the glow plugs in a pre-glow situation. After the engine starts and de-

HIGHEST CYLINDER	LOWEST CYLINDER
Maximum kPa (PSI)	Minimum kPa (PSI)
1792 (260)	1344 (195)
1929 (280)	1447 (210)
2067 (300)	1551 (225)
2205 (320)	1654 (240)
2343 (340)	1757 (255)
2481 (360)	1860 (270)
2619 (380)	1964 (285)
2756 (400)	2067 (300)
2894 (420)	2171 (315)
3032 (440)	2274 (330)

CA7414-1B

Figure 24–27. 6.9L Diesel engine typical compression pressure readings. [Courtesy of Ford Motor Company]

pending upon its coolant temperature, an after-glow situation will exist with the glow plugs.

The major function of the glow plugs is to preheat the incoming air in the combustion chamber to reduce the ignition delay period of the injected fuel when the engine coolant is lower than normal operating temperature.

The length of time that the glow plugs are energized depends upon how cold the engine coolant is. When energized, the glow plugs will heat up similar to a heating element on a kitchen stove. The glow plugs will not operate if the engine coolant temperature is higher than 165°F (91°C). A glow plug is used in each cylinder and controlled by:

1. A glow plug controller cycling switch.
2. An after-glow timer.
3. A power relay.
4. A circuit breaker.
5. Two fusible links connected between the power relay and the glow plug harness (one for each bank of glow plugs).
6. Glow plug indicator lamp.
7. A wiring harness with eight fusible links located between the wiring harness and the individual glow plugs.

Figure 24–28 illustrates a typical glow plug wiring harness along with its control switch, which is located on the rear of the left-hand cylinder head. The control switch senses engine coolant temperature and determines how long the glow plugs will remain energized—which varies between 4 and 10 seconds during the initial cycle.

Most diesel engine manufacturers today employ 6-volt glow plugs energized from a 12-volt battery source in order to provide rapid heating of the glow plug element and reduce start-up time when the engine is cold. A glow plug cycling controller ensures that the glow plugs will not overheat by shutting off battery current to them after a given time period depending upon engine coolant temperature.

Glow Plug Operation

Refer to Figure 24–29, which is a wiring circuit diagram for the system. Turning the ignition switch to *run* allows battery current to flow through the glow plug controller and on to the glow plug power relay where a set of contacts close and power flows to:

1. The glow plug indicator lamp
2. The eight glow plugs
3. The controller cycling switch

Power flows to the glow plugs for between 0 and 9 seconds, depending upon the engine coolant temperature; and then the cycling switch will open, cutting off current to the glow plug power relay. The relay contacts open and the glow plug indicator lamp goes out along with a power loss to the eight glow plugs.

The after-glow timer controls the cycling time of the current flow to the glow plugs through the controller cycling switch. Cycling of the glow plug current supply depends upon how long the after-glow timer switch remains closed. This timer remains closed for up to two minutes or sooner if the engine reaches approximately 165°F (74°C).

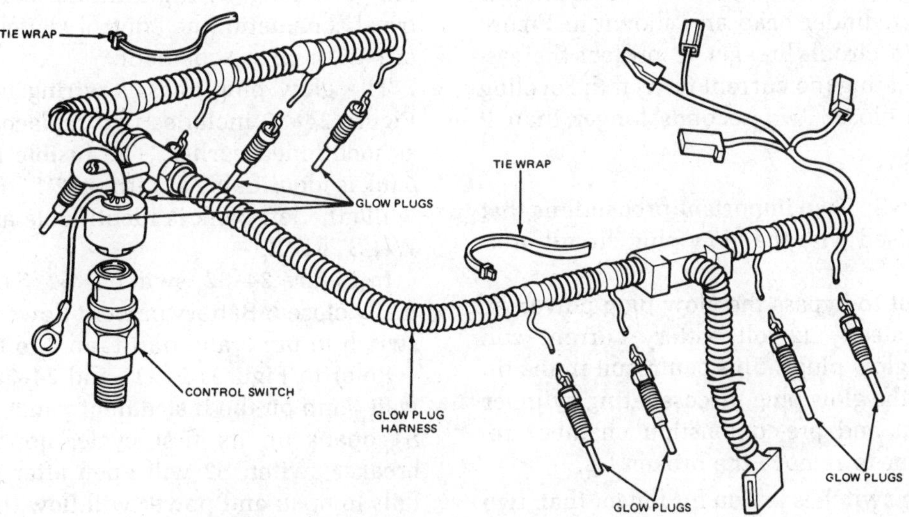

Figure 24–28. 6.9L Diesel glow plug wiring harness. (Courtesy of Ford Motor Company)

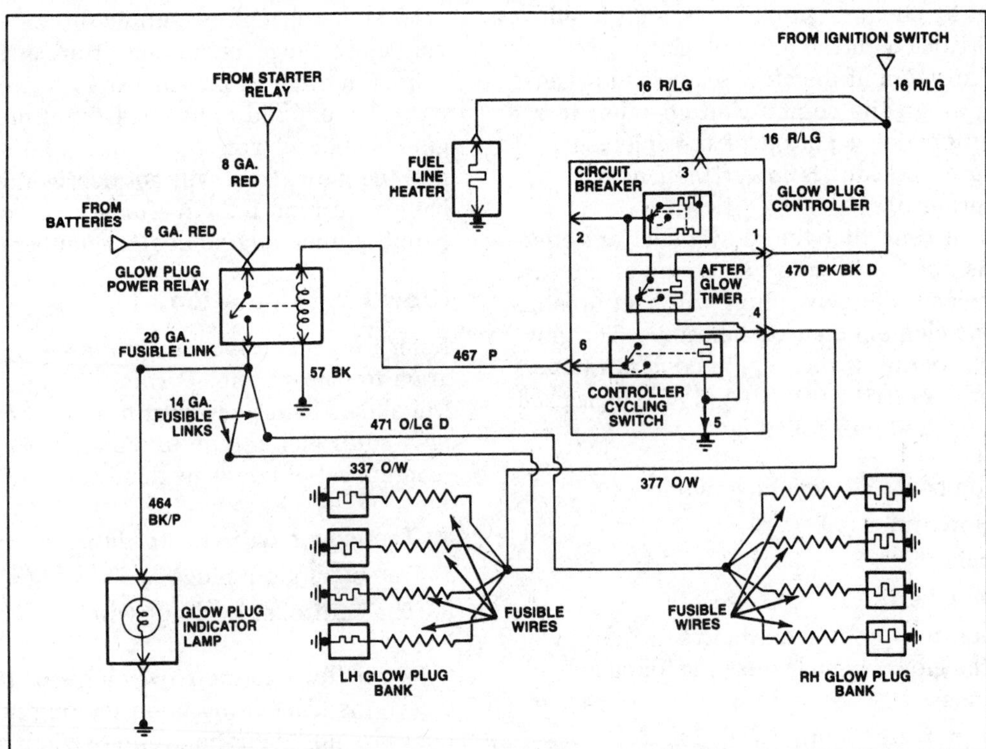

Figure 24–29. Engine wiring system schematic. (Courtesy of Ford Motor Company)

While the glow plug after-glow timer is heating up, the controller cycling switch turns the glow plugs on/off to keep them hot during the period immediately after engine start-up to ensure smoother warm-up cycle and to minimize exhaust smoke.

The glow plug indicator lamp will also cycle on/off during this time period.

The glow plug controller switch located in the rear of the left-hand cylinder head and shown in Figure 24–28 includes a circuit breaker to protect the system from overheating (no current flow) if the cycling switch remains closed two seconds longer than it normally would.

PRECAUTIONS: Two important precautions that should be exercised with the glow plug circuit are:

1. Never attempt to bypass the glow plug power relay because steady 12-volt battery current will burn out the glow plugs. This can result in the tip falling off of the glow plug, necessitating cylinder head removal and pre-combustion chamber removal in order to remove the broken tip.
2. If the ignition switch is left on for longer than two minutes prior to attempting to start the engine, hard starting will result because the after-glow switch will have opened, stopping the cycling action of power flow to the glow plugs. To remedy the situation, it will be necessary to turn the ignition key *off* for one or two minutes to allow the glow plug timer to cool down sufficiently so that it will close and allow battery power to again flow to the glow plugs.

To assist you in understanding the glow plug circuit further, Figures 24–30 through 24–33 illustrate the glow plug wiring harness, control switch electrical schematic, the control switch circuit, and the circuit breaker schematic.

The glow plug chassis wiring harness shown in Figure 24–30 includes two replaceable fusible links as mentioned earlier. The fusible link for the right bank is identifiable as circuit 471 OR/LT GRN DOTS, while the left bank is identifiable as circuit 337 OR/WH STRIPE.

In Figure 24–32, switches S2, S1, and S3 are normally closed. Battery current flows from the ignition switch to pin 3 and on to pin 6 to the power relay.

Refer to Figures 24–31 and 24–33. The glow plug wait lamp on the instrument panel will go out when S1 opens on its first cycle; however, the circuit breaker switch S2 will open after 14 seconds if S1 fails to open and power will flow through R4 (to pin 6 of the power relay), which will reduce the current to prevent activation of the power relay while current continues to heat switch S2 keeping it open.

Pin No. 5 acts as a ground for the current flow from

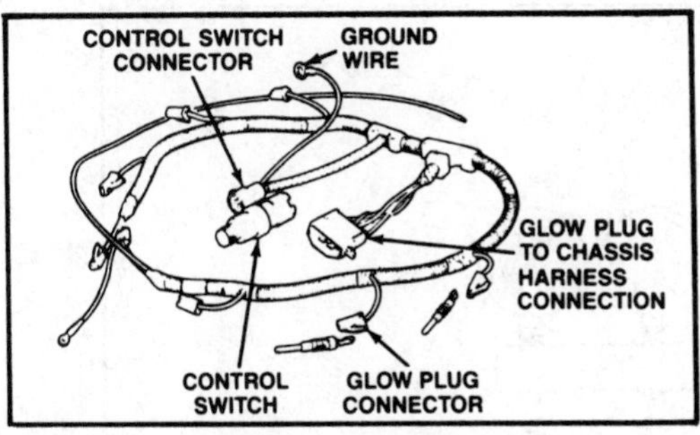

Figure 24-30. Glow plug wiring harness and connectors. (Courtesy of Ford Motor Company)

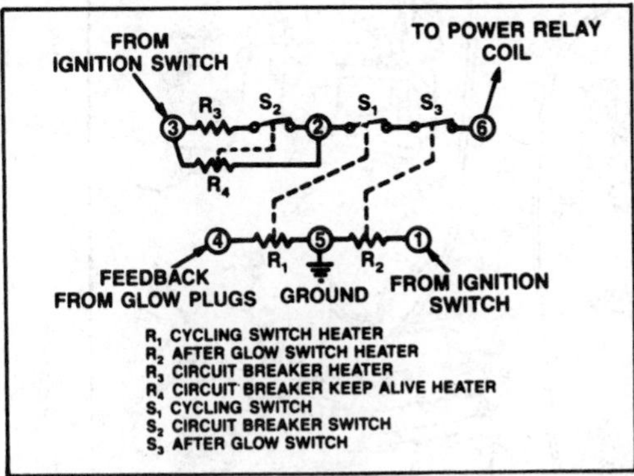

Figure 24-31. Glow plug control switch electrical schematic. (Courtesy of Ford Motor Company)

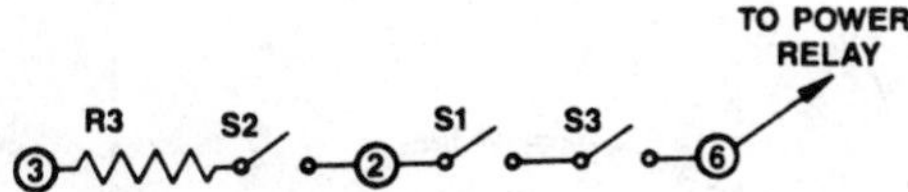

Figure 24-32. Glow plug control switch circuit. (Courtesy of Ford Motor Company)

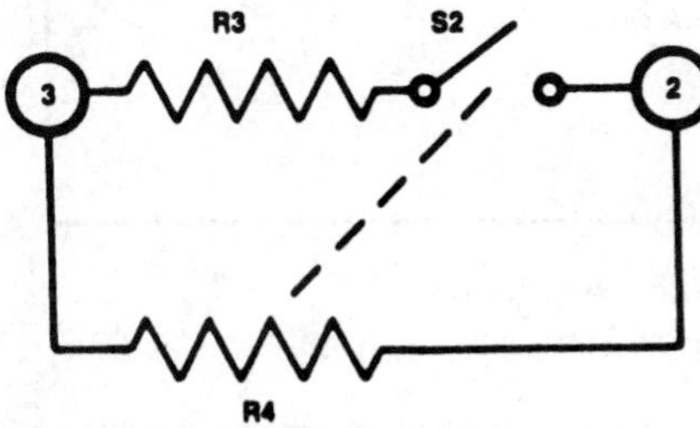

Figure 24-33. Glow plug circuit breaker schematic. (Courtesy of Ford Motor Company)

the glow plugs through pin No. 4 of the control switch through resistor R1. S1 is a *bimetallic* switch that will cut off power to the glow plugs when it is heated by current flow through R1.

Once the engine is running, power will flow to ground through pin No. 5 after the ignition switch closes the relay switch and current flows to pin No. 1 and then resistor R2.

Another *bimetallic* switch is S3, which will stop current flow to the power relay within 20–90 seconds and the alternator is charging when S3 is heated by R2.

Glow Plug Troubleshooting

Glow plug circuit troubleshooting should start with a check of the circuit to establish if battery voltage is present at the glow plugs with the engine coolant temperature below 165°F (74°C).

To do this, remove all leads from the glow plugs and connect a 12-volt test lamp between the power relay output (green leads) (to glow plugs) and ground. Turn the key *on* and note the voltage at each glow plug lead with a voltmeter when the test lamp is lit, which should not be less than 11 volts. If no voltage is present, check the chassis harness fusible links as well as the engine harness fusible links.

Glow plugs should also be checked for resistance between the glow plug terminal and its metal body, which should be less than 2 ohms. To do this, make sure that the ignition key is *off* and that the engine wiring harness is disconnected at the glow plugs. A systematic check of the circuit will confirm where the problem area is. If detailed analysis of the glow plug circuit is required, refer to the Ford Truck Shop Manual P/N FPS 365-326-84HT, Emission Diagnosis and Engine Electronics.

6.9L Engine Performance Diagnostic Procedure

A number of special tools and test equipment is available from either Ford or Rotunda to successfully troubleshoot the 6.9L diesel engine operation.

A pressure test kit Rotunda P/N 19-0002 can be hooked up to various engine test points as shown in Figure 24–34 to monitor systems operation.

Fuel System Checks

When the engine performance is poor—such as sluggish throttle response, low horsepower, and hard starting or stumble at idle—check the color of the exhaust smoke as a guide to where the problem might be. Black exhaust smoke generally indicates either an air starvation situation or excess fuel delivery.

Although the air cleaner element can be inspected visually for signs of plugging, this in itself is not conclusive proof that the element is in fact offering a high air inlet restriction. Use a water (H_2O) manometer or a vacuum gauge connected to the test port on the air cleaner to establish the amount of restriction to air flow. Run the engine at its maximum no-load rpm and note the reading which should not exceed

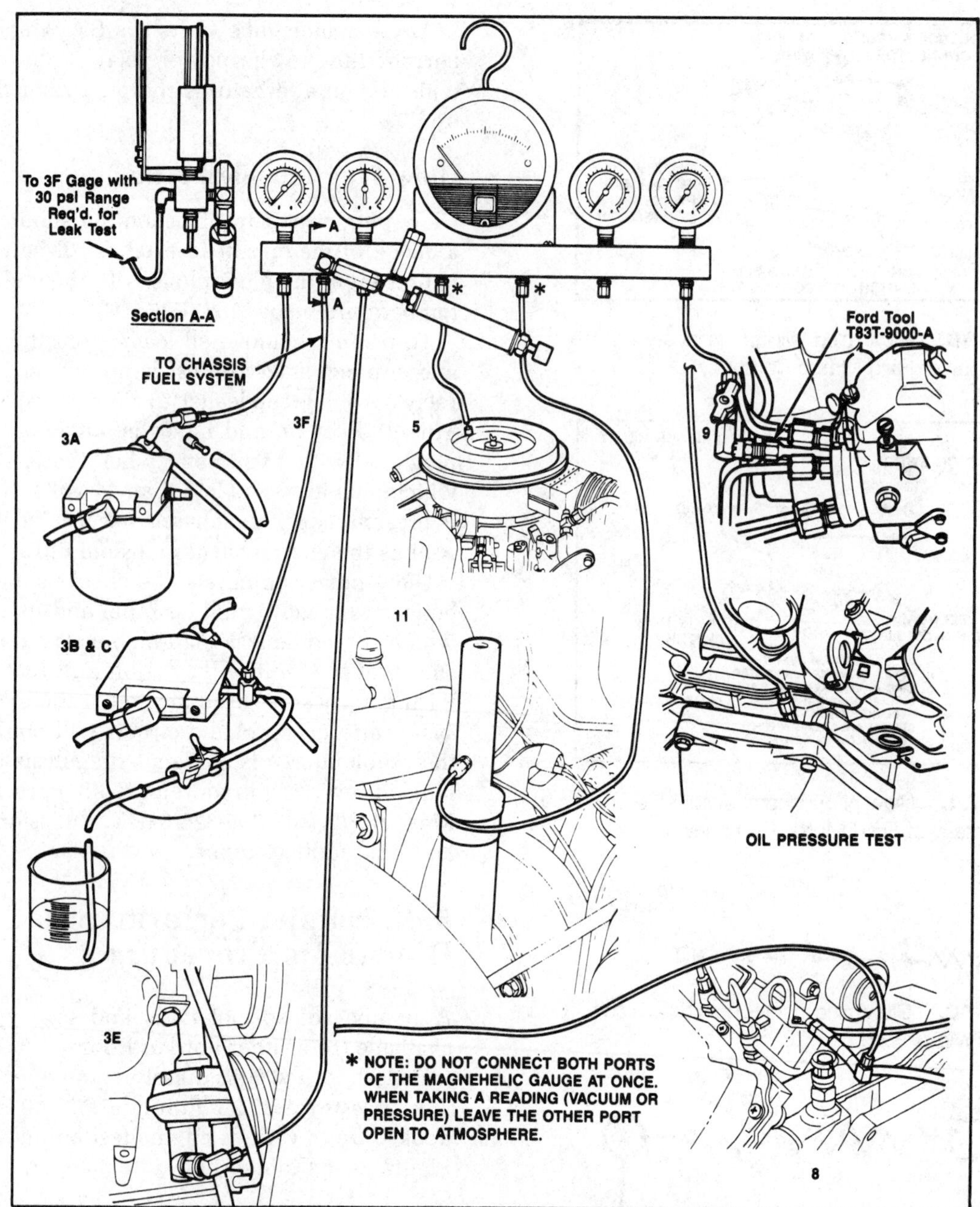

Figure 24-34. Pressure test kit hookup—engine performance diagnosis. (Courtesy of Ford Motor Company)

25″ H_2O (maximum). Since 25″ is the maximum reading, anything close to this reading would be reason enough to change the filter element.

In the majority of cases, the air starvation problem is more common. When poor engine performance is evident with no black smoke or very little at the exhaust, this is usually indicative of a fuel starvation problem. Faulty injection pump timing and faulty injection nozzles can also be a cause of black exhaust smoke as well as high exhaust back pressure.

To check the fuel system for fuel starvation, the following checks would be performed: The suction side of the system is between the fuel tank(s) and the inlet side of the lift pump. From the outlet side of the lift pump, the fuel is under low pressure until it enters the vane-type transfer pump in the injection pump. Fuel from the injection pump is at high pressure (2000–2150 psi, or 13790–14824 kPa) and will spray fuel over the engine compartment if leaks exist.

1. Check fuel supply lines for kinks or damage (restriction) and also the return fuel line for restriction problems.

2. Check fuel inlet (suction) lines for loose fittings, which would cause air to be drawn into the system.
3. Check the pressure side of the system for fuel leaks.
4. Check the condition of the fuel, especially the Cetane rating with the test kit as well as for conditions such as dirt and water.
5. Check the fuel supply (lift) pump pressure as well as its delivery capacity.
6. Check the fuel filter outlet pressure.
7. Check for fuel tank restriction.
8. Check the fuel return pressure.
9. Check injection pump transfer pressure.

Procedural Checks

Part 1 of Item 1 above can be checked visually while Part 2 of Item 1 can be checked by teeing in a vacuum gauge to the suction side of the lift pump and running the engine at 3300 rpm (wheels jacked clear of the ground for safety reasons) with the transmission in *neutral* (standard) or in *park* (automatic). Maximum restriction to the suction side of the system should not exceed 6″ Hg.

Item 2 can be confirmed by using soap suds on the fittings while the engine is running to determine if suction leaks exist.

Item 3 can be checked by visual means, while Item 4 can be checked both visually and with the aid of the Cetane rating test kit tool, shown in Figure 24–35.

Part 1 of Item 5 can be checked by removing the vacuum purge valve from the top of the fuel filter and installing a suitable low pressure gauge, or use test kit P/N 19-0002 as shown in Figure 24–34 with the clamp closed on the sampling hose (3B and C) as shown. Start and run the engine at idle rpm no-load and note the pressure gauge reading, which should be 2 psi (13.8 kPa) or higher.

Part 2 of Item 5 can be checked by referring to Figure 24–34 (3B and C) and running the engine at an idle rpm no-load. With the clamp open on the sample hose, allow fuel to flow into a graduated container for 30 seconds and measure the volume returned. Minimum volume delivery should be one pint; otherwise the lift pump is not operating correctly. Do not immediately change the lift pump unless you are satisfied that Items 1 and 2 above are satisfactory as well as ensuring that the fuel/water separator is not plugged.

Item 6 can be checked by removing the air-bleed orifice hose from the fuel filter and installing a pressure gauge. Run the engine to 3300 rpm no-load and note the pressure reading, which should be at least 1 psi.

Item 7 can be checked by using a test container of fuel to draw from to confirm if indeed there is a restriction at the fuel tank.

Item 8 can be checked by removing the fuel return line located at the junction of the left rear of the engine and installing the pressure test kit or a suitable low pressure gauge. Run the engine at 3300 rpm and note the fuel return pressure, which should be no more than 2 psi (13.8 kPa).

Item 9 can be checked by removing the screw from the transfer pump port cover and installing special tool P/N T83T-9000-A (shown in Figure 24–35) into and through the cover and O-ring port. Connect this adapter to the pressure test kit 19-0002 or equivalent unit and run the engine to 3300 rpm no-load and note the fuel pressure reading, which should be 90–110 psi (620–758 kPa).

Having checked out the above causes, if engine performance complaints still exist, check the static and dynamic timing of the injection pump. Other areas that can be checked would of course involve removal and testing of the fuel injectors and also a compression check of each cylinder—both of which are discussed in detail in other areas of this chapter and book.

Hard starting would always involve a check of the glow plug circuit as well as the injection pump static and dynamic timing checks discussed earlier in this chapter.

Lack-of-power complaints in hot weather operation can also be attributed to both air inlet and fuel inlet temperatures. An increase in air inlet temperatures beyond 90°F (32°C) will affect the horsepower of the engine. This will vary between different engines from as low as 0.15 to 0.7 horsepower per cylinder for every 10°F rise beyond 90°F (32°C).

When the fuel temperature increases beyond the optimum of 90°F (32°C), there is a 1% loss of the total gross horsepower for every 10°F beyond 90°F. For example, if the fuel temperature was to increase to 130°F (54.4°C), the fuel expansion means that it becomes less dense and will affect the horsepower. This temperature of 130°F (54.4°C) is 40°F above the optimum fuel temperature and will decrease the engine horsepower by a total of 4%, or 4 horsepower for every 100 that the engine produces.

Both of these conditions of air and fuel temperature rise will adversely affect the engine's operation.

Other areas that can adversely affect the engine's performance are high crankcase pressure and high exhaust back pressure. The crankcase pressure can be checked with a water manometer or a vacuum gauge at the breather attachment, while the exhaust back pressure can be monitored at the exhaust manifold companion flange.

The following test equipment (Figs. 1 through 5) is required for adjusting idle speed and timing.

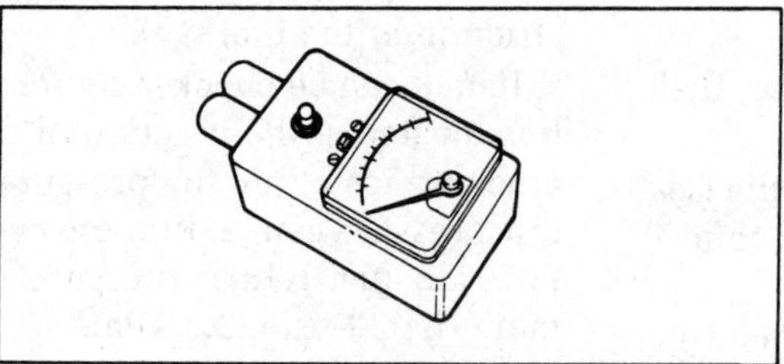

Figure 1 Rotunda 99-0001 Photoelectric Tachometer (For Engine RPM Checking Only)

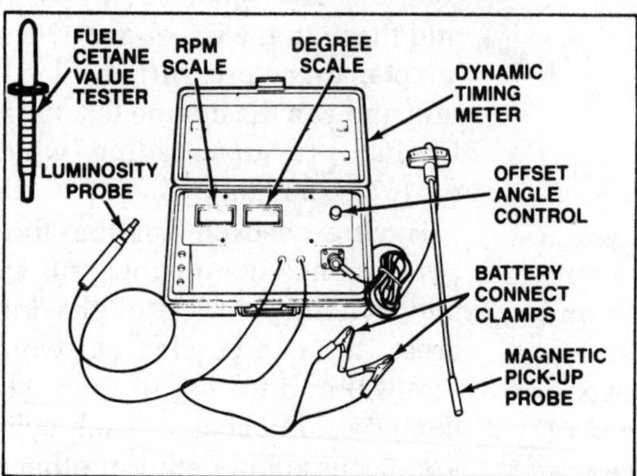

Figure 2 Rotunda 78-0100 Dynamic Timing Meter

Figure 3 T83T-9000-B Injection Pump Mounting Wrench

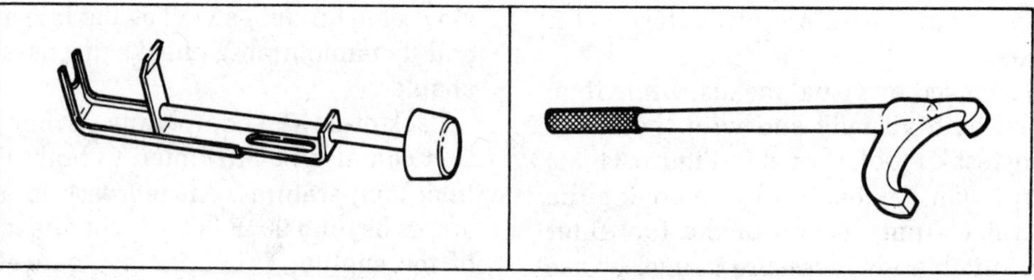

Figure 4 Rotunda 14-0302 Throttle Control

Figure 5 T83T-9000-C Injection Pump Rotating Tool

The following test equipment (Figs. 6 and 7) is required for performing the Engine Performance Diagnostic Procedure.

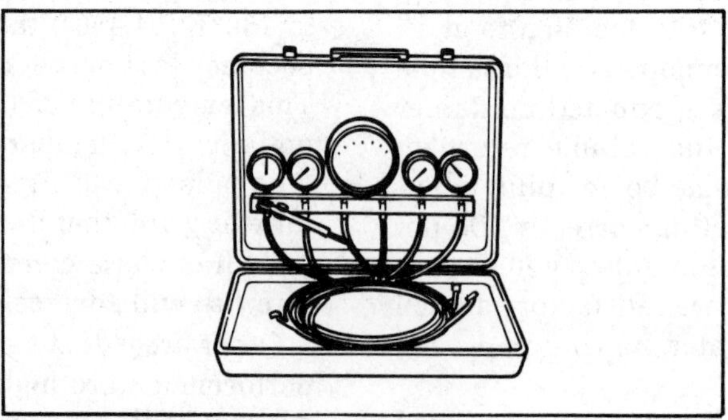

Figure 6 Rotunda 19-0002 Pressure Test Kit

Figure 24–35. Special tools and equipment. (Courtesy of Ford Motor Company)

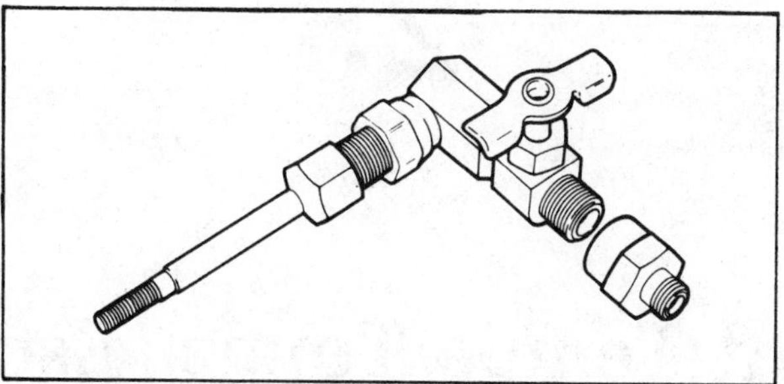

Figure 7 T83T-9000-A Fuel Transfer Pump Pressure Adapter

The following test equipment (Figs. 8 and 9) is required for performing the WAIT TO START Lamp Diagnostic Procedure and the Fast Start Glow Plug System Diagnostic Procedure.

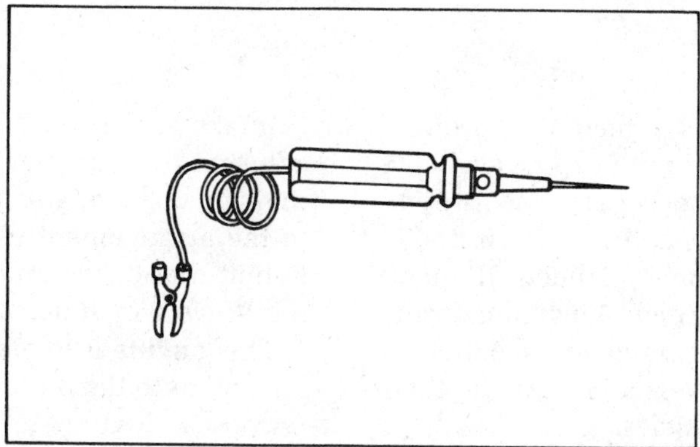

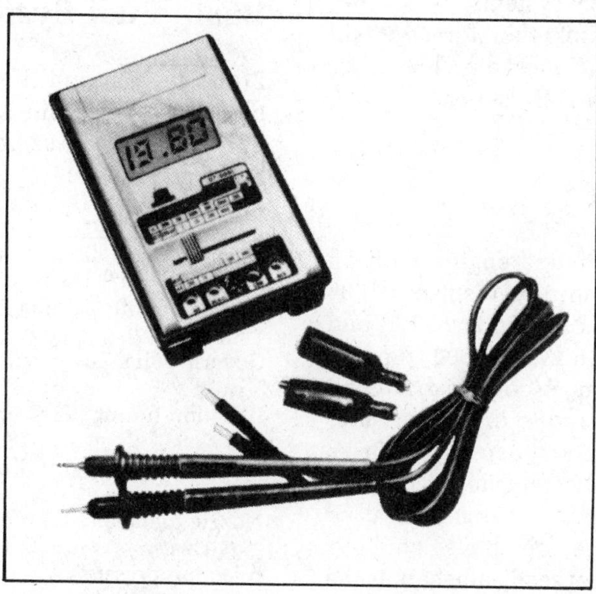

Figure 24-35. *(Continued)*

25

Ford Tempo/Topaz, Escort/ Lynx 2.0L Diesel Engine

Ford offers four diesel engines in their 1984 product lines: the 2.0L (122 cu.in.) 4-cylinder engine available as an option in the Tempo/Topaz, Escort/Lynx models; the 2.2L (132 cu.in.) 4-cylinder Toyo Kogyo (Mazda) engine in their Ranger/Bronco II small pickup units; the 2.4L BMW/Steyr 6-cylinder turbocharged engine in the Continental; and the 6.9L (421 cu.in.) V8 International-Harvester engine in their F350 and Econoline E350 vehicles.

From this offering it would appear that Ford is reasonably well committed to diesel engine power as an optional powerplant in their line of existing vehicles!

The 2.4L and 6.9L engine fuel systems are covered in other sections in this textbook; therefore, we will concentrate on the two 4-cylinder diesel engines here, namely the 2.0L and the 2.2L models.

Basic Engine Design

The 2.0L diesel is a 4-cycle in-line engine with an overhead camshaft arrangement that displaces 1998 cc (122 cu.in.). The compression ratio is 22.7:1, and the engine produces 53 hp (39.5 kW) at 4000 rpm and 82 lb.ft. of torque at 2750 rpm. A cogged drive belt at the front of the engine is used to drive the water pump and the camshaft, while another cogged drive belt driven off of the rear of the camshaft drives the fuel injection pump similar to the arrangement used on the Volvo D24 diesel engine, which also employs the same type and layout of overhead camshaft as the 2.0L Ford engine.

The 2.0L engine is mounted transversely in the vehicle frame to accommodate the front-wheel drive layout used with this vehicle.

Ford claims that fuel economy with this engine can be as high as 64 mpg (Imperial gallon) or 51 mpg (US gallon).

Intake and exhaust valves on the 2.0L engine are adjusted by changing valve shims located on top of the cam followers as shown in Figure 25–1 similar to the arrangement used by VW in their 4-cylinder Rabbit diesel, the Audi 5000 5-cylinder diesel, and the Volvo 6-cylinder 2.4L diesel engines.

This engine also employs oil cooling jets to spray lube oil onto the underside of the pistons for cooling purposes. Figure 25–2 illustrates the combustion chamber design, glow plug, fuel injector, and overhead cam design of this engine.

Basic Fuel System Design

Both the 2.0L and the 2.2L engines employ distributor-type rotary injection pumps. The basic operation of each is the same with the main characteristics being as follows:

2.0L Engine	*2.2L Engine*
Pump plunger diameter 8.0 mm (0.3150″)	Pump plunger diameter 10 mm (0.394″)
Pump cam lift: 2.2 mm (0.0866″)	Pump cam lift: 2.2 mm (0.0866″)
Governor: half/all speed	Governor: hydraulic/mechanical type
Injection timing: TDC hot	Injection timing: 2 Degrees ATDC
Injection nozzle: throttling pintle type	Injection nozzle: throttling pintle type
Nozzle diameter: 1.0 mm (0.0394″)	Nozzle diameter: 0.8 mm (0.031″)
Injection pressure: New: 1990–2105 psi (14000–14800 kPa) Used: 1849–1990 psi (13000–14000 kPa)	Injection pressure: 1957–2030 psi (13500–13997 kPa)

As you can see from the general comparison of the 2.0L and 2.2L fuel systems, the fuel systems are very similar in both design and operation.

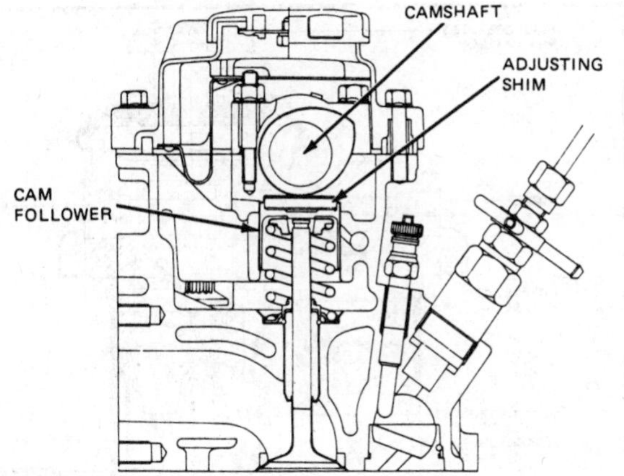

Figure 25-1. Camshaft/valve arrangement 2.0L engine. (Courtesy of Ford Motor Company)

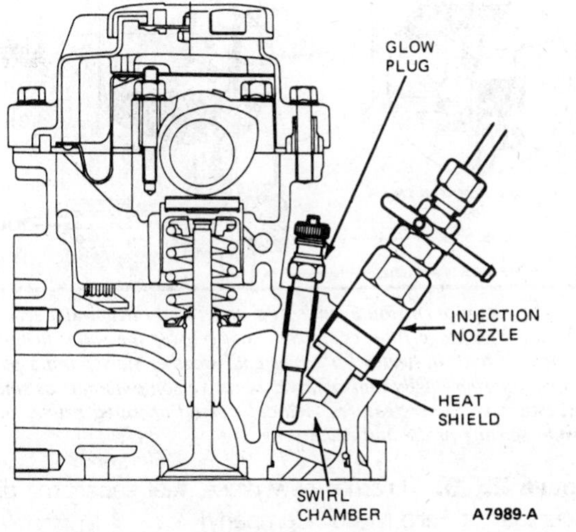

Figure 25-2. Pre-combustion chamber design—2.0L engine. (Courtesy of Ford Motor Company)

2.0L Fuel System

The fuel system on the 2.0L diesel engine is illustrated in Figure 25–3. The fuel injection pump used on this engine is a Robert Bosch design VE type distributor unit; therefore, a detailed explanation of the operation, maintenance and repair of this pump can be found in Chapter 12, Robert Bosch diesel fuel system, VE pump section. The fuel system is arranged so that an integral mounted mechanical fuel pump (vane-type) within the injection pump assembly draws fuel from the tank up to a fuel filter/water separator illustrated in Figure 25–4. Water and dirt in the fuel is removed in the filter/conditioner; the fuel then passes to the fuel feed pump. The fuel feed pump then pushes the fuel under pressure to the injection pump where the fuel is placed under high pressure and delivered to the injector nozzles in engine firing order sequence.

Throttle position determines how much fuel will be metered and sent to the injectors. This is achieved by the position of a control sleeve as shown in Figure 12–17 that is connected through linkage to the accelerator.

Fuel at feed pump pressure (100 psi, or 7 bar, approximately) is delivered to a plunger that is rotating as well as moving backward and forward (reciprocating). This rotating/reciprocating plunger increases the fuel pressure and distributes the fuel in firing order sequence to the injectors. Pressure from the injection pump plunger must be high enough to cause the injector needle valve to be lifted off of its seat against spring pressure. In the 2.0L engine, 1920 psi (135 kg/cm^2) minimum is required to open the injector and allow fuel to be sprayed directly into the

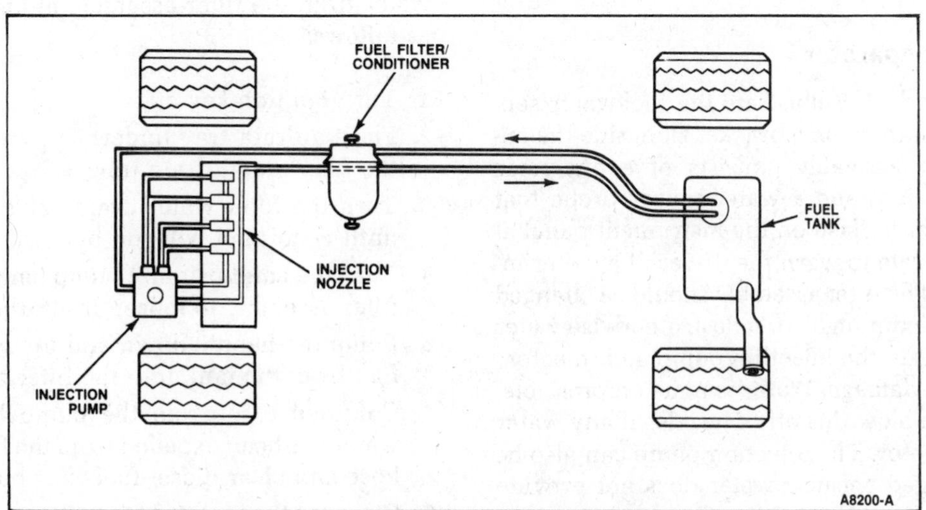

Figure 25-3. Fuel system component layout. (Courtesy of Ford Motor Company)

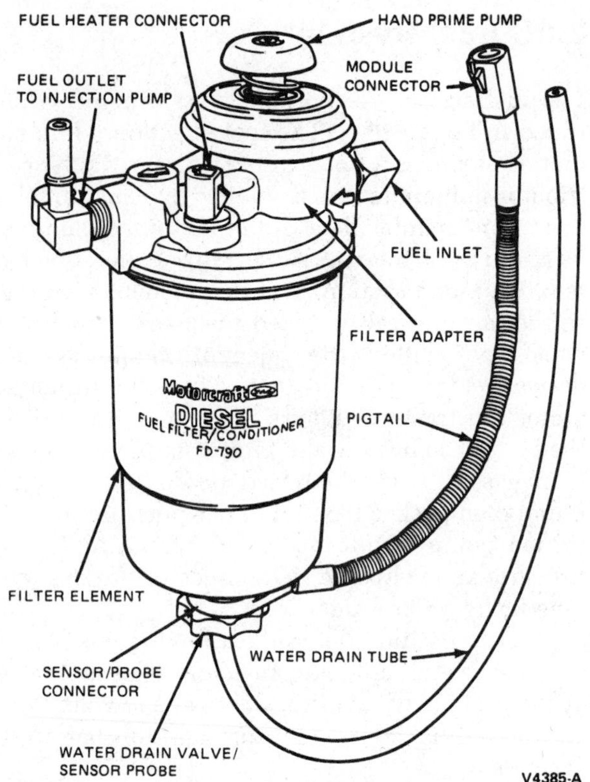

Figure 25–4. Fuel filter/water separator assembly. [Courtesy of Ford Motor Company]

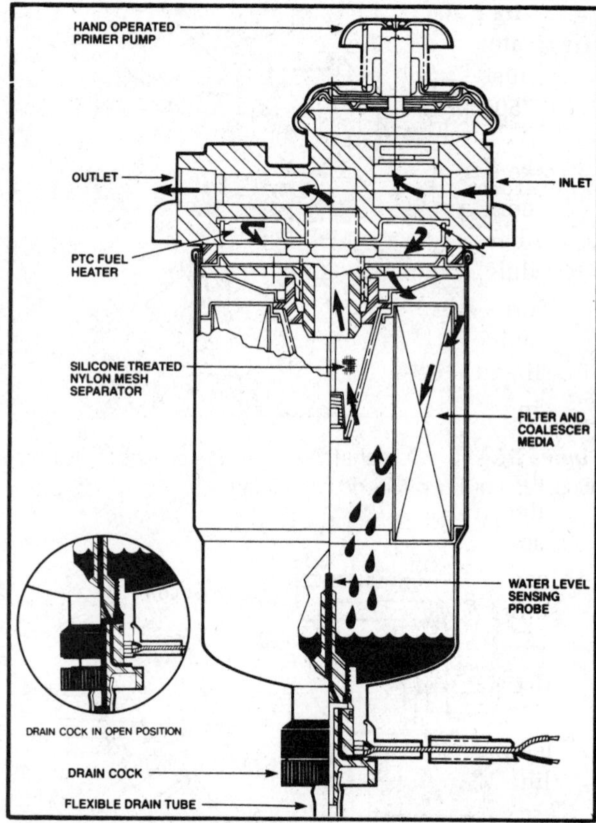

The key components on Fram's new diesel fuel filter/water separator include: two-stage filter coalescer that removes the water in the fuel system; a built-in heater for easier cold weather starts; and a pleated paper filtering media that removes abrasive contaminants as small as 1.5 microns. Other features include: a hand-operated primer pump; water sensing probe and drain tube.

Figure 25–5. Fram's new diesel fuel separator design. [Courtesy of Ford Motor Company]

pre-combustion chamber. When fuel pressure drops off, the injector needle valve spring closes it and injection stops.

Some fuel is used to lubricate the internal parts of the injector; this excess fuel not used at the injectors returns through a series fuel line connected to all injectors, back to the fuel injection pump, and then back to the fuel tank.

Filter/Water Separator

Figures 25–4 and 25–5 illustrate the fuel/water separator that is mounted on a bracket alongside the engine. The filter assembly consists of a one-piece replaceable element and a water sensor probe that will cause a light to flash on the instrument panel at a 1 to 3 second rate to warn the driver of a water-in-fuel condition. The filter assembly should be changed immediately (maximum of 10 miles), otherwise water may be forced into the injection pump and injectors causing serious damage. Water is non-compressible; therefore, it can blow tips off of nozzles if any water reaches the injector. The injection pump can also be seriously damaged because water does not provide the necessary lubrication of the closely fitted components, resulting in scuffing/overheating and serious injection pump damage.

Draining Water from Filter Assembly. To drain water from the filter assembly at any time, proceed as follows:

1. Turn ignition key *off*.
2. Place a drain tray underneath the vehicle below the filter water drain tube.
3. Turn the filter water drain valve 2 to 3 turns or until water/fuel will run out of the drain hose.
4. Turn the hand-priming pump handle on top of the filter assembly to loosen it off, then pull it up.
5. Pump the handle down and up, which will draw fuel from the tank into the filter bowl.
6. Continue to operate the pump handle until all water has been expelled from the filter drain valve hose and clear diesel fuel runs out of the line.
7. Close up the water drain valve securely.
8. Start and run the engine while inspecting the filter assembly for signs of leaks at the drain valve.

Changing Fuel/Water Element. To change the fuel filter/water separator, the complete filter element shell must be replaced as a unit at approximately every 7500 miles, or 12070 km. Proceed as follows:

1. Make sure that the ignition key is turned off.
2. Disconnect the wiring harness pigtail located on the base of the filter element by depressing the module connector clip.
3. Drain some fuel by opening the drain valve and allowing fuel to run out of the drain tube into a drain tray underneath.
4. Using a filter or strap-type wrench, unscrew the filter element shell assembly from its adapter.
5. With the filter assembly removed, unscrew the water drain valve/sensor probe assembly from its base.
6. Unsnap the sensor probe pigtail connector from the filter base.

Installation of the Filter Assembly.

1. Clean off the sensor probe pigtail assembly with a lint-free rag and then snap it into position at the base of the new filter element.
2. Lightly lubricate both O-ring seals on the water sensor probe with 10W motor oil, then securely tighten the probe into the base of the filter element.
3. Use a lint-free rag and clean off the gasket sealing surface of the filter adapter.
4. Lightly coat the captive seal ring on the top of the filter with clean 10W engine oil.
5. Screw the filter onto the adapter until the gasket just makes contact, then hand tighten it another 1/2 to 5/8 of a turn.
6. Reconnect the water module sensor connector wire.
7. Prime the fuel system by unscrewing the filter hand-pump, pulling it up and then pumping the handle until clear fuel is observed flowing through the transparent plastic line from the filter to the injection pump housing.
8. Pump the handle until an increase in effort is required and push the handle down to its normal closed position.
9. Start the engine and check for signs of fuel leakage at the filter assembly.

NOTE: It may be necessary to rebleed the system if hard starting occurs; it may also be necessary to individually loosen off the injector fuel lines about 1/2 turn at the nozzle end to adequately bleed the system of trapped air. Wrap a rag around the fuel line while you do this to prevent fuel spraying into your eyes or over the vehicle. Tighten each line once you have bled it.

Troubleshooting Water Sensor. If the water-in-fuel warning light on the instrument panel fails to illuminate when water is in fact in the system, proceed as shown in Figure 25–6.

Ford 2.0L Fuel Heater

Contained within the fuel/water filter separator is a fuel heater, which is illustrated in Figure 25–5. This fuel heater is necessary to prevent fuel filter plugging due to waxing of the fuel in cold weather operation.

The fuel heater is automatically controlled through the use of a bimetallic snap switch located below the connector. Power is available to the fuel heater when the engine is running so that the snap switch will relay power to the fuel heater anytime that the fuel temperature is between 30–55°F (−1–13°C).

When the snap switch moves into the *on* position, it supplies current to two (PTC) positive temperature coefficient heaters that are self-regulating due to the fact that as their temperature goes up there is an increase in the resistance to electrical flow that will automatically limit the maximum temperature of the two heaters. Because of the self-regulating feature of the heaters, no control module is required.

Glow Plug System

All pre-combustion chamber diesel engines require the use of a glow plug in order to preheat the incoming air to promote rapid start up, especially in cold weather. The glow plug is screwed into each pre-combustion chamber as shown in Figures 25–2 and 25–7.

The wiring circuit for the glow plugs is illustrated in Figure 25–8 and it contains the following major items:

1. Four glow plugs
2. A control module
3. Two relays
4. A glow plug resistor assembly
5. Coolant temperature switch
6. Clutch and neutral switches
7. Connecting wiring

Both the relay power and the feedback circuits are fuse protected by the insertion of fusible links in-

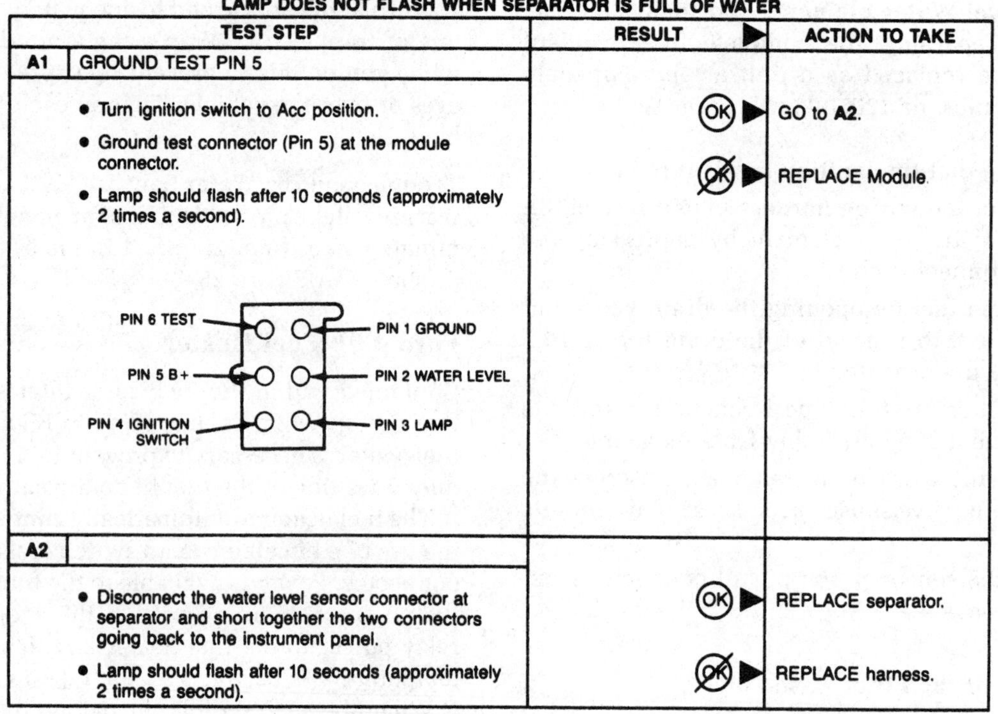

LAMP DOES NOT FLASH WHEN SEPARATOR IS FULL OF WATER

	TEST STEP	RESULT ▶	ACTION TO TAKE
A1	GROUND TEST PIN 5		
	• Turn ignition switch to Acc position. • Ground test connector (Pin 5) at the module connector. • Lamp should flash after 10 seconds (approximately 2 times a second).	OK ▶	GO to **A2**.
		Not OK ▶	REPLACE Module.
A2			
	• Disconnect the water level sensor connector at separator and short together the two connectors going back to the instrument panel. • Lamp should flash after 10 seconds (approximately 2 times a second).	OK ▶	REPLACE separator.
		Not OK ▶	REPLACE harness.

LAMP FLASHES CONSTANTLY OR LAMP ON CONTINUOUSLY

	TEST STEP	RESULT ▶	ACTION TO TAKE
B1			
	• Disconnect module connector. • Lamp should stop flashing.	OK ▶	GO to **B2**.
		Not OK ▶	SERVICE harness wiring.
B2			
	• Disconnect connector at separator. • Lamp should stop flashing.	OK ▶	REPLACE separator.
		Not OK ▶	GO to **B3**.
B3			
	• Connect harness at separator. Disconnect Pin 2 at water in diesel fuel module connector. • Lamp should stop flashing.	OK ▶	REPLACE harness between module and separator.
		Not OK ▶	REPLACE module.

CV4384-A

Figure 25-6. Troubleshooting chart for water-in-fuel lamp circuit. (Courtesy of Ford Motor Company)

serted into the wiring harness. A 10-amp fuse in the fuse block protects the control module.

As soon as the driver turns the ignition key *on*, the instrument panel glow plug lamp will illuminate. The voltage signal closes the No. 1 relay, allowing full battery voltage to be applied to all of the glow plugs. The No. 2 relay will only close if the engine coolant temperature is lower than 86°F (30°C).

The glow plug lamp will go *out* after approximately 3 seconds to allow starting of the engine. If the engine for some reason is not started by the driver after the glow plug lamp goes out, within three more seconds the control module will open No. 1 relay, cutting current to the glow plugs.

If, however, the engine coolant is below 86°F (30°C), when No. 1 relay opens, No. 2 relay will re-

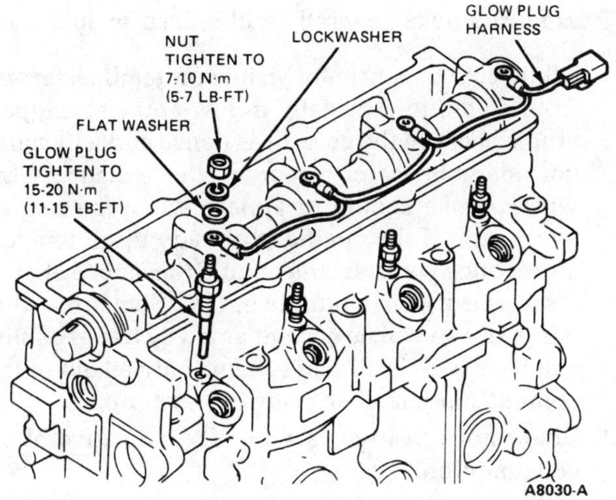

Figure 25-7. Glow plug arrangement. (Courtesy of Ford Motor Company)

main closed to allow a reduced electrical flow to the glow plugs via the resistor until the ignition key is turned *off.*

SPECIAL NOTE: If the ignition key is left *on* without attempting to start the engine, the battery will eventually be drained completely.

Cranking of the engine to start it causes the control module assembly to cycle the No. 1 relay off/on. Because of this feature, the voltage to the glow plugs varies between 12 and 4 volts while cranking the engine with No. 2 relay closed, and between 12 volts and zero volts if No. 2 relay is open.

Upon engine start-up, the alternator output sends a voltage signal to the control module that prevents cycling of the No. 1 relay and allows the after-glow feature to come into play. The after-glow feature is

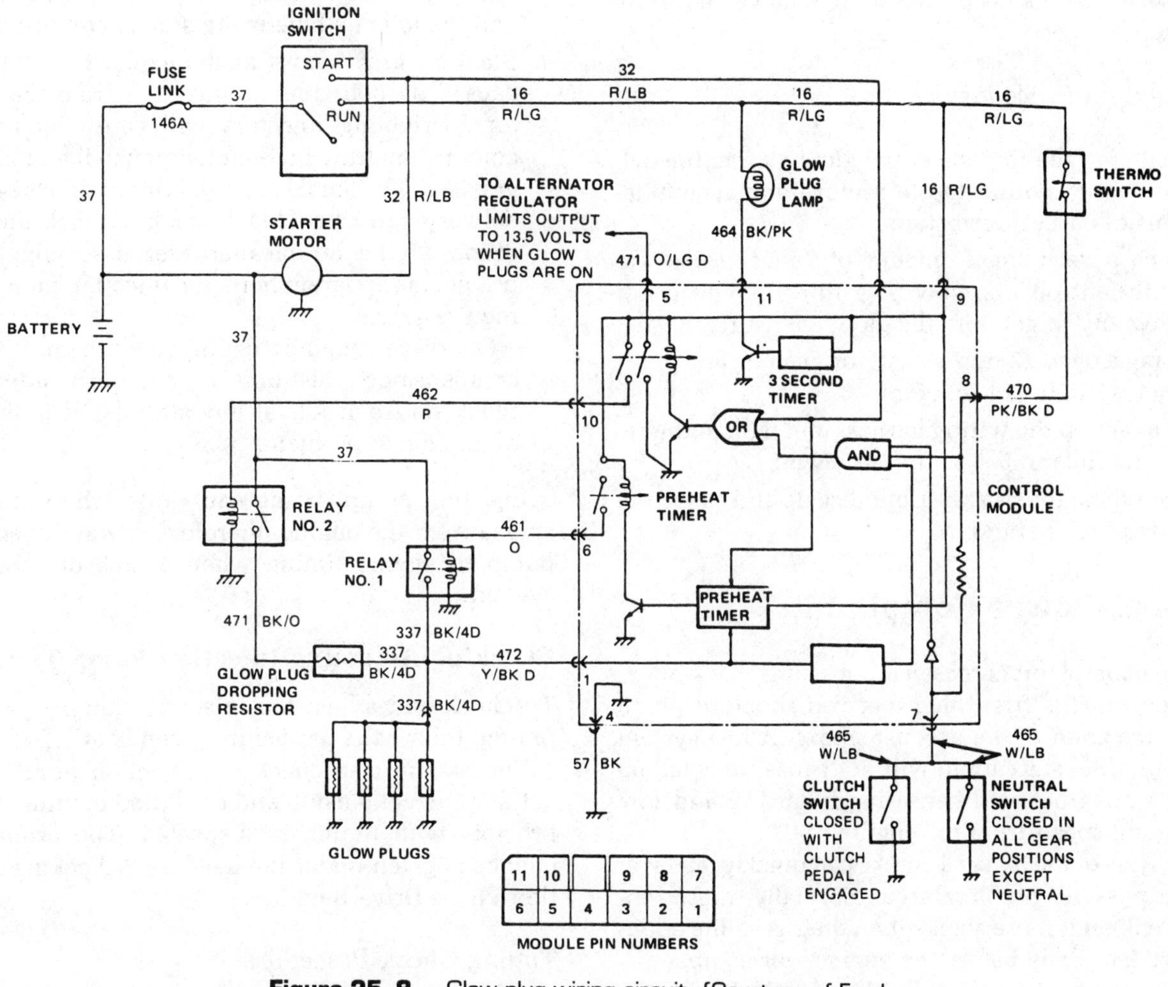

Figure 25-8. Glow plug wiring circuit. (Courtesy of Ford Motor Company)

to allow smoother and faster engine warm-up by allowing the No. 2 relay to supply between 4.2 and 5.3 volts to the glow plugs via the glow plug resistor. This reduced voltage will only exist until either the vehicle is put in gear and driven off (clutch and neutral switches closed) or until the coolant temperature rises above 30°C (86°F) with the vehicle parked and the engine running. The control module will send a signal to open the No. 2 relay, thereby cutting off current to the glow plugs.

Glow Plug Removal and Installation

Refer to Figure 25–7, which illustrates the location of the glow plugs in the cylinder head.

Removal Procedure

1. Always disconnect the battery ground cable located in the luggage compartment.
2. Remove the glow plug wiring harness connecting the glow plugs together.
3. Use a 12-mm deep socket to remove the glow plugs.

Installation Procedure

1. Clean out the threads of the glow plug in the cylinder head with a tap to remove any accumulations of carbon formation.
2. Place a very small amount of "never seize" or equivalent on the glow plug threads but do not allow any to get onto the glow plug surface.
3. Using a deep 12-mm socket, torque the glow plugs to 11–15 lb.ft. (15–20 N·m).
4. Connect up the wiring harness and tighten the retaining nuts to 5–7 lb.ft. (7–10 N·m).
5. Reconnect the battery ground cable and check the glow plug operation.

Loss-of-Power Complaint

When a complaint is received regarding a lack of engine power, the first thing that you should do is to check the color of the exhaust smoke. A fuel system problem (fuel starvation) will not cause unusual exhaust smoke but will cause hard starting and low maximum speed from the engine.

Analysis of the exhaust smoke can quickly lead you to the possible problem area. Normally, diesel engines will emit more smoke than their gasoline counterpart especially before the engine warms up.

Exhaust smoke is generally classified by its density or opacity. Diesel exhaust smoke with concentrations lower than 5% is barely visible to the naked eye.

Exhaust smoke generally will appear as follows:

1. White smoke is usually visible on engine start-up from cold and especially in low ambient temperatures because the cold air is dense and will cause considerable water vapor at the exhaust. This white smoke should disappear when the engine warms up. If the engine is already up to temperature, once you start the engine the smoke should disappear within a minute or two. Sometimes this white smoke will be accompanied by traces of blue smoke, which is oil that is being carried out of the combustion chamber prior to warm up.
2. Blue smoke in itself is generally indicative of oil consumption.
3. A combination of blue/white smoke can occur sometimes when the engine is up to operating temperature if the engine is allowed to idle for long periods because very little fuel is injected at a low engine speed. Little heat is developed within the engine cylinder, so consequently the engine will cool down leading to incomplete combustion and the tell-tale blue/white smoke condition.
4. Black or gray smoke at the exhaust is generally caused by poor fuel atomization from the injectors. Dribbling injectors will cause incomplete combustion (raw fuel—non-atomized) and gray or grayish/black smoke. Air starvation or excess fuel delivery can cause black smoke. Check the condition of the air cleaner/filter assembly. If no problem is apparent here, the injection pump may require service.

   On diesel engines equipped with an altitude compensator, this unit may require attention. Black smoke at idle at low altitudes is indicative of an engine problem.

Injection pump timing can cause white or black smoke from the engine; therefore, always check the pump to engine timing when a lack of power is evident.

Checking/Adjusting Injection Pump Timing

To check and adjust the injection pump to engine timing, follow the procedure given below.

You should also check the injection pump drive belt for correct tension and condition anytime that a problem with timing is suspected. The proper sequence is given under the heading "Checking Injection Pump Drive Belt."

Timing Check Procedure

1. Always make sure that the engine is at its normal operating temperature before performing an in-

jection pump-to-engine timing check. Failure to do this will result in minor variations in the actual timing because of heat expansion of the components. The timing will be close, but not as accurate when set cold as it will be when set at the engine's normal operating temperature. The coolant temperature should be above 176°F (80°C) before the timing check.

2. Isolate the engine starter by disconnecting the battery ground cable. The battery on these vehicles is located in the luggage compartment.
3. Carefully remove the large bolt in the center of the injection pump hydraulic head.
4. Refer to Figures 12–30 and 12–31; install special tool Rotunda 14-0303, which is a static timing gauge adapter that has a plunger that is inserted through the hole in the hydraulic head and actually comes into contact with the injection pump plunger inside the pump.
5. Refer to Figure 25–9, which illustrates the timing cover removed from the transmission housing.
6. Gently rotate the crankshaft clockwise until the flywheel TDC mark is in perfect alignment with the stationary pointer on the engine rear cover plate.
7. Read the dial indicator; it should show 1 ± 0.02 mm ($0.040 \pm 0.0008''$); if the reading is not within these specifications, the injection pump-to-engine timing is incorrect and must be adjusted as follows:
8. Loosen off the injection pump retaining bolt and nuts slightly so that you can nudge the pump into its correct position.
9. Grasp the injection pump with your hands and move the pump towards the engine slightly to "advance" the timing, or pull it away from the engine to "retard" the timing. The degree that the pump must be moved will depend on how far out it is. Move the pump until the specification stated in Step 7 above is obtained.

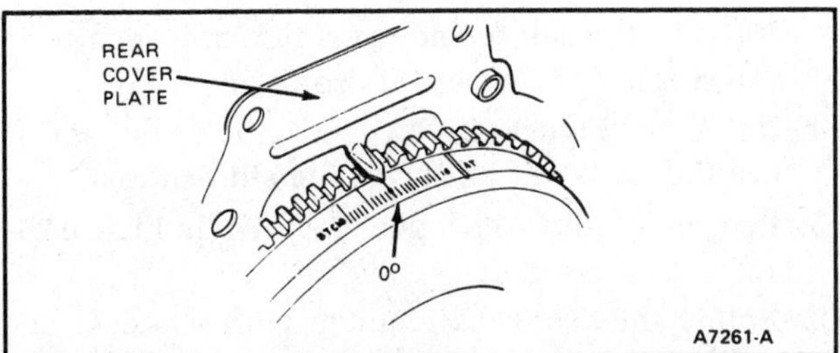

Figure 25–9. Flywheel timing mark. (Courtesy of Ford Motor Company)

10. Tighten up the retaining nuts and bolt to 13–20 lb.ft. (18–27 N·m), then recheck the timing by repeating the procedure stated above in Steps 6 and 7.
11. When the timing has been rechecked, remove the dial gauge and its adapter bracket.
12. Install the center plug and its washer in the hydraulic head and torque it to 10–14.5 lb.ft. (13.5–19.5 N·m).
13. Reconnect the battery ground strap.
14. Start and run the engine; check for fuel leaks at the hydraulic head bolt.
15. Check and adjust the engine idle speed to specs.

Idle Speed Adjustment

To adjust the engine idle speed refer to Figure 25–10 and proceed as follows:

1. Ensure that the engine is up to its normal operating temperature of 176°F (80°C) coolant temperature.
2. Stop the engine and place the transmission in *neutral.*
3. An accurate digital or photoelectric magnetic sensitive-type tachometer such as the Rotunda P/N 99-0001 unit, shown in Figure 24–35, can be used by placing a piece of reflective tape on a clean

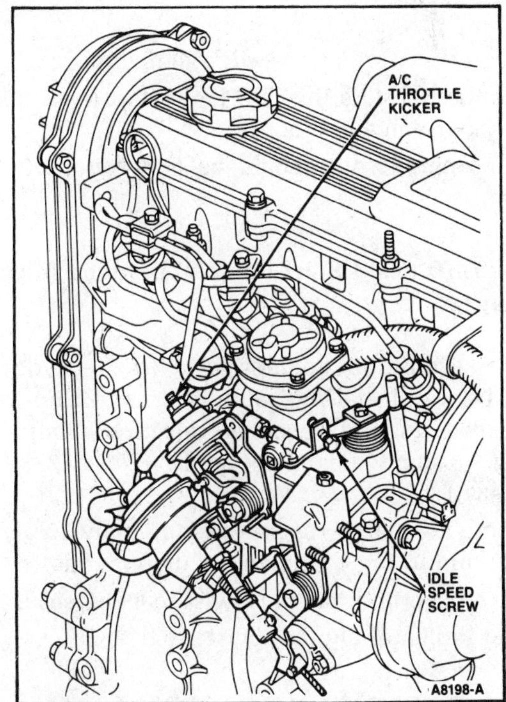

Figure 25–10. Idle speed adjustment setting. (Courtesy of Ford Motor Company)

flywheel surface after removing the timing hole cover.

4. Start and run the engine and monitor the idle speed.
5. The recommended vehicle idle rpm can be found on the VECI (vehicle emissions control information) decal.
6. To adjust the idle speed, loosen off the idle screw lock nut and turn the screw clockwise to increase the rpm, and counterclockwise to decrease the rpm, then tighten the lock nut.

NOTE: On automatic transmission equipped vehicles, the transmission should be in *drive (D)* with the brakes applied to check the "curb idle speed." On standard transmissions, place the gear lever in *neutral.*

7. After an adjustment, place the transmission in *neutral* and rev up the engine. Allow the engine to return to idle, place the automatic transmission equipped vehicle back into *drive (D)* and recheck the idle rpm.
8. On vehicles equipped with air conditioning, switch the A/C on and check the idle rpm. To adjust the idle speed now, refer to Figure 25–10, loosen off the nut on the A/C throttle kicker, and turn the screw until the idle speed is correct.

Check Cold Idle RPM. To check the cold idle rpm proceed as follows:

1. Engine at normal operating temperature.
2. Manually rotate the CSD (cold start device) lever clockwise against its stopper.
3. The engine rpm should be between 1450–1550 rpm.

Check CSD (Cold Start Device) Advance Injection Timing.

1. Turn crankshaft counterclockwise 30–50 degrees BTDC until dial indicator returns to zero as described under "Checking Injection Pump Timing." (Gauge hookup shown in Figure 12–30 and 12–31.)
2. Using a suitable vacuum pump, apply 20" Hg or 500 mm to the CSD vacuum diaphragms.
3. Turn the crankshaft clockwise to 8 degrees BTDC.
4. Dial indicator should register 0.039".

Check CSD (Cold Start Device) Diaphragms.

1. Apply 20 Hg or 138 kPa to the CSD diaphragm.
2. Measure the distance that the CSD lever moves, which should be 1.10" or 28 mm.

Check CSD Water Thermo Valve.

1. Remove the thermo valve from the thermostat housing.
2. Apply 20" Hg or 138 kPa to the valve and check its operation as per the following:

Temperature	*Stage 1*	*Stage 2*
Below 45°F (7°C)	O	O
45–68°F (7–20°C)	O	X
over 68°F (20°C)	X	X

O pass X does not pass

Timing Belt Adjustment

As mentioned earlier in the introduction to the 2.0L diesel engine, an overhead belt-driven camshaft is used with this unit. The camshaft is belt driven from the front of the crankshaft, while the injection pump is belt driven from the rear of the camshaft.

A sprocket is used with a toothed belt to transfer the drive from the crankshaft to the camshaft and then to the injection pump. Since belts are subject to stretch, it may become necessary to periodically adjust these two belts and also inspect their condition. Belt stretch can affect injection pump timing; therefore, both the camshaft and injection pump drive belts may require attention.

Figure 25–11 illustrates the cogged belt arrangement used to drive the camshaft from the crankshaft, while Figure 25–12 illustrates the location of the injection pump drive sprocket belt.

Front Timing Belt Adjustment

Procedure

1. Remove the flywheel timing mark cover.
2. Remove the engine front timing belt upper cover assembly.
3. Remove the belt tension spring that is located in the storage pocket in the front cover.
4. Refer to Figure 25–13 and hook the spring around the lever and over the stud shown.
5. Loosen off the large lockbolt shown in Figure 25–13.
6. Rotate the crankshaft pulley with a socket and ratchet two revolutions until the flywheel TDC mark aligns with the stationary pointer as shown in Figure 25–9 (timing check).

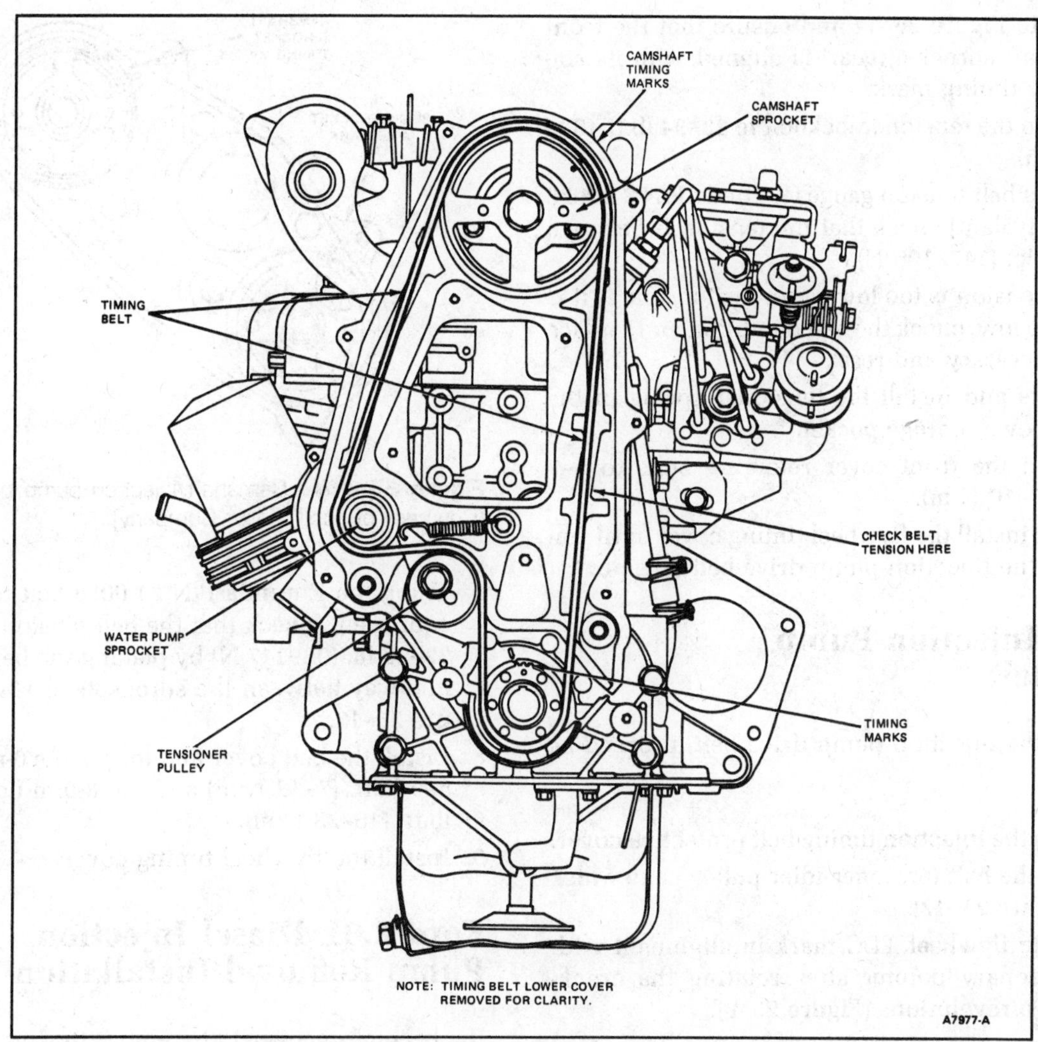

Figure 25-11. Crankshaft-to-camshaft timing belt arrangement 2.0L Ford diesel engine. [Courtesy of Ford Motor Company]

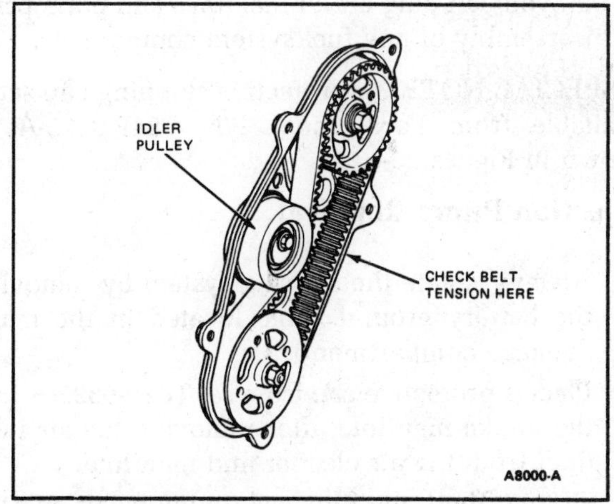

Figure 25-12. Injection pump drive belt and tensioner. [Courtesy of Ford Motor Company]

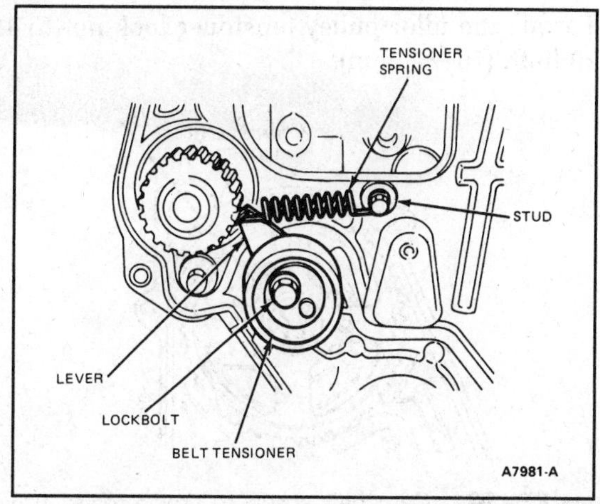

Figure 25-13. Front belt tensioner spring installation. [Courtesy of Ford Motor Company]

7. Refer to Figure 25–14 and ensure that the front camshaft sprocket (gear) is aligned with its stationary timing mark.
8. Tighten the tensioner lockbolt to 23–34 lb.ft. (32–47 N·m).
9. Using a belt tension gauge (Rotunda P/N 21-0028 or equivalent) check that the tension is between 33–44 lb. (147–196 N).
10. If the tension is too low, reset the tension; if it is still too low, check the belt condition and replace it if necessary and reset.
11. Remove and install the tensioner spring in the front cover storage pocket.
12. Tighten the front cover retaining bolts to 5–7 lb.ft. (7–10 N·m).
13. Do not install the flywheel timing cover until you check the injection pump drive belt tension.

Adjust Injection Pump Drive Belt

To adjust the injection pump drive belt, proceed as follows:

1. Remove the injection timing belt protective cover.
2. Loosen the belt tensioner idler pulley center lock nut (Figure 25–12).
3. Place the flywheel TDC mark in alignment with the stationary pointer after rotating the crankshaft two revolutions (Figure 25–9).
4. Make certain that both the camshaft and injection pump sprockets (gears) are properly aligned with their stationary timing marks (Figure 25-15).
5. Loosen or tighten the idler pulley tensioner with the aid of a large screwdriver as shown in Figure 25–12.
6. Torque the idler pulley tensioner lock nut to 15–20 lb.ft. (20–27 N·m).

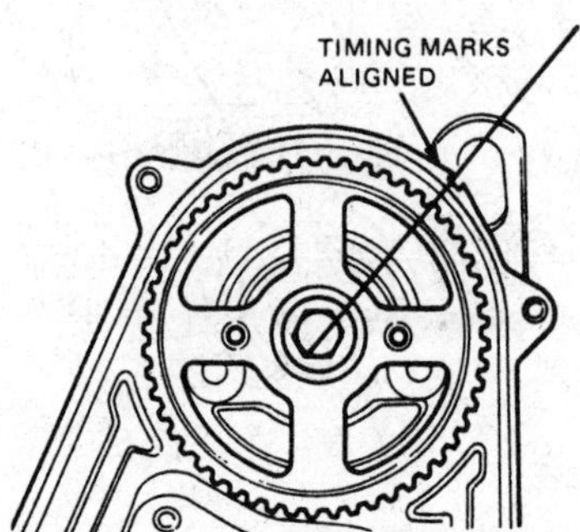

Figure 25–14. Camshaft sprocket timing mark. (Courtesy of Ford Motor Company)

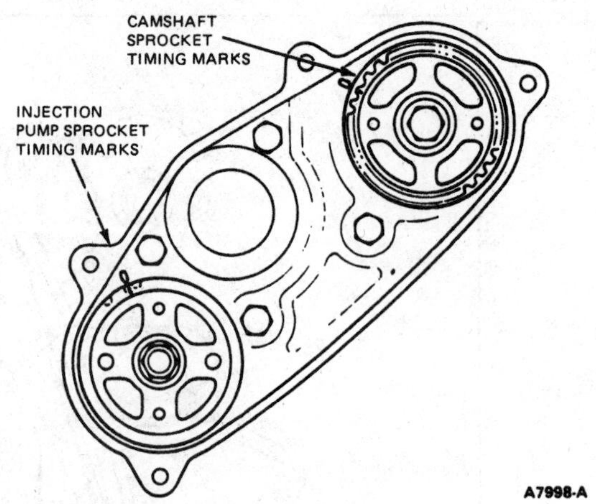

Figure 25–15. Camshaft/injection pump timing marks. (Courtesy of Ford Motor Company)

7. Using the Rotunda P/N 21-0028 belt tensioner or equivalent, check that the belt tension is between 22–23 lb. (98–147 N) by placing the belt tensioner mid-way between the sprockets as shown in Figure 25–12.
8. Install the belt cover and torque the 6-mm bolts to 5–9 lb.ft. (7–12 N·m) and the 8-mm bolt to 12–16 lb.ft. (16–23 N·m).
9. Install the flywheel timing cover.

Ford 2.0L Diesel Injection Pump Removal/Installation

If it becomes necessary to remove the injection pump from the engine for repair or during a major engine overhaul, care must be exercised to ensure that no dirt is allowed to enter any injection components.

To avoid this situation, always have an adequate supply of male/female plastic shipping caps on hand so that you can plug off all fuel lines and ports prior to disassembly of any fuel system components.

SPECIAL NOTE: A protective shipping cap set is available from Ford under P/N T84P-9395-A as shown in Figure 25–16.

Injection Pump Removal

1. Always isolate the starting system by removing the battery ground cable located in the trunk (luggage compartment).
2. Place a protective cap from set T84P-9395-A into the intake manifold after removing the air inlet duct from the air cleaner and manifold.
3. Take off both the flywheel timing cover and the injection pump belt cover.

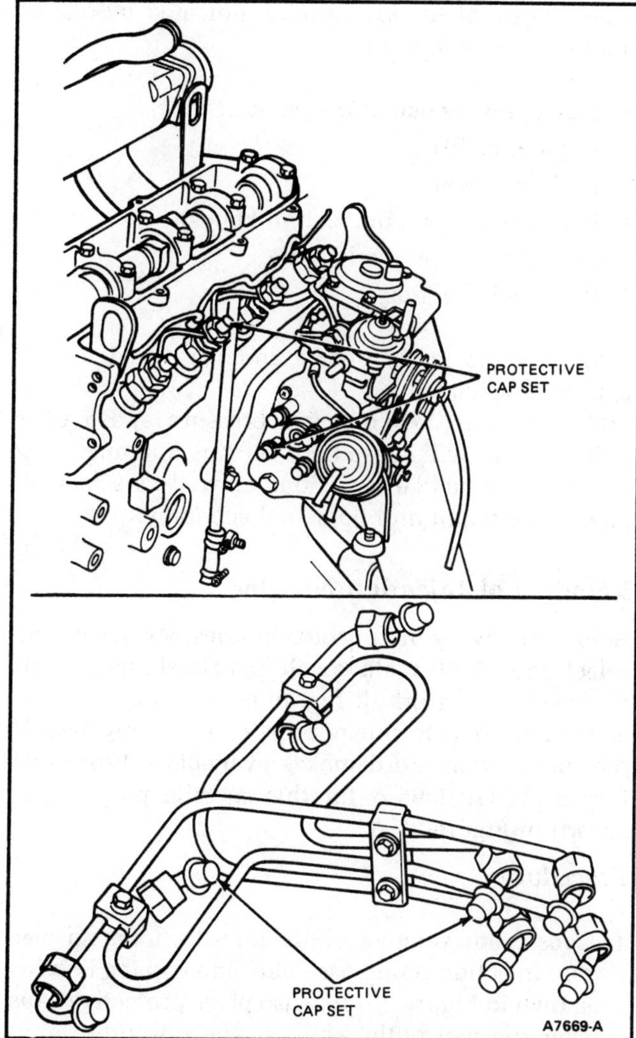

Figure 25-16. a. Protective caps over injectors and injection pump fuel studs. b. Protective caps over injector fuel lines. (Courtesy of Ford Motor Company)

4. Remove the injection pump belt by loosening off the idler pulley bolt and rotating the idler hub with a screwdriver to relieve the belt tension.
5. Disconnect the throttle cable and speed control cable if so equipped.
6. Remove the vacuum hoses from the altitude compensator and cold start diaphragm.
7. Unhook the wire from the fuel cutoff solenoid.
8. Disconnect all fuel hoses and injector high pressure lines from the injection pump.
9. Be sure to cap all lines and fittings as suggested earlier with plastic shipping caps from kit T84P-9395-A or equivalent.
10. Refer to Figure 25-15 and rotate the injection pump drive sprocket until the timing marks are aligned.
11. Install two M8 × 1.25 bolts in the threaded sprocket holes to prevent rotation of the unit when you are attempting to pull it from its mounting as shown in Figure 25-17. Remove the sprocket retaining nut.
12. Use a gear puller (T77F-4220-B1) and adapter (D80L-625-4) to remove the sprocket.
13. Remove the bolt from the pump housing to front support bracket.
14. Remove the pump rear bracket bolts.

Injection Pump Installation

Installation of the injection pump is basically the reverse of removal. However, ensure that you install and adjust the pump drive belt as explained earlier under "Belt Adjustment."

Check and adjust the injection pump timing as explained earlier also. Bleed all air from the fuel system as described herein, and start and run the engine to check for leaks as well as ensuring that the curb idle speed is set to specifications on the VECI decal.

Fuel Injectors

The fuel injection nozzles are screwed into the cylinder head as shown in Figure 25-18. High pressure steel lines connect the nozzles with the outlet fuel studs on the hydraulic head of the distributor injection pump.

The nozzles are of the single hole (pintle) type employing a needle valve held on its seat with a spring. Fuel pressure developed in the injection pump is sent through the high-pressure lines in engine firing order sequence to the individual injectors.

Normal injector nozzle opening pressure is between 1957-2030 psi (13,500-14,000 kPa). At this point, the fuel pressure acting upon a tapered face of the nozzle needle valve will lift it up against the

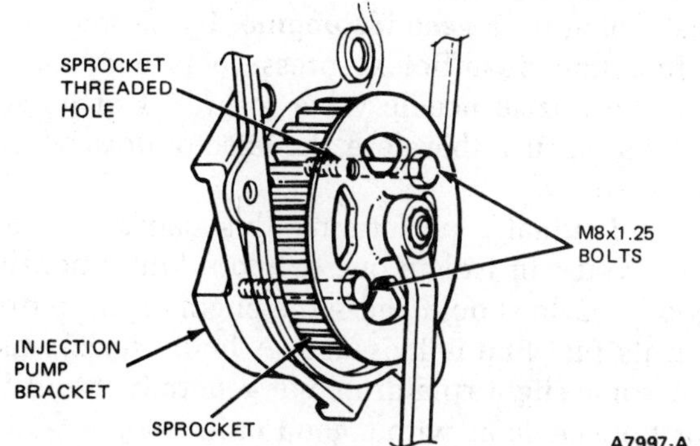

Figure 25-17. Injection pump timing sprocket removal. (Courtesy of Ford Motor Company)

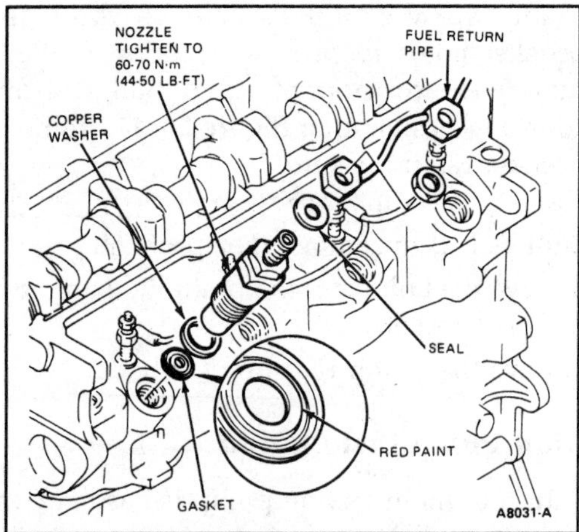

Figure 25–18. Location of injection nozzle in cylinder head. (Courtesy of Ford Motor Company)

force of its return spring. Fuel is then forced through the small hole in the nozzle tip directly into the pre-combustion chamber. The nozzle is located in the pre-combustion chamber as shown in Figure 25–2. In this position, fuel can be sprayed towards the hot glow plug to facilitate rapid vaporization of the fuel on cold starts thereby ensuring quick start-up in cold ambients.

A small percentage of the fuel delivered to each injector will leak past the internal components to act as a lubricant. This fuel is returned from each injector through a return fuel line back to the injection pump.

Troubleshooting Faulty Nozzles

To isolate a suspected faulty nozzle, bring the engine up to normal operating temperature (coolant at least 176°F, or 80°C). Run the engine at an idle rpm and slacken off one of the injector fuel lines at the nozzle end while you place a rag around the fuel line nut to prevent fuel leakage.

This sequence is somewhat similar to shorting out a spark plug on a gasoline engine. By loosening off the fuel line, insufficient pressure is available to cause the nozzle needle valve to lift off of its seat against its spring; therefore, the injector does not fire (injects no fuel).

By individually performing this check on each nozzle, you can isolate one or more faulty nozzles. A good nozzle should cause the engine rpm to drop when its fuel line is loosened. A faulty nozzle may cause some slight rpm drop, but generally it will not be as noticeable as with a good unit.

If this test is not positive at an idle speed, run the engine at the rpm at which the misfire or rough operation occurs and repeat the same test procedure.

Nozzle problems are usually indicated by one of the following conditions:

1. Black/gray smoke at the exhaust pipe
2. Engine misfire
3. Lack of power
4. Unusual knock (combustion knock)
5. Possible engine overheating
6. Poor fuel economy

Remove the faulty nozzle and pop test it as explained in Chapter 29.

Nozzles with low opening pressure and a poor spray pattern will cause hard starting and unacceptable exhaust smoke emissions as well as a lack-of-power complaint and poor fuel economy.

Removal of Injection Nozzles

Before removing the injector nozzles, always have a selection of both male/female plastic shipping caps on hand to plug off all fuel lines after they are disconnected so as to prevent dirt entering any area of the fuel system. Ford makes available a Protective Cap Set T84P-9395-A for this specific purpose, as shown in Figure 25–16.

Procedure

1. Loosen and remove all fuel lines from the nozzles and injection pump assembly and cap the lines as shown in Figure 25–16. Also place protective caps over the fuel outlet studs of the injection pump and over the ends of the injectors as shown in Figure 25–16.
2. Remove the fuel return line from the top of the injectors along with its seals.
3. To remove the nozzles, always use a deep 27-mm socket.
4. Always remove the nozzle gasket and washer from the seat area of the nozzles since these should always be replaced anytime that the nozzle is removed for any purpose. This can be successfully accomplished by using O-ring pick tool T71P-19703-C. If this tool is unavailable, make up a small pointed pick with a small bend on its end to facilitate the nozzle gasket removal from the injector bore. The nozzle gasket is clearly shown in Figure 25–18 (nozzle location).

Installation of Injector Nozzles

Once the nozzles have been cleaned externally, proceed as follows:

1. Refer to Figure 25–18 (nozzle location in cylinder head) and insert a new nozzle gasket into the bore of the cylinder head with the *red* painted surface facing up.
2. Refer to Figure 25–18 and place a new copper washer into the nozzle bore.
3. Screw the nozzles into their bores and hand tighten; using the deep 27-mm socket, torque the injectors to 44–51 lb.ft. (60–70 N·m).
4. Using new seals, install the injector fuel return lines and tighten the nuts to 10 lb.ft. (14 N·m).
5. Install the high pressure fuel lines between the injection pump fuel outlet studs and the nozzles, and torque them to 18–22 lb.ft. (25–29 N·m).
6. Bleed air from the fuel system as explained under "Bleed Fuel System."
7. Start and run the engine until it attains normal operating temperature, and check and correct any fuel leaks.

Fuel System and Engine Pressure Tests

To check the fuel system pressure and other areas of the engine, refer to Figure 25–19, which shows the necessary hookup of test lines and gauges.

When checking for sufficient fuel delivery, disconnect the fuel inlet at the injection pump and install

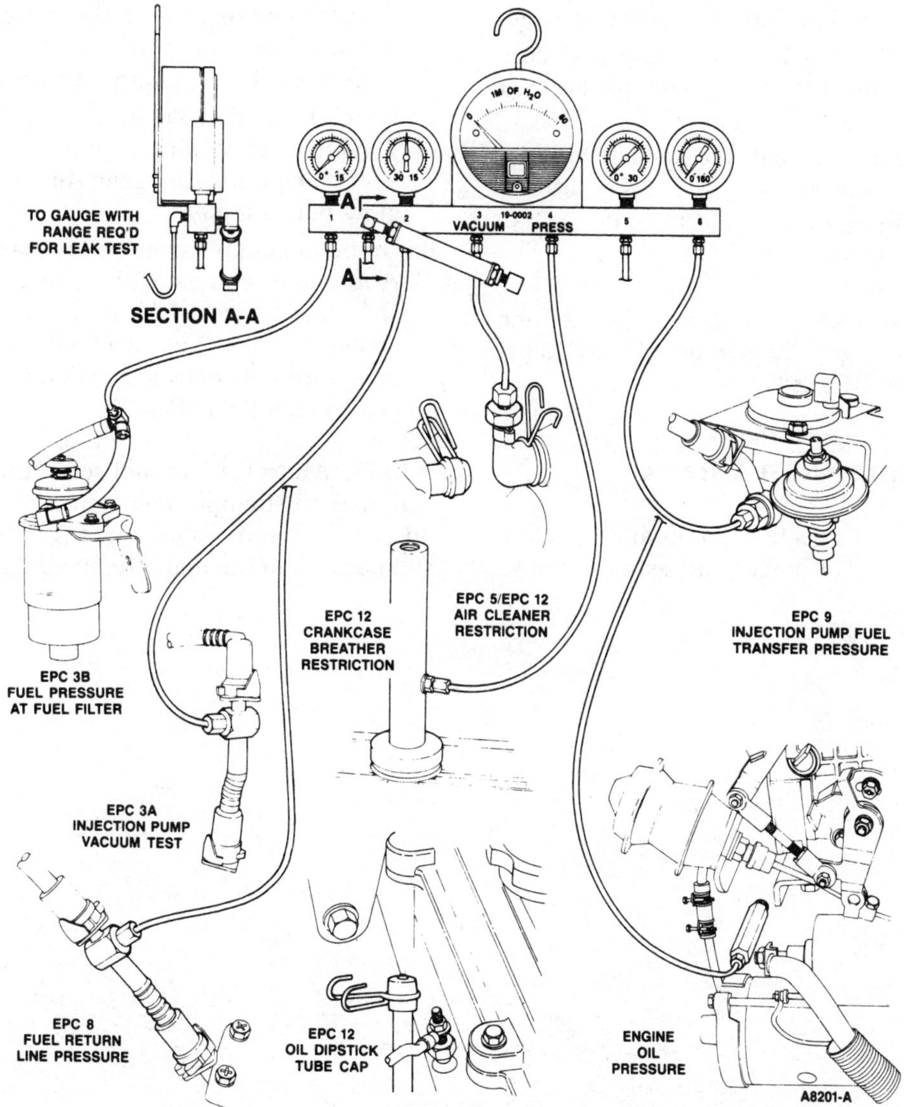

Figure 25–19. Fuel system and engine pressure test kit hookup—Rotunda model 19-0002. (Courtesy of Ford Motor Company)

an adapter tee onto the pump, then reconnect the fuel line onto the installed tee fitting. Connect the other end of the tee fitting to a 30″ Hg gauge or manometer. Run the engine at 4000 rpm no-load and note the vacuum reading, which should not be less than 3″ Hg (20.6 kPa), otherwise, the vane-type transfer pump is at fault or there is a suction leak between the pump and the fuel tank.

When checking the fuel filter pressure drop, connect the lines as shown in Figure 25–19 and run the engine at 4000 rpm. Maximum reading on the gauge should not exceed 6″ Hg. Fuel return line pressure should not exceed 2 psi (14 kPa) at 4000 rpm.

The injection pump transfer pressure can be checked by:

1. Removing the banjo bolt from the fuel return line.
2. Installing adapter 10326 banjo fitting and connecting the adapter line to the 0–160 psi (0–1103 kPa) gauge.
3. Running the engine to 4000 rpm no-load with the transmission in *neutral*.
4. Transfer pump pressure should be between 95–110 psi (655758 kPa).

Fuel Supply Line Check. Disconnect the fuel line at the injection pump and the fuel tank; apply 15 psi to the line and check for leaks.

Engine Compression Check

Often the engine fuel injection pump and injectors are blamed for a lack of power and exhaust smoke complaint when the cause may be low compression in one or more cylinders.

To conduct a compression check on the 2.0L diesel engine, you have to first of all run the engine until it is at normal coolant temperature of 176°F (80°C).

Procedure

1. Stop the engine and remove all four glow plugs.
2. Disconnect the wire from the fuel cutoff solenoid on the injection pump to prevent any fuel delivery when cranking the engine.
3. Also disconnect the glow plug harness (blue/red stripe) from the engine wiring harness to prevent any possibility of an electrical short when the ignition/starter is engaged.
4. Spin the engine over on the starter to establish if, in fact, it is being rotated fast enough (200 rpm minimum). If not, charge the battery.
5. Install Rotunda 5634 glow plug adapter into No. 1 cylinder glow plug hole and connect Rotunda 19-0001 compression gauge or equivalent onto the glow plug adapter.
6. Crank the engine over on the starter about 4 to 6 revolutions and note the gauge reading, which should be between 384 psi (2700 kPa) minimum to an average of 427 psi (3000 kPa); maximum allowable variation between cylinders should not exceed 43 psi (300 kPa).

SPECIAL NOTE: Do not squirt oil into the engine cylinders to attempt to obtain a wet reading because oil is non-compressible and engine damage can result because of the high pressures that are produced.

26

Ford Ranger/Bronco II Diesel Fuel System

The 2.2L (2209 cc/135 cu.in.) diesel engine is a 4-cylinder, 4-stroke cycle water-cooled overhead valve design unit that has a compression ratio of 22:1 and develops 59 hp (44 kW) at 4000 rpm and 90 lb.ft. of torque at 2500 rpm. The engine is of Toyo-Kogyo (Mazda) design and manufacture, and is also used in the Mazda 2200 series diesel pickup truck.

The camshaft is supported by bores in the cylinder block and is held in place by a thrust plate. The valve mechanism employs a rocker arm actuated by a solid tappet/lifter operating a pushrod; therefore, periodic valve adjustment through a conventional screw and lock nut in the end of the rocker arm is required.

The cylinder head is a cross-flow design with the intake ports on the left side and the exhaust ports on the right side. A heat-resisting alloy pre-combustion chamber is designed to produce excellent fuel swirl. The design of the pre-combustion chamber is similar to that used with the 2.0L engine offered in the Tempo/Topaz and Escort/Lynx vehicles.

Briefly, the Ranger/Bronco II uses a conventional valve train arrangement whereas the 2.0L Tempo/Topaz, Escort/Lynx models offer an overhead camshaft arrangement.

This engine employs a distributor-type fuel injection pump of Robert Bosch VE design. A detailed explanation of the operation, maintenance and repair of this fuel pump can be found under Chapter 12, Robert Bosch diesel fuel pump.

The fuel systems used with the Ford 2.0L and 2.2L diesel engines are very similar; the major differences are highlighted under the 2.0L engine fuel system introductory comments.

The injection nozzles used with both engines are very similar as well as the glow plug arrangement. Injection pump timing varies slightly between these engines with the 2.0L being timed to TDC while the 2.2L is timed at 2 degrees ATDC.

The service procedures for both engines are very similar; therefore, only the apparent differences between these will be discussed here. No fuel injector test sequence is given for the 2.2L because the information already given for the 2.0L engine can be followed.

Basic Fuel System Arrangement of the 2.2L Ranger/Bronco II Diesel

Figure 26–1 illustrates the basic fuel system arrangement of the 2.2L which differs from that of the 2.0L engine in that both a primary fuel filter and a secondary fuel filter are employed with the 2.2L engine.

The primary filter is located between the fuel tank and the secondary filter as shown. The primary filter or sedimenter, as it is often called, is used to remove water in the fuel, which would of course cause serious damage to the injection pump and nozzles. The sedimenter is equipped with a water-in-fuel sensor similar to that used on the 2.0L diesel engine and will activate a dash-mounted warning light to warn the driver to drain the sedimenter unit as soon as possible.

Water volume trapped in the fuel sedimenter depends upon fuel storage conditions/quality and the actual amount of vehicle usage. Water should normally be drained from the sedimenter every 5000 miles (8046 km), or anytime that the water-in-fuel warning lamp goes on.

The water-in-fuel sensing unit will activate the wiring circuit to the dash light anytime that 0.5 quart (0.57 liter) of water has collected in the sedimenter bowl.

Draining Water from Sedimenter. To drain water from the sedimenter, refer to Figure 26–2, which il-

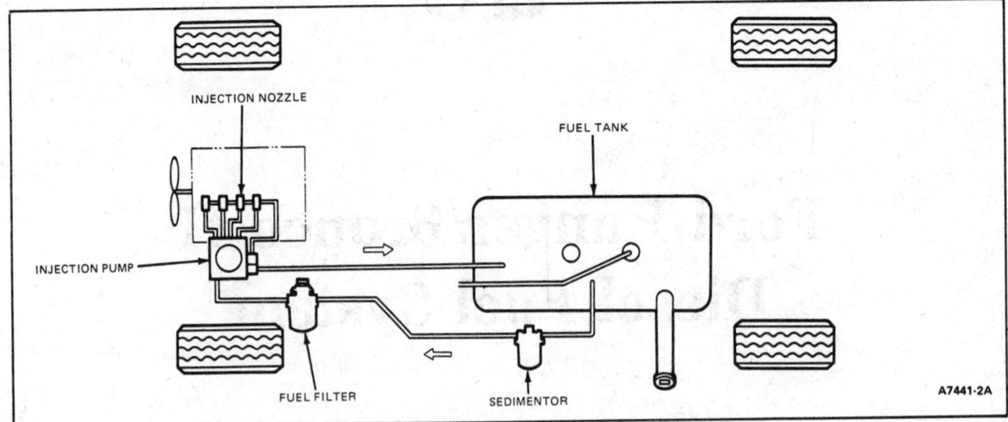

Figure 26–1. 2.2L Diesel fuel system arrangement. (Courtesy of Ford Motor Company)

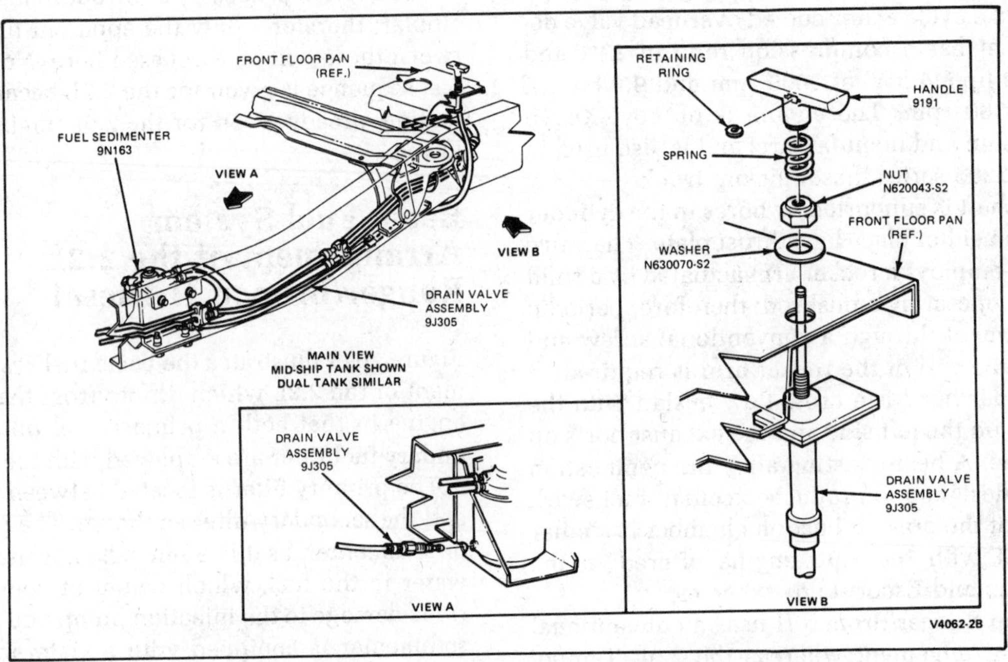

Figure 26–2. Ranger/Bronco II fuel sedimenter drain system. (Courtesy of Ford Motor Company)

lustrates a T-handle located on the vehicle cab floor behind the driver's seat. Pull up on this T-handle and turn the ignition key to the *on* position so that the warning lamp (water-in-fuel) glows; hold the T-handle up until the warning lamp goes out, which usually takes about 45 seconds.

NOTE: Be sure to place a drain tray under the sedimenter on the driver's side where the sedimenter is located inside the frame rail to catch drained water/fuel.

After 45 seconds or after the water-in-fuel light goes out, release the T-handle and crawl under the vehicle to see that the fluid has stopped draining from the sedimenter.

The other major difference between the 2.0L and the 2.2L engines is that the 2.0L engine is transversely mounted while the 2.2L engine is mounted in the conventional manner, that is, longitudinally.

Fuel leaving the sedimenter is passed to the secondary fuel filter before it passes on into the injection pump.

The fuel transfer (feed) pump is enclosed within the injection pump in the same manner as that used in the 2.0L engine system. The feed pump is a vane-type design that is driven from the injection pump drive shaft.

Dual Fuel Tanks

On vehicles equipped with dual fuel tanks, a six-port motorized fuel valve is energized by a small electric motor in order to open and close the selected valve ports for each fuel tank when desired by the driver. The system is arranged so that when the driver switches tanks with the dash-mounted selector valve, the supply and return fuel ports for both tanks are opened and closed at the same time.

Figure 26–3 illustrates the wiring circuitry for the dual fuel tank arrangement. Terminals 1 and 2 in the wiring circuit diagram are used in two operating modes as follows:

1. Positive battery voltage (12 V) is applied to terminal 2 and ground to terminal 1 when the valve is shifted to the front position.
2. Positive battery voltage (12 V) is applied to terminal 1 and ground to terminal 2 when the valve is in the rear mode; this is accomplished by an internal switch that opens the circuit and stops the motor when the valve reaches the front mode position; returning the valve to the rear mode is accomplished by reversing power.

Terminals 3 and 5 are connected to the front and rear fuel tank sending units, while terminal 4 is connected to the fuel gauge.

Injection Pump Layout

The fuel injection pump used with the 2.2L diesel engine uses an altitude/boost compensator to control exhaust emissions. This device is mounted on the top of the injection pump as shown at the bottom right-hand side of Figure 12–17 and identified as an LDA (manifold pressure compensator).

When an altitude/boost compensator is used, the injection pump overflow valve is mounted immediately below the compensator. In Figure 12–17, it would be the line below and to the right of the compensator. A description of the operation of the altitude/boost compensator can be found in Chapter 12, Robert Bosch diesel fuel pump, VE section, and in Chapter 27.

Fuel Filter Service

Filter service is similar to that on the 2.0L engine.

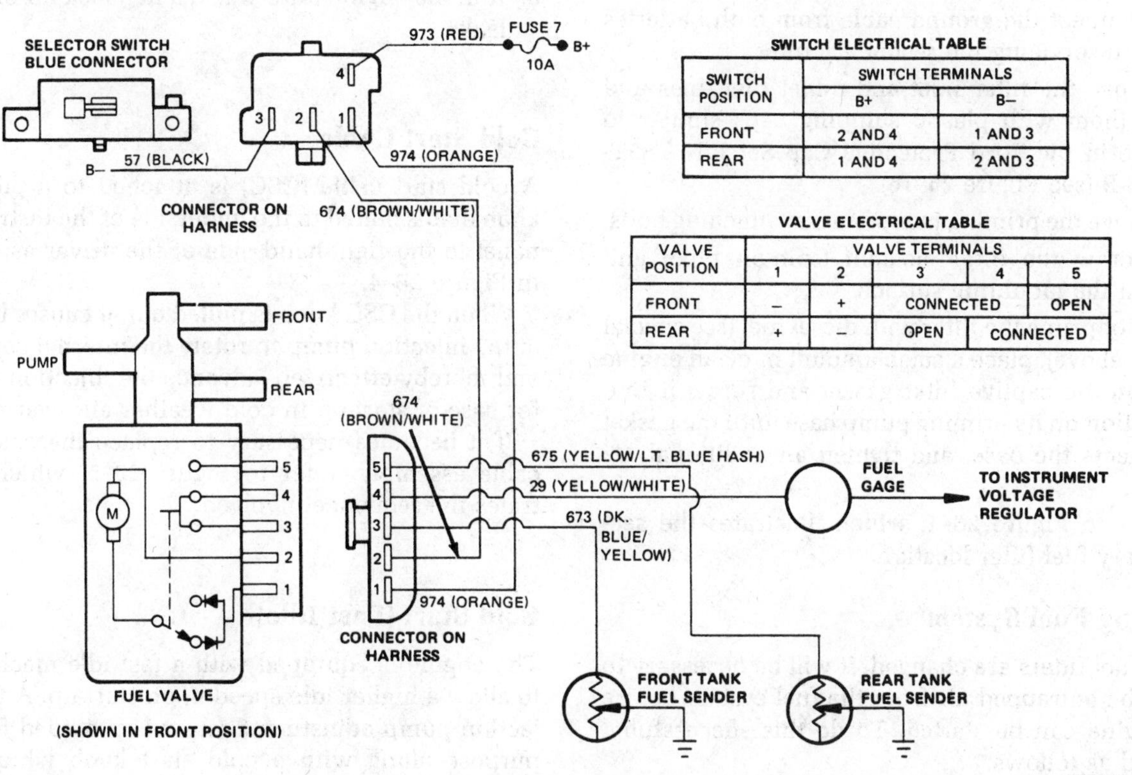

SWITCH ELECTRICAL TABLE

SWITCH POSITION	SWITCH TERMINALS B+	B–
FRONT	2 AND 4	1 AND 3
REAR	1 AND 4	2 AND 3

VALVE ELECTRICAL TABLE

VALVE POSITION	VALVE TERMINALS 1	2	3	4	5
FRONT	–	+	CONNECTED		OPEN
REAR	+	–	OPEN	CONNECTED	

Figure 26–3. Dual fuel tank wiring circuitry. (Courtesy of Ford Motor Company)

Procedure

1. Replace the primary filter (sedimenter) (see Figure 25–4 and 25–5, Ford 2.0L Diesel) by rotating the body counterclockwise with a filter wrench or your hands, and when removed, discard it since it is a throwaway type unit.
2. Clean the filter mounting surface.

   SPECIAL NOTE: Do not fill the primary filter with diesel fuel to assist in priming it (unless the diesel fuel is prefiltered) because some of the fuel will not pass through the filter element. Fuel in the center passage goes directly to the secondary filter where any water or dirt can damage the secondary element.
3. Place (coat) a small amount of clean engine oil on the captive spin-on filter assembly gasket and screw it into position. Once the filter makes contact with the gasket surface, tighten it an additional 1/2 turn.
4. Priming of the system can be done by using the priming pump located at the secondary filter assembly.

Secondary Filter Replacement.

1. Disconnect the ground cable from both batteries before servicing the secondary filter.
2. Remove the filter inlet and outlet fuel lines and cap them with plastic shipping caps similar to those in the Ford Protective Cap Set P/N T83T-9395-B (see Figure 25–16).
3. Remove the priming pump bracket attaching bolts.
4. Unscrew the filter element (spin-on type) and clean the mounting surface.
5. Do not prime the filter with diesel fuel (see special note above); place a small amount of clean engine oil on the captive filter gasket and screw it into position on its priming pump base until the gasket contacts the base, and tighten an additional 1/2 turn.
6. Refer to Figure 26–1, which illustrates the secondary fuel filter location.

Bleeding Fuel System

When fuel filters are changed, it will be necessary to bleed the entrapped air from the fuel system before the engine can be started. To do this successfully, proceed as follows:

1. Loosen off the filter air vent plug (secondary filter, not the sedimenter).
2. Pull up on the priming pump handle shown in Figures 25–4 and 25–5, and push it down/up to draw fuel from the fuel tank.
3. Keep pumping the priming handle until fuel free of air bubbles vents from the filter plug.
4. Push the handle down and keep it there as you tighten up the filter vent plug.

If the fuel system has run out of fuel, or if major work has been done to any of the fuel system components, proceed as follows:

1. Bleed the system as stated above.
2. Disconnect the fuel injection pump return hose.
3. Operate the filter priming pump until clear fuel flows from the return port on the injection pump.
4. Reconnect the fuel return line to the pump and start and run the engine.
5. If the engine still runs rough, you may have to bleed fuel to the injectors, especially if the fuel lines had been removed for a service operation. To do this, simply loosen off the nut at the injector end of the fuel line and crank the engine until fuel flows from the lines, then tighten them up. See Figure 12–25.
6. Run the engine until warm and check for any fuel leaks.

Cold Start Cable

A cold start cable (CSC) is attached to a pull-type knob that is bolted to the underside of the instrument panel to the right-hand side of the driver as shown in Figure 26–4.

When the CSC knob is pulled out, it causes linkage at the injection pump to rotate the internal cam ring and thereby effectively advance the injection timing for ease of start-up in cold weather and warm-up.

If it becomes necessary to replace the cold start cable assembly, refer to Figure 26–5, which illustrates the sequence involved.

Cold Start (Fast Idle)

The engine is equipped with a fast idle mechanism to allow a higher idle speed upon start-up. A fuel injection pump adjustment screw is provided for this purpose along with a cold start knob which was shown in Figure 26–4. Figure 26–6 illustrates the cold start (fast idle) adjustment screw location on the injection pump assembly.

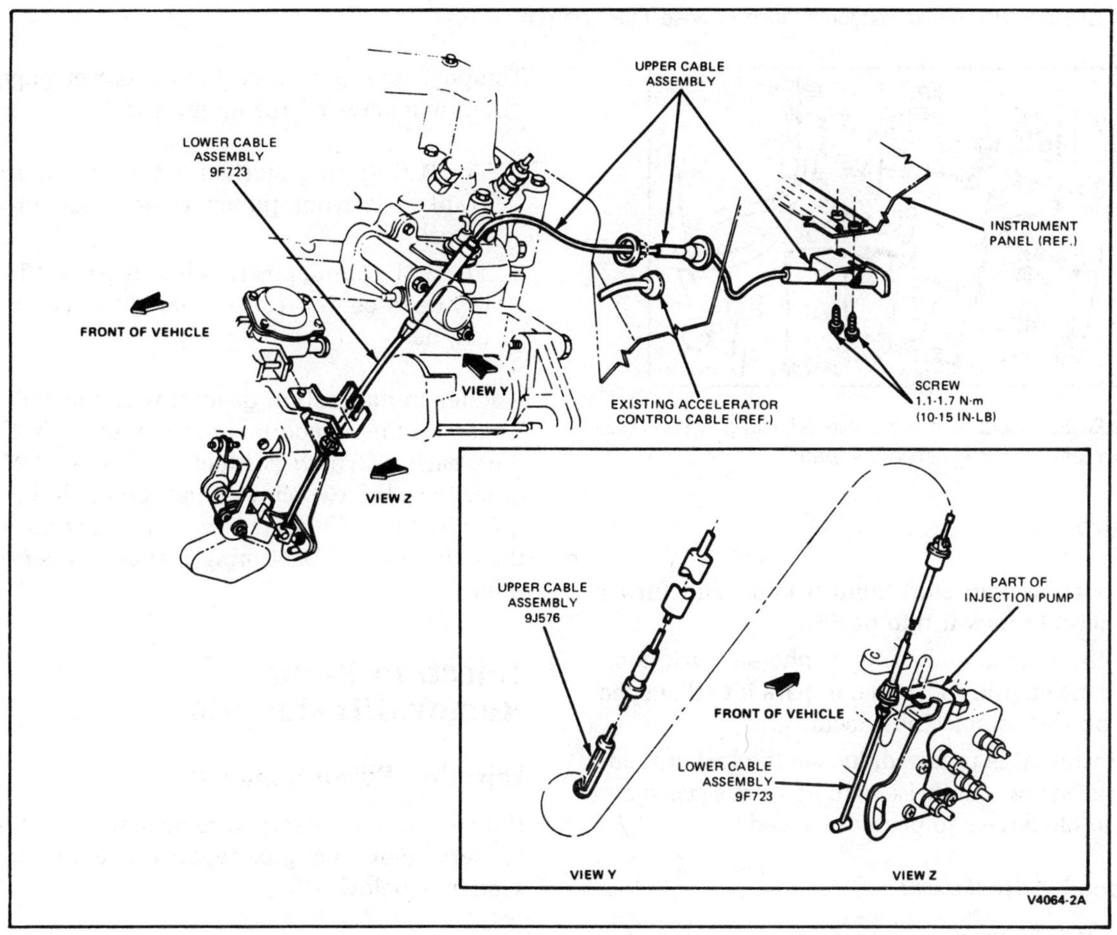

Figure 26-4. Cold start cable linkage arrangement. (Courtesy of Ford Motor Company)

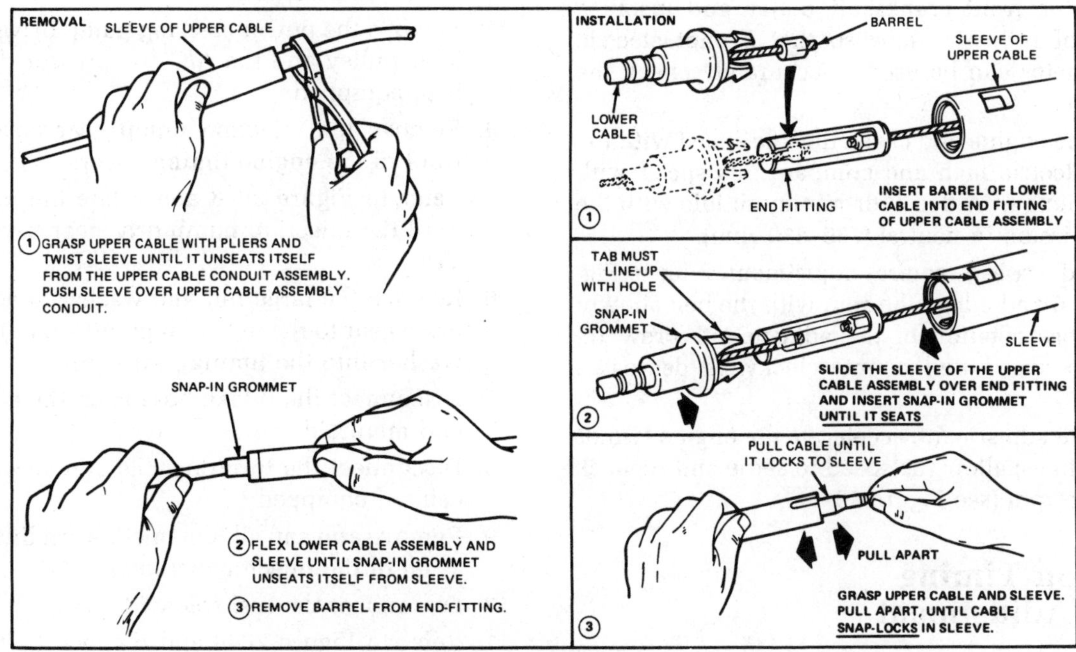

Figure 26-5. Removal/installation—Cold start cable. (Courtesy of Ford Motor Company)

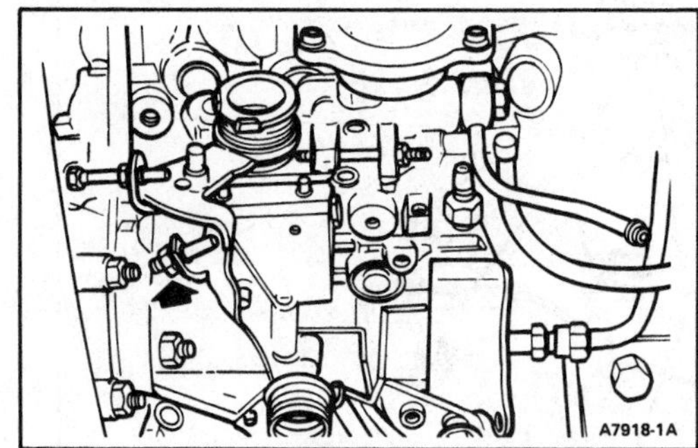

Figure 26-6. Cold start/fast idle adjusting screw location. (Courtesy of Ford Motor Company)

Procedure

1. Pull out the cold start control knob and turn it clockwise to lock it into position.
2. Start the engine and, using a photoelectric tach, check the engine rpm and compare it to that specified on the vehicle EPA decal.
3. If adjustment is required, loosen the lock nut and turn the screw clockwise to increase the speed and counterclockwise to decrease speed.

Idle Speed Adjustment

Proceed as follows:

1. Run the engine until it attains normal operating temperature, then shut it off.
2. Clean the front crankshaft pulley and install a piece of reflective tape so that a photoelectric tachometer can be used to accurately check the idle rpm.
3. Start the engine and check the idle speed with the photoelectric tach and compare this speed with the vehicle EPA decal for minimum idle with the transmission in neutral (780–830 rpm).
4. If the idle speed requires adjustment, refer to Figure 26–6 and adjust the rpm with the bolt shown; clockwise rotation of the adjustment screw increases speed and counterclockwise decreases speed.
5. After an adjustment, accelerate the engine two or three times; allow the speed to settle and recheck the idle rpm (see Figure 26–6).

Injection Timing Check/Adjustment

The injection pump timing check/adjustment is similar to that given for the 2.0L engine used with the Tempo/Topaz and Escort/Lynx diesel engine. The difference between the engines is:

1. The 2.2L timing marks for the engine are on the crankshaft front pulley rather than on the flywheel.
2. The 2.2L timing position is 2 degrees ATDC (after top dead center) rather than TDC as on the 2.0L engine.

Refer to the section dealing with the 2.0L Tempo/Topaz engine/injection pump timing check procedure earlier, which is identical for the 2.2L engine other than the two conditions stated above.

Figure 26–7 illustrates the location of the TDC and the 2 degrees ATDC timing marks on the crankshaft pulley.

Injection Pump Removal/Installation

Injection Pump Removal

If it becomes necessary to remove the injection pump for service or an engine repair procedure, follow the sequence below.

1. Always isolate the starting system by disconnecting the ground cable from both batteries.
2. Drain the cooling system and remove the radiator, fan, and shroud.
3. Remove the power steering pump drive belt and idler pulley and the A/C compressor drive belt, if so equipped.
4. Remove the injection pump gear cover at the front of the engine timing cover.
5. Refer to Figure 26–8 and rotate the crankshaft until the injection pump drivegear keyway is at TDC.
6. Remove the large nut and washer attaching the drive gear to the injection pump (do not drop the washer into the timing gear case).
7. Disconnect the intake hose from the air cleaner and manifold.
8. Disconnect the throttle cable and speed control cable if equipped.
9. Remove and cap all fuel inlet/return lines as well as all of the injector fuel lines.
10. Disconnect the fuel solenoid wire.
11. Refer to Figure 26–9 and remove the two front attaching nuts and the single rear attaching bolt.
12. In order to pull the injection pump drive gear

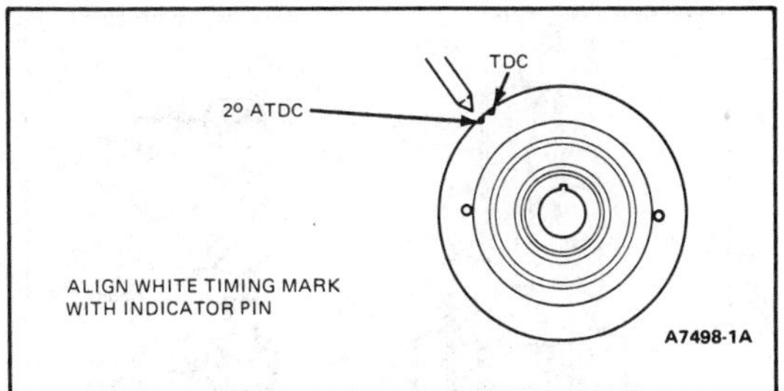

Figure 26-7. 2.2L Ranger/Bronco II crankshaft pulley timing marks. (Courtesy of Ford Motor Company)

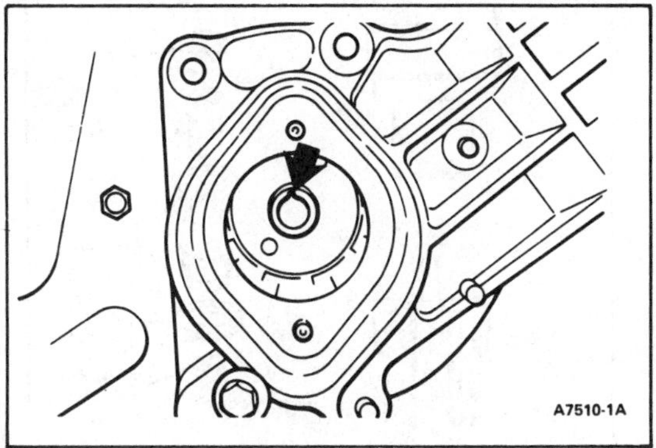

Figure 26-8. Injection pump gear opening and shaft key slot at TDC. (Courtesy of Ford Motor Company)

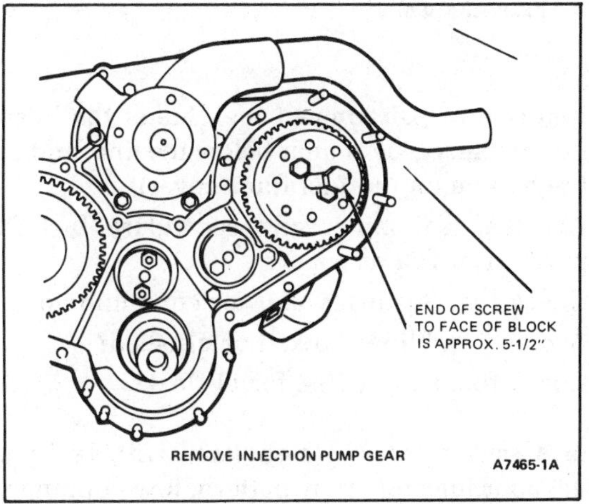

Figure 26-9. Removing injection pump drive gear. (Courtesy of Ford Motor Company)

from the pump shaft, refer to Figure 26–9 and install puller tool P/N T83T-6306-A into the drive gear cover and attach it to the injection pump drive gear. Rotate the puller center screw until the gear is withdrawn from the shaft and pull off the injection pump assembly as a unit. Take care that the key on the injection pump shaft does not drop into the timing cover.

Injection Pump Installation

The installation of the injection pump is basically a reversal of the removal procedure; however, the following information indicates those areas that require attention.

1. Connect up the cold start cable to the injection pump prior to its installation.
2. Refer to Figure 26–8 shown under removal procedure and install the injection pump into its mounting position on the back side of the engine timing cover, ensuring that the key and keyway in the drive gear are aligned.
3. Install the injection pump retaining nuts and rear mounting bolt, and lightly snug them up.
4. Install and torque the pump drive gear to 29–51 lb.ft. (36–69 N·m).
5. Install the injection pump drive gear cover with a new gasket onto the front of the timing gear cover and tighten the bolts to 12–17 lb.ft. (16–24 N·m).
6. Check and adjust the injection pump/engine timing as described earlier in this section.
7. Install the remaining drive belts, hoses, and all fuel lines.
8. Bleed air from the fuel system as stated earlier.
9. Install the radiator, shroud, and fan and fill the coolant system.
10. Connect up both battery ground cables.
11. Start and run the engine and check for any oil or fuel leaks.

Engine Timing Gear Marks

If it becomes necessary at any time to check that the engine itself is correctly timed, refer to Figure 26–10, which shows the respective gear train arrangement as well as the timing gear alignment marks in behind the engine front timing cover.

Injection Nozzles

The injection nozzles used with the 2.2L engine are of the same basic design and operation as those used with the 2.0L diesel engine offered in the Tempo/Topaz, Escort/Lynx models. Both are Bosch DN pintle nozzles mounted in a KC type holder (body), illustrated in Figure 29–4.

The difference between them is that the nozzle diameter of the 2.0L engine is 1.0 mm (0.0394"), while the nozzle diameter of the 2.2L unit is 0.8 mm (0.031"). Injector popping (release) pressure on the

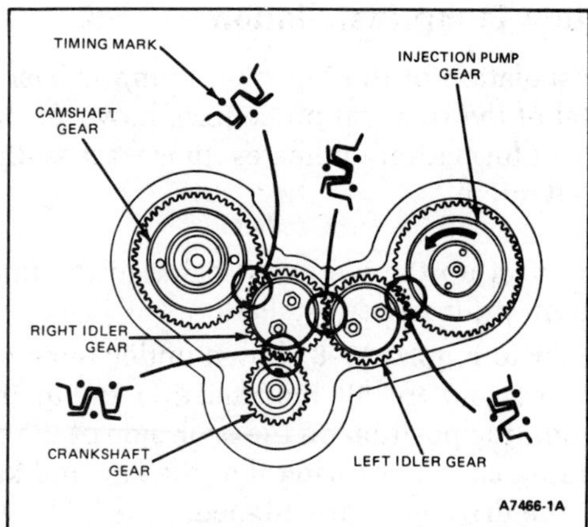

Figure 26–10. Engine timing gear alignment marks. [Courtesy of Ford Motor Company]

2.0L engine is between 1990–2105 psi (14,000–14,800 kPa) on new nozzles and 1849–1990 psi (13,000–14,000 kPa) with used units.

Release pressure on the 2.2L engine nozzles is 1957–2030 psi (13,500–14,000 kPa), new or used.

Other than the above changes, the removal, inspection, cleaning, testing, and installation of these injector nozzles is the same as that described for the 2.0L engine; therefore, follow the information under the 2.0L engine section when service is required on these engines.

Figure 26–11 illustrates the assembled and disassembled injector unit used with the 2.2L diesel engine.

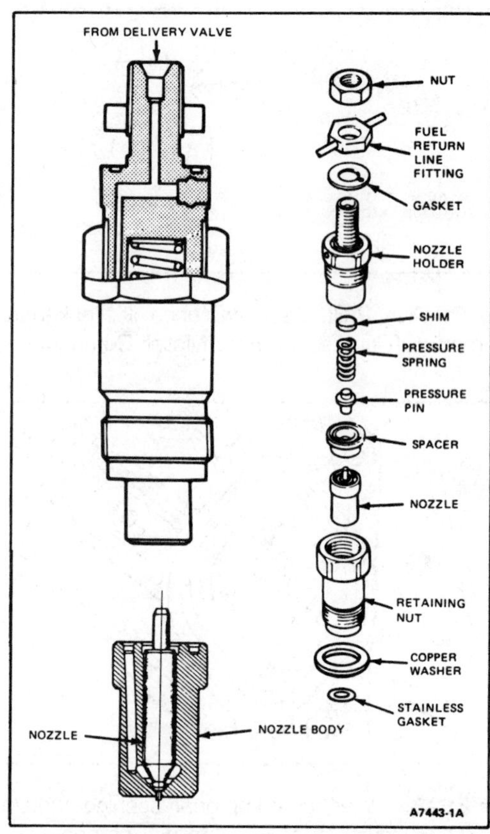

Figure 26–11. 2.2L Engine injection nozzle. [Courtesy of Ford Motor Company]

Engine Fuel System Problems

When it is suspected that the engine fuel system is at fault, follow the same sequence given under that for the 2.0L diesel used with the Tempo/Topaz and Escort/Lynx models.

Typical areas to check are:

1. Is there current to the fuel shutoff solenoid?
2. Is there enough fuel in the tank (dual fuel tanks)? Check fuel return to selector valve by disconnecting the fuel return hose from the engine at the selector valve and placing a container below the line. Turn the ignition key to the *run* position and move the tank selector switch to the problem tank. Start the engine and see if a steady flow of fuel occurs. Also check the return fuel flow to the problem tank by disconnecting the return line and placing it in a container.
3. Is fuel flowing to the transfer pump at the injection pump? (Disconnect fuel line at the injection pump, place a container below the line, and pump the handle on the fuel filter cover).
4. Are the sedimenter and secondary fuel filters plugged or full of water?
5. Check fuel for dirt, water, or contaminates.
6. Are any fuel lines kinked or crushed?
7. Check fuel supply line for air leaks.

Hard starting can be associated with faulty glow plugs, poor injector spray pattern, low popping pressure, low cylinder compression, faulty cold start device cable/linkage.

Generally, the engine will smoke when it is cold if one or more glow plugs are faulty, as well as missing when cold. If this condition clears up when the engine is at operating temperature, it probably is the glow plug circuit.

Faulty injectors can be checked once the engine is at operating temperature by placing a rag around the fuel line nut at the injector end and loosening the nut off about 1/2 a turn with the engine running at an idle rpm. The engine speed should decrease when you do this (similar to shorting out a spark plug) because the cylinder will not receive fuel.

Do this individually to each injector to establish if one or more are at fault. If the miss occurs at a specific engine speed, run the engine up to that speed and repeat the same procedure.

Faulty injectors should be removed and tested.

Engine Compression Test

Often the engine fuel system and injectors are blamed for a poor performance condition when in fact the cause may be low cylinder compression. To perform a compression check on the engine, refer to the 2.0L diesel engine section that covers this procedure. The 2.0L and 2.2L engine compression check is performed with the same special tools. Both engines should show the readings given under the 2.0L test for pressures and allowable variations between cylinders.

Glow Plug Control Circuit

Figure 26–12 illustrates the glow plug circuit for the 2.2L diesel engine, which is similar to that used on the 2.0L Tempo/Topaz and Escort/Lynx models described earlier in detail.

The troubleshooting charts below indicate the test procedure that should be followed to check out the glow plug circuit when a problem is suspected, which is Ford's recommended practice.

Glow Plug Troubleshooting Charts

Glow plug troubleshooting charts are found in Figure 26–13.

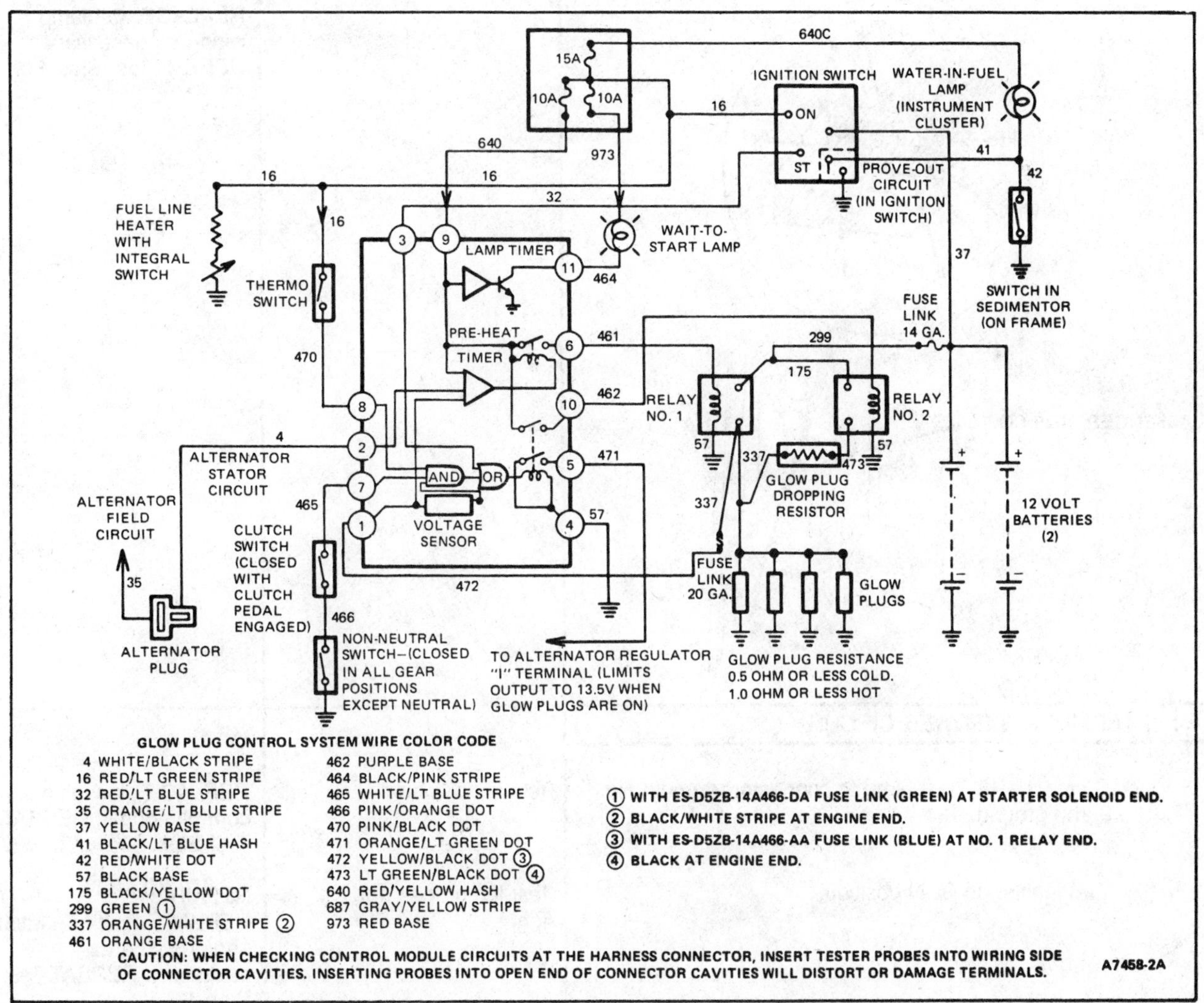

Figure 26–12. 2.2L Diesel engine (glow plug circuit) quick start control system schematic. (Courtesy of Ford Motor Company)

WAIT-TO-START LAMP DIAGNOSTIC GUIDE
(REFER TO QUICK START CONTROL SYSTEM SCHEMATIC)

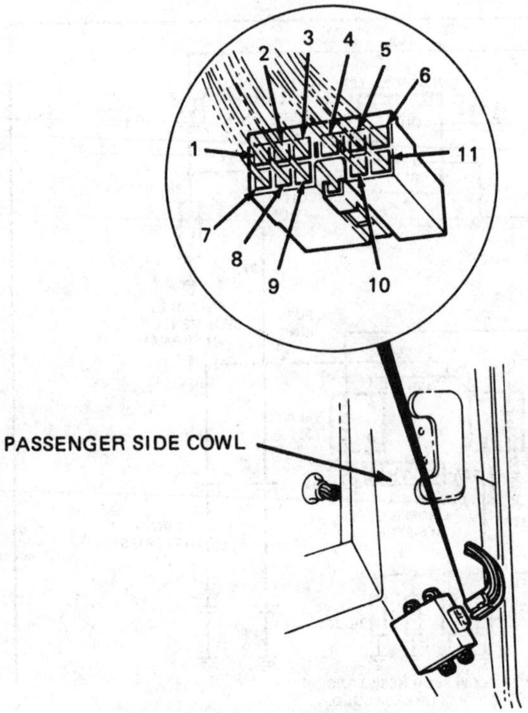

TEST STEP	RESULT ▶	ACTION TO TAKE
E0 WAIT LAMP		
• Turn ignition to RUN position. Wait lamp should stay on for 3 seconds then go out.	OK ▶	GO to Glow Plug Control System Diagnosis Guide in this Section.
	Lamp does not light ▶	GO to **E1**.
	Lamp lights, but does not go out ▶	REPLACE glow plug control module and REPEAT Test Step **E0**.
E1 WAIT LAMP BULB		
• Connect a jumper wire between connector terminal No. 11 and ground.	Wait lamp lights ▶	GO to **E2**.
• Turn ignition to RUN position.	Wait lamp does not light ▶	REPLACE wait lamp bulb or REPAIR or REPLACE wait lamp wiring as necessary. REPEAT Test Step **E0**.
1 2 3 4 5 6 7 8 9 10 11 PASSENGER SIDE COWL		
E2 TERMINAL 9 (POWER CIRCUIT)		
• Connect a 12V test lamp to connector terminal No. 9 and ground.	Test lamp lights ▶	REPLACE glow plug control module. REPEAT Test Step **E0**.
• Turn ignition to RUN position.	Test lamp does not light ▶	REPAIR and/or REPLACE ignition switch and/or wiring as necessary. REPEAT Test Step **E0**.

CA7452-2A

Figure 26-13. Glow plug diagnostic troubleshooting chart. (Courtesy of Ford Motor Company)

GLOW PLUG CONTROL SYSTEM DIAGNOSTIC GUIDE (REFER TO QUICK START CONTROL SYSTEM SCHEMATIC)

	TEST STEP	RESULT ▶	ACTION TO TAKE
F0	CHECK VOLTAGE TO EACH GLOW PLUG		
	• Place transmission gear selector in NEUTRAL.	Voltage OK ▶	REMOVE jumper from coolant thermoswitch. GO to **F13**.
	NOTE: If engine coolant temperature is above 30°C (86°F), jumper connections at coolant thermoswitch.	No Voltage ▶	GO to **F1**.
	• Turn ignition switch to RUN position.	No voltage at 3 or less glow plugs ▶	REPLACE glow plug harness. REPEAT Test Step **F0**.
	• Using a voltmeter, check voltage at each glow plug lead. Minimum of 11 volts at each lead for 6 seconds, then drops to 4.2 to 5.3 volts.	Voltage is OK for 6 seconds, then drops to zero (0) ▶	GO to **F6**.
F1	ENGINE HARNESS TO GLOW PLUG HARNESS		
	• Disconnect glow plug harness from engine harness and glow plugs.	Test lamp lights ▶	RECONNECT glow plug harness. GO to **F2**.
	• Connect a self-powered test lamp between glow plug harness connector and each glow plug terminal.	Test lamp does not light ▶	REPAIR or REPLACE glow plug harness. REPEAT Test Step **F0**.
F2	TERMINAL 9 (POWER CIRCUIT)		
	• Connect a 12 volt test lamp between glow plug control module terminal No. 9 and ground.	Test lamp lights ▶	GO to **F3**.
	• Turn ignition switch to RUN position.	Test lamp does not light ▶	REPAIR and/or REPLACE ignition switch and/or wiring as necessary. REPEAT Test Step **F0**.
F3	TERMINAL 6 (NO. 1 GLOW PLUG RELAY SIGNAL)		
	• Connect a 12 volt test lamp between glow plug control module terminal No. 6 (signal) and ground.	Test lamp lights for 6 seconds ▶	GO to **F4**.
	• Turn ignition switch to RUN position.	Test lamp does not light ▶	REPLACE quick start control unit. REPEAT Test Step **F3**.
F4	NO. 1 GLOW PLUG RELAY WIRING		
	• Connect a 12 volt test lamp between No. 1 glow plug relay signal terminal and ground.	Test lamp lights for 6 seconds ▶	GO to **F5**.
	• Turn ignition to RUN position.	Test lamp does not light ▶	REPAIR or REPLACE wiring between quick start control unit terminal 6 and No. 1 glow plug relay. REPEAT Test Step **F4**.

CA7454-2A

Figure 26-13. [*Continued*]

GLOW PLUG CONTROL SYSTEM DIAGNOSTIC GUIDE (Cont'd.)

TEST STEP		RESULT	▶	ACTION TO TAKE
F5	NO. 1 GLOW PLUG RELAY			
	• Connect a volt meter between No. 1 glow plug relay output terminal (to glow plugs) and ground.	11 volts or more for 6 seconds	▶	GO to **F12**.
	• Turn ignition switch to RUN position.	Less than 11 volts	▶	REPLACE No. 1 glow plug relay. REPEAT Test Step **F5**.
F6	TERMINAL NO. 10 (NO. 2 GLOW PLUG RELAY SIGNAL)			
	• Connect a 12 volt test lamp between glow plug control module terminal No. 10 (signal) and ground.	Test lamp lights	▶	GO to **F8**.
	• Turn ignition switch to RUN position.	Test lamp does not light	▶	GO to **F7**.
F7	CLUTCH SWITCH/NEUTRAL SWITCH			
	• Using a self-powered test lamp, check the functioning of clutch and neutral switch in both open and closed positions.	OK	▶	GO to **F8**.
	• With transmission in gear and clutch pedal released, both switches should be closed.	NOT OK (OK crossed out)	▶	REPLACE malfunctioning clutch or neutral switch. REPEAT Test Step **F7**.
	• With transmission in Neutral and clutch pedal depressed, both switches should be open.			
F8	NO. 2 GLOW PLUG RELAY WIRING			
	• Connect a 12 volt test lamp between No. 2 glow plug relay signal terminal and ground.	Test lamp lights	▶	GO to **F9**.
	• Place transmission gear selector in Neutral. • Turn ignition switch to RUN position.	Test lamp does not light	▶	REPAIR or REPLACE wiring between glow plug control module terminal No. 10 and No. 2 glow plug relay. REPEAT Test Step **F8**.
F9	NO. 2 GLOW PLUG RELAY			
	• Connect a 12 volt test lamp between No. 2 glow plug relay output terminal (to glow plugs) and ground.	Test lamp lights	▶	GO to **F10**.
	• Turn ignition to RUN position.	Test lamp does not light	▶	REPLACE No. 2 glow plug relay. REPEAT Test Step **F9**.

CA7455-2A

Figure 26-13. [*Continued*]

GLOW PLUG CONTROL SYSTEM DIAGNOSTIC GUIDE (Cont'd.)

TEST STEP		RESULT ▶	ACTION TO TAKE
F10	DROPPING RESISTOR WIRING		
	• Disconnect dropping resistor from wiring harness.	Test lamp lights ▶	GO to **F11**.
	• Connect a 12 volt test lamp between the dropping resistor input terminal on wiring harness and ground. • Turn ignition to RUN position.	Test lamp does not light ▶	REPAIR or REPLACE wiring between No. 2 glow plug relay and dropping resistor. REPEAT Test Step **F10**.
F11	DROPPING RESISTOR		
	• Connect an ohmmeter to the connector terminals on the resistor.	(OK) ▶	RECONNECT dropping resistor to wiring harness. GO to **F12**.
	• Set multiply by knob to X1. • Ohmmeter should indicate less than 1 ohm.	(~~OK~~) ▶	REPLACE dropping resistor. REPEAT Test Step **F11**.
F12	GLOW PLUG HARNESS		
	• Connect a 12 volt test lamp between any glow plug terminal and ground.	Test lamp lights ▶	GO to **F0**.
	• Turn ingition to RUN position.	Test lamp does not light ▶	REPAIR or REPLACE wiring from No. 1 glow plug relay to glow plug harness. REPEAT Test Step **F12**.
F13	GLOW PLUGS		
	• Disconnect leads from each glow plug.	Meter indicates less than one ohm ▶	Problem is not in glow plug system. REFER to Engine Performance Diagnostic Guide in this Section.
	• Connect one lead of ohmmeter to glow plug terminal and one lead to a good ground. • Set ohmmeter multiply by knob to X1. • Test each glow plug.	Meter indicates one ohm or more ▶	REPLACE glow plug. REPEAT Test Step **F13**.

CA7456-2A

Figure 26-13. [*Continued*]

27

Ford 2.4L Diesel Engine

Ford has contracted with BMW/Steyr in Austria to have them produce a 2.4L (149 cu.in.) overhead camshaft in-line 6-cylinder turbocharged diesel engine for use in the Lincoln Continental, Mark VII, Thunderbird, and Cougar models starting in the 1984 model year.

Maximum power is quoted as 85 kW (112 bhp) at 4800 rpm with 155 lb.ft. of torque at 2400 rpm thus providing adequate power/performance for mid-size and full-size cars. The engine, which is of indirect injection design and designated the M 105, was developed at BMW AG in West Germany, but it will be produced and marketed by BMW/STEYR Motoren Ges.m.b.H, which is a joint company between BMW AG and Steyr-Diamler-Puch AG.

The engine block is cast iron, and the cylinder head is cast aluminum, which keeps the engine weight to 430 pounds (195 kg). Engine dimensions and specifications are shown in Table 27–1.

The engine has many of the common dimensions found on BMW's existing line of gasoline 6-cylinder engines and is of similar design with the engine being inclined 20 degrees from a vertical mounting position to ensure a lower overall hood line.

A pre-combustion chamber using the swirl concept is used in an off-center position to allow the use of large valve diameters to ensure high power output.

Because the engine is turbocharged, the pistons are oil cooled by nozzles fed from the main oil gallery, and each has a check valve set at 1.5 bar (approximately 22 psi).

The cylinder head follows the BMW cross-flow design, and the pre-combustion chamber, which is positioned on the inlet side of the engine, contains 50% of the compression volume. The glow plugs are mounted off-center at each cylinder, and the injection nozzles are tangential to the pre-combustion chamber.

Table 27–1. *Specifications*

ENGINE	
Type	6-cylinder, in-line, 4-cycle, overhead valve, water-cooled
Bore	3.150 in. (80 mm)
Stroke	3.189 in. (81 mm)
Displacement	149 cu. in. (2442.9 cc)
Compression ratio	22:1
Horsepower	114 (85 kW) at 4800 rpm
Minimum torque	150 lb. ft. (210 N·m) at 2400 rpm
Compression pressure	348 psi (2400 kPa)
Valve clearance (cold engine)	Intake: 0.010 in. (0.3 mm)
	Exhaust: 0.010 in. (0.3 mm)
Cam Timing	
Intake valve opens	6°BTDC
Intake valve closes	34°ABDC
Exhaust valve opens	46°BBDC
Exhaust valve closes	6°ATDC
Intake valve lift	0.374 in. (9.5 mm)
Exhaust valve lift	0.376 in. (9.55 mm)
Weight	433 lb. (196.4 kg) dry

The overhead camshaft is belt driven, and the rocker arms have a mechanical adjustment procedure similar to the BMW gasoline engines. The Robert Bosch VE6/10 distributor-type injection pump is also driven from the same cogged drive belt as the overhead camshaft.

The Robert Bosch throttling pintle-type injection nozzles are mounted in holders with integral heat shields and a nozzle opening pressure of 150 bar (2175 psi approximately).

The turbocharger is of Garrett/AiResearch manufacture (Model T 03) with an integral swing-valve waste gate controlled by boost pressure to open at 11.6 psi, see Chapter 31.

Exhaust gases from cylinders 1 through 3 and 4 through 6 follow separate paths to the turbocharger.

Service procedures for the fuel system are very similar to that for the Volvo D24 engine featured in this textbook because both engines employ the Robert Bosch rotary/distributor VE6 model injection pump and similar injection nozzles.

Maintenance of the nozzles (see Chapter 29) follows the same routine as that for the Ford 2.0L/2.2L diesel engines, the VW Rabbit, Volvo D24, Datsun/Nissan, Toyota, Isuzu/GMC diesel engines.

2.4L Ford/BMW Basic Engine Compartment Layout

The major components and their location underhood are illustrated in Figure 27–1. Due to the fact that this engine is turbocharged and uses EGR (exhaust gas recirculation), it is one of the most heavily electronically monitored diesel engines used by Ford Motor Company.

Basic Fuel System Arrangement

Since the 2.4L diesel engine employs a Robert Bosch-type VE distributor fuel injection pump, all information dealing with the pump operation, maintenance, and adjustment can be found in Chapter 12, the VE pump section.

The fuel system used with the 2.4L diesel engine is very similar in arrangement to that used with both the Ford 2.0L and 2.2L 4-cylinder diesel engines described in this book.

A schematic of the basic fuel system arrangement is shown in Figure 12–17, with the main difference between this illustration and the fuel system used on the 2.4L diesel engine being that an electric fuel transfer pump is located between the fuel filter/water separator and the vane-type transfer pump mounted inside the injection pump housing. This pump lifts fuel from the fuel tank and pulls it through the fuel filter/water separator (fuel conditioner) and delivers it to the vane-type transfer pump.

Since the 2.4L diesel engine is turbocharged, an altitude/boost compensator similar to that (LDA) shown at the bottom right-hand side of Figure 12–17 is used.

The vane-type transfer pump pressurizes the diesel fuel for delivery to the hydraulic head of the injection pump. Vane-type pump transfer pressure is about 36 psi (250 kPa) at an idle rpm and as high as 116 psi (800 kPa) at 4500 rpm.

Contained within the fuel tank is a sending unit

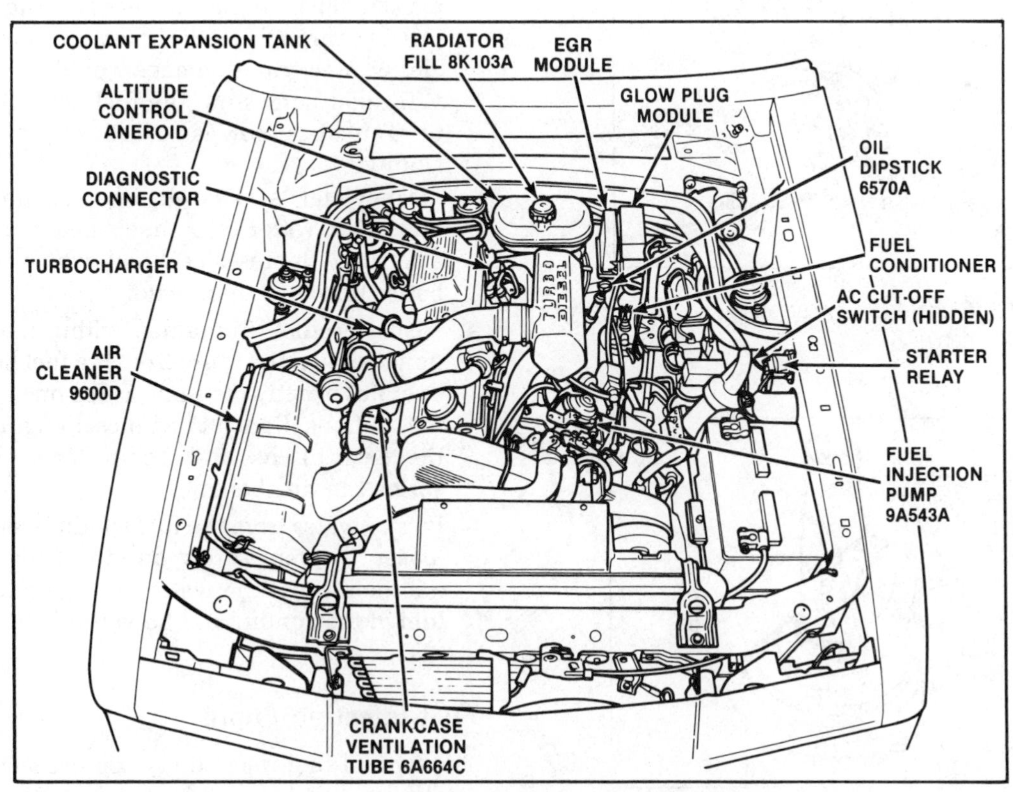

Figure 27–1. 2.4L Turbocharged diesel engine underhood component layout. (Courtesy of Ford Motor Company)

that is a unique Ford designed system; it is illustrated in Figures 27–2 and 27–3. On the bottom of the fuel intake are a series of nibs that are in contact with the base of the fuel tank at all times. These nibs set up a restriction in the fuel flow to create a venturi-type effect on the fuel flow with the result that when fuel passes these nibs it accelerates and therefore creates a pressure drop. The main reason for this design is to allow the diesel fuel intake to actually draw any water in the fuel tank towards it so that this water will pass through to the fuel filter/water separator assembly. Here it will be trapped in the base of the filter where it can then be drained instead of either lying in the tank to create rust or being drawn through to the fuel pump where it would create serious damage to the injection pump and nozzles.

The fuel system used with the 2.4L engine is of *high flow* design in that only 10% of the fuel that enters the injection pump is actually used for injection purposes. The remaining 90% is used for cooling and lubrication purposes for both the injection pump and nozzles; therefore, warm fuel is always returned to the fuel tank—which is helpful in cold weather operation in preventing wax crystal formation in the fuel which would plug the fuel lines.

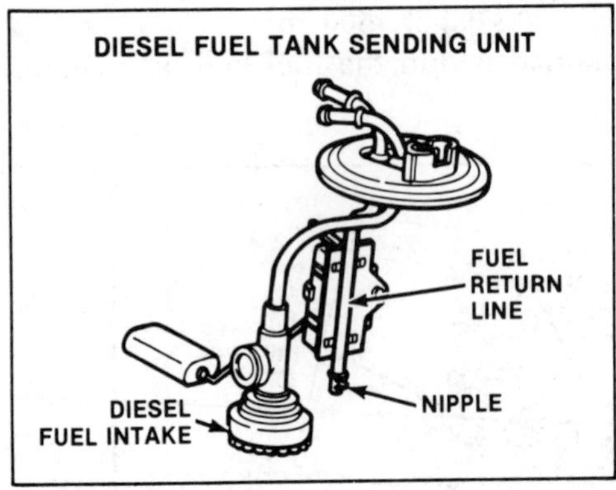

Figure 27–2. Diesel fuel tank sending unit. (Courtesy of Ford Motor Company)

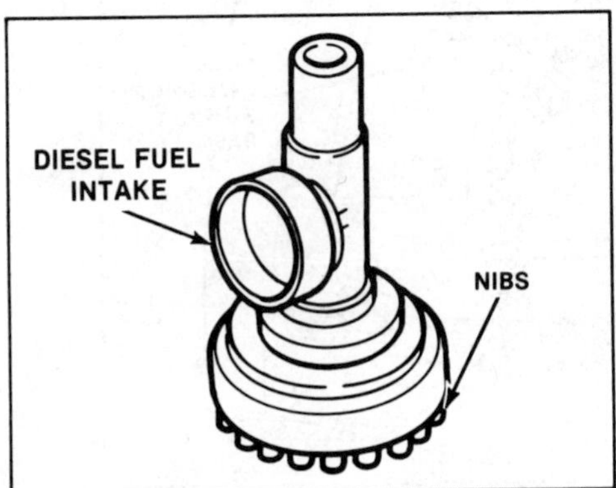

Figure 27–3. Diesel fuel intake. (Courtesy of Ford Motor Company)

Fuel Filter Conditioner

The fuel conditioner used with the 2.4L diesel engine has a capacity of 600 cc (6/10 of a liter) and is generally replaced at 30,000-mile (48279-km) intervals or sooner if necessary. The fuel conditioner performs five different functions, which are:

1. It filters the fuel.
2. It separates any water from the diesel fuel. Since water has a heavier specific gravity than diesel fuel, the water is dumped at the bottom of the conditioner bowl as it passes through the conditioner assembly.
3. Senses the water level in the fuel conditioner. Contained within the base of the fuel conditioner is a water level sensor that provides an electrical ground to the water-in-fuel light on the vehicle instrument panel when approximately 82 cc of water collect in the conditioner base.

   With this amount of water, the WIF light will flash on and off. The maximum quantity of retained water within the conditioner should not exceed 350 cc because there is a danger of water passing through to the injection pump and nozzles where serious damage would result. When the WIF light comes on, the fuel conditioner should be drained as soon as possible to prevent this possibility.

   The water level sensor operates upon the principle that water is a much better conductor of electricity than is diesel fuel, which tends to be a better insulator than water.
4. Heats the fuel. Contained within the fuel conditioner shown in Figure 27–4 is a fuel heater, which operates exactly the same as the one described for the 2.0L 4-cylinder Ford diesel engine used with the Escort/Lynx and Tempo/Topaz models described in this book.
5. Purges water from the fuel conditioner. To do this, loosen the vent screw on top of the conditioner and then press the water drain button at the bottom of the conditioner bowl until clear diesel fuel comes out.

Fuel Injection Pump

The type of fuel injection pump used with the 2.4L 6-cylinder diesel engine is of Robert Bosch manufacture and therefore operates as explained in detail in several chapters of this book, namely Chapter 12,

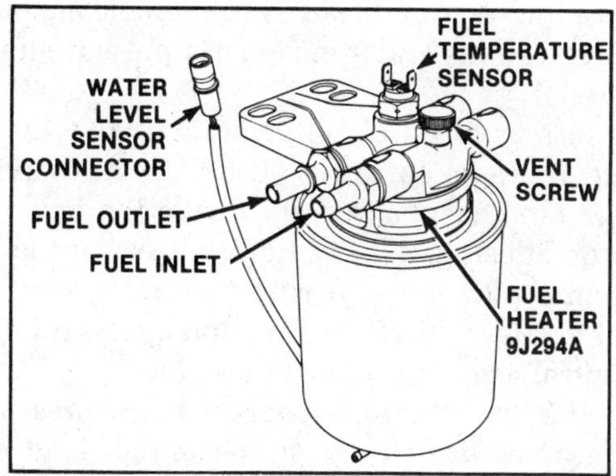

Figure 27-4. Fuel conditioner [9155A]. (Courtesy of Ford Motor Company)

Robert Bosch, as well as in the VW Rabbit, Volvo D24, and both the 2.0L and 2.2L Ford 4-cylinder engine chapters.

Note, however, that the vehicle certified in 49 states uses a VE injection pump that is a mechanically controlled unit whereas the California model employs a VP-20 injection pump. The VP-20 pump operates the same as the VE unit with the exception being that the VP-20 pump is electronically controlled. Identification of these two pumps can be done by looking at the pump ID plate on its side which will indicate the type of pump as shown in Figure 27–5.

The VP-20 injection pump is controlled electronically by a vehicle trunk-mounted fuel flow computer that receives engine operating information data from the engine coolant sensor, the engine speed sensor, and from an instrumented fuel injector in cylinder No. 4 on the engine.

The sensing arrangement is illustrated in Figure 27–6. The trunk-mounted computer, after receiving these voltage signals from each sensor, compares these readings to a loop controller that has been calibrated at manufacture within the computer. A corrected voltage signal is then relayed to a fuel control

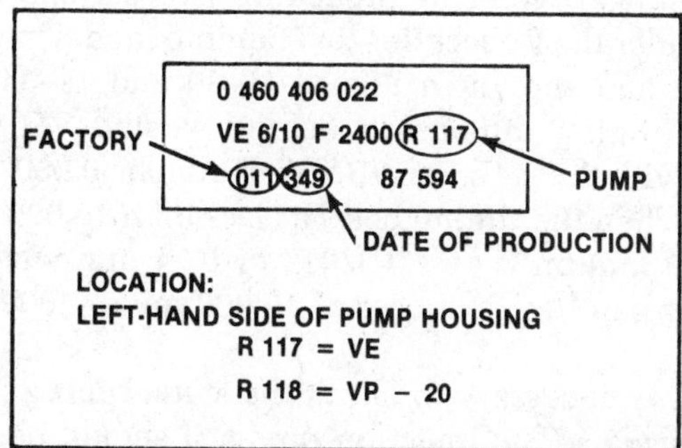

Figure 27-5. Fuel injection pump identification tag. (Courtesy of Ford Motor Company)

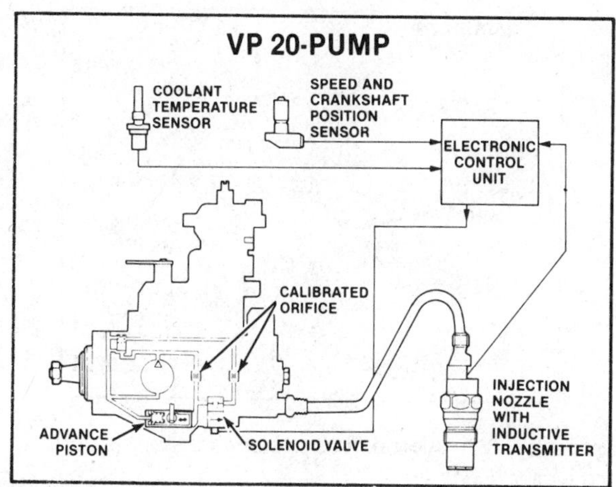

Figure 27-6. VP-20 Injection pump. (Courtesy of Ford Motor Company)

solenoid located in the injection pump, which is cycled opened and closed. This action will vary the fuel pressure from the vane-type transfer pump that acts upon the timing advance piston by providing an additional fuel bleed passage whenever it is open. Since the automatic timing advance of the fuel injection pump is created by the action of the vane-type transfer pump fuel pressure acting upon the advance piston, the constant cycling on/off of the fuel control solenoid thereby effectively controls the fuel injection timing to the engine.

Boost Compensator

Since the 2.4L 6-cylinder diesel engine is turbocharged it is necessary to employ an altitude/boost compensator to control exhaust emissions. One function is to limit the amount of fuel injected in relation to atmospheric pressure, which will decrease with an increase in altitude. The second function is to increase or decrease the amount of fuel injected in relation to the amount of boost pressure provided by the turbocharger. The altitude/boost compensator is illustrated in Figure 27–7.

The altitude/boost compensator is located on top of the injection pump and is used in conjunction with a reference vacuum that is modulated by an altitude control aneroid mounted on the vehicle cowl and shown earlier in Figure 27–1. Connections to this altitude aneroid control are shown in Figure 27–8.

Operation. The basic function of the altitude/boost compensator is to ensure that the amount of fuel injected into the engine combustion chambers is in relation to both the atmospheric and turbocharger boost pressures so as to prevent incomplete combus-

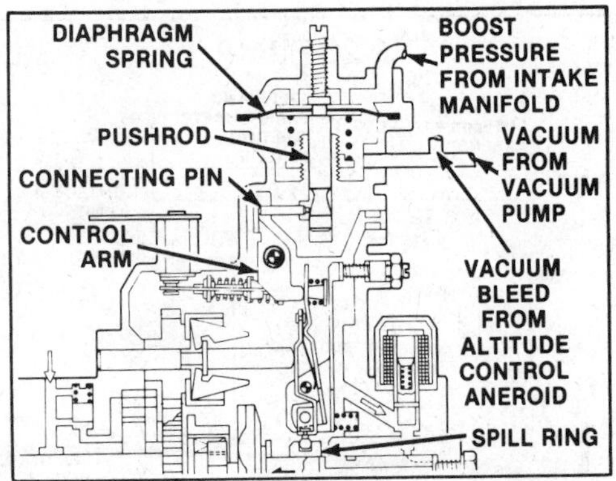

Figure 27–7. Altitude/boost compensator. (Courtesy of Ford Motor Company)

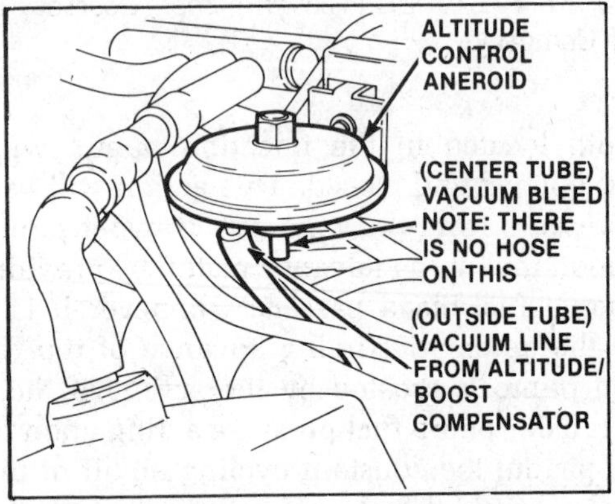

Figure 27–8. Altitude control aneroid. (Courtesy of Ford Motor Company)

tion in the engine cylinder due to low air pressure and flow.

It achieves this simply by the action of the boost compensator spring forcing a tapered pushrod up and down inside the housing, which is attached to a connecting pin and control arm that will move the injection pump spill ring (control sleeve) back and forth to vary the effective pump plunger stroke. This action will automatically increase or decrease the volume of fuel injected in direct relationship to both altitude and turbocharger boost pressure.

Specific operation of the boost compensator is as follows. A vacuum pump that is located under the valve rocker cover is driven from an eccentric lobe on the overhead camshaft. This vacuum pump is required on a diesel engine since the air to the engine is unthrottled (no butterfly valve such as is found on a gasoline engine).

The vacuum created by this pump is used for vehicle power brake operation and also to operate the altitude/boost compensator. Vacuum is admitted to the underside of the diaphragm as shown in Figure 27–7 and is modulated by the action of the vehicle cowl mounted altitude control aneroid, which will increase the vacuum bleed as the vehicle encounters higher elevations and atmospheric pressure tends to decrease.

The loss or bleed off of vacuum allows the force of the diaphragm spring to push both the diaphragm and the tapered pushrod up inside the boost compensator housing. This action will cause the injection pump plunger spill ring (control sleeve) to be moved to a fuel decrease direction by the rotation of the control arm and connecting pin.

Anytime that the turbocharger boost pressure increases as a result of high speed/high load conditions, maximum fuel delivery is required in order to produce maximum horsepower. Therefore the turbo boost pressure is applied to the top of the diaphragm, which will push down on the diaphragm and its spring as well as the tapered pushrod. This action will result in a longer injection period because the injection pump spill ring or control sleeve will be moved to an increased fuel position through the linkage.

Fuel Injection Pump Service

Repair procedures for the VE and VP-20 injection pumps are basically the same and are discussed in detail in Chapter 12 of this book along with a detailed explanation of its operation.

The most common injection pump service that will be performed by the automotive mechanic is a check of the pump-to-engine timing. Always refer to the vehicle emissions label/decal for information on specific timing degrees.

NOTE: Dynamic timing of VE pumps that are certified in 49 states should be 2.5 ± 1 degree BTDC at 750 rpm, while the California certified engine (VP-20) injection pump should be timed to 6 ± 1.5 degrees BTDC at 2000 engine rpm.

Static injection pump timing with the engine stopped requires the use of a special adapter and dial indicator gauge. This procedure is described in detail under the Robert Bosch VE pump model in Chapter 12 and shown in Figure 12–30 and 12–31. The static timing dimension on the 49-state-VE pump used with the 2.4L Ford/BMW should be 0.030″ (0.76 mm) when the timing belt has less than 10,000 miles (16093 km) on it and 0.0291″ or 0.74 mm when the timing belt has in excess of 10,000 miles (16093 km) on it.

If it is necessary to check the static timing on the California VP-20 injection pump, it should be set to 0.0256 ± 0.0015″ (0.65 ± 0.04 mm) with less than 10,000 miles (16093 km) on the timing belt and to 0.0248 ± 0.0015″ (0.63 ± 0.04 mm) when the timing belt has in excess of 10,000 miles (16093 km) on it. Once the injection pump timing has been adjusted, restart the engine and check the timing, then retighten the pump retaining bolts in the following order.

1. Tighten the nut on the inside of the engine at the front flange.
2. Tighten the nut on the outside of the engine at the front flange.
3. Tighten both rear bolts.

Checking Dynamic Timing (Engine Running)

The 2.4L diesel engine is equipped with a diagnostic connector that is located between the engine valve rocker cover and the air intake plenum chamber, which is shown in Figure 27–1.

This diagnostic connector can be used to check both the engine rpm and the dynamic (engine running) injection pump-to-engine timing specification. Connected to this diagnostic connector is Rotunda dynamic timing meter P/N 078-00100, along with a special adapter P/N 078-00117, designed to be used specifically with the 2.4L diesel engine. The timing meter and adapter can be obtained under P/N 078-00116.

By using the diagnostic connector and the timing meter, the use of a pre-combustion chamber luminosity probe such as is used with the 6.9L V8 pickup truck diesel engine is not necessary.

Always set the dynamic timing when the engine is at normal operating temperature. If the pump timing requires adjustment, follow the same procedure as that used when changing static pump timing by loosening off the injection pump retaining bolts and rotating the pump towards the engine in order to advance the timing or away from the engine in order to retard timing.

Engine Speed Settings

Idle and maximum speed settings are discussed in Chapter 12, dealing with Robert Bosch VE model injection pumps. The recommended idle speed setting for the 2.4L engine is 750–800 rpm and the maximum speed setting is 5350 ± 100 rpm.

Adjusting the Injection Pump Operating Lever

Perform this adjustment after completing the idle and maximum speed settings. Ensure that the engine is at normal operating temperature and that the partial load enrichment is cancelled or the cable clamp has been disconnected to provide play between the knurled nut and operating lever (as shown in Figure 27–14).

Procedure

1. Refer to Figure 27–9 and measure distance "A," then push the operating lever against the full-load stop screw and measure distance "B."
2. Subtract measurement "B" obtained in Step 1 above from "A."
3. The difference between these two readings should be noted and compared to the adjusting table shown in Table 27–2.
4. If, for example, dimension "A" was 3.5″ (88.9 mm) and dimension "B" was 1.65″ (41.91 mm), then the difference would be 1.85″ (46.99 mm). This dimension would be known as dimension "Y."
5. Refer to the Adjusting Tables shown in Table 27–2 and determine what dimension "C" should be. In this case "C" would be 2.64."
6. Refer to Figure 27–10 and disconnect the linkage to establish what dimension "C" is. If "C" is not as per the adjusting table, then adjust the linkage to obtain the dimension "C" which in this example is 2.64."
7. Check and adjust the injection pump operating lever setting shown as dimension "X" in the idle position. The correct dimension is 2.68″ (68 mm) and is shown in Figure 27–11.
8. Also check the injection pump operating lever setting (Z) in the wide open throttle (WOT) position as shown in Figure 27–12. This dimension should be 1.14 ± 0.020″ (29 ± 0.5 mm). If dimension "Z" is incorrect, repeat Steps 1 through 5.
9. Check and adjust the throttle cable setting if necessary.

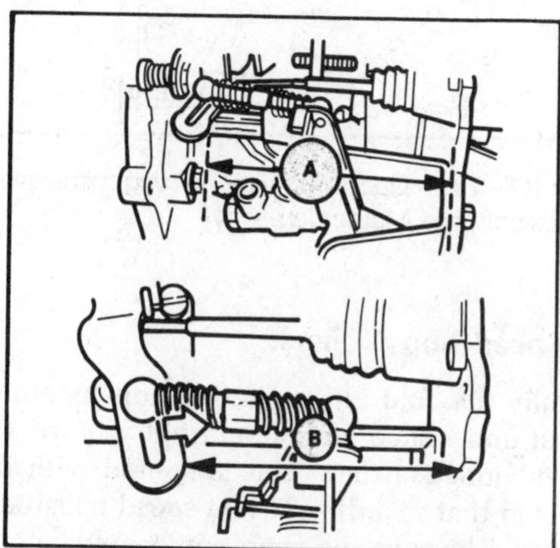

Figure 27–9. Operating lever setting, dimension "A" and "B." (Courtesy of Ford Motor Company)

Adjusting Table: English

Y (in.)	1.61	1.63	1.65	1.67	1.69	1.71	1.73	1.75	1.77	1.79	1.81	1.83
C (in.)	3.07	3.03	3.00	2.94	2.90	2.87	2.83	2.79	2.76	2.73	2.70	2.66
Y (in.)	1.85	1.87	1.89	1.91	1.93	1.95	1.97	1.99	2.01	2.03	2.05	2.07
C (in.)	2.64	2.61	2.58	2.55	2.53	2.50	2.47	2.45	2.42	2.40	2.38	2.35
Y (in.)	2.09	2.11	2.13	2.15	2.17	2.19	2.20					
C (in.)	2.33	2.31	2.29	2.27	2.26	2.24	2.22					

Adjusting Table: Metric

Y (mm)	41	41.5	42	42.5	43	43.5	44	44.5	45	45.5	46	46.5
C (mm)	78.1	77.0	76.0	74.9	73.9	73.0	72.0	71.1	70.3	69.4	68.6	67.8
Y (mm)	47	47.5	48	48.5	49	49.5	50	50.5	51	51.5	52	52.5
C (mm)	67.0	66.3	65.6	64.9	64.2	63.5	62.9	62.3	61.6	61.0	60.5	59.9
Y (mm)	53	53.5	54	54.5	55	55.5	56					
C (mm)	59.4	58.8	58.3	57.8	57.3	56.8	56.4					

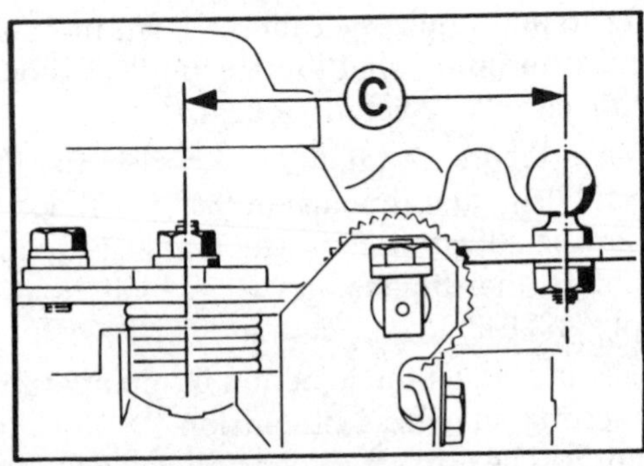

Figure 27-10. Operating lever setting, dimension "C." (Courtesy of Ford Motor Company)

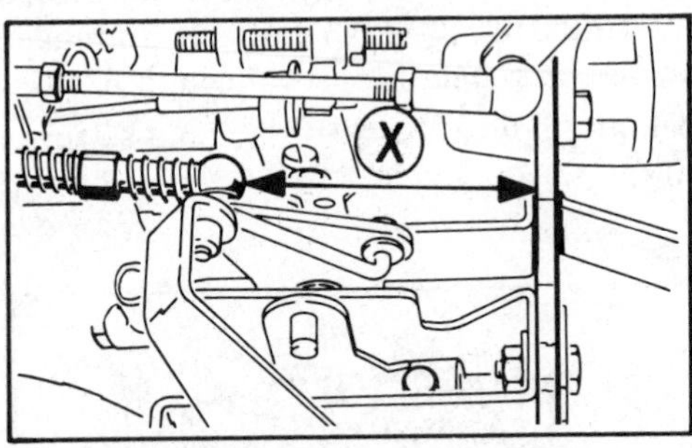

Figure 27-11. Operating lever setting, dimension "X." (Courtesy of Ford Motor Company)

Idle Speed Boost Check

Normally the cold idle operation adjustment is factory set and sealed and should not require any adjustment unless it has been tampered with or it is suspected that insufficient idle speed boost is being achieved. The cold idle operation should be checked as per the following procedure.

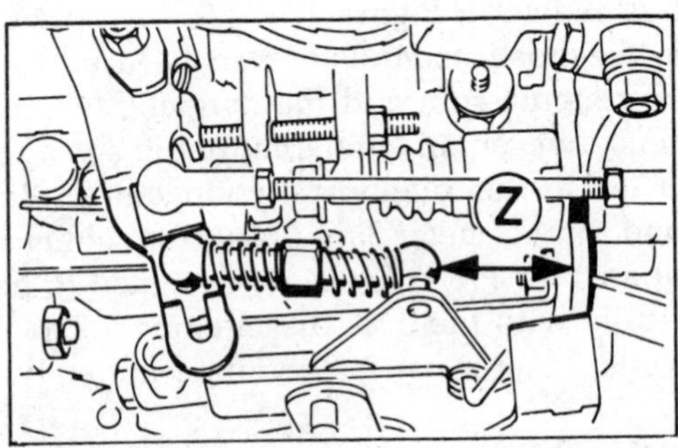

Figure 27-12. Operating lever setting, dimension "Z." (Courtesy of Ford Motor Company)

1. Refer to Figure 27-13 and remove the rear thermostat housing so that you can measure the dimension "A" between the lock nut and the holding bracket which should have a measurement of 0.216±0.015" (5.5±0.4 mm).
2. Loosen the front and rear clamps shown in Figure 27-13 (above).
3. Adjust dimension "A" by moving the clamps; always tighten the rear clamp first.
4. Tighten the pinch screw on the front clamp.
5. Recheck dimension "A" and seal the nut at the rear clamp with yellow paint.

Knurled Nut Adjustment

This adjustment should be done with the engine coolant operating temperature above 77°F (25°C) and the correct engine idle rpm as shown in Figure 27-14.

Procedure

1. Measure distance "B" between the lever and

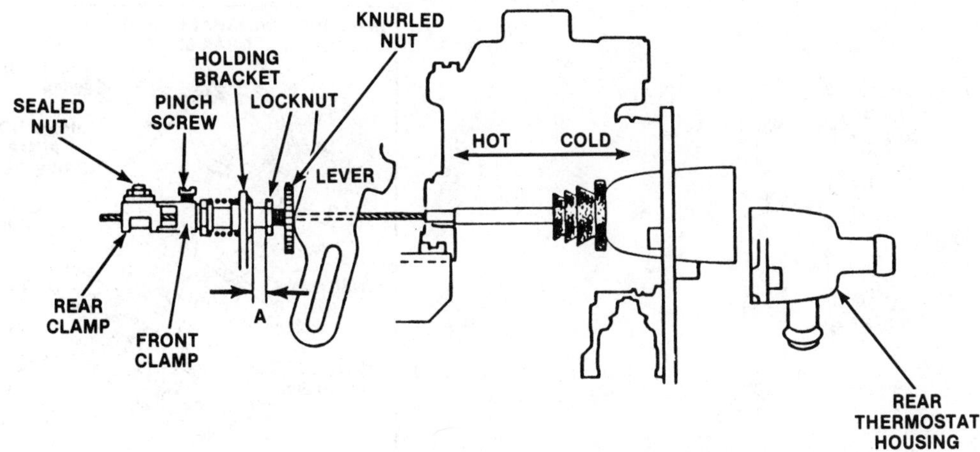

Figure 27–13. Idle speed boost adjustment. (Courtesy of Ford Motor Company)

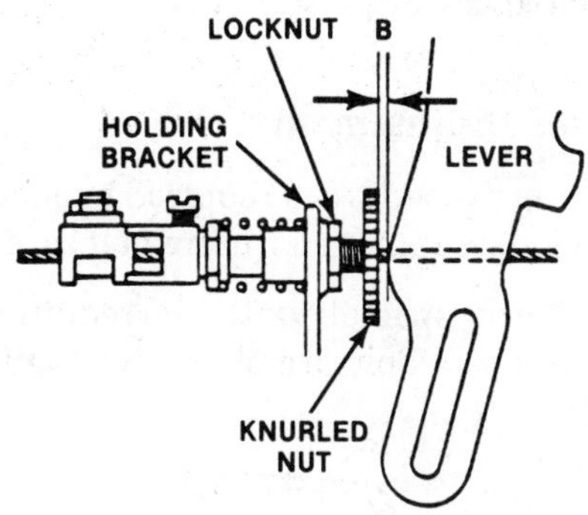

Figure 27–14. Knurled nut adjustment. (Courtesy of Ford Motor Company)

knurled nut, which should be 0.020±0.011″ (0.5±0.3 mm).

2. If adjustment is necessary, loosen the lock nut and move the knurled nut to change the setting to specifications, then tighten the lock nut.

Fuel Injection Nozzles

Detailed information on the maintenance, service, and repair of the fuel injectors can be found under Chapter 29 dealing with nozzle operation, service, and overhaul.

The fuel injection nozzles used with the 2.4L diesel engine are of Robert Bosch manufacture and are similar in appearance and design to those used in the 2.0L and 2.2L Ford passenger car and light truck diesel engines. Because of minor differences in release pressures, do not interchange nozzles from one engine type to another.

These nozzles are of the throttling-pintle-type since they are used with a pre-combustion chamber-type cylinder design. Figure 27–15 illustrates a cutaway view of the nozzle while Figure 27–16 illustrates the instrumented injector used on this engine, which is slightly longer than the other units.

NOTE: 49-state-certified vehicles have one instrumented injector located in No. 5 cylinder that sends signals to the trip-minder computer located beneath the package tray portion of the instrument panel. In California certified vehicles, both injectors No. 4 and No. 5 are instrumented. Injector No. 4 sends voltage signals to a trunk-mounted computer that monitors fuel flow and engine timing. The trunk-mounted computer then sends out any necessary corrected signals to change engine timing by opening and closing (cycling) a fuel control solenoid in the VP-20 injection pump only. This causes fuel pressure acting upon the injection pump advance piston to be bled off anytime that the fuel solenoid is open (energized)

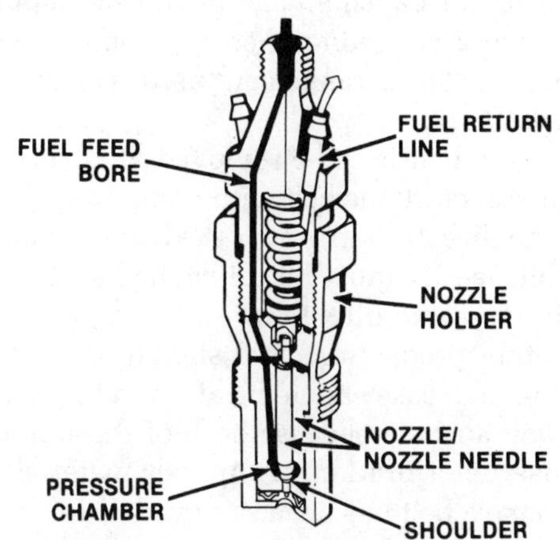

Figure 27–15. 2.4L Fuel injection nozzle (9E527A). (Courtesy of Ford Motor Company)

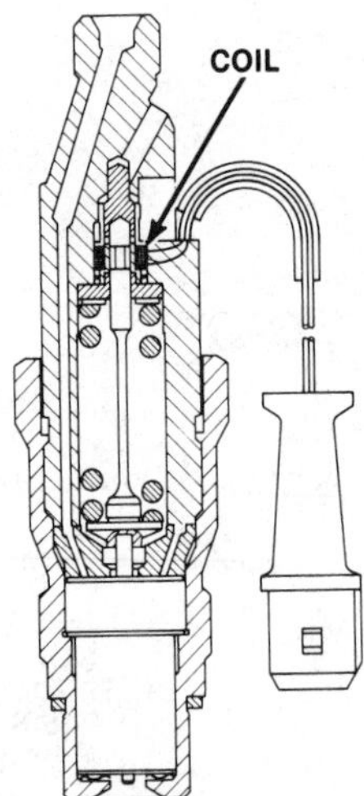

Figure 27–16. Instrumented injector (9E527B). (Courtesy of Ford Motor Company)

thereby varying injection pump timing automatically. The fully instrumented injector shown in Figure 27–16 consists of an internal solenoid/coil and has two wire outlets protruding from it.

Movement of the internal nozzle needle valve is transmitted to the trip-minded computer, which takes this voltage signal along with voltage signals from the engine coolant sensor and engine speed and crankshaft position sensors which are sent to the trunk-mounted computer in the California certified vehicles to determine the beginning of injection and its duration.

The trip-minded computer calculates the fuel economy and displays it on the instrument panel readout.

Timing Belt Arrangement

Both the engine camshaft and the injection pump are driven from a cogged/toothed type belt at the front of the engine. The arrangement used is clearly shown in Figure 27–17.

If a timing belt is to be removed, it is very important that you mark the belt with paint or a felt marker so that the direction in which the belt is rotating can be established. It must be reinstalled so that it will rotate in the same direction.

Two other precautions that should be exercised is that no oil or grease should be allowed to accumulate on the belt surface because both of these lubricants can cause deterioration of the belt material. In addition, once a belt has been removed or loosened off, it must be correctly adjusted once it has been reinstalled; otherwise the valves could contact the pistons and cause serious engine damage.

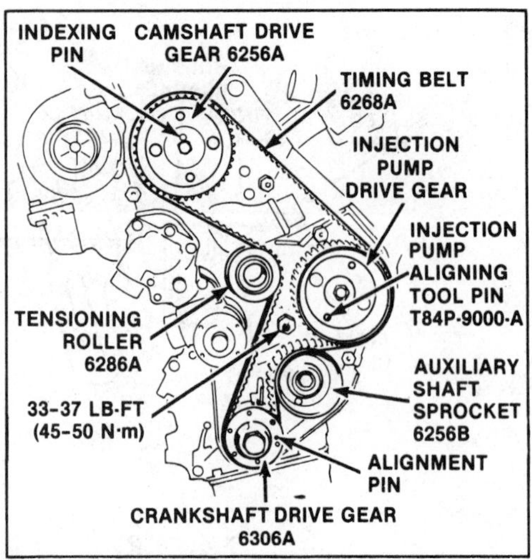

Figure 27–17. Timing belt—2.4L Diesel engine. (Courtesy of Ford Motor Company)

Timing Belt Adjustment

The following procedure is required to accurately install a previously removed timing belt or a new one.

NOTE: Three special tools are required for this procedure. These tools are shown in Figures 27–18 and 27–19.

CAM POSITIONING TOOL
T84P-6256-A

This precision ground tool is required to index the camshaft to the crankshaft for timing the valves. Used when installing the cogged-timing belt. Mounts over the front of the camshaft and is locked in position to the cylinder head.

Figure 27–18.

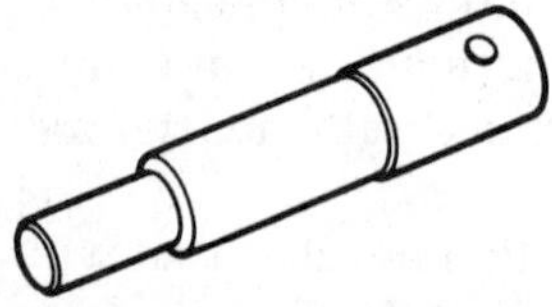

INJECTION PUMP ALIGNING PIN
T84P-9000-A

This pin is required to index the injection pump sprocket in its precise position during timing belt installation. Used in conjunction with the TDC Aligning Pin (T84P-6400-A) when static timing the injection pump.

Figure 27–19.

Procedure

1. Manually rotate the engine crankshaft over to place No. 1 piston at TDC. This can be accurately established by inserting the timing pin P/N T84P-6400-A through the hole in the block and into the hole in the flywheel as shown in Figure 27–20.

   TDC can be confirmed by the fact that the alignment pin on the front crankshaft damper flange will be at the 3 o'clock position and the indexing pin on the camshaft sprocket (gear) shown in Figure 27–17 should be at about the 12 o'clock position.
2. Using special tool P/N T84P-6256-A, Figure 27–18, hold the camshaft at the TDC position by placing this tool over the front of the camshaft and locking it into position on the cylinder head.

   NOTE: On a new belt—up to 10,000 miles (15,000 km)—place a 0.100" (2.5-mm) feeler gauge/spacer between the valve cover sealing surface and the special tool P/N T84P-6256-A on the engine exhaust valve side.
3. Install and torque all drive gears shown in Figure 27–17.
4. Insert special tool P/N T84P-9000-A into the hole through the injection pump drive gear as shown in Figures 27–17 and 27–19 to prevent movement of this gear.
5. Install the camshaft sprocket (gear) and tighten its retaining bolt by hand, then move the sprocket in the direction of normal engine rotation towards the pin and against the stop.
6. Install the timing belt tensioning roller and tighten its bolt and nuts by hand.

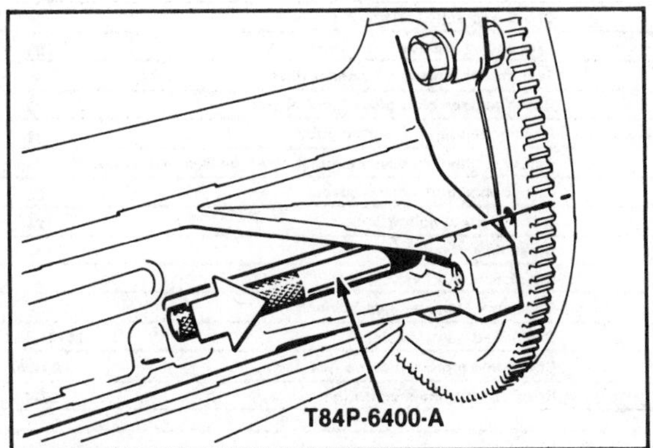

Figure 27–20. Fixing TDC on cylinder No. 1. (Courtesy of Ford Motor Company)

7. To install the belt proceed as follows: Slip the belt over the crankshaft gear, then around the auxiliary shaft and injection pump sprocket, over the cam sprocket, and then over the belt tensioner.
8. Remove the holding alignment pin on the injection pump sprocket, shown in Figures 27–17 and 27–19.
9. Force the tensioning roller against the back of the belt using a dial-type torque wrench rather than a click type between 33–37 lb.ft. (45–50 N•m) if the belt is new, or between 22–26 lb.ft. (30–35 N•m) for a used belt. This action can be done on the roller by the nut shown in Figure 27–17, bottom/right.

   CAUTION: If the belt jumps out of position, repeat the previous procedure Steps 1 through 9.
10. Torque the baseplate of the tensioning roller to between 15–18 lb.ft. (20–24 N·m).
11. Torque the camshaft sprocket to between 41–48 lb.ft. (55–65 N•m), then remove the special tools from the cylinder head and crankshaft.
12. Manually rotate the engine over at least 720 degrees and recheck the belt setting by putting the engine at No. 1 TDC with special tool T84P-6400-A, shown in Figure 27–20. The camshaft holding tool in Figure 27–18 should slide into place and the injection pump aligning tool should also be lined up.

Valve Adjustment

The valve adjustment used with the Ford/BMW diesel engine is the same arrangement used by BMW on their engines which is shown in Figure 27–21. Valve adjustments should be made with the engine *cold*, meaning that the engine coolant temperature should be below 95°F (35°C).

Refer to Figure 27–21 and measure the clearance between the rocker arm sliding surface and the camshaft when the piston is at TDC on its compression stroke (valve closed). Valves should be set to 0.012" (0.30 mm) for both the intake and exhaust.

To adjust the valve clearance, you have to place a 12-mm wrench over the flats machined onto the back side of the rocker arm to hold it while you slacken off the retaining bolt that passes through the eccentrically mounted roller. Move the roller towards the

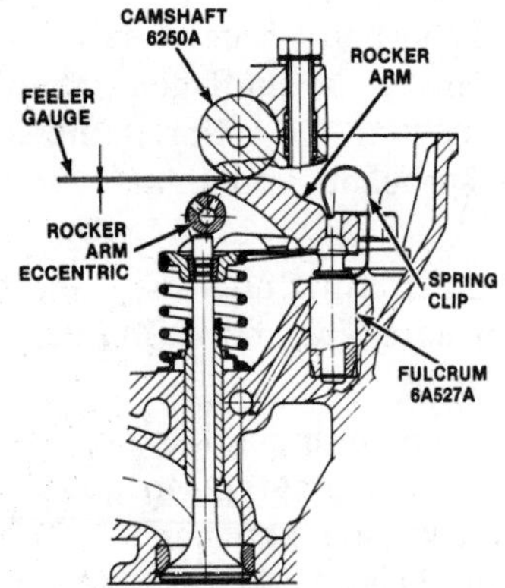

Figure 27-21. Measuring valve clearance. [Courtesy of Ford Motor Company]

valve stem or away from it until you have obtained the correct clearance, then tighten the roller adjusting bolt nut.

NOTE: If it becomes necessary at any time to replace a rocker arm, its fulcrum shown in Figure 27–21 must also be replaced. This fulcrum can be removed using special tool D84P-6564-A along with a slide hammer. When installing a new fulcrum, make certain that the bore in the cylinder head is completely clean and dry, then coat the fulcrum shaft with Loctite, and tap it into position using a plastic hammer.

Figure 27–22 provides a diagnosis troubleshooting guide for the 2.4L turbo-diesel.

DIAGNOSIS GUIDE 2.4L TURBO-DIESEL

1. Engine does not start or starts poorly in warm condition.
2. Engine does not start or starts poorly in cold condition.
3. Irregular idling (hunting).
4. Erratic idling when engine is warm.
5. Engine misfires while driving car.
6. Engine power output insufficient.
7. Fuel consumption too high.
8. Engine is running on.
9. Engine is running rough, black smoke from exhaust in full-load range poss. loss of power.
10. Smoke from exhaust (white/blue).
11. Improper warm idle speed and deviation in maximum speed.
12. Engine does not rev up in cold condition.
13. Injection pump is overheating.
14. Smoke at full load higher speeds only.

NOTE:
The use of this troubleshooting guide requires, that the basic engine adjustment, the boost pressure equipment, the electric and vacuum system are correct.

1	2	3	4	5	6	7	8	9	10	11	12	13	14	CAUSE:	CORRECTION:	REFERENCE MANUAL PAGE
●	●			●										Tank empty	Fill up fuel. Bleed fuel system.	31
●	●		●	●	●						●			Air in fuel system	Bleed/seal fuel system.	31
●	●						●							Electric shut off valve without current, or faulty	Check electric power supply or replace fuel shutoff.	24
●	●			●	●						●			Fuel filter clogged	Replace fuel filter.	20 & 46
●	●				●						●			Injection line clogged (diameter reduced)	Check, clean or replace fuel lines.	31
●	●				●						●			Fuel supply line clogged (diameter reduced)	Check, clean or replace fuel lines.	19 & 46
●	●			●	●									Connections loose, lines leaking, broken	Test and/or tighten/seal connections, replace fuel lines.	19 & 46
	●				●						●			Extreme paraffine deposits on filter	Test and/or replace fuel filter, use winter fuel.	19 & 46
●	●	●	●	●	●	●		●	●		●			Injection, timing not correct	Check and adjust injection timing.	26
●	●		●		●			●	●					Fuel injector faulty	Test and/or replace fuel injector.	33
					●	●		●					●	Engine air cleaner clogged	Test and/or replace air cleaner filter.	46
●	●										●			Glow plug system faulty	Check/replace glow plugs, control unit.	43
●	●		●											Injection order does not correspond with firing order	Fit injection lines in correct order.	31
			●							●				Engine idle speed not correct	Adjust engine idle speed (setting screw on injection pump)	28
					●					●				Engine maximum speed not correct	Adjust maximum engine speed.	28
		●			●				●	●	●			Wrong hollow bolt on return line	Install correct hollow bolt.	24
				●								●		Hollow bolt clogged	Clean hollow bolt.	24
	●										●			Cold start advance faulty	Check cold start valve.	24
●	●		●			●		●			●			Engine compression low or different	Check engine (see Shop Manual).	48
●	●	●	●	●	●	●	●	●	●	●	●			Injection pump faulty or maladjusted	Check and adjust/replace.	26-31,39,47
	●				●						●			Electric fuel pump faulty	Check and replace electric fuel pump.	20 & 47
													●	EGR system operating improperly	Check EGR system operation.	37-39

Figure 27-22. Diagnosis troubleshooting guide, 2.4L turbo-diesel. [Courtesy of Ford Motor Company]

28

Lucas CAV Fuel Systems

Lucas CAV is a subsidiary of Lucas Industries of Birmingham, England. The world headquarters for Lucas CAV Limited is in West London, England (PO Box 36, Warple Way).

CAV is presently the world's largest manufacturer of diesel fuel injection equipment, with over one-third of total world production of diesel engines fitted with CAV equipment.

In Britain, the rotary pump product center is located at Medway, Kent; production capacity at Medway is almost 1 million injection pumps per year.

The in-line injection pumps are manufactured at Finchley in North London; the third product center in the United Kingdom is at Sudbury, Suffolk. This is the principal location for the manufacture of injectors and filters, with current production capacity for 8 million injectors and nozzles, and substantially more filters and filter elements.

In France, a subsidiary company, CAV RotoDiesel, supplies two-thirds of the fuel injection systems for the French diesel industry. In Spain, associate company Condiesel supplies two-thirds of Spain's diesel fuel-injection systems.

In Brazil, subsidiary company CAV do Brazil makes rotary pumps, injectors, and filters, and supplies one in every three diesel engines made in Brazil with fuel injection equipment.

In Japan, subsidiary company Lucas CAV Kk produces rotary fuel-injection pumps for the Japanese market, along with injectors and filters to Japanese customers. In Mexico, associate company CAV InyecDiesel makes injectors and filters, producing more than half of the country's total requirement.

In the United States, Lucas CAV is based in Greenville, South Carolina. The first phase of a major development program has been completed, and the company is producing rotary pumps for North American customers.

In Korea, Lucas CAV has established a partnership company, CAV Korea Limited, to manufacture injection nozzles for the country's rapidly expanding diesel engine industry.

In addition, Lucas CAV has established licensee agreements with MEFIN in Romania, IPM in Yugoslavia, and WSK in Poland for the production of rotary pumps.

In the United Kingdom, three quarters of their output is either directly exported, or exported as part of a complete diesel engine from a British manufacturer.

After sales, service support for all CAV fuel injection equipment is provided through the Lucas Service network, covering 4500 outlets worldwide in more than 130 countries.

Whenever fuel injection equipment is discussed, Lucas CAV remains as one of the truly great leaders. Lucas CAV is the parent company of Lucas, Simms, and Bryce fuel injection companies: therefore, fuel injection equipment made by these companies or made under their names is similar in design and operation to those with the CAV name.

Lucas CAV manufactures injection pumps in both in-line and rotary configurations, with the Minimec, Majormec, and Maximec being the best known of the in-line pumps, and the legendary DPA, DPC and DPS rounding out their distributor-type pumps. The latest improvement to the DP model pumps is the "epic" (electronic pump injection control) system. The fuel injectors manufactured by CAV operate on the same basic principles as those produced by both Robert and American Bosch companies (see Chapter 29).

Rotary Fuel Injection Pumps (DPA)

In 1956 CAV signed an agreement with Roosa Master in the USA that enabled CAV to manufacture their now legendary DPA diesel fuel injection pump. This pump was based upon the basic design of the Roosa Master "A" Model injection pump. Over the years of course, CAV has redesigned and improved the DPA pump so that few similarities exist now with the old Roosa Master model pump on which it was based.

The letters DPA stand for distributor pump assembly, and it is available with either a mechanical or hydraulic governor assembly. Although a similar pump known as the DP15 was released in 1975–76, this later model was discontinued in favor of improvements to the long used DPA model.

The DPA pump is capable of being adapted to diesel engines with cylinder capacities up to 2 liters (122 in.^{3}) and is generally found on 2-, 3-, 4-, and 6-cylinder engines. It can be supplied with pumping plungers of 5-, 6-, 7-, 8-, 9-, or 10-mm diameter with a normal maximum fuel output of 100 mm^3/stroke. These pumps do not require phasing or calibrating.

DPA Fuel Injection Pump

The pump derives its name from the fact that its main shaft is driven and runs through the center of the pump housing lengthwise. Fuel is in turn distributed from a single cylinder opposed plunger control somewhat similar to a rotating distributor rotor in a gasoline engine. The pump can be hub mounted or gear driven because its shaft is very stiff to eliminate torsional oscillation and ensure constant accuracy of injection.

Figure 28–1 shows a cutaway view of a typical DPA fuel-injection pump with a mechanical governor. Figure 28–2 shows a DPA pump with a hydraulic governor.

All internal parts are lubricated by fuel oil under pressure from the delivery pump. The pump can be fitted with either a mechanical or hydraulic governor, depending on the application; a hydraulically operated automatic advance mechanism controls the start of injection in relation to engine speed.

The operation of the fuel distribution is similar to that found in Roosa Master distributor pumps in that a central rotating member forms the pumping and distributing rotor driven from the main drive shaft on which is mounted the governor assembly.

Fuel Flow

Mounted on the outer end of the pumping and distributing rotor is a sliding vane-type transfer pump that receives fuel under low pressure from a lift pump mounted and driven from the engine. This lift pump pressure enters the vane-type pump through the fitting on the injection pump end plate opposite

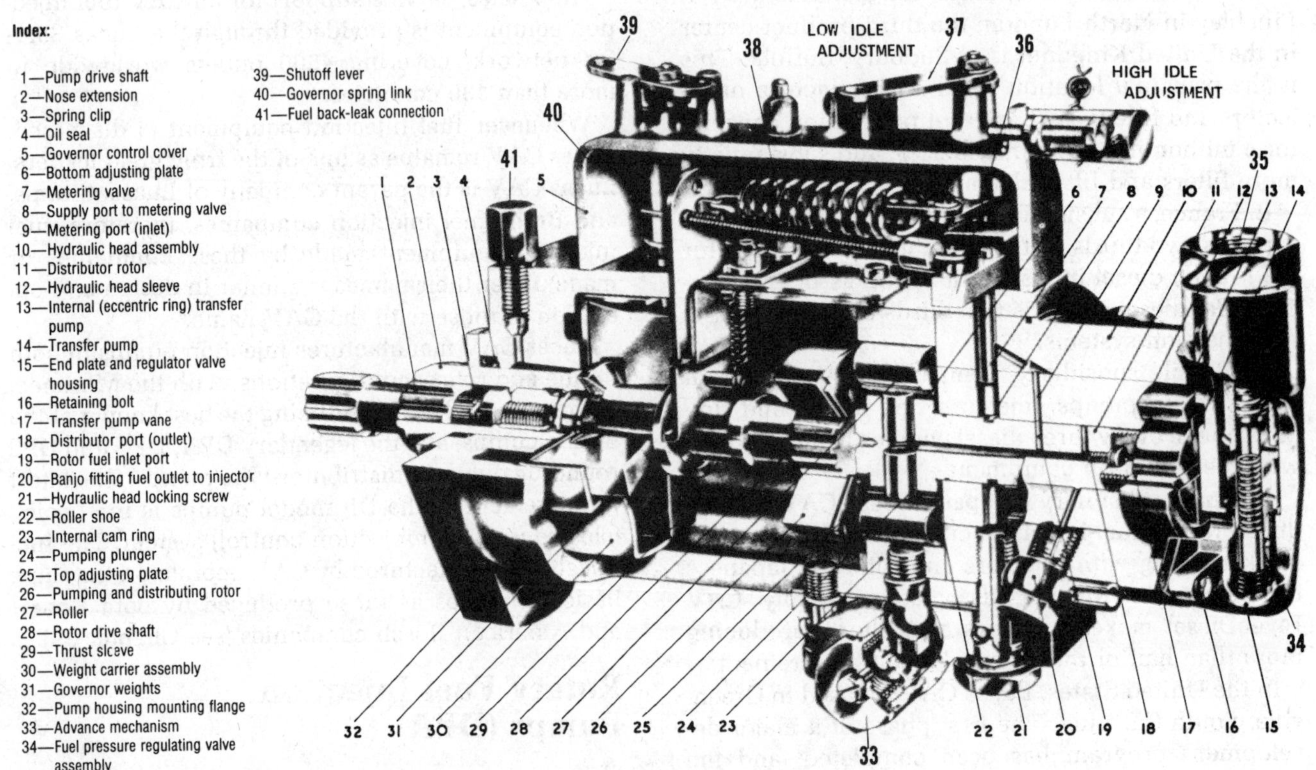

Figure 28–1. DPA pump with mechanical governor. [Courtesy of Lucas CAV Ltd.]

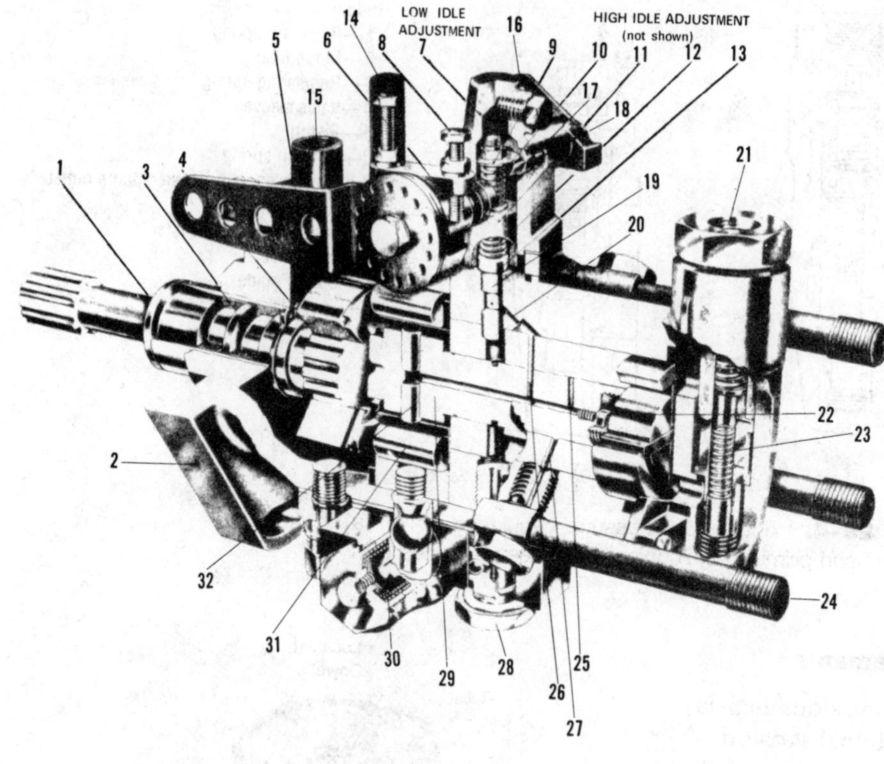

Index:
1—Pump drive shaft
2—Pump mounting flange
3—Outer seal
4—Circlip
5—Throttle lever
6—Pinion shaft
7—Governor housing
8—Idling stop screw
9—Shutoff washer
10—Idling spring
11—Rack
12—Governor spring
13—Damping washers
14—Maximum speed stop screw
15—Back leakage mounted on inspection cover
16—Shutoff lever
17—Half-round cam end
18—Shutoff spindle
19—Metering valve
20—Metering port
21—Fuel inlet
22—Transfer pump
23—Pressure regulator assembly
24—To fuel injector
25—Distributor port (outlet)
26—Rotor fuel inlet port
27—Sleeve outlet port
28—Hydraulic head locking screw
29—Rotor
30—Advance mechanism
31—Roller
32—Plunger

Figure 28-2. DPA pump with hydraulic governor. (Courtesy of Lucas CAV Ltd.)

its drive end and passes through a fine nylon gauze filter.

The vane-type pump has the capability of delivering more fuel than the injection pump will need; therefore, a pressure-regulating valve housed in the injection pump end plate allows excess fuel to be bypassed back to the suction side of the vane transfer pump. This valve is shown in Figure 28-3.

In addition to regulating fuel flow, the pressure-regulating valve also provides a means of bypassing fuel through the outlet of the transfer pump on into the injection pump for priming purposes. As seen in Figure 28-3, the regulating valve is round and contains a small *free piston* whose travel is controlled by two light springs. During priming of the injection pump, fuel at lift pump pressure enters the central port of the regulating valve sleeve and causes the free piston to move against the retaining spring pressure, thereby uncovering the priming port at the lower end of the sleeve, which connects by a passage in the end plate to the delivery side of the vane-type transfer pump, which leads to the injection pump itself.

Once the engine starts, we now have the vane-type transfer pump producing fuel under pressure, which enters the lower port of the regulating valve and causes the free piston to move up against the spring.

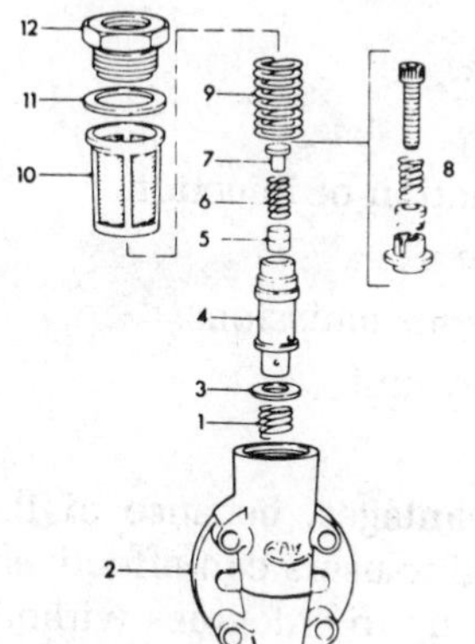

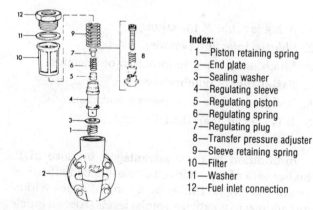

Index:
1—Piston retaining spring
2—End plate
3—Sealing washer
4—Regulating sleeve
5—Regulating piston
6—Regulating spring
7—Regulating plug
8—Transfer pressure adjuster
9—Sleeve retaining spring
10—Filter
11—Washer
12—Fuel inlet connection

Figure 28-3. Pressure regulating valve. (Courtesy of Lucas CAV Ltd.)

As the engine is accelerated, fuel pressure increases, allowing the free piston to progressively uncover the regulating port, thereby bypassing fuel from the outlet side of the vane-type pump. This action automatically controls the fuel requirements of the injection pump. Figure 28-4 shows the fuel pressure-regulating valve assembled into its bore and its action during *priming* and *regulating*.

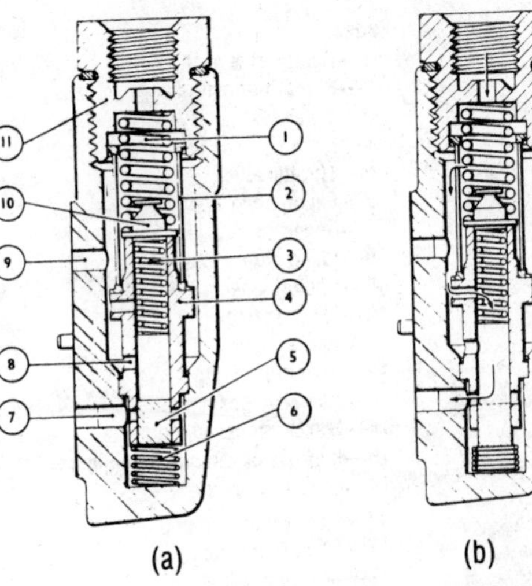

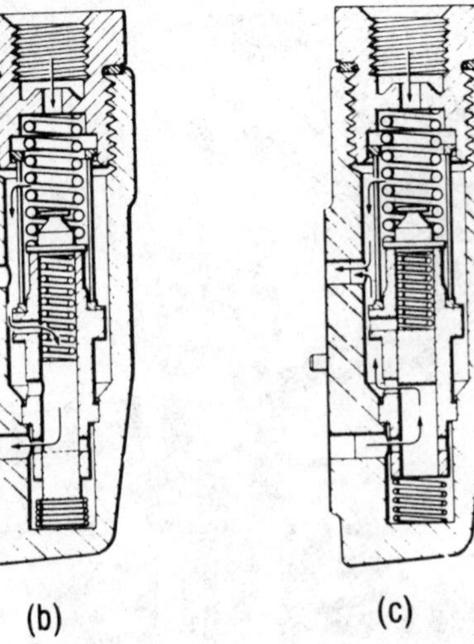

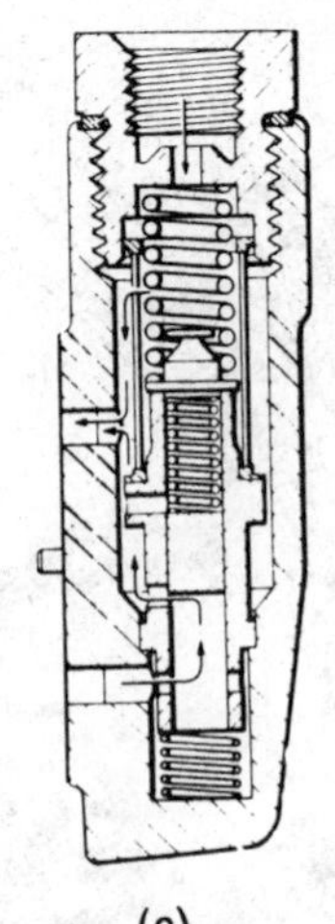

Index:
1—Retaining spring
2—Nylon filter
3—Regulating spring
4—Valve sleeve
5—Piston
6—Priming spring
7—Fuel passage to transfer pump outlet
8—Regulating port
9—Fuel passage to transfer pump inlet
10—Spring guide
11—Fuel inlet connection

Figure 28-4. Action of pressure regulating valve. a) Rest; b) hand priming; c) running. [Courtesy of Lucas CAV Ltd.]

DPA Distributor Pump Improvements

With the advent of diesel engine emissions legislation, Lucas CAV introduced an optional version of the legendary DPA injection pump, known as DPS which uses four pumping plungers rather than the conventional two of the standard DPA pump. The four-plunger DPS unit has the following inherent advantages:

1. A higher rate of injection.
2. Shorter injection periods.
3. Much more rapid termination of injection.
4. Faster rate of combustion.
5. Reduced smoke and gaseous emissions.
6. It is shorter than the DPA unit.

In addition to these advantages, because of the higher rate of injection, engine users can effectively retard the engine timing by up to 6 degrees without any increase in exhaust smoke levels. Also, a quieter engine results owing to the later start to injection, with a shorter ignition delay period. With the four-plunger design, cam contact stresses are within an acceptable level even at the peak injection pressures. Figure 28–5 shows a 6-cylinder version of the new four-plunger CAV DPA pump.

The four-plunger pump is available on engines ranging from 8 to 19 kW/cylinder (10 to 25 bhp/cylinder), which gives an application range of engines rated from 30 to 112 kW (40 to 150 bhp) in the premium performance and mildly turbocharged types. The first application for the four-plunger DPA is the Perkins turbocharged T6.354 for industrial and agricultural use, with automotive applications coming in the near future.

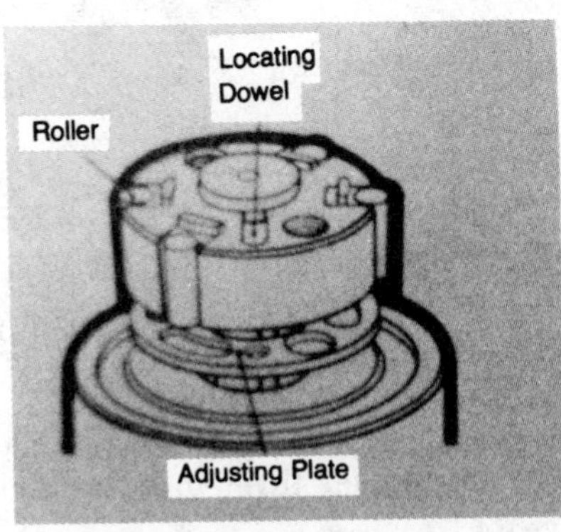

Roller and Shoe Assemblies

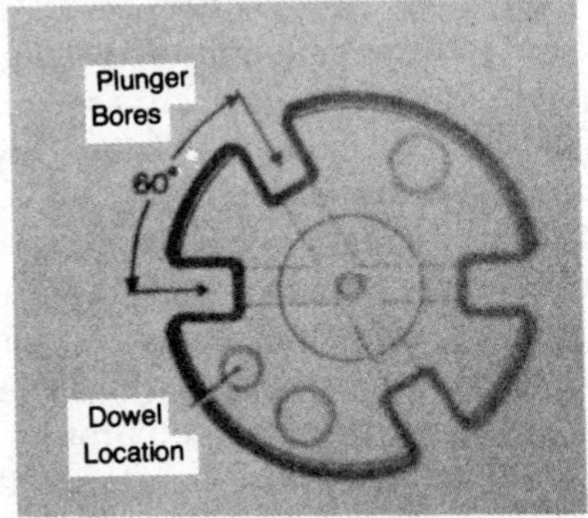

End View of Rotor

Figure 28-5. Six cylinder version of the four plunger DPA pump. [Courtesy of Lucas CAV Ltd.]

In addition to the Perkins contract, Volkswagen (Germany) has contracted with Lucas CAV to use the DPA injection pump on their VW model Golf diesel car sold in Europe, the equivalent of which is the diesel Rabbit and recently the Golf in North America fitted with the Robert Bosch VE model injection pump. The DPA injection pump being supplied for the VW Golf car includes the following features:

1. A cold-start device to advance injection timing for quick starting in low ambient temperatures without blue smoke.
2. Automatic excess fuel to further ensure rapid starting under all cold conditions.
3. Ignition key to fuel solenoid fuel shutoff (stop and start).
4. Two-speed hydraulic governing for smooth accelerator response throughout the engine speed range.
5. Light-load advance to give optimum timing for minimum noise and smoke emissions.
6. A 5600 rpm maximum speed limit to avoid a sharp cutoff at the top of the engine performance range.
7. Use of new CAV low spring injectors.

Other diesel car manufacturers using Lucas CAV model DPA pumps and injectors include Peugeot, Citroen, and Renault in France and Chrysler in Spain, supplied by CAV plants in those countries. The DPA pump on the Perkins engine is also used on the VW model LT van. The DPA injection pump is the world's best selling rotary fuel injection pump, with more than 1 million a year being produced from Lucas CAV subsidiaries, associates, and licensees abroad.

DPA Excess Fuel Starting

The use of the excess fuel starting device cuts engine cranking time in half at −7°C in tests conducted on a typical 100-bhp (75-kW) automotive diesel. Both battery and starter life are prolonged, and a quicker run up to normal operating rpm is obtained, with elimination of initial hesitancy.

The excess fuel device is so arranged that an additional pair of plungers in the rotor head is brought into operation by an automatic valve when the engine is stopped. When the engine fires and reaches adequate rpm to continue running, the excess fuel starting valve automatically cuts off fuel supply to these two plungers, allowing the pump to operate as a standard DPA pump. The point of excess fuel cutoff can be adjusted for each particular engine, thereby offering good starting quality with minimum exhaust smoke. The CAV valve mechanism prevents reengagement of the excess fuel plungers with the engine running.

Figure 28–6 shows a typical flow diagram for a DPA pump using a mechanical governor. Let us study the action of the fuel under pressure once it leaves the vane-type pump and flows to the injection pump. The pumping and distributor rotor, which is driven from the drive on the engine, rotates within the stationary hydraulic head, which contains the ports leading to the individual injectors. The number of ports varies with the number of engine cylinders. Figure 28–7 shows the rotor during the charging cycle and delivery cycle. In Figure 28–7, fuel from the vane-type transfer pump passes through a passage in the hydraulic head to an annular groove surrounding the rotor and then to a metering valve (see Figure 28–1), which is controlled by the throttle position.

The flow of fuel into the rotor (volume) is controlled by the vane-type pump's pressure, which depends on the speed of the engine and hence throttle or governor position. Fuel flowing into the rotor [Figure 28–7(a)] comes from the inlet or metering port in the hydraulic head. These inlet ports are equally spaced around the rotor; therefore, as the rotor turns, these are aligned successively with the hydraulic head inlet port.

The distributor part of the rotor has a centrally drilled axial passage that connects the pumping space between the plungers with the inlet ports (the number depending on the number of engine cylinders) and single distributing port drilled radially in the rotor. As the rotor turns around, the single outlet port will successfully distribute fuel to the outlet ports of the hydraulic head and on to its respective injector.

The pumping section of the rotor has a cross-drilled bore that contains the twin opposed plungers, which are operated by means of a cam ring (internal) carried in the pump housing, through rollers and shoes that slide in the rotor. The internal cam ring has as many lobes as there are engine cylinders. For example, a 4-cylinder engine would have four internal lobes operating in diagonally opposite pairs.

The opposed plungers have no return springs and are moved outward by fuel pressure, the amount being controlled by throttle position, metering valve, and the time during which an inlet port in the rotor is exposed to the inlet port of the hydraulic head. As a result, the rollers that operate the plungers do not follow the contour of the internal cam ring entirely, but they will contact the cam lobes at points that will vary according to the amount of plunger displacement.

The maximum amount of fuel delivered to an injector is therefore controlled by limiting the maximum outward movement of the plungers. Refer to

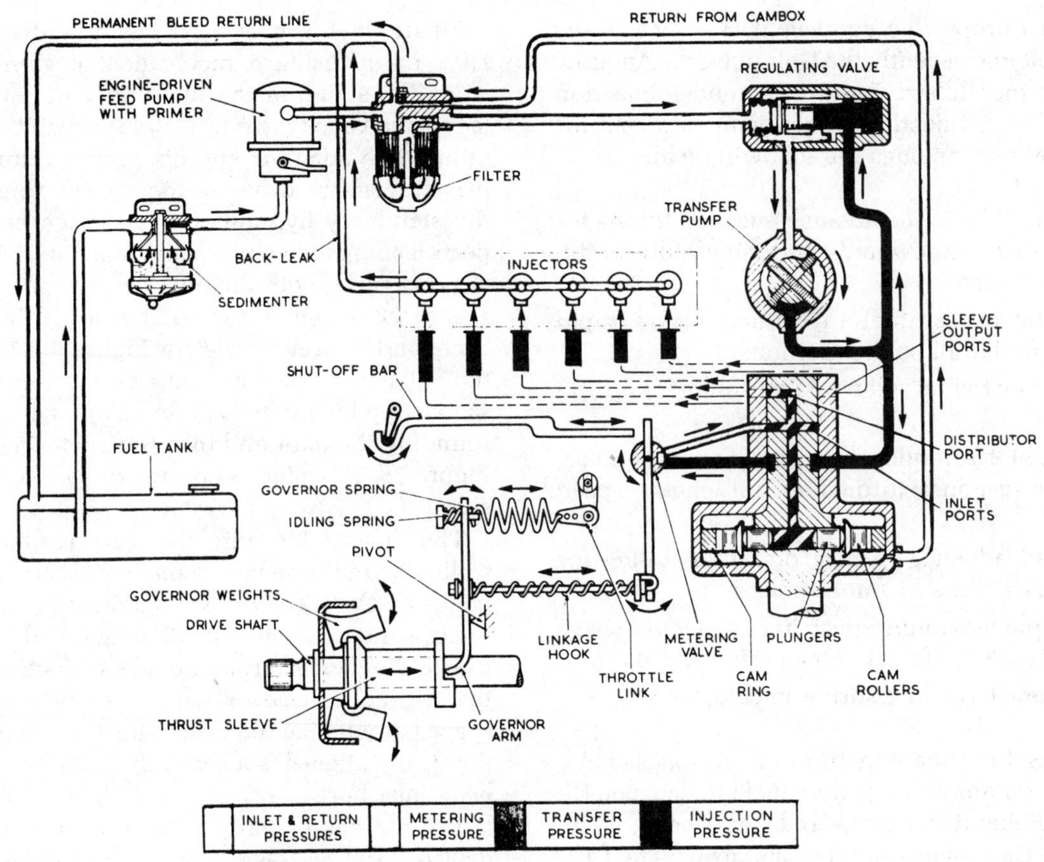

Figure 28-6. Flow diagram for a DPA pump using a mechanical governor. [Courtesy of Lucas CAV Ltd.]

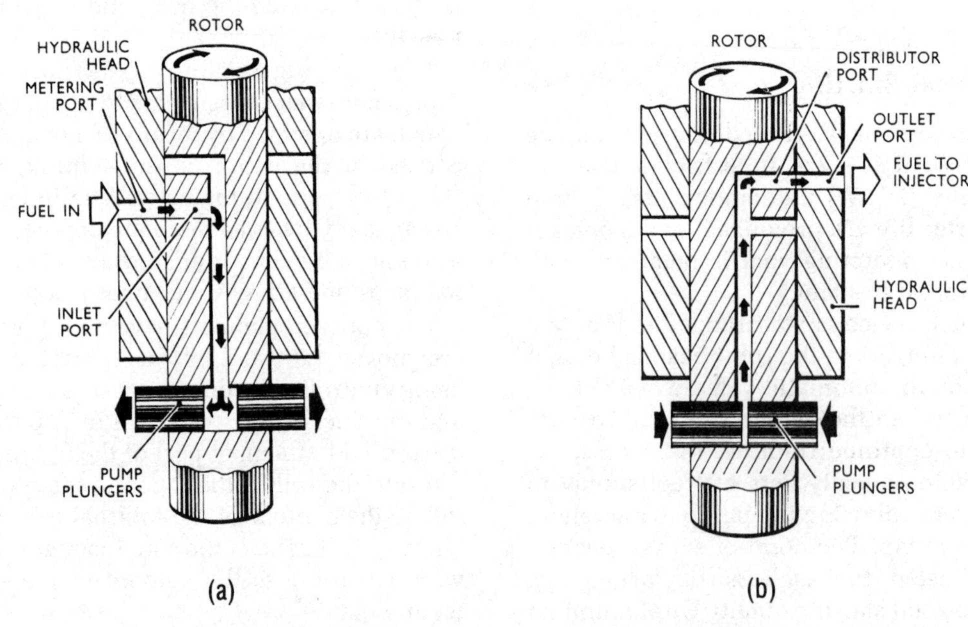

Figure 28-7. Rotor charging and delivery cycle. a] Inlet stroke; b] injection stroke. [Courtesy of Lucas CAV Ltd.]

Figure 28–8, which shows an end-on view of the rotor with the cam rollers carried in shoes bearing against the plunger ends. These roller shoes sliding in slots in the rotor have projecting ears that engage eccentric slots in the top and bottom adjusting plates located against one another by two lugs in slots.

The top adjusting plate is clamped to the rotor by the drive plate; however, the adjusting plate is cut away at the locating screws to allow adjustment of the plates by rotation. In effect, then, the ears of the roller shoes coming into contact with the curved slot sides of the adjusting plates limit the maximum outward travel of the plungers. The slots are eccentric for adjustment purposes.

Control of DPA Pump Delivery

Figure 28–9 shows the method used to control injection pump delivery. The roller-to-roller dimension is checked with an outside micrometer; however, the stirrup pipe tool (Item 1), P/N 7144-262A, must be connected to two of the hydraulic head high-pressure outlet ports along with a suitable relief valve (Item 2), P/N 7144-155.

Connect an injector nozzle testing pump unit to the relief valve and raise the fuel pressure to the outlet ports until 30 atmospheres, or that specified in the pump test data sheet, is reached. By manually turning the pump rotor, the fuel pressure internally will force the pumping plungers and rollers to their maximum outward (full fuel) position. At this point, using the micrometer, measure the roller-to-roller dimension. To adjust the roller-to-roller dimension, the adjusting plates can be moved, then the drive plate screws torqued to specifications.

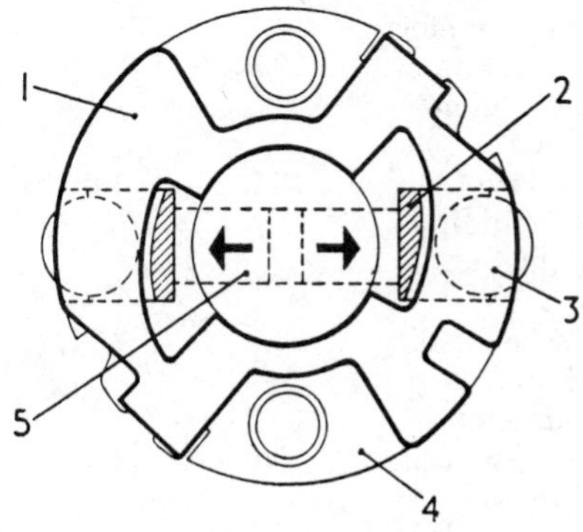

Index:
1—Top adjusting plate
2—Roller shoe ear
3—Roller
4—Pumping end of rotor
5—Pumping plunger

Figure 28–8. End view of rotor. (Courtesy of Lucas CAV Ltd.)

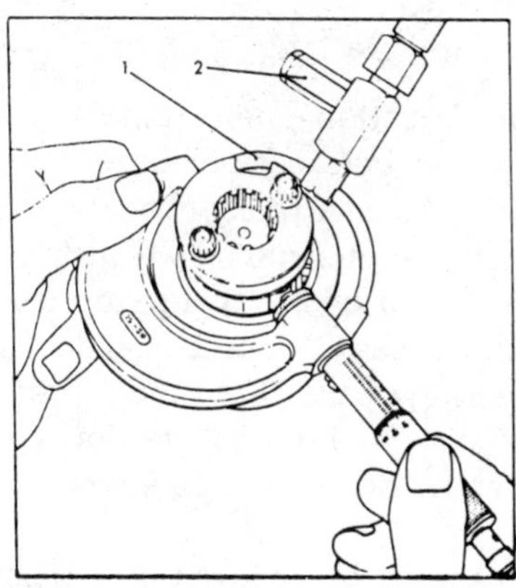

Figure 28–9. Measuring roller to roller dimension. (Courtesy of Lucas CAV Ltd.)

All necessary pump information can be found by consulting the proper test data sheet for your particular pump. This can be readily obtained from Lucas CAV or your local fuel injection repair company, most of which are members of the Association of Diesel Specialists.

As indicated earlier, the maximum amount of fuel delivered can be regulated either by limiting the outward travel of the plungers or by limiting the stroke of a shuttle valve according to pump type, as shown in Figure 28–10.

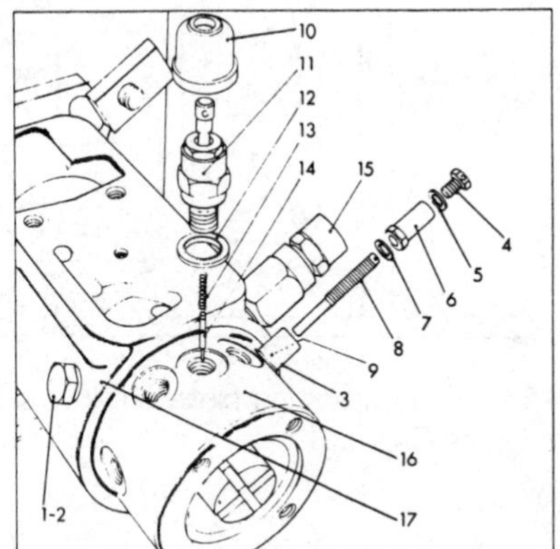

Index:
1 and 2—Hydraulic head locking screw and washer
3—Shuttle valve body
4—Shuttle plug screw
5—Screw plug washer
6—Shuttle tubular nut
7—Tubular nut washer
8—Shuttle stop screw
9—Shuttle valve piston
10—Cover (excess fuel)
11—Excess fuel selector assembly
12—Washer
13—Latch pin spring
14—Latch pin
15—Proportional pressurizing valve and washer
16—Hydraulic head and rotor assembly
17—Pump housing

Figure 28–10. Shuttle valve assembly used on an injection pump equipped with an excess fuel device and external maximum fuel adjustment. (Courtesy of Lucas CAV Ltd.)

Maximum Fuel Setting

Although the injection pump roller-to-roller dimension must be set before assembling the pump, the pump can only be accurately checked for fuel delivery when mounted and run on a test stand. The specifications are listed on the fuel pump test data sheet available from Lucas CAV or a local dealer. The maximum fuel setting on the DPA pump fitted with either the mechanical or hydraulic governor is basically the same, and would be done as follows:

1. The pump must be checked at a specified speed with both the throttle and shutoff controls fully open, and the maximum fuel delivery rate compared to specifications.
2. If the delivery rate is low, then the pump inspection cover, as shown in Figures 28–11 and 28–12, must be removed to gain access to the pump drive plate screws.
3. A special adjustment tool, P/N 7144-875, is available, which is used to engage the slot located in the periphery of the pump adjusting plate.
4. With the pump drive plate screws slackened off as mentioned in Step 2, the end of the tool can be lightly tapped to adjust the internal pump plate. Although the direction of fuel increase and decrease will vary, depending on the type of adjusting plate fitted, the following holds true. Viewing the pump, from its drive end, if the top adjusting plate has a shallow slot 3 mm in depth (Figure 28–11), then turning the plate in the counterclockwise direction will increase fuel delivery, whereas if the plate is turned in a clockwise direction, the fuel delivery rate will be reduced. If, however, the top adjusting plate has a deeper slot 5.5 mm deep (Figure 28–12), turning the top plate clockwise will increase fuel delivery, and counterclockwise will decrease it.
5. It is extremely important that the drive plate screws be torqued up evenly as listed in Table 28–1, using special adaptor tool, P/N 7144-482, spanner tool, P/N 7144-511A, and a torque wrench.
6. Install the pump inspection cover, prime the pump with fuel, and recheck the maximum fuel delivery rate. If necessary, repeat the previous procedure until the maximum fuel rate is as specified in the pump test data sheet.
7. On pumps with external maximum fuel adjustment, the shuttle stop screw is turned inward to reduce fuel delivery or outward to increase delivery. After adjustment, re-tighten shuttle tubular nut and replace shuttle plug screw.

Figure 28–11. Top adjusting plate shallow slot (3mm). (Courtesy of Lucas CAV Ltd.)

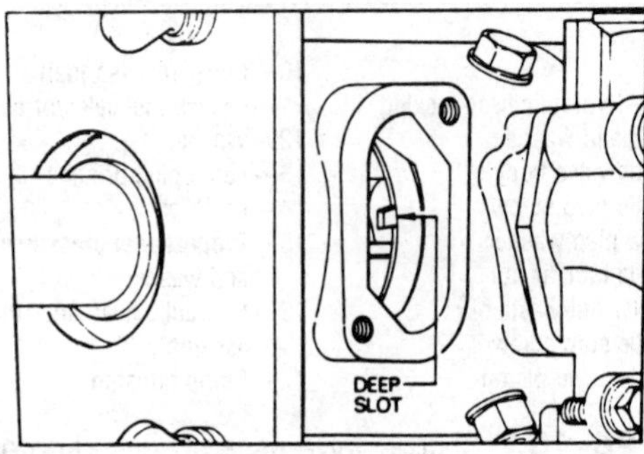

Figure 28–12. Top adjusting plate deep slot (5.5mm). (Courtesy of Lucas CAV Ltd.)

Table 28–1. *Pump Drive Plate Screw Torques*

	N•M	kg/m	lb.in.
Direct torque, plungers up to and including 7.5-mm diameter	18.1	1.85	160
Direct torque, plungers 8.00-mm diameter and above	28.4	2.90	250
Drive plate screw (A),[1] up to and including 7.5-mm diameter plungers	15.7	1.60	140
Drive plate screw (A),[1] 8 mm and above plunger diameter	24.3	2.48	215
Drive plate screw (B),[2] up to and including 7.5-mm plunger diameter	13.0	1.33	115
Drive plate screw (B),[2] 8.00 mm and above plunger diameter	20.6	2.10	180

[1](A): obtained with spanner 7144/511 and adapter 7144/482.
[2](B): obtained with spanner 7144/511A and adapter 7144/482.
Adapter and spanner *must* form a straight line when torquing up screws with a 127-mm (5.0-m) adapter center to ring spanner.

Pump Timing

The correct timing in degrees for any DPA pump is given in the test data sheet. All pumps must be correctly timed. The sequence of events for DPA pumps with either a mechanical or hydraulic governor is basically the same. To check the timing, the inspection cover immediately above the pump advance unit must be removed.

When checking the timing, the pump should be removed from the test stand, and drained of fuel. You must then connect tool no. 7144-262A along with tool no. 7144-155, which was shown in Figure 28–9. This can be done by connecting the stirrup pipe tool to the fuel outlet specified on the test data sheet and to the diametrically opposite outlet. Bring the fuel pressure up to 30 atm by use of a nozzle testing pump unit, and proceed as follows:

1. Manually turn the pump drive shaft in its normal direction until resistance to further movement is felt, which is the *timing position.*
2. The timing ring is visible through the inspection cover.
3. Refer to Figure 28–13, and move the timing ring until the straight edge of the timing circlip, or the line scribed on the ring as is the case with the older-style clips, aligns with the mark on the drive plate as specified in the test plan sheet.
4. Circlips with two straight ears are only for spacing, and the circlip ends are positioned remote from the inspection hole aperture. Refit the inspection cover when timing is complete.
5. Pumps fitted with excess fuel devices have the timing marks machined on the drive plate and the cam ring and are not adjustable. The pump is timed by removing the timing inspection cover and turning the rotor until the marks are in alignment.

A flange marking gauge tool, P/N 7244-27, is available for use in remarking pump flanges at any time a change is required.

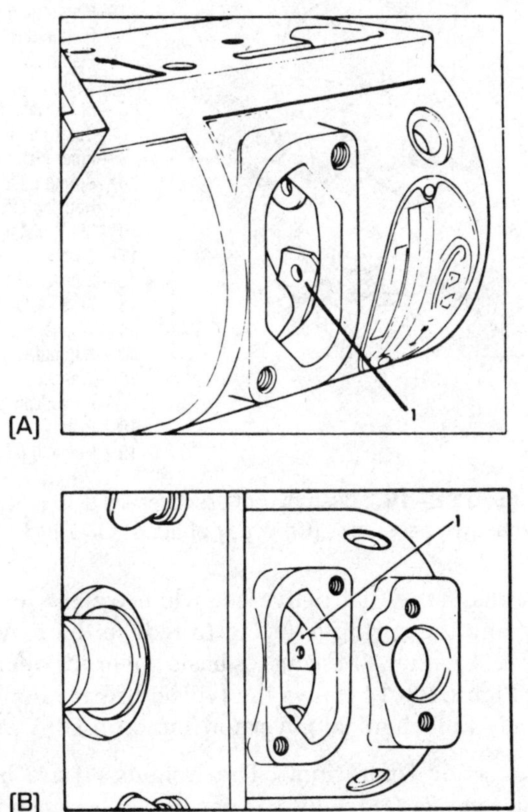

Figure 28–13. a) DPA pump mechanical governor housing; b) DPA pump hydraulic governor housing. (Courtesy of Lucas CAV Ltd.)

Automatic Light-Load Advance Mechanism

The light-load advance device automatically varies the start of injection in relation to the speed of the engine. Reference to Figure 28–14 will assist you in understanding its principle of operation. Rotation of the internal cam ring in which the rotor revolves affects the point at which the discharge port of the rotor will align with the hydraulic head outlet to the injector. Its action is similar to that of the advance mechanism used with an ignition distributor in a gasoline engine.

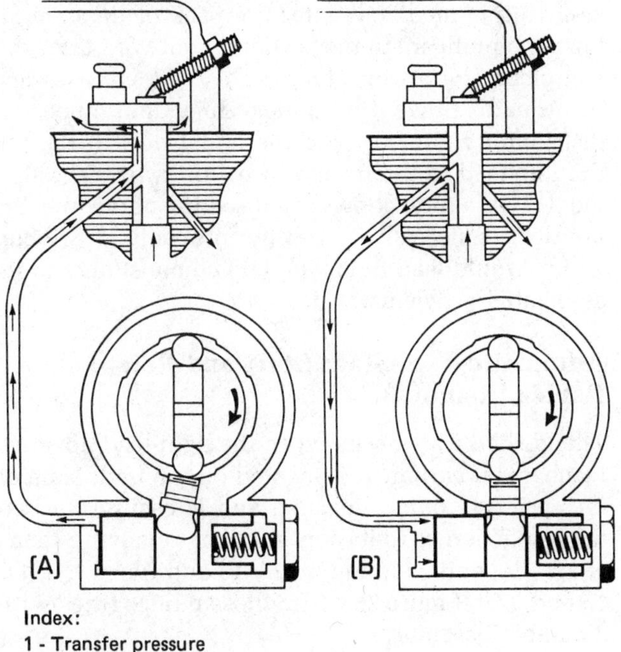

Index:
1 - Transfer pressure
2 - To pumping element

Figure 28–14. Automatic light-load advance mechanism. a) Full load; b) Low load. (Courtesy of Lucas CAV Ltd.)

Screwed into the cam ring is a lever with a ball end on it, one side of which is acted upon by a spring-loaded piston, while the other side is subjected to fuel pressure or drain pressure, depending on the engine load and throttle position, from the metering valve, which feeds through a hollow locating bolt and port in the housing.

Operation: Full Load. With the engine running under a situation of full load, the helical groove of the fuel metering valve will be aligned with the hydraulic head timing port, as shown in Figure 28–14(a); therefore, fuel at drain pressure is bled away from the advance piston, allowing the spring pressure to hold the internal cam ring in its fully retarded position.

Operation: Light Load. With the engine running under light load, the fuel metering valve would be in a position whereby the flat at the lower end of the metering valve would now be aligned with the timing port to the piston. The transfer pump pressure would now flow to the advance piston, forcing it back against the spring and allowing the cam ring to move to its fully advanced position.

When starting the engine, due to the tension of the governor spring, the metering valve would be in the full-load position with the advance piston subjected to fuel at drain pressure; therefore, the internal cam ring would remain in the fully retarded position.

A retraction curve machined on each lobe of the cam ring immediately after the peak of the cam allows the plungers to move slightly outward after the completion of an injection cycle. This is necessary in order to prevent secondary injection, since the distributor rotor port and the outlet port in the hydraulic head to the injector are still in partial alignment. This action allows rapid seating of the injector nozzle needle valve, thereby preventing dribble, which would lead to incomplete combustion and its associated problems.

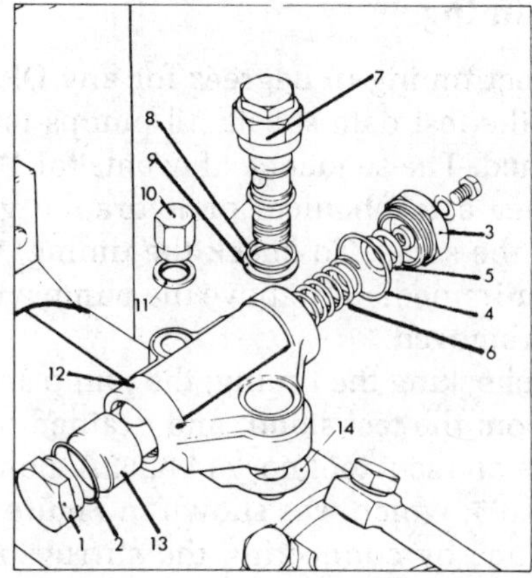

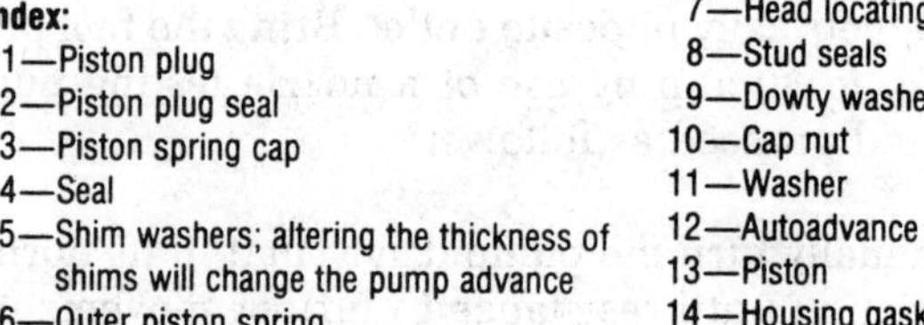

Index:
1—Piston plug
2—Piston plug seal
3—Piston spring cap
4—Seal
5—Shim washers; altering the thickness of shims will change the pump advance
6—Outer piston spring
7—Head locating screw
8—Stud seals
9—Dowty washer
10—Cap nut
11—Washer
12—Autoadvance housing
13—Piston
14—Housing gasket

Figure 28–15. Removal of start retard device. (Courtesy of Lucas CAV Ltd.)

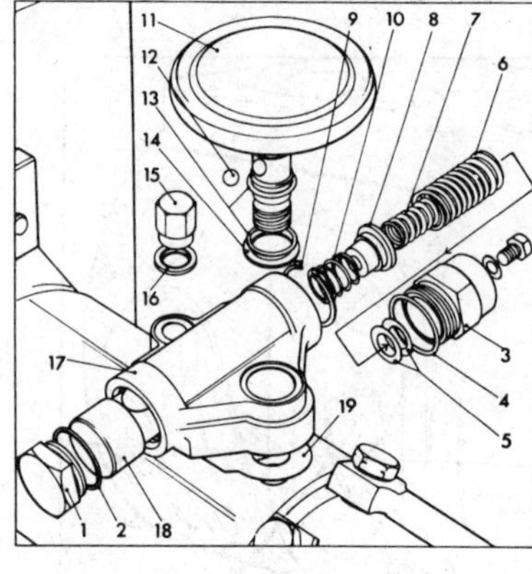

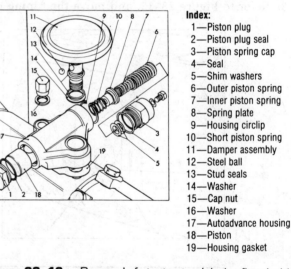

Index:
1—Piston plug
2—Piston plug seal
3—Piston spring cap
4—Seal
5—Shim washers
6—Outer piston spring
7—Inner piston spring
8—Spring plate
9—Housing circlip
10—Short piston spring
11—Damper assembly
12—Steel ball
13—Stud seals
14—Washer
15—Cap nut
16—Washer
17—Autoadvance housing
18—Piston
19—Housing gasket

Figure 28–16. Removal of start retard device fitted with a damper assembly. (Courtesy of Lucas CAV Ltd.)

Automatic Two-Stage Start and Retard Device (Removal)

This device was discussed as the light-load advance mechanism earlier. It is located on the underside of the injection pump housing, and is easily removed for inspection as to its condition, by removing Items 1, 3, 7, 10, and 12 in Figure 28–15 and Items 1, 3, 11, 15 and 17 in Figure 28–16 on those pumps fitted with a damper assembly.

Mechanical Governor

The governor functions in the same manner as all mechanical units in that spring pressure is trying to increase the fuel delivered, while weight force opposing the spring is trying to reduce the fuel delivered. Chapter 11 discusses basic governor operation.

Figure 28–17 shows the typical linkage hookup of such a mechanical governor found on DPA pumps.

Governor Operation. The weights (1) are held in the retainer (15), which is positioned between the injection pump drive hub and drive shaft (13) and thus rotates with the shaft as a unit. The weights will move outward under the influence of centrifugal force, the distance being related to engine rpm.

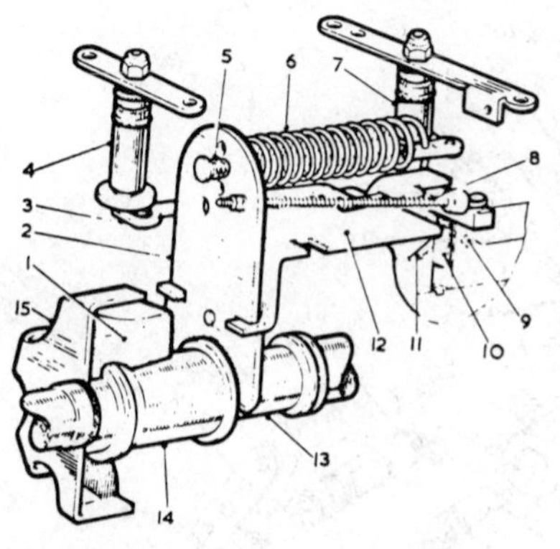

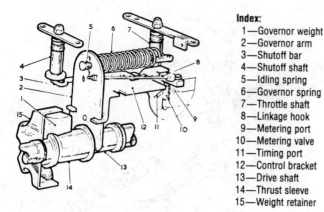

Index:
1—Governor weight
2—Governor arm
3—Shutoff bar
4—Shutoff shaft
5—Idling spring
6—Governor spring
7—Throttle shaft
8—Linkage hook
9—Metering port
10—Metering valve
11—Timing port
12—Control bracket
13—Drive shaft
14—Thrust sleeve
15—Weight retainer

Figure 28-17. Typical linkage of a mechanical governor. (Courtesy of Lucas CAV Ltd.)

A thrust sleeve (14), which is hollow, is free to move lengthwise along the extension nose of the injection pump drive shaft under the influence of the governor weight force. Any such movement would be transmitted by the governor arm (2) and the spring loaded linkage hook (8) to cause rotation of the fuel metering valve (10). The governor arm (2) pivots around a fulcrum on the control bracket (12) and is held in contact with the thrust sleeve (14) by spring tension. The governor arm (2) and throttle arm (7) and shaft assembly are connected through the governor spring (6) and the idling spring (5) and its guide.

A shutoff bar (3) connected to an external shutdown control rotates the fuel metering valve to close off the metering port when it is desired to stop the engine. In automotive applications, a fuel solenoid cuts off fuel when the ignition key is turned off.

Both throttle and governor action directly affect the rotation of the fuel metering valve and therefore the speed of the engine. Movement of the throttle arm (7) toward a fuel increase position causes the light idling spring (5) to be compressed as its guide is drawn through the governor arm (2). As the engine accelerates, the centrifugal force developed by the rotating flyweights will cause the thrust sleeve to move the governor arm (2) and therefore the metering valve (10) to a decreased fuel position.

However, as with all mechanical governors, within the governed range, for any fixed throttle position, the centrifugal force of the flyweights will be balanced by the force of the governor spring. Anytime that this occurs, the engine will run at a steady speed. This state of balance can occur at any speed range as long as the rpm of the engine is capable of carrying the load placed upon it.

This state of balance can be upset either by the operator or driver moving the throttle or by a change in load applied or removed from the engine. Therefore, an increase in load for a fixed throttle position will cause a reduction in speed, lowering the centrifugal force of the weights and upsetting the state of balance in favor of the spring, causing it to increase the fuel by rotation of the fuel metering valve. As the engine speed climbs again and a state of balance is reached, the engine will again run at a steady rpm. If the load were decreased, the rpm would increase, upsetting the state of balance in favor of the weights and causing a fuel decrease and engine rpm loss, until once again a corrected state of balance existed.

With the throttle at the idle position, governor action is controlled by the light idle spring only.

Mechanical Governor DPA Pump Linkage Adjustment

With the governor cover removed (see Figure 28-18), set the link length as shown with the use of a vernier caliper. This dimension is listed on the correct test data sheet available from CAV or a local dealer, and is taken between the larger base diameters of the governor control cover stud (1) and the metering valve

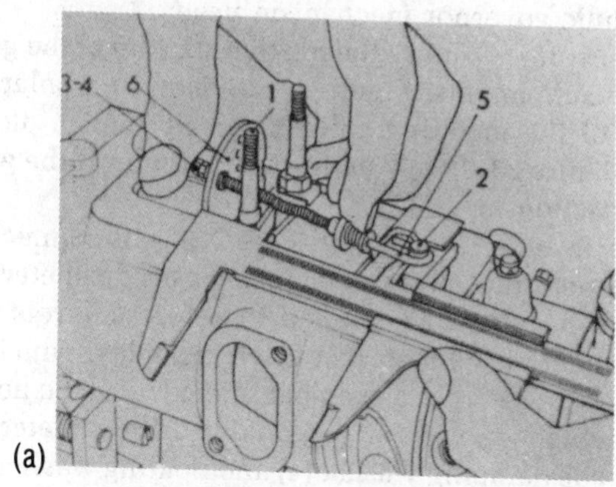

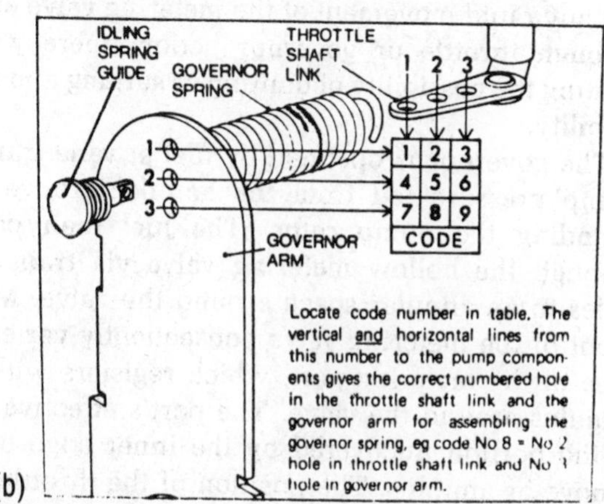

Figure 28-18. a) Setting governor linkage; b) application of setting code. (Courtesy of Lucas CAV Ltd.)

lever pin (2). To adjust the linkage length, back off the lock nut (3), and turn nut (4). Hold light finger pressure against the control arm (6) to hold the metering valve in the fully open position when setting the desired length.

When installing the governor control spring to the governor control arm and throttle shaft link, ensure that it is inserted into the proper hole as shown above. This information is also given on the pump test data sheet.

Hydraulic Governor

The DPA pump fitted with a hydraulic governor operates on the same principle as the pump with a mechanical governor. However, with the hydraulic governor, there are no governor flyweights, elimination of the pump drive hub, and use of a sliding piston-type metering valve in place of the slotted semirotary type. The hydraulic governor is contained within a smaller housing than that of the mechanical, but it is also located on top of the pump housing. Since the basic injection pump components remain the same, we will only concern ourselves with the hydraulic governor mechanism itself.

Figure 28–19 shows the relative stackup of the governor components. Figure 28–20 shows an enlargement of the metering valve assembly, which along with Figure 28–19 will assist in explaining of the governor action.

The throttle lever (17) in Figure 28–19 connected to the metering valve pinion (2) engages the metering rack (6 in Figure 28–20) and therefore controls the movement of the metering valve assembly, which is free to move within a chamber of the hydraulic head, into which opens the diagonally drilled metering port. The damping washer (4) and floating washer (3) shown in Figure 28–20 act as a dashpot to dampen out any rapid movement of the metering valve either through throttle or governor action, thereby preventing the possibility of hunting or surging and poor stability.

The governor is operated by fuel at vane transfer pump pressure fed from the annular groove surrounding the pump rotor. The fuel then passes through the hollow metering valve via transverse holes to an annular space around the valve. Movement of the metering valve consequently varies the area of the metering port, which registers with the annulus around the valve. The port's effective area is the portion uncovered by the inner edge of the groove or annulus. The position of the throttle arm will vary the compressive force of the metering valve spring (5 in Figure 28–20).

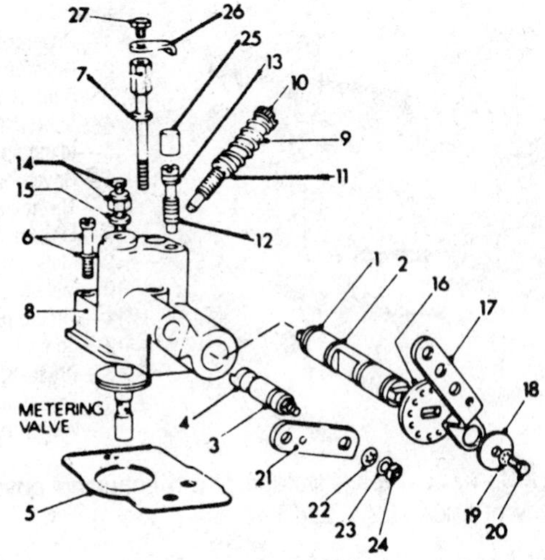

Index:
1—O ring seal
2—Metering valve pinion
3—O seals
4—Shutoff shaft
5—Cover gasket
6—Short screw
7—Long screw
8—Governor housing
9—Idling spring
10—Idle adjusting screw
11—O seal
12—O seal
13—Maximum speed adjusting screw
14—Vent screw
15—Washer
16—Locking plate
17—Throttle arm
18—Washer
19—Lock washer
20—Lock screw
21—Shutoff lever
22—Washer
23—Plain washer
24—Nut
25—Rubber plug
26—Locking plate
27—Screw

Figure 28-19. Stackup of hydraulic governor components. [Courtesy of Lucas CAV Ltd.]

Speed Increase. When the throttle is moved to an increased fuel position, the metering valve will be forced toward the maximum delivery position by the action of the spring (5). This allows the engine speed to increase; in so doing, the vane-type transfer pump pressure also increases thereby forcing the metering valve back against the spring pressure until a balance is reached, which can occur at any engine speed as long as the rpm is capable of carrying the load placed upon the engine.

Speed Decrease. Movement of the throttle toward the idling position allows the idling spring (7 in Figure 28–20) to be compressed. A state of balance will exist when the forces exerted by the idling spring and fuel pump transfer pressure equal that of the metering valve spring (5). In a deceleration situation, the compression of the metering valve spring will become proportionately lower as the throttle approaches the idling stop screw. This allows the reducing transfer pump fuel pressure at low speeds to operate the metering valve and therefore perform the governing function throughout the idling range.

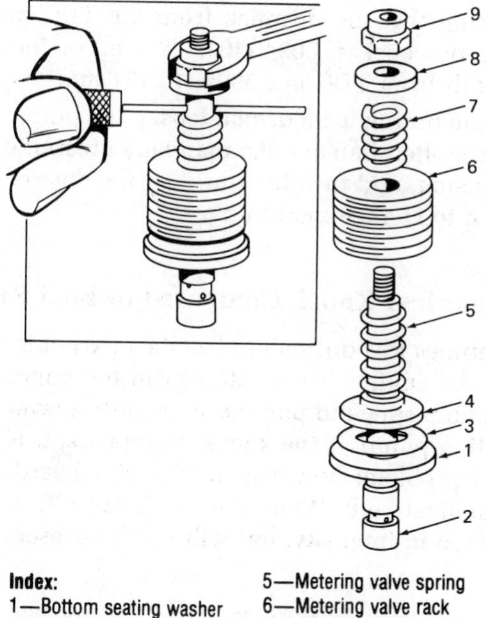

Index:

1—Bottom seating washer
2—Metering valve stem
3—Floating washer
4—Top damper washer
5—Metering valve spring
6—Metering valve rack
7—Idling spring
8—Shutoff washer
9—Nut

Figure 28–20. Enlargement of metering valve assembly. [Courtesy of Lucas CAV Ltd.]

Shut-down Device. The shutoff shaft (4 in Figure 28–19) has a half-round end on it that contacts the underside of the shutoff washer (8 in Figure 28–20). When the shutoff lever (21 in Figure 28–19) is rotated, it lifts the metering valve to a position where the metering port is blanked off, thereby stopping the engine.

Reversible Hydraulic Governor

The reversible-type governor is identical in operation to the standard unit, with the only exception being that movement of the metering valve is obtained by using an eccentric formed onto the end of the throttle control shaft. Figures 28–21 and 28–22 show the basic component part layout.

Rotation of the throttle shaft (11) causes the eccentric to move the control sleeve of the metering valve against the spring (6), thus loading the metering valve. The shutdown is the same as for the other hydraulic governor.

Perkins Diesel Engines

Perkins engines presently use either the DPA, DP15, or Simms/CAV fuel injection pumps on their range of diesel engines. Older model engines used in-line CAV pumps; however, on current in-line engines, the DPA style of rotary distributor pump is used, mounted either horizontally or vertically and driven from the front of the engine.

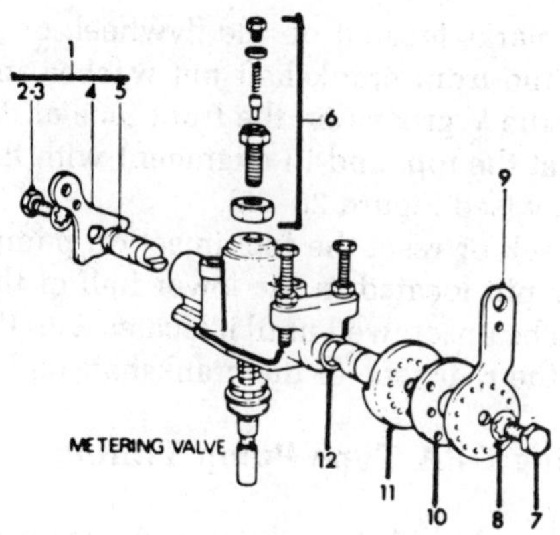

Index:

1—Shutoff shaft assembly
2—Screw
3—Washer
4—Shutoff lever
5—Shutoff shaft and seal
6—Antistall and idling stop assembly
7—Lock screw
8—Washer (shakeproof)
9—Throttle lever
10—Vernier plate
11—Throttle shaft and plate assembly
12—O ring seal

Figure 28–21. Component parts layout of reversible hydraulic governor. [Courtesy of Lucas CAV Ltd.]

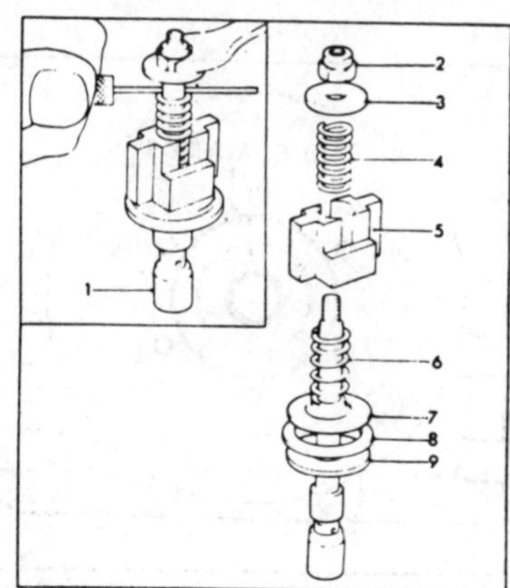

Index:

1—Metering valve
2—Metering valve nut
3—Shutoff washer
4—Idling spring
5—Guided control sleeve
6—Metering valve spring
7—Top damper washer
8—Center floating washer
9—Bottom sealing washer

Figure 28–22. Metering valve. [Courtesy of Lucas CAV Ltd.]

The V8-510 and 540 engines use the in-line Minimec pump along with the newer V8-640.

DPA Pump Installation

Top dead center (TDC) can be determined on Perkins engines by either (1) looking through the inspection hole provided on the engine flywheel housing at the scribed marks located on the flywheel, or (2) by removing the front crankshaft nut washer and ensuring that the V groove on the front face of the crankshaft is at the top, and in alignment with the dot on the pulley (see Figure 28–23).

To check or reset the fuel injection pump timing, a timing pin located in the lower half of the timing case can be unscrewed until it locates with the drilled hole in the rear face of the crankshaft pulley.

Checking DPA Type Pump Timing

1. Remove the side inspection cover from the pump housing shown in Figure 28–24.
2. Turn the engine over manually until the scribed line in the pump inspection window marked H on the pump rotor coincides with the scribed line of the snap ring. (This letter can be different on other model engines.) See Figure 28–24.
3. Check that the flywheel timing mark is aligned at 29 degrees BTDC. This can also be checked by ensuring that the distance from the TDC mark on the flywheel is 4.02″ (102.108 mm) or the piston depth from TDC is 0.399″ (10.134 mm).
4. If the timing is incorrect, loosen the pump retaining slotted flange nuts; manually grasp the pump housing, and turn the pump in the desired direction to change the timing.

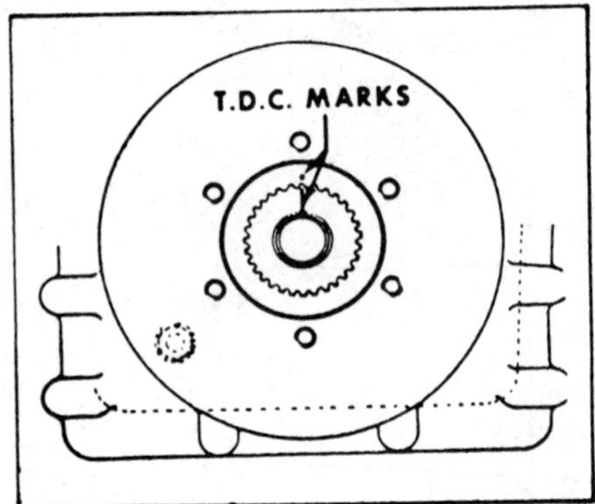

Figure 28–23. Determining TDC. (Courtesy of Perkins Engines)

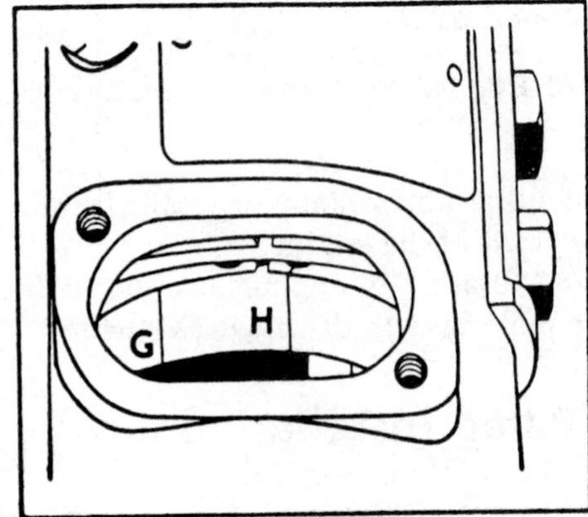

Figure 28–24. Checking DPA pump timing. (Courtesy of Perkins Engines)

Mechanical Knock Compared to Fuel Knock

Sometimes it is difficult to isolate a particular knock from the engine. To isolate it, run the engine at its maximum rpm and pull the stop control lever on the injection pump. If the knock disappears, it is due to a fuel problem; however, if it is still heard, it is a mechanical fault. When the fuel is cut off, the noise will fade in intensity, but will still be present.

Injection Pump Installation

A machined slot located in the hub of the worm gear on the injection pump drive shaft mates with a slot approximately 1/8″ wide (3.175 mm) in the adapter plate. With these slots aligned and piston 1 at TDC on the compression stroke, install the pump with the scribed line on its mounting flange in line with the scribed line on the pump adapter plate.

If, however, the machined slots are not in alignment, proceed as follows:

1. Place piston 1 at TDC on compression.
2. Remove the small auxiliary drive gear cover from the front cover timing case.
3. Remove the auxiliary drive gear by taking out the three retaining bolts.
4. Turn the auxiliary drive shaft until the machined slots are in alignment; then reinstall the auxiliary drive gear and bolt it into position, making sure that the drive shaft does not move during this operation.
5. Install the auxiliary drive gear cover back onto the timing case.

Engine Firing Orders

1. All 3-cylinder clockwise-rotation engines: 1–2–3.
2. All 4-cylinder clockwise-rotation engines: 1–3–4–2.
3. All 6-cylinder clockwise-rotation engines: 1–5–3–6–2–4.
4. All V8 clockwise rotation engines: 1–8–7–5–4–3–6–2.

Injection Pump Overhaul

All fuel injection pumps, are precision manufactured pieces of equipment; therefore, unless you have the knowledge and expertise along with the necessary special tools and test equipment, *do not attempt* to overhaul or tinker with the pump. Send it to your local fuel injection dealer.

The test procedure for any CAV injection pump can be found in specific test literature relating to the particular model of pump, along with the necessary test readings. The following procedure therefore is of a general nature and gives an idea of the kinds of tests that are required during testing and overhauling of a pump. It is not meant to replace the manufacturer's service information, which is the best available.

These tests must be carried out under spotlessly clean surroundings, preferably in a fuel test lab, to prevent the possibility of problems associated with the entrance of dust or dirt.

Pump Testing

Some of the major factors to be considered before testing a pump are as follows:

1. During testing, only recommended fuel test oils should be used.
2. Carry out all tests in the sequence given by the manufacturer.
3. The pump *must* only be rotated in its normal direction of rotation; otherwise severe damage can result.
4. Do not run the pump for long periods without having the fuel inlet valve to the pump open.
5. Do not at any time run the pump with either of the advance piston plugs loose or removed.
6. Oil leakage while testing the pump at the various adjusting screws is a normal occurrence. When finished, these will be corrected by the insertion of sealing caps.
7. Ensure that the fuel *test oil* temperature does not exceed that recommended by the manufacturer.
8. Do not run a cold pump on the test machine. Submerge the pump assembly in test oil for 20 minutes before mounting on the machine.
9. Ensure that the pump is mounted correctly and square to the drive, as well as being secure.
10. Make sure that you are familiar with the test machine controls before undertaking any tests. Do you know where the emergency control switch is in the electrical panel feeding the machine?
11. Make sure that you do not have any loose clothing on while testing the pump, especially around the drive shaft area.
12. Attach the pump specifications and test procedure to a suitable place on the machine so that you can read it during testing.
13. Do you have all the necessary tools and fittings handy?
14. Do you have a set of ear protectors or head phones available?
15. Do you have on a set of safety glasses?
16. Do you have enough wiping rags (lint-free) available?
17. Is the immediate floor area around the test machine free of obstructions and spilled test oil?

Testing of Advance Devices

The checking of the DPA injection pump advance devices is similar on pumps with mechanical or hydraulic governors. Figures 28–25 and 28–26 show each pump type with the cam advance checking tool installed.

The special advance tool P/N 7244-50 (hydraulic governor) or P/N 7244-59 (mechanical governor) is inserted into the hole in the spring cap, after first unscrewing the small screw normally located in the spring cap. A feeler pin tool P/N 7244-70 used with both types of governors is inserted into the hole in the spring cap. The scale on the timing gauge gen-

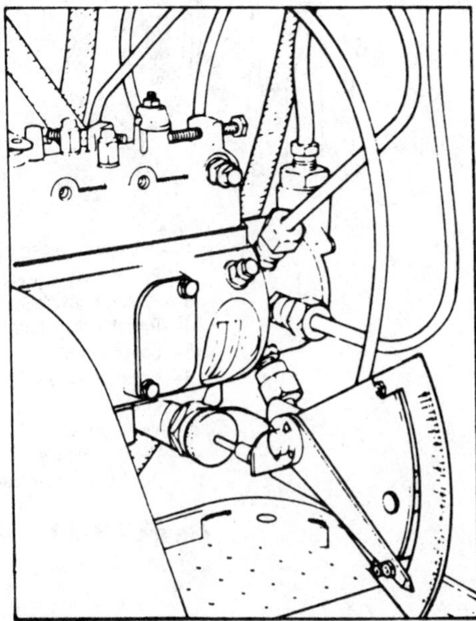

Figure 28–25. Mechanical governor with cam advance checking tool installed. (Courtesy of Lucas CAV Ltd.)

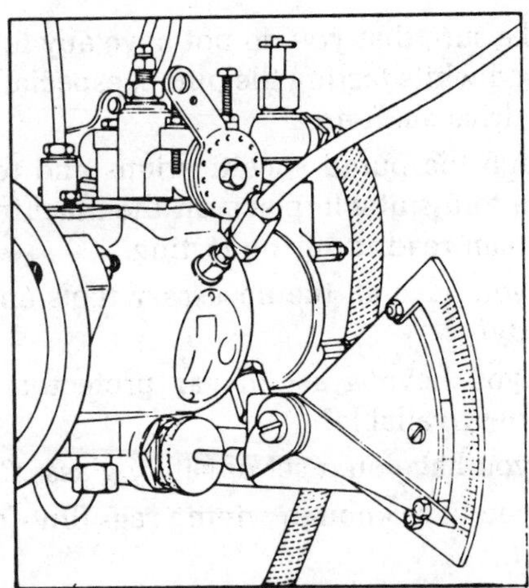

Figure 28-26. Hydraulic governor with cam advance checking tool installed. (Courtesy of Lucas CAV Ltd.)

erally covers 0 to 18 degrees. Zero the gauge by moving the scale relative to the pointer.

The pump should be reprimed with fuel after fitting the advance tool; then with the pump running on the test stand at 100 rpm, operate the throttle and press in and release the advance gauge pin several times. Reference to the test data sheet for the particular pump on test will indicate the advance at a given rpm.

The speed advance device on both the mechanical and hydraulic governor types is altered by changing the shims behind the piston spring and cap. On those units employing the combined load and speed advance, alter the shims beneath the inner and outer piston springs.

Fuel Transfer (Lift)

The common type of fuel lift pump used with CAV in-line and distributor-type injection pumps is a diaphragm-type unit manufactured by AC. It has a delivery pressure of between 2.75 and 4.25 psi (18.96 and 29.3 kPa). Such a pump is shown in Figure 28–27. This pump can be used to prime the fuel system by actuating the primer (18) on the side of the pump. The pump is normally engine driven from the camshaft. However, on in-line injection pumps, it can be mounted and driven from the injection pump camshaft.

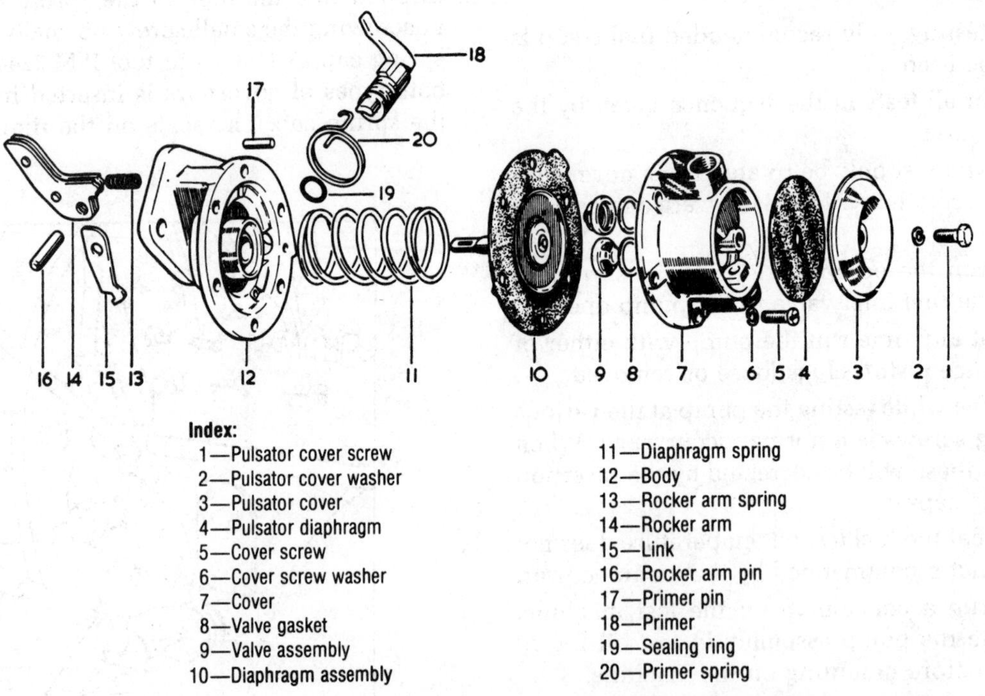

Index:

1—Pulsator cover screw
2—Pulsator cover washer
3—Pulsator cover
4—Pulsator diaphragm
5—Cover screw
6—Cover screw washer
7—Cover
8—Valve gasket
9—Valve assembly
10—Diaphragm assembly
11—Diaphragm spring
12—Body
13—Rocker arm spring
14—Rocker arm
15—Link
16—Rocker arm pin
17—Primer pin
18—Primer
19—Sealing ring
20—Primer spring

Figure 28-27. AC fuel lift pump, series VP. (Courtesy of Lucas CAV Ltd.)

29

Injection Nozzles—Types and Operation

This chapter is designed to familiarize you with the concept of operation and the various design types of nozzles currently in use on automotive passenger car and light truck diesel engines.

At the present time there are no mass-produced diesel passenger cars or light pickup trucks that use the direct-injection principle—which is where the injector sprays fuel directly into the combustion chamber formed by the piston crown and the underside of the cylinder head.

For reasons of quietness and the ability to meet the EPA smoke emissions regulations, manufacturers have chosen to use the swirl type pre-combustion chamber whereby the injector sprays fuel into a small ante-chamber above the cylinder. The fuel is then ignited and enters the main combustion chamber through a passageway cast at an angle leading to the main combustion chamber area.

Pre-combustion chambers are located in the cylinder head and can be removed from the cylinder head during overhaul for servicing purposes. Both the pre-combustion chamber type and direct injection type of engines are illustrated in Chapter 3, dealing with combustion chamber types. The reasons for using one type of combustion chamber over another is discussed in some detail in combustion chamber designs, Chapter 3.

Types of Injection Nozzles

Contained within this book are chapters dealing with the majority of well-known passenger car and light truck diesel engine fuel systems now in use worldwide. With the exception of the Mercedes-Benz, Toyota and Datsun/Nissan pickup trucks, which use an in-line-type fuel injection pump; all the other diesel engines employ a distributor/rotary-type injection pump.

The three most commonly used distributor-type injection pumps in use today are (1) the Robert Bosch VE model, which is also manufactured under a licencing agreement with both Diesel Kiki and Nippondenso in Japan; (2) the Roosa Master/Stanadyne DB2-type pump used by General Motors on their 4.3L, 5.7L, and 6.2L V6 and V8 passenger car and light truck engines, Ford/IHC on their 6.9L V8 and (3) the legendary DPA pump model manufactured by Lucas CAV and also built under licence by RotoDiesel in France and several others worldwide.

Because the pre-combustion style swirl chamber is in use on all passenger cars and many light truck diesel engines, the type of fuel injector in use is very similar in both physical appearance and actual internal design and operation, regardless of the manufacturer. Within this chapter, therefore, the operation, maintenance, and repair of these injectors can be considered common to most diesel engines.

Four basic types of fuel injectors are used in the diesel engines discussed in this textbook:

1. The CAV/GM Microjector which is used in the 4.3L and 5.7L General Motors V6 and V8 diesel engines (1980 and later).
2. The Roosa Master/Stanadyne pencil-type nozzle used in pre-1980 General Motors 4.3L and 5.7L diesel engines.
3. The throttling pintle-type nozzle manufactured by Robert Bosch and its Japanese licensees Diesel Kiki and Nippondenso and found in the following vehicles discussed in this book:
   a. VW Rabbit
   b. Volvo D24 (VW designed/manufactured engine)
   c. Audi 5000 (VW designed/manufactured engine)
   d. Peugeot

e. Dodge Ram 50/Mitsubishi
f. Isuzu/GMC
g. Ford 2.0L Tempo/Topaz, Escort/Lynx
h. Ford/Mazda 2.2L Ranger/Bronco II
i. Ford/BMW 2.4L Continental/Thunderbird
j. Datsun/Nissan
k. Toyota Landcruiser/passenger cars
l. Mercedes-Benz
m. GMC 6.2L V8 light pickup truck engine (GMC/Chevrolet)

4. Lucas CAV low spring 17 mm and United Technologies Diesel Systems group 17-mm injectors used by Ford/International-Harvester in their 6.9L V8 light truck diesel engine.

Injection Nozzle Design

Direct-injection-type engines employ an injector commonly referred to as a hole-type nozzle, which can employ either one or more holes in the spray tip. When the nozzle spray tip contains more than one hole, the nozzle is known as a multi-hole nozzle. Hole-type nozzles can have either a single hole with a spray-in angle of from 4 to 15 degrees when used with turbulence chamber type engines, or from 3 to 18 holes with orifice diameters ranging from 0.006" to 0.033" (0.076 to 0.838 mm), depending upon the engine. Fuel release pressures of between 2300–3300 psi (15858–22753 kPa) are common on direct-injection engines.

The indirect-injection or pre-combustion chamber engine uses what is commonly called a throttling pintle-type nozzle, which only employs one hole (the exception is the pre-1980 5.7L GMC pencil nozzles which had 4 spray holes) and has fuel release pressures in the 1800–2100 psi (12411–14480 kPa) range.

Pintle nozzles usually have an orifice (hole) that can range from 0.030" to 0.060" average on most automotive applications, although some can be smaller than this depending on the particular engine make.

When a pre-combustion chamber-type engine is used, the use of glow plugs is mandatory in order to facilitate ease of starting in cold weather. These glow plugs are required because of the greater overall surface area of the combustion chamber that is exposed to loss of heat radiation, plus the use of lower injection pressure nozzles that do not atomize the fuel as readily as in the multi-hole nozzle.

Two types of nozzles, namely the inward and outward opening style, are shown in Figure 29–1.

The inward opening nozzle is also referred to as a closed nozzle owing to the fact that it opens away from the combustion chamber during fuel injection and closes towards it when injection is complete.

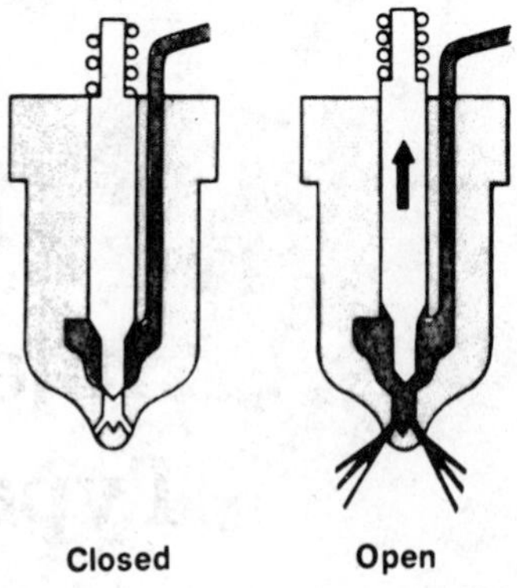

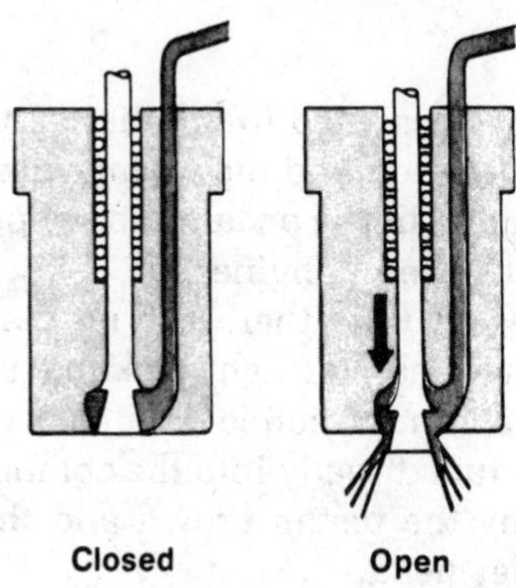

Figure 29–1. Inward opening nozzle and outward opening nozzle.

The outward opening nozzle is also referred to as an open nozzle because it moves toward the combustion chamber during injection and away from it when injection is complete.

The component parts of typical multi-hole and pintle-type nozzles are shown in Figures 29–2 and 29–3. The nozzle fuel spray-in angle is shown in the figures which is an "included spray-in angle." For example, if the spray-in angle was given as 160 degrees, this is the dimension of the included angle (total width of fuel spray from one side to the other).

Many nozzles may appear to be the same physical size as another model; however, never intermix nozzles of different models or manufacture in an engine because they can have different spray-in angles as well as different fuel release pressures. Serious engine damage can result if this is attempted.

Nozzle Function

The purpose of an injection nozzle is to receive from the injection pump a metered amount of fuel under

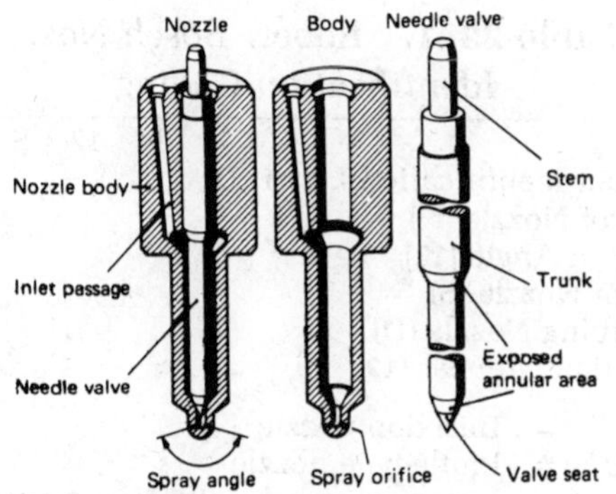

Figure 29–2. Multi-hole injection nozzle.

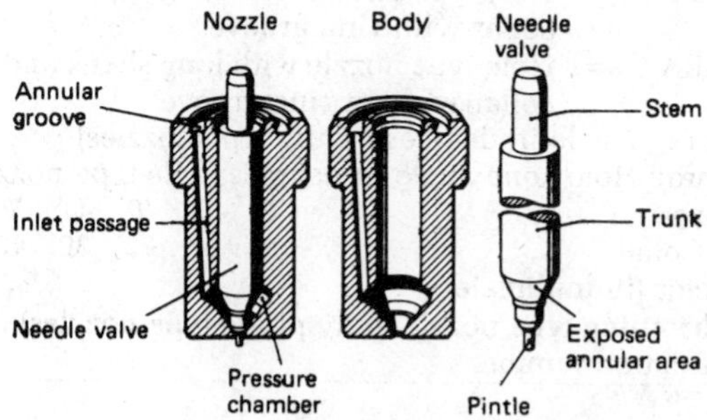

Figure 29–3. Pintle-type injection nozzle.

high pressure that will then be delivered by the injector nozzle into the combustion chamber. The fuel leaving the nozzle spray tip must do so at a high enough pressure to penetrate the compressed air charge in the combustion chamber in an atomized spray pattern so as to produce ignition with a minimum of delay, thereby producing optimum combustion and a smooth running, economical engine. To achieve this, the injector assembly must be made to extremely close and exacting tolerances.

As shown in Figure 29–4, the nozzle employs a tapered face type of needle valve which is held on a lapped nozzle seat in the spray tip by the action of a coil spring. Fuel under high pressure from the injection pump delivered to the nozzle through an internal fuel passage acts upon the tapered needle valve face causing the valve to lift upwards or move downwards depending upon whether the nozzle is an inward- or outward-opening type.

When the fuel pressure from the injection pump decreases, the needle valve is returned rapidly to its seat by the action of the coil spring above the needle valve. This action effectively ends fuel injection to that cylinder. The action of the fuel pressure on the needle valve causes the term *closed differential hy-*

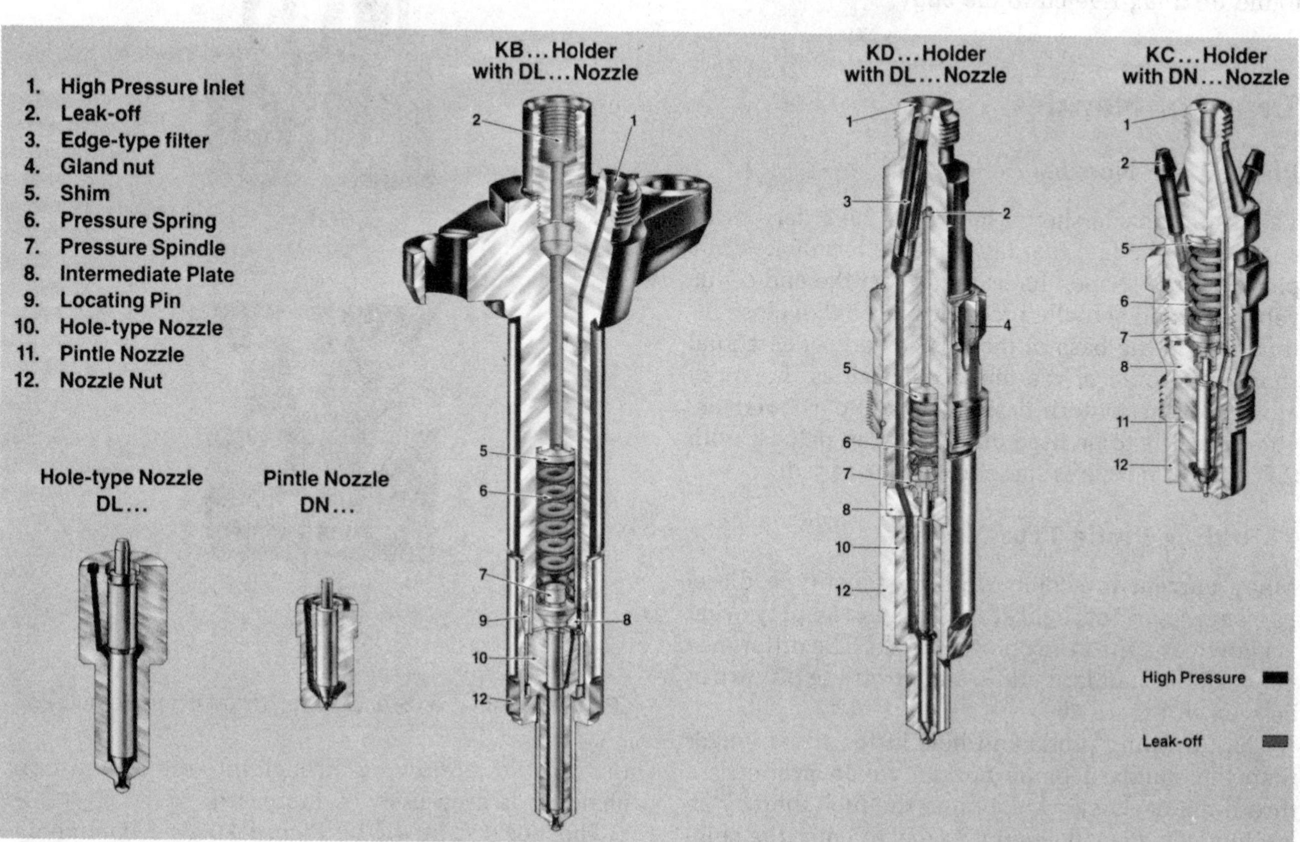

Figure 29–4. Holders and nozzles. [Courtesy of Robert Bosch Corporation]

draulically operated type to be used in describing these types of nozzles.

The nozzle and its spray tip are matched to one another at the time of manufacture and should not be intermixed when overhauling or repairing the injector.

The conical area at the base of the nozzle needle valve is ground to a slightly different angle with respect to the valve seat, which results in line contact seating thereby creating a high pressure sealing area to prevent leakage that could cause an increase in fuel consumption, unburned fuel, and thus smoke at the exhaust pipe as well as carbon buildup around the nozzle tip which can cause the nozzle to hang up or stay open. Plugging of the tip is also a possibility.

Similar nozzles can be used with various types of nozzle holders depending upon the application and make of engine. Examples of Robert Bosch nozzles and holders are illustrated in Figure 29–4.

An example of the coding used to identify Robert Bosch fuel injection nozzles is shown in Table 29–1. This is typical of the type of coding employed by most nozzle manufacturers.

The nozzle code number is stamped or etched on the body of the injector, or in some cases can be found on a tag riveted to the body.

Table 29–1. *Robert Bosch Nozzle Identification Coding*

D N 12 S D 12

1 Product Identification Letter (D)
2 Type of Nozzle (N)
3 Spray-in Angle (12)
4 Size of Nozzle (S)
5 Throttling Nozzle (D)
6 Variation Number (12)

1	"D"	=	Injection nozzle
2	"L"	=	Pintle-type nozzle
	"L"	=	Hole-type nozzle with ring groove
	"LA"	=	Hole-type nozzle without ring groove
	"LF"	=	Hole-type nozzle liquid cooled
	"LL"	=	Hole-type nozzle with long shank and long collar with ring groove
	"LLA"	=	Hole-type nozzle with long shank and long collar without ring groove

3 Spray Angle in degrees (pintle-type nozzles)
Spray Hole cone angle in degrees (hole-type nozzles)

4 Sizes:	R	S	T	U	V	W
Body Collar Diameter (in millimeters)	16	17	22	30	42	50

5 Throttling-type nozzle (only pintle-type nozzles)
6 Variation number

Types of Nozzles

Pintle-Type Nozzles

The pintle nozzle shown in Figure 29–3 derives its name from the fact that the valve tip terminates in a pin, or pintle shape, that extends past the end of the valve seat and actually protrudes through a close fitting hole in the base of the nozzle body. The actual shape or profile of the pintle determines the spray angle and the pattern desired for the particular engine that it is to be used on. The spray pattern with this type of nozzle is therefore cone-shaped.

Throttling Pintle-Type Nozzles

Many current pre-combustion chamber-type diesel passenger car and light truck engines employ what is known as a throttling pintle nozzle. The difference between the standard pintle and throttling pintle can be seen in Figure 29–5.

The throttling pintle and hole in the tip is longer than the standard pintle nozzle, which produces a throttling, or delaying, action to the fuel, thereby allowing only a small amount of fuel to enter the combustion chamber at the beginning of injection. However, as the valve continues to lift (by fuel pressure), the rate of fuel flowing through into the combustion chamber is progressively increased.

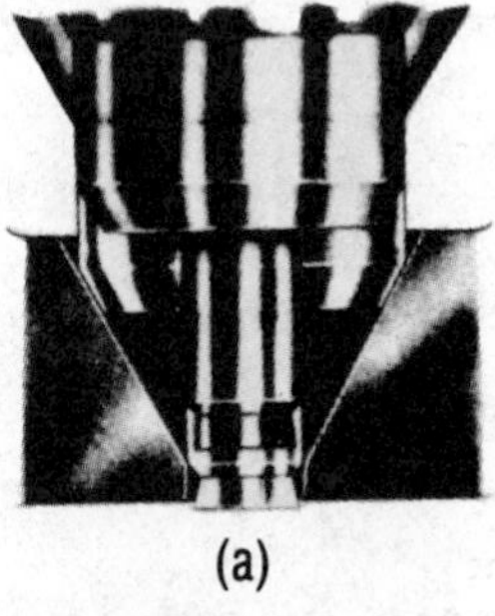

(a)

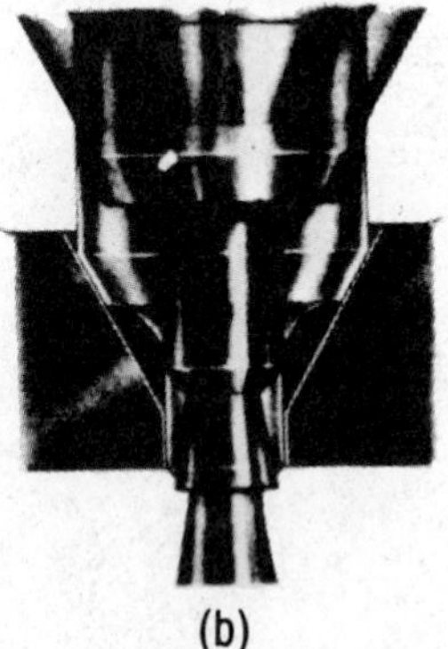

(b)

Figure 29–5. a. Standard. b. Throttling pintle nozzles.

The nozzle shown in Figure 29–6 is commonly known as a pintaux pintle nozzle, which is similar to the standard pintle type, but with an additional

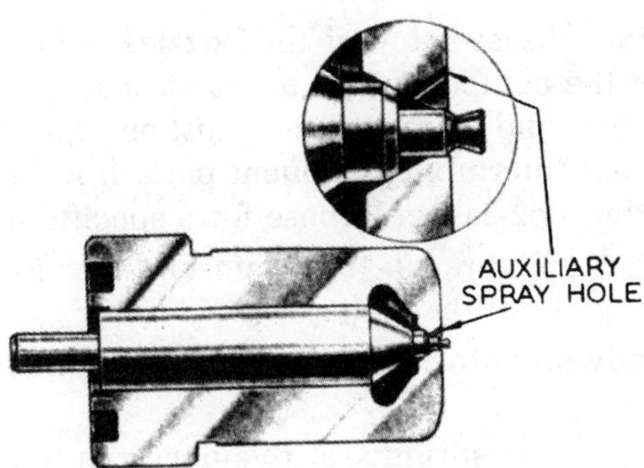

Figure 29-6. Pintaux pintle nozzle.

small plain orifice located so as to direct a finely atomized spray of fuel into the hottest part of the turbulence chamber during the starting period to assist cold weather starting of the engine.

During engine operation, when the fuel pressure delivered to the nozzle is higher, the nozzle valve will be withdrawn from the pintle hole, allowing the major portion of the fuel to enter the combustion chamber through the main orifice.

Nozzle Problems

Service life of injection nozzles is directly attributable to the following conditions:

1. Proper control of engine operating temperature to ensure complete combustion of injected fuel.
2. Water- and dirt-free fuel supply.
3. Correct grade of fuel for ambient temperature conditions encountered.

Injection nozzle problems are usually indicated when one of the following conditions exists:

1. Black smoke at the exhaust.
2. Poor performance and a lack of power.
3. Hard starting.
4. Rough idle and misfire.
5. Increased fuel consumption.
6. Combustion knock.
7. Engine overheating.

A quick check of the nozzle operation when it is still in the engine can normally be performed by running the engine at the speed that the problem is most noticeable. Loosen off a high-pressure fuel line at each nozzle one at a time between one-half to one full turn. Cover the fuel line with a rag to prevent fuel spraying onto you or the engine compartment.

With the fuel line loose, the injector will not be able to inject because insufficient fuel pressure will be present to lift the internal nozzle valve against the return spring. Under such a situation, the cylinder will receive no fuel. The engine speed should decrease, and its sound should change indicating that it is running on one less cylinder.

If one nozzle is found where loosening the high pressure fuel line makes little or no difference either in the misfiring condition or visible black smoke concentrations at the exhaust pipe, then that nozzle should be removed and checked on a pop tester for release pressure, spray pattern, and holding pressure. If the nozzle passes these tests, then the problem is either in the injection pump itself or low compression in that cylinder.

Before testing nozzles, *do not* clean them, especially at the spray tip, because if you remove any carbon that was affecting the nozzle spray pattern or release pressure, you will have removed this evidence. Test the nozzle just as it is when it is removed from the engine.

Cleaning/Testing Microjectors

Current 4.3L (262 cu.in.) V6 and 5.7L (350 cu.in.) V8 model General Motors automotive diesel engines employ a miniature fuel injector that is screwed into the cylinder head. This nozzle was first released in 1980 and was produced at that time by Lucas CAV and termed the "Microjector"; it is also commonly known as a poppet nozzle due to its design and operating characteristics. The nozzle is shown in Figure 29–7. Since that time, the GM/DED (Diesel Equipment Division) has undertaken manufacture of these nozzles.

Identification of the CAV versus the GM/DED injection nozzle can be seen in Figure 29–8, which shows that the GM/DED nozzle has hash marks on

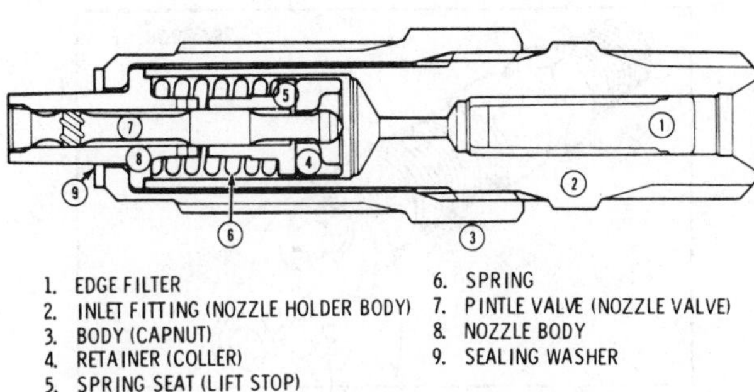

1. EDGE FILTER
2. INLET FITTING (NOZZLE HOLDER BODY)
3. BODY (CAPNUT)
4. RETAINER (COLLER)
5. SPRING SEAT (LIFT STOP)
6. SPRING
7. PINTLE VALVE (NOZZLE VALVE)
8. NOZZLE BODY
9. SEALING WASHER

Figure 29-7. Construction of a poppet nozzle. [Courtesy of Buick Motor Car Division, GMC]

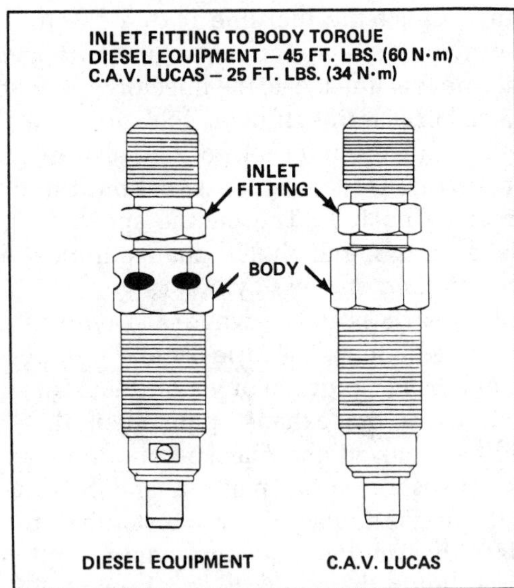

Figure 29-8. Identification of GM/DED and Lucas CAV injection nozzles.

the flats of the nozzle body hex, while the Lucas CAV unit does not.

Removal and installation of these nozzles can be found in Chapter 22, dealing with the 4.3L and 5.7L diesel fuel systems.

Cleaning of these nozzles requires that the old copper sealing washer on the end of the nozzle be removed prior to cleaning with the use of a pair of side cutter-type pliers as shown in Figure 29–9.

Cleaning.

1. Always wash the exterior of the nozzle assembly with a brass bristle brush so as not to scratch the nozzle surfaces. Clean diesel fuel can be used for this purpose or one of the readily available commercial cleaners. Do not attempt at this time to remove accumulated hard carbon because this will be done later. Remove only loose dirt or carbon.

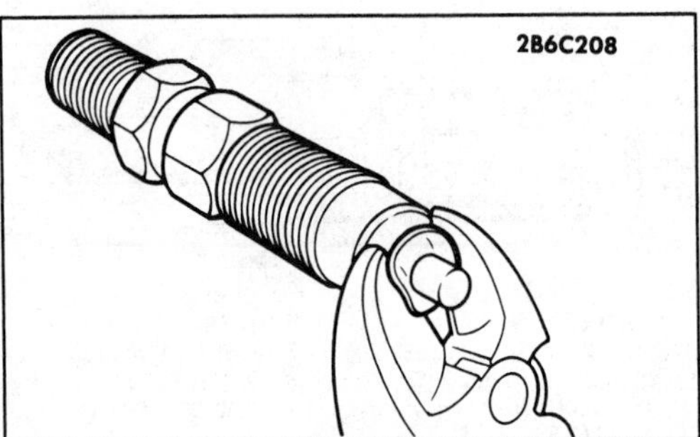

Figure 29-9. Removal of the nozzle copper washer. (Courtesy of Buick Motor Car Division, GMC)

2. Before disassembly of the nozzle, keep in mind that the component parts are all matched to one another and therefore they *must* be kept together. Do not intermix component parts from other injection nozzles with those for a specific nozzle.
3. Refer to Figure 29–10 and unscrew the long inlet fitting out of the body.
4. Gently press the actual nozzle spray tip out of the body.
5. Remove the spring seat retainer from the assembly by sliding the retainer sideways.
6. Remove the spring seat and its valve.
7. Press the pintle valve from the spray tip.

Refer to Figure 29–11 and insert all of the matched components for the nozzle into a common rack slot so that they will not be intermixed with those parts from other nozzles out of the engine.

The best method used to effectively clean nozzle parts and ensure removal of accumulated carbon is to use an ultrasonic bath cleaner such as a BT-8041 or Kent-Moore (GM) J-29653 unit or other such make

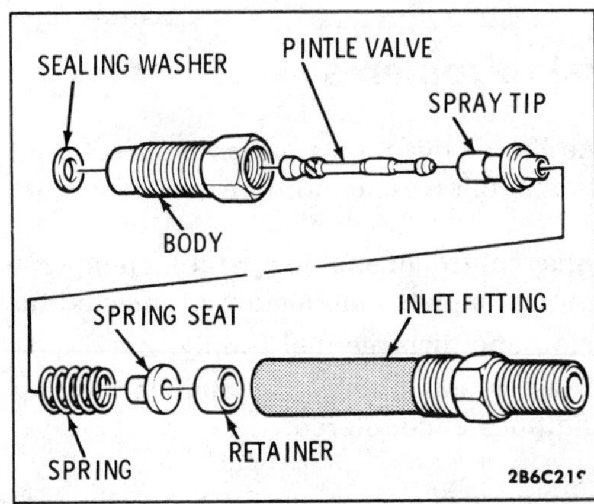

Figure 29-10. Parts of injection nozzle. (Courtesy of Buick Motor Car Division, GMC)

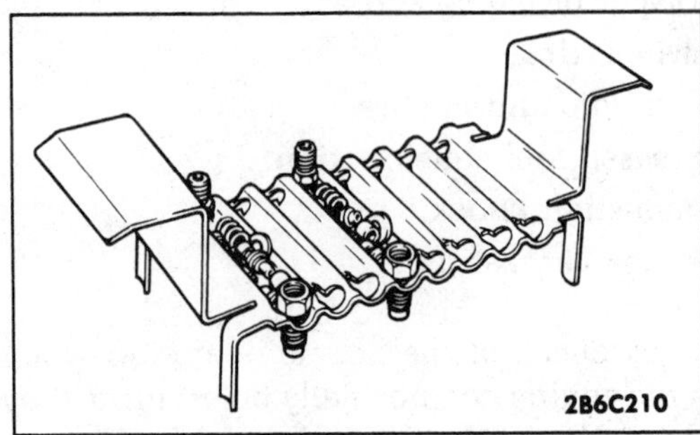

Figure 29-11. Ultrasonic cleaning tray/rack. (Courtesy of Buick Motor Car Division, GMC)

that will perform the same job. Ultrasonic cleaning uses a hot liquid along with high pressure sound waves to break down the carbon accumulations and is very effective in doing this without causing physical damage to the injector parts.

Hartridge Equipment offer a variety of nozzle test machines for successful cleaning and servicing of all types of injectors and nozzles. These special tools and equipment can be procured through local dealers. Visual inspection of cleaned components can be assisted by using a magnifying glass or Hartridge's Nozzle Viewer equipment.

Component Assembly. If all components are satisfactory, they can be assembled as follows:

1. Dip all components in clean calibrating fluid such as ISO 4113 or SAE J967D during assembly so that they are not assembled dry.
2. Insert the pintle valve into the spray tip.
3. Install the spring, its seat, and retainer.
4. To effectively compress the spring and lock the retainer into position, special tool J-29653-3 or BT-8018 shown in Figure 29–12 can be used; if this is unavailable, using a piece of steel as a base, compress the spring and slide the retainer to a centered position putting pressure on it. Check that the retainer has seated correctly by rotating the retainer by hand.
5. Carefully insert the nozzle tip into the body, then screw the inlet fitting into the body and tighten it to 35 lb.ft. (47 N•m).
6. After assembling any injection nozzle, perform the checks and tests to the nozzle described below.

Testing Injection Nozzles

Testing of the injection nozzles follows the same basic pattern as that for other diesel engine nozzles in that extreme cleanliness must be exercised as well as using calibrating fluid or special test oil such as SAE J967D at a room temperature of approximately 70°F (20°C) or ISO 4113 test oil.

Diesel fuel should never be used unless the situation precludes the use of available calibrating fluid. Since, diesel fuel is unstable especially if it has been in storage for any length of time. It tends to absorb moisture and will therefore corrode internal injector components as well as test equipment and is also a fire hazard when compared to calibrating fluid—and can create possible skin problems.

The use of a Hartridge Testmaster or similar test machine will prove to be satisfactory when testing these nozzles.

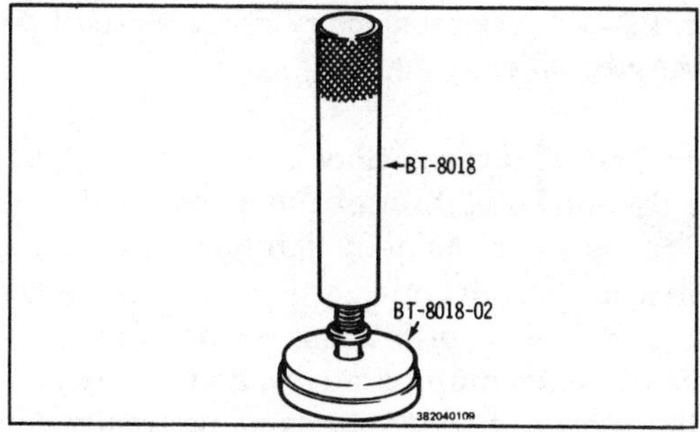

Figure 29–12. Special tool BT-8018/J-29653-3. (Courtesy of Buick Motor Car Division, GMC)

Nozzle Identification

Injection nozzles used on the current 4.3L and 5.7L diesel engines are identified by a color coded band around the injector body as shown in Figures 29–13 and 29–14.

V6 engines use either a red or green color band, while the V8 engines use a blue color band. In addition, stamped on the nozzle body hex nut are the numbers "403," which signify that this nozzle is for a V6 4.3L engine, while the numbers "404" indicate that the nozzle is for a V8 5.7L engine.

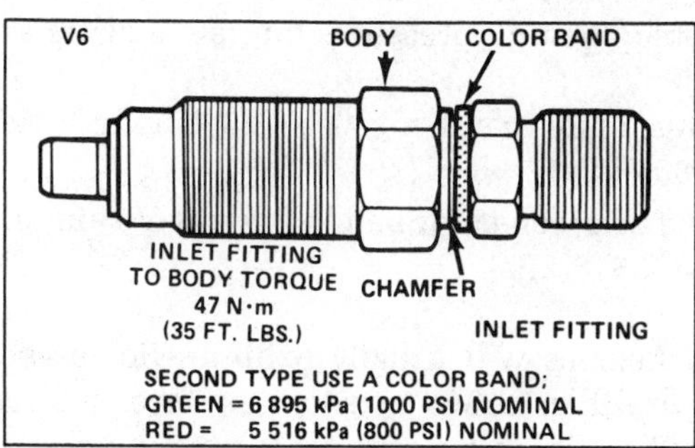

Figure 29–13. VIN T with aluminum heads nozzle installation.

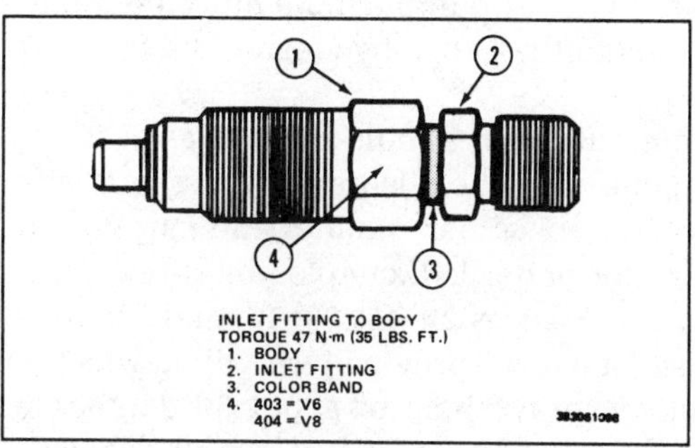

Figure 29–14. Injection nozzle.

Test Procedures

A nozzle pop tester similar to Figure 29–35 is required in order to determine the spray pattern, leakage, and opening pressure of each nozzle. Various nozzle testers are available and are shown throughout this textbook in sections dealing with injector overhaul and individual sections dealing with the test and service of other diesel engine nozzles.

CAUTION: When testing injection nozzles, be certain to wear eye protection and keep your hands away from all areas of fuel spray because penetration of the skin by fuel can occur leading to serious personal injury such as blood poisoning. Nozzles should always be tested within a clear plastic receptacle to protect the operator and also to see the pattern and condition of the spray.

Opening Pressure Check. To perform the opening pressure check, mount the nozzle to the pop tester machine and with the control valve on the machine open slightly, manipulate the tester pump handle until the injector releases fuel. Normally the rate at which the tester handle must be stroked will vary slightly between the type of nozzle under test, but generally anywhere from one full stroke in a period of from 3 to 5 seconds is considered acceptable.

Carefully note the gauge needle and establish at what pressure the nozzle actually releases fuel. Acceptable opening pressures on new nozzles are:

1. Blue color band nozzles—V8 engines: 1000 psi (6895 kPa)
2. Red or green color band nozzles—V6 engines: 800 psi (5516 kPa)

Used nozzles will usually exhibit a lower popping pressure than a new nozzle; therefore, subtract 150 psi (1034 kPa) as an acceptable opening pressure on used nozzles.

Spray Pattern. In order to protect the pop tester gauge against high pressure pulsations that are caused by rapid pump handle movement during this test, isolate the gauge by closing the pressure gauge valve.

Pump the tester handle slightly faster than you did when checking the release (popping) pressure (about 20 to 30 strokes per minute) and carefully note the fuel spray pattern. Examples of spray patterns are shown in Figures 29–15, 29–16, and Figure 29–17.

A satisfactory spray pattern will have a maximum included spray angle of about 30 degrees with acceptable atomization of fuel. Small droplets of fuel that form near the spray tip make the nozzle unacceptable.

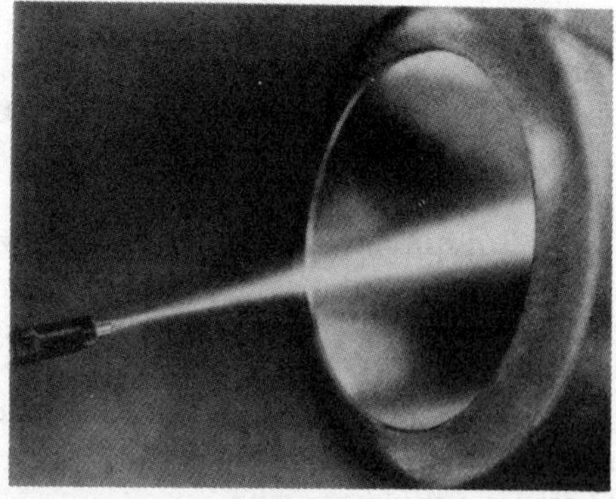

Figure 29–15. Satisfactory spray formation.

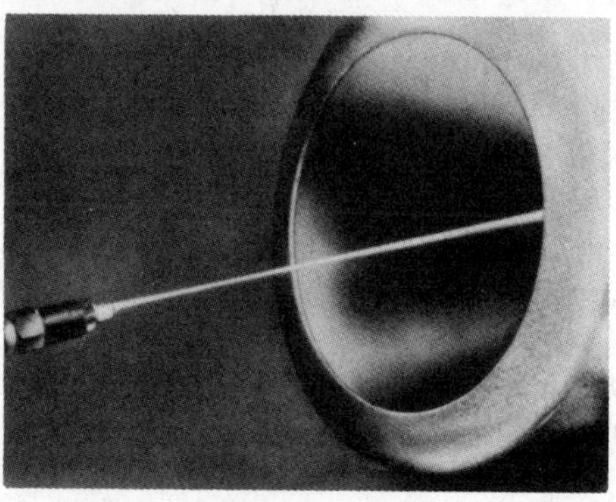

Figure 29–16. Unsatisfactory spray: hosing, no atomization.

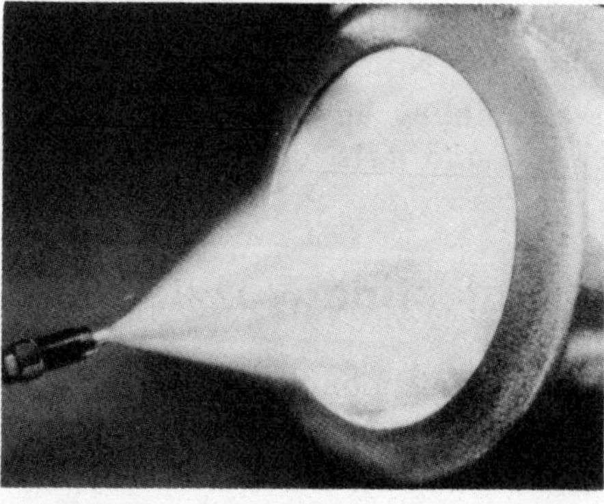

Figure 29–17. Unsatisfactory spray: unequal distribution over wide angle.

Nozzle Seat Tightness Check. Refer to Figure 29–18 for the interpretation of this check; with the pop tester gauge valve opened slightly wait until the fuel line pressure has dropped at least 290 psi (2000 kPa) lower than the recorded nozzle opening pressure. Dry the tip with compressed air, then pump the tester handle until the fuel pressure gauge reads approximately 150 psi below the nozzle opening pressure; hold this pressure for about 5 seconds.

Compare the state of the fuel droplets or formation

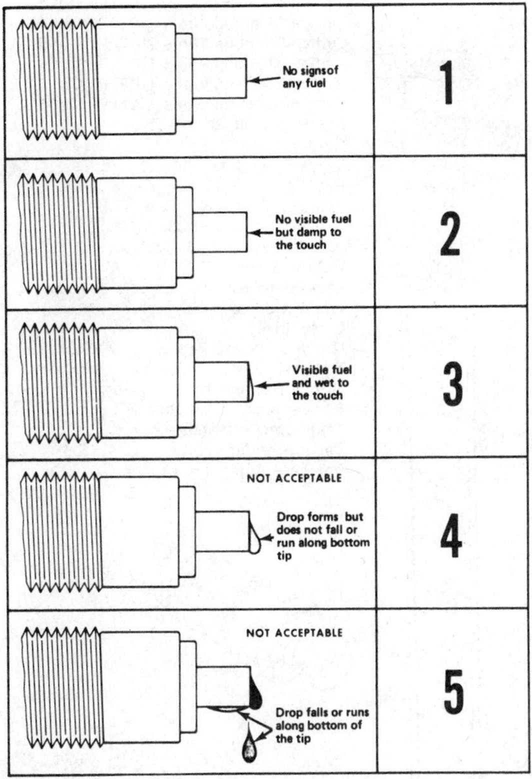

Figure 29–18. Microjector nozzle seat tightness check. (Courtesy of Buick Motor Car Division, GMC)

at the end of the nozzle tip with those shown in Figure 29–18; patterns 1, 2, and 3 are acceptable, while patterns 4 and 5 are *not*.

Nozzle Installation

When installing the nozzles, check to make sure that no foreign objects have dropped into the bore in the cylinder head and that all possible signs of carbon have been removed from the bore seat and thread area.

Always install new copper washers on the nozzles before screwing the injector into the cylinder head. This usually requires the use of a pair of pliers to pull the old copper washer off as shown in Figure 29–9; however, be careful not to damage the nozzle tip.

On aluminum cylinder head V6 engines, you *must* apply GM lubricant P/N 1052771 or its equivalent to the threads of the nozzle prior to installing it into a VIN T engine.

Install the nozzle into the threaded bore of the head and torque it to 25 lb.ft. (34 N•m) applying force only to the large hex on the nozzle body.

Connect up all of the fuel lines and torque them to the same figure as that for the nozzle while using a back-up wrench on the upper injection nozzle hex.

Roosa Master Pencil Nozzles

The basic operation and removal and installation of the Roosa Master pencil nozzle can be found under the 4.3L and 5.7L GMC diesel engine fuel system (Chapter 22).

Cleaning and testing of this nozzle is given below. Figure 29–19 illustrates the recommended Roosa Master Nozzle Cleaning Kit P/N 16494 along with a number of fittings and tools required for successful nozzle service. Cleaning of the pencil nozzle follows the same basic routine as that for other nozzles described in this chapter.

Testing Procedure

Before performing a nozzle test, refer to Figure 29–20 and using a sharp utility knife, remove the carbon dam seal from the lower end of the nozzle taking care not to scratch the seal ring groove.

A brass bristle brush can be used to remove loose carbon/dirt from the nozzle body and tip; however, do not attempt to dislodge hard carbon accumulations because tip damage may result. Using one part Brulin 815Q cleaner to ten parts of water, heat this solution to 200°F (93°C) and suspend the nozzles in this solution to a depth that just covers the carbon dam seal ring groove as shown in Figure 29–21. Usually 20 to 30 minutes will be required to dislodge heavy carbon accumulations.

Once the nozzles have been cleaned of all hard carbon accumulations, they should be flushed on a nozzle pop tester to remove all traces of the cleaning solution.

Testing Pencil Nozzles

Assemble the nozzle to a pop tester as shown in Figures 29–22 and 29–23 and close the pop tester gauge valve to protect it from damage while rapidly pumping the handle to flush/prime the nozzle.

Nozzle opening pressure for the 5.7L Oldsmobile/GMC diesel engine is 1800 + 100 or − 50 psi (12411 + 690 or − 345 kPa), which can be determined by slowly manipulating the pop tester handle. Do not concern yourself at this time with either nozzle spray pattern or chatter because these cannot be effectively checked until the needle valve "lift" has been set, which is explained below.

SPECIAL NOTE: Some pencil nozzles have threaded caps on the upper end; these must be removed when pop testing the nozzle, otherwise a hydrostatic lock will occur within the nozzle due to the

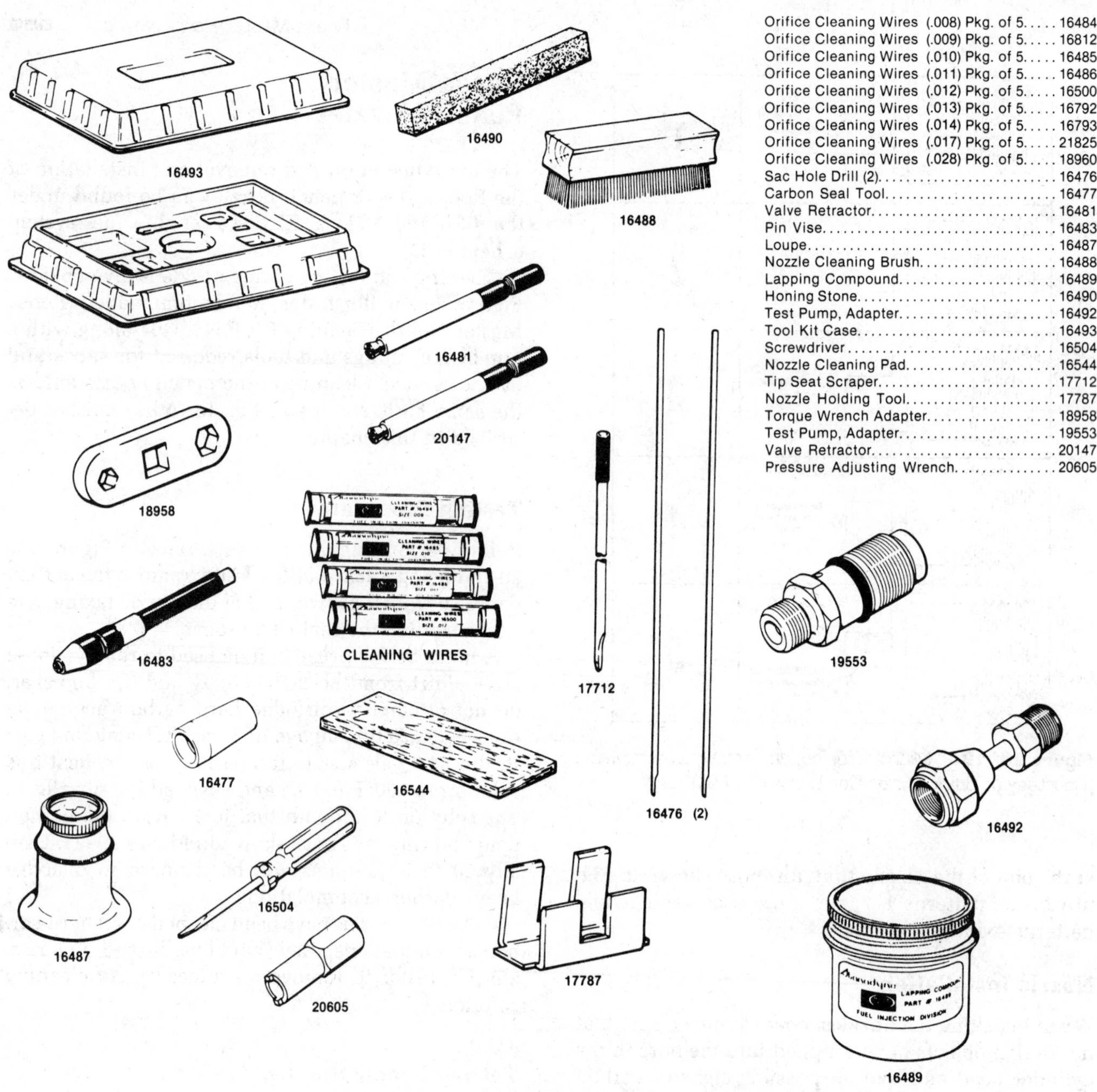

Orifice Cleaning Wires (.008) Pkg. of 5..... 16484
Orifice Cleaning Wires (.009) Pkg. of 5..... 16812
Orifice Cleaning Wires (.010) Pkg. of 5..... 16485
Orifice Cleaning Wires (.011) Pkg. of 5..... 16486
Orifice Cleaning Wires (.012) Pkg. of 5..... 16500
Orifice Cleaning Wires (.013) Pkg. of 5..... 16792
Orifice Cleaning Wires (.014) Pkg. of 5..... 16793
Orifice Cleaning Wires (.017) Pkg. of 5..... 21825
Orifice Cleaning Wires (.028) Pkg. of 5..... 18960
Sac Hole Drill (2)........................ 16476
Carbon Seal Tool.......................... 16477
Valve Retractor........................... 16481
Pin Vise.................................. 16483
Loupe..................................... 16487
Nozzle Cleaning Brush..................... 16488
Lapping Compound.......................... 16489
Honing Stone.............................. 16490
Test Pump, Adapter........................ 16492
Tool Kit Case............................. 16493
Screwdriver............................... 16504
Nozzle Cleaning Pad....................... 16544
Tip Seat Scraper.......................... 17712
Nozzle Holding Tool....................... 17787
Torque Wrench Adapter..................... 18958
Test Pump, Adapter........................ 19553
Valve Retractor........................... 20147
Pressure Adjusting Wrench................. 20605
16493
16490
16488
16481
20147
18958
CLEANING WIRES
16483
17712
16476 (2)
19553
16492
16477
16544
16504
16487
20605
17787
16489

RECOMMENDED NOZZLE CLEANING FLUID MANUFACTURERS

Brulin 815QR
Brulin & Company, Inc.
2920 Martindale Avenue
P.O. Box 270-B
Indianapolis, Indiana 46206

Penetone 423
Penetone Products
74 Hudson Avenue
Tenafly, New Jersey 07670

K-21 Carbon Removal Chemical
Hydrac Oy
Liisankatu 27A7
Henlsinki, Finland

Amway Industrial Cleaner
Amway Corporation
7575 East Fulton Road
Ada, Michigan 49355

Figure 29-19. Contents of 16494 nozzle cleaning kit. [Courtesy of Stanadyne Diesel Systems]

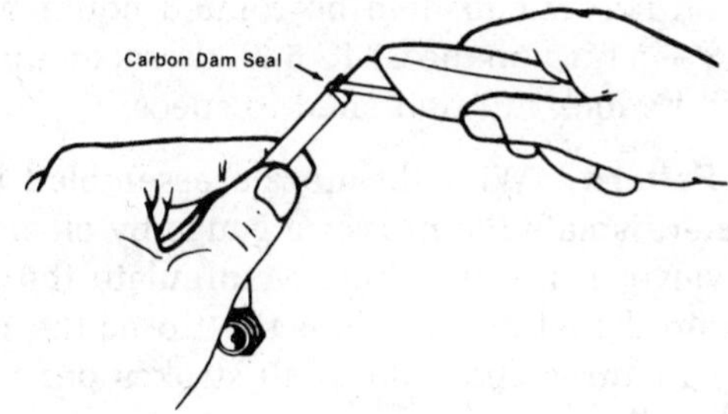

Figure 29–20. Removing carbon dam deal. [Courtesy of Stanadyne Diesel Systems]

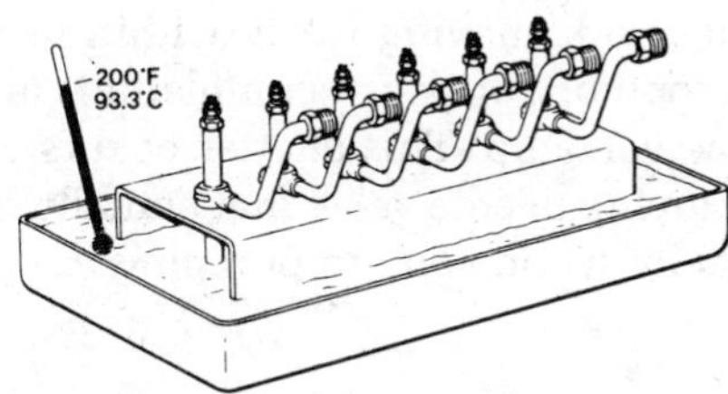

Figure 29–21. Removing carbon from nozzle tips. [Courtesy of Stanadyne Diesel Systems]

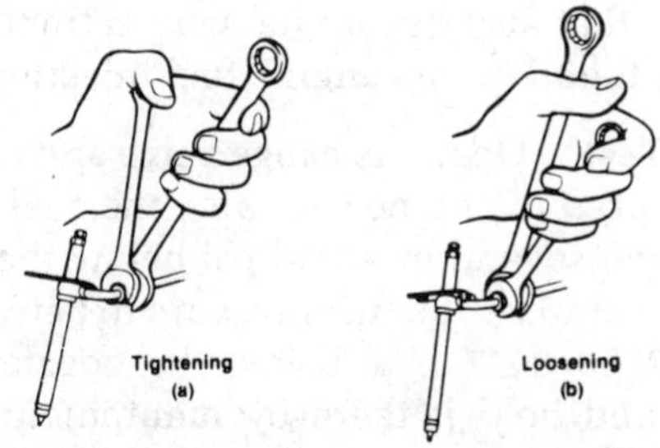

Figure 29–22. Tightening/loosening nozzle-to-pop tester fuel line. [Courtesy of Stanadyne Diesel Systems]

Figure 29–23. Nozzle assembled ready for testing. [Courtesy of Stanadyne Diesel Systems]

fact that there is no reduction in line pressure on the pop tester such as there is on an injection pump used on an engine.

Nozzle Adjustment. If the nozzle fails to open within the specified opening pressure, it can be assembled into special holding fixture P/N 17787 as shown in Figure 29–24 and its lock nut loosened off. The nozzle would be reassembled into the pop tester again and the lift adjusting screw shown in Figure 29–25 loosened off about 2 to 3 turns so that no interference will exist while adjusting the nozzle opening pressure adjusting screw shown in Figure 29–26. Pump up the fuel pressure on the pop tester and note at what pressure the nozzle now opens. Turning the adjustment screw in a clockwise direction will increase the force on the internal needle valve spring and therefore increase the opening pressure. Turning the screw in a counterclockwise direction will decrease the pressure of the internal spring, and consequently it will decrease the opening pressure of the nozzle.

Gently snug up the adjustment screw lock nut; remove the nozzle assembly from the pop tester and place it back into the special holding fixture P/N 17787 and torque the lock nut to spec using a torque wrench adapter P/N 18958 as shown in Figure 29–27.

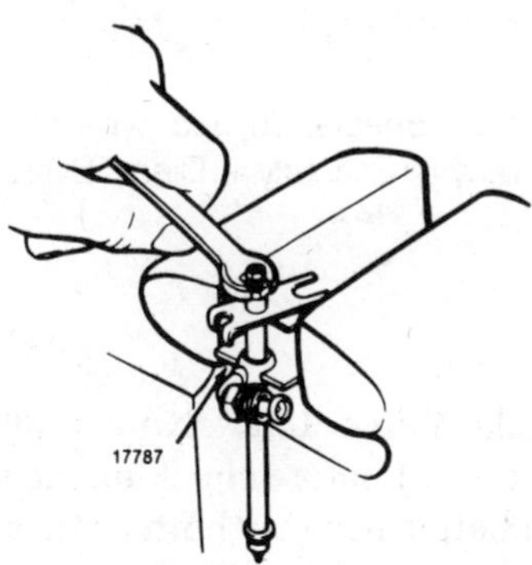

Figure 29–24. Holding fixture P/N 17787. [Courtesy of Stanadyne Diesel Systems]

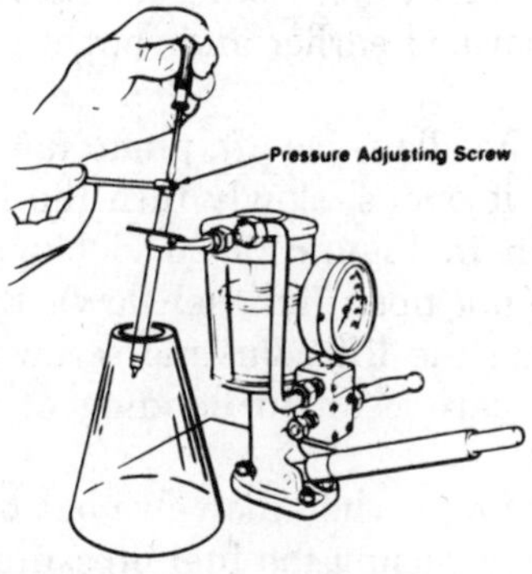

Figure 29–25. Lift adjusting screw. [Courtesy of Stanadyne Diesel Systems]

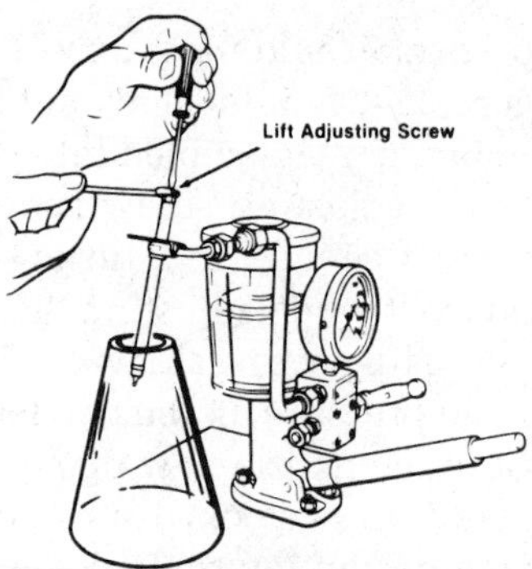

Figure 29-26. Nozzle pressure adjusting screw. (Courtesy of Stanadyne Diesel Systems)

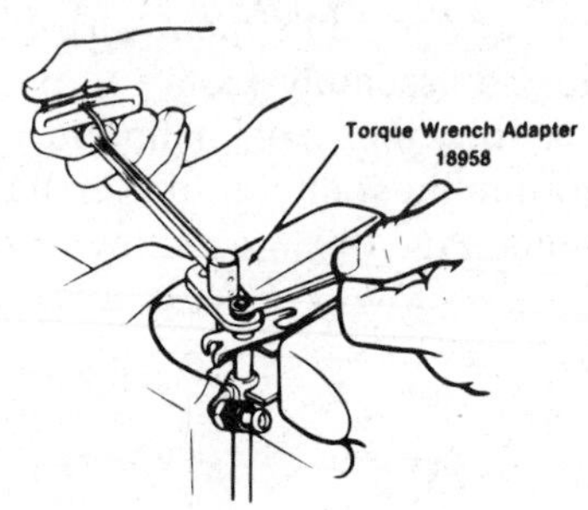

Figure 29-27. Special torque wrench adapter P/N 18958. (Courtesy of Stanadyne Diesel Systems)

Setting Nozzle Valve Lift. Note that the needle valve lift *must* be set before checking the nozzle spray pattern and chatter because both of these characteristics are affected by this setting.

An adjustment to the lift adjusting screw is required on these nozzles to set just how far the internal needle valve will move up when fuel pressure acts upon the tapered face of the internal needle valve. This is achieved by connecting the nozzle to the pop tester as described earlier and shown in Figure 29–23.

To set the needle valve lift, pump fuel through the nozzle until it opens; slowly turn the lift adjusting screw shown in Figure 29–25 clockwise until the nozzle does not open (no fuel flow). Exercise care when rotating the lift adjusting screw because extreme force can result in bending of the internal needle valve.

Confirmation of when the valve has *bottomed* can be obtained by raising the fuel pressure to between 200–500 psi (1379–3447 kPa) above the normal nozzle opening pressure with no fuel flow, although a slight dribble may appear at the tip.

The lift screw can then be rotated counterclockwise 5/8 of a turn on the GMC 5.7L diesel engine nozzles and its lock nut tightened to specs.

Spray Pattern. With the nozzle assembled to the pop tester, isolate the pressure gauge by closing the gauge valve in the fuel line. Manipulate the pump handle until fuel sprays from the tip of the nozzle (usually requires about 30 to 40 strokes per minute of the handle).

The spray pattern should be similar to that shown in Figure 29–28. At the same time, listen for an audible chatter from the nozzle—which is an indication that the needle is moving freely within its body and that no carbon or varnish accumulations are present to hang the valve up. This chatter occurs as a result of a free moving needle valve that actually opens and closes rapidly while you are pumping the pop tester handle.

NOTE: Chatter of a nozzle does not occur when the nozzle is in the engine due to the limited volume of fuel per injection stroke versus the large volume of fuel delivered in rapid succession when using a pop tester that also has no delivery retraction valve such as is found on the engine fuel injection system.

Leakage Test. Open the gauge valve and dry off the spray tip area of the nozzle after the fuel pressure has dropped several hundred psi below the opening pressure. Pump up the fuel pressure to between 400–500 psi (2758–3447 kPa) below the nozzle opening pressure and hold it there by manipulation of the tester handle.

A slight dampness or sweat is acceptable during this test; however, if a drop or dribble of fuel occurs at the tip within a 10-second period, then the nozzle requires servicing.

If the nozzle is installed into the engine with a drip or dribble of fuel from the spray tip, this raw, non-

Figure 29-28. Typical 4-hole nozzle spray pattern. (Courtesy of Stanadyne Diesel Systems)

atomized fuel will cause incomplete combustion, black smoke in the exhaust emissions, and poor fuel economy.

Return Fuel Check. On those nozzles using a return fuel line, loosen the connector nuts and place the nozzle into the position shown in Figure 29–29 just off the horizontal plane. Retighten the pop tester connector nuts and pump up the fuel pressure to below the opening pressure (see particular vehicle nozzle spec sheet) and carefully note the amount of return fuel. If the amount of return fuel is too much, then the nozzle requires servicing.

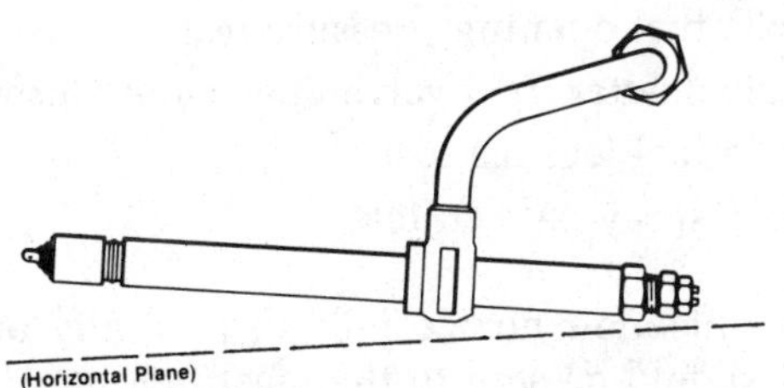

Figure 29–29. Checking return fuel rate. (Courtesy of Stanadyne Diesel Systems)

Troubleshooting Pencil Nozzles. Figure 29–30 illustrates typical problems that may occur with a pencil type nozzle.

Robert Bosch Pintle Nozzle Testing/Repair

Robert Bosch pintle and throttling pintle-type nozzles are used extensively by most of the world's current manufacturers of automotive and light pickup trucks.

Typical users of Bosch nozzles is listed at the start of this chapter under "Types of Injection Nozzles," Item 3, A through M.

Although these various engines all use Robert Bosch pintle-type nozzles, there are minor differences between these nozzles such as release pressures and the method in which the fuel is returned from the nozzle back to the fuel tank, etc.

Basic operation of all Bosch pintle-type nozzles is shown in Figure 29–31, which illustrates the fuel flow into, through, and out of the nozzle body holder or injector. Figure 29–32 shows that the fuel pressure

CAUSE Numbers in problem column indicate order in which to check possible cause of problem.	PROBLEM: Incorrect opening pressure	Nozzle will not open	Poor spray pattern	Poor atomization	Inconsistent chatter	No chatter	Seat leakage	High leak-off	Low leak-off	CORRECTION
Plugged or chipped orifices		1	1	1						1. Clean orifices 2. Discard if holes are chipped or eroded.
Spring components misaligned					1	1				1. Readjust opening pressure while flushing nozzle.
Varnish on valve					2	2			1	1. Clean guide area with solvent.
Deposits in seat area					3	3	1			1. Disassemble and clean.
Bent valve					4	4				1. Replace valve.
Broken spring	2									1. Replace.
Bottomed lift screw		2								1. Readjust.
Valve seat eroded or pitted						5	2			1. Lap valve to tip seat 2. Replace valve.
Tip seat pitted						6	3			1. Lap valve to tip seat *and* replace valve 2. Replace assembly.
Seat interference angle worn						7				1. Replace valve 2. Replace assembly.
Wrong adjustment	1									1. Readjust.
Wear or scratches at guide								1		1. Lap old valve in guide and replace valve.
Cracked tip			2	3			6			1. Replace assembly.
Valve not free				2			4			1. See "Problem - Inconsistent or no Chatter".
Insufficient clearance									2	1. Clean 2. Lap valve to guide 3. Replace valve.
Distorted body					5	8	5			1. Replace assembly.

Figure 29–30. Pencil nozzle troubleshooting. (Courtesy of Stanadyne Diesel Systems)

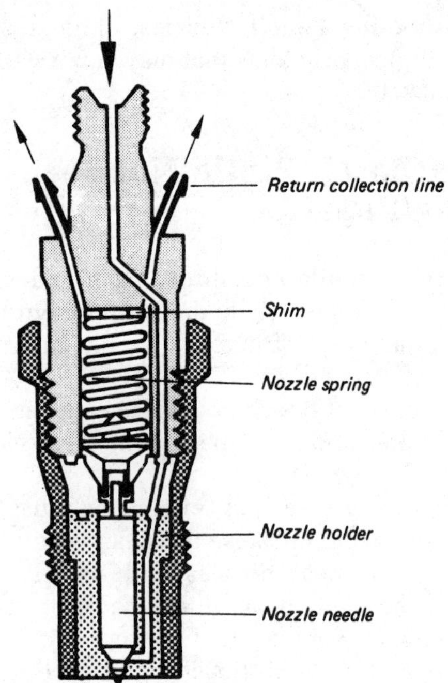

Figure 29-31. Cross section of a pintle nozzle. [Courtesy of Robert Bosch Corporation]

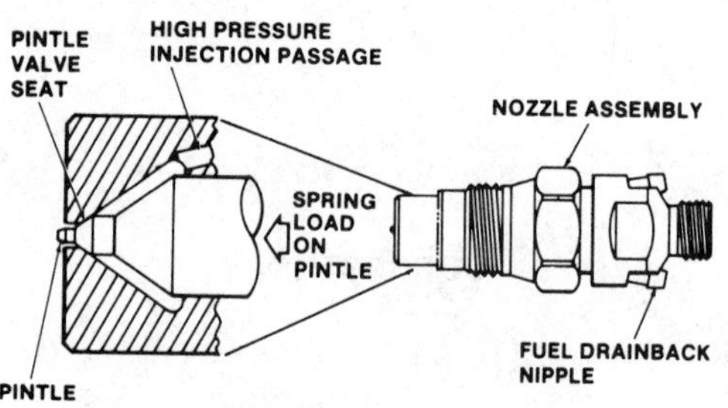

Figure 29-32. Pintle nozzle. [Reprinted with permission © 1984 Society of Automotive Engineers]

inside the nozzle acts upon the pintle valve tapered surface behind the pintle seat. When this pressure is high enough, the pintle valve will be lifted against the force of the spring inside the nozzle holder to allow fuel to flow out past the pintle valve and into the engine combustion chamber. When the fuel pressure in the line between the injection pump and the nozzle decreases, the spring pressure will close the pintle valve and injection ends.

The nozzle body holder or injector may differ slightly in outward appearance and the injector installation torque to the cylinder head may also differ; however, the testing, inspection, and overhaul of these nozzles can be considered common for all engines that use a Bosch KC holder and a DN or DNO type nozzle such as is shown in Figure 29-4.

Each manufacturer specifies that a particular injector nozzle pop tester be used with specific fittings for checking their engine's nozzles; however, there are a variety of nozzle pop testers available on the market that can be used to check any number of different nozzles since all that is required is to adapt the correct fitting to the injector body for the various tests.

Testing Nozzles for Performance

Testing of Robert Bosch nozzles follows the same basic routine as that for other manufacturers' nozzles. Tests that would be conducted to the nozzles with a pop tester are as follows:

1. Nozzle fuel opening pressure test
2. Nozzle chatter (fuel valve opens and closes) test
3. Nozzle fuel leakage test
4. Nozzle spray pattern test

If the injection nozzle fails to pass any of these tests, it should be sent to the local fuel injection repair shop in your area for repair or exchanged for a new or rebuilt one.

CAUTION: The fuel pressure build-up required to cause the injector nozzle to release fuel is commonly known as the *popping pressure,* or release pressure. This fuel is forced out of the nozzle tip in a finely atomized spray due to the high pressure build-up within the injector body and at the nozzle tip.

This fuel pressure is high enough to cause penetration of the skin leading to blood poisoning—therefore *never* place your hands or fingers into this spray area.

Pop testing machines come equipped with a protective receptacle, usually manufactured from a heavy transparent plastic that protects you from the high pressure spray while still allowing you to see the spray pattern of the fuel similar to Figure 29-23.

Test Equipment

Most pop testers such as that shown in Figure 29-33 come equipped with a variety of fittings and lines to allow a number of different nozzles to be tested.

NOTE: Use only calibrating fluid as per ISO specification ISO 4113, or SAE J968D or SAE No. 208629 when checking a nozzle because it is more stable than diesel fuel, which can contain traces of water and sediment.

Do not smoke or allow an open flame when testing nozzles and always wear safety goggles.

Nozzle cleaning kits and special tools are available from the nozzle manufacturer to facilitate the nec-

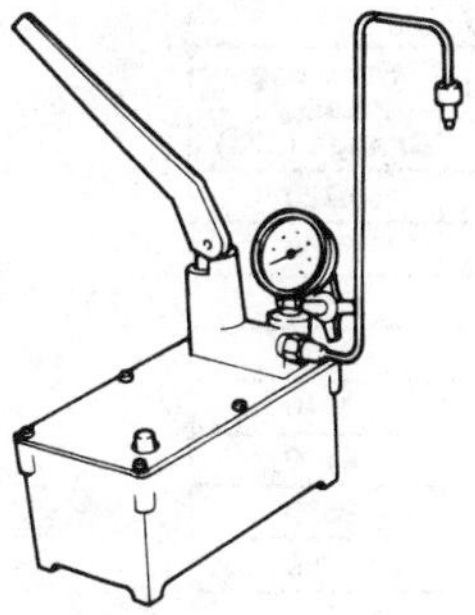

Figure 29-33. Rotunda 14-6300 injector nozzle tester. [Courtesy of Ford Motor Company, Dearborn]

essary repairs to the nozzle. An example of such a cleaning/repair kit was shown earlier in Figure 29-19 for the Roosa Master pencil-type nozzles. Figure 29-34 illustrates a similar type of kit that can be used for service/repair of Bosch pintle-type nozzles.

Nozzle Test Sequence

Popping Pressure.
Procedure

1. Connect the nozzle and holder to a suitable pop tester such as shown in Figure 29-35 and place clear plastic tubes to the injector overflow connections (fuel return lines) so that return fuel is not confused with injector leakage.
2. Leave the tester-to-nozzle holder fuel line loose and manually pump the tester handle until clear fuel free of air flows from the nozzle end of the tester fuel line, then tighten this fuel line nut.
3. Open the tester gauge shutoff valve if it is closed, and pump the tester handle fairly rapidly (about 45-55 strokes per minute) and note at what pressure the nozzle pops or releases fuel. Compare the nozzle opening or release pressure with the specifications shown in the enclosed chart in this chapter for your particular make of engine.
4. Low popping/release pressure necessitates injector replacement or disassembly and repair. Nozzle opening pressure can be adjusted by adding or removing shims from above the internal nozzle valve spring. Figure 29-36 shows the location of the spring and shims, while Figure 29-37 illustrates the resulting pressure change to a typical Bosch nozzle when the shim thickness is altered. This particular chart shows shim thickness changes of 0.04 mm (0.00157″). These shim thicknesses can vary between different Bosch nozzles and the make of engine that it is used with.

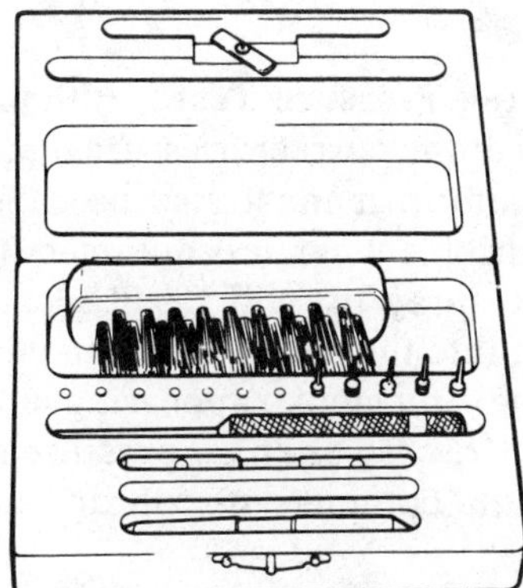

Figure 29-34. Rotunda 14-0301 injector nozzle cleaning kit. [Courtesy of Ford Motor Company, Dearborn]

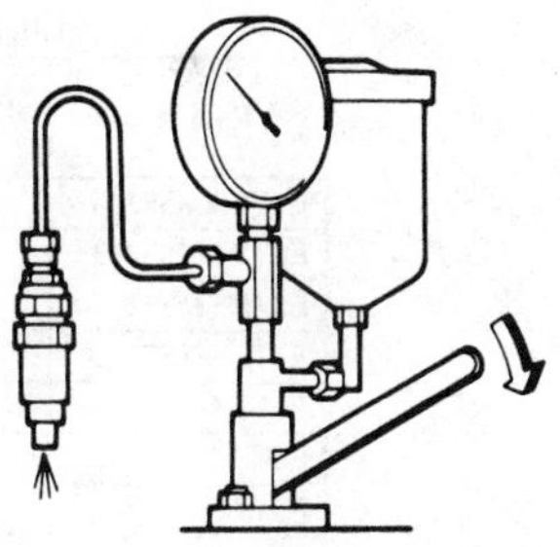

Figure 29-35. Checking injection pressure. [Courtesy of Chrysler Corporation]

Nozzle Chatter Test. This term is in reference to the fact that the nozzle needle valve will open and close as you pump the tester handle up and down. Some nozzles may chatter more than others due to minor variations in seating angles, carbon deposits, and gum from combustion blow-by.

Lack of a nozzle chatter is no positive indication in itself that the nozzle is faulty as long as it passes the other tests. While checking the release pressure of the nozzle, listen for a chatter or hissing sound, which is an indication that the internal nozzle valve is in fact free and moving within the nozzle bore. No chatter is generally accompanied by a poor spray pattern and/or low release pressure.

Lack of a good chatter is usually caused by carbon or varnish build-up within the nozzle, which can re-

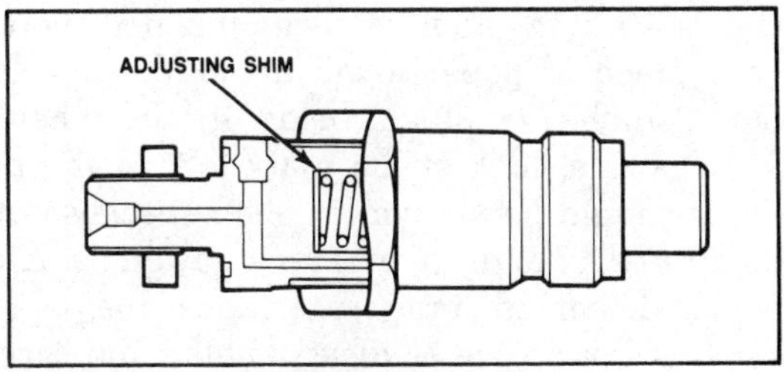

Figure 29-36. Adjusting shim location. [Courtesy of Ford Motor Company, Dearborn]

QUICK REFERENCE NOZZLE SHIM ADJUSTMENT CHART

Part No.	Thickness (mm)	Resulting Pressure Change (kPa)	Part No.	Thickness (mm)	Resulting Pressure Change (kPa)
E3TZ-9M557-A	0.50	Base	E3TZ-9M557-Q	1.06	6720
E3TZ-9M557-B	0.54	480	E3TZ-9M557-R	1.10	7200
E3TZ-9M557-C	0.58	960	E3TZ-9M557-S	1.14	7680
E5TZ-9M557-D	0.62	1440	E3TZ-9M557-T	1.18	8160
E3TZ-9M557-E	0.66	1920	E3TZ-9M557-U	1.22	8640
E3TZ-9M557-F	0.70	2400	E3TZ-9M557-V	1.26	9120
E3TZ-9M557-G	0.74	2880	E3TZ-9M557-W	1.30	9600
E3TZ-9M557-H	0.78	3360	E3TZ-9M557-X	1.34	10,080
E3TZ-9M557-J	0.82	3840	E3TZ-9M557-Y	1.38	10,560
E3TZ-9M557-K	0.86	4320	E3TZ-9M557-Z	1.42	11,040
E3TZ-9M557-L	0.90	4800	E3TZ-9M557-AA	1.46	11,520
E3TZ-9M557-M	0.94	5280	E3TZ-9M557-BB	1.50	12,000
E3TZ-9M557-N	0.98	5760	E3TZ-9M557-CC	1.54	12,480
E3TZ-9M557-P	1.02	6240			

Figure 29–37. Adjusting shim chart. (Courtesy of Ford Motor Company, Dearborn)

strict flow and act with a cushion effect on the moving nozzle valve.

Procedure

1. Close the shutoff valve to the pressure gauge to protect it.
2. Pump the handle quickly up and down until a chatter, hiss or squeal is detected.

Nozzle Spray Pattern. Spray patterns for Bosch pintle-type nozzles are very similar regardless of the make of engine that the nozzle is used in. Generally, pintle nozzles are designed to provide a concentrated spray angle that is fairly narrow or cone shaped.

To give you a sampling of what is required from the nozzle spray pattern, Figure 29–38 illustrates the spray pattern recommended by Ford in their 2.0L, 2.2L, and 2.4L engines, which is similar to that recommended by Nissan/Datsun; while Figure 29–39 shows the pattern recommended by Mitsubishi/Dodge in the Ram 50 pickup truck.

As you can see from the diagrams, both figures are basically the same. To accurately determine if the nozzle is spraying fuel as recommended, you can place a piece of paper towel under the nozzle tip (about 12 inches or 30 cm below it) and when you pop the nozzle, look at the mark left on the paper towel. It should be circular in shape because of the concentrated 4 degree or greater spray angle. A wide or poorly dispersed spray pattern over the paper requires that the nozzle be disassembled and serviced or replaced.

NOTE: Some pop testers will not deliver fuel at a great enough velocity to obtain the correct spray pattern for proper analysis. Therefore, if the nozzle cannot be properly tested for spray pattern yet it passes the other tests, then it should be considered acceptable unless performance problems occur with it in the engine.

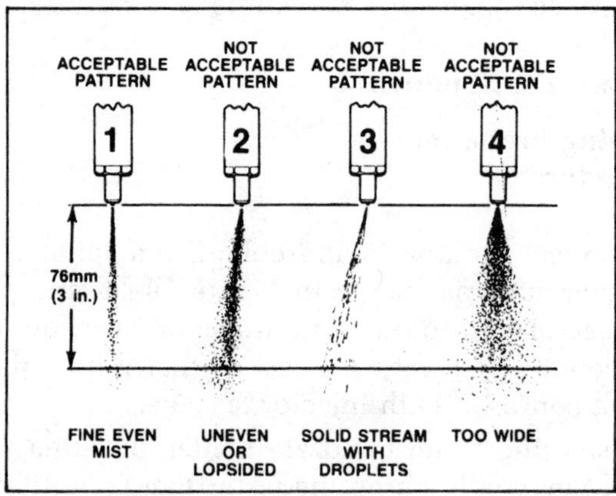

Figure 29–38. Injection nozzle spray pattern. (Courtesy of Ford Motor Company, Dearborn)

Nozzle Holding Pressure Test. Although the nozzle may open at the correct pressure and have a suitable spray pattern, it must also be capable of preventing fuel dribble when it is not injecting fuel. Fuel dribble at the spray tip will result in non-atomized fuel dribbling into the combustion chamber resulting in unburnt fuel and smoke appearing at the exhaust.

Using a lint-free rag or an air pressure nozzle, wipe or blow dry the complete nozzle tip and its holder (injector body).

This condition is best checked by bringing the nozzle pop tester up to a point usually 150–200 psi (1034–1379 kPa) below the nozzle popping or release

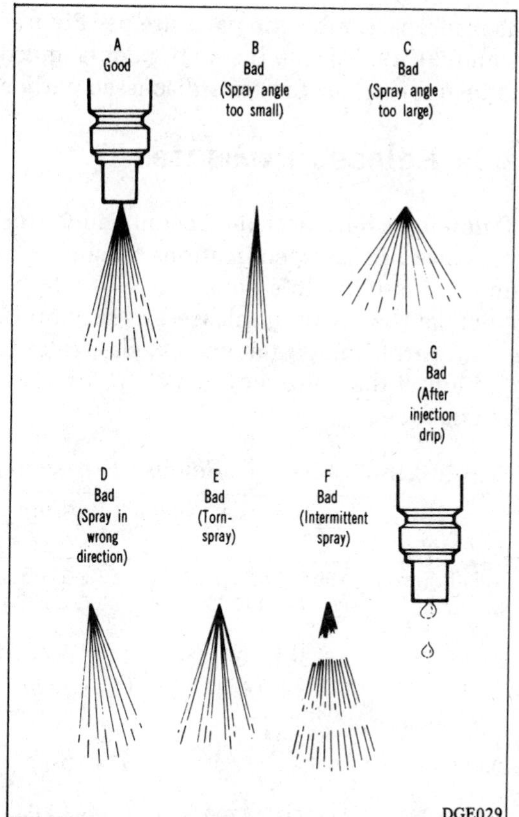

Figure 29–39. Injection nozzle spray condition. (Courtesy of Chrysler Corporation)

pressure and while keeping pressure on the pump handle, inspect the condition of the spray tip as shown in Figure 29–39 for signs of raw fuel leakage or a bubble/dribble of fuel at the spray tip, although a slight sweat is acceptable at the tip after 5 seconds as long as no fuel droplets appear.

Inspect the other sealing surfaces on the nozzle holder (injector body) for signs of external leakage.

To check for fuel leak-off (internal leakage past internal parts), quickly operate the tester handle while looking at the nozzle return fuel outlet—which can be a single outlet at the top center of the nozzle holder, or the nozzle may have two separate fuel return lines such as is shown in Figure 29–31. A few drops per pump handle stroke is acceptable, but a steady flow of return fuel indicates that there is wear between the internal injector parts and the nozzle should be replaced.

Nozzle Disassembly/Assembly

Extreme cleanliness must be exercised when repairing fuel injection nozzles as well as having access to the special tools and equipment to successfully complete a proper repair procedure.

To properly clean and decarbonize nozzles/holders, place them into a parts holder similar to that shown in Figure 29–11. Both cold and hot cleaning solutions are available for cleaning purposes—handle these with care; always wear eye protection.

If special cleaners are unavailable to you, clean solvent or diesel fuel can be used with a small brass bristle brush similar to the one shown in the injector cleaning kit in Figures 29–19 and 29–34.

Do not use a hand-held steel wire brush or a bench grinder wire buffing wheel to clean up the injector parts because these can scratch and damage injector components.

Procedure

1. Always obtain a suitable container(s) prior to disassembly so that each nozzle and its components can be kept together. Do not intermix components between nozzles and holders.
2. Wash the exterior of the injector body first to remove all dirt and loose carbon formation.
3. Refer to Figure 29–40 and place the injector nozzle holder in a soft jaw vise if the manufacturer's special tools are not available. Do not overtighten the vise, otherwise nozzle damage can result. Loosen off the nozzle holder nut with a deep socket.
4. Disassemble the injector/nozzle components and lay them out in a tray or individual container per injector. Figure 29–41 illustrates typical components of a pintle nozzle with a fuel return line fitting at the top of the nozzle body, while Figure 29–42 shows the parts found with a nozzle using two separate return fuel lines on the upper nozzle body retainer.
5. Clean all disassembled parts in a cleaning solution.
6. Inspect all components under a lighted magnifying glass or a lighted microscope. Check for signs

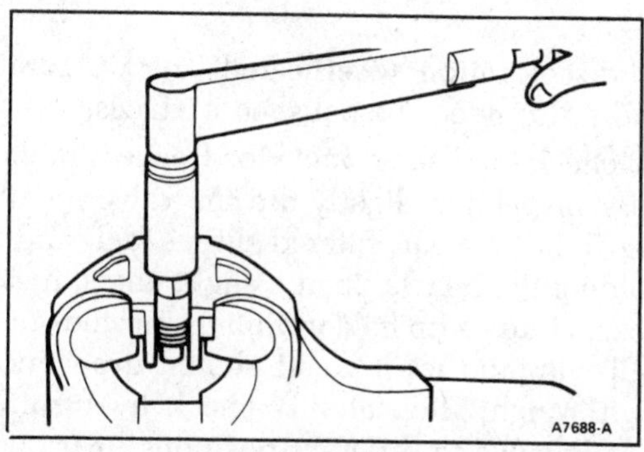

Figure 29–40. Nozzle holder removal. (Courtesy of Ford Motor Company, Dearborn)

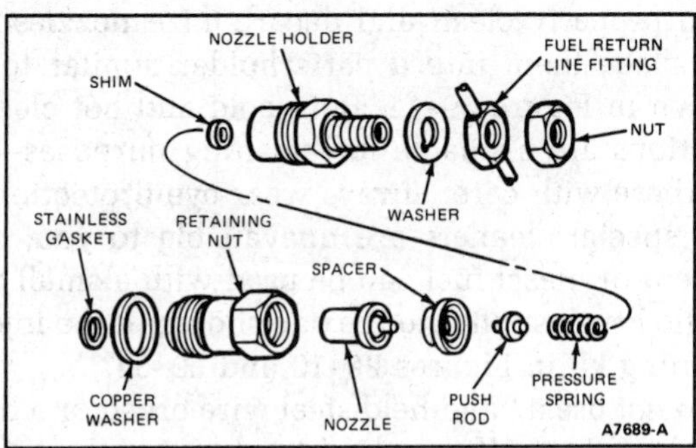

Figure 29-41. Injection nozzle assembly components. (Courtesy of Ford Motor Company)

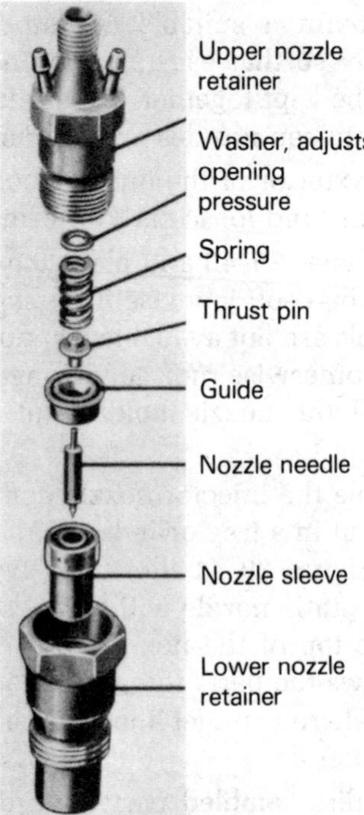

Figure 29-42. Component parts of a Bosch pintle nozzle. (Courtesy of Volvo of America Corporation)

of discoloration (overheated), nicks, scratches, and scuffing on the polished surfaces.

7. To check the needle valve for freeness in its body after inspection, lightly dip the valve in calibrating fluid or clean filtered diesel fuel, and while holding the nozzle tip at a slight angle, insert the nozzle into its tip holder. Pull the nozzle out about half way and let it go. It should drop under its own weight. Repeat this check by turning the needle valve to different positions. If it does not drop under its own weight, replace the nozzle and tip (sleeve) assembly.
8. When all parts have been cleaned, inspected and checked, reassemble the parts are per Figures 29–41 and 29–42. Torque the components and retest the injector as per the tests discussed earlier.

Nozzle Release Pressures

The following chart lists the recommended nozzle fuel pressure release specifications for all the diesel engines discussed in this book.

To change the opening release pressure on Bosch pintle and throttling pintle-type nozzles, refer to the information at the bottom of this chart for your particular engine.

Table 29–2. *Nozzle Release Pressure*

	Nozzle Opening Pressure	
Vehicle Make	*New*	*Used*
Mercedes-Benz 1978–84 240D	1564–1706 psi (10784–11763 kPa)	1422–1706 psi (9805–11763 kPa)
1978–81 300D, 300CD and 1979–80 300TD	1635–1750 psi (11273–12066 kPa)	1422 psi (9805 kPa)
1978–83 300SD Turbo	1958–2074 psi	1740 psi
1981–84 300TD Turbo	(13500–14300 kPa)	(11997 kPa)
1982–84 300D Turbo	Same	Same
1982–84 300CD Turbo	Same	Same
**Difference in injector nozzle opening pressures should not vary more than 71 psi (490 kPa)		
1979–81 Audi 5000	1706–1849 psi (11763–12749 kPa)	
1982–84 Audi 5000 Turbo	2139–2306 psi (14748–15900 kPa)	
Volkswagen Rabbit		
Non Turbocharged	1885 psi (12997 kPa)	1706 psi (11763 kPa)
Turbocharged	2306 psi (15900 kPa)	2139 psi (14748 kPa)
Volvo 80–84 D24	1700–1845 psi (11722–12721 kPa)	1775–1920 psi (12239–13238 kPa) When Servicing Nozzle
Volvo 83–84 D24 Turbo	2062–2318 psi (14217–15983 kPa)	2205–2318 psi (15203–15983 kPa) When Servicing Nozzle
Peugeot 1978–84 XD2, XD2C	1740–1813 psi (11997–12500 kPa)	1668–1813 psi (11500–12500 kPa)
1981–84 XD2S Turbo	1813 psi (12500 kPa)	NA

(Continued)

Table 29–2. *(Continued)*

Vehicle Make	*Nozzle Opening Pressure* New	Used
1983 Turbodiesel	2175 psi (14997 kPa)	NA
1984 Dodge Ram 50 (Mitsubishi 2.3L)	1707–1849 psi (11770–12749 kPa)	1565 psi (10791 kPa)
Mitsubishi 4.0L 6-Cylinder	1706–1949 psi (11763–13438 kPa)	
1984 Datsun/Nissan 2.4L SD25 Pickup	1493–1607 psi 10294–11080 kPa)	1422–1493 psi (9805–10294 kPa)
CD17 1.7L Engine	1920–2033 psi (13238–14018 kPa)	NA
LD28 2.8L Engine	Same	
Toyota 1981–84 2.2L Engine	1636–1778 psi (11280–12259 kPa)	1492–1777 psi (10287–12252 kPa)
Toyota 1984 2.4L Landcruiser engine	Same	Same
Isuzu 1981–84 1.8L	1707 psi (11770 kPa)	
Isuzu 1981–84 2.2L	1493 psi (10294 kPa)	
General Motors 5.7L/V8	1750–1900 psi (12066–13100 kPa)	
Roosa Master Pencil Nozzle General Motors 4.3L/5.7L		
Blue Band Nozzles V8	1000 psi (6895 kPa)	850 psi (5861 kPa)
Red Band Nozzles V6	800 psi (5516 kPa)	N
General Motors 6.2L V8	1750–1850 psi (12066–12756 kPa)	1600 psi (11032 kPa)
Ford/IHC 6.9L V8	2000–2150 psi (13790–14824 kPa)	1600–1850 psi (11032–12756 kPa)
1984 Ford/BMW 2.4L Turbo	2175 psi (14966 kPa)	
Ford/Mazda 2.0L Tempo/Topaz/ Escort/Lynx	1990–2105 psi (13721–14514 kPa)	1849–1990 psi (12749–13721 kPa)
Ford/Mazda 2.2L Ranger/Brono II	1957–2030 psi (13493–13997 kPa)	1957–2030 psi (13493–13997 kPa)

Injector Release Pressure Adjustment

Injector nozzle release pressures can be altered by adding or removing shims from the injector spring. Figure 29–36 illustrates the location of this spring inside the Robert Bosch-type pintle or throttling pintle nozzle.

The 6.9L Ford diesel engine can be equipped with either United Technologies Diesel Systems Group nozzles or CAV nozzles. Both are 17-mm shank diameter injector bodies of similar design and operation with the test procedure and removal and installation from the engine being the same for both.

Refer to the 6.9L diesel engine fuel systems (Chapter 24) for information dealing with injection nozzle removal and installation.

The nozzles should be removed from the engine and checked for correct opening pressure and spray pattern every 60,000 miles (96,500 km). Normal opening pressure of the nozzles is between 2000–2150 psi (13790–14824 kPa), although an acceptable service level on a used nozzle is 1600 psi (11032 kPa) as long as it passes all of the other checks/tests.

Cleaning of the nozzles can be done in the same manner as that for other nozzles using ultrasonic cleaning for best results. A nozzle cleaning test kit contains all of the necessary tools for service and repair of the nozzles. Several tool kits which can be considered common for cleaning of most nozzles can be found in this chapter.

Use calibrating fluid rather than diesel fuel for testing the nozzles based upon the reasons given in this chapter for other nozzles. The same safety precautions should be exercised regarding eye protection and hand protection as for other nozzle test procedures.

Perform the nozzle checks and tests in the same manner as you would for other nozzles in this chapter checking the following conditions:

1. Nozzle release pressure (popping pressure)
2. Nozzle spray pattern
3. Nozzle chatter
4. Nozzle tip leakage
5. Nozzle fuel leak-off

Use a suitable pop testing machine such as a Rotunda P/N 14-6300 unit or equivalent shown earlier in the Bosch nozzle section, Figure 29–33.

To check the nozzle opening pressure, bleed fuel to the nozzle, then tighten the retaining nut. Slowly operate the pump handle and note the pressure at which the nozzle opens and releases fuel.

To check the fuel leak-off, operate the tester handle quickly and note how much fuel flows from the injector fuel return port at the top. One to two drops per pump stroke is considered acceptable. If, however, the fuel squirts from the nozzle return port, the nozzle is faulty and should be repaired.

To check the spray pattern refer to Figure 29–43 and compare the spray pattern with the examples shown. Table 29–3 lists possible causes of faulty diesel nozzles.

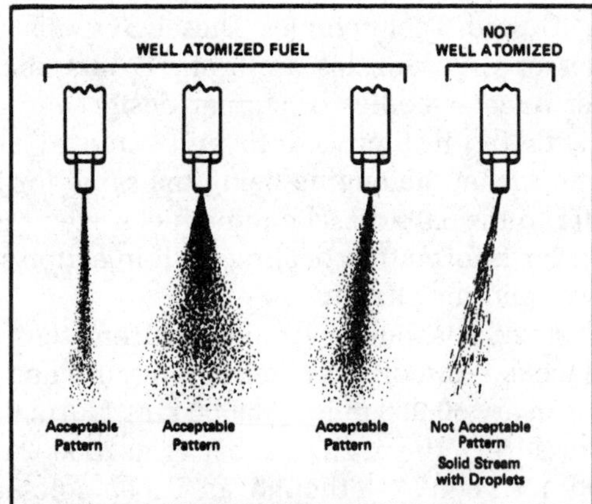

Figure 29–43. 6.9L Ford diesel nozzle spray patterns. (Courtesy of Ford Motor Company, Dearborn)

Testing Bosch "K" Gasoline Fuel Injectors

Robert Bosch produces a D, L, Motronic and a K type of gasoline fuel injection system for use on a large number of vehicles produced worldwide. For specific information on these systems, refer to Chapter 6, dealing with Robert Bosch gasoline fuel systems.

The fuel injectors used with the K system are mechanically operated, whereas the D, L and Motronic systems have electronically controlled fuel injectors.

Because the K system employs a mechanically operated injector, it can be tested similar to a diesel engine fuel injection nozzle. The sequence to check the Robert Bosch K system fuel injector is as follows: Maintain absolute cleanliness when testing gasoline fuel injectors just as you would when testing diesel fuel injectors.

CAUTION: *Never* use gasoline as a fluid medium when testing the injectors because of possible fire danger and also because of the atomized gasoline fumes that can be dangerous to your health.

NOTE: Use Shell Mineral Spirits 135 if available or a suitable alternative with a viscosity of 1.8–2.8 cSt (centistokes), a density of 0.77–0.81 g/cm³, a boiling point of 140–200°C (392°F) and self-ignition temperature of at least 245°C (473°F).

Procedure for Contamination Test

1. Prime the injector by pouring clean test fluid into the inlet port, then attach it loosely to the pop tester as shown in Figure 29–44.
2. Vent all air from the pop tester and line by pumping the tester handle until clear fuel flows from the end of the line attached to the injector.

Table 29–3. *Troubleshooting Faulty Nozzles*

Fault	*Possible Cause*	*Remedy*
Excessive leak-off	Dirt between pressure face of nozzle, spring retainer, or plate and nozzle holder.	Clean nozzle.
	Loose nozzle retainer nut.	Inspect lapped faces and tighten retainer nut.
	Defective nozzle.	Replace nozzle.
Nozzle blueing	Faulty installing or tightening.	Replace nozzle.
	Insufficient cooling.	Correct cooling system.
Nozzle opening pressure too high	Incorrect shim adjustment.	Replace nozzle.
	Nozzle valve dirty or sticky or opening is clogged.	Clean nozzle.
	Seized nozzle.	Replace nozzle.
Nozzle opening pressure too low	Incorrect shim adjustment.	Readjust nozzle.
	Nozzle valve spring broken.	Replace spring and readjust pressure.
	Nozzle seat worn.	Install new or reconditioned nozzle.
Nozzle drip	Nozzle leaks because of carbon deposit or sticking nozzle valve.	Clean nozzle.
	Defective nozzle.	Replace nozzle.
Spray pattern distorted	Carbon deposit on tip of nozzle valve.	Clean nozzle.
	Nozzle hole partially blocked.	Clean nozzle.
	Defective nozzle.	Replace nozzle.

3. Tighten the injector fuel line connection.
4. Open the pop tester gauge valve and slowly pump the handle at about one stroke every two seconds without exceeding 1.5 bar (21.75 psi).
5. If pressure fails to build up, or if the injector leaks raw fuel at the tip, the injector is defective.

Before proceeding further, close the tester valve and pump the handle until no bubbles are evident at the nozzle in order to confirm that no air has been trapped in the system.

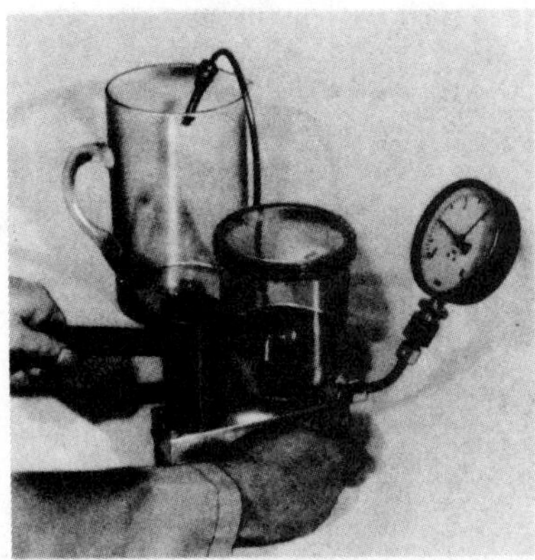

Figure 29–44. Pop testing a gasoline fuel injector (K-Jetronic). (Courtesy of Robert Bosch Corporation)

Procedure for Opening Pressure Test

1. Open the pop tester gauge valve.
2. Pump the handle until the injector actually sprays test fuel, and note the gauge reading.
3. The injector opening pressure must be within 2.5–3.6 bar (36.3–52.3 psi).

Procedure for Leakage Test

1. Pump up the pop tester handle until the gauge reads 0.5 bar (7.26 psi) below the opening pressure, but not less than 2.3 bar (33.4 psi). Hold the pressure here for 15 seconds and carefully inspect the injector to make sure that there are no fuel drips/drops at the end of the tip.

Procedure for Spray Pattern

1. Isolate the pop tester valve to avoid damage to it during this test by closing the tester valve.
2. Pump the handle at about 1 stroke per second while carefully watching the spray pattern.
3. A good fuel injector will emanate a discernible chatter, which indicates that all internal valve components are free and not hanging up due to carbon or vanish formations.
4. No drops should form on the tip during this time.
5. The spray pattern of the injector should appear as shown in Figure 29–45. If the injector spray pattern appears as shown under incorrect spray patterns, the injector should be replaced.
6. Always install new seal rings at the base of the injector before installing them back into the engine.
7. New and old injectors are interchangeable, and new and old injectors can be mixed in a set as long as they pass all of the tests described above.

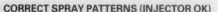

INCORRECT SPRAY PATTERNS
(INJECTOR FAULTY)

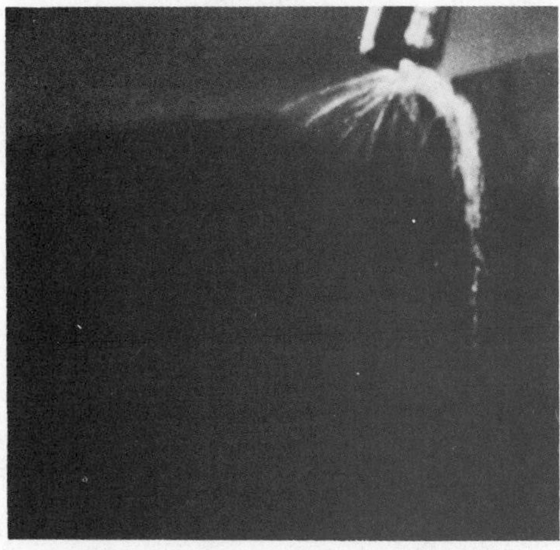

Figure 29–45. Acceptable and unacceptable injector spray patterns for Robert Bosch K injectors. (Courtesy of Robert Bosch Corporation)

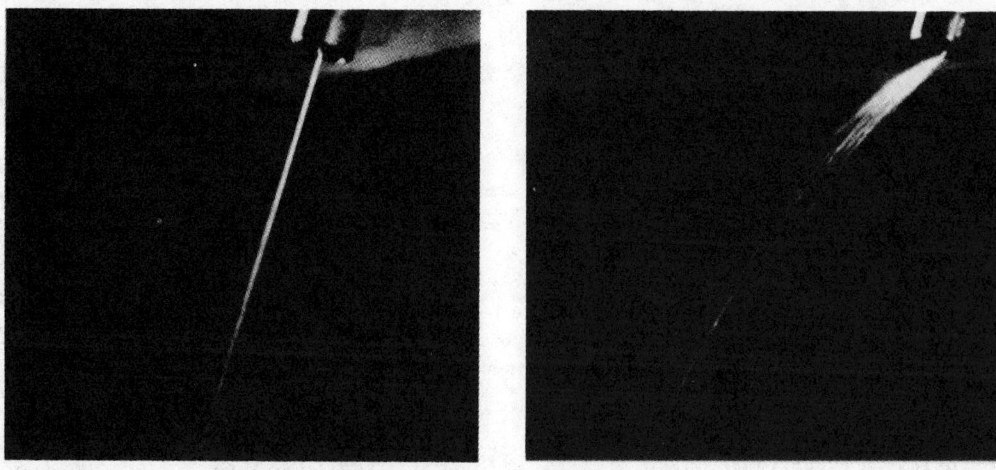

Figure 29-45. (*Continued*)

30

General Troubleshooting

The ability to be able to effectively and quickly diagnose an engine problem is basically related to the following:

1. A thorough understanding of the basic fundamentals of what actually goes on within an internal combustion engine.
2. The amount of experience of the mechanic involved.
3. How familiar the mechanic is with a particular make of engine; also, how up-to-date you are.
4. The ability to be analytical.
5. The ability to control one's temper when an irrate customer or operator is pushing you for an answer.
6. Don't second guess yourself; if in doubt, check it out.
7. If necessary, do not hesitate to refer to the manufacturer's specifications or troubleshooting charts in the respective engine service manual.

People often refer to someone as being a really good mechanic. How do you think that person was able to achieve such respect? Obviously, experience is in many instances a series of a great many mistakes that were learned on the way up through the apprenticeship stage and early journeyman years. But one must have a genuine desire to want to succeed and be nothing less than the best in the field of mechanics. Certainly, in this ever-changing technological era, and especially with high labor costs and overhead, it is easy to become simply the proverbial parts replacer instead of a highly skilled and dedicated craftsman. Company policy can, of course, dictate to what degree they desire their maintenance personnel to replace component parts.

In many instances, a new part may be required; however, there are many, many instances when a new component part is installed with the assumption that this will in fact cure the problem at hand. Within a short time, the same problem exists, leading you to scratch your head and ask why.

Don't accept at face value, unless a part shows particular excessive wear or damage, that it is nonserviceable. Learn to accept where possible nothing less than the best; place yourself in the position on every job that you do that the engine or equipment is your own. It is hard work to stay abreast of the many changes that are constantly going on in this business, and that is one of the reasons that it is such a challenge to you as an individual and as a skilled craftsman.

People will remember your abilities as a first-class mechanic only as long as you continue to produce first-class work. Foul up once, and that is the job that stays in their minds, regardless of how many jobs you completed successfully for them previously.

With these few considerations and thoughts, tackle a troubleshooting problem with an open and keen mind. Don't panic, take it easy, and eventually you will find that the majority of problems are in many instances of a minor nature. Walk before you run.

The problems that can relate to the fuel system are obviously somewhat diversified in nature. The method that you choose to effectively pinpoint the particular problem will depend upon how familiar you are with the make of engine. However, if you systematically collect all the information as to what led up to the problem at hand, you should be able to analyze on a step-by-step basis the reason for the existing problem. Remember, satisfactory operation of the engine primarily depends upon the following items:

1. An adequate supply of clean, relatively cool air, which once in the cylinder can be compressed to

a high enough pressure to effect proper combustion.

2. The injection of the correct amount of fuel at the proper time during the compression stroke.
3. Use of the proper grade of fuel for the environment in which the engine operates.
4. The ability to maintain the fuel oil if possible at an optimum temperature range of 90 to 95°F (32 to 36°C) for high speed diesel operation (maximum allowable of 150°F (65°C).
5. Clean, filtered water and sediment-free fuel.
6. Maintenance of the proper engine water temperature. Satisfactory water treatment.
7. Maintenance of exhaust back pressure within specifications.
8. Use of the proper grade of oil with proper service intervals.
9. Proper selection and application of the engine for what it was intended.

When collecting information before analyzing a problem, keep an open mind. There are always those who are more than ready to tell you what the problem is. Listen to their suggestions by all means, but remember that it is you who are the trained and skilled mechanic; therefore, proceed with this in mind. It is very easy to become side-tracked and led into believing that what an operator happens to say is in fact the cure for the problem at hand. Maybe it is! However, think for a minute before jumping in.

In these days of high labor costs, it is worth every penny that the customer pays if, after 5 or 10 minutes of basic checks and collecting your thoughts, you are able to effectively arrive at a solution to the problem, rather than going off on a haphazard guess and idea that you may just be lucky enough to stumble upon the problem. With the high costs involved in purchasing equipment, most companies have a reasonably good maintenance program that in most instances reflects itself in minimum engine failures and downtime. When a problem occurs then, you will find that it is of a minor nature in very many instances. Therefore, don't automatically suspect and look for a major reason for failure before considering some of the foregoing thoughts.

There are many types and shapes of internal combustion engines on the world market today found in a wide variety of applications. Obviously, the more familiar you are with a particular make of engine, the more confident you will be when faced with a problem on that engine. However, if you take into account the seven items listed earlier, you should have little trouble in effectively finding the problem. After all, every engine has pistons, crankshafts, and pumps and injectors, which are all very similar in basic operation.

Most manufacturers supply special tools that makes working on their engine that much easier; however, you can in some troubleshooting situations resort to the use of a few locally and readily available tools to reach the same conclusions.

Diesel Engine Troubleshooting

This chapter deals primarily with troubleshooting of automotive passenger car and light pickup truck diesel engines because the chapters dealing with gasoline-powered vehicles within this book deal with typical areas that would be checked when problems exist with the engine.

Contained within this book are various diesel engine fuel system chapters that discuss the specific techniques and procedures for that particular engine; however, keep in mind that the techniques used will always follow the same basic pattern regardless of the particular make of engine.

An example of this would be when different engines use the same injection pump—such as the 4.3L, 5.7L, and 6.2L General Motors V6 and V8 diesel engines and the Ford 6.9L (IHC) diesel V8, which all use the Roosa Master/Stanadyne model DB2 fuel injection pump.

Engines using the Robert Bosch VE model injection pump are the VW Rabbit, the Audi 5000, the Volvo D24, Peugeot, Dodge Ram 50 (Mitsubishi), Isuzu/GMC, the BMW/Ford 2.4L 6-cylinder turbocharged diesel engine, the 2.0L Ford Tempo/Topaz, Escort/Lynx, and 2.2L Ranger/Bronco II models.

Some Peugeot, Renault, and Citroen diesel engines along with some VW Rabbit/Jetta engines are also equipped with the CAV DPA distributor injection pump. Information on these pumps can be found under Chapter 28 dealing with the DPA pump.

Datsun/Nissan SD25 pickup trucks, Toyota Landcruisers, and Mercedes-Benz cars use an in-line type of fuel injection pump on their engines at present, but the operating and troubleshooting principles for all can be considered common in most instances.

Injection nozzles may vary in their release pressures and spray patterns; however, the testing and inspection of all diesel injection nozzles can be considered common. Typical test procedures for a variety of widely used nozzles can be found under Chapter 29 herein dealing with Injection Nozzles.

With the exception of the pre-1980 GMC 5.7L V8

diesel engine, which employed Roosa Master pencil nozzles, and the current Ford/IHC 6.9L V8 pickup truck diesel engine, which uses either United Technologies or CAV 17-mm nozzles, all other engines discussed in this book use Robert Bosch-type pintle or throttling pintle nozzles with KC holders. Therefore, when troubleshooting any diesel engine fuel system, information contained in one chapter may very well be applied to that of another system.

Special Tools and Equipment

To successfully troubleshoot as well as service/repair a particular engine or fuel system, each engine manufacturer makes available through their dealers a number of special tools that are not only required in many instances but will also save time and allow you to effectively isolate and confirm a problem area. In many cases an alternate equivalent tool can be used to successfully perform a given repair.

Examples of special tools and equipment can be found at the end of each fuel system chapter for that particular engine/fuel system as well as in the chapters dealing with fuel injection test equipment and nozzles. Where possible, use these tools to assist you in successfully servicing and repairing the fuel system.

Problems that occur with a diesel engine regardless of its model and manufacturer can be considered common in most instances. When a gasoline engine has a problem, the model and manufacturer may indicate a particular problem area to the experienced mechanic; however, in most cases, the sequence used is generally similar in other to pinpoint the possible problem area.

One area that is often overlooked when troubleshooting a diesel engine for a lack of power complaint is the throttle linkage adjustment. Always check to ensure that, when the throttle pedal is placed into the full-fuel position, the throttle lever on the injection pump housing is also at full-fuel. If not, adjust the linkage to allow full-fuel.

Another area that can contribute to the performance complaint is the grade of fuel oil being used. Refer to Chapter 4 dealing with fuel oils for information on fuel selection criteria.

Winter versus Summer Operation

The grade of fuel used in the summer months versus that required in the winter months can have a drastic effect upon the engine's performance. Summer fuel that is used in the winter can lead to hard starting, exhaust smoke, and a rough running engine. Similarly a winter blend used in the summer can also create engine problems. These specific problems can be found in Chapter 4 dealing with fuel oils.

One common problem in the winter months that affects diesel engine operation in areas subjected to low ambient temperatures is one of *fuel waxing*. This is a condition that occurs at the temperature at which wax crystals form in the diesel fuel; it restricts or even completely prevents fuel from flowing to the injection pump. This can be improved with certain additives; however, it is much better to use a thermostatically controlled fuel heater to prevent this problem. Several types of fuel heaters are discussed in the various chapters dealing with passenger car and light truck diesel engines.

In the summer months, a lack of power complaint is often associated only with very warm ambient temperatures. Generally speaking, it is felt that the optimum diesel fuel temperature should not exceed 90–95°F (32–35°C), and that for every 10°F rise beyond this temperature the engine will lose 1% of its gross horsepower (1.5% if turbocharged) through expansion of the fuel, which makes it less dense and therefore produces less horsepower per injection. This problem can become worse if fuel filters are located very close to an area of exhaust heat radiation, etc.

Also, most fuel injection systems in use today use the diesel fuel as a lubricant for internal components as well as acting as a cooling agent. This fuel is then returned to the fuel tank where it will then be picked up and pumped back through the system.

Fuel coolers are not required on automotive applications; however, if the vehicle is equipped with dual tanks, it is better to pull fuel from one tank and return the warm fuel to the other with a balance line between the tanks.

Detonation

Do not confuse the normal combustion sound within the engine for this complaint. Some engines do run louder than others, and many of them have a peculiar sound common to that particular engine or application. Pressure pulsations within the engine cylinder create the condition often referred to as *diesel knock;* it is an inherent characteristic of all diesel engines.

Experience will tune your ear to pick up sounds other than the normal combustion pressure sounds. However, it is often helpful even to an experienced mechanic to isolate any irregular noises with the use of an engine stethoscope, which amplifies any sounds remarkably well. Often a piece of welding rod or even a lead pencil placed on the engine with the other end at your ear can magnify sounds reasonably well.

If detonation occurs, check for the following:

1. Lube oil picked up by the air intake stream to the engine, which cannot only cause detonation but engine overspeed.
2. Low coolant temperature caused by excessive periods of idling and light-load operation. Or cold weather operation without proper attention to maintaining operating temperatures.
3. Faulty injectors: leaking fuel, fuel spray-in pressure low.

Checking the Fuel System

If the engine is misfiring, running rough, and lacking power, you can start and run it at an idle rpm. On 4-cycle engine with a high pressure fuel system, you can loosen the fuel line retaining nut at the injector one-half turn and note if there is any change in the sound of the engine. If the injector is firing properly, there should be a positive change as you loosen the line; otherwise, the injector is not firing properly. Exhaust smoke analysis is very helpful in determining the possible causes of poor performance. See Figure 30–1.

If the engine fails to reach its maximum governed speed and generally seems to be starving for fuel, install a fuel pressure gauge into the secondary filter, run the engine, and check the fuel pressure with the engine manufacturer's specifications. Some engines employ a small filter screen located just under the cover of the fuel transfer pump; check that this is not plugged.

If a fuel strainer or fuel/water separator is used, check them for plugging and excessive amounts of water. Check that all fuel lines are free of sharp bends and kinks. Check the tightness of all fittings and connections from the suction side of the transfer pump back to the fuel tank. Install a clear pipe into the suction line to check for air bubbles.

You may have to undertake a restriction check to the fuel flow with a mercury manometer plumbed into the suction side of the primary filter.

Check the fuel pump drive for security and proper engagement. Ensure that there are no external fuel leaks, especially at the pump or injectors. Also, if more than one fuel tank is employed, check to see that the balance line valve is open between them; if a three-way valve is employed, check that it is in the correct position.

In certain instances you may also find that there is a restriction to fuel flow from inside the fuel tank caused by sediment or some foreign object that has dropped into the tank either during filling or maintenance checks.

One complaint that you may occasionally come across is that the engine runs well in the early part of a shift, but stalls and lacks power as the day wears on. This could be caused by debris, which only creates a restriction to fuel flow as the level in the fuel tank drops and the debris is drawn over the suction line.

If the engine has recently been overhauled or the injection pump or injectors serviced, doublecheck the injection pump timing, injection release pressure, or injector timing.

If the engine has a considerable amount of hours or miles on it, it very well may be in need of a tune-up; however, this alone may not be the cause of the problem. Too many people immediately assume that if an engine is lacking power the answer is to tune it up. Although many large companies have developed a sequence of maintenance checks to be carried out at certain intervals of time, a tune-up should be done only if other checks show that everything else is according to specifications.

When doing a tune-up, do not back off all adjustments and start from scratch. Check each adjustment first, and if necessary readjust. One of the first checks that should be carried out is to disconnect the throttle linkage and manually hold the speed control lever on the governor to the full-fuel position and accurately record the maximum governed engine rpm. Reconnect the throttle linkage, place it in the full-fuel position, and compare the readings. If not the same, adjust the linkage to correctly obtain the maximum engine rpm. Similarly, the maximum governor speed setting may require adjustment. Ensure that there is no binding anywhere in the fuel control linkage.

Fuel Temperature

On high speed diesel engines, fuel temperature can adversely affect the horsepower output of the engine. The optimum fuel temperature should be kept between 90 and 95°F (32 and 35°C). With each 10°F temperature rise beyond this figure, there is approximately a 1% loss in horsepower due to expansion in the fuel and 1.5% on a turbocharged engine. Therefore, if you were running at a fuel temperature of 135 to 140°F (57 to 60°C), theoretically your engine would be producing approximately 4% less horsepower on a naturally aspirated engine, and about 6% on a turbocharged engine.

Black or Dark Gray Smoke

SYMPTOM	PROBABLE CAUSE	ACTION TO TAKE	COMMENTS
Smoke at full load, particularly at high and medium speeds; engine quieter than normal	Pump timing retarded	Check and adjust timing to specifications	Pump timing changes only if moved intentionally, or if mountings are not properly tightened
Smoke at full load, particularly at low and medium speeds; engine noisier than normal	Pump timing advanced	Check and adjust timing to specifications	
Smoke at full load, particularly at high and medium speeds, probably with loss of power	Injection nozzle(s) discharge hole fully or partly blocked	Clean or replace injection nozzles as required	
Smoke at full load, at higher speeds only	Air cleaner filter restricted; EGR system	Replace air cleaner filter; Check EGR system	Replace filter at scheduled intervals or more often under severe service conditions
Intermittent or puffy smoke, sometimes with white or bluish tinge, usually accompanied by engine knocking	Injector nozzle valve sticks open intermittently	Check injection nozzles for sticking valve, broken spring, or very low opening pressure; also check for cross-threading in head	May be caused by dirt or water in fuel; injection nozzle should screw into head freely and must not be over-tightened
Smoke at all speeds at high loads, mostly at low and medium speeds and probably accompanied by hard starting	Loss of cylinder compression because of stuck rings, bore wear, burning, sticking valves, or incorrect valve setting	Engine requires overhaul if wear indications are present; check valve setting Check compression	May be caused by improper crankcase oil or incorrect valve clearance
Smoke at full load, mostly at medium and high speeds, and probably accompanied by low power	Injection fuel lines clogged or restricted because of damage	Check injection fuel lines; clean or replace as required	Fuel lines must be clear and unrestricted

Blue, Bluish Gray or Grayish White Smoke

SYMPTOM	PROBABLE CAUSE	ACTION TO TAKE	COMMENTS
Blue or whitish smoke particularly when cold, and at high speeds and light load, but reducing or changing to black when hot and at full load, with loss of power at least at high speeds	Pump timing retarded	Check and adjust pump timing	Some engines show this symptom for less retard than gives rise to black smoke, but usually substantial retard is required to produce blue smoke when running hot and under load
Blue or whitish smoke when cold, particularly at light loads but persisting when hot, probably accompanied by knocking	Injection nozzle valve stuck open or nozzle tip damaged and leaking	Examine nozzle for valve stuck open or nozzle tip damage	
Blue smoke at all speeds and loads, hot or cold	Engine oil being passed by piston rings because of sticking rings or bore wear	Engine reconditioning as required	May be caused by improper crankcase oil; will be associated with high oil consumption
Blue smoke, particularly when accelerating from period of idling, tending to clear with running	Engine oil being passed by worn inlet valve guides, or valve stem umbrella seals worn or missing	Engine reconditioning as required	
Light blue smoke at high-speed light loads, or running downhill, usually accompanied by sharp odor	Engine running cold; thermostat stuck open	Replace thermostat	Low temperatures also increase bore wear

Figure 30-1. Exhaust smoke diagnosis. [Courtesy of Ford Motor Company, Dearborn]

Compression Checks

A compression check may be necessary to determine the condition of the valves and rings. On many engines, this check is taken with the use of a dummy injector and the engine running while on others, all injectors must be removed. Some engine manufacturers recommend glow plug removal only. Detail on compression checking can be found in the respective engine chapters. Each make of engine will have some variation in the sequence of events required to do this; therefore, check the respective engine manufacturer's service manual for the routine and specifications.

A crankcase pressure check taken with the use of a water manometer could tip you off to worn rings, as can an exhaust smoke analysis, hard starting, and low power.

Engine Will Not Start

Although there can be a variety of reasons for not starting, these generally fall into one of the following areas:

1. Cranking problem: A low battery high circuit resistance in the electrical system caused by corroded terminals, loose terminals, faulty wiring, and so on. It could also be a solenoid failing to engage or starter drive or internal starting problems, to name but a few.
2. Lack of air: See air inlet restriction checks.
3. Lack of fuel: See fuel system checks and the troubleshooting charts.

Engine Timing

Improper engine timing can lead to actual physical mechanical damage, such as valves hitting pistons. However, this will depend on just how far out the timing is and is more acute on some engines than on others.

If the injection pump timing or injector release pressure timing is off, problems of smoking exhaust, low power, high fuel consumption, and internal engine damage can result; therefore, always ensure that the engine is timed as per manufacturer's specifications and that the injection pump is timed for the particular application for which the engine is being employed.

There is also a noticeable change in the engine's exhaust temperature with a timing change. The degree of change will vary; however, most heavy-duty on- and off-highway trucks and equipment employ dash-mounted pyrometers, which constantly monitor the exhaust gas temperatures and are readily visible at a glance. One pyrometer is employed per engine bank on Vee engines. If these are not installed, the individual cylinder temperature can be read by turning a numbered switch that is connected through thermocouples at each cylinder. This type is more common to larger engines used in stationary industrial and marine applications. Automotive and light truck engines do not have either of these, a hand-held contact-type pyrometer can be used to individually check cylinder exhaust temperatures.

Exhaust temperatures will vary between engines; therefore, always check the manufacturer's specifications regarding normal and maximum allowable temperatures. The size of the injectors, fuel pump delivery rate, and engine speed and load will all contribute to exhaust temperature.

An engine timing light that operates off fuel pressure through a transducer pickup can be used on engines that employ a high pressure fuel system.

Use of Troubleshooting Charts

Contained within this chapter are a number of troubleshooting charts that can be used to pinpoint a possible trouble source in either the fuel injection pump system or with the cranking and glow plug systems as well as general diesel engine problem areas.

In addition to these troubleshooting charts, specific chapters in this book such as the Volkswagen Rabbit (Chapter 14), the Robert Bosch in-line injection pump (Chapter 12), and the Ford/BMW 2.4L 6-cylinder turbocharged diesel engine (Chapter 27) also contain troubleshooting charts relating to these particular pumps or vehicles.

For example, the Robert Bosch VE pump is used extensively on a variety of passenger cars and light trucks, namely:

1. VW Rabbit
2. Volvo D24
3. Audi 5000
4. Datsun/Nissan 1.7L 4-cylinder and 2.8L 6-cylinder engines
5. Isuzu 1.8L and 2.2L 4-cylinder engines
6. Peugeot 2.3L 4-cylinder engine
7. Dodge Ram 50 (Mitsubishi) 2.3L 4-cylinder engine
8. Ford 2.0L Tempo/Topaz, Escort/Lynx and 2.2L (Mazda) Ranger/Bronco II 4-cylinder engines

9. Ford/BMW 2.4L 6-cylinder turbocharged Continental/Thunderbird cars
10. Toyota 2.2L 4-cylinder light truck engine

Therefore, refer to various troubleshooting charts throughout this book when you run into a problem that may be considered common to more than one engine type.

Diesel Engine Diagnosis

Diesel Engine Mechanical Diagnosis such as noisy lifters, rod bearings, main bearings, valves, rings, and pistons is the same as for a gasoline engine. This diagnosis covers only those conditions that are different for the diesel engine.

Condition	*Possible Cause*	*Correction*
Engine will not crank	1. Loose or corroded battery cables	Check connections at batteries, engine block, and starter solenoid.
	2. Discharged batteries	Check generator output and generator belt adjustment.
	3. Starter inoperative	Check voltage to starter and starter solenoid. If OK, remove starter for repair.
Engine cranks slowly—will not start (Minimum engine cranking speed—100 rpm cold, 240 rpm hot)	1. Battery cable connections loose or corroded	Check connections at batteries, engine block and starter.
	2. Batteries undercharged	Check charging system
	3. Wrong engine oil	Drain and refill with oil of recommended viscosity.
Engine cranks normally—will not start	1. Incorrect starting procedure	Use recommended starting procedure.
	2. Glow plugs inoperative	Electrical problem.
	3. Glow plug control system inoperative	Electrical problem.
	4. No fuel into cylinders	Remove any one glow plug. Depress the throttle part way and crank the engine for 5 seconds. If no fuel vapors come out of the glow plug hole, go to step 5. If fuel vapors are noticed, remove the remainder of the glow plugs and see if fuel vapors come out of each hole when the engine is cranked. If fuel comes out of one glow plug hole only, replace the injection nozzle in that cylinder. Crank the engine and check to see that fuel vapors are coming out of all glow plug holes. If fuel is coming from each cylinder, go to Step 11.
	5. No fuel to injection pump	Loosen the line coming out of the filter. Crank the engine. The fuel should spray out of the fitting—use care to direct fuel away from sources of ignition. If fuel sprays from the fitting go to Step 10.
	6. Restricted fuel filter	Loosen the line going to the filter. If fuel sprays from the fitting, the filter is plugged and should be replaced. Use care to direct the fuel away from sources of ignition.
	7. Fuel pump inoperative	Remove inlet hose to fuel pump. Connect a hose to the pump from a separate container that contains fuel. Loosen the line going to the filter. If fuel does not spray from the fitting, replace the pump. Use care to direct the fuel away from source of ignition.
	8. Restricted fuel tank filter	Remove fuel tank and check filter.
	9. Plugged fuel return system	Disconnect fuel return line at injection pump and route hose to a metal container. Connect a hose to the injection pump connection, route it to the metal container. Crank the engine. If it starts and runs, correct restriction in fuel return lines. If it does not start, remove the

(Continued)

Condition	Possible Cause	Correction
		ball check connector from the top of the injection pump and make sure that it is not plugged.
	10. No voltage to fuel solenoid	Connect a volt meter to the wire at the injection pump solenoid and ground. The voltage should be a minimum of 9 volts. If there is inadequate voltage, there is an electrical problem.
	11. Incorrect or contaminated fuel	Flush fuel system and install correct fuel.
	12. Pump timing incorrect	Make certain that pump timing mark is aligned with mark on adapter.
	13. Low compression	Check compression to determine cause.
	14. Injection pump malfunction	Remove injection pump for repair.
Engine starts but will not continue to run at idle	1. Slow idle incorrectly adjusted	Adjust idle screw to specification.
	2. Fast idle solenoid inoperative	With engine cold, start engine; solenoid should move to hold injection pump lever in "fast idle position." If solenoid does not move, it is an electrical problem.
	3. Restricted fuel return system	Disconnect fuel return line at injection pump and route hose to a metal container. Connect a hose to the injection pump connection; route it to the metal container. Crank the engine and allow it to idle. If engine idles normally, correct restriction in fuel return line. If engine does not idle normally, remove the return line check valve fitting from the top of the pump and make sure it is not plugged.
	4. Glow plugs turn off too soon	Electrical problem.
	5. Pump timing incorrect	Make certain that timing mark on injection pump is aligned with mark on adapter.
	6. Limited fuel to injection pump	Test the engine fuel pump, check for plugged filters; check fuel lines. Replace or repair as necessary.
	7. Incorrect or contaminated fuel	Flush fuel system and install correct fuel.
	8. Low compression	Check compression to determine cause.
	9. Fuel solenoid closes in run position	Ignition switch out of adjustment. If OK, it is an electrical problem.
	10. Injection pump malfunction	Remove injection pump for repair.
Engine starts, idles rough, without abnormal noise or smoke	1. Slow idle incorrectly adjusted	Adjust slow idle screw to specification.
	2. Injection line leaks	Wipe off injection lines and connections. Run engine and check for leaks. Correct leaks.
	3. Restricted fuel return system	Disconnect fuel return line at injection pump and route hose to a metal container. Connect a hose to the injection pump connection; route it to the metal container. Start the engine and allow it to idle; if engine idles normally, correct restriction in fuel return lines. If engine does not idle normally, remove the return line check valve fitting from the top of the pump and make sure it is not plugged.
	4. Air in system	Install a section of clear plastic tubing on the fuel return fitting from the engine. Evidence of bubbles in fuel when cranking or running indicates the presence of an air leak in the suction fuel line. Locate and correct.

Condition	*Possible Cause*	*Correction*
	5. Incorrect or contaminated fuel	Flush fuel system and install correct fuel.
	6. Nozzle(s) malfunction	Remove and clean or replace.
Engine cold starts and idles rough *with* excessive noise and/or smoke but clears up after warm-up	1. Injection pump timing incorrect	Be sure timing mark on injection pump is aligned with mark on adapter.
	2. Insufficient engine break-in time	Break-in engine 2000 or more miles.
	3. Air in system	Install a section of clear plastic tubing on the fuel return fitting from the engine. Evidence of bubbles in fuel when cranking or running indicates the presence of an air leak in the suction fuel line. Locate and correct.
	4. Nozzle(s) malfunction	Remove and clean or replace.
Engine misfires above idle but idles correctly	1. Plugged fuel filters	Replace filters.
	2. Incorrect injection pump timing	Be sure that timing mark on injection pump and adapter are aligned.
	3. Incorrect or contaminated fuel	Flush fuel system and install correct fuel.
Engine will not return to idle	1. External linkage binding or misadjusted	Free up linkage. Adjust or replace as required.
	2. Fast idle malfunction	Check fast idle adjustment.
	3. Internal injection pump malfunction	Remove injection pump for repair.
Fuel leaks on ground—no engine malfunction	1. Loose or broken line or connection	Examine complete fuel system, including tank, and injection lines. Determine source and cause of leak and repair.
	2. Injection pump internal seal leak	Remove injection pump for repair.
Noticeable loss of power	1. Restricted air intake	Check air cleaner element.
	2. EGR malfunction	Refer to service manual.
	3. Restricted or damaged exhaust system	Check system and replace as necessary.
	4. Plugged fuel filter	Replace filter.
	5. Plugged fuel tank vacuum vent in fuel cap	Remove fuel cap. If loud "hissing" noise is heard, vacuum vent in fuel cap is plugged. Replace cap. (Slight hissing sound is normal.)
	6. Restricted fuel supply from fuel tank to injection pump	Examine fuel supply system to determine cause of restriction. Repair as required.
	7. Restricted fuel tank filter	Remove fuel tank and check filter.
	8. Pinched or otherwise restricted return system	Examine system for restriction and correct as required.
	9. Incorrect or contaminated fuel	Flush fuel system and install correct fuel.
	10. External compression leaks	Check for compression leaks at all nozzles and glow plugs, using "Leak-Tec" or equivalent. If leak is found, tighten nozzle or glow plug.
	11. Plugged nozzle(s)	Remove nozzles. Have them checked for plugging and repair or replace.
	12. Low compression	Check compression to determine cause.

(Continued)

Condition	Possible Cause	Correction
Noise—"Rap" from one or more cylinders (sounds like rod bearing knock)	1. Nozzle(s) sticking open or with very low nozzle opening pressure 2. Mechanical problem	Remove nozzle for test and replace as necessary. Refer to service manual.
Excessive black smoke and/or objectionable overall combustion noise	1. Timing not set to specification 2. EGR malfunction 3. Injection pump housing pressure out of specifications 4. Injection pump internal problem	Make certain that timing mark on injection pump is aligned with mark on adapter. Refer to emission diagnosis. Check housing pressure. Remove injection pump for repair.
Engine noise—internal or external	1. Engine fuel pump generator, water pump, valve train, vacuum pump, bearings, etc.	Repair or replace as necessary. If noise is internal, see Diagnosis for Noise—Rap From One or More Cylinders and Engine Starts and Idles Rough With Excessive Noise and/or Smoke.
Engine overheats	1. Coolant system leak, oil cooler system leak or coolant recovery system not operating 2. Belt slipping or damaged 3. Thermostat stuck closed 4. Head gasket leaking	Check for leaks and correct as required. Check coolant recovery jar, hose, and radiator cap. Replace or adjust as required. Check and replace if required. Check and repair as required.
Oil warning lamp *on* at idle	1. Oil cooler or oil or cooler line restricted 2. Oil pump pressure low	Remove restriction in cooler or cooler line. See oil pump repair procedures (service manual).
Engine will not shut off with key NOTE: With engine at idle, pinch the fuel return line at the flexible hose to shut off engine.	1. Injection pump fuel solenoid does not return fuel valve to *off* position	Electrical problem.

Vaccum Pump Diagnosis

Condition	Possible Cause	Correction
Excessive noise or clattering noise	1. Loose screws between pump assembly and drive assembly 2. Loose tube on pump assembly	1. a. Tighten screws to spec. b. Replace O-ring. c. Replace pump assembly. 1. Replace pump assembly.
Hooting noise	Valves not functioning properly	Replace pump assembly.
Pump assembly loose on drive assembly	Stripped threads	Replace pump assembly.
Oil around end plug	Loose plug	1. Seat plug. 2. Replace drive assembly.
Oil leaking out crimp	Bad crimp	Replace pump assembly
Install hose and vacuum gauge to pump, engine running, gauge should have reading of 20 inches vacuum minimum. With engine off, vacuum level loss should not drop from 20 inches to 19 inches in less than 1½ seconds.	1. Defective valves 2. Defective diaphragm 3. Worn push rod seal 4. Loose tube	Replace pump assembly.

Engine Vibration

Misfiring cylinders due to low compression or faulty injectors, improper timing of individual pumping units or injector setting, valves set too tight, improperly balanced cylinder banks on vee type engines, water in the fuel, and plugged fuel filters are some of the typical causes of engine vibration. However, it may be caused by accessory items on the engine, necessitating a more thorough analysis with a vibration meter.

Exhaust Smoke Analysis

Always make exhaust smoke checks with a minimum water outlet temperature as specified by the engine manufacturer, usually around 160°F (70°C); it is also advisable if possible to check the opacity of the exhaust smoke at both no load and full load.

Although there are smoke meters readily available on the market, not everyone has such a device; therefore, a Ringelmann-type smoke chart can be used to approximate the density of the exhaust smoke emanating from the stack (see Figure 30–2).

Less than 5% opacity is hardly visible to the naked eye. The acceptable standards for exhaust smoke opacity are controlled by the EPA. Each engine produced by a manufacturer must meet certain limits of maximum allowable exhaust smoke opacity under a variety of conditions, including acceleration and lug down conditions.

Many states in the United States are very strict on exhaust gas emissions; therefore, it is encumbent upon the maintenance personnel to ensure that all engines meet these specificiations; otherwise, costly fines can result. Also, if the engine is producing abnormal amounts of smoke, it is an indication of poor performance and is reflected as a direct economic loss to the user.

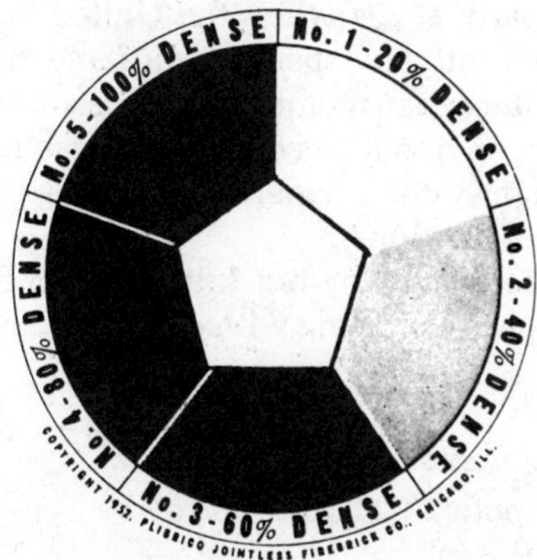

Figure 30–2. Ringelmann-type smoke chart. (Courtesy of Detroit Diesel Allison Division, GMC)

A Ringelmann smoke scale enables the user to observe conveniently the approximate density of the smoke coming out of the engine's exhaust stack. The scale should be held at arm's length, at which distance the shaded areas on the chart can be compared to the shade or density of the smoke coming from the exhaust stack. The observer's line of observation should be at right angles to the direction of smoke travel and not be less than 100 ft (30.48 m) or more than a $\frac{1}{4}$ mile (0.4 km) from the stack. The background directly beyond the top of the exhaust stack should be free of buildings or other dark objects and direct sunlight. By recording the changes in smoke density, the average *percentage of smoke density* for any period of time can be determined.

One of the quickest methods that can be used to pinpoint a problem with a diesel engine is the color of the exhaust smoke. There are basically four colors of exhaust smoke that you will come across, namely white, black, gray, and blue.

A number of conditions can contribute to exhaust smoke as well as its density. Smoke concentrations less than 5% opacity are generally not visible to the naked eye. Exhaust smoke meters are used increasingly today to monitor the condition and opacity (concentration/density) of all types of internal combustion engines.

White. This color of smoke usually occurs during cold starts due to tiny droplets of unburned fuel. The smoke is a result of condensed fuel particles. When starting from cold, the air temperature is low, which leads to a low temperature at the end of compression before the fuel is sprayed into the combustion chamber. For example, a reduction in intake air temperature from 80°F (26.6°C) in the summer to −20°F in the winter can result in a reduction in air temperature at the end of the compression stroke of as much as 230–300°F (110–149°C). The total reduction depends upon such things as piston compression ratio, piston design, combustion chamber design, and speed of cranking. In addition, when the cylinder wall temperature is cold, air will condense, usually at wall temperatures below 195°F (90.5°C).

With low compression temperatures, the speed of combustion is reduced, which increases the ignition delay period once the fuel has been injected, allowing the fuel particles a longer time to condense. The result is that the engine sounds noisier and runs rougher along with the smoke condition. This is why in a diesel engine with a higher compression ratio and fuel that is not taken in on the intake stroke white smoke is generally more noticeable than it is on a gasoline engine. This smoke generally will disappear after a short time as the engine warms up. However, continuous white smoke is an indication that a problem exists, especially if white smoke still exists when the engine is at operating temperature.

The following list summarizes general causes of white smoke.

1. Insufficient fuel delivery (cold start device faulty)
2. Injection timing late (retarded)
3. Faulty injectors
4. Glow plugs not operating correctly
5. Low compression pressure/temperature
6. Low engine coolant operating temperature (faulty thermostat, etc.)
7. Water leak into combustion chamber/cylinder (head gasket)
8. Wrong grade/blend of fuel

Black/Gray Smoke. Gray smoke can often turn to black smoke if the problem is not found and corrected. Generally some gray smoke will emanate from the diesel exhaust under low speed/idling conditions; however, large quantities of grayish-black smoke are cause for concern.

Driving and traffic conditions can contribute to this condition and frequent acceleration in city traffic after long periods of idling will cause large quantities of exhaust smoke due to the low cylinder temperatures that exist at idle speed (minute quantities of fuel sprayed in). When the throttle is suddenly moved to a large fuel delivery situation, there is an incomplete combustion phase for a short time until cylinder temperature becomes warm enough to allow a minimum ignition time delay period.

General causes of black/gray smoke are:

1. Air starvation (dirty air cleaner or air inlet restriction)
2. High exhaust back pressure (plugged muffler/pipe, crushed piping)
3. Too much fuel (check cold start device)
4. Excessive injector delivery (raw fuel)
5. Worn injectors
6. Low compression temperature
7. Wrong grade of fuel

Exhaust temperatures should be checked with a contact type pyrometer to isolate a faulty cylinder. Cylinder temperatures should generally be within 50°F (10°C) of one another on a well-balanced engine (compression pressures and injector fuel delivery the same).

Excess fuel delivery will cause an increase in exhaust temperatures and combustion temperatures with possible turbocharger overspeeding on turbocharged engines without any appreciable horsepower gain.

Faulty fuel injectors can cause the engine to knock fairly loudly. Do not confuse cylinder combustion knock with a bearing knock. To isolate a knock, you can loosen off each injector fuel line nut at the injector and listen for a change in the engine speed and sound where the problem is most pronounced.

If one nozzle is found where loosening the fuel line makes no difference in the misfire, or the puffing black smoke at the exhaust disappears, then that is the faulty nozzle or cylinder.

Faulty injectors can cause:

1. An overheated engine
2. Reduced power output
3. Uneven idle (rough) or misfiring
4. Black exhaust smoke
5. High fuel consumption
6. Combustion knock

Blue Smoke. Blue smoke, although it can be caused by the engine burning oil, can also be caused by incomplete combustion and condensed fuel particles. Some gray smoke can accompany a burning oil condition based upon the degree of oil being burned.

1. Retarded (late) injection timing
2. Faulty injectors
3. Low engine coolant temperature
4. High fuel consumption
5. Leaking injector heat shield gasket under nozzle body

Exhaust Emissions/Smokemeters

Since the early 1970s when the United States Congress passed into law specific rules and regulations governing the exhaust emissions of gasoline and diesel engines, engine and vehicle manufacturers have had to design systems capable of meeting these strict emissions regulations.

In Europe, the following International Standards have been passed dealing with engine exhaust smoke emissions:

1. ECE Regulation 24, Annex 8
2. EEC Directive 72/306, Annex 7
3. ISO DIS 3173

We are all familiar with the numerous engineering

changes that have taken place since that time and know that this will be an on-going process in the future.

Exhaust smoke in any form is harmful to the atmosphere in which we live as well as being an indication that the engine is not running properly. Exhaust smoke color and its interpretation is discussed earlier in this section.

Several major manufacturers produce exhaust smoke meters or analyzers in both fixed and portable form to quickly allow the service technician to take a measurement of exhaust smoke opacity or density. Commonly used smokemeters are manufactured by:

1. Hartridge Equipment Company, Diesel Division
2. Robert Bosch Corporation
3. Berkeley Controls
4. Robert H. Wager, Co., Inc.

Most of these smokemeters can be operated from either a 12/24-volt battery source or a main supply.

Hartridge Smokemeter Mk3. Hartridge has had available a smokemeter for many years; however, until the early to mid-1970s, its use was limited because most manufacturers of internal combustion engines were not controlled by governmental legislation as to limits of exhaust smoke.

Hartridge has developed their line of smokemeters to a point that they now have a model known as the Mark 3 unit that has been certified to meet all the British and International standards.

Studies have established that the regular use of a smokemeter followed up by action on the engine problem contributing to abnormal smoke emissions can create a fuel savings of at least 5% on a conservative basis.

Smokemeters are used in the following situation tests:

1. Steady state engine operation
2. Controlled accleration
3. Free acceleration
4. Mixed mode cycle
5. Lug down

Figure 30–3 illustrates the Hartridge Smokemeter Mk3 model in mobile unit form. This portable unit would be used at roadside test stations or where such a feature was required; however, the Smokemeter can be purchased in simpler form without Items 3 and 4 shown in the diagram. A smaller portable form, Figure 30–4, illustrates the base smokemeter that can

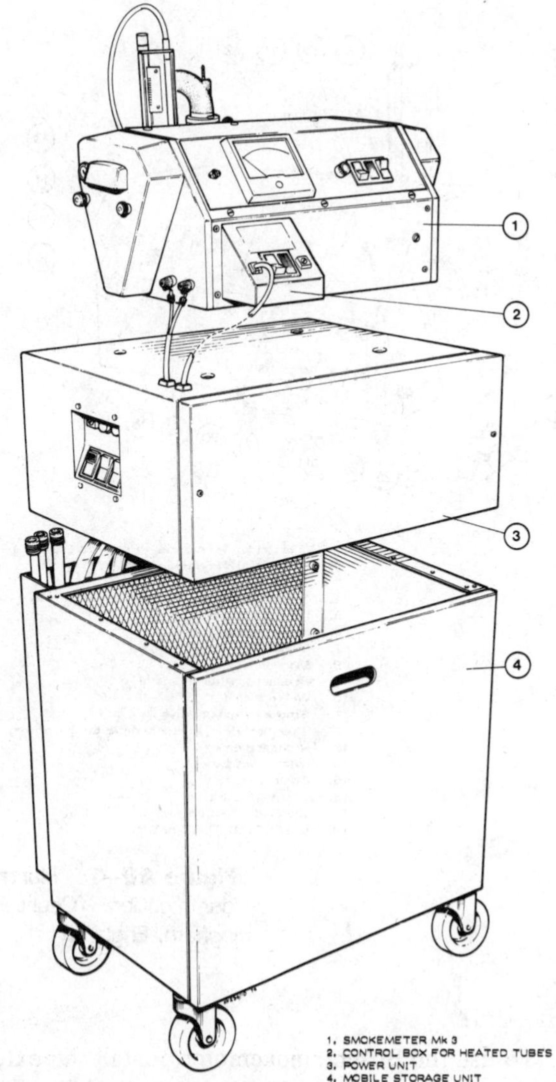

Figure 30–3. Hartridge Mk3 Smokemeter and mobile units. (Courtesy of Hartridge Equipment, Buckingham, England)

be mounted on a bench or on the vehicle during testing.

The operation of the smokemeter is explained in detail courtesy of Hartridge Equipment along with its operational diagram shown in Figure 30–5.

Robert Bosch Smokemeter Operation. The Robert Bosch smokemeter is shown in Figures 30–6 through 30–12 connected to a Volvo D24 6-cylinder diesel engine exhaust system.

The Volvo diesel engine is similar in design to the 4-cylinder Volkswagen Rabbit and the Audi 5000 diesel engines; therefore, use of this smokemeter sequence would apply equally to those engines and any other diesel engine that it was hooked up to. Figure 30–6 illustrates the Bosch smokemeter kit.

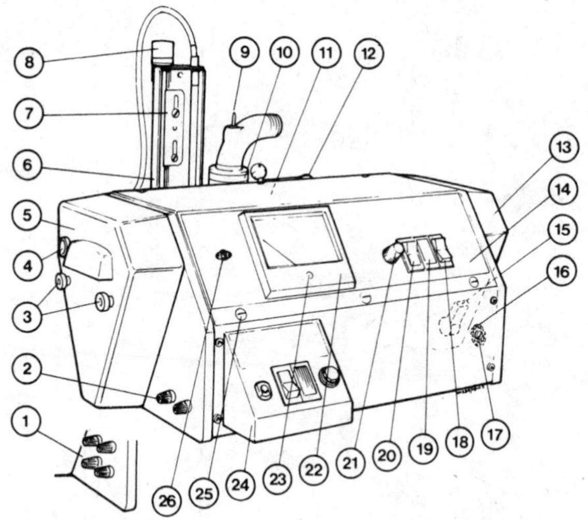

1. Electrical Terminals - Pre Mod Std G
2. Electrical Terminals - Mod Std G
3. Cover securing screws
4. Thermometer (smoke tube temperature)
5. Left end cover
6. Manometer
7. Manometer scale
8. Manometer filler cap
9. Pressure relief indicator
10. Water trap
11. Smoke by-pass valve
12. Thermometer (smoke sample temperature)
13. Right end cover
14. Instrument panel
15. Control knob
16. Clean air inlet
17. Clean air inlet control
18. V TEST / S TEST switch
19. WARNING LIGHT
20. ON/OFF switch
21. ZERO RESET knob
22. Meter
23. Needle reset screw
24. Control box for heated tubes
25. Fastener for instrument panel (six off)
26. Chart recorder output socket
27. Manometer mounting screws
28. Relief valve outlet
29. By-pass closed position (smoke to outlet 32)
30. Smoke inlet
31. By-pass open position (smoke into meter)
32. Smoke by-pass outlet
33. Air and sample smoke outlet
34. Housing pressure connection
35. Water trap drain plug

Figure 30–4. Hartridge Smokemeter Mk3 component identification. (Courtesy of Hartridge Equipment, Buckingham, England)

Procedure

1. To use the Bosch smokemeter, install the extension pipe shown in Figure 30–7 to the vehicle exhaust, then refer to Figure 30–8 and install the sampling pump by inserting the probe so that it enters the vehicle exhaust in the center of the pipe and is at least 8 inches (200 mm) into the pipe; otherwise, a false reading may result.
2. To check the sampling pump operation, refer to Figure 30–9 and manually push in the pump plunger knob fully (rubber flexible hose shown immediately below white arrow in the figure; push it towards the exhaust pipe).
3. Squeeze the rubber bladder by hand to release the sampling pump plunger, which should move back as shown in Figure 30–9 in the same direction as the arrow. Make sure that the pump plunger does move freely. The pump plunger will not release if you place your thumb or hand over the vent hole at the top of the rubber bladder. Push the handle of the sampling pump all the way in again and leave it there.
4. Refer to Figure 30–10 and screw out the sampling pump cover opposite the pump handle end. Check and clean the contact surfaces. Insert and position the new clean white filter into the opening on the end of the sampling pump cover. Install and tighten the cover by hand only, then fold over the rubber seal in order to prevent any moisture or dirt from entering the filter which could affect the exhaust smoke reading.
5. It is necessary to route the rubber bladder and its hose up into the vehicle passenger compartment. Tape the hose in place and run it in through the rear side window on a passenger car and in through the tailgate on station wagons.
6. The engine must be at normal operating temperature prior to conducting the test.
7. On manual transmission equipped vehicles, drive in second gear at 30 mph (50 km/hr). This speed must be maintained while pushing the accelerator to the floor and applying the brakes to maintain this speed (Figure 30–11).

A. SMOKEMETER

A1. Description

The smokemeter contains two dimensionally and optically similar tubes (1) and (18). The reference tube (18) is connected to the clean air blower (17). Air is drawn through a mesh along a tube containing a screw controlled valve (15) from the clean air inlet (14). The smoke tube (1) is connected to the smoke inlet (8) via the smoke by pass valve (9). This inlet tract contains a thermometer (6) and water trap (4) and has a pressure relief valve (5) controlled outlet, a pick off for the manometer connection and a vane (3) at the tube entry.

The light (21) and photoelectric cell (11) are each mounted on an arm pivoted on each end of a spring located control shaft (19) which is operated by a control knob (13).

The cabinet is constructed to enable easy access for servicing yet remain airtight. Hence the main access panel is located and secured by six Dzeus type fasteners and sealed by a composite rubber gasket. The meter, warning light and main controls are mounted on this access panel. If supplied, the tube heater control box is mounted at the lower left front of the smokemeter and all electrical interconnects and fittings are leakproofed. End casings for access to the light and photoelectric cell are cast alloy and are retained by knurled screws.

A2. Operation

The smokemeter operates on a comparative basis. The opacity of the smoke sample is compared to a clean air sample by first taking a meter reading through the clean air tube (18) to set the zero, then moving the light (21) and photoelectric cell (11) to the smoke tube (1) for an instantaneous comparative reading of the smoke sample. For repetative accurate results, the pressure of the clean air sample and the pressure of the smoke sample must be accurately controlled. Also, limiting factors, such as condensation, temperature, even smoke distribution and a constant intensity light source must be held within practical limits. Sooting up of the photoelectric cell and the lamp is prevented by a flow of clean air from the blower which after passing through the clean air tube (18), enters the end covers and passes across the photoelectric cell (11) and the lamp (21) to carry away any smoke particles into the instrument casing with the sampled smoke. This flow of cool air also reduces the surface temperature of the photoelectric cell.

To achieve the required accuracy and ensure a representative sample, the smoke pressure at the inlet is controlled by careful positioning of the exhaust probe. Inlet smoke temperature can be monitored by thermometer (6) for excessive temperature and should not normally exceed 210°C. For specialised smoke readings the temperature of the sample in the smoke tube is checked on thermometer (22) and should be within the range 70°C to 115°C, and the pressure is checked on manometer (2) which should be between 40mm and 65mm of water. If the smoke sample is at the correct temperature, condensation should not be formed in the smoke tube (1). The blow off valve (5) above the water trap is a safety relief for excessive smoke pressure. Vane (3) ensures an equal distribution of the smoke sample in the smoke tube. The air pressure in the clean air tube (18) is adjusted by the clean air inlet control (16) and is set using Kit HV009.

Adjustable stops ensure the accurate positioning of the photoelectric cell and light in the ZERO CHECK (clean air tube) and (SMCKE CHECK) smoke tube positions.

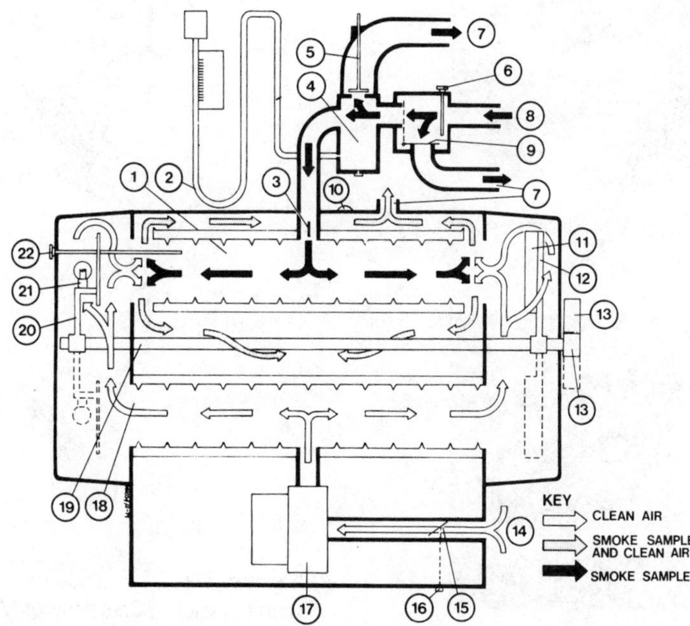

Figure 30–5. Smokemeter—schematic layout, shown in smoke check position, zero check position dotted. (Courtesy of Hartridge Equipment, Buckingham, England)

8. On automatic transmission equipped vehicles, place the gear selector in position 2 and maintain a speed of 30 mph (50 km/hr) without allowing the kickdown to come into engagement by applying the brakes as you would do with a manual transmission equipped vehicle (Figure 30–11).
9. Maintain 30 mph (50 km/hr) for several seconds while squeezing the rubber bladder hard several times (do not cover the vent hole at the top) to ensure that the sampling pump will obtain its full stroke.
10. Bring the vehicle to a stop and remove the filter from the sampling pump and repeat the same procedure over again.
11. When complete, refer to Figure 30–12 and follow the Robert Bosch instructions that come with the smokemeter to calibrate the test analyzer. Use the second sample from the test to evaluate the condition of the engine exhaust gases.

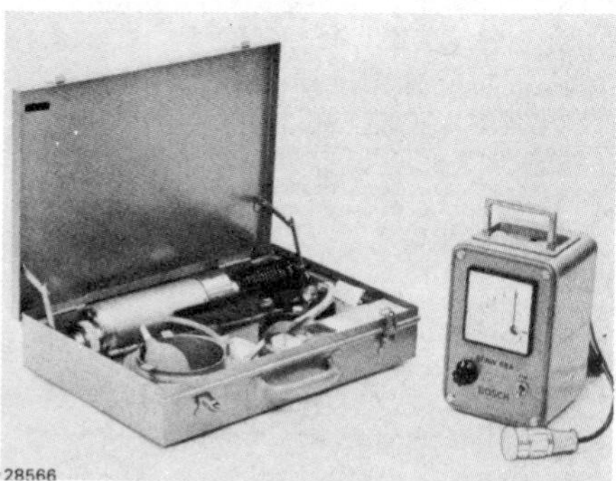

Figure 30-6. Robert Bosch smokemeter. [Courtesy of Volvo Corporation of America]

Figure 30-7. Extension pipe installation. [Courtesy of Volvo of America Corporation]

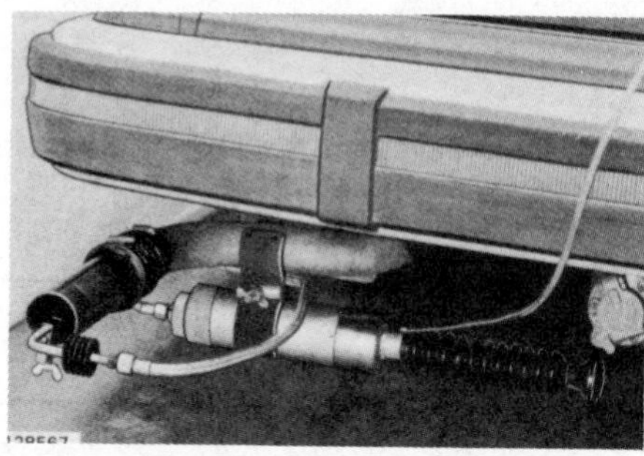

Figure 30-8. Sampling probe installation. [Courtesy of Volvo of America Corporation]

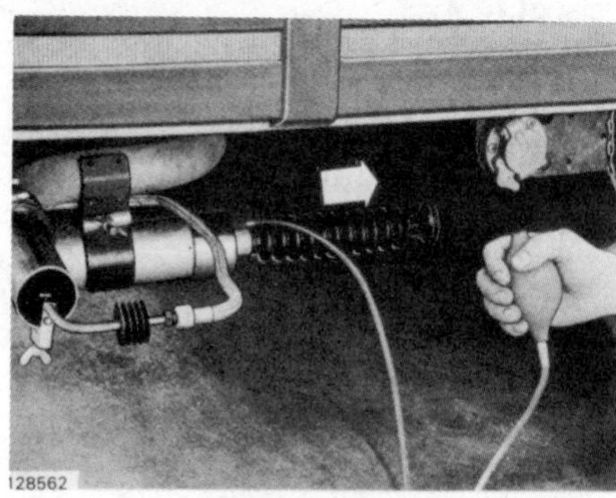

Figure 30-9. Checking sampling pump operation. [Courtesy of Volvo of America Corporation]

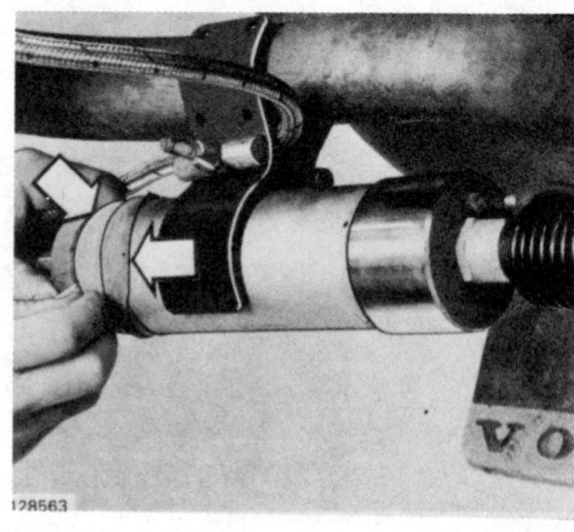

Figure 30-10. Installing filter in sampling pump. [Courtesy of Volvo of America Corporation]

Figure 30-11. Test procedure, manual and automatic transmission. [Courtesy of Volvo of America Corporation]

Figure 30-12. Calibrating smokemeter and sampling smoke filter. [Courtesy of Volvo of America Corporation]

31

Turbochargers

Turbochargers have been in use for many years on internal combustion chamber engines with their use, until recently, limited to high speed heavy-duty truck diesel engines and larger slow speed diesel engines.

Gasoline engines have used turbochargers in the past also, but their use was mainly limited to high performance racing engines. However, General Motors did offer a turbocharged version of the Oldsmobile F-85 Jetfire in 1962 and also a turbocharged Chevrolet Corvair with its air-cooled flat six-cylinder engine.

The word turbocharger is briefly defined in the dictionary as whirlwind, spinning top, etc.; the word is actually an abbreviation for turbo-super-charger. Gas turbines and steam turbines have been in use for many years on stationary applications and for ship propulsion.

In 1897, Sir Charles Parsons, who had spent considerable time in developing the turbine principle, astounded skeptics by installing a turbine in his yacht "Turbinia," which had excellent speed characteristics.

The turbocharger was taken a stage further in 1905 by Alfred J. Buchi whose name and company still produce turbochargers in a variety of sizes for many different engine manufacturers.

The first commercial turbocharged diesel marine engine was used in 1923; however, it wasn't until 1952 that the first turbocharged diesel engine was applied to a race car, which was the Cummins diesel used in the Indianapolis 500. The first turbocharged gasoline engine to use a turbocharger in a USAC (United States Auto Club) race car occurred in 1966, and it was two years later before a turbocharged gasoline engine actually won in the Indianapolis 500.

The turbocharger was used fairly extensively for many years on aircraft piston-driven engines, especially between the years 1922 to 1950. The first operational aircraft squadron with turbos was formed in 1922.

The jet engine (gas turbine) idea was patented by Sir Frank Whittle in 1930 with the first British jet engine being run on April 12, 1937. The first British jet-propelled airplane flew on May 15, 1941, from which subsequent British and U.S. jet engines were developed.

The concept of both the jet engine and the automotive turbocharger depends upon the expansion of gases against vaned wheels.

Since 1976, production passenger cars have been offered with turbocharger options as a means of improving both the horsepower output of the engine and the overall vehicle performance with the use of smaller displacement engines, namely in-line 4- and V6-cylinder arrangements.

The first manufacturer to produce what can be considered "new wave" turbocharging was Porsche in 1976 with the 930 Model Carrera. This vehicle was capable of 0–60 mph in 5.0 seconds, 0–100 mph in 11.9 seconds, and a standing start quarter-mile time of 13.7 seconds at 106.5 mph—truly an outstanding performer.

This was followed (also in 1976) by the Saab 99 turbo released in Europe in 1976 and in the USA in 1977.

Mercedes-Benz introduced their 5-cylinder 300SD (diesel) in the 1978 model year as did Buick with their 3.8L (232 cu.in.) V6 gasoline engine. Ford introduced their 2.3L 4-cylinder engine option for the Mustang Cobra and Capri Turbo R/S in 1979. Porsche extended their turbocharging to the 924 in 1979, while Pontiac introduced the Trans Am 301 cu.in. V8 in 1980.

This chapter will deal with the operating and design characteristics of turbochargers that are now in use on automotive engines—both gasoline and diesel powered. Several types and vehicle models will be studied so that you can assimilate the parallels that exist between turbocharged gasoline and diesel engines.

Most vehicle manufacturers obtain their turbochargers from major suppliers such as:

1. Garrett/AiResearch
2. Schwitzer Division of Wallace Murray Corporation
3. Holset
4. Rajay
5. Hitachi
6. KKK (Kuhnle, Kopp and Kausch)
7. Brown Boveri
8. IHI (Warner-Ishi) (IHI—Ishikawajima-Harima Heavy Industries)

The above turbocharger manufacturers represent a good cross section of manufacturers in North America, Britain, Europe, and Japan. There are a large number of after-market suppliers and custom engine tuneup and performance companies—too many to name here; however, such companies are capable of installing turbocharger kits to a wide variety of engines, both gasoline and diesel powered.

Workable Engine Formulas

Volumetric Efficiency

The term *volumetric efficiency* (VE) is used to express the amount or weight of air that an engine can take into the cylinders during the intake stoke. On a naturally aspirated engine, air is forced into the cylinder by the action or force contained in atmospheric air, which is 14.7 psi at sea level. The power that a given displacement engine can produce is directly related to the quantity or volume of air that it can inhale during its intake stroke.

Since air density decreases with an increase in temperature, the final power determining factor rests with the density or weight of air that the engine can retain in the cylinder. Although colder air can be denser than warm air, the other consideration here is one of heat loss. If the air temperature is low enough, problems of starting and smooth combustion can be affected, especially in a diesel engine, which relies upon the heat of the compressed air charge to stimulate combustion.

The engine's VE can be considered as the weight of dry air contained in the engine cylinder when the piston is at BDC (bottom dead center) with the engine stopped versus the weight of dry air contained in the cylinder with the piston at BDC when the engine is running. The weight of air in the cylinder when the engine is running on a naturally aspirated engine is generally about 80–85% of atmospheric.

Since energy must be expended to get the air into the engine, due to air-flow restrictions and the very short time period during which the intake valves are open, this pressure will always be less than atmospheric unless the engine is equipped with a gear driven blower or exhaust gas driven turbocharger.

Consider that an engine running at about 2500 rpm has its intake valves open for less than 0.017 second; therefore, very little time is available for air to enter the cylinder.

The greater the weight of air that can be inhaled or packed into the engine cylinder, the greater the volume of fuel that can be drawn or injected into the air mass, resulting in higher horsepower from the greater heat energy released.

In theory, most engines should be capable of drawing in a volume of air that is equal to their cylinder displacement. Therefore, if we consider a 2.0L (122 cu.in.) engine, it would basically draw in a 122 cu. in. supply of air. However, on a naturally aspirated engine, the engine would be incapable of doing this due to restrictions to air flow caused by air cleaners and the short time that the intake valves are open.

The formula for Volumetric Efficiency can be shown as follows:

$$\text{VE} = \frac{\text{Weight of air in the cylinder at BDC/Engine stopped}}{\text{Weight of air in the cylinder at BDC/Engine running}} \times 100\%$$

VE is generally measured in cubic feet per minute of air flow. Therefore, to calculate how much air an engine should draw in on one crankshaft revolution, we simply convert the cubic inch displacement of the engine into cubic feet first. To do this we can remember that there are 1728 cubic inches in a cubic foot; therefore, if we divide an engine's displacement by 1728 we can arrive at an engine's displacement in cubic feet.

Using the following formula we can find what we need to know:

Formula for Displacement in Cubic Inches.

1728 cubic inches = cubic feet in one crankshaft revolution

Example: An engine with a displacement of 2.0L

(2000 cc) divided by 16.4 equals 122 cu. in.; therefore

$$\frac{122 \text{ cu.in.}}{1728 \text{ cu.in.}} = 0.0706 \text{ cu.ft.}$$

On each revolution, our 122 cu.in. engine displaces 0.0706 cu.ft. of air; therefore, if we assume that the engine is red lined at 6000 rpm, how much air will flow through that engine at its maximum rpm point?

To determine this, we multiply 0.0706 cu.ft. by 6000 rpm and divide our result by 2 (since we only have one intake stroke for every two crankshaft revolutions in a 4-stroke cycle engine). Therefore the formula becomes

$$\frac{\text{Displacement} \times \text{Engine rpm}}{2} = \text{CFM}$$

Solution to our example would become

$$\frac{0.0706 \times 6000}{2} = 211.8 \text{ cu.ft./min.}$$

Based upon this formula, this engine would have an ideal air-flow rate of 211.8 CFM.

If we were to consider that the engine had an actual flow rate (by measurement) of 180 CFM, its VE would be 85% arrived at by

$$\frac{\text{Flow Rate CFM Actual}}{\text{Flow Rate CFM Ideal}} = \text{Volumetric Efficiency (VE)}$$

By inserting the figures arrived at in the example above we have

$$\frac{180}{211.8} = 84.98\%$$

This particular engine is therefore capable of a volumetric efficiency of 85%, which is typical of that found on naturally aspirated engines.

Thermal Efficiency

The thermal efficiency (TE) is the heat efficiency that the engine is actually able to produce. An easier way to think about this is just how much work value is returned to the flywheel for every dollar of fuel that is burned in the engine. For example, if this figure were 35 cents in every dollar, then the TE would be 35%.

To determine just how thermally efficient any engine is, we have to take several things into account. First we know that no engine today is 100% thermally efficient and that a lot of the heat that is released into the combustion chamber is lost by friction, radiation, cooling, and exhaust systems. Generally, the cooling and exhaust systems can absorb about 27–30% at each one, plus another 7–10% to friction and radiation. The net result is that what is left is actually the TE of the engine, or just how efficiently the engine has consumed and used the BTU heat content from the burning fuel source.

Since any internal combustion engine is using either a liquid (gasoline, diesel, or methanol) or a gaseous fuel in the form of LPG or CNG (propane or compressed natural gas/methane), the fuel is rated as to its heat value per pound or per gallon in British Thermal Units (BTU's). Each BTU is capable of releasing 778 foot pounds of energy, which is commonly known as Joule's Equivalent.

Chapter 4 provides each type and grade of fuel's heat value in BTU's per gallon.

Common thermal efficiencies of gasoline-powered vehicles average anywhere between 23–28%, while diesel-powered vehicles average anywhere from about 33–40% and even greater in some instances.

Example

An example of how to calculate the thermal efficiency of a given engine is as follows. Find out the following information:

1. Rated engine horsepower
2. Fuel consumption in lb./bhp/hr.
3. BTU content of the fuel being used

Items 1 and 2 above can be obtained from a manufacturer's sales information bulletin/brochure, while the heat value of the specific diesel fuel can be found from the table dealing with diesel/gasoline fuels in Chapter 4.

Let's assume for instructional purposes that a diesel engine produces 135 horsepower at 3500 rpm and that its quoted fuel consumption by the manufacturer is 0.350 lb./hp/hr. while it is running on a No. 1 grade diesel fuel.

The TE can be calculated as follows. Multiply the horsepower by the fuel rate, which in this case would be 135 × 0.350 = 47.25 pounds; this is the amount of fuel (weight) that this engine would consume in one hour while running at 3500 rpm, producing 135 horsepower.

If we now assume that the particular fuel being used had an API gravity of 40, then as per the chart shown in Chapter 4, this fuel would contain 135,800

BTU's per gallon high heat value (HHV) or 19,750 BTU's per pound.

By multiplying 47.25 pounds × 19,750 we would establish that this engine has consumed 933,187.5 BTU's of heat in one hour while producing its 135 horsepower. A perfect internal combustion engine requires 2545 BTU's to produce one horsepower in one hour, or in metric equivalency, it requires 3413 BTU's to produce one kilowatt per hour of power.

Since we now know that we require 2545 BTU's to produce one horsepower in an hour, if we divide the total heat that our engine used by the number of horsepower that it produced, we will be able to establish how much heat our engine consumed versus that of a perfect engine.

Therefore divide 933,187.5 by 135 equals 6912.5 BTU's; our engine consumed 6912.5 versus 2545 in a perfect engine, which means that we have an engine that is less than perfect. How efficient our engine is can be arrived at by simply dividing 2545 by 6912.5 × 100% = 36.82%, or 37%, which is the thermal efficiency of this particular engine.

The engine actually required 6912.5 BTU's to produce one horsepower over a one-hour time period; therefore, where did the rest of this heat go? Heat losses in the engine such as coolant system, exhaust, and radiation accounted for the remainder of the BTU heat losses. If we assume that this engine gave up 28% to both the cooling system and the exhaust, then this would account for 56% with the remaining 7.19% being taken up with friction and radiation of heat.

If we multiply the total BTU's used by each one of these percentages of heat lost, we can establish that the cooling system and exhaust with a 28% heat loss each would account for 6912.5 × .28 = 1935.5 BTU's each, while the friction and radiation accounted for 6912.5 × 7.19 = 497 BTU's.

Therefore this engine is returning approximately 37 cents in every dollar as useful work at the engine flywheel, which is useable work output.

Another method that can be used to establish an engine's thermal efficiency is to simply use the following formula:

$$\text{Brake Thermal Efficiency} = \frac{\text{BHP/hr} \times 100\%}{\text{Fuel Energy in BTU's input/hr.}}$$

For example, if a gasoline engine were to produce 135 horsepower over a one-hour period at a steady throttle position similar to the diesel engine example given above with a fuel consumption of 9.31 gallons per hour, what would its brake thermal efficiency be?

The grade of gasoline used will determine the actual heat value released and used by the engine to produce its 135 horsepower over a one-hour period. A mean average weight for a U.S. gallon of gasoline is about 5.8 lb., and it contains approximately 113,800 BTU's. If we divide the BTU's per gallon by its weight, we establish that each pound of gasoline will produce 19,621 BTU's.

To calculate this engine's TE, we simply insert the known information into the formula given above; therefore, it becomes as follows:

$$\text{TE} = \frac{\text{BHP} \times 2545 \times 100\%}{\text{BTU's} \times \text{lb. of Gas/hr.}} =$$

$$\frac{135 \times 2545 \times 100\%}{19{,}621 \times 53.99} = 32.43\% \text{ Thermal Efficiency}$$

Gasoline engine thermal efficiency is similar to that of a diesel engine with respect to the fact that many variables can come into play. However, on an average basis, the gasoline engine will return between 23–28% TE with some of the newer style engines, especially the turbocharged units, producing thermal efficiencies in the 33–34% range. Overall operation of the gasoline engine versus its diesel counterpart results in a TE difference in favor of the diesel in the region of 7–10%. Again however, in some cases this can be as high as 10–15% or greater depending upon operating conditions such as load and speed of the vehicle, geographical location, and ambient air conditions.

Basic Turbocharger Operation/Design

The turbocharger concept is basically derived from the days of the steam engine or waterwheel, whereby a supply of water flowed from a river or stream down against paddles connected to a wheel rotating on an axis.

No direct mechanical connection from the water to the paddle wheel existed; however, the force of the falling water hitting the paddle wheel caused it to rotate and thereby produce work at the shaft of the waterwheel.

The turbocharger operates on a similar principle in that, instead of using water or steam pressure to drive the wheel, exhaust gases are used. The exhaust gases are under pressure when they leave the cylinder and have considerable heat energy. This expansion of the gases as they leave the engine is what drives the turbocharger assembly. Since approximately 27–30% of the heat generated within the engine cylinder is lost out of the exhaust stream, the use of a turbocharger allows these gases to give up their energy to driving the turbine portion of the rotating members, without any direct mechanical connection to the engine, which would require a horsepower loss to drive it such as you have on a gear driven supercharger or blower unit.

Figure 31–1 shows the basic arrangement of an exhaust gas driven turbocharger assembly, which contains in its simplest form a shaft that supports two vaned wheels, one at either end. One of these vaned wheels is known as the "turbine" since hot exhaust gases expanding against it cause it to rotate. Since the turbine is solidly connected to its support shaft, the shaft will also rotate and spin the vaned wheel at its other end; this wheel is called the "compressor."

The compressor assembly draws air into it through the air cleaner where the air is then pressurized inside the turbocharger housing and directed through piping to the cylinders.

The housing that encompasses the turbine end of the turbocharger is generally referred to as the "hot end" since exhaust gases flow through it, while the air inlet end is commonly called the "cold end" since fresh air flows through it on its way to the engine cylinders.

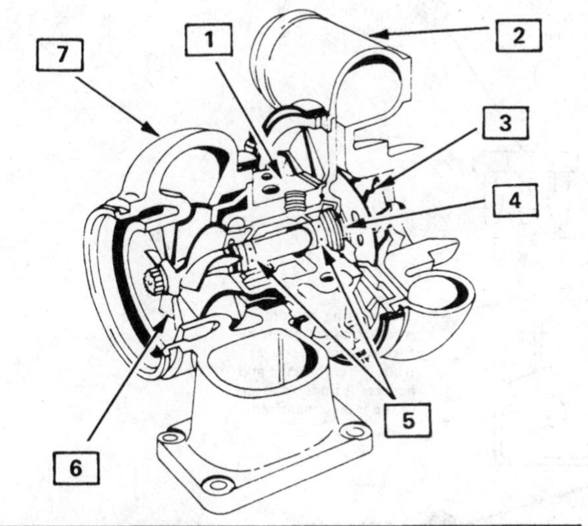

1–CENTER HOUSING
2–COMPRESSOR HOUSING
3–COMPRESSOR WHEEL
4–THRUST BEARING
5–FULL FLOATING SHAFT BEARING
6–TURBINE WHEEL
7–TURBINE HOUSING

420001-6J

Figure 31–1. Basic components-turbocharger assembly. (Courtesy of Pontiac Motor Division)

The rotating components shown in Figure 31–1 are mounted in a housing that has pressurized lube oil from the engine directed into the center support section of the housing for lubrication of the turbocharger turbine/compressor support shaft bearings/bushings. This is necessary because current turbochargers routinely spin at speeds of 80,000 to 140,000 rpm with some high-output special application models running even higher. Figure 31–2 illustrates a typical oil supply diagram for a turbocharger assembly. A schematic describing the basic air and exhaust flow into and out of the turbocharger is shown in Figure 31–3.

Performance Improvements Using a Turbocharger

The use of a turbocharger increases the engine power output and vehicle performance generally from an average of 10–30% on gasoline engines, although higher specific vehicle ratings have been achieved such as:

1. Ford 2.3L T/C in-line 4-cylinder engine displays a 40% horsepower and 30% torque increase over its non-turbocharged version.
2. Buick's 1976 Indianapolis 500 V6 pace car developed 315 horsepower over the then standard engine's 105, for a horsepower increase of 300%. Manifold boost pressures of 22 psi (152 kPa) gauge were used that resulted in BMEP's (brake mean effective pressures) of 240 psi (1655 kPa).

   The current non-turbocharged engine produces 110 while the turbo engine produces 180, for a horsepower increase of 39%.
3. British Leylands Mini-Metro car with a 1.3L engine was capable of a horsepower increase of 43% and a torque increase of 32% when fitted with a turbocharger.
4. Datsun/Nissan's 280ZX turbo engine produces 24% more power and 30% more torque than the non-turbocharged version.
5. Chrysler Laser/Daytona 2.2L turbocharged engines produce 142 horsepower compared to the 99 horsepower of the non-turbocharged version for an increase of almost 50% along with a 35% torque increase.
6. Mercedes 5-cylinder 300D (diesel) produces 83 horsepower in standard trim while the T/C version produces 120 for an increase of 31%.
7. Pontiac's 4.9L (301 cu.in.) V8 in standard trim

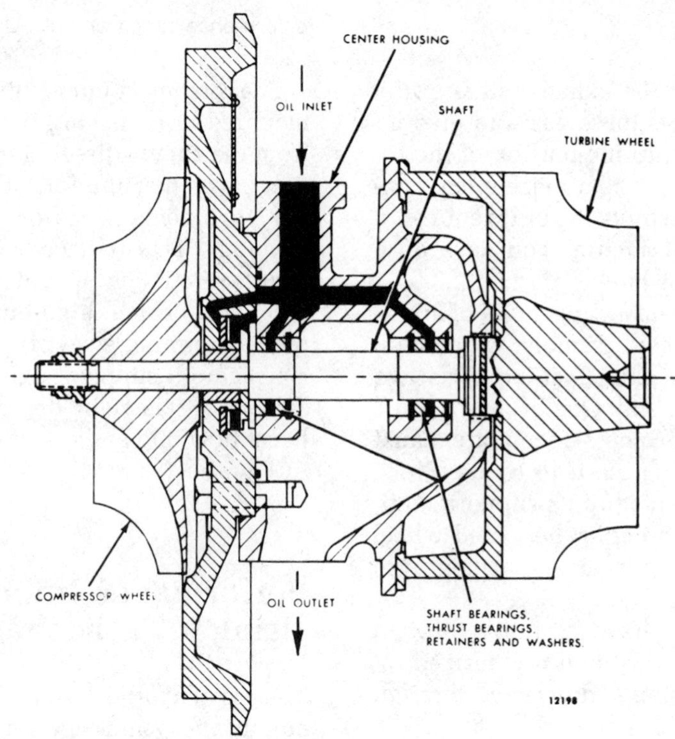

Figure 31-2. Turbocharger lube oil flow diagram. (Courtesy of Detroit Diesel Allison Division of GMC)

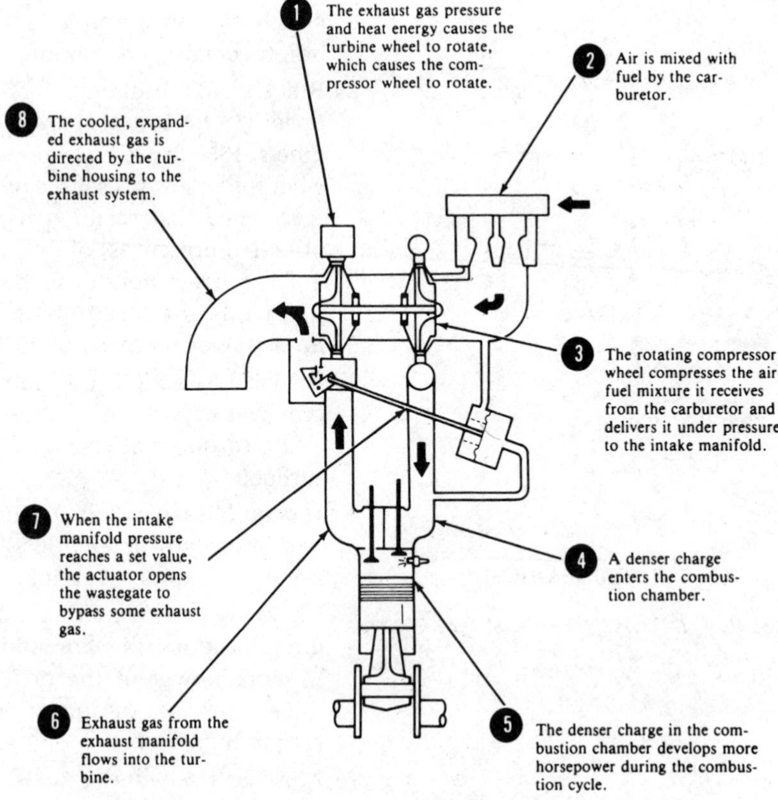

1 The exhaust gas pressure and heat energy causes the turbine wheel to rotate, which causes the compressor wheel to rotate.
2 Air is mixed with fuel by the carburetor.
3 The rotating compressor wheel compresses the air-fuel mixture it receives from the carburetor and delivers it under pressure to the intake manifold.
4 A denser charge enters the combustion chamber.
5 The denser charge in the combustion chamber develops more horsepower during the combustion cycle.
6 Exhaust gas from the exhaust manifold flows into the turbine.
7 When the intake manifold pressure reaches a set value, the actuator opens the wastegate to bypass some exhaust gas.
8 The cooled, expanded exhaust gas is directed by the turbine housing to the exhaust system.

Figure 31-3. Turbocharged engine system air/exhaust flow schematic. (Reprinted with permission 1985, Society of Automotive Engineers, Inc.)

produces 150 horsepower while the T/C version produces 210 for a 29% increase.

8. The 924 Porsche produces 95 horsepower non-turbo, while the turbo version produces 154 for a horsepower increase of 38%.
9. The Audi 5000 produces 103 horsepower while the turbo produces 130 for an increase of 21%.

These examples show just a few of the many engine/vehicle manufacturers now offering turbocharging as an optional source of better fuel economy and vehicle performance improvement.

Performance improvements on automotive and light duty diesel engines by turbocharging are even more dramatic than that for gasoline engines. Improvements of 40–60% are possible on diesel engines versus the mean average of 20–30% of gasoline engines.

Turbocharger Lag/ Response Time

Although a turbocharger can improve the performance of a vehicle by increasing the engine volumetric efficiency, all turbochargers have what is commonly called "turbo lag," which is the condition that occurs when the driver pushes the accelerator or throttle towards an increase fuel position.

This lag time is basically a short delay period before the added turbocharger boost pressure flows to the engine. The reason for this is due to the fact that the turbocharger is not mechanically connected to the engine; therefore, before the turbocharger turbine/compressor wheels can speed up, they must first of all receive an increase in exhaust gas flow from the engine. Therefore, when the driver accelerates, there is a short delay period before the additional fuel flow can be transformed into increased exhaust pressure and flow rate.

The turbo lag point or rpm will vary between different engines and is directly related to such items as:

1. The exhaust system gas flow volume
2. The shape/contour of exhaust manifolds and pipes
3. The exhaust system gas flow restriction
4. The intake system gas flow restriction/volume and shape
5. Engine valve timing (camshaft design)
6. The rotating inertia (size and weight/mass) of the turbocharger turbine/compressor and shaft
7. The turbine A/R ratio (explained herein)
8. Turbocharger wastegate calibration
9. The aerodynamic performance of the turbine/compressor assembly
10. Vehicle transmission and axle ratios

Turbocharged engines with excessive lag characteristics are normally found on a high-horsepower/high-rpm operation. Because of the large diameter turbochargers required with large displacement high speed engines, a large induction and exhaust system volume exists; therefore these large volumes, especially in the induction system, create excessive delays in turbocharger transient response.

The trend today is to smaller turbochargers capable of responding rapidly to exhaust flow rate variations so that a minimum of "lag," or transient response time, exists.

T/C Tuned Intake System

A method that is gaining increased attention today with the use of a turbocharger is the design of a tuned intake and exhaust system. A tuned intake manifold system can improve the engine breathing performance (volumetric efficiency) at the peak torque speed, improve the cylinder-to-cylinder air distribution, as well as improving turbocharger response time. This increased air flow improves combustion efficiency and reduces exhaust smoke emissions.

An example of one such tuned intake system is shown in Figure 31–4 where a damping reservoir or chamber is placed between the T/C compressor outlet and the intake manifold. Tuned intake pipes connect the front and rear intake manifolds with the engine cylinders.

The diagram shows a 6-cylinder engine with a firing order of 1-5-3-6-2-4, which lends itself to the adoption of two separate intake manifolds since 1-2-3 and 4-5-6 can be interconnected to the same manifold. With irregular firing orders such a system can-

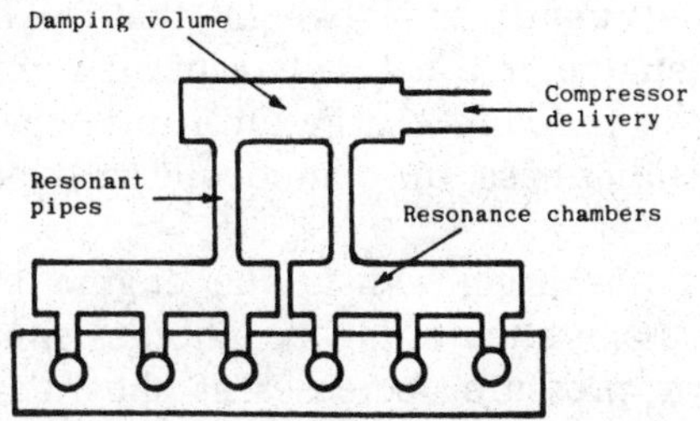

Figure 31–4. Tuned turbocharger intake manifold system. [Reprinted with permission 1985, Society of Automotive Engineers, Inc.]

not be used, but instead individual organ style pipes must be employed.

Volumetric Efficiency Increase

To understand why turbocharging can increase the power and torque of a given internal combustion engine, we must consider the term *volumetric efficiency* and how it relates to an engine's power output capabilities.

Any piston engine that does not use a turbocharger or supercharger is classified as a naturally aspirated engine. What this means is that the pistons as they move up and down in their cylinders act like an air pump. When the intake valves are open, pressure in the cylinder is less than atmospheric due to the short time period that the valve is open and the resulting restriction to air flow through air cleaners and manifolding. When the exhaust valves are open, the pistons simply push the burnt gases out of the cylinder.

Engine valve timing does affect the cross-flow effect that exists between the opening inlet and closing exhaust valve. When the inlet valve opens BTDC, the exhaust valve is already open (BBDC—before bottom dead center). This action causes a scavenging effect on the escaping exhaust gases, which are under pressure, with the result that when the intake valve opens, a suction effect occurs so that additional air/fuel mixture can be drawn in to the cylinder on a carburetor-type engine and produce better air flow on a fuel injected engine.

The power that a given displacement engine can produce is directly related to the quantity or volume of air that it can inhale during the valve overlap period and the remaining intake stroke.

Since air density decreases with an increase in air temperature, the final power determining factor rests with the denseness, or weight, of air that the engine can retain within the engine cylinder. The heavier (denser or colder) the air surrounding the engine, the greater the quantity of air (weight) that will be drawn into the engine on a naturally aspirated unit, or the greater the pressure charge on a turbocharged engine. In either case, the engine will produce greater power.

The turbocharger assists the engine especially when it is operating at higher altitudes since the atmospheric pressure decreases as shown in Figure 31–5. As a rule of thumb, engine horsepower will decrease by approximately 1% for an increase in altitude of every 328 feet (100 meters).

The degree of cylinder charge (pressure at the end of the intake stroke versus ambient air pressure) establishes the engine's power output. Combustion efficiency decreases in cold thin air as a result of reduced fuel evaporation and lower compression temperatures, which lead to sluggish combustion or an increase in the ignition delay period in a diesel engine.

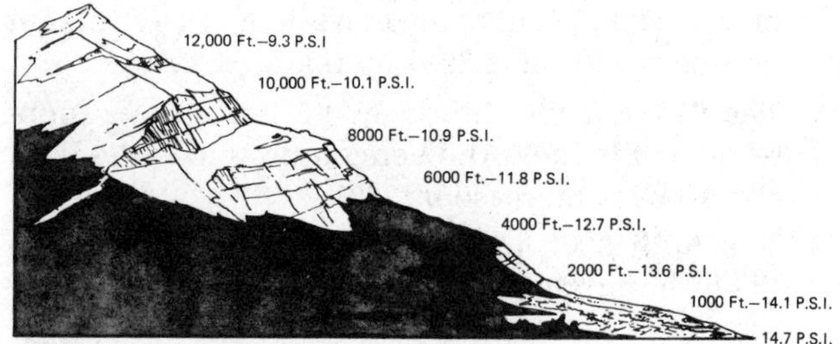

Figure 31–5. Typical atmospheric pressures at various altitudes. (Courtesy of Reston Publishing Co; Inc.)

Combustion chamber temperatures in an engine operating at 80°F above zero versus one at −30°F can result in lowered combustion chamber temperatures of as much as 230–300°F (110–149°C). The difference in the temperature spread is related to engine compression ratio, combustion chamber design, and air turbulence. Should the air be preheated in the intake manifold, then dependent upon engine design, the air can become rarefied and the engine will lose power. Of course, glow plugs are required on pre-combustion chamber diesel engines to facilitate starting.

Air humidity can also affect engine power since it contains less oxygen than dry air and results in noticeable power losses especially when operating in tropical conditions.

The turbocharger under both hot and cold ambients will improve engine performance although an intercooler or aftercooler may be required in hot climates to gain an appreciable advantage. Obviously, on a carbureted engine the fuel flow is directly tied to the air flow; and on current fuel injected gasoline engines, an air flow and air temperature sensor establishes the pulse time of the fuel injectors. The longer the injectors are electrically energized, the greater will be the fuel flow rate into the engine.

The engine's volumetric efficiency can be considered as the weight of dry air contained in the cylinder when the piston is at BDC when it is stopped, versus the weight of air contained within the cylinder with the piston at BDC when the engine is running. The weight of air in the cylinder when the engine is running on a naturally aspirated engine is generally about 80–85% of atmospheric. Since atmospheric pressure is 14.7 psi (101.35 kPa) at sea level, we can say that the air pressure in the cylinder is less than atmospheric (vacuum) when the engine is running. With a volumetric efficiency of 80%, cylinder pressure would be about 11.76 psi (81 kPa) while at 85% VE it would be about 12.5 psi (86.15 kPa).

Refer to the earlier heading *Workable Engine Formulas* which describe how to determine an engine's volumetric efficiency and air flow rate.

Engine Air Flow Requirements

The amount of extra air that an engine requires when it is turbocharged can be established by referring to Figure 31–6 dealing with the CFM air requirements for a naturally aspirated 4-cycle engine versus a turbocharged version. These air flow rates are based upon a VE of 80%.

For example, we can see that if we were to consider a 2.0L (122 cu.in.) engine running at a speed of 3000 rpm, it would require approximately 84 CFM; at 6000 rpm, it would require approximately double this, or 168 CFM.

Engine air flow requirements can be calculated by using the formula

$$\frac{\text{Displacement in cubic feet} \times \text{Revs/Min.}}{2} = \text{Cubic feet/minute}$$

Therefore

$$\frac{D \times RPM}{2} = CFM$$

Example: If we were to consider the 2.0L (122 cu.in), engine discussed earlier when we were considering

4-CYCLE ENGINES

CUBIC INCH DISPLACEMENT	ENGINE RPM 1250	1500	1750	2000	2250	2500	2750	3000
100	28	35	41	46	52	58	64	69
120	35	42	49	56	63	69	75	84
140	41	49	57	65	73	81	89	97
160	46	56	65	74	83	93	102	111
180	52	63	73	83	94	104	114	125
200	58	69	81	93	104	116	127	139
220	64	76	89	102	114	127	140	153
240	70	84	98	111	125	139	153	167
260	75	90	105	120	135	150	165	180
280	82	98	114	130	146	162	176	194
300	87	104	121	139	156	173	191	208
320	92	111	130	148	167	185	204	222
340	98	118	138	158	177	197	217	236
360	104	125	146	167	187	208	229	250
380	110	132	154	176	198	220	242	264
400	115	139	162	185	208	232	255	278
420	121	146	170	194	218	244	268	292
440	127	152	178	202	229	255	280	305
460	133	160	186	213	240	266	293	320
480	138	166	194	221	250	278	306	334
500	144	173	202	231	260	289	318	347
520	150	180	210	240	270	300	330	360
540	156	187	218	250	281	312	344	375
560	162	194	226	259	292	324	356	389
580	168	201	235	269	302	335	369	401
600	174	208	244	277	312	346	382	416
620	179	215	251	287	323	359	395	431
640	185	222	259	296	334	371	408	435
660	191	229	267	305	343	382	420	458
680	197	237	276	315	354	394	434	473
700	201	243	284	325	365	406	445	486
720	209	251	292	334	375	417	459	500
740	214	257	300	343	385	429	472	514
760	219	264	308	352	396	440	484	528
780	226	272	316	361	408	453	497	542
800	232	278	324	371	418	484	510	556
850	246	296	345	384	444	493	542	591
900	260	312	365	417	469	521	574	625
950	275	330	385	440	495	550	605	660
1000	280	348	406	463	521	580	638	695
1050	302	362	425	487	547	608	669	730
1100	319	382	446	510	574	638	702	765
1150	333	400	467	534	600	667	734	800
1200	348	418	487	557	626	696	765	835
1250	362	485	587	579	632	724	796	868
1300	376	453	528	609	678	754	829	904
1350	390	469	548	625	704	782	860	938

Figure 31–6. Air flow requirements for four-cycle engines. (Courtesy of Farr Air Cleaner Company)

how to determine volumetric efficiency/air flow, this engine had a cubic foot displacement of 0.0706 cu. ft.

Solution

$$\frac{0.0706 \times 6000}{2} = 211.8 \text{ Cubic feet/minute}$$

Based upon the above, this engine would have an ideal air/fuel flow rate of 211.8 CFM.

If, for example, we were to consider that the engine had an actual flow rate by measurement of 180 CFM, its volumetric efficiency would therefore be 85% arrived at by

$$\frac{\text{flow rate CFM actual}}{\text{flow rate CFM ideal}} \times 100\% = \text{Volumetric Efficiency}$$

Inserting the figures arrived at above we have

$$\frac{180}{211.8} \times 100\% = 84.98\%$$

This particular engine is therefore capable of a volumetric efficiency of 85%, which is very good.

The amount of extra air that an engine requires when it is turbocharged can be established by referring to Figure 31–6 dealing with the CFM air requirements for a naturally aspirated 4-cycle engine versus a turbocharged version. These air flow rates are based upon a VE of 80%.

For example, we can see that if we were to consider a 3.0L (183 cu.in.) engine running at a speed of 3000 rpm, it would require approximately 125 CFM; at 6000 rpm, it would require approximately double this, or 250 CFM.

Engine air-flow requirements can be calculated by using the formula explained above under the heading Volumetric Efficiency.

If we were to turbocharge the engine, we would have to double the air flow requirements to 500 CFM at 6000 rpm. The general rule of thumb is that you require twice as much air when you add a turbocharger.

NOTE: Volumetric Efficiency on a naturally aspirated 4 cycle gasoline engine will decrease with an increase in engine speed in the following approximate values:

Volumetric Efficiency = 0.80 up to 2499 RPM
= 0.75 between 2500 and 2999 RPM
= 0.70 between 3000 and 4000 RPM
= 0.65 between 4000 and 5000 RPM
= 0.60 between 5000 and 6000 RPM

The main reason for this is due to the very short time that the intake valves are open.

In an engine running at 2500 RPM for example, the intake valves are only open for approximately 0.017 seconds, therefore very little time is available for air to enter the cylinder.

Naturally Aspirated Versus Supercharged Engine

The term *supercharged* is often misunderstood when discussing engines. Theoretically speaking, any engine that receives air at a pressure higher than atmospheric and then compresses it is classified as a supercharged engine. However, although some engines may *receive* their air at pressures higher than atmospheric, if they don't actually start to compress this air before it returns to atmospheric (because of the exhaust valve timing, e.g. Detroit Diesel—some engines), then the engine will not in effect be supercharged.

Valve timing therefore plays an important part in ensuring that any air delivered to the engine cylinder is retained and compressed by the upward moving piston.

Using a gear-driven type of supercharger, such as a Rootes unit that is found on Detroit Diesel Allison 2-cycle engines and often employed on high-performance gasoline drag racing engines, requires that the engine power the blower through a belt arrangement. On the diesel engine, this blower is gear driven. Horsepower is required to drive the blower, and valve timing ensures that the cylinders will always have air at a greater pressure than atmospheric.

The turbocharger, on the other hand, doesn't require a belt or gear drive mechanism since it is driven by the expanding hot exhaust gases; therefore, the turbocharger does not rob the engine of horsepower.

Whether a gear-driven blower or turbocharger is used on the engine, the reason for its use is to increase the volumetric efficiency.

From our earlier discussion we know that any increase in volumetric efficiency will cause an increase in the engine's developed horsepower.

Many after-market turbocharger kits are readily available for non-turbocharged engines that generally include changes to the air and exhaust systems and often cylinder head changes to receive the maximum benefit from the turbocharger.

Compression Ratio

Compression ratio is the ratio of the volume of air above the piston when it is at BDC versus the volume of air above the piston when it is at TDC. Compression ratio can be expressed as:

$$CR = \frac{\text{Total volume in a cylinder/piston at BDC}}{\text{Total volume in a cylinder/piston at TDC}}$$

or it can be calculated by using the formula

$$CR = \frac{\text{Piston displacement + Clearance volume}}{\text{Clearance volume}}$$

For example, if a cylinder contained 100 cu.in. of space above the piston when it was at BDC and only 10 cu.in. of volume above the piston when it is at TDC, then the compression ratio would in fact be 10:1 since the air in the cylinder has been compressed into an area or volume 1/10 what it was at BDC.

Naturally aspirated engines generally have higher compression ratios than a turbocharged engine. The turbocharged engine uses a lower compression ratio because this allows the BMEP (brake mean effective pressure), which is the average pressure on the piston crown during the power stroke, to be kept within the designed limits for that particular engine.

If the compression ratio is not lowered, problems with pre-ignition can occur due to the greater air turbulence/pressure and temperature from the use of the turbocharger. In addition, shorter engine life may result as a consequence of the greater air/fuel ratio leading to higher peak cylinder pressures and temperatures during the power stroke.

Some examples of the changes in the compression ratio between the N.A. versus the T/C engines of some popular vehicles are as follows:

Vehicle Make	*N.A. Engine CR*	*T/C Engine CR*
Audi 5000	8:1	7:1
Buick 3.8L (231 cu.in.)	8.3:1	8:1
Chevrolet 3.8L	8.6:1	8:1
Chrysler Laser/Daytona	9:1	8.2:1
Datsun 280ZX	8.3:1	7.4:1
Ford 2.3L (140 cu.in.)	9:1	8:1
Ford 1.6L (97 cu.in.)	9.5:1	8:1
Pontiac 5.0L (301 cu.in.)	8:1	7.6:1
Porsche 924	8:1	8:1
Mercedes-Benz 300D (Diesel)	21:1	21.5:1

As you can see from the above figures, some manufacturers have decreased their compression ratio when a turbocharger is added, while others have left it the same. Usually the ignition timing is different between the N.A. engine and the T/C unit, plus the T/C unit employs a knock-sensor on some engines to retard ignition timing to prevent piston damage through detonation.

Sometimes different camshafts are used, although many current N.A. and T/C engines use the same valve timing specifications.

Intake piping is often increased in diameter along with a high-flow, low-restriction air cleaner assembly with about 50% more capacity than that used on the naturally aspirated engine.

Exhaust pipe diameters are often enlarged on the T/C engine, plus the exhaust flow requirements/mufflers are usually different than on the N.A. engine to provide lower exhaust back pressure and suitable flow rates to ensure the turbocharger will respond within the best parameters of its designed boost pressure and engine exhaust gas flow rates.

The use of a turbocharger generally helps to silence exhaust noise in the turbine section; therefore, tuned mufflers are used for the appropriate noise reduction and reduced exhaust gas back pressure.

Generally, a different radiator is also required with the turbocharged engine because of the higher heat load on the engine (higher horsepower). This is generally accomplished by using a heavy-duty-type radiator with increased thickness and fin density along with a higher capacity fan. For example, on Ford's 1.6L T/C engine, two 10-inch four-bladed fans are mounted ahead of the radiator to push cooling air through the fins to ensure adequate cooling under the most adverse conditions.

Cylinder head design and gasket material are often changed for the T/C engine, along with intake and exhaust valve metals, piston and ring package, piston-to-bore clearances, and main and rod bearings. The oil pump is generally of a higher flow rate and has a higher relief valve setting. Many turbocharged engines also employ intercoolers (see below) and engine oil coolers to offset the higher heat load.

Generally, the oil cooler uses engine coolant flowing around it to maintain oil temperatures within safe operating levels. Such an arrangement is used on Ford's new 1.6L T/C powerplant in the EXP and Escort/Lynx models.

Turbocharger Arrangements

The location of the turbocharger will vary slightly on the engine between different makes of vehicles; however, this in itself does not alter the operation of the turbocharger. Two basic arrangements of turbocharger/carburetor are found on gasoline engines, namely:

1. A draw-through system whereby the carburetor or throttle body (TB1) fuel injection system would

be placed "upstream" of the turbocharger compressor such as shown in Figure 31–7.

2. A blow-through system whereby the carburetor or fuel injection system would be placed "downstream" of the turbocharger compressor such as shown in Figure 31–8.

NOTE: When a multi-point fuel injection system is used (individual cylinder fuel injectors) it is located after the T/C compressor.

In the upstream arrangement of turbocharger mounting, the carburetor or throttle body injection system would be located between the air cleaner and the turbocharger compressor. On this system, the carburetor is able to function in its usual manner by metering fuel before it enters the turbocharger. Once the fuel enters the compressor, the air/fuel under pressure will be properly atomized and delivered evenly to each cylinder.

A disadvantage of an upstream arrangement, however, occurs in that longer piping between the carburetor and cylinders can cause some throttle lag. Other problem areas are the condensing of fuel droplets within the intake manifold passages and the possible icing of the carburetor throttle plate (butterfly valve), which could necessitate the use of heated manifolds.

An example of a throttle body attachment bolted directly to the turbocharger compressor mounting flange is shown in Figures 31–9 and 31–10. In this

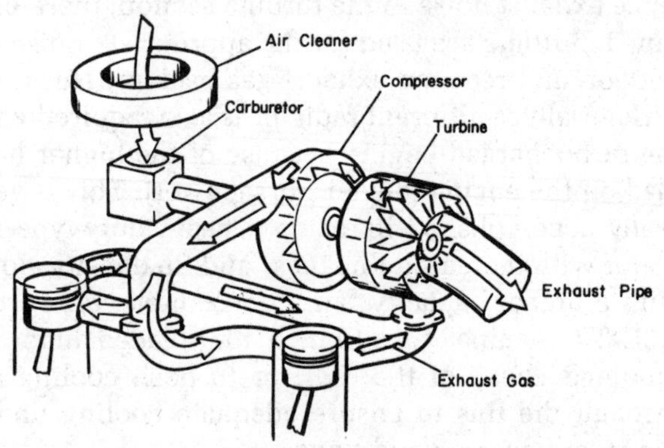

Figure 31–7. Downstream turbocharger arrangement with upstream carburetor or TBI unit [Air is pulled through the Carburetor or TBI unit]. (Courtesy of Reston Publishing Co; Inc.)

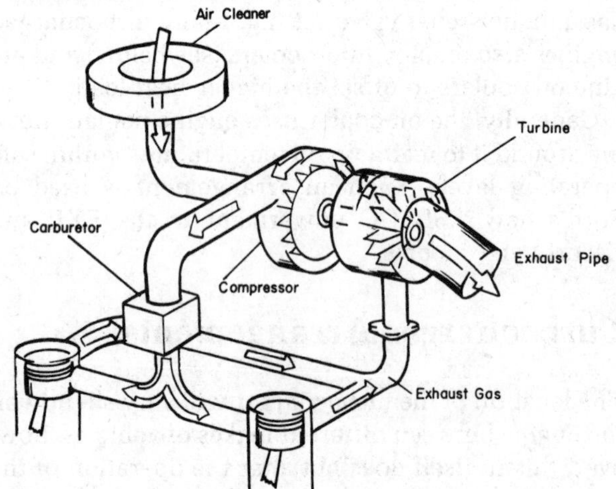

Figure 31–8. Upstream turbocharger arrangement with a downstream carburetor or TBI unit [Air is blown through the carburetor or TBI unit]. (Courtesy of Reston Publishing Co; Inc.)

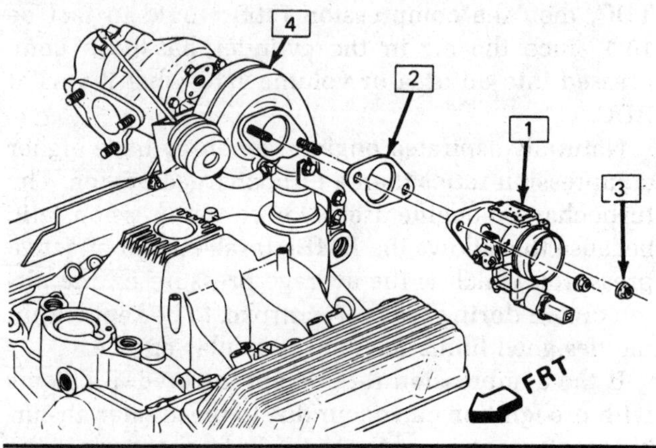

Figure 31–9. Throttle body injection (TBI) system attachment-G series vehicles. (Courtesy of Buick Motor Division)

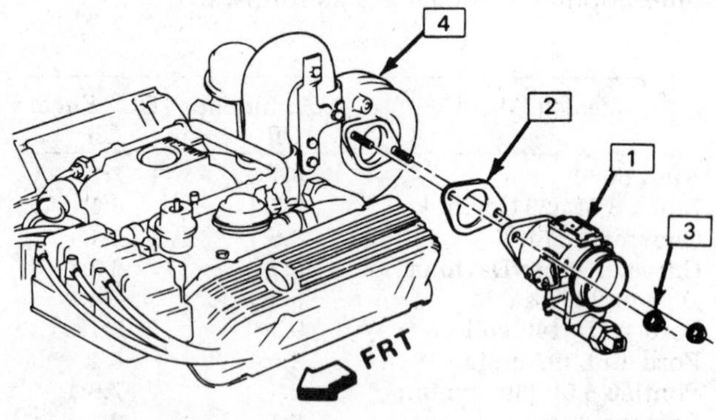

Figure 31–10. Throttle body injection system (TBI) attachment-E series vehicles. (Courtesy of Buick Motor Division)

Buick system used with the 3.8L (232 cu.in.) V6 engine, individual fuel injectors are used at each cylinder. Therefore only fresh air is drawn through the throttle body to the turbocharger. The throttle body uses a TPS (throttle position sensor) to send a voltage signal to the ECM (electronic control module), which establishes the pulse time (on/off) of the fuel injectors to determine fuel flow into the engine cylinders.

On the downstream arrangement of turbocharger to carburetor or throttle body system, the routing of the exhaust pipes is simpler. It is necessary, however, with this system to ensure that the carburetor or throttle body and manifold passages are in fact airtight to preven the loss of air/fuel under pressure to the engine cylinders.

Air/Fuel Mixture Control—Turbocharger

There are basically three types of air/fuel mixture control available with a turbocharged engine:

1. Electronic or mechanical port-type fuel injection such as Robert Bosch L or K respectively or equivalent, typically located after the compressor close to the inlet valve port area.
2. A carburetor on the outlet side of the turbocharger compressor.
3. A carburetor on the inlet side of the turbocharger compressor.

A carburetor on the pressure side of the compressor can generally relate to a lower installation cost although the carburetor must be hermetically sealed against the atmosphere during charging operation.

Specific advantages of the carburetor mounted on the intake side of the compressor or "upstream" are as follows:

1. There is no restriction as to the particular make of carburetor used.
2. The only modification to the carburetor is "rejetting."
3. The naturally aspirated engine fuel supply system can be used.
4. Carburetor tuning is easier for some operating ranges than the tuning of the carburetor when mounted on the outlet side (downstream) of the turbocharger compressor.
5. Excellent fuel distribution to all engine cylinders.
6. The evaporation heat can be used for charge cooling purposes.
7. The dynamic behavior of the T/C piston engine combination is improved.

Disadvantages however of the "upstream" or draw-through system are as follows:

1. A two-stage carburetor must be used because of the significant difference between idle and rated air-mass-flow for optimum power and driveability.
2. A special gasket must be used on the turbocharger compressor in order to avoid oil leakage from the turbocharger bearings during part load operation.
3. The gasoline wetted surfaces of the intake manifold are larger than those in other systems.

Turbocharger Boost

The pressure of intake air delivered to the engine cylinders with a turbocharger is more or less related to both engine speed and engine load characteristics. Although an engine may be accelerated to a high rpm, if there is little load on the vehicle, the engine will not be required to produce its maximum horsepower output.

Since the turbocharger is exhaust gas driven, it will only receive its highest gas pressure/flow rate when the engine is producing its rated horsepower unless the engine/turbo matching has been designed to produce peak torques at low to mid-range speeds. The pressure delivered by the turbocharger is commonly referred to as *boost pressure,* or simply *turbo-boost.*

Each turbocharger is designed to produce a given boost psi for a particular engine and rated exhaust gas flow rate. The ability of the turbocharger to provide a given air flow rate (CFM) and boost pressure is controlled by what is commonly referred to as the A/R ratio. This simply means that the ratio of the turbine inlet (nozzle) area to the radius of the centroid of that area will determine the turbocharger's performance.

Reducing the A/R ratio with a given turbine wheel size and trim will result in an increase in turbine rotating speed for the same exhaust gas flow, resulting in a higher boost pressure unless the maximum boost pressure is regulated by the use of a wastegate such as explained later in this chapter. In this case, *trim* is the variation in the actual machined shape of the outer contour of the turbine wheel, which will produce variations in flow.

Increased turbine rotating speeds will result in higher manifold pressures coupled with increased engine torque at the low and mid-range speeds. If the

A/R ratio is increased, then the opposite situation will occur.

Wastegate Control Mechanism

In order to control engine pressures, stresses, and temperatures, the maximum allowable turbocharger boost pressure must be controlled. Several methods can be used to limit the maximum boost from the turbocharger; these are:

1. A variable area turbine
2. A compressor restrictor
3. A pressure relief valve
4. A turbine outlet restrictor
5. An exhaust gas wastegate assembly to bypass exhaust gases

Many current turbocharged engines use method 5 above as a desirable method of controlling the turbocharger maximum boost pressure. An example of a wastegate actuator is shown in Figure 31–3 earlier and identified as Item No. 7.

The operation of the wastegate is controlled by turbocharger compressor pressure flowing through a hose to the wastegate actuator. The actuator consists of a spring-loaded diaphragm connected through external linkage to a wastegate control valve as shown in Figure 31–11.

The actual location of the wastegate valve is more clearly shown in the exploded view in Figure 31–12.

When the turbocharger boost pressure acts upon the actuator diaphragm, the diaphragm and its linkage are forced to move so that the internal wastegate valve inside the turbine end of the T/C is opened. The degree of opening is related to the boost from the intake manifold.

The wastegate valve therefore controls the exhaust flow through an orifice that connects the turbine inlet nozzle area and the discharge housing to effectively control the amount of exhaust gases that will be bypassed.

Another example of a wastegate control mechanism is shown in Figures 31–13 and 31–14 for the V6 3.8L (232 cu.in.) Buick engine and Ford/BMW 2.4L diesel engine overhead cam 6 respectively. This system operates in the same basic manner as that used by other manufacturers.

By using a wastegate control mechanism, Ford maintains their 2.3L engine T/C boost to a maximum of 7.2 ± 0.5 psi (50 ± 3.5 kPa), while the 1.6L engine boost peaks at about 8 psi (55 kPa) and the 2.4L diesel is limited to 12 psi with an intake manifold mounted safety release valve opening at 13 psi. The Buick V6 maintains peak T/C boost at approximately 8.8 psi or 61 kPa at 2400 engine rpm when the wastegate control valve begins to bypass the appropriate amount of exhaust gases to maintain that peak manifold pressure. In the Buick V6, BMEP levels of 175 psi (1207 kPa) occur from 2400–3200 engine rpm compared to the 116 psi (800 kPa) for the naturally aspirated V6. The BMEP is the average pressure exerted on the piston crown during the engine power stroke.

Datsun/Nissan maintains a maximum turbocharger boost pressure of 8.5 psi (58.6 kPa) on their 300ZX V6 engine while the earlier 280ZX in-line 6 engine was limited to a boost pressure of 7 psi (48.2 kPa). Both engines employ a wastegate control mechanism plus a safety blow-off valve located on the intake manifold.

The Dodge/Plymouth Conquest turbo model limits its boost to 8.7 psi or 60 kPa in its 2.6L (158 cu.in.) 4-cylinder aluminum hemi-head fuel injected engine.

As you can see, most current production turbocharged engines limit boost pressures to between 7 and 9 psi (48–62 kPa) maximum on their engines.

When problems of low or high turbocharger boost pressures occur, always check that the wastegate control mechanism is free and operating correctly. Air leaks from the pressure side of the T/C compressor to the wastegate control diaphragm or a damaged diaphragm can cause high T/C boost pressures while a seized wastegate valve or bent or misadjusted linkage can also cause high boost to occur. Low boost can be caused by an air inlet restriction or a wastegate valve that is stuck/seized in the open position.

Ford 2.3L Turbocharger

The turbocharger used on the Ford 2.3L (140 cu.in.) and 1.6L engines is a blow-through arrangement mounted as shown on the right-hand side of the engine in Figure 31–15. In the blow-through arrangement, the fuel is introduced downstream of the compressor.

The turbocharger used with the 2.3L engine is employed with multi-point Bosch/Ford EFI (electronic fuel injection), which has an individual fuel injector for each engine cylinder. Although the naturally aspirated 2.3L and T/C versions of this engine look the same, parts and equipment are not interchangeable between these two engines.

Turbocharger boost starts to come on at around 1700 rpm, is more noticeable at 2000 rpm, and from 2400 rpm and up is progressively higher.

Lubrication of the turbocharger bearings on this engine is accomplished as shown in Figure 31–16 in the following manner:

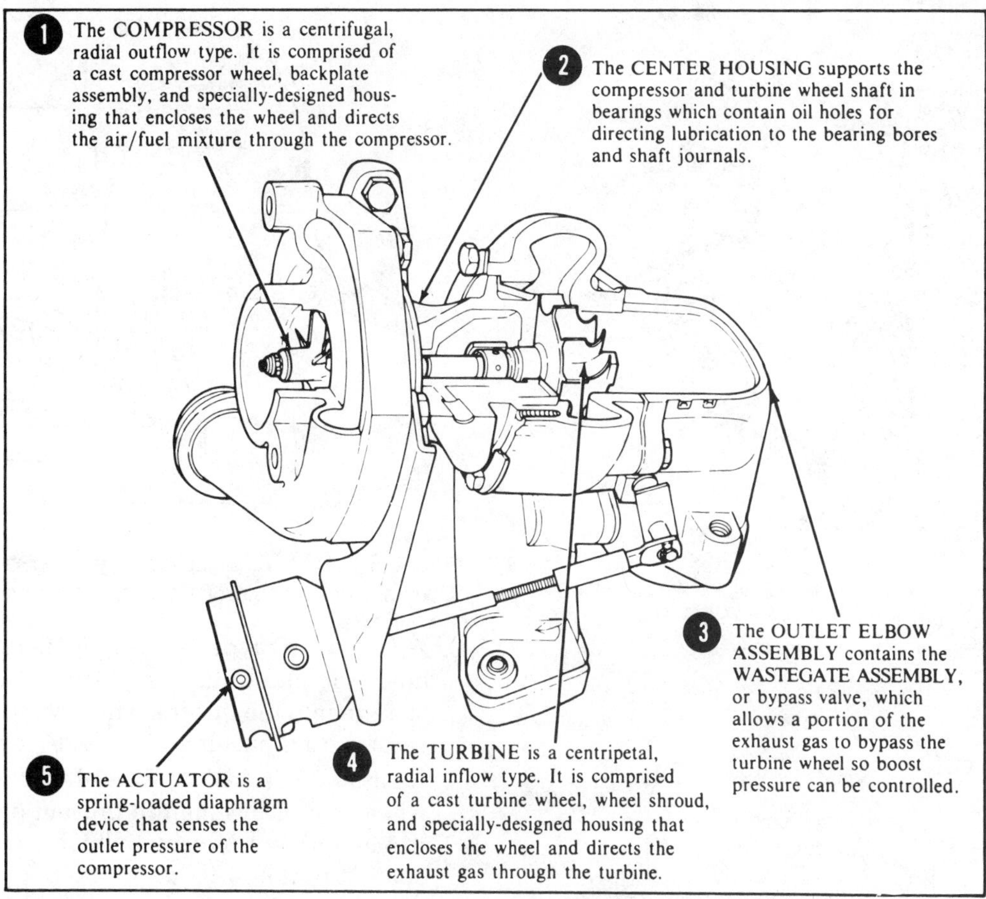

Figure 31-11. 2.3L Ford turbocharger and wastegate actuator mechanism. [Reprinted with permission 1985, Society of Automotive Engineers, Inc.]

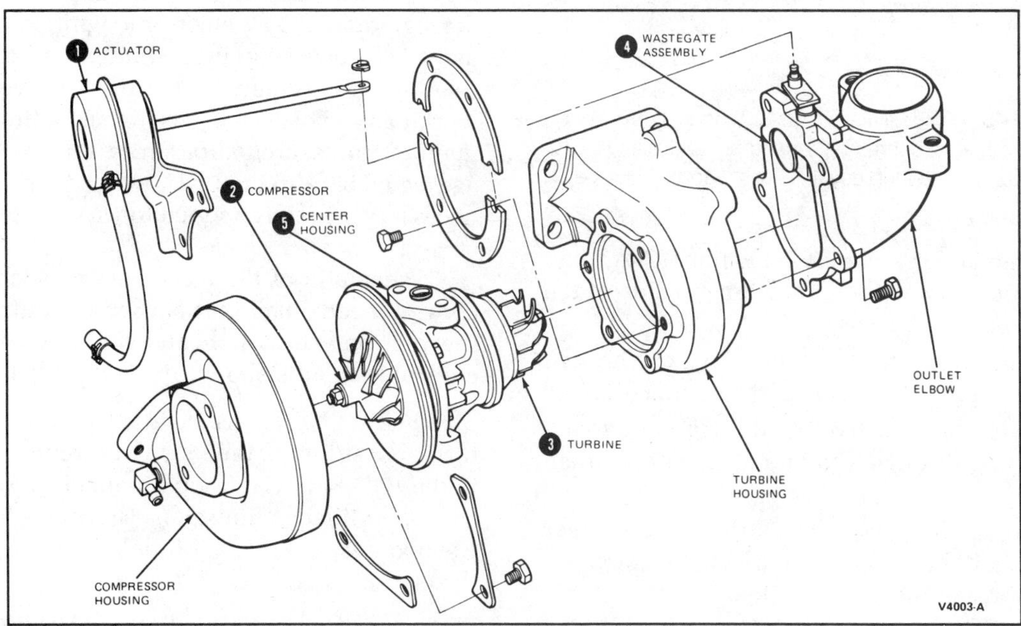

Figure 31-12. 2.3L engine turbocharger-exploded view. [Courtesy of Ford Motor Company, Dearborn]

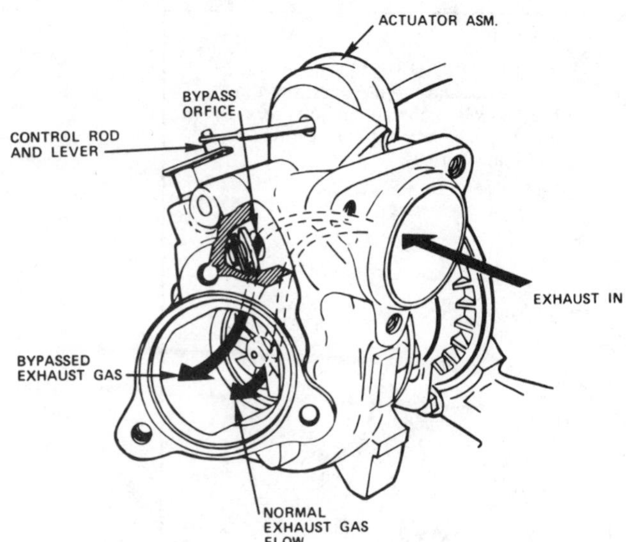

Figure 31–13. Buick V6 turbocharger wastegate control valve. (Reprinted with permission 1985, Society of Automotive Engineers, Inc.)

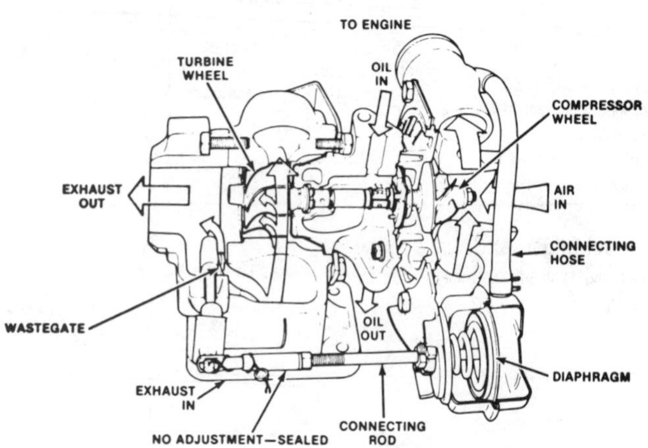

Figure 31–14. Ford/BMW 2.4L overhead CAM 6 cylinder diesel engine turbocharger assembly and waste gate control mechanism. (Courtesy of Ford Motor Company, Dearborn)

1. Engine main oil pressure is fed to the T/C through a "tee" fitting located on the left-hand side of the engine (item 1).
2. Oil pressure from the "tee" fitting passes through a crossover pipe at the rear of the engine (item 2).
3. Oil under pressure is fed into the T/C center support housing through a top mounted fitting (item 3).
4. The main rotating assembly (turbine/compressor support shaft) bearings are lubricated through oil passages inside the housing (item 4).
5. The support bearings (5) are drilled to improve oil circulation.

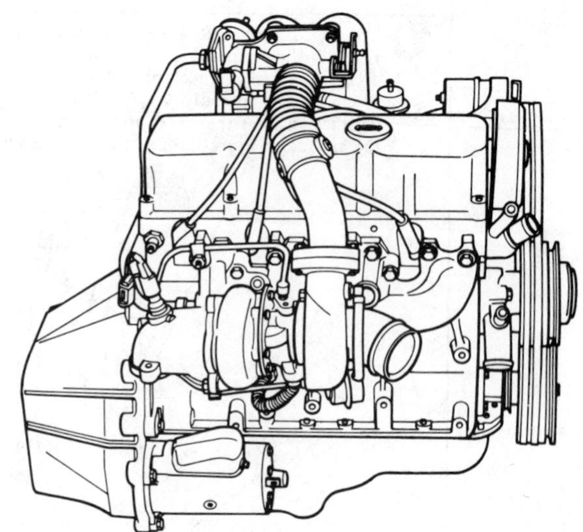

Figure 31–15. 2.3L turbocharged engine-right side view. (Courtesy of Ford Motor Company, Dearborn)

6. A piston ring type seal ring (6) is used at the turbine (hot) end of the support shaft to prevent oil leakage into the turbine wheel housing, otherwise carbon from burning oil would result.
7. A carbon face seal (7) is used on the compressor wheel end of the shaft to prevent oil leakage into the compressor housing, which would cause detonation in the cylinder.
8. Oil is drained from the bottom (8) of the center support.
9. Oil drains back to the oil pan through line No. 9.

An electronic boost control that is activated through the EEC-1V electronic ignition system modulates T/C boost, ignition timing, and air charge in relation to barometric pressure to give a constant working performance at higher altitudes. Intake manifold pressure/temperature sensors feed a voltage signal back to the ECU, which then sends out the necessary corrective signal to vary the timing, boost, etc.

A schematic of the electronic turbocharger boost control mechanism that is used in conjunction with the fuel octane switch and the EEC-1V electronic control system is shown in Figure 31–17.

SPECIAL NOTE. A detailed pinpoint test of the turbo boost/fuel octane switch circuit can be found in the 1984 Ford Car Shop Manual P/N FPS 365–126–84HC, entitled "Emission Diagnosis and Engine Electronics, All Car Models" on pages 25–188 through 25–194.

Used in conjunction with the electronic boost control is a premium/regular fuel selector switch on the

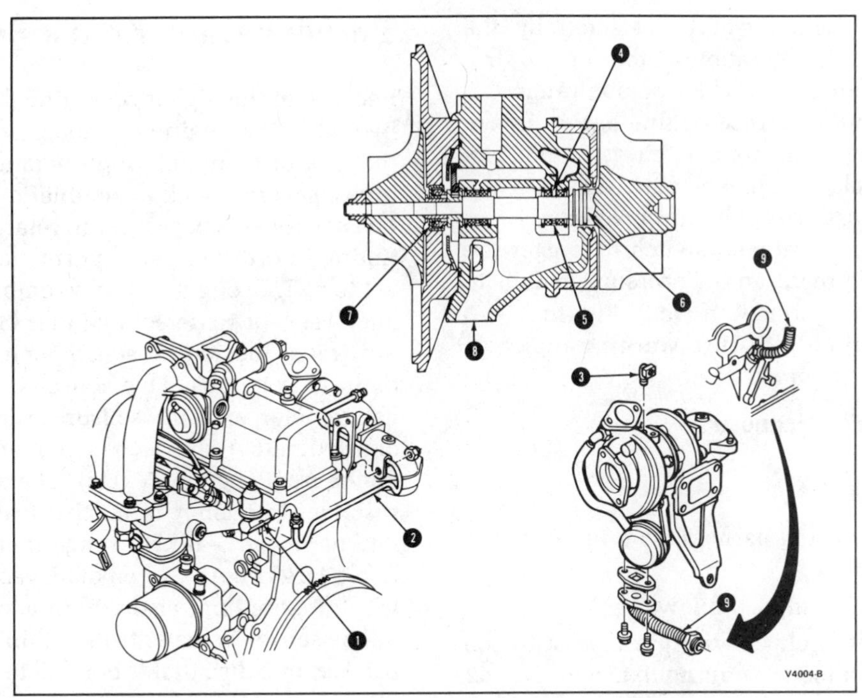

Figure 31–16. 2.3L engine turbocharger oil supply and return circuit. (Courtesy of Ford Motor Company, Dearborn)

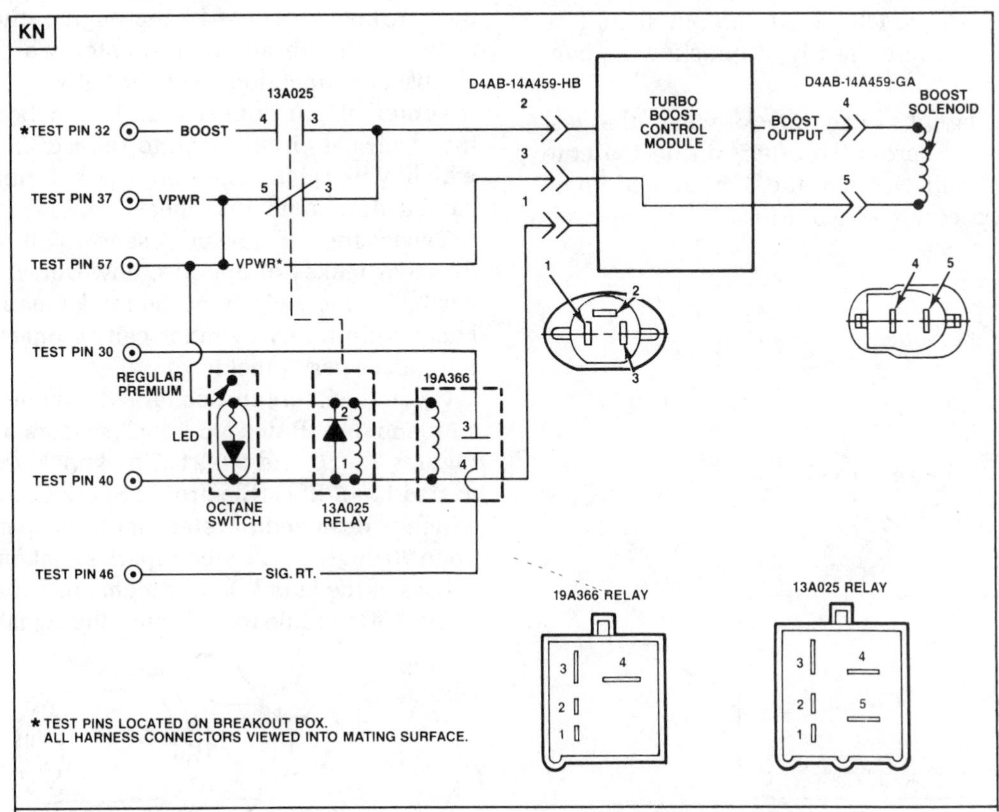

Figure 31–17. 2.3L turbocharged engine, electronic boost/fuel octane circuit. (Courtesy of Ford Motor Company, Dearborn)

instrument panel, which can be selected by the driver. This allows recalibration of the fuel control system for operation over a wider octane range.

The Ford 2.3L turbocharged engine has an intake manifold "warning indicator switch assembly" located on the vehicle instrument panel to warn the driver of "over boost" from the turbocharger.

The unit employs a vacuum switch to trigger the turbo boost lamp (green) and a pressure switch to trigger an over-boost condition (red). Figure 31–18 illustrates the manifold pressure warning indicator switch assembly arrangement.

Turbo Boost Switch Diagnosis.

Procedure

1. Disconnect the wiring harness from the warning switch.
2. Use a 12-volt test lamp as follows.
3. Boost lamp switch check—the pin closest to the notch on the connector running to circuit No. 42 should not be connected to ground (light shouldn't come on) when a vacuum signal of 11.8 kPa (3.5″ Hg or 1.71 psi) or more is applied to the switches. Below 8.4 kPa (2.5″ Hg or 1.22 psi) and up to a pressure of 6.3 psi (43.4 kPa), the pin should be connected to ground (test light on, showing continuity exists).
4. Over-boost lamp switch check—when the test lamp is placed across the other pin on the connector to circuit No. 160, the light should come on when a pressure of 6.8 psi (47 kPa) or greater is applied.

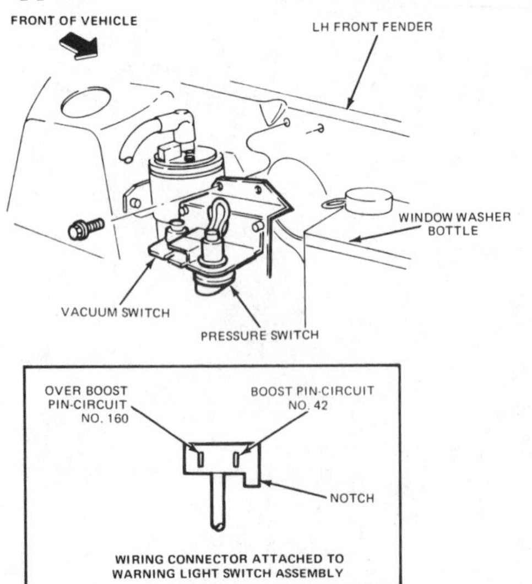

Figure 31–18. 2.3L turbocharged engine manifold pressure warning indicator switch assembly. [Courtesy of Ford Motor Company, Dearborn]

Turbocharger Knock Sensor

Because of the higher operating (boost) pressures delivered to the engine cylinders on a T/C engine and the lack of a suitably high octane fuel supply at all times, severe knock or detonation can result. When this occurs, it is necessary to retard the ignition spark timing in order to avoid permanent engine damage.

Most T/C engines today employ some form of knock sensor to detect cylinder knock. Figure 31–19 illustrates the knock sensor employed by Ford on their T/C engines. This sensor is a piezoelectric accelerometer with the sensor designed to resonate at approximately the same frequency as the engine knock frequency (5–6 kHz). The sensor employs the resonant frequency to mechanically amplify the variations in the 5–6 kHz range thereby allowing relatively large signals to be achieved without electrical amplification and also within a small package size.

The sensor is made from a thin circular piezoelectric ceramic disc that is bonded to a metal diaphragm with electrical connections made through a two-pin integral connector.

The sensor is designed to vibrate when a given level of resonance is attained; therefore, when a combustion knock occurs in the engine, the metal diaphragm will vibrate (or resonate), which causes an electrical connection to occur between the disc and the integral pin connectors. This action completes the electrical circuit back to the EEC/MCU system, which will retard ignition spark timing until the combustion knock disappears.

The location of the knock sensor will vary between different makes of engines; however, it is generally located in the vicinity of the intake manifold so that knock within any cylinder can resonate back up to the sensor very quickly.

General Motors turbocharged engines such as the 1.8L and 3.8L Buick V6 knock sensors are shown in Figures 31–20 and 31–21. The knock sensor is connected to an ESC (Electronic Spark Control) system, which is designed to retard ignition spark timing by up to 20 degrees to reduce spark knock in the engine.

Loss of the knock sensor signal or a loss of ground at the ESC module would cause the signal at the ECM

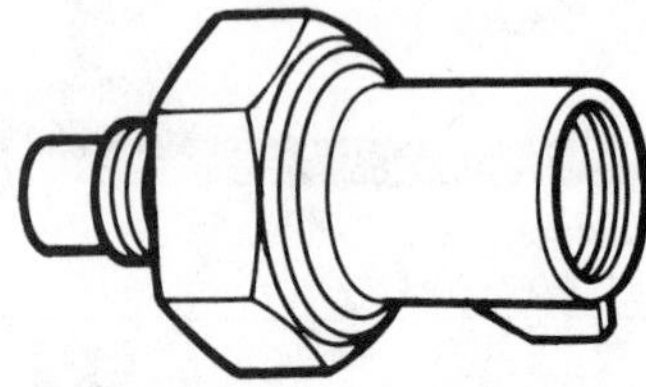

Figure 31–19. 2.3L turbocharged engine knock sensor. [Courtesy of Ford Motor Company, Dearborn]

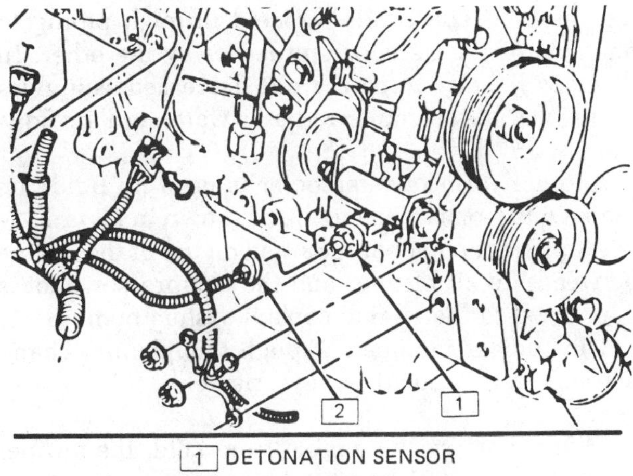

Figure 31-20. 1.8L turbocharged engine knock sensor location. [Courtesy of Buick Motor Division]

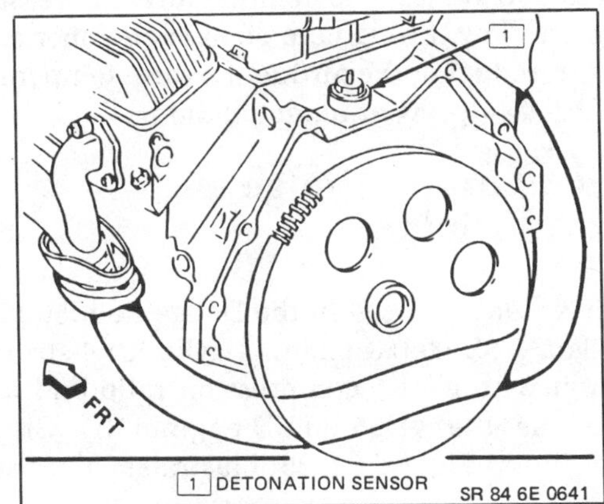

[A]

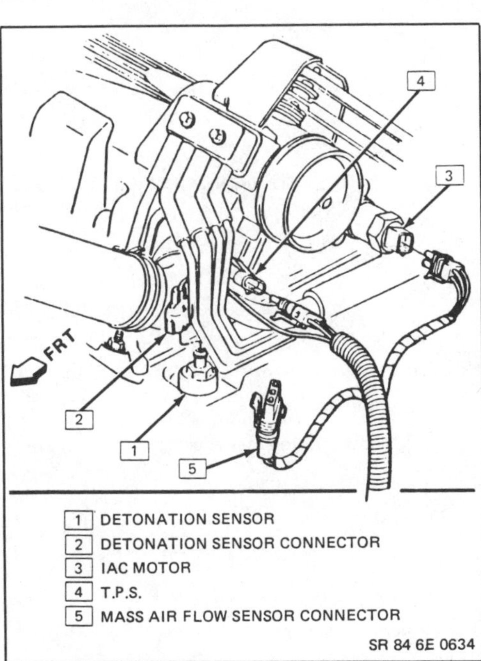

[B]

(electronic control module) to remain high and therefore no ignition retard would occur. Under such a condition, spark knock would become severe under heavy throttle/load conditions and serious engine damage would occur.

On the other hand, the loss of a voltage signal from the ESC to the ECM would cause the ECM to constantly retard the ignition spark timing, resulting in very poor performance. On General Motors vehicles, a Code 43 would be set into the ECM memory bank along with the "check engine light" flashing on the instrument panel.

Turbocharger Intercooler

Several high performance factory equipped turbocharged engines are now fitted with an "air intercooler" to reduce the temperature of the charge air being delivered to the engine from the turbocharger compressor wheel.

Toyota now uses these on their 6-cylinder M 2.0L engine in Japan on their Supra and Soarer model cars, so expect to see this eventually on the North American models. In addition, Mitsubishi's Starion and Lancer turbo models have an intercooler installed. Nissan/Datsun's 300ZX may also have an intercooler soon.

Isuzu has also decided to use an intercooler on their J-car turbo diesel, and since General Motors has an interest in Isuzu and uses their diesel engines, we may soon see an intercooled turbo diesel on future General Motors cars and pickup trucks.

An example of the intercooler used on the Ford 2.3L 4-cylinder EFI engine is shown in Figure 31-22.

The intercooler is an "air-to-air" design and is lo-

Figure 31-22. Ford 2.3L turbocharged engine air-to-air intercooler. [Reprinted with permission 1985, Society of Automotive Engineers, Inc.]

Figure 31-21. [a] 3.8L V6 turbocharged engine [E and G series vehicles] knock sensor location; [b] 3.8L V6 turbocharged engine [a series vehicles] knock sensor location.

cated between the T/C compressor outlet and the throttle body air inlet of the fuel injection system. The function of any intercooler or aftercooler is to reduce the air inlet temperature after it has been compressed by the T/C thereby removing heat and increasing the density of the air for a higher charge or boost pressure.

The intercooler reduces the air temperature from 300°F (149°C) to about 175°F (79.5°C) and thereby allows a horsepower increase of about 20% to 175 bhp.

Another concept of turbocharger intercooler is shown in Figure 31–23; it is used on the Volvo 760 model vehicle. This intercooler is built into the normal radiator in a wafer fashion. Compressed air leaving the turbocharger is directed through piping to one side of the radiator and leaves by the other. In this way the air temperature is decreased as it gives up its heat to the cooler radiator water and air flowing around it.

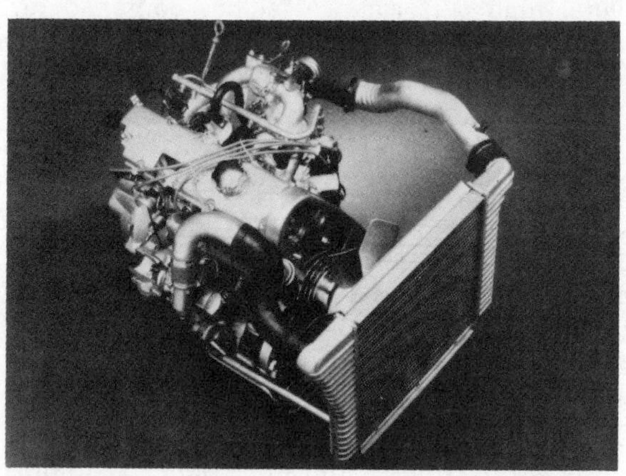

Figure 31–23. 2.3L turbocharged engine intercooler location/arrangement. (Courtesy of Volvo of America Corporation)

Another form of intercooler is used by Buick on their V6 turbocharged engine as shown in Figure 31–24. A plenum chamber mounted on top of the engine between the air cleaner and the compressor inlet is connected to the engine coolant system by hoses.

The coolant plumbed through this plenum chamber acts in two positive ways, namely:

1. When starting the engine from cold, the air/fuel mixture on TBI (throttle body injection) systems will pick up heat from the warm coolant passing through the plenum chamber, which will assist vaporization of the fuel.
2. As the engine warms up and ambient temperatures and underhood temperatures increase, the coolant flow through the plenum chamber can assist in reducing the air/fuel mixture temperatures to assist in preventing detonation.

Turbocharged Dodge Ram 50 Pickup

The diesel engine used in the Dodge Ram 50 pickup truck is a 2.3L (140.3 cu.in.) 4-cylinder 4-stroke design unit with a 21:1 compression ratio and is designated as the 4D55 (TBD-engine) by Chrysler Corporation. Figure 31–25 illustrates a cross-sec-

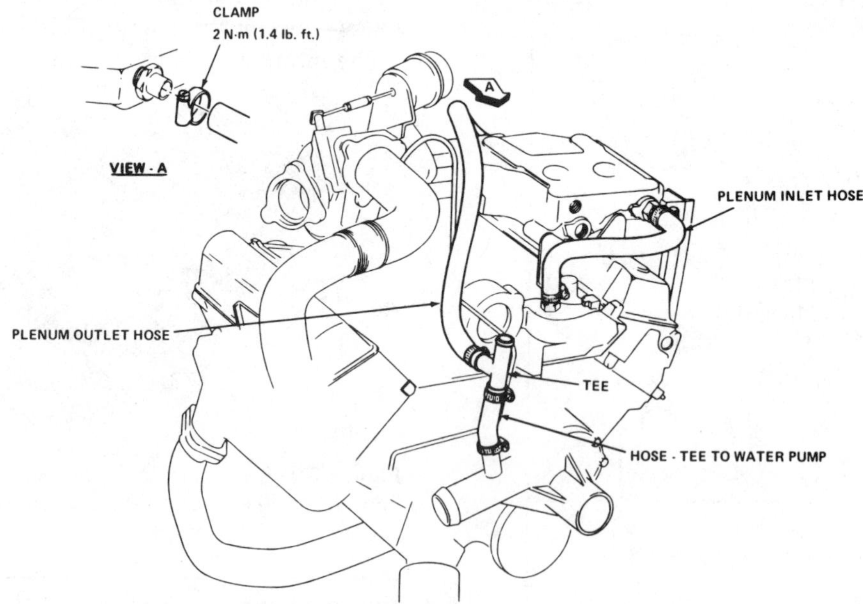

Figure 31–24. 3.8L V6 engine turbocharger plenum chamber to control air/fuel temperatures. (Courtesy of Buick Motor Division)

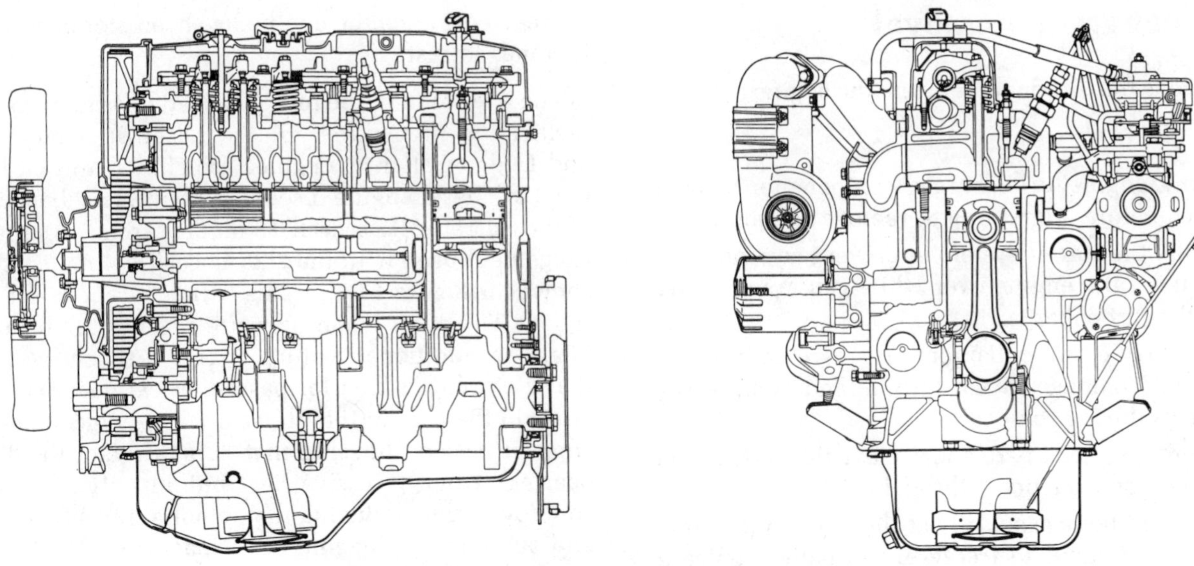

Figure 31–25. Dodge Ram 50 pickup truck diesel engine and turbocharger location. [Courtesy of Chrysler Corporation]

tional end view of the engine with the turbocharger location shown on the left-hand side of the engine block.

The turbocharger used with this engine employs the conventional wastegate design to limit the maximum "boost" pressure of the turbocharger unit by bypassing exhaust gases around the turbine end of the T/C to the exhaust system.

Lubrication of the turbocharger is via a line from the engine main oil pressure feed on the side of the block to the top center of the turbocharger. Oil drain is via a larger pipe from the bottom center of the T/C center support housing.

A schematic of the turbocharger used with the Dodge Ram 50 pickup is illustrated in Figure 31–26.

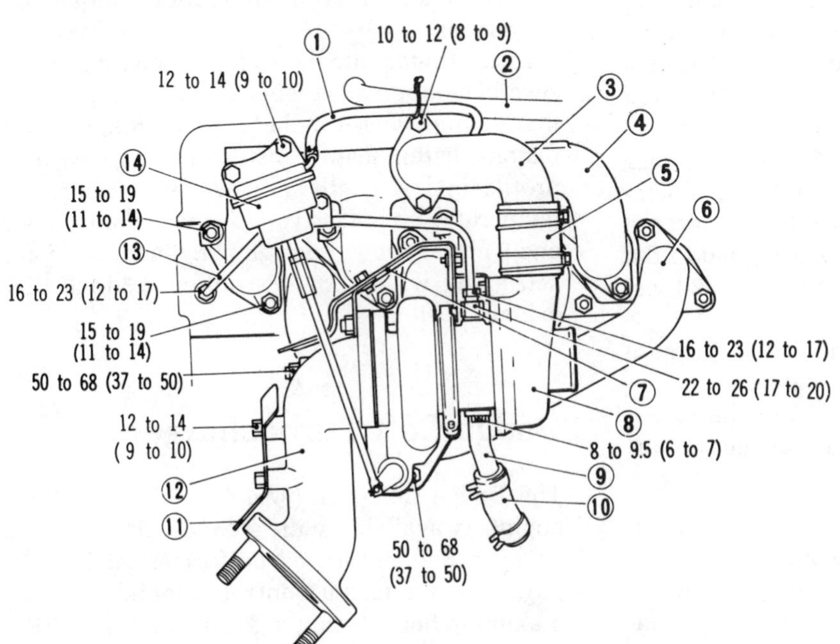

(1) Boost hose
(2) Breather hose
(3) Inlet fitting
(4) Inlet manifold
(5) Air hose
(6) Exhaust manifold
(7) Heat protector
(8) Turbocharger
(9) Oil return pipe
(10) Oil hose
(11) Heat protector
(12) Exhaust fitting
(13) Oil pipe
(14) Waste gate actuator

Figure 31–26. Dodge Ram 50 pickup truck diesel engine turbocharger inlet and exhaust manifold components. [Courtesy of Chrysler Corporation]

Turbocharger Removal

To remove the turbocharger from the engine proceed as follows:

1. Allow engine/turbocharger components to cool down before attempting to remove assembly.
2. Isolate the starter system to prevent anyone from cranking the engine over while you are working on the turbocharger.
3. Handle the turbocharger with care to prevent possible damage to the turbine and compressor wheels during removal.
4. Refer to Figure 31–26 and remove the parts in the numerical sequence 1 through 14.
5. Nuts and bolts especially at the hot end (turbine) of the T/C tend to freeze and usually require a permeable lubricant applied to them to loosen them off.
6. When the T/C oil lines are removed, plug the T/C connections with plastic shipping caps to prevent the entrance of foreign matter, which could damage bearings and the shaft.

Turbocharger Installation

Once the turbocharger assembly is installed back on to the engine, always pour clean oil into the top center oil hole prior to installing the lubrication line. Manually rotate the turbocharger by placing your hand into the compressor end and turning it slowly as you pour between 1 pint to 1 quart of oil through the turbo unit.

To ensure adequate oil supply to the turbocharger bearings, especially on a new or rebuilt T/C assembly, leave the oil supply line loose while you start the engine to ensure that pressure oil is being fed to the turbo assembly. If oil has not appeared within 20 seconds, shut the engine down.

If oil appears, stop the engine and tighten the supply line. When the engine is restarted, do not allow it to be accelerated to a high rpm until the turbocharger warms up and oil has had a chance to circulate through the turbo.

CAUTION: After an engine has been running under load, the turbocharger components are very hot. Before stopping the engine, the engine should be idled for several minutes to let the turbocharger components cool down. If the T/C is stopped abruptly, bearing damage will generally result because of overheated bearings. The turbocharger will continue to freewheel once the engine has been stopped until it comes to rest.

The operation of the turbocharger, its removal and installation procedures for the Laser/Daytona 2.2L and Dodge Colt 1.6L engines are similar to that for the 2.3L diesel engine discussed above. A schematic of the Garrett AiResearch T3 model turbocharger arrangement used with the 2.2L Laser/Daytona cars is shown in Figure 31–27. This turbocharger is set to relieve T/C boost when pressure reaches 7.2 psi (50 kPa). An additional feature of this turbocharger is that it employs a first for passenger car gasoline engines in the U.S. in that it has a water-cooled center support bearing housing that helps protect the T/C bearings from excessive heat buildup after engine shutdown, thus reducing the potential for oil coking and subsequent clogging of oil passages. The bearings are still oil lubricated, but the design of the center housing is such that engine coolant is routed around the bearing casting area of the T/C for increased cooling. This arrangement has been in use on heavy-duty diesel engines for a number of years.

The turbocharged 2.2 Laser/Daytona engines are equipped with a "knock sensor" mounted on the intake manifold between the No. 2 and No. 3 cylinders. The sensor detects any combustion detonation and sends a signal to the on-board vehicle computer which retards spark timing until the knock disappears. This is necessary to prevent the internal components, such as the piston, from being damaged due to low-octane fuel or operation under extreme conditions.

Incorporated into the 2.2L turbocharged engine are several components that have been designed only for that engine. These include high strength deep dish pistons, better sealing piston rings, a new camshaft profile, high strength intake valves and springs, high temperature exhaust valves and exhaust manifold, and a more efficient design of cylinder head and gasket, along with a low-restriction 2.5-inch diameter exhaust system.

Dodge Colt Turbocharger

The Dodge Colt 1.6L (97.4 cu.in.) in-line 4-cylinder engine is available with a Model TC04–09B–5 turbocharger for improved performance. The T/C employs the wastegate control principle to limit the maximum boost from the T/C to 8.5 psi (19 kPa). Figure 31–28 shows the basic T/C arrangement on the engine.

Removal of the turbocharger and installation fol-

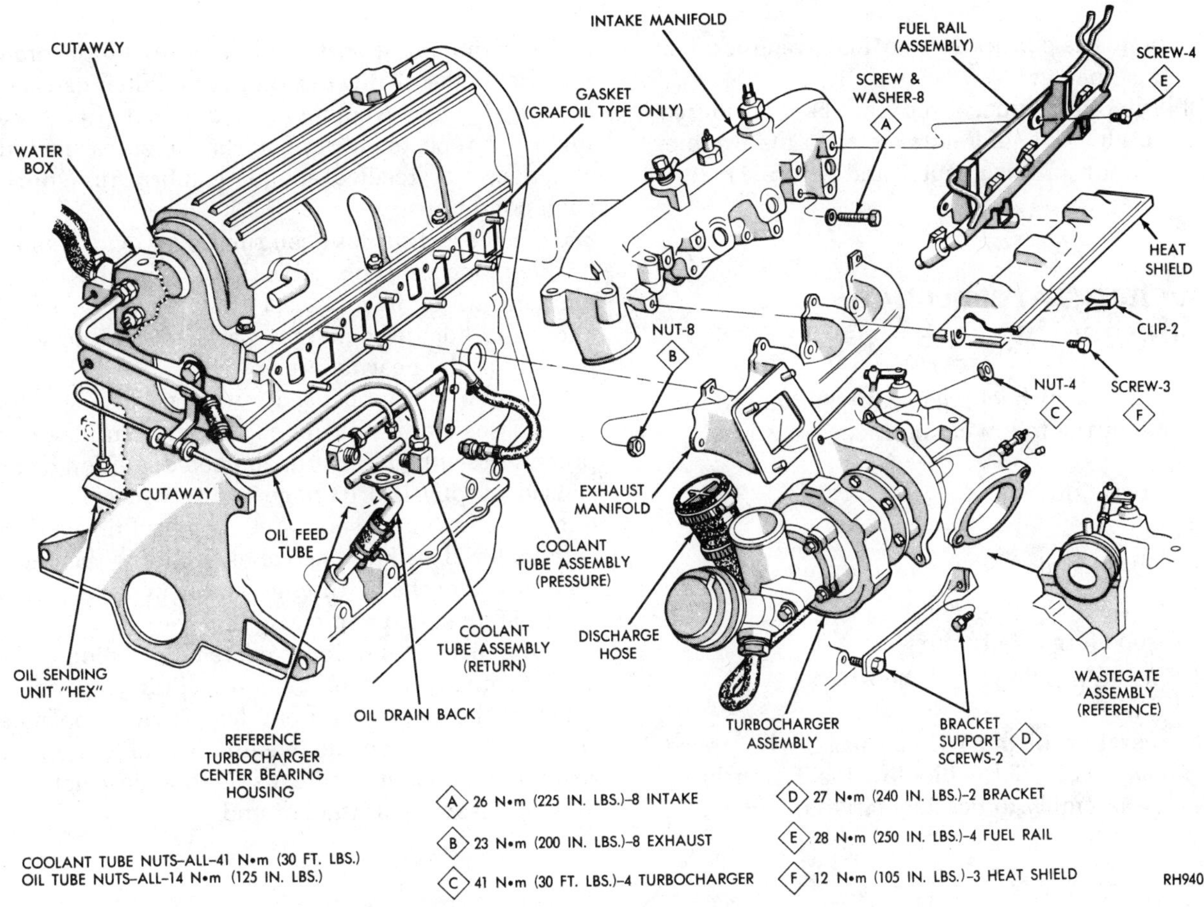

Figure 31-27. 2.2L Laser/Daytona turbocharged engine components. [Courtesy of Chrysler Corporation]

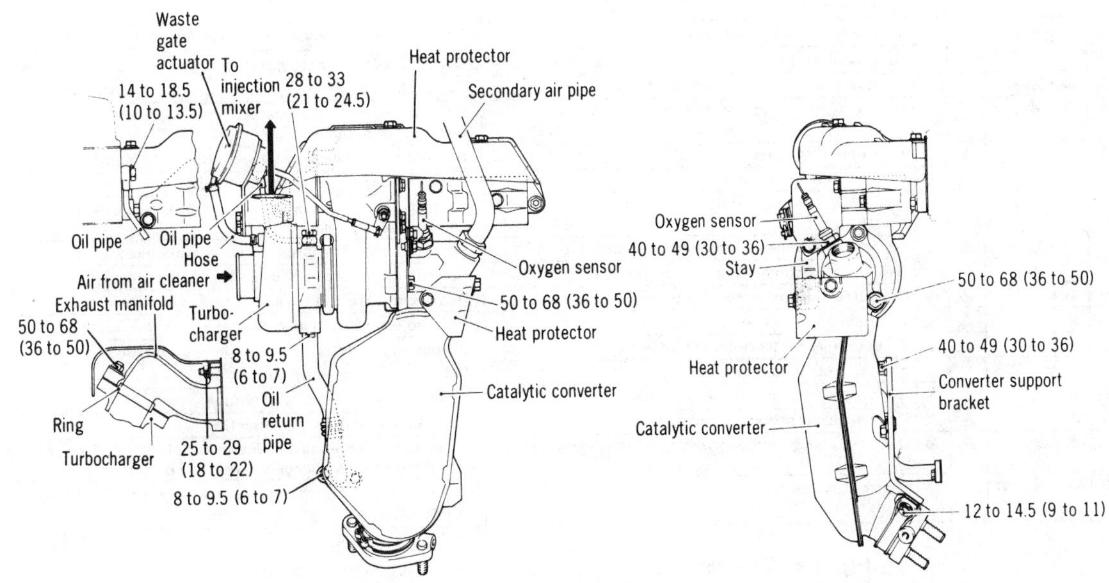

Figure 31-28. Dodge Colt 1.6L engine turbocharger system. [Courtesy of Chrysler Corporation]

lows the same basic procedure as that given for other T/C's in this chapter.

Troubleshooting procedures can be considered common to all T/C's; therefore, refer to the troubleshooting section in this chapter and Figure 31–29.

Turbocharger Inspection/ Troubleshooting

Problems with a turbocharger can generally be related to one of the following problem areas:

1. Noisy operation
2. Vibration
3. Damaged oil seals and engine burning oil
4. High pitched whistling sound
5. Low turbocharger output
6. High T/C output (too much boost)

Noisy operation of the T/C can usually be traced to worn bearings, which will allow the T/C turbine/compressor assembly to possibly strike the housing. Dirty engine oil or sustained operation with foreign material in the oil due to a plugged oil filter can cause bearing damage. Another condition that affects bearing life is repeated abrupt starts and stops since the T/C does not receive adequate lubrication during such conditions.

Once the engine has been run under load, it should be idled for a long enough period to allow the coolant temperature to drop at least 10 degrees F. This will allow the hot turbocharger parts to cool down especially in the bearing areas.

Vibration can be caused by worn bearings or in some cases by an unbalanced rotating turbine/compressor assembly. This unbalanced condition can be caused by physical damage to the vanes/blades on either the turbine or compressor wheel through foreign material striking the vanes. Running the engine without either the air filter assembly in place or without a safety shield can cause this.

At the turbine end, incomplete combustion of fuel can eventually lead to carbon building up on the vanes of the unit, which can lead to an unbalanced condition. Vibration can eventually lead to the vanes striking the housing and causing total destruction of the T/C assembly at the hot end.

TURBOCHARGER TROUBLESHOOTING

PROBLEM: Engine lacks power	Black exhaust smoke	Excessive engine oil consumption	Blue exhaust smoke	Turbocharger noisey	Cyclic sound from turbocharger	Oil leak from compressor seal	Oil leak from turbine seal	CAUSE	REMEDY
								Clogged air filter element	Replace element
	▲	▲	▲	▲	▲	▲		Obstructed air intake duct to turbo compressor	Remove obstruction or replace damaged parts as required
▲	▲			▲				Obstructed air outlet duct from compressor to intake manifold	Remove obstruction or replace damaged parts as required
▲	▲			▲				Obstructed intake manifold	Refer to engine mechanical section & remove obstruction
				▲				Air leak in duct from air cleaner to compressor	Correct leak by replacing seals or tightening fasteners as required
▲	▲	▲	▲	▲				Air leak in duct from compressor to intake manifold	Correct leak by replacing seals or tightening fasteners as required
▲	▲	▲	▲	▲				Air leak at intake manifold to engine joint	Refer to engine mechanical section & replace gaskets or tighten fasteners as required
▲	▲	▲	▲	▲		▲		Obstruction in exhaust manifold	Refer to engine mechanical section & remove obstruction
▲	▲					▲		Obstruction in exhaust system	Remove obstruction or replace faulty components as required
▲	▲					▲		Gas leak in exhaust manifold to engine joint	Refer to engine mechanical section & replace gaskets or tighten fasteners as required
▲	▲			▲		▲		Gas leak in turbine inlet to exhaust manifold joint	Replace gasket or tighten fasteners as required
				▲				Gas leak in ducting after the turbine outlet	Refer to engine mechanical section & repair leak
		▲	▲			▲	▲	Obstructed turbocharger oil drain line	Remove obstruction or replace line as required
		▲	▲			▲	▲	Obstructed engine crankcase ventilation	Refer to engine mechanical section, clear obstruction
		▲	▲			▲	▲	Turbocharger center housing sludged or coked	Change engine oil & oil filter, overhaul or replace turbo as required
▲	▲							Engine camshaft timing incorrect	Refer to engine mechanical section
▲	▲	▲	▲			▲	▲	Worn engine piston rings or liners (blowby)	Refer to engine mechanical section
▲	▲	▲	▲			▲	▲	Internal engine problem (valves, pistons)	Refer to engine mechanical section.
▲	▲	▲	▲	▲	▲	▲	▲	Dirt caked on compressor wheel and/or diffuser vanes	Clean using a Non-Caustic cleaner & Soft Brush. Find & correct source of unfiltered air & change engine oil & oil filter
▲	▲	▲	▲	▲	▲	▲	▲	Damaged turbocharger	Analyze failed turbocharger, find & correct cause of failure, overhaul or replace turbocharger as required

4B6J14

Figure 31–29. Buick turbocharger troubleshooting chart. (Courtesy of Buick Motor Division)

Leaking oil seals will allow oil to be consumed and burnt at the hot end of the T/C causing carbon buildup on the turbine vanes and an unbalanced condition. Leaking seals at the compressor end can lead to an external source of fuel that will cause severe detonation within the cylinder causing piston damage, burning oil at the exhaust pipe, and a tendency for the engine to overspeed.

Worn bearings, caused by dirty oil or overheated oil which will coke, can starve the bearings of lubrication and cause a serious condition such as scoring of the turbine/compressor support shaft. When this occurs, the shaft will wobble and again either the turbine or compressor vanes can strike the housing.

A high pitched whistling sound from the turbocharger can be caused by a leak between the compressor (cold end) and engine intake manifold or throttle body unit since pressurized air will escape at these points. A leak at the turbine (hot end) can also cause a noise, but it is generally not as high pitched as that at the compressor end.

Low T/C output or boost pressure when the condition of the T/C appears satisfactory from a mechanical point of view can be traced to:

1. Exhaust gas leaks
2. High exhaust gas back pressure from a plugged muffler or pipe, or an exhaust pipe that has been crushed
3. Air leaks on the discharge side of the compressor
4. A plugged air filter element
5. A stuck open wastegate valve

If the turbine or compressor wheel does not rotate smoothly when turned by hand, replace or overhaul the T/C assembly after checking both the axial and radial clearances as explained below.

High T/C boost pressures can usually be traced to a wastegate valve that has stuck in the closed or partially closed position.

Exhaust Smoke Analysis

Analysis of the exhaust smoke color on both a gasoline and diesel engine can allow you to quickly pinpoint the cause of the problem. Four basic colors of exhaust smoke are generally visible under a variety of operating conditions. These colors are black, gray, white, or blue. Exhaust smoke opacity can be determined by the use of a smokemeter; refer to Chapter 30, dealing with engine troubleshooting for information on smokemeters.

Exhaust smoke color should always be checked when the engine is at its normal operating temperature since cold starts can cause smoke due to a number of reasons, the least of which is usually cold ambient air that causes a lack of gasoline vaporization and in a diesel can increase the ignition lag. In both cases, this condition will disappear as the engine warms up. Detonation in a gasoline engine can be caused by the wrong grade of gasoline, EGR system defect, or too much boost pressure.

Black/Gray Smoke. This is usually caused by incompletely burned fuel, excessive fuel or irregular fuel distribution as well as the possibility of the wrong grade of fuel (diesel); high air inlet restriction caused by plugged filters, crushed piping, etc., can also cause black/gray smoke due to air starvation.

Black smoke is usually accompanied by a complaint of the engine lacking power and can be caused by the conditions shown in the table.

Possible Problem	*Possible Cause*
Air Inlet Restriction	Plugged air filter or dirty Air inlet duct obstruction Air inlet tube air leaks
Turbocharger	Exhaust leaks Wastegate actuator problem T/C problem internally
Exhaust	High exhaust back pressure—check for a collapsed muffler or piping, plugged or restricted piping.

White Smoke. This is usually the result of faulty injectors, low compression, or the use of low Cetane fuel in a diesel engine. In a gasoline engine, white smoke is generally related to leakage of oil from the T/C into the exhaust pipe or intake pipe. Therefore check for a possible plugged or crushed oil return pipe or leaking T/C oil seals.

Blue Smoke. This is usually caused by lube oil being burned in the cylinder during the exhaust stroke and can be related to worn piston rings, inlet valve guides and in some cases to T/C oil seals.

Figure 31–29, turbocharger troubleshooting chart, will assist you when attempting to pinpoint a turbocharger problem.

Axial and Radial T/C Bearing Check

Before removing a suspected faulty turbocharger from an engine, both the axial and radial bearing clearances should be checked with the use of a dial indicator gauge as shown in Figures 31–30 and 31–31. It may be necessary to remove the T/C assembly from the engine to gain suitable access to the shaft for the radial check.

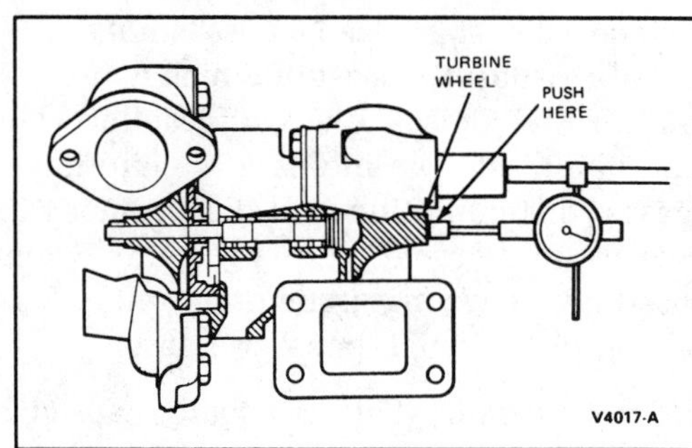

Figure 31–30. Turbocharger bearing axial clearance check. (Courtesy of Ford Motor Company, Dearborn)

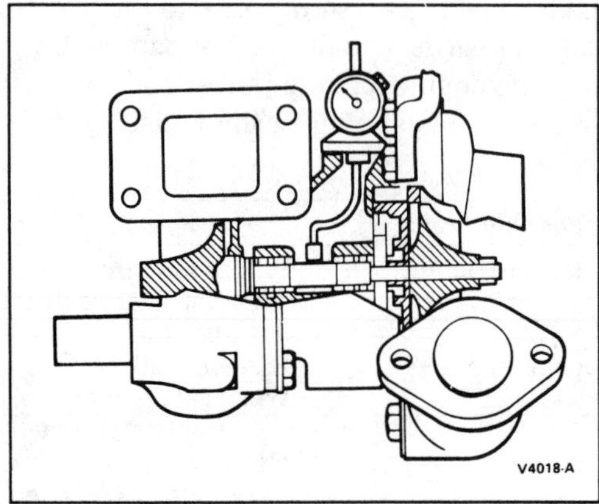

Figure 31–31. Turbocharger bearing radial clearance check. (Courtesy of Ford Motor Company, Dearborn)

NOTE: Before checking the axial/radial clearances, closely inspect the turbine/compressor vanes and housing for signs of contact. Any sign of contact between these rotating members and the housing would require T/C disassembly and repair/replacement.

Procedure for Axial Clearance Check

1. Isolate the starter system before starting this check.
2. Disconnect the air inlet piping from the compressor (cold end) of the turbocharger so that you can gain access to the compressor wheel.
3. Allow the turbine (hot end) to cool down sufficiently to allow you to remove the exhaust piping from this end of the T/C.
4. Select a dial indicator assembly that can be mounted to the T/C unit as shown in Figure 31–30.
5. Mount the dial indicator so that its pointer contacts the end of the turbine/compressor support shaft as illustrated in Figure 31–30.
6. Force the T/C shaft away from the dial indicator and hold it there while zeroing in the gauge needle (set gauge to zero).
7. Manually force the turbine towards the dial gauge and push it away again to make sure that the dial gauge is still zeroed.
8. Repeat Step 7 and carefully note the gauge reading.
9. On the Ford 2.3L turbocharger unit, the reading should not exceed 0.003″ (0.076 mm) or be less than 0.001″ (0.025 mm); otherwise the T/C requires overhaul/replacement; the specification for T/C's used by other manufacturers may differ from the spec given here; therefore, always consult the Service Manual.

Procedure for Radial Clearance Check

NOTE: A special dial gauge extension is required to contact the shaft for this check. It is an offset adapter as you can see from the diagram. Ford Tool Number TOOL–4201–C and T79L–4201–A or equivalent should be used for this purpose.

1. Remove the wastegate actuator rod retaining clip and remove the rod.
2. Remove the turbine oil outlet line to the center support housing.
3. Attach the dial indicator as shown in Figure 31–31 so that the dial indicator shaft pointer contacts the bearing support shaft of the turbine/compressor.
4. Place your hand through the turbine/compressor outlet/inlets and push the vaned wheels down; while keeping one hand on one of the wheels, zero-in the dial gauge onto the support shaft.
5. Grasp both the turbine/compressor vaned wheels and force the assembly up towards the dial gauge and carefully note the reading.
6. If bearing radial clearance is more than 0.006″ (0.152 mm) or less than 0.003″ (0.076 mm), the turbocharger requires overhaul/replacement.
7. Repeat Steps 4 and 5 to double-check your readings.

Turbocharger Overhaul

Overhaul of the turbocharger requires the use of certain special tools and equipment to successfully complete this task. If the specific Service Repair Manual is available to you, the repair of the turbocharger can be successfully completed with some care and attention on your part.

An example of the retaining bolts for the turbocharger on the Pontiac 1.8L engine is shown in Figure 31–32.

Turbocharger removal involves the same basic procedure as that indicated earlier, namely the removal of the oil feed and drain pipes, air inlet tubes and hoses, and air cleaner and intake ducting along with the exhaust manifold piping to the turbocharger.

Once the turbocharger has been removed after confirmation of damage, it can be systematically disassembled unless you are going to install an exchange rebuilt unit or a new one.

If either the turbine or compressor vaned wheels are damaged or bent and are in need of repair, special balancing equipment is necessary to ensure that the rotating members are correctly balanced. Attempting to straighten bent vanes, although appearing successful, may cause T/C destruction when the engine is restarted because they are, in fact, out of balance.

If the only problem with the turbocharger appears to be bearings/seals, this job can be undertaken without the need for balancing equipment. In most cases, however, it is usually cheaper to have the complete T/C unit exchanged through a local supplier/repair specialist unless your company policy is to repair it in-house.

If you are repairing a T/C, refer to Figure 31–1, which illustrates the main parts that make up the T/C assembly, namely the center housing, turbine, and compressor housing.

Once the T/C assembly has been split into these four main components as shown in the diagram, the center support housing containing the turbine/compressor rotating assembly must be disassembled by loosening off the retaining nut on the compressor end of the shaft in order to withdraw the compressor wheel. Once the compressor wheel has been pulled from the shaft, the turbine and shaft are withdrawn as an assembly from the hot end of the housing. Bearings and seals can then be inspected and replaced, and the support shaft, turbine and compressor wheels, and the center housing should also be closely inspected.

Wastegate Boost Pressure Test Procedures

Two examples of the recommended test procedure for the turbocharger wastegate operation are contained within this section.

The Pontiac and Buick Divisions of General Motors Corporation both employ turbochargers on their 1.8L in-line 4-cylinder engine and 3.8L V6 engines respectively. Control of both of these turbocharger wastegates are through linkage, which is shown in Figures 31–33 and 31–34.

Basic Wastegate Operation

The wastegate is normally held in a closed condition during normal engine operation in order to allow all

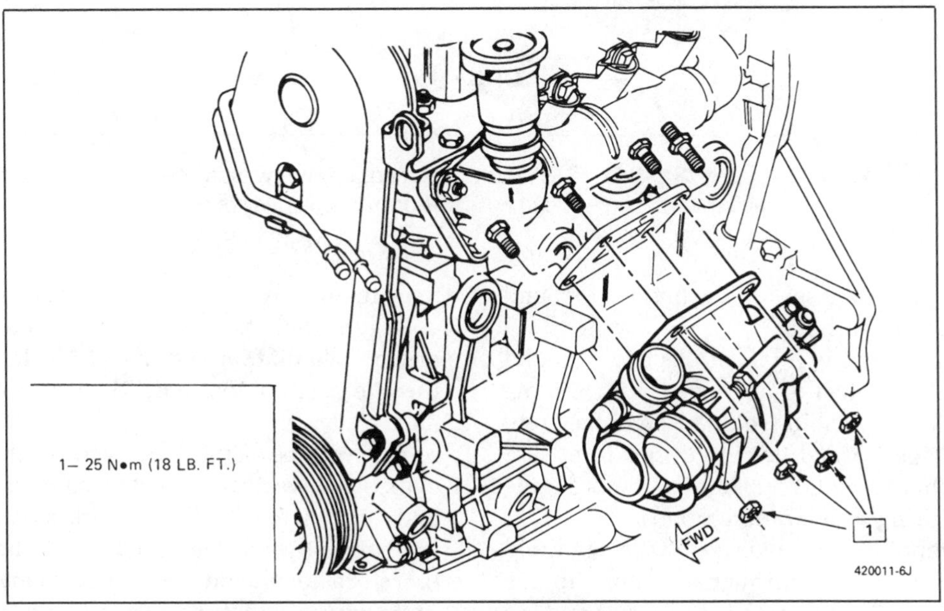

Figure 31–32. 1.8L engine-turbocharger retaining bolts location. (Courtesy of Pontiac Motor Division)

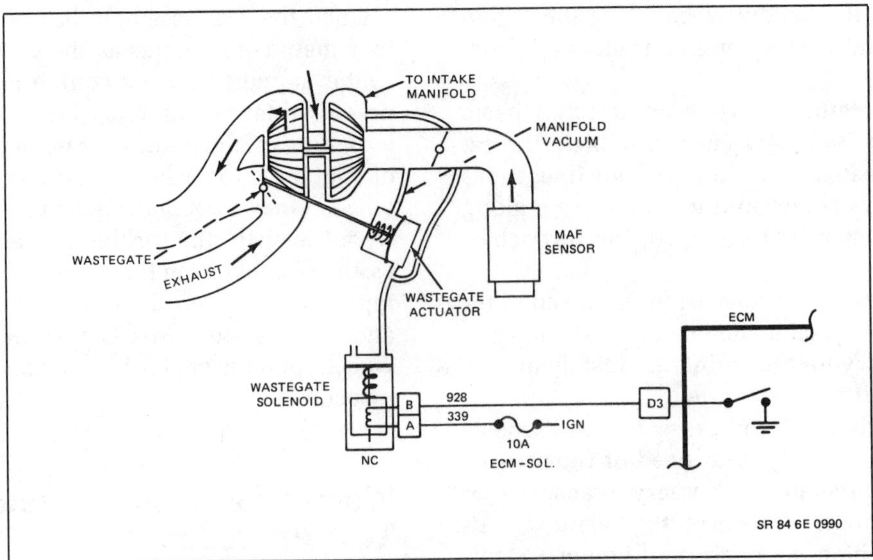

Figure 31–33. 3.8L V6 turbocharged engine wastegate control arrangement and wiring diagram. (Courtesy of Buick Motor Division)

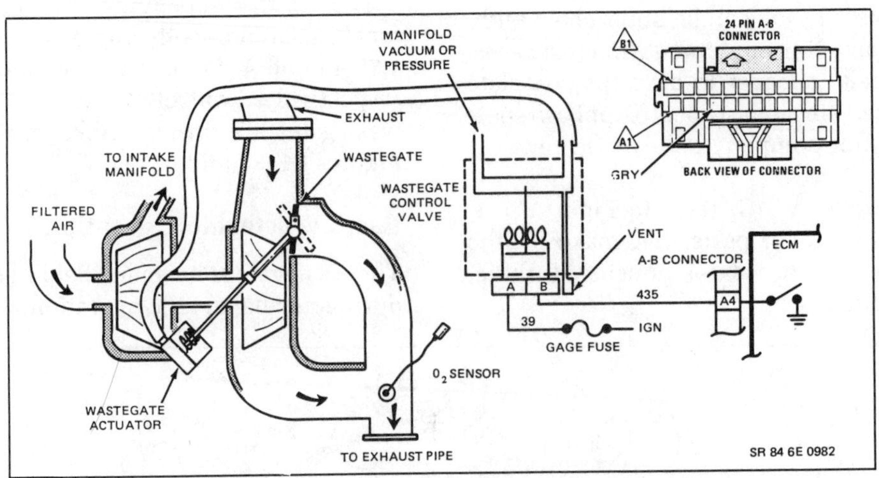

Figure 31–34. 1.8L turbocharged engine wastegate control arrangement and wiring diagram. (Courtesy of Pontiac Motor Division)

of the engine's exhaust gases to flow to the "hot" end of the turbocharger (turbine).

Figure 31–34 illustrates that an electric solenoid (wastegate control valve) is employed with the actuator system. This solenoid would be energized (on) at all times during normal driving cycles to ensure that the turbocharger will receive maximum engine exhaust flow during acceleration modes.

With the solenoid energized, no intake manifold pressurized air can flow through the hose to the wastegate actuator. As the engine accelerates and the turbocharger boost increases, the MAP (manifold absolute pressure) sensor responds to the change in intake manifold pressure. The MAP sensor relays a voltage signal that varies between 1 to 1.5 volts at engine idle up to between 4–4.5 volts at a wide open throttle position to the ECM (electronic control module).

The voltage signal received at the MAP sensor causes the wastegate electric solenoid to pulse "on" and "off," which will pulse pressurized intake manifold air to the wastegate actuator. The wastegate will therefore be opened and closed to effectively bypass some of the exhaust gases and thereby control the turbocharger boost pressure.

When the boost pressure decreases due to operating conditions, the ECM will close the control solenoid and the remaining pressurized air from the wastegate actuator will bleed off through the vent in the solenoid control valve.

When the wastegate mechanism is at fault, this is generally reflected by either an over- or under-boost situation that is noticeable by the driver of the vehicle.

With a low boost condition, a noticeable loss of power and slow acceleration from a low speed is evident due to the wastegate valve being in a partially or fully open position, which bypasses the majority of the pressurized exhaust gases leaving the engine cylinders. With little exhaust gas flow into the turbine (hot) end of the turbocharger, low boost will exist. During an over-boost condition, the wastegate remains in a closed position and the possibility of engine and turbocharger damage can result through an overspeed situation. Under such a situation, the MAP sensor voltage signal causes the ECM to reduce the duration of its voltage signal to the fuel injector thereby decreasing both fuel and the flow of exhaust gases to the turbocharger.

Since GMC cars employ an ECM (electronic control module) that is fed engine and vehicle operating condition information from a variety of sensors, a "check engine light" will be activated on the vehicle dash or in the memory bank, which will store a "Code 31" on the ECM.

ECM Solenoid Actuator

On the Pontiac 1.8L turbocharged engine, the trouble code 31 will appear anytime that the intake manifold boost pressure exceeds approximately 15 psi (199 kPa). This pressure is sensed by the MAP sensor (manifold absolute pressure) and if this pressure is maintained for at least 2 seconds, code 31 and the *check engine light* will illuminate on the instrument panel for about 10 seconds, then the code 31 will be stored in the ECM memory bank.

If the MAP sensor fails, the ECM will substitute a fixed MAP value and use the throttle position sensor (TPS) to effectively control fuel injector delivery.

The 1.8L engine uses both a MAP and a MAT (manifold air temperature) sensor. The MAT sensor is a thermistor unit that is mounted in the intake manifold. Low temperatures produce a high resistance (100,000 ohms at −40°C or −40°F) in the sensor, while higher temperatures cause a low resistance reading (70 ohms at 130°C or 266°F). This results in a high voltage signal when the manifold air is cold and a low voltage signal when the air is hot. The voltage signal is fed to the ECM for fuel control purposes. Should the MAT sensor fail, a code 23 or 25 will be stored in the memory bank.

Figure 31–35 shows the MAT sensor location while Figure 31–36 shows the MAP sensor unit, which is generally mounted on the intake manifold.

On the Buick V6 turbocharged engine, a MAF (mass air flow) sensor on the suction side of the turbocharger is used to measure the flow of air entering the engine; this signal is used by the ECM to control the fuel injectors. Figures 7–35 and 31–37 illustrates the mass air flow sensor used with the 3.8L Buick V6 engine.

A code 31 is set on the Buick V6 3.8L turbocharged engine when the ECM is commanding a wastegate duty cycle between 5 and 95%, no voltage pulses are received on the wastegate monitor, and both of the above conditions have been met for more than 5 seconds.

Causes of possible over-boost conditions are:

1. A sticking wastegate actuator or wastegate valve itself.

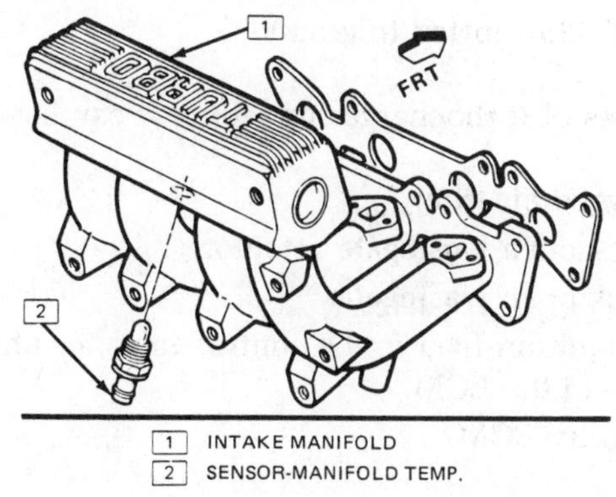

Figure 31–35. 1.8L turbocharged engine manifold air temperature (MAT) sensor location. (Courtesy of Pontiac Motor Division)

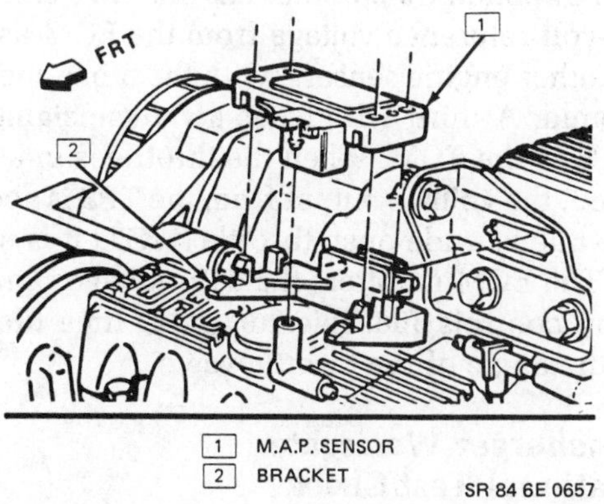

Figure 31–36. 1.8L turbocharged engine manifold absolute pressure (MAP) sensor location. (Courtesy of Pontiac Motor Division)

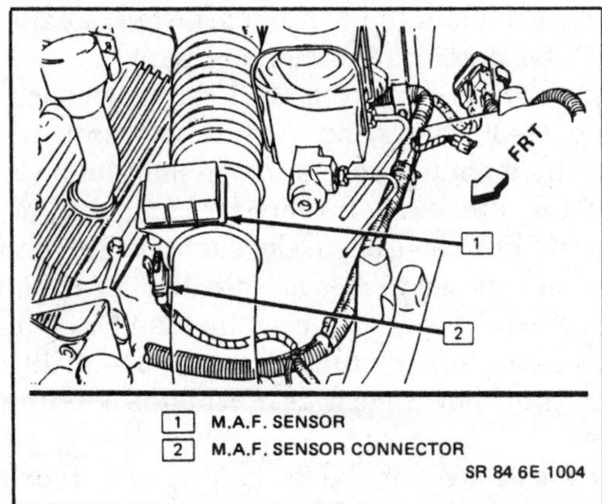

Figure 31-37. 3.8L V6 turbocharged engine mass air flow (MAF) sensor location. (Courtesy of Buick Motor Division)

2. A cut or pinched hose from the intake manifold to the wastegate actuator.
3. A faulty ECM.
4. CKT 435 shorted to ground.

Causes of turbocharger under-boost can be:

1. Plugged air filter.
2. Stuck open wastegate actuator.
3. Stuck open wastegate.
4. No ignition feed to the control valve, or an open wire to the ECM.
5. A faulty ECM.

The throttle position sensors (TPS) used with both the 1.8L and 3.8L turbocharged engines are shown in Figures 31-38 and 31-39.

The throttle position sensor is a variable resistance switch or potentiometer that has one wire connected to a 5-volt reference voltage from the ECM as do all of the other engine sensors, with the other end going to ground. A third wire feeds a voltage signal from the TPS to the ECM. When the throttle is in a closed position, the voltage output from the TPS is about 0.4 volt, while at wide open throttle (WOT) it is about 5 volts. The ECM monitors the TPS voltage signal and thereby controls fuel injector pulse time (delivery) from the angle of the throttle valve.

Turbocharger Wastegate Operational Test/Check

An operational check of the turbocharger wastegate should be performed as follows for the Buick V6 3.8L unit and Pontiac 1.8L unit:

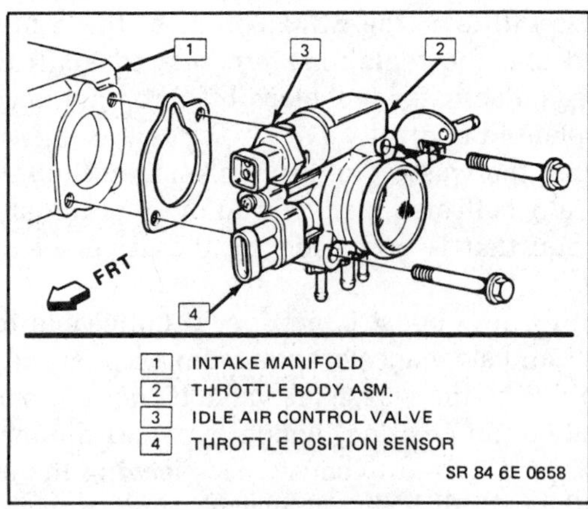

Figure 31-38. 1.8L engine throttle position sensor (TPS) location. (Courtesy of Buick Motor Division)

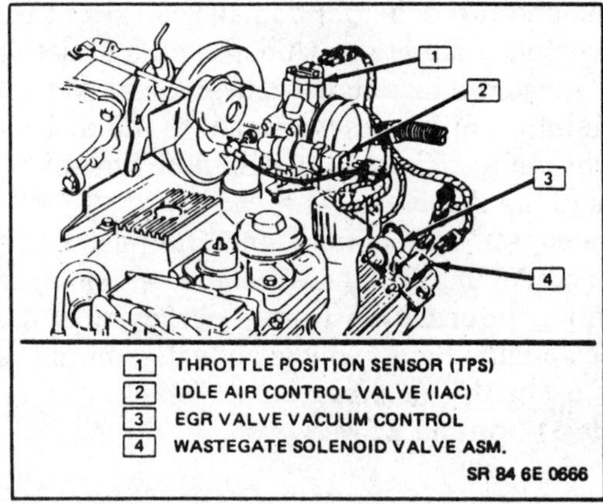

Figure 31-39. 3.8L V6 turbocharged engine throttle position sensor (TPS) location. (Courtesy of Buick Motor Division)

1. Check the mechanical linkage to the wastegate for possible linkage wear or damage.
2. Visually check and manually inspect the condition of the hose assembly that runs from the throttle body to the wastegate solenoid and also from the solenoid to the actuator.
3. To check the operation of the wastegate actuator linkage, apply a hand-operated vacuum/pressure pump such as Kent-Moore P/N J-23738 (Figure 7-56) in series with J-28474 component gauge to the actuator assembly, replacing the wastegate solenoid to actuator assembly hose.
4. Apply pressure to the actuator; between 7.5-8.5 psi (52-58.6 kPa) on the V6 Buick and between 3.5-4.5 psi (24-31 kPa) on the Pontiac 1.8L 4-cylinder unit to the actuator rod end, which should move 0.015" and move the wastegate linkage. If this does not occur, replace the actuator assembly and recheck the operation.

Turbocharger Solenoid Valve Check

1. Ignition key off and apply 15 psi (103 kPa) to the actuator valve through the intake manifold hose, which should cause movement.
2. Remove the pressure source; the actuator should move back closing the wastegate. If the pressure doesn't bleed off, then the electric solenoid vent line could be plugged.
3. Turn ignition key *on* and ground the diagnostic terminal; the solenoid should receive battery power, which will close off the intake manifold to wastegate actuator.
4. With the ignition key *on* and the engine stopped, the solenoid control valve should not be energized. With the key *on* and the diagnostic terminal grounded, the solenoid should be energized.

SPECIAL NOTE: For vehicle diagnostic procedure terms, refer to Chapter 7 dealing with General Motors fuel injection systems.

Turbocharger Seal Leakage Troubleshooting

Seal leakage from a turbocharger can occur at either the compressor end (cold end) or the turbine (hot end). Seal leakage can lead to oil being drawn into the engine, which can cause detonation.

Oil leakage on the outlet or hot end of the turbocharger can cause heavy carbon buildup on the rotating turbine vanes, and a possible imbalance situation can arise. In addition, blue smoke will generally be visible from the exhaust pipe.

Although visual inspection of the seals can be undertaken, a more positive approach has been developed whereby a fluorescent tracer material can be added to the engine oil. When an ultraviolet light is directed towards the oil, it will glow white-yellow. This method is used often on heavy-duty high-speed truck diesels to determine the difference between oil and diesel fuel slobber—which will be dark blue.

High-intensity, long-wave ultraviolet lamps are available from a variety of tool manufacturers for this purpose since this type of lamp is also used for zyglow inspection. The fluorescent tracer material is available from a number of sources with directions as to its use and mixture quantity.

For example, the Cummins Engine Company recommends that one package of their tracer material P/N 3376891 be added to each 10 gallons of engine oil. Obviously, if you were to use this tracer material to pinpoint a leak on a turbocharged engine such as a small 4-cylinder gasoline engine used in a passenger car, the quantity of tracer would be proportional to the amount of oil required in that particular engine.

When a turbocharger turbine seal leak is suspected, the proper quantity of tracer can be added to the engine oil through the oil filler tube or opening. The engine should be started and allowed to run at an idle speed for approximately 10 minutes maximum, which is sufficient time to allow the tracer to mix with the engine oil and also gives the leaking seal enough time to allow oil to leak if this is the problem.

Once the engine is stopped, allow enough time for the hot turbine outlet pipe to cool so that you can effectively remove it from the turbocharger. Refer to Figure 31–40 and shine the ultraviolet light into the turbine outlet area. Oil seal leakage will be confirmed by the fact that the tracer material will glow yellow.

On diesel engines, fuel slobber through a faulty injector or incomplete combustion (raw or unburned fuel) will glow dark blue under the ultraviolet light. If this is the case, then the engine or fuel system should be checked for the cause of raw fuel appearing at the turbocharger hot end.

Another method that can be employed when oil or fuel seems to be emanating from the turbine end of the turbocharger is to simply wipe the excess drips from the exhaust pipe flange with a clean shop towel once tracer material has been added to the engine oil as described above. Using the ultraviolet light, if the material glows dark blue, then the leakage is caused by fuel, while a yellow-white glow indicates a turbo seal leakage problem.

Turbocharger compressor seal leakage is usually confirmed by a wet oil film at the compressor housing outlet as shown in Figure 31–41. Compressor seal leakage can be caused by air filter, crankcase breather, and air inlet piping problems since these can cause an increase in the pressure difference at the compressor seal ring leading to oil leakage.

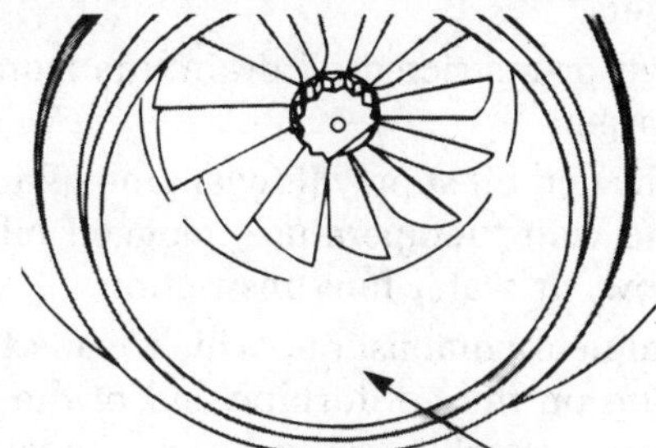

Figure 31–40. Inspecting turbocharger turbine end for seal leakage. (Courtesy of Cummins Engine Company, Inc; Columbus, Indiana)

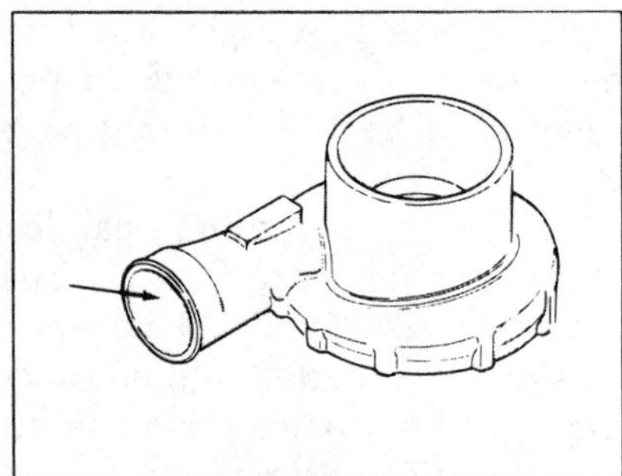

Figure 31–41. Oil leakage location-compressor housing outlet. [Courtesy of Cummins Engine Company, Inc; Columbus, Indiana]

Turbocharger Failure Analysis

Failure analysis of any engine component requires that you gather as much information as you possibly can dealing with the engine's mode of operation. Obtain all service records so that you can check on previous turbocharger or engine problems that may have led to the current turbocharger failure.

Generally, turbocharger service life will be directly affected by the following considerations and conditions:

1. Cleanliness of the air entering the compressor end of the turbocharger.
2. Rapid acceleration of the engine immediately after startup, especially on cold mornings since this will starve the T/C for oil and bearing/shaft damage will result.
3. Shutting the engine down immediately after a hard high speed run without allowing several minutes for cool down of the turbocharger and the cylinder heads and hot exhaust manifolds.
4. Dirty oil supply (poor engine oil change procedures).
5. Low oil pressure from engine.
6. Kinked or leaking oil supply line to the turbocharger.
7. Kinked or restricted oil drain line from the turbocharger.
8. Overheated oil supply (low engine oil level, high engine water temperature, plugged oil cooler—air flow, or water flow restriction.
9. Incomplete combustion, which leads to carbon buildup on the hot turbine end of the balanced rotating assembly leading to an imbalance and the rotating blades striking the housing.
10. Bearing damage caused by a number of the above items, which can lead to imbalance and destruction.

Do not simply remove a turbocharger and replace it without giving any thought as to what might have caused the failure. Replacing a turbocharger assembly onto the engine without checking to ensure that the problem that caused the last failure has been rectified could lead to another costly repair bill.

Check all of the listed items above, and from the following information, establish that the reason for the failure has been fixed.

SPECIAL NOTE: When a new turbocharger has been installed, make certain that the oil supply and drain lines are clean and kink free. With the battery disconnected, pour a small amount of clean engine oil into the inlet feed port of the turbocharger with a funnel while manually rotating the compressor wheel by hand.

Connect up the battery and isolate the ignition circuit; with the turbocharger return oil line loose or disconnected (use a funnel or drain tray to catch the oil or route it back to a receptacle), crank the engine over until engine oil flows freely from the T/C drain line.

Tighten up the drain line and after starting the engine, check for signs of oil leakage with the engine running at an idle rpm for several minutes to allow adequate lubrication to the T/C bearings and allow gradual expansion of the internal components before loading the engine.

Typical examples of turbocharger component part damage after a failure can be seen in Figures 31–42 through 31–57. Figure 31–58 is a T/C bearing failure troubleshooting chart.

Typical examples of turbocharger component part damage after a failure can be seen in Figures 31–42 through 31–57. Figure 31–58 is a T/C bearing failure troubleshooting chart, while 31–59 is a T/C inspection/troubleshooting chart.

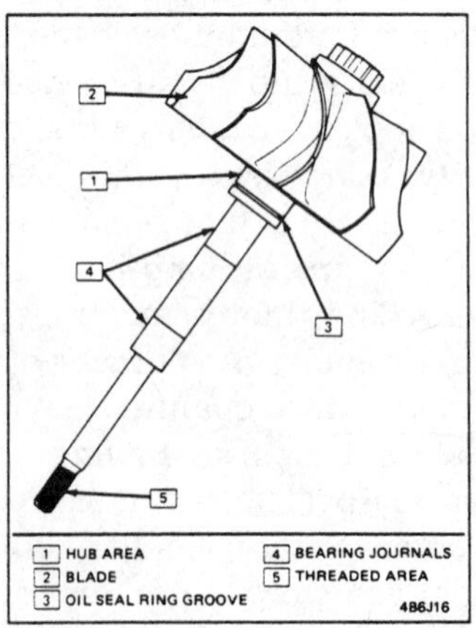

Figure 31–42. [Courtesy of Buick Motor Division]

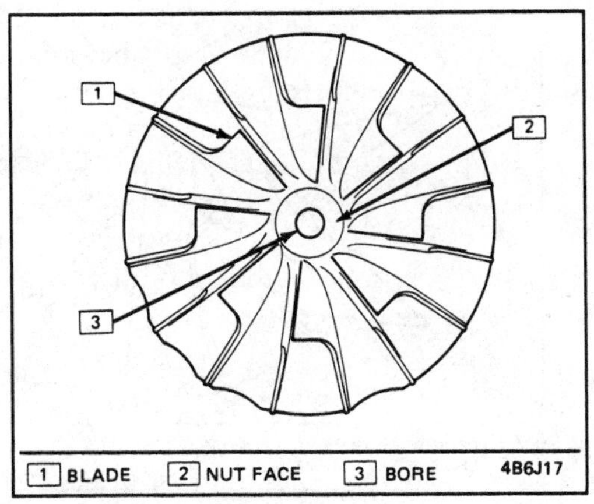

Figure 31-43. (Courtesy of Buick Motor Division)

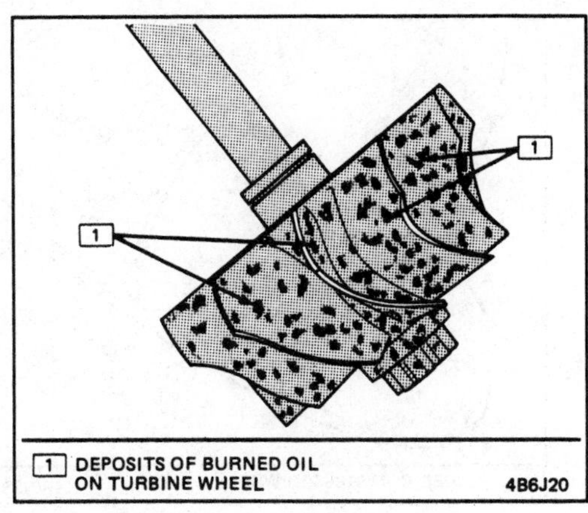

Figure 31-46. (Courtesy of Buick Motor Division)

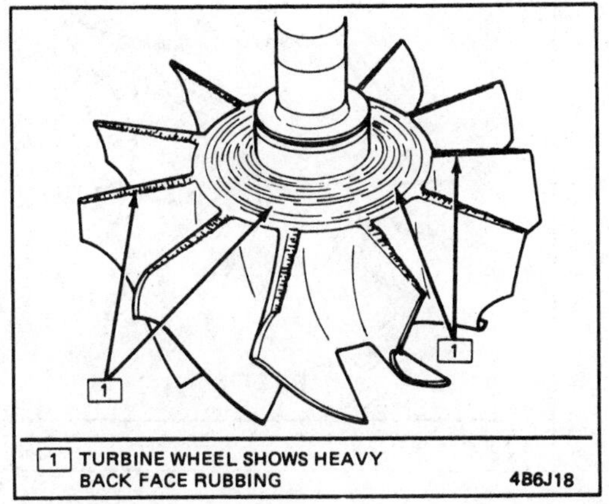

Figure 31-44. (Courtesy of Buick Motor Division)

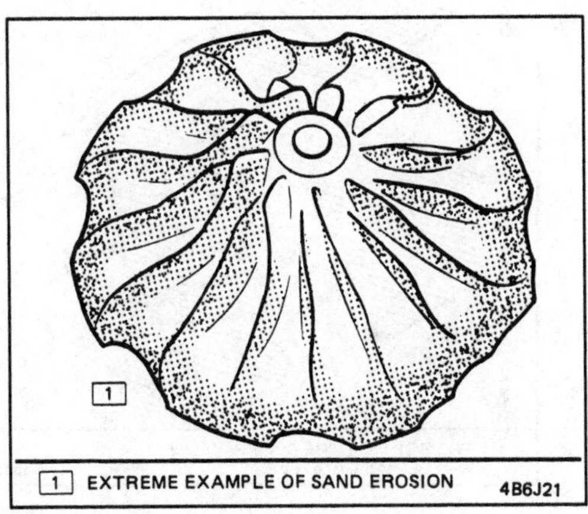

Figure 31-47. (Courtesy of Buick Motor Division)

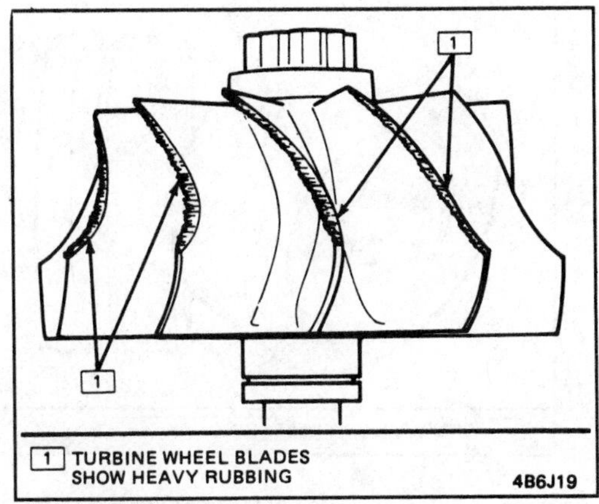

Figure 31-45. (Courtesy of Buick Motor Division)

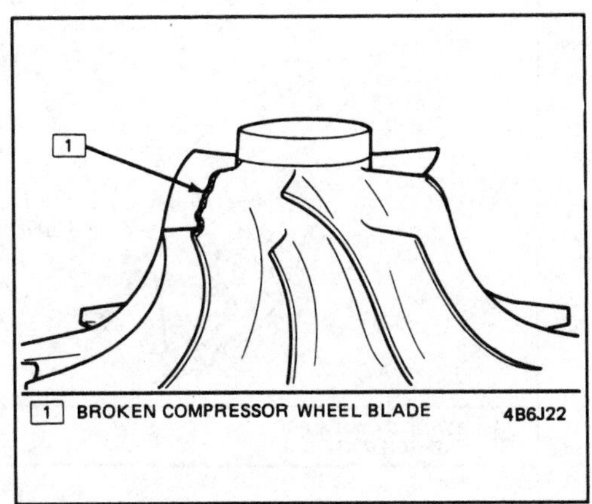

Figure 31-48. (Courtesy of Buick Motor Division)

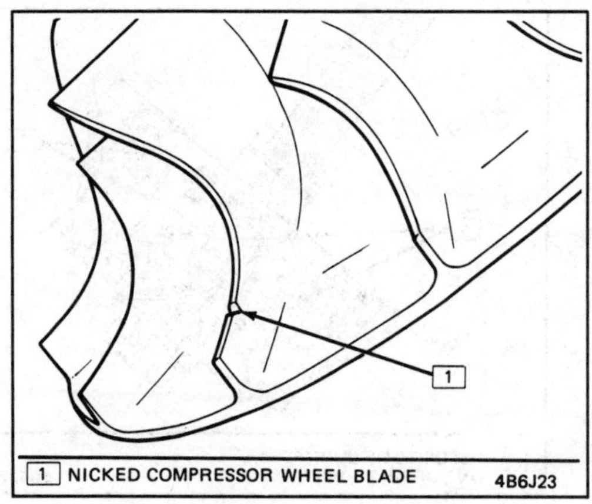

Figure 31–49. (Courtesy of Buick Motor Division)

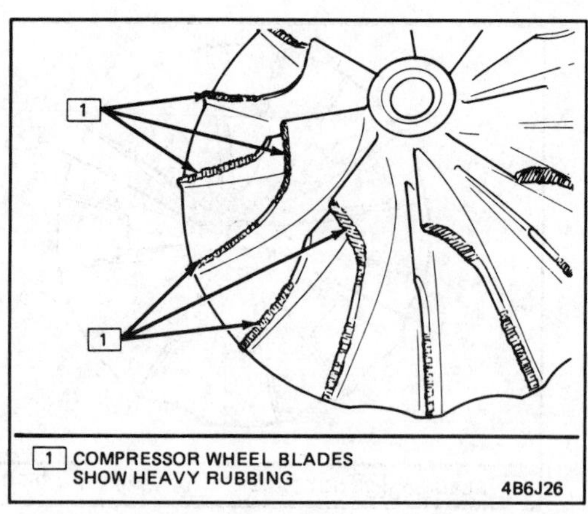

Figure 31–52. (Courtesy of Buick Motor Division)

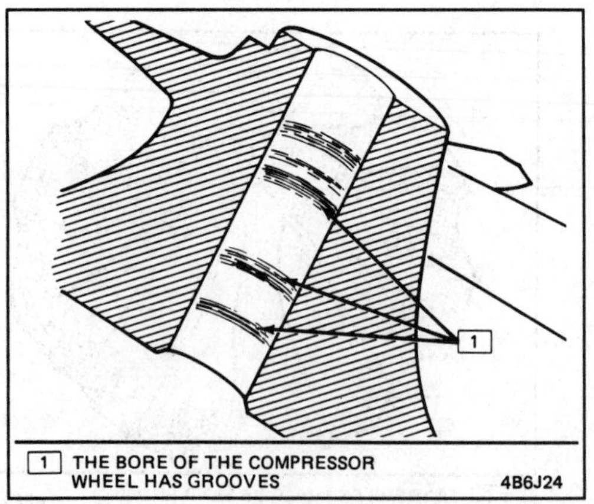

Figure 31–50. (Courtesy of Buick Motor Division)

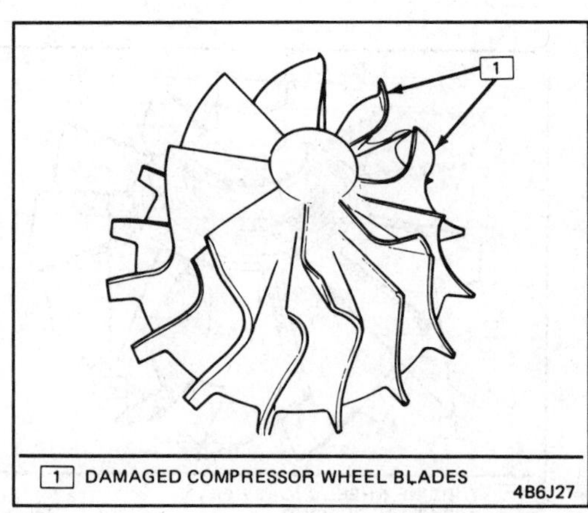

Figure 31–53. (Courtesy of Buick Motor Division)

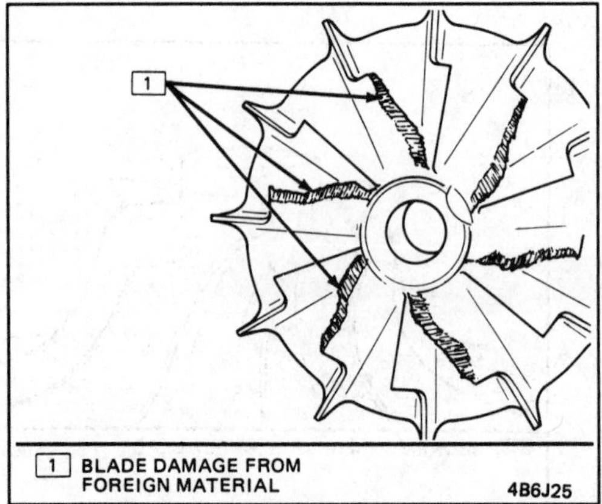

Figure 31–51. (Courtesy of Buick Motor Division)

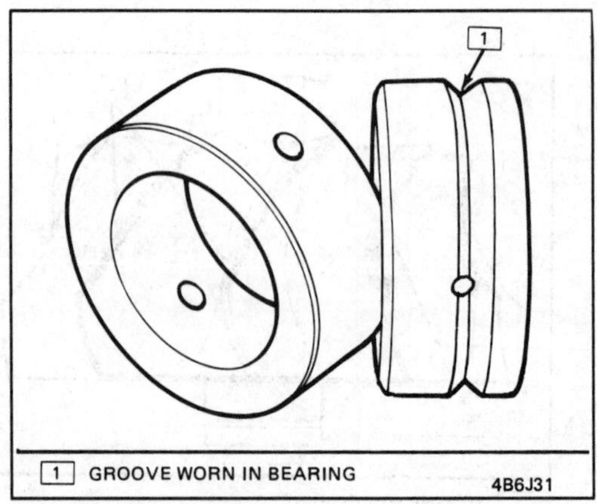

Figure 31–54. (Courtesy of Buick Motor Division)

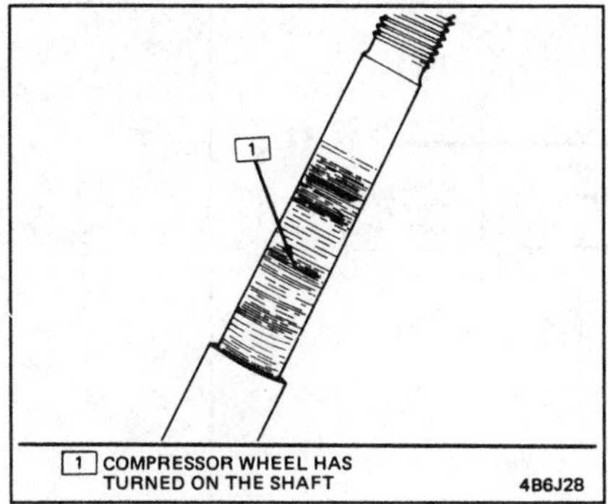
1
1 COMPRESSOR WHEEL HAS TURNED ON THE SHAFT
4B6J28

Figure 31–55. (Courtesy of Buick Motor Division)

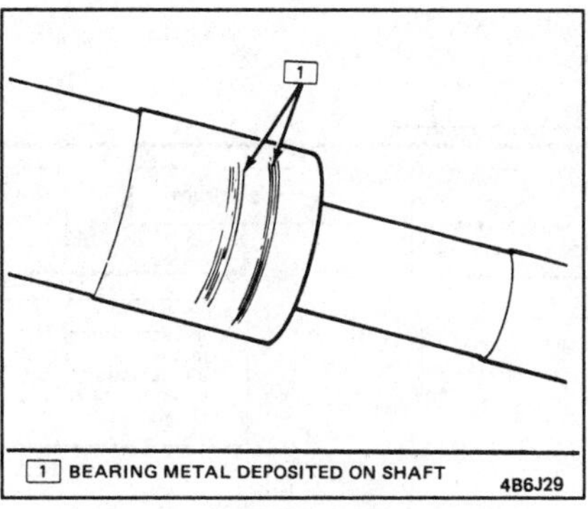
1
1 BEARING METAL DEPOSITED ON SHAFT
4B6J29

Figure 31–56. (Courtesy of Buick Motor Division)

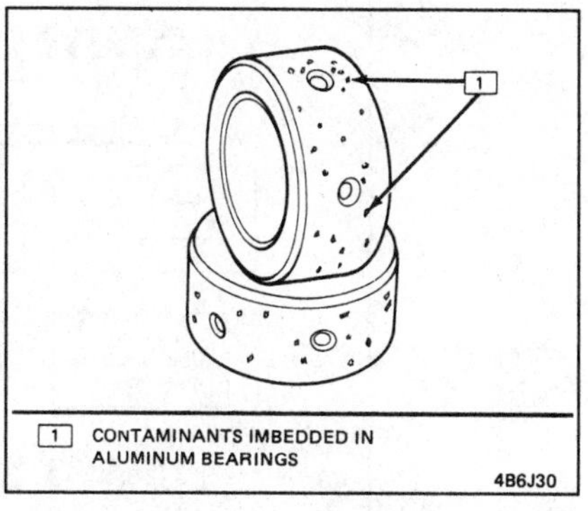
1
1 CONTAMINANTS IMBEDDED IN ALUMINUM BEARINGS
4B6J30

Figure 31–57. (Courtesy of Buick Motor Division)

4B6J15

BEARING TROUBLESHOOTING

APPEARANCE OF BEARING

Slight wear or scratches	Moderate to heavy grooving on O.D. only	Moderate to heavy grooving on O.D. & I.D.	Extruded, or pounded. (May be stuck in ctr. hsg.)	Smooth undersized O.D.	Cracked or broken	Deep groove around center of O.D.	Oil holes fully or partially closed	Oil Holes plugged with carbon	Polished looking O.D.	I.D. Polished & worn oversize	Melted (aluminum bearing)	CONDITION	PROBABLE CAUSE
●												Normal use	Acceptable operating & maintenance procedures
	●											Contaminated oil (dirt in oil)	Engine oil & oil filter(s) not changed frequently enough, unfiltered air entering engine intake, malfunction in prelube valve or oil filter bypass valve
		●				●						severely contaminated (dirty oil)	
			●		●		●					Pounded by eccentric shaft motion	Foreign object damage, coked or loose housing, excessive bearing clearance due to lube problem
				●					●			Center housing bearing bores, rough finish	Incorrect cleaning of ctr. hsg. during overhaul of turbo. (wrong chemicals, bores sand or bead blasted)
						●						Metal or large particle oil contamination	Severe engine wear. i.e., bearing damage, camshaft or lifter wear, broken piston
							●				●	Lack of lube, oil lag, insufficient lube	Low oil level, high speed shutdowns, lube system failure, turbo plugged
								●				Coking	Hot shutdowns, engine overfueled, restricted or leaking air intake
				●					●	●		Fine particles in oil (contaminated oil)	See contaminated oil
										●		Rough bearing journals on shaft	Bearing journals not protected from sand or bead blast cleaning during overhaul

Figure 31–58. Turbocharger bearing failure troubleshooting chart. (Courtesy of Buick Motor Division)

TROUBLE SHOOTING CHARTS

CHART 1

drag on turbine wheel

- carbon build-up behind turbine wheel
  - coked oil or combustion deposits
- dirt build-up behind compressor wheel
  - air intake leaks
- bearing seizure
- dirty or worn bearings
  - (bearing seizure / dirty or worn bearings): excessive temperatures, imbalance, dirty oil, oil starvation, insufficient lubrication, normal wear
- foreign object damage to either wheel

CHART 2

oil on compressor wheel or in compressor housing

oil being pushed or pulled through from center housing

- clogged engine oil breathers
- restricted exhaust flow—excessive back pressure
- air-intake restriction
- clogged air cleaner element or collapsed inlet line

low power

excessive exhaust temperatures

cracked housings — overheated bearings

turbocharger failure

CHART 3

oil in manifolds or dripping from housings

- seal failure
  - damaged or worn journal bearings
    - unbalance of rotating assembly
      - damage to turbine or compressor wheel or blade; dirt or carbon build-up on wheel or wheels
      - bearings appear pounded out or extruded. Float bearings become fixed or expanded into center housing bearing boxes
    - bearing wear
      - shaft journal is worn and grooved, but not blued
    - oil starvation or insufficient lube
      - bearing journal and shaft are blued. In early stages, aluminum bearing material can be found on journals.
- seal leakage
- flooding due to excessive cranking or idling for long periods

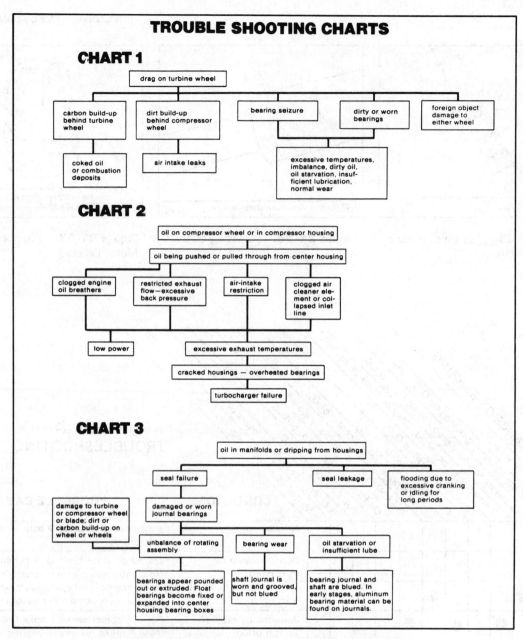

Figure 31-59. Turbocharger inspection/troubleshooting chart. (Courtesy of Detroit Diesel Allison Division of GMC)

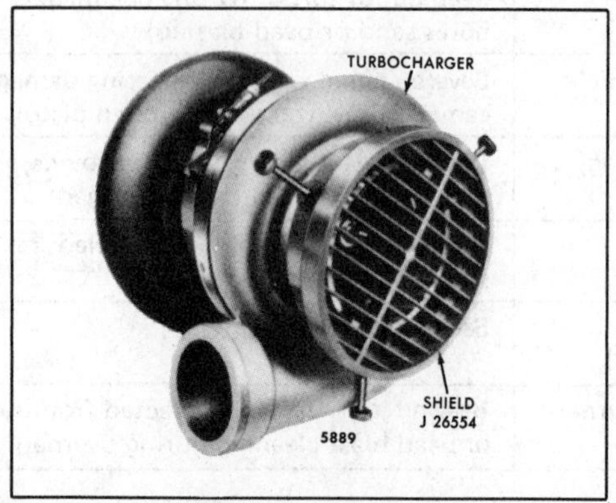

Figure 31-60. Turbocharger safety protective inlet shield. (Courtesy of Detroit Diesel Allison Division of GMC)

SAFETY NOTE: Prior to running a turbocharged engine during test purposes, always install a protective screen similar to that illustrated in Figure 31-60 over the inlet side of the turbocharger assembly. This will prevent possible serious personal injury when working around a running engine as well as ensuring that no foreign materials such as shop towels and cleaning rags or loose tools can fall or be sucked into the compressor section of the turbocharger!

Appendix

METRIC BOLT AND NUT IDENTIFICATION

Common metric fastener strength property classes are 9.8 and 10.9 with the class identification embossed on the head of each bolt. Customary (inch) strength classes range from grade 2 to 8 with line identification embossed on each bolt head. Markings correspond to two lines less than the actual grade (i.e. grade 7 bolt will exhibit 5 embossed lines on the bolt head). Some metric nuts will be marked with single digit strength identification numbers on the nut face. The following figure illustrates the different strength markings.

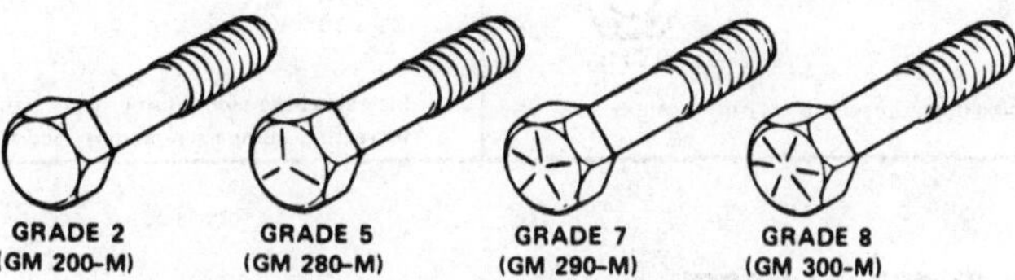

Customary (inch) bolts - Identification marks correspond to bolt strength - Increasing numbers represent increasing strength.

Metric Bolts - Identification class numbers correspond to bolt strength - Increasing numbers represent increasing strength.

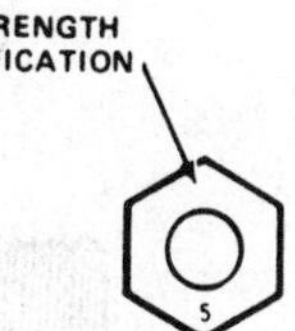

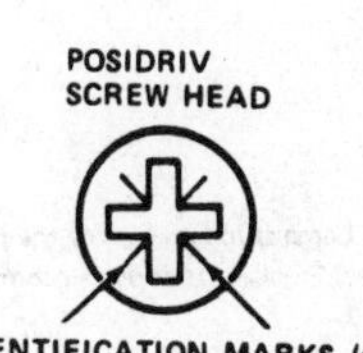

NOMENCLATURE FOR BOLTS

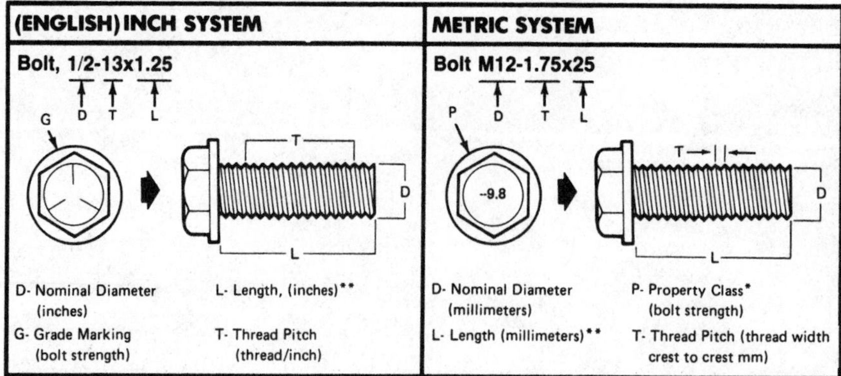

(ENGLISH) INCH SYSTEM		METRIC SYSTEM	
D- Nominal Diameter (inches)	L- Length, (inches)**	D- Nominal Diameter (millimeters)	P- Property Class* (bolt strength)
G- Grade Marking (bolt strength)	T- Thread Pitch (thread/inch)	L- Length (millimeters)**	T- Thread Pitch (thread width crest to crest mm)

*The property class is an Arabic numeral distinguishable from the slash SAE English grade system.
**The length of all bolts is measured from the underside of the head to the end.

HEX NUT STRENGTH IDENTIFICATION

(ENGLISH) INCH SYSTEM		METRIC SYSTEM	
Grade	Identification	Class	Identification
Hex Nut Grade 5	3 Dots	Hex Nut Property Class 9	Arabic 9
Hex Nut Grade 8	6 Dots	Hex Nut Property Class 10	Arabic 10
Increasing dots represent increasing strength.		May also have blue finish or paint daub on hex flat. Increasing numbers represent increasing strength.	

OTHER TYPES OF PARTS

Metric identification schemes vary by type of part, most often a variation of that used of bolts and nuts. Note that many types of English and metric fasteners carry no special identification if they are otherwise unique.

—Stamped "U" Nuts

9

—Tapping, thread forming and certain other case hardened screws

2 MET

—Studs, Large studs may carry the property class number. Smaller studs use a geometric code on the end.

CLASS 9.8 CLASS 8.8

3

REUSE OF PREVAILING TORQUE NUT(S) AND BOLT(S)

PREVAILING TORQUE NUTS ARE THOSE NUTS WHICH INCORPORATE A SYSTEM TO DEVELOP AN INTERFERENCE BETWEEN NUT AND BOLT THREADS INTERFERENCE IS MOST COMMONLY ACHIEVED BY DISTORTING TOP OF ALL-METAL NUT, BUT ALSO MAY BE ACHIEVED BY DISTORTING AT MIDDLE OF HEX FLAT, BY NYLON PATCH ON THREADS, BY NYLON WASHER INSERT AT TOP OF NUT AND BY NYLON INSERT THROUGH NUT.

PREVAILING TORQUE BOLTS ARE THOSE BOLTS WHICH INCORPORATE A SYSTEM TO DEVELOP AN INTERFERENCE BETWEEN BOLT AND NUT OR TAPPED HOLE THREADS. INTERFERENCE IS ACHIEVED BY DISTORTING SOME OF THE THREADS (SEVERAL METHODS EXIST), BY APPLYING A NYLON PATCH OR STRIP OR BY ADHESIVE COATING ON THREADS.

PREVAILING TORQUE NUTS

TOP LOCK MANY TYPES

CENTER LOCK

NYLON INSERT

NYLON PATCH

NYLON WASHER INSERT

PREVAILING TORQUE BOLTS

DRY ADHESIVE COATING

OUT OF ROUND THREAD AREA

NYLON STRIP OR PATCH

THREAD PROFILE DEFORMED

RECOMMENDATIONS FOR REUSE

A. CLEAN, UNRUSTED PREVAILING TORQUE BOLTS AND NUTS MAY BE REUSED AS FOLLOWS:

1. CLEAN DIRT AND OTHER FOREIGN MATERIAL OFF NUT AND BOLT.
2. INSPECT BOLT AND NUT TO ASSURE THERE ARE NO CRACKS, ELONGATION OR OTHER SIGNS OF ABUSE OR OVERTIGHTENING. LIGHTLY LUBRICATE THREADS. (IF ANY DOUBT, REPLACE WITH NEW PREVAILING TORQUE FASTENER OF EQUAL OR GREATER STRENGTH.)
3. ASSEMBLE PARTS AND START BOLT OR NUT.
4. OBSERVE THAT BEFORE FASTENER SEATS, IT DEVELOPS PREVAILING TORQUE PER CHART BELOW. (IF ANY DOUBT, INSTALL NEW PREVAILING TORQUE FASTENER OF EQUAL OR GREATER STRENGTH).
5. TIGHTEN TO TORQUE SPECIFIED IN SERVICE MANUAL.

B. BOLTS AND NUTS WHICH ARE RUSTY OR DAMAGED SHOULD BE REPLACED WITH NEW PARTS OF EQUAL OR GREATER STRENGTH.

METRIC SIZES

		6 & 6.3	8	10	12	14	16	20
NUTS AND ALL METAL BOLTS	N•m	0.4	0.8	1.4	2.2	3.0	4.2	7.0
	In. Lbs.	4.0	7.0	12	18	25	35	57
ADHESIVE OR NYLON COATED BOLTS	N•m	0.4	0.6	1.2	1.6	2.4	3.4	5.6
	In. Lbs.	4.0	5.0	10	14	20	28	46

INCH SIZES

		.250	.312	.375	.437	.500	.562	.625	.750
NUTS AND ALL METAL BOLTS	N•m	0.4	0.6	1.4	1.8	2.4	3.2	4.2	6.2
	In. Lbs.	4.0	5.0	12	15	20	27	35	51
ADHESIVE OR NYLON COATED BOLTS	N•m	0.4	0.6	1.0	1.4	1.8	2.6	3.4	5.2
	In. Lbs.	4.0	5.0	9.0	12	15	22	28	43

SI METRIC-CUSTOMARY CONVERSION TABLE

Multiply	by	to get equivalent number of:
LENGTH		
Inch	25.4	millimeters (mm)
Foot	0.304 8	meters (m)
Yard	0.914 4	meters
Mile	1.609	kilometers (km)
AREA		
$Inch^2$	645.2	millimeters2 (mm2)
	6.45	centimeters2 (cm2)
$Foot^2$	0.092 9	meters2 (m2)
$Yard^2$	0.836 1	meters
VOLUME		
$Inch^3$	16 387.	mm^3
	16.387	cm^3
	0.016 4	liters (1)
Quart	0.946 4	liters
Gallon	3.785 4	liters
$Yard^3$	0.764 6	meters3 (m3)
MASS		
Pound	0.453 6	kilograms (kg)
Ton	907.18	kilograms (kg)
Ton	0.907	tonne (t)
FORCE		
Kilogram	9.807	newtons (N)
Ounce	0.278 0	newtons
Pound	4.448	newtons
TEMPERATURE		
Degree Fahrenheit	(°F-32) ÷ 1.8	degree Celsius (C)

°F: -40, 0, 32, 40, 80, 98.6, 120, 160, 200, 212 °F

°C: -40, -20, 0, 20, 37, 40, 60, 80, 100 °C

Multiply	by	to get equivalent number of:
ACCELERATION		
$Foot/sec^2$	0.304 8	meter/sec2 (m/s2)
$Inch/sec^2$	0.025 4	meter/sec2
TORQUE		
Pound-inch	0.112 98	newton-meters (N·m)
Pound-foot	1.355 8	newton-meters
POWER		
Horsepower	0.746	kilowatts (kW)
PRESSURE OR STRESS		
Inches of mercury	3.377	kilopascals (kPa)
Pounds/sq. in.	6.895	kilopascals
ENERGY OR WORK		
BTU	1 055.	joules (J)
Foot-pound	1.355 8	joules
Kilowatt-hour	3 600 000. or 3.6×10^6	joules (J = one W's)
LIGHT		
Foot candle	10.764	lumens/meter2 (lm/m2)
FUEL PERFORMANCE		
Miles/gal	0.425 1	kilometers/liter (km/1)
Gal/mile	2.352 7	liters/kilometer (1/km)
VELOCITY		
Miles/hour	1.609 3	kilometers/hr. (km/h)

DECIMAL AND METRIC EQUIVALENTS

Fractions	Decimal In.	Metric MM.	Fractions	Decimal In.	Metric MM.
1/64	.015625	.39688	33/64	.515625	13.09687
1/32	.03125	.79375	17/32	.53125	13.49375
3/64	.046875	1.19062	35/64	.546875	13.89062
1/16	.0625	1.58750	9/16	.5625	14.28750
5/64	.078125	1.98437	37/64	.578125	14.68437
3/32	.09375	2.38125	19/32	.59375	15.08125
7/64	.109375	2.77812	39/64	.609375	15.47812
1/8	.125	3.1750	5/8	.625	15.87500
9/64	.140625	3.57187	41/64	.640625	16.27187
5/32	.15625	3.96875	21/32	.65625	16.66875
11/64	.171875	4.36562	43/64	.671875	17.06562
3/16	.1875	4.76250	11/16	.6875	17.46250
13/64	.203125	5.15937	45/64	.703125	17.85937
7/32	.21875	5.55625	23/32	.71875	18.25625
15/64	.234375	5.95312	47/64	.734375	18.65312
1/4	.250	6.35000	3/4	.750	19.05000
17/64	.265625	6.74687	49/64	.765625	19.44687
9/32	.28125	7.14375	25/32	.78125	19.84375
19/64	.296875	7.54062	51/64	.796875	20.24062
5/16	.3125	7.93750	13/16	.8125	20.63750
21/64	.328125	8.33437	53/64	.828125	21.03437
11/32	.34375	8.73125	27/32	.84375	21.43125
23/64	.359375	9.12812	55/64	.859375	21.82812
3/8	.375	9.52500	7/8	.875	22.22500
25/64	.390625	9.92187	57/64	.890625	22.62187
13/32	.40625	10.31875	29/32	.90625	23.01875
27/64	.421875	10.71562	59/64	.921875	23.41562
7/16	.4375	11.11250	15/16	.9375	23.81250
29/64	.453125	11.50937	61/64	.953125	24.20937
15/32	.46875	11.90625	31/32	.96875	24.60625
31/64	.484375	12.30312	63/64	.984375	25.00312
1/2	.500	12.70000	1	1.00	25.40000

LIST OF AUTOMOTIVE ABBREVIATIONS WHICH MAY BE USED IN THIS MANUAL

A - Ampere(s)
A-6 - Axial 6 Cyl. A/C Compressor
A/C - Air Conditioning
ACC - Automatic Climate Control
Adj. - Adjust
A/F - Air Fuel (As in Air Fuel Ratio)
AIR - Air Injection Reaction System
ALC - Automatic Level Control
ALCL - Assembly Line Communications Link
Alt. - Altitude
APT - Adjustable Part Throttle
AT - Automatic Transmission
ATC - Automatic Temperature Control
ATDC - After Top Dead Center

BARO - Barometric Absolute Pressure Sensor
Bat. - Battery
Bat. + - Positive Terminal
Bbl. - Barrel
BHP - Brake Horsepower
BP - Back Pressure
BTDC - Before Top Dead Center

Cat. Conv. - Catalytic Converter
CC - Catalytic Converter
- Cubic Centimeter
- Converter Clutch
CCC - Computer Command Control
C-4 - Computer Controlled Catalytic Converter
CB - Citizens Band (Radio)
CCOT - Cycling Clutch (Orifice) Tube
CCP - Controlled Canister Purge
C.E. - Check Engine
CEAB - Cold Engine Airbleed
CEMF - Counter Electromotive Force
CID - Cubic Inch Displacement
CLOOP - Closed Loop
CLCC - Closed Loop Carburetor Control
CLTBI - Closed Loop Throttle Body Injection
Conv. - Converter
CP - Canister Purge
Cu. In. - Cubic Inch
CV - Constant Velocity
Cyl. - Cylinder(s)

DBB - Dual Bed Bead
DBM - Dual Bed Monolith
DEFI - Digital Electronic Fuel Injection
DFI - Digital Fuel Injection
Diff. - Differential
Distr. - Distributor

EAC - Electric Air Control Valve
EAS - Electric Air Switching Valve
ECC - Electronic Comfort Control
ECM - Electronic Control Module
ECS - Emission Control System
ECU - Engine Calibration Unit
EEC - Evaporative Emission Control
EEVIR - Evaporator Equalized Valves in Receiver
EFE - Early Fuel Evaporation
EFI - Electronic Fuel Injection
EGR - Exhaust Gas Recirculation
ELC - Electronic Level Control
EMF - Electromotive Force
EMR - Electronic Module Retard
EOS - Exhaust Oxygen Sensor
ESC - Electronic Spark Control
EST - Electronic Spark Timing
ETC - Electronic Temperature Control
ETCC - Electronic Touch Comfort Control
ETR - Electronically Tuned Receiver
Exh. - Exhaust

FMVSS - Federal Motor Vehicle Safety Standards
Ft. Lb. - Foot Pounds (Torque)
FWD - Front Wheel Drive
- Four Wheel Drive
4 x 4 - Four Wheel Drive

HD - Heavy Duty
HEI - High Energy Ignition
Hg. - Mercury
Hi. Alt. - High Altitude
HVAC - Heater-Vent-Air Conditioning
HVACM - Heater-Vent-Air Conditioning Module
HVM - Heater-Vent-Module

IAC - Idle Air Control
IC - Integrated Circuit
ID - Identification
- Inside Diameter
ILC - Idle Load Compensator
I/P - Instrument Panel
ISC - Idle Speed Control

km - Kilometers
km/hr - Kilometers Per Hour
KV - Kilovolts (Thousands of Volts)
km/L - Kilometers/Liter (mpg)
kPa - Kilopascals

L - Liter
L-4 - Four Cylinder In-Line (Engine)
L-6 - Six Cylinder In-Line (Engine)
LF - Left Front
LR - Left Rear

Man. Vac. - Manifold Vacuum
MAP - Manifold Absolute Pressure
MAT - Manifold Air Temperature Sensor
M/C - Mixture Control
MPG - Miles Per Gallon
MPH - Miles Per Hour
MT - Manual Transmission

N·m - Newton Metres (Torque)

OD - Outside Diameter
OHC - Overhead Cam
OL - Open Loop
OXY - Oxygen

PAIR - Pulse Air Injection Reaction System
P/B - Power Brakes
PCV - Positive Crankcase Ventilation
PECV - Power Enrichment Control Valve
P/N - Park, Neutral
PROM - Programmable, Read Only Memory
P/S - Power Steering
PSI - Pounds Per Square Inch
Pt. - Pint
PTO - Power Takeoff

Qt. - Quart

R - Resistance
R-4 - Radial Four Cyl. A/C Compressor
RF - Right Front
RPM - Revolutions Per Minute
RR - Right Rear
RTV - Room Temperature Vulcanizing (Sealer)
RVR - Response Vacuum Reducer
RWD - Rear Wheel Drive

SAE - Society of Automotive Engineers
SI - System International
Sol. - Solenoid

TAC - Thermostatic Air Cleaner
TACH - Tachometer
TBI - Throttle Body Injection
TCC - Transmission Converter Clutch
TCS - Transmission Controlled Spark
TDC - Topdead Center
TPS - Throttle Position Sensor
TURB - Turbocharger
T/V - Throttle Valve
TVBV - Turbocharger Vacuum Bleed Valve
TVRS - Television & Radio Suppression
TVS - Thermal Vacuum Switch

UJT - Universal Joint

V - Volt(s)
V-6 - Six Cylinder Engine - Arranged in a "V"
V-8 - Eight Cylinder Engine - Arranged in a "V"
Vac. - Vacuum
VATS - Vehicle Anti-Theft System
VIN - Vehicle Identification Number
VIR - Valves in Receiver
VSS - Vehicle Speed Sensor
VMV - Vacuum Modulator Valve

W/ - With
W/B - Wheel Base
W/O - Without
WOT - Wide Open Throttle

X-Valve - Expansion Valve

Glossary

Glossary of Abbreviations and Terms (Courtesy of Ford Motor Co.)

Many abbreviations appear throughout this book that deal with terminology used by various engine/vehicle manufacturers for their particular fuel systems. In addition to these abbreviations, a number of terms that may be new to you appear for reference purposes.

The following information contains a list of automotive abbreviations that are common to the industry. Also included is an "A" to "Z" glossary of terms relating to "Emissions Related Components," which basically applies to gasoline fuel injected engines; typical terms used relating to diesel engines are also provided.

A number of these terms are courtesy of the Society of Automotive Engineers.

A Ampere(s).

A-6 Axial 6-cylinder A/C compressor.

A/C Air conditioning.

ACC Automatic climate control.

Acceleration Smoke Limiter A device that limits the smoke of a diesel engine during acceleration by temporarily limiting the amount of fuel injected into the engine cylinders during speed and/or load transients below the steady-state limit.

Accelerator Switch (Courier) Located on accelerator pedal. Activates anti-afterburn valve to reduce rich mixture on deceleration.

A/CL Bi Met (Air Cleaner Bimetal Sensor).

A/CL CWM See CWM.

A/CL DV (Air Cleaner Duct and Valve Vacuum Motor).

ACT (Air Charge Temperature Sensor) Air Charge Temperature Sensor or its signal line. Thermistor Sensor can be located in intake manifold (EFI Systems) or air cleaner.

Active Circuit A circuit consisting of some active (semiconductor) elements.

ACV (Air Control Valve) A vacuum controlled valve in the Thermactor system that diverts air pump air to either the upstream (exhaust manifold) or downstream (underbody catalyst) air injection points as required.

Adj. Adjust.

A/F Air/fuel (as in air/fuel ratio).

Afterglow The period of time the glow plug system on a diesel engine remains on after the engine is started.

AIR Air injection reaction system.

AIR BPV See BPV.

ALC Automatic level control.

ALCL Assembly line communications link.

Alt. Altitude.

Altitude Compensating Modulator Device to control transmission shift spacing and engine timing. Improves performance at higher altitudes by changing the vacuum signals.

Ambient Temperature Temperature of air surrounding an object.

Anti-Afterburn Valve (Courier) Prevents abnormal combustion in exhaust system by increasing fresh air in intake manifold on sudden deceleration. This tends to reduce vacuum and fuel flow.

ANTI-BFV (Anti-Backfire Valve).

APT Adjustable part throttle.

AT Automatic transmission.

ATC Automatic temperature control.

ATDC After top dead center.

Automotive Emissions Gaseous and particulate compounds that are emitted from a car's crankcase, exhaust, carburetor, and fuel tank (hydrocarbons, nitrogen oxides, and carbon monoxide).

Axis A real or imaginary straight line on which an object rotates or is regarded as rotating.

BARO Barometric absolute pressure sensor.

Bat. Battery.

Bat. + Positive terminal.

Bbl. Barrel.

BHP Brake horsepower.

BHS (Bimetal Heat Sensor) A unit using a metallic part, consisting of 2 layers with different metals, to sense temperature changes. The layers expand at different rates as temperature changes, causing the part to bend in a predetermined manner, thereby causing a changing signal.

BMAP (Barometric and Manifold Absolute Pressure Sensor) Sensor housing containing both MAP and BP sensors. The barometric pressure sensor monitors atmospheric pressure and allows adjustment of spark advance, EGR flow, and air/fuel ratio as a function of altitude. The manifold absolute pressure sensor monitors engine vacuum and provides engine load information that is used to calculate spark advance, EGR flow, and air/fuel ratio. See BP and MAP.

BP (Barometric Pressure Sensor) An active circuit sensor (one-half of BMAP sensor) or its signal line. Sensor signal is proportional to the barometric pressure at the sensor location.

BP Back pressure.

B/P (EGR Back Pressure Valve).

BSV (Backfire Suppressor Valve) A device used in conjunction with the early design "Thermactor" exhaust emission system. Its primary function is to lean-out the excessively rich fuel mixture that follows closing of the throttle after acceleration. Allows additional air into the induction system whenever intake manifold vacuum increases.

BPV (Bypass Valve) A vacuum controlled valve in the Thermactor system that allows air from the air pump to be injected into the exhaust system or bypassed to the atmosphere when not required; it also provides a pressure relief function to protect the air pump from high pressure. These valves may be mounted on the air pump (pump mounted) or mounted in the Thermactor hose line (remote mounted). The two basic types are: normally closed (bypasses air with no vacuum signal applied) and normally open (allows air to flow through the valve to the exhaust system with no vacuum signal applied).

BTDC Before top dead center.

BV (Bowl Vent Port).

Calibration (1) *Balancing:* The setting of the delivery of an injection system or the setting of the rack pointer on a single unit pump in relation to predetermined positions of a quantity control member. (2) *Adjustment:* Fixing fuel delivery and speed adjustments to specified engine requirements.

Calibration Assembly See ECA.

Camshaft Pump An injection pump containing a camshaft to operate the pumping element or elements. It can be classified as "in-line," "distributor," "submerged," etc.

Camshaft Pump Mountings (1) *Base mounted:* A pump, mounted on a surface of the engine, parallel to the axis of the pump camshaft. (2) *Cradle mounted:* A special form of a base mount in which the base is contoured to permit rotation of the pump around the axis of the pump camshaft. (3) *Flange mounted:* A pump, mounted on a surface of the engine, at a right angle to the axis of the pump camshaft.

Canister A container, in an evaporative emission control system, that contains charcoal to trap vapors from the fuel system.

CANP (Canister Purge Solenoid) Electrical solenoid or its control line. Solenoid opens valve from fuel vapor canister line to intake manifold when energized. Controls flow of vapors between carburetor bowl vent and carbon canister.

Catalyst A muffler-like device located in the exhaust system where hot exhaust gas comes in contact with special metals (i.e., platinum or palladium) that promote more complete combustion of unburned hydrocarbons and reduction of carbon monoxide.

Cat. Conv. Catalytic converter.

CB Citizens band (radio).

CC (Catalytic Converter) A muffler-like assembly placed in the exhaust system and containing a catalyst to change hydrocarbons and carbon monoxide into water vapor and carbon dioxide.

CC Catalytic converter.
Cubic centimeter.
Converter clutch.

CCC Computer command control.

C-4 Computer controlled catalytic converter.

CCOT Cycling clutch (orifice) tube.

CCP Controlled canister purge.

CDR (Crankcase Depression Regulator) A device on a diesel engine used to regulate the amount of crankcase gases admitted into the intake manifold to be burned to lower engine emissions.

C.E. Check engine.

CEAB Cold engine airbleed.

CEMF Counter electromotive force.

CFI (Central Fuel Injection) See EFI.

Chamfer A beveled edge or corner.

Check Valve A one-way valve that allows a liquid or gas to flow in one direction only—prevents backflow.

CID Cubic inch displacement.

CLCC Closed-loop carburetor control.

CLOOP Closed loop.

Closed Nozzle A nozzle incorporating either a poppet valve or a needle valve, loaded in order to open at some predetermined pressure. (1) *Poppet nozzle:* A closed nozzle provided with an outward opening, spring-loaded poppet valve. (2) *Differential nozzle:* A closed nozzle provided with a spring-loaded needle valve. (3) *Pintle nozzle:* A closed nozzle provided with a spring-loaded needle valve. The body of the nozzle has a single large orifice into which enters a projection from the lower end of the needle, this projection being so formed as to influence the rate and shape of the fuel spray. (4) *Hole-type nozzle:* A closed nozzle provided with one or more orifices through which the fuel issues. Nozzles with more than one orifice are known as multihole nozzles.

CLTBI Closed-loop throttle body injection.

CO (Carbon Monoxide) A colorless, odorless gas that is a by-product of incomplete combustion of carbon. This gas is poisonous in confined spaces when it is at high concentration.

Coasting Richer Solenoid Valve (Courier) Adds additional fuel to lean mixture caused by deceleration. Controlled by speedometer switch and accelerator switch.

COC (Conventional Oxidation Catalyst) Acts on two of the major pollutants: HC and CO.

Combination Thermactor Air Bypass and Air Diverter Valve Combines the function of a normally closed air bypass valve and an air control valve in one integral valve.

Computed Timing (CT) Relationship of Spark Plug Firing to CP pulse, expressed in crankshaft degrees. Since CP pulse is at 10 BTDC, a Computed Timing reading of 10 would be equal to a Timing reading of 20 BTDC.

Control Pinion (Control Sleeve) A collar engaging the plunger and having a segment of gear teeth, integral or attached, which mesh with the control rack. By this means, linear motion of the control rack is transformed into rotary movement of the plunger to regulate the amount of fuel delivered by the pump.

Control Rack (Control Rod) The rack or rod by which the fuel delivery is regulated.

Conv. Converter.

CP Canister purge.

CP (Crankshaft Position Sensor) Permanent magnet with coil. Sensor produces AC pulse when magnetic field or sensor is disrupted by a gear tooth or notch. Senses pulse ring lobes. Resultant signal enables the ECA to determine when the ignition module should be triggered to produce a spark. Also, provides RPM information that is used to calculate spark advance and EGR flow, and air/fuel ratio (replaces distributor centrifugal advance weights and timing pickup). One pulse is produced for each cylinder every other engine rotation (one for each spark plug firing). Example: 4 pulses are produced for each V8 engine rotation. The CP pulse is generated at 10 crankshaft degrees Before Top Dead Center (BTDC).

CP+ CP sensor signal line, Pin No. 18, TSS A-10.

CP− CP sensor return line. Connects to ground (Pin No. 8) in ECA processor. Pin No. 2, TSS B-7.

CPRV (Canister Purge Regulator Valve).

Crankshaft Timing See Timing.

CRK Cranking signal. Battery voltage that is switched to ECA by ignition switch during cranking (start) only. Pin No. 1, TSS A-2.

CSC (Coolant Spark Control).

CSSA (Cold Start Spark Advance System) Added to the distributor spark control on some engines. System momentarily traps spark port vacuum on the distributor spark advance diaphragm when engine coolant temperature is below 53°C (128°F). Used in conjunction with a DRCV and a CSSA PVS.

CSSH (Cold Start Spark Hold System) Provides improved cold engine acceleration. When engine coolant is less than 53°C (128°F), the CSSH PVS is closed and the distributor vacuum signal travels through a restrictor providing a modified vacuum advance during the initial stage of acceleration.

CT See Computed Timing.

CTAV (Cold Temperature Actuated Vacuum Switch) Used with MCU system to signal open loop or closed loop operating mode.

Cu.In. Cubic inch.

CV Constant velocity.

CV (Control Valve).

CWM (Cold Weather Modulator) A vacuum modulator located in the carburetor air cleaner on some models. Prevents the air cleaner duct door from opening to non-heated intake air when fresh air is below 13°C (55°F).

Cyl. Cylinder(s).

Darlington A two-transistor switch with transistors connected so that their collectors are common and gains are multiplied.

DBB Dual bed bead.

DBM Dual bed monolith.

Deceleration Valve See DVAC.

DEFI Digital electronic fuel injection.

Delivery Valve Assembly A valve installed in a pump, interposed between the pumping chamber and outlet, to control residual line pressures and which may or may not have an unloading or retraction function.

Delivery Valve Holder A device that retains the delivery valve assembly within the pump.

DFI Digital fuel injection.

Diff. Differential.

Differential Angle The difference between the angles of the seat face of the valve and that of the seat in the body provided to ensure its effective sealing.

Differential Ratio The ratio between the guide diameter of the needle valve and the effective diameter of the needle valve seat.

Distr. Distributor.

Distributor Pump An injection pump where each metered delivery is directed to the appropriate engine cylinder by a distributing device.

DMS (Distributor Modulator System).

DRCV (Distributor Retard Control Valve) Used on engines equipped with the CSSA cold start spark advance system. Used in combination with a CSSA PVS.

Dribble Insufficiently atomized fuel issuing from the nozzle at or immediately following the end of main injection.

Dual Catalytic Converter See TWC.

Duty Cycle (DC) The measurement of the off (high voltage) time of the IMS pulse, expressed in percent.

DVAC (Distributor Vacuum Advance Control Valve) A device used in conjunction with the dual diaphragm vacuum advance unit to advance timing under deceleration conditions.

DVCV See DV/DV.

DV/DV (Differential Valve/Delay Valve) Installed in series with VDV. Used to delay sudden drops in manifold vacuum as during hard acceleration. Permits instant signal when deceleration occurs.

DVOM Digital volt ohmmeter.

DV-TW (Delay Valve-Two Way).

DVVV (Distributor Vacuum Vent Valve) Required by some engines to prevent fuel migration to distributor advance diaphragm and to act as a spark advance delay valve.

EAC Electric air control valve.

Earth Refer to Ground.

EAS Electric air switching valve.

ECA (Electronic Control Assembly) A vehicle computer consisting of a calibration assembly containing the computer memory and thus its control program and a processor assembly, which is the computer hardware. The calibration assembly plugs into the processor. The processor plugs into the vehicle harness.

ECC Electronic comfort control.

ECM Electronic control module.

ECS Emission control system.

ECT (Engine Coolant Temperature Sensor) Refers to thermistor sensor or its signal line. Sensor is immersed in engine coolant fluid. Provides engine coolant temperature information that is used to alter spark advance and EGR flow during warm-up or overheat conditions (replaces cooling and EGR PVS).

ECU Electronic control unit.

ECU Engine calibration unit.

EDM (Electronic Distributor Modulator).

EEC (Electronic Engine Control) A computer directed system of engine control. EEC-I—Control of engine timing. EEC-II—Control of engine timing and fuel (with FBC system). EEC-III—FBC-descendent of EEC-II EEC-III—EFI-control of engine timing and fuel (with EFI system).

EEC Evaporative emission control.

EESS (Evaporative Emission Shed System) A system for containment of evaporative fuel vapors, introduced in 1978. Annual improvements have modified this system.

EEVIR Evaporator equalized valves in receiver.

EFE Early fuel evaporation.

EFI (Electronic Fuel Injection) A system using two injectors, that are computer controlled, spraying fuel into a divided throttle body and thence to a dual plane intake manifold.

EFM See EFI.

EGC (Exhaust Gas Check Valve) Allows thermactor air to enter exhaust manifold but prevents reverse flow in event of improper operation of other components.

EGO (Exhaust Gas Oxygen Sensor) Exhaust Oxygen Sensor or its signal line. Sensor changes its output voltage as exhaust gas oxygen content changes when compared to oxygen content of atmosphere. Constantly changing voltage signal is sent to the ECA and MCU for analysis and adjustment of the air/fuel ratio.

EGOR (EGO Signal Return) EGO signal return with separate ground on engine block.

EGR (Exhaust Gas Recirculation) A procedure where a small amount of exhaust gas is readmitted to the combustion chamber to reduce peak combustion temperatures and thus reduce NOx emissions.

EGR-ACT EGR Solenoid pressure valve assembly.

EGR Cooler Assembly Heat exchanger using engine coolant to reduce exhaust gas temperature.

EGRC (EGR Control Solenoid) Electrical solenoid or its control line. Solenoid switches engine manifold vacuum to operate EGR valve. Vacuum opens EGR valve when solenoid is energized.

EGRV (EGR Vent Solenoid) Electrical solenoid or its control line. Solenoid normally vents EGRC vacuum line. When EGRV is energized, EGRC can open EGR valve.

EGR VA (EGR Valve Actuator).

ELC Electronic level control.

EMF Electromotive force.

EMR Electronic module retard.

EOS Exhaust oxygen sensor.

EPA (Environmental Protection Agency) Federal agency having responsibility for administrating congressional programs relating to the protection of the environment.

ERTN See EGOR.

ESC (Electronic Spark Control System) An electronic distributor vacuum control system aiding in more complete combustion through controlled spark retard under certain temperature and speed conditions.

EST Electronic spark timing.

ETC Electronic temperature control.

ETCC Electronic touch comfort control.

ETR Electronically tuned receiver.

EVP (Exhaust Valve Position Sensor) Potentiometric sensor or its signal line. Sensor wiper position is proportional to EGR valve pintle position. This allows ECA to determine actual EGR flow at any point in time.

Excess Fuel Device Any device provided for giving an increased fuel setting for starting only, generally designed to automatically restore action of the normal full load stop after starting.

Exh. Exhaust.

Exhaust Gas Analyzer An instrument for determining the amount of HC and CO emitted to the atmosphere. These instruments usually incorporate a "wheatstone bridge" to analyze the thermal conductivity of the mixture. A means of determining the efficiency with which the engine is then burning fuel.

Exhaust Heat Control Valve A valve that routes hot exhaust gases to the intake manifold heat riser during cold engine operation. The valve can be thermostatically controlled, or vacuum operated.

FBC (Feedback Carburetor System) A system of fuel control employing a computer controlled stepper motor that varies the carburetor air/fuel mixture.

FBCA (Feedback Carburetor Actuator) Controls carburetor air/fuel ratio. A system of fuel control employing a computer controlled stepper motor that varies the carburetor air/fuel mixture.

FCV (Fuel Decel Valve) A valve that, during deceleration, adds additional air-fuel to the intake manifold to enrich the air/fuel mixture and thus provides for more complete combustion. See DVAC.

FED Federal.

FMVSS Federal Motor Vehicle Safety Standards.

FP (Fuel Pump Relay) Relay or its control line. Relay is controlled by ECA processor. Supplies power to electric fuel pump of EFI system.

Fuel Injection Tubing The tube connecting the injection pump to the nozzle holder assembly.

Fuel Pump Housing The main casing into or to which are assembled all the components of the injection pump. It may accommodate the camshaft in the case of camshaft pumps, or the camshaft, or driveshaft in the case of distributor type pumps.

Fuel-Vacuum Separator Used to filter waxy hydrocarbons from carburetor ported vacuum to protect vacuum delay valve and distributor vacuum controls.

Fulcrum The pivot point of a valve rocker arm on an engine.

Full Load Stop A device that limits the maximum amount of fuel injected into the engine cylinders at the rated load and speed specified by the engine manufacturer.

FWD Front wheel drive.
Four wheel drive.

4 × 4 Four wheel drive.

Glow Plug An electronically operated device used to heat up the combustion chamber in a diesel engine to aid in starting.

GND., or Ground Common line for all vehicle power, vehicle chassis ground and engine block ground, connected to vehicle battery negative terminal.

HC (Hydrocarbon) Any compound composed of carbon and hydrogen, such as petroleum products. Excessive amounts in the atmosphere are considered undesirable contaminants and a major contributor to air pollution.

HCV (Exhaust Control Valve).

HD Heavy duty.

Header The portion of a diesel fuel injection pump that distributes fuel to the individual cylinders.

HEI High energy ignition.

Helix Hand The hand of the helix in plungers is designated right or left, the same as a thread.

Helix Lead The axial advance of the helix edge in one revolution.

Helix (Scroll) A term used to describe the control edge of a spill groove provided on the plunger, usually of helical form. The helices may be upper or lower or both and may be the same hand or opposite. They can also be duplicated on both sides of the plunger.

Hg Mercury.

Hi. Alt. High altitude.

HIC (Hot Idle Compensator) A thermostatically controlled carburetor valve that opens whenever inlet air temperatures are high. Additional air is allowed to discharge below the throttle plates at engine idle. This feature improves idle stability and does not allow the rich fuel mixture normally associated with increased fuel vaporization of a hot engine.

HVAC Heater-vent-air conditioning.

HVACM Heater-vent-air conditioning module.

HVM Heater-vent module.

Hydraulic Governor A mechanical governor having a hydraulic servo-booster to increase output force.

Hydraulic Head Assembly The assembly containing the pumping, metering, and distributing elements (and may include the delivery valve) for distributor-type pumps.

IAC Idle air control.

IAT (Intake Air Temperature Sensor) See ACT.

IC Integrated circuit.

ID Identification.
Inside diameter.

Idle Limiter A device to control minimum and maximum idle fuel richness of the carburetor. Aids in preventing unauthorized persons from making overly rich idle adjustments. The limiters are of two distinct types: the external plastic limiter caps installed on the head of idle mixture adjustment screws or the internal needle-type located in the idle channel.

Idle Vacuum Valve This device may be used in conjunction with other vacuum controls to dump thermactor air during extended period of idle. Provides protection to the catalyst.

IGN Ignition.

Ignition Injection A small charge of fuel used to ignite the main gas charge in dual fuel engines.

Ignition Pressure Switch Assembly Pressure switches used on turbocharged engines in conjunction with dual mode ignition to retard spark timing during boost to prevent engine damage.

ILC Idle load compensator.

IMS (Ignition Module Signal) The signal produced by the ECA that controls the ignition module ON and OFF time and therefore the spark plug timing and coil dwell.

Inertia Switch EFI system switch. Designed to shut off electric fuel pump if vehicle receives severe rear end impact.

Injection Lag The time interval (usually expressed in degrees of crank angle) between the nominal start of injection pump delivery and the actual start of injection at the nozzle.

Injection Pump The device that meters the fuel and delivers it under pressure to the nozzle and holder assembly.

Injection Pump Assembly A complete assembly consisting of the fuel pump proper, together with additional units such as governor, fuel supply pump, and additional optional devices, when these are assembled with the fuel injection pump to form a unit. (1) *Right-hand mounted:* When the pump is mounted on the right-hand side of the engine commonly viewed from the engine flywheel end. (2) *Left-hand mounted:* When the pump is mounted on the left-hand side of the engine commonly viewed from the engine flywheel end.

Injection Timing The matching of the pump timing mark, or the injector timing mechanism, to some index mark on an engine component, such that injection will occur at the proper time with reference to the engine cycle. Injection advance or retard is respectively an earlier, or later, injection pump delivery cycle in reference to the injection cycle.

Injector EFI system electrical solenoid (one of two) or its control line. Solenoid, when energized, allows fuel flow into throttle body and thence into one plane of dual-plane intake manifold. Injector No. 1 (driver's side)—Pin No. 28. Injector No. 2 (passenger's side)—Pin No. 12.

Inlet Metering A system of metering fuel delivery by controlling the amount of fuel entering the pumping chamber during the filling or charging portion of the pump's cycle.

Inlet Valve A valve used to admit fuel to the pump barrel.

In-Line Pump An injection pump with two or more pumping elements arranged in line, each pumping element serving one engine cylinder only. A pump that has the elements arranged in line and in more than one bank, for instance, in two banks forming a "V," is a specific case of an in-line pump.

I/P Instrument panel.

IPS (Ignition Pressure Switch).

ISC Idle Speed Control.

ITVS (Ignition Timing Vacuum Switch).

IVV (Idle Vacuum Valve).

km Kilometers.

km/hr Kilometers per hour.

km/L Kilometers per liter (mpg).

kPa Kilopascals.

kV Kilovolts (thousands of volts).

L Liter.

LCV (Load Control Valve).

Leakdown Rate The period of time it takes for a valve tappet to collapse when being tested on a leakdown tester.

Leak-Off Fuel that escapes between the nozzle valve and its guide. (This term is also used to describe the leakage past the plunger of a fuel pump.)

LF Left front.

LR Left rear.

L-4 Four cylinder in-line (engine).

L-6 Six cylinder in-line (engine).

Linear Similar to straight line. Directly proportional. Example: A linear circuit would produce 1X out with 1Y in, 3X out with 2Y in, a 9X out with 3Y in.

Load-Sensing Governor An engine speed control device for use on engine-generator sets to control engine fuel settings as a function of electrical load to anticipate resulting changes in engine speed. It may or may not incorporate a mechanical speed-sensing device as well.

LOS (Limited Operational Strategy) Certain types of computer malfunctions will place the ECA into the LOS mode. Output commands are cut off the important control solenoids.

MAF Mass air flow.

Man. Vac. Manifold vacuum.

MAP Manifold absolute pressure.

MAP (Manifold Absolute Pressure Sensor) An active circuit sensor (one-half of BMAP sensor) or its signal line. Sensor signal is proportional to the absolute pressure of intake manifold.

MAT Manifold air temperature sensor.

Maximum-Minimum Governor Any one of the above varieties, which exerts control only at the upper and lower limits of the designed engine speed range, intermediate speeds being controlled by the operator setting the fuel delivery directly by throttle action.

M/C Mixture control.

MCT (Manifold Charge Temperature Sensor) See ACT.

MCU (Microprocessor Control Unit) Integral part of electronically controlled feedback carburetor system using TWC catalyst. Various sensors monitor mode conditions. MCU is now widely used on Ford built vehicles for the control of air/fuel ratios.

MCV (Manifold Control Valve) A thermostatically operated valve in the exhaust manifold for varying heat to intake manifold with engine temperature.

Mechanical Governor A speed sensitive device of the centrifugal type, which controls the injection pump delivery solely by mechanical means.

MFI Multiport fuel injection.

Monolithic Substrate The ceramic honeycomb structure used as a base to be coated with a metallic catalyst material for use in the converter.

MPG Miles per gallon.

MPH Miles per hour.

MT Manual transmission.

N.C. Normally closed.

Needle Valve (In a Closed Nozzle) A needle valve

has two diameters, the smaller at the valve seat. The fuel injection pressure acting on a portion of the total valve area lifts the valve at the predetermined pressure, then acts on the total area. The end opposite the valve seat is never subjected to injection pressure.

N•m Newton meters (torque).

N.O. Normally open.

NOx (Nitrous Oxides) Any of various compounds formed during the engine's combustion process when oxygen in the air combines with nitrogen in the air to form nitrous oxides, which are agents in photochemical smog.

Nozzle The assembly of parts employed to atomize and deliver fuel to the engine.

Nozzle Body That part of the nozzle that serves as a guide for the valve and in which the actual spray openings may be formed. These two parts, the body and the valve, are considered as a unit for replacement purposes.

Nozzle and Holder Assembly The complete apparatus that injects the pressurized fuel into the combustion chamber.

Nozzle Holder Assembly The assembly of all parts of the nozzle and holder assembly other than those included in the nozzle.

Nozzle Holder Cap A cap nut or other type of closure that covers the outer end of the nozzle holder.

Nozzle Holder Shank Length The distance from the top of the cylindrical shank to the seating face of the nozzle holder.

Nozzle Opening Pressure The pressure needed to unseat the nozzle valve.

Nozzle Retaining Nut The nozzle holder part that secures the nozzle or nozzle tip to the other nozzle holder parts.

Nozzle Tip The extreme end of the nozzle body containing the spray holes (may be a separate part).

O2 Oxygen sensor.

OD Outside diameter.

OHC Overhead cam.

OL Open loop.

Open Nozzle A nozzle incorporating no valve.

Overspeed Governor A mechanical speed-sensitive device that, through mechanical or electrical action (operation of a switch), acts to shut down the engine and limit the speed by cutting off fuel and/or air supply if the engine speed exceeds a preset maximum.

OXY Oxygen.

PAIR Pulse air injection reaction system.

Passive Circuit A circuit consisting of all passive elements. Resistors, capacitors, or inductors.

P/B Power brakes.

P.C.V. (Positive Crankcase Ventilation) A valve that controls the flow of vapors from the crankcase into the engine intake system where they are burned in the engine cylinders rather than being discharged unburned into the air from the crankcase.

Peak Injection Pressure The maximum fuel pressure attained during the injection period (not to be confused with opening pressure).

PECV Power enrichment control valve.

PFI Port fuel injection.

Pilot Injection A small initial charge of fuel delivered to the engine cylinder in advance of the main delivery of fuel.

Pintle Valve (In a Closed Nozzle) A special type of a "needle valve" wherein an integral projection from the lower end of the needle is so formed as to influence the rate and/or shape of the fuel spray during operation.

Plunger and Barrel Assembly (Or Plunger and Bushing Assembly) The combination of a pump plunger and its barrel constituting a pumping element. The plunger and barrel assembly may also perform the additional functions of timing and metering.

Plunger Control Arm A lever attached to a collar or sleeve engaging the plunger, or attached directly to the plunger, its other end engaging possibly adjustable fittings on the control rod. This transforms linear motion of the control rod to rotary motion of the plunger to regulate the amount of fuel delivered by the pump.

P/N Park, neutral.

Pneumatic Governor (1) *Vacuum or suction governor:* One operated by a change in pressure created by the air actually consumed by the engine. (2) *Air governor:* One operated by air displaced by a device provided for this particular purpose and driven by the engine.

Poppet Valve An outwardly opening valve used with certain forms of closed nozzles.

Port Closing A term referring to the fuel injection pump of the port and helix or sleeve metering type in which timing is determined by the point of the closing of the port by the metering member, corresponding to the nominal start of pump delivery.

Port and Helix Metering A system of metering fuel delivery by means of one or more helical cuts in the plunger and one or more ports in the barrel. Axial rotation of the plunger alters the effective portion of the stroke by changing the points at which the helices close and/or open the port or ports.

Port Opening A term referring to a fuel injection pump of the port and helix or sleeve metering type in which timing is determined by the point of the opening of the port by the metering member, corresponding to the nominal end of pump delivery.

Positive Valve Rotator A device at the base of the valve spring assembly used to rotate the valve slightly each time the valve is opened. This action prevents the buildup of carbon on the valve face and valve seat.

Potentiometer (POT) A variable registor with three connections. The third connection (wiper) moves physically up and down the resistive element, which has each end attached to one of the other two connections.

Potentiometric Processor Like a potentiometer.

PPS (Ported Pressure Switch).

Pressure Adjusting Screw (Shims) The screw (shims) by means of which the spring load on the nozzle valve is adjusted to obtain the prescribed opening pressures.

Processor Assembly See ECA.

PROM Programmable read only memory.

P/S Power steering.

PSI Pounds per square inch.

Pt. Pint.

PTO Power takeoff.

PTC (Positive Temperature Coefficient Heater) A temperature sensitive electric assist choke system activated when ambient temperature is above 16°C (60°F). Permits earlier opening of the choke plates to reduce fuel consumption.

Pump Rotation (1) *Clockwise:* The rotation of the pump camshaft or driveshaft is clockwise when viewed from the pump drive end. (2) *Counterclockwise:* The rotation of the pump camshaft or driveshaft is counterclockwise when viewed from the pump drive end.

Purge CV (Purge Control Valve).

PVS (Ported Vacuum Switch) A temperature-actuated switch that changes vacuum connections when the coolant temperature changes.

Qt. Quart.

R Resistance.

Rated RPM The maximum RPM of a diesel engine run at wide open throttle.

Ratio The relation between two similar magnitudes in respect of quantity determined by the number of times one contains the other (integrally or fractionally).

Ratiometric As used in EEC. A voltage divided by the VREF multiplied by 10.

Ratiometrically Similar to a Ratio. See Ratiometric for specific EEC meaning.

Reed Valve (Courier) Check valve used on some engine applications. Prevents reverse flow of air from exhaust manifold to intake air cleaner.

Relay A switching device operated by a low-current circuit, which controls the opening and closing of another circuit of higher current capacity.

Relief Valve A pressure limiting valve located in the exhaust chamber of the air supply pump. Functions to relieve part of the exhaust air flow if the pressure exceeds a pre-determined value.

Retraction Volume The volume of fuel retracted from the high-pressure delivery line by action of the delivery valve's retraction piston in the process of the delivery valve returning to its seat following the end of injection.

R-4 Radial four-cylinder A/C compressor.

RF Right front.

Rotary Injection Pump A type of diesel injection pump that rotates and distributes fuel to the individual cylinders at precisely timed intervals.

RPM Revolutions per minute.

RR Right rear.

RTV Room temperature vulcanizing (sealer).

RVR Response vacuum reducer.

RWD Rear wheel drive.

Sac Hole The recess immediately within the nozzle tip and acting as a feeder to the spray hole(s) of a hole-type nozzle.

SAE Society of Automotive Engineers.

SA-FV (Separator Assembly-Fuel Vacuum).

SC (Signal Conditioner).

SDV (Spark Delay Valve) A valve that delays spark vacuum advance during rapid acceleration from idle or from speeds below 15 mph, and cuts off spark advance immediately on deceleration. Has internal sintered orifice to slow air in one direction, a check valve for free air flow in opposite direction, and a filter.

Seating Face The face upon which the nozzle and holder assembly seats to make a gas-tight seal with the cylinder head. Commonly, this face is on the nozzle retaining nut.

Secondary Injection The fuel discharged from the nozzle as a result of a reopening of the nozzle valve after the main discharge.

SFI Sequential fuel injection.

SHED (Sealed Housing Evaporative Determination System) See EESS.

SI System International.

SILN (Silencer-Exhaust Air Supply).

Sleeve Metering A system of metering fuel delivery by incorporating a movable sleeve with which port opening and/or port closing is controlled.

Spill Valve A valve used to terminate injection at a controllable point on the pumping stroke by allowing fuel to escape from the pumping chamber.

Spindle A spindle transmits the load from the spring to the valve.

Spray Angle The included angle of the cone embracing the axis of the several spray holes of a multihole nozzle. In the case of nozzles for large engines, more than one spray angle may be needed to embrace all the sprays; for example, an inner and an outer spray angle.

Spray Dispersal Angle The included angle of the cone of fuel leaving any single orifice in the nozzle or tip including pintle type.

Spray Inclination Angle The angle that the axis of a cone of spray holes makes with the axis of the nozzle holder.

Spray Orifice/Orifices The opening or openings in the end of the nozzle through which the fuel is sprayed into the cylinder.

Spring Retainer The spring retainer encloses the spring and carries the adjusting screw or shims.

Sol. Solenoid.

Solenoid A wire coil with a movable core that changes position by means of electromagnetism when current flows through the coil.

Sol V (EGR Solenoid Vacuum Valve Assembly).

Spacer Entry EGR System Exhaust gases are routed directly from exhaust manifold through stainless steel tube to carburetor base.

SPK or Spark See IMS.

SRTN (Sensor Return Line) A return line for all sensors, except EGO and CP.

Submerged Pump A pump with the mounting flange raised to limit pump projection above the mounting face.

Supply Pump A pump for transferring the fuel from the tank and delivering it to the injection pump.

SV-CBV (Solenoid Valve-Carburetor Bowl Vent).

SVV (Solenoid Vent Valve) Energized by ignition switch to control fuel vapor flow to canister. Ignition off opens valve.

TAB (Thermactor Air Bypass Solenoid) Electrical solenoid or its control line. Solenoid switches engine manifold vacuum. Vacuum bypasses thermactor air to atmosphere.

TAC Thermostatic air cleaner.

TACH Tachometer.

TAD (Thermactor Air Diverter Solenoid) Electrical solenoid or its control line. Solenoid switches engine manifold vacuum. Vacuum switches thermactor air from downstream (past EGO sensor in exhaust system) to upstream (before EGO sensor) when solenoid is energized.

Tailshaft Governor A mechanical speed-sensitive device commonly mounted on an engine driven torque convertor to monitor its tailshaft speed. It is mechanically connected to the normal engine governor such that engine output will be governed to maintain a constant tailshaft speed regardless of torque load.

TAV (Temperature Actuated Vacuum).

TBI Throttle body injection.

TCC Transmission converter clutch.

TCS Transmission controlled spark.

TDC (Top Dead Center) The point of rotation of the crankshaft when the piston is at the top of its stroke.

THERMAC Thermostatic air cleaner.

Thermactor An air injection type of exhaust emission control system. Similar to exhaust emission control systems used by GM and American Motors vehicles. A product identification name of Ford Motor Company.

Thermactor II This is a pulse air system that utilizes the natural exhaust pulses in a tuned pipe. A reed-type check valve responds to the negative pressure pulses so generated and permits air to be drawn into the exhaust system.

Thermistor A resistor that changes its resistance with temperature.

Timing Relationship between spark plug firing and piston position, usually expressed in crankshaft degrees before or after top dead center (TDC) of compression stroke. Example: 15 degrees BTDC.

Timing Device A device responsive to engine speed and/or load to control the timed relationship between injection cycle and engine cycle.

TIV (Thermactor Idle Vacuum Valve) Used in some evaporative emission systems to improve engine idling after a hot start.

TKA (Throttle Kicker Actuator) See TKS.

TKS (Throttle Kicker Solenoid) Electrical solenoid or its control line. Solenoid, when energized, supplies manifold vacuum to throttle kicker actuator as directed by the ECA. Actuator moves throttle linkage to increase idle rpm. TKS vents the actuator vacuum line when de-energized.

Torque Control A device that modifies the maximum amount of fuel injected into the engine cylinders at speeds below rated speed to obtain the desired torque output.

TP (Throttle Position Sensor) Potentiometric sensor or its signal line. Sensor wiper position is proportional to throttle position. Defines engine operation mode: closed, part, or wide open throttle. (Replaces carburetor spark/EGR ports.)

TPS Throttle position sensor.

Transducer An electrically actuated vacuum regulator.

TRS or TRS+1 (Transmission Regulated Spark Control System) A distributor vacuum control system aiding in more complete fuel combustion by regulating spark retard when the outside air temperature is above 60°F. The system uses a switch responding to the gear range, or hydraulic pressure in the transmission, to regulate a vacuum control valve.

TSP (Throttle Solenoid Positioner) A device for providing variable throttle stops. A higher position for maintaining the specified idle rpm, and a lower position, when the ignition is turned OFF, to prevent dieseling. Can be used to increase throttle opening when air conditioning is turned on.

TSP-VOTM (TSP-Vacuum Operated Throttle Modulator).

TURB Turbocharger.

T/V Throttle valve.

TVBV Turbocharger vacuum bleed valve.

TVRS Television and radio suppression.

TVS (Thermal Vacuum Switch) Used with some ported EGR systems. Located in the carburetor air cleaner to control vacuum to the EGR valve, instead of a PVS valve. Responds to temperature of the inlet air heated by the exhaust manifold.

TVV (Thermal Vent Valve) Temperature sensitive valve assembly located in the canister vent line. Closes when engine is cold and opens when hot. Prevents fuel tank vapors from being vented through the carburetor fuel bowl when fuel tank heats up before engine compartment.

TWC (Three-Way Catalyst) Sometimes referred to as a dual catalytic converter. Combines two converters in one shell. Controls NOx, HC, and CO.

UJT Universal joint.

Unit Fuel Injector An assembly that receives fuel under supply pressure and is then actuated by an engine mechanism to meter and inject the charge of fuel to the combustion chamber at high pressure and at the proper time.

Unit Pump An injection pump containing no actuating mechanism to operate the pumping element or elements. It can be classified as in-line, distributor, submerged, etc.

V Volt(s).

V-6 Six-cylinder engine—arranged in a "V."

V-8 Eight-cylinder engine—arranged in a "V."

Vac. Vacuum.

Vacuum A term used to describe a pressure that is less than atmospheric pressure; hence, a partial vacuum. Used for control purposes throughout the automotive industry.

Vacuum Advance Advances ignition timing with relation to engine load conditions. This is achieved by using engine intake manifold vacuum to operate the distributor diaphragm.

Vacuum Regulator Provides constant vacuum output from manifold when vehicle is at idle. Switches to engine vacuum at off idle.

Vacuum Regulator/Solenoid Regulator valve provides vacuum to feedback carburetor. Solenoid is controlled by ECA to provide feedback carburetor control.

Vacuum Retard Delay Valve Delays the loss of vacuum signal during some vehicle deceleration modes.

VACVV-D (Distributor Vacuum and Vent Control Valve).

VACVV-T (Thermactor Vacuum and Vent Control Valve).

VACW-D (Vent Valve Vacuum-Delay) Delays vacuum to the distributor through a sintered disk.

VATS Vehicle anti-theft system.

VBAT Vehicle battery, battery voltage, Pin No. 24, TSS A-1.

VCKV (Vacuum Control Check Valve).

VCS (Vacuum Controlled Switch).

VCS-CT (Vacuum Controlled Switch-Cold Temperature).

VCS-DI (Vacuum Controlled Switch-Decel Idle).

VCTS (Vacuum Control Temperature Sensing Valve) A valve that connects manifold vacuum to the distributor advance mechanism under hot idle conditions.

VCV (Vacuum Check Valve) Used to retain vacuum signal after vacuum source is gone.

VDV (Vacuum Differential Valve) A valve used in the Thermactor system with catalyst to cut off vacuum to the air bypass valve during deceleration.

VECI (Vehicle Emission Control Information Decal) Critical specifications for servicing emission systems. Decal located in engine compartment.

Vehicle Speed Switch (Courier) Attached to speedometer to help control air fuel mixture at cruising speed.

VIN Vehicle identification number.

VIR Valves in receiver.

Viscosity The rating of an engine oil's resistance to flow.

VMV Vacuum modulator valve.

VOTM (Vacuum Operated Throttle Modulator).

VRDV (Vacuum Retard Delay Valve) A valve that delays a decrease in vacuum at the distributor vacuum advance unit when the source vacuum decreases.

VREF (Reference Voltage) A power supplied by ECA to some sensors regulated at a specific voltage between 8 and 10.

VRESER (Vacuum Reservoir).

VREST (Vacuum Restrictor).

VR/S (Vacuum Regulator/Solenoid).

VRV (Vacuum Regulator Valve) Three-port valve.

VSS Vehicle speed sensor.

VVA (Venturi Vacuum Amplifier) Used with some EGR systems so that carburetor venturi vacuum can control EGR valve operation. Venturi vacuum is desirable because it is proportional to the air flow through the carburetor.

VVV (Vacuum Vent Valve).

W/ With.

W/B Wheel base.

WIPER See Potentiometer.

W/O Without.

WOT (Wide Open Throttle Valve) In some EGR applications, it is desirable to cut off EGR flow at wide-open throttle, where a richer fuel mixture is needed for performance. Valve is vacuum actuated when "full throttle" signal is received from the carburetor.

X-Valve Expansion valve.

Index